Monographs in Electrical and Electronic Engineering

Series editors: P. Hammond, T. J. E. Miller, and T. Kenjo

Monographs in Electrical and Electronic Engineering

25. *Electrical machines and drives: a space-vector theory approach* (1992) Peter Vas
26. *Spiral vector theory of a.c. circuits and machines* (1992) Sakae Yamamura
27. *Parameter estimation, condition monitoring, and diagnosis of electrical machines* (1993) Peter Vas
28. *An introduction to ultrasonic motors* (1993) T. Sashida and T. Kenjo
29. *Ultrasonic motors: theory and applications* (1993) S. Ueha and Y. Tomikawa
30. *Linear induction drives* (1993) J. F. Gieras
31. *Switched reluctance motors and their control* (1993) T. J. E. Miller
32. *Numerical modelling of eddy currents* (1993) Andrzej Krawczyk and John A. Tegopoulos
33. *Rectifiers, cycloconverters, and a.c. controllers* (1994) Thomas M. Barton
34. *Stepping motors and their microprocessor controls. Second edition* (1994) T. Kenjo and A. Sugawara
35. *Inverse problems and optimal design in electricity and magnetism* (1995) P. Neittaanmäki, M. Rudnicki, and A. Savini
36. *Electric drives and their control* (1995) Richard M. Crowder
37. *Design of brushless permanent magnet motors* (1995) J. R. Hendershott and T. J. E. Miller
38. *Reluctance synchronous machines and drives* (1995) I. Boldea
39. *Geometry of electromagnetic systems* (1996) D. Baldomir and P. Hammond
40. *Permanent-magnet d.c. linear motors* (1996) Amitava Basak
41. *Vector control and dynamics of a.c. drives* (1996) D. W. Novotny and T. A. Lipo
42. *Sensorless vector and direct torque control* (1998) Peter Vas

Sensorless Vector and Direct Torque Control

Peter Vas

Professor of Electrical Engineering
University of Aberdeen

OXFORD NEW YORK TOKYO
OXFORD UNIVERSITY PRESS

This book has been printed digitally and produced in a standard specification in order to ensure its continuing availability

OXFORD
UNIVERSITY PRESS

Great Clarendon Street, Oxford OX2 6DP

Oxford University Press is a department of the University of Oxford.
It furthers the University's objective of excellence in research, scholarship, and education by publishing worldwide in

Oxford New York

Auckland Bangkok Buenos Aires Cape Town Chennai
Dar es Salaam Delhi Hong Kong Istanbul Karachi Kolkata
Kuala Lumpur Madrid Melbourne Mexico City Mumbai Nairobi
Sáo Paulo Shanghai Taipei Tokyo Toronto

Published in the United States
by Oxford University Press Inc., New York

Reprinted 2003

ISBN 0-19-856465-1

Printed in Great Britain by

Antony Rowe Ltd., Eastbourne

Dedicated to my wife Zsuzsanna, my sons Peter and Victor
and also
to the memory of my father and identical twin brother

Foreword

Professor Peter Lawrenson FEng, FRS

It is a pleasure to welcome this new book from Professor Vas, and I congratulate him on his immense grasp of his subject.

Everyone involved in or with the subject area of this book, whether primarily intellectually, or in the advancement of practical engineering capability or for commercial advantage, knows that the last few years have been particularly fast changing and exciting – and the future promises to be no less so. The principal enabling influence in this has been the striking progress in electronic components, firstly in the power switches, with wide ranging quality improvements, including speed of operation, ease of use, thermal and mechanical robustness and packaging, and secondly in the intelligence level components, particularly ASICs and digital signal processors, with their massive increases in speed and power and with major reductions in cost.

These developments, together with a steady build-up of experience and understanding, have led to significant all-round improvements in the basic performance of electronically-fed motor drives; and they have opened the way to more modern and sophisticated means of controlling these drives – and the larger systems in which they are embedded. Additionally, newer motor types, some using permanent magnets, and particularly others based on reluctance principles, have started to make an impact following their successful introduction into some specially large and demanding markets. At the operational level these advances have led to more efficient use of energy and resources, to products and services of higher quality, to lower costs, and to improved safety.

The general pattern of these changes may be illuminated by mention of four specific features:

(i) The basic all-round performance capabilities of induction motor drives has been advanced significantly, not least in robustness and dynamic capability. One result is that the induction motor is now increasingly displacing the d.c. motor, the historic benchmark for demanding control applications.

(ii) Cost reductions and performance improvements are opening the way to using electronically-controlled drives, with wide-ranging advantages, in situations where previously only simple single-speed motors were viable. Important examples are in the massive markets for household appliances and automotive auxiliaries, traditionally conceived as supporting only the simplest, low cost motors.

(iii) The introduction of electronic controllers able to supply different types of motor and, more generally, the development of drives, not least of switched reluctance type, with a single motor type and a single controller type, able to cover widely different duties previously requiring separate designs and possibly different types of motor and controller. The significance of some such truly 'universal machine', conceptually and commercially (through aggregation of production), is great.

(iv) Integration and design, not only of the electronics, but of the motor with the electronics and, indeed, of the whole drive with the equipment or system of which it is a part.

There have been many advances over the past five years, and the challenge of understanding, modelling, designing and successfully applying these modern types of motor drives and their containing systems is substantial. It embraces, of necessity, the electromechanical and electromagnetic functions of the motor, the electronic power switching, a full range of traditional and modern control strategies, and the now extensive computing, communication and software requirements of the drive and the system.

From his extensive personal researches and from his wide range of contacts, Peter Vas is excellently equipped to meet this challenge and, indeed, has already so demonstrated in his three earlier books. This new book is in one sense a replacement of his 1990 *Vector control of ac machines* (second edition now out of print), but it is very much more than that. In structure and style it is very similar, using space-phasors to model and unify the treatment of the motors, proceeding with a control theory view of the overall drive systems, but including numerous helpful comments on the relationship to, and implementation of, the physical realisations, and leading for a vast number of drives to 'solution' equations and block diagrams in 'user ready' form. But the range of this book is vastly increased. It includes most of the earlier book and is improved generally and expanded in major ways (to more than twice the size) by the incorporation of significant new material on (a) direct-torque-control techniques; (b) a variety of 'sensorless' techniques; (c) artificial-intelligence-based systems more generally, including a chapter on analysis and estimators; (d) a further new chapter on switched reluctance motors; and (e) another on self-commissioning. The book is therefore a major contribution to the literature. Its somewhat encyclopedic nature may deter the casual reader, and it will not be accessible to most undergraduate readers until their third year of study, but those engaged in its field – in research, development, design or application, and whether in industry or university – will undoubtedly wish to study it closely, and they are likely to find themselves returning to it often.

Preface

The first book of the present author, *Vector control of a.c. machines*, was published in 1990 by Oxford University Press, and the manuscript was completed in 1989. The book was based on the important publications available at that time and also on the results obtained by the author, who was the first in the UK to perform research in the field of vector-controlled drives. This was the first book in the literature which was fully devoted to vector-controlled drives (which are one particular type of instantaneous torque-controlled drive) and which gave a unified treatment of the three main types of vector control: stator-flux-oriented, rotor-flux-oriented, and magnetizing-flux-oriented control of a.c. machines. It also incorporated new material on the vector control of saturated a.c. machines, and discussed various parameter determination techniques. The book was well received and according to the publisher was a best seller of its type. In 1995 the second impression was published (the printing errors of the first edition have been corrected), but no major changes were made.

Based on the success of the first edition of *Vector control of a.c. machines*, the author has published *Electrical machines and drives: a space-vector theory approach* (Oxford University Press 1992), which contains the space-vector analysis and simulation of not only vector drives but many other types of drives as well. This has found use in industry and universities worldwide. However, since the design, analysis, simulation, and implementation of drives relies heavily on machines parameters, a third book was also published by the author: *Parameter estimation, condition monitoring and diagnosis of electrical machines* (Oxford University Press 1993). This book contains numerous on-line and off-line parameter estimation techniques and real-time implementation issues and, where possible, it uses space-vector theory. However, due to recent trends and developments the need for the present book has also emerged, as is discussed below.

In recent years vector-controlled a.c. drives have become increasingly popular for many high-performance applications in which, traditionally, only d.c. drives were employed. These include drive applications for machine tools, spindles, rolling mills, high-speed elevators, mine winders, etc., where control of speed, position, electromagnetic torque, etc. is required. Due to rapid advances in power electronics, microprocessors, and digital signal processors, further replacement of d.c. drives is expected in various applications, and fully digital drives will become increasingly competitive. Although d.c. drives are manufactured at a declining rate, and further decline is expected, they are not dead, and some manufacturers are even marketing some improved d.c. drives. Recently various speed and position sensorless vector drives have emerged (these are usually referred to as 'sensorless drives', although these drives still contain some sensors, e.g. current sensors) and at least two large manufacturers have introduced high-performance drives which incorporate artificial intelligence. Although several excellent books have been published since 1990 which discuss vector control, none of these discusses the various forms of sensorless control techniques and latest trends and developments (e.g. sensorless control, intelligent control, etc.). Therefore it was decided to publish the present book, which discusses the latest developments as well. Furthermore, since in 1995 a major manufacturer marketed the first industrial version of direct-torque-controlled induction motor drive, which is also a torque-controlled drive (some publications even categorize it as a specific type of vector drive, e.g. it is sometimes referred to as the torque-vector-controlled drive), it was decided to discuss also various types of direct-torque-controlled drives in the present book. For this purpose again a unified treatment is given by using space-vector theory and emphasis is laid on detailed mathematical and physical presentations.

Pioneering work on switched reluctance motors has been performed in the UK by Professor P. J. Lawrenson and Professor T. J. E. Miller. Recently switched reluctance motors (SRMs) have gained in interest, mainly due to the simple technology involved. Furthermore, they have high efficiency over a wide speed range and they require a reduced number of switching devices. They also have superior fault tolerance characteristics. However, most of the difficulty in understanding the operation and design of switched reluctance motor drives originates from the fact that the SRM is a doubly salient, highly non-linear machine. Because of their importance, switched reluctance motor drives have also been included in the present book. It is expected that these drives will find an increasing number of applications in the future.

There are many excellent publications in the literature where a fuzzy-logic or artificial-neutral-network-based controller (e.g. speed controller) is used in a drive system, but in most cases such a controller merely replaces an existing classical one (e.g. a PI controller). The main benefits of this replacement could be improved performance and reduced tuning effort. However, in addition to replacing conventional controllers by artificial-intelligence-based controllers, there are many other possibilities for the application of artificial intelligence in drives. For example, in contrast to the conventional steady-state and dynamic analysis of electrical machines and drives, which is based on mathematical models (e.g. space-vector models of a.c. machines and drives), it is possible to perform such an analysis without using any mathematical models. For this purpose artificial intelligence (AI)-based techniques (artificial neutral networks, fuzzy logic, fuzzy-neutral networks, etc.) can also be used. Such an approach has many advantages over the conventional techniques (e.g. it does not rely on the many simplifying assumptions related to the conventional mathematical models, it can be robust to parameter variations, it offers the possibility of reduced computation times, etc.). This approach can be used not only for simulation purposes but also for implementing AI-based virtual speed, position, flux, torque, etc. 'sensors'. The transient simulation of a.c. and d.c. machines using AI-based techniques is also introduced in the present book, and it is believed that this is the first time that the details of such simulation results have been presented. Furthermore, AI-based virtual (rotor speed, electromagnetic torque, etc.) sensor implementations are also discussed. It is believed that such AI-based techniques will find increasing industrial applications in the future.

Comparing this book with the first by the author, *Vector control of a.c. machines*, the completely new sections in this book are as follows:

- General introduction to various torque-controlled drives; comparison of vector and direct torque control techniques
- Vector control of sensorless permanent magnet synchronous motor drives
- New sections relating to the vector control of synchronous reluctance motors, including sensorless synchronous reluctance motor drives
- New flux models for induction motors and synchronous motors
- Sensorless vector control techniques for induction motors
- Direct torque control of VSI- and CSI-fed induction motors, including sensorless drives
- Direct torque control of VSI- and CSI-fed electrically excited, permanent-magnet synchronous and synchronous reluctance motors, including sensorless drives
- Switched reluctance motor drives; torque control of switched reluctance motors
- Artificial-intelligence-based drive control, including simulation of electrical machines and drives using artificial neural networks, fuzzy logic, and fuzzy-neural networks

The reference list given at the end of each chapter attempts to represent the state-of-the-art up to December 1996, when the manuscript was completed. It was the author's aim to provide a list of the most important publications, but in view of the continually growing number of excellent publications it is very difficult to provide a complete list.

Aberdeen, UK
December 1996

P. V.

Acknowledgements

The author wishes to acknowledge the continuous support and help of a former colleague, Dr J. E. Brown, starting from the time the author was at the University of Newcastle upon Tyne, UK. The author started to work on instantaneous torque-controlled drives at the University of Newcastle upon Tyne more than 20 years ago. The author also wishes to thank his colleague, Dr A. F. Stronach, at the Department of Engineering, University of Aberdeen, for his help and stimulating discussions involving drives with intelligent control.

Acknowledgements are also due to Ferranti (UK), NEI Electronics (UK) (now part of Rolls Royce Industrial Power Group), ASEA (now ABB) (Sweden), and ABB Industry Oy (Finland) for their support and help during various stages of the research work performed by the author, to Professor Ueda for his help, and to Professor K. Ohnishi, Professor P. J. Lawrenson, and Professor J. Robert for their earlier help. Thanks are also due to Professor H. Grotstollen, Professor A. Vagati, and Professor W. Leonhard who have provided the author with some valuable material. The author also wishes to acknowledge the strong support and help of Dr W. Drury, Technical Director, Control Techniques plc (UK) and Dr L. Taufiq, Chief Engineer, GEC Traction (UK). The support of Dr S. Beierke, Director of Motion Control, Texas Instruments (Germany), EPSRC (UK), and the University of Aberdeen is greatly appreciated.

Furthermore, the author wishes to thank various universities and organisations for inviting him to deliver courses, seminars, and tutorials on different aspects of high-performance drives: these have considerably helped the author to write the present book. It is not possible to give a full list, but thanks are also due to the University of Tampere (Finland), University of Lappenranta (Finland), University of Palermo (Italy), University of Padova (Italy), University of Milano (Italy), University of Seville (Spain), PCIM (Germany), IEE, EPE, etc.

The author wishes to emphasize that many engineers have contributed to this book through their valuable publications, which are referenced throughout.

Expressions of gratitude and apology are directed to the author's wife, Zsuzsanna, and children Peter and Victor, who patiently endured the long working hours dedicated to this book.

Finally the author wishes to thank the staff at Oxford University Press for their effective cooperation, and special thanks are given to Mr R. Lawrence, Engineering Editor. Suggestions for future improvements will be gratefully received by the author (e.g. by e-mail: p.vas@eng.abdn.ac.uk).

Contents

Glossary of symbols

a	complex spatial operator ($e^{j2\pi/3}$)	
$\mathbf{C}_2$	commutator transformation matrix	
f	magnetomotive force (m.m.f.)	At
i_a	instantaneous value of the armature current	A
i_D	instantaneous value of the d.c. link current	A
$\bar{i}_{ms}$ $(\bar{i}_{ms})$	space phasor of the stator magnetizing currents expressed in the magnetizing-flux-oriented reference frame	A
$\bar{i}_{mr}$ $(\bar{i}_{mr})$	space phasor of the rotor magnetizing currents expressed in the magnetizing-flux-oriented reference frame	A
$\bar{i}_r$ $(\bar{i}_r)$	space phasor of the rotor currents expressed in the rotor reference frame	A
$\bar{i}_r'$ $(\bar{i}_r')$	space phasor of the rotor currents expressed in the stator reference frame	A
$\bar{i}_{rg}$ $(\bar{i}_{rg})$	space phasor of the rotor currents expressed in the general reference frame	A
$\bar{i}_{rm}$ $(\bar{i}_{rm})$	space phasor of the rotor currents expressed in the magnetizing-flux-oriented reference frame	A
$\bar{i}_{r\psi s}$ $(\bar{i}_{r\psi s})$	space phasor of the rotor currents expressed in the stator-flux-oriented reference frame	A
i_{ra}, i_{rb}, i_{rc}	instantaneous values of the rotor currents in rotor phases ra, rb, rc respectively	A
i_{rd}, i_{rq}	instantaneous values of direct- and quadrature-axis rotor currents expressed in the stator reference frame	A
$i_{r\alpha}$, $i_{r\beta}$	instantaneous values of direct- and quadrature-axis rotor current components respectively and expressed in the rotor reference frame	A
i_{rx}, i_{ry}	instantaneous values of the direct- and quadrature-axis rotor current components respectively and expressed in the general or special reference frames	A
$\bar{i}_s$ $(\bar{i}_s)$	space phasor of the stator current expressed in the stator reference frame	A
$\bar{i}_s'$ $(\bar{i}_s')$	space phasor of the stator currents expressed in the rotor reference frame	A
$\bar{i}_{sg}$ $(\bar{i}_{sg})$	space phasor of the stator currents expressed in the general reference frame	A
$\bar{i}_{sm}$ $(\bar{i}_{sm})$	space phasor of the stator currents expressed in the magnetizing-flux-oriented reference frame	A
$\bar{i}_{s\psi r}$ $(\bar{i}_{s\psi r})$	space phasor of the stator currents expressed in the rotor-flux-oriented reference frame	A

$\bar{i}_{s\psi s}$ ($\bar{i}_{s\psi s}$)	space phasor of the stator currents expressed in the stator-flux-oriented reference frame	A
i_{sA}, i_{sB}, i_{sC}	instantaneous values of the stator currents in stator phases sA, sB, sC respectively	A
i_{sD}, i_{sQ}	instantaneous values of direct- and quadrature-axis stator current components respectively and expressed in the stationary reference frame	A
i_{sd}, i_{sq}	instantaneous values of the direct- and quadrature-axis stator current components respectively and expressed in the rotor reference frame	A
i_{sx}, i_{sy}	instantaneous values of the direct- and quadrature-axis stator current components respectively and expressed in the general or special reference frames (flux and torque producing stator current components respectively)	A
i_{s0}	instantaneous value of zero-sequence stator current component	A
J	polar moment of inertia	kgm^2
L	dynamic inductance	H
L_{DQ}	cross-magnetizing coupling inductance	H
L_F	inductance of filter	H
L_m	magnetizing inductance	H
L_s, L_{sl}	self- and leakage inductances of the stator respectively	H
L_r, L_{rl}	self- and leakage inductances of the rotor respectively	H
L'_s, L'_r	stator and rotor transient inductances respectively	H
$p = \mathrm{d}/\mathrm{d}t$	differential operator	
p_s, p_r	instantaneous powers of the stator and rotor respectively	W
P	number of pole-pairs	
R_F	resistance of filter	Ohm
R_s, R_r	resistances of a stator and rotor phase winding respectively	Ohm
t	time	s
t_e	instantaneous value of the electromagnetic torque	Nm
t_1	load torque	Nm
T	time lag	s
T_s, T_r	stator and rotor time constants respectively	s
T'_s, T'_r	stator and rotor transient time constants respectively	s
u_D	instantaneous output voltage of rectifier	V
u_{dx}, u_{dy}	instantaneous values of the direct- and quadrature-axis decoupling voltage components expressed in the special reference frames	V
$\bar{u}_s$	space phasor of the stator voltages expressed in the stator reference frame	V
$\bar{u}_{sg}$	space phasor of the stator voltages expressed in the general reference frame	V

$\bar{u}_{sm}$	space phasor of the stator voltages expressed in the magnetizing-flux-oriented reference frame	V
$\bar{u}_{s\psi s}$	space phasor of the stator voltages expressed in the stator-flux-oriented reference frame	V
$\bar{u}_{s\psi r}$	space phasor of the stator voltages expressed in the rotor-flux-oriented reference frame	V
u_{sA}, u_{sB}, u_{sC}	instantaneous values of the stator voltages in phases sA, sB, sC respectively	V
u_{sD}, u_{sQ}	instantaneous values of direct- and quadrature-axis stator voltage components respectively and expressed in the stationary reference frame	V
u_{sx}, u_{sy}	instantaneous values of the direct- and quadrature-axis stator voltage components respectively and expressed in the general or special reference frames	V
$\bar{u}_r$	space phasor of the rotor voltages expressed in the rotor reference frame	V
$\bar{u}_r'$	space phasor of the rotor voltages expressed in the stator reference frame	V
$\bar{u}_{rg}$	space phasor of the rotor voltages expressed in the general reference frame	V
$\bar{u}_{r\psi s}$	space phasor of the rotor voltages expressed in the stator-flux-oriented reference frame	V
u_{rd}, u_{rq}	instantaneous values of the direct- and quadrature-axis rotor voltages respectively and expressed in the stator reference frame	V
u_{rx}, u_{ry}	instantaneous values of the direct- and quadrature-axis rotor voltages respectively and expressed in the general or special reference frames	V
$u_{r\alpha}$, $u_{r\beta}$	instantaneous values of the direct- and quadrature-axis rotor voltages respectively and expressed in the rotor reference frame	V
u_{r0}	instantaneous value of the zero-sequence rotor voltage	V
W_{mech}	instantaneous value of mechanical output energy	J
$z = e^{sT}$	discrete Laplace variable	
δ	load angle	rad
θ_r	rotor angle	rad
μ_m	phase angle of the magnetizing flux linkage space phasor with respect to the direct-axis of the stator reference frame	rad
ρ_r	phase angle of the rotor flux linkage space phasor with respect to the direct-axis of the stator reference frame	rad
ρ_s	phase angle of stator flux linkage space phasor with respect to the direct-axis of the stator reference frame	rad
σ	resultant leakage constant	
σ_r	rotor leakage constant	

ψ_f	excitation flux linkage	Wb
$\bar{\psi}'_m$	space phasor of magnetizing flux linkages expressed in the stator reference frame	Wb
$\bar{\psi}_{m\psi r}$	space phasor of magnetizing flux linkages expressed in the rotor-flux-oriented reference frame	Wb
$\bar{\psi}_{m\psi s}$	space phasor of magnetizing flux linkages expressed in the stator-flux-oriented reference frame	Wb
ψ_{md}, ψ_{mq}	instantaneous values of the direct- and quadrature-axis magnetizing flux linkage components expressed in the rotor reference frame	Wb
$\bar{\psi}_r$	space phasor of rotor flux linkages expressed in the rotor reference frame	Wb
$\bar{\psi}'_r$	space phasor of the rotor flux linkages expressed in the stator reference frame	Wb
$\bar{\psi}_{rg}$	space phasor of the rotor flux linkages expressed in the general reference frame	Wb
$\bar{\psi}_{r\psi r}$	space phasor of the rotor flux linkages expressed in the rotor-flux-oriented reference frame	Wb
ψ_{ra}, ψ_{rb}, ψ_{rc}	instantaneous values of the flux linkages in rotor phases ra, rb, rc respectively	Wb
ψ_{rx}, ψ_{ry}	instantaneous values of the direct- and quadrature-axis rotor flux linkage components respectively and expressed in the general or special reference frames	Wb
$\bar{\psi}_s$	space phasor of stator flux linkages expressed in the stator reference frame	Wb
$\bar{\psi}'_s$	space phasor of the stator flux linkages expressed in the rotor reference frame	Wb
$\bar{\psi}_{sg}$	space phasor of the stator flux linkages expressed in the general reference frame	Wb
$\bar{\psi}_{sm}$	space phasor of the stator flux linkages expressed in the magnetizing-flux-oriented reference frame	Wb
$\bar{\psi}_{s\psi s}$	space phasor of the stator flux linkages expressed in the stator-flux-oriented reference frame	Wb
ψ_{sA}, ψ_{sB}, ψ_{sC}	instantaneous values of stator flux linkages in stator phases sA, sB, sC respectively	Wb
ψ_{sd}, ψ_{sq}	instantaneous values of the direct- and quadrature-axis stator flux linkages expressed in the rotor reference frame	Wb
ψ_{sD}, ψ_{sQ}	instantaneous values of the direct- and quadrature-axis stator flux linkage components expressed in the stator reference frame	Wb
ψ_{sx}, ψ_{sy}	instantaneous values of the direct- and quadrature-axis stator flux linkage components respectively and expressed in the general or special reference frames	Wb
ω_m	angular speed of the magnetizing-flux-oriented reference frame	rad s^{-1}

ω_{mr}	angular speed of the rotor-flux-oriented reference frame	rad s^{-1}
ω_{ms}	angular speed of the stator-flux-oriented reference frame	rad s^{-1}
ω_r	angular rotor speed	rad s^{-1}
ω_{sl}	angular slip frequency	rad s^{-1}

Subscripts

A, B	upper and lower cages respectively
b	base
g	general reference frame
i	induced
m	magnetizing
n	normalized
r	rotor
ra, rb, rc	rotor phases
ref	reference
s	stator
sA, sB, sC	stator phases
x	direct-axis component in general reference frame or in special reference frames (fixed to the stator flux-linkage, rotor flux-linkage, or magnetizing flux-linkage space phasors respectively)
y	quadrature-axis component in general reference frame or in special reference frames (fixed to the stator flux-linkage, rotor flux-linkage, or magnetizing flux-linkage space phasors respectively)

Mathematical symbols

$\times$	cross vector product
$*$	complex conjugate

1 Introduction to torque-controlled drives

1.1 General introduction

In the past, d.c. motors were used extensively in areas where variable-speed operation was required, since their flux and torque could be controlled easily by the field and armature current. In particular, the separately excited d.c. motor has been used mainly for applications where there was a requirement of fast response and four-quadrant operation with high performance near zero speed. However, d.c. motors have certain disadvantages, which are due to the existence of the commutator and the brushes. That is, they require periodic maintenance; they cannot be used in explosive or corrosive environments and they have limited commutator capability under high-speed, high-voltage operational conditions. These problems can be overcome by the application of alternating-current motors, which can have simple and rugged structure, high maintainability and economy; they are also robust and immune to heavy overloading. Their small dimension compared with d.c. motors allows a.c. motors to be designed with substantially higher output ratings for low weight and low rotating mass.

Variable-speed a.c. drives have been used in the past to perform relatively undemanding roles in applications which preclude the use of d.c. motors, either because of the working environment or commutator limits. Because of the high cost of efficient, fast switching frequency static inverters, the lower cost of a.c. motors has also been a decisive economic factor in multi-motor systems. However, as a result of the progress in the field of power electronics, the continuing trend is towards cheaper and more effective power converters, and single motor a.c. drives compete favourably on a purely economic basis with the d.c. drives.

Among the various a.c. drive systems, those which contain the cage induction motor have a particular cost advantage. The cage motor is simple and rugged and is one of the cheapest machines available at all power ratings. Owing to their excellent control capabilities, variable speed drives incorporating a.c. motors and employing modern static converters and torque control can well compete with high-performance four-quadrant d.c. drives.

Vector control techniques incorporating fast microprocessors and DSPs have made possible the application of induction-motor and synchronous-motor drives for high-performance applications where traditionally only d.c. drives were applied. In the past such control techniques would have not been possible because of the complex hardware and software required to solve the complex control problem. As for d.c. machines, torque control in a.c. machines is achieved by controlling the motor currents. However, in contrast to a d.c. machine, in an a.c. machine, both the phase angle and the modulus of the current has to be controlled, or in other words, the current vector has to be controlled. This is the reason for the terminology 'vector control'. Furthermore, in d.c. machines, the orientation of the field flux and armature m.m.f. is fixed by the commutator and the brushes, while in a.c. machines the field flux and the spatial angle of the armature m.m.f. require external control. In the absence of this control, the spatial angles between the various fields in a.c. machines vary with the load and yield unwanted oscillating dynamic response. With vector control of a.c. machines, the torque- and flux-producing current components are decoupled and the transient response characteristics are similar to those of a separately excited d.c. machine, and the system will adapt to any load disturbances and/or reference value variations as fast as a d.c. machine. It is the primary aim of the present book to give a unified and detailed treatment of the various forms of instantaneous torque-controlled drives. Vector-controlled drives are one particular type of torque-controlled drive. The other type of high-performance torque-controlled drive is the so-called direct-torque-controlled drive. In this drive, direct torque control is achieved by direct and independent control of the flux linkages and electromagnetic torque through the selection of optimal inverter switching modes which give fast torque response, low inverter switching frequency, and low harmonic losses.

It is expected that with the rapid developments in the field of microelectronics, torque control of various types of a.c. machines will become a commonly used technique when, even though high dynamic performance is not required, servo-like high performance plays a secondary role to reliability and energy (efficiency) savings. It is possible to contribute to the energy savings by the application of intelligent control of the flux- and torque-producing components of the stator currents.

In the case of d.c. drives, the power circuits are relatively uniform and in most cases contain a line-commutated thyristor converter or a transistorized chopper for low power applications. However, for a.c. drives, there is much greater variety, due to the different types of converters (voltage source, current source, natural commutation, forced commutation, d.c. link, cycloconverter) which can be combined with various types of a.c. machines.

Following the early works of Blaschke and Hasse, and largely due to the pioneering work of Professor Leonhard, vector control of a.c. machines has become a powerful and frequently adopted technique worldwide. In recent years, numerous important contributions have been made in this field by contributors from many countries, including Canada, Germany, Italy, Japan, the UK, and the USA. Many industrial companies have marketed various forms of induction-motor and synchronous-motor drives using vector control. At present direct-torque-controlled drives are receiving great attention worldwide, although they were first introduced by German and Japanese researchers more than 10 years ago. Presently only one large manufacturer is marketing one form of direct-torque-controlled induction-motor drive.

In the present book the very wide field of torque control of smooth-air-gap and salient-pole a.c. machines is discussed. Four types of vector control are considered: rotor-oriented control, rotor-flux-oriented control, stator-flux-oriented control, and magnetizing-flux-oriented control. Great emphasis is laid on presenting a unified and detailed physical and mathematical treatment which relies on space-phasor theory. The book contains a step-by-step, physical and mathematical development of space-phasor theory and its applications to the various forms of vector control of a.c. machines. The theory developed also covers the operation of smooth-air-gap and salient-pole machines under non-linear magnetic conditions.

The book contains many novel features. The details of a very large number of 'sensorless' (speed and/or position sensorless) drive schemes for different types of permanent-magnet synchronous motors, synchronous reluctance motors, and induction motors are presented for the first time in a textbook. Furthermore, the mathematical and physical details of numerous direct torque control (DTC) schemes of synchronous motors and induction motors are also discussed. For this purpose four types of DTC schemes for a VSI-fed permanent-magnet synchronous motor are presented. The book also discusses two types of DTC schemes of a VSI-fed synchronous reluctance motor and also the DTC of a VSI-fed and a CSI-fed electrically excited synchronous motor. In addition, the DTC of a VSI-fed induction motor is also discussed and predictive and non-predictive schemes are also considered. However, the DTC scheme of a CSI-fed induction motor is also presented. Due to the increasing importance of switched reluctance motors (SRM), the fundamentals, main techniques of position-sensorless implementations, and the torque control of SRMs are also discussed. The book also covers some drive applications using artificial intelligence (fuzzy logic, artificial neural networks, fuzzy-neural networks). In addition, artificial-intelligence-based steady-state and transient analysis of electrical machines is also presented. Finally, self-commissioning of induction machines is discussed.

This book is intended to cover the needs of students, academics, and industrial users who require a deep understanding of the various aspects of torque control.

The book contains almost 220 figures and 530 references on the above topics. The author has attempted to give a most up-to-date list of references published before December 1996, when the complete manuscript was released for printing.

1.2 Basic torque control schemes; sensorless drives

There are basically two types of instantaneous electromagnetic torque-controlled a.c. drives (briefly, torque-controlled drives) used for high-performance applications: vector- and direct-torque-controlled (DTC) drives. Vector-controlled drives were introduced more than 20 years ago in Germany by Blaschke, Hasse, and Leonhard. They have achieved a high degree of maturity and have become increasingly popular in a wide range of applications. They have established a substantial and continually increasing worldwide market. Direct-torque-controlled drives were introduced in Japan by Takahashi and also in Germany by Depenbrock more than 10 years ago. However, so far only one form of a direct-torque-controlled induction-motor drive has been marketed by an industrial company, but it is expected that other manufacturers will soon follow.

In this section a brief description is given of the fundamentals of vector- and direct-torque-controlled drives. This will help the reader to have a good understanding of these drives and also to recognize the main differences between these two high-performance drive schemes. The required knowledge of space-vector theory (which is also referred to as space-phasor theory in the literature) will be kept to a minimum, but details of space-vector theory are described in Chapter 2. However, readers familiar with this theory can completely skip Chapter 2. Details of all the vector drives and direct-torque-controlled drives described in this section are given later in the book.

1.2.1 FUNDAMENTALS OF VECTOR DRIVES

1.2.1.1 D.C. machine control

Due to the stationary orthogonal field axes, the control structure of d.c. machines is relatively simple, but their mechanical construction is complicated. In a separately excited d.c. machine, the instantaneous electromagnetic torque, t_e, is proportional to the product of the field current, i_f (flux-producing current), and the armature current, i_a (torque-producing current),

$$t_e = c i_f i_a = c_1 \psi_f i_a. \tag{1.2-1}$$

In eqn (1.2-1), c and c_1 are constants and ψ_f is the field flux. The field flux can be established either by a stationary d.c. excited field winding, or by permanent magnets. Torque control can be achieved by varying the armature current, and quick torque response is obtained if the armature current is changed quickly and the field current (field flux) is constant. This principle can also be used for the instantaneous torque control of a.c. machines (both induction and synchronous, the latter can be of the electrically excited, reluctance, or the permanent-magnet excited type), as discussed below.

Conventional d.c. motor drives continue to take a large part of the variable-speed drive market. However, it is expected that this share will slowly decline. Similarly to a.c. drives, efforts are being made to reduce their costs and to increase their reliability and thus new and improved d.c. drives are being introduced by various manufacturers.

1.2.1.2 Induction machine control

Squirrel-cage induction machines are simple and rugged and are considered to be the 'workhorses' of industry. At present, induction motor drives dominate the world market. However, the control structure of an induction motor is complicated since the stator field is revolving, and further complications arise due to the fact that the rotor currents or rotor flux of a squirrel-cage induction motor cannot be directly monitored.

The mechanism of torque production in an a.c. machine and in a d.c. machine is similar. Unfortunately this similarity was not emphasized before the 1970s, and this is one of the reasons why the technique of vector control did not emerge earlier. The formulae given in many well-known textbooks on machine theory (which do not discuss space-vector theory) have also implied that, for the monitoring of the instantaneous electromagnetic torque of an induction machine, it is also necessary to monitor the rotor currents and the rotor position. Even in the 1980s some publications seemed to strengthen this false conception, which only arose because the complicated formulae derived for the expression of the instantaneous electromagnetic torque have not been simplified. However, by using fundamental physical laws and/or space-vector theory, it is easy to show that, similarly to the expression of the electromagnetic torque of a separately excited d.c. machine, the instantaneous electromagnetic torque of an induction motor can be expressed as the product of a flux-producing current and a torque-producing current, if a special, flux-oriented reference frame is used, i.e. if flux-oriented control is employed. In this case, the stator current components (which are expressed in the stationary reference frame) are transformed into a new rotating reference frame, which rotates together with a selected flux-linkage space vector. There are in general three possibilities for the selection of the flux-linkage vector, so that the chosen vector can be either the stator-flux-linkage vector, rotor-flux-linkage vector or magnetizing-flux-linkage vector. Hence the terminology: stator-flux-, rotor-flux-, and magnetizing-flux-oriented control. In these three cases the instantaneous electromagnetic torque can be expressed as follows:

$$t_e = c_{1s}|\bar{\psi}_s| i_{sy}^s \quad \text{for stator-flux-oriented control} \qquad (1.2\text{-}2)$$

$$t_e = c_{1r}|\bar{\psi}_r| i_{sy}^r \quad \text{for rotor-flux-oriented control} \qquad (1.2\text{-}3)$$

$$t_e = c_{1m}|\bar{\psi}_m| i_{sy}^m \quad \text{for magnetizing-flux-oriented control.} \qquad (1.2\text{-}4)$$

These expressions are similar to eqn (1.2-1), and for linear magnetic conditions c_{1s}, c_{1r}, c_{1m} are constants, $|\bar{\psi}_s|$, $|\bar{\psi}_r|$, and $|\bar{\psi}_m|$ are the modulus of the stator-, rotor-, and magnetizing-flux-linkage space vectors, respectively. Furthermore, the torque-producing stator currents in the stator-, rotor-, and magnetizing-flux-oriented reference frame are denoted by i_{sy}^s, i_{sy}^r, and i_{sy}^m respectively (the superscript indicates the special reference frame used). It should be noted that in the present book, the real and imaginary axes of a rotating reference frame fixed to a flux-linkage space vector are denoted by x and y, respectively. The torque-producing stator currents in eqns (1.2-2)–(1.2-4) take the role of the armature current in eqn (1.2-1). The application of eqn (1.2-3) will be shown in the drive scheme of Fig. 1.1(a) below. Equations (1.2-2)–(1.2-4) can be derived from a single equation, according to which the electromagnetic torque for an induction machine can be expressed as the cross vectorial product of the stator flux linkage and current space vectors (in every reference frame). The derivation is not given here, but the details are shown in Chapter 2. It follows from eqn (1.2-2) that when the stator-flux-linkage modulus is constant ($|\bar{\psi}_s|$=const.), and the torque-producing stator current is changed quickly, quick torque response is obtained. Similar considerations hold for eqns (1.2-3) and (1.2-4). This physical picture forms the basis of vector-controlled drives. It should be noted that for simplicity, later in the book, the superscripts in the notation of the torque-producing stator currents will be omitted, but it will always be made clear in which reference frame the stator currents are established. Furthermore, it is important to note that the $t_e = c|\bar{\psi}|i_{sy}$ type of expression for the instantaneous electromagnetic torque also holds for smooth-air-gap synchronous motors, as discussed below, and this equation is utilized in the vector control implementations of smooth-air-gap synchronous motors.

Since the expression for the electromagnetic torque contains the transformed stator currents, it is obvious that in a vector-controlled drive, the stator currents must be transformed into the required special reference frame (e.g. for a stator-flux-oriented controlled drive, the stator current components in the stationary reference frame must be transformed into the stator current components in the stator-flux-oriented reference frame). It is a common feature of all vector-controlled drives that the modulus and phase angle of the a.c. excitation are controlled, and this is the reason why this type of control is called vector control. However, since the reference frame is aligned with the selected flux-linkage space vector (e.g. stator-flux-linkage space vector), the transformation contains the angle of the flux-linkage space vector (e.g. stator flux angle, with respect to the real axis of the stationary reference frame). Thus, the implementation requires the flux angle and also the flux-linkage modulus, and a major task for an implementation is to have an accurate flux-linkage estimation. As shown below in Section 1.2.2, this is also a major task in a drive employing direct torque control.

When the selected flux-linkage is the stator flux-linkage, it can easily be obtained by using terminal voltages and currents, since in the stationary reference frame, the stator flux-linkage is the integral of the terminal voltage minus the ohmic stator loss, $\bar{\psi}_s = \int(\bar{u}_s - R_s\bar{i}_s)\,dt$ (where $\bar{\psi}_s$, $\bar{u}_s$, and $\bar{i}_s$ are the space vectors of the stator flux-linkage, stator voltage, and stator current respectively and they are expressed in the stationary reference frame). However, at low frequencies some problems arise when this technique is applied, since the stator voltages become very small and the ohmic voltage drops become dominant, requiring very accurate knowledge of the stator resistance and very accurate integration. The stator resistance can vary due to temperature changes; this effect can also be taken into consideration by using a thermal model of the machine. Drifts and offsets can greatly influence the precision of integration. The overall accuracy of the estimated flux-linkage vector will also depend on the accuracy of the monitored voltages and currents. At low frequencies more sophisticated techniques must be used to obtain the stator flux-linkages. The monitoring of the stator voltages can be eliminated in a voltage-source inverter-fed machine, since they can be reconstructed by using the monitored value of the d.c. link voltage and also the switching states of the switching devices (of the inverter). The components of the stator flux-linkage space vector can also be used to obtain the rotor speed signal in a speed sensorless drive, i.e. by utilizing the speed of the stator flux-linkage space vector (which is equal to the rate of change of the angle of the stator flux-linkage space vector).

When the selected flux-linkage space vector is the rotor flux-linkage vector ($\bar{\psi}'_r$), it is also possible to obtain it from terminal quantities by first obtaining the stator flux-linkage vector ($\bar{\psi}_s$), and then by applying the appropriate modifications, $\bar{\psi}'_r=(L_r/L_m)(\bar{\psi}_s-L'_s\bar{i}_s)$, where L_r, L_m, and L'_s are the rotor, magnetizing, and stator transient inductances respectively and $\bar{i}_s$ is the space vector of the stator currents expressed in the stationary reference frame. Thus an accurate knowledge of the machine inductances is required and this is a difficult problem, since they can vary with the saturation level, and the conventional tests (no-load and blocked rotor tests) do not give accurate values for an induction machine with closed rotor slots. Furthermore, at low frequencies the same problems arise as with the stator flux-linkage estimation, and to avoid these, various rotor flux models can be used. Some of these models also use the rotor speed (or rotor position), which can be monitored, but in 'sensorless' applications, the speed or position is not directly monitored, but is estimated by using advanced control (observers, intelligent systems, etc.) or other techniques.

When the selected flux-linkage is the magnetizing flux-linkage space vector, in principle it is possible to obtain it by using Hall sensors or search coils, and such schemes were used in the first type of vector-controlled drives. However, Hall sensors increase the cost and size of the motor. They are also temperature-sensitive and limit the operation of the motor (i.e. below 75 °C). Schemes using Hall sensors and search coils have been outdated by schemes in which the magnetizing flux-linkage vector is obtained by using monitored values of the terminal voltages and currents. In principle, the magnetizing flux-linkage vector ($\bar{\psi}_m$) can be obtained similarly to the stator flux-linkage vector from terminal quantities, since it is equal to the stator flux-linkage vector minus the stator leakage flux vector, the latter being equal to the product of the stator leakage inductance and the stator current vector ($\bar{\psi}_m=\bar{\psi}_s-L_{s1}\bar{i}_s$). It can also be obtained by using other types of machine models, which may also contain the rotor speed or rotor position, in a non-sensorless implementation. Furthermore, it can be obtained by using special sensorless techniques. Details of various sensorless drive implementations are given in this book.

There are basically two different types of vector control techniques: direct and indirect techniques. The direct implementation relies on the direct measurement or estimation of the rotor-, stator-, or magnetizing-flux-linkage vector amplitude and position. The indirect method uses a machine model, e.g. for rotor-flux-oriented control, it utilizes the inherent slip relation. In contrast to direct methods, the indirect methods are highly dependent on machine parameters. Traditional direct vector control schemes use search coils, tapped stator windings, or Hall-effect sensors for flux sensing. This introduces limitations due to machine structural and thermal requirements. Many applications use indirect schemes, since these have relatively simpler hardware and better overall performance at low frequencies, but since these contain various machine parameters, which may vary with temperature, saturation level, and frequency, various parameter adaptation schemes have been developed. These include self-tuning controller applications, Model Reference Adaptive System (MRAS) applications, applications of observers, applications of intelligent controllers (fuzzy, neuro, fuzzy-neuro controllers), etc. To obtain a solution, in the development of the theory it is sometimes assumed that the mechanical time-constant is much greater than the electrical time-constants, but this becomes an invalid assumption if the machine inertia is low. If incorrect modulus and angle of the flux-linkage space vector are used in a vector control scheme, then flux and torque decoupling is lost and the transient and steady-state responses are degraded. Low frequency response, speed oscillations, and loss of input–output torque linearity are major consequences of detuned operation, together with decreased drive efficiency.

To illustrate a vector-controlled drive, Fig. 1.1(a) shows the schematic drive scheme of an induction motor with impressed stator currents employing direct

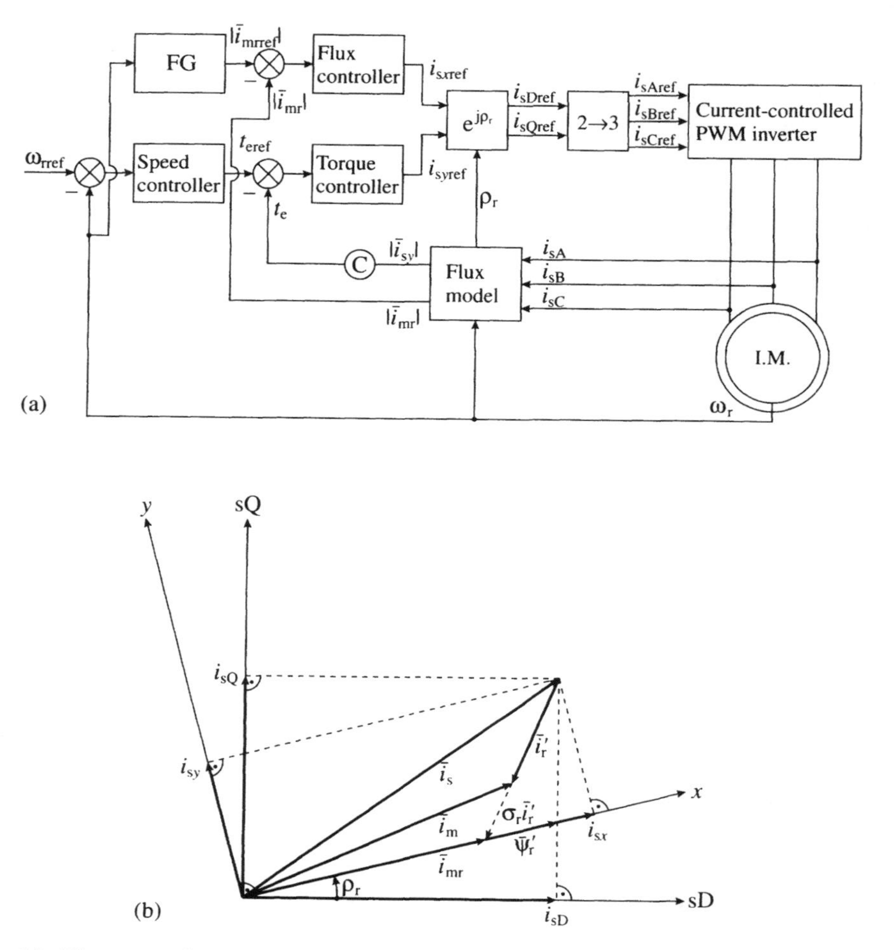

Fig. 1.1. Direct rotor-flux-oriented control of induction motor with impressed stator currents. (a) Drive scheme; (b) vector diagram.

rotor-flux-oriented control. In the drive scheme of Fig. 1.1(a) the expression for the electromagnetic torque given by eqn (1.2-3) has been used, but as mentioned above, the notation has been simplified, thus $c=c_{1r}$, $i_{sy}=i_{sy}^{r}$ and hence $t_e=c_{1r}|\bar{\psi}_r|i_{sy}$. Furthermore, the modulus of the rotor flux-linkage space vector can be expressed as $|\bar{\psi}_r|=L_m|\bar{i}_{mr}|$, where L_m is the magnetizing inductance of the induction machine and $|\bar{i}_{mr}|$ is the modulus of the rotor magnetizing-current space vector shown in Fig. 1.1(b). The flux- and torque-producing stator current references (i_{sxref}, i_{syref}) are obtained on the outputs of the flux and torque controllers respectively, and are then transformed to the stator currents in the stationary reference frame (i_{sDref}, i_{sQref}), where all the current components are also shown in Fig. 1.1(b). For this purpose $\bar{i}_{sref}=i_{sDref}+ji_{sQref}=(i_{sxref}+ji_{syref})\exp(j\rho_r)$ is used, where ρ_r is the position of the rotor flux-linkage space vector (with respect to the real axis of the stationary reference frame). For illustration purposes Fig. 1.1(b) shows all the stator current components and also the angle ρ_r. In the drive scheme of Fig. 1.1(a), the torque-producing stator current (i_{sy}) and the modulus of the rotor magnetizing current are obtained on the output of a rotor flux-linkage model. In the simple case shown, the rotor flux-linkage model uses the monitored rotor speed and also the monitored stator currents as inputs. Such a model can be obtained by considering the rotor space-vector voltage equation of the induction motor (see Section 4.1.1.4). This yields the modulus and angle of the rotor flux-linkage space vector (hence the name flux model), but since the rotor flux modulus is the product of L_m and $|\bar{i}_{mr}|$, thus $|\bar{i}_{mr}|$ can also be obtained. Furthermore, since the position of the rotor flux-linkage space vector (ρ_r) is also determined in the flux model, and since the stator currents are also inputs to the flux model (in practice only two stator currents are required), the torque-producing stator current i_{sy} can also be obtained. For this purpose the relationship between the stator current space vector expressed in the rotor-flux-oriented reference frame ($\bar{i}_s'$) and the stator current space vector expressed in the stationary reference frame ($\bar{i}_s$) is considered; thus $\bar{i}_s'=i_{sx}+ji_{sy}=\bar{i}_s\exp(-j\rho_r)$, where $\bar{i}_s$ is the space vector of the stator currents in the stationary reference frame $\bar{i}_s=(2/3)(i_{sA}+ai_{sB}+a^2i_{sC})$. In Fig. 1.1(a), FG is a function generator, from the output of which the modulus of the rotor magnetizing-current reference ($|\bar{i}_{mrref}|$) is obtained, and its input is the monitored speed. This can be used for field-weakening purposes. When the conventional field-weakening technique is used, then below base speed $|\bar{i}_{mrref}|$ is constant, and above base speed it is reduced, i.e. it is inversely proportional to the rotor speed (see Section 6.1.4.1). However, when a non-conventional field-weakening technique is used, then FG is implemented in a different way (see Section 6.1.4.2). In Fig. 1.1(a) the electromagnetic torque reference is obtained from the output of the speed controller. This can be a conventional PI controller, a fuzzy logic controller, etc. It can be seen that the drive scheme shown in Fig. 1.1(a) also contains two current controllers and i_{sy} is obtained (in the flux model) by utilizing the transformation $\exp(-j\rho_r)$, but the transformation $\exp(j\rho_r)$ is also used (to obtain i_{sDref}, i_{sQref}). However, it will be shown below in Section 1.2.2 that the current controllers and these transformations are eliminated when direct torque control is used.

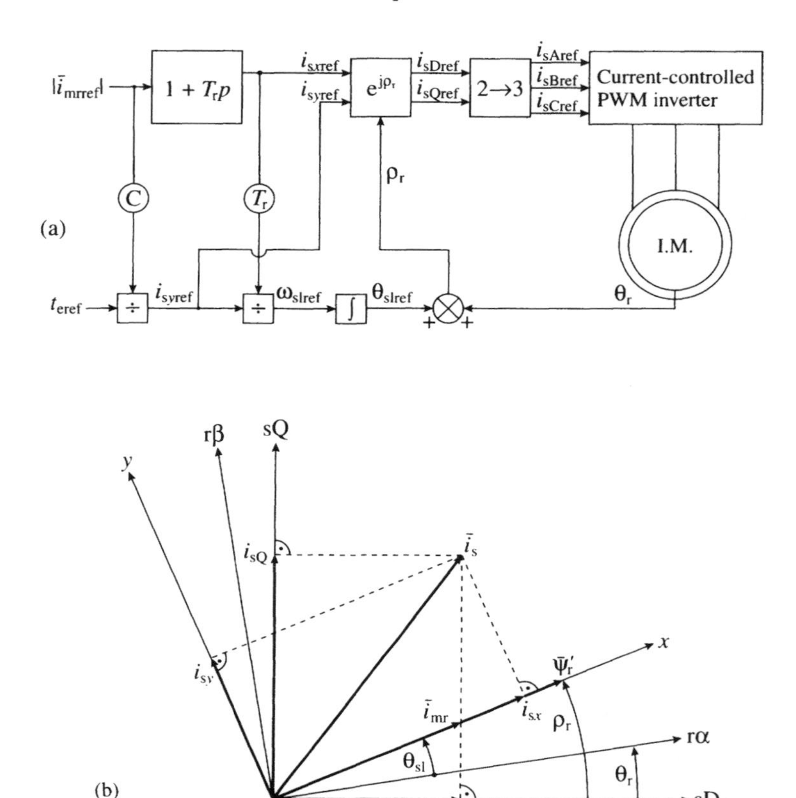

Fig. 1.2. Indirect rotor-flux-oriented control of induction motor with impressed stator currents. (a) Drive scheme; (b) vector diagram.

Figure 1.2 shows the schematic drive scheme of an induction motor with impressed stator currents employing indirect rotor-flux-oriented control. In Fig. 1.2(a) the torque- and flux-producing stator current references (i_{sxref}, i_{syref}) and also the angular slip frequency reference (ω_{slref}) are generated from the reference electromagnetic torque (t_{eref}) and reference rotor magnetizing-current modulus ($|\bar{i}_{mrref}|$). The rotor position (θ_r) is monitored, and it is added to the reference value of the slip angle (θ_{slref}), to yield the position of the rotor magnetizing-current space vector (or rotor flux-linkage space vector). These angles are also shown in Fig. 1.2(b). The transformation of i_{sxref} and i_{syref} into the three-phase stator reference currents is similar to that shown in the direct scheme above (see Fig. 1.1(a)). The expressions used for the estimation of i_{sxref} and ω_{slref} follow from the rotor voltage equation of the induction machine. The details are shown in Section 4.1.2; these expressions are $\omega_{slref}=i_{syref}/(T_r i_{sxref})$ and $i_{sxref}=(1+T_r p)|\bar{i}_{mrref}|$ [see eqns (4.1-131) and (4.1-135)] where T_r is the rotor time-constant and $p=\mathrm{d}/\mathrm{d}t$.

1.2.1.3 Permanent-magnet synchronous machine and synchronous reluctance machine control

Inverter-fed synchronous motors are widely used in high-performance variable-speed drive systems. If the synchronous machine is supplied by a current-controlled voltage-source PWM inverter, then the stator currents are decided by the reference speed or reference electromagnetic torque and the inverter drives the synchronous motor so that the instantaneous stator currents follow their reference values. For high-performance drives it is possible to use various rotor configurations: rotors with permanent magnets, reluctance type, and electrically excited rotors.

There are basically three types of permanent-magnet synchronous machine (the permanent magnets are on the rotor). In the permanent-magnet synchronous machine with surface-mounted magnets, the (polar) magnets are located on the surface of the rotor and the machine behaves like a smooth-air-gap machine (the direct- and quadrature-axis synchronous inductances are equal, $L_{sd}=L_{sq}$) and there is only magnet torque produced. In the permanent-magnet synchronous machine with inset magnets, the (subpolar) magnets are inset into the rotor (directly under the rotor surface), and this is a salient-pole machine, ($L_{sd}\neq L_{sq}$) and both magnet torque and reluctance torque are produced. In the permanent-magnet synchronous machine with buried magnets (interior magnets), the magnets are buried in the rotor, ($L_{sd}\neq L_{sq}$) and again both magnet and reluctance torques are produced. There are basically three types of permanent magnet machines with buried magnets, depending on how the magnets are buried into the rotor: the magnets can be radially placed, axially placed, or they can be inclined.

A synchronous machine with reluctance rotor (SYRM) can have various rotor configurations. In earlier constructions, rotor saliency was achieved by removing certain teeth from the rotor of a conventional squirrel cage. Such synchronous reluctance machines with low output power have been used for a long time and their inferior performance combined with their relatively high price have resulted in a limited use. However, as a result of recent developments, more reliable and robust new constructions exist; these have basically three types of rotors: segmential, flux barrier, and axially laminated rotors. In the SYRM with segmental rotor, saliency ratios (L_{sd}/L_{sq}) of 6–7 have been obtained. If the number of rotor segments is very large, then a distributed anisotropic structure is obtained, which is similar to the various axially laminated structures used in the past. By using multiple segmental structures, the saliency ratio can be increased. In the SYRM with axially laminated rotor, the rotor is made of conventional axial laminations bent into U or V shapes and stacked in the radial direction. With this structure it is possible to produce very high saliency ratios, and saliency ratios of 9–12 have been obtained. This also leads to fast torque responses. A synchronous reluctance machine is a salient-pole machine which produces reluctance torque.

Permanent-magnet machines

In the past, ferrite and cobalt–samarium magnets have been used in permanent-magnet machines. However, recently, magnetic materials with higher energy density (e.g. 35 $\mathrm{J\,cm^{-3}}$) and coercitivity have become available (e.g. neodymium–iron–boron magnets). This has opened up new possibilities for a larger-scale application of permanent-magnet synchronous machines with high torque density, low loss/torque, high power factor, fast torque and speed response. However, the high price, operating temperature limitations, and danger of demagnetization of these permanent magnets can be restrictive for many applications. It would appear that the limitations imposed by the high cost might be overcome in the near future, since there are some countries where there is a large reservoir of these magnets, and their cost may be reduced.

In a permanent-magnet synchronous motor with surface-mounted magnets, torque control can be achieved very simply, since the instantaneous electromagnetic torque can be expressed similarly to that of the d.c. machine given by eqn (1.2-1):

$$t_e = c_F \psi_F i_{sq}, \tag{1.2-5}$$

where c_F is a constant, ψ_F is the magnet flux (rotor flux in the rotor reference frame), and i_{sq} is the torque-producing stator current component (quadrature-axis stator current in the rotor reference frame). Both ψ_F and i_{sq} are shown in Fig. 1.3(b) below. Equation (1.2-5) can be obtained in many ways. However, according to simple physical considerations, the electromagnetic torque must be maximum when the torque angle (δ) is 90°, and the torque varies with the sine of the torque angle, which is the angle between the space vector of the stator currents and the magnet flux (it is also shown in Fig. 1.3(b)). It follows from Fig. 1.3(b) that the quadrature-axis stator current is $|\bar{i}_s| \sin \delta$. The torque expression is also similar to those shown in eqns (1.2-2)–(1.2-4) for the induction motor, and this is an expected physical feature. It can also be seen that the electromagnetic torque can be controlled by controlling the stator currents. The torque expression also follows directly from the fact that for all singly salient electrical machines the instantaneous electromagnetic torque can be considered to be the cross vectorial product of the stator flux-linkage space vector and stator current space vector:

$$t_e = c_F(\bar{\psi}_s' \times \bar{i}_s') = c_F(L_s \bar{i}_s' + \psi_F) \times \bar{i}_s' = c_F \psi_F \times \bar{i}_s' = c\psi_F i_{sq}, \tag{1.2-6}$$

where $\bar{\psi}_s'$ and $\bar{i}_s'$ are the space vectors of the stator flux linkage and stator current space vectors respectively in the rotor reference frame, and L_s is the stator inductance. It can be seen that the electromagnetic torque is proportional to the magnet flux (rotor flux) and the quadrature-axis stator current (torque-producing stator current). If the magnet flux is constant, and the quadrature-axis current is changed rapidly (e.g. by a current-controlled PWM inverter), a quick torque response is obtained.

However, in contrast to the vector control scheme of the induction motor, the vector control scheme of the permanent-magnet induction motor is simpler. This follows from the fact that the magnet flux is fixed to the direct axis of the rotor reference frame and the space angle between the magnet flux and the direct axis of the stator reference frame is equal to the rotor angle, θ_r, which can be monitored by using a position sensor. In contrast to this, in the induction motor, e.g. if rotor-flux-oriented control is performed, it is also necessary to know the position of the rotor flux-linkage space vector with respect to the direct axis of the stator reference frame, and in general this angle is not equal to the rotor angle and its determination requires the use of a flux model. The vector control scheme for the permanent-magnet synchronous motor is simple, and the quadrature-axis stator current (in the rotor reference frame) can be obtained simply by considering that when the stator current vector is expressed in the rotor reference frame, the following transformation holds: $\bar{i}'_s = \bar{i}_s \exp(-j\theta_r)$. In this expression the rectangular form of the stator current vector in the rotor reference frame is $\bar{i}'_s = i_{sd} + ji_{sq}$, and the stator current vector in the stator (stationary) reference frame is $\bar{i}_s = i_{sD} + ji_{sQ} = |\bar{i}_s| \exp(j\alpha_s)$, where, as shown in Fig. 1.3(b) α_s is the angle of the stator current vector (with respect to the direct axis of the stationary reference frame). Thus finally $\bar{i}'_s = i_{sd} + ji_{sq} = |\bar{i}_s| \exp[j(\alpha_s - \theta_r)]$ is obtained, yielding

$$i_{sq} = |\bar{i}_s| \sin(\alpha_s - \theta_r) = |\bar{i}_s| \sin(\delta) \tag{1.2-7}$$

in agreement with the discussion above, and hence the electromagnetic torque can be expressed as

$$t_e = c_F \psi_F |\bar{i}_s| \sin(\alpha_s - \theta_r) = c_F \psi_F i_{sq}. \tag{1.2-8}$$

It follows that if the stator current modulus and angle are known (e.g. by using a rectangular-to-polar conversion of the measured direct- and quadrature-axis stator current components i_{sD}, i_{sQ}) and the rotor angle (θ_r) is measured (by a suitable position sensor), then the quadrature-axis stator current can be simply determined according to eqn (1.2-7). As stated above, eqn (1.2-8) is physically expected, since it shows that the electromagnetic torque varies with the sine of the torque angle (torque angle $\delta = \alpha_s - \theta_r$). It also follows that maximum torque per stator current is obtained when the torque angle is 90°.

Equation (1.2-8) forms the basis of rotor-oriented vector control schemes, which use a position sensor. Figure 1.3(a) shows a possible implementation of a vector-controlled (rotor-oriented-controlled) permanent-magnet synchronous machine with surface-mounted magnets, where the machine is supplied by a current-controlled PWM inverter.

In Fig. 1.3(a) the electromagnetic torque reference (t_{eref}) is proportional to i_{sqref}, since $t_{eref} = c\psi_F i_{sqref}$. Below base speed, $i_{sdref} = 0$, since in this case the torque angle is 90° and maximum torque/ampere is obtained. However, above base speed, in the field-weakening range, where the inverter voltage limit is reached, a negative value of i_{sdref} is used and in this case the torque angle is larger than 90°. Since the stator current modulus $|\bar{i}_s|$ cannot exceed its maximum value (i_{smax}) thus $|\bar{i}_s| = \sqrt{(i_{sd}^2 + i_{sq}^2)} \leqslant i_{smax}$, so the quadrature-axis stator current i_{sq} may have to be

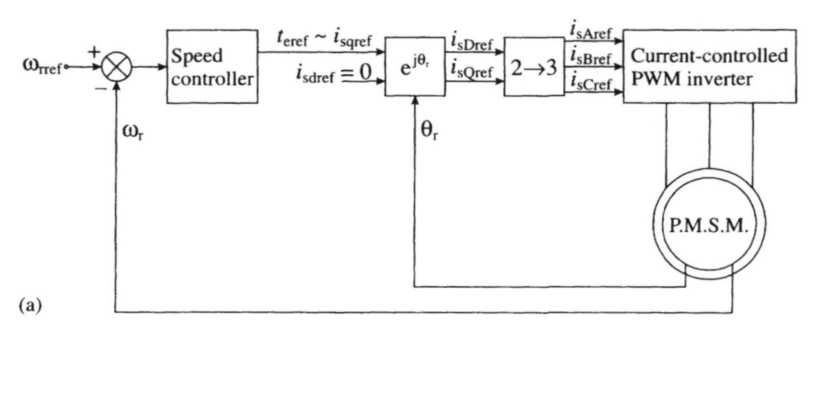

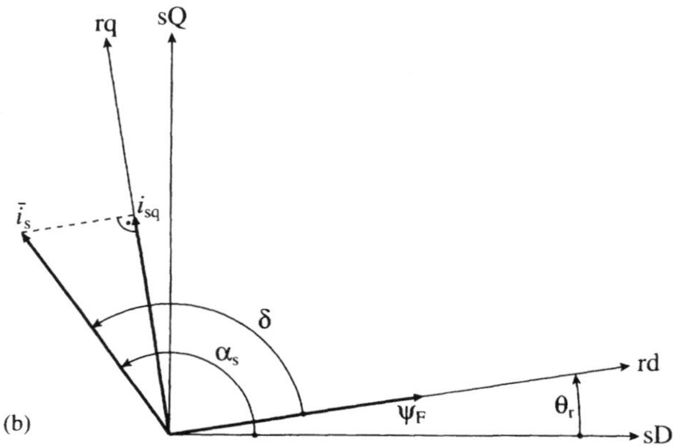

Fig. 1.3. Rotor-oriented control of a permanent-magnet synchronous motor with surface-mounted magnets supplied by a current-controlled PWM inverter. (a) Drive scheme; (b) vector diagram.

reduced, thus t_{eref} may also have to be limited. The stator current components in the stationary reference frame (i_{sDref}, i_{sQref}) are obtained by using the transformation $\exp(j\theta_r)$, and this is followed by the application of the two-phase-to-three-phase transformation. It should be noted that this scheme is similar to that shown in Fig. 1.2(a). However, for the permanent-magnet synchronous motor, in Fig. 1.3(a) the rotor position (θ_r) is used (rotor-oriented control) instead of the position (ρ_r) of the rotor flux-linkage space vector (rotor-flux-oriented control) and the angular slip frequency reference is zero. Furthermore, below base speed $i_{sdref}=0$.

It is also possible to have a speed-sensorless implementation, where the rotor speed is obtained by utilizing the speed of the stator flux-linkage space vector (rate of change of the angle of the stator flux-linkage space vector). Although such a drive can be used in a wide speed range, it will not give full controlled-torque operation down to zero speed. It is expected that permanent-magnet machines will be used more widely when the price of magnet materials is reduced.

One main difference between the vector control of a synchronous machine and the vector control of an induction machine is their cross-magnetizing behaviour. In an electrically excited synchronous machine, the stator currents produce a rotor flux which is in space-quadrature to the flux produced by the field winding. Thus, similarly to the cross-magnetizing armature reaction of a d.c. machine, a reduction of the field flux can be caused at high values of torque. In a synchronous machine with surface-mounted magnets, this demagnetization effect is normally small, due to the large air-gap associated with the magnets, but at high currents it can cause magnet demagnetization. This effect must be considered when designing the motor. However, in an induction machine subjected to vector control (e.g. rotor-flux-oriented control), the quadrature-axis rotor flux linkage (in the rotor-flux-oriented reference frame) is zero. Thus there is no demagnetization effect caused by the torque-producing stator current.

It should be noted that the $i_{sd}=0$ control method has been a popular technique for a long time and has been used to avoid the demagnetization of the magnet material. The demagnetization of the permanent magnet is non-reversible when a large d-axis current is applied. However, recent developments in permanent-magnet technology have resulted in magnet materials with large coercitivity (e.g. NdFeB). In permanent-magnet synchronous motors using these materials, a direct-axis stator current can be applied without the danger of demagnetizing the magnets. In general, in a permanent-magnet synchronous motor, where the direct- and quadrature-axis stator inductances (L_{sd}, L_{sq}) are not equal, e.g. for a machine with interior magnets, independent control of i_{sd} (flux-producing current) and, i_{sq} (torque-producing current) is possible by considering that the direct-axis stator flux linkage is $\psi_{sd}=\mathrm{Re}(\bar{\psi}_s')=\psi_F+L_{sd}i_{sd}$ [see also the definition used in eqn (1.2-6)] and the electromagnetic torque is $c[\psi_F i_{sq}+(L_{sd}-L_{sq})i_{sd}i_{sq}]$. This last equation follows from the fact that $t_e=c\bar{\psi}_s'\times\bar{i}_s'$, where $\bar{\psi}_s'=\psi_{sd}+j\psi_{sq}$, and the direct-axis stator flux-linkage, $\psi_{sd}=\psi_F+L_{sd}i_{sd}$ and the quadrature-axis stator-flux linkage component is $\psi_{sq}=L_{sq}i_{sq}$.

Synchronous reluctance machines

Due to their low cost, simplicity of control, absence of rotor losses, and field-weakening capability, research efforts have recently been increased towards various industrial and automotive applications of synchronous reluctance motors, although at present no large manufacturer produces this type of drive. This increased interest is the main reason for incorporating new material in the present book on the torque control of this drive. These types of motors were developed in the UK by Professor Lawrenson more than 30 years ago and by Honsinger in the USA (Professor Lawrenson is also the main developer of switched reluctance motors). The synchronous reluctance motor (SYRM) is a greatly robust singly-salient machine (similarly to the stator of an induction machine, the stator bore is smooth but slotted and there is a symmetrical three-phase stator winding carrying balanced three-phase currents; however, there is a salient-pole rotor). In a high-performance synchronous reluctance motor the rotor is axially laminated: this results in high saliency ratios (high L_{sd}/L_{sq}, where L_{sd} and L_{sq} are the direct- and quadrature-axis synchronous inductances respectively). The high saliency ratio results in high power density, increased torque, high power factor, and increased efficiency. Reluctance torque is created by the alignment of the minimum reluctance path of the rotor with the rotating magnetizing m.m.f. which is produced by the stator currents (there are no rotor currents in a cageless design).

The electromagnetic torque of a synchronous reluctance motor can also be expressed similarly to that of the induction motor or the permanent-magnet synchronous motor. By considering that the electromagnetic torque is developed by the interaction of the stator flux-linkage space vector and stator current space vector, and this relationship is valid in all reference frames, thus $t_e=(3/2)P\bar{\psi}'_s \times \bar{i}'_s$, where P is the number of pole-pairs and the stator flux-linkage and current space vectors in the rotor reference frame are $\bar{\psi}'_s=\psi_{sd}+j\psi_{sq}$ and $\bar{i}'_s=i_{sd}+ji_{sq}$ respectively. The subscripts d and q denote the direct- and quadrature-axis quantities respectively in the rotor (synchronous) reference frame. Due to the absence of rotor currents, the stator flux linkages are established by the stator currents, thus $\psi_{sd}=L_{sd}i_{sd}$, and $\psi_{sq}=L_{sq}i_{sq}$, where L_{sd} and L_{sq} are the direct- and quadrature-axis synchronous inductances respectively. It follows that

$$t_e=\tfrac{3}{2}P\bar{\psi}'_s \times \bar{i}'_s=\tfrac{3}{2}P(\psi_{sd}i_{sq}-\psi_{sq}i_{sd})=\tfrac{3}{2}P(L_{sd}-L_{sq})i_{sd}i_{sq}. \tag{1.2-9}$$

It is important to note that in eqn (1.2-9) the stator current components (i_{sd}, i_{sq}) are expressed in the rotor reference frame, and can be obtained from the stator currents in the stationary reference (i_{sD}, i_{sQ}) frame by considering that

$$\bar{i}'_s=i_{sd}+ji_{sq}=\bar{i}_s\exp(-j\theta_r)=(i_{sD}+ji_{sQ})\exp(-j\theta_r), \tag{1.2-10}$$

where θ_r is the rotor angle. This is why in general the rotor position is required (e.g. in a control scheme using rotor-oriented control) for the transformation of the measured stator currents (i_{sD}, i_{sQ}). Figure 1.4(b) shows the stator current vector, the stator flux-linkage vector, and also the various reference frames.

Equation (1.2-9) together with eqn (1.2-10) can be used for the implementation of the rotor-oriented control of the synchronous reluctance motor. However, eqn (1.2-9) can also be put into the following form, if $\psi_{sd}=L_{sd}i_{sd}$ is utilized:

$$t_e=\tfrac{3}{2}P(L_{sd}-L_{sq})i_{sd}i_{sq}=[\tfrac{3}{2}P(1-L_{sq}/L_{sd})]\psi_{sd}i_{sq}=c\psi_{sd}i_{sq}, \tag{1.2-11}$$

where $c=(3/2)P(1-L_{sq}/L_{sd})$ and it should be noted that in general it is saturation dependent. Equation (1.2-11) resembles the torque expression of a separately excited d.c. motor, that of a vector-controlled induction motor and also that of a permanent-magnet synchronous motor with surface magnets; see eqns (1.2-1), (1.2-2), (1.2-3), and (1.2-5). In a vector-controlled synchronous reluctance motor drive, where the motor is supplied by a current-controlled PWM inverter, and where rotor-oriented control is performed, independent control of the torque and flux (torque-producing stator current and flux-producing stator current) can be achieved as shown in Fig. 1.4(a).

The gating signals of the six switching devices of the inverter are obtained on the output of hysteresis current controllers. On the inputs of these the difference

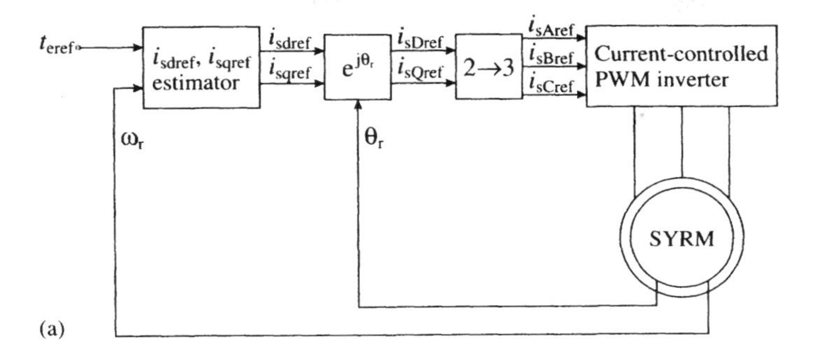

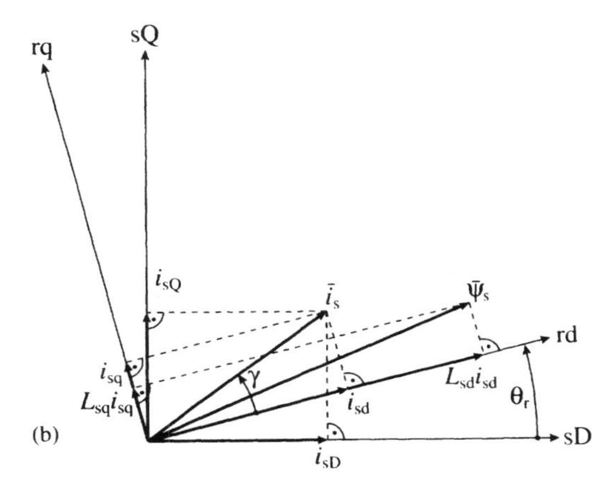

Fig. 1.4. Rotor-oriented control of synchronous reluctance motor supplied by a current-controlled PWM inverter. (a) Drive scheme; (b) vector diagram.

between the actual (measured) and reference stator line currents are present. The three-phase stator current references are obtained from their two-axis components (i_{sD}, i_{sQ}) by the application of the two-phase-to-three-phase ($2 \rightarrow 3$) transformation, and these are obtained from the reference values of i_{sd}, i_{sq} by the application of the $\exp(j\theta_r)$ transformation. These reference values are obtained in the appropriate estimation block from the reference value of the torque (t_{eref}) and the monitored rotor speed (ω_r). This estimation block can be implemented in various ways, according to the required control strategy. In general there are three different constant-angle control strategies: fastest torque control, maximum torque/ampere control, and maximum power factor control. It will be proved in Section 3.2.2.1 that, for example, the fastest torque response can be obtained by a controller where $\gamma = \tan^{-1}(L_{sd}/L_{sq})$, where γ is the angle of the stator current space vector with respect to the real axis of the rotor reference frame (d-axis), as shown in Fig. 1.4(b). It should be noted that the rotor-oriented control scheme of the synchronous reluctance machine shown in Fig. 1.4(a) is very similar to the rotor-oriented control scheme of the permanent-magnet synchronous machine shown in Fig. 1.3(a).

1.2.1.4 Switched reluctance machine control

A switched reluctance motor (SRM) is a singly-excited doubly-salient d.c. motor. In contrast to a variable reluctance stepper motor, the SRM produces continuous motion. The concentrated stator windings on diametrically opposite poles (teeth) are connected in series to form a stator phase. The rotor is of the reluctance type and is made of steel laminations.

In the SRM, the electromagnetic torque is developed by the tendency of the magnetic circuit to adopt a position with minimum reluctance of the magnetic paths and is independent of the direction of the current flow. Thus the stator currents are unidirectional currents and one stator phase conducts at a time. In a variable-speed SRM drive a sequence of current pulses is applied to each stator phase by using the appropriate power converter. Due to the unidirectional stator currents the topology of the power converter is simple. The operation of the SRM is also simple: a pair of stator poles is excited and a pair of rotor poles aligns itself to these. When a stator pole-pair is energized, the corresponding rotor pole-pair is attracted toward the energized stator pole-pair to minimize the reluctance of the magnetic path. Thus by energizing the consecutive stator phases in succession, it is possible to develop constant torque in either direction of rotation. In a conventional SRM drive, a position sensor is used to switch the stator currents at the appropriate instants, but in more recent implementations, the position sensor is eliminated by estimating the rotor position. In contrast to other types of a.c. and d.c. motors, the SRM cannot run directly from a.c. or d.c. (the flux is not constant), but the flux must be established from zero at every step and the converter must supply unipolar current pulses, which are timed accurately by using information on the rotor position. In the SRM the direction of rotation is controlled by the stator phase excitation sequence, e.g. the sequence sA, sB, sC, sA ... gives clockwise rotation while sA, sC, sB, sA ... yields counter-clockwise rotation. The speed of the SRM drive can be changed by varying the stator frequency. If the fundamental switching frequency is f, then $f=\omega_r N_r$, where ω_r is the angular rotor speed and N_r is the number of rotor poles. The non-uniform torque production causes torque ripples (and noise).

In an idealized SRM where saturation and fringing effects are neglected, for constant phase current, in the non-aligned rotor positions the electromagnetic torque is constant (since i_s=constant and $dL_s/d\theta_r$ is linear, thus $t_e=(1/2)i_s^2 dL_s/d\theta_r$=constant, and, for example, for motoring operation this constant is positive). However, in a real machine, due to saturation effects and field fringing, the electromagnetic torque produced for a constant phase current is a non-linear function of saturation. Thus to produce constant torque, profiled stator currents are required. The required current profile is a non-linear function of the rotor position and also of the electromagnetic torque. It should be noted that the general expression for the electromagnetic torque in an SRM is $t_e=t_{esA}+t_{esB}+t_{esC}$, where the three torque components are the electromagnetic torque components produced by each stator phase respectively. For example, the electromagnetic torque produced by stator phase sA can be expressed as

$$t_{esA} = \frac{\partial W_c(\theta_r, i_{sA})}{\partial \theta_r}, \tag{1.2-12}$$

where W_c is the coenergy, which is the area under the corresponding magnetization curve. It follows from eqn (1.2-12) that in general it is not possible to express the electromagnetic torque as a product of the flux- and torque-producing stator current components, as in various types of vector-controlled a.c. drives. When the effects of magnetic saturation are neglected, $W_c = (1/2)L_{sA}i_{sA}^2$, and the expression for the torque becomes

$$t_{esA} = \frac{1}{2} i_{sA}^2 \frac{dL_{sA}}{d\theta_r}, \tag{1.2-13}$$

where L_{sA} is the stator self-inductance. It follows that the electromagnetic torque is zero when the stator inductance is maximum. This corresponds to the aligned position. However, when the rotor is in a non-aligned position, the electromagnetic torque is not zero. Since the torque is not zero in the unaligned positions, this torque causes the rotor to align with stator phase sA in agreement with the above discussion. The direction of the torque is always towards the nearest aligned position. Thus positive torque can only be produced if the rotor is between misaligned and aligned positions in the forward direction. It follows that to obtain positive (motoring) torque, the stator phase current has to be switched on during the rising inductance region of L_{sA}. To obtain negative (braking) torque, it has to be switched on during the decreasing part of the corresponding stator inductance region. However, it should be noted that to obtain maximal motoring torque, the current in a stator phase should be switched on during the constant inductance region so it can build up before the region of increasing inductance starts. In a conventional SRM drive, the detection of these regions is made by using a position sensor. In a 'sensorless' drive the position information is obtained without using a position sensor (e.g. by using an observer).

When designing stator-current-reference waveforms for constant torque operation, one of the main difficulties is to maintain constant torque over a wide speed-range by considering in addition the variation in the d.c. link voltage. This is due to the fact that the non-linear machine characteristics make it difficult to take into account the rate of change of the flux linkage which is required for a given current reference. The maximum rate of change of the stator current is a non-linear function of the stator current and rotor position. Calculation of the phase current references for constant torque, which the current controllers are able to track, is made difficult by the non-linear variation of the current slew rate. However, if instead of current controllers, flux-linkage controllers are used, then the maximum rate of change of the stator flux linkage with respect to rotor position can be simply determined from the maximum available supply voltage

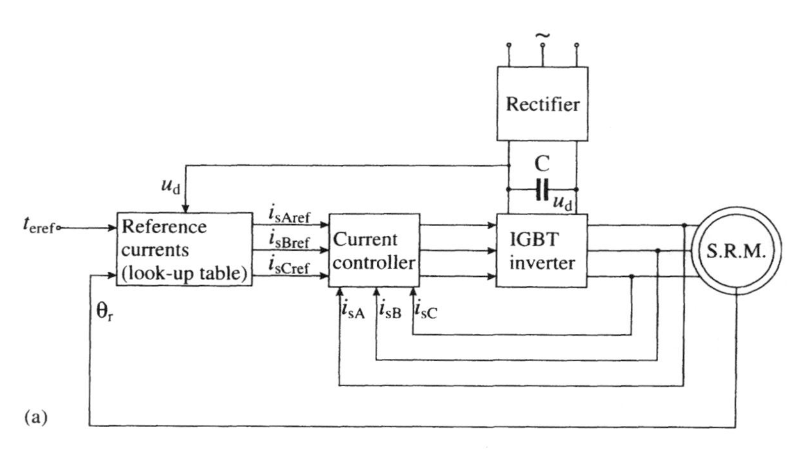

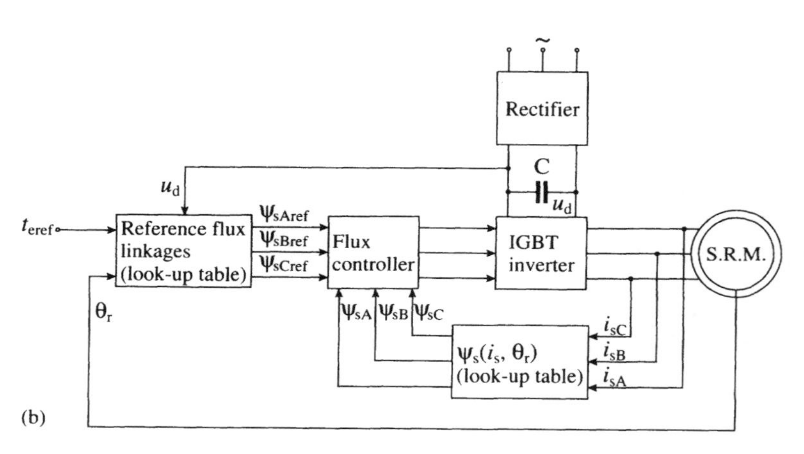

Fig. 1.5. Torque-controlled SRM drive. (a) SRM drive with a current controller; (b) SRM drive with a flux controller.

and rotor speed. When such a scheme is used, a required stator flux-linkage reference waveform (as a function of the rotor position) contains a series of linear ramps, each with a gradient not higher than the greatest actually achievable for a given speed and supply voltage. However, by neglecting the stator ohmic drop, each linear flux ramp can be realized by the application of a constant voltage to the stator phase. The gradient of each flux ramp determines the maximum speed at which it is possible to track the constant torque-flux characteristic for a given d.c. link voltage. If the constant-torque operation has to be extended to the high-speed operation, the stator flux-linkage reference waveform has to be planned over the cycle as a whole, and cannot be performed on a point-by-point basis. By using linear flux-linkage ramps, it is possible to predict the operation later in the cycle, as the time taken to reach a new stator flux-linkage value can be simply determined.

Thus the two main schemes for instantaneous torque-controlled switched reluctance motors will now be presented. The first solution uses profiled stator currents, and thus the drive contains a current controller, but the second solution uses profiled stator flux-linkages, and the drive scheme contains a stator flux-linkage controller. It is shown that the second scheme is simpler to implement. Figure 1.5 shows the schematics of the two torque-controlled SRM drive systems. Figure 1.5(a) shows the drive scheme where a current controller is used, and Fig. 1.5(b) corresponds to the drive with a flux controller.

In Fig. 1.5(a) the input is the torque reference. This is used together with the rotor position (θ_r) and d.c. link voltage (U_d) for the determination of the stator current references. The d.c. link voltage is obtained on the output of a three-phase uncontrolled rectifier. The current references are inputs to the current controller (e.g. hysteresis controller), which also requires the measured values of the stator currents. The current controller outputs the required switching signals for an IGBT converter.

In Fig. 1.5(b) the torque-controlled SRM drive scheme is shown, but instead of the current controller there is a flux-linkage controller (a dead-beat controller, where the desired flux linkage is reached in one sampling time without overshoot). The flux-linkage controller also requires the actual stator flux linkages, since the actual values are compared with the reference flux-linkage values. The actual flux linkages can be obtained from the stator currents and the rotor position, but the information required can be stored in a look-up table. By using the flux linkage error ($\Delta\psi_s$), the d.c. link voltage and also the stator ohmic voltage-drop correction (since the stator ohmic drop is neglected in the stator flux-reference calculation stage), the inverter switching times can be simply determined.

The torque-controlled SRM can operate in a wide speed-range, where the torque is constant (chopping mode) and also where the torque is reduced (constant power region). The torque-controlled SRM is an ideal candidate for traction purposes or electric vehicles. It is expected that in the future SRMs will be more widely used.

1.2.2 FUNDAMENTALS OF DIRECT-TORQUE-CONTROLLED DRIVES

In addition to vector control systems, instantaneous torque control yielding fast torque response can also be obtained by employing direct torque control. Direct torque control was developed more than a decade ago by Japanese and German researchers (Takahashi and Noguchi 1984, 1985; Depenbrock 1985). Drives with direct torque control (DTC) are being shown great interest, since ABB has recently introduced a direct-torque-controlled induction motor drive, which according to ABB can work even at zero speed. This is a very significant industrial contribution, and it has been stated by ABB that 'direct torque control (DTC) is the latest a.c. motor control method developed by ABB' (Tiitinen 1996). It is expected that other manufacturers will also release their DTC drives and further developments are underway for speed-sensorless and artificial-intelligence-based implementations.

In a DTC drive, flux linkage and electromagnetic torque are controlled directly and independently by the selection of optimum inverter switching modes. The selection is made to restrict the flux-linkage and electromagnetic torque errors within the respective flux and torque hysteresis bands, to obtain fast torque response, low inverter switching frequency and low harmonic losses. The required optimal switching-voltage vectors can be selected by using a so-called optimum switching-voltage vector look-up table. This can be obtained by simple physical considerations involving the position of the stator flux-linkage space vector, the available switching vectors, and the required torque and flux linkage.

Figure 1.6 shows the schematic of one simple form of DTC induction motor drive, which uses the stator flux linkages. Thus it will be referred to as a stator-flux-based DTC induction motor drive. However, it should be noted that other forms are also possible, which are based on the rotor flux linkages or magnetizing flux linkages.

In Fig. 1.6 the induction motor is supplied by a VSI inverter and the stator flux-linkage and electromagnetic torque errors are restricted within their respective hysteresis bands. For this purpose a two-level flux hysteresis comparator and a three-level torque hysteresis comparator are used, respectively. The outputs of the flux and torque comparators ($d\psi, dt_e$) are used in the inverter optimal switching table, which also uses information on the position (ρ_s) of the stator flux-linkage space vector. It should be noted that it is not the actual flux-linkage vector position which has to be determined, but only the sector where the flux linkage is located. It can be seen that the drive scheme requires stator flux-linkage and electromagnetic torque estimators. Similarly as is done in one form of a stator-flux-oriented vector control scheme, the stator flux linkages can be estimated by integrating the terminal voltage reduced by the ohmic losses. However, at low frequencies large errors can occur due to the variation of the stator resistance, integrator drift, and noise. Therefore instead of using open-loop

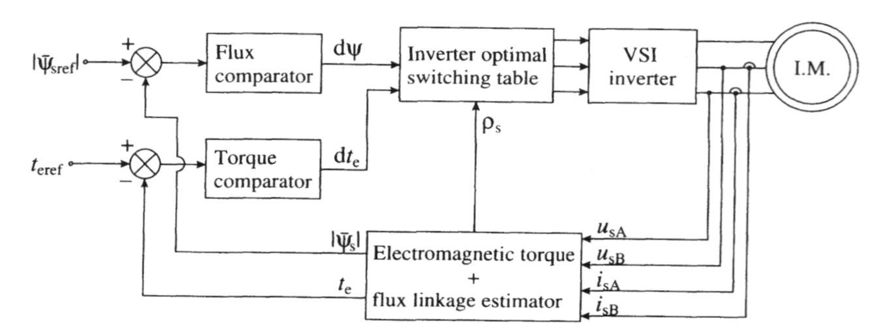

Fig. 1.6. Schematic of a stator-flux-based DTC induction motor drive.

flux-linkage estimators, other techniques should be used (e.g. a flux observer). Observers have reduced sensitivity to parameter variations, but the accuracy of a stator flux-linkage observer can be increased by also using on-line parameter estimators (e.g. for the estimation of the stator resistance), or a thermal model of the machine the outputs of which can be used for accurate stator resistance estimation. However, it is also possible to use a joint state and parameter observer which estimates the stator flux linkages, the stator resistance or temperature (required to estimate the 'hot' value of the stator resistance), and also the rotor speed. It should be noted that for stator flux-linkage estimation it is not necessary to monitor the stator voltages since they can be reconstructed by using the inverter switching functions and the monitored d.c. link voltage together, with a model for the voltage drops across the inverter switches.

It is also possible to have such a DTC scheme which contains a predictive switching-voltage vector estimation. For this purpose various predictive schemes will be discussed in the present book. For example, such a scheme will be discussed which can be used when there is a transient in the reference electromagnetic torque (which is the most practical situation), whilst maintaining dead-beat control of the stator flux linkage. A simplified predictive scheme will also be discussed, which, although it does not yield dead-beat control of the stator flux linkage, is computationally non-intensive and the performance degradation is minimal.

The main features of DTC are: direct control of stator flux and electromagnetic torque; indirect control of stator currents and voltages; approximately sinusoidal stator fluxes and stator currents; reduced torque oscillations; excellent torque dynamics; inverter switching frequency depending on flux and torque hysteresis bands. The main advantages of DTC are: absence of coordinate transformations (which are required in most of the vector-controlled drive implementations); absence of separate voltage modulation block (required in vector drives); absence of voltage decoupling circuits (required in voltage-source inverter-fed vector drives); reduced number of controllers (e.g. only a speed controller is required if the drive contains a speed loop); the actual flux-linkage vector position does not have to be determined, but only the sector where the flux linkage is located, etc. However, in general, the main disadvantages of a conventional DTC can be: possible problems during starting and low-speed operation and during changes in torque command; requirement for flux-linkage and electromagnetic torque estimators (the same problem exists for vector drives); variable switching frequency.

In the ABB DTC induction motor drive, torque response times typically better than 2 ms have been claimed by ABB together with high torque-control linearity even down to low frequencies including zero speed. It has also been claimed by ABB that the new a.c. drive technology rests chiefly on a new motor model which enables computation of the motor states to be performed without using a speed or position sensor. The motor model used by ABB is a conventional type of mathematical model (using various machine parameters, i.e. the stator resistance, mutual inductance, etc.), and not a model based on artificial intelligence techniques (which does not require a conventional mathematical model). However, it is believed that it is also possible to implement DTC drives which utilize intelligent control techniques.

Other types of DTC drives utilize principles similar to those discussed above, and in the present book direct-torque-controlled permanent-magnet synchronous machine drives, direct-torque-controlled synchronous reluctance-motor drives, and direct-torque-controlled electrically excited synchronous motor drives are also discussed. Both the voltage-source and current-source inverter-fed drives (VSI, CSI) are considered. Thus the eleven types of DTC drives discussed in this book are as follows:

1. DTC of induction motor
   - (a) DTC of VSI-fed induction motor
     - Scheme 1: Drive with non-predictive switching-vector selection
     - Scheme 2: Drive with predictive switching-vector selection
   - (b) DTC of CSI-fed induction motor
2. DTC of synchronous motor
   - (a) DTC of VSI-fed permanent-magnet synchronous motor
     - Scheme 1: Direct control of the electromagnetic torque and stator flux using optimal voltage switching look-up table
     - Scheme 2: Direct control of the electromagnetic torque and d-axis stator current using optimal voltage switching look-up table
     - Scheme 3: Direct control of the electromagnetic torque and reactive torque using optimal voltage switching look-up table
     - Scheme 4: Direct torque control using a predictive algorithm for switching voltage vector selection
   - (b) DTC of VSI-fed synchronous reluctance induction motor
     - Scheme 1: Direct control of the electromagnetic torque and stator flux
     - Scheme 2: Direct control of the electromagnetic torque and stator current
   - (c) DTC of VSI-fed electrically excited synchronous motor
   - (d) DTC of CSI-fed electrically excited synchronous motor

It is expected that in the future DTC drives will have an increased role and predictive-algorithm-based DTC schemes will be more widely used.

1.2.3 SPEED- AND POSITION-SENSORLESS A.C. DRIVES

In the past few years great efforts have been made to introduce speed- and/or shaft position-sensorless torque-controlled (vector- and direct-torque-controlled) drives. These drives are usually referred to as 'sensorless' drives, although the terminology 'sensorless' refers to only the speed and shaft sensors: there are still other sensors in the drive system (e.g. current sensors), since closed-loop operation cannot be performed without them.

Sensorless vector drives have become the norm for industry and almost every large manufacturer (Control Techniques plc, Siemens, Hitachi, Yaskawa,

Eurotherm, etc.) has introduced a sensorless induction motor drive. However, it is a main feature of almost all of these industrial drives, that they cannot operate at very low frequencies without speed or position sensors. Only one large manufacturer (ABB) has developed one form of a direct-torque-controlled induction motor drive, which (according to the claims made by the manufacturer) can work very close to zero frequency, and no speed or position estimator is used. In general, to solve the problems which occur at low frequencies, a number of special techniques have also been proposed by various investigators, e.g. techniques which deliberately introduce asymmetries in the machine, or in which extra signals are injected into the stator. However, so far these techniques have not been accepted by industry, due to their undesirable side effects and other problems. It would appear that, in general, industrially acceptable sensorless solutions must be applicable to off-the-shelf motors, unless the drive system is an integrated drive, in which the motor, inverter, and controllers are part of a single system provided by the same manufacturer. In such a case special techniques are also possible.

Conventionally the speed of an electrical machine can be measured conveniently by d.c. tachogenerators, which are nowadays brushless d.c. tachogenerators. Rotor position can be measured by using electromagnetic resolvers or digitally by using incremental or absolute encoders. Optical encoders are one of the most widely used position sensors. Electromagnetic resolvers are popular for measuring the rotor position because of their rugged construction and higher operating temperature. Obviously if the rotor position is monitored, the speed (which is the first derivative of the position) can be estimated directly from the position, but the speed resolution is limited by the resolution of the position transducer and also the sampling time.

To reduce total hardware complexity and costs, to increase the mechanical robustness and reliability of the drive, and to obtain increased noise immunity, it is desirable to eliminate these sensors in vector-controlled and direct-torque-controlled drives. Furthermore, an electromechanical sensor increases the system inertia, which is undesirable in high-performance drives. It also increases the maintenance requirements. In very small motors it is impossible to use electromechanical sensors. In a low-power torque-controlled drive the cost of such a sensor can be almost equal to the other costs. In drives operating in hostile environments, or in high-speed drives, speed sensors cannot be mounted. As real-time computation (DSP) costs are continually and radically decreasing, speed and position estimation can be performed by using software-based state-estimation techniques where stator voltage and/or current measurements are performed. It is also possible to use other types of solutions, e.g. the stator phase third harmonic, rotor slot harmonics, etc.

In summary the main objectives of sensorless drive control are:

- reduction of hardware complexity and cost
- increased mechanical robustness and overall ruggedness
- operation in hostile environments
- higher reliability

- decreased maintenance requirements
- increased noise immunity
- unaffected machine inertia
- applicability to off-the-shelf motors.

The main techniques of sensorless control for induction motor drives are:

1. Open-loop estimators using monitored stator voltages/currents and improved schemes;
2. Estimators using spatial saturation stator-phase third harmonic voltages;
3. Estimators using saliency (geometrical, saturation) effects;
4. Model reference adaptive systems (MRAS);
5. Observers (Kalman, Luenberger);
6. Estimators using artificial intelligence (neural network, fuzzy-logic-based systems, fuzzy-neural networks etc.).

The details of these techniques will be discussed in Sections 4.5 and 4.6 and Chapter 7 for both vector drives and direct-torque-controlled drives.

The main techniques of sensorless control of permanent-magnet synchronous motor drives are:

1. Open-loop estimators using monitored stator voltages/currents;
2. Stator-phase third harmonic voltage-based position estimators;
3. Back e.m.f.-based position estimators;
4. Observer-based (Kalman, Luenberger) position estimators;
5. Estimators based on inductance variation due to geometrical and saturation effects;
6. Estimators using artificial intelligence (neural network, fuzzy-logic-based systems, fuzzy-neural networks etc.).

The details of these will be discussed in Sections 3.1.3 and 3.3.2 for both vector-controlled and direct-torque-controlled drives.

The position-sensorless control of synchronous reluctance motors is discussed in Sections 3.2.2.2 and 3.3.3 respectively for vector- and direct-torque-control applications. The main techniques for sensorless control of synchronous reluctance motors are:

1. Estimators using stator voltages and currents, utilizing speed of stator flux-linkage space vector.
2. Estimators using spatial saturation third-harmonic voltage component.
3. Estimators based on inductance variation due to geometrical effects:
   (a) Indirect position estimation using the measured rate of change of the stator current (no test-voltage vector used).
   (b) Indirect flux detection by using measured rate of change of the stator currents and applying test-voltage vectors (on-line reactance measurement (INFORM) method); low speed application.

(c) Indirect position estimation using the rate of change of the measured stator current vector by using zero test-voltage vector (stator is short-circuited); high speed application.

4. Estimators using observers (e.g. extended Kalman filter).
5. Estimators using artificial intelligence (neural network, fuzzy-logic-based systems, fuzzy-neural networks etc.).

The main techniques for sensorless control of switched reluctance motors are:

1. Position estimation from the monitored stator currents;
2. Position estimation using an observer (extended Kalman filter, extended Luenberger observer);
3. Stator-flux- and stator-current-based position estimation;
4. Artificial-intelligence-based position estimation.

These techniques are discussed in detail in Chapter 5.

It is expected that in the future further speed- and position-sensorless drives will emerge and find wide industrial applications. It is believed that the most significant recent industrial contributions to variable-speed drives are due to Control Techniques plc, who introduced in 1995 the first universal drive (Unidrive) in the world, and also due to ABB who introduced the first industrial direct-torque-controlled drive (also in 1995). Unidrive is a radically new drive, which combines sensorless vector open-loop, closed-loop flux vector and high-performance brushless servo technologies as a single fully enclosed product. Figure 1.7 shows the Unidrive product family.

Unidrive is available in five frame sizes from 0.75 kW to 1 MW. It can be directly controlled from a control module panel or remotely through a serial communications interface. Universal drives offer distinct advantages to the user in that the control method, parameter listing, and user interface are the same for each method of control, eliminating the need for different training programmes. For the manufacturer of the drive, a universal drive can lead to significant rationalization of the manufacturing process and hence to lower production costs.

The ABB DTC induction motor drive contains the ACS 600 frequency converter (inverter); this is shown in Fig. 1.8. The inverter switchings directly control the motor flux linkages and electromagnetic torque. The ACS 600 product family suits many applications and operating environments, with a large selection of a.c. voltage, power, and enclosure ratings, combined with highly flexible communication capabilities. Details of the ABB DTC drive are given in Section 4.6.2.11.

It is expected that in the future further sensorless universal and direct-torque-controlled drives will emerge. However, it is possible that these will take the form of integrated drives. At present there exist several integrated drives, but they are not universal drives and do not contain direct torque control. Integrated, sensorless, universal drives with some type of intelligent control also justify the material and unified analysis presented in this book. Together with further developments in semiconductor technologies, analogue and digital signal electronics,

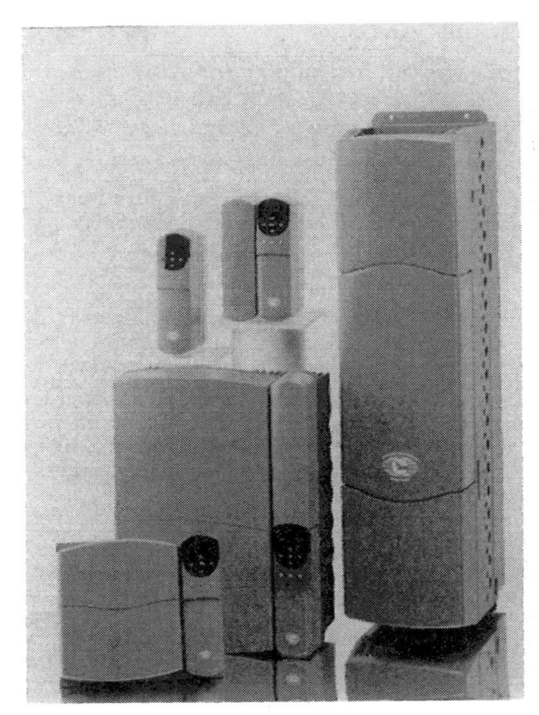

Fig. 1.7. Unidrive (Courtesy of Control Techniques plc, UK).

Fig. 1.8. ACS 600 frequency converter (Courtesy of ABB Industry Oy, Helsinki).

advances in soft-computing based intelligent control techniques, and designs yielding higher efficiency, these topics offer much further research work and have the potential to introduce revolutionary changes in the drive industry.

Bibliography

Barrass, P. and Mecrow, B. C. (1996). Torque control of switched reluctance motor drives. *ICEM, Vigo*, 254–9.

Beierke, S., Vas, P., Simor, B., and Stronach, A. F. (1997). DSP-controlled sensorless a.c. vector drives using the extended Kalman filter. *PCIM, Nurnberg*, 31–42.

Blaschke, F. (1972). The principle of field-orientation as applied to the new Transvektor closed-loop control system for rotating machines. *Siemens Review* **34**, 217–20.

Boldea, L. and Nasar, S. A. (1988). Torque vector control (TVC) – A class of fast and robust torque-speed and position digital controllers for electric drives. *Electric Machines and Power Systems* **15**, 135–47.

Boldea, I., Fu, Z., and Nasar, S. A. (1991). Digital simulation of a vector controlled axially laminated anisotropic (ALA) rotor synchronous motor servo-drive. *Electrical Machines and Power Systems* **19**, 415–24.

Crowder, R. M. (1995). *Electric drives and their controls*. Clarendon Press, Oxford.

Damiano, A., Vas, P., Marongiu, I., and Stronach, A. F. (1997). Comparison of speed sensorless DTC drives. *PCIM, Nurnberg*, 1–11.

Depenbrock, M. (1985). Direkte Selbstregelung (DSR) für hochdynamische Drehfeld-antriebe mit Stromrichterschaltung. *ETZ A* **7**, 211–18.

Du, T., Vas, P., and Stronach, A. F. (1995). Design and application of extended observers for joint state and parameter estimation in high-performance a.c. drives. *IEE Proc. Pt. B* **142**, 71–8.

Fratta, A. and Vagati, A. (1987). A reluctance motor drive for high dynamic performance applications. *IEEE IAS Meeting*, pp. 295–302.

Habetler, T. G. and Divan, D. M. (1991). Control strategies for direct torque control using discrete pulse modulation. *IEEE Transactions on Ind. Applications* **27**, 893–901.

Harris, M. R., Finch, J. W., Mallick, J. A., and Miller, T. J. (1986). A review of the integral horsepower switched reluctance drive. *IEEE Transactions of Ind. Applications* **IA-22**, 716–21.

Hasse, K. (1972). Drehzahlregelverfahren für schnelle Umkehrantriebe mit stromrichter-gespeisten Asynchron-Kurzschlusslaufermotoren. *Regelungstechnik* **20**, 60–6.

Hofman, H., Sanders, S. R., and Sullivan, C. (1995). Stator-flux-based vector control of induction machines in magnetic saturation. *IEEE IAS Meeting, Orlando*, 152–8.

Holtz, J. and Bube, E. (1991). Field-oriented asynchronous pulse-width modulation for high-performance a.c. drives operating at low switching frequency. *IEEE Transactions on Ind. Applications* **27**, 574–9.

Hopper, E. (1995). The development of switched reluctance motor applications. *PCIM Europe* **7**, 236–41.

Kazmierkowski, M. P. and Sulkovski, W. (1991). A novel vector control scheme for transistor PWM inverter-fed induction motor drives. *IEEE Transactions on Ind. Electronics* **38**, 41–7.

Lawrenson, P. J. (1965). Two-speed operation of salient-pole reluctance machines. *Proc. IEE Pt. B* **112**, 545–51.

Lawrenson, P. J. and Gupta, S. K. (1967). Developments in the performance and theory of segmented-rotor reluctance machines. *Proc. IEE* **114**, 645–53.

Lawrenson, P. J. and Vamaruju, S. R. (1978). New 4/6 pole reluctance motor. *Electric Machines and Electromechanics*, 311–23.

Lawrenson, P. J., Stephenson, J. M., Blenkinshop, P. T., Corda, J., and Fulton, N. N. (1980). Variable speed reluctance motors. *Proc. IEE Pt. B* **127**, 253–65.

Leonhard, W. (1985, 1996). *Control of electrical drives.* Springer-Verlag, Berlin.

Lorenz, R. D. and Yang, S. M. (1992). Efficiency-optimised flux trajectories for closed-cycle operation of field-orientation induction machine drives. *IEEE Transactions on Ind. Applications* **28**, 574–9.

Miller, T. J. E. (1985). Converter volt-ampere requirements of the switched reluctance motor drive. *IEEE Transactions on Ind. Applications* **IA-21**, 1136–44.

Miller, T. J. E. (1989). *Brushless permanent magnet and reluctance motor drives.* Clarendon Press, Oxford.

Miller, T. J. E. (1993). *Switched reluctance motors and their control.* Magna Physics Publishing and Clarendon Press, Oxford.

Miller, T. J. E. and McGilp, M. (1990). Nonlinear theory of the switched reluctance motor for rapid computer-aided design. *Proc. IEE Pt. B* **137**, 337–47.

Stronach, A. F. and Vas, P. (1995). Variable-speed drives incorporating multi-loop adaptive controllers. *IEE Proc. Pt. D* **142**, 411–20.

Stronach, A. F. and Vas, P. (1997). Application of artificial-intelligence-based speed estimators in high-performance electromechanical drives. *PCIM, Nurnberg* (in press).

Stronach, A. F., Vas, P., and Neuroth, M. (1996). Implementation of intelligent self-organising controllers in DSP-controlled electromechanical drives. *IEE Proc. Pt. D* **144**, 1–7.

Takahashi, I. and Noguchi, T. (1984). Quick torque response control of an induction motor using a new concept. *IEEE J. Tech. Meeting on Rotating Machines*, paper RM84-76, 61–70.

Takahashi, I. and Noguchi, T. (1985). A new quick response and high efficiency control strategy of an induction motor. *IEEE IAS Annual Meeting*, pp. 496–502.

Tiitinen, P. (1996). The next generation motor control method, DTC, direct torque control. *PEDES*, 37–43.

Vas, P. (1990). *Vector control of a.c. machines.* Oxford University Press.

Vas, P. (1992). *Electrical machines and drives: a space-vector theory approach.* Oxford University Press.

Vas, P. (1993). *Parameter estimation, condition monitoring and diagnosis of electrical machines.* Oxford University Press.

Vas, P. (1995). Application of space-vector techniques to electrical machines and variable-speed drives. *EPE tutorial, Seville*, pp. 1–130.

Vas, P. (1995). Artificial intelligence applied to variable-speed drives. *PCIM Seminar, Nurnberg*, pp. 1–149.

Vas, P. and Drury, W. (1996). Future developments and trends in electrical machines and variable-speed drives. *Energia Elettrica* **73**, January, 11–21.

Vas, P. and Stronach, A. F. (1996). Design and application of multiple fuzzy controllers for servo drives. *Archiv für Elektrotechnik* **79**, 1–12.

Vas, P., Stronach, A. F., and Drury, W. (1996). Artificial intelligence in drives. *PCIM Europe* **8**, 152–5.

Vas, P., Stronach, A. F., and Neuroth, M. (1997). Full fuzzy control of a DSP-based high performance induction motor drive. *IEE Proc. Pt. D* **144**, pp. 361–8.

Vas, P., Stronach, A. F., Neuroth, and Neuroth, M. (1998). Application of conventional and AI-based techniques in sensorless high-performance torque-controlled induction motor drives. *IEE Colloquium*, **23** February.

2 The space-phasor model of a.c. machines

As discussed in Chapter 1, the principles of the various forms of vector control and direct torque control of a.c. machines can be well understood by comparing the production of electromagnetic torque in d.c. and a.c. machines. However, for this purpose first the space phasors of various quantities (m.m.f.s, currents, flux linkages, etc.) will be introduced by utilizing physical and mathematical considerations. With the help of these, it is possible to formulate the space-phasor models of a.c. machines and to give a detailed explanation of the fundamental principles involved in vector-controlled and direct-torque-controlled machines.

2.1 The space-phasor model of smooth-air-gap machines

For the purpose of understanding and designing torque controlled drives, it is necessary to know the dynamic model of the machine subjected to control. The machine models which are needed to design control loops are very different to those used for designing the machine. Machine designers must have tolerance levels (e.g. for power levels) which are less than 1%, while control designs are only rough approximations, where even a 10% error can be considered as acceptable. This is because every control scheme must absorb the changes of the plant parameters, which are due to the changes in the temperature, supply, non-linearity etc. and the effects of the load are only approximately considered. However, a model of the electrical machine which is adequate for designing the control system must preferably incorporate all the important dynamic effects occurring during steady-state and transient operation. It should also be valid for any arbitrary time variation of the voltages and currents generated by the converter which supplies the machine.

Such a model, valid for any instantaneous variation of voltage and current and adequately describing the performance of the machine under both steady-state and transient operation, can be obtained by the utilization of space-phasor theory. This is very closely related to the two-axis theory of electrical machines, but the simplicity and compactness of the space-phasor equations and the very clear physical pictures obtained by its application can yield further advantages. For better understanding, the relationship of the space-phasor equations to the two-axis equations will be emphasized throughout the book.

First the space-phasor quantities (voltages, currents, m.m.f.s, flux densities, flux linkages, etc.) will be introduced by using both mathematical and physical considerations. For simplicity, a smooth-air-gap a.c. machine is considered with symmetrical two-pole, three-phase windings. Figure 2.1 shows the cross-sectional view of the machine under consideration; the effects of slotting have been neglected. It is also assumed that the permeability of the iron parts is infinite and the flux density is radial in the air-gap. The effects of iron losses and end-effects

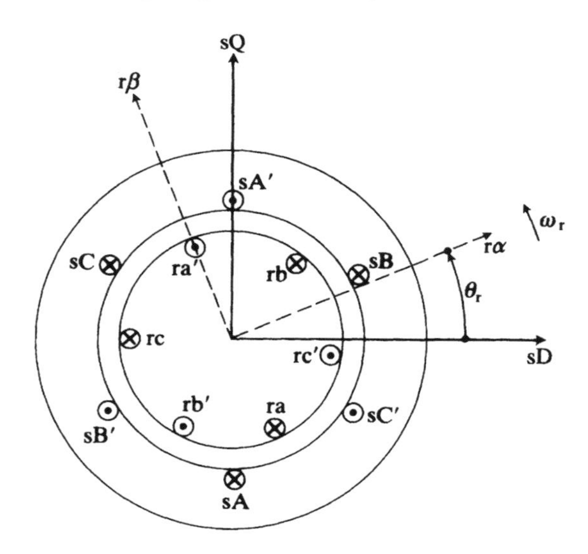

Fig. 2.1. Cross-section of an elementary symmetrical three-phase machine.

are also neglected. In Fig. 2.1 the stator and rotor windings are shown as single, multiple-turn full pitch coils situated on the two sides of the air-gap; these, however, represent distributed windings, which at every instant produce sinusoidal m.m.f. waves centred on the magnetic axes of the respective phases. The phase windings are displaced by 120 electrical degrees from each other. In Fig. 2.1 θ_r is the rotor angle, the angle between the magnetic axes of stator winding sA and rotor winding ra. In general, the speed of the rotor is $\omega_r = d\theta_r/dt$, and its positive direction is also shown in Fig. 2.1.

2.1.1 THE SPACE PHASOR OF THE STATOR M.M.F.s AND STATOR CURRENTS

If the stator windings are supplied by a system of three-phase currents $i_{sA}(t)$, $i_{sB}(t)$, and $i_{sC}(t)$, which can vary arbitrarily in time, but the neutral point is isolated, there can be no zero-sequence stator currents. Then

$$i_{s0}(t) = i_{sA}(t) + i_{sB}(t) + i_{sC}(t) = 0, \tag{2.1-1}$$

where $i_{s0}(t)$ is the instantaneous value of the zero-sequence stator currents. If it is assumed that the stator windings have an equal number of effective turns $N_{se} = N_s k_{ws}$, where N_s and k_{ws} are the number of turns and the winding factor respectively of a stator winding, then the resultant m.m.f. distribution $f_s(\theta, t)$ produced by the stator is as follows: if θ is the angle around the periphery with reference to the axis of stator winding sA, which coincides with the real axis of the stator denoted by sD in Fig. 2.1, then

$$f_s(\theta, t) = N_{se}[i_{sA}(t)\cos\theta + i_{sB}(t)\cos(\theta - 2\pi/3) + i_{sC}(t)\cos(\theta - 4\pi/3)]. \tag{2.1-2}$$

By using complex notation, it is possible to put eqn (2.1-2) into the following form:

$$f_s(\theta, t) = \tfrac{3}{2} N_{se}\,\mathrm{Re}\{\tfrac{2}{3}[i_{sA}(t) + a i_{sB}(t) + a^2 i_{sC}(t)\,e^{-j\theta}\}. \tag{2.1-3}$$

In eqn (2.1-3), $e^{-j\theta}$ is multiplied by the following quantity:

$$\bar{i}_s(t) = \tfrac{2}{3}[1 i_{sA}(t) + a i_{sB}(t) + a^2 i_{sC}(t)] = |\bar{i}_s|\,e^{j\alpha_s}, \tag{2.1-4}$$

which is the complex space phasor of the three-phase stator currents in the complex plane in the stationary reference frame fixed to the stator. Furthermore, in eqn (2.1-4) 1, a, and a^2 are spatial operators, $a = e^{j2\pi/3}$ and $a^2 = e^{j4\pi/3}$. Although these spatial operators are formally the same as the time operators in the theory of complexors used for the steady-state analysis of sinusoidal voltages and currents, it is very important that the two should not be confused. From this point of view, it would be justifiable to use a different notation for these operators, but historically they have been introduced by using the same notation, which is therefore used throughout this book.

In eqn (2.1-4) $|\bar{i}_s|$ is the modulus of the stator current space phasor and α_s is its phase angle with respect to the real axis of the stationary reference frame fixed to the stator. The real axis of the stator is denoted by sD, corresponding to the terminology: Direct-axis of the stator. It should be noted that, since in general all the currents vary in time, both the modulus and the phase angle vary with time. Physically the space phasor of the stator currents determines the instantaneous magnitude and spatial displacement of the peak of the sinusoidal stator m.m.f. distribution produced by the three stator windings. Thus the space phasor of the stator m.m.f.s is defined as follows:

$$\bar{f}_s(t) = N_{se}\bar{i}_s(t) = \bar{f}_{sA}(t) + \bar{f}_{sB}(t) + \bar{f}_{sC}(t), \tag{2.1-5}$$

where $\bar{f}_{sA}(t)$, $\bar{f}_{sB}(t)$, and $\bar{f}_{sC}(t)$ are the space phasors of the individual phase m.m.f.s.

It follows from eqn (2.1-4) that the three-phase space phasor of the stator currents can be obtained by the addition of the space phasors corresponding to each phase current, $\bar{i}_{sA} = 1 i_{sA}(t)$; $\bar{i}_{sB} = a i_{sB}(t)$; $\bar{i}_{sC} = a^2 i_{sC}(t)$. In the symmetrical steady state, when the currents are sinusoidal and form a three-phase balanced system,

$$i_{sA} = I_s\cos(\omega_1 t - \phi_s); \qquad i_{sB} = I_s\cos(\omega_1 t - \phi_s - 2\pi/3);$$
$$i_{sC} = I_s\cos(\omega_1 t - \phi_s - 4\pi/3).$$

Thus the space phasor of the stator currents is $\bar{i}_s = I_s e^{j(\omega_1 t - \phi_s)}$ and since $I_s =$ constant, this corresponds to a circle in the complex plane. Thus the locus of the stator current phasor is a circle which the tip of this space phasor runs around, in the positive direction in space; the angular speed of this space phasor is constant and equal to the synchronous speed. In this case the space phasor of the stator currents is identical to the positive-sequence complexor (complex phasor) of the three-phase currents and this is the reason why in this case the time-vector diagrams are identical to the space-phasor diagrams. However, in other cases, the space phasors and complexors should not be confused. If the currents form an asymmetrical system, in the steady state the locus of the stator-current space phasor will become an ellipse or a straight line. In the transient state the space-phasor locus can have an arbitrary shape.

It should be noted that it is an important consequence of the assumed sinusoidal m.m.f. distribution that the space phasor m.m.f.s of the three phases can be added together. Furthermore, it should also be noted that whilst the m.m.f. waves are measurable, physically existing real quantities, the m.m.f. space phasors are only convenient mathematical abstractions, which of course can be displayed (e.g. in real time, on monitors, oscilloscopes, etc.).

It should be emphasized that it is also possible to introduce the space phasors by utilizing two-axis theory; this was the method originally followed by Park. The space phasor of the stator currents can be defined as a phasor whose real part is equal to the instantaneous value of the direct-axis stator current component, $i_{sD}(t)$, and whose imaginary part is equal to the quadrature-axis stator current component, $i_{sQ}(t)$. Thus, the stator-current space vector in the stationary reference frame fixed to the stator can be expressed as

$$\bar{i}_s = i_{sD}(t) + j i_{sQ}(t). \tag{2.1-6}$$

In the technical literature the notation $s\alpha$ and $s\beta$ is sometimes used instead of the notation sD, sQ.

In symmetrical three-phase machines, the direct- and quadrature-axis stator currents i_{sD}, i_{sQ} are fictitious, quadrature-phase ('two-phase') current components, which are related to the actual, three-phase stator currents as follows:

$$i_{sD} = c[i_{sA} - \tfrac{1}{2} i_{sB} - \tfrac{1}{2} i_{sC}] \tag{2.1-7}$$

and

$$i_{sQ} = c \frac{\sqrt{3}}{2} (i_{sB} - i_{sC}), \tag{2.1-8}$$

where c is a constant. For the so-called classical, non-power-invariant form of the phase transformation from three-phase to quadrature-phase components, $c = 2/3$. However, for the power-invariant form $c = \sqrt{2/3}$. It follows from the definition of the space phasor of the stator currents (eqn 2.1-4) that, if $c = 2/3$, its real-axis component yields

$$\mathrm{Re}(\bar{i}_s) = \mathrm{Re}[\tfrac{2}{3}(i_{sA} + a i_{sB} + a^2 i_{sC})] = \tfrac{2}{3}(i_{sA} - \tfrac{1}{2} i_{sB} - \tfrac{1}{2} i_{sC}) = i_{sD}, \tag{2.1-9}$$

which agrees with eqn (2.1-7) if the non-power-invariant form of the phase transformation is used, and its imaginary-axis component yields

$$\mathrm{Im}(\bar{i}_s) = \mathrm{Im}[\tfrac{2}{3}(i_{sA} + a i_{sB} + a^2 i_{sC})] = (i_{sB} - i_{sC})/\sqrt{3} = i_{sQ}, \tag{2.1-10}$$

which is in agreement with eqn (2.1-8). In a quadrature-phase machine i_{sD} and i_{sQ} are actual, not transformed, currents, which flow in the two stator windings sD and sQ which are in space quadrature.

It should be emphasized that the space phasor does not contain the zero-sequence component and thus if there is a zero-sequence component, an additional definition must be utilized. In general, similarly to eqn (2.1-1), the instantaneous zero-sequence current component of the stator currents is defined as

$$i_{s0} = c_1[i_{sA}(t) + i_{sB}(t) + i_{sC}(t)], \tag{2.1-11}$$

where c_1 is a constant; $c_1 = 1/3$ for the non-power-invariant form, and $c_1 = 1/\sqrt{3}$ for the power-invariant form.

If the non-power-invariant form of the transformations is used, it is a useful consequence that if there are no zero-sequence components, the projections of a space-phasor quantity on the corresponding phase axes directly yield the instantaneous values of the phase variables of the same quantity. This is shown in Fig. 2.2 for the case of the space phasor of the stator currents.

Mathematically this means that, by utilizing $i_{s0} = 0$, and by using the non-power-invariant forms of the zero-sequence current component and the space phasor of the stator currents, the following equations are obtained:

$$\mathrm{Re}(\bar{i}_s) = \mathrm{Re}[\tfrac{2}{3}(i_{sA} + ai_{sB} + a^2 i_{sC})] = \tfrac{2}{3}[i_{sA} - \tfrac{1}{2}i_{sB} - \tfrac{1}{2}i_{sC}] = i_{sA} \tag{2.1-12}$$

$$\mathrm{Re}(a^2\bar{i}_s) = \mathrm{Re}[\tfrac{2}{3}(a^2 i_{sA} + i_{sB} + ai_{sC})] = i_{sB} \tag{2.1-13}$$

$$\mathrm{Re}(a\bar{i}_s) = \mathrm{Re}[\tfrac{2}{3}(ai_{sA} + a^2 i_{sB} + i_{sC})] = i_{sC}. \tag{2.1-14}$$

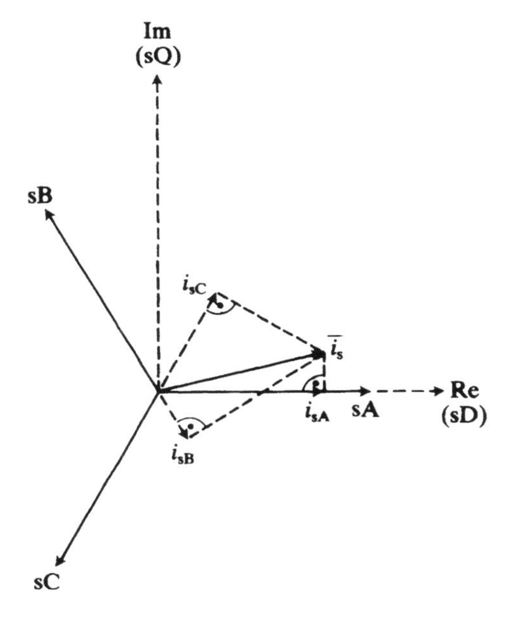

Fig. 2.2. Projections of the stator-current space phasor.

If, however, the zero-sequence current is not zero, the phase-variable stator currents are obtained as

$$i_{sA}=\mathrm{Re}(\bar{i}_s)+i_{s0} \tag{2.1-15}$$

$$i_{sB}=\mathrm{Re}(a^2\bar{i}_s)+i_{s0} \tag{2.1-16}$$

$$i_{sC}=\mathrm{Re}(a\bar{i}_s)+i_{s0}, \tag{2.1-17}$$

and it follows that the zero-sequence current components are added to the corresponding projections of the space phasor on the corresponding axes.

2.1.2 THE SPACE PHASOR OF ROTOR M.M.F.s AND ROTOR CURRENTS

Similar considerations to those for the resultant stator m.m.f. hold for the resultant m.m.f. produced by the symmetrical three-phase sinusoidally distributed rotor windings but the rotor phase ra is displaced from stator phase sA by an angle θ_r and α is the angle around the periphery (also shown in Fig. 2.1) with respect to the axis of rotor winding ra. If it is assumed that the rotor windings have an equal number of effective turns $N_{re}=N_r k_{wr}$, where N_r is the number of turns and k_{wr} the winding factor of a rotor winding, then similarly to eqn (2.1-2), if there is no zero-sequence rotor current, the resultant rotor m.m.f. distribution $f_r(\theta,t)$ produced by the rotor windings carrying currents $i_{ra}(t)$, $i_{rb}(t)$, and $i_{rc}(t)$ is as follows:

$$f_r(\theta,t)=N_{re}[i_{ra}(t)\cos\alpha+i_{rb}(t)\cos(\alpha-2\pi/3)+i_{rc}(t)\cos(\alpha-4\pi/3)]. \tag{2.1-18}$$

By introducing complex notation, it is possible to express eqn (2.1-18) as follows:

$$f_r(\theta,t)=\tfrac{3}{2}N_{re}\,\mathrm{Re}\{\tfrac{2}{3}[i_{ra}(t)1+ai_{rb}(t)+a^2i_{rc}(t)]e^{-j\alpha}\}. \tag{2.1-19}$$

In eqn (2.1-19) the complex quantity multiplied by $e^{-j\alpha}$ is the rotor current space phasor i_r,

$$\bar{i}_r=\tfrac{2}{3}[i_{ra}(t)+ai_{rb}(t)+a^2i_{rc}(t)]=|\bar{i}_r|e^{-j\alpha_r}, \tag{2 1-20}$$

expressed in the reference frame fixed to the rotor (the real axis of this reference frame is denoted by rα and its imaginary axis by rβ, as shown in Fig. 2.1). The speed of this reference frame is $\omega_r=d\theta_r/dt$, where θ_r, is the rotor angle (also shown in Fig. 2.1). This definition is similar to that of the stator-current space phasor expressed in the stationary reference frame (eqn 2.1-4). The rotor-current space phasor determines the instantaneous magnitude and angular displacement of the peak of the sinusoidally distributed rotor m.m.f. produced by the sinusoidally distributed rotor windings.

Thus it is possible to generalize, and it follows that the definitions of the space-phasor quantities in their 'own' reference frames ('natural' reference frames) are similar. Furthermore, it should be noted that it would again be possible to introduce the definition of the rotor-current space phasor by the direct application of two-axis theory. Thus let $i_{r\alpha}$ and $i_{r\beta}$ be the instantaneous values of the direct- and quadrature-axis rotor-current components of the three-phase rotor currents which, similarly to eqns (2.1-7) and (2.1-8), are related to the instantaneous values of the actual three-phase currents by

$$i_{r\alpha}=c[i_{ra}-\tfrac{1}{2}i_{rb}-\tfrac{1}{2}i_{rc}] \tag{2.1-21}$$

$$i_{r\beta}=c\sqrt{3}/2(i_{rb}-i_{rc}), \tag{2.1-22}$$

where $c=\frac{2}{3}$ for the non-power-invariant, classical form of the transformation. Then the definition of the space phasor of the rotor currents in the reference frame fixed to the rotor is as follows

$$\bar{\imath}_r=i_{r\alpha}+ji_{r\beta}. \tag{2.1-23}$$

Thus the real part of $\bar{\imath}_r$ yields the transformed direct-axis rotor current and the imaginary part gives the transformed quadrature-axis rotor current and all the currents can vary arbitrarily in time. For a machine with quadrature-phase rotor winding, $i_{r\alpha}$ and $i_{r\beta}$ are non-transformed, actual rotor currents which flow in the rotor windings rα and rβ respectively.

It follows from eqns (2.1-19) and (2.1-20) that the instantaneous variation of the rotor m.m.f. wave, which rotates at speed ω_r, can be expressed in terms of the rotor-current space phasor and the angles θ and θ_r as follows, if it is considered that in accordance with Fig. 2.1, $\alpha=\theta-\theta_r$:

$$f_r(\theta,\theta_r,t)=\tfrac{3}{2}N_{re}\,\mathrm{Re}[\bar{\imath}_r\,e^{-j(\theta-\theta_r)}]=\tfrac{3}{2}N_{re}\,\mathrm{Re}(\bar{\imath}_r'\,e^{-j\theta}) \tag{2.1-24}$$

where

$$\bar{\imath}_r'=\bar{\imath}_r\,e^{-j\theta_r}=|\bar{\imath}_r|\,e^{j\alpha_r'} \tag{2.1-25}$$

is the space phasor of the rotor currents expressed in the stationary reference frame. Substitution of eqn (2.1-20) into eqn (2.1-25) finally gives

$$\bar{\imath}_r'=|\bar{\imath}_r|\,e^{j\alpha_r'}=|\bar{\imath}_r|\,e^{j(\alpha_r+\theta_r)}. \tag{2.1-26}$$

The fact that $\bar{\imath}_r'$ is the space phasor of the rotor currents expressed in the stationary reference frame can be proved as follows. The relationship between the reference frame fixed to the stator (whose direct and quadrature axes are denoted by sD and sQ respectively) and the reference frame fixed to the rotor, which rotates at angular speed ω_r, is shown in Fig. 2.3.

It follows from eqn (2.1-20) that the space phasor of rotor currents expressed in the rotating reference frame is $\bar{\imath}_r=|\bar{\imath}_r|\,e^{j\alpha_r}$, where α_r is the angle of the current space phasor with respect to the rα-axis, which is stationary with respect to the moving rotor. This angle is also shown in Fig. 2.3. Here $\bar{\imath}_r'$ is the space phasor of the rotor currents expressed in the reference frame, which is stationary with respect to the stator, and it follows from Fig. 2.3 that $\bar{\imath}_r'=|\bar{\imath}_r|\,e^{j\alpha_r'}$, where $\alpha_r'=\alpha_r+\theta_r$, which is in full agreement with eqn (2.1-26).

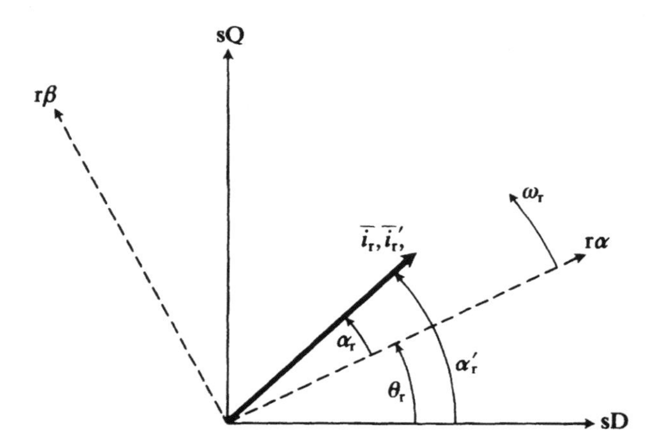

Fig. 2.3. The relationship between the stationary and rotating reference frames.

2.1.3 THE MAGNETIZING-CURRENT SPACE PHASOR

Due to the combined effects of stator and rotor excitations, the resultant m.m.f. wave is equal to the sum of the stator and rotor m.m.f. distributions,

$$f(\theta, \theta_r, t) = f_s(\theta, t) + f_r(\theta, \theta_r, t) \tag{2.1-27}$$

and by considering eqns (2.1-3), (2.1-4), and (2.1-24), it follows that

$$f(\theta, \theta_r, t) = \tfrac{3}{2} N_{se}[\mathrm{Re}(\bar{i}_s \mathrm{e}^{-j\theta}) + N_{re}/N_{se} \mathrm{Re}(\bar{i}_r' \mathrm{e}^{-j\theta})]. \tag{2.1-28}$$

Thus

$$f(\theta, \theta_r, t) = \tfrac{3}{2} N_{se} \mathrm{Re}[(\bar{i}_s + N_{re}/N_{se} \bar{i}_r') \mathrm{e}^{-j\theta}] \tag{2.1-29}$$

is obtained. In eqn (2.1-29), the complex current in the brackets multiplied by $\mathrm{e}^{-j\theta}$ is the magnetizing-current space phasor expressed in the stationary reference frame (which is fixed to the stator) and therefore it is defined as

$$\bar{i}_m = \bar{i}_s + (N_{re}/N_{se})\bar{i}_r'. \tag{2.1-30}$$

This is the sum of the stator-current space phasor and the rotor-current space phasor expressed in the stationary reference frame fixed to the stator. However, if the rotor-current space phasor is referred to the stator, the reference factor is $N_{se}/N_{re} = (N_s k_{ws})/(N_r k_{wr})$. When $\bar{i}_r'$ is divided by this factor, then $(N_{re}\bar{i}_r)/N_{se}$ is obtained, which is the rotor-current space phasor referred to the stator and expressed in the stationary reference frame. By a special selection of the turns ratio, e.g. if the reference factor is equal to the ratio of the magnetizing inductance and the rotor self inductance, eqn (2.1-30) yields the so-called rotor magnetizing-current space phasor expressed in the stationary reference frame, $\bar{i}_{mr}$ (for more detailed discussion see Section 4.1.1). The rotor magnetizing-current space phasor is extensively utilized when the so-called rotor-flux-oriented control of a.c. machines is employed (e.g. see Sections 1.2, 2.1.8 and 4.1).

2.1.4 FLUX-LINKAGE SPACE PHASORS

In this section the space phasors of the stator and rotor flux linkages will be obtained in various reference frames.

2.1.4.1 The stator flux-linkage space phasor in the stationary reference frame

Similarly to the definitions of the stator-current and rotor-current space phasors, it is possible to define the space phasor of the stator flux linkages $\bar{\psi}_s$ in terms of the instantaneous values of the flux linkages of the three stator windings. Thus in the stationary reference frame fixed to the stator, the total flux-linkage space phasor can be expressed as follows

$$\bar{\psi}_s = \tfrac{2}{3}(\psi_{sA} + a\psi_{sB} + a^2\psi_{sC}), \tag{2.1-31}$$

where the instantaneous values of the phase-variable flux-linkage components are

$$\begin{aligned}\psi_{sA} = {} & \bar{L}_s i_{sA} + \bar{M}_s i_{sB} + \bar{M}_s i_{sC} + \bar{M}_{sr} \cos\theta_r i_{ra} \\ & + \bar{M}_{sr}\cos(\theta_r + 2\pi/3) i_{rb} + \bar{M}_{sr}\cos(\theta_r + 4\pi/3) i_{rc}\end{aligned} \tag{2.1-32}$$

$$\begin{aligned}\psi_{sB} = {} & \bar{L}_s i_{sB} + \bar{M}_s i_{sA} + \bar{M}_s i_{sC} + \bar{M}_{sr} \cos(\theta_r + 4\pi/3) i_{ra} \\ & + \bar{M}_{sr}\cos\theta_r i_{rb} + \bar{M}_{sr}\cos(\theta_r + 2\pi/3) i_{rc}\end{aligned} \tag{2.1-33}$$

$$\begin{aligned}\psi_{sC} = {} & \bar{L}_s i_{sC} + \bar{M}_s i_{sB} + \bar{M}_s i_{sA} + \bar{M}_{sr} \cos(\theta_r + 2\pi/3) i_{ra} \\ & + \bar{M}_{sr}\cos(\theta_r + 4\pi/3) i_{rb} + \bar{M}_{sr}\cos\theta_r i_{rc}.\end{aligned} \tag{2.1-34}$$

In these equations $\bar{L}_s$ is the self-inductance of a stator phase winding, $\bar{M}_s$ is the mutual inductance between the stator windings, and $\bar{M}_{sr}$ is the maximal value of the stator–rotor mutual inductance. It can be seen that the phase-variable flux linkages contain six flux-linkage terms, a self-flux-linkage component produced by the stator currents in the stator winding under consideration, two mutual stator flux-linkage components due to the other two stator currents and three stator–rotor mutual flux-linkage components, which are due to the three rotor currents. Substitution of eqns (2.1-32), (2.1-33), and (2.1-34) into eqn (2.1-31) yields the following space-phasor equation for the stator flux linkages if eqns (2.1-4), (2.1-20), and (2.1-25) are also considered:

$$\bar{\psi}_s = L_s \bar{i}_s + L_m \bar{i}_r' = L_s \bar{i}_s + L_m \bar{i}_r e^{j\theta_r}, \tag{2.1-35}$$

where $L_s = \bar{L}_s - \bar{M}_s$ is the total three-phase stator inductance and L_m is the so-called three-phase magnetizing inductance, $L_m = (\tfrac{3}{2})\bar{M}_{sr}$.

The stator flux-linkage space phasor describes the modulus and phase angle of the peak of the sinusoidal stator flux distribution in the air-gap. In eqn (2.1-35) there are two flux-linkage space phasor components. The first component, which is equal to $L_s \bar{i}_s$, is the self-flux-linkage space phasor of the stator phases, which is caused by the stator currents. The second component, $L_m \bar{i}_r'$, is a mutual flux-linkage space phasor, which is due to the rotor currents and is expressed in the stationary reference frame. It is important to note that eqn (2.1-35) is general, and holds even under non-linear magnetic conditions. Thus it is also valid when the leakage or main flux paths are saturated. In this case L_s and L_m are not constant, but also depend on the currents of the machine. This will be discussed in detail in Chapter 6.

It is possible to give the definition of the stator flux linkages in terms of the direct- and quadrature-axis flux-linkage components ψ_{sD} and ψ_{sQ}:

$$\bar{\psi}_s = \psi_{sD} + j\psi_{sQ}, \tag{2.1-36}$$

where it follows by considering eqn (2.1-35) that the direct-axis stator flux-linkage component is defined as

$$\psi_{sD} = L_s i_{sD} + L_m i_{rd} \tag{2.1-37}$$

and the quadrature-axis stator flux-linkage component is given by

$$\psi_{sQ} = L_s i_{sQ} + L_m i_{rq}. \tag{2.1-38}$$

The relationship between the instantaneous values of the direct- and quadrature-axis flux-linkage components ψ_{sD} and ψ_{sQ} and the instantaneous values of the three-phase stator flux-linkage components ψ_{sA}, ψ_{sB}, and ψ_{sC} is similar to the relationship between the two-axis currents and the three-phase currents described by eqns (2.1-9) and (2.1-10).

In the equations above, i_{sD}, i_{sQ} and i_{rd}, i_{rq} are the instantaneous values of the direct- and quadrature-axis stator and rotor currents respectively, and it is important to note that all four currents are defined in the stationary reference frame fixed to the stator and they can vary arbitrarily in time. The rotor currents i_{rd} and i_{rq} are related to the rotor currents $i_{r\alpha}$ and $i_{r\beta}$ by eqn (2.1-25), and the latter current components, defined by eqns (2.1-21) and (2.1-22), are the two-axis components of the rotor currents in the reference frame fixed to the rotor. Thus it follows from eqn (2.1-25) that

$$\bar{i}'_r = i_{rd} + j i_{rq} = \bar{i}_r e^{j\theta_r} \tag{2.1-39}$$

and this yields the following transformational relationship between the d, q and α, β components of the rotor currents, if for convenience the matrix form is used, as is usually the case in the generalized theory of electrical machines:

$$\begin{bmatrix} i_{rd} \\ i_{rq} \end{bmatrix} = \begin{bmatrix} \cos\theta_r & -\sin\theta_r \\ \sin\theta_r & \cos\theta_r \end{bmatrix} \begin{bmatrix} i_{r\alpha} \\ i_{r\beta} \end{bmatrix} = \mathbf{C}_2^{-1} \begin{bmatrix} i_{r\alpha} \\ i_{r\beta} \end{bmatrix}. \tag{2.1-40}$$

In eqn (2.1-40) the inverse of the so-called commutator transformation matrix ($\mathbf{C}_2$) appears. It is also an advantage of the application of the space phasors that, in contrast to the matrix forms used in generalized machine theory, where the various machine models corresponding to different reference frames are obtained by the application of matrix transformations, here complex transformations—e.g. $e^{j\theta_r}$ in eqn (2.1-39)—are used, which result in more compact, more easily manipulable equations. Furthermore, by expressing the space phasor of a given quantity in various reference frames, all the matrix transformations of generalized machine theory can be obtained [Vas 1992].

2.1.4.2 The rotor flux-linkage space phasor in the rotating reference frame fixed to the rotor

The space phasor of the rotor flux linkages expressed in its own (natural) reference frame, i.e. in the reference frame fixed to the rotor, and rotating at speed ω_r, is defined as follows

$$\bar{\psi}_r = \tfrac{2}{3}[\psi_{ra}(t) + a\psi_{rb}(t) + a^2\psi_{rc}(t)], \tag{2.1-41}$$

where $\psi_{ra}(t)$, $\psi_{rb}(t)$, and $\psi_{rc}(t)$ are the instantaneous values of the rotor flux linkages in the rotor phases ra, rb, and rc respectively. In terms of the instantaneous values of the stator and rotor currents they can be expressed as

$$\psi_{ra} = \bar{L}_r i_{ra} + \bar{M}_r i_{rb} + \bar{M}_r i_{rc} + \bar{M}_{sr}\cos\theta_r i_{sA} + \bar{M}_{sr}\cos(\theta_r + 4\pi/3)i_{sB} + \bar{M}_{sr}\cos(\theta_r + 2\pi/3)i_{sC} \tag{2.1-42}$$

$$\psi_{rb} = \bar{L}_r i_{rb} + \bar{M}_r i_{ra} + \bar{M}_r i_{rc} + \bar{M}_{sr}\cos(\theta_r + 2\pi/3)i_{sA} + \bar{M}_{sr}\cos\theta_r i_{sB} + \bar{M}_{sr}\cos(\theta_r + 4\pi/3)i_{sC} \tag{2.1-43}$$

$$\psi_{rc} = \bar{L}_r i_{rc} + \bar{M}_r i_{ra} + \bar{M}_r i_{rb} + \bar{M}_{sr}\cos(\theta_r + 4\pi/3)i_{sA} + \bar{M}_{sr}\cos(\theta_r + 2\pi/3)i_{sB} + \bar{M}_{sr}\cos\theta_r i_{sC}, \tag{2.1-44}$$

where $\bar{L}_r$ is the self-inductance of a rotor winding and $\bar{M}_r$ is the mutual inductance between two rotor phases. It can be seen that all three rotor flux-linkage components $\psi_{ra}(t)$, $\psi_{rb}(t)$, and $\psi_{rc}(t)$, contain three flux-linkage components produced by the rotor currents and three mutual flux-linkage components produced by the stator currents. A considerable simplification is achieved if eqns (2.1-42), (2.1-43), and (2.1-44) are substituted into eqn (2.1-41) and thus the space phasor of rotor flux linkages in the rotor reference frame is obtained as

$$\bar{\psi}_r = L_r\bar{i}_r + L_m\bar{i}_s', \tag{2.1-45}$$

where $L_r = \bar{L}_r - \bar{M}_r$ is the total three-phase rotor inductance and $\bar{i}_s'$ is the space phasor of the stator currents expressed in the reference frame fixed to the rotor. Equation (2.1-45) contains two terms: (i) the space phasor $L_r\bar{i}_r$ is the rotor self-flux-linkage space phasor expressed in the rotor reference frame and is solely due to the rotor currents and; (ii) the space phasor $L_m\bar{i}_s'$ is a mutual flux-linkage space phasor, produced by the stator currents and expressed in the same reference frame.

Instead of defining the rotor flux-linkage space phasor in terms of the rotor flux-linkage components corresponding to the three phases, it is possible to define it in terms of its two-axis components ($\psi_{r\alpha}$, $\psi_{r\beta}$),

$$\bar{\psi}_r = \psi_{r\alpha} + j\psi_{r\beta}, \tag{2.1-46}$$

where it follows from eqn (2.1-45) that the direct-axis rotor flux-linkage component can be defined as

$$\psi_{r\alpha}=L_r i_{r\alpha}+L_m i_{sd} \tag{2.1-47}$$

and the quadrature-axis rotor flux-linkage component can be given as

$$\psi_{r\beta}=L_r i_{r\beta}+L_m i_{sq}. \tag{2.1-48}$$

In eqns (2.1-47) and (2.1-48) $i_{r\alpha}$, $i_{r\beta}$, i_{sd}, and i_{sq} are the direct- and quadrature-axis rotor and stator current components respectively, and all the current components are expressed in the reference frame fixed to the rotor. The relationship of the stator current components i_{sd}, i_{sq} and the stator current components i_{sD}, i_{sQ} will be shown in the next section.

2.1.4.3 The rotor flux-linkage space phasor in the stationary reference frame

The rotor flux-linkage components in the reference frame fixed to the rotor ($\psi_{r\alpha}, \psi_{r\beta}$) are related to the rotor flux-linkage components expressed in the stationary reference frame (ψ_{rd}, ψ_{rq}) by the same transformation $e^{j\theta_r}$ as given by eqn (2.1-39) for the rotor currents. Thus the following equation holds

$$\bar{\psi}'_r=\psi_{rd}+j\psi_{rq}=\bar{\psi}_r e^{j\theta_r}=(\psi_{r\alpha}+j\psi_{r\beta})e^{j\theta_r} \tag{2.1-49}$$

and this can be put into the following matrix form:

$$\begin{bmatrix}\psi_{rd}\\ \psi_{rq}\end{bmatrix}=\begin{bmatrix}\cos\theta_r & -\sin\theta_r\\ \sin\theta_r & \cos\theta_r\end{bmatrix}\begin{bmatrix}\psi_{r\alpha}\\ \psi_{r\beta}\end{bmatrix}. \tag{2.1-50}$$

In eqn (2.1-50) the transformation matrix is the inverse of the commutator transformation matrix given in eqn (2.1-40).

By the substitution of eqn (2.1-45) into eqn (2.1-49) and by considering eqn (2.1-39), the space phasor of the rotor flux linkages in the stationary reference frame can be expressed as

$$\bar{\psi}'_r=(L_r\bar{i}'_r+L_m\bar{i}'_s e^{j\theta_r})=L_r\bar{i}'_r+L_m\bar{i}_s. \tag{2.1-51}$$

This contains two flux-linkage components, a self flux linkage produced by the rotor currents but expressed in the stationary reference frame ($L_r\bar{i}'_r$) and a mutual flux-linkage component produced by the stator currents and also expressed in the stationary reference frame ($L_m\bar{i}_s=L_m\bar{i}'_s e^{j\theta_r}$). Thus the stator currents in the stationary reference frame are related to the stator current components in the rotating reference frame fixed to the rotor by the following complex transformation:

$$\bar{i}_s=\bar{i}'_s e^{j\theta_r}, \tag{2.1-52}$$

where $\bar{i}_s$ and $\bar{i}'_s$ are expressed in terms of their two-axis components as follows:

$$\bar{i}_s=i_{sD}+ji_{sQ}$$

$$\bar{i}'_s=i_{sd}+ji_{sq}. \tag{2.1-53}$$

It follows from eqn (2.1-52) that the transformed stator-current space phasor in the rotating reference frame fixed to the rotor can be obtained from the space phasor of the stator currents, expressed in the stationary reference frame, as

$$\bar{\imath}_s' = \bar{\imath}_s e^{-j\theta_r}. \tag{2.1-54}$$

The corresponding two-axis form can be obtained by the substitution of eqn (2.1-53) into eqn (2.1-54),

$$\begin{bmatrix} i_{sd} \\ i_{sq} \end{bmatrix} = \begin{bmatrix} \cos\theta_r & \sin\theta_r \\ -\sin\theta_r & \cos\theta_r \end{bmatrix} \begin{bmatrix} i_{sD} \\ i_{sQ} \end{bmatrix} \tag{2.1-55}$$

and these are the stator currents used in eqns (2.1-47) and (2.1-48). The transformation described by eqn (2.1-54) can also be obtained by considering Fig. 2.4.

It follows from eqn (2.1-4) that the stator-current space phasor expressed in the stationary reference frame ($\bar{\imath}_s$) can be expressed in terms of its modulus ($|\bar{\imath}_s|$) and its phase angle (α_s), which are also shown in Fig. 2.4, as $\bar{\imath}_s = |\bar{\imath}_s| e^{j\alpha_s}$. However, when the space phasor of the stator currents is expressed in the reference frame rotating with the rotor speed ω_r, then it follows from Fig. 2.4 that $\bar{\imath}_s' = |\bar{\imath}_s| e^{j\alpha_s'}$, where the angle α_s' is the angle between $\bar{\imath}_s'$ and the real axis (sd) of the rotating reference frame. However, since it follows from Fig. 2.4 that $\alpha_s' = \alpha_s - \theta_r$, $\bar{\imath}_s' = |\bar{\imath}_s| e^{j(\alpha_s - \theta_r)}$ is obtained, which can be expressed as $\bar{\imath}_s' = \bar{\imath}_s e^{-j\theta_r}$, and this is in agreement with eqn (2.1-54).

2.1.4.4 The stator flux-linkage space phasor in the reference frame fixed to the rotor

It should be noted that if, for example, it is necessary to know the space phasor of the stator flux linkages expressed in the rotating reference frame ($\bar{\psi}_s'$) in terms of the stator flux-linkage space phasor expressed in the stationary reference

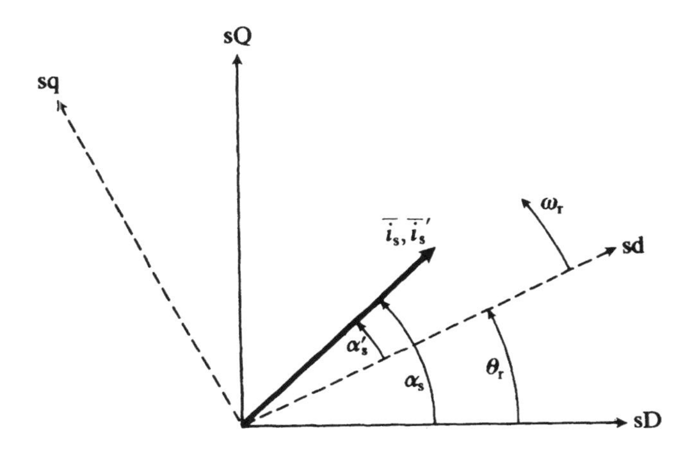

Fig. 2.4. Transformation of the stator-current space phasor.

frame ($\bar{\psi}_s$), then similarly to eqn (2.1-54) the equation

$$\bar{\psi}_s' = \bar{\psi}_s e^{-j\theta_r} \tag{2.1-56}$$

holds, or in two-axis form, by considering $\bar{\psi}_s' = \psi_{sd} + j\psi_{sq}$, $\bar{\psi}_s = \psi_{sD} + j\psi_{sQ}$,

$$\begin{bmatrix} \psi_{sd} \\ \psi_{sq} \end{bmatrix} = \begin{bmatrix} \cos\theta_r & \sin\theta_r \\ -\sin\theta_r & \cos\theta_r \end{bmatrix} \begin{bmatrix} \psi_{sD} \\ \psi_{sQ} \end{bmatrix}, \tag{2.1-57}$$

where the transformation matrix is the commutator transformation $\mathbf{C}_2$.

Incidentally, the substitution of eqn (2.1-35) into eqn (2.1-56) yields $\bar{\psi}_s'$ as follows, if eqns (2.1-39) and (2.1-54) are also considered:

$$\bar{\psi}_s' = (L_s \bar{i}_s + L_m \bar{i}_r') e^{-j\theta_r} = L_s \bar{i}_s' + L_m \bar{i}_r, \tag{2.1-58}$$

where $\bar{i}_s'$ and $\bar{i}_r$ are the space phasors of the stator and rotor currents respectively and are expressed in the reference frame fixed to the rotor. By resolving into real and imaginary parts, and using the notation introduced in eqns (2.1-23) and (2.1-53), the direct- and quadrature-axis stator flux linkages in the rotor reference frame are obtained as

$$\psi_{sd} = L_s i_{sd} + L_m i_{r\alpha} \tag{2.1-59}$$

$$\psi_{sq} = L_s i_{sq} + L_m i_{r\beta}, \tag{2.1-60}$$

where i_{sd}, i_{sq}, $i_{r\alpha}$, and $i_{r\beta}$ are the two-axis stator and rotor currents respectively in the reference frame fixed to the rotor.

2.1.5 THE SPACE PHASORS OF THE STATOR AND ROTOR VOLTAGES

The space phasors of the stator and rotor voltages can be defined similarly to the space-phasor quantities defined in the earlier sections [e.g. eqn (2.1-4)]. Thus the stator-voltage space phasor in the stationary reference frame is

$$\bar{u}_s = \tfrac{2}{3}[u_{sA}(t) + a u_{sB}(t) + a^2 u_{sC}(t)] = u_{sD} + j u_{sQ} \tag{2.1-61}$$

and the rotor-voltage space phasor in the reference frame fixed to the moving rotor is

$$\bar{u}_r = \tfrac{2}{3}[u_{ra}(t) + a u_{rb}(t) + a^2 u_{rc}(t)] = u_{r\alpha} + j u_{r\beta}. \tag{2.1-62}$$

In eqns (2.1-61) and (2.1-62) $u_{sA}(t)$, $u_{sB}(t)$, $u_{sC}(t)$ and $u_{ra}(t)$, $u_{rb}(t)$, and $u_{rc}(t)$ are the instantaneous values of the stator and rotor phase voltages respectively and u_{sD}, u_{sQ}, $u_{r\alpha}$ and $u_{r\beta}$ are the corresponding direct- and quadrature-axis components. The relationship between the three-phase and quadrature-phase voltages immediately follows from these equations. For example, for the stator voltages,

$$u_{sD} = \mathrm{Re}\{\tfrac{2}{3}[u_{sA}(t) + a u_{sB}(t) + a^2 u_{sC}(t)]\} = \tfrac{2}{3}(u_{sA} - \tfrac{1}{2} u_{sB} - \tfrac{1}{2} u_{sC}) \tag{2.1-63}$$

$$u_{sQ} = \mathrm{Im}\{\tfrac{2}{3}[u_{sA}(t) + a u_{sB}(t) + a^2 u_{sB}(t)]\} = (u_{sB} - u_{sC})/\sqrt{3}. \tag{2.1-64}$$

Similarly, the following equations hold for the rotor voltage components:

$$u_{r\alpha}=\tfrac{2}{3}(u_{ra}-\tfrac{1}{2}u_{rb}-\tfrac{1}{2}u_{rc}) \quad (2.1\text{-}65)$$

and

$$u_{r\beta}=(u_{rb}-u_{rc})/\sqrt{3}. \quad (2.1\text{-}66)$$

It has been emphasized in Section 2.1.1 that the space phasor does not contain the zero-sequence components. Thus if there are zero-sequence voltages, they have to be considered separately in terms of the following stator and rotor zero-sequence components

$$u_{s0}=\tfrac{1}{3}[u_{sA}(t)+u_{sB}(t)+u_{sC}(t)] \quad (2.1\text{-}67)$$

$$u_{r0}=\tfrac{1}{3}[u_{ra}(t)+u_{rb}(t)+u_{rc}(t)]. \quad (2.1\text{-}68)$$

It should be noted that if eqns (2.1-63), (2.1-64), and (2.1-67) are put into matrix form,

$$\begin{bmatrix} u_{s0} \\ u_{sD} \\ u_{sQ} \end{bmatrix} = \frac{2}{3}\begin{bmatrix} 1/2 & 1/2 & 1/2 \\ 1 & -1/2 & -1/2 \\ 0 & \sqrt{3}/2 & -\sqrt{3}/2 \end{bmatrix}\begin{bmatrix} u_{sA} \\ u_{sB} \\ u_{sC} \end{bmatrix}, \quad (2.1\text{-}69)$$

where the transformation matrix is the inverse of the so-called phase transformation matrix ($\mathbf{C}_1$). A similar transformation matrix applies for the rotor voltages. It should also be pointed out that similarly to eqns (2.1-12)–(2.1-14), in the absence of the zero-sequence components, the projections of the space phasor of voltages on the corresponding axes yield the phase voltages, i.e.

$$u_{sA}=\mathrm{Re}(\bar{u}_s) \quad (2.1\text{-}70)$$

$$u_{sB}=\mathrm{Re}(a^2\bar{u}_s) \quad (2.1\text{-}71)$$

$$u_{sC}=\mathrm{Re}(a\bar{u}_s). \quad (2.1\text{-}72)$$

The stator-voltage space phasor expressed in the stationary reference frame ($\bar{u}_s$) can be transformed into the stator space phasor expressed in the reference frame fixed to the rotor ($\bar{u}_s'$) similarly to eqn (2.1-54). Thus

$$\bar{u}_s'=\bar{u}_s e^{-j\theta_r}=u_{sd}+ju_{sq}. \quad (2.1\text{-}73)$$

If this is resolved into real- and imaginary-axis components, the same transformation matrix ($\mathbf{C}_2$) will appear as in eqn (2.1-57).

The rotor-voltage space phasor expressed in the reference frame fixed to the rotor ($\bar{u}_r$) can be expressed in the reference frame fixed to the stator ($\bar{u}_r'$), and in this case the complex transformation to be used is the same as the one used in eqn (2.1-39). Thus

$$\bar{u}_r'=\bar{u}_r e^{j\theta_r}=u_{rd}+ju_{rq}. \quad (2.1\text{-}74)$$

When this is put into the two-axis form, the same transformation matrix ($\mathbf{C}_2^{-1}$) will appear as in eqn (2.1-40).

If, for example, the relationship between u_{sA}, u_{sB}, u_{sC} and u_{sd}, u_{sq} is required, it follows from eqns (2.1-73) and (2.1-61) that

$$u_{sd}+\mathrm{j}u_{sq}=\tfrac{2}{3}[u_{sA}(t)+au_{sB}(t)+a^2u_{sC}(t)](\cos\theta_r-\mathrm{j}\sin\theta_r). \quad (2.1\text{-}75)$$

Resolution of eqn (2.1-75) into real and imaginary parts yields

$$u_{sd}=\tfrac{2}{3}[u_{sA}\cos\theta_r+u_{sB}\cos(\theta_r-2\pi/3)+u_{sC}\cos(\theta_r-4\pi/3)] \quad (2.1\text{-}76)$$

$$u_{sq}=-\tfrac{2}{3}[u_{sA}\sin\theta_r+u_{sB}\sin(\theta_r-2\pi/3)+u_{sC}\sin(\theta_r-4\pi/3)], \quad (2.1\text{-}77)$$

which define the so-called Park transformation. Incidentally, if the zero-sequence stator voltage component defined by eqn (2.1-67) is combined with eqns (2.1-76) and (2.1-77), the following matrix form is obtained,

$$\begin{bmatrix} u_{s0} \\ u_{sd} \\ u_{sq} \end{bmatrix} = \frac{2}{3} \begin{bmatrix} 1/2 & 1/2 & 1/2 \\ \cos\theta_r & \cos(\theta_r-2\pi/3) & \cos(\theta_r-4\pi/3) \\ -\sin\theta_r & -\sin(\theta_r-2\pi/3) & -\sin(\theta_r-4\pi/3) \end{bmatrix} \begin{bmatrix} u_{sA} \\ u_{sB} \\ u_{sC} \end{bmatrix}. \quad (2.1\text{-}78)$$

2.1.6 THE MECHANISM OF ELECTROMAGNETIC TORQUE PRODUCTION

In this section, the mechanism of electromagnetic torque production in d.c. and a.c. machines will be discussed and for simplicity the effects of magnetic non-linearities will be neglected. To enhance the analogy between the mechanism of torque production in d.c. and a.c. machines, the space-phasor formulation of the electromagnetic torque will be presented for both types of machines.

2.1.6.1 Electromagnetic torque production in d.c. machines

Figure 2.5 shows the schematic diagram of a compensated d.c. machine, which for simplicity has a smooth air-gap.

On the stator of the machine there are the field (f) and compensating windings (c) and on the rotor there is the armature winding (a). The current in the field winding i_f produces an excitation flux linkage ψ_f. If current i_a flows in the armature winding, the interaction of the armature current and the excitation flux-linkage will result in forces (F) acting on the conductors, as shown in Fig. 2.5. Since the excitation flux linkage is in space quadrature to the armature current, maximal forces are applied to the shaft and therefore the position of the armature winding is optimal for electromagnetic torque production.

It follows from Fig. 2.5 that the armature winding also produces a field, which is superimposed on the field produced by the field winding, but it is in space quadrature with respect to the excitation flux. Thus the resultant field will be displaced from its optimal position. However, this effect can be cancelled by the application of the compensating winding (c), which carries current i_c which is equal to $-i_a$. The currents in the compensating winding and the field winding produce an electromagnetic torque which acts against the armature.

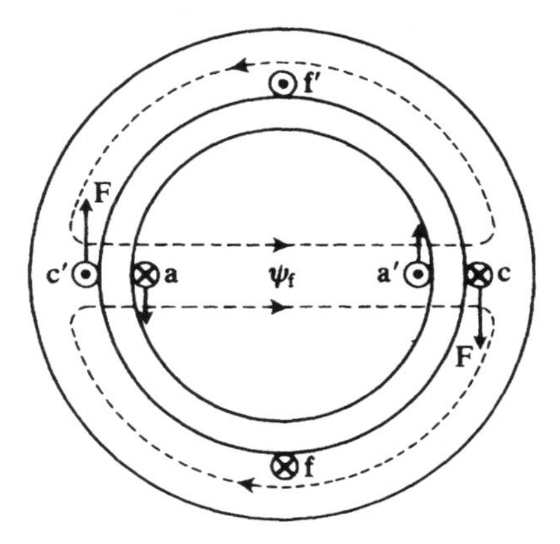

Fig. 2.5. Electromagnetic torque production in a d.c. machine.

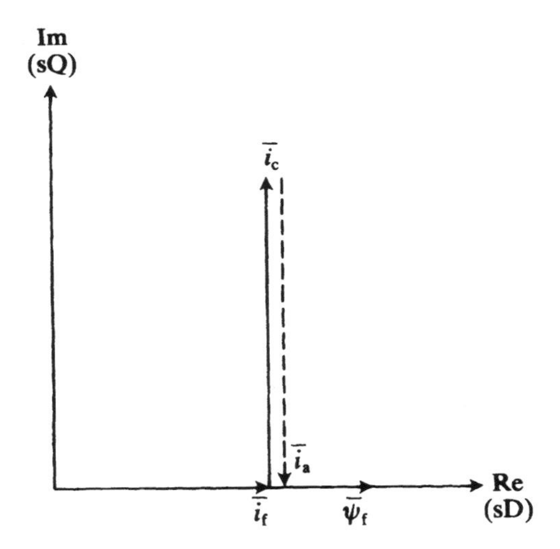

Fig. 2.6. The space phasors in a d.c. machine.

Figure 2.6 shows the space phasors of the currents in the field winding, compensating winding, and armature winding respectively and also the excitation flux-linkage. These are denoted by $\bar{i}_f$, $\bar{i}_c$, $\bar{i}_a$, and $\bar{\psi}_f$ respectively and it should be noted that the stationary reference frame with direct (sD) and quadrature (sQ) axes has been used, so that these space phasors are expressed in the stationary reference frame fixed to the stator of the machine.

It follows that in the d.c. machine described above, the current in the field winding creates the excitation flux-linkage, and the interaction of the excitation flux with the currents in the armature winding and compensating winding produce the electromagnetic torque under both steady-state and transient conditions.

Under linear magnetic conditions it is possible to express the instantaneous value of the electromagnetic torque (t_e) as a vector (cross) product of the excitation flux-linkage and armature-current space phasors, and thus the following expression is obtained:

$$t_e = c\bar{\psi}_f \times \bar{i}_a. \tag{2.1-79}$$

In eqn (2.1-79) c is a constant and $\times$ denotes the vector product. Since the two space phasors are in space quadrature, eqn (2.1-79) can be put into the following form:

$$t_e = c\psi_f i_a, \tag{2.1-80}$$

where ψ_f and i_a are the moduli of the respective space phasors and are equal to the instantaneous values of the excitation flux-linkage and armature current respectively. If the excitation flux is maintained constant, the electromagnetic torque can be controlled by varying the armature current and a change in the armature current will result in a rapid change in the torque. It is the purpose of vector control of a.c. machines to have a similar technique of rapid torque control, as discussed in the next section (see also Section 1.2).

2.1.6.2 Electromagnetic torque production in a.c. machines

In this section the analogy between the electromagnetic torque production in d.c. and a.c. machines is described. It will be shown that in general, in symmetrical three-phase or quadrature-phase smooth-air-gap a.c. machines, the developed instantaneous electromagnetic torque can be put into the following vectorial form, which is similar to eqn (2.1-79),

$$t_e = c\bar{\psi}_s \times \bar{i}_r', \tag{2.1-81}$$

where under linear magnetic conditions c is a constant and $\bar{\psi}_s$ and $\bar{i}_r'$ are the space phasors of the stator flux linkages and rotor currents respectively, expressed in the stationary reference frame. Thus the electromagnetic torque is the cross product of a flux linkage and a current space phasor. In eqn (2.1-81) the stator flux-linkage space phasor and the rotor-current space phasors can be expressed in reference frames other than the stationary reference frame. However, in this case other transformed space phasors will be present in the expression for the electromagnetic torque, e.g. $\bar{\psi}_s'$ and $\bar{i}_r$, which correspond to the space phasors of the stator flux linkages and rotor currents expressed in the rotor reference frame.

By utilizing the Euler forms of vector quantities, it is possible to express eqn (2.1-81) as

$$t_e = c|\bar{\psi}_s|\,|\bar{i}_r|\sin\gamma, \tag{2.1-82}$$

where $|\bar{\psi}_s|$ and $|\bar{i}_r|$ are the moduli of the stator flux-linkage and rotor-current space phasors respectively and γ is the so-called torque angle. It follows that when $\gamma = 90\,^\circ$, eqn (2.1-82) will take the form of eqn (2.1-80) and maximum torque is obtained.

Thus the similarity between the production of the electromagnetic torque in a compensated d.c. machine and in symmetrical, smooth-air-gap a.c. machines has been established. However, it should be noted that whilst in the d.c. machine, where the armature current and main flux distributions are fixed in space—the former is due to the action of the commutator—and where torque control can be established by independently controlling the excitation flux (ψ_f) and armature current (i_a) and where, due to the fixed relationship between the stator field and armature current distribution, maximal torque is produced for given field and armature currents, in an a.c. machine, it is much more difficult to realize this principle, because these quantities are coupled and are stationary with respect to the stator and rotor respectively; they also depend on the modulus, frequency, and phase angles of the stator currents. It is a further complication that in machines with a squirrel-cage rotor, it is not possible to monitor the rotor currents by simple means. Of course, there are ways of preparing the rotor cage in advance and using certain transducers which yield signals proportional to the instantaneous values of the rotor currents, but these are only suitable for laboratory work or under very special conditions. The search for simple control schemes, similar to those used for d.c. machines, has led to the development of the so-called vector-controlled schemes, where similarly to d.c. motors, it is possible to obtain two current components, one of which is a flux-producing current component and the other is the torque-producing current component. It will be shown in later sections (e.g. Section 2.1.8.2) how the expression for the instantaneous electromagnetic torque can be formulated in terms of these two currents. There will also be a discussion on how they can be independently controlled so as to achieve the required decoupled control.

For completeness, a proof of eqn (2.1-81) will now be presented and this will be discussed in the remainder of the present section. This proof will be based on energy considerations, but for better understanding, in Sections 2.1.7 and 2.1.8 other derivations will also be given which utilize other concepts. It is possible to obtain an expression for the electromagnetic torque by putting the rate of change of the mechanical output energy ($\mathrm{d}W_{mech}/\mathrm{d}t$) equal to the mechanical power p_{mech}; the latter is equal to the product of the instantaneous rotor speed and electromagnetic torque,

$$p_{mech} = \frac{\mathrm{d}W_{mech}}{\mathrm{d}t} = t_e \omega_r. \qquad (2.1\text{-}83)$$

For simplicity a two-pole machine is considered, and the expression for the electromagnetic torque is derived from eqn (2.1-83). For this purpose, it is first necessary to formulate an equation for the differential mechanical energy $\mathrm{d}W_{mech}$.

Since for every machine the principle of conservation of energy must be valid during motion, it follows that the input electrical energy (W_e) has to cover the energies related to the stator and rotor losses (W_{loss}), the magnetic energy stored in the field (W_{field}) and the mechanical output energy (W_{mech}), and thus

$$W_e = W_{loss} + W_{field} + W_{mech}, \qquad (2.1\text{-}84)$$

and the differential mechanical output energy can be obtained as

$$\mathrm{d}W_{\mathrm{mech}} = \mathrm{d}W_{\mathrm{e}} - \mathrm{d}W_{\mathrm{loss}} - \mathrm{d}W_{\mathrm{field}}. \tag{2.1-85}$$

The three different energy components given on the right-hand side of eqn (2.1-85) can be obtained as follows. In general, for a doubly-fed polyphase machine, the differential input electrical energy can be expressed as

$$\mathrm{d}W_{\mathrm{e}} = \tfrac{3}{2}\,\mathrm{Re}(\bar{u}_{\mathrm{s}}\bar{\imath}_{\mathrm{s}}^{*} + \bar{u}_{\mathrm{r}}'\bar{\imath}_{\mathrm{r}}'^{*})\,\mathrm{d}t, \tag{2.1-86}$$

where the stator voltage and current space phasors ($\bar{u}_{\mathrm{s}}, \bar{\imath}_{\mathrm{s}}$) are expressed in the stationary reference frame and the rotor voltage and current space phasors ($\bar{u}_{\mathrm{r}}', \bar{\imath}_{\mathrm{r}}'$) are expressed in the reference frame fixed to the stator. Equation (2.1-86) follows from the physical fact that the total instantaneous power is the sum of the instantaneous power of the stator

$$p_{\mathrm{s}} = \tfrac{3}{2}\,\mathrm{Re}(\bar{u}_{\mathrm{s}}\bar{\imath}_{\mathrm{s}}^{*}) \tag{2.1-87}$$

and the instantaneous power of the rotor

$$p_{\mathrm{r}} = \tfrac{3}{2}\,\mathrm{Re}(\bar{u}_{\mathrm{r}}'\bar{\imath}_{\mathrm{r}}'^{*}) = \tfrac{3}{2}\,\mathrm{Re}(\bar{u}_{\mathrm{r}}\bar{\imath}_{\mathrm{r}}^{*}), \tag{2.1-88}$$

where the asterisk denotes the complex conjugate. Furthermore, the rate of change of the electrical energy must be equal to the total power $p_{\mathrm{s}} + p_{\mathrm{r}}$. Equations (2.1-87) and (2.1-88) can be proved by expressing the stator and rotor space-phasor voltages and currents in terms of the instantaneous phase-variable components as defined by eqns (2.1-4), (2.1-20), (2.1-25), (2.1-61), (2.1-62), and (2.1-74), and also by utilizing the assumption that there are no zero-sequence stator and rotor voltages and currents.

The losses are due to heat dissipation across the stator and rotor winding resistances, hysteresis and eddy-current losses within the magnetic material, friction losses between moving parts and either their bearings or the surrounding air, and dielectric losses in the electric fields. However, if only the winding losses are considered, the differential energy related to the stator and rotor losses can be expressed as

$$\mathrm{d}W_{\mathrm{loss}} = \tfrac{3}{2}\,(R_{\mathrm{s}}|\bar{\imath}_{\mathrm{s}}|^{2} + R_{\mathrm{r}}|\bar{\imath}_{\mathrm{r}}|^{2})\,\mathrm{d}t, \tag{2.1-89}$$

where R_{s} and R_{r} are the stator and rotor resistances respectively and the terms $\tfrac{3}{2}R_{\mathrm{s}}|\bar{\imath}_{\mathrm{s}}|^{2}$ and $\tfrac{3}{2}R_{\mathrm{r}}|\bar{\imath}_{\mathrm{r}}|^{2}$ correspond to the ohmic losses across the stator and rotor windings respectively.

The third differential energy component, the differential field energy, can be obtained by considering that the rate of change of the magnetic energy stored in the field must be equal to the input power minus the sum of the stator and rotor losses and the mechanical power. Thus

$$\mathrm{d}W_{\mathrm{field}}/\mathrm{d}t = \tfrac{3}{2}\,\mathrm{Re}(\bar{u}_{\mathrm{si}}\bar{\imath}_{\mathrm{s}}^{*} + \bar{u}_{\mathrm{ri}}'\bar{\imath}_{\mathrm{r}}'^{*}), \tag{2.1-90}$$

where $\bar{u}_{\mathrm{si}}$ and $\bar{u}_{\mathrm{ri}}'$ are the space phasors of the induced stator and rotor transformer e.m.f.s respectively, both expressed in the stationary reference frame fixed to the stator. The induced stator e.m.f. is due to the rate of change of the stator flux-linkage space phasor

$$\bar{u}_{si} = \frac{d\bar{\psi}_s}{dt} \tag{2.1-91}$$

and similarly the induced rotor e.m.f., which is due to the rate of change of the rotor flux linkages, is defined as

$$\bar{u}'_{ri} = \frac{d\bar{\psi}'_r}{dt}. \tag{2.1-92}$$

Substitution of eqns (2.1-91) and (2.1-92) into eqn (2.1-90) yields the following expression for the differential magnetic energy stored in the field:

$$dW_{field} = \tfrac{3}{2}\,\mathrm{Re}\left(\bar{i}_s^*\frac{d\bar{\psi}_s}{dt} + \bar{i}_r'^*\frac{d\bar{\psi}'_r}{dt}\right)dt \tag{2.1-93}$$

and it follows that the stored magnetic energy can be expressed as

$$W_{field} = \tfrac{3}{2}\,\mathrm{Re}(\bar{i}_s^*\bar{\psi}_s + \bar{i}_r'^*\bar{\psi}'_r)$$

which is the physically expected result. It can be put into the more familiar non-space-phasor form if the space-phasor quantities are expressed in terms of their phase-variable components.

Thus, by the substitution of eqns (2.1-86), (2.1-89), and (2.1-93) into eqn (2.1-85), the following equation is obtained for the differential mechanical energy:

$$dW_{mech} = dW_{mechs} + dW_{mechr} \tag{2.1-94}$$

where dW_{mechs} and dW_{mechr} are the mechanical energies due to the stator and rotor respectively,

$$dW_{mechs} = \tfrac{3}{2}[\mathrm{Re}(\bar{u}_s\bar{i}_s^*) - R_s|\bar{i}_s|^2 - \mathrm{Re}(\bar{i}_s^* d\bar{\psi}_s/dt)]dt \tag{2.1-95}$$

and

$$dW_{mechr} = \tfrac{3}{2}[\mathrm{Re}(\bar{u}'_r\bar{i}_r'^*) - R_r|\bar{i}'_r|^2 - \mathrm{Re}(\bar{i}_r'^* d\bar{\psi}'_r/dt)]\,dt. \tag{2.1-96}$$

Since in the stationary reference frame, the stator-voltage space phasor $\bar{u}_s$ can only be balanced by the stator ohmic drop ($R_s\bar{i}_s$) plus the rate of change of the stator flux linkages ($d\bar{\psi}_s/dt$)—see also Section 2.1.7—it follows from eqn (2.1-95) that dW_{mechs} must be equal to zero. Furthermore, in the same reference frame, the rotor-voltage space phasor $\bar{u}'_r$ must be balanced by the sum of the rotor ohmic voltage drop ($R_r\bar{i}'_r$) plus the rate of change of the rotor flux linkages ($d\bar{\psi}'_r/dt$) and a rotational voltage $-j\omega_r\bar{\psi}'_r$. (See also Section 2.1.7 where it is shown that in the stationary reference frame, the rotor voltage space-phasor equation must also contain this rotational e.m.f. term, because the rotor is rotating in the direction from stator phase A to stator phase B as shown in Fig. 2.1.) Therefore, by considering eqns (2.1-83), (2.1-94), (2.1-95), and (2.1-96), the mechanical power is obtained as

$$p_{\text{mech}}=\tfrac{3}{2}\,\text{Re}(-\text{j}\omega_r\bar{\psi}_r'\bar{i}_r'^*)=\tfrac{3}{2}\,\omega_r\,\text{Re}(-\text{j}\bar{\psi}_r'\bar{i}_r'^*)=-\tfrac{3}{2}\,\omega_r\bar{\psi}_r'\times\bar{i}_r'. \tag{2 1-97}$$

According to eqn (2.1-97), the mechanical power is proportional to the instantaneous rotor speed and to the vectorial product of the rotor flux-linkage and current space phasors, and it follows by considering eqn (2.1-83) that the electromagnetic torque can be expressed as

$$t_e=-\tfrac{3}{2}\bar{\psi}_r'\times\bar{i}_r'. \tag{2.1-98}$$

For a machine with P pole pairs this has to be multiplied by P (for a two-pole machine $P=1$, for a four-pole machine $P=2$, etc.). It should be noted that in eqn (2.1-98) the rotor flux-linkage and current space phasors $\bar{\psi}_r'$ and $\bar{i}_r'$ are expressed in the stationary reference frame, but since the torque is invariant to the change of the reference frame, the expression $-\tfrac{3}{2}\bar{\psi}_r\times\bar{i}_r$ is also valid, where $\bar{\psi}_r$ and $\bar{i}_r$ are the space phasors of the rotor flux linkages and currents respectively, but are expressed in the reference frame fixed to the rotor.

It is possible to put eqn (2.1-98) into many other forms and eqn (2.1-81) can be obtained as follows. From eqns (2.1-51) and (2.1-52) the rotor flux-linkage space phasor expressed in the stationary reference frame must contain two flux-linkage components, one of which, $L_r\,\bar{i}_r'$, is produced by the rotor currents only, where L_r is the self-inductance of a rotor winding, and the other $L_m\bar{i}_s$ is a mutual flux-linkage component produced by the stator currents, where L_m is the magnetizing inductance and $\bar{i}_s$ is the space phasor of the stator currents in the stationary reference frame. Thus eqn (2.1-98) can be put into the following form

$$t_e=-\tfrac{3}{2}(L_r\bar{i}_r'+L_m\bar{i}_s)\times\bar{i}_r'=-\tfrac{3}{2}L_m\bar{i}_s\times\bar{i}_r', \tag{2.1-99}$$

where the property that a vector product of a vector with itself is zero has been used. Expanding eqn (2.1-99), the following expression is obtained:

$$t_e=-\tfrac{3}{2}L_m\bar{i}_s\times\bar{i}_r'=-\frac{3L_m}{2L_s}(L_s\bar{i}_s+L_m\bar{i}_r')\times\bar{i}_r', \tag{2.1-100}$$

since the vector product $\bar{i}_r'\times\bar{i}_r'$ gives zero. In eqn (2.1-100) L_s is the self-inductance of a stator winding and according to eqn (2.1-35), the term $L_s\bar{i}_s+L_m\bar{i}_r'$ is equal to the space phasor of the stator flux linkages expressed in the stationary reference frame ($\bar{\psi}_s$) and contains a self-flux-linkage component produced by the stator currents ($L_s\bar{i}_s$) and a mutual flux linkage produced by the rotor currents ($L_m\bar{i}_r'$). Thus the electromagnetic torque produced by the two-pole machine can be expressed as

$$t_e=-\frac{3L_m}{2L_s}\bar{\psi}_s\times\bar{i}_r'. \tag{2.1-101}$$

If the effects of the main flux and leakage flux saturation are neglected, the magnetizing inductance (L_m) and the total stator self-inductance (L_s) are constant and eqn (2.1-101) yields eqn (2.1-81). Further expressions for the electromagnetic torque will be given in Section 2.1.8.

Finally, it should be noted that the general expressions derived for the instantaneous value of the electromagnetic torque are valid for symmetrical three-phase or quadrature-phase machines and no restriction has been made on the time variation of the stator and rotor currents, apart from the fact that there can be no zero-sequence stator or rotor currents. The space phasors used in this book rely on the assumption that the spatial distributions of the flux density and current density are sinusoidal. This is an important assumption frequently adopted in the generalized theory of electrical machines. It would be possible to define harmonic space phasors which correspond to non-sinusoidal flux density and current density distributions, but in this book the effects of space harmonics are incorporated only in the parameters of the machine under consideration and not in extra harmonic equations.

2.1.7 THE VOLTAGE EQUATIONS OF SMOOTH-AIR-GAP MACHINES IN VARIOUS REFERENCE FRAMES

In this section the stator- and rotor-voltage differential equations, which are valid in both the steady-state and transient operation of smooth-air-gap machines, will be described by using three-phase variables, two-axis variables, and space phasors. The application of space-phasor theory results in a drastic simplification of the voltage equations compared with the phase-variable forms and, while it is possible to give both physical and mathematical analyses of the dynamics concerned in terms of the phase-variable equations, this would lead to unnecessary complications.

For better understanding and to enable a direct comparison to be made between the non-space-phasor and space-phasor forms, the three-phase equations are first briefly introduced and then the two-axis models are presented together with the corresponding matrix forms of the equations. Finally, the space-phasor equations are presented.

2.1.7.1 The phase-variable voltage equations in the natural and other reference frames

The phase-variable forms of the three-phase stator and rotor voltage equations are first formulated in their natural reference frames. Thus the stator voltage equations are formulated in the stationary reference frame fixed to the stator and the rotor voltage equations are formulated in the rotating reference frame fixed to the rotor.

The three-phase model: Here the symmetrical three-phase two-pole smooth-air-gap machine with sinusoidally distributed windings, discussed earlier, is considered and the effects of m.m.f. space harmonics are neglected. The schematic of the machine has been shown in Fig. 2.1. It is assumed that the stator and rotor voltages and currents can vary arbitrarily in time. The phase-variable form of the voltage equations is as follows.

In the stationary reference frame the stator voltage equations can be expressed as

$$u_{sA}(t)=R_s i_{sA}(t)+\mathrm{d}\psi_{sA}(t)/\mathrm{d}t \tag{2.1-102}$$

$$u_{sB}(t)=R_s i_{sB}(t)+\mathrm{d}\psi_{sB}(t)/\mathrm{d}t \tag{2.1-103}$$

$$u_{sC}(t)=R_s i_{sC}(t)+\mathrm{d}\psi_{sC}(t)/\mathrm{d}t, \tag{2.1-104}$$

where $u_{sA}(t)$, $u_{sB}(t)$, and $u_{sC}(t)$, $i_{sA}(t)$, $i_{sB}(t)$, and $i_{sC}(t)$ are the instantaneous values of the stator voltages and currents respectively (in the stationary reference frame) and R_s is the resistance of a stator winding. Here $\psi_{sA}(t)$, $\psi_{sB}(t)$, and $\psi_{sC}(t)$ are the instantaneous values of the stator flux linkages in phases sA, sB, and sC respectively as defined by eqns (2.1-32), (2.1-33), and (2.1-34).

Similar expressions hold for the rotor voltage equations expressed in the reference frame fixed to the rotor,

$$u_{ra}(t)=R_r i_{ra}(t)+\mathrm{d}\psi_{ra}(t)/\mathrm{d}t \tag{2.1-105}$$

$$u_{rb}(t)=R_r i_{rb}(t)+\mathrm{d}\psi_{rb}(t)/\mathrm{d}t \tag{2.1-106}$$

$$u_{rc}(t)=R_r i_{rc}(t)+\mathrm{d}\psi_{rc}(t)/\mathrm{d}t, \tag{2.1-107}$$

where $u_{ra}(t)$, $u_{rb}(t)$, and $u_{rc}(t)$, $i_{ra}(t)$, $i_{rb}(t)$, and $i_{rc}(t)$ are the instantaneous values of the rotor phase voltages and currents respectively, R_r is the resistance of a rotor winding and $\psi_{ra}(t)$, $\psi_{rb}(t)$, and $\psi_{rc}(t)$ are the instantaneous values of the rotor flux linkages in rotor phases ra, rb, and rc respectively, as defined by eqns (2.1-42), (2.1-43), and (2.1-44).

For convenience, the stator and rotor voltage equations [eqns (2.1-102)–(2.1-107), eqns (2.1-32)–(2.1-34)], and eqns (2.1-42)–(2.1-44) of the three-phase machine can be combined into a single matrix equation:

$$\begin{bmatrix} u_{sA} \\ u_{sB} \\ u_{sC} \\ u_{ra} \\ u_{rb} \\ u_{rc} \end{bmatrix} = \begin{bmatrix} R_s+p\bar{L}_s & p\bar{M}_s & p\bar{M}_s & p\bar{M}_{sr}\cos\theta & p\bar{M}_{sr}\cos\theta_1 & p\bar{M}_{sr}\cos\theta_2 \\ p\bar{M}_s & R_s+p\bar{L}_s & p\bar{M}_s & p\bar{M}_{sr}\cos\theta_2 & p\bar{M}_{sr}\cos\theta & p\bar{M}_{sr}\cos\theta_1 \\ p\bar{M}_s & p\bar{M}_s & R_s+p\bar{L}_s & p\bar{M}_{sr}\cos\theta_1 & p\bar{M}_{sr}\cos\theta_2 & p\bar{M}_{sr}\cos\theta \\ p\bar{M}_{sr}\cos\theta & p\bar{M}_{sr}\cos\theta_2 & p\bar{M}_{sr}\cos\theta_1 & R_r+p\bar{L}_r & p\bar{M}_r & p\bar{M}_r \\ p\bar{M}_{sr}\cos\theta_1 & p\bar{M}_{sr}\cos\theta & p\bar{M}_{sr}\cos\theta_2 & p\bar{M}_r & R_r+p\bar{L}_r & p\bar{M}_r \\ p\bar{M}_{sr}\cos\theta_2 & p\bar{M}_{sr}\cos\theta_1 & p\bar{M}_{sr}\cos\theta & p\bar{M}_r & p\bar{M}_r & R_r+p\bar{L}_r \end{bmatrix} \begin{bmatrix} i_{sA} \\ i_{sB} \\ i_{sC} \\ i_{ra} \\ i_{rb} \\ i_{rc} \end{bmatrix}, \tag{2.1-108}$$

where $p=\mathrm{d}/\mathrm{d}t$, which operates on the inductances, since in general they can vary with current. The angles θ, θ_1 and θ_2 are defined as $\theta=\theta_r$, $\theta_1=\theta_r+2\pi/3$, and $\theta_2=\theta_r+4\pi/3$, where θ_r is the rotor angle shown in Fig. 2.1. The stator self-inductance of one stator phase winding $\bar{L}_s$ can be expressed as the sum of the stator leakage inductance L_{sl} and the stator magnetizing inductance L_{sm}, $\bar{L}_s=L_{sl}+L_{sm}$, and it should be noted that this relationship holds even when the m.m.f. distribution around the periphery is not sinusoidal. For the sinusoidal distribution, the mutual inductance between two stator windings $\bar{M}_s$ can be expressed in terms of the stator magnetizing inductance L_{sm} as

$$\bar{M}_s = L_{sm}\cos(2\pi/3) = -L_{sm}/2.$$

Similarly, it is possible to express the rotor self-inductance of one rotor phase winding $\bar{L}_r$ as the sum of the leakage inductance of a rotor phase winding L_{rl} and the rotor magnetizing inductance L_{rm}, $\bar{L}_r = L_{rl} + L_{rm}$. For a sinusoidal winding distribution the mutual inductance between the rotor windings is $\bar{M}_r = L_{rm}\cos(2\pi/3) = -L_{rm}/2$.

By simple physical considerations of the winding inductances it can be shown that the magnetizing inductance of the stator L_{sm} is related to the maximal value of the mutual inductance between the stator and rotor $\bar{M}_{sr}$ as $L_{sm} = (N_{se}/N_{re})\bar{M}_{sr}$, where N_{se} and N_{re} are the effective number of stator and rotor turns (see Section 2.1.2). Furthermore, the magnetizing inductance of the rotor is related to $\bar{M}_{sr}$ as $L_{rm} = (N_{re}/N_{se})\bar{M}_{sr}$. It also follows that $\bar{M}_{sr}^2 = L_{sm}L_{rm}$.

If the resultant three-phase magnetizing inductance L_m, first introduced in eqn (2.1-35), is utilized, where $L_m = \frac{3}{2}\bar{M}_{sr}$, then it follows from above that the total three-phase stator inductance, L_s (first used in eqn (2.1-35)), takes the form

$$L_s = \bar{L}_s - \bar{M}_s = L_{sl} + L_{sm} + \tfrac{1}{2}L_{sm} = L_{sl} + \tfrac{3}{2}L_{sm}$$

and the total three-phase rotor inductance (first introduced in eqn (2.1-45)) can be expressed as

$$L_r = \bar{L}_r - \bar{M}_r = L_{rl} + L_{rm} + \tfrac{1}{2}L_{rm} = L_{rl} + \tfrac{3}{2}L_{rm}.$$

There are smooth-air-gap machines—e.g. the three-phase squirrel cage induction machine—where instead of coil-wound rotor windings, the rotor currents flow in bars embedded in the rotor slots and connected to end-rings. There are various types of cage rotors, but a so-called single-cage rotor has uniformly distributed bars: it can be considered as a multi-phase winding system, where the number of rotor phases depends on the number of rotor bars and pole-pairs and for a symmetrical rotor construction it is possible to represent this by an equivalent symmetrical three-phase rotor winding and the three-phase model described above can be used for this machine too. If the machine has a so-called double-cage rotor, with an inner and an outer cage, then for modelling purposes, it is still possible to use an equivalent three-phase rotor winding where the rotor parameters (resistance, inductance) contain the effects related to the double cage. However, for certain forms of double cage, or machines with deep bar rotor, it can be advantageous to use a more accurate model. The space-phasor equations for double-cage induction machines or induction machines with deep rotor bars will be discussed in detail in Section 4.3.5.

It is important to note that in eqn (2.1-108) the stator voltages and currents are expressed in the stationary reference frame fixed to the stator, and thus these quantities can be directly monitored on the stator. However, the rotor voltages and currents are expressed in the reference frame fixed to the rotor, so they can be measured on the rotating rotor which of course is almost impossible in the case of a squirrel-cage rotor. Thus, two reference frames have been used for establishing the hybrid model represented by eqn (2.1-108); these are the so-called natural reference frames.

It follows from eqn (2.1-108) that the system of voltage differential equations is non-linear, and contains variable, time-dependent coefficients, since in general the rotor angle is a non-linear function of time and, furthermore, some other parameters can also be variable. For example, under saturated conditions some of the inductance parameters vary with the currents. However, even when all the machine parameters are considered to be constant, the voltage differential equations will contain variable coefficients, since in general the rotor angle varies with time. Furthermore, in eqn (2.1-108), there are 36 terms in the impedance matrix of the machine and the windings are fully coupled.

The complete performance of the machine under transient conditions can be determined by using eqn (2.1-108) together with the equation of motion

$$t_e - t_1 = J\frac{d\omega_r}{dt} + D\omega_r, \qquad (2.1\text{-}109)$$

where t_e is the developed electromagnetic torque of the machine, usually given in Newton meters (N m), t_1 is the load torque (N m), ω_r is the rotor speed (rad s^{-1}) and is equal to the first time derivative of the electrical rotor angle, $d\theta_r/dt$. The relationship between the electrical and mechanical rotor angles is $\theta_r = P\theta_{rm}$, where θ_{rm} is the mechanical rotor angle, and P is the number of pole-pairs. J is the inertia of the rotor, the unit of which is usually given either as kilogram metre^2 (kg m^2) or Joules second^2 (J s^2), or as a quantity called WR^2 expressed in units of pound mass feet^2 (lb m ft^2). The quantity $J\omega_r$ is called the inertia constant and is closely related to the kinetic energy of the rotating masses, which can be expressed as $W = \frac{1}{2}J\omega_r^2$. The quantity $D\omega_r$ is the damping torque and D is the damping constant which represents dissipation due to windage and friction.

In the voltage equations, significant simplification can be performed if the three-phase variables are replaced by their two-phase equivalents expressed in the same reference frame, since in this case, in the absence of zero-sequence components, there will only be four voltage equations, corresponding to direct- and quadrature-axis stator and rotor voltage equations respectively. Thus there will only be 16 elements in the corresponding new (transformed) impedance matrix. Further simplification can be achieved by using other than the natural reference frames. For example, if instead of expressing the rotor quantities (voltages, currents, flux linkages) in their natural reference frames, they are expressed in the stationary reference frame, the resulting voltage differential equations will contain constant coefficients if the parameters of the machine are considered to be constant. Because of the advantages of these simplifications, they will be briefly discussed below.

The quadrature-phase slip-ring model: To describe the so-called quadrature-phase slip-ring model of smooth-air-gap machines, first the number of phase variables is reduced in the three-phase model by assuming that there are no zero-sequence voltages and currents on the stator or rotor. Physically this corresponds to using a quadrature-phase machine model instead of the three-phase model. In this so-called quadrature-phase slip-ring model, shown in Fig. 2.7, there is a quadrature-phase stator winding (sD, sQ) and a quadrature-phase rotor winding ($r\alpha$, $r\beta$); rotor winding $r\alpha$ is displaced from stator winding sD by the angle θ_r.

The relationship between the two-axis stator voltage and current components and the corresponding three-phase components can be obtained by considering eqns (2.1-63)–(2.1-66):

$$u_{sD}=\tfrac{2}{3}(u_{sA}-\tfrac{1}{2}u_{sB}-\tfrac{1}{2}u_{sC}) \tag{2.1-110}$$

$$u_{sQ}=(u_{sB}-u_{sC})/\sqrt{3} \tag{2.1-111}$$

and

$$i_{sD}=\tfrac{2}{3}(i_{sA}-\tfrac{1}{2}i_{sB}-\tfrac{1}{2}i_{sC}) \tag{2.1-112}$$

$$i_{sQ}=(i_{sB}-i_{sC})/\sqrt{3}. \tag{2.1-113}$$

Similar transformations hold for the rotor voltages and currents:

$$u_{r\alpha}=\tfrac{2}{3}(u_{ra}-\tfrac{1}{2}u_{rb}-\tfrac{1}{2}u_{rc}) \tag{2.1-114}$$

$$u_{r\beta}=(u_{rb}-u_{rc})/\sqrt{3} \tag{2.1-115}$$

and

$$i_{r\alpha}=\tfrac{2}{3}(i_{ra}-\tfrac{1}{2}i_{rb}-\tfrac{1}{2}i_{rc}) \tag{2.1-116}$$

$$i_{r\beta}=(i_{rb}-i_{rc})/\sqrt{3}. \tag{2.1-117}$$

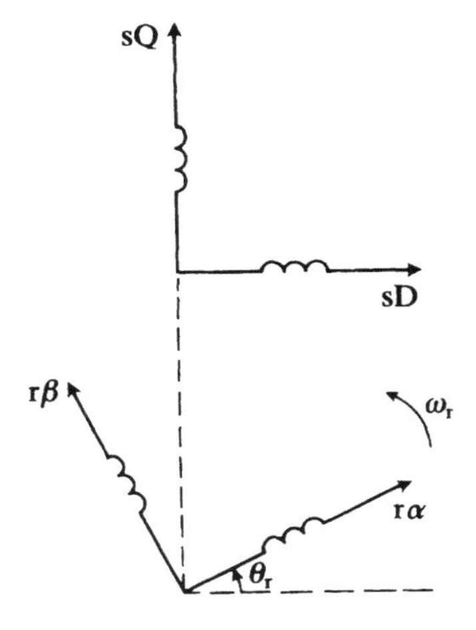

Fig. 2.7. Schematic of the quadrature-phase slip-ring model.

Thus with eqns (2.1-110)–(2.1-117), after some algebraic manipulation, eqn (2.1-108) takes the following form:

$$\begin{bmatrix} u_{sD} \\ u_{sQ} \\ u_{r\alpha} \\ u_{r\beta} \end{bmatrix} = \begin{bmatrix} R_s + pL_s & 0 & pL_m \cos\theta_r & -pL_m \sin\theta_r \\ 0 & R_s + pL_s & pL_m \sin\theta_r & pL_m \cos\theta_r \\ pL_m \cos\theta_r & pL_m \sin\theta_r & R_r + pL_r & 0 \\ -pL_m \sin\theta_r & pL_m \cos\theta_r & 0 & R_r + pL_r \end{bmatrix} \begin{bmatrix} i_{sD} \\ i_{sQ} \\ i_{r\alpha} \\ i_{r\beta} \end{bmatrix}, \qquad (2.1\text{-}118)$$

where L_s and L_r are the stator and rotor inductances, introduced in eqns (2.1-35) and (2.1-45) respectively and $L_m = \frac{3}{2}\bar{M}_{sr}$ is the magnetizing inductance. In eqn (2.1-118) the stator and rotor variables are expressed in their natural reference frames and, even if the machine parameters are constant, in general the system of voltage differential equations will be time-dependent, since the equations contain the rotor angle θ_r which changes with time. If the inductances are constant, the differential operator $p = \mathrm{d}/\mathrm{d}t$ can be moved after the inductance elements.

In the so-called generalized theory of electrical machines, where the various models are obtained by using matrix transformations, the model described by eqn (2.1-118) is usually obtained from the three-phase model described by eqn (2.1-108) by the application of the so-called phase transformation matrix $\mathbf{C}_1$. Thus if $\mathbf{Z}_s$ is the impedance matrix of the three-phase machine described by eqn (2.1-108), the impedance matrix of the quadrature-phase slip-ring model described by eqn (2.1-118) can be obtained by performing $\mathbf{Z}_s' = \mathbf{C}_t \mathbf{Z}_s \mathbf{C}$, where $\mathbf{C} = \mathrm{diag}(\mathbf{C}_1; \mathbf{C}_1)$. The inverse of the phase transformation matrix has been defined by eqn (2.1-69). It follows from eqn (2.1-118) that as a consequence of the phase transformation, there are four zero elements in the impedance matrix of the machine and in contrast to the impedance matrix of the three-phase model, which contains 36 non-zero elements, the impedance matrix of the quadrature-phase slip-ring machine contains 12 non-zero elements. However, it is possible to achieve a further reduction in the elements of the impedance matrix, and this is discussed in the following section.

The quadrature-phase commutator model: If the stationary-axis stator quantities of the quadrature-phase slip-ring model are unchanged, but the rotor voltages and currents $u_{r\alpha}$, $i_{r\alpha}$, $u_{r\beta}$, $i_{r\beta}$ are transformed (from the rotor reference frame) into a new reference frame fixed to the stator, the so-called quadrature-phase commutator model with pseudo-stationary rotor windings is obtained. The schematic of this machine is shown in Fig. 2.8; on the stator there are the direct- and quadrature-axis windings denoted by sD, sQ and on the rotor there are the windings denoted by rd and rq respectively.

The following transformations can be obtained from eqns (2.1-40) and (2.1-74) for the rotor voltages and currents:

$$u_{r\alpha} = \cos\theta_r u_{rd} + \sin\theta_r u_{rq} \qquad (2.1\text{-}119)$$

$$u_{r\beta} = -\sin\theta_r u_{rd} + \cos\theta_r u_{rq} \qquad (2.1\text{-}120)$$

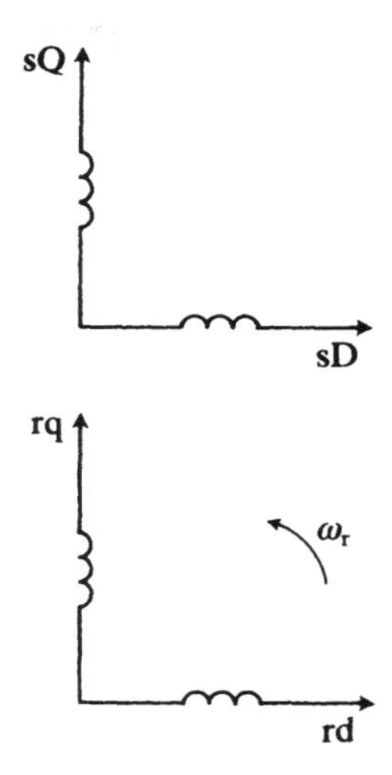

Fig. 2.8. Schematic of the quadrature-phase commutator model.

and similarly

$$i_{r\alpha}=\cos\theta_r i_{rd}+\sin\theta_r i_{rq} \tag{2.1-121}$$

$$i_{r\beta}=-\sin\theta_r i_{rd}+\cos\theta_r i_{rq}. \tag{2.1-122}$$

Thus by considering eqns (2.1-118)–(2.1-122), the following transformed set of voltage equations of the commutator model (sometimes also referred to as the commutator primitive model) is obtained:

$$\begin{bmatrix} u_{sD} \\ u_{sQ} \\ u_{rd} \\ u_{rq} \end{bmatrix} = \begin{bmatrix} R_s+pL_s & 0 & pL_m & 0 \\ 0 & R_s+pL_s & 0 & pL_m \\ pL_m & \omega_r L_m & R_r+pL_r & \omega_r L_r \\ -\omega_r L_m & pL_m & -\omega_r L_r & R_r+pL_r \end{bmatrix} \begin{bmatrix} i_{sD} \\ i_{sQ} \\ i_{rd} \\ i_{rq} \end{bmatrix}. \tag{2.1-123}$$

In this model the cosine and sine functions of the rotor angle are not present in the impedance matrix, but the rotor speed ω_r is present in the rotor equations. In the generalized theory of electrical machines, this model is usually obtained from the quadrature-phase slip-ring model by performing $\mathbf{C}_t\mathbf{Z}'_s\mathbf{C}$, where $\mathbf{C}=\mathrm{diag}(\mathbf{I}_2;\mathbf{C}_2)$, $\mathbf{I}_2$ is a 2×2 identity matrix, the matrix $\mathbf{C}_2^{-1}$ has been defined in eqn (2.1-40), and $\mathbf{Z}'_s$ is the impedance matrix of the quadrature-phase slip-ring machine described by eqn (2.1-118). It is also possible to obtain eqn (2.1-123) by physical considerations only, e.g. by considering that in the direct-axis rotor winding, rd, there must be voltages induced by transformer effects, i.e. $p(L_m i_{sD}+L_r i_{rd})$ and rotational voltages due to the rotation of the rotor, i.e.

$\omega_r(L_m i_{sQ} + L_r i_{rq})$. Under linear magnetic conditions, the operator p can be moved after the inductances in eqn (2.1-123). Some aspects of the effects of magnetic saturation will be discussed in Chapter 6 and in Section 6.1.1 the quadrature-phase commutator model will be described, where the effects of the saturation of the main flux paths are incorporated in the equations; eqn (2.1-123) is significantly modified.

2.1.7.2 The space-phasor form of the equations

In the present section the space-phasor forms of the voltage equations of three-phase and quadrature-phase smooth-air-gap machines will be presented. The equations will be expressed both in the stationary reference frames and in a general rotating reference frame, which rotates at a general speed ω_g. The relationship between the space-phasor and matrix forms, given in the previous section, will also be shown.

The space-phasor voltage equations in the stationary reference frame: The stator- and rotor-voltage differential equations of the smooth-air-gap three-phase machine have been given above in their natural reference frames. By utilizing the definitions of the voltage space-phasors [eqns (2.1-61), (2.1-62)], current space-phasors [eqns (2.1-4), (2.1-20)] and flux-linkage space phasors [eqns (2.1-31), (2.1-41)], the stator and rotor voltage equations [eqns (2.1-102)–(2.1-104) and eqns (2.1-105)–(2.1-107)] can be put into the following space-phasor form:

$$\bar{u}_s = R_s \bar{i}_s + \frac{d\bar{\psi}_s}{dt} \tag{2.1-124}$$

and

$$\bar{u}_r' = R_r \bar{i}_r' + \frac{d\bar{\psi}_r'}{dt} - j\omega_r \bar{\psi}_r'. \tag{2.1-125}$$

In eqns (2.1-124) and (2.1-125) the stator and rotor space-phasor flux linkages $\bar{\psi}_s$ and $\bar{\psi}_r'$ are present and these have been defined in terms of the machine inductances and the space phasors of the stator and rotor currents in eqns (2.1-35) and (2.1-51) respectively. The quantity ω_r is the instantaneous angular speed of the rotor.

For the sake of a better overview, the definitions of all the space-phasor quantities will be repeated below using the definitions of the space phasors of the three-phase quantities. The space phasors of the stator voltages, currents, and flux linkages in the stationary reference frame fixed to the stator are [see eqns (2.1-61), (2.1-4), (2.1-31), and (2.1-35)]:

$$\bar{u}_s = \tfrac{2}{3}[u_{sA}(t) + a u_{sB}(t) + a^2 u_{sC}(t)] = u_{sD} + j u_{sQ} \tag{2.1-126}$$

$$\bar{i}_s = \tfrac{2}{3}[i_{sA}(t) + a i_{sB}(t) + a^2 i_{sC}(t)] = i_{sD} + j i_{sQ} \tag{2.1-127}$$

$$\bar{\psi}_s = \tfrac{2}{3}[\psi_{sA}(t) + a\psi_{sB}(t) + a^2 \psi_{sC}(t)] = \psi_{sD} + j\psi_{sQ} = L_s \bar{i}_s + L_m \bar{i}_r' \tag{2.1-128}$$

and similarly the space phasors of the rotor voltages, currents, and flux linkages in the reference frame fixed to the rotor are [see eqns (2.1-62), (2.1-20), (2.1-41), and (2.1-45)]:

$$\bar{u}_r = \tfrac{2}{3}[u_{ra}(t) + au_{rb}(t) + a^2u_{rc}(t)] = u_{r\alpha} + ju_{r\beta} \tag{2.1-129}$$

$$\bar{i}_r = \tfrac{2}{3}[i_{ra}(t) + ai_{rb}(t) + a^2i_{rc}(t)] = i_{r\alpha} + ji_{r\beta} \tag{2.1-130}$$

$$\begin{aligned}\bar{\psi}_r &= \tfrac{2}{3}[\psi_{ra}(t) + a\psi_{rb}(t) + a^2\psi_{rc}(t)] = L_r\bar{i}_r + L_m\bar{i}_s' \\ &= L_r\bar{i}_r + L_m\bar{i}_s e^{-j\theta_r} = \psi_{r\alpha} + j\psi_{r\beta}.\end{aligned} \tag{2.1-131}$$

The rotor quantities defined above are, in the reference frame fixed to the stator [see eqns (2.1-74), (2.1-39), and (2.1-51)],

$$\bar{u}_r' = \bar{u}_r e^{j\theta_r} = u_{rd} + ju_{rq} \tag{2.1-132}$$

$$\bar{i}_r' = \bar{i}_r e^{j\theta_r} = i_{rd} + ji_{rq} \tag{2.1-133}$$

$$\bar{\psi}_r' = \bar{\psi}_r e^{j\theta_r} = L_r\bar{i}_r' + L_m\bar{i}_s = L_r\bar{i}_r e^{j\theta_r} + L_m\bar{i}_s = \psi_{rd} + j\psi_{rq}. \tag{2.1-134}$$

For completeness, the space phasors of the stator voltages, currents, and flux linkages are also given in the rotating reference frame fixed to the rotor [see eqns (2.1-73), (2.1-54), and (2.1-56)] as

$$\bar{u}_s' = \bar{u}_s e^{-j\theta_r} = u_{sd} + ju_{sq} \tag{2.1-135}$$

$$\bar{i}_s' = \bar{i}_s e^{-j\theta_r} = i_{sd} + ji_{sq} \tag{2.1-136}$$

$$\bar{\psi}_s' = \bar{\psi}_s e^{-j\theta_r} = \psi_{sd} + j\psi_{sq}. \tag{2.1-137}$$

The very compact form of eqns (2.1-124) and (2.1-125) makes their application extremely convenient. The first term on the right-hand side of these equations is the space-phasor form of the ohmic losses, the second term is a transformer e.m.f., which is the first time derivative of the flux-linkage space phasor of the stator and rotor respectively. Finally, in eqn (2.1-125), the term $-j\omega_r\bar{\psi}_r'$ represents a rotational e.m.f., which is due to the rotation of the rotor and contributes to electromechanical energy conversion. Equations (2.1-124) and (2.1-125) together with the flux-linkage equations, eqns (2.1-128), (2.1-131), and (2.1-134), are also valid under saturated conditions.

If the flux-linkage space phasors defined by eqns (2.1-128) and (2.1-134) are substituted into eqns (2.1-124) and (2.1-125), then the space-phasor voltage equations take the following form:

$$\bar{u}_s = R_s\bar{i}_s + d(L_s\bar{i}_s)/dt + d(L_m\bar{i}_r')/dt \tag{2.1-138}$$

and

$$\bar{u}_r' = R_r\bar{i}_r' + d(L_r\bar{i}_r')/dt + d(L_m\bar{i}_s)/dt - j\omega_r(L_r\bar{i}_r' + L_m\bar{i}_s). \tag{2.1-139}$$

These equations can also be put into matrix form:

$$\begin{bmatrix}\bar{u}_s \\ \bar{u}_r'\end{bmatrix} = \begin{bmatrix}R_s & 0 \\ 0 & R_r\end{bmatrix}\begin{bmatrix}\bar{i}_s \\ \bar{i}_r'\end{bmatrix} + \frac{d}{dt}\begin{bmatrix}L_s & L_m \\ L_m & L_r\end{bmatrix}\begin{bmatrix}\bar{i}_s \\ \bar{i}_r'\end{bmatrix} - j\omega_r\begin{bmatrix}0 & 0 \\ L_m & L_r\end{bmatrix}\begin{bmatrix}\bar{i}_s \\ \bar{i}_r'\end{bmatrix}. \tag{2.1-140}$$

The simplicity of these equations should be contrasted with eqn (2.1-108), but while in eqn (2.1-108) the phase-variable voltages and currents appear, in eqn (2.1-140) the space phasors of the voltages and currents are present. If these space phasors are resolved into their real- and imaginary-axis components in accordance with eqns (2.1-126), (2.1-127), (2.1-132), and (2.1-133), then eqn (2.1-140) yields eqn (2.1-123), which corresponds to the quadrature-phase commutator model. Thus it can be seen that if the space-phasor voltage equations are established in the stationary reference frame, they can be used to obtain the equations of the commutator model, without having to perform any matrix transformation. The ability to obtain various models of a machine, in various reference frames, without using matrix transformations, is also an advantage of the application of space-phasors over the application of the conventional generalized matrix theory of electrical machines.

Furthermore, from eqns (2.1-132) and (2.1-133), $\bar{u}_r = \bar{u}_r' e^{-j\theta_r}$ and $\bar{i}_r = \bar{i}_r' e^{-j\theta_r}$, and the rotor quantities can be expressed in the reference frame fixed to the rotor and eqn (2.1-140) becomes

$$\begin{bmatrix} \bar{u}_s \\ \bar{u}_r \end{bmatrix} = \begin{bmatrix} R_s & 0 \\ 0 & R_r \end{bmatrix} \begin{bmatrix} \bar{i}_s \\ \bar{i}_r \end{bmatrix} + \frac{d}{dt} \begin{bmatrix} L_s & L_m e^{j\theta_r} \\ L_m e^{-j\theta_r} & L_r \end{bmatrix} \begin{bmatrix} \bar{i}_s \\ \bar{i}_r \end{bmatrix}. \tag{2.1-141}$$

If all the space-phasor quantities are expressed in terms of their real- and imaginary-axis components, i.e. from eqns (2.1-126), (2.1-127), (2.1-129), and (2.1-130), eqn (2.1-141) yields eqn (2.1-118), which corresponds to the quadrature-phase slip-ring model. Another model can therefore be obtained from the space-phasor equations without utilizing matrix transformations. Of course, the transformation procedure is now 'hidden' in the complex transformations, which contain the terms $e^{j\theta_r}$ and $e^{-j\theta_r}$.

For slip-ring induction machines with short-circuited rotor windings, or induction machines with squirrel-cage rotor, $\bar{u}_r = 0$.

The space-phasor voltage equations in the general reference frame: Here the voltage space-phasor equations will be formulated in a general reference frame, which rotates at a general speed ω_g.

It follows from the analysis presented earlier, e.g. from eqn (2.1-136) or Fig. 2.4, that the space phasor of the stator currents in the rotor reference frame is $\bar{i}_r' = \bar{i}_s e^{-j\theta_r}$ where $\bar{i}_s$ is the stator-current space phasor in the stator reference frame. Similarly it follows from eqn (2.1-133) that the space phasor of the rotor currents in the stationary reference frame is $\bar{i}_r' = \bar{i}_r e^{j\theta_r}$ where $\bar{i}_r$ is the space phasor of the rotor currents in the rotor reference frame. However, if instead of a reference frame fixed to the rotor, a general reference frame, with direct and quadrature axes x, y rotating at a general instantaneous speed $\omega_g = d\theta_g/dt$, is used, as shown in Fig. 2.9, where θ_g is the angle between the direct axis of the stationary reference frame sD fixed to the stator and the real axis (x) of the general reference frame, then the following equation defines the stator-current space phasor in the general

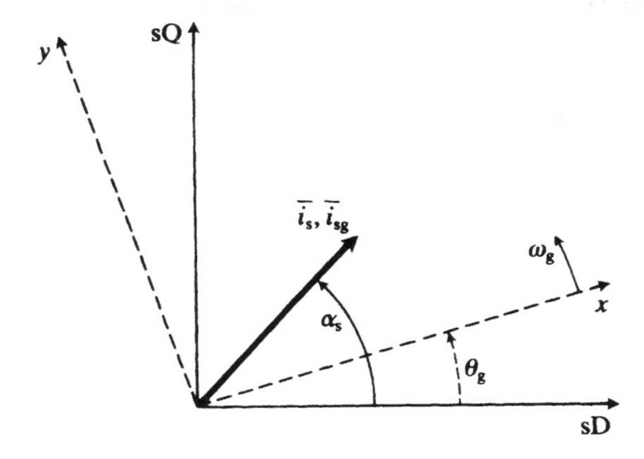

Fig. 2.9. Application of the general reference frame; transformation of the stator quantities.

reference frame

$$\bar{i}_{sg} = \bar{i}_s \, e^{-j\theta_g} = i_{sx} + j i_{sy}. \tag{2.1-142}$$

By the substitution of $\theta_g = \theta_r$, eqn (2.1-136) is obtained, which gives the space phasor of the stator currents expressed in the rotor reference frame. Equation (2.1-142) can be proved mathematically by considering that, from Fig. 2.9, in the stator reference frame $\bar{i}_s = |\bar{i}_s| \, e^{j\alpha_s}$ and in the general reference frame $\bar{i}_{sg} = |\bar{i}_s| \, e^{-j(\alpha_s - \theta_g)}$, thus

$$\bar{i}_{sg} = |\bar{i}_s| \, e^{j\alpha_s} e^{-j\theta_g} = \bar{i}_s \, e^{-j\theta_g}.$$

The stator voltage and flux-linkage space phasors can be similarly obtained in the general reference frame,

$$\bar{u}_{sg} = \bar{u}_s \, e^{-j\theta_g} = u_{sx} + j u_{sy} \tag{2.1-143}$$

$$\bar{\psi}_{sg} = \bar{\psi}_s \, e^{-j\theta_g} = \psi_{sx} + j\psi_{sy}, \tag{2.1-144}$$

where $\bar{u}_s$ and $\bar{\psi}_s$ are the space phasors of the stator voltages and stator flux linkages respectively in the stationary reference frame.

Similar considerations hold for the space phasors of the rotor voltages, current, and flux linkages. In Fig. 2.10 three reference frames are shown, the reference frame fixed to the rotor, the reference frame fixed to the stator, and the general reference frame. The magnetic axes of the rotor winding are rα, rβ and the rotor axis rα is displaced from the direct axis of the stator reference frame by the rotor angle θ_r.

It follows from Fig. 2.10 that in the rotor reference frame, the space phasor of the rotor currents can be expressed in terms of its modulus and phase angle α_r as $\bar{i}_r = |\bar{i}_r| \, e^{j\alpha_r}$. However, since it can be seen that the angle between the real axis (x) of the general reference frame and the real axis of the reference frame

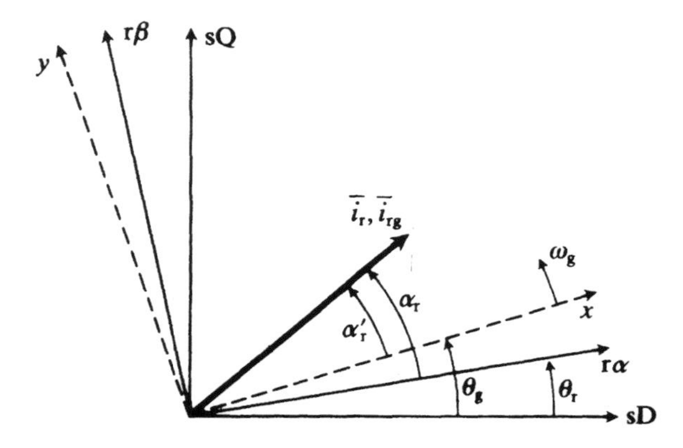

Fig. 2.10. Application of the general reference frame; transformation of the rotor quantities.

rotating with the rotor (rα) is $\theta_g - \theta_r$, in the general reference frame the space phasor of the rotor currents can be expressed as $\bar{i}_{rg} = |\bar{i}_s| e^{j\alpha'_r}$. where $\alpha'_r = \alpha_r - (\theta_g - \theta_r)$. Thus

$$\bar{i}_{rg} = |\bar{i}_r| e^{j\alpha_r} e^{-j(\theta_g - \theta_r)} = \bar{i}_r e^{-j(\theta_g - \theta_r)} = i_{rx} + j i_{ry}. \tag{2.1-145}$$

In the reference frame fixed to the stator, $\theta_g = 0$ and $\bar{i}_{rg} = \bar{i}'_r = \bar{i}_r e^{j\theta_r}$ in accordance with eqn (2.1-133). Similarly, the space phasors of the rotor voltages and rotor flux linkages in the general reference frame can be expressed as

$$\bar{u}_{rg} = \bar{u}_r e^{-j(\theta_g - \theta_r)} = u_{rx} + j u_{ry} \tag{2.1-146}$$

and

$$\bar{\psi}_{rg} = \bar{\psi}_r e^{-j(\theta_g - \theta_r)} = \psi_{rx} + j\psi_{ry}. \tag{2.1-147}$$

Substitution of eqns (2.1-142)–(2.1-147) into eqns (2.1-124), (2.1-125), (2.1-128), and (2.1-131) yields the following stator and rotor space-phasor voltage equations in the general reference frame:

$$\bar{u}_{sg} = R_s \bar{i}_{sg} + \frac{d\bar{\psi}_{sg}}{dt} + j\omega_g \bar{\psi}_{sg} \tag{2.1-148}$$

$$\bar{u}_{rg} = R_r \bar{i}_{rg} + \frac{d\bar{\psi}_{rg}}{dt} + j(\omega_g - \omega_r)\bar{\psi}_{rg}, \tag{2.1-149}$$

where the stator and rotor flux linkages in the general reference frame can be expressed in terms of the stator and rotor current space phasors as

$$\bar{\psi}_{sg} = L_s \bar{i}_{sg} + L_m \bar{i}_{rg} \tag{2.1-150}$$

and

$$\bar{\psi}_{rg} = L_r \bar{i}_{rg} + L_m \bar{i}_{sg}. \tag{2.1-151}$$

Equations (2.1-148)–(2.1-151) can be combined into the following two complex voltage equations:

$$\bar{u}_{sg}=R_s\bar{i}_{sg}+\frac{d}{dt}(L_s\bar{i}_{sg})+\frac{d}{dt}(L_m\bar{i}_{rg})+j\omega_g(L_s\bar{i}_{sg}+L_m\bar{i}_{rg}) \tag{2.1-152}$$

$$\bar{u}_{rg}=R_r\bar{i}_{rg}+\frac{d}{dt}(L_r\bar{i}_{rg})+\frac{d}{dt}(L_m\bar{i}_{sg})+j(\omega_g-\omega_r)(L_r\bar{i}_{rg}+L_m\bar{i}_{sg}), \tag{2.1-153}$$

which can be put into the following matrix form

$$\begin{bmatrix}\bar{u}_{sg}\\ \bar{u}_{rg}\end{bmatrix}=\begin{bmatrix}R_s & 0\\ 0 & R_r\end{bmatrix}\begin{bmatrix}\bar{i}_{sg}\\ \bar{i}_{rg}\end{bmatrix}+\frac{d}{dt}\begin{bmatrix}L_s & L_m\\ L_m & L_r\end{bmatrix}\begin{bmatrix}\bar{i}_{sg}\\ \bar{i}_{rg}\end{bmatrix}+j\omega_g\begin{bmatrix}L_s & L_m\\ L_m & L_r\end{bmatrix}\begin{bmatrix}\bar{i}_{sg}\\ \bar{i}_{rg}\end{bmatrix}$$
$$-j\omega_r\begin{bmatrix}0 & 0\\ L_m & L_r\end{bmatrix}\begin{bmatrix}\bar{i}_{sg}\\ \bar{i}_{rg}\end{bmatrix}. \tag{2.1-154}$$

On the right-hand side of eqn (2.1-153) there are the voltage terms due to the ohmic losses, plus the transformer e.m.f.s, plus the voltages due to the rotation of the general reference frame, plus the true rotational voltages. This last component is only present in the rotor equations and it is this term $-j\omega_r\bar{\psi}_{rg}=-j\omega_r(L_r\bar{i}_{rg}+L_m\bar{i}_{sg})$ which contributes to electromechanical energy conversion. It is also possible to utilize the expression for the rotational voltage to obtain the expressions for the electromagnetic torque in various reference frames, as shown in the next section. In eqns (2.1-152)–(2.1-154), under linear magnetic conditions all the inductance elements can be considered to be constant, and thus the operator d/dt can be moved after the inductances, so it directly operates on the space phasors of the currents.

It is also possible to express all the space-phasor quantities in terms of their real- and imaginary-axis components and thus by utilizing eqns (2.1-142), (2.1-143), (2.1-145), and (2.1-146), eqn (2.1-154) can be put into the following two-axis form:

$$\begin{bmatrix}u_{sx}\\ u_{sy}\\ u_{rx}\\ u_{ry}\end{bmatrix}=\begin{bmatrix}R_s+pL_s & -\omega_g L_s & pL_m & -\omega_g L_m\\ \omega_g L_s & R_s+pL_s & \omega_g L_m & pL_m\\ pL_m & -(\omega_g-\omega_r)L_m & R_r+pL_r & -(\omega_g-\omega_r)L_r\\ (\omega_g-\omega_r)L_m & pL_m & (\omega_g-\omega_r)L_r & R_r+pL_r\end{bmatrix}\begin{bmatrix}i_{sx}\\ i_{sy}\\ i_{rx}\\ i_{ry}\end{bmatrix}, \tag{2.1-155}$$

where again $p=\mathrm{d}/\mathrm{d}t$. If $\omega_g=0$, eqn (2.1-155) yields eqn (2.1-123), which corresponds to the quadrature-phase commutator model. If $\omega_g=\omega_1$, where ω_1 is the so-called synchronous speed, and where, say, for an induction machine, $\omega_1-\omega_r=s\omega_1$ is the slip speed and s is the slip, the two-axis equations of the induction machine in the synchronously rotating reference frame are

$$\begin{bmatrix} u_{sx} \\ u_{sy} \\ u_{rx} \\ u_{ry} \end{bmatrix} = \begin{bmatrix} R_s+pL_s & -\omega_1 L_s & pL_m & -\omega_1 L_m \\ \omega_1 L_s & R_s+pL_s & \omega_1 L_m & pL_m \\ pL_m & -s\omega_1 L_m & R_r+pL_r & -s\omega_1 L_r \\ s\omega_1 L_m & pL_m & -s\omega_1 L_r & R_r+pL_r \end{bmatrix} \begin{bmatrix} i_{sx} \\ i_{sy} \\ i_{rx} \\ i_{ry} \end{bmatrix}, \tag{2.1-156}$$

where u_{sx}, u_{sy}, and i_{sx}, i_{sy} are the two-axis components of the stator voltages and currents respectively in the synchronously rotating reference frame. The rotor voltage and current components in the same reference frame are $u_{rx}=u_{ry}=0$ and i_{rx}, i_{ry}. Under linear magnetic conditions, the operator p can be moved after the inductance terms.

The voltage equations are valid under both transient and steady-state conditions. However, under transient conditions the equation of motion [eqn (2.1-109)],

$$t_e - t_l = J\frac{d\omega_r}{dt} + D\omega_r, \tag{2.1-157}$$

is also required, where for a machine with P pole-pairs, similarly to eqn (2.1-99), the electromagnetic torque can be expressed as follows, in terms of the stator and rotor current space phasors and their two-axis components formulated in the general reference frame [defined by eqns (2.1-142) and (2.1-145)]:

$$t_e = -\tfrac{3}{2} P L_m \bar{i}_{sg} \times \bar{i}_{rg} = -\tfrac{3}{2} P L_m (i_{sx} i_{ry} - i_{sy} i_{rx}). \tag{2.1-158}$$

In eqn (2.1-157) the rotor speed can be expressed as

$$\omega_r = \frac{d\theta_r}{dt}, \tag{2.1-159}$$

where θ_r is the rotor angle.

Equation (2.1-158) can also be proved by considering eqns (2.1-98) and (2.1-99). However, in eqn (2.1-98) $\bar{\psi}_r'$ and $\bar{i}_r'$ are the rotor flux linkage and current space phasors formulated in the stationary reference frame fixed to the stator and by considering $\theta_g=0$, it follows from eqns (2.1-145) and (2.1-147) that they are defined as $\bar{i}_r' = \bar{i}_r e^{j\theta_r}$ and $\bar{\psi}_r' = \bar{\psi}_r e^{j\theta_r}$, where $\bar{i}_r$ and $\bar{\psi}_r$ are space phasors in the rotor reference frame. Similarly, in the general reference frame they are defined as

$$\bar{i}_{rg} = \bar{i}_r e^{-j(\theta_g-\theta_r)} = \bar{i}_r e^{j\theta_r} e^{-j\theta_g} \text{ and } \bar{\psi}_{rg} = \bar{\psi}_r e^{-j(\theta_g-\theta_r)} = \bar{\psi}_r e^{j\theta_r} e^{-j\theta_g}.$$

Thus by also considering eqn (2.1-98), the electromagnetic torque is

$$t_e = -\tfrac{3}{2} \bar{\psi}_r' \times \bar{i}_r' = -\tfrac{3}{2} \bar{\psi}_{rg} e^{j\theta_g} \times \bar{i}_{rg} e^{j\theta_g} = -\tfrac{3}{2} \bar{\psi}_{rg} \times \bar{i}_{rg}, \tag{2.1-160}$$

where $\bar{\psi}_{rg}$ has been defined in eqn (2.1-151). Substitution of eqn (2.1-151) into eqn (2.1-160) gives $t_e = -\frac{3}{2} L_m \bar{i}_{sg} \times \bar{i}_{rg}$ for the two-pole machine and this agrees with eqn (2.1-158) when $P=1$.

Another derivation of the electromagnetic torque can be based on the technique used in the conventional generalized matrix theory of electrical machines, where the electromagnetic torque is obtained from the so-called torque matrix,

$$t_e = \tfrac{3}{2} P \mathbf{i}_t^* \mathbf{G} \mathbf{i} \qquad (2.1\text{-}161)$$

where $\mathbf{G}$ is the torque matrix and $\mathbf{i}$ is the column vector of the currents in the general reference frame. From eqn (2.1-156), $\mathbf{i}_t$ is defined as $\mathbf{i}_t = [i_{sx}, i_{sy}, i_{rx}, i_{ry}]$, where t denotes the transpose, and the torque matrix is defined as

$$\mathbf{G} = \begin{bmatrix} 0 & 0 & 0 & 0 \\ 0 & 0 & 0 & 0 \\ 0 & L_m & 0 & L_r \\ -L_m & 0 & -L_r & 0 \end{bmatrix}, \qquad (2.1\text{-}162)$$

since in eqn (2.1-156) $s\omega_1 = \omega_1 - \omega_r$ and $\mathbf{G}$ contains the terms multiplied by ω_r.

It follows from eqns (2.1-161) and (2.1-162) that

$$t_e = \tfrac{3}{2} P (i_{rx}\psi_{ry} - i_{ry}\psi_{rx}), \qquad (2.1\text{-}163)$$

which can be expressed as a vector product,

$$t_e = -\tfrac{3}{2} P \bar{\psi}_{rg} \times \bar{i}_{rg}$$

which agrees with eqn (2.1-160) if a two-pole machine ($P=1$) is considered.

It follows that the stator and rotor voltage equations (two complex or four real) expressed in the general reference frame, together with eqns (2.1-157) and (2.1-158) are suitable for the simulation of the transient operation of smooth-air-gap a.c. machines and for designing various forms of control of a.c. drives. With different constraints in the rotor circuit, the general model is applicable to smooth-air-gap synchronous, asynchronous, and doubly-fed machines and valid for an arbitrary instantaneous variation of the voltages and currents; the only restriction is that there can be no zero-sequence line-to-neutral voltages and phase currents. Thus the general model is suitable for the analysis of a large number of converter-fed machines, where different voltage or current waveforms are generated by the converter under consideration. If the neutral point of the machine is isolated, but the line-to-neutral voltages are unbalanced and contain a zero-sequence component, the space-phasor equations can still be used, since the voltage asymmetry means only that the neutral of the stator windings has a different potential with respect to the neutral of the supply voltages.

2.1.8 VARIOUS EXPRESSIONS FOR ELECTROMAGNETIC TORQUE

Here different expressions for the electromagnetic torque are obtained, by utilizing various space-phasor quantities (voltages, currents, flux linkages) expressed in different reference frames. It will also be shown that in special reference frames fixed to the rotor flux, the stator flux, or the magnetizing flux-linkage space phasor, the expression of the electromagnetic torque is similar to the expression for the electromagnetic torque produced by a separately excited d.c. machine. This analogy serves as a basis for various forms of vector control, where the torque control of the a.c. machine is similar to the torque control of the separately excited d.c. machine.

2.1.8.1 Application of the general reference frame

Further to the expressions for the electromagnetic torque given in Section 2.1.6, it is possible to obtain various other useful expressions for the electromagnetic torque. To give a better physical insight into the processes involved, it is first shown that eqn (2.1-149) can be used directly to obtain a general expression for the electromagnetic torque.

It follows from eqn (2.1-149) that the space-phasor form of the rotational e.m.f. in the rotor windings is $-j\omega_r\bar{\psi}_{rg}=-j\omega_r(\psi_{rx}+j\psi_{ry})$. Thus the real-axis component of this e.m.f. is $u_{rotx}=\omega_r\psi_{ry}$ and the imaginary-axis rotational e.m.f. is $u_{roty}=-\omega_r\psi_{rx}$. It follows that the mechanical power can be expressed as

$$P_{mech}=\tfrac{3}{2}(u_{rotx}i_{rx}+u_{roty}i_{ry})=\tfrac{3}{2}\,\omega_r(\psi_{ry}i_{rx}-\psi_{rx}i_{ry})=-\tfrac{3}{2}\,\omega_r\bar{\psi}_{rg}\times\bar{i}_{rg}$$

and when this is divided by the rotor speed ω_r, it yields an expression for the electromagnetic torque which is in agreement with eqn (2.1-160).

It follows from eqn (2.1-160) that the expression for the electromagnetic torque is similar in all reference frames, which is an expected result. For a machine with P pole-pairs, eqn (2.1-160) takes the following form:

$$t_e=-\tfrac{3}{2}P\bar{\psi}'_r\times\bar{i}'_r=-\tfrac{3}{2}P\bar{\psi}_{rg}\times\bar{i}_{rg}, \qquad (2.1\text{-}164)$$

according to which the torque is produced by the interaction of the rotor flux-linkage space phasor and the rotor current space phasor. By a similar argument, it follows that eqn (2.1-99) can be written as

$$t_e=-\tfrac{3}{2}PL_m\bar{i}_s\times\bar{i}'_r=-\tfrac{3}{2}PL_m\bar{i}_{sg}\times\bar{i}_{rg}, \qquad (2.1\text{-}165)$$

according to which the electromagnetic torque is proportional to the vector product of the stator and rotor current space phasors. It is also possible to obtain eqn (2.1-165) by the substitution of eqn (2.1-151) into eqn (2.1-164). Similarly, eqn (2.1-101) can be written as

$$t_e=-\tfrac{3}{2}P(L_m/L_s)\bar{\psi}_{sg}\times\bar{i}_{rg}, \qquad (2.1\text{-}166)$$

which could also be obtained from eqns (2.1-150) and (2.1-165).

By the application of the principle of action–reaction, it is possible to write eqn (2.1-164) as

$$t_e=\tfrac{3}{2}P\bar{\psi}_{sg}\times\bar{i}_{sg}. \qquad (2.1\text{-}167)$$

There are many other ways of giving a rigorous mathematical proof of eqn (2.1-167), e.g. by considering eqn (2.1-165), the equation

$$t_e = \tfrac{3}{2} P L_m \bar{i}_{rg} \times \bar{i}_{sg} \tag{2.1-168}$$

is obtained, which can be expanded as follows, by considering that the vector product $\bar{i}_{sg} \times \bar{i}_{sg} = 0$ and using eqn (2.1-150):

$$t_e = \tfrac{3}{2} P(L_s \bar{i}_{sg} + L_m \bar{i}_{rg}) \times \bar{i}_{sg} = \tfrac{3}{2} P \bar{\psi}_{sg} \times \bar{i}_{sg}, \tag{2-1-169}$$

which agrees with eqn (2.1-167).

It is also possible to split the total stator flux linkages into the sum of a leakage ($\bar{\psi}_{slg}$) and a magnetizing flux linkage ($\bar{\psi}_{mg}$); thus from eqn (2.1-150)

$$\bar{\psi}_{sg} = \bar{\psi}_{slg} + \bar{\psi}_{mg} \tag{2.1-170}$$

where

$$\bar{\psi}_{slg} = L_{sl} \bar{i}_{sg} \tag{2.1-171}$$

is the space phasor of the stator leakage flux linkages in the general reference frame and L_{sl} is the leakage inductance of a stator winding. If the stator and rotor windings have equal numbers of effective turns,

$$\bar{\psi}_{mg} = L_m \bar{i}_{mg}, \tag{2.1-172}$$

which gives the space phasor of the magnetizing flux linkages in the general reference frame, where $\bar{i}_{mg}$ is the magnetizing-current space phasor in the general reference frame,

$$\bar{i}_{mg} = \bar{i}_{sg} + \bar{i}_{rg}. \tag{2.1-173}$$

Thus substitution of eqns (2.1-170)–(2.1-173) into eqn (2.1-169) yields

$$t_e = \tfrac{3}{2} P(L_{sl} \bar{i}_{sg} + L_m \bar{i}_{mg}) \times \bar{i}_{sg} = \tfrac{3}{2} P L_m \bar{i}_{mg} \times \bar{i}_{sg} = \tfrac{3}{2} P \bar{\psi}_{mg} \times \bar{i}_{sg}, \tag{2.1-174}$$

according to which the electromagnetic torque can be expressed as the vector product of the magnetizing-current and stator-current space phasors. As physically expected, it also follows that the stator leakage fluxes do not contribute to the production of the electromagnetic torque.

By resolving all the space phasors into their real- (x) and imaginary-axis (y) components, it is possible to obtain the electromagnetic torque in terms of the two-axis components of the general reference frame. Thus from eqn (2.1-174), with $\bar{\psi}_{mg} = \psi_{mx} + j\psi_{my}$ and $\bar{i}_{sg} = i_{sx} + j i_{sy}$,

$$t_e = \tfrac{3}{2} P(\psi_{mx} i_{sy} - \psi_{my} i_{sx}). \tag{2.1-175}$$

Expansion of eqn (2.1-168) yields the following equation for the electromagnetic torque, if $\bar{i}_{sg} \times \bar{i}_{sg} = 0$ and eqn (2.1-151) is utilized:

$$t_e = \tfrac{3}{2} P L_m \bar{i}_{rg} \times \bar{i}_{sg} = \tfrac{3}{2} P \frac{L_m}{L_r} (L_r \bar{i}_{rg} \times \bar{i}_{sg})$$

$$= \tfrac{3}{2} P \frac{L_m}{L_r} (L_r \bar{i}_{rg} + L_m \bar{i}_{sg}) \times \bar{i}_{sg} = \tfrac{3}{2} P \frac{L_m}{L_r} \bar{\psi}_{rg} \times \bar{i}_{sg}. \tag{2.1-176}$$

By the resolution into direct-axis (x) and quadrature-axis (y) components of the general reference frame and using $\bar{\psi}_{rg}=\psi_{rx}+j\psi_{ry}$, $\bar{i}_{sg}=i_{sx}+ji_{sy}$,

$$t_e=\frac{3}{2}P\frac{L_m}{L_r}(\psi_{rx}i_{sy}-\psi_{ry}i_{sx}), \tag{2.1-177}$$

where ψ_{rx} and ψ_{ry} are the two-axis components of the rotor flux-linkage space phasor in the general reference frame, and which from eqn (2.1-151) can be expressed as $\psi_{rx}=L_r i_{rx}+L_m i_{rx}$ and $\psi_{ry}=L_r i_{ry}+L_m i_{sy}$.

2.1.8.2 The application of special reference frames and the basis for vector control

In Chapter 1 and Section 2.1.6 one aim of the application of vector-controlled a.c. motors is given, which is to be able to implement control schemes which yield high dynamic performance and are similar to those used to control d.c. machines. For this purpose it must first be proved that, under transient conditions, the electromagnetic torque in smooth-air-gap machines is proportional to the product of a flux-producing current component and a torque-producing current component which are in space quadrature. It will be shown in this section that there are three similar ways of achieving this. These are to use reference frames fixed to the stator flux-linkage space phasor, to the rotor flux-linkage space phasor, or to the magnetizing flux-linkage space phasor (in smooth-air-gap machines the latter is equivalent to fixing the reference frame to the magnetizing-current space phasor) to derive an expression for the electromagnetic torque in a form which enables independent control of the torque- and flux-producing current components. The three cases will be discussed in turn.

The application of a special reference frame fixed to the magnetizing flux-linkage space phasor: The expression for the electromagnetic torque will be obtained in terms of the modulus of the magnetizing flux-linkage space phasor and the quadrature-axis stator current component, expressed in the reference frame fixed to the magnetizing flux-linkage space phasor.

An induction machine with a symmetrical multiphase, short-circuited rotor winding is assumed, where for simplicity the three-phase stator windings (sA, sB, sC) are replaced by their equivalent quadrature-phase stator windings (sD, sQ). In contrast to Fig. 2.5, which was used to explain the mechanism of torque production in a d.c. machine, with armature winding on the stator, and field and compensating windings on the rotor, in Fig. 2.11 is shown the schematic of an induction machine, with stator windings sD and sQ and a multi-phase short-circuited rotor winding system.

The direct-axis stator current i_{sD} produces the magnetizing flux ψ'_m. If the quadrature-axis current i_{sQ} is suddenly injected into the stator winding sQ—the compensating winding for the d.c. machine—rotor current i'_r will flow in the rotor windings, which at the first instant of time (t_0) is $i'_r=-i_{sQ}$. Thus the space-phasor

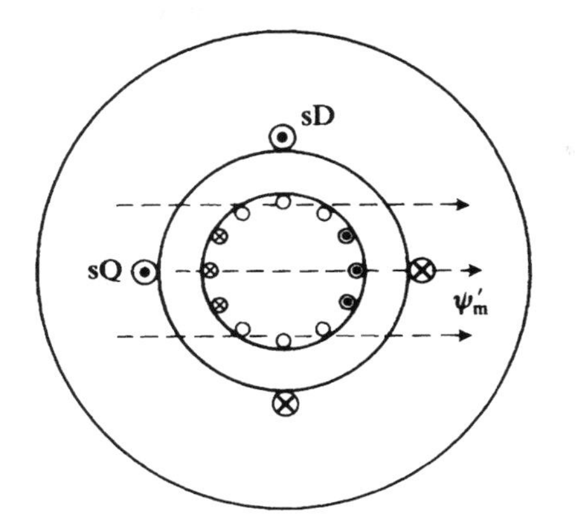

Fig. 2.11. Schematic of the induction machine with quadrature-phase stator windings.

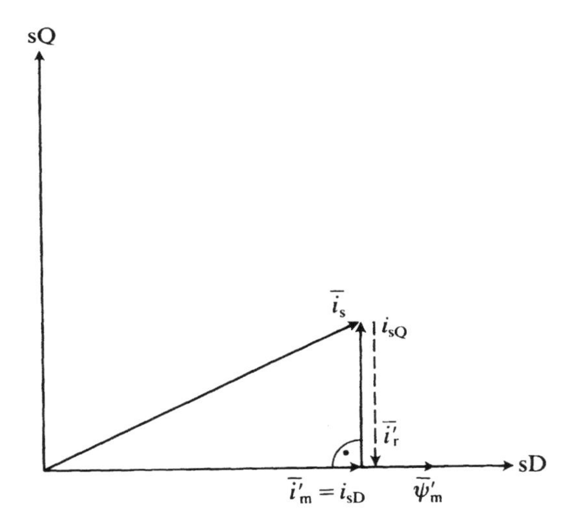

Fig. 2.12. Space phasors of $\bar{i}_s$, $\bar{i}_r'$, and $\bar{\psi}_m'$ at instant t_0.

diagram of the magnetizing flux-linkage and stator and rotor currents will be similar to the space-phasor diagram shown in Fig. 2.6 for a d.c. machine. This is shown in Fig. 2.12, where $\bar{i}_s = i_{sD} + ji_{sQ}$ is the space phasor of the stator currents, and $\bar{i}_r'$ is the space phasor of the rotor currents in the stationary reference frame.

From eqn (2.1-173), the sum of the stator and rotor current space phasors gives the magnetizing-current phasor if the stator and rotor windings are assumed to have the same number of effective turns. Furthermore, it follows from eqn (2.1-172) that the magnetizing flux-linkage space phasor in the stationary reference frame is $\bar{\psi}_m' = L_m \bar{i}_m'$, which is proportional to the magnetizing-current space phasor if the magnetizing inductance L_m is constant, and coaxial with the direct axis of the stationary reference frame. However, at $t>t_0$, the space-phasor diagram must be changed, since the field must change to induce the rotor currents. For simplicity, it is assumed that the rotor is locked in the same position as in Fig. 2.11. The new space-phasor diagram is shown in Fig. 2.13, with the new (changed) magnetizing flux and rotor-current space phasors.

It follows from Fig. 2.13 that, in contrast to Fig. 2.12 or Fig. 2.5 (which corresponds to a d.c. machine), the magnetizing flux-linkage space phasor $\bar{\psi}'_m = L_m \bar{i}'_m$ is not coaxial with the direct axis of the stationary reference frame and is not in space quadrature to the quadrature axis of the same reference frame, but it is rotated by angle μ_m, with respect to the direct axis sD. Thus $\bar{\psi}'_m$ is not along the direct axis sD.

If however, there is a fictitious rotation of the stator through angle μ_m in the counter-clockwise direction, the magnetizing flux-linkage space phasor $\bar{\psi}'_m = L_m \bar{i}'_m$ will again be coaxial with the direct-axis stator current and will be in space quadrature to the quadrature-axis stator current, as shown in Fig. 2.14, which is similar to the conditions shown in Fig. 2.5 for the d.c. machine.

In order to keep $\bar{\psi}'_m$ along the direct-axis, if $\bar{\psi}'_m$ is not allowed to rotate, the stator has to be continuously rotated by the appropriate value of μ_m, until the required orientation is achieved. The conditions are then similar to those for the compensated d.c. machine discussed in Section 2.1.6. In reality, of course, the stator is not rotated and the stator windings remain stationary, and to ensure separation of the stator currents into two current components, one of which produces the magnetizing flux linkage and the other of which produces the electromagnetic torque, a special reference frame (with direct and quadrature axes x, y) has to be used. There are several ways to choose such a special reference frame. One is to select

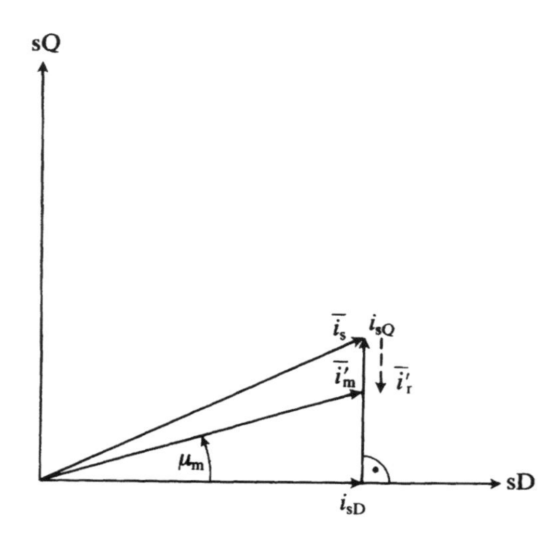

Fig. 2.13. Space phasors of $\bar{i}_s$, $\bar{i}'_r$, and $\bar{\psi}'_m$ at instant $t>t_0$.

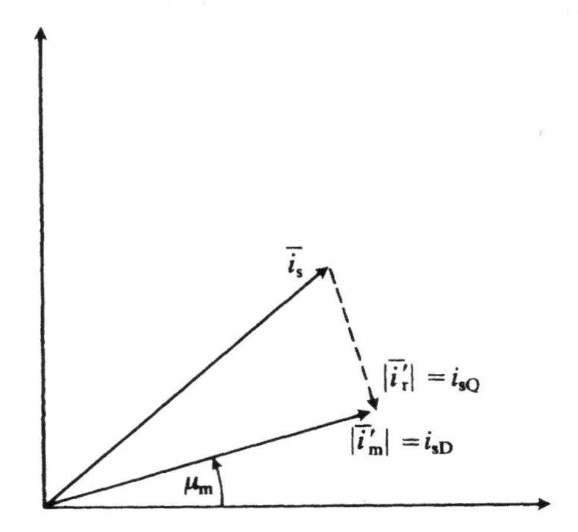

Fig. 2.14. Space phasors of the currents and magnetizing flux linkage when the stator is rotated by the angle μ_m.

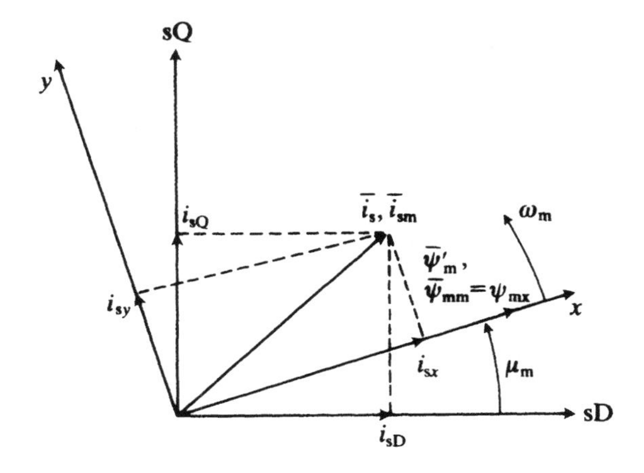

Fig. 2.15. Relationship of stationary (sD, sQ) and special (x, y) reference frames.

a reference frame which rotates at the speed of the magnetizing flux-linkage space phasor and has its real-axis coaxial with the magnetizing flux-linkage space phasor, as shown in Fig. 2.15.

It follows from Fig. 2.15 that, if in the stationary reference frame (sD, sQ) the magnetizing flux-linkage space phasor is defined as

$$\bar{\psi}'_m = |\bar{\psi}_m| e^{j\mu_m} = \psi_{mD} + j\psi_{mQ} \tag{2.1-178}$$

where $|\bar{\psi}_m|$ and μ_m are the modulus and space angle respectively of the magnetizing space phasor and ψ_{mD} and ψ_{mQ} are its two-axis components in the special reference frame, the space phasor of the stator currents in the special reference frame rotating at speed

$$\omega_m = \frac{d\mu_m}{dt} \quad (2.1\text{-}179)$$

is

$$\bar{i}_{sm} = \bar{i}_s e^{-j\mu_m} = i_{sx} + ji_{sy}, \quad (2.1\text{-}180)$$

where $\bar{i}_s$ is the space phasor of the stator currents in the stationary reference frame. Equation (2.1-180) also follows from eqn (2.1-142), by using $\theta_g = \mu_m$ and replacing the subscript g (used to denote the general reference frame) by the subscript m (to denote the reference frame fixed to the magnetizing flux-linkage space phasor). Since the two-axis components of $\bar{i}_{sm}$ are i_{sx} and i_{sy} and the two-axis components of $\bar{i}_s$ are i_{sD} and i_{sQ} (see Fig. 2.15), eqn (2.1-180) can be written as

$$\bar{i}_{sm} = i_{sx} + ji_{sy} = (i_{sD} + ji_{sQ}) e^{-j\mu_m} \quad (2.1\text{-}181)$$

which can be put into the matrix form

$$\begin{bmatrix} i_{sx} \\ i_{sy} \end{bmatrix} = \begin{bmatrix} \cos\mu_m & \sin\mu_m \\ -\sin\mu_m & \cos\mu_m \end{bmatrix} \begin{bmatrix} i_{sD} \\ i_{sQ} \end{bmatrix}. \quad (2.1\text{-}182)$$

Equation (2.1-182) shows how the direct- and quadrature-axis stator current components in the special reference frame can be obtained from the direct- and quadrature-axis stator current components of the stationary reference frame if the angle μ_m, which is continuously changing in time, is known. In the transient state all the current components vary arbitrarily in time. In the special reference frame the direct axis is coaxial with the magnetizing flux-linkage space phasor, and thus the quadrature-axis component of the magnetizing flux-linkage space phasor is zero. Therefore it follows from eqn (2.1-174) or eqn (2.1-175) that, in the special magnetizing flux-linkage-oriented reference frame, the electromagnetic torque is proportional to the product of the modulus of the magnetizing flux-linkage space phasor $|\bar{\psi}_m|$ and the quadrature-axis component of the stator current space phasor i_{sy},

$$t_e = \tfrac{3}{2} P |\bar{\psi}_m| i_{sy}, \quad (2.1\text{-}183)$$

It is possible to expand eqn (2.1-183). From eqns (2.1-172) and (2.1-174), in the special reference frame the magnetizing flux-linkage space phasor $\bar{\psi}_{mm}$, which contains only a direct-axis (x-axis) component ($\bar{\psi}_{mm} = \psi_{mx}$), can be expressed in terms of the magnetizing-current space phasor and the stator and rotor currents as

$$\psi_{mx} = \bar{\psi}_{mm} = L_m \bar{i}_{mm} = L_m(\bar{i}_{sm} + \bar{i}_{rm}) = L_m i_{mx} = |\bar{\psi}_m| = L_m |\bar{i}_{mm}|, \quad (2\text{-}1\text{-}184)$$

where $\bar{i}_{sm}$ has been defined in eqn (2.1-181) and the space phasor of the rotor currents in the special reference frame ($\bar{i}_{rm}$) is obtained from eqn (2.1-145) with $\theta_g = \mu_m$ as

$$\bar{i}_{rm} = \bar{i}_r e^{-j(\mu_m - \theta_r)} = i_{rx} + ji_{ry}. \quad (2.1\text{-}185)$$

Thus by the substitution of eqn (2.1-184) into eqn (2.1-183), the electromagnetic torque can be expressed as

$$t_e = \tfrac{3}{2} P \psi_{mx} i_{sy} = \tfrac{3}{2} P L_m i_{mx} i_{sy}, \tag{2.1-186}$$

where i_{mx} is the main flux-producing current component (direct-axis magnetizing-current component) and i_{sy} is the torque-producing stator current component.

Eqn (2.1-183) forms the basis of the so-called magnetizing-flux-oriented control of smooth-air-gap machines, which can be implemented in many different ways, as will be discussed in later sections.

The application of a special reference frame fixed to the rotor flux-linkage space phasor: The expression for the electromagnetic torque will be obtained in terms of the modulus of the rotor flux-linkage space phasor and the quadrature-axis stator current component, expressed in a special reference frame fixed to the rotor flux-linkage space phasor.

From eqn (2.1-176), or its two-axis form eqn (2.1-177), in the general reference frame, if the effects of saturation are neglected, and thus L_m and L_r are constant, the torque is proportional to the vector product of the rotor flux linkage and the stator-current space phasors expressed in the general reference frame. Thus, if such a rotating reference frame is used where the quadrature-axis component of the rotor flux-linkage space phasor is zero ($\psi_{ry}=0$), the electromagnetic torque will be produced by the interaction of the rotor flux linkage in the direct axis (of the special reference frame) and the quadrature-axis component of the stator currents (expressed in the same reference frame). Mathematically this can be described as follows.

According to eqn (2.1-147) or (2.1-134), the rotor flux-linkage space phasor in the stationary ($\omega_g=0$) reference frame can be expressed as

$$\bar{\psi}'_r = \bar{\psi}_r e^{j\theta_r} = \psi_{rd} + j\psi_{rq} = |\bar{\psi}_r| e^{j\rho_r}, \tag{2.1-187}$$

where $|\bar{\psi}_r|$ and ρ_r, are the modulus and phase angle of the rotor flux-linkage space phasor in the stationary reference frame. They are shown in Fig. 2.16, which illustrates the relationship between the stator current components in the stationary reference frame and the special reference frame fixed to the rotor flux-linkage space phasor.

The special x, y reference frame shown in Fig. 2.16 rotates at the speed of the rotor flux-linkage space phasor,

$$\omega_{mr} = \frac{d\rho_r}{dt}. \tag{2.1-188}$$

From eqn (2.1-142) with $\theta_g=\rho_r$, the stator-current space phasor in the special reference frame is

$$\bar{i}_{s\psi r} = \bar{i}_s e^{-j\rho_r} = i_{sx} + j i_{sy}, \tag{2.1-189}$$

where $\bar{i}_s$ is the space phasor of the stator currents in the stationary reference frame. It also follows mathematically from eqn (2.1-147) and (2.1-187) that in the

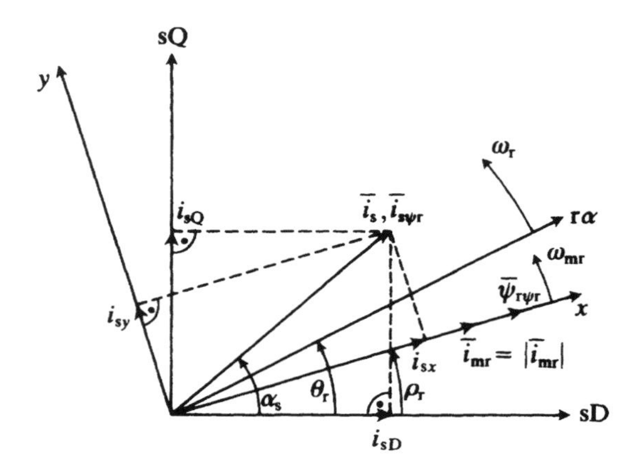

Fig. 2.16. Stator-current and rotor flux-linkage space phasors in the stationary reference frame and in the special reference frame fixed to the rotor flux-linkage space phasor.

special reference frame ($\theta_g = \rho_r$) the rotor flux-linkage space phasor has only a direct-axis component,

$$\bar{\psi}_{r\psi r} = \bar{\psi}_r e^{-j(\rho_r - \theta_r)} = \bar{\psi}_r e^{j\theta_r} e^{-j\rho_r} = \bar{\psi}_r' e^{-j\rho_r} = |\bar{\psi}_r| e^{j\rho_r} e^{-j\rho_r} = |\bar{\psi}_r| = \psi_{rx}, \quad (2.1\text{-}190)$$

which is also shown in Fig. 2.16.

Thus by substituting eqns (2.1-189) and (2.1-190) into eqns (2.1-176) or (2.1-177), the electromagnetic torque is given by

$$t_e = \tfrac{3}{2} P \frac{L_m}{L_r} \psi_{rx} i_{sy}, \quad (2.1\text{-}191)$$

where $\psi_{rx} = |\bar{\psi}_r|$ is the modulus of the rotor flux-linkage space phasor and i_{sy} is the quadrature-axis stator current in the reference frame fixed to the rotor flux-linkage space phasor.

The relationship between the stator current components in the stationary reference frame (i_{sD}, i_{sQ}) and the stator current components in the special reference frame (i_{sx}, i_{sy}) can be obtained by considering eqn (2.1-189) as

$$\begin{bmatrix} i_{sx} \\ i_{sy} \end{bmatrix} = \begin{bmatrix} \cos\rho_r & \sin\rho_r \\ -\sin\rho_r & \cos\rho_r \end{bmatrix} \begin{bmatrix} i_{sD} \\ i_{sQ} \end{bmatrix}, \quad (2.1\text{-}192)$$

which can also be proved by considering Fig. 2.16.

The rotor flux-linkage space phasor in the general reference frame (defined by eqn (2.1-190)) can also be expressed in terms of the stator and rotor currents by considering eqn (2.1-151):

$$\bar{\psi}_{r\psi r} = \psi_{rx} = |\bar{\psi}_r| = L_r \bar{i}_{r\psi r} + L_m \bar{i}_{s\psi r}, \quad (2.1\text{-}193)$$

where the rotor current space phasor in the special reference frame is obtained from eqns (2.1-145) and (2.1-133),

$$\bar{\imath}_{r\psi r}=i_{rx}+ji_{ry}=\bar{\imath}_r e^{-j(\rho_r-\theta_r)}=\bar{\imath}_r e^{j\theta_r}e^{-j\rho_r}=\bar{\imath}_r' e^{-j\rho_r}=(i_{rd}+ji_{rq})e^{-j\rho_r}, \qquad (2.1\text{-}194)$$

where $\bar{\imath}_r'$ is the rotor-current space phasor in the stationary reference frame and $\bar{\imath}_r$ is the rotor-current space phasor in the reference frame fixed to the rotor. It follows from eqn (2.1-194) that in the special reference frame, both of the rotor current components (i_{rx}, i_{ry}) are non-zero. This is of course a physically expected result, since in the special reference frame the quadrature-axis component of the rotor flux-linkage component is zero. Thus the quadrature-axis rotor current in the same reference frame can only be zero if there is no mutual rotor flux component due to the quadrature-axis stator current in the special reference frame, i.e. if the quadrature-axis stator current is zero ($i_{sy}=0$), which in general is not the case.

From eqn (2.1-193), the so-called rotor magnetizing current in the special reference frame is defined in terms of the stator- and rotor-current space phasors given by eqns (2.1-189) and (2.1-194) respectively, as

$$\bar{\imath}_{mr}=\frac{\bar{\psi}_{r\psi r}}{L_m}=\frac{L_r}{L_m}\bar{\imath}_{r\psi r}+\bar{\imath}_{s\psi r}=\bar{\imath}_{s\psi r}+(1+\sigma_r)\bar{\imath}_{s\psi r}, \qquad (2.1\text{-}195)$$

where $\sigma_r = L_{rl}/L_m$ is the rotor leakage factor (L_{rl} and L_r are the rotor leakage inductance and self-inductance respectively). In Fig. 2.16 the rotor magnetizing-current space phasor is also shown. From eqn (2.1-195), the rotor magnetizing current in the special reference frame must also have a component, but only along the real axis of the special reference frame:

$$\bar{\imath}_{mr}=i_{mrx}+ji_{mry}=i_{mrx}=|\bar{\imath}_{mr}|=\frac{|\bar{\psi}_r|}{L_m}=\bar{\imath}_{s\psi r}+(1+\sigma_r)\bar{\imath}_{r\psi r}. \qquad (2.1\text{-}196)$$

Thus, if $\psi_{rx}=|\bar{\psi}_r|$, and $\sigma_r=L_{rl}/L_m$, substitution of eqn (2.1-196) into eqn (2.1-191) yields

$$t_e=\tfrac{3}{2}P\frac{L_m^2}{L_r}|\bar{\imath}_{mr}|i_{sy}=\tfrac{3}{2}P\frac{L_m}{1+\sigma_r}|\bar{\imath}_{mr}|i_{sy}. \qquad (2.1\text{-}197)$$

It is a very important feature of eqn (2.1-197) that it shows that the electromagnetic torque can be controlled by independently controlling the flux-producing current component $|\bar{\imath}_{mr}|$ and the torque-producing stator current component i_{sy}. Under linear magnetic conditions L_m, L_r, and the term $\frac{3}{2}PL_m/(1+\sigma_r)$ are constant, and the expression for the torque is similar to that of the separately excited, compensated d.c. machine.

For an induction machine, under linear magnetic conditions, when $|\bar{\imath}_{mr}|$ is kept constant, $|\bar{\imath}_{mr}|=i_{sx}=$constant. This will be proved in Section 4.1.1. Thus the rotor flux is solely determined by $|\bar{\imath}_{mr}|=$constant and the torque is proportional to i_{sy}. That $|\bar{\imath}_{mr}|=$constant leads to the expression $|\bar{\imath}_{mr}|=i_{sx}$ can also be proved by considering that the rotor voltage of the x-axis short-circuited rotor winding expressed in the special reference frame must be balanced by a rotor ohmic voltage drop ($R_r i_{rx}$) plus the rate of change of the rotor flux linkage expressed in the special reference frame ($d\psi_{rx}/dt$), and that there is no rotational voltage component due to the quadrature-axis rotor flux linkage (ψ_{ry}), since this component is zero due to the special selection of the reference frame. Under linear conditions L_m = constant and, if $|\bar{i}_{mr}|$ = constant, it follows that $\psi_{rx} = L_m|\bar{i}_{mr}|$ = constant (and thus its rate of change is zero). It follows that $i_{rx} = 0$ and thus $|\bar{i}_{mr}| = i_{sx} + (1+\sigma_r)i_{rx} = i_{sx}$.

Torque control schemes of smooth-air-gap a.c. machines based on eqn (2.1-197) have so far found the most widespread applications and, despite the fact that in this case the special reference frame is aligned with the rotor flux-linkage space phasor, and rotates synchronously with the rotor flux-linkage space phasor, and therefore is not fixed to the magnetizing flux-linkage space phasor (as discussed in the previous section), this type of control is often referred to as field-oriented control. However, this terminology can be justified by considering that the rotor flux-linkage space phasor is coaxial with the rotor magnetizing-current space phasor and 'field-orientation' in this sense has the meaning of rotor magnetizing-current orientation. It would be more accurate to use the general terminology of vector control and to distinguish between magnetizing-flux (or magnetizing-current)-oriented control (for the control ideas presented in the previous section) and rotor-flux-oriented control or rotor-magnetizing-current-oriented control for the control using the principles developed in the present section. There is also the possibility of stator-flux-oriented control (or stator-magnetizing-current-oriented control), which will be discussed in the next section.

The application of a special reference frame fixed to the stator flux-linkage space phasor: It has already been shown [eqn (2.1-174)] that the electromagnetic torque can be expressed in terms of the cross product of the stator flux linkage and current space phasors. From eqn (2.1-167), if a special reference frame attached to the rotating stator flux-linkage space phasor is used, i.e. where the quadrature-axis stator flux-linkage component is zero, in both the steady-state and transient states the electromagnetic torque of a two-pole machine is

$$t_e = \tfrac{3}{2}\psi_{sx} i_{sy}. \tag{2.1-198}$$

Here ψ_{sx} is the real-axis component of the stator flux-linkage space phasor in the special reference frame rotating at the speed of the stator flux-linkage space phasor (ω_{ms}), and i_{sy} is the imaginary-axis component of the stator-current space phasor in the same reference frame. The components of the different space phasors in the special reference frame are related to their corresponding components expressed in the stationary reference frame as follows.

It follows from eqn (2.1-128) that in the stationary reference frame the space phasor of the stator flux linkages can be expressed as

$$\bar{\psi}_s = \psi_{sD} + j\psi_{sQ} = |\bar{\psi}_s|\, e^{j\rho_s}, \tag{2.1-199}$$

where ψ_{sD} and ψ_{sQ} are the stationary axis components of the stator flux-linkage space phasor, $|\bar{\psi}_s|$ is its modulus, and ρ_s is its phase angle—a spatial angle—with respect to the real axis of the stationary reference frame (sD). The speed of the special reference frame can be expressed as

$$\omega_{ms} = \frac{d\rho_s}{dt}. \qquad (2.1\text{-}200)$$

Figure 2.17 shows the relationship between the special and the stationary reference frames.

From eqns (2.1-142) and (2.1-144), in the special reference frame shown in Fig. 2.17 ($\theta_g = \rho_s$), the space phasors of the stator current and stator flux linkages are

$$\bar{i}_{s\psi s} = \bar{i}_s e^{-j\rho_s} = (i_{sD} + ji_{sQ})e^{-j\rho_s} = i_{sx} + ji_{sy} \qquad (2.1\text{-}201)$$

and

$$\bar{\psi}_{s\psi s} = \bar{\psi}_s e^{-j\rho_s} = (\psi_{sD} + j\psi_{sQ})e^{-j\rho_s} = \psi_{sx} + j\psi_{sy}, \qquad (2.1\text{-}202)$$

where the two-axis stator current and the stator flux-linkage components in the special reference frame are i_{sx}, i_{sy} and ψ_{sx}, ψ_{sy} respectively and in the stationary reference frame are i_{sD}, i_{sQ} and ψ_{sD}, ψ_{sQ} respectively. However, substitution of eqn (2.1-199) into eqn (2.1-202) yields

$$\bar{\psi}_{s\psi s} = |\bar{\psi}_s| e^{j\rho_s} e^{-j\rho_s} = \psi_{sx} + j0 = \psi_{sx} = |\bar{\psi}_s|, \qquad (2.1\text{-}203)$$

which shows mathematically that in the special reference frame the space phasor of the stator flux linkages contains only a direct-axis component, which is equal to the modulus of the space phasor (since it has been deliberately aligned with the stator flux-linkage space phasor). For convenience, eqn (2.1-201) can be put into matrix form:

$$\begin{bmatrix} i_{sx} \\ i_{sy} \end{bmatrix} = \begin{bmatrix} \cos\rho_s & \sin\rho_s \\ -\sin\rho_s & \cos\rho_s \end{bmatrix} \begin{bmatrix} i_{sD} \\ i_{sQ} \end{bmatrix}, \qquad (2.1\text{-}204)$$

which gives the relationship of i_{sD}, i_{sQ} and the transformed currents i_{sx}, i_{sy}.

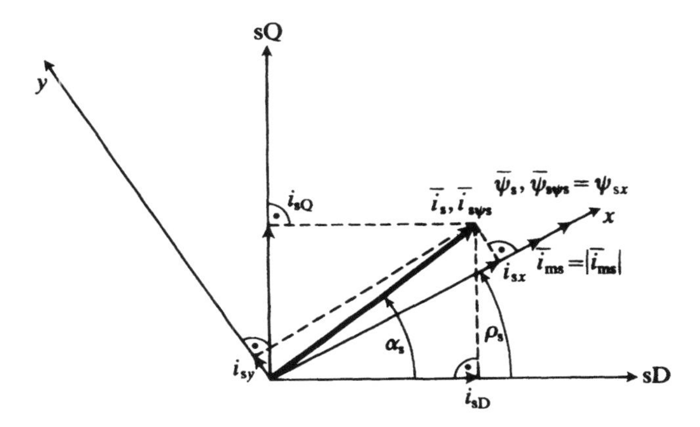

Fig. 2.17. Relationship between the stationary reference frame and the special reference frame fixed to the stator flux-linkage space phasor.

It is possible to put the equation for the torque in such a form that the stator flux linkage ψ_{sx} is expressed in terms of the currents. For this purpose the space phasor of the stator flux linkages in the special reference frame [defined by eqn (2.1-201)] is expressed in terms of the stator and rotor currents and from eqn (2.1-150):

$$\bar{\psi}_{s\psi s}=L_s\bar{i}_{s\psi s}+L_m\bar{i}_{r\psi s}, \tag{2.1-205}$$

where $\bar{i}_{s\psi s}$ is defined in eqn (2.1-201) and $\bar{i}_{r\psi s}$ is the space phasor of the rotor currents in the special reference frame. It follows from eqns (2.1-133), (2.1-145), and $(\theta_g=\rho_s)$ that

$$\bar{i}_{r\psi s}=i_{rx}+ji_{ry}=\bar{i}_r e^{-j(\rho_s-\theta_r)}=\bar{i}_r e^{j\theta_r}e^{-j\rho_s}=\bar{i}_r' e^{-j\rho_r}=(i_{rd}+ji_{rq})e^{-j\rho_s}, \tag{2.1-206}$$

where $\bar{i}_r$ is the space phasor of the rotor currents in the rotor reference frame and $\bar{i}_r'$ is the space phasor of the rotor currents in the stationary reference frame fixed to the stator. It is now possible to define the so-called stator magnetizing current $\bar{i}_{ms}$ in the special reference frame, which rotates at speed ω_{ms}. From eqns (2.1-205) and (2.1-206)

$$\bar{i}_{ms}=\frac{\bar{\psi}_{s\psi s}}{L_m}=\frac{L_s}{L_m}\bar{i}_{s\psi s}+\bar{i}_{r\psi s}=(1+\sigma_s)\bar{i}_{s\psi s}+\bar{i}_{r\psi s}, \tag{2.1-207}$$

where $\bar{i}_{r\psi s}$ and $\bar{i}_{s\psi s}$ are defined in eqns (2.1-206) and (2.1-201) respectively. Since it follows from the choice of the special reference frame that $\bar{\psi}_{s\psi s}$ is aligned with the real axis, and thus has only one component [see eqn (2.1-203)], from eqn (2.1-207), the stator magnetizing-current space phasor will be coaxial with $\bar{\psi}_{s\psi s}$, as shown in Fig. 2.17. From eqn (2.1-207)

$$\bar{i}_{ms}=i_{msx}+ji_{msy}=i_{msx}=|\bar{i}_{ms}|=\frac{\psi_{sx}}{L_m}=\bar{i}_{s\psi s}(1+\sigma_s)+\bar{i}_{r\psi s} \tag{2.1-208}$$

and, if eqn (2.1-208) is substituted into eqn (2.1-198) and in general $P\neq 1$, finally the instantaneous electromagnetic torque can be expressed as

$$t_e=\tfrac{3}{2}PL_m|\bar{i}_{ms}|i_{sy}. \tag{2.1-209}$$

$|\bar{i}_{ms}|$ is the stator flux-producing current component and i_{sy} is the torque-producing stator current component and this equation is again similar to the equation of the electromagnetic torque developed by the separately excited d.c. machine. Thus the torque can be controlled by independently controlling these two currents. It should be noted that in the steady state, these currents are d.c. values.

2.2 Electromagnetic torque production and the basis for vector control in salient-pole machines

The mechanism of electromagnetic torque production in salient-pole a.c. machines will be discussed here and it will be assumed that the saliency exists only on one member of the machine, the rotor. For simplicity, a two-pole synchronous machine is assumed. It will be shown that the mechanism of electromagnetic torque production is similar to that shown for smooth-air-gap machines, but owing to the saliency, reluctance torque will also occur.

2.2.1 THE ELECTROMAGNETIC TORQUE IN THE ROTOR REFERENCE FRAME

A two-pole ($P=1$) salient-pole machine is assumed with symmetrical three-phase, sinusoidally distributed stator windings (sA, sB, sC). Figure 2.18 shows the cross section of the machine; on the rotor there is a field winding (rF) and there are also quadrature-phase damper windings ($r\alpha, r\beta$), which are short-circuited. θ_r is the rotor angle, the angle between the magnetic axis of stator phase sA (the direct axis of the stationary reference frame sD) and the direct axis of the rotor ($r\alpha$). It is possible to prove that the equation for the electromagnetic torque in the salient-pole machine can be put into a form similar to eqn (2.1-167):

$$t_e = \tfrac{3}{2} P(\psi_{sx} i_{sy} - \psi_{sy} i_{sx}), \tag{2.2-1}$$

where P is the number of pole pairs, ψ_{sx}, ψ_{sy} and i_{sx}, i_{sy} are the direct- (x) and quadrature-axis (y) components of the stator flux linkages and currents respectively in the general reference frame, rotating at the instantaneous speed ω_g. If the reference frame fixed to the rotor ($\omega_g = \omega_r = d\theta_r/dt$) is used

$$t_e = \tfrac{3}{2} P(\psi_{sd} i_{sq} - \psi_{sq} i_{sd}), \tag{2.2-2}$$

where ψ_{sd}, ψ_{sq} and i_{sd}, i_{sq} are the direct- and quadrature-axis stator flux-linkage and current components expressed in the rotor reference frame.

From eqn (2.1-142) the stator currents in the reference frame fixed to the rotor (i_{sd}, i_{sq}) are related to the two-axis currents expressed in the stationary reference

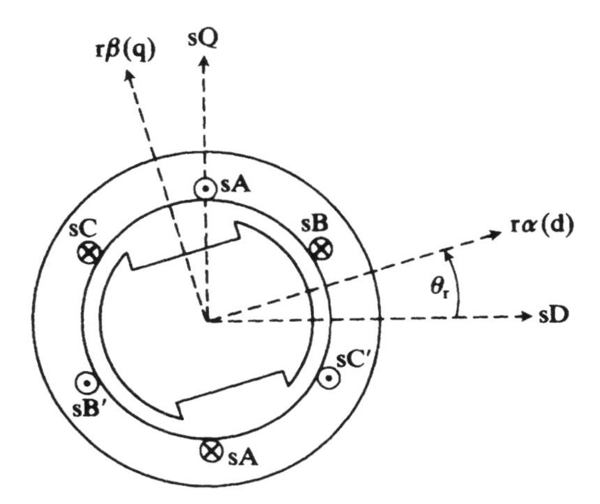

Fig. 2.18. Cross section of the salient-pole machine.

frame (i_{sD}, i_{sQ}) as

$$i_{sd}+ji_{sq}=(i_{sD}+ji_{sQ})\,e^{-j\theta_r}, \tag{2.2-3}$$

The matrix form of this is

$$\begin{bmatrix} i_{sd} \\ i_{sq} \end{bmatrix} = \begin{bmatrix} \cos\theta_r & \sin\theta_r \\ -\sin\theta_r & \cos\theta_r \end{bmatrix} \begin{bmatrix} i_{sD} \\ i_{sQ} \end{bmatrix}. \tag{2.2-4}$$

From physical considerations, the stator flux linkages in the rotor reference frame can be defined in terms of the currents as

$$\psi_{sd}=L_{sd}i_{sd}+L_m i_{rF}+L_{md}i_{r\alpha} \tag{2.2-5}$$

and

$$\psi_{sq}=L_{sq}i_{sq}+L_{mq}i_{r\beta}, \tag{2.2-6}$$

where L_{sd}, L_{sq} are the self-inductances of the stator winding along the d and q axes, L_{md}, L_{mq} are the magnetizing inductances along the same two axes, and owing to saliency $L_{sd}\neq L_{sq}$ and $L_{md}\neq L_{mq}$. The currents i_{rF} and $i_{r\alpha}$, $i_{r\beta}$ are those currents in the field winding and the direct- and quadrature-axis damper windings respectively and are expressed in the reference frame fixed to the rotor.

Substitution of eqns (2.2-5) and (2.2-6) into eqn (2.2-2) yields the electromagnetic torque,

$$t_e=\tfrac{3}{2}P[(L_{sd}-L_{sq})i_{sd}i_{sq}+L_{md}(i_{rF}+i_{r\alpha})i_{sq}-L_{mq}i_{r\beta}i_{sd}], \tag{2.2-7}$$

where $(L_{sd}-L_{sq})i_{sd}i_{sq}$ is the so-called reluctance torque, which is solely due to the saliency and disappears for a smooth-air-gap machine, where $L_{sd}=L_{sq}=L_s$. The expression $L_{md}i_{rF}i_{sq}$ is the field torque, $L_{md}i_{r\alpha}i_{sq}$ is the direct-axis damper-torque component, and $L_{mq}i_{r\beta}i_{sd}$ is the quadrature-axis damper-torque component. In the steady-state the damper currents are zero and all the other currents are constant.

It is possible to rearrange the torque equation in terms of the magnetizing flux linkages. Since the total inductances along the direct and quadrature axes can be decomposed into leakage and magnetizing inductances, $L_{sd}=L_{sl}+L_{md}$, $L_{sq}=L_{sl}+L_{mq}$, the flux-linkage components defined by eqns (2.2-5) and (2.2-6) can be written as

$$\psi_{sd}=L_{sl}i_{sd}+\psi_{md} \tag{2.2-8}$$

$$\psi_{sq}=L_{sl}i_{sq}+\psi_{mq}. \tag{2.2-9}$$

The first term on the right-hand side of eqns (2.2-8) and (2.2-9) is a leakage flux component and ψ_{md}, ψ_{mq} are the direct- and quadrature-axis magnetizing flux-linkage components respectively in the rotor reference frame:

$$\psi_{md}=L_{md}i_{md}=L_{md}(i_{sd}+i_{rF}+i_{r\alpha}) \tag{2.2-10}$$

$$\psi_{mq}=L_{mq}i_{mq}=L_{mq}(i_{sq}+i_{r\beta}), \tag{2.2-11}$$

where i_{md} and i_{mq} are the magnetizing currents along the two axes. The leakage flux components do not contribute to electromechanical energy conversion, and indeed this can also be seen, if eqns (2.2-8)–(2.2-11) are substituted into eqn (2.2-2) to yield

$$t_e = \tfrac{3}{2} P(\psi_{md} i_{sq} - \psi_{mq} i_{sd}) \tag{2.2-12}$$

or

$$t_e = \tfrac{3}{2} P[(L_{md} - L_{mq}) i_{sd} i_{sq} + L_{md}(i_{rF} + i_{r\alpha}) i_{sq} - L_{mq} i_{r\beta} i_{sd}], \tag{2.2-13}$$

which does not contain the leakage flux linkages. According to eqn (2.2-12) the electromagnetic torque is caused by

(i) the interaction of the direct-axis magnetizing flux-linkage component and the quadrature-axis stator current component; and

(ii) by the interaction of the quadrature-axis magnetizing flux-linkage component and the direct-axis stator current component;

the same mechanism exists in smooth-air-gap machines. In eqn (2.2-13) the term $(L_{md} - L_{mq}) i_{sd} i_{sq}$ is reluctance torque and is equal to the reluctance torque component given in eqn (2.2-7). It follows from eqn (2.2-13) that if a smooth air gap is assumed ($L_{md} = L_{mq}$), and if on the rotor only the field winding is excited ($i_{r\alpha} = 0$; $i_{r\beta} = 0$), the electromagnetic torque can be expressed as

$$t_e = \tfrac{3}{2} P L_m i_{rF} i_{sq}. \tag{2.2-14}$$

This expression is similar to the equation for the electromagnetic torque of a synchronous machine with permanent-magnet excitation (see Section 3.1.1).

It can also be seen that in the transient state, if the direct-axis components of the stator currents expressed in the rotor reference frame are absent ($i_{sd} = 0$ or $\gamma = 0$, where γ is the torque angle), there is no reluctance torque and no torque produced by the quadrature-axis damper current. If $i_{sd} = 0$ and $i_{r\alpha} = 0$, the field current (i_{rF}) and the stator current (which lies along the q-axis) are in space quadrature, similarly to a separately excited d.c. machine, and 'field-orientation' is achieved, where the stator and field m.m.f.s are in space quadrature. Furthermore, if $i_{sd} = 0$ and the field winding is supplied by a d.c. voltage, then the current in the field winding—expressed in the reference frame fixed to the rotor—must be constant ($i_{rF} = I_F$) and it follows from eqn (2.2-13) that the electromagnetic torque will be

$$t_e = \tfrac{3}{2} P L_{md} I_F i_{sq}. \tag{2.2-15}$$

This expression is again similar to the equation for the torque of a synchronous machine with permanent-magnet excitation (Section 3.1) and it is also similar to the expression for the electromagnetic torque of the compensated, separately excited d.c. machine.

It is an important consequence that, in contrast to an induction machine, where the electromagnetic torque production requires that the stator currents should contain both the excitation (flux-producing) and torque-producing stator current components, in a synchronous machine it is not necessary for the stator currents to contain the excitation component, since excitation is provided by the field currents and torque is produced by the interaction of the field and quadrature-axis (torque-producing) stator current.

If the d-axis magnetizing reactance $X_{md} = \omega L_{md}$, in the steady-state $i_{rF} = I_{rF}$ and the e.m.f. $jX_{md}I_{rF}$ is the so-called internal voltage $\bar{E}$, which lies along the q-axis, as shown in Fig. 2.19.

In the conventional steady-state theory, the time-phase angle between the q axis, where $\bar{E}$ is positioned, and the vector of the steady-state currents $(I_{sd} + jI_{sq})$ expressed in the rotor reference frame, is denoted by γ. In the steady state the torque component $X_{md}I_{rF}I_{sq} = X_{md}I_{rF}|\bar{I}'_s| \cos \gamma$ is produced as a result of the interaction of the field current and the quadrature-axis stator current, where $|\bar{I}'_s|$ is the modulus of the stator currents. In the transient state, if $i_{r\alpha} = 0$, the torque component $L_{md}i_{rF}i_{sq}$ is present, caused by the interaction of the flux linkage $L_{md}i_{rF}$ and the stator current i_{sq}. If in the transient state the space angle γ is defined as the angle between the stator current space phasor $(i_{sd} + ji_{sq})$ and the quadrature-axis, then the expression $L_{md}i_{rF}i_{sq} = L_{md}i_{rF}|I'_s| \cos\gamma$ is obtained for the torque.

If $\psi_{md} + j\psi_{mq}$ is defined to be the magnetizing flux-linkage space phasor, then as a result of saliency, this vector will not be coaxial with $i_{md} + ji_{mq}$, which could be defined as the magnetizing-current space phasor.

2.2.2 THE ELECTROMAGNETIC TORQUE IN SPECIAL REFERENCE FRAMES

It has been shown in Section 2.1.8 that for smooth-air-gap machines, by suitable orientation of the reference frame, it is possible to obtain expressions for the torque which resemble the expression for the torque of the separately excited, compensated d.c. machine. It is possible to use similar concepts for the salient-pole machine, as will now be briefly discussed.

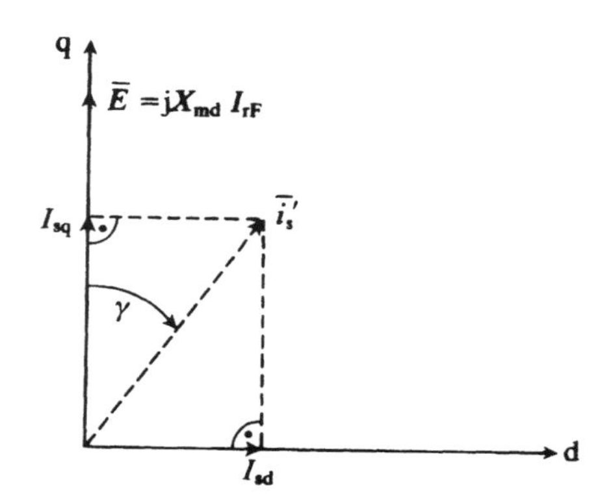

Fig. 2.19. Steady-state vector diagram of the stator currents and internal voltage in the synchronously rotating reference frame (d, q) fixed to the rotor.

Application of a special reference frame fixed to the magnetizing flux-linkage space phasor

By considering eqn (2.2-12), where the current and flux components are expressed in the rotor reference frame, it can be shown that in the general reference frame (x,y) rotating at the speed ω_g, for the salient-pole machine discussed in Section 2.2.1, the equation of the torque will be similar:

$$t_e = \tfrac{3}{2} P(\psi_{mx} i_{sy} - \psi_{my} i_{sx}), \tag{2.2-16}$$

where ψ_{mx}, ψ_{my} and i_{sx}, i_{sy} are the direct- and quadrature-axis components of the magnetizing flux linkages and currents respectively in the general reference frame.

If, as in Section 2.1.8, a special reference frame is used, whose real axis is coaxial with the magnetizing flux-linkage space phasor and which rotates with the speed of the magnetizing flux-linkage space phasor (ω_m), then in the special reference frame, the magnetizing flux-linkage space phasor can be expressed as

$$\bar{\psi}_{mm} = \psi_{mx} + j0 = |\bar{\psi}_m| \tag{2.2-17}$$

since $\psi_{my} = 0$ in the special reference frame. Thus substitution of eqn (2.2-17) into eqn (2.2-16) yields

$$t_e = \tfrac{3}{2} P |\bar{\psi}_m| i_{sy}, \tag{2.2-18}$$

where $|\bar{\psi}_m|$ is the modulus of the magnetizing flux-linkage space phasor and i_{sy} is the quadrature-axis component of the stator current in the special reference frame. If the angle μ_m is the space angle of the magnetizing flux-linkage space phasor with respect to the real axis of the stationary reference frame (sD), then from eqn (2.1-142), i_{sy} can be defined in terms of the two-axis current components (i_{sD}, i_{sQ}) defined in the stationary reference frame as

$$i_{sy} = -\sin \mu_m i_{sD} + \cos \mu_m i_{sQ}, \tag{2.2-19}$$

which also follows from eqn (2.1-182). i_{sy} is in space quadrature to $\bar{\psi}_{mm}$. Equation (2.2-18) serves as the basis of the magnetizing-flux-oriented control of salient-pole synchronous machines.

If the magnetizing-current space phasor in the special reference frame is defined as

$$\bar{i}_{mm} = i_{mx} + j i_{my} \tag{2.2-20}$$

then, in contrast to the smooth-air-gap machine, where the magnetizing flux linkage and magnetizing current space phasors are coaxial, as a result of saliency, $\bar{\psi}_{mm}$ and $\bar{i}_{mm}$ will not be coaxial. Thus whilst for the smooth-air-gap machine the application of a $\bar{\psi}_{mm}$- or $\bar{i}_{mm}$-oriented reference frame has the same effect, this is not the case for the salient-pole machine.

Bibliography

Blaschke, F. (1972). The principle of field-orientation as applied to the new Transvektor closed-loop control system for rotating-field machines. *Siemens Review* **34**, 217–20.

Jones, C. V. (1967). *The unified theory of electrical machines.* Butterworths, London.

Kovács, K. P. and Rácz, I. (1959). *Transiente Vorgänge in Wechselstrommaschinen*, I–II. Akadémia Kiadó, Budapest.

Leonhard, W. (1985). *Control of electrical drives.* Springer-Verlag, Berlin.

Naunin, D. (1979). The calculation of the dynamic behaviour of electrical machines by space phasors. *Electrical Machines and Electromechanics* **4**, 33–45.

Stronach, A. F., Vas, P., and Neuroth, M. (1997). Implementation of intelligent self-organising controllers in DSP-controlled electromechanical drives. *Proceedings of the IEEE, Pt. D* **144**, 324–30.

Vas, P. and Brown, J. E. (1985). Real-time monitoring of the electromagnetic torque of multi-phase a.c. machines. *IEEE IAS Meeting, Toronto, 6–11 October*, pp. 732–7.

Vas, P., Brown, J. E., and Shirley, A. (1984). The application of *n*-phase generalized rotating-field theory to induction machines with arbitrary stator winding connection. *IEEE Transactions on Power Apparatus and Systems* **PAS-103**, 1270–7.

Vas, P., Willems, J. L., and Brown, J. E. (1987). The application of space-phasor theory to the analysis of electrical machines with space harmonics. *Archiv für Elektrotechnik* **69**, 359–63.

Vas, P. (1992). *Electrical machines and drives: a space-vector theory approach.* Oxford University Press.

Vas, P. and Alakula, M. (1990). Field-oriented control of saturated induction machines. *IEEE Transactions on Energy Conversion* **5**, 218–24.

Vas, P. (1993). *Parameter estimation, condition monitoring and diagnosis of electrical machines.* Oxford University Press.

Vas, P. (1995). Application of space-vector techniques to electrical machines and variable-speed drives. *EPE Tutorial, Seville*, 1–130.

Vas, P. (1995). Artificial intelligence applied to variable-speed drives. *PCIM Seminar, Nurnberg*, 1–149.

Vas, P., Stronach, A.F., and Neuroth, M. (1997). Full fuzzy control of a DSP-based high performance induction motor drive. *IEE Proc. Pt. D.* **144**, 361–8.

3 Vector and direct torque control of synchronous machines

In this chapter vector control and direct torque control of different types of synchronous machines are described. For this purpose permanent-magnet synchronous machines (PMSM), the electrically excited synchronous machine, and synchronous reluctance machines (SYRM) are considered. Various 'sensorless' control schemes are also discussed.

3.1 Vector control of permanent-magnet synchronous machines

Synchronous machines with an electrically excited rotor winding have a three-phase stator winding (called the armature) and an electrically excited field winding on the rotor which carries a d.c. current. There can also be damper windings on the rotor. The armature winding is similar to the stator of induction machines. The synchronous machine is a constant-speed machine which always rotates at a synchronous speed which depends on the supply frequency and on the number of poles. The electrically excited field winding can be replaced by permanent magnets. The use of permanent magnets has many advantages including the elimination of brushes, slip-rings, and rotor copper losses in the field winding, which leads to higher efficiency. Since the copper and iron losses are then concentrated in the stator, cooling of the machine, through the stator, is more easily achieved. The higher efficiency allows a reduction in the machine frame size. Another advantage of using permanent magnets is that for the same-size machine, the characteristics of the machine can be varied greatly according to the type of magnets chosen and the way they are arranged. Permanent-magnet synchronous machines (PMSM) can be categorized into two main groups:

(i) the brushless d.c. machines fed by current source inverters and having trapezoidal flux distribution; and

(ii) synchronous machines having approximately sinusoidal air-gap flux-density distribution, fed by sinusoidal stator currents.

Permanent-magnet machines are extensively used in servo drives containing machines with low power for machine-tool (e.g. spindle motors, positioning drives etc.) and robotic applications, but large machines up to 1 MW have also been built; for example, a 1 MW machine has been used in a ship propulsion drive. It is also possible to use such machines in the fields of electricity generation, solar pumping, wind-energy applications, etc. In the present book, however, only servo applications are discussed. The design criteria for synchronous servomotors, to be used in machine-tool feed drives, manipulators, and industrial robots differ from that of conventional synchronous machines since the following requirements must be met:

- High air-gap flux density;
- High power/weight ratio (greatest possible power/motor mass);
- Large torque/inertia ratio (to enable high acceleration);
- Smooth torque operation (small torque ripples) even at very low speeds (to achieve high positioning accuracy);
- Controlled torque at zero speed;
- High speed operation;
- High torque capability (quick acceleration and deceleration for short time);
- High efficiency and high $\cos\phi$ (low expense for the power supply);
- Compact design.

These requirements can be met well by the permanent-magnet synchronous machine employing vector control. Either the rotor construction can be such that a slim-drum rotor is used with large length/diameter ratio, resulting in a low mechanical time constant (such rotors are sometimes also referred to as hot-dog rotors) or a light-weight, aluminium disc rotor (pancake rotor) is used. Sometimes the disc is manufactured from high-strength plastic material, and the magnets are embedded and encapsulated in this using resin. These machines are also referred to as radial-field machines (they have a long cylindrical rotor) and axial-field machines (with disc rotor). The machines with the disc rotor are usually employed in robotic applications, whilst in servo drives for machine tools the slim-drum rotor is usually applied. The machine with the pancake rotor can yield the same high dynamic performance as the machine with the drum-type rotor, but its power factor, maximum torque, and efficiency can be lower. Furthermore, the drum rotor leads to a low-inertia (low mechanical time constant) design, which makes possible a more rapid acceleration. The manufacturing costs of axial-field machines are higher than those for the radial-field machines.

There are various practical implementations of mounting the magnets. In one solution, rare-earth magnets, e.g. samarium–cobalt ($SmCo_5$, Sm_2Co_{17}) or neodymium–iron–boron (NdFeB), which have a high energy product, and thus for which the effects of demagnetization are avoided, are mounted on the surface of the solid steel rotor, by using high-strength adhesive material. To obtain high mechanical strength, which is especially important for high-speed operation, either the gaps between the magnets are filled with non-magnetic material and the rotor can be wrapped with high-strength material, e.g. fibreglass, or sometimes screwed pole-shoes are used. At present a disadvantage of using rare-earth magnets is their relatively high price, but with new developments this may change in the future. Furthermore, it is also expected that the characteristics of the NdFeB magnets will improve (at present there can be problems because the field strength decreases with increasing temperature). When the surface-magnet synchronous motor is used for variable frequency operation there are no damper windings on the rotor. In another construction, buried or so-called interior magnets are used.

In the case where the magnet bars are mounted on the rotor surface, the incremental permeability of the magnets relative to the external fields is 1.02–1.2. They have high resistivity, and thus the machine can be considered to have a large effective air gap, which makes the effects of saliency negligible (thus the direct-axis magnetizing inductance is equal to the quadrature-axis magnetizing inductance, $L_{md}=L_{mq}=L_m$). Furthermore, because of the large air-gap, the synchronous inductance ($L_s=L_{sl}+L_m$) is small and therefore the effects of armature reaction are negligible. A further consequence of the large air gap is that the electrical time constant of the stator winding is small. The magnets can have various shapes. Bar-shaped magnets and circumferential segments with angles up to 90° and thickness of a few millimetres are available. Radially magnetized circumferential segments produce a smoother air-gap flux density distribution and fewer torque ripples. These torque ripples can also be reduced by the appropriate design of the stator windings.

When the magnets are buried inside the rotor, a mechanically robust construction is obtained which can be used for high-speed applications since the magnets are physically contained and protected. However, with the interior permanent-magnet construction, the machine cannot be considered to have a uniform air gap. In this case the magnets are mounted inside the steel rotor core and physically there seems to be almost no variation of the surface of the magnet geometry. However, since each magnet is covered by a steel pole-piece, this significantly changes the magnetic circuits of the machine, since owing to these iron pole-pieces, high-permeance paths are created for the magnetic flux across these poles and also in space quadrature to the magnet flux. Thus there are saliency effects which significantly alter the torque production mechanism of the machine. As described in Section 2.2, in addition to the so-called magnet (or field-alignment) torque, which arises from the interaction of the magnet flux and the quadrature-axis stator current component, there is a reluctance torque component which is due to saliency. It is also a feature of these type of machines that in contrast to conventional, electrically excited salient-pole machines, where the quadrature-axis reactance is smaller than the direct-axis reactance, in synchronous machines with interior permanent magnets on the rotor, the quadrature-axis reactance is larger than the direct-axis reactance. In synchronous machines with buried (interior) permanent magnets, the magnet cost is minimized by the low magnet-weight requirements of the particular design.

One of the requirements to be met for servo applications has been smooth torque operation. Therefore both the so-called cogging torques (which are due to the slotting) and the pulsating torques, due to space harmonics and time harmonics, must be reduced. The reduction of the cogging torque can be achieved in a number of ways, such as: shaping of the magnets, skewing of the magnets on the rotor with respect to the rotor axis, skewing of the stator, coordinating the design of the number of stator slots, slot opening and magnet dimensions.

However, the skewed rotor construction is costly in the case of high-strength ceramic magnets and rare-earth magnets, because it is difficult to manufacture the magnets in complex geometric shapes. Furthermore, the required magnetizing techniques can result in lower flux levels. However, any method used for minimizing the cogging torque must be evaluated on the basis of manufacturing considerations from the point of view of cost, assembly, and the equipment required for the process. Under normal circumstances one or two per cent of the rated torque is an acceptable value for the cogging torque in servo applications. However, most of the effects of cogging torque can be eliminated in a converter-fed permanent-magnet synchronous machine drive by a high-performance converter and controller containing an accurate speed or position monitoring device.

It should be noted that whilst in conventional permanent-magnet machines the stator contains teeth, it is possible to construct such permanent-magnet machines where there are no teeth on the stator. In this case the stator windings are assembled outside the machine and are then inserted and fixed in the stator. Such a machine will not cog at low speed and the iron losses are reduced. Furthermore, in this construction more space is available for the stator windings than in the conventional construction. Thus larger size conductors can be used and the current rating of the windings can be higher. Since there is more space peripherally, the radial dimension of the stator windings can be smaller than in the case of a toothed construction. Thus for a toothless machine the rotor diameter can be larger than for a conventional machine of the same frame size. The air gaps of machines with toothless stators are usually several times larger than in the machines which have teeth. This has an adverse effect on the air-gap flux density, which can be partly offset by the application of a rotor with larger diameter and a larger magnet surface.

The reduction of the pulsating torques (ripple torques) is also an important aspect of achieving smooth torque operation. This can be obtained in a number of ways including the minimization of the time harmonic content of the stator currents.

With rare-earth magnets air-gap flux densities up to 1 T (Tesla) can be obtained. However, because of the saturation of the teeth and iron losses, it is not always best to use the strongest magnet material. The teeth can be highly saturated, especially with armature reaction during overload. Despite the large air gap, a field distortion occurs due to saturation, and this can reduce the flux and therefore the induced e.m.f. At high speed, the iron losses are the limiting factor.

3.1.1 VECTOR CONTROL OF PMSMs WITH SURFACE-MOUNTED MAGNETS

3.1.1.1 General introduction

It is assumed that the permanent magnets are located on the rotor surface and are of the rare-earth magnet type. The effects of magnetic saturation are neglected and thus for modelling purposes the permanent magnets can be considered as fictitious equivalent constant-current sources (I_{rF}=const.). Thus in the reference frame fixed to the rotor, the rotor-current space phasor is

$$\bar{i}_r = cI_{rF} = I_{rf} = \text{constant} \tag{3.1-1}$$

if there are no damper windings on the rotor. The magnets are positioned at angle θ_r relative to the direct axis of the stator, which coincides with the magnetic axis of stator winding sA. In the d, q reference frame fixed to the rotor the flux linkage with the stator windings due to the permanent magnets is

$$\psi_F = L_m \bar{i}_r = L_m I_{rf}. \tag{3.1-2}$$

Since the synchronous reactance is small, ψ_F is almost equal to the modulus of the magnetizing flux-linkage space vector and thus the angle between the magnetizing flux-linkage space vector and ψ_F is small; this angle is the so-called load angle and is denoted by δ. Thus the induced stator magnetizing e.m.f., which is equal to the rate of change of the magnetizing flux linkage, will in the steady state almost be positioned along the quadrature-axis. As a consequence, the terminal voltage will also be almost co-phasal with the quadrature-axis.

It should be noted that the magnet flux linkage ψ_F can be obtained by performing simple tests. If the stator currents are zero and the rotor is rotated at a constant speed (e.g. by a d.c. motor coupled to the shaft), the induced e.m.f.s in the stator windings are proportional to the magnet flux. Thus the magnet flux linkage can be obtained by dividing the terminal voltage by the rotor speed. It follows from eqn (3.1-2) that the equivalent fictitious rotor current can also be determined if the magnetizing inductance and the magnet flux linkage are known. The fictitious rotor current is proportional to the number of pole pairs, the residual flux density of the magnet material (approximately 1.2 T for NdFeB), the radial thickness of the magnets, and the sine of half of the electrical angle span of a magnet. Furthermore it is inversely proportional to the recoil permeability of the magnet material (near to unity for NdFeB) and to the equivalent number of sinusoidally distributed stator turns per phase.

The electromagnetic torque of the permanent-magnet synchronous machine with surface-mounted magnets and with a symmetrical three-phase stator winding can be obtained physically by considering that the torque is produced by the tendency of the magnets to align with the axis of the stator m.m.f. and now, owing to the lack of saliency, there is no reluctance torque component. Thus the magnet torque has to vary sinusoidally with the angle between the stator m.m.f. axis and the direct axis of the rotor, to which the excitation m.m.f. is frozen. This can also be seen by considering eqns (2.2-14) and (3.1-2) and thus the electromagnetic torque is

$$t_e = \tfrac{3}{2} P L_m I_{rf} i_{sq} = \tfrac{3}{2} P \psi_F i_{sq}, \tag{3.1-3}$$

where i_{sq} is the quadrature-axis component of the stator-current space vector ($\bar{i}_s'$) expressed in the reference frame fixed to the rotor, as shown in Fig. 3.1.

It follows from Fig. 3.1 that the space angle of the stator-current space phasor relative to the magnetic axis of stator winding sA is α_s, and since its space angle

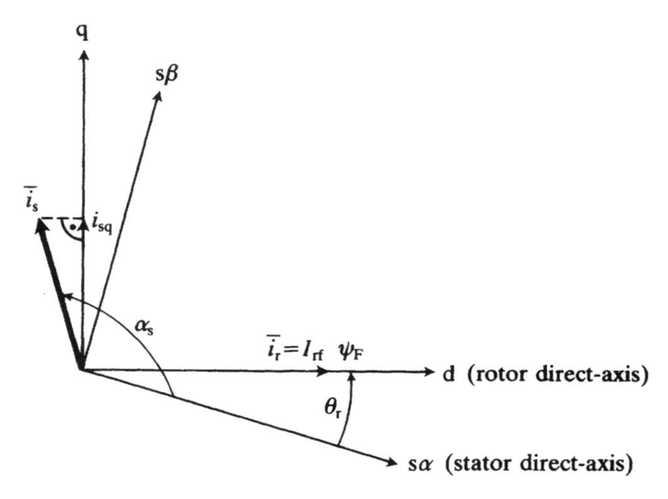

Fig. 3.1. The stator- and rotor-current space phasors and the excitation flux of the PMSM.

relative to the direct axis of the reference frame fixed to the rotor is $\alpha_s - \theta_r$, eqn (3.1-3) can be put into the form

$$t_e = \tfrac{3}{2} P \psi_F |\bar{i}_s| \sin(\alpha_s - \theta_r), \tag{3.1-4}$$

where $|\bar{i}_s|$ is the modulus of the stator-current space phasor.

Thus the physically expected result has been obtained and the torque varies with the sine of the angle $(\alpha_s - \theta_r)$, the so-called torque angle (β). Since the flux produced by the permanent magnets has been assumed to be constant, the electromagnetic torque can be varied by changing the quadrature-axis stator current expressed in the rotor reference frame (i_{sq}). Thus a constant torque is obtained if i_{sq}=constant. It also follows from eqn (3.1-4) that the maximum torque per stator current is obtained if the torque angle is 90°. Since the electromagnetic torque is a constant times the quadrature-axis stator current component (i_{sq}), it immediately follows that a quick torque response is obtained if the quadrature-axis stator current is changed quickly, e.g. by the application of a current-controlled pulse-width-modulated inverter. It is this feature which makes possible the rotor-oriented control of permanent-magnet synchronous machines.

3.1.1.2 Control scheme of the rotor-oriented controlled PMSM

The control scheme of the rotor-oriented permanent-magnet synchronous machine discussed above is relatively simple. The excitation flux is frozen to the direct axis of the rotor and thus its position can be obtained directly from the rotor shaft by monitoring the rotor angle θ_r or the rotor speed ω_r.

In practice the rotor speed can be monitored by, say, the application of an analogue tachometer and the rotor angle can be monitored by, say, a resolver.

Analogue tachometers have about 0.1% accuracy and at low-speed operation of a servo drive, very high-speed resolution is necessary to obtain accurate results; for feed drives and machine tools, for example, a speed resolution of approximately $0.2\,\text{min.}^{-1}$ is required. This problem can be solved by using the same digital encoder for sensing the rotor position and rotor speed. It should be noted that at high speeds and high-bandwidth speeds, optical encoders can have limited accuracy and temperature susceptibility, since the monitoring device has to be mounted within the motor enclosure. Resolvers are inherently accurate, but they must be combined with high-resolution digital circuits to achieve and maintain high accuracy over a wide speed-range and give no errors due to acceleration. Furthermore, it is also possible to utilize the fact that the rotor speed is equal to the first time derivative of the rotor position, $\omega_r = d\theta_r/dt$. In Section 3.1.3 various rotor speed and rotor position estimators are described for sensorless PMSM drives.

As shown above, in order to produce the largest torque for a given stator current, an optimally efficient operation is achieved by stator current control which ensures that the stator-current space phasor contains only a quadrature-axis component (i_{sq}) when expressed in the reference frame fixed to the rotor. This is analogous to the separately excited d.c. machine, where this is achieved by the consecutive switching of the armature coils through the commutator. In the base speed region, for motor operation, the space phasor of the stator currents expressed in the rotor reference frame is therefore $\bar{i}_s' = ji_{sq}$ [see Fig. 3.2(a)], and for braking, it is $\bar{i}_s' = -ji_{sq}$ as shown in Fig. 3.2(b).

This optimal mode of operation is suitable below the base rotor speed, where sufficient voltage is available from the inverter which supplies the stator windings of the machine. However, at higher speeds, above the base speed—constant-horsepower range—the induced e.m.f. increases directly with the rotor speed (the excitation flux is constant due to the permanent magnets), and if a given speed

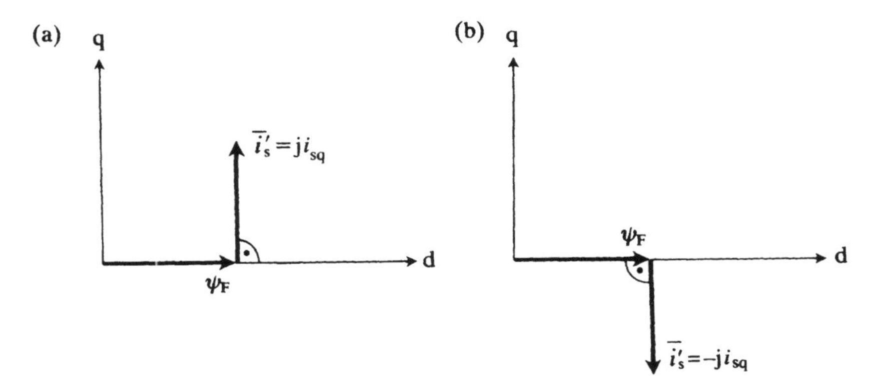

Fig. 3.2. Stator-current space phasors in the case of optimal operation in the base speed region: (a) motor operation; (b) braking operation.

is to be reached, the terminal voltage must also be increased to match the increased stator e.m.f. The increased stator terminal voltage would require an increase in the voltage rating of the inverter used. However, with a given inverter, there is a ceiling voltage which cannot be exceeded. Thus to limit the terminal voltage of the machine to the ceiling voltage of the inverter, field weakening has to be introduced.

In smaller drives, there is no need for field weakening, but for larger drives since, owing to the permanent-magnet construction, it is not possible to achieve direct field weakening, the effect of field weakening can be obtained by controlling the stator currents in such a way that the stator-current space phasor in the rotor reference frame should contain a direct-axis component i_{sd} along the negative direct-axis of the rotor reference frame, in addition to the quadrature-axis stator current component i_{sq}. This is shown in Fig. 3.3, where the same i_{sq} has been maintained as for operation below base speed.

The limiting amplitude of the stator voltages can be obtained by utilizing the steady-state stator equations of the PMSM. It follows from eqn (2.1-141) that in the stationary reference frame, when the effects of magnetic saturation are neglected, the stator voltage equation is

$$\bar{u}_s = R_s \bar{i}_s + L_s \frac{d\bar{i}_s}{dt} + L_m \frac{d}{dt}(\bar{i}_r e^{j\theta_r}). \tag{3.1-5}$$

By the substitution of ψ_F defined in eqn (3.1-2), this will take the following form:

$$\bar{u}_s = R_s \bar{i}_s + L_s \frac{d\bar{i}_s}{dt} + \frac{d}{dt}(\psi_F e^{j\theta_r}). \tag{3.1-6}$$

By performing the required differentiation, but keeping the excitation flux constant,

$$\bar{u}_s = R_s \bar{i}_s + L_s \frac{d\bar{i}_s}{dt} + j\omega_r \psi_F e^{j\theta_r}. \tag{3.1-7}$$

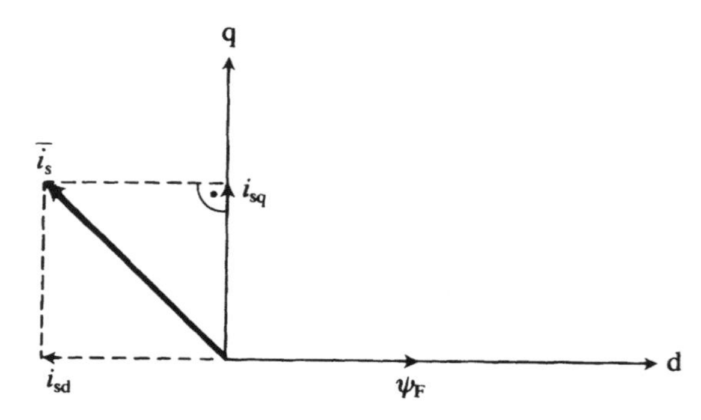

Fig. 3.3. Space phasors of PMSM in the field-weakening range (overspeed range).

In the steady state, where $\theta_r = \omega t$, ω=constant, the space phasor of the stator currents in the stationary reference frame is $\bar{i}_s = c_1 \bar{I}_s e^{j\omega t}$, the voltage space phasor in the same reference frame is $\bar{u}_s = c_1 \bar{U}_s e^{j\omega t}$, where c_1 is a constant (see the discussion which follows eqn (2.1-5)). Equation (3.1-7) thus takes the form

$$\bar{U}_s = R_s \bar{I}_s + j\omega L_s \bar{I}_s + \bar{U}_p = Z_s \bar{I}_s + \bar{U}_p, \tag{3.1-8}$$

where

$$\bar{U}_p = \frac{j\omega\psi_F}{c_1} \tag{3.1-9}$$

is the phasor of the induced e.m.f.s in the stator winding, $Z_s = R_s + j\omega L_s = R_s + jX_s$ is the impedance of a stator winding, and X_s is the synchronous reactance.

By utilizing eqns (3.1-8) and (3.1-9), the two steady-state phasor diagrams shown in Fig. 3.4 can be plotted. The steady-state phasor diagram shown in Fig. 3.4(a) corresponds to the case where there is no direct-axis stator current component ($I_{sd} = 0$), thus $\bar{I}_s = jI_{sq}$ and

$$\bar{U}_s = (R_s + j\omega L_s) jI_{sq} + \bar{U}_p, \tag{3.1-10}$$

which holds for operation below base speed (constant-torque region). The angle between $\bar{U}_p$ and $\bar{U}_s$ is the load angle, and in this case agrees with the displacement angle (ϕ) of the stator current.

In Fig. 3.4(b) the phasors are shown for operation above the base speed (constant-output-power region) and for stator currents where $I_{sd} \neq 0$ ($I_{sd} < 0$). Thus

$$\bar{U}_s = (R_s + j\omega L_s) jI_{sq} + \bar{U}_p + (R_s + j\omega L_s) I_{sd}. \tag{3.1-11}$$

Figure 3.4(b) corresponds to the case where minimum stator voltage ($\bar{U}_{smin}$) is obtained. It follows from Fig. 3.4(b) that the minimum stator voltage is obtained if the stator voltage is adjusted to be in quadrature to the voltage phasor corresponding to the voltage drop along the d-axis, which is equal to $(R_s + j\omega L_s) I_{sd}$.

It is then possible to utilize Fig. 3.4 to obtain the necessary direct-axis stator current component to achieve speeds in the field-weakening range (above base speed). It should be noted that because of the extra current along the d axis, the stator currents are increased, but since the stator currents are limited by the inverter, a corresponding reduction of the maximum quadrature-axis stator current has to be performed, which of course leads to a smaller torque. In other words, when the rotor speed is increased the torque angle is increased for the purposes of field weakening and thus the torque per unit of stator current is reduced.

Owing to the increased stator currents, the copper losses will be increased and the efficiency of the drive will deteriorate. However, field weakening is restricted to short transients at light load. The maximum speed is determined by the current rating of the inverter. It is possible to achieve high-speed operation with light

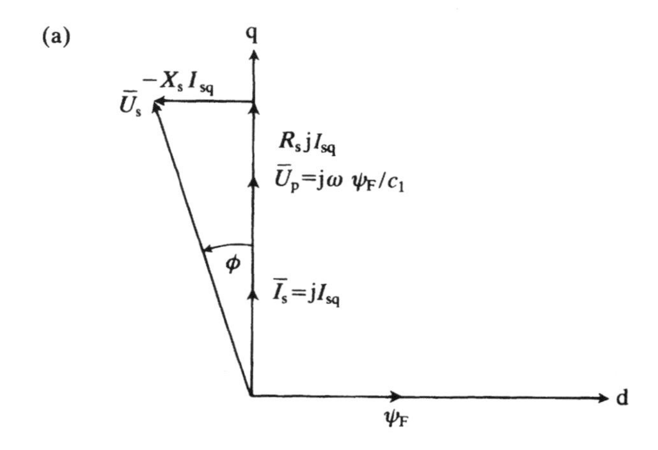

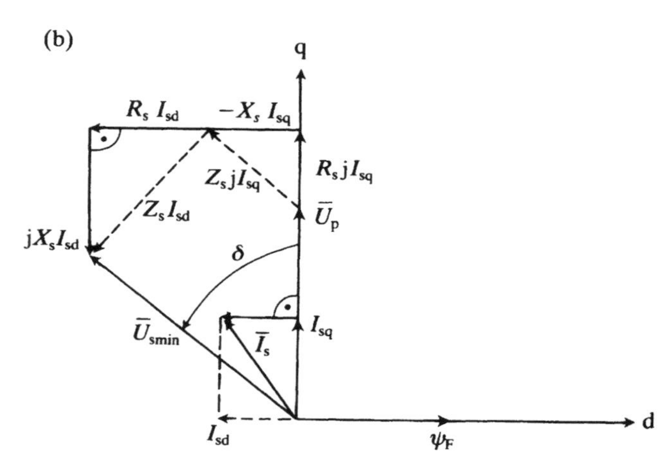

Fig. 3.4. Steady-state phasor diagrams of the PMSM: (a) operation below base speed; (b) operation above base speed.

loads and high values of the demagnetizing stator current component. However, in practice, the maximum speed does usually not exceed twice the base speed.

It follows from the equations and vector diagrams presented above that the degree of field weakening with a PMSM depends on the synchronous reactance of the machine, i.e. on the armature reaction of the machine. With the magnets mounted on the surface of the rotor, the air gap can be considered to be relatively large, as discussed earlier. So the synchronous reactance is small and therefore the field-weakening range is limited.

In the following discussion it is assumed that the stator windings of the machine are supplied by impressed stator currents obtained from a current-controlled switched transistor inverter with a high switching frequency. This is similar to the application of armature current control in a converter-fed d.c. drive. The main advantage of this type of operation is that because of the impressed stator currents, the effects of stator resistance and leakage inductance on the drive dynamics are eliminated and the motor interactions are simplified. Figure 3.5 shows the schematic diagram of a Pulse-Width-Modulated (PWM) transistorized inverter, using fast current-control loops, which supplies the stator windings of the permanent-magnet synchronous machine. Figure 3.5(a) shows a simplified circuit and Fig. 3.5(b) shows a more detailed circuit.

An uncontrolled rectifier produces the constant voltage U across the filter capacitor C_F. The filter establishes a steady d.c. voltage. The constant d.c. voltage source could also be a battery bank. With an appropriately high U, fast current-control loops can be designed, which will yield machine currents (i_{sA}, i_{sB}, i_{sC}) which coincide with the demanded currents $(i_{sAref}, i_{sBref}, i_{sCref})$. Thus the stator windings can be considered to be supplied by fictitious current sources. This approximation is valid if the ceiling voltage of the inverter is sufficient.

It should be noted that inverters used for small drives, under a few kilowatts, are almost always pulse-width-modulated inverters, due to the fact that they avoid the need for a variable d.c. source voltage. Furthermore, they yield superior waveforms and at these power levels it is easy to obtain controlled-turn-off power devices.

The current controllers can be hysteresis or PWM controllers. When a hysteresis current controller is used, upper and lower hysteresis levels are defined relative to the reference value of the current and the inverter is used to ensure that the actual stator current is limited to between the hysteresis bands. By defining narrow hysteresis bands, it is possible to ensure tight control, but this demands a higher switching frequency from the inverter. The actual currents contain harmonics which produce high-frequency torque ripples, but these are filtered out by the inertia of the machine so the speed of the machine can almost be free from the effects of these ripples. The hysteresis current controller reacts almost instantaneously to changes in the reference values of the current, and thus the time lag in the model of this controller is very small. When a PWM current controller is used, the actual current is compared with the reference current, and an error signal is generated. The error signal is compared to a sawtooth wave, and if the resulting error is larger than the sawtooth, the phase voltage is switched positively. However, if it is smaller than the sawtooth wave, it is switched negatively. The advantage of this controller over the hysteresis current controller is that the switching frequency of the inverter is preset and it is easy to determine the time delay between the time the reference changes and when corrective switching action is performed by the current controller.

In Fig. 3.5(b), the current control is implemented by the appropriate firing of the six bipolar transistors $T_1 \dots T_6$ of a full-bridge inverter which produces adjustable frequency three-phase excitation from the d.c. source. The bipolar transistors can be replaced by any other bipolar or metal-oxide-semiconductor (MOS) power switch devices, e.g. by Metal-Oxide Semiconductor Field-Effect

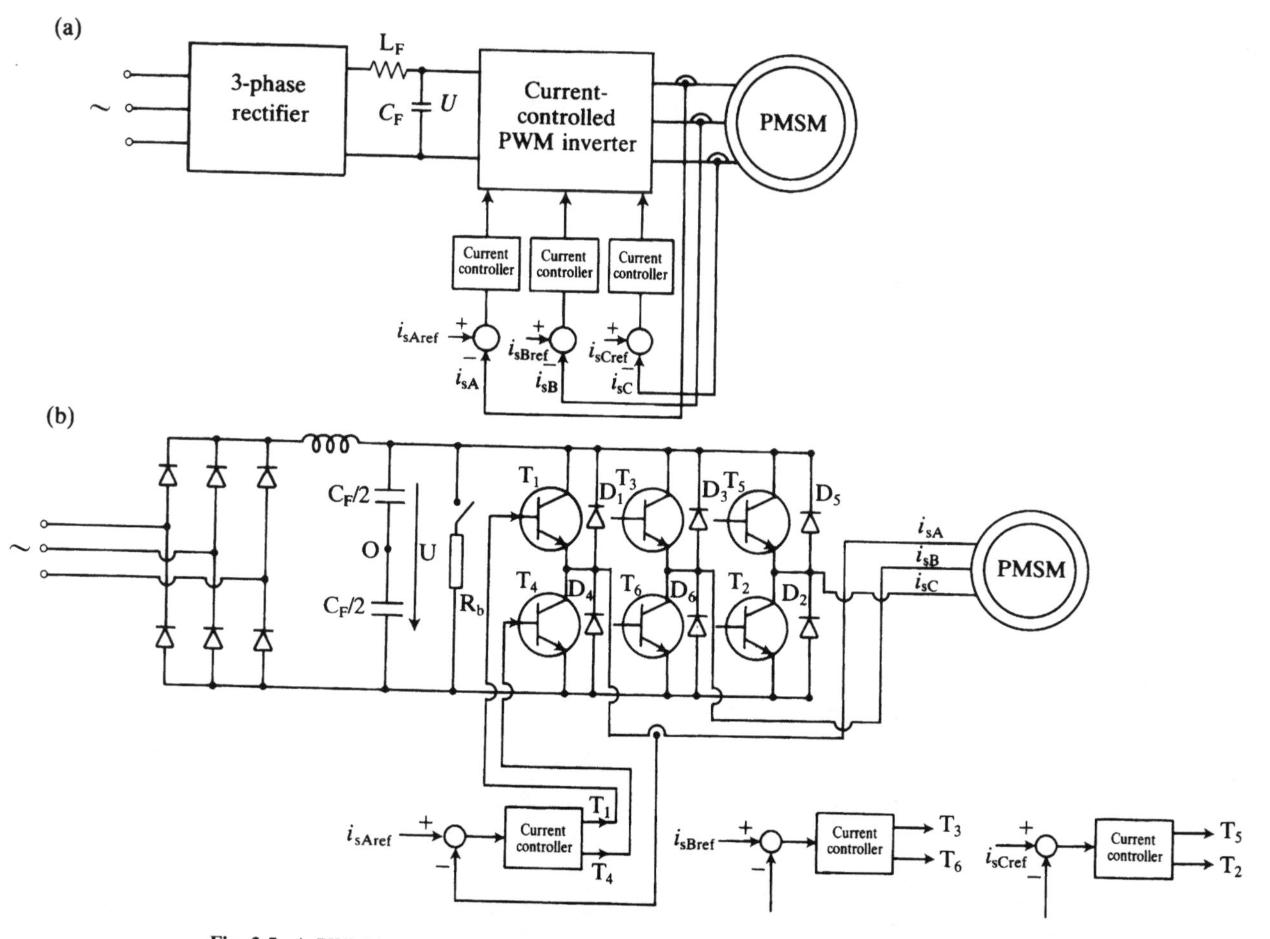

Fig. 3.5. A PWM inverter-fed PMSM (a) simplified circuit; (b) more detailed circuit.

(MOSFET) transistors, which are significantly faster than the bipolar transistors, or by Metal-Oxide-Semiconductor Insulated Gate Transistors (MOSIGT), all of which can be turned both off and on with low-level gating signals (which can be gate voltages or gate currents depending on the specific type of power transistor used). The three-phase full-bridge inverter consists of three half-bridge units (which are phase shifted by 120°). In Fig. 3.5(b) all the transistors are combined with a parallel freewheeling diode to provide circulation paths for the flow of reverse stator currents of the synchronous machine during reactive power flow and regeneration; they also clamp the load voltage to the input level.

For braking, the motor is operated in the regenerating mode and the kinetic energy stored in the system inertia is converted into electrical energy. This can be dissipated across a resistor R_b shown in Fig. 3.5(b). During braking, active power flows from the machine to the d.c. link. The braking resistor can be switched on and off within the hysteresis band of the d.c. link voltage to dissipate the braking power.

Implementation of rotor-oriented control of PMSM in polar coordinates: The block diagram of the rotor-oriented controlled permanent-magnet synchronous machine in polar coordinates is shown in Fig. 3.6, where the stator currents (i_{sA}, i_{sB}, i_{sC}) are impressed by fast current loops, e.g. by utilizing a transistorized PWM inverter as discussed above. When the control structure is put into polar form and a microprocessor implementation is used, for algorithmic reasons it can have advantages over an implementation utilizing the Cartesian form.

Since the sum of the stator currents is assumed to be zero, it is sufficient to monitor only two stator currents, $i_{sA}(t)$ and $i_{sB}(t)$. When a microprocessor is used, these currents are converted into their digital forms by using A/D converters. By using rectangular-to-polar (R$\rightarrow$P) conversion, the modulus ($|\bar{i}_s|$) and the phase angle (α_s) of the stator-current space phasor expressed in the stationary reference frame are obtained as shown in Fig. 3.6. It should be noted that the block labelled '3$\rightarrow$2' performs the transformation of the three-phase currents into the quadrature-phase current components, in accordance with eqns (2.1-9) and (2.1-10).

The rotor speed ω_r is measured and the rotor angle θ_r is obtained by integration (denoted by '$1/p$'). It is also possible to monitor the rotor angle directly and to use this to obtain the rotor speed. However, whatever method is used, it is very important to have an extremely accurate knowledge of the rotor position, since rotor-oriented control is based on the accurate knowledge of this quantity. The rotor angle is subtracted from the angle α_s of the stator-current space phasor to obtain $\beta=\alpha_s-\theta_r$, which is the angle of the stator-current space phasor in the rotor reference frame. The sine of this angle is obtained by using a function generator (FG1) and it is then multiplied by the modulus of the stator-current space phasor and thus the quadrature-axis stator current component (i_{sq}) is obtained. This is multiplied by the excitation flux (ψ_F) and a signal proportional to the instantaneous value of the electromagnetic torque (t_e) is obtained, in agreement with eqns (3.1-3) and (3.1-4).

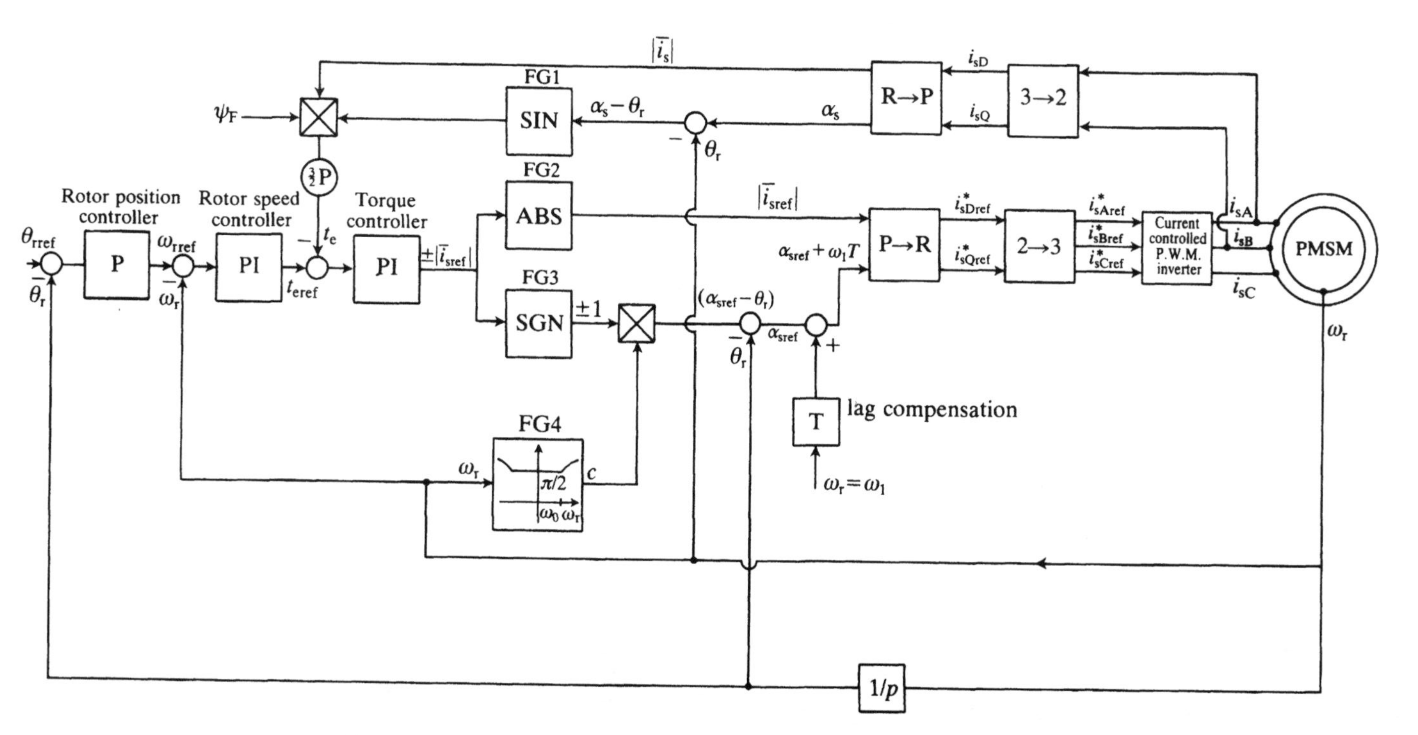

Fig. 3.6. Block diagram of a rotor-oriented controlled PMSM using polar coordinates.

The necessary current references to the PWM inverter are generated as follows. A cascaded control structure is shown with superimposed torque, rotor-speed, and rotor-position control loops. The difference between the reference value of the rotor angle and the monitored rotor angle serves as an input signal to the rotor position controller, which is a proportional (P) controller. The output of this controller is the rotor-speed reference value. The difference between this and the actual monitored rotor speed is the input to the speed controller, which is a proportional-integrating (PI) controller, the output of which is the torque reference signal. The actual value of the electromagnetic torque (t_e) is subtracted from the reference value (t_{eref}) and this is the input signal to the torque controller, which is a PI controller. The output of the torque controller is the modulus of the reference value of the stator-current space phasor but multiplied by $+1$ or -1, indicated in Fig. 3.6 by $\pm|\bar{i}_{sref}|$. The reference values of the modulus and phase angle of the stator-current space phasor in the rotor reference frame are $|\bar{i}_{sref}|$ and $(\alpha_{sref}-\theta_r)$ respectively, and they are obtained by using three function generators and by utilizing the signal for the rotor speed ω_r and the output signal of the torque controller. The signal $\pm|\bar{i}_{sref}|$ is fed into a function generator (FG2) which takes its absolute value, $|\bar{i}_{sref}|$. The angle $(\alpha_{sref}-\theta_r)$ must be positive or negative, corresponding to a positive or negative output from the torque controller, and therefore a sign function generator (FG3) is used. Thus when $+|\bar{i}_{sref}|$ is the output of the torque controller, the output of FG3 is $+1$ and when $-|\bar{i}_{sref}|$ is the output of the torque controller, the output of FG3 is -1. The factor ±1 is then multiplied by another factor which is obtained with the fourth function generator FG4. This is equal to $\pi/2$ if the rotor speed is less than the base speed (ω_b), or to a factor greater than $\pi/2$ if the rotor speed is greater than the base speed, in order to obtain a stator-current phasor with a negative i_{sd} component for field-weakening purposes in accordance with the technique described above.

The actual rotor angle (θ_r) is subtracted from the reference value of the angle of the stator-current space phasor in the rotor reference frame, and thus the angle α_{sref} of the stator-current space phasor in the stationary reference frame is obtained. Thus the outputs of the P→R converter are the two-axis stator-current references i^*_{sQref} and i^*_{sDref}, which are transformed into the three-phase stator-current references i^*_{sAref}, i^*_{sBref}, and i^*_{sCref}. This two-phase to three-phase transformation is indicated by the block labelled '2→3'. The three-phase reference currents are the inputs to the inverter which supplies the permanent-magnet synchronous motor.

The role of the lag compensation showed in Fig. 3.6 will be described in the following section, and it should be noted that i^*_{sQref} and i^*_{sDref} are the reference values of the two-axis stator currents, which are obtained when the time lag of the control loops (T) is not negligible ($T\neq0$). Similarly, i^*_{sAref}, i^*_{sBref}, and i^*_{sCref} are the reference values of the stator currents when $T\neq0$.

It follows from the block diagram shown in Fig. 3.6 that there is a substantial amount of signal processing required when the scheme is implemented. With analogue techniques, it would be problematic and uneconomical to realize accurately the required multiplications, to implement the four function generators used to obtain the required modulations by the rotor angle. However, the use of a microprocessor simplifies the required hardware significantly. It is believed that in the future dedicated signal processors will perform the tasks set by the rotor-oriented control scheme described above.

Implementation of rotor-oriented control of PMSM in Cartesian coordinates: The block diagram of the rotor-oriented controlled permanent-magnet synchronous machine using Cartesian coordinates is shown in Fig. 3.7. It is again assumed that the stator currents of the machine are impressed by fast current loops, e.g. by using a current-controlled PWM inverter.

Since there are no zero-sequence stator currents, again only two stator currents are monitored, $i_{sA}(t)$ and $i_{sB}(t)$. These are transformed into direct- and quadrature-axis stator current components $i_{sD}(t)$ and $i_{sQ}(t)$, which are the real- and imaginary-axis components of the stator-current space phasor in the stationary reference frame. The rotor speed is measured and the rotor angle is obtained by integration. However, it is again possible to have another implementation, where the rotor angle is directly monitored. The rotor angle is used to transform the space phasor of the stator currents in the stationary reference frame into the space phasor of the stator currents in the rotor reference frame ($\bar{i}_s' = \bar{i}_s e^{-j\theta_r} = i_{sd} + j i_{sq}$, see eqn (2.1-136)) and thus the current components i_{sd} and i_{sq} are obtained. The excitation flux (ψ_F) is multiplied by the quadrature-axis stator current component (i_{sq}) expressed in the rotor reference frame, and thus the electromagnetic torque (t_e) is obtained in agreement with eqn (3.1-3).

The current references to the inverter are generated as follows, by again using a cascaded control with superimposed torque, speed, and rotor control and also using field-weakening.

The difference between the limit stator voltage (U_{smax}) and the peak of the actual stator voltages ($|\bar{U}_s|$) serves as input to the field-weakening controller, which is an integrator and the output of which is the reference value of the stator current in the rotor d-axis (i_{sdref}). The stator current reference in the rotor q-axis (i_{sqref}) is obtained at the output of a PI torque controller, whose input is the difference between the torque reference (t_{eref}) and the actual torque (t_e). The torque reference is obtained at the output of the rotor speed controller, a PI controller, whose output is the difference between the reference rotor speed (ω_{rref}) and the actual rotor speed (ω_r). The reference value of the rotor speed is obtained by the application of a rotor position controller, the input of which is the difference between the rotor angle reference and the actual monitored rotor angle.

The function generator (FG) shown in Fig. 3.7 ensures that if i_{smax} is the limiting value of the stator currents, i_{sq} may have to be reduced and this is achieved by utilizing $(i_{sd}^2 + i_{sq}^2)^{1/2} < i_{smax}$, and thus the function generator performs the operation $(i_{smax}^2 - i_{sd}^2)^{1/2}$.

It has been assumed that the PWM inverter which supplies the machine is current controlled and that the current controllers react instantaneously to the changes in the current commands, i.e. there is no time lag in the current control loops. Thus the transformations cancel and the control is extremely simple and is similar to the control of a converter-fed d.c. machine. However, when this

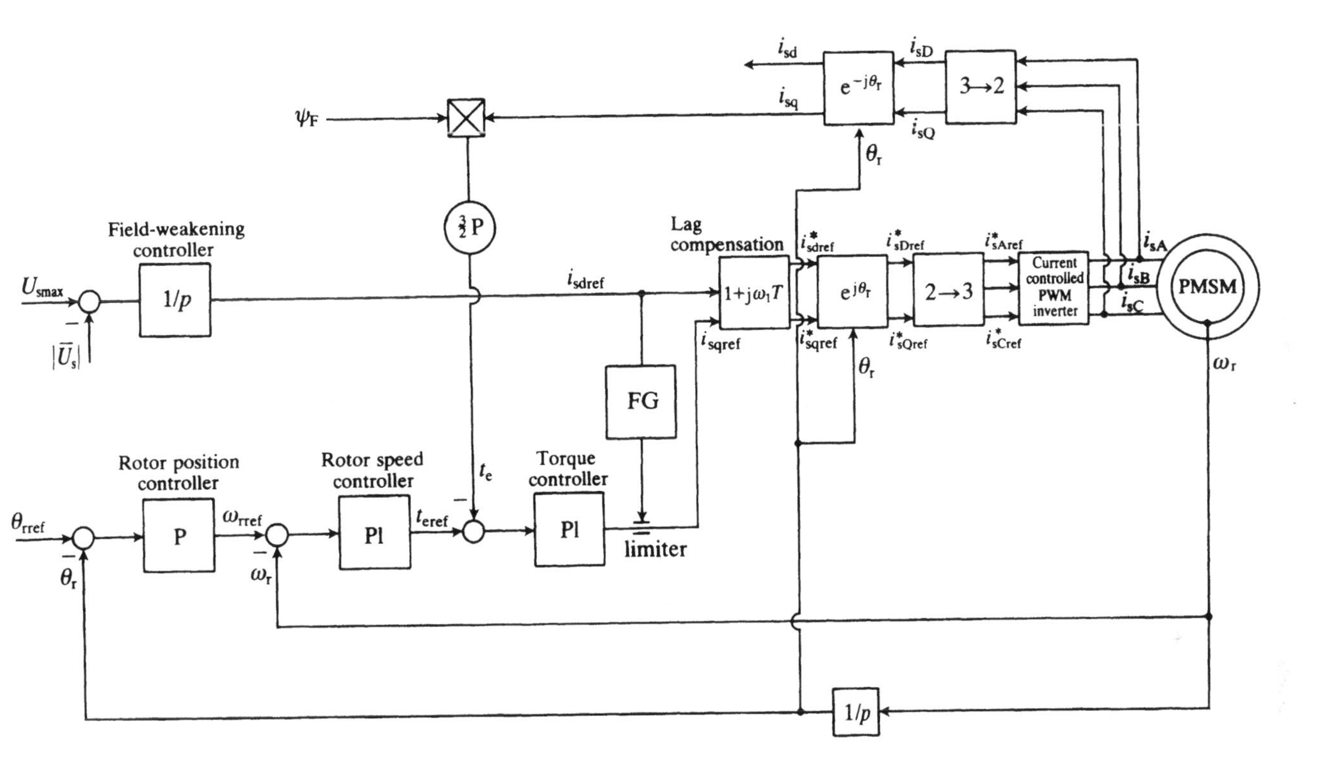

Fig. 3.7. Block diagram of a rotor-oriented controlled PMSM using Cartesian coordinates.

assumption is not valid, that is when the reference value of the stator-current space phasor ($i_{sDref}+ji_{sQref}$) is not equal to the actual stator-current space phasor ($\bar{i}_s=i_{sD}+ji_{sQ}$), owing to the time lag in the current control loops, the following space phasor equation holds

$$\bar{i}_{sref}=T\frac{d\bar{i}_s}{dt}+\bar{i}_s, \tag{3.1-12}$$

where T is the time lag in the control loops. In two-axis form, this is

$$i_{sDref}=T\frac{di_{sD}}{dt}+i_{sD} \tag{3.1-13}$$

$$i_{sQref}=T\frac{di_{sQ}}{dt}+i_{sQ}. \tag{3.1-14}$$

For the rotor-oriented control to work well, it is essential that the required decoupling of the stator current components is achieved, even when this lag is present. A solution to this can be obtained easily if it is assumed that the unwanted coupling caused by the lag is due only to the phase shift of the stator-current space phasor and that its modulus is unchanged. In this case, in Fig. 3.7, in front of the block labelled '$e^{j\theta_r}$', which represents the transformation of the reference values of the two-axis stator current components in the rotor-coordinates into the two-axis stator current components of the stationary reference frame, an extra block labelled '$1+j\omega_1 T$' is added, as shown in Fig. 3.7.

This can also be proved by considering Fig. 3.8. When there is no lag ($T=0$), the relationship between the actual and reference values of the stator-current space phasors in the stationary reference frame is $\bar{i}_s=\bar{i}_{sref}$, and of course a similar relationship must hold in the rotor reference frame $\bar{i}_s'=\bar{i}_{sref}'$, since it follows from eqn (2.1-142) that $\bar{i}_s'=\bar{i}_s e^{-j\theta_r}$, and $\bar{i}_{sref}'=\bar{i}_{sref}e^{-j\theta_r}$. However, when there is a lag, if a steady state, where the modulus of the stator-current space phasor is constant ($|\bar{i}_s'|=\text{const.}$), is assumed, the reference stator-current phasor in the rotor reference frame ($\bar{i}_{sref}'^{*}$) will be in a new position, displaced by angle γ from $\bar{i}_{sref}'$, where $\gamma=\omega_1 T$, but its length is the same as the length of $\bar{i}_s'$. By assuming small values for the lag, and thus of γ,

$$\bar{i}_{sref}'^{*}=\bar{i}_{sref}'e^{j\gamma}=\bar{i}_{sref}'e^{j\omega_1 T}=\bar{i}_{sref}'(\cos\omega_1 T+j\sin\omega_1 T)\approx\bar{i}_{sref}'(1+j\omega_1 T). \tag{3.1-15}$$

Thus in front of the $e^{j\theta_r}$ block in Fig. 3.7, the block $(1+j\omega_1 T)$ must be inserted, the outputs of which are the two-axis components of $\bar{i}_{sref}'^{*}(i_{sdref}^{*}, i_{sqref}^{*})$ and whose inputs are the two-axis components of $\bar{i}_{sref}(i_{sdref}, i_{sqref})$. Thus, in two-axis form, the equations which define the lag compensation are

$$i_{sdref}^{*}=i_{sdref}-\omega_1 T i_{sqref} \tag{3.1-16}$$

$$i_{sqref}^{*}=i_{sqref}+\omega_1 T i_{sdref}. \tag{3.1-17}$$

Of course, it is possible to combine the transformation for the lag compensation with the transformation $e^{j\theta_r}$, so that a compound transformation $e^{j\theta_r}e^{j\gamma}=e^{j(\theta_r+\gamma)}$ is

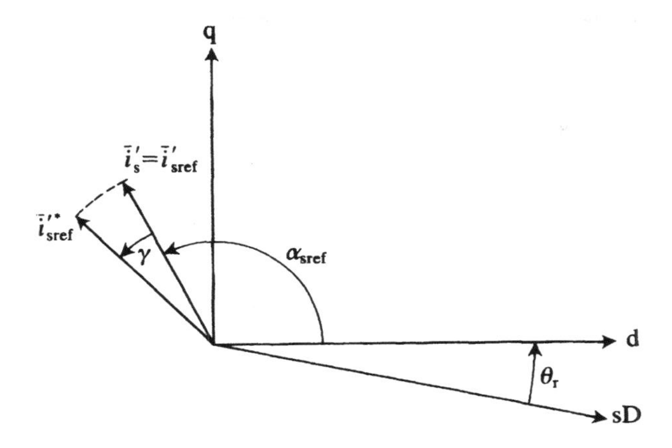

Fig. 3.8. The effects of controller lag.

present. It can thus be seen that the removal of the unwanted coupling caused by the lag of the converter can be implemented by replacing the angle θ_r by $\theta_r + \gamma = \theta_r + \omega_1 T$ in the $e^{j\theta_r}$ block. This polar form can be utilized directly in Fig. 3.6, if the lag of the inverter is not negligible. In this case, since the angle of $\bar{i}_{sref}$ relative to the direct axis of the stator (sD) is α_{sref} (as shown in Fig. 3.8), and furthermore, since due to the $e^{j\gamma}$ displacement, the angle of $\bar{i}'^{*}_{sref}$ relative to the sD axis is $\alpha_{sref} + \gamma$, in front of the polar-to-rectangular transformation block, $\gamma = \omega_1 T$ has to be added to α_{sref}. Finally, it should be noted that at higher speeds, the lag of the current control loops can increase, but in this case T can be adapted in the blocks which contain T. In Section 4.1.1 some other aspects of the effects of time delay are discussed.

3.1.2 VECTOR CONTROL OF PMSMs WITH INTERIOR MAGNETS

3.1.2.1 General introduction

Interior-permanent-magnet synchronous machines (IPMSM) are mechanically robust and thus allow high-speed applications. As discussed in Section 3.1.1, in these machines the effective air gap is small and therefore the effects of armature reaction are significant. This allows the control of the synchronous machine in the constant-torque region as well as in the field-weakening (constant-power) region up to a high speed. Furthermore, in accordance with the discussion presented in Section 3.1.1, in the interior-permanent-magnet machine the quadrature-axis stator inductance (L_{sq}) is smaller than the direct-axis stator inductance (L_{sd}) and this is a distinct difference from the conventional electrically excited salient-pole synchronous machine. In practice there can exist inductance ratios (L_{sq}/L_{sd}) as high as five. The extra inductance can be useful as it allows the use of lower switching frequencies for the IPM synchronous machine than those used with other inverter-fed machines. The relation $L_{sq} < L_{sd}$ has a direct effect on the electromagnetic torque production and also on the excitation flux requirements of the IPM synchronous machine. Both of these aspects will also be discussed here.

An IPMSM allows economical machine design, because in accordance with eqn (2.2-13), in contrast to the machine with surface-mounted permanent magnets, where the torque is produced by the interaction of the magnet flux and the quadrature-axis stator current, now the electromagnetic torque contains two components. One of these is due to the interaction of the magnet flux and the quadrature-axis stator current and the other one is the so-called reluctance torque component, which is proportional to the difference of the stator inductances in the two axes, $L_{sd} - L_{sq} = L_{md} - L_{mq}$ where L_{md} and L_{mq} are the direct- and quadrature-axis magnetizing inductances respectively. Thus rotor saliency provides a way of reducing the volume of the magnets, which would otherwise be high in order to obtain the desired rating of the synchronous machine.

The system performance characteristics of the IPM synchronous machine drive can be influenced by the proper adjustment of the rotor parameters of the machine to control the two torque components described above. For example, overexcitation in a permanent-magnet synchronous machine drive endangers the drive electronics when the induced stator voltages (back e.m.f.s) generated by the magnets significantly exceed the source voltage at high speeds. However, the rotor saliency can be utilized to reduce the excitation flux linkage due to the permanent magnets, in order to obtain large speed ranges while reducing the amplitudes of overexcitation.

In the IPM synchronous machine, when the m.m.f. components created by the two-axis stator currents and the rotor magnets are added, because of magnetic saliency, the resultant m.m.f. distribution will be non-sinusoidal, even under linear magnetic conditions. Under linear magnetic conditions, the resultant flux-density distribution is equal to the sum of the component flux-density distributions produced by the permanent magnets (B_m) and the direct- and quadrature-axis components of the stator currents (B_d, B_q). These component distributions, together with the resultant distribution, are shown in Fig. 3.9. The flux-density waveforms shown correspond to operation as a motor and are for rated torque. With a light load or operation as a generator other waveforms apply. In Fig. 3.9, the direct and quadrature axes of the rotor (d, q) are also shown and τ_p is the pole pitch and θ is the angle around the periphery.

Due to the higher permeance along the rotor direct-axis, high flux densities can arise at the edges of the steel pole-pieces. Thus, the stator teeth which are opposite to the leading edges of these poles tend to saturate as the current excitation level is increased. For given stator currents, saturation of these parts of the stator teeth reduces the magnitude of the fundamental component of the air-gap flux-density distribution around the periphery and also shifts this fundamental component in the direction of the centre of the rotor pole. When viewed from the stator terminals, this reduction of the air-gap flux appears as a reduction in the stator inductances, especially along the quadrature-axis (q). Thus, for a given applied terminal voltage, the maximum torque capability of the machine is greater than

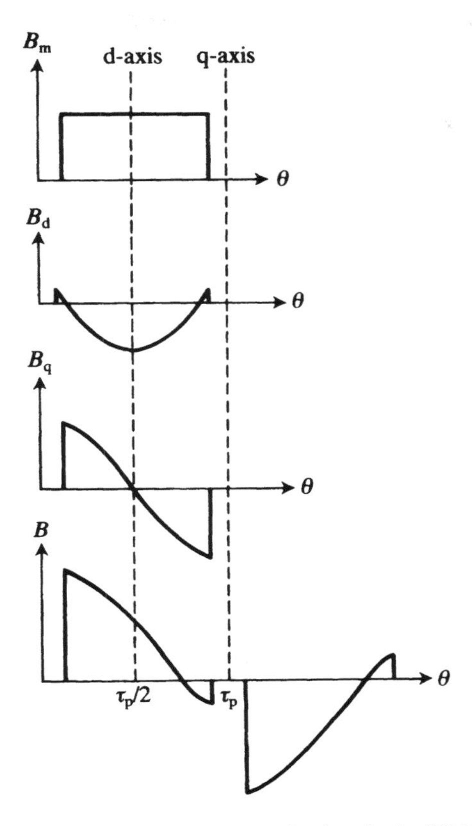

Fig. 3.9. Component and resultant flux-density distributions in the IPM synchronous machine.

the one which could be predicted by the application of a linear model of the machine, which neglects the effects of magnetic saturation.

In the following discussion, the main assumptions (apart from magnetic saliency) are the same as those used for the permanent-magnet synchronous machine with surface-mounted magnets. However, it should be noted that when the effects of magnetic saturation cannot be neglected, there will be a cross-coupling between the direct and quadrature axes not only due to the inherent physical saliency, but also due to the effects of main-flux saturation. This is the phenomenon of cross-saturation. Some aspects of magnetic saturation will be discussed in Chapter 6.

Since the flux-density distribution in the IPM synchronous machine is non-sinusoidal, the most convenient way to produce a smooth, pulsation-free electromagnetic torque is to produce a stator-current m.m.f. distribution which rotates synchronously and which is fixed in space relative to the rotor surface. This requires symmetrical three-phase stator currents, when sinusoidally distributed stator windings are assumed. It should be noted that in contrast to the synchronous motor with surface-mounted magnets, which can be designed for both sinusoidal and square-wave excitation, in the IPM synchronous motor with square-wave excitation, smooth torque operation cannot be obtained since the square waves produce an m.m.f. distribution which discretely shifts along the air-gap periphery only at the instants of switching.

It follows from eqn (2.2-13) that in the absence of damper windings, the electromagnetic torque produced in the IPMSM is

$$t_e = \tfrac{3}{2}P[\psi_F i_{sq} + (L_{sd} - L_{sq}) i_{sd} i_{sq}], \tag{3.1-18}$$

where ψ_F is the constant flux linkage produced by the permanent magnets. However, it should be noted that when neodymium–iron–boron magnets are used, the magnet material has approximately 0.1% per °C negative temperature sensitivity, such that with the rise of the rotor temperature, the magnet flux linkage decreases. If the rotor temperature is monitored, it is possible to compensate this flux linkage for the variations of the rotor temperature. There is no simple method to monitor the rotor temperature directly, but, among other methods, it is possible to estimate this parameter by utilizing the monitored stator temperature and the dynamic thermal model of the machine. The control of the instantaneous stator currents provides the possibility of directly controlling the instantaneous electromagnetic torque. In the absence of damper windings, the torque immediately responds to changes in the direct- and quadrature-axis stator currents.

It is possible to put eqn (3.1-18) into the following form,

$$t_e = \tfrac{3}{2}P(\psi_{sd} i_{sq} - \psi_{sq} i_{sd}) = \tfrac{3}{2}P|\bar{\psi}_s| i_{sy}, \tag{3.1-19}$$

where ψ_{sd} and ψ_{sq} are the stator flux linkages in the direct and quadrature axes of the rotor, which in the absence of damper circuits can be expressed in terms of the stator currents and the magnet flux as

$$\psi_{sd} = L_{sd} i_{sd} + \psi_F,$$
$$\psi_{sq} = L_{sq} i_{sq}. \tag{3.1-20}$$

In eqn (3.1-19), $|\bar{\psi}_s|$ is the modulus of the stator flux-linkage space phasor,

$$|\bar{\psi}_s| = (\psi_{sd}^2 + \psi_{sq}^2)^{1/2}, \tag{3.1-21}$$

and i_{sd}, i_{sq} are the direct- and quadrature-axis components of the stator currents in the rotor reference frame. i_{sy} is the quadrature-axis component of the stator currents expressed in a special reference frame (x, y) which rotates at the speed of the stator flux-linkage space phasor ($\bar{\psi}_s$). The real axis of this reference frame is aligned with the stator flux-linkage space phasor.

The phasor diagram of the IPMSM in the steady state is shown in Fig. 3.10, where θ_r is the rotor angle, which is the angle between the direct axis of the stator (sD) and the direct axis of the rotor (d). Both the magnet flux-linkage (ψ_F) and the stator flux-linkage ($\bar{\psi}_s$) space phasors are shown; δ is the load angle and $\bar{U}_s$ is the phasor of the terminal voltage.

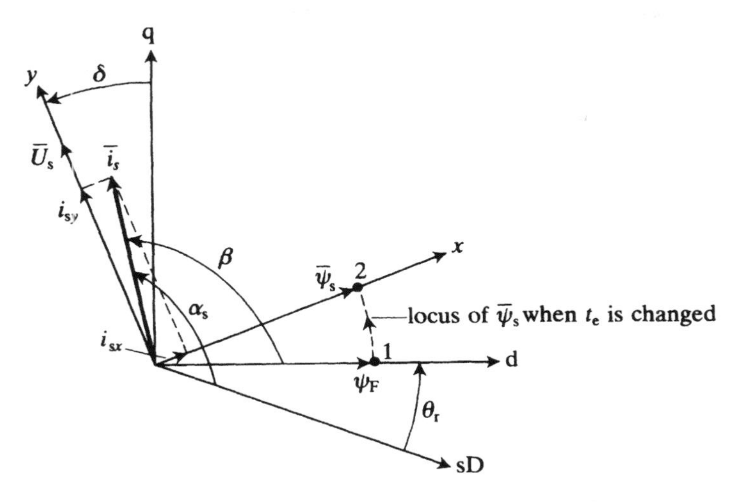

Fig. 3.10. Steady-state phasor diagram of the IPMSM.

In the special (x,y) reference frame, the two-axis components of the stator-current space phasor are denoted by i_{sx} and i_{sy} respectively. By utilizing Fig. 3.10, it is possible to prove that the electromagnetic torque is proportional to the quadrature-axis stator current component i_{sy}, since it follows from Fig. 3.10 that

$$\sin\delta=\frac{\psi_{sq}}{|\bar{\psi}_s|}$$

$$\cos\delta=\frac{\psi_{sd}}{|\bar{\psi}_s|}. \tag{3.1-22}$$

Thus by the substitution of eqn (3.1-22) into eqn (3.1-19), the electromagnetic torque can be expressed as

$$t_e=\tfrac{3}{2}P|\bar{\psi}_s|(i_{sq}\cos\delta-i_{sd}\sin\delta)=\tfrac{3}{2}P|\bar{\psi}_s|i_{sy}. \tag{3.1-23}$$

The relationship $i_{sy}=i_{sq}\cos\delta-i_{sd}\sin\delta$ follows from Fig. 3.10, but it also follows from eqn (2.1-142), since by considering that $\theta_g=\delta+\theta_r$, the transformation

$$i_{sx}+ji_{sy}=\bar{i}_s e^{-j(\theta_r+\delta)}=(i_{sD}+ji_{sQ})e^{-j(\theta_r+\delta)}=(i_{sd}+ji_{sq})e^{-j\delta} \tag{3.1-24}$$

is obtained. In eqn (3.1-24) the relationship between the stator current phasor in the synchronously rotating reference frame $(i_{sd}+ji_{sq})$ and in the stationary reference frame $(i_{sD}+ji_{sQ})$ has been obtained from eqn (2.1-142) by the substitution of $\theta_g=\theta_r$. Equation (3.1-24) yields the following transformation between the stator current components of the special and rotor reference frames

$$i_{sx}=i_{sd}\cos\delta+i_{sq}\sin\delta, \tag{3.1-25}$$

$$i_{sy}=-i_{sd}\sin\delta+i_{sq}\cos\delta. \tag{3.1-26}$$

Equation (3.1-23) will also be used for the development of the vector-control system of the IPM synchronous machine.

3.1.2.2 Implementation of the stator-flux-oriented control of the IPMSM in Cartesian coordinates

Similarly to that shown in Section 3.1.1, a current-controlled PWM inverter provides the necessary control of the instantaneous currents of the IPMSM. The simplified schematic of the vector-controlled IPMSM drive is shown in Fig. 3.11.

In Fig. 3.11 a three-phase uncontrolled rectifier produces a d.c. voltage across the filter capacitor C_F. The rectifier can also be replaced by a battery bank. The d.c. voltages supply power to the current-controlled PWM inverter which supplies the IPM synchronous machine. Similarly to the PWM inverter discussed in Section 3.1.1, the inverter shown in Fig. 3.11 is a full-bridge three-phase inverter with six switching devices, which can be, for example, bipolar transistors. Each transistor is combined with a parallel freewheeling diode to provide a circulation path for the reactive stator currents of the machine. The drive system operates in all four quadrants.

The stator currents are monitored and because there are no zero-sequence currents, it is sufficient to monitor only two of these. The machine shaft is connected to a rotor position encoder, which supplies information on the rotor angle (θ_r) to the drive control system. The drive control electronics are also provided with information from the stator currents. The inverter/motor control also receives the reference value of the electromagnetic torque (t_{eref}) signal, and by utilizing the feedback signals of the currents and the rotor position, it generates the transistor base-drive signals. It should be noted that sinusoidal excitation of the IPMSM requires rotor-angle feedback with sufficient resolution to synchronize the sinusoidal references properly with the rotor position. This requirement is more demanding than for the comparable six-step square-wave current excitation configuration, for which the rotor angle information is necessary only in 60° increments.

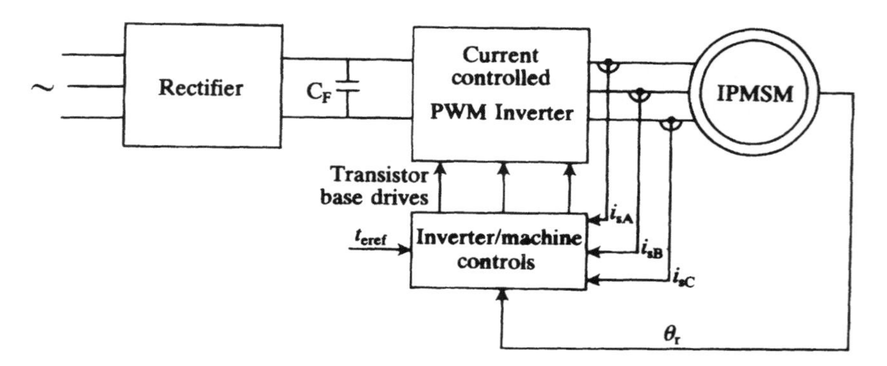

Fig. 3.11. Simplified schematic of the vector-controlled IPMSM drive system.

For simplicity, in the first implementation of the drive system, it is assumed that the drive operates in the constant-torque mode, below base speed. In the constant-torque region, the stator flux is constant with the speed and the inverter is current controlled in the PWM mode to maintain the required flux–torque relationship. Thus the inverter has the characteristics of a current-source inverter. The stator flux can be controlled by the reactive stator currents, which can be magnetizing or demagnetizing.

If it is assumed that the drive can enter the constant-power operation (over-speed) region, then there are some special aspects which have to be considered. At high speeds the d.c. supply voltage is responsible for limiting the torque-speed envelope of the drive. This limit can be explained by considering that for any given direct- and quadrature-axis currents, the amplitude of the stator-voltage space phasor is dependent on and nearly proportional to the rotor speed. When the speed increases, the resulting terminal line-to-line voltages approach the d.c. voltage and the voltage which is necessary to force the stator currents to their reference values decreases to zero. Under these conditions, at higher speed, the inverter saturates, the pulses in the phase-voltage waveforms drop out, the current control is lost and the inverter will generate constant-amplitude, six-step square-wave voltages. As the rotor speed increases in the constant-power region, the stator flux decreases inversely (with the speed) and the motor demands demagnetizing reactive currents to reduce the magnet flux (ψ_F). The saturation of the inverter requires the IPMSM control to change from current to voltage control. The transition between the PWM and square-wave modes must be fast and smooth under all conditions of operation of the drive. Such a transition involves some degradation in the torque control characteristics, since only the phase-angle of the voltage phasor can be adjusted during six-step voltage excitation; its magnitude cannot be adjusted.

From eqn (3.1-23) it follows that the torque control of the interior permanent-magnet synchronous machine can be performed similarly to the torque control of the separately excited d.c. machine. In the constant-torque region (below base speed) the torque can be controlled by controlling the quadrature-axis stator current is i_{sy}. It is possible to control the stator flux by controlling the direct-axis stator current i_{sx}.

In Fig. 3.12 the block diagram is shown of the torque control of the IPM synchronous machine in the constant-torque region, utilizing stator-flux-oriented control. The inverter is current controlled, using the principles described in Section 3.1.1. The drive system is designed with an outer torque control loop, but it is possible to add speed and position control loops. In the form which is discussed, it is directly applicable to electrical vechicles.

In Fig. 3.12 the difference between the reference torque (t_{eref}) and the actual torque (t_e) serves as the input to the torque controller, which is a PI controller. Its output is the reference value of the torque-producing stator current components (i_{syref}). The reference value of the flux-producing stator current component (i_{sxref}) is generated from i_{syref} through the function generator FG2 as follows. The reference value of the modulus of the stator flux linkage $|\bar{\psi}_{sref}|$ is obtained by

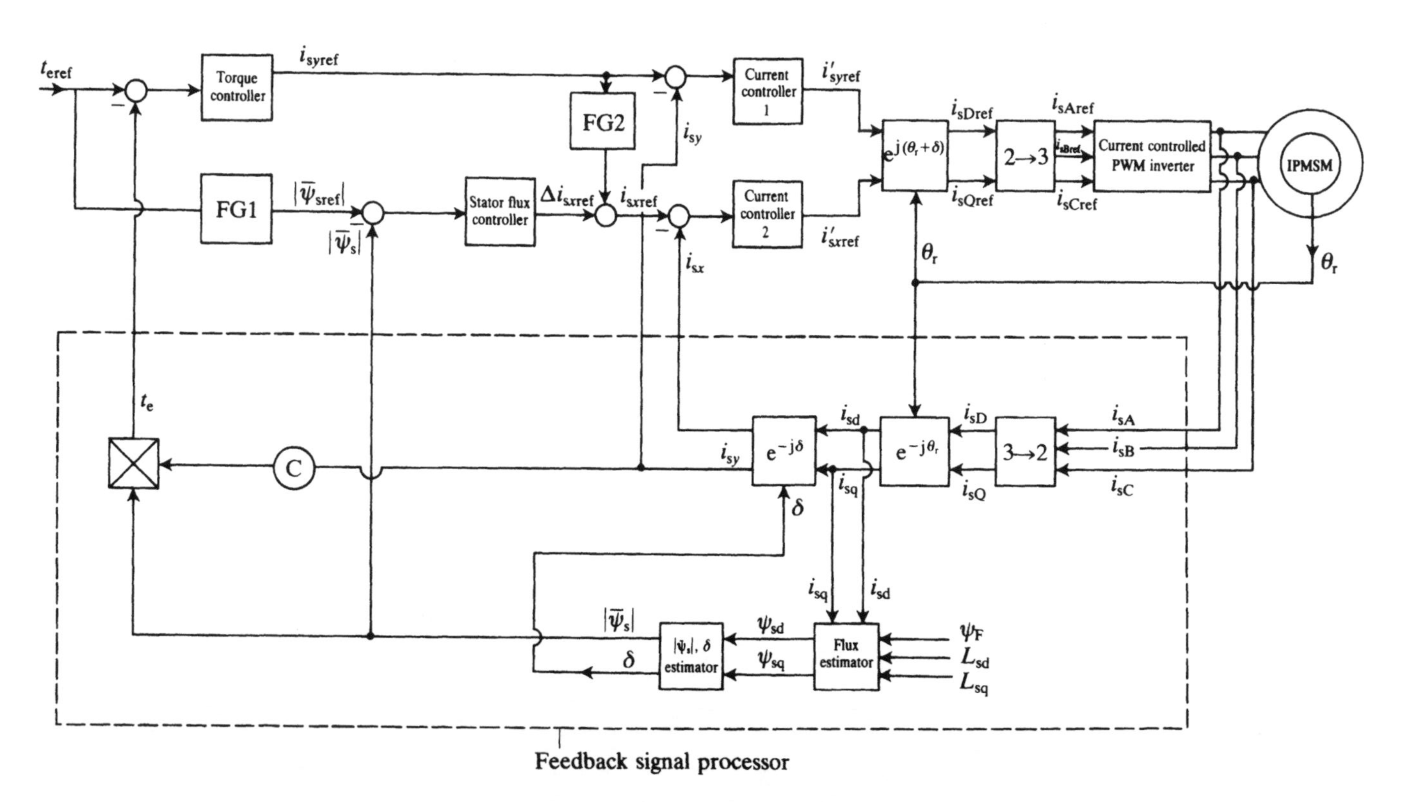

Fig. 3.12. Control block diagram of the stator-flux-oriented control of the IPMSM.

utilizing the function generator FG1. This serves to optimize the core losses in order to improve overall drive efficiency. Figure 3.13 shows the non-linear characteristics implemented by function generator FG1. Point 1 corresponds to zero electromagnetic torque, since at this point the stator flux is equal to the magnet flux (ψ_F) and thus the load angle is zero. Point 2 corresponds to rated stator flux at rated torque.

The drive system incorporates the $|\bar{\psi}_s|$ flux control loop to prevent flux drift due to the variation of machine parameters. The difference between the reference stator flux $|\bar{\psi}_{sref}|$ and the actual value of the modulus of the stator flux $|\bar{\psi}_s|$ appears as the input of the stator flux controller, which is a PI controller, the output of which is Δi_{sxref}. The modulus of the actual stator flux is estimated as described below. The output signal of the flux controller is added to the output signal of function generator FG2 whose input is i_{syref}.

The role of FG2 can be understood by considering the phasor diagram shown in Fig. 3.10. In the constant-torque region, the electromagnetic torque can be controlled by i_{sy} and $|\bar{\psi}_{sref}|$ can be maintained at a value determined by FG1.

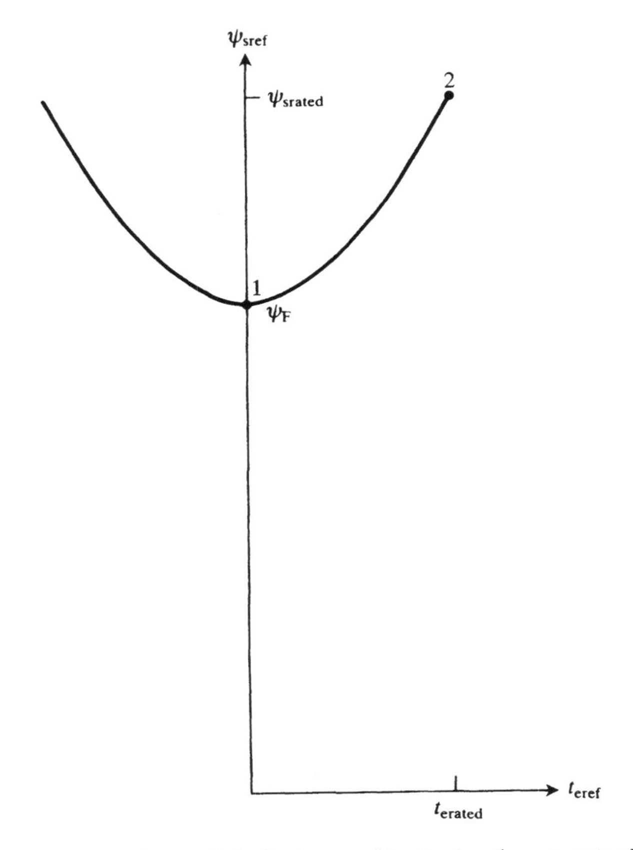

Fig. 3.13. The characteristics implemented by the function generator FG1.

In Fig. 3.10, the locus of the stator flux-linkage space phasor is also shown when the electromagnetic torque is increased from zero to its rated value in the constant-torque region. At point 1, $t_e=0$, $i_{sx}=i_{sy}=0$ and $|\bar{\psi}_s|=\psi_F$. When the electromagnetic torque is increased, $|\bar{\psi}_s|$ will also increase, but i_{sx} and i_{sy} also have to increase until t_e is equal to its rated value. The function generator FG2 determines the relationship between i_{sxref} and i_{syref}, shown in Fig. 3.14.

In an electrically excited synchronous machine, $|\bar{\psi}_s|$ is controlled by the field current and thus the machine can always operate at unity power factor. However, in the IPMSM, $|\bar{\psi}_s|$, is controlled by the lagging stator current component i_{sx} and thus the machine operates at lagging power factor. At the boundary between the constant-torque region and the constant-power region, the control of i_{sx} and i_{sy} is lost as a result of the saturation of the inverter as described above.

The reference signals i_{sxref}, i_{syref} are processed by current control loops, which allow vector control to be effective in partial saturation of the inverter, and if there is also a constant power region, help the smooth transition between the PWM and square-wave modes. The inputs to the two current controllers, which are PI controllers, are $i_{syref}-i_{sy}$ and $i_{sxref}-i_{sx}$ is respectively. Below base speed, the operation of these current loops is redundant. The output currents of these controllers (i'_{sxref}, i'_{syref}) are fed into the block containing the complex transformation $e^{j(\theta_r+\delta)}$. This transformation is used to align i'_{sxref} with $\bar{\psi}_s$ and i'_{syref} with $\bar{U}_s$, as shown in Fig. 3.10, since this ensures the high dynamic performance of the system. Mathematically, it can be seen that it is this transformation which is required, since it follows from eqn (3.1-24) that

$$i_{sD}+ji_{sQ}=(i_{sx}+ji_{sy})e^{j(\theta_r+\delta)}. \tag{3.1-27}$$

The resolution of eqn (3.1-27) into its real- and imaginary-axis components gives

$$i_{sD}=i_{sx}\cos(\theta_r+\delta)-i_{sy}\sin(\theta_r+\delta),$$

$$i_{sQ}=i_{sx}\sin(\theta_r+\delta)+i_{sy}\cos(\theta_r+\delta), \tag{3.1-28}$$

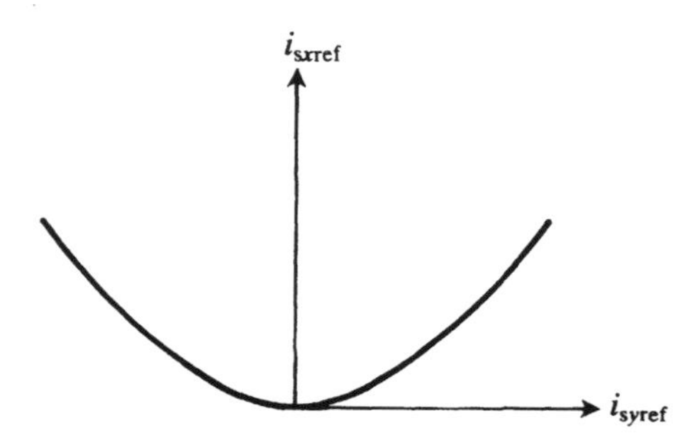

Fig. 3.14. The characteristic described by FG2.

and the unit-vector signals necessary for the transformation can be obtained as

$$\cos(\theta_r+\delta)=\cos\theta_r\sin\delta-\sin\theta_r\sin\delta,$$

$$\sin(\theta_r+\delta)=\sin\theta_r\cos\delta+\cos\theta_r\sin\delta. \quad (3.1\text{-}29)$$

The three-phase reference stator currents are obtained from the stationary-axis stator reference currents i_{sDref}, i_{sQref} by the application of the two-to-three-phase transformation. It should be noted that in the absence of zero-sequence quantities, it is only necessary to obtain two current references. The reference currents are inputs to the inverter as described earlier. The actual three-phase stator currents (i_{sA}, i_{sB}, i_{sC}) are monitored, but it is necessary to monitor only two of these in the absence of the zero-sequence currents.

The monitored values of the stator currents are used to obtain the electromagnetic torque (t_e) load angle (δ), modulus of the stator flux ($|\bar{\psi}_s|$), and the two-axis stator current components i_{sx} and i_{sy}, but it is also necessary to know the magnet flux (ψ_F), which can be determined by preliminary tests or can be calculated from the dimensions of the magnets and knowledge of the magnet material, as described in Section 3.1.

It is possible to obtain a fully digital implementation of the feedback signal processing, but to achieve high performance of the drive, it is necessary to compute all feedback signals ($i_{sx}, i_{sy}, \delta, |\bar{\psi}_s|, t_e$) with high precision in real time. Thus a large-wordsize microcomputer with small sampling intervals is required. This could be achieved, for example, by the application of the Texas 32010 digital signal processor.

In this case the two-axis currents i_{sx} and i_{sy} are obtained from the monitored currents i_{sA}, i_{sB}, i_{sC} by utilizing the transformation $e^{-j(\theta_r+\delta)}$. Furthermore, the flux modulus $|\bar{\psi}_s|$ can be obtained from the two-axis stator flux-linkage components ψ_{sd} and ψ_{sq} as $|\bar{\psi}_s|=(\psi_{sd}^2+\psi_{sq}^2)^{1/2}$, where the flux-linkage components are obtained by considering eqn (3.1-20) and the torque can be obtained by considering eqn (3.1-19), according to which it is proportional to the product of $|\bar{\psi}_s|$ and i_{sy}.

Figure 3.12 also shows how these signals are generated. Thus the monitored stator currents are inputs to the three-phase to two-phase transformation block, the outputs of which are the stationary-axis currents i_{sD} and i_{sQ}. These are transformed into the two-axis components of the rotor reference frame (i_{sd}, i_{sq}) by the application of the transformation $e^{-j\theta_r}$, and finally i_{sd}, i_{sq} are transformed into the stator current components of the stator flux-oriented reference frame by the aid of transformation $e^{-j\delta}$. The inputs to the flux computation block are the current components i_{sd}, i_{sq}, ψ_F, L_{sd}, L_{sq} and its outputs ψ_{sd} and ψ_{sq}. These are inputs to the block which computes $|\bar{\psi}_s|$ and δ, as described above.

It is possible to have many other implementations of the vector control of the IPMSM, and rotor-oriented control of the IPMSM will be discussed in the next section.

3.1.2.3 Implementation of the rotor-oriented control of the IPMSM

Here rotor-oriented control of the interior-permanent-magnet synchronous machine is described in both constant-torque and flux-weakening regions.

Implementation in the constant-torque region: It is assumed that the direct-axis of the special reference frame is aligned with the magnet flux (rotor-oriented control) and the stator current components expressed in the rotor reference frame (i_{sd} and i_{sq}) are controlled to control the torque. Figure 3.15 shows the block diagram of an implementation for the constant-torque region. The stator currents i_{sd} and i_{sq} are controlled in an open loop in such a way that the torque per stator ampere is maximum. All the other assumptions are the same as used in the previous section and it also follows that there are no damper circuits.

In Fig. 3.15, the reference torque serves as the input to the two function generators FG1 and FG2. Function generator FG1 gives the relationship between the torque and the direct-axis stator current component i_{sd}, and function generator FG2 gives the relationship between the torque and the quadrature-axis stator current i_{sq} for maximum torque per stator current ampere, if the effects of magnetic saturation are neglected. In a permanent-magnet machine, which operates at a given speed and torque, optimal efficiency can be obtained by the application of an optimal voltage that minimizes the electrical losses. This

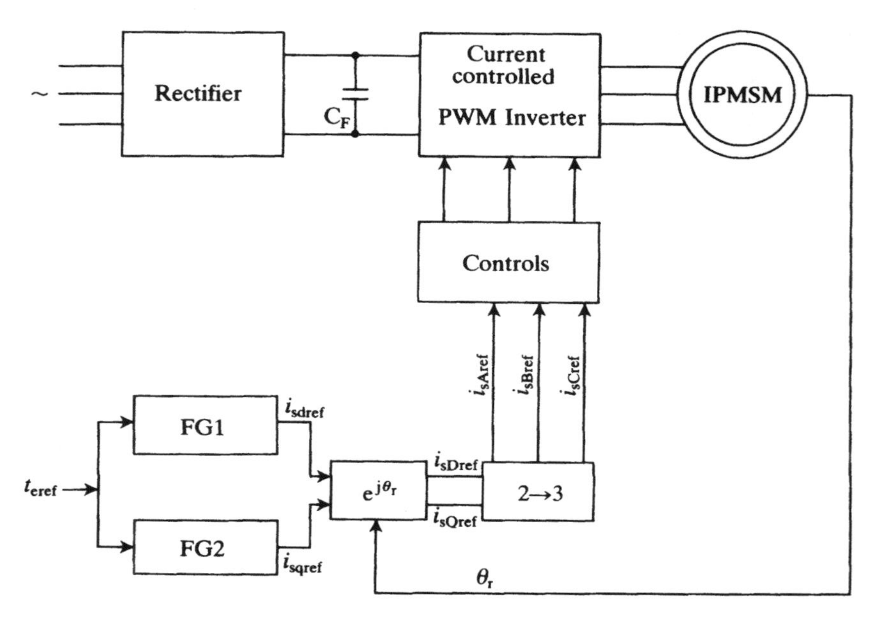

Fig. 3.15. Torque control of the IPMSM.

optimum will not coincide with the condition of maximum torque per stator ampere owing to the presence of core losses and the difference is more significant in the case of high-speed operation. In a permanent-magnet machine with no core losses, the machine which operates at maximum torque per stator ampere is optimally efficient. Such operation also leads to minimal copper, inverter, and rectifier losses. Furthermore, minimization of the stator currents for a given torque leads to higher current rating of the inverter and rectifier, and thus the overall cost of the system is reduced.

In the synchronous machine with surface-mounted magnets, the electromagnetic torque does not contain a reluctance torque term, since there is only a very small difference between the reluctances of the direct and quadrature axes. It has been shown that in this case, the torque is produced by the interaction of the stator currents with the air-gap flux and maximum torque is obtained when the space phasor of the stator currents $\bar{i}_s$ is in space quadrature to the magnet flux, and thus when the direct-axis stator current component (i_{sd}) is zero. However, in the synchronous machine with interior permanent magnets, there is a significant variation of the reluctance between the direct and quadrature axes and as discussed before, $L_{sd}<L_{sq}$. As a consequence, there also exists a reluctance torque component, whose sign is negative compared with the reluctance torque of an electrically excited (wound rotor) salient-pole synchronous machine. Thus when this component is added to the torque component produced by the stator currents and the air-gap flux, maximum torque will be produced at the stator current angle $\beta>0$, where this angle is the phase angle of the stator-current space phasor relative to the direct-axis of the rotor reference frame, as shown in Fig. 3.10. Thus if $\beta=0$ is chosen for the IPMSM, maximal torque per stator ampere cannot be obtained. In an IPMSM higher torque per ampere can be produced than in the PMSM with surface-mounted magnets. The improved torque capability of the IPMSM is an advantage, but this must also be compared with the higher manufacturing costs.

To obtain generally applicable functions for the description of function generators FGl and FG2, for any IPMSM, a normalized torque–current relationship is obtained and used in the function generators shown in Fig. 3.15. By selecting the following base value of the electromagnetic torque:

$$t_{eb}=\tfrac{3}{2}P\psi_F i_b, \tag{3.1-30}$$

where ψ_F is the magnet flux and i_b is the base value of the currents

$$i_b=\frac{\psi_F}{L_{sq}-L_{sd}}, \tag{3.1-31}$$

the normalized value of the torque can be expressed as

$$t_{en}=\frac{t_e}{t_{eb}}. \tag{3.1-32}$$

Substitution of eqns (3.1-18), (3.1-30), and (3.1-31) into eqns (3.1-32) yields

$$t_{en}=\left(1-\frac{i_{sd}}{i_b}\right)\frac{i_{sq}}{i_b}=i_{sqn}(1-i_{sdn}), \tag{3.1-33}$$

where i_{sdn} and i_{sqn} are normalized values of the two-axis currents in the rotor reference frame,

$$i_{sdn} = \frac{i_{sd}}{i_b}; \qquad i_{sqn} = \frac{i_{sq}}{i_b}. \tag{3.1-34}$$

Thus it is possible to plot the constant-torque loci of the IPM synchronous machine in terms of the normalized values of the two-axis stator currents i_{sdn}, i_{sqn}. This is shown in Fig. 3.16, where on every locus the point which is nearest to the (0, 0) point (e.g. point A) corresponds to the minimum stator current, so that the maximum torque per stator ampere curve is obtained by connecting points CBAA′B′C′.

The maximum torque per stator ampere curve shown in Fig. 3.16 can be used to obtain the normalized stator current components as a function of the normalized torque for maximum torque per stator ampere. Thus the functions $i_{sdn}=f_1(t_{en})$ and $i_{sqn}=f_2(t_{en})$ shown in Fig. 3.17 are obtained and these are implemented by function generators FG1 and FG2 respectively.

In Fig. 3.15, the outputs of function generators FG1 and FG2 are the reference signals i_{sdref} and i_{sqref} respectively. These are rotated by utilizing the complex transformation $e^{j\theta_r}$ to yield the two-axis stator references in the stationary reference frame. Finally the two-to-three-phase transformation is used to obtain the three-phase reference currents which are used to control the current-controlled PWM inverter which supplies the IPMSM. An absolute rotor-position transducer is used to obtain the rotor angle and this angle is utilized during the $e^{j\theta_r}$ transformation. The combination of the current-controlled PWM inverter and rotor transducer yields a current-source type of control, where both the amplitude and the phase angle of the stator current are regulated.

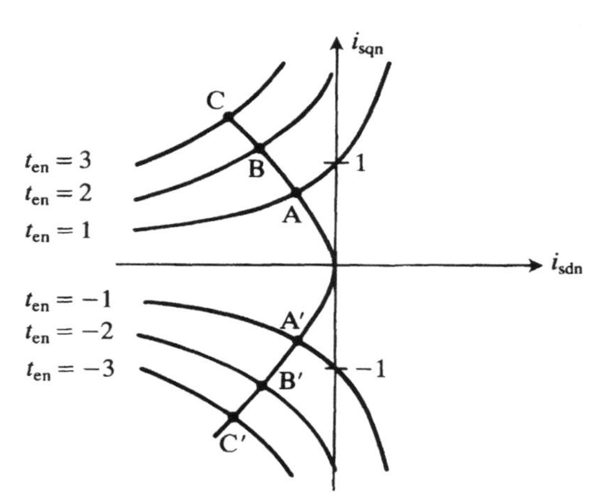

Fig. 3.16. Constant-torque loci for the IPMSM.

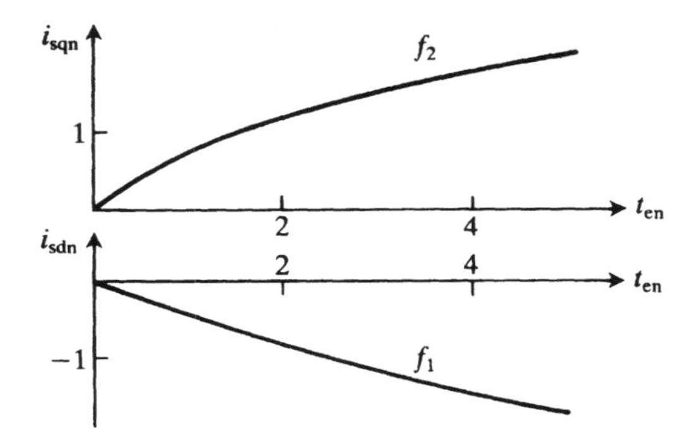

Fig. 3.17. The functions f_1 and f_2.

It is only possible to obtain satisfactory operating characteristics of the drive described above at low frequencies. However, it is possible to achieve torque control at high speeds, in the flux-weakening mode, by commanding the demagnetizing component of the stator currents to keep the PWM controllers from saturating completely. This operating mode is different from the maximum torque per stator current condition, but is required because of the voltage limitations of the inverter. A possible implementation will now be described.

Implementation in the flux-weakening region: As discussed above, saturation of the current controllers of the current-controlled PWM inverter occurs at higher rotor speeds when the terminal voltages of the motor approach the ceiling voltage of the inverter. The rotor speed is an important factor which determines the saturation of the current controllers, since both the motor reactances and the induced stator e.m.f.s are proportional to the excitation frequency, which is synchronized with the rotor speed in order to yield smooth responsive torque control.

The maximum available per-phase fundamental stator voltage ($|\bar{u}_s|$) is determined by the d.c. link voltage, and can be expressed as

$$|\bar{u}_s| = (u_{sd}^2 + u_{sq}^2)^{1/2}, \tag{3.1-35}$$

where u_{sd} and u_{sq} are the direct- and quadrature-axis components of the stator voltages in the rotor reference frame and can be obtained as follows.

In the synchronously rotating reference frame the direct- and quadrature-axis stator voltage equations of a salient-pole synchronous machine are

$$u_{sd} = R_s i_{sd} + \frac{d\psi_{sd}}{dt} - \omega_1 \psi_{sq} \tag{3.1-36}$$

and

$$u_{sq} = R_s i_{sq} + \frac{d\psi_{sq}}{dt} + \omega_1 \psi_{sd}, \tag{3.1-37}$$

where R_s is the stator resistance and for the IPMSM the stator flux linkages ψ_{sd} and ψ_{sq} are defined by eqn (3.1-20). Thus in the steady state

$$U_{sd} = R_s I_{sd} - X_{sq} I_{sq}, \tag{3.1-38}$$

$$U_{sq} = R_s I_{sq} + X_{sd} I_{sd} + \omega_1 \psi_F, \tag{3.1-39}$$

where $X_{sd} = \omega_1 L_{sd}$ and $X_{sq} = \omega_1 L_{sq}$ are the direct- and quadrature-axis stator reactances respectively and ω_1 is the excitation frequency (electrical rad/sec).

The maximum limits for the steady-state current components i_{sd} and i_{sq}, which correspond to maximal stator voltage $|\bar{u}_s|$, can thus be obtained by considering eqns (3.1-35), (3.1-38), and (3.1-39):

$$|\bar{u}_s|^2 = (R_s i_{sd} - X_{sq} i_{sq})^2 + (R_s i_{sq} + X_{sd} i_{sd} + \omega_1 \psi_F)^2. \tag{3.1-40}$$

In the complex d–q plane, for specific values of the rotor speed, d.c. voltage, and machine parameters, eqn (3.1-40) defines the locus of the stator current phasor $\bar{i}_s' = i_{sd} + ji_{sq}$. If, for simplicity, the ohmic voltage drops are neglected ($R_s = 0$), eqn (3.1-40) takes the recognizable form of an ellipse,

$$|\bar{u}_s|^2 = (X_{sq} i_{sq})^2 + \left[X_{sd} \left(i_{sd} + \frac{\omega_1 \psi_F}{X_{sd}} \right) \right]^2 \tag{3.1-41}$$

or

$$\left(\frac{|\bar{u}_s|}{X_{sq}} \right)^2 = i_{sq}^2 + \left(\frac{X_{sd}}{X_{sq}} \right)^2 \left(i_{sd} + \frac{\omega_1 \psi_F}{X_{sd}} \right)^2. \tag{3.1-42}$$

For any value of the rotor speed, the stator current phasor $\bar{i}_s' = i_{sd} + ji_{sq}$ is directed away from the origin of the complex plane to a point on the ellipse, but because there is a maximal voltage, it must be always located inside the ellipse and not outside it. Such an ellipse will be referred to as a voltage-limit ellipse. For increasing rotor speeds, there is a group of ellipses, where the axes of the ellipses are inversely proportional to the rotor speed.

In comparison to the drive control shown in Fig. 3.15, in Fig. 3.18 a stator-current feedback is shown for the identification of current-regulator saturation. In Fig. 3.18, the actual currents are first transformed into their direct- and quadrature-axis components (i_{sD}, i_{sQ}) in the stationary reference frame by the application of the three-phase to two-phase transformation and these components are transformed into the stator currents i_{sd}, i_{sq} formulated in the rotor reference frame, by utilizing the monitored rotor angle (θ_r). The direct-axis stator current reference i_{sdref} is obtained from the torque reference by utilizing function generator FG1, described above. The difference between the d-axis stator current i_{sd} and the reference current i_{sdref} is Δi_{sd} and this is used to identify the saturation of the current controllers of the inverter. When these are not saturated, this error is very small, since their gains are set high. The current difference Δi_{sd} serves as an input to the flux-weakening control.

Of course it is possible to perform the three-to-two-phase transformation and the $e^{-j\theta_r}$ transformation in one step and, similarly to the d-axis component in eqn (2.1-78),

$$i_{sd} = \tfrac{2}{3}[i_{sA} \cos \theta_r + i_{sB} \cos(\theta_r - 2\pi/3) + i_{sC} \cos(\theta_r - 4\pi/3)]. \tag{3.1-43}$$

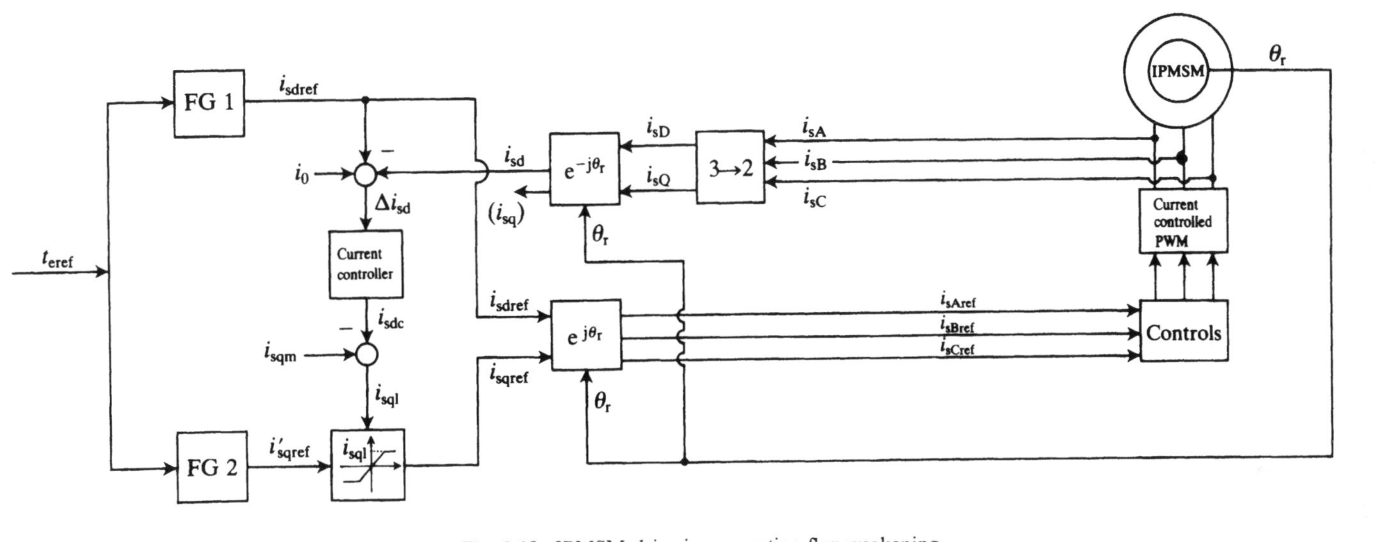

Fig. 3.18. IPMSM drive incorporating flux weakening.

In Fig. 3.18 the quadrature-axis stator current reference i'_{sqref} is obtained from the torque reference by the application of function generator FG2 described above and i'_{sqref} is reduced in response to the increasing difference Δi_{sd}, which corresponds to the saturation of the current controllers of the inverter. Owing to the reduction of i'_{sqref}, at a given rotor speed, the reference-current space phasor $\bar{i}'_{sref} = i_{sdref} + ji_{sqref}$ will move from outside the so-called voltage-limit ellipse—which represents the locus of the stator-current space phasor—to inside this ellipse and therefore a correspondence is forced between the reference and resultant current space phasors, and therefore the current controllers of the inverter will work again.

The difference Δi_{sd} is fed into a PI current controller, the output of which is the current i_{sdc}. The quadrature-axis current reference i'_{sqref} is limited and the adjustable limiter shown in Fig. 3.18 ensures that if $i'_{sqref} < i_{sql}$, where i_{sql} is the limiting value of i_{sq} and is obtained as $i_{sqm} - i_{sdc}$, the output of the limiter is $i_{sqref} = i'_{sqref}$; otherwise the output signal is limited to $\pm i_{sql}$. For all speed and load conditions this provides the desired coupling of the flux-weakening operation from the normal current-controlled constant-torque operation until saturation of the current controllers of the inverter is reached. In Fig. 3.18 the values i_{sqm} and i_0 are adjusted to ensure adequate decoupling of the current and flux-weakening modes, while yielding high dynamic performance during the transition from one mode of operation to the other.

When the flux-weakening operation is inactive, the dynamic response of the drive is limited by the response characteristics of the current controllers of the inverter. However, during flux-weakening the dynamic response depends on the feedback loop of the direct-axis stator current (i_{sd}), and on the saturating current controllers of the inverter.

With the flux-weakening control described above, the upper speed limit is approximately equal to the overexcitation threshold speed of the interior-permanent-magnet synchronous machine, which is the speed at which the induced stator e.m.f. $\omega_1 \psi_F$ is equal to the maximum available stator phase voltage $|\bar{u}_s|$.

It should be noted that, in contrast to the synchronous machine with surface-mounted magnet, the synchronous machine with interior permanent magnets will move into the flux-weakening mode in the underexcitation speed mode, when $\omega_1 \psi_F < |\bar{u}_s|$. This is an advantage, since problems are caused if the motor back e.m.f. is allowed to exceed the d.c. source voltage. For example, d.c. link-voltage and current surges can arise if the inverter gating is suddenly removed at high speeds.

3.1.3 VECTOR CONTROL OF PERMANENT-MAGNET SYNCHRONOUS MACHINES WITHOUT SENSORS

3.1.3.1 General introduction

It has been discussed in Section 1.2.1.3 and Sections 3.1.1 and 3.1.2 that the torque control of a permanent-magnet synchronous machine (PMSM) requires knowledge of the rotor position to perform an effective stator current control. Furthermore, for speed control, the speed signal is also required. The fundamentals of vector control schemes for permanent-magnet synchronous machines employing position sensors have been discussed in Sections 1.2.1.3 and a detailed analysis has been presented in Sections 3.1.1 and 3.1.2 for machines with surface-mounted magnets and interior magnets respectively. However, as discussed in Section 1.2.3, to reduce total hardware complexity and costs, to increase the mechanical robustness and reliability of the drive, to reduce the maintenance requirements, to ensure that the inertia of the system is not increased, and to have noise immunity, it is desirable to eliminate these sensors in vector-controlled drives. For this purpose indirect sensoring techniques are used. It has been shown in Section 1.2.3 that the main techniques of sensorless control of PMSM drives are as follows:

1. Open-loop estimators using monitored stator voltages/currents;
2. Stator phase third-harmonic voltage-based position estimators;
3. Back e.m.f.-based position estimators;
4. Observer-based (e.g. extended Kalman filter) speed and position estimators;
5. Estimators based on inductance variation due to geometrical and saturation effects;
6. Estimators using artificial intelligence (neural network, fuzzy-logic-based systems, fuzzy-neural networks, etc.).

Some of these techniques will be discussed in the following sections, although not all of them are suitable for high-dynamic-performance drives. Furthermore, for better understanding, drive schemes for both the permanent-magnet synchronous machine with sinusoidal back e.m.f.s and the permanent-magnet synchronous machine with trapezoidal back e.m.f.s (the nomenclatures 'brushless d.c. machine' or 'electronically commutated machine' are also widely used in the literature) will be discussed, although vector control providing high-dynamic performance is only possible with the machine with sinusoidal back e.m.f.s. This is due to the fact that in the brushless d.c. machine, only two stator phases are excited at any time instant.

It should be noted that in general, when a position-sensorless control scheme is used, the PMSM is not self-starting, since around standstill it is difficult to estimate the stator-flux position by the estimators listed above. To avoid this problem, several starting strategies have been developed, and some of these will also be discussed in the following sections together with their corresponding sensorless schemes. For example, when open-loop stator flux-linkage estimators are used, at low speeds it is impossible to get accurate flux-linkage estimates and to ensure acceleration from standstill to a speed where the stator flux-linkage can be accurately estimated, and open-loop control (with the speed control loop opened) can be performed. For this purpose ramp acceleration can be used (the reference speed signal is a ramp). This will only produce a smooth start (no oscillations in the speed) if the initial rotor position is chosen adequately (it is a small value). Similar problems exist with back e.m.f.-based estimators, where the back e.m.f. can only be estimated after the motor has been first started and brought to a certain speed. In such a case it is possible to have a starting scheme in which two or three stator windings are energized, and the rotor will then align itself to the desired rotor position to yield the required accelerating torque. However, such a solution leads to low dynamic performance and does not provide the rotor position at standstill. It is also possible to utilize saturation or other saliency effects, which result in different inductances in two axes, and indirect position estimation is performed by using the inductance variations (which can be obtained directly by direct inductance measurements or indirectly by using measured currents even at standstill). Such schemes can offer high dynamic performance even at standstill.

3.1.3.2 Vector control using open-loop flux and speed estimators using monitored stator voltages/currents

In the present section a simple position-sensorless control system is described for the vector control of a PMSM (with sinusoidal back e.m.f.) supplied by a current-controlled PWM voltage source inverter. The goal is to control the phase angles of the stator currents to maintain near unity power factor over a wide range of speed and torque. For this purpose, in the first implementation, the monitored stator voltages and currents are used to estimate the position of the stator flux-linkage space vector through which the phase angles of the stator currents can be controlled. However, in an alternative position-sensorless scheme, only the stator currents are monitored together with the d.c. link voltage. Because these are the first two position-sensorless drives discussed in the present book, both schemes will be discussed in great detail. This also allows practical implementation.

Many types of open-loop and closed-loop flux-linkage, position, and speed estimators have been discussed in detail in a recent book [Vas 1993]. The position-sensorless estimators discussed first are open-loop estimators and can be used in a wide range of a.c. drive applications, including induction motor drives. Such an estimator will yield the angle of the stator flux-linkage space vector with respect to the real-axis of the stator reference frame (ρ_s), together with the modulus of the flux space vector. These quantities are shown in Fig. 3.19. In general, for a synchronous machine in the steady-state, the first time-derivative of this angle gives exactly the rotor speed, $\omega_r = d\rho_s/dt$. However, in the transient state, in a drive where there is a change in the reference electromagnetic torque, the stator flux-linkage space vector moves relative to the rotor (to produce a new torque level), and this influences the rotor speed. This effect can be neglected if the rate of change of the electromagnetic torque is limited.

In general the angle between the stator-current space vector and stator flux-linkage space vector is not $90°$, but if the stator power factor is to be unity, then the stator-current space vector should lead the stator flux-linkage space vector by $90°$, and this case is shown in Fig. 3.19, since this will be used in a

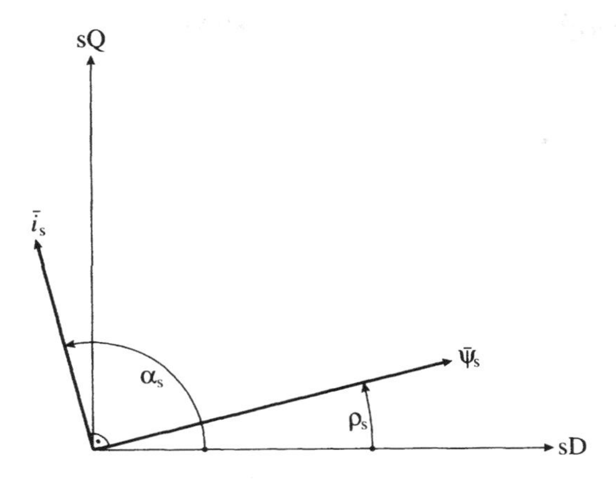

Fig. 3.19. Stator current and flux-linkage space vector (for unity power factor).

particular implementation of the control of the permanent-magnet synchronous machine discussed below.

3.1.3.2.1 Open-loop flux estimators, voltage reconstruction, drift compensation

In general, the stator-flux space vector can be obtained by integration of the terminal voltage minus the stator ohmic drop (this also follows from eqn (2.1-124)):

$$\bar{\psi}_s = \int (\bar{u}_s - R_s \bar{i}_s)\,dt. \tag{3.1-44}$$

Thus the following direct- and quadrature-axis stator flux-linkage components in the stator reference frame are obtained (by considering $\bar{u}_s = u_{sD} + ju_{sQ}$, $\bar{i}_s = i_{sD} + ji_{sQ}$, and $\bar{\psi}_s = \psi_{sD} + j\psi_{sQ}$):

$$\psi_{sD} = \int (u_{sD} - R_s i_{sD})\,dt, \tag{3.1-45}$$

$$\psi_{sQ} = \int (u_{sQ} - R_s i_{sQ})\,dt. \tag{3.1-46}$$

Equations (3.1-45) and (3.1-46) contain the two-axis stator voltages and currents. These can be obtained from the measured line voltages and currents as follows:

$$u_{sD} = \tfrac{1}{3}(u_{BA} - u_{AC}) \tag{3.1-47}$$

$$u_{sQ} = -(1/\sqrt{3})(u_{AC} + u_{BA}) \tag{3.1-48}$$

$$i_{sD} = i_{sA} \tag{3.1-49}$$

$$i_{sQ} = (1/\sqrt{3})(i_{sA} + 2i_{sB}). \tag{3.1-50}$$

Equations (3.1-47) and (3.1-48) can be obtained from $\bar{u}_s = u_{sD} + ju_{sQ} = (2/3)(u_{sA} + au_{sB} + a^2u_{sC})$ and by also considering

$$u_{CB} = u_{sB} - u_{sC}, \tag{3.1-51}$$

$$u_{AC} = u_{sC} - u_{sA}, \tag{3.1-52}$$

$$u_{BA} = u_{sA} - u_{sB}, \tag{3.1-53}$$

$$u_{sA} + u_{sB} + u_{sC} = 0. \tag{3.1-54}$$

Equations (3.1-49) and (3.1-50) follow from $\bar{i}_s = i_{sD} + ji_{sQ} = (2/3)(i_{sA} + ai_{sB} + a^2 i_{sC})$ and by also considering $i_{sA} + i_{sB} + i_{sC} = 0$.

The angle of the stator flux-linkage space vector (which is also shown in Fig. 3.19) can be obtained from the two-axis stator flux-linkage components as

$$\rho_s = \tan^{-1}(\psi_{sQ}/\psi_{sD}). \tag{3.1-55}$$

It is important to note that the performance of a PMSM drive using eqn (3.1-55) depends greatly on the accuracy of the estimated stator flux-linkage components and these depend on the accuracy of the monitored voltages and currents, and also on an accurate integration technique. Errors may occur in the monitored voltages and currents due to the following factors: phase shift in the measured values (because of the sensors used), magnitude errors because of conversion factors and gain, offset in the measurement system, quantization errors in the digital system, etc. Furthermore, an accurate value has to be used for the stator resistance. For accurate flux-linkage estimation, the stator resistance must be adapted to temperature changes. The integration can become problematic at low frequencies, where the stator voltages become very small and are dominated by the ohmic voltage drops. At low frequencies the voltage drop of the inverter must also be considered. This is a typical problem associated with open-loop flux estimators used in other a.c. drives as well, which use measured terminal voltages and currents.

Drift compensation is also an important factor in a practical implementation of the integration since drift can cause large errors of the position of the stator flux-linkage space vector. In an analog implementation the source of drift is the thermal drift of analog integrators. However, a transient offset also arises from the d.c. components which result after a transient change. An incorrect flux angle will cause phase modulation in the control of the currents at fundamental frequency, which, however, will produce an unwanted fundamental frequency oscillation in the electromagnetic torque of the machine. Furthermore, since in the open-loop speed estimator which utilizes the stator flux-linkage components, the rotor speed is determined from the position of the stator flux-linkage space vector, a drift in the stator flux-linkage vector will cause incorrect and oscillatory speed values. In a speed control loop, this drift error will cause an undesirable fundamental frequency modulation of the modulus of the reference stator-current space vector ($|\bar{i}_{sref}|$). The open-loop stator flux-linkage estimator can work well down to 1–2 Hz, but not below this, unless special techniques are used.

It should also be noted that in addition to the stator flux estimation based on eqns (3.1-45) and (3.1-46) [which were obtained from eqn (2.1-124)], and which is shown in Fig. 3.20(a), it is also possible to construct another stator flux-linkage estimator, where the integration drifts are reduced at low frequency. For this purpose, instead of open-loop integrators, closed-loop integrators are introduced. In general the space vector of the stator flux linkages can be expressed in the stationary reference frame as

$$\bar{\psi}_s = |\bar{\psi}_s| \exp(j\rho_s|) = \psi_{sD} + j\psi_{sQ},$$

where $|\bar{\psi}_s|$ is the modulus and ρ_s is the angle of the stator flux-linkage space vector (with respect to the direct axis of the stationary reference frame). Thus if $|\bar{\psi}_s|$ and ρ_s are known, the direct- and quadrature-axis flux linkages can be obtained as $\psi_{sD} = |\bar{\psi}_s| \cos \rho_s$ and $\psi_{sD} = |\bar{\psi}_s| \sin \rho_s$ respectively. However, the modulus and the angle of the stator flux-linkage space vector can be obtained by considering the stator voltage equation in the stator-flux-oriented reference frame. Thus it follows from eqn (2.1-148) that if the stator-voltage equation is expressed in the

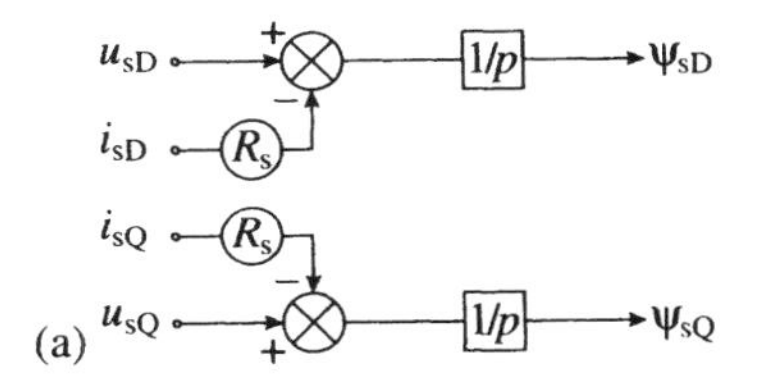

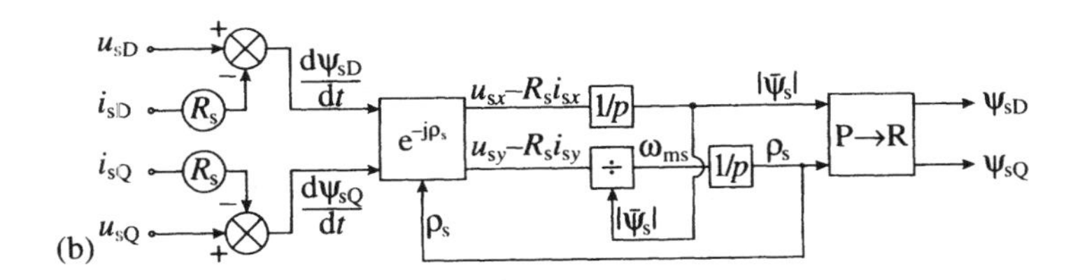

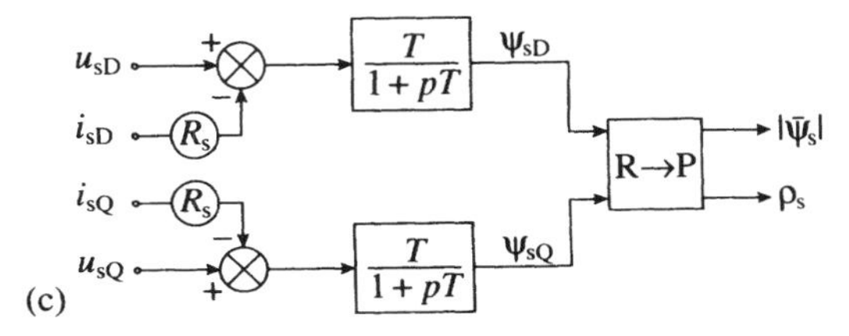

Fig. 3.20. Stator flux-linkage estimators. (a) Estimation in the stationary reference frame; (b) estimation in the stator-flux-oriented reference frame; (c) estimation in the stationary reference frame using quasi-integrators.

stator-flux-oriented reference frame, which rotates at the speed of the stator flux-linkage space vector ($\omega_{ms} = d\rho_s/dt$),

$$\bar{u}_{s\psi s} = R_s \bar{i}_{s\psi s} + \frac{d|\bar{\psi}_s|}{dt} + j\omega_{ms}|\bar{\psi}_s| \tag{3.1-56}$$

is obtained. It follows by resolution into real and imaginary components that the rate of change of the stator flux modulus is

$$\frac{d|\bar{\psi}_s|}{dt} = u_{sx} - R_s i_{sx} \tag{3.1-57}$$

and the speed of the stator-flux space vector is obtained as

$$\omega_{ms} = \frac{u_{sy} - R_s i_{sy}}{|\bar{\psi}_s|}. \tag{3.1-58}$$

In these equations x and y denote the direct and quadrature axes of the stator-flux-oriented reference frame respectively. Equations (3.1-57) and (3.1-58) can be used to estimate ω_{ms} and $|\bar{\psi}_s|$ from u_{sD}, u_{sQ}, i_{sD}, and i_{sQ} as shown in Fig. 3.20(b). The angle ρ_s is then obtained by the integration of ω_{ms}. This angle is fed back into a transformation block, but it is not necessary to use two transformations to transform u_{sD} and u_{sQ} into u_{sx} and u_{sy} and also i_{sD} and i_{sQ} into i_{sx} and i_{sy}. This follows from the fact that in the two voltage equations above, $u_{sx} - R_s i_{sx}$ and $u_{sy} - R_s i_{sy}$ are present, and thus it is possible to obtain these in a single step from $u_{sD} - R_s i_{sD}$ and $u_{sQ} - R_s i_{sQ}$ by the transformation $\exp(-j\rho_s)$ as shown in Fig. 3.20(b).

It is also possible to estimate the stator flux-linkage components by using low-pass filters instead of pure integrators. In this case $1/p$ is replaced by $T/(1+pT)$ where T is a suitable time constant [see also eqns (3.3-5)–(3.3-7)]. Such a flux-linkage estimation scheme is shown in Fig. 3.20(c). To obtain accurate flux estimates at low stator frequency, the time constant T has to be large and the variation of the stator resistance with the temperature has also to be considered. It is an advantage of using the flux-linkage estimation scheme shown in Fig. 3.20(c) that the effects of initial conditions are damped by the time constant T. Obviously there will be a phase shift between the actual and estimated flux linkages, but increased T decreases the phase shift. However, an increased T decreases the damping.

The estimation of the stator flux-linkage components described above requires the stator terminal voltages. However, it is possible to have a scheme in which these voltages are not monitored, but they are reconstructed from the d.c. link voltage (U_d) and the switching states (S_A, S_B, S_C) of the six switching devices of a six-step voltage-source inverter. The switching functions (S_A for stator phase sA, S_B for stator phase sB, and S_C for stator phase sC) are defined as follows:

$S_A = 1$ when upper switch in phase sA of inverter (S1) is ON and lower switch (S4) is OFF

$S_A = 0$ when upper switch in phase sA of inverter (S1) is OFF and lower switch (S4) is ON

$S_B=1$ when upper switch in phase sB of inverter (S3) is ON and lower switch (S6) is OFF

$S_B=0$ when upper switch in phase sB of inverter (S3) is OFF and lower switch (S6) is ON

$S_C=1$ when upper switch in phase sC of inverter (S5) is ON and lower switch (S2) is OFF

$S_C=0$ when upper switch in phase sC of inverter (S5) is OFF and lower switch (S2) is ON

Figure 3.21 shows the schematic of the six switches of the voltage-source inverter and the values of the switching functions are also shown together with the appropriate switch positions.

The stator-voltage space vector (expressed in the stationary reference frame) can be obtained by using the switching states and the d.c. link voltage U_d, as

$$\bar{u}_s=\tfrac{2}{3}U_d(S_A+aS_B+a^2S_C). \tag{3.1-59}$$

A simple proof of eqn (3.1-59) can be obtained in various ways, but only one technique will now be discussed, which is a direct consequence of the physical fact that the stator line-to-line voltages of the induction motor (output voltages of the inverter) can be expressed as $u_{ab}=U_d(S_A-S_B)$, $u_{bc}=U_d(S_B-S_C)$, and $u_{ca}=U_d(S_C-S_A)$. However, the stator phase voltages (line-to-neutral voltages) can be obtained from the line-to-line voltages as $u_{sA}=(u_{ab}-u_{ca})/3$, $u_{sB}=(u_{bc}-u_{ab})/3$, $u_{sc}=(u_{ca}-u_{bc})/3$. Substitution of the expressions for the line-to-line voltages into the stator phase voltages gives

$$u_{sA}=(1/3)U_d(2S_A-S_B-S_C), \qquad u_{sB}=(1/3)U_d(-S_A+2S_B-S_C),$$
$$u_{sC}=(1/3)U_d(-S_A-S_B+2S_C).$$

Thus by considering that $\bar{u}_s=(2/3)(u_{sA}+au_{sB}+a^2u_{sC})$, finally eqn (3.1-59) is obtained. These results are now summarized:

$$u_{sA}=\mathrm{Re}(\bar{u}_s)=\tfrac{1}{3}U_d(2S_A-S_B-S_C), \tag{3.1-60}$$

$$u_{sB}=\mathrm{Re}(a^2\bar{u}_s)=\tfrac{1}{3}U_d(-S_A+2S_B-S_C), \tag{3.1-61}$$

$$u_{sC}=\mathrm{Re}(a\bar{u}_s)=\tfrac{1}{3}U_d(-S_A-S_B+2S_C). \tag{3.1-62}$$

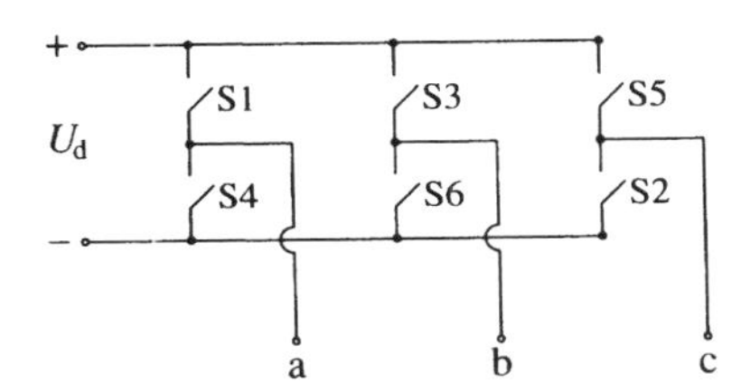

$S_A=1$	S_1 ON	S_4 OFF
$S_A=0$	S_1 OFF	S_4 ON
$S_B=1$	S_3 ON	S_6 OFF
$S_B=0$	S_3 OFF	S_6 ON
$S_C=1$	S_5 ON	S_2 OFF
$S_C=0$	S_5 OFF	S_2 ON

Fig. 3.21. Inverter switches and the switching functions.

Thus by the substitution of eqn (3.1-59) into (3.1-44), the stator flux-linkage space vector can be obtained as follows for a digital implementation. The kth sampled value of the estimated stator flux-linkage space vector (in the stationary reference frame) can be obtained as

$$\bar{\psi}_s(k)=\bar{\psi}_s(k-1)+\tfrac{2}{3}U_d T[S_A(k-1)+aS_B(k-1)$$
$$+a^2 S_C(k-1)]-R_s T\bar{i}_s(k-1), \tag{3.1-63}$$

where T is the sampling time (flux control period) and $\bar{i}_s$ is the stator-current space vector. Resolution of eqn (3.1-63) into its real- and imaginary-axis components gives ψ_{sD} and ψ_{sQ}:

$$\psi_{sD}(k)=\psi_{sD}(k-1)+\tfrac{2}{3}U_d T\,\mathrm{Re}[S_A(k-1)+aS_B(k-1)$$
$$+a^2 S_C(k-1)]-R_s T i_{sD}(k-1) \tag{3.1-64}$$
$$\psi_{sQ}(k)=\psi_{sQ}(k-1)+\tfrac{2}{3}U_d T\,\mathrm{Im}[S_A(k-1)+aS_B(k-1)$$
$$+a^2 S_C(k-1)]-R_s T i_{sQ}(k-1). \tag{3.1-65}$$

It is important to note that eqns (3.1-64) and (3.1-65) are sensitive to:

- voltage errors caused by dead-time effects (e.g. at low speeds, the pulse widths become very small and the dead time of the inverter switches must be considered);
- the voltage drop in the power electronic devices;
- the fluctuation of the d.c. link voltage;
- the variation of the stator resistance (but this resistance variation sensitivity is also a feature of the method using monitored stator voltages).

Due to the finite turn-off times of the switching devices (power transistors) used in a three-phase PWM VSI, there is a need to insert a time delay (dead time, t_{dead}) after switching a switching device off and the other device on in one inverter leg. This prevents the short-circuit across both switches in the inverter leg and the d.c. link. Due to the presence of the dead time, in a voltage-controlled drive system, the amplitude of the output voltages of the inverter (stator voltages of the machine) will be reduced and will be distorted. This effect is very significant at low speeds. In space-vector terms this means that at very low speeds, the stator-voltage space-vector locus produced by the inverter, without using any dead-time effect compensation scheme, becomes distorted and the amplitude of the stator-voltage space vector is unacceptably small. Due to the presence of the dead time, the output-voltage space vector of the inverter is not equal to the desired (reference) voltage space vector, but it is equal to the sum of the desired voltage space vector and an error-voltage space vector ($\Delta\bar{u}_k$).

The error-voltage space vector can be estimated by utilizing information available on the switching times of the power transistors. By using the known value of the dead time (t_{dead}, which in practice can be several microseconds) together with the various switching times supplied by the manufacturer of the switching device: ON delay time (t_{don}), rise time (t_r), OFF delay time (t_{doff}), fall time (t_f), it is possible to determine the control time (t_c). Good approximation is obtained by considering $t_c = t_{dead} + t_{don} + t_r/2 - t_{doff} - t_f/2$. When neither the upper nor the lower transistor conduct in a particular inverter leg, then the stator currents also have an effect on the error voltage, since they determine which diodes conduct (in that inverter leg). For example, if in the first inverter leg, the upper switch (S1) and lower switch (S2) are not conducting, then either diode D1 (across S1) or diode D2 (across S2) conducts, depending on the sign of the stator phase current i_{sA}. It follows that this inverter leg can be connected either to the positive d.c. rail, or to the negative d.c. rail. In general the error-voltage space vector also depends on the angle of the stator current space vector, α_s, where $\alpha_s = \tan^{-1}(i_{sQ}/i_{sD})$ and e.g. $i_{sD} = i_{sA}$, $i_{sQ} = (i_{sB} - i_{sC})/\sqrt{3} = i_{sA}/\sqrt{3} + 2i_{sB}/\sqrt{3}$. However, it is not necessary to compute this angle by using trigonometric functions, since we only need to know in which of six sectors the stator-current space vector is located, and this can be determined by using comparators. Due to the six-step inverter each of the sectors are 60 electrical degrees wide, the first sector spans the region $-30°$ to $30°$, the second one spans $30°$ to $90°$, . . . , the sixth sector spans the region from $270°$ to $330°$. For example, when $i_{sA} > 0$, $i_{sB} < 0$, $i_{sC} < 0$ then the stator-current space vector is in the first sector; when $i_{sA} > 0$, $i_{sB} > 0$, $i_{sC} < 0$ the stator-current space vector is in the second sector, etc. If the stator-current space vector is located in the kth sector, then the error stator-voltage space vector can be determined by using $\Delta\bar{u}_k = [4U_d t_c/(3T)]\exp[j(k-1)\pi/3]$, where U_d is the d.c. link voltage, T is the switching period, and $k = 1, 2, \ldots, 6$. It should be noted that it is possible to implement such dead-time compensation schemes where the existing space vector PWM strategy is not changed but only the reference stator-voltage space vector ($\bar{u}_{sref}$) is modified by adding the error stator-voltage space vector. However, it is also possible to have other solutions, which do not require knowledge of the position of the stator-current space vector (see also Section 8.2.6).

Recent trends are such that in position-sensorless drives, the stator-voltage sensors are also eliminated and only current sensors are used, but the d.c. link voltage is also monitored, so in eqns (3.1-64) and (3.1-65) the actual value of the d.c. link voltage is used. It is possible to have a position-sensorless drive implementation based on eqns (3.1-64) and (3.1-65), where the dead-time effects are also considered and the thermal variation of the stator resistance is also incorporated into the control scheme, which will even work at very low speeds. Obviously there are other possibilities as well to solve the problem of integration at low frequencies, e.g. cascaded low-pass filters can be used with programmable time constants, etc. In the voltage-source permanent-magnet synchronous motor drive, it is also possible to use appropriate voltage references instead of actual voltages if the switching frequency of the inverter is high compared to the electrical time-constant of the motor. In this case the appropriate reference voltages force the actual currents to follow their references. Such a solution has the additional advantage of further reducing the required number of sensors, and since no filtering of the voltages is required, no delay is introduced due to filtering. In general, a delay introduced by a low-pass filter (acting on the voltages) is a function of the motor speed (since the modulus of the voltage is a direct function of the speed, e.g. at low speeds it is small); thus in a vector-controlled drive, it is not possible to keep the stator current in the desired position as speed changes. This adversely affects the torque/ampere capability and motor efficiency, unless filter-delay compensation is introduced (e.g. by using a lead-lag network).

It has been emphasized above that drift-error compensation is important. Any d.c. offset arising in the voltage- and current-measurement circuits of a drive must be minimized. One solution is to calibrate these sensors every time the drive is started. For this purpose, during the calibration stage (prior to starting up the drive system), average offset values of the voltages and currents can be obtained (by taking hundreds of readings). During normal operation of the drive, these average values are then subtracted from the measured values.

It has also been emphasized that if analog integrators are used, a source of drift is the thermal drift of these integrators and a transient offset also arises from the d.c. components which result after a transient change. Drift will cause an error in the flux position calculation, whose 'drifted' position is ρ'_s which differs from ρ_s (thus the position error due to drift is $\Delta\rho_s=\rho'_s-\rho_s$). Drift compensation of the direct- and quadrature-axis stator flux-linkage components can be achieved by subtracting the respective drift flux-linkage components. If in the stationary reference frame the stator flux-linkage space vector without drift is $\bar{\psi}_s=\psi_{sD}+j\psi_{sQ}=|\bar{\psi}_s|\exp(j\rho_s)$, and the drift is characterized by a drift vector $\bar{d}$, then the 'drifted' stator flux-linkage space vector is $\bar{\psi}_s^{drift}=\bar{\psi}_s+\bar{d}=\psi_{sD}^{drift}+j\psi_{sQ}^{drift}=|\bar{\psi}_s^{drift}|\exp(j\rho'_s)$. These vectors are also shown in Fig. 3.22. Thus the stator flux components which are required for accurate position and speed estimation are

$$\psi_{sD}=\psi_{sD}^{drift}-a \tag{3.1-66}$$

$$\psi_{sQ}=\psi_{sQ}^{drift}-b, \tag{3.1-67}$$

where a and b are the real- and imaginary-axis components of the drift vector (d) as shown in Fig. 3.22.

In the steady state, the locus of the stator flux-linkage space vector is a circle and during the transient state it is almost a circle, since the main component of the stator flux is the magnet flux component. Equations (3.1-66) and (3.1-67) simply mean that the corresponding drift components of the drift vector are subtracted. In other words, a simple physical property has been utilized: due to the drift of the analog integrators, the space-vector locus of the stator flux-linkage space vector is shifted (drifted) by an amount characterized by the drift vector, ($\bar{d}=a+jb$). It is interesting to note that if digital integrators are used, this shift is not present. However, a similar shift of the stator flux-linkage space vector will arise if the stator flux is estimated by using an incorrect value of the stator resistance. This can be simply proved since it follows from eqn (3.1-44) that the actual value of the stator flux-linkage space vector is $\bar{\psi}_s=\int(\bar{u}_s-R_s\bar{i}_s)\,dt$, where R_s is the actual value of the stator resistance, and similarly the estimated value of the stator flux-linkage space vector is $\bar{\psi}_{se}=\int(\bar{u}_s-R_{se}\bar{i}_s)\,dt$, where R_{se} is an estimated value of the stator resistance. Thus by introducing the stator resistance error,

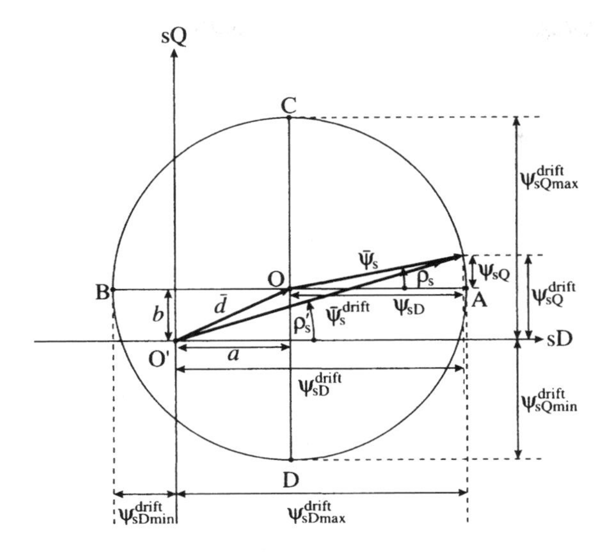

Fig. 3.22. Space-vector locus of the drifted stator flux-linkage space vector.

$\Delta R_s = R_{se} - R_s$, the expression of the estimated flux-linkage space vector becomes $\bar{\psi}_{se} = \bar{\psi}_s - \Delta R_s \int \bar{i}_s \, dt = \bar{\psi}_s + \Delta\bar{\psi}_s$, where $\Delta\bar{\psi}_s$ is the drift flux-linkage space vector due to the error in the estimated stator resistance. This can be expressed as $\Delta\bar{\psi}_s = -\Delta R_s \int \bar{i}_s \, dt$. If $\Delta R_s = 0$, it follows that $\Delta\bar{\psi}_s = 0$, which is the expected result. It should also be noted that this drift can be used for the determination of the actual value of the stator resistance, since it can be expressed as $R_s = R_{se} + \Delta R_s = R_{se} - \Delta\bar{\psi}_s / \int \bar{i}_s \, dt$, and it follows that if the drift due to the stator resistance error is known, R_s can be estimated by using this deviation. If the estimated stator resistance is deliberately chosen to be zero, then the stator resistance is obtained simply from $R_s = -\Delta\bar{\psi}_s / \int \bar{i}_s \, dt$ (and of course it is incorrect to deduce that $R_s = 0$ if $\Delta\bar{\psi}_s = 0$, since due to the deliberate selection of $R_{se} = 0$, $\Delta\bar{\psi}_s \neq 0$).

For accurate position estimation the required position angle of the stator flux-linkage space vector is not ρ_s' but it is ρ_s, which can be expressed as

$$\rho_s = \tan^{-1}\left(\frac{\psi_{sQ}}{\psi_{sD}}\right) = \frac{\psi_{sQ}^{drift} - b}{\psi_{sD}^{drift} - a}. \tag{3.1-68}$$

It should be noted that it is eqn (3.1-68) which must be used to obtain the correct value of the flux-linkage space vector position, but this contains the parameters a and b. However, these parameters can be obtained from the space-vector locus of the drifted stator flux-linkage space vector, by considering its four extreme points shown in Fig. 3.22. These correspond to ψ_{sDmax}^{drift} (point A, where the direct-axis stator flux-linkage component of the drifted flux linkage is maximal), ψ_{sDmin}^{drift} (point B where the direct-axis stator flux-linkage component of the drifted flux linkage is minimal), ψ_{sQmax}^{drift} (point C, where the quadrature-axis stator flux-linkage component of the drifted flux linkage is maximal), and finally to $\psi_{\mathrm{sQmin}}^{\mathrm{drift}}$ (point D, where the direct-axis stator flux-linkage component of the drifted flux linkage is minimal). Thus parameters a and b can be determined from

$$a=\tfrac{1}{2}(\psi_{\mathrm{sDmax}}^{\mathrm{drift}}+\psi_{\mathrm{sDmin}}^{\mathrm{drift}}) \tag{3.1-69}$$

$$b=\tfrac{1}{2}(\psi_{\mathrm{sQmax}}^{\mathrm{drift}}+\psi_{\mathrm{sQmin}}^{\mathrm{drift}}). \tag{3.1-70}$$

Obviously, the minimum values must be substituted by considering their negative sign (and when a maximal value is equal to the minimal value, then the corresponding drift value is zero, as expected).

3.1.3.2.2 Rotor position estimation using stator flux linkages

In addition to the techniques described above to determine the stator flux linkages, it is also possible to use another technique, which also allows rotor position detection over a wide speed-range, including acceleration from standstill. For simplicity, only a permanent-magnet synchronous machine with surface-mounted magnets is considered, but the technique can also be used for a machine with interior permanent magnets. For this purpose it is again considered that the stator voltage equation in the stationary reference frame is $\bar{u}_s=R_s\bar{i}_s+\mathrm{d}\bar{\psi}_s/\mathrm{d}t$. However, for the machine with surface-mounted magnets, the stator flux-linkage space vector in the stationary reference frame can be expressed as $\bar{\psi}_s=L_s\bar{i}_s+\psi_F\exp(\mathrm{j}\theta_r)$, where ψ_F is the magnet flux defined in eqn (3.1-2) and θ_r is the rotor position. It should be noted that when the expression of the stator flux-linkage space vector is substituted into the stator voltage equation, and linear magnetic conditions are assumed, then eqn (3.1-5) is obtained, but in general $\bar{u}_s=R_s\bar{i}_s+\mathrm{d}[L_s\bar{i}_s+\psi_F\exp(\mathrm{j}\theta_r)]/\mathrm{d}t$. If the stator flux linkages are known (e.g. by integrating the terminal voltage minus the ohmic voltage drop), then it is possible to obtain the rotor position information by utilizing the expression which defines the stator flux-linkage space vector. Thus by using the known value of $\bar{\psi}_s$ in the expression $\bar{\psi}_s=L_s\bar{i}_s+\psi_F\exp(\mathrm{j}\theta_r)$, and if the stator currents are also measured and L_s and ψ_F are known, it is possible to estimate θ_r. There are various possibilities for implementing this technique, and only one specific solution is now briefly discussed, and the algorithmic steps are summarized here.

Step 1: *Estimation of stator flux-linkage space vector* by using

$$\bar{\psi}_{se}=\int(\bar{u}_s-R_s\bar{i}_s)\,\mathrm{d}t. \tag{3.1-71}$$

When a numerical integration routine is used, this contains the initial value of the stator flux-linkage space vector, $\bar{\psi}_{se}(0)$, which is defined by the position of the magnet. Thus if the rotor is brought to a known initial position (θ_{r0}), the initial value in the integration is defined. Of course this estimated stator flux-linkage space vector may not be equal to the actual one, and will be corrected later in the algorithm. Incorrect stator flux linkage leads to incorrect position estimation, i.e. to position error, which will also be estimated (in Step 4).

Step 2: *Estimation of stator currents* (first estimation) by using the initial rotor position and the estimated stator flux-linkage space vector by using the definition given above:

$$\bar{\psi}_s = L_s \bar{i}_{s1} + \psi_F \exp(j\theta_r). \tag{3.1-72}$$

Thus the estimated stator-current space vector can be expressed as

$$\bar{i}_{se1} = \frac{\bar{\psi}_s - \psi_F \exp(j\theta_{r0})}{L_s} \tag{3.1-73}$$

(the subscript 1 appears due to the fact that this is a first current estimate, and later the subscript 2 will be used for a second current estimate). Of course, in a practical implementation the line currents or the direct- and quadrature-axis stator currents are determined, e.g. $\bar{i}_{sAe1} = [\psi_{sA} - \psi_F \cos(j\theta_{r0})]/L_s$.

Step 3: *Stator current error estimation* by using

$$\Delta\bar{i}_{s1} = \bar{i}_s - \bar{i}_{se1}, \tag{3.1-74}$$

where $\bar{i}_s$ is the actual stator-current space vector ($\bar{i}_{se1}$ is the estimated stator-current space vector determined in Step 2). Again, it should be noted that in a practical implementation, the line current errors, or the direct- and quadrature-axis stator current errors, are determined, e.g. $\Delta i_{sA1} = i_{sA} - i_{sAe1}$.

Step 4: *Rotor position error estimation* from the line current errors. This uses the fact that the stator flux linkage is a function of the stator currents and also a function of the rotor position, and thus the rotor position error can be obtained by using the current estimation errors. For example, for stator phase sA, mathematically this follows from the following expression: $\Delta\psi_{sA} = (\partial\psi_{sA}/\partial i_{sA1})\Delta i_{sA1} + (\partial\psi_{sA}/\partial\theta_r)\Delta\theta_{rsA}$, where by considering the real part of eqn (3.1-72), it follows that the first term gives $L_s \Delta i_{sA1}$ and the second term gives $-\psi_F \sin(\theta_{r0})\Delta\theta_{rsA}$. Thus if it is assumed that $\Delta\psi_{sA} = 0$, then the position error for stator phase sA is obtained as

$$\Delta\theta_{rsA} = -\frac{L_s \Delta i_{sA1}}{\psi_F \sin(\theta_{r0})}. \tag{3.1-75}$$

The expressions of the position errors due to the current errors in the other two stator phases ($\Delta\theta_{rsB}, \Delta\theta_{rsC}$) can be similarly obtained and are similar to eqn (3.1-75), but contain the corresponding current error, and also the initial rotor-angle values are displaced by 120° and 240° respectively. The three position errors are then used to obtain a single average value,

$$\Delta\theta_r = \tfrac{1}{3}(\Delta\theta_{rsA} + \Delta\theta_{rsB} + \Delta\theta_{rsC}). \tag{3.1-76}$$

Step 5: *Rotor position estimation correction* using the rotor position error and the previous rotor position estimate:

$$\theta_r = \theta_{r0} + \Delta\theta_r. \tag{3.1-77}$$

Step 6: *Stator line current estimation* using the corrected rotor position:

$$\bar{i}_{se2}=\frac{\bar{\psi}_s-\psi_F\exp(j\theta_r)}{L_s} \tag{3.1-78}$$

(again in a practical implementation the line currents or the direct- and quadrature-axis stator currents are obtained).

Step 7: *Stator current error estimation* (second estimation) using

$$\Delta\bar{i}_{s2}=\bar{i}_s-\bar{i}_{se2} \tag{3.1-79}$$

Step 8: *Flux-linkage error estimation* using the current errors, assuming that the flux-linkage error arises only due to the current errors:

$$\Delta\bar{\psi}_{se}=L_s\Delta\bar{i}_{s2}. \tag{3.1-80}$$

Step 9: *Flux-linkage space vector estimation correction* using the flux-linkage error,

$$\bar{\psi}_s=\Delta\bar{\psi}_{se}+\bar{\psi}_{se}. \tag{3.1-81}$$

Thus the stator flux linkage obtained by integration, $\bar{\psi}_{se}$, is updated by adding $\Delta\bar{\psi}_{se}$.

It should be noted that the rotor estimation technique described above is problematic at very low speeds (the problems are associated with the flux estimation). Parameter variations (due to saturation and temperature effects) also influence the accuracy of the estimation. Since the rotor-position estimation technique described with this algorithm is based on flux-linkage estimation, the technique can be extended to other machines, including the brushless d.c. machine and the synchronous reluctance machine. By using the determined rotor position, it is also possible to determine the rotor speed and then a drive without position and speed sensors can be implemented. Since the technique is computationally intensive, for practical implementation, the application of a DSP is required, e.g. for this purpose the TMS320C30 can be used.

3.1.3.2.3 *Open-loop speed estimators*

For the synchronous machine, the first derivative of the angle ρ_s is equal to the rotor speed,

$$\omega_r=\frac{d\rho_s}{dt}, \tag{3.1-82}$$

and in the speed control loop of the permanent-magnet synchronous motor drive, this relationship can be directly utilized. The estimation of the rotor speed based on the derivative of the position of the stator flux-linkage space vector can be slightly modified, by considering that the analytical differentiation $d\rho_s/dt$, where $\rho_s=\tan^{-1}(\psi_{sQ}/\psi_{sD})$, gives

$$\omega_r=\frac{\psi_{sD}\frac{d\psi_{sQ}}{dt}-\psi_{sQ}\frac{d\psi_{sD}}{dt}}{\psi_{sD}^2+\psi_{sQ}^2}. \tag{3.1-83}$$

In eqn (3.1-83) the derivatives of the flux-linkage components can be eliminated, since they are equal to the respective terminal voltages minus the corresponding ohmic drops [this also follows from eqn (2.1-124)]. Thus

$$\omega_r = \frac{\psi_{sD}(u_{sQ} - R_s i_{sQ}) - \psi_{sQ}(u_{sD} - R_s i_{sD})}{\psi_{sD}^2 + \psi_{sQ}^2}. \tag{3.1-84}$$

In eqn (3.1-84) the denominator is the square of the stator flux-vector modulus ($|\bar{\psi}_s|^2$) and the stator flux-linkage components can be obtained by using eqns (3.1-45) and (3.1-46).

For digital implementation, eqn (3.1-84) can be transformed into the following form:

$$\omega_r(k) = \frac{\psi_{sD}(k-1)\psi_{sQ}(k) - \psi_{sQ}(k-1)\psi_{sD}(k)}{T_s(\psi_{sD}^2(k) + \psi_{sQ}^2(k))} \tag{3.1-85}$$

where $\omega_r(k)$ is the value of the speed at the kth sampling time, and T_s is the sampling time. It should be noted that eqn (3.1-85) is only one specific sampled-data form of eqn (3.1-84) and contains a modelling error; thus it yields an error in the estimated speed. However, this can be removed by using a low-pass filter.

The rotor speed estimation based on the position of the stator flux-linkage space vector can also be conveniently used in other schemes, i.e. stator-flux-oriented control of the permanent-magnet synchronous machine or in vector-controlled or direct-torque-controlled induction motor drives (e.g. in the case of the induction motor see Section 4.5.3.1). However, in all of these applications, for operation at low speeds, the same limitations hold as above. Furthermore, for the induction motor, $d\rho_s/dt$ will not give the rotor speed, but will only give the speed of the stator m.m.f.s, which is the synchronous speed (ω_s), and the rotor speed is the difference between the synchronous speed and the slip speed ($\omega_r = \omega_s - \omega_{sl}$).

3.1.3.2.4 *Stator current-controlled inverter-fed PMSM drive scheme*

In the present section a stator current-controlled inverter-fed PMSM drive control scheme is discussed which uses an open-loop position and speed estimator. It follows from Fig. 3.19 that in general, in a drive which contains a current-controlled voltage-source PWM inverter, the reference stator-current space vector (expressed in the stator reference frame) takes the following form when expressed in polar coordinates:

$$\bar{\imath}_{sref} = |\bar{\imath}_{sref}| \exp(\alpha_s), \tag{3.1-86}$$

where α_s is the position of the stator-current space vector with respect to the direct-axis of the stator reference frame. For the special case of unity power factor, the stator flux-linkage space vector is in quadrature to the stator-current space vector (as shown in Fig. 3.19), thus

$$\alpha_s = \rho_s + \frac{\pi}{2}. \tag{3.1-87}$$

In general, by knowing the reference stator-current space vector, the required stator current references can be obtained in two steps by employing a polar-to-rectangular conversion to give the stator current components (i_{sDref}, i_{sQref}), and these can be transformed into the three-phase stator current references by using the phase transformation. However, it is simpler to obtain the reference stator currents in one step, since it directly follows from eqn (3.1-86) that

$$i_{sAref} = |\bar{i}_{sref}| \cos \alpha_s \tag{3.1-88}$$

$$i_{sBref} = |\bar{i}_{sref}| \cos\left(\alpha_s - \frac{2\pi}{3}\right), \tag{3.1-89}$$

and by considering that no zero-sequence stator currents are present:

$$i_{sCref} = -(i_{sAref} + i_{sBref}) \tag{3.1-90}$$

is obtained. These currents are used in the current-control loop of the drive, the control scheme of which is shown in Fig. 3.23. Since $\alpha_s = \rho_s + \pi/2$, eqns (3.1-88)–(3.1-90) depend on the angle ρ_s, which can be estimated by using the monitored stator voltages and currents, and also by using the known value of the stator resistance (R_s) as shown above.

As discussed above, around standstill, it is difficult to estimate the position of the stator flux-linkage space vector by the open-loop flux-linkage estimator described above, since the terminal voltages minus the ohmic drops are too small.

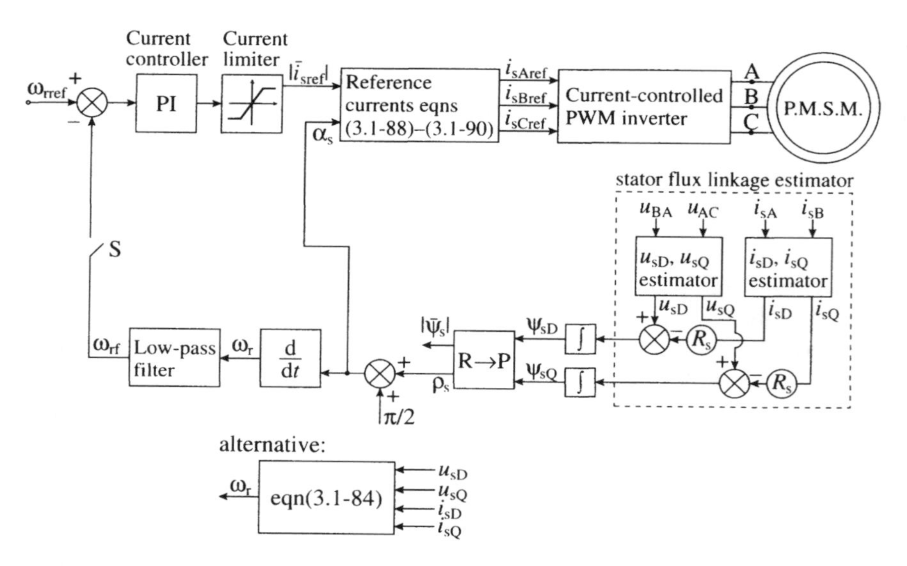

Fig. 3.23. Position-sensorless vector-control of a PM synchronous machine supplied by a current-controlled PWM voltage-source inverter.

However, to ensure acceleration from standstill to a speed where the stator flux-linkage space vector can be accurately estimated an open-loop control (with the speed control loop opened) can be performed. For this purpose ramp acceleration can be used (the reference speed signal is a ramp). This will only produce a smooth start (no oscillations in the speed) if the initial rotor position is chosen adequately (it is a small value).

In Fig. 3.23 the machine is supplied by a current-controlled PWM voltage-source inverter, and the current controllers are hysteresis controllers. As discussed above, the stator flux linkages are obtained by an open-loop flux estimator surrounded by the dashed lines. In Fig. 3.23 this is obtained by using the monitored values of the stator voltages and currents, but, as shown in Section 3.1.3.2.1, the stator flux-linkage components can also be obtained by using the monitored stator currents and the monitored d.c. link voltage together with the switching signals (since in this case it is possible to reconstruct the stator voltages). The obtained flux components (ψ_{sD}, ψ_{sQ}) are then used for the calculation of the flux angle ρ_s and then the current angle $\alpha_s = \rho_s + \pi/2$ is obtained. This is the phase angle of the reference stator currents and this is controlled to maintain near-unity power factor. The first time-derivative of this angle gives the unfiltered speed signal ω_r, and this is the input to a low-pass filter on the output of which the filtered speed signal ω_{rf} is present. This is then compared to the speed reference. However, as discussed above, it is also possible to obtain the speed signal by using eqn (3.1-84) and in this case, there is no need for the differentiation block.

In Fig. 3.23, there is a speed control loop, which is the outer loop, and there is also a current loop, which is the inner loop. The speed error is fed into a PI controller, the output of which is the input to a current limiter which outputs the modulus of the reference stator-current space vector. As discussed above, the limiter limits the initial acceleration to ensure a small value of acceleration during starting to yield minimal speed oscillations during start-up. During start-up, the flux-linkage estimator is inactive and switch S is open, and ω_{rref} is the reference value of the ramp speed. The modulus and position of the reference stator-current space vector are then input to a circuit which uses eqns (3.1-88)–(3.1-90) and outputs the three reference stator currents. These are then used by the hysteresis current controllers, which control the PWM inverter.

In the PWM-fed drive the line-to-line stator voltages change very rapidly and have high-frequency content (modulation noise). The stator currents also contain high frequency components. It follows that any differentiation used in the differentiation block shown in Fig. 3.23 would amplify these. Thus a low-pass filter is used to reduce this noise. If the filter is a digital filter, then its algorithm can be obtained simply as follows. If the output of the filter is ω_{rf}, and the input to the filter is ω_r (see Fig. 3.23), then the filter transfer function in the Laplace domain is

$$F(s) = \frac{\omega_{rf}(s)}{\omega_r(s)} = \frac{c}{(s+c)}, \tag{3.1-91}$$

where s is the Laplace operator. By the substitution $s=2(z-1)/[T(z+1)]$, where T is the sampling time, the pulse transfer function $F(z)$ is obtained as

$$F(z)=\frac{\omega_{\text{rf}}(z)}{\omega_{\text{r}}(z)}=\frac{cT(z+1)}{(2+cT)z-(2-cT)}. \tag{3.1-92}$$

Thus the filtered speed signal at the kth sampling instant, $\omega_{\text{rf}}(k)$, can be obtained from the unfiltered speed signal and also by using the previous sampled value of the filtered speed signal, $\omega_{\text{rf}}(k-1)$, as follows:

$$\omega_{\text{rf}}(k)=\frac{2-cT}{2+cT}\omega_{\text{rf}}(k-1)+\frac{cT}{2+cT}[\omega_{\text{r}}(k)+\omega_{\text{r}}(k+1)]. \tag{3.1-93}$$

It has been emphasized above that if analog integrators are used to obtain the stator flux-linkage components (by integrating the terminal voltages reduced by the ohmic voltage drops), then it is important to eliminate the effects of stator flux-linkage drift. For this purpose eqns (3.1-69) and (3.1-70) can be used. The importance of this can be simply proved by considering that the rotor speed which would result by using the stator flux vector with drift is

$$\omega_{\text{r}}^{\text{drift}}=\frac{\mathrm{d}\rho_{\text{s}}'}{\mathrm{d}t}=\frac{\mathrm{d}\{\tan^{-1}[(\psi_{\text{sQ}}+b)/(\psi_{\text{sD}}+a)]\}}{\mathrm{d}t}, \tag{3.1-94}$$

which is similar to eqn (3.1-82), but now the position ρ_{s}' of $\bar{\psi}_{\text{s}}^{\text{drift}}$ is used, where ρ_{s}' is also shown in Fig. 3.22 above. The drifts a and b are also shown in Fig. 3.22. In the steady-state, the stator flux linkage (without drift) can be expressed as $|\bar{\psi}_{\text{s}}|\exp(\mathrm{j}\omega_1 t)=|\bar{\psi}_{\text{s}}|\exp(\mathrm{j}\rho_{\text{s}}^{\text{steady}})$, thus $\psi_{\text{sD}}=|\bar{\psi}_{\text{s}}|\cos\rho_{\text{s}}^{\text{steady}}$, $\psi_{\text{sQ}}=|\bar{\psi}_{\text{s}}|\sin\rho_{\text{s}}^{\text{steady}}$, and it follows by considering eqn (3.1-94) that

$$\omega_{\text{r}}^{\text{drift}}=\frac{(\psi_{\text{sD}}+a)\dfrac{\mathrm{d}\psi_{\text{sQ}}}{\mathrm{d}t}-(\psi_{\text{sQ}}+b)\dfrac{\mathrm{d}\psi_{\text{sD}}}{\mathrm{d}t}}{(\psi_{\text{sD}}+a)^2+(\psi_{\text{sQ}}+b)^2}$$

$$=\frac{\omega_1(|\bar{\psi}_{\text{s}}|+a\cos\rho_{\text{s}}^{\text{steady}}+b\sin\rho_{\text{s}}^{\text{steady}})}{|\bar{\psi}_{\text{s}}|+(a^2+b^2)/|\psi_{\text{s}}|^2+2(a\cos\rho_{\text{s}}^{\text{steady}}+b\sin\rho_{\text{s}}^{\text{steady}})}. \tag{3.1-95}$$

By introducing per-unit drift values

$$a_{\text{pu}}=\frac{a}{|\bar{\psi}_{\text{s}}|} \tag{3.1-96}$$

$$b_{\text{pu}}=\frac{b}{|\bar{\psi}_{\text{s}}|}, \tag{3.1-97}$$

eqn (3.1-95) can be put into the following form:

$$\omega_{\text{r}}^{\text{drift}}=\frac{\omega_1[1+a_{\text{pu}}\cos\rho_{\text{s}}^{\text{steady}}+b_{\text{pu}}\sin\rho_{\text{s}}^{\text{steady}}]}{1+a_{\text{pu}}^2+b_{\text{pu}}^2+2(a\cos\rho_{\text{s}}^{\text{steady}}+b\sin\rho_{\text{s}}^{\text{steady}})}. \tag{3.1-98}$$

From eqn (3.1-98) the ratio $\omega_{\text{r}}^{\text{drift}}/\omega_1$ can be determined, which is the ratio of the incorrect speed (when drift is present in the flux-linkage estimation) to the actual speed in the steady state. This ratio is plotted in Fig. 3.24, for various values of a_{pu} and b_{pu}, as a function of ρ_s^{steady} (which is the stator flux-linkage position in the steady state). It has been assumed that the flux-linkage drift in the direct and quadrature axes is equal, $a_{pu}=b_{pu}$.

It follows from eqn (3.1-98) that if there is no drift ($a_{pu}=b_{pu}=0$), and $\omega_r^{drift}=\omega_1$ as expected, but in the presence of drift (when $a_{pu}\neq 0$, $b_{pu}\neq 0$), the speed error increases with the drift significantly. This increase can be well visualized in Fig. 3.24 where the largest error in the examined cases occurs when $a_{pu}=b_{pu}=0.4$, and even for $a_{pu}=b_{pu}=0.1$, the error is more than 15%. It is interesting to note that for all the curves shown, the maxima occur at the same flux-linkage position, which is 225°. This is physically due to the fact that the maximum point in the drifted speed signal must be present at a position where the position error is a minimum and the minima are at the same place for all the cases considered (but where $a_{pu}=b_{pu}$). A simple physical proof of this is now presented.

The position of the drifted flux-linkage space vector is ρ_s' and thus the drifted speed is the rate of change of this position, $\omega_r^{drift}=d\rho_s'/dt$, in agreement to that

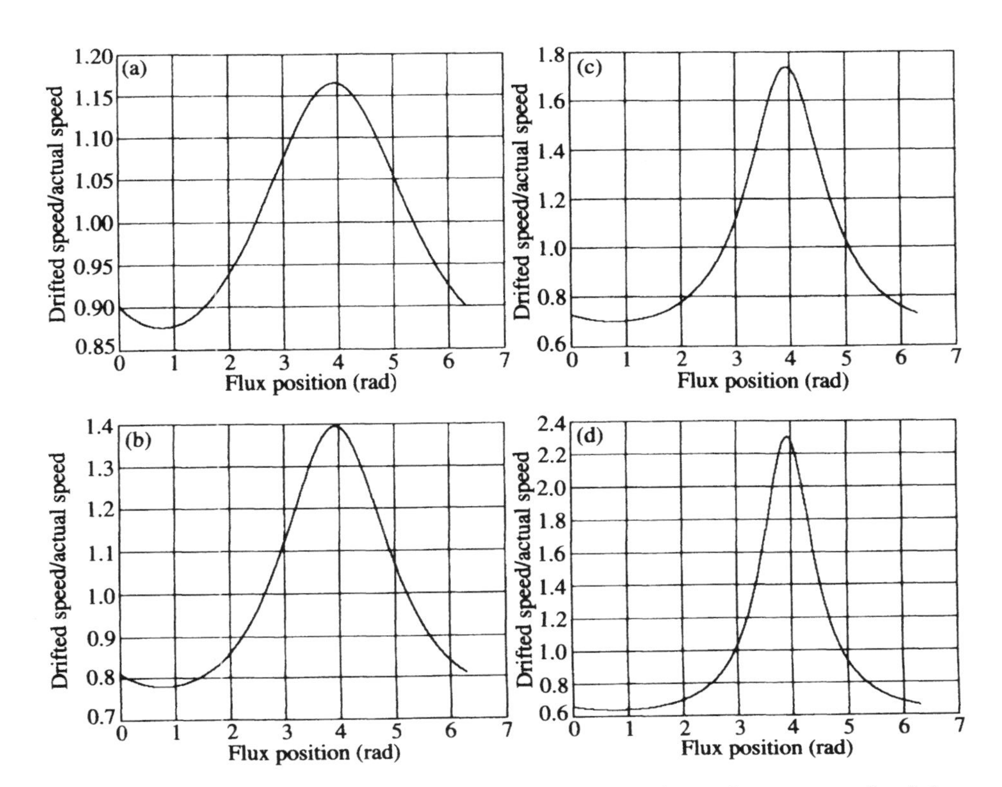

Fig. 3.24. Ratio of incorrect rotor speed to actual speed versus the steady-state stator flux-linkage position. (a) $a_{pu}=b_{pu}=0.1$; (b) $a_{pu}=b_{pu}=0.2$; (c) $a_{pu}=b_{pu}=0.3$; (d) $a_{pu}=b_{pu}=0.4$.

shown above. However, $\rho_s' = \Delta\rho_s + \rho_s$, thus the drifted speed must be maximum, where the position error $\Delta\rho_s$ is minimum. When $a_{pu} = b_{pu} \neq 0$ (this holds for all the cases considered above) the drift vector has a position of 45 ° (see also Fig. 3.22 above), and if the resultant drifted stator flux-linkage vector ($\bar{\psi}_s^{drift}$) is coaxial with this, then the position error is minimal. A minimum position error occurs when the position of the undrifted flux vector is $\rho_s = 45$ ° (since the resultant drifted flux-linkage vector will then have a position angle $\rho_s' = 45$ ° and $\Delta\rho_s = \rho_s' - \rho_s = 0$) or $\rho_s = 180 + 45 = 225$ ° (since the resultant drifted flux-linkage vector will then have a position angle $\rho_s' = 225$ ° and $\Delta\rho_s = \rho_s' - \rho_s = 0$). For better understanding, the drift vector and the drifted stator flux-linkage space vector and the various angles are shown for these two cases in Fig. 3.25. Thus both of these cases result in zero position errors. However, the rate of change of the position error is larger in the neighbourhood of 225 ° than around 45 ° so the 225 ° is the position at which the maximum value of the drifted speed must occur.

A mathematical proof of the two positions (45 ° and 225 °) corresponding to the minimum position error can be obtained by considering that in the steady state, the position error can be expressed as follows (for simplicity no superscript is used for the steady-state value of the stator flux-linkage position):

$$\Delta\rho_s = \rho_s' - \rho_s = \tan^{-1}\left[\frac{\psi_{sQ} + b}{\psi_{sD} + a}\right] - \rho_s$$

$$= \tan^{-1}\left[\frac{|\bar{\psi}_s|\sin\rho_s + b}{|\bar{\psi}_s|\cos\rho_s + a}\right] - \rho_s = \tan^{-1}\left[\frac{\sin\rho_s + b_{pu}}{\cos\rho_s + a_{pu}}\right] - \rho_s \qquad (3.1\text{-}99)$$

It is this expression which can be used to find the zero-crossings of the $\Delta\rho_s(\rho_s)$ function and if $a_{pu} = b_{pu}$, then, as expected, $\rho_s = 45$ ° and 225 ° result. It should be noted that if $a_{pu} \neq b_{pu}$ then the minima of the flux-linkage position error will not be at 45 ° and 225 °.

A similar technique can be used to obtain the positive and negative maximum values of the flux position error. For this purpose eqn (3.1-99) can be used. It can be shown, by differentiation of $\Delta\rho_s$ and then by equating the time-derivative $d(\Delta\rho_s)/dt$ to zero, that these occur where

$$a_{pu}\cos\rho_s + b_{pu}\sin\rho_s + a_{pu}^2 + b_{pu}^2 = 0. \qquad (3.1\text{-}100)$$

For example, for $a_{pu} = b_{pu} = 0.4$, the negative maximum occurs at 166.2 ° and the positive maxima occurs at 280.7 °.

It is also interesting to note that for all the cases shown in Fig. 3.24, the drifted speed is equal to the undrifted speed in two points of the curve. However, these two points must occur at the positions where the position error due to drift is maximal, so e.g. for $a_{pu} = b_{pu} = 0.4$, these must occur at 166.2 ° and 280.7 ° respectively, and it can be seen from Fig. 3.24 ($a_{pu} = b_{pu} = 0.4$) that these are indeed the angles where $\omega_r^{drift}/\omega_1 = 1$. Physically this is due to the fact that the drifted speed is equal to the rate of change of the position of the resultant flux-linkage space vector due to drift. However, the angle of this vector (ρ_s') is the sum of the position error due to drift ($\Delta\rho_s$) and the position of the undrifted flux-linkage

(a)

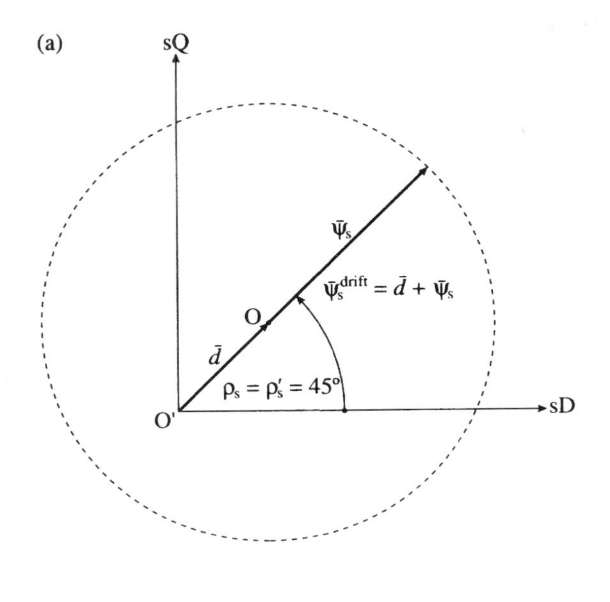

(b)

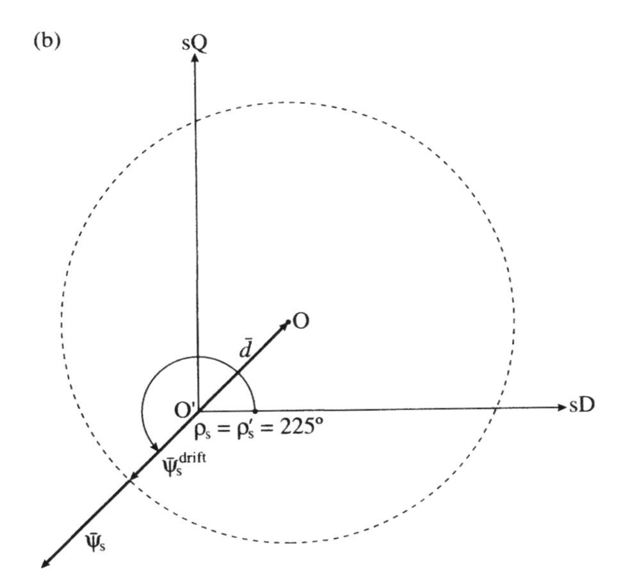

Fig. 3.25. Illustration of minimum position errors for two cases when $a_{pu} = b_{pu} = 0.4$. (a) $\Delta\rho_s = \rho_s' - \rho_s = 0$; $\rho_s = \rho_s' = 45°$; (b) $\Delta\rho_s = \rho_s' - \rho_s = 0$; $\rho_s = \rho_s' = 225°$;

vector (ρ_s). In the steady state, the rate of change of the position of the undrifted vector is ω_1, and thus it follows that the drifted speed can only be equal to ω_1 if the rate of change of the position error is zero. This occurs where the oscillatory position error has its extreme values (for $a_{pu} = b_{pu} = 0.4$, this is at 166.2° and

280.7° respectively). Finally it should also be noted that when the drift increases, the two points will move closer, where $\omega_r^{drift}/\omega_1 = 1$.

3.1.3.3 Vector control using a stator phase third-harmonic voltage-based position estimator

In the present section the monitored stator phase third-harmonic voltage is utilized to obtain indirectly the rotor position of a permanent-magnet synchronous machine. This information can also be used to obtain the rotor flux-linkage space vector. The rotor position can be used to control the switching functions of the inverter which supplies the motor.

For a PMSM with trapezoidal flux distribution (brushless d.c. machine), only two of the stator phases are excited at any instant of time, so that constant current flows into one of the excited windings and out of the other. The stator currents of the machine are square waves with 120 electrical degree conduction periods. The back e.m.f.s (open-circuit stator voltages induced in the stator windings due to the magnets) are approximately trapezoidal, where a back e.m.f. waveform contains two constant parts, each of which is 120 electrical degrees wide. The amplitude of the back e.m.f. is proportional to the speed. To continually synchronize the stator excitation with the m.m.f. wave produced by the magnets, the six inverter switches are switched at every 60 electrical degrees and switching is initiated at the beginning and end of the flat part of the corresponding trapezoidal back e.m.f. waveform. As a result of the synchronization, the developed electromagnetic torque of the machine is proportional to the phase winding current. The polarity of the torque is reversed by reversing the direction of the current flow through the two excited stator windings. Thus the machine can operate as a motor or a generator in both directions of rotation, providing the basis for four-quadrant operation.

It follows that in the drive when a position sensor is not used, the measurement of the back e.m.f.s is required. However, basically all back e.m.f.-based techniques utilize measured phase voltages, which however contain modulation noise (high-frequency noise). This can be eliminated by using a low-pass filter which, however, causes a phase delay. This delay is a function of the speed, and thus it is not possible to achieve optimum control (stator current is in space quadrature to the rotor flux linkage), which affects the torque/ampere capability and efficiency of the motor adversely as the speed changes. Although this problem can be reduced by applying a filter phase-delay compensator (lead-lag network), maximum performance can still not be achieved at low speeds. However, it is possible to extract the rotor position by using the monitored third-harmonic stator voltage due to the trapezoidal shape. When this simple technique is used, lower speed operation can be achieved than with the conventional method of back e.m.f. sensing, since the third harmonic voltage has a frequency three times higher than the fundamental back e.m.f. (but zero speed operation cannot be achieved). The derivation of the inverter switching functions using the monitored third harmonic stator voltage is discussed below.

As shown in Fig. 3.26 the inverter contains six switches (Sl, S2, ..., S6) and the goal is to obtain the switching strategy (six switching functions). Due to the approximately trapezoidal resultant air-gap flux-density distribution, in the stator windings of the brushless d.c. machine with surface-mounted magnets, trapezoidal back e.m.f.s., e_{sA}, e_{sB}, e_{sC} (open-circuit stator voltages due the magnets) are induced. The idealized back e.m.f. waveforms are shown in Fig. 3.27. However, a trapezoidal e.m.f. contains the fundamental component plus higher-order harmonics, and in particular it contains a dominant third-harmonic component. The harmonic content depends only on the magnets and the stator winding configuration, and saturation does not influence the back e.m.f.s. The stator pole-pitch and rotor pitch have to be different to 2/3, otherwise no third-harmonic voltages are induced in the stator windings.

When the three stator phase voltages are added, the obtained zero-sequence stator voltage will contain a dominant third-harmonic component (u_{s3}) and also high frequency zero-sequence components (u_{sh}),

$$u_{s0}=u_{sA}+u_{sB}+u_{sC}=u_{s3}+u_{sh}. \tag{3.1-101}$$

The third-harmonic voltage component keeps a constant phase relationship with the fundamental air-gap voltage for all speed and load conditions. It is practically free of inverter modulation noise, and only a small amount of filtering is required to eliminate the switching frequency and its side bands. The main disadvantage is that it has a relatively low value at low speed. In a star-connected machine

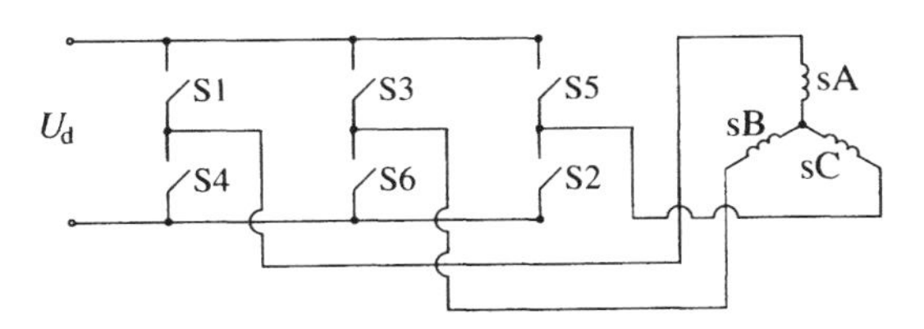

Fig. 3.26. Schematic of inverter supplyig the brushless d.c. machine.

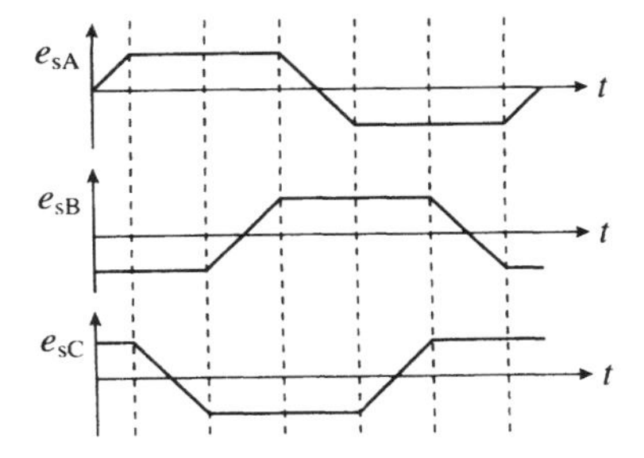

Fig. 3.27. Idealized trapezoidal e.m.f.s in a brushless d.c. machine.

where zero-sequence stator currents cannot flow, the third-harmonic voltage is an air-gap voltage component, and thus it is equal to the rate of change of the rotor flux. It follows that integration of the third-harmonic voltage gives the third-harmonic rotor flux

$$\psi_{r3}=\int u_{s3}\,dt. \tag{3.1-102}$$

It follows that zero-crossings of the third-harmonic rotor flux occur at 60 electrical degrees, which are exactly the desired switching instants (current commutations instants). Thus the third-harmonic rotor flux (integrated third-harmonic stator voltage) is input to a zero-crossing detector and the output of the zero-crossing detector determines the switching functions. To get maximum torque/ampere, the stator-current space vector is kept at 90 electrical degrees with respect to the rotor flux ψ_r. The complete derivation of the switching functions is shown in Fig. 3.28, where for completeness the time variations of e_{sA}, e_{sB}, e_{sC} are also shown in addition to u_{s3}, ψ_{r3}, rotor flux ψ_r, stator currents i_{sA}, i_{sB}, i_{sC} and the six switching functions. It can be seen that, as discussed above, the back e.m.f. waveforms are trapezoidal and contain two constant parts, each of which is 120 degrees wide. The stator currents are square waves with the appropriate 120 degree conduction periods.

It follows from Fig. 3.28 that for the implementation of the required switching functions, it is necessary to know the positive zero-crossing of the back e.m.f. in stator phase sA, and for this purpose the zero-crossing of one of the stator phase voltages has to be detected. The zero-crossing of the back e.m.f. in stator phase sA is possible, since at that instant the stator phase current i_{sA} is zero, and thus $e_{sA}=u_{sA}$. When the zero-crossing of e_{sA} is detected, the control algorithm waits for the next zero-crossing of the integrated third-harmonic stator voltage (third-harmonic rotor flux), in order to turn on the current i_{sA} and to turn off the current i_{sC}. The six zero-crossings of the third-harmonic rotor flux are numbered in Fig. 3.28 (by 1, 2, ..., 6). In the second zero-crossing of the third-harmonic rotor flux, stator current i_{sB} is turned off, and i_{sC} is turned on. At the third zero-crossing i_{sA} is turned off and i_{sB} is turned on, at the fourth zero-crossing i_{sA} is turned on, i_{sC} is turned off. In the fifth zero-crossing i_{sC} is ON, i_{sB} is OFF, and it can be seen from Fig. 3.28 that in the sixth zero-crossing i_{sB}, is ON, i_{sA} is OFF. This completes the entire period.

To obtain the zero-sequence stator voltage, it is necessary to know the stator phase voltages. This requires access to the neutral point of the stator phases. In this case the zero-sequence voltage can be obtained by adding the phase voltages of the stator (e.g. by using potential transformers, or by using operational amplifiers) and then u_{s3} can be obtained by eliminating the high-frequency voltage component (by using an analog or digital filter). When three potential transformers are used, their secondary windings are connected in series as shown in Fig. 3.29.

However, it is also possible to monitor the zero-sequence voltage by the scheme shown in Fig. 3.30, where three identical resistors are connected in star, and this

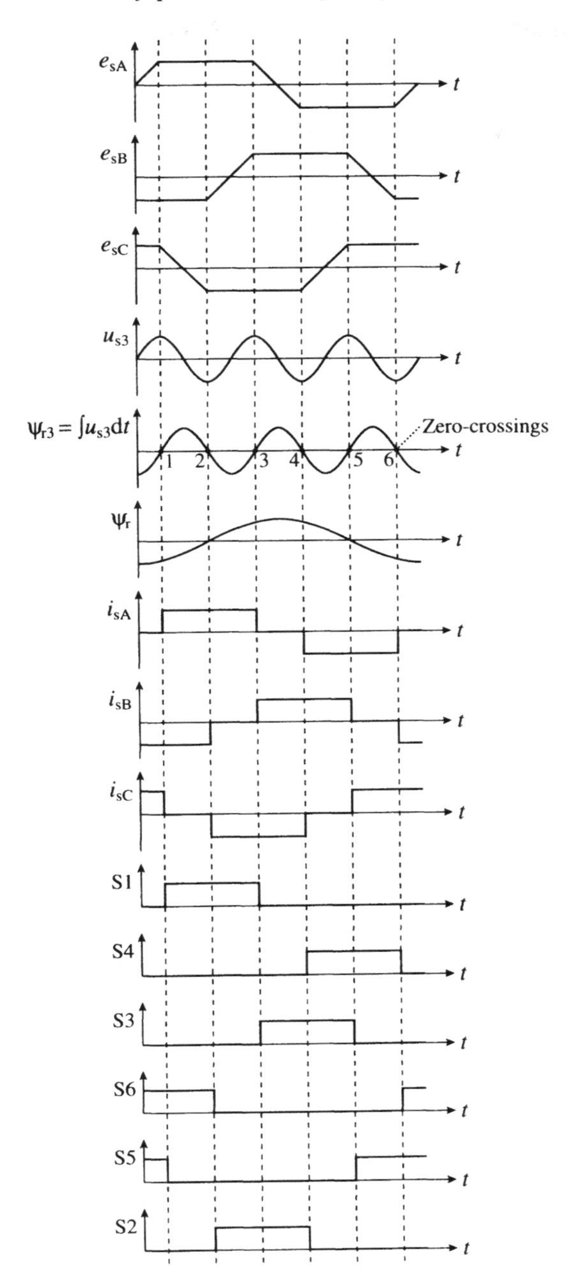

Fig. 3.28. Time variation back e.m.f.s, u_{s3}, ψ_{r3}, ψ_r, i_{sA}, i_{sB}, i_{sC} and the switching functions.

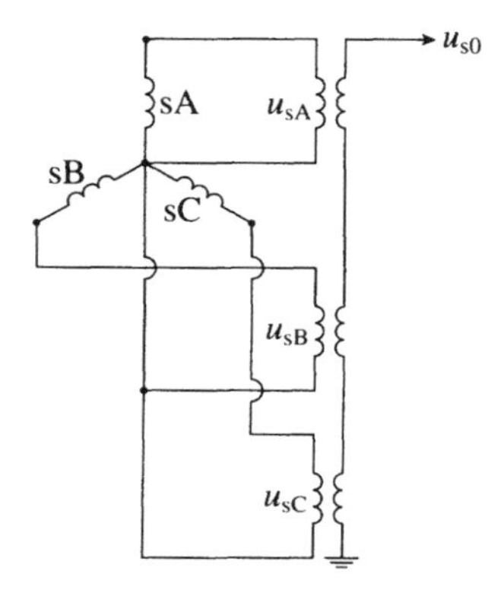

Fig. 3.29. Zero-sequence stator voltage monitoring using three potential transformers.

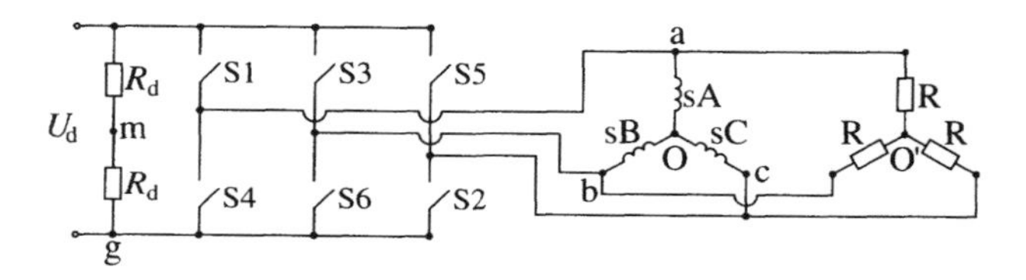

Fig. 3.30. Zero-sequence stator voltage monitoring using three identical external resistors.

arrangement is connected in parallel with the motor windings. The zero-sequence stator voltage is obtained across the neutral point of the stator winding (point 0) and the pseudo-neutral point of the three external resistors (0′). This must then be isolated. This can be simply proved by using the following three voltage equations which can be obtained by inspection of the appropriate loops in Fig. 3.30:

$$u_{sA}+u_{00'}+u_{0'a}=0 \tag{3.1-103}$$

$$u_{sB}+u_{00'}+u_{0'b}=0 \tag{3.1-104}$$

$$u_{sC}+u_{00'}+u_{0'c}=0. \tag{3.1-105}$$

By adding these three equations, and considering that the sum of the three voltages across the star-connected resistor network is zero ($u_{0'a}+u_{0'b}+u_{0'c}=0$),

$$u_{sA}+u_{sB}+u_{sC}=-3u_{00'} \tag{3.1-106}$$

is obtained. Substituting eqn (3.1-101) into eqn (3.1-106) and considering that $u_{0'0}=-u_{00'}$, finally

$$u_{0'0}=\tfrac{1}{3}(u_{s3}+u_h) \tag{3.1-107}$$

is obtained. Thus it can be seen that the voltage between the two neutral points can be used directly to obtain $(1/3)(u_{s3}+u_h)$, from which u_{s3} can be obtained by filtering out the voltage u_h. When this technique is used, it is an advantage that the required third-harmonic stator voltage can be obtained without having to add the three stator voltages.

It is also possible to use an alternative monitoring scheme to obtain the third-harmonic stator voltage, without the need to access the neutral point of the stator windings. This is now discussed. For this purpose the voltage between the neutral of the resistor network and the mid-point reference (point m) of the d.c. bus can be used. This mid-point is also shown in Fig. 3.30, between the two resistors R_d, which are connected across the d.c. rails of the d.c. link. It will be shown below that when only two switches of the inverter are conducting at one time (e.g. S1 and S2; or S2 and S3; or S3 and S4; or S5 and S6; or S6 and S1) then this voltage can be expressed as

$$u_{m0'}=-\tfrac{1}{3}(u_{s3}+u_h). \tag{3.1-108}$$

During commutation, when two switches are connected to the positive rail of the d.c. bus (e.g. when the switching changes from S1 and S2 to S2 and S3) then

$$u_{m0'}=-\tfrac{1}{3}(u_{s3}+u_h)-\frac{U_d}{6}. \tag{3.1-109}$$

However, during commutation when two switches are connected to the negative rail of the d.c. bus,

$$u_{m0'}=-\tfrac{1}{3}(u_{s3}+u_h)+\frac{U_d}{6}, \tag{3.1-110}$$

where U_d is the d.c. link voltage. It follows from eqns (3.1-108)–(3.1-110) that the voltage between the neutral of the resistor network and the d.c. bus mid-point can be used to obtain the third-harmonic stator voltage, and the high-frequency component u_h can be eliminated by a filter, but during commutation, on the third-harmonic voltage there is a superimposed a.c. voltage, which varies from $-U_d$ to $+U_d$. This superimposed voltage is of three times the fundamental frequency, since six commutations occur in any given period of the fundamental inverter output voltage. The superimposed voltage introduces notches in the third-harmonic voltage and they are at the same frequency as the third-harmonic component for the case when the motor is driven by a six-stepped (not PWM) waveform. The presence of the third-harmonic component is not so clear when PWM is used, because a commutation notch of $\pm U_d$ is generated at the PWM frequency.

Finally a proof of eqns (3.1-108)–(3.1-110) is now given. It follows from Fig. 3.30 that

$$u_{ag}+u_{g0'}+u_{0'0}+u_{0a}=0 \tag{3.1-111}$$

$$u_{bg}+u_{g0'}+u_{0'0}+u_{0b}=0 \tag{3.1-112}$$

$$u_{cg}+u_{g0'}+u_{0'0}+u_{0c}=0, \tag{3.1-113}$$

where g denotes the negative rail of the d.c. bus. If only two switches of the inverter conduct, e.g. S1 and S2, then $u_{ag}=U_d$, $u_{cg}=0$, $u_{0c}=U_d/2$, $u_{0a}=-U_d/2$ and it follows from eqn (3.1-113) that $u_{g0'}=-u_{0'0}-U_d/2$. The same expression holds for all the other cases when only two switches conduct. However, the d.c. term in $u_{g0'}$ can be eliminated if instead of point g, the mid-point m is used and in this case by using eqn (3.1-107) for $u_{0'0}$ given above,

$$u_{m0'}=-u_{0'0}=-\tfrac{1}{3}(u_{s3}+u_h) \qquad (3.1\text{-}114)$$

is obtained which agrees with eqn (3.1-108).

The proof of eqn (3.1-109) is as follows. During commutation, when e.g. the switching sequence changes from S1 and S2 to S2 and S3, the following conditions hold: $u_{ag}=U_d$, $u_{bg}=U_d$, $u_{cg}=0$. When these are substituted into eqns (3.1-111)–(3.1-113) and by using the mid-point m instead of the point g [similarly to the operation performed for the derivation of eqn (3.1-114)], then

$$u_{m0'}=-\tfrac{1}{3}(u_{s3}+u_h)-\frac{U_d}{6} \qquad (3.1\text{-}115)$$

is obtained, which agrees with eqn (3.1-109). The proof of eqn (3.1-110) is similar, and therefore is not discussed further.

3.1.3.4 Vector control using back e.m.f.-based position estimators

It has also been discussed in Section 3.1.3.2 that in a PMSM the magnitude of a back e.m.f. (open-circuit voltage induced in a stator winding due to the magnets) is position dependent. Thus if this can be accurately monitored, the rotor position can be accurately determined in real-time and this can be used to control the switching pattern of the inverter. Although not all the back e.m.f.-based techniques can be used for high-dynamic-performance applications, they are briefly reviewed here for better understanding and to help the reader when developing a position-sensorless drive system. The four main methods are as follows:

1. The zero-crossing method, where the instant (switching point) is detected at which the back e.m.f. of an unexcited stator phase crosses zero or reaches a predetermined level (this is the simplest technique, but is only suitable for steady-state operation).
2. The phase-locked loop method, where the position signals are locked on to the back e.m.f. in the unexcited stator phase during each sixty-degree interval (there is automatic adjustment to inverter switching instants to changes in rotor speed).
3. The back e.m.f. integration method, where a switching pulse is obtained when the absolute value of the integrated back e.m.f. reaches a pre-set threshold value (integrator results in reduced sensitivity to high-frequency modulation noise caused by the inverter and also provides automatic adjustment to inverter switching instants to changes in rotor speed).
4. The indirect estimation of the back e.m.f.s by detecting the conduction interval of free-wheeling diodes connected in anti-parallel with the power transistors of

the inverter (allows detection of position at very low speeds as well, but not zero speed).

However, all of these methods are problematic at low speeds. Since the zero-crossing method is the simplest, it will now be first discussed, although it cannot be used in high-dynamic-performance applications. This will be followed by the discussion of the back e.m.f. integration method and finally the indirect estimation of back e.m.f.s will be described briefly, a method which uses the detected conduction interval of the free-wheeling diodes.

3.1.3.4.1 *Application of the zero-crossing method*

The zero-crossing method can be used to obtain equivalent rotor position information in an inverter-fed brushless d.c. motor (PMSM with trapesoidal m.m.f.) drive. The zero-crossing method is based on detecting the instant (switching point) where the back e.m.f. in the unexcited phase crosses zero. This zero-crossing is then used to obtain the switching sequence. The PMSM with trapesoidal m.m.f. distribution is an ideal candidate for this technique since, as discussed above, only two of the three stator phases are excited at any time instant, and thus the back e.m.f. can be conveniently measured.

The zero-crossing method can be simply proved by considering that if the inverter is connected to a brushless d.c. machine with star-connected stator windings, as shown in Fig. 3.31, then at the instants of the zero-crossings of a back e.m.f., the corresponding terminal voltage is equal to the neutral voltage. In Fig. 3.31 there is also shown the equivalent circuit of the machine when switches S3 and S2 are ON.

For example, it follows from Fig. 3.31 that $u_a = u_N + e_{sA}$, where u_a is the terminal voltage of stator phase sA and u_N is the neutral voltage of the machine with respect to the negative d.c. bus. Due to the fact that S3 and S2 are ON, the terminal voltages u_b and u_c can be expressed as $u_b = U_d$ and $u_c = 0$, as shown in the equivalent circuit of Fig. 3.31, and physically the unexcited stator phase winding sA sees an infinite impedance and conducts no current despite the continued presence of its back e.m.f. e_{sA} voltage source, which is therefore also shown in the equivalent circuit. Thus at the instant of the zero-crossing of the back e.m.f. e_{sA}, $u_a = u_N$ as expected. Furthermore it will now be shown below that $u_N = U_d/2$, where U_d is the d.c. link voltage, and therefore $u_a = u_N = U_d/2$. Figure 3.32 shows

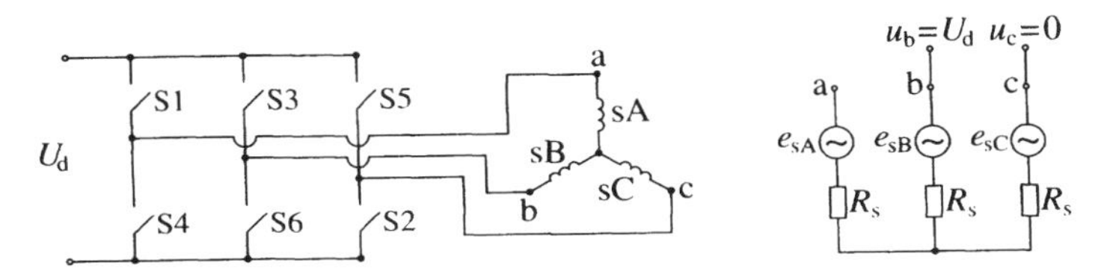

Fig. 3.31. Inverter-fed machine and the equivalent circuit when switches S3 and S2 are ON.

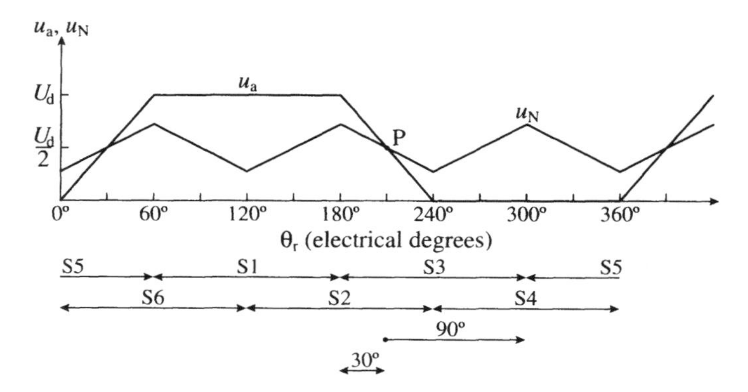

Fig. 3.32. Detection of switching point P from the crossing of the neutral voltage and terminal voltage.

the variation of the neutral voltage and terminal voltage u_a, which are used to detect the switching point P (at the position of 210°). Since $u_a = U_N$, the intersection of the two voltage curves gives point P (when $e_{sA} = 0$). The position defined by point P is independent of the load current. In a practical implementation the required intersection is performed by a comparator and the neutral voltage is obtained from the terminal voltages by connecting three equal resistors in star, and by connecting the thus-obtained network in parallel with the stator windings (see also Section 3.1.3.3). However, since the terminal voltages contain higher-frequency components due to the PWM inverter, the terminal voltages are filtered (by using low-pass filters).

By also considering Fig. 3.28 in Section 3.1.3.3, it can be seen that to use the switching point P (which is at position 210°) to obtain the switching signals, this point has to be phase shifted by 30° (e.g. it also follows from Fig. 3.28 that if this point is shifted 30° to the left (to the 180° position), this is where the 120°-wide conduction period of switch S1 ends or the 120° conduction period of switch S3 begins, etc.). The conduction of the six switches is also shown in Fig. 3.32. In a practical implementation, to achieve the 30° shift, the terminal voltages are first converted into triangular waveforms, where the peak of the triangle is at point P, and these triangles are then intersected (compared) with the neutral voltage.

The proof of $u_N = U_d/2$ is as follows. By inspection of the equivalent circuit in Fig. 3.31, $u_N = U_d - e_{sB} - R_s i_{sB}$, and $U_d = 2R_s i_{sB} + e_{sB} - e_{sC}$. When an expression for i_{sB} is obtained from the second equation, and is substituted into the first equation, $u_N = U_d/2 - (1/2)(e_{sB} + e_{sC})$ is obtained, but when $e_{sA} = 0$, the equality $e_{sB} + e_{sC} = 0$ holds (since the back e.m.f.s are symmetrical, and therefore $e_{sA} + e_{sB} + e_{sC} = 0$) and thus $u_N = U_d/2$ is obtained.

Since the back e.m.f.s are zero at zero speed, the technique cannot be used at zero speed, and the machine is not self-starting (see also the first part in Section 3.1.3). Ideal applications for such a system are steady-state applications. Since the modulation noise is eliminated by using low-pass filters (as discussed above), this can cause a phase delay which is a function of the speed, and thus it is not possible to achieve optimum control (the stator current is in space quadrature to the rotor flux), which affects the torque/ampere capability and efficiency of the motor adversely as the speed changes. Although this problem can be reduced by applying a filter phase-delay compensator (lead-lag network), maximum performance can still not be achieved at low speeds.

3.1.3.4.2 Application of the back e.m.f. integration method

The back e.m.f. integration technique can be used for the detection of the rotor position of the brushless d.c. machine, which is supplied by an inverter. When this technique is used, first the phase-to-neutral voltage of the unexcited stator winding of the brushless d.c. machine is selected. This is equal to the desired back e.m.f. required for position sensing as soon as the residual inductive current flowing in the unexcited winding, immediately following the removal of excitation, decays to zero. The modulus of this back e.m.f. is integrated as soon as the back e.m.f. crosses zero. The integrated voltage (u_i) contains pulses whose shapes can be described by $u_i = E_0 t^2/(2k)$, since the instantaneous back e.m.f. varies approximately linearly with time in the neighbourhood of zero-crossing, so $e_s(t) = E_0 t$, where E_0 is the amplitude of the back e.m.f., and thus its integrated value becomes $E_0 t^2/(2k)$, where k is the gain of the integrator. When the integrated voltage reaches a pre-set threshold value, the next switching instant occurs.

The integrator has also the advantage of reducing the switching noises. Furthermore, since the back e.m.f. amplitude is proportional to the speed, the conduction intervals automatically scale inversely with the speed, so there is an automatic adjustment of the inverter switching instants to changes in the speed. The values of the threshold voltage and integrator gain depend on the motor and also on the alignment of the phase-current excitation waveform with the back e.m.f. Varying the threshold value or integrator gain has the effect of varying the current–voltage alignment. A reduction of the gain can be used to change this alignment and to force faster current build-up to develop extra torque at high speeds.

Similarly to other back-e.m.f-based control schemes, accuracy problems arise at low speed and in addition the machine is not self-starting. Control of the motor currents is important in a high-dynamic-performance drive, since this provides the basis for instantaneous torque control. It is possible to utilize the back e.m.f. integration method in a drive to have high quality current and instantaneous torque control in all operating modes [Becerra *et al.* 1991].

3.1.3.4.3 Application of the indirect estimation of the back e.m.f.s by detecting the conduction interval of free-wheeling diodes

Indirect position estimation of the back e.m.f.s can be based on detecting the conduction state of the free-wheeling diodes in the unexcited stator phase of a brushless d.c. machine. The approach can be used over a wide speed-range, even at very low speeds (but not zero speed). The free-wheeling diodes are connected in anti-parallel with the power transistors, and provide a current flow path to dissipate inductive energy stored in a winding.

In an inverter-fed brushless d.c. motor drive, the inverter-gate drive signals are chopped during each 120 electrical degree operation (conduction interval is 120 electrical degrees). Thus only two transistors, i.e. a positive-side transistor in one phase and a negative-side transistor in another phase, are ON, at one particular instant. The other phase, in which no active drive signal is given to the positive-side and negative-side transistors, is called the open-phase.

The open-phase current under chopped operation results from the back e.m.f.s. produced in the stator windings. To continue producing the maximum torque, the inverter switching has to be performed every 60 electrical degrees, so the rectangular motor line current is in phase with the back e.m.f. The position information is obtained at every 60 degrees by detecting whether the free-wheeling diodes are conducting, or not. Since the detected position signal leads the next switching by 30 electrical degrees, the switching signals of the inverter are shifted by a phase-shifter.

Since the amplitude of the back e.m.f. is proportional to the rotor speed, similarly to the other back-e.m.f.-based techniques, the motor is not self-starting and a special starting technique must be applied. On the other hand, it is an advantage that it allows the detection of the position at very low speeds as well, but the technique requires the inverter to operate in the chopping mode [Ogasawara and Akagi 1991].

3.1.3.5 Vector control using observer-based speed and position estimators

By using monitored stator voltages and currents and Kalman or Luenberger observers, it is possible to implement high-dynamic-performance PMSM drives without position and speed sensors. In the present section, the extended Kalman filter will be used for this purpose.

3.1.3.5.1 Extended Kalman filter application: general introduction

An Extended Kalman Filter (EKF) is a recursive optimum-state estimator which can be used for the joint state and parameter estimation of a non-linear dynamic system in real-time by using noisy monitored signals that are distributed by random noise. This assumes that the measurement noise and disturbance noise are uncorrelated. The noise sources take account of measurement and modelling inaccuracies. In a first stage of the calculations, the states are predicted by using a mathematical model (which contains previous estimates) and in the second stage, the predicted states are continuously corrected by using a feedback correction scheme. This scheme makes use of actual measured states, by adding a term to the predicted states (which is obtained in the first stage). The additional term contains the weighted difference of the measured and estimated output signals. Based on the deviation from the estimated value, the EKF provides an optimum output value at the next input instant. In a PMSM drive the EKF can be used for the real-time estimation of the rotor position and speed. This is possible since a mathematical dynamical model of the machine is sufficiently well known. For this purpose the stator voltages and currents are measured and for example, the speed and position of the machine can be obtained by the EKF quickly and accurately.

To be more specific, the goal of the Kalman filter is to obtain unmeasurable states (e.g. speed and rotor position) by using measured states, and also statistics of the noise and measurements (i.e. covariance matrices $\mathbf{Q}$, $\mathbf{R}$, $\mathbf{P}$ of the system noise vector, measurement noise vector, and system state vector ($\mathbf{x}$) respectively). In general, by means of the noise inputs, it is possible to take account of computational inaccuracies, modelling errors, and errors in the measurements. The filter estimation ($\hat{\mathbf{x}}$) is obtained from the predicted values of the states ($\mathbf{x}$) and this is corrected recursively by using a correction term, which is the product of the Kalman gain ($\mathbf{K}$) and the deviation of the estimated measurement output vector and the actual output vector ($\mathbf{y}-\hat{\mathbf{y}}$). The Kalman gain is chosen to result in the best possible estimated states.

Thus the filter algorithm contains basically two main stages, a prediction stage and a filtering stage. During the prediction stage, the next predicted values of the states $\mathbf{x}(k+1)$ are obtained by using a mathematical model (state-variable equations) and also the previous values of the estimated states. Furthermore, the predicted state covariance matrix ($\mathbf{P}$) is also obtained before the new measurements are made and for this purpose the mathematical model and also the covariance matrix of the system ($\mathbf{Q}$) are used. In the second stage, which is the filtering stage, the next estimated states, $\hat{\mathbf{x}}(k+1)$, are obtained from the predicted estimates $\mathbf{x}(k+1)$ by adding a correction term $\mathbf{K}(\mathbf{y}-\hat{\mathbf{y}})$ to the predicted value. This correction term is a weighted difference between the actual output vector ($\mathbf{y}$) and the predicted output vector ($\hat{\mathbf{y}}$), where $\mathbf{K}$ is the Kalman gain. Thus the predicted state estimate (and also its covariance matrix) is corrected through a feedback correction scheme that makes use of the actual measured quantities. The Kalman gain is chosen to minimize the estimation error variances of the states to be estimated. The computations are realized by using recursive relations.

The algorithm is computationally intensive, and the accuracy also depends on the model parameters used. A critical part of the design is to use correct initial values for the various covariance matrices. These can be obtained by considering the stochastic properties of the corresponding noises. Since these are usually not known, in most cases they are used as weight matrices, but it should be noted that sometimes simple qualitative rules can be set up for obtaining the covariances in the noise vectors (see also Section 3.1.3.5.2). With advances in DSP technology, it is possible to implement an EKF conveniently in real time.

3.1.3.5.2 Application of an EKF to a PMSM with sinusoidal flux distribution

The main design steps for a speed- and position-sensorless PMSM drive implementation using a discretized EKF algorithm are as follows:

1. Selection of the time-domain machine model;
2. Discretization of the machine model;

3. Determination of the noise and state covariance matrices **Q**, **R**, **P**;
4. Implementation of the discretized EKF algorithm; tuning.

These main design steps are now discussed.

1. Selection of the time-domain machine model It is possible to have EKF implementations using time-domain machine models expressed in the stationary reference frame, or expressed in the rotor reference frame. Obviously the selection of the reference frame has an important effect on the execution time, which is a crucial factor, especially in view of the computational intensity of the EKF filter. First, the EKF using equations expressed in the rotor reference frame is discussed, and this will form a firm basis for comparison with the other solution, which uses the machine model obtained in the stationary reference frame.

The voltage equations of the PMSM in the rotor reference frame are obtained as follows. In the stationary reference frame the stator voltage equation is similar to that of the induction machine and

$$\bar{u}_s = R_s \bar{i}_s + \frac{d\bar{\psi}_s}{dt}, \tag{3.1-116}$$

where for the PMSM with surface-mounted magnets the stator flux-linkage space vector can be defined as

$$\bar{\psi}_s = L_s \bar{i}_s + \psi_F \exp(j\theta_r). \tag{3.1-117}$$

In eqns (3.1-116) and (3.1-117) R_s and L_s are the stator resistance and stator inductance respectively, ψ_F is the magnet flux [also defined in eqn (3.1-2)], and θ_r is the rotor position. The stator voltage equation in the rotor reference frame is obtained by considering that the transformed stator voltage, stator current, and stator flux-linkage space vectors are defined as $\bar{u}_s' = \bar{u}_s \exp(-j\theta_r)$, $\bar{i}_s' = \bar{i}_s \exp(-j\theta_r)$, and $\bar{\psi}_s' = \bar{\psi}_s \exp(-j\theta_r)$, and thus by considering eqns (3.1-116) and (3.1-117),

$$\bar{u}_s' = R_s \bar{i}_s' + \frac{d\bar{\psi}_s'}{dt} + j\omega_r \bar{\psi}_s' \tag{3.1-118}$$

is obtained. In eqn (3.1-118)

$$\bar{\psi}_s' = L_s \bar{i}_s' + \psi_F \tag{3.1-119}$$

is the stator flux-linkage space vector in the rotor reference frame. Resolution of eqns (3.1-118) and (3.1-119) into their real- and imaginary-axis components gives

$$u_{sd} = R_s i_{sd} + \frac{d\psi_{sd}}{dt} - \omega_r \psi_{sq} \tag{3.1-120}$$

$$u_{sq} = R_s i_{sq} + \frac{d\psi_{sq}}{dt} + \omega_r \psi_{sd}, \tag{3.1-121}$$

where

$$\psi_{sd}=L_s i_{sd}+\psi_F \quad (3.1\text{-}122)$$

$$\psi_{sq}=L_s i_{sq}. \quad (3.1\text{-}123)$$

In these equations, i_{sd} and i_{sq} are the stator current components in the rotor reference frame. To obtain the state-space model of the PMSM, the voltage equations are arranged into their state-variable form, where i_{sd} and i_{sq} are used as state-variables, but these are augmented with the quantities which have to be estimated, i.e. ω_r and θ_r. The extra two equations required are $\omega_r=d\theta_r/dt$, and by assuming that the rotor speed derivative is negligible, $d\omega_r/dt=0$. This last equation corresponds to infinite inertia, which in practice is not true, but the required correction is performed by the Kalman filter. For the derivation of the state-variable equations, it will be assumed that the effects of saturation of the magnetic paths of the machine can be neglected, thus the stator inductance is assumed to be constant. This assumption is justified, since it can be shown that the EKF is not sensitive to changes in the stator inductance, since any changes in the stator parameters are compensated by the current loop inherent in the EKF. Thus from eqns (3.1-120)–(3.1-123) and the extra two equations given above, and also considering the component equations arising from $\bar{u}_s'=u_{sd}+ju_{sq}=\bar{u}_s\exp(-j\theta_r)$ $=(u_{sD}+ju_{sQ})(\cos\theta_r-j\sin\theta_r)$ and $\bar{i}_s'=i_{sd}+ji_{sq}=\bar{i}_s\exp(-j\theta_r)=(i_{sD}+ji_{sQ})(\cos\theta_r-j\sin\theta_r)$, finally

$$\dot{\mathbf{x}}=\mathbf{A}\mathbf{x}+\mathbf{B}\mathbf{u} \quad (3.1\text{-}124)$$

is obtained, and the output vector is

$$\mathbf{y}=\mathbf{C}\mathbf{x}. \quad (3.1\text{-}125)$$

In these equations, $\mathbf{x}$ is the state vector, $\mathbf{x}=[i_{sd},i_{sq},\omega_r,\theta_r]^T$, $\mathbf{y}$ is the output vector, $\mathbf{y}=[i_{sD},i_{sQ}]^T$, $\mathbf{u}$ is the input vector, $\mathbf{u}=[u_{sD},u_{sQ},u_p]^T$, where u_p is the voltage induced in the stator windings by the magnet flux, $u_p=\omega_r\psi_F$ and in the EKF, rated rotor speed is used in the expression of the induced voltage (thus u_p=constant in the EKF). Furthermore, in eqns (3.1-124) and (3.1-125) **A** is the state matrix, **B** is the input matrix, and **C** is the transformation matrix,

$$\mathbf{A}=\begin{bmatrix}-R_s/L_s & \omega_r & 0 & 0\\ -\omega_r & -R_s/L_s & 0 & 0\\ 0 & 0 & 0 & 0\\ 0 & 0 & \omega_r & 0\end{bmatrix} \quad (3.1\text{-}126)$$

$$\mathbf{B}=\begin{bmatrix}\cos\theta_r/L_s & \sin\theta_r/L_s & 0\\ -\sin\theta_r/L_s & \cos\theta_r/L_s & -1/L_s\\ 0 & 0 & 0\\ 0 & 0 & 0\end{bmatrix} \quad (3.1\text{-}127)$$

$$\mathbf{C}=\begin{bmatrix}\cos\theta_r & -\sin\theta_r & 0 & 0\\ \sin\theta_r & \cos\theta_r & 0 & 0\end{bmatrix} \tag{3.1-128}$$

Equations (3.1-124) and (3.1-125) are non-linear. It can be seen from eqn (3.1-128) that $\mathbf{C}$ contains a sub-matrix that transforms the two-axis stator current components in the stationary reference frame into the two-axis stator current components of the rotor reference frame. Equations (3.1-124) and (3.1-125) describe the time-domain model of the PMSM and can be visualized by the block-diagram shown Fig. 3.33.

It should be noted that so far ω_r and θ_r have been treated as known quantities, although they have to be estimated. To allow a digital implementation of the Extended Kalman Filter, the time-domain equations of the PMSM will now be discretized.

2. Time-discrete state-space model of the PMSM The time-discrete state-space model of the PMSM can be obtained from eqns (3.1-124) and (3.1-125) as follows

$$\mathbf{x}(k+1)=\mathbf{A}_d\mathbf{x}(k)+\mathbf{B}_d(k)\mathbf{u}(k) \tag{3.1-129}$$

$$\mathbf{y}(k)=\mathbf{C}_d(k)\mathbf{x}(k). \tag{3.1-130}$$

It should be noted that for the sampled value of $\mathbf{x}$ in the t_k instant, the notation $\mathbf{x}(t_k)$ could be used, or the corresponding more simple $\mathbf{x}(k)$, which strictly means $\mathbf{x}(kT)$, which corresponds to sampling at the kth instant, and the sampling time is $T(T=t_{k+1}-t_k)$. In eqns (3.1-129) and (3.1-130) $\mathbf{A}_d$ and $\mathbf{B}_d$ are the discretized system matrix and input matrix respectively,

$$\mathbf{A}_d=\exp[\mathbf{A}(k)]\approx\mathbf{I}+\mathbf{A}(k)T \tag{3.1-131}$$

$$\mathbf{B}_d=\mathbf{B}(k)T, \tag{3.1-132}$$

where $\mathbf{I}$ is an identity matrix and $\mathbf{C}_d$ is the discrete transformation matrix, and is obtained from eqn (3.1-129) as

$$\mathbf{C}_d=\mathbf{C}(k). \tag{3.1-133}$$

It should be noted that when eqns (3.1-131) and (3.1-132) are used, they require very short sampling times to give a stable and accurate discretized model. However, a better approximation is obtained with a second-order series expansion, and in this case

$$\mathbf{A}_d=\exp[\mathbf{A}(k)]\approx\mathbf{I}+\mathbf{A}(k)T+[\mathbf{A}(k)T]^2 \tag{3.1-134}$$

$$\mathbf{B}_d\approx\mathbf{B}(k)T+\mathbf{A}(k)\mathbf{B}(k)\frac{T^2}{2} \tag{3.1-135}$$

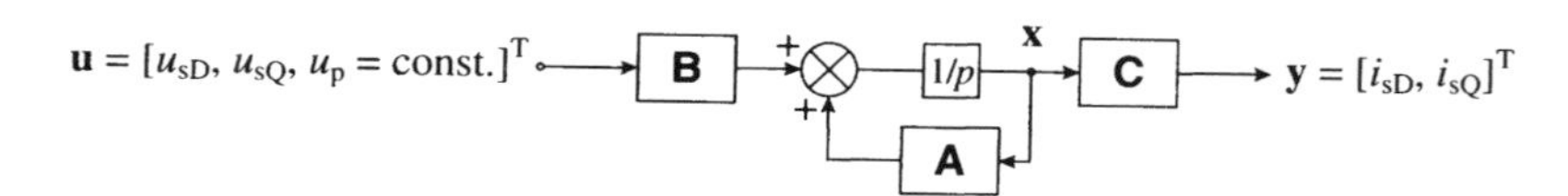

Fig. 3.33. Block-diagram of the time-domain state-space model of the PMSM machine.

In general, to achieve adequate accuracy, the sampling time should be appreciably smaller than the characteristic time-constants of the machine. The final choice for this should be based on obtaining adequate execution time of the full EKF algorithm, satisfactory accuracy, and stability. The second-order technique obviously increases the computational time.

By using eqns (3.1-129) and (3.1-130), the time-discrete state-space model of the PMSM is obtained as shown in Fig. 3.34. The obtained time-discrete model is a fourth-order model, it is non-linear, and is time variant. This model is used in the EKF algorithm for the estimation of the state vector, if the stator voltages and currents are monitored (and sampled).

The discrete EKF utilizes the state-variable equations of the time-discrete model of the PMSM, given by eqns (3.1-129) and (3.1-130). However, the state vector is disturbed by the noise vector $\mathbf{v}$ (system noise vector), thus

$$\mathbf{x}(k+1)=\mathbf{A}_d(k+1)\mathbf{x}(k)+\mathbf{B}_d(k)\mathbf{u}(k)+\mathbf{v}(k) \tag{3.1-136}$$

and the output vector (measurement vector) is disturbed by the noise vector $\mathbf{w}$ (measurement noise):

$$\mathbf{y}(k)=\mathbf{C}_d(k)\mathbf{x}(k)+\mathbf{w}(k). \tag{3.1-137}$$

The addition of noise is required, since the equations without noise (deterministic model) define an ideal system. However, in practice, the machine cannot be modelled perfectly due to the various assumptions used for deriving the ideal model, and also due to errors arising from measurements. A more realistic model (stochastic model) is obtained by adding the noise vectors. To summarize: by means of noise inputs, it is possible to take account of computational inaccuracies, modelling errors, and errors in the measurements (see also the section discussing discretized EKF implementation and tuning).

The system noise vector $\mathbf{v}$ is a zero-mean, white Gaussian noise, independent of the initial state vector, and its covariance matrix is $\mathbf{Q}$. White noise is the terminology used for a noise source that is perfectly random from one time instant to the next. Thus white noise possesses a constant power spectral density at all frequencies. A Gaussian noise refers to the fact that the probability density function of the amplitudes of a Gaussian noise source has the familiar bell-shaped (Gaussian) curve. The noise vector is also a zero-mean white Gaussian noise that is independent of the initial state vector and also of $\mathbf{w}$, and has a covariance matrix $\mathbf{R}$. The covariance matrices (their elements) are assumed to be known.

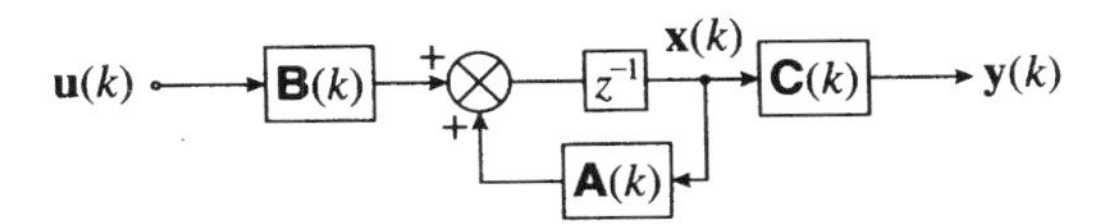

Fig. 3.34. Block-diagram of the time-discrete state-space model of the PMSM.

By using eqns (3.1-136) and (3.1-137) the PMSM model of the EKF is shown in Fig. 3.35.

3. Determination of the noise and state covariance matrices **Q**, **R**, **P** A critical part of the design of the EKF is to use correct initial values for the various covariance matrixes, **Q**, **R**, and **P**. These have important effects on the filter stability and convergence time. The system noise covariance **Q** accounts for the model inaccuracy, the system disturbances, and the noise introduced by the voltage measurements (sensor noise, A/D converter quantization). The noise covariance **R** accounts for measurement noise introduced by the current sensors and A/D quantization.

The noise covariance matrices **Q** and **R** have to be obtained by considering the stochastic properties of the corresponding noises. However, since these are usually not known, in most cases they are used as weight matrices, but it should be noted that sometimes simple qualitative rules can be set up for obtaining the covariance values. It is even possible to design fuzzy-logic-assisted covariance estimation (see also the discussion at the end of the section below on the implementation and tuning of the discretized EKF). In many applications, trial and error is used for the initial estimates of the elements of **Q**, **R** and the state covariance matrix **P**, and diagonal covariance matrices are assumed, due to the lack of statistical information to evaluate the off-diagonal elements. Sometimes these are set to be identity matrices, or they are identity matrices multiplied by a constant.

In general, **Q** is a 4 by 4 matrix, **R** is a 2 by 2 matrix, and **P** is a 4 by 4 matrix. This means that in total there are 36 covariance elements to be determined. However, since the noise signals are not correlated, a reduction of the required elements results, and e.g. **Q** and **R** will only contain 6 unknown diagonal elements. Similarly, it can be assumed that **P** is diagonal, and it contains 4 elements. However, a further reduction arises by considering that the covariance matrix elements do not depend on the axes d and q. Thus the first two elements of **Q** are equal, the only two elements of **R** are equal and the first two elements of **P** are equal, i.e.

$$\mathbf{Q}=\mathbf{Q}_0=\mathrm{diag}(a,a,b,c)\quad \mathbf{P}_0=\mathrm{diag}(e,e,f,\mathrm{g})\quad \mathbf{R}=\mathbf{R}_0=\mathrm{diag}(m,m).$$

Among the various covariance elements, the elements of the initial-state covariance matrix have the smallest influence in the initial tuning procedure. To obtain the best estimated values of the states, the EKF has to be tuned, and tuning involves an iterative search for the covariance matrices which yield the best estimates.

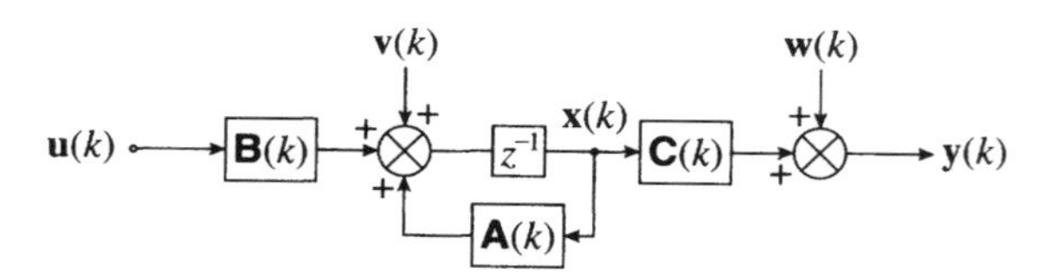

Fig. 3.35. System model of the EKF.

4. Implementation of the discretized EKF algorithm; tuning As discussed above, the EKF is an optimal, recursive state-estimator, and it contains basically two main stages, a prediction stage and a filtering stage. During the prediction stage, the next predicted values of the states $\mathbf{x}(k+1)$ and the predicted state covariance matrix ($\mathbf{P}$) is also obtained. For this purpose the state-variable equations of the machine are used, and also the system covariance matrix ($\mathbf{Q}$). During the filtering stage, the filtered states ($\hat{\mathbf{x}}$) are obtained from the predicted estimates by adding a correction term to the predicted value ($\mathbf{x}$), this correction term is $\mathbf{Ke}=\mathbf{K}(\mathbf{y}-\hat{\mathbf{y}})$, where $\mathbf{e}=(\mathbf{y}-\hat{\mathbf{y}})$ is an error term, and it uses measured values. This error is minimized in the EKF. The structure of the EKF is shown in Fig. 3.36.

The state estimates are obtained in the following steps:

Step 1: *Initialization of the state vector and covariance matrices* Starting values of the state vector $\mathbf{x}_0=\mathbf{x}(t_0)$ and the starting values of the noise covariance matrixes $\mathbf{Q}_0$ and $\mathbf{R}_0$ are set, together with the starting value of the state covariance matrix $\mathbf{P}_0$, where $\mathbf{P}$ is the covariance matrix of the state vector (the terminology 'covariance matrix of prediction', 'error covariance matrix' is also used in the literature).

Step 2: *Prediction of the state vector* Prediction of the state vector at sampling time $(k+1)$ from the input vector $\mathbf{u}(k)$ and the state vector at previous sampling time $\mathbf{x}(k)$, by using $\mathbf{A}_d(k)$ and $\mathbf{B}_d(k)$, is obtained by performing

$$\mathbf{x}(k+1|k)=\mathbf{x}(k+1)=\mathbf{A}_d(k)\mathbf{x}(k)+\mathbf{B}_d(k)\mathbf{u}(k). \tag{3.1-138}$$

The notation $\mathbf{x}(k+1|k)$ means that this is a predicted value at the $(k+1)$th instant, and it is based on measurements up to the kth instant. However, to simplify the notation, it has been replaced by $\mathbf{x}(k+1)$. Similarly, $\mathbf{x}(k|k)$ has been replaced $\mathbf{x}(k)$.

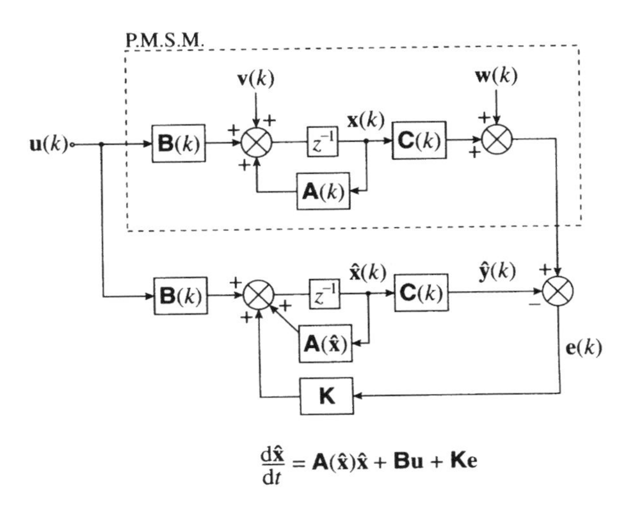

Fig. 3.36. Structure of the EKF.

Step 3: *Covariance estimation of prediction* The covariance matrix of prediction is estimated as

$$\mathbf{P}(k+1)=\mathbf{f}(k+1)\mathbf{P}(k)\mathbf{f}^{\mathrm{T}}(k+1)+\mathbf{Q}, \tag{3.1-139}$$

where (to have simpler notation), $\mathbf{P}(k+1|k)$ has been replaced by $\mathbf{P}(k+1)$, $\mathbf{P}(k|k)$ has been replaced by $\mathbf{P}(k)$ [where $k|k$ denotes prediction at time k based on data up to time k], and $\mathbf{f}(k+1|k)$ has been replaced by $\mathbf{f}(k+1)$, where $\mathbf{f}$ is the following gradient matrix:

$$\mathbf{f}(k+1)=\left.\frac{\partial[\mathbf{A}_{\mathrm{d}}(k)\mathbf{x}+\mathbf{B}_{\mathrm{d}}(k)\mathbf{u}(k)]}{\partial\mathbf{x}}\right|_{\mathbf{x}=\hat{\mathbf{x}}(k+1)}. \tag{3.1-140}$$

Step 4: *Kalman filter gain computation* The Kalman filter gain (correction matrix) is computed as

$$\mathbf{K}(k+1)=\mathbf{P}(k+1)\mathbf{h}^{\mathrm{T}}(k+1)[\mathbf{h}(k+1)\mathbf{P}(k+1)\mathbf{h}^{\mathrm{T}}(k+1)+\mathbf{R}]^{-1}. \tag{3.1-141}$$

For simplicity of notation $\mathbf{P}(k+1|k)$ has been replaced by $\mathbf{P}(k+1)$, and $\mathbf{h}(k+1)$ is a gradient matrix, defined as

$$\mathbf{h}(k+1)=\left.\frac{\partial[\mathbf{C}_{\mathrm{d}}(k)\mathbf{x}]}{\partial\mathbf{x}}\right|_{\mathbf{x}=\hat{\mathbf{x}}(k+1)}. \tag{3.1-142}$$

Step 5: *State vector estimation* The state vector estimation ('corrected state vector estimation', 'filtering') at time $(k+1)$ is performed as

$$\hat{\mathbf{x}}(k+1)=\mathbf{x}(k+1)+\mathbf{K}(k+1)[\mathbf{y}(k+1)-\hat{\mathbf{y}}(k+1)], \tag{3.1-143}$$

where for simplicity of notation, $\hat{\mathbf{x}}(k+1|k+1)$ has been replaced by $\hat{\mathbf{x}}(k+1)$, $\mathbf{x}(k+1|k)$ has been replaced by $\mathbf{x}(k+1)$, and

$$\hat{\mathbf{y}}(k+1)=\mathbf{C}_{\mathrm{d}}(k+1)\mathbf{x}(k+1). \tag{3.1-144}$$

Step 6: *Covariance matrix of estimation error* The error covariance matrix can be obtained from

$$\hat{\mathbf{P}}(k+1)=\mathbf{P}(k+1)-\mathbf{K}(k+1)\mathbf{h}(k+1)\mathbf{P}(k+1), \tag{3.1-145}$$

where for simplicity of notation $\hat{\mathbf{P}}(k+1|k+1)$ has been replaced by $\hat{\mathbf{P}}(k+1)$, and $\hat{\mathbf{P}}(k+1|k)$ has been replaced by $\mathbf{P}(k+1)$.

Step 7: $k=k+1$, $\mathbf{x}(k)=\mathbf{x}(k-1)$, $\mathbf{P}(k)=\mathbf{P}(k-1)$ and go to Step 1.

For the realization of the EKF algorithm it is very convenient to use a signal processor because of the large number of multiplications required and also because of the fact that all of the computations have to be performed fast—within one sampling interval. The extended Kalman filter described above can be used under both steady-state and transient conditions for the estimation of the rotor speed and rotor angle. By using this filter in the drive system, it is possible to implement a PWM inverter-fed PMSM drive without the need of extra position and speed sensors. It should be noted that accurate speed and position sensing is obtained in a very wide speed-range, down to very low values of speed (but not zero speed). However, care must be taken in the selection of the machine parameters and covariance values used. The tuning of the filter involves an iterative modification of these parameters in order to yield the best estimates of the states. Changing the covariance matrices **Q** and **R** affects both the transient duration and steady-state operation of the filter. Increasing **Q** corresponds to stronger system noises, or larger uncertainty in the machine model used. The filter gain matrix elements will also increase and thus the measurements will be more heavily weighted and the filter transient performance will be faster. If the covariance **R** is increased, this corresponds to the fact that the measurements of the currents are subjected to a stronger noise, and should be weighted less by the filter. Thus the filter gain matrix elements will decrease and this results in slower transient performance. Finally it should be noted that in general, the following qualitative tuning rules can be obtained:

Rule 1: if **R** is large then **K** is small (and the transient performance is faster);

Rule 2: if **Q** is large then **K** is large (and the transient performance is slower).

However, if **Q** is too large or if **R** is too small, instability can arise. It is possible to derive similar rules to these rules and to implement a fuzzy-logic-assisted system for the selection of the appropriate covariance elements.

3.1.3.5.3 An alternative machine model and EKF

It has been mentioned above that it is possible to speed up the execution time of the EKF algorithm if another machine model is used. For this purpose it is possible to use the machine model expressed in the stationary reference frame. In this case, the output vector $\mathbf{y}=[i_{sD}, i_{sQ}]^T$ is obtained simply from the state variables, and the computation time is reduced.

The space vector form of the stator voltage equation of the PMSM in the stationary reference frame has been given by eqns (3.1-116) and (3.1-117). The corresponding two-axis voltage equations are

$$\frac{di_{sD}}{dt}=-\left(\frac{R_s}{L_s}\right)i_{sD}+\left(\frac{\psi_F}{L_s}\right)\omega_r\sin\theta_r+\frac{u_{sD}}{L_s} \tag{3.1-146}$$

$$\frac{di_{sQ}}{dt}=-\left(\frac{R_s}{L_s}\right)i_{sQ}-\left(\frac{\psi_F}{L_s}\right)\omega_r\cos\theta_r+\frac{u_{sQ}}{L_s}. \tag{3.1-147}$$

If the state variables are chosen to be the stator currents (i_{sD}, i_{sQ}), the rotor speed (ω_r), and the rotor position (θ_r), then the state vector is $\mathbf{x}=[i_{sD}, i_{sQ}, \omega_r, \theta_r]^T$; this contains both ω_r and θ_r which have to be estimated. The input is defined as $\mathbf{u}=[u_{sD}, u_{sQ}]^T$, the output vector as $\mathbf{y}=[i_{sD}, i_{sQ}]^T$. The state-variable form of the equations will now be obtained by also assuming that the rotor inertia has an infinite value (thus the derivative of the rotor speed is negligible, $d\omega_r/dt=0$) and $\omega_r=d\theta_r/dt$. Although in practice the rotor inertia is not infinite, the required correction is performed by the Kalman filter algorithm, as discussed above. Thus the following state-variable equation is obtained:

$$\frac{d\mathbf{x}}{dt}=\mathbf{f}(\mathbf{x})+\mathbf{Bu} \tag{3.1-148}$$

and the output vector is

$$\mathbf{y}=\mathbf{Cx}. \tag{3.1-149}$$

In eqns (3.1-148) and (3.1-149) $\mathbf{f}(\mathbf{x})$ is defined as:

$$\mathbf{f}(\mathbf{x})=\begin{bmatrix} -(R_s/L_s)i_{sD}+(\psi_F/L_s)\omega_r\sin\theta_r \\ -(R_s/L_s)i_{sQ}-(\psi_F/L_s)\omega_r\cos\theta_r \\ 0 \\ \omega_r \end{bmatrix}. \tag{3.1-150}$$

Furthermore, the input matrix $\mathbf{B}$ is defined as

$$\mathbf{B}=\begin{bmatrix} 1/L_s & 0 \\ 0 & 1/L_s \\ 0 & 0 \\ 0 & 0 \end{bmatrix} \tag{3.1-151}$$

and the transformation matrix is

$$\mathbf{C}=\begin{bmatrix} 1 & 0 & 0 & 0 \\ 0 & 1 & 0 & 0 \end{bmatrix}. \tag{3.1-152}$$

In contrast to the transformation matrix of the model expressed in the rotor reference frame, which was given by eqn (3.1-128), the transformation matrix in eqn (3.1-152) contains only constant elements, and this also results in simplifications in the implementation of the EKF. The system described by eqns (3.1-148) and (3.1-149) is a non-linear system. It is this non-linearity which ensures that the EKF has to be used, and not the conventional, linearized Kalman filter (unextended version). It is important to note that this model cannot be used for the simulation of the performance analysis of the machine, but it can be used in the EKF algorithm for the estimation of the augmented state-vector, if the stator voltages and currents are monitored. To allow digital implementation of the EKF, the system equations of the PMSM have to be first discretized, and then the discretized EKF algorithm can be applied. This will not be discussed in detail, but it should be noted that the time-discrete model is now put into the following form (where the noise vectors have also been added to obtain a system model required by the EKF):

$$\mathbf{x}(k)=\mathbf{f}[\mathbf{x}(k),k]+\mathbf{B}(k)\mathbf{u}(k)+\mathbf{v}(k) \tag{3.1-153}$$

$$\mathbf{y}(k)=\mathbf{C}(k)\mathbf{x}(k)+\mathbf{w}(k). \tag{3.1-154}$$

For example, in eqn (3.1-153) $\mathbf{B}(k)=\mathbf{B}T$, where T is the sampling time ($T=t_{k+1}-t_k$), similarly to that shown above in eqn (3.1-132). Furthermore, $\mathbf{C}(k)=\mathbf{C}$. Equation (3.1-153) is in a different form compared to eqn (3.1-136), since now $\mathbf{f}[\mathbf{x}(k)]$ is present instead of $\mathbf{A}_d(k+1)\mathbf{x}(k)$. Thus the discretized EKF algorithm given above has to be slightly changed and is now summarized. The covariance matrices can be chosen to be diagonal, and $\mathbf{Q}$ is a 4 by 4 matrix, $\mathbf{R}$ is a 2 by 2 matrix, and $\mathbf{P}_0=\mathbf{P}(0)$ is a 4 by 4 matrix. In general they are assumed to take the form

$$\mathbf{Q}=\mathbf{Q}_0=\mathrm{diag}(a,a,b,c)\quad \mathbf{P}_0=\mathrm{diag}(e,e,f,g)\quad \mathbf{R}=\mathbf{R}_0=\mathrm{diag}(m,n).$$

The steps of the discretized EKF algorithm are as follows:

Step 1: *Initialization of the state vector and covariance matrices* Starting values of the state vector $\mathbf{x}_0=\mathbf{x}(t_0)$ and the starting values of the noise covariance matrixes $\mathbf{Q}_0$ and $\mathbf{R}_0$ are set, together with the starting value of the state covariance matrix $\mathbf{P}_0$. The initial value of the state vector can be a zero vector (with all elements zero), since if the motor starts from standstill, the initial values of the stator currents are zero and the initial value of the speed is zero. For simplicity the initial value of the rotor position can be assumed to be zero as well. It is important to note that in general, if incorrect initial values are used, the EKF algorithm will not converge to the correct values.

Step 2: *Prediction of the state vector* Prediction of the state vector at sampling time $(k+1)$ from the input $\mathbf{u}(k)$, the state vector at previous sampling time $\mathbf{x}(k)$, by using $\mathbf{f}$ and $\mathbf{B}$, is obtained by performing

$$\begin{aligned}\mathbf{x}(k+1|k)&=\mathbf{x}(k+1)=\mathbf{x}(k|k)+T\mathbf{f}[\mathbf{x}(k|k)+\mathbf{B}\mathbf{u}(k)]\\&=\mathbf{x}(k)+T\mathbf{f}[\mathbf{x}(k)+\mathbf{B}\mathbf{u}(k)].\end{aligned}\tag{3.1-155}$$

On the right-hand side of this equation, the simplified notation has been used, and it should be noted that this is different from the expression given in Step 2 of the earlier algorithm, since integration of $\mathbf{f}[\mathbf{x}(k)+\mathbf{B}\mathbf{u}(k)]$ yields the new expression (when first-order Euler integration is used). In eqn (3.1-155) a prediction of the states is performed by using the previous state estimate $\mathbf{x}(k|k)$, and also the mean voltage vector $\mathbf{u}(k)$ which is applied to the motor in the period t_k to t_{k+1}.

Step 3: *Covariance estimation of prediction* The covariance matrix of prediction is estimated as

$$\begin{aligned}\mathbf{P}(k+1|k)=\mathbf{P}(k|k)+T[&\mathbf{F}(k)\mathbf{P}(k+1|k+1)\\&+\mathbf{P}(k+1|k+1)\mathbf{F}^{\mathrm{T}}(k)]+\mathbf{Q},\end{aligned}\tag{3.1-156}$$

where $\mathbf{F}$ is the following system gradient matrix (Jacobian matrix):

$$\mathbf{F}(k)=\left.\frac{\partial\mathbf{f}(\mathbf{x})}{\partial\mathbf{x}}\right|_{\mathbf{x}=\hat{\mathbf{x}}(k|k)}.\tag{3.1-157}$$

Step 4: *Kalman filter gain computation* The Kalman filter gain (correction matrix) is computed as

$$\mathbf{K}(k+1)=\mathbf{P}(k+1|k)\mathbf{C}^{\mathrm{T}}[\mathbf{CP}(k+1|k)\mathbf{C}^{\mathrm{T}}+\mathbf{R}]^{-1}. \tag{3.1-158}$$

Step 5: *State vector estimation* The state vector estimation ('corrected state vector estimation', 'filtering') at time $(k+1)$ is performed through the feedback correction scheme, which makes use of the actual measured quantities ($\mathbf{y}$):

$$\hat{\mathbf{x}}(k+1|k+1)=\mathbf{x}(k+1|k)+\mathbf{K}(k+1)[\mathbf{y}(k+1)-\mathbf{Cx}(k+1|k)]. \tag{3.1-159}$$

Step 6: *Covariance matrix of estimation error* The error covariance matrix can be obtained from

$$\hat{\mathbf{P}}(k+1|k+1)=\mathbf{P}(k+1|k)-\mathbf{K}(k+1)\mathbf{CP}(k+1|k). \tag{3.1-160}$$

Step 7: $k=k+1$, $\mathbf{x}(k)=\mathbf{x}(k-1)$, $\mathbf{P}(k)=\mathbf{P}(k-1)$ and go to Step 1.

It should be noted that this scheme requires the measurement of the stator voltages and currents. However, in a PWM voltage-source inverter-fed PMSM drive, it is also possible to reconstruct the stator voltages from the d.c. link voltage and switching states as discussed in connection with eqn (3.1-59). This requires the measurement of the d.c. link voltage, which can vary. However, it is also possible to use such a scheme in which it is not necessary to monitor any voltage at all, but instead of using the actual stator voltages, they can be replaced by the voltage references, which are inputs to the PWM modulator used. This is possible if the PWM inverter switching period is small with respect to the stator electrical time constant ($T_s=L_s/R_s$) of the machine. Such an approach has the advantage of not only leading to a further reduction of the number of sensors required, but it does not require the low-pass filtering used with the usual method of stator voltage monitoring (to eliminate the modulation noise). The reference voltages can be determined by using the estimated values of the currents, position, and speed obtained by the EKF. The need for the presence of the estimated currents is obvious in the reference voltages, since these voltages must contain an ohmic voltage component, which contains these currents (and the currents are multiplied by an ohmic type of gain to give the ohmic reference voltage). However, the presence of the estimated speed and position is also required, because the reference voltages must also contain back e.m.f. components, which depend on the speed and position in the stator reference frame. This can also be seen from eqns (3.1-146) and (3.1-147), where the direct-axis back e.m.f. contains $\psi_F\omega_r\sin\theta_r$, and the quadrature-axis back e.m.f. contains $\psi_F\omega_r\cos\theta_r$. Thus a direct-axis stator reference voltage will contain the estimated direct-axis stator current, the estimated value of the speed, and the estimated value of the sine of the rotor angle. Similarly, the quadrature-axis stator reference voltage will contain the estimated quadrature-axis stator current, the estimated value of the speed, and the estimated value of the cosine of the rotor angle. Thus the following reference voltages can be used in the interval $t(k)$ to $t(k+1)$, and they can be used in a PWM inverter-fed

PMSM drive with a space vector modulator [Bado *et al.* 1992]:

$$u_{sDref}(k+1)=\mathbf{K}[i_{sDref}(k+2)-\exp(-T/T_s)\hat{\imath}_{sD}(k+1|k)]$$
$$-\psi_F\hat{\omega}_r(k+1|k)\sin[\hat{\theta}_r(k+1|k)+T\hat{\omega}_r(k+1|k)/2] \quad (3.1\text{-}161)$$
$$u_{sQref}(k+1)=\mathbf{K}[i_{sQref}(k+2)-\exp(-T/T_s)\hat{\imath}_{sQ}(k+1|k)]$$
$$+\psi_F\hat{\omega}_r(k+1|k)\cos[\hat{\theta}_r(k+1|k)+T\hat{\omega}_r(k+1|k)/2], \quad (3.1\text{-}162)$$

where T_s is the stator time constant and T is the sampling time, $\hat{\imath}_{sD}(k+1|k)$, $\hat{\imath}_{sQ}(k+1|k)$, $\hat{\omega}_r(k+1|k)$, and $\hat{\theta}_r(k+1|k)$ are the estimated stator current components, speed and rotor position respectively, which are obtained by the EKF. The second part of the voltage references is the back e.m.f. component. In the next sampling period, $t(k+1)$ to $t(k+2)$, these reference voltages to the space-vector modulator force the actual stator currents $i_{sD}(k+1)$, $i_{sQ}(k+1)$ to match their reference values $i_{sDref}(k+2)$, $i_{sQref}(k+2)$ at the sampling instant $t(k+2)$. The technique described is one type of predictive current control, which ensures that the reference currents match the ones estimated by the EKF.

An EKF-based position- and speed-sensorless vector-controlled PMSM drive can work in a very wide speed-range, down to very low speeds. However, around zero speed, the speed control system loses its control properties. This is due to the fact that at lower speeds the stator voltages become smaller and the measurement errors and the inaccuracies in the machine model become significant and lead to estimation errors of the states.

3.1.3.6 Vector control using position estimators based on inductance variation due to geometrical and saturation effects

In vector-controlled permanent-magnet synchronous motors (PMSM) the rotor position can also be estimated by using inductance variations due to saturation and geometrical effects. These two techniques are discussed below.

3.1.3.6.1 Scheme 1, utilizing saturation effects: machine with surface-mounted magnets

In a PMSM, because of saturation effects, the stator inductances are a function of the rotor position. This variation can be used for the estimation of the rotor position. The variation arises even in a PMSM with surface-mounted magnets. In such a machine, due to saturation in the stator teeth the direct-axis synchronous inductance is smaller than the quadrature-axis synchronous inductance and these inductances are functions of the rotor position. It can be shown that, in general, these inductances depend on 2β, where β is the angle of the magnet flux (with respect to the stator-voltage space vector) [Schroedl 1991]. It is possible to determine β even at standstill by applying test stator voltages with different directions (angles, with respect to the stator reference frame) and by measuring the resulting changes of the stator-current space vector. This can be proved by considering the stator voltage equation at standstill. Thus by neglecting the stator resistance, from the stator voltage equation we find that $d\bar{\imath}_s/dt = \bar{u}_s/L$ holds, where $\bar{\imath}_s$ and $\bar{u}_s$ are the space vectors of the stator voltages and currents respectively, and L is a complex inductance, which depends on 2β. This technique can be combined with the open-loop flux estimator techniques described in Section 3.1.3.2 and could be used for the detection of the stator flux in the zero-speed region. It is interesting to note that a similar technique could also be used for an induction motor for the estimation of the angle of the magnetizing flux (or the rotor flux) with respect to the real-axis of the stator reference frame. This is based on the fact that in the induction motor, due to saturation of the stator and rotor teeth, the stator inductances depend not only on the level of saturation but also on the position of the main flux. At standstill $d\bar{\imath}_s/dt = \bar{u}_s/L$ holds, where L is a complex stator transient inductance of the induction machine and, by applying appropriate stator voltage test vectors ($\bar{u}_s$), $d\bar{\imath}_s/dt$ call be measured. The angle of the magnetizing flux can then be obtained since the locus of the modulus of the complex transient inductance is an ellipse and the minimum of this ellipse is in the direction of the magnetizing flux.

3.1.3.6.2 Scheme 2, utilizing geometrical saliency effects: machine with interior magnets

A PMSM with interior magnets behaves like a salient-pole machine, and the inductances are different in the direct and quadrature axes. Furthermore, it should be noted that $L_{sq} > L_{sd}$ is obtained (in contrast to a salient-pole synchronous machine with wound rotor and damper windings, where $L_{sd} > L_{sq}$). The inductance variation can again be used for the estimation of the rotor position and the inductances can be obtained by using the monitored stator voltages and currents. This will be discussed for the case when the PMSM with interior magnets is supplied by a current-controlled PWM inverter and the switching frequency is assumed to be high.

In a PMSM with interior magnets, the different inductances are physically due to the fact that each magnet is covered by a steel pole-piece, and thus high-permeance paths are produced for the magnetic fluxes across these poles and also in space quadrature to the magnet flux. However, in the magnetic circuit on the quadrature-axis of the rotor there is only iron, and in the direct-axis of the rotor a part of the magnetic circuit consists of the magnet whose permeability is approximately equal to that of air. Thus the reluctance in the direct-axis is increased and the inductance is decreased, and since the reluctance in the direct axis is much larger than in the quadrature axis, $L_{sq} > L_{sd}$ arises.

As mentioned above, the inductances of the PMSM with internal permanent magnets can be obtained by using monitored stator voltages and currents. In general, due to the saliency, all the stator inductances vary with $(2\theta_r)$, where θ_r is the rotor angle. For example, the stator phase sA self-inductance can be expressed as

$$L_{AA} = L_{sl} + L_{s0'} + L_{s2}\cos(2\theta_r) = L_{s0} + L_{s2}\cos(2\theta_r), \tag{3.1-163}$$

where $L_{s0} = L_{sl} + L_{s0'}$. The stator self-inductances in stator phases sB and sC can be expressed similarly, but are functions of $(2\theta_r + 2\pi/3)$ and $(2\theta_r - 2\pi/3)$ respectively. The mutual inductances between the stator phases are also functions of twice the rotor angle, and for example the inductance L_{AB} between stator phase sA and stator phase sB can be expressed as

$$L_{AB} = -\frac{L_{s0'}}{2} + L_{s2}\cos(2\theta_r - 2\pi/3). \tag{3.1-164}$$

Similarly L_{BC} varies with $(2\theta_r)$ and L_{AC} varies with $(2\theta_r + 2\pi/3)$. In these expressions L_{sl} is the leakage inductance of a stator winding, $L_{s0'}$ is the magnetizing inductance due to the fundamental air-gap flux, and L_{s2} is the self-inductance due to the position-dependent flux. If the direct- and quadrature-axis synchronous inductances are known (L_{sd}, L_{sq}), then L_{s2} and $L_{sl} + (3/2)L_{s0'}$ can be expressed in terms of L_{sd} and L_{sq} by considering [Vas 1993]

$$L_{sd} = L_{sl} + \tfrac{3}{2}(L_{s0'} - L_{s2}) \tag{3.1-165}$$

$$L_{sq} = L_{sl} + \tfrac{3}{2}(L_{s0'} + L_{s2}), \tag{3.1-166}$$

since it follows from eqns (3.1-165) and (3.1-166) that

$$L_{s2} = \frac{L_{sq} - L_{sd}}{3} \tag{3.1-167}$$

$$L_{sl} + \tfrac{3}{2}L_{s0'} = \frac{L_{sd} + L_{sq}}{2}. \tag{3.1-168}$$

In general the stator voltage equation in the stationary reference for stator phase sA can be expressed as $u_{sA} = R_s i_{sA} + d\psi_{sA}/dt$, where the stator flux linkage ψ_{sA} contains the stator flux-linkage components due to the three stator currents and also a component due to the magnet flux. However, if it is assumed that the switching frequency of the PWM inverter supplying the motor is high (inductance variation with rotor position can be neglected in one switching period) then the stator voltage equation takes the form

$$u_{sA} = R_s i_{sA} + L_{sA}\frac{di_{sA}}{dt} + e_{sA}, \tag{3.1-169}$$

where R_s and L_{sA} are the resistance and synchronous inductance of stator phase sA respectively, and e_{sA} is the back e.m.f. of stator phase sA. In general the back e.m.f is proportional to the motor speed. In eqn (3.1-169) the inductance L_{sA} can be expressed as follows:

$$L_{sA} = L_{AA} - L_{AB}. \tag{3.1-170}$$

Before performing the measurement of L_{sA}, first the variation of L_{sA} with the rotor angle is obtained using the known values of L_{sd} and L_{sq}. For this purpose eqns (3.1-163) and (3.1-164) are substituted into eqn (3.1-170), and then in the resulting equation, the inductances L_{s2} and $L_{sl}+(3/2)L_{s0'}$ are substituted by eqns (3.1-167), (3.1-168) respectively. Thus

$$L_{sA}=\frac{L_{sd}+L_{sq}}{2}+\left[\frac{L_{sq}-L_{sd}}{3}\right][\cos(2\theta_r)-\cos(2\theta_r-2\pi/3)] \qquad (3.1\text{-}171)$$

is obtained. Similar expressions can be obtained for the synchronous inductances of the other two stator phases, L_{sB}, L_{sC} respectively:

$$L_{sB}=\frac{L_{sd}+L_{sq}}{2}+\left[\frac{L_{sq}-L_{sd}}{3}\right][\cos(2\theta_r-2\pi/3)-\cos(2\theta_r-4\pi/3)] \qquad (3.1\text{-}172)$$

$$L_{sC}=\frac{L_{sd}+L_{sq}}{2}+\left[\frac{L_{sq}-L_{sd}}{3}\right][\cos(2\theta_r-4\pi/3)-\cos(2\theta_r-2\pi)]. \qquad (3.1\text{-}173)$$

The obtained variation of the three synchronous inductances with the rotor angle is shown in Fig. 3.37.

By using eqns (3.1-171)–(3.1-173), a look-up table can be created which contains corresponding values of the synchronous inductances and the rotor angle. This table can then be used for the estimation of the rotor angle by using measured values of these inductances. The measurements will be described below, but it

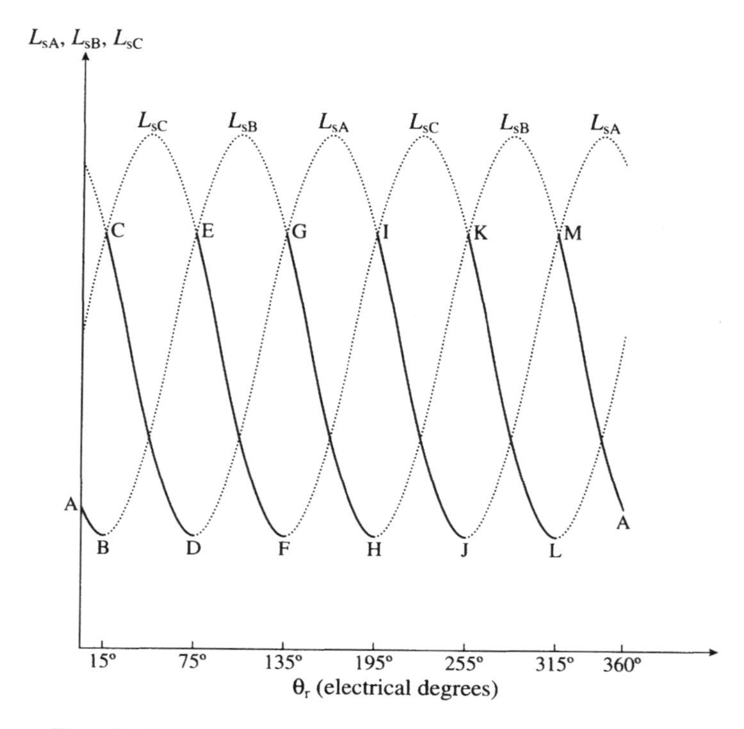

Fig. 3.37. Variation of synchronous inductances with the rotor angle.

should be noted that to obtain accurate values of the rotor position, L_{sd} and L_{sq} have to be known with great accuracy. The reason for using three synchronous inductances is now discussed. In theory, the rotor position can be estimated from the measured inductance L_{sA}, but it must be considered that since it varies with twice the rotor angle, in every electrical cycle it goes through two cycles and a specific value of the inductance corresponds to four different rotor positions. For example, in Fig. 3.37, the value of L_{sA} is the same at four points C, G, I, and M. Therefore to obtain an unambiguous rotor-position estimation for a specific value of the inductance, the information on all three inductances (L_{sA}, L_{sB}, and L_{sC}) has to be utilized by calculating them during different intervals of each electrical cycle. A simple solution is obtained if the appropriate inductances are determined in the AB, CD, EF, GH, IJ, KL, and MA intervals shown in Fig. 3.37. For this purpose, L_{sB} is determined from point A to point B in Fig. 3.37 (0 to 15 electrical degrees interval), L_{sA} is determined from point C to point D (15 to 75 electrical degrees), and L_{sC} is determined from point E to point F (75 to 135 electrical degrees). L_{sB} is then determined from point G to point H (135 to 195 electrical degrees), L_{sA} is determined from point I to point J (195 to 255 electrical degrees), L_{sC} is determined from point K to point L (255 to 315 electrical degrees), and finally L_{sB} is determined from point M to point A (315 to 15 electrical degrees). The determined inductance values are stored in the look-up table, e.g. for every 0.05 electrical degree steps. However, since only parts of each inductance variation are utilized to estimate the rotor position, the look-up table has to be only 60 electrical degrees long and a position estimator using this scheme with an accuracy of 0.1 electrical degrees requires only 1200 stored elements (memory locations).

By using the monitored voltage u_{sA} and monitored current i_{sA}, the inductance L_{sA} can be determined from eqn (3.1-169), but again it should be emphasized that it is assumed that the PMSM with interior magnets is supplied by a current-controlled PWM inverter with a hysteresis controller and the switching frequency is high (thus in one switching period the variation of the inductance can be neglected). By using eqn (3.1-169), it is possible to estimate the synchronous inductance L_{sA} from

$$L_{sA}=\frac{u_{sA}-R_s i_{sA}-e_{sA}}{\mathrm{d}i_{sA}/\mathrm{d}t}, \tag{3.1-174}$$

where the back e.m.f is assumed to be constant (during a switching period); it is proportional to the rotor speed (ω_r):

$$e_{sA}=c\omega_r=c\frac{\mathrm{d}\theta_r}{\mathrm{d}t}. \tag{3.1-175}$$

In eqn (3.1-175) c is a constant, and the derivative $\mathrm{d}\theta_r/\mathrm{d}t$ can be approximated by using the rotor positions of two previous time instants, t_{i-1} and t_{i-2}. Thus the rate of change of the rotor angle is

$$\frac{\mathrm{d}\theta_r}{\mathrm{d}t}\approx\frac{\Delta\theta_r}{\Delta t}=\frac{\theta_r(t_{i-1})-\theta_r(t_{i-2})}{t_{i-1}-t_{i-2}}. \tag{3.1-176}$$

At lower speeds the back e.m.f. is smaller, hence the assumption of constant back e.m.f. at lower speed results in a smaller error. Similarly, in eqn (3.1-174) the rate of change of the stator current can be approximated as

$$\frac{di_{sA}}{dt} \approx \frac{\Delta i_{sA}}{\Delta t} = \frac{i_{sA}(t_{i-1}) - i_{sA}(t_{i-2})}{t_{i-1} - t_{i-2}}. \tag{3.1-177}$$

It should be noted that a similar technique can be used for the estimation of the synchronous inductances in phases sB and sC respectively (these are L_{sB} and L_{sC} respectively). The synchronous inductances obtained by measurements can then be used in the look-up table described above to obtain the corresponding rotor position. However, it should also be considered that a measured inductance may not be contained in the look-up table and in this case the rotor position can be determined by considering the inductance in the table whose value is closest to the measured one.

The position estimation using the real-time estimates of the inductances can be conveniently performed by using a DSP. The position sensing error can be minimized by using a higher switching frequency (in this case the assumption of the constancy of the inductances and the back e.m.f during a switching period is more valid).

3.1.3.6.3 Scheme 3, utilizing geometrical saliency effects: machine with interior magnets

For the estimation of the rotor position and rotor speed of an interior-permanent-magnet synchronous machine, even at zero and low speeds, it is also possible to use saliency effects in another way to that described above. In order to be able to track the saliency, high-frequency voltages are injected into the stator (the amplitude and angular frequency of these are U_{si} and ω_i respectively). These produce high-frequency currents which vary with the rotor position. By detecting these currents and by using appropriate processing, a signal is produced, which is proportional to the difference between the actual rotor position and the estimated rotor position. The position error signal is then input into a controller which outputs the speed estimates and when this is integrated the rotor position estimates are obtained.

First, it is shown below that in the permanent-magnet synchronous machine with interior magnets, the position dependency of the stator inductances causes position-dependent current responses when the stator is supplied by high-frequency voltages. By measuring these stator currents, it is then possible to extract the information on the rotor position.

In the interior-permanent-magnet synchronous machine the stator inductances (L_{sd}, L_{sq}) in the direct and quadrature axes of the rotor reference frame are different ($L_{sd} < L_{sq}$) and the injected stator flux-linkage components in the d, q axes are $\psi_{sdi} = L_{sd} i_{sdi}$, $\psi_{sqi} = L_{sq} i_{sqi}$. However, in the stator reference frame, the injected stator flux linkages (ψ_{sDi}, ψ_{sQi}) contain rotor position (θ_r) dependent inductances, $L_{sD} = L_l + \Delta L \cos(2\theta_r)$, $L_{sQ} = L_1 - \Delta L \cos(2\theta_r)$, $L_{sDQ} = L_{sQD} = \Delta L \sin(2\theta_r)$, where

$L_1=(L_{sd}+L_{sq})/2$ and $\Delta L=(L_{sd}-L_{sq})/2$. A simple proof of this can be obtained, for example, by performing the transformation of the injected stator-flux space vector expressed in the rotor reference frame, $\bar{\psi}'_{si}=L_{sd}i_{sdi}+\mathrm{j}L_{sq}i_{sqi}$, into the injected stator flux-linkage space vector expressed in the stationary reference frame, $\bar{\psi}_{si}=\bar{\psi}'_{si}\exp(\mathrm{j}\theta_r)=\psi_{sDi}+\mathrm{j}\psi_{sQi}$, and then by resolving the obtained flux linkages into their real- and imaginary-axis components. For this purpose the expressions of the stator-current space vector in the rotor reference frame, $\bar{i}'_{si}=i_{sdi}+\mathrm{j}i_{sqi}$, and also in the stator reference frame, $\bar{i}_{si}=i_{sDi}+\mathrm{j}i_{sQi}=\bar{i}'_{si}\exp(\mathrm{j}\theta_r)$, are also introduced. Thus

$$\begin{aligned}\bar{\psi}_{si}&=\bar{\psi}'_{si}\exp(\mathrm{j}\theta_r)=(L_{sd}i_{sdi}+\mathrm{j}L_{sq}i_{sqi})\exp(\mathrm{j}\theta_r)\\&=\left(\frac{L_{sd}+L_{sq}}{2}\right)\bar{i}'_{si}\exp(\mathrm{j}\theta_r)+\left(\frac{L_{sd}-L_{sq}}{2}\right)\bar{i}'^{*}_{si}\exp(\mathrm{j}\theta_r)=L_1\bar{i}_{si}+\Delta L\bar{i}^{*}_{si}\exp(2\mathrm{j}\theta_r)\\&=\psi_{sDi}+\mathrm{j}\psi_{sQi}\end{aligned} \tag{3.1-178}$$

is obtained, and resolution into its components (ψ_{sDi}, ψ_{sQi}) yields the injected stator flux-linkage components (in the stationary reference frame) which contain the inductances defined above. However, due to the appropriate injected high-frequency stator voltages, the injected stator flux-linkage components in the stationary reference frame are obtained from $\psi_{sDi}+\mathrm{j}\psi_{sQi}=\bar{\psi}_{si}=\bar{\psi}'_{si}\exp(\mathrm{j}\hat{\theta}_r)=(\psi_{sdi}+\mathrm{j}\psi_{sqi})\exp(\mathrm{j}\hat{\theta}_r)$, where $\hat{\theta}_r$ is the estimated value of the rotor position, $\psi_{sdi}=0$, and $\psi_{sqi}=\int u_{si}(t)\,\mathrm{d}t$, $u_{si}(t)=U_{si}\cos(\omega_i t)$ (due to the high frequency, the stator ohmic voltage drop can be neglected), thus $\psi_{sqi}=(U_{si}/\omega_i)\sin(\omega_i t)$. It is important to note that the direct-axis injected stator flux is zero and only a quadrature-axis stator flux is injected. It follows that

$$\begin{aligned}\bar{\psi}_{si}&=\psi_{sDi}+\mathrm{j}\psi_{sQi}=-\psi_{sqi}\sin\hat{\theta}_r+\mathrm{j}\psi_{sqi}\cos\hat{\theta}_r\\&=-(U_{si}/\omega_i)\sin(\omega_i t)\sin\hat{\theta}_r+\mathrm{j}(U_{si}/\omega_i)\sin(\omega_i t)\cos\hat{\theta}_r.\end{aligned} \tag{3.1-179}$$

It follows that in the steady-state the stator currents due to the injected stator voltages can be obtained from eqns (3.1-178) and (3.1-179) as

$$i_{si}=\{I_{si0}\exp(\mathrm{j}\hat{\theta}_r)+I_{si1}\exp[\mathrm{j}(2\theta_r-\hat{\theta}_r)]\}\sin(\omega_i t)=i_{sDi}+\mathrm{j}i_{sQi}, \tag{3.1-180}$$

where the high-frequency current amplitudes are

$$I_{si0}=\frac{(U_{si}/\omega_i)L_1}{L_1^2-\Delta L^2} \tag{3.1-181}$$

and

$$I_{si1}=\frac{(U_{si}/\omega_i)\Delta L}{L_1^2-\Delta L^2}. \tag{3.1-182}$$

It follows from eqn (3.1-180) that the high-frequency stator current in the rotor reference frame is

$$\bar{i}'_{si}=\bar{i}_{si}\exp(-\mathrm{j}\hat{\theta}_r)=\{I_{si1}\exp[2\mathrm{j}(\theta_r-\hat{\theta}_r)]\}\sin(\omega_i t)=i_{sdi}+\mathrm{j}i_{sqi}. \tag{3.1-183}$$

Thus the direct-axis stator current in the rotor reference frame due to the injected high-frequency stator voltages is obtained as

$$i_{sdi}=\{I_{si1}\sin[2(\theta_r-\hat{\theta}_r)]\}\sin(\omega_i t). \tag{3.1-184}$$

It follows that the frequency of this stator current is equal to the frequency of the injected high-frequency voltages and is amplitude-modulated by the error between the actual rotor position and the estimated rotor position. It is this signal from which the rotor position can be extracted. For this purpose, this signal is low-pass filtered and demodulated and thus the d.c. signal $\varepsilon_f=I_{si1}\sin[2(\theta_r-\hat{\theta}_r)]$ is obtained, which contains the position error $\theta_r-\hat{\theta}_r$. The demodulation can be performed by multiplication of the high-frequency signal by another signal at the injected frequency.

The obtained demodulated error signal (ε_f) is then fed into a linear controller which updates the estimated rotor position and drives the error to zero (in other words the d-axis stator current at the injected frequency is forced to zero). Thus the filtered and demodulated signal drives a controller, K_1+K_2/p, where K_1 and K_2 are gains of the controller (linear observer), thus the rotor position is obtained as $\hat{\theta}_r=\int\hat{\omega}_r\,dt$, where $\hat{\omega}_r=(K_1+K_2/p)\varepsilon_f$ is the estimated rotor speed. The estimated electromagnetic torque can be used to obtain the dynamic accuracy of the estimation. For this purpose the estimated electromagnetic torque is used as feedforward which drives an estimated mechanical model of the machine [Corley and Lorenz 1996]. It is a great advantage that the steady-state tracking ability of the estimator does not depend on machine parameters. Although the current amplitude I_{si1} depends on the machine inductances, it is only a scaling term and does not affect the accuracy of the rotor position estimates since its spatial angle is tracked and not the amplitude. Furthermore, I_{si1} is independent of the speed, thus the observer eigenvalues are independent of the speed if a linear observer is used. The estimation scheme relies on the total decoupling of the direct and quadrature-axes in the rotor reference frame. However, under saturated magnetic conditions (at load) a position error will arise due to saturation, but this error is only an offset [Corley and Lorenz 1996] which can be compensated for. The estimator can be used in a wide speed-range, including zero speed.

The voltage command feedforward signals which are required to ensure that $\bar{\psi}'_{si}=\psi_{sdi}+j\psi_{sqi}=j(U_{si}/\omega_i)\sin(\omega_i t)$ can be obtained by considering the stator voltage equation expressed in the rotor reference frame. However, due to the high injected frequency, the stator ohmic drop can be neglected, and it follows from eqn (2.1-148) that $\bar{u}'_{si}=d\bar{\psi}'_{si}/dt+j\hat{\omega}_r\bar{\psi}'_{si}$ holds (all the primed quantities are expressed in the rotor reference frame and $\hat{\omega}_r$ is the estimated rotor speed). By substitution of $\bar{\psi}'_{si}=j(U_{si}/\omega_i)\sin(\omega_i t)$, finally

$$\bar{u}'_{si}=u_{sdi}+ju_{sqi}=-\hat{\omega}_r(U_{si}/\omega_i)\sin(\omega_i t)+jU_{si}\cos(\omega_i t) \tag{3.1-185}$$

is obtained. In a vector-controlled drive, the command stator voltages in the rotor reference frame can then be obtained by considering that the drive contains synchronous current regulators on the outputs of which u_d and u_q are present, and thus the reference voltages u_{sdref}, u_{sqref} in the rotor reference frame can be obtained from

$$\bar{u}_{sref} = u_{sdref} + ju_{sqref} = u_d + u_{sdi} + j(u_q + u_{sqi}), \quad (3.1\text{-}186)$$

where u_{sdi}, u_{sqi} are given by eqn (3.1-185). Thus the command voltages in the stationary reference frame can be obtained from

$$\bar{u}_{sref} = u_{sDref} + ju_{sQref} = \bar{u}_{sref} \exp(-j\hat{\theta}_r). \quad (3.1\text{-}187)$$

It should be noted that in Sections 4.5.3.3.3 and 4.5.3.3.4 a similar rotor speed estimator is discussed for an induction machine, but in the induction motor saliency due to magnetic saturation or deliberately introduced rotor asymmetry is utilized.

3.1.3.6.4 Scheme 4, utilizing geometrical saliency effects: machine with interior magnets

In the final part of the present section the initial rotor-position estimation of an interior-permanent-magnet machine is described, by utilizing saliency effects. As discussed above, in a PMSM with interior magnets, there exists saliency and the direct and quadrature axis inductances (L_{sd}, L_{sq}) are different, $L_{sd} < L_{sq}$. When symmetrical three-phase sinusoidal stator voltages are applied to this machine at standstill, due to the saliency, in the steady-state, the locus of the space vector of the stator currents in the stationary reference frame becomes an ellipse (in a PWM VSI-fed drive system, the currents are not perfectly sinusoidal and this locus is not perfect ellipse). If the stator currents are monitored, and thus the locus of the space vector of the stator currents is known, then the position of the major axis of the ellipse can be used to obtain information on the rotor angle (θ_r). By definition, the direct axis of the rotor (rd) is in the direction of the north pole of the rotor magnet. If the major axis of the ellipse is located at angle δ from the real-axis of the stationary reference frame (sD), and the direct-axis of the rotor (rd) is located at angle γ from the major axis of the ellipse, then the rotor angle can be obtained from $\theta_r = \delta + \gamma$. The displacement angle, γ, is caused by the armature impedance. If the frequency of the armature current is constant, then γ is constant. The frequency of the applied stator voltage can be selected to be high. This generates a very small torque so the rotor cannot turn and stays still during the initial rotor-position estimation procedure.

When the machine is at standstill, in the absence of saliency ($L_s = L_{sd} = L_{sq}$), in the steady-state, the locus of the stator-current space vector is a circle when the stator voltages are symmetrical three-phase voltages. This fact also follows from the space-vector voltage equation of the machine at standstill, $\bar{u}_s' = R_s \bar{i}_s' + d(L_s \bar{i}_s')/dt$ (the magnet flux, ψ_F, is not present in this equation, since because $\omega_r = 0$, the rotational voltage component $j\omega_r \psi_F \exp(j\theta_r) = 0$). Thus $i_{sd} + ji_{sq} = |\bar{i}_s| \exp[(\omega t - \phi)]$ is obtained, where the phase angle is $\phi = \tan^{-1}(X_s/R_s)$ and the radius of the circle is equal to $|\bar{i}_s| = U_s/(R_s^2 + X_s^2)^{1/2}$, where $X_s = \omega L_s$. However, for the machine with saliency, from the direct- and quadrature-axis voltage equations of the machine at standstill ($u_{sd}=R_s i_{sd}+d(L_{sd}i_{sd})/dt$, $u_{sq}=R_s i_{sq}+d(L_{sq}i_{sq})/dt$) it follows that $i_{sd}=|i_{sd}|\cos(\omega t-\phi_d)$ and $i_{sq}=|i_{sq}|\cos(\omega t-\phi_q)$. In these expressions $|i_{sd}|=U_s/(R_s^2+X_{sd}^2)^{1/2}$, $|i_{sq}|=U_s/(R_s^2+X_{sq}^2)^{1/2}$, $\phi_d=\tan^{-1}(X_{sd}/R_s)$, and $\phi_q=\tan^{-1}(X_{sq}/R_s)$. Thus the space-vector locus of the stator currents is an ellipse. The major axis of the stator-current space-vector locus can be obtained from the monitored stator currents (i_{sD}, i_{sQ}). Since in the interior-permanent-magnet machine $L_{sq} \gg L_{sd}$, it follows that the modulus of i_{sd} is larger than the modulus of i_{sq} and thus the major axis of the ellipse is close to the rd-axis. If the effects of magnetic saturation are neglected then the ellipse is described by the second-order equation: $Ai_{sD}^2+2Bi_{sD}i_{sQ}+Ci_{sD}^2+1=0$ and the angle of the major axis can be obtained from $\delta=(1/2)\tan^{-1}[2B/(A-C)]$.

It has been stated above that due to the PWM inverter the stator-current space-vector locus is not a perfect ellipse. When the d.c. link voltage is large, the ripple in the currents is larger and the deviation from the ellipse is larger. In the case when there are current ripples, the coefficients A, B, and C can be determined in such a way that the square error between the ellipse described by the equation given above and the current space-vector locus should be minimal. For this purpose it is possible to use the Newton–Raphson method. Since the number of data used to obtain the stator-current space-vector locus coefficients is large, the estimation is very accurate and is robust against current ripples and noise.

Since δ is estimated by using the inverse tangent function, there is an ambiguity of π degrees in the estimation. Physically this means that it is not known if this angle is the angle between the north pole of the magnet and the sD-axis or the south pole of the magnet and the sD-axis. However, it is possible to identify the direction of the north pole by using the fact that the centre of the stator-current space-vector locus will shift due to magnetic saturation [Kondo *et al.* 1995].

3.1.3.7 Vector control using artificial-intelligence-based estimators

The application of two types of artificial-intelligence-based estimators is briefly discussed below; these use an artificial neural network (ANN) or a fuzzy-neural network. However, for further details, the reader is referred to Section 4.4 and Chapter 7.

It is possible to train a supervised multi-layer feedforward ANN with back-propagation training for the estimation of the rotor position and rotor angle. By using the back-propagation algorithm, the square of the error between the required and actual ANN output is minimized. The trained ANN can then be used in real-time applications. Such an ANN contains an input layer, an output layer, and the hidden layers. However, the number of hidden layers to be used is not known in advance: this has to be determined by trial and error, although it should be noted as a guideline that in electrical engineering applications the number of hidden layers is usually one or two. Furthermore, the number of hidden nodes in the hidden layers is also not known in advance and again this has to be obtained by trial and error. The number of input nodes depends on the training data used, and various possibilities exist for this purpose. This also depends on the type of PMSM (machine with surface-mounted magnets or machine with interior magnets). However, it is possible to construct such a neural network which also uses at its inputs the stator currents of the machine (i_{sD}, i_{sQ}), but for each of the stator currents used, there are two inputs, corresponding to a present and also to a past input [$i_{sD}(k)$, $i_{sD}(k-1)$, $i_{sQ}(k)$, $i_{sQ}(k-1)$]. It is an advantage of such an approach that, in contrast to other conventional techniques described in the previous sections, it does not require a mathematical model of the machine, since a well-trained ANN which uses supervised learning is capable of approximating any non-linear function (by using the examples during the learning stage). It is also an advantage that the training can be automized.

It is possible to overcome some of the difficulties of the ANN-based approach described above by using a fuzzy-neural estimator instead of the conventional neural estimator. This is basically a neural network with fuzzy features. As discussed above, a conventional neural network using supervised learning uses a fixed topology and back-propagation learning. However, it is difficult to relate the structure of the network to the physical processes and there are no guidelines for the selection of the number of hidden layers and nodes. A fuzzy-neural system combines the advantages of fuzzy-logic and neural networks. In a fuzzy-neural system the structure of the network is based on a fuzzy-logic system. It is another advantage of a fuzzy-neural network that the number of layers and also the number of nodes is known. In a conventional fuzzy-logic-based system, the number of rules, the rules themselves, the number of membership functions, and the membership functions themselves have to be known *a priori*. However, in an adaptive fuzzy-neural network this is not the case, and thus it is possible to obtain automated design and tuning of the fuzzy-neural estimator.

There are many possibilities for implementing an adaptive fuzzy-neural estimator, e.g. there exist Mamdani-type and Sugeno-type fuzzy-neural networks (see the reference list at the end of Chapter 7). In a Mamdani-type fuzzy-neural system, there is an input layer, followed by a fuzzification layer, and there can be three more layers corresponding to the fuzzy AND, fuzzy OR, and defuzzification. Each layers contain well-defined node functions. The design of such a network can use a two-stage learning process [Stronach 1997], a first-stage clustering by supervised learning and a second-stage fine-tuning by supervised learning. The first-stage learning algorithm can be the competitive (instar) learning algorithm and after the first stage, the number of rules, the rules themselves, the number of input and output membership functions, and initial shapes of these are known. In the second-stage final tuning is achieved. There are again many possibilities for the selection of the input data, and e.g. it is possible to train a fuzzy-neural network where the measured stator currents are also used (as with the ANN described above) for position and speed estimation. It should be noted that when a fuzzy-neural network is used, the network architecture can be more complicated than the required architecture with a 'pure' (non-fuzzy) ANN (e.g. it could contain an increased number of nodes), and this could be a disadvantage of this technique.

Although some large manufacturers (Yaskawa, Hitachi) have already incorporated artificial intelligence into their induction motor drives, it is expected that in the future artificial-intelligence-based estimators will have a larger role in other drive applications as well.

3.2 Vector control of synchronous reluctance machines and synchronous machines with electrically excited rotor

3.2.1 MAGNETIZING-FLUX-ORIENTED CONTROL OF A SYNCHRONOUS MACHINE WITH ELECTRICALLY EXCITED ROTOR

3.2.1.1 General introduction

It is possible to design high-performance drive systems employing salient-pole synchronous machines with electrically excited rotor. In this section, the principle of the drive control will be based on field-oriented or more precisely, magnetizing-flux-oriented control. The control method can be applied to voltage-source inverter-fed synchronous machines, cycloconverter-fed synchronous machines, and force-commutated current-source inverter-fed synchronous machines.

Vector control is not suitable for the load-commutated inverter-fed synchronous machine. In this drive, there is a phase-controlled rectifier on the supply side of the d.c. current link, which has a current control loop and operates as a controlled current source. The controlled d.c. current is passed through a d.c. link inductor to the thyristor-commutated three-phase inverter which supplies the synchronous machine with quasi-square-wave line currents. The role of the d.c. link inductor is to smooth the d.c. link current and to allow the rectifier and inverter to operate independently. Load commutation is ensured by overexcitation of the synchronous machine and thus the machine operates at leading power factor. Since the inverter is commutated by the induced voltages of the synchronous machine, load commutation will not function at a very low rotor speed, since in this case the induced voltage is low. The torque and the speed of the machine are controlled by the d.c. link current and the frequency of the inverter respectively. Vector control is not suitable for this drive because the stator currents of the synchronous machine cannot be freely controlled as, for example, in the case of the current-controlled PWM inverter, since the firing angle of the machine-side inverter must be formed to ensure safe commutation of the inverter and to yield a maximum link voltage and minimal link current for a given torque to achieve a higher power factor of the machine-side inverter. The control of the firing must be performed to give a minimal value of the turn-off angle (which is equal to the inverter advance angle minus the overlap angle), to give maximal motor power factor and efficiency, but it should be large enough not to produce commutation failure. Thus at any speed the turn-off angle must have a value such that it will allow a safe recovery of the outgoing thyristor. Furthermore, as a result of operation at leading power factor, decoupling of the torque- and flux-producing components of the stator currents is not possible. Since this drive requires a smaller number of switching devices (e.g. thyristors) than those found in cycloconverter drives, and since higher frequencies and speeds can be obtained than with the drive using a cycloconverter, the load-commutated inverter-fed synchronous machine has found extensive high-power, high-speed applications. At such high power levels (several megawatts) there are no d.c. machines available. However, in these load-commutated inverter-fed synchronous drives high dynamic performance is usually not required, and thus the fact that vector control cannot be employed does not represent a great disadvantage.

Similarly to that described for smooth-air-gap machines, it is a main feature of the magnetizing-flux-oriented control of salient-pole synchronous machines that the stator current is divided into two components, a direct-axis component i_{sx} and a quadrature-axis component i_{sy}, which determine the magnetization and the electromagnetic torque independently of each other. The quadrature-axis stator current component is in space quadrature to the magnetizing flux-linkage space phasor $\bar{\psi}_m$, and the electromagnetic torque is proportional to the product of the modulus of the magnetizing flux-linkage space phasor $|\bar{\psi}_m|$ and the quadrature-axis stator current i_{sy}. Figure 3.38 shows the steady-state phasor diagram for the salient-pole synchronous machine.

In Fig. 3.38 ψ_{md} and ψ_{mq} are the magnetizing flux-linkage components in the reference frame fixed to the rotor and the space phasor of the stator currents $\bar{i}_s$ is also shown for the case of unity power factor. By using eqns (2.2-10) and (2.2-11), these flux components can be expressed in terms of the stator and rotor currents as

$$\psi_{md} = L_{md}(i_{sd} + i_{r\alpha} + i_{rF}) = L_{md} i_{md} \quad (3.2\text{-}1)$$

$$\psi_{mq} = L_{mq}(i_{sq} + i_{r\beta}) = L_{mq} i_{mq}, \quad (3.2\text{-}2)$$

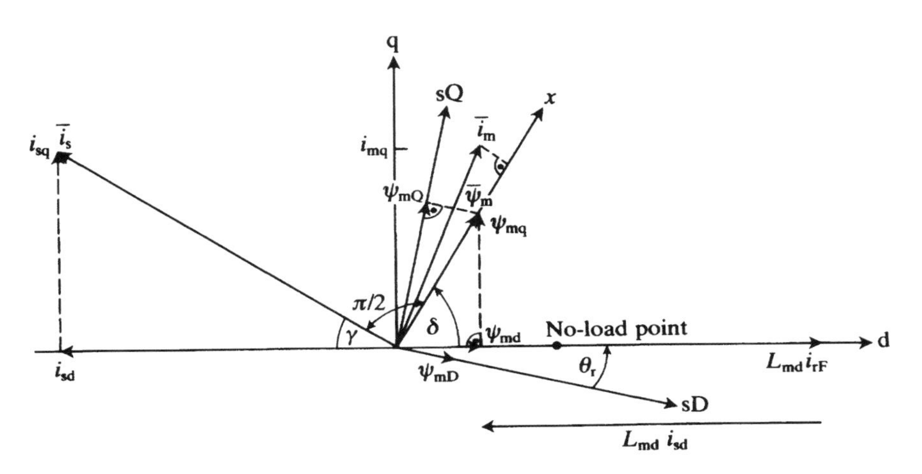

Fig. 3.38. Steady-state phasor diagram of the salient-pole synchronous machine.

where L_{md} and L_{mq} are the magnetizing inductances along the direct (d) and quadrature (q) axes fixed to the rotor, i_{sd} and i_{sq} are the direct- and quadrature-axis stator currents in the rotor reference frame, $i_{r\alpha}$ and $i_{r\beta}$ are the damper currents along the d and q axes respectively, which are zero in the steady state, i_{rF} is the field current and i_{md}, i_{mq} are the magnetizing current components in the rotor reference frame. The magnetizing flux-linkage space phasor is defined as $\bar{\psi}_m = \psi_{md} + j\psi_{mq}$ and is also shown in Fig. 3.38. As previously, a special (x, y) reference frame is chosen, where the direct axis (x) is aligned with the magnetizing flux-linkage space phasor and the special reference frame rotates at the speed of the magnetizing flux-linkage space phasor. In Fig. 3.38 the magnetizing-current space phasor is also shown and is defined as $\bar{i}_m = i_{md} + ji_{mq}$. As a result of saliency this is not coaxial with the magnetizing flux-linkage space phasor.

3.2.1.2 Implementation for a cycloconverter-fed salient-pole synchronous machine

Here the control scheme for the magnetizing-flux-oriented control of a salient-pole synchronous machine with electrically excited rotor winding is discussed. The excitation winding on the rotor of the synchronous machine is supplied by a controllable three-phase rectifier. The stator windings are supplied by a cycloconverter, which utilizes the current control described in Section 3.1.1.

The cycloconverter is a frequency converter which converts power directly (without an intermediate d.c. link) from a fixed frequency to a lower frequency. In one specific implementation, the cycloconverter consists of three sets of anti-parallel six-pulse bridge rectifiers without circulating currents. Figure 3.39 shows the schematic of the cycloconverter-fed synchronous machine drive.

Each of the motor phases is supplied through a three-phase transformer and an anti-parallel thyristor bridge and the field winding is supplied by another three-phase transformer and a three-phase rectifier using the bridge connection. The synchronous machine is equipped with rotor-position (θ_r) and rotor-speed (ω_r) transmitters and the stator currents (i_{sA}, i_{sB}, i_{sC}) and the excitation current (i_{rF}) are also monitored. The control circuits utilize the rotor speed, rotor position, and monitored currents to control the relationship between the magnetizing flux and currents of the machine by modifying the voltages of the cycloconverter and the excitation current in the field winding.

Since each of the anti-parallel converters carries only one direction of the output current, and the output voltage of each converter can have positive or negative polarity, the cycloconverter is suitable for four-quadrant operation and can supply inductively or capacitively excited synchronous machines without the need for any extra equipment. The output currents of the cycloconverter have low harmonic content. The excellent current waveforms enable the synchronous machine to produce a uniform torque at standstill and during starting. In practice there are several power-circuit configurations to connect the cycloconverter to the supply. In one of the applications there is an isolation transformer (or three separate transformers) between the supply network and the cycloconverter phases as

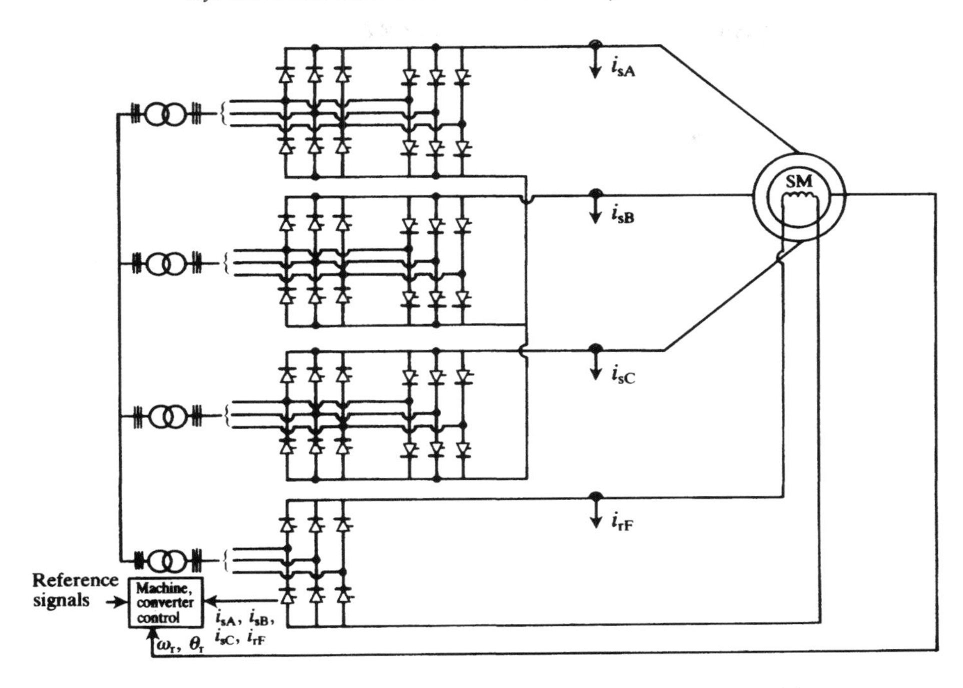

Fig. 3.39. Schematic of the cycloconverter-fed synchronous machine drive.

shown in Fig. 3.39. This is necessary to ensure that the three sets of anti-parallel bridge rectifiers can operate independently.

It should be noted that since the bridge rectifiers are line-commutated, the frequency range is limited to approximately half the supply frequency and thus with a line frequency of 50 Hz, the output frequency of a 6-pulse cycloconverter can vary between approximately 0 and 25 Hz. However, this limited output frequency is of no importance in low-speed drives.

At higher voltages and frequencies the utilization of the cycloconverter can be improved by changing the output voltages of the cycloconverter from sinusoidal to trapezoidal waveforms since the amplitude of the fundamental voltage component exceeds the maximum instantaneous value by 15%. However, it should be ensured that air-gap harmonics, which are uneven multiples of 3, should be as low as possible, to give the best utilization of the drive system. This can be achieved by appropriate machine design.

In the past, for large-power and low-speed applications, d.c. drives have been used. However, owing to the problems associated with d.c. machines discussed earlier and also because in large drives, owing to the limited output power of d.c. machines, it would be necessary to employ a multi-motor d.c. drive, it is more economical to use a single-motor a.c. drive. Since the cycloconverter has practically unlimited output power, it is very suitable for driving low-speed and large-horsepower a.c. motors, e.g. mine-shaft winders, rolling mill drives used in the steel industry, or gearless cement-mill drives. Conventionally low speeds could be obtained by, for example, the use of a four-pole machine together with a gearset having a large gear ratio. However, the maintenance and cost of these large gears is a disadvantage which can be avoided by using the cycloconverter-fed synchronous machine. Cycloconverter-fed synchronous machines are also used in ship propulsion drives, large compressors, large excavators, and large belt conveyor systems. In such drives it is preferable to use synchronous machines rather than induction machines, because of their high power factor and large torque capability at low speeds. Furthermore, a synchronous machine draws less reactive power from the supply. In comparison with converter-fed d.c. drives, in the cycloconverter-fed synchronous machine drive, commutation problems do not exist, the drive has higher efficiency and higher overload capacity, and there are lower maintenance and ventilation requirements. The rotor diameter of the synchronous machine can be kept smaller in comparison with an equivalent d.c. machine and thus the inertia is reduced. The physical size of the synchronous machine fed by a cycloconverter is smaller than that of the converter-fed d.c. machine of equal power rating.

The control scheme described in the present section is suitable for, say, a rolling mill application. For the control system of a rolling mill, the following characteristics are required:

- high torque and power control response;
- quick acceleration/deceleration;
- frequent start/stop operation;
- four-quadrant operation including field-weakening mode;
- high overload capacity from zero to maximum speed;
- production of constant torque at zero speed;
- small torque oscillations.

It should be noted that similar characteristics are required for propulsion systems, where until recently d.c. drives have played a dominant role. However, in the cycloconverter-fed propulsion drive it can be a great disadvantage to employ isolation transformers which are bulky and heavy. Therefore these can be eliminated by isolating the three stator phases of the synchronous machine and by supplying all the stator phases separately from a separate phase of the cycloconverter (cycloconverter with isolated output phases).

All the characteristics described above are almost perfectly combined in the drive containing a salient-pole synchronous machine fed by the line-commutated cycloconverter and subjected to field-oriented control. With the implementation shown in Fig. 3.40, operation is possible in both constant-flux and flux-weakened modes.

Derivation of the reference signals: In Fig. 3.40 the reference rotor speed (ω_{rref}) and the actual monitored value of the rotor speed (ω_r) are compared and their difference is fed into the speed controller, which is a PI controller. The output voltage of the speed controller is proportional to the developed electromagnetic

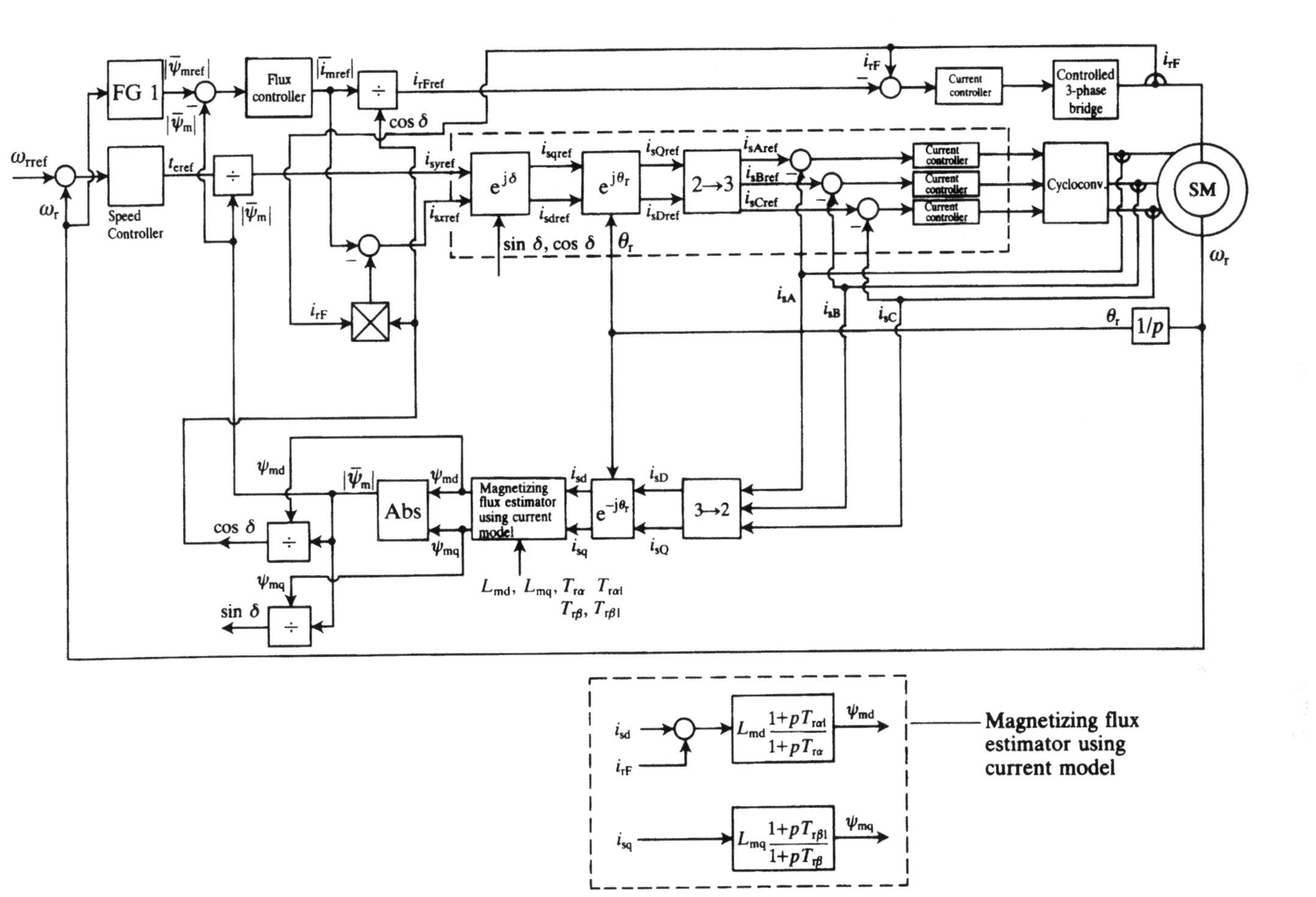

Fig. 3.40. Block diagram of field-oriented control of a cycloconverter-fed salient-pole synchronous machine.

torque (t_e) of the synchronous machine, and thus the reference torque (t_{eref}) is obtained. This is divided by the modulus of the magnetizing flux-linkage space phasor ($|\bar{\psi}_m|$) to yield the reference value of the torque-producing stator current component i_{syref}.

The monitored rotor speed serves as input to function generator FG1, which below base speed yields a constant value of the magnetizing flux reference $|\bar{\psi}_{mref}|$; above base speed this flux is reduced. The flux-linkage reference $|\bar{\psi}_{mref}|$ is compared with the actual value of the magnetizing flux linkage $|\bar{\psi}_m|$ and their difference is fed into the flux controller, which is also a PI-controller. The flux controller maintains the magnetizing flux linkage in the machine at a preset value, independent of the load. To ensure optimal utilization of the motor, the magnetizing flux-linkage reference $|\bar{\psi}_{mref}|$ is raised to a very high value during starting to yield a high breakdown torque, but for normal operating conditions it is set to the rated value. The output of the flux controller is the reference value of the magnetizing current $|\bar{i}_{mref}|$.

In Fig. 3.40 the reference value of the field current (i_{rFref}) and the reference values of the magnetizing and torque-producing stator current components (i_{sxref}, i_{syref}) are obtained as follows. The general expression for the magnetizing-current space phasor in the rotor reference frame is obtained from eqns (3.2-1) and (3.2-2), as

$$\bar{i}_m = i_{md} + ji_{mq} = \frac{\psi_{md}}{L_{md}} + \frac{j\psi_{mq}}{L_{mq}} = i_{sd} + i_{r\alpha} + i_{rF} + j(i_{sq} + i_{r\beta}) \tag{3.2-3}$$

and in accordance with Fig. 3.38, it follows from this equation that because of magnetic saliency, the magnetizing flux-linkage space phasor and the magnetizing-current space phasor are not coaxial. In the steady state the damper currents $i_{r\alpha}$ and $i_{r\beta}$ are zero.

The control strategy is defined in such a way that in the steady state there is no reactive power drawn from the stator. In this case the power factor is maximum ($\cos\phi_s = 1$, where ϕ_s is the phase angle between the stator-current space phasor and the stator-voltage space phasor) and the stator currents are optimal, i.e. at the minimal level. The zero-reactive-power condition can be fulfilled by controlling the stator component i_{sx} to be zero and thus from Fig. 3.38 and eqn (3.2-3), the following expression is obtained for the reference value of the magnetizing current (reference of the modulus of the magnetizing-current vector), if it is assumed that the space angle of the magnetizing-current space phasor and the space angle of the magnetizing flux-linkage space phasor with respect to the direct-axis of the rotor are approximately equal,

$$|\bar{i}_{mref}| = i_{rFref}\cos\delta = (i_{rFref}^2 - i_{syref}^2)^{1/2}, \tag{3.2-4}$$

where i_{rFref} and i_{syref} are the reference values of the field current and quadrature-axis stator current respectively, and the latter current component is formulated in the reference frame fixed to the special reference frame. Thus it follows from eqn (3.2-4) that the reference value of the field current can be obtained as

$$i_{rFref} = \frac{|\bar{i}_{mref}|}{\cos\delta} \tag{3.2-5}$$

or as

$$i_{\mathrm{rFref}}=(|\bar{i}^{2}_{\mathrm{mref}}|^{2}+i^{2}_{\mathrm{syref}})^{1/2}. \tag{3.2-6}$$

Because the field winding has a large inductance, it is sometimes necessary to keep the magnetizing flux linkage constant by keeping the field voltage transiently high. It is therefore effective to provide an additional magnetizing-current component i_{sx} from the stator in the transient state. It should be noted that i_{sx} is the direct-axis stator current component in the special reference frame, and thus it is collinear with the magnetizing flux-linkage space phasor. Thus in the transient state the modulus of the magnetizing-current space phasor is not equal to the magnetizing current component produced by the field current in the real axis of the special reference frame ($i_{\mathrm{rF}}\cos\delta$), but is the sum of the current components $i_{\mathrm{rF}}\cos\delta$ and i_{sx}. It follows that the reference value of the stator current component in the special reference frame can be obtained as

$$i_{\mathrm{sxref}}=|\bar{i}_{\mathrm{mref}}|-i_{\mathrm{rF}}\cos\delta. \tag{3.2-7}$$

This reference value becomes zero in the steady-state when the required magnetizing current is produced only by the field winding.

The stator current components i_{sxref}, i_{syref} are first transformed into the stator current components i_{sdref}, i_{sqref} established in the rotor reference frame by utilizing eqn (3.1-24):

$$i_{\mathrm{sdref}}=\cos\delta\, i_{\mathrm{sxref}}-\sin\delta\, i_{\mathrm{syref}} \tag{3.2-8}$$

$$i_{\mathrm{sqref}}=\sin\delta\, i_{\mathrm{sxref}}+\cos\delta\, i_{\mathrm{syref}}. \tag{3.2-9}$$

These components are transformed into the stationary-axes stator current components i_{sDref}, i_{sQref} by a similar transformation, but taking into account that the phase displacement between the stator direct axis (sD) and the direct axis of the rotor (d) is θ_{r} in accordance with Fig. 3.19:

$$i_{\mathrm{sDref}}=\cos\theta_{\mathrm{r}}\, i_{\mathrm{sdref}}-\sin\theta_{\mathrm{r}}\, i_{\mathrm{sqref}} \tag{3.2-10}$$

$$i_{\mathrm{sQref}}=\sin\theta_{\mathrm{r}}\, i_{\mathrm{sdref}}+\cos\theta_{\mathrm{r}}\, i_{\mathrm{sqref}}. \tag{3.2-11}$$

Of course it is possible to combine these two transformations into a compound transformation and thus the following reference values of the currents are obtained:

$$i_{\mathrm{sDref}}=\cos(\theta_{\mathrm{r}}+\delta)i_{\mathrm{sxref}}-\sin(\theta_{\mathrm{r}}+\delta)i_{\mathrm{syref}} \tag{3.2-12}$$

$$i_{\mathrm{sQref}}=\sin(\theta_{\mathrm{r}}+\delta)i_{\mathrm{sxref}}+\cos(\theta_{\mathrm{r}}+\delta)i_{\mathrm{syref}}. \tag{3.2-13}$$

These equations are similar to the eqns (3.1-28). It should be noted that in eqns (3.2-12) and (3.2-13) the angle $\theta_{\mathrm{r}}+\delta$ appears; this is the space angle between the magnetizing flux-linkage space phasor and the real-axis of the stationary reference frame fixed to the stator. By utilizing the inverse of the transformation matrix defined in eqn (2.1-69), the obtained two-axis stator current references are transformed into the three-phase stator current references by the application of the three-phase to two-phase transformation indicated by the block containing the symbol '2→3'. The reference stator currents are compared with their respective monitored values, and their differences are fed into the respective stator current controllers, which are PI-controllers. The output signals from these current controllers are used to generate the firing pulses of the cycloconverter which supplies the salient-pole synchronous machine.

Similarly, the reference value of the field current and the actual value of the field current are compared and their difference serves as input to the field current controller. This supplies the necessary signal to a controlled three-phase bridge rectifier.

To obtain the reference values of the three-phase stator currents and the reference value of the field current, it is necessary to utilize the rotor speed and the modulus and phase angle of the magnetizing flux-linkage space phasor in various reference frames. Thus it is necessary to obtain $|\bar{\psi}_m|$, the load angle δ, the rotor speed ω_r, and the rotor angle θ_r. For this purpose the rotor speed is monitored, e.g. by a d.c. tachogenerator. In Fig. 3.40 the rotor angle is obtained by using the integration block. In practice this can be obtained by using a resolver, or by other techniques described in earlier sections. The load angle, the components and the modulus of the magnetizing-flux space phasor can be obtained as described in the next section.

Determination of the components of the magnetizing flux-linkage space phasor using a current model: The two-axis components of the magnetizing flux-linkage space phasor ψ_{md}, ψ_{mq} can be determined in many ways. It is possible to use a method similar to the one described in the case of the interior-permanent-magnet machine. Thus these flux linkages can be determined from the monitored values of the machine currents and some machine parameters. However, in contrast to the technique described in Section 3.1.2 for the IPMSM, there are now damper circuits as well and it is not possible to monitor the damper currents directly. Therefore it is not possible to use directly the flux linkage equations, eqns (3.2-1) and (3.2-2), but when the voltage equations of the machine are considered and the damper currents are eliminated from the rotor voltage equations, it is possible to obtain expressions for the required magnetizing flux components in terms of the stator currents and the field current. Such a current model is now described.

It follows from eqns (3.1-36) and (3.1-37) that the stator voltage equations of the salient-pole synchronous machine in the reference frame fixed to the rotor can be put into the form,

$$u_{sd}=R_s i_{sd}+\frac{d\psi_{sd}}{dt}-\omega_r\psi_{sq} \qquad (3.2\text{-}14)$$

$$u_{sq}=R_s i_{sq}+\frac{d\psi_{sq}}{dt}+\omega_r\psi_{sd}, \qquad (3.2\text{-}15)$$

where the two-axis components of the stator flux linkages are

$$\psi_{sd}=L_{sl}i_{sd}+\psi_{md} \qquad (3.2\text{-}16)$$

and

$$\psi_{sq} = L_{sl} i_{sq} + \psi_{mq}. \tag{3.2-17}$$

R_s and L_{sl} are the resistance and leakage inductance respectively of a stator winding and the magnetizing flux-linkage components are defined by eqns (3.2-1) and (3.2-2). The voltage equation for the field winding can be put into the following form in the reference frame fixed to the rotor:

$$u_{rF} = R_{rF} i_{rF} + \frac{d\psi_{rF}}{dt}, \tag{3.2-18}$$

where R_{rF} is the resistance of the field winding. ψ_{rF} is the flux linking the field winding and can be expressed as:

$$\psi_{rF} = L_{F1} i_{rF} + \psi_{md}, \tag{3.2-19}$$

where L_{F1} is the leakage inductance of the field winding. Similar equations hold for the damper circuits:

$$0 = R_{r\alpha} i_{r\alpha} + \frac{d\psi_{r\alpha}}{dt} \tag{3.2-20}$$

$$0 = R_{r\beta} i_{r\beta} + \frac{d\psi_{r\beta}}{dt}, \tag{3.2-21}$$

where $R_{r\alpha}$ and $R_{r\beta}$ are the resistances of the damper windings along the direct and quadrature axes respectively, $i_{r\alpha}$ and $i_{r\beta}$ are the currents in the two damper windings, and $\psi_{r\alpha}$ and $\psi_{r\beta}$ are the flux components linking the damper windings. These can be expressed as

$$\psi_{r\alpha} = L_{r\alpha 1} i_{r\alpha} + \psi_{md} \tag{3.2-22}$$

$$\psi_{r\beta} = L_{r\beta 1} i_{r\beta} + \psi_{mq}, \tag{3.2-23}$$

where $L_{r\alpha 1}$ and $L_{r\beta 1}$ are the leakage inductances of the damper windings in the direct and quadrature axes of the rotor respectively.

Expressions for the magnetizing flux linkages which contain the stator and field current components and not the damper currents can be obtained from the damper voltage equations by eliminating the damper currents. Thus by considering eqns (3.2-l), (3.2-20), and (3.2-22), the voltage equation for the α-axis damper winding can be put into the following form

$$T_{r\alpha} \frac{d\psi_{md}}{dt} + \psi_{md} = L_{md}(i_{sd} + i_{rF}) + L_{md} T_{r\alpha 1} \frac{d(i_{rF} + i_{sd})}{dt}, \tag{3.2-24}$$

where $T_{r\alpha}$ is the time constant of the damper winding along the α-axis, $T_{r\alpha} = (L_{md} + L_{r\alpha 1})/R_\alpha$, and $T_{r\alpha 1}$ is the leakage time constant of the same damper winding, $T_{r\alpha 1} = L_{r\alpha 1}/R_\alpha$.

Similarly by considering eqns (3.2-2), (3.2-21), and (3.2-23), the voltage equation of the damper winding along the β axis can be obtained as

$$T_{r\beta} \frac{d\psi_{mq}}{dt} + \psi_{mq} = L_{mq} i_{sq} + L_{mq} T_{r\beta 1} \frac{di_{sq}}{dt}, \tag{3.2-25}$$

where $T_{r\beta}$ is the time constant of the β-axis damper winding, $T_{r\beta}=(L_{mq}+L_{r\beta1})/R_{\beta}$, and $T_{r\beta1}$ is the leakage time constant of the damper winding in the β-axis, $T_{r\beta1}=L_{r\beta1}/R_{\beta}$.

Equations (3.2-24) and (3.2-25) can be arranged as

$$\psi_{md}=L_{md}(i_{sd}+i_{rF})\frac{1+pT_{r\alpha1}}{1+pT_{r\alpha}} \tag{3.2-26}$$

$$\psi_{mq}=L_{mq}i_{sq}\frac{1+pT_{r\beta1}}{1+pT_{r\beta}}, \tag{3.2-27}$$

where $p=\mathrm{d}/\mathrm{d}t$. When the magnetizing inductances and the time constants used in these equations are known (e.g. by utilizing the results of measurements performed prior to starting up the drive system) and the currents i_{sd}, i_{sq}, i_{rF} are obtained from monitored currents, eqns (3.2-26) and (3.2-27) can be used to obtain the magnetizing flux linkages ψ_{md}, ψ_{mq} in the direct and quadrature axes of the rotor. For this purpose it is necessary to obtain i_{sd} and i_{sq} from the monitored stator currents i_{sA}, i_{sB}, i_{sC}, and therefore to use the inverse of the above transformations. This is also shown in Fig. 3.40, where the monitored stator currents are first transformed into their stationary two-axis components i_{sD} and i_{sQ} by the application of the three-phase to two-phase transformation, and these components are then transformed into the i_{sd} and i_{sq} currents by the application of the block containing the $e^{-j\theta_r}$ transformation. The currents i_{sd}, i_{sq}, and i_{rF} serve as inputs to the block which contains the magnetizing flux-linkage estimator, which is based on eqns (3.2-26) and (3.2-27).

To achieve successful field-oriented control, it is extremely important to obtain accurate values of the flux-linkage components. However, this requires accurate values of the machine parameters in the current model described by eqns (3.2-26) and (3.2-27).

In contrast to the current model, it is also possible to use a voltage model, where the magnetizing flux-linkage components in the reference frame fixed to the stator (ψ_{mD}, ψ_{mQ}) are directly obtained from the terminal voltages. In this case the monitored terminal voltages of the salient-pole synchronous machine are reduced by their respective ohmic and leakage voltage drops and the resulting magnetizing voltages are integrated. However, when such a scheme is implemented, at low speeds problems can arise concerning the accuracy of the determined flux-linkage components. It is possible to have a hybrid implementation, where at lower speeds (say below 10% of the rated speed), the current model is used for the estimation of the flux-linkage components and at higher speed the voltage model is used.

Derivation of the load angle and modulus of the magnetizing flux-linkage space phasor: By considering the vector diagram shown in Fig. 3.38, the load angle can be obtained from

$$\cos\delta=\frac{\psi_{md}}{|\bar{\psi}_m|} \tag{3.2-28}$$

or

$$\sin\delta = \frac{\psi_{mq}}{|\bar{\psi}_m|} \tag{3.2-29}$$

and the modulus of the magnetizing flux-linkage space phasor is obtained as

$$|\bar{\psi}_m| = (\psi_{md}^2 + \psi_{mq}^2)^{1/2}. \tag{3.2-30}$$

Equations (3.2-28) and (3.2-29) are also utilized in Fig. 3.40, and the block labelled 'Abs' yields the absolute value of the magnetizing flux-linkage space phasor in accordance with eqn (3.2-30).

In the control scheme described in Fig. 3.40 it has been necessary to use the monitored rotor angle in the blocks containing the transformation $e^{j\theta_r}$ and its inverse. However, when the space phasor of the magnetizing flux linkages is directly determined in terms of its stationary-axes components, ψ_{mD} and ψ_{mQ}, which are obtained from the monitored stator voltages and currents in the way described above, it is not necessary to use the monitored value of the rotor angle for the determination of the two-axis stator current references i_{sDref}, i_{sQref} since it follows from Fig. 3.38 that

$$\cos(\delta + \theta_r) = \frac{\psi_{mD}}{|\bar{\psi}_m|} \tag{3.2-31}$$

or

$$\sin(\delta + \theta_r) = \frac{\psi_{mQ}}{|\bar{\psi}_m|}, \tag{3.2-32}$$

and thus the required angle in the compound transformation $e^{j(\theta_r + \delta)}$ is obtained from ψ_{mD} and $|\bar{\psi}_m|$ or ψ_{mQ} and $|\bar{\psi}_m|$ by using eqn (3.2-31) or eqn (3.2-32).

It should be noted that by using high-speed microprocessors or signal processors it is possible to perform the computation of the direct- and quadrature-axis magnetizing flux linkages, the modulus of the magnetizing flux-linkage space phasor and the load angle, as well as all the required transformations.

Current control with e.m.f. compensation: The control system described in Fig. 3.40 is similar to the control of a d.c. drive with a bridge converter. However, at higher output frequencies of the cycloconverter, even in the steady state, owing to a.c. induced stator voltages, which disturb the current control system shown in Fig. 3.40, there will be a phase error and also an amplitude (gain) error which the current controllers cannot eliminate. There are several ways of eliminating these errors and in the present section a current control scheme with e.m.f. compensation is described.

In the current control system shown in Fig. 3.41, which replaces the components surrounded with a broken line in Fig. 3.40, feedforward voltages are added to the outputs of the current controllers and the cycloconverter is mainly controlled by the feedforward stator voltage references u_{isAref}, u_{isBref}, and u_{isCref}.

These voltages are obtained by utilizing the rotor speed and the direct- and quadrature-axis components of the stator flux linkages. It follows from eqns (3.2-14) and (3.2-15) that the direct- and quadrature-axis components of the feedforward e.m.f.s in the reference frame fixed to the rotor are

$$u_{isd} = -\omega_r \psi_{sq} \tag{3.2-33}$$

and

$$u_{isq} = \omega_r \psi_{sd}, \tag{3.2-34}$$

where the direct- and quadrature-axis stator flux linkages have been defined by eqns (3.2-16) and (3.2-17) respectively. Thus to obtain the corresponding two-axis components u_{isD} and u_{isQ} in the stationary reference frame, by considering eqn (2.1-143), it is necessary to perform the transformation $u_{isD}+ju_{isQ}=(u_{isd}+ju_{isq})e^{j\theta_r}$, and the transformation $e^{j\theta_r}$ is also shown in Fig. 3.41. This transformation is followed by the two-phase to three-phase transformation and thus the induced voltages u_{isA}, u_{isB}, and u_{isC} are obtained.

The induced stator voltages can be obtained from eqns (3.2-33) and (3.2-34), together with eqns (3.2-16) and (3.2-17). Thus the block containing the induced voltage estimator uses as its inputs the monitored rotor speed and the stator currents i_{sd}, i_{sq}. These stator current components are obtained from the monitored three-phase stator currents by utilizing the three-phase transformation and the complex transformation $e^{-j\theta_r}$, as shown in Fig. 3.40.

Current control with no e.m.f. compensation, but utilizing controllers operating in the magnetizing-flux-oriented reference frame: It is possible to have an implementation of the current control loops where there is no e.m.f. compensation but the drive will still have excellent control characteristics. In such a control system integrating current controllers operating in the magnetizing-flux-oriented reference frame are used, since the deviations in the steady state between the reference stator currents and the actual values of the stator currents will be zero when i_{sx} and i_{sy} are controlled by integrating (I) controllers. Figure 3.42 shows a possible implementation and this should replace the components surrounded with a broken line in Fig. 3.40.

In Fig. 3.42 the differences $i_{sxref}-i_{sx}$ and $i_{syref}-i_{sy}$ serve as inputs to the respective I controllers. The output signals of these controllers are transformed into the three-phase components of the stationary reference frame by the successive application of the transformation $e^{j(\theta_r+\delta)}$ and the three-phase to two-phase transformation, since the first transformation transforms the real- and imaginary-axis components of the magnetizing-flux-oriented reference frame into the real- and imaginary-axis components of the stationary reference frame. The stator currents i_{sx} and i_{sy} are obtained from the monitored values of the three-phase stator currents by utilizing the three-phase to two-phase transformation and also the transformation $e^{-j(\theta_r+\delta)}$.

However, when only the currents i_{sx} and i_{sy} are controlled, any imbalance and non-linearity of the gain of the $e^{-j(\theta_r+\delta)}$ and $2\rightarrow3$ coordinate converters, gate

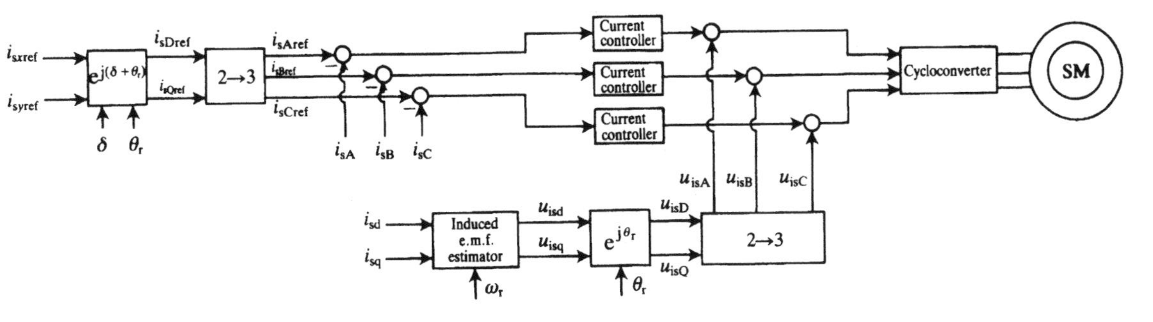

Fig. 3.41. Block diagram of field-oriented control of the salient-pole synchronous machine incorporating stator current control with e.m.f. compensation.

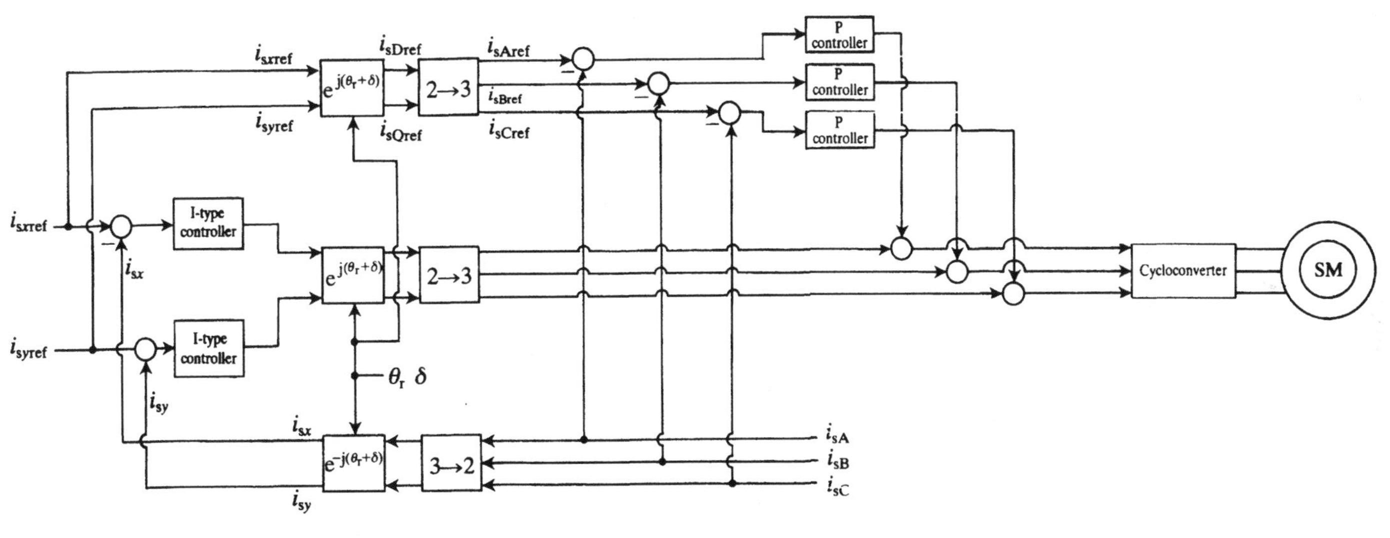

Fig. 3.42. Current control using current controllers operating in the magnetizing-flux-oriented reference frame.

pulse generators of the cycloconverter, etc. are amplified by the gain of the stator circuit of the synchronous machine under consideration. It is possible to eliminate this deficiency by controlling the three-phase stator currents, which are a.c. quantities, using proportional (P) controllers. For this purpose it is necessary to obtain the signals $i_{sAref}-i_{sA}$, $i_{sBref}-i_{sB}$, and $i_{sCref}-i_{sC}$ respectively, and in Fig. 3.42 the reference stator currents i_{sAref}, i_{sBref}, and i_{sCref} are obtained from i_{sxref} and i_{syref} by the application of the transformation $e^{j(\theta_r+\delta)}$ and the two-phase to three-phase transformation. The output signals of the P controllers are added to the output signals of the transformed signals of the I controllers and the signals obtained are the commands for the gate control of the cycloconverter. With the implementation shown in Fig. 3.42 a high-response current control is achieved.

3.2.2 VECTOR CONTROL OF SYNCHRONOUS RELUCTANCE MOTORS (SYRM)

In the present section the general aspects of synchronous reluctance motors are first described. This is followed by a description of implementations using position sensors and then position-sensorless implementations.

3.2.2.1 General introduction; rotor-oriented control using a position sensor

The synchronous reluctance motor (SYRM) is a singly-salient synchronous motor where the symmetrical three-phase sinusoidally distributed stator windings are excited with balanced a.c. currents and there is a reluctance rotor. The three-phase stator windings are situated in the smooth but slotted stator bore. The reluctance rotor is made of steel laminations, it is salient-pole, but does not have any windings or magnets. The rugged simple structure, low-cost manufacturing, possibility of high torque per unit volume, and the absence of rotor windings resulting in simple control schemes and decreased losses make this motor an attractive candidate for numerous industrial and automotive applications. However, the torque density, power factor, and efficiency of the SYRM is only high if the saliency ratio (L_{sd}/L_{sq}) is high, and if $L_{sd}-L_{sq}$ is high. Furthermore, in general, a SYRM is characterized by torque pulsations, vibration and acoustic problems, but these are not as severe as for variable reluctance, doubly-salient machines. Both vector control and direct torque control techniques can be applied to synchronous reluctance motors.

It is interesting to note that, despite the fact that synchronous reluctance motors were discussed as early as 1923 (the early versions with conventional reluctance rotor are characterized by high torque pulsations, low torque density, very low power factor and efficiency) and axially laminated versions appeared in the 1960s (which improved the performance), and more advanced axially laminated versions have been available for many years, at present there are only a few companies who manufacture these motors. However, it is expected that the practical applications of the SYRM will increase in the future especially if reliable position- and speed-sensorless drives are industrially implemented.

In earlier constructions, rotor saliency was achieved by removing certain teeth from the rotors of conventional squirrel cages. Such synchronous reluctance machines with low output power have been used for a long time. Their inferior performance combined with their relatively high price have resulted in their limited use. However, as a result of recent developments, more reliable and robust constructions exist. In the new type of synchronous reluctance machines there are basically three types of rotors: segmential, flux barrier, and axially laminated rotors. In the SYRM with segmental rotor, saliency ratios of 6–7 have been obtained. If the number of rotor segments is very large, then a distributed anisotropic structure is obtained, which is similar to the various axially laminated structures used in the past. By using multiple segmental structures, the saliency ratio can be increased. In the SYRM with axially laminated rotor, the rotor is made of conventional axial lamination bent into U or V shapes and stacked in the radial direction. With this structure it is possible to produce very high saliency ratios, and L_{sd}/L_{sq} ratios of 9–12 have been obtained. This also leads to fast torque responses. However, it should be noted that there is a physical limit to the maximum value of L_{sd}/L_{sq}, since the maximum value of L_{sd} is the synchronous inductance, and the minimum value of L_{sq} is the stator leakage inducance. Synchronous reluctance motors with high rotor anisotropy are suited for high-performance applications, such as machine-tool drives, robotics, electric vehicles, electric traction, etc., and can become strong competitors for existing variable-speed brushless high-performance drives. The new type of synchronous reluctance machines have higher output power which is comparable to that of corresponding induction machines. However, it is a distinct advantage over induction machines that reluctance machines run at exactly the synchronous speed—which is solely determined by the stator excitation frequency and the number of poles—and there are no rotor currents. These are important aspects in many industrial applications.

It is possible to operate reluctance motors without a rotor cage. Conventionally, it is necessary to have a cage winding in a reluctance machine, since the machine does not have a starting torque when excited from a constant frequency supply. By using the cage winding, the machine runs up to synchronous speed by induction motor action where the rotor locks into synchronism with the field produced by the stator. For sinusoidal stator excitation, at synchronous speed the only role of the cage winding is to damp the oscillations in the rotor speed. However, if variable stator excitation is used, it is not necessary to have a rotor cage for starting purposes, since the excitation can be controlled in such a way that the motor is always kept in synchronism. Reluctance machines with cageless rotors lead to a reduction of the rotor losses, improved efficiency, higher power factor, and higher torque/weight ratio.

The operation of the SYRM is now briefly discussed. Similarly to the switched reluctance motor discussed in Chapter 5, the synchronous reluctance motor utilizes the principle that electromagnetic torque is produced to minimize the reluctance of the magnetic paths. The three-phase stator currents carry balanced three-phase currents, and thus a rotating air-gap flux is produced. When the rotor rotates in synchronism with the air-gap flux, torque is produced which tries to align the minimum reluctance path of the rotor with the rotating air-gap flux. When a load torque is present, the rotor begins to lag the rotating air-gap flux producing a misalignment of the minimum reluctance path and the rotating air-gap flux. Thus electromagnetic torque is produced to minimize the reluctance which tries to maintain alignment. When this torque is equal and opposite to the load torque on the rotor, the rotor will again rotate with synchronous speed.

For better understanding, in Table 3.1 the SYRM is compared with the switched reluctance motor (SRM). It can be seen that the SYRM is a singly-excited, singly-salient rotating field motor, in contrast to the SRM, which is also a singly-excited, but doubly-salient motor. In the SYRM the stator currents are polyphase a.c. currents and in the SRM the currents are unidirectional (see also Chapter 5). The stator windings of the SYRM are polyphase windings, whilst the stator windings of the SRM are concentrated windings. The stator windings of the SYRM are similar to those of a three-phase induction motor. A converter-fed SYRM does not require a starting cage. Thus the rotor can be designed purely for synchronous performance.

Similarly to the SRM drive, the SYRM drive also requires the information on the rotor position for closed-loop control, but this information is in addition required for starting purposes as well. This is a disadvantage of SYRM drives; however, speed and position information can be obtained without using speed and position sensors, as discussed below. Both vector control and direct torque control techniques can be applied to the SYRM, but it is possible to have implementations of these drives where there is no conventional position sensor present ('sensorless' drives). The anisotropic rotor of the SYRM offers an advantage for rotor position estimation since, due to anisotropy, information on the rotor position can be simply extracted even at low speeds including zero speed, e.g. from the monitored stator currents.

It was shown in Chapter 1 that the electromagnetic torque of a synchronous reluctance motor can also be expressed similarly to that of the induction motor or the permanent-magnet synchronous motor. For convenience eqn (2.1-9) is

Table 3.1 Main features of switched reluctance motors (SRM) and synchronous reluctance motors (SYRM)

	SRM	SYRM
Stator	salient pole	smooth bore (but slotted)
Rotor	salient pole	salient pole
Stator winding	concentrated	multiphase
Rotor winding	—	—
Stator currents	unidirectional	multiphase balanced
Control	position feedback	position feedback

repeated here:

$$t_e = \tfrac{3}{2}P\bar{\psi}'_s \times \bar{i}'_s = \tfrac{3}{2}P(\psi_{sd}i_{sq} - \psi_{sq}i_{sd}) = \tfrac{3}{2}P(L_{sd} - L_{sq})i_{sd}i_{sq}, \qquad (3.2\text{-}35)$$

where L_{sd} and L_{sq} are the direct- and quadrature-axis synchronous inductances respectively (sometimes these are denoted in the literature by L_d and L_q). In eqn (3.2-35) the stator current components (i_{sd}, i_{sq}) are expressed in the rotor reference frame, and can be obtained from the stator currents (i_{sD}, i_{sQ}) in the stationary reference frame by considering

$$\bar{i}'_s = i_{sd} + \mathrm{j}i_{sq} = \bar{i}_s \exp(-\mathrm{j}\theta_r) = (i_{sD} + \mathrm{j}i_{sQ})\exp(-\mathrm{j}\theta_r), \qquad (3.2\text{-}36)$$

where θ_r is the rotor angle, and it is this transformation which implies the use of a position sensor, which most often is a resolver. Equation (3.2-35) can also be put into the following form, if $\psi_{sd} = L_{sd}i_{sd}$ is utilized:

$$t_e = \tfrac{3}{2}P(L_{sd} - L_{sq})i_{sd}i_{sq} = [\tfrac{3}{2}P(1 - L_{sq})/L_{sd}]\psi_{sd}i_{sq} = c\psi_{sd}i_{sq}, \qquad (3.2\text{-}37)$$

where $c = (3/2)P(1 - L_{sq})/L_{sd}$ and in general it is saturation dependent. The direct-axis synchronous inductance L_{sd} depends strongly on i_{sd}, and at very high torques on i_{sq} as well. The direct-axis stator flux is also saturation dependent, in general it also depends on i_{sq} and decreases when the quadrature-axis current is increased (cross-saturation effect). In general there is cross-saturation coupling between the d and q axes, due to saturation. Equation (3.2-37) greatly resembles the electromagnetic torque expression of a separately excited d.c. motor, that of the vector-controlled induction motor, and also that of the permanent-magnet synchronous motor (PMSM) with surface magnets (see also Section 1.2). When the torque expression of the PMSM motor [$t_e = c_F\psi_F i_{sq}$ as shown in eqn (2.1-5)] is compared with that for the synchronous reluctance motor given by eqn (3.2-37), it follows that the torque of the synchronous reluctance motor is lower than that of the PMSM, since eqn (3.2-37) contains the factor $(1 - L_{sq})/L_{sd} < 1$ and also because the quadrature-axis stator current of the synchronous reluctance motor is smaller than that of the PMSM, due to the magnetizing requirement. However, at high speeds the synchronous reluctance motor can be superior to the PMSM, because the iron losses can be consistently reduced by some type of flux weakening.

For rotor-oriented control, eqn (3.2-35) [or eqn (3.2-37)] together with eqn (3.2-36) can be used. Two control strategies will be discussed below: current control and combined current–voltage control.

3.2.2.1.1 Current-type control

In a vector-controlled synchronous reluctance motor drive, where the motor is supplied by a current-controlled PWM inverter, and where rotor-oriented control is performed, independent control of the torque and flux (torque-producing stator current and flux-producing stator current) can be achieved by the current control scheme shown in Fig. 3.43.

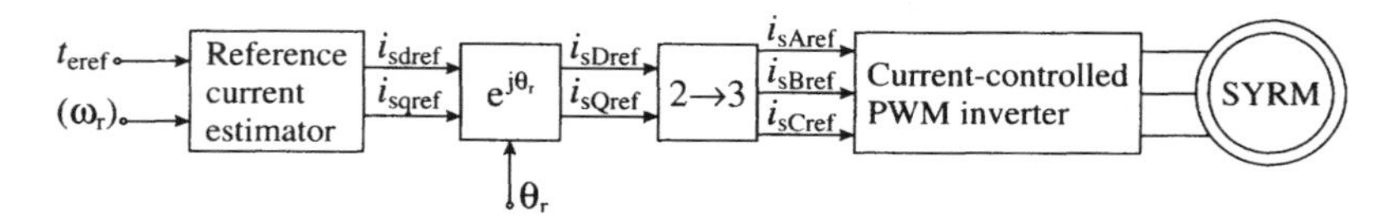

Fig. 3.43. Schematic of rotor-oriented current control of a synchronous reluctance motor (supplied by a current-controlled PWM inverter) using a position sensor.

The gating signals of the six switching devices of the inverter are obtained on the output of hysteresis current controllers. On the inputs of these the difference between the actual (measured) and reference stator line currents are present. The three-phase stator current references are obtained from their two-axis components (i_{sD}, i_{sQ}) by the application of the two-phase to three-phase ($2 \rightarrow 3$) transformation, and these are obtained from the reference values of i_{sd}, i_{sq} (which are i_{sdref}, i_{sqref}) by the application of the $\exp(j\theta_r)$ transformation. These reference values are obtained in the appropriate estimation block from the reference value of the torque (t_{eref}) and sometimes the monitored rotor speed (ω_r) is also used (e.g. when there is field weakening). This estimation block can be implemented in various ways, according to the required control strategy. In general there are two main types of control strategies:

- constant angle (γ) control;
- control with constant i_{sd} and controlled i_{sq}.

The second type of control is more suitable below base speed; it gives the highest rate of change of torque at low speeds. The expressions for i_{sdref} below and above base speed are now given.

Expression for i_{sdref} below base speed Below base speed, i_{sd} is kept constant:

$$i_{sdref} = K = \text{constant} \tag{3.2-38}$$

where

$$K = \frac{\psi_{smaxref}}{L_{sd}\sqrt{2}}. \tag{3.2-39}$$

In eqn (3.2-39) the maximal value of the reference stator flux linkage can be obtained from the reference value of the electromagnetic torque as

$$\psi_{smaxref} = \left[\frac{4|t_{eref}|L_{sd}L_{sq}}{3P(L_{sd}-L_{sq})}\right]^{1/2}. \tag{3.2-40}$$

This expression follows from eqn (3.2-44) which is given below. It should be noted that in general the inductance L_{sd} is constant, it depends greatly on i_{sd} (and due to cross-saturation, on i_{sq} as well). When the cross-saturation effect is neglected, L_{sd} depends only on i_{sd} and the non-linear function $L_{sd}(i_{sd})$ can be stored in a look-up table and can be used for the generation of i_{sdref}.

Expression for i_{sdref} above base speed Above base speed i_{sdref} is decreased with the speed:

$$i_{\mathrm{sdref}}=\frac{K\omega_{\mathrm{base}}}{|\omega_{\mathrm{r}}|}. \tag{3.2-41}$$

Expression for i_{sqref} By considering eqn (3.2-35), the quadrature-axis stator current reference can be obtained as

$$i_{\mathrm{sqref}}=\frac{t_{\mathrm{eref}}}{3P(L_{\mathrm{sd}}-L_{\mathrm{sq}})i_{\mathrm{sdref}}}. \tag{3.2-42}$$

When i_{sd} varies, L_{sd} also varies and t_{e} is not proportional to i_{sq}, and thus to eliminate the effects of parameter detuning, more sophisticated techniques (application of self-tuning controllers, or model reference adaptive controllers etc.) must be used to produce i_{sqref}.
There are three types of different constant-angle control strategies:

- fastest torque control (gives the fastest torque response);
- maximum torque/ampere control (for this control, the inverter runs out of volts at rated speed);
- maximum power factor control (operates the machine at highest power factor).

The estimation block in Fig. 3.43 contains the estimation of the reference currents i_{sdref}, i_{sqref} and as shown below, these depend on $\tan\gamma$ for the three constant-angle control strategies. In the next three subsections the estimation of the reference currents will be discussed and the expressions for $\tan\gamma$ will be obtained for the three control strategies.

Fastest torque control strategy It will now first be proved that fastest torque response can be obtained by a controller where $\gamma=\tan^{-1}(L_{\mathrm{sd}}/L_{\mathrm{sq}})$ and γ is the angle of the stator-current space vector with respect to the real-axis of the rotor reference frame (d-axis) as shown in Fig. 1.4(b) in Section 1.2. This is a so-called constant-angle control strategy. Fastest torque response can be obtained for maximum torque for a given stator flux. From $\bar{\psi}_{\mathrm{s}}'=\psi_{\mathrm{sd}}+\mathrm{j}\psi_{\mathrm{sq}}=L_{\mathrm{sd}}i_{\mathrm{sd}}+\mathrm{j}L_{\mathrm{sq}}i_{\mathrm{sq}}$ the flux vector modulus is $|\bar{\psi}_{\mathrm{s}}|=[(L_{\mathrm{sd}}i_{\mathrm{sd}})^2+(L_{\mathrm{sq}}i_{\mathrm{sq}})^2]^{1/2}$, it follows by substitution into eqn (3.2-35) that the electromagnetic torque for a given flux modulus is

$$t_{\mathrm{e}}=\tfrac{3}{2}P[(L_{\mathrm{sd}}-L_{\mathrm{sq}})/L_{\mathrm{sq}}]i_{\mathrm{sd}}[|\bar{\psi}_{\mathrm{s}}|^2-L_{\mathrm{sd}}^2i_{\mathrm{sd}}^2]^{1/2}. \tag{3.2-43}$$

Thus the maximum torque for a given stator flux (modulus) can be obtained from eqn (3.2-43) by considering $\partial t_{\mathrm{e}}/\partial i_{\mathrm{sd}}=0$. This yields $i_{\mathrm{sd}}=|\bar{\psi}_{\mathrm{s}}|/[L_{\mathrm{sd}}\sqrt{2}]$ and also $i_{\mathrm{sq}}=|\bar{\psi}_{\mathrm{s}}|/[L_{\mathrm{sq}}\sqrt{2}]$, and thus the maximum torque can be obtained from eqn (3.2-43) as

$$t_{\mathrm{emax}}=\frac{\tfrac{3}{2}P(L_{\mathrm{sd}}-L_{\mathrm{sq}})|\bar{\psi}_{\mathrm{s}}|^2}{2L_{\mathrm{sd}}L_{\mathrm{sq}}}. \tag{3.2-44}$$

It follows that for fastest torque response, the torque is maximum for a given stator flux if the following condition holds:

$$\frac{i_{sq}}{i_{sd}}=\frac{|\bar{\psi}_s|/[L_{sq}\sqrt{2}]}{|\bar{\psi}_s|/[L_{sd}\sqrt{2}]}=\frac{L_{sd}}{L_{sq}}. \tag{3.2-45}$$

However it follows from Fig. 1.4(b) in Section 1.2 that $\tan\gamma=i_{sq}/i_{sd}$, thus by also considering eqn (3.2-45),

$$\gamma=\tan^{-1}(L_{sd}/L_{sq}) \tag{3.2-46}$$

is obtained. The estimation block in Fig. 3.43 contains the estimation of the reference currents as follows. Substitution of $|\bar{\psi}_s|=[L_{sd}\sqrt{2}]i_{sd}$ into eqn (3.2-44), rearranging for i_{sd}, and replacing the actual torque by its reference torque value (t_{eref}), gives

$$i_{sdref}=\left[\frac{2t_{eref}\tan\gamma}{3P(L_{sd}-L_{sq})}\right]^{1/2}. \tag{3.2-47}$$

By utilizing eqns (3.2-45) and (3.2-46), the quadrature-axis stator current reference can be obtained from

$$i_{sqref}=\frac{i_{sdref}\,\mathrm{sgn}(t_{eref})}{\tan\gamma}. \tag{3.2-48}$$

Maximum torque/ampere control strategy The second control strategy considered is maximum torque/ampere control. It will now be proved that for this case $\gamma=\pi/4$ is required. It follows from Fig. 1.4(b) that $\bar{i}_s=|\bar{i}_s|\exp[j(\gamma+\theta_r)]$, where γ is the torque angle, thus the transformed stator current is

$$\bar{i}_s'=\bar{i}_s\exp(-j\theta_r)=|\bar{i}_s|\exp[j(\gamma+\theta_r)]\exp(-j\theta_r)=|\bar{i}_s|\exp(j\gamma), \tag{3.2-49}$$

and substitution into eqn (3.2-35) gives

$$t_e=\tfrac{3}{2}P(L_{sd}-L_{sq})|\bar{i}_s|^2\frac{\sin(2\gamma)}{2}. \tag{3.2-50}$$

Thus maximum torque/ampere is obtained with $\gamma=\pi/4$. The reference stator current components can be determined by using eqns (3.2-47) and (3.2-48) with $\gamma=\pi/4$.

Maximum power factor control strategy The third control strategy is maximum power factor control. It will now be proved that this requires $\gamma=\tan^{-1}[(L_{sd}/L_{sq})^{1/2}]$ and the reference stator current components can be determined by using eqns (3.2-47) and (3.2-48) with this γ value.

If the stator winding losses and core losses are neglected, the power factor can be obtained as follows:

$$\cos\phi=\frac{t_e\omega_1/P}{(3/2)|\bar{u}_s||\bar{i}_s|}. \tag{3.2-51}$$

Substitution of the torque expression from eqn (3.2-35) into eqn (3.2-51) and by also considering $|\bar{u}_s| = \omega_1 |\bar{\psi}_s|$ where $|\bar{\psi}_s| = [(L_{sd} i_{sd})^2 + (L_{sq} i_{sq})^2]^{1/2}$, and $|\bar{i}_s| = (i_{sd}^2 + i_{sq}^2)^{1/2}$,

$$\cos\phi = \frac{(L_{sd} - L_{sq}) i_{sd} i_{sq}}{|\bar{\psi}_s||\bar{i}_s|} = \frac{[(L_{sd}/L_{sq}) - 1] i_{sd} i_{sq}}{[(i_{sd} L_{sd}/L_{sq})^2 + i_{sq}^2]^{1/2} (i_{sd}^2 + i_{sq}^2)^{1/2}} \tag{3.2-52}$$

is obtained. It can be seen that the power factor depends on the ratio of the synchronous inductances. This equation can be manipulated into a form such that the ratio of the stator currents (i_{sq}/i_{sd}) is present and thus the maximum power factor can be obtained from eqn (3.2-52) by considering $\partial(\cos\phi)/\partial(i_{sq}/i_{sd}) = 0$. This gives

$$\frac{i_{sq}}{i_{sd}} = \left(\frac{L_{sd}}{L_{sq}}\right)^{1/2}. \tag{3.2-53}$$

3.2.2.1.2 Combined current–voltage control

In this section the rotor-oriented control of the synchronous reluctance machine is discussed, but a scheme using combined current and voltage control is described. When such a scheme is used, delays in the a.c. currents and saturation at high speeds can be avoided. The scheme is similar to the corresponding one developed later in the book for an induction machine. The machine is supplied by a PWM voltage source inverter and is current-controlled along the direct (d) and quadrature (q) axes of the rotating reference frame fixed to the rotor. It is assumed that the drive operates in both constant-torque and constant-power regions. Figure 3.44 shows the block diagram of the drive system.

In Fig. 3.44 the rotor speed (ω_r) is monitored and the rotor angle (θ_r) is also obtained. However, it should be noted that the integration block is only symbolic; there are many ways of obtaining the variation of the rotor angle in real time, by the use of a resolver. It is also possible to obtain the rotor speed from the rotor angle from the expression

$$\omega_r = \cos\theta_r \frac{d(\sin\theta_r)}{d\theta_r} - \sin\theta_r \frac{d(\cos\theta_r)}{d\theta_r}. \tag{3.2-54}$$

The monitored rotor speed is fed into the function generator FG1, the output of which is the direct-axis stator flux reference (ψ_{sdref}). During field weakening, this reference signal is a function of the rotor speed.

The direct-axis stator flux linkage ψ_{sd} is obtained with the stator flux-estimation circuit surrounded with the dashed lines. Below base speed, the direct-axis stator flux loop is current controlled. The difference between ψ_{sdref} and ψ_{sd} is fed into a flux controller, a PI controller, and the output is the direct-axis stator current reference i_{sdref}. The error $\Delta i_{sd} = i_{sdref} - i_{sd}$ is fed into a current controller, also a PI controller, and the output signal of this controller is added to the rotational voltage component, $\omega_r \psi_{sq}$, where ψ_{sq} is the quadrature-axis stator flux linkage

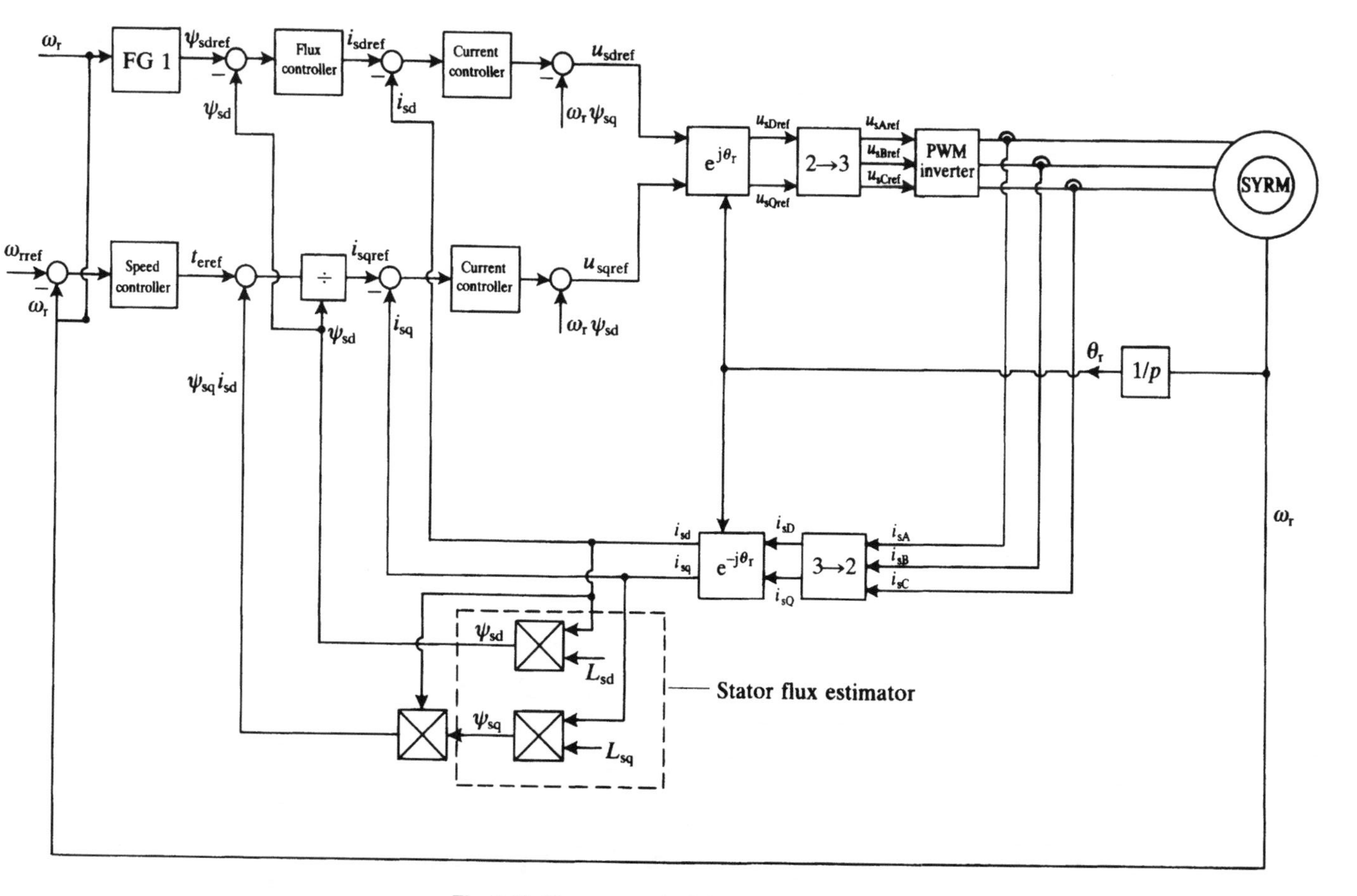

Fig. 3.44. Vector control of the reluctance machine.

expressed in the reference frame fixed to the rotor. Thus d-axis stator voltage reference u_{sdref} is obtained and it should be noted that it is expressed in the rotor reference frame.

The speed error $\omega_{rref}-\omega_r$ is fed into a speed controller to produce the reference signal for the electromagnetic torque (t_{eref}). From eqn (3.2-35), it is possible to obtain the reference torque as

$$t_{eref}=\psi_{sd}i_{sqref}-\psi_{sq}i_{sd} \tag{3.2-55}$$

and thus the quadrature-axis stator current reference i_{sqref} is obtained as

$$i_{sqref}=\frac{t_{eref}+\psi_{sq}i_{sd}}{\psi_{sd}}. \tag{3.2-56}$$

The error $\Delta i_{sq}=i_{sqref}-i_{sq}$ is fed into the quadrature-axis current controller (a PI controller) and the output signal together with the rotational voltage component $\omega_r\psi_{sd}$ is added to yield the q-axis stator voltage reference u_{sqref}, which is expressed in the reference frame fixed to the rotor.

With eqn (2.1-143), the voltage components u_{sdref} and u_{sqref} are transformed into their stationary-axes components u_{sDref}, u_{sQref} via the complex transformation $e^{j\theta_r}$, and these components are finally transformed into the three-phase stator voltage references u_{sAref}, u_{sBref}, and u_{sCref}, by the application of the two-phase to three-phase transformation shown in the block labelled '2→3'. These are the reference voltages for the inverter.

It has been necessary to use the stator current components i_{sd}, i_{sq} which correspond to the rotor reference frame. In Fig. 3.44 these are obtained from the monitored three-phase stator currents (i_{sA}, i_{sB}, i_{sC}) by the application of the three-phase to two-phase and $e^{-j\theta_r}$ complex transformations. Because of the absence of the zero-sequence currents, it is possible to obtain i_{sd} and i_{sq} by utilizing only two monitored stator currents.

The stator flux components are obtained by the stator flux estimator circuit shown in Fig. 3.44, which uses $\psi_{sd}=L_{sd}i_{sd}$ and $\psi_{sq}=L_{sq}i_{sq}$ and assumes L_{sd} and L_{sq} to be known; they can even vary with i_{sd} and i_{sq}. It is also possible to implement other flux estimators, e.g. those which involve monitored machine terminal voltages and currents.

3.2.2.2 Position-sensorless and speed-sensorless implementations

In a high-performance speed-sensorless and position-sensorless SYRM drive, the information on the rotor position and/or rotor speed and/or flux position and flux speed can be obtained by using one of the following techniques:

1. Estimation using stator voltages and currents, utilizing the speed of the stator flux-linkage space vector;
2. Estimation using the spatial saturation third-harmonic voltage component;

3. Estimators based on inductance variation due to geometrical and saturation effects;
4. Estimators using observers (e.g. extended Kalman filter);
5. Estimators using artificial intelligence (neural networks, fuzzy-logic-based systems, fuzzy-neural networks, etc.).

If the SYRM is supplied by a PWM inverter and is subjected to rotor-oriented vector control (see also Sections 1.2.1.3, 3.2.2.1) then it can operate at maximum torque per ampere. However, the SYRM requires the rotor position information to synchronize the inverter output voltages with the rotor position for starting and closed-loop control of the motor. In a vector-controlled rotor-oriented-controlled SYRM drive, the position information is used for transformation purposes [e.g. to transform the stator currents of the stationary reference frame (i_{sD}, i_{sQ}) into stator currents in the synchronously rotating (rotor-oriented) reference frame (i_{sd}, i_{sq})], $\bar{i}_s' = i_{sd} + ji_{sq} = (i_{sD} + ji_{sQ})\exp(-j\theta_r)$, and it is also used to obtain the speed, since this is required for speed control purposes.

In a direct-torque controlled SYRM drive with a speed-control loop, the speed signal has also to be known; this can also be obtained by using the monitored stator voltages and currents and using three machine parameters (R_s, L_{sd}, L_{sq}). An accurate estimation requires accurate values of R_s and for this purpose a thermal model of the machine can be used. Furthermore, in general, the inductances are not constant, and the saturated values have to be used (see also below). However, if there is no speed-control loop in the direct-torque-controlled SYRM drive, then it is very simple to implement the required control scheme, which does not require L_{sd} and L_{sq} since the flux linkage and electromagnetic torque estimator uses only R_s and the monitored stator voltages and currents. It should be noted that in a torque-controlled SYRM drive, it is also possible to obtain a stator flux-linkage estimator (flux model) which uses a position estimator and the monitored values of the stator currents. In this case, by using the monitored rotor position, first the stator currents expressed in the stationary reference frame (i_{sD}, i_{sQ}) are transformed into the stator current components in the rotor reference frame (i_{sd}, i_{sq}), $i_{sd} + ji_{sq} = (i_{sD} + ji_{sQ})\exp(-j\theta_r)$ (see also Fig. 3.45), and then the flux-linkage components are obtained as $\psi_{sd} = L_{sd} i_{sd}$, $\psi_{sq} = L_{sq} i_{sq}$. However, due to saturation the appropriate $L_{sd}(i_{sd})$ function has to be used. This gives the stator flux modulus as $|\bar{\psi}_s| = (\psi_{sd}^2 + \psi_{sq}^2)^{1/2}$. The angle of the stator flux-linkage space vector, ρ_s (with respect to the direct-axis of the stator reference frame), which is also shown in Fig. 3.45, can also be obtained, but this again uses the monitored rotor position, since the stator flux-linkage space vector in the stationary reference frame can be obtained from that in the rotor reference frame as $\psi_{sD} + j\psi_{sQ} = (\psi_{sd} + j\psi_{sq})\exp(j\theta_r)$. Since the angle (δ) of the stator flux-linkage space vector with respect to the real-axis of the rotor reference frame can be determined from the flux linkages ψ_{sd} and ψ_{sq} by using $\delta = \sin^{-1}(\psi_{sq}/|\bar{\psi}_s|)$ (see also Fig. 3.45; mathematically this follows from $\psi_{sd} + j\psi_{sq} = |\bar{\psi}_s|\exp(j\delta)$), thus

$$\psi_{sD} + j\psi_{sQ} = (\psi_{sd} + j\psi_{sq})\exp(j\theta_r) = |\bar{\psi}_s|\exp j(\delta + \theta_r) = |\bar{\psi}_s|\exp(j\rho_s). \qquad (3.2\text{-}57)$$

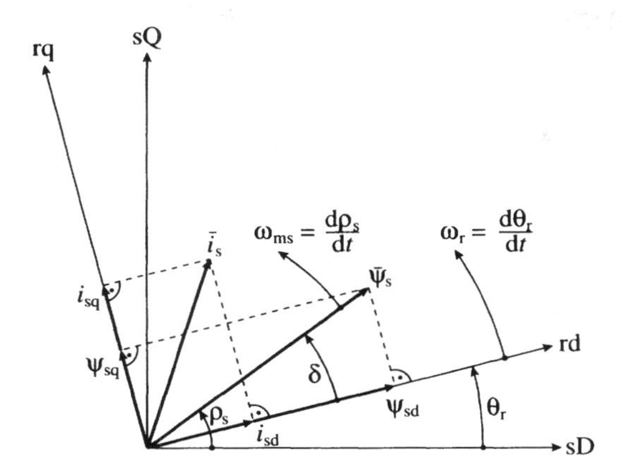

Fig. 3.45. Stator flux-linkage and stator current space vectors in the SYRM.

Hence it can be seen that $\rho_s = \sin^{-1}(\psi_{sq}/|\bar{\psi}_s|) + \theta_r$ is the angle of the stator flux-linkage space vector with respect to the real-axis of the stator reference frame, where $\psi_{sq} = L_{sq}(i_{sq})i_{sq}$. Thus

$$\rho_s = \sin^{-1}\left[\frac{L_{sq}i_{sq}}{|(L_{sd}i_{sd})^2 + (L_{sq}i_{sq})^2|}\right] + \theta_r. \tag{3.2-58}$$

The various 'sensorless' schemes are described below, but it should be noted that due to the rotor anisotropy, in the SYRM it is possible to extract accurate information on the rotor position by utilizing simple concepts (e.g. inductance variation with rotor speed), and the rotor position can be accurately estimated even at zero speed.

3.2.2.2.1 Estimation using stator voltages and currents (utilizing speed of stator flux-linkage space vector)

It is possible to estimate the rotor angle and the speed in both the transient and steady state of the SYRM by using the estimated stator flux-linkage components. In the steady state the speed of the stator flux-linkage space vector (ω_{ms}) is equal to the rotor speed (ω_r), since in this case the angle (δ) between the stator flux-linkage space vector and the direct-axis of the rotor is constant. However, in the transient state, e.g. when there is a change in the torque reference, the stator flux-linkage space vector moves relative to the rotor to produce the desired new torque. For example, if an increased torque is required, the stator flux-linkage space vector will move in the opposite direction of speed rotation.

In Fig. 3.45, θ_r is the rotor angle (the angle between the real axis of the stationary reference frame and the real axis of the rotor reference frame), and ρ_s is the angle of the stator flux-linkage space vector with respect to the real axis of the stationary reference frame. It can be seen that δ is the angle between the stator flux-linkage space vector, $\bar{\psi}_s$ (expressed in the stationary reference frame), and the real axis of the rotor reference frame. As discussed above, the angle δ is constant in the steady-state. It follows from Fig. 3.45 that the rotor angle can be expressed as

$$\theta_r = \rho_s - \delta. \tag{3.2-59}$$

Thus the rotor speed can be expressed as

$$\omega_r = \frac{d\theta_r}{dt} = \frac{d\rho_s}{dt} - \frac{d\delta}{dt} = \omega_{ms} - \omega_d, \tag{3.2-60}$$

where $\omega_{ms} = d\rho_s/dt$ is the speed of the stator flux-linkage space vector (relative to the stator) and $\omega_d = d\delta/dt$ is the speed of the stator flux-linkage space vector relative to the rotor. It is important to note that in the steady-state $\omega_d = 0$, but in the transient state ω_d is not zero (e.g. when there is a change in the torque demand, as discussed above). In a drive where the rate of change of the torque is limited, ω_d is small. However, in general, ρ_s, ω_{ms}, δ, and ω_d can be obtained as follows, by utilizing measured stator voltages and currents.

Determination of ρ_s and ω_{ms} Since in the stationary reference frame, the stator flux-linkage space vector can be expressed as

$$\bar{\psi}_s = |\bar{\psi}_s| \exp(\rho_s) = \psi_{sD} + j\psi_{sQ}, \tag{3.2-61}$$

thus

$$\omega_{ms} = \frac{d\rho_s}{dt} = \frac{d[\tan^{-1}(\psi_{sQ}/\psi_{sD})]}{dt} = \frac{[\psi_{sD}(d\psi_{sQ}/dt) - \psi_{sQ}(d\psi_s/dt)]}{|\bar{\psi}_s|^2}, \tag{3.2-62}$$

where $|\bar{\psi}_s| = (\psi_{sD}^2 + \psi_{sQ}^2)^{1/2}$ is the modulus of the stator flux-linkage space vector. The stator flux components can be obtained by considering that the stator voltage equation of the SYRM in the stationary reference frame is

$$\bar{u}_s = R_s \bar{i}_s + \frac{d\bar{\psi}_s}{dt}. \tag{3.2-63}$$

Thus the stator flux-linkage components can be obtained from the measured direct- and quadrate-axis stator voltages and currents (expressed in the stationary reference frame) by using the following integration.

$$\psi_{sD} = \int (u_{sD} - R_s i_{sD})\, dt \tag{3.2-64}$$

$$\psi_{sQ} = \int (u_{sQ} - R_s i_{sQ})\, dt. \tag{3.2-65}$$

Furthermore, the angle of the stator flux-linkage space vector can be obtained by using the known stator flux-linkage components as

$$\rho_s = \tan^{-1}\left(\frac{\psi_{sQ}}{\psi_{sD}}\right). \tag{3.2-66}$$

It should be noted that the obtained stator flux-linkage estimator is much simpler than the one which uses the rotor position and the monitored stator currents i_{sD}, i_{sQ} (see the introductory part of Section 3.2.2.2), in which the stator flux-linkage modulus is obtained as $|\bar{\psi}_s|=[(L_{sd}i_{sd})^2+(L_{sq}i_{sq}^2)]^{1/2}$ and the angle of the stator flux linkage is obtained by using $\rho_s=\sin^{-1}[L_{sq}i_{sq}/|(L_{sd}i_{sd})^2+(L_{sq}i_{sq})^2|]+\theta_r$, where the stator currents are obtained from $i_{sd}+ji_{sq}=(i_{sD}+ji_{sQ})\exp(-j\theta_r)$.

Estimation of δ and ω_d The angle δ [which was introduced in eqn (3.2-59)] and the speed $\omega_d=d\delta/dt$ can be determined as follows. The angle δ can be determined from the estimated electromagnetic torque (t_e), or directly from the stator currents. The electromagnetic torque has to be estimated in both vector- and direct-torque-controlled drives, so if this is known, it is possible to estimate δ from it.

By considering that in the rotor reference frame the stator flux-linkage space vector can be expressed as

$$\bar{\psi}_s'=L_{sd}i_{sd}+jL_{sq}i_{sq}, \tag{3.2-67}$$

where L_{sd} and L_{sq} are the direct- and quadrature-axis synchronous inductances and i_{sd}, i_{sq} are the stator current components in the rotor reference frame (they are also shown in Fig. 3.45), thus

$$t_e=\tfrac{3}{2}P\bar{\psi}_s'\times\bar{i}_s'=\tfrac{3}{2}P(L_{sd}i_{sd}+jL_{sq}i_{sq})\times(i_{sd}+ji_{sq})=\tfrac{3}{2}P(L_{sd}-L_{sq})i_{sd}i_{sq}$$
$$=\frac{\tfrac{3}{4}P(L_{sd}-L_{sq})|\bar{\psi}_s|^2\sin(2\delta)}{L_{sd}L_{sq}}, \tag{3.2-68}$$

where it has also be utilized that $i_{sd}=|\bar{\psi}_s|\cos\delta/L_{sd}$ and $i_{sq}=|\psi_s|\sin\delta/L_{sq}$ (since it follows from Fig. 3.45 that $\psi_{sd}=|\bar{\psi}_s|\cos\delta=L_{sd}i_{sd}$ and $\psi_{sq}=|\psi_s|\sin\delta=L_{sq}i_{sq}$). However, since it is also possible to express the electromagnetic torque using the stator flux linkages expressed in the stationary reference frame, $t_e=(3/2)\,P\bar{\psi}_s\times\bar{i}_s=(3/2)P(\psi_{sD}i_{sQ}-\psi_{sQ}i_{sD})$, thus by substitution of this torque expression into eqn (3.2-68), it follows that

$$\sin(2\delta)=\frac{(4/3P)t_e}{|\bar{\psi}_s|^2(L_{sd}-L_{sq})/(L_{sd}L_{sq})}, \tag{3.2-69}$$

where

$$t_e=\tfrac{3}{2}P(\psi_{sD}i_{sQ}-\psi_{sQ}i_{sD}). \tag{3.2-70}$$

In eqn (3.2-70) the stator flux components are obtained by using eqns (3.2-64) and (3.2-65). It follows that

$$\delta=\tfrac{1}{2}\sin^{-1}\left[\frac{(4/3P)(L_{sd}L_{sq})t_e}{|\bar{\psi}_s|^2(L_{sd}-L_{sq})}\right] \tag{3.2-71}$$

and

$$\omega_d=d\delta/dt. \tag{3.2-72}$$

However, it is also possible to obtain the angle δ by considering Fig. 3.45, and thus $\psi_{sd}=|\bar{\psi}_s|\cos\delta=L_{sd}i_{sd}$. It follows that $\cos\delta=L_{sd}i_{sd}/|\bar{\psi}_s|$, and by also considering $|\bar{i}_s|^2 = i_{sd}^2 + i_{sq}^2$, and from Fig. 3.45 $i_{sq} = \psi_{sq}/L_{sq} = |\bar{\psi}_s|\sin\delta/L_{sq}$, finally

$$\cos^2\delta = \frac{1 - L_{sq}^2|\bar{i}_s|^2/|\bar{\psi}_s|^2}{1 - L_{sq}^2/L_{sd}^2} \tag{3.2-73}$$

is obtained. Hence by using $\sin^2\delta + \cos^2\delta = 1$,

$$\delta = \sin^{-1}\left\{\left[\frac{L_{sq}^2|\bar{i}_s|^2/|\bar{\psi}_s|^2 - L_{sq}^2/L_{sd}^2}{1 - L_{sq}^2/L_{sd}^2}\right]^{1/2}\right\}. \tag{3.2-74}$$

It also follows that since ρ_s and δ have been determined, the rotor angle can be obtained as $\rho_s - \delta$ [see eqn (3.2-59)].

The estimation of the stator flux linkage based on eqns (3.2-64) and (3.2-65) requires the measured stator voltages and currents and also the stator resistance. For accurate estimation, accurate measurements of the voltages and currents are required. Furthermore, at low frequencies, the stator resistance has to be known very accurately. However, pure integration will cause some drift, but this can be avoided by using special techniques described in other Sections. (e.g. see Sections 3.1.3.2.1 and 4.5.3.1). Furthermore, it should also be noted that in a voltage-source inverter-fed SYRM, the stator voltages can be reconstructed by utilizing the switching signals of the inverter and the monitored d.c. link voltage (e.g. see details in Sections 3.1.3.2.1 and 4.5.3.1). The estimation of the angle δ also requires the parameters L_{sd}, L_{sq}. By considering eqn (3.2-74), it can be seen that it contains L_{sq}^2 and also L_{sq}^2/L_{sd}^2. In an SYRM with axially laminated rotor it can be assumed that L_{sq} is constant due to the large air-gap, although in some axially laminated rotors, non-linearity is introduced due to ribs which are present (which connect the various segments). Furthermore, L_{sd} varies strongly with saturation (e.g. if cross-coupling effects are neglected, it varies strongly with i_{sd} and, for example, the saturated value can even be 30% less than the unsaturated value). However, the ratio L_{sq}^2/L_{sd}^2 is very small for an SYRM with high L_{sd}/L_{sq} saliency ratio, and thus

$$\delta \approx \sin^{-1}\left[\left(\frac{L_{sq}^2|\bar{i}_s|^2}{|\bar{\psi}_s|^2}\right)^{1/2}\right]. \tag{3.2-75}$$

In general, the variation of L_{sd} with saturation can be considered by starting the estimation with a certain value of L_{sd}, and in the next computation step (sampling instant), this is changed to the value which can be obtained from

$$L_{sd} = \frac{|\bar{\psi}_s|\cos\delta}{i_{sd}} = \frac{|\bar{\psi}_s|\cos\delta}{(|\bar{i}_s|^2 - i_{sq}^2)^{1/2}} = \frac{|\bar{\psi}_s|\cos\delta}{(|\bar{i}_s|^2 - |\bar{\psi}_s|^2\sin^2\delta/L_{sq}^2)^{1/2}}. \tag{3.2-76}$$

All the equations used for the derivation of eqn (3.2-76) have been given above. In eqn (3.2-76), the stator flux-linkage space-vector modulus is estimated from the stator flux-linkage components by using $|\bar{\psi}_s| = (\psi_{sD}^2 + \psi_{sQ}^2)^{1/2}$, where the flux-linkage components are obtained from the voltages and currents by using eqns (3.2-64) and (3.2-65). The modulus of the stator-current space vector is obtained from the measured stator currents as $|\bar{i}_s| = (i_{sD}^2 + i_{sQ}^2)^{1/2}$. Thus by knowing the constant L_{sq}, the modulus of the stator flux-linkage space vector, $|\bar{\psi}_s|$, and also knowing the angle δ (estimated by using the value of L_{sd} of the previous time step), the new value of L_{sd} can be estimated by using eqn (3.2-76). By using a small sampling time, which is much smaller than the time constant of the machine, the computational error (due to the fact that the δ-estimation always uses a previous value of L_{sd}), will be very small.

Finally it should be noted that the estimation scheme described above can only be used for the low-speed operation of a vector-controlled SYRM drive, or a direct-torque-controlled SYRM drive, with a speed-control loop, if the stator flux-linkage components are estimated very accurately. For this purpose it is not adequate to use the open-loop integrators based on eqns (3.2-64) and (3.2-65), but other techniques have to be used (see also Section 3.1.3.2.1). Furthermore, as mentioned above, when eqns (3.2-64) and (3.2-65) are used, this requires an accurate value of the stator resistance. However, for this purpose a thermal model of the machine can also be used, which yields the 'hot' value of the stator resistance. It is also possible to use a Kalman filter or a fuzzy-neural estimator to obtain this resistance. If the stator voltages are not monitored, but they are reconstructed from the switching signals of the inverter and the d.c. link voltage, then it is also necessary to compensate the effects of dead time and also the voltage drops across the inverter switches (see also Sections 3.1.3.2.1 and 8.2.6).

3.2.2.2.2 Estimation using spatial saturation third-harmonic voltage component

It is possible to estimate the magnitude, position, and speed of the magnetizing-flux space vector of the SYRM by utilizing the spatial saturation third-harmonic voltage. In general the magnetizing voltage ($\bar{u}_m$) and magnetizing flux-linkage space vector ($\bar{\psi}_m$) are related by

$$\bar{u}_m = \frac{d\bar{\psi}_m}{dt}, \tag{3.2-77}$$

where

$$\bar{\psi}_m = \bar{\psi}_s - L_{sl}\bar{i}_s \tag{3.2-78}$$

and $\bar{\psi}_s$ is the stator flux-linkage space vector, L_{sl} is the stator leakage inductance, and $\bar{i}_s$ is the space vector of the stator currents (all space vectors are expressed in the stationary reference frame). Thus if the space vector $\bar{\psi}_m$ is known, $\bar{\psi}_s$ can be estimated by using eqn (3.2-78). By using these vectors, it is possible to implement various vector- or direct-torque-controlled drives, which do not use speed or position sensors. The details of this type of estimator using spatial saturation third-harmonic voltage are discussed in Section 4.5.3.2, where this technique is used for an induction machine. Since the technique only uses measured stator quantities, and the stator of the SYRM is similar to that of the induction machine, the approach for the SYRM is identical to that used for the induction machine, so only the main aspects are summarized here.

In the SYRM with stator windings without a neutral point, the sum of the stator voltages is monitored. This is a spatial saturation third-harmonic voltage component and is due to saturation effects. Due to the wye-connection, no third-harmonic stator currents exist and thus no third-harmonic voltage drops exist in the stator. Thus the stator third-harmonic saturation voltage is always in phase with the third-harmonic component of the magnetizing voltage for any load condition of the motor. The amplitude of the third-harmonic voltage is a function of the saturation level and is determined by the fundamental component of the magnetizing flux. Thus saturation function relating the third-harmonic stator voltage and magnetizing flux exists and it can be used to estimate the fundamental magnetizing flux linkage of the machine. It follows that if the third-harmonic voltage is integrated, then the third-harmonic flux is obtained and the fundamental component of the magnetizing flux ($|\bar{\psi}_{m1}|=|\bar{\psi}_m|$) is then determined by using a saturation function which gives the relationship of the fundamental magnetizing flux and third-harmonic flux. However, the integration can be problematic due to offsets and noise. This integration can be avoided, e.g. by estimating the flux position directly from the position of the third-harmonic voltage and by introducing a constant phase shift of $\pi/2$.

Furthermore, it is also possible to obtain the angle of the magnetizing flux space vector by utilizing the monitored third-harmonic voltage. For this purpose the current in stator phase sA is also monitored and the displacement angle (γ) between this stator current maximum and the fundamental magnetizing flux-linkage maximum is determined. It should be noted that the fundamental magnetizing flux-linkage component does not have to be monitored, since only the location of its maximum point is required and this can be obtained from the third-harmonic stator voltage (the suitable zero-crossing point of this voltage). By knowing the angle of the fundamental magnetizing-current space vector with respect to the stator-current space vector (γ), and also by measuring the angle of the stator-current space vector with respect to the direct-axis of the stationary reference frame (α_s), it is possible to obtain the angle of the magnetizing flux-linkage space vector with respect to the direct-axis of the stationary reference frame (μ_m).

By using the modulus of the fundamental magnetizing flux-linkage space vector ($|\bar{\psi}_m|$) and also its phase angle with respect to the stationary reference frame, it follows that $\bar{\psi}_m=|\bar{\psi}_m|\exp(j\mu_m)$ is known, and this can also be used to construct the stator flux-linkage space vector by using eqn (3.2-78). The stator flux-linkage space vector can be used in a vector control scheme or in a direct-torque control scheme. The rotor speed can be obtained, for example, by also using the first derivative of the angle of the stator-flux space vector with respect to the direct-axis of the stationary reference frame (e.g. see Section 3.2.2.2.1).

It should be emphasized that this scheme requires the stator windings to be wye-connected and also there has to be access to the neutral point of the stator windings. At low speeds problems arise due to the distorted third-harmonic voltage.

3.2.2.2.3 Estimators based on inductance variations due to geometrical and saturation effects

Due to rotor saliency, the inductances of the SYRM depend on the rotor position. Thus in a SYRM drive, the rotor position can be estimated by using inductance variations due to geometrical effects. Two techniques are discussed below. When the first method is used, measurement of stator current ripples in a hysteresis current-controlled inverter-fed SYRM is performed and the rate of change of current ripples is converted into the rotor position signal. However, it will be shown that accurate position estimation can only be performed at zero and low speeds. When the second technique is used for the estimation of the rotor angle at low speeds, suitable test stator-voltage space vectors are applied and again the rate of change of the stator current space vectors ($d\bar{i}_s/dt$) is measured. By using the locus of the complex inductance, which describes the relationship between the applied test stator-voltage vectors and the measured rate of change of the stator currents, it is then possible to extract the information on the rotor position. This technique can also be used at zero speed. However, at higher speeds, to obtain higher accuracy, the position information is extracted from the measured rate of change of the stator-current space vector, when the machine terminals are short-circuited for a short time (this is the case of applying a zero test stator-voltage space vector).

Indirect position estimation using the measured rate of change of the stator currents (in a SYRM drive with a hysteresis current controller) In a synchronous reluctance motor (SYRM), which is a singly-salient machine, due to the rotor saliency, the stator self- and mutual inductances are position dependent. Thus, for example, in a SYRM drive, where the SYRM is supplied by a hysteresis current-controlled inverter, the rate of change of the stator currents (which is related to the inverter switchings) is also position dependent. The rotor position can thus be estimated from the monitored rate of change of the stator currents. The estimation scheme can be used at low rotor speed, including zero speed, but the estimation accuracy decreases radically at higher speeds.

In a singly-salient synchronous machine, due to saliency of the rotor the inductances are different in the direct and quadrature axes ($L_{sd} > L_{sq}$). Furthermore, due to the saliency, all the stator inductances vary with ($2\theta_r$), where θ_r is the rotor angle. For example, by neglecting higher-order harmonics, the stator phase sA self-inductance can be expressed as

$$L_{AA} = L_{s1} + L_{s0'} + L_{s2}\cos(2\theta_r) = L_{s0} + L_{s2}\cos(2\theta_r). \tag{3.2-79}$$

The sB-phase and sC-phase stator self-inductances can be expressed similarly, but are functions of ($2\theta_r + 2\pi/3$) and ($2\theta_r - 2\pi/3$) respectively:

$$L_{BB} = L_{s1} + L_{s0'} + L_{s2}\cos(2\theta_r + 2\pi/3) = L_{s0} + L_{s2}\cos(2\theta_r + 2\pi/3) \tag{3.2-80}$$

$$L_{CC} = L_{s1} + L_{s0'} + L_{s2}\cos(2\theta_r - 2\pi/3) = L_{s0} + L_{s2}\cos(2\theta_r - 2\pi/3). \tag{3.2-81}$$

The mutual inductances between the stator phases are also functions of twice the rotor angle, and the inductance L_{AB} between stator phase sA and stator phase sB can be expressed as

$$L_{AB} = -\frac{L_{s0'}}{2} + L_{s2}\cos(2\theta_r - 2\pi/3). \qquad (3.2\text{-}82)$$

Similarly L_{BC} varies with $(2\theta_r)$ and L_{AC} varies with $(2\theta_r + 2\pi/3)$,

$$L_{BC} = -\frac{L_{s0'}}{2} + L_{s2}\cos(2\theta_r) \qquad (3.2\text{-}83)$$

$$L_{AC} = -\frac{L_{s0'}}{2} + L_{s2}\cos(2\theta_r + 2\pi/3). \qquad (3.2\text{-}84)$$

In these expressions L_{s1} is the leakage inductance of a stator winding, $L_{s0'}$ is the magnetizing inductance due to the fundamental air-gap flux, and L_{s2} is the self-inductance due to the position-dependent flux. If the direct- and quadrature-axis synchronous inductances are known (L_{sd}, L_{sq}), then L_{s2} and $L_{s1} + (3/2)L_{s0'}$ can be expressed in terms of L_{sd} and L_{sq} by considering [Vas 1992]

$$L_{sd} = L_{s1} + L_{md} = L_{s1} + \tfrac{3}{2}(L_{s0'} + L_{s2}) \qquad (3.2\text{-}85)$$

$$L_{sq} = L_{s1} + L_{mq} = L_{s1} + \tfrac{3}{2}(L_{s0'} - L_{s2}), \qquad (3.2\text{-}86)$$

since it follows from eqns (3.2-85) and (3.2-86) that

$$L_{s2} = \frac{L_{sq} - L_{sd}}{3} \qquad (3.2\text{-}87)$$

$$L_{s1} + \tfrac{3}{2} L_{s0'} = \frac{L_{sd} + L_{sq}}{2}. \qquad (3.2\text{-}88)$$

Thus from eqns (3.2-85) and (3.2-86),

$$L_{s2} = \frac{L_{md} - L_{mq}}{3} \qquad (3.2\text{-}89)$$

$$L_{s0'} = \frac{L_{md} + L_{mq}}{3}, \qquad (3.2\text{-}90)$$

where in the SYRM $L_{md} > L_{mq}$, and in general the magnetizing inductances vary due to saturation. Thus substitution of eqns (3.2-89) and (3.2-90) into eqns (3. 2-79)–(3.2-84) yields the following stator self- and mutual inductances:

$$L_{AA} = L_{s1} + \frac{L_{md} + L_{mq}}{3} + \frac{(L_{md} - L_{mq})}{3}\cos(2\theta_r) \qquad (3.2\text{-}91)$$

$$L_{BB} = L_{s1} + \frac{L_{md} + L_{mq}}{3} + \frac{(L_{md} - L_{mq})}{3}\cos(2\theta_r - 2\pi/3) \qquad (3.2\text{-}92)$$

$$L_{CC}=L_{s1}+\frac{L_{md}+L_{mq}}{3}+\frac{(L_{md}-L_{mq})}{3}\cos(2\theta_r+2\pi/3) \quad (3.2\text{-}93)$$

$$L_{AB}=-\frac{(L_{md}+L_{mq})}{6}+\frac{(L_{md}-L_{mq})}{3}\cos(2\theta_r-2\pi/3) \quad (3.2\text{-}94)$$

$$L_{BC}=-\frac{(L_{md}+L_{mq})}{6}+\frac{(L_{md}+L_{mq})}{3}\cos(2\theta_r) \quad (3.2\text{-}95)$$

$$L_{AC}=-\frac{(L_{md}+L_{mq})}{6}+\frac{(L_{md}+L_{mq})}{3}\cos(2\theta_r+2\pi/3). \quad (3.2\text{-}96)$$

All the inductances shown above vary with $2\theta_r$; this is due to the rotor saliency. By knowing the values of the stator leakage inductance and the direct- and quadrature-axis magnetizing inductances, it is possible to plot the variation of all the inductances versus the rotor angle. For simplicity constant magnetizing inductances can be used (which can also correspond to a given saturation level). When a voltage-source inverter supplies the SYRM, there are six conduction modes. It is then possible to manipulate the machine voltage equations in all conduction modes into a form in which the rates of change of the stator currents (di_{sA}/dt, di_{sB}/dt, di_{sC}/dt) are obtained. The resulting expressions contain the known inductance variations defined by eqns (3.2-91)–(3.2-96). However, the simplest form of these expressions is obtained if the stator resistances and also the speed voltages are neglected (which is only valid at zero speed). In this case, the resulting expressions for the rate of change of the stator currents will only contain the inductances defined above, and thus the rate of change of the currents as a function of the rotor angle can be plotted. Thus if the measured values of the rate of change of the stator currents are known, then it is possible to use these curves to obtain indirectly the rotor position. Since there are three stator phases, three independent signals are available at every rotor position for the estimation of the rotor angle. It follows that the position estimator described above depends on the motor self- and mutual inductances and to obtain accurate estimation, the dependence of saturation must also be considered. Furthermore, in general, a compensation scheme must be applied to improve the accuracy of the estimation, since the rotor speed and the initial value of the stator currents also influence the rate of change of the stator currents. However, since the equations are obtained with the assumption of zero speed, the most accurate results can be obtained at zero speed and at low speeds. The resolution of the estimated rotor angle depends on the resolution of the current measurements and the A/D converters used for the currents, and is also related to the controller sampling time and controller bandwidth.

For accurate rotor-angle estimation in a wide speed-range, including zero speed, a hybrid scheme could be used. At zero speed and low speeds the scheme discussed above could be used, but at higher speeds the scheme discussed in the previous section could be used (where the position and speed information is obtained by using the estimated stator flux-linkage components).

Indirect flux detection by the on-line reactance measurement (INFORM) method; low-speed application As discussed above, in a SYRM, due to geometrical effects the stator inductances are a function of the rotor position. This variation can be used for the estimation of the rotor position. In general, these inductances depend on $2\theta_r$, where θ_r is the rotor angle. It is possible to determine the rotor angle even at standstill by applying test stator voltages with different directions (angles, with respect to the stator reference frame) and by measuring the resulting changes of the stator current space vector [Schroedl 1988]. This can be proved by considering the stator voltage equation at standstill. In the model used the effects of core losses will be neglected. The space vector equation of the SYRM in the rotor reference frame can be obtained from

$$\bar{u}_s' = R_s \bar{i}_s' + \frac{d\bar{\psi}_s'}{dt} + j\omega_r \bar{\psi}_s', \tag{3.2-97}$$

where the space vector of the stator flux linkages in the rotor reference frame is

$$\bar{\psi}_s' = \psi_{sd} + j\psi_{sq}. \tag{3.2-98}$$

The stator flux-linkage components can be defined in terms of the machine currents and inductances as

$$\psi_{sd} = L_{sd} i_{sd} \tag{3.2-99}$$

$$\psi_{sq} = L_{sq} i_{sq}. \tag{3.2-100}$$

As discussed in the previous section, the direct-axis synchronous inductance depends strongly on the direct-axis stator current (cross-saturation coupling between the d and q axes is neglected), but due to the large air-gap, the quadrature-axis synchronous inductance can be assumed to be constant. At standstill, and neglecting the ohmic drops, for a test voltage space vector $\bar{u}_s' = u_{sd} + ju_{sq}$, the stator voltage components are

$$u_{sd} = \frac{d\psi_{sd}}{dt} = \left(\frac{d\psi_{sd}}{di_{sd}}\right)\frac{di_{sd}}{dt} = L_{dyn}\frac{di_{sd}}{dt} \tag{3.2-101}$$

$$u_{sq} = L_{sq}\frac{di_{sq}}{dt}, \tag{3.2-102}$$

and they determine the rate of change of the stator current components. In eqn (3.2-101)

$$L_{dyn} = \frac{d\psi_{sd}}{di_{sd}} \tag{3.2-103}$$

is a dynamic (incremental) inductance of the stator in the direct-axis. Figure 3.46 shows the rate of change of the current space vector ($d\bar{i}_s/dt$) at standstill, if the test stator-voltage space vector $\bar{u}_s$ is applied at a given rotor position (θ_r) and in the example shown, it is coaxial with the real-axis of the stator reference frame. The position of $d\bar{i}_s/dt$ with respect to the direct-axis of the stationary reference

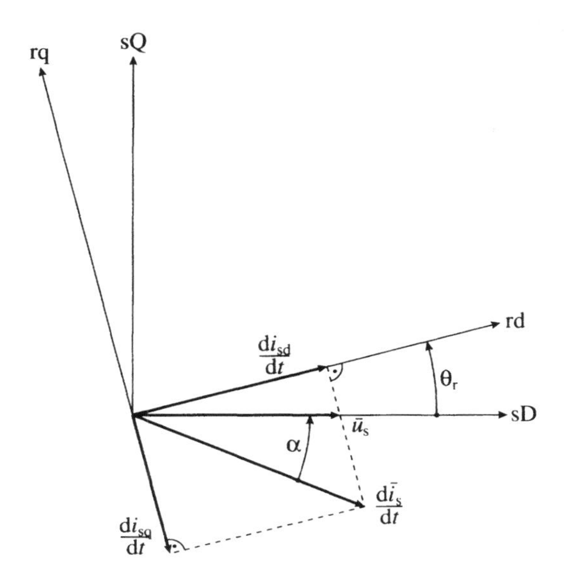

Fig. 3.46. Rate of change of stator-current space vector for a test voltage space vector (at standstill).

frame is described by the angle α, and the rates of change of the stator current components in the rotor reference frame (di_{sd}/dt, di_{sq}/dt) are also shown.

By measuring the rates of change of the stator current components (in the stator reference frame) and by using the known test stator-voltage space vector, the angle α can be determined. However, since the inductances vary with $2\theta_r$, the angle α must be a function of $2\theta_r$, thus $\alpha=\alpha(2\theta_r)$. Hence $2\theta_r$ can be estimated from a given α. In some special cases α is zero, e.g. for $\theta_r=k\pi/2$ $(k=0,1,2,\ldots)$.

When the rotor speed of the SYRM is not zero, it follows from eqns (3.2-97)–(3.2-100) that in an operating point $\bar{i}_{s0'}=i_{sd0}+ji_{sq0}$, the stator voltage equations in the rotor reference frame are obtained as

$$u_{sd0}=R_s i_{sd0}+L^0_{dyn}\frac{di_{sd}}{dt}-\omega_r L_{sq} i_{sq0} \tag{3.2-104}$$

$$u_{sq0}=R_s i_{sq0}+\omega_r L_{sd} i_{sd0}, \tag{3.2-105}$$

where

$$L^0_{dyn}=\left[\frac{d\psi_{sd}}{di_{sd}}\right]_{i_{sd0}} \tag{3.2-106}$$

is the dynamic inductance in the d-axis in the operating point. It follows from eqns (3.2-104) and (3.2-105) that the rotational voltages influence the rate of change of the currents. To eliminate this effect, two measurements are made, measurement 1 and measurement 2, where the applied stator-voltage space vectors are different (e.g. their modulus is the same, but their phase angle is different).

The measuring time must be short, so the operating point currents (i_{sd0}, i_{sq0}) do not change (thus the ohmic drops are unchanged and also L_{dyn}^{0} does not change), and also the rotor speed does not change. Thus it follows from eqns (3.2-104) and (3.2-105) that

$$u_{sd1}-u_{sd2}=\Delta u_{sd}=L_{dyn}^{0}\left(\frac{di_{sd1}}{dt}-\frac{di_{sd2}}{dt}\right) \tag{3.2-107}$$

$$u_{sq1}-u_{sq2}=\Delta u_{sq}=L_{sq}\left(\frac{di_{sq1}}{dt}-\frac{di_{sq2}}{dt}\right), \tag{3.2-108}$$

where u_{sd1} and u_{sq1} are the direct- and quadrature-axis stator voltages in the rotor reference frame for test space vector 1 ($\bar{u}'_{s1}=u_{sd1}+ju_{sq1}$), and similarly u_{sd2} and u_{sq2} are the direct- and quadrature-axis stator voltages in the rotor reference frame for test space vector 2 ($\bar{u}'_{s2}=u_{sd2}+ju_{sq2}$). The space vector form of eqns (3.2-107) and (3.2-108) in the stationary reference frame is as follows if it is considered that the first test stator-voltage space vector is $\bar{u}_{s1}=|\bar{u}_s|\exp(j\beta_1)$ and the second test stator-voltage space vector is $\bar{u}_{s2}=|\bar{u}_s|\exp(j\beta_2)$:

$$\frac{d\bar{i}_{s1}}{dt}-\frac{d\bar{i}_{s2}}{dt}=L^{-1}(2\varepsilon)(\bar{u}_{s1}-\bar{u}_{s2}). \tag{3.2-109}$$

It is important to note that, as expected by physical considerations in eqn (3.2-109) 2ε is present, where the angle ε is defined as

$$\varepsilon=\beta_1-\beta_2-\theta_r. \tag{3.2-110}$$

Furthermore, in eqn (3.2-109) L is a complex inductance, which depends on the operating point of the machine and is a function of 2ε, where ε contains the rotor angle. Equation (3.2-109) can also be expressed as

$$\frac{d\Delta\bar{i}_s}{dt}=L^{-1}(2\varepsilon)\Delta\bar{u}_s, \tag{3.2-111}$$

where $\Delta\bar{u}_s=\bar{u}_{s1}-\bar{u}_{s2}$ is the difference of the two test stator-voltage space vectors (expressed in the stationary reference frame) and $\Delta\bar{i}_s=\bar{i}_{s1}-\bar{i}_{s2}$ is the difference of the resulting stator-current space vectors (expressed in the stationary reference frame). However, by considering eqns (3.2-107) and (3.2-108) the inverse complex inductance can be expressed as follows

$$L^{-1}(2\varepsilon)=g-\Delta g\exp(-2j\varepsilon), \tag{3.2-112}$$

where

$$g=\tfrac{1}{2}\left(\frac{1}{L_{dyn}^{0}}+\frac{1}{L_{sq}}\right) \tag{3.2-113}$$

and

$$\Delta g = \tfrac{1}{2}\left(\frac{1}{L_{sq}} - \frac{1}{L^0_{dyn}}\right). \tag{3.2-114}$$

Thus it follows that if $\Delta\bar{u}_s$ is a voltage space vector which is coaxial with the direct-axis of the stationary reference frame, and the measured $d\Delta\bar{i}_s/dt$ is plotted, then the space-vector locus of the inverse complex inductance $L^{-1}(2\varepsilon)$ is obtained. This is shown in Fig. 3.47. It can be seen that the angle ε is the angle between the direct-axis of the rotor reference frame and the real-axis of the inverse complex inductance (which is coaxial with g). The interpretation of the locus is as follows: the rate of change of the stator-current vector is obtained, which depends on ε, which depends on the rotor position ($\varepsilon = \beta_1 - \beta_2 - \theta_r$).

It follows that if the machine parameters L_{sq}, L^0_{dyn} are known, it is possible to determine the angle 2ε (from the space-vector locus or from the corresponding equations). However, the determination of these parameters can be avoided by imposing two or three test voltage space vectors ($\Delta\bar{u}_s$) in different directions (voltage vectors with different positions). Thus 2ε can be determined with the following parameter independent tests (for this purpose two tests, A and B, will be performed). For the first test (test A) the angle of $\Delta\bar{u}_s$ is $2\pi/3$; thus it is in the direction of stator phase sB. In this case it follows from eqn (3.2-110) that

$$\varepsilon_A = \beta_1 - \beta_2 - \theta_r = 2\pi/3 - \theta_r \tag{3.2-115}$$

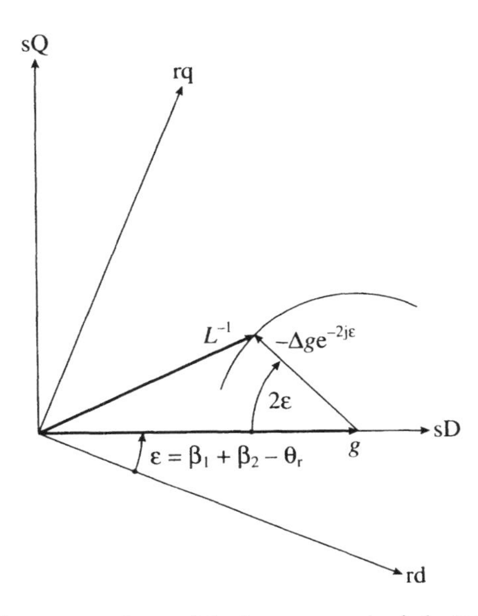

Fig. 3.47. Space-vector locus of the inverse complex inductance $L^{-1}(2\varepsilon)$.

and from eqn (3.2-109)

$$L_{\mathrm{A}}^{-1}(2\varepsilon_{\mathrm{A}})=|\Delta\bar{u}_{\mathrm{s}}^{-1}|\left[\frac{\mathrm{d}\bar{i}_{\mathrm{s}1}}{\mathrm{d}t}-\frac{\mathrm{d}\bar{i}_{\mathrm{s}2}}{\mathrm{d}t}\right]_{\mathrm{A}}\exp(-\mathrm{j}2\pi/3). \tag{3.2-116}$$

Similarly, for test B, the angle of $\Delta\bar{u}_{\mathrm{s}}$ is $4\pi/3$, thus it is collinear with stator phase sC. In this case it follows from eqn (3.2-110) that

$$\varepsilon_{\mathrm{B}}=\beta_1-\beta_2-\theta_{\mathrm{r}}=4\pi/3-\theta_{\mathrm{r}}, \tag{3.2-117}$$

and from eqn (3.2-109)

$$L_{\mathrm{B}}^{-1}(2\varepsilon_{\mathrm{B}})=|\Delta\bar{u}_{\mathrm{s}}^{-1}|\left[\frac{\mathrm{d}\bar{i}_{\mathrm{s}1}}{\mathrm{d}t}-\frac{\mathrm{d}\bar{i}_{\mathrm{s}2}}{\mathrm{d}t}\right]_{\mathrm{B}}\exp(-\mathrm{j}4\pi/3). \tag{3.2-118}$$

By using the real and imaginary parts of the known complex inverse inductances L_{A}^{-1} and L_{B}^{-1}, the rotor angle can be extracted as follows by using eqn (3.2-112):

$$x_1=\mathrm{Re}[L_{\mathrm{A}}^{-1}(2\varepsilon_{\mathrm{A}})]-\mathrm{Re}[L_{\mathrm{B}}^{-1}(2\varepsilon_{\mathrm{B}})]=-\Delta g\sqrt{3}\sin(2\theta_{\mathrm{r}}) \tag{3.2-119}$$

$$x_2=\mathrm{Im}[L_{\mathrm{A}}^{-1}(2\varepsilon_{\mathrm{A}})]-\mathrm{Im}[L_{\mathrm{B}}^{-1}(2\varepsilon_{\mathrm{B}})]=-\Delta g\sqrt{3}\cos(2\theta_{\mathrm{r}}). \tag{3.2-120}$$

Thus by considering eqns (3.2-119) and (3.2-120) the rotor angle is finally obtained as

$$\theta_{\mathrm{r}}=\tfrac{1}{2}\tan^{-1}\left(\frac{x_1}{x_2}\right). \tag{3.2-121}$$

As expected the modulus $|\Delta\bar{u}_{\mathrm{s}}^{-1}|$ and the parameters g and Δg do not influence the estimation. Thus no voltage measurement or machine parameters are required for the estimation of the rotor angle. The technique is simple, and gives an accurate estimate of the rotor position even at low speeds. To summarize: appropriate stator test voltages are applied to the motor and the rate of change of the stator currents is measured. Due to geometrical saliency the voltage space vector and the rate of change of the current space vector are related via a complex inductance, which is a function of twice the rotor angle. By conducting two tests (A and B), and using the obtained two inverse complex inductances (L_{A}^{-1} and L_{B}^{-1}), the rotor angle can be determined by using eqn (3.2-121), where $x_1=\mathrm{Re}(L_{\mathrm{A}}^{-1})-\mathrm{Re}(L_{\mathrm{B}}^{-1})$ and $x_2=\mathrm{Im}(L_{\mathrm{A}}^{-1})-\mathrm{Im}(L_{\mathrm{B}}^{-1})$. It should be noted that if there were no error, the locus $x_1+\mathrm{j}x_2$ would be a circle, but in practice there will be some deviation from the circle. It is possible to obtain the rotor position more accurately by using a hybrid technique. For this purpose the estimated rotor position can be used as an input to a Kalman filter, and the Kalman filter can accurately determine the rotor position.

It should be noted that in a permanent-magnet synchronous machine (PMSM) a similar technique can be used for the estimation of the rotor position, even if the machine has surface-mounted magnets, and thus when geometrical saliency can be neglected. This is due to the fact that because of saturation effects, in the PMSM with surface-mounted magnets, the stator inductances are a function of the rotor position and this variation can be used for the estimation of the rotor position. In this machine, due to saturation in the stator teeth, the direct-axis synchronous inductance will be smaller than the quadrature-axis synchronous inductance and these inductances are functions of the rotor position. In general, these inductances depend on 2β, where β is the angle of the magnet flux (with respect to the stator-voltage space vector). It is then possible to determine β even at standstill by applying test stator voltages with different directions (angles, with respect to the stator reference frame) and by measuring the resulting rates of changes of the stator-current space vector. This can be simply proved by considering the stator voltage equation at standstill. Thus by neglecting the stator resistance, it follows from the stator voltage equation of the PMSM expressed in the stationary reference frame that $d\bar{i}_s/dt = \bar{u}_s/L$, where $\bar{i}_s$ and $\bar{u}_s$ are the space vectors of the stator voltages and currents respectively and L is a complex inductance, which depends on 2β, $L = L(2\beta)$.

It is interesting to note that a similar technique can also be used for a three-phase squirrel-cage induction motor for the estimation of the angle of the magnetizing flux with respect to the real-axis of the stator reference frame. This is based on the fact that in the induction motor, due to saturation of the stator and rotor teeth, the stator inductances depend not only on the level of saturation but also on the position of the main flux. The technique can be used even at standstill. This can be proved by considering that (similarly to that shown for the PMSM machine) at standstill $d\bar{i}_s/dt = \bar{u}_s/L$ holds, where L is a complex stator transient inductance which also depends on the position of the magnetizing flux, and by applying appropriate test stator-voltage vectors $d\bar{i}_s/dt$ can be measured, thus the locus of the complex inductance can be determined. The angle of the magnetizing flux can be obtained since the locus of the modulus of the complex transient inductance is an ellipse and the minimum of this ellipse is in the direction of the magnetizing flux.

The rotor angle estimator discussed above for the SYRM can be used in a inverter-fed SYRM drive. As discussed earlier in Section 1.2 (Fig. 1.4(a)) and Section 3.2.2.1.1 (Fig. 3.43), the drive contains a stator reference-current estimator (first the stator current references are obtained in the rotor reference frame), which is obtained from the reference value of the electromagnetic torque and estimated value of the rotor speed. The current references are then transformed into the stator reference frame by using the estimated rotor angle. The stator current references are then inputs to three current controllers, which output the required switching signals. In the drive with current control, the method described above is executed while the current control algorithm is inactive. However, this is a considerably long time (approx. 100 μs), and it will reduce the quality of the control at higher speed. Thus another simple scheme is proposed for the rotor position estimation, which can be used at higher speeds. This is discussed below.

Indirect position estimation using the rate of change of the measured stator-current vector if the stator is short-circuted; high-speed application It is possible to extract the rotor position information when the zero inverter switching states (0 0 0) or (1 1 1) are applied for a short time by measuring the rate of change of the stator currents. In the first case, all the lower switches of the inverter are connected to the negative d.c. rail, and in the second case, all the upper inverter switches are connected to the positive d.c. rail. This means that the stator windings are short-circuited. The stator voltage equations of the short-circuited SYRM will now be used to obtain an expression for the rotor angle by using the angle of the measured $\mathrm{d}\bar{i}_s/\mathrm{d}t$ vector.

If the stator ohmic voltage drops are neglected, then the stator voltage equation of the SYRM with short-circuited stator windings can be obtained from eqn (3.2-97) and is as follows in the rotor reference frame:

$$0=\frac{\mathrm{d}\bar{\psi}_s'}{\mathrm{d}t}+\mathrm{j}\omega_r\bar{\psi}_s'. \tag{3.2-122}$$

Thus resolution into real and imaginary parts, and also considering eqns (3.2-99) and (3.2-100), gives

$$0=\frac{\mathrm{d}\psi_{sd}}{\mathrm{d}t}-\omega_r\psi_{sq}=L_{dyn}\frac{\mathrm{d}i_{sd}}{\mathrm{d}t}-\omega_r L_{sq}i_{sq} \tag{3.2-123}$$

$$0=\frac{\mathrm{d}\psi_{sq}}{\mathrm{d}t}+\omega_r\psi_{sd}=L_{sq}\frac{\mathrm{d}i_{sq}}{\mathrm{d}t}+\omega_r L_{sd}i_{sd}. \tag{3.2-124}$$

It follows that

$$\frac{\mathrm{d}i_{sd}}{\mathrm{d}t}=\frac{\omega_r L_{sq}i_{sq}}{L_{dyn}} \tag{3.2-125}$$

$$\frac{\mathrm{d}i_{sq}}{\mathrm{d}t}=\frac{-\omega_r L_{dyn}i_{sd}}{L_{sq}}. \tag{3.2-126}$$

By considering that the stator-current space vector in the rotor reference frame is obtained as $\bar{i}_s'=i_{sd}+\mathrm{j}i_{sq}$, the space vector form of eqns (3.2-125) and (3.2-126) is

$$\frac{\mathrm{d}\bar{i}_s'}{\mathrm{d}t}=\omega_r\left(\frac{L_{sq}i_{sq}}{L_{dyn}}-\frac{\mathrm{j}L_{dyn}i_{sd}}{L_{sq}}\right). \tag{3.2-127}$$

It can be seen that if the rate of change of the stator currents is measured, then at zero speed it is not possible to use eqn (3.2-127), since $\mathrm{d}\bar{i}_s'/\mathrm{d}t=0$. Problems also occur if the speed is very small and also the stator current modulus is small. The rotor position is determined from the angle of $\mathrm{d}\bar{i}_s'/\mathrm{d}t$. It follows from eqn (3.2-127) that in the rotor reference frame, the angle of the rate of change of the stator-current space vector is equal to the argument of the speed (which for positive speeds is π, and for negative speeds is $-\pi$) minus $\tan^{-1}(L_{dyn}^2 i_{sd}/L_{sq}^2 i_{sq})$, thus the angle of $\mathrm{d}\bar{i}_s'/\mathrm{d}t$ is

$$\alpha_s'=\pm\pi-\tan^{-1}\left(\frac{L_{dyn}^2 i_{sd}}{L_{sq}^2 i_{sq}}\right). \tag{3.2-128}$$

It should be noted that eqn (3.2-127) gives the rate of change of the currents expressed in the rotor reference frame, if the stator windings are short-circuited, and for this purpose it is possible to use the inverter supplying the SYRM by using the 000 or alternatively the 111 zero switching states for a short time. However, if the rate of change of the stator currents is measured in the stator reference frame, since $\bar{i}_s' = \bar{i}_s \exp(-j\theta_r)$, it follows from eqn (3.2-127) that $d\bar{i}_s/dt = d[\bar{i}_s' \exp(j\theta_r)]/dt$ can be obtained. Thus the angle of $d\bar{i}_s/dt$ ($\alpha_s = \angle d\bar{i}_s/dt$) is equal to the angle of $d\bar{i}_s'/dt$ plus the rotor angle. Thus the rotor angle can be obtained as the angle of the measured rate of change of $d\bar{i}_s/dt$ minus the angle of $d\bar{i}_s'/dt$, and hence

$$\theta_r = \alpha_s - \alpha_s' = \frac{\angle d\bar{i}_s}{dt} - \left[\pm\pi - \tan^{-1}\left(\frac{L_{dyn}^2 i_{sd}}{L_{sq}^2 i_{sq}}\right)\right]. \tag{3.2-129}$$

In the current-controlled SYRM drive, it is possible to use the reference values of i_{sd} and i_{sq} in eqn (3.2-129).

To summarize: if the inverter supplying the SYRM is forced for a short time into one of the zero switching states (where the stator windings are short-circuited), then it is possible to obtain indirectly the rotor position information by measuring the rate of change of the stator currents ($d\bar{i}_s/dt$), and by using eqn (3.2-129). Thus, first the angle of $d\bar{i}_s/dt$ is obtained, this gives α_s, and from this α_s' is deducted, which can be obtained from $[\pm\pi - \tan^{-1}(L_{dyn}^2 i_{sdref}/L_{sq}^2 i_{sqref})]$. It should be noted that this technique cannot be used at zero and low speed and at low values of the stator reference current. However, if this technique is combined with the INFORM technique described in the previous section, which is applicable at low speeds, then a hybrid scheme is obtained which can be used in a wide speed-range for the estimation of the rotor position. It also follows from eqn (3.2-179) that for rotor position estimation, it is necessary to know the values of the quadrature-axis synchronous inductance and also the dynamic inductance (in the direct-axis). The quadrature-axis inductance can be assumed to be constant (due to the large air-gap), but the dynamic inductance varies with the direct-axis current and to obtain an accurate estimate, its current-dependent value should be used. However, the accuracy of the position estimator can be increased if the position determined by the hybrid estimator is used as an input to a Kalman filter, as described in the next section.

3.2.2.2.4 Estimators using observers (e.g. extended Kalman filters)

It is possible to estimate the rotor position together with other quantities (e.g. the load torque, speed) by using an extended Kalman filter (EKF) or an extended Luenberger observer. The extended Kalman filter algorithm was discussed in great detail in Section 3.1.3.5.1, so only the most important aspects of the algorithm will be discussed here. It should be noted that this is a computationally intensive algorithm, and a DSP must be used for this purpose.

It has been discussed in the previous section that the accuracy of the hybrid position estimator can be improved, if the estimated rotor angle of the hybrid estimator is input to a Kalman filter. This will optimally reduce the effects of stochastic disturbances, and will also yield an accurate estimate of the load torque and rotor speed. The load torque (t_L) can be used in the inverter-fed SYRM drive, since the stator reference-current estimator uses at its inputs the electromagnetic torque reference and also the estimated rotor speed (which is obtained by the Kalman filter), but the electromagnetic torque reference can be estimated accurately by taking the difference of the output of the speed controller and the estimated load torque.

The digital EKF requires the discrete form state-variable equations of the SYRM (augmented with the estimation noise vector). Furthermore it also requires the discretized output equations (augmented with the estimation noise vector). By means of noise inputs, it is possible to take account of computational inaccuracies, modelling errors, and errors in the measurements. The matrix forms of these two sets of equations will now be derived. However, to obtain a simple EKF, only the mechanical parts of the SYRM are modelled. Thus the input to the model is the electromagnetic torque computed by using eqn (3.2-68):

$$t_{ei}=\tfrac{3}{2}P(L_{sd}-L_{sq})i_{sdref}i_{sqref}. \tag{3.2-130}$$

The difference between the actual torque and the reference torque is modelled by a stochastic term w_1 (noise term). Thus the state model of the SYRM considered contains the states ω_r and θ_r, and the equation of motion gives

$$\frac{d\omega_r}{dt}=\frac{t_{ei}}{J}+\frac{t_L}{J}, \tag{3.2-131}$$

where J is the inertia. Furthermore,

$$\frac{d\theta_r}{dt}=\omega_r. \tag{3.2-132}$$

The discretized form of eqns (3.2-131) and (3.2-132) are as follows, with the torque estimation noise considered:

$$\omega_r(k+1)=\omega_r(k)+\frac{T_s}{J}t_{ei}+\frac{T_s}{J}t_L+\frac{T_s}{J}w_1(k) \tag{3.2-133}$$

$$\theta_r(k+1)=T_s\omega_r(k)+\theta_r(k)+\frac{T_s^2}{2J}t_{ei}+\frac{T_s^2}{2J}t_L+\frac{T_s^2}{2J}w_1(k), \tag{3.2-134}$$

where T_s is the sampling time.

Since the hybrid estimator gives noisy measurements of the rotor angle, the output equations of the model will consist of two equations. The first output equation corresponds to the output when the first scheme of the hybrid estimator is used (which is the INFORM scheme described in the previous section), and which gives $2\theta_r$, and the measurement noise is v_1. The second output equation corresponds to the second scheme of the hybrid estimator, which gives θ_r, and where the measurement noise is v_2. Thus

$$y_1(k)=2\theta_r(k)+v_1(k) \tag{3.2-135}$$

$$y_2(k)=\theta_r(k)+v_2(k). \tag{3.2-136}$$

The noises w_1 and v_1, v_2 are assumed to be uncorrelated white Gaussian noises. Since the load torque is unknown, it can be modelled as

$$t_L(k+1)=t_L(k)+w_2(k), \tag{3.2-137}$$

where $w_2(k)$ is an estimation noise of the load torque, and it is also a white Gaussian noise (it gives the expected change of load torque in a sampling period), and it is independent of w_1 and w_2. Thus the matrix form of the state-variable equations can be obtained from eqns (3.2-133), (3.2-134), and (3.2-137) as follows:

$$\mathbf{x}(k+1)=\mathbf{A}\mathbf{x}(k)+\mathbf{B}\mathbf{u}(k)+\mathbf{C}\mathbf{w}(k), \tag{3.2-138}$$

where $\mathbf{x}$ is the state vector $\mathbf{x}(k+1)=[\omega_r(k+1), \theta_r(k+1), t_L(k+1)]^T$, $\mathbf{A}$ is the state matrix

$$\mathbf{A}=\begin{bmatrix} 1 & 0 & T_s/J \\ T_s & 1 & T_s^2/2J \\ 0 & 0 & 1 \end{bmatrix}, \tag{3.2-139}$$

$\mathbf{u}$ is the input vector, $\mathbf{u}=[t_{ei}(k)]$, $\mathbf{B}$ is the input matrix, $\mathbf{B}=[T_s/J, T_s^2/2J, 0]^T$, $\mathbf{w}(k)$ is the estimation noise vector, $\mathbf{w}(k)=[w_1(k), w_2(k)]^T$. In eqn (3.2-138) the weighting matrix of estimation noise is defined as

$$\mathbf{C}=\begin{bmatrix} T_s/J & 0 \\ T_s^2/2J & 0 \\ 0 & 1 \end{bmatrix} \tag{3.2-140}$$

The matrix form of the output equation is obtained from eqns (3.2-135) and (3.2-136) as

$$\mathbf{y}(k)=\mathbf{F}\mathbf{x}(k)+\mathbf{v}(k), \tag{3.2-141}$$

where the output vector is $\mathbf{y}=[y_1, y_2]^T$ and $\mathbf{v}(k)$ is the measurement noise vector, $\mathbf{v}(k)=[v_1(k), v_2(k)]^T$. If in the hybrid model, the first scheme (INFORM scheme) is used, then $\mathbf{y}(k)=y_1(k)$, $\mathbf{v}=v_1(k)$, and the output matrix is $\mathbf{F}=[0\ 2\ 0]$. However, when the hybrid system uses the second scheme (where the stator windings are short-circuited) then $\mathbf{y}(k)=y_2(k)$, $\mathbf{v}=v_2(k)$, and the output matrix is $\mathbf{F}=[0\ 1\ 0]$.

In the prediction stage the EKF estimates the state variables by using the state-variable equations of the system given above. Thus the prediction of the state vector at sampling time $(k+1)$ is obtained from the input $\mathbf{u}(k)$ and state vector at previous sampling time $\mathbf{x}(k)$ by performing

$$\hat{\mathbf{x}}(k+1|k)=\mathbf{A}\hat{\mathbf{x}}(k|k)+\mathbf{B}\mathbf{u}(k), \tag{3.2-142}$$

where $\hat{\mathbf{x}}$ is the predicted state-vector at sampling time $(k+1)$. The notation $\mathbf{x}(k+1|k)$ means that it is a predicted value at the $(k+1)$-th instant, and it is based on measurements up to the kth instant. In the filtering stage, the predicted state is corrected by comparing the actual measured outputs $\mathbf{y}(k+1)$ with the expected measured values $\hat{\mathbf{y}}(k+1)=\mathbf{F}\hat{\mathbf{x}}(k+1|k)$, and using the error to minimize the state error. Thus

$$\hat{\mathbf{x}}(k+1|k+1)=\hat{\mathbf{x}}(k+1|k)+\mathbf{K}(k+1)[\mathbf{y}(k+1)-\hat{\mathbf{y}}(k+1|k], \qquad (3.2\text{-}143)$$

where $\mathbf{K}$ is the Kalman gain, which is computed iteratively. The main steps of the EKF algorithm are now summarized.

Step 1: *Initialization of the covariance matrices* $(\mathbf{P}_0, \mathbf{R}_0, \mathbf{Q}_0)$

$\mathbf{P}_0$ is the covariance matrix of the state vector, $\mathbf{R}_0$ is the covariance matrix of the measurement noise, $\mathbf{Q}_0$ is the covariance matrix of the estimation noise.

Step 2: *Prediction of the state vector*

Prediction of the state vector at sampling time $(k+1)$ from the input $\mathbf{u}(k)$ and also the state vector at previous sampling time $\hat{\mathbf{x}}(k)$ by using the known matrices $\mathbf{A}$ and $\mathbf{B}$:

$$\mathbf{x}^*(k+1|k)=\mathbf{A}\hat{\mathbf{x}}(k|k)+\mathbf{B}\mathbf{u}(k).$$

Step 3: *Covariance estimation of prediction*

The covariance matrix of prediction is estimated as

$$\mathbf{P}^*(k+1|k)=\mathbf{A}\hat{\mathbf{P}}(k|k)\mathbf{A}^{\mathrm{T}}+\mathbf{Q},$$

where $k|k$ denotes prediction at time k based on data up to time k, and $\hat{\mathbf{P}}(k|k)$ is the covariance matrix of the state estimation error after the correction is made by using the actual measurements.

Step 4: *Kalman filter gain computation*

The Kalman filter gain (correction matrix) is computed as

$$\mathbf{K}(k+1)=\mathbf{P}^*(k+1|k)\mathbf{F}^{\mathrm{T}}[\mathbf{F}\mathbf{P}^*(k+1|k)\mathbf{F}^{\mathrm{T}}+\mathbf{R}]^{-1}.$$

Step 5: *State vector estimation*

The state vector estimation (corrected state vector estimation, filtering) at time $(k+1)$ is performed through the feedback correction scheme, which makes use of the actual measured quantities ($\mathbf{y}$):

$$\hat{\mathbf{x}}(k+1|k+1)=\mathbf{x}^*(k+1|k)+\mathbf{K}(k+1)[\mathbf{y}(k+1)-\mathbf{C}\mathbf{x}^*(k+1)|k].$$

Step 6: *Covariance matrix of estimation error*

The error covariance matrix can be obtained from

$$\hat{\mathbf{P}}(k+1|k+1)=\mathbf{P}^*(k+1|k)-\mathbf{K}(k+1)\mathbf{F}\mathbf{P}^*(k+1|k).$$

It should be noted that the Kalman gain matrix ($\mathbf{K}$) does not depend on the system state and input quantities, and thus it can be computed off-line. Furthermore, since the Kalman gain matrix of the SYRM reaches it steady-state value within a few hundred milliseconds, the steady-state Kalman gain matrix can be used in the EKF. When the second model in the hybrid position estimator is used, then the filter coefficients should not be changed (although the measurement noise v_2 is a function of the speed), since this will considerably reduce the computational requirements, because the state estimator will only contain constant coefficients. Although this will give a pessimistic weight of the measurement vector y_2 at high speeds, according to experimental results, satisfactory estimates are obtained. By using a DSP, it is possible to have a high-dynamic-performance SYRM drive, which gives excellent responses in a wide speed-range without any speed or position sensor, including zero speed [Schroedl 1994].

It should be noted that it is also possible to have another implementation of the EKF, where another model of the SYRM is used, which also uses the stator voltage equations and does not use the hybrid model to generate inputs to the EKF. However, this EKF would be computationally more demanding than the one described above. Such an EKF is described in the present book for an induction machine and also for permanent-magnet synchronous machines.

3.2.2.2.5 Estimators using artificial intelligence

So far no applications of using artificial intelligence (fuzzy logic, artificial neural networks, fuzzy-neural networks, etc.) for the estimation of the speed and position of a SYRM have been reported in the literature. However, since it is possible to approximate any non-linear function with great accuracy by using these techniques (see also Chapter 7), it is believed that such implementations will also emerge in the future for both vector- and direct-torque-controlled SYRM drives.

It is believed that advances in the field of position-sensorless SYRM will significantly contribute to the industrial acceptance of various types of SYRM drives. In this case the SYRM drive could become a strong competitor to variable-speed brushless a.c. drives.

3.3 Direct torque control (DTC) of synchronous motors

3.3.1 GENERAL INTRODUCTION

In general, the direct torque control (DTC) of a synchronous motor involves the direct control of the flux linkages (e.g. stator flux linkages, stator transient flux linkages, etc.) and electromagnetic torque by applying optimum current or voltage switching vectors of the inverter which supplies the motor. In the case of a CSI-fed motor, the optimum current switching vectors are selected, but in the case of a VSI-fed motor, the optimal voltage switching vectors are selected. A brief description of the general aspects of direct torque control has been given in Section 1.2.2 and a very detailed description of the DTC of induction motors is presented in Section 4.6.

In the present section, four main DTC control schemes of synchronous machines are described:

1. DTC of a permanent-magnet synchronous machine supplied by a voltage-source inverter. Four schemes are considered; the first three use an optimum switching voltage vector look-up table while the fourth one uses a predictive technique.

Scheme 1: Direct control of the electromagnetic torque and stator flux linkages using an optimal voltage switching look-up table.
Scheme 2: Direct control of the electromagnetic torque and direct-axis stator current using an optimal voltage switching look-up table.
Scheme 3: Direct control of the electromagnetic torque and reactive torque using an optimal voltage switching look-up table.
Scheme 4: Direct torque control using a predictive algorithm for the switching voltage vector selection.

2. DTC of a synchronous reluctance machine supplied by a voltage-source inverter.
3. DTC of a synchronous machine supplied by a voltage-source inverter.
4. DTC of a synchronous machine supplied by a current-source inverter.

3.3.2 DTC OF THE VSI-FED PERMANENT-MAGNET SYNCHRONOUS MOTOR

Vector-controlled PM synchronous motor drives have been discussed in Sections 3.1.2, 3.1.3, and also a brief introduction has been given in Sections 1.2.1 and 1.2.2 on vector control and direct torque control techniques. In the present section four types of DTC schemes are discussed for the PM synchronous machine. In the first scheme, there is direct control of the stator flux linkages and electromagnetic torque. In the second scheme, direct control of the direct-axis stator current (expressed in the rotor reference frame) and the electromagnetic torque is performed. In the third scheme there is direct control of the electromagnetic torque and the reactive power. The first three schemes use an optimal switching vector look-up table. However, a fourth scheme is also discussed, which uses predictive algorithms for the selection of the appropriate switching voltage vectors.

3.3.2.1 Direct control of the electromagnetic torque and stator flux, using optimal voltage switching look-up table (Scheme 1)

3.3.2.1.1 Control scheme

In this first scheme, a direct-torque-controlled PM synchronous motor supplied by a voltage-source inverter is present and the stator flux linkage and the electromagnetic torque are controlled directly by applying optimum voltage switching vectors of the inverter. It is a principal goal to select those voltage switching vectors which yield the fastest electromagnetic torque response. The six active switching vectors $(\bar{u}_1, \bar{u}_2, \ldots, \bar{u}_6)$ and the control scheme of the DTC PM synchronous motor supplied by a VSI are shown in Fig. 3.48.

In Fig. 3.48(b), the electromagnetic torque error and the stator flux-linkage error are inputs to the respective flux-linkage and torque hysteresis comparators. The flux-linkage comparator is a three-level comparator and the electromagnetic

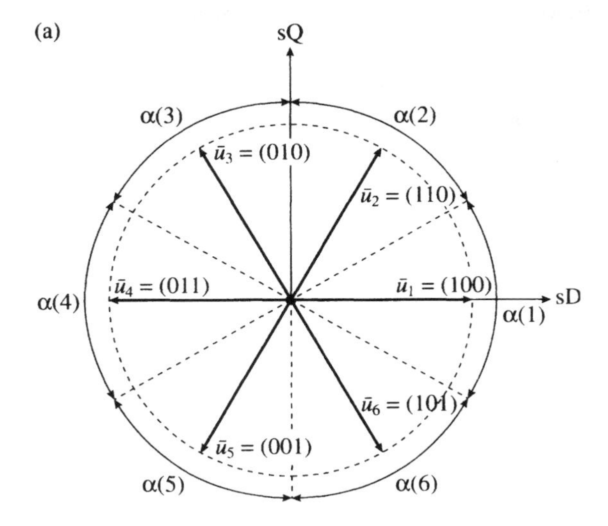

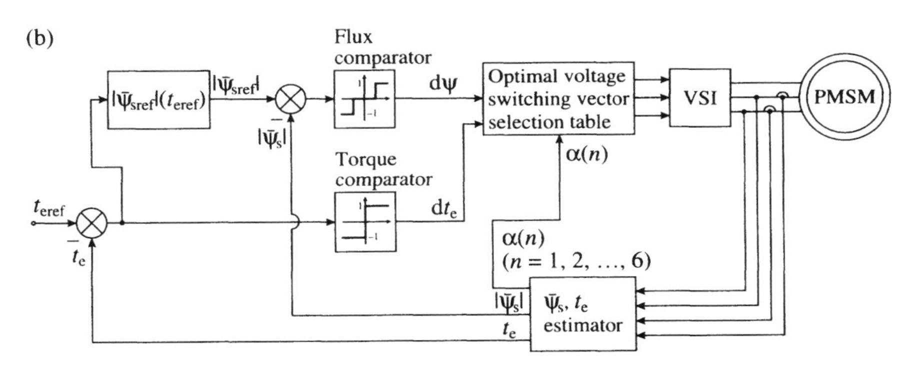

Fig. 3.48. Direct control of the flux linkages and electromagnetic torque of a PM synchronous motor supplied by VSI. (a) Six active switching vectors ($\bar{u}_1, \bar{u}_2, \ldots, \bar{u}_6$); (b) control scheme.

torque comparator is a two-level comparator. The discretized outputs of the hysteresis comparators ($d\psi, dt_e$) are inputs to the optimum voltage switching selection look-up table. However, the information on the position of the stator flux-linkage space vector (sector number) is also an input to the look-up table. Figure 3.48(a) shows the six voltage switching vectors together with the six sectors; these sectors cover the angles $\alpha(1), \alpha(2), \ldots, \alpha(6)$ respectively. It can be seen that each sector spans 60 electrical degrees (e.g. sector 1 spans from -30° to 30°) and for a six-step inverter, the minimal number of sectors is six.

The drive scheme also contains an estimator, which provides estimates of the stator flux-linkage space vector, $\bar{\psi}_s = |\bar{\psi}_s| \exp(j\rho_s)$ (its modulus, $|\bar{\psi}_s|$, and its position, ρ_s) and also the estimate of the electromagnetic torque (t_e). The estimation of the stator flux-linkage space vector is briefly discussed in the next section.

In a speed-sensorless DTC drive, the rotor speed can also be estimated by using the techniques presented in Section 3.1.3. If the drive scheme contains a speed control loop, then the reference electromagnetic torque is present on the output of the speed controller (which can be a PI controller).

3.3.2.1.2 Flux-linkage estimation

The stator voltage equation of the PM synchronous machine in the stationary reference frame is

$$\bar{u}_s = R_s \bar{i}_s + \frac{d\bar{\psi}_s}{dt}, \tag{3.3-1}$$

where R_s is the stator resistance and $\bar{u}_s$, $\bar{i}_s$, $\bar{\psi}_s$ are the space vectors of the stator voltage, stator current, and stator flux-linkage respectively. Thus by considering eqn (3.3-1), the stator flux-linkage space vector can be obtained from the measured stator voltages and currents as

$$\bar{\psi}_s = \int (\bar{u}_s - R_s \bar{i}_s)\, dt. \tag{3.3-2}$$

Thus the direct- and quadrature-axis stator flux-linkage components in the stationary reference frame are obtained as

$$\psi_{sD} = \int (u_{sD} - R_s i_{sD})\, dt \tag{3.3-3}$$

$$\psi_{sQ} = \int (u_{sQ} - R_s i_{sQ})\, dt. \tag{3.3-4}$$

It can be seen that this technique requires only one machine parameter, which is the stator resistance. However, an open-loop integrator will cause initial value and drift problems, especially at low stator frequencies, and to avoid this, various techniques can be used (see also Section 3.1.3.2.1). For example, one possibility is to use a low-pass filter instead of the pure integrator and thus $1/p$ is replaced by $T/(1+pT)$ and hence

$$\bar{u}_s = R_s \bar{i}_s + \left[\frac{1+pT}{T}\right] \bar{\psi}_s. \tag{3.3-5}$$

In eqn (3.3-5) $p = d/dt$, and T is a suitably chosen time constant, which gives a low cut-off frequency and thus allows eqn (3.3-5) to approximate a pure integration in the widest speed-range. It follows from eqn (3.3-5) that the stator flux-linkage components can be estimated by using

$$\psi_{sD} = \frac{u_{sD} - R_s i_{sD}}{p + 1/T} \tag{3.3-6}$$

$$\psi_{sQ} = \frac{u_{sQ} - R_s i_{sQ}}{p + 1/T}. \tag{3.3-7}$$

However, to obtain accurate flux estimates at low stator frequencies, the time constant has to be large, and the variation of the stator resistance with the temperature has to be considered. For this purpose it is also possible to use a thermal model of the machine.

It is also possible to use such a stator flux-linkage estimator where the drift problems associated with 'pure' open-loop integrators at low frequency are avoided by a band-limited integration of the high-frequency components, and where the inaccurate flux-linkage estimation at frequencies below $1/T$ is replaced by its reference value in a smooth transition. For this purpose a first-order delay element, $1/(1+pT)$, is used; thus the stator flux-linkage space vector is obtained from

$$\bar{\psi}_s = \frac{T(\bar{u}_s - R_s\bar{i}_s) + \bar{\psi}_{sref}}{1+pT}, \tag{3.3-8}$$

where $\bar{\psi}_{sref}$ is the stator flux-linkage space vector in the stationary reference frame, $\bar{\psi}_{sref} = |\bar{\psi}_{sref}| \exp(j\rho_s)$. The inputs to this stator flux estimator are the measured values of the stator-voltage space vector ($\bar{u}_s$) and space-current space vector ($\bar{i}_s$), expressed in the stationary reference frame. However, there is also a third input, which is the reference value of the stator flux-linkage space vector ($\bar{\psi}_{sref}$), which is also expressed in the stationary reference frame. It should be noted that since the stationary reference frame is used, $\bar{\psi}_{sref}$ contains two components, ψ_{sDref} and ψ_{sQref}. In eqn (3.3-8) the space vector of the induced stator voltages is $\bar{u}_{si} = \bar{u}_s - R_s\bar{i}_s$ and in an open-loop stator flux estimator using a 'pure' integrator, its integrated value ($\int \bar{u}_{si}\,dt$) yields the stator flux-linkage space vector $\bar{\psi}_s$. However, in eqn (3.3-8) $\bar{u}_{si}$ is multiplied by T and the reference stator flux-linkage space vector is added to $T\bar{u}_{si}$ yielding $T\bar{u}_{si} + \bar{\psi}_{sref}$. This is then the input to the first-order delay element, $1/(1+pT)$, on the output of which the estimated value of the stator flux-linkage space vector is obtained. In the permanent-magnet synchronous machine it is possible to have such an implementation of the flux-linkage estimator, where the reference stator flux-linkage space vector is obtained as $\bar{\psi}_{sref} \approx \psi_F \exp(j\theta_r)$ (this assumes $i_{sd} = 0$; see below), where ψ_F is the magnet flux (in the reference frame fixed to the rotor). In this case $\bar{\psi}_F \exp(j\theta_r)$ is the magnet flux in the reference frame fixed to the stator and θ_r is the rotor angle. It follows that this solution requires the use of a position sensor (to obtain the rotor angle θ_r), and adequate results can be obtained for a PM synchronous machine with surface-mounted magnets, but at very low frequencies (speeds), other techniques are required.

To obtain greater accuracy, it is possible to use observers (Luenberger, Kalman observers), see also Sections 3.1.3.5, 3.2.2.4, 4.5.3.5. When an extended Kalman filter is used, it is possible to estimate the rotor speed in addition to the stator flux linkages and some machine parameters (joint state and parameter estimation). The stator flux-linkage components can also be estimated by using a model reference adaptive control (MRAS) system, e.g. see Section 4.5.3.4. It should also be noted that it is possible to obtain accurate stator flux-linkage estimates by using an artificial neural network, or a fuzzy-neural estimator (see Chapter 7). Furthermore, the fuzzy-logic techniques can be combined with conventional observer-based techniques.

3.3.2.1.3 Electromagnetic torque estimation

The estimator shown in the drive scheme of Fig. 3.48(b) also estimates the electromagnetic torque. For instantaneous torque control it is necessary to have a very accurate electromagnetic torque estimator. There are many possibilities for estimating the electromagnetic torque, e.g. it can be obtained by using the estimated flux linkages and monitored stator currents and in this case, it follows from eqn (2.1-167) that

$$t_e = \tfrac{3}{2}P(\psi_{sD}i_{sQ} - \psi_{sQ}i_{sD}), \tag{3.3-9}$$

where the estimation of the stator flux linkages has been discussed in the previous section. However, it can also be obtained by using the stator currents in the rotor reference frame (i_{sd}, i_{sq}), and it follows from eqn (3.1-18) that

$$t_e = \tfrac{3}{2}P[\psi_F i_{sq} + (L_{sd} - L_{sq})i_{sd}i_{sq}]. \tag{3.3-10}$$

However, this requires the synchronous inductances L_{sd} and L_{sq}, and also the rotor position has to be monitored (to obtain i_{sd}, i_{sq} from the measured stator currents i_{sD}, i_{sQ}). If this technique is used, in general the saturation dependency of the inductances has also to be considered. It is also possible to estimate the electromagnetic torque by using a third technique, where the stator currents and rotor position are measured during the operation of the drive, but L_{sd} and L_{sq} are not required. In this case the electromagnetic torque versus rotor position versus current characteristics of the machine are derived (this is a technique also used in switched reluctance motor drives; see Chapter 5). These characteristics can then be stored in a look-up table, and during the operation of the drive, by monitoring the rotor position and the stator currents, the electromagnetic torque can be estimated. For a given machine, the characteristics of the look-up table can be obtained in a self-commissioning stage by considering the fact that the electromagnetic torque can be estimated from the rate of change of the coenergy:

$$t_e = \partial W_c / \partial \theta_r. \tag{3.3-11}$$

In eqn (3.3-11) the coenergy versus current (i) characteristic can be obtained from the total flux linkage (ψ) versus current versus rotor position characteristics by integration

$$W_c = \int_0^i \psi \, dt. \tag{3.3-12}$$

The total flux linkage versus current versus rotor position characteristics can be obtained by utilizing monitored stator voltages and currents at various rotor positions. For example, by performing blocked rotor tests, and by applying a step voltage at specific rotor positions, it is possible to obtain the data required for the flux linkage versus current versus rotor position characteristics. For this purpose it is also considered that the flux linking a stator phase can be obtained by integrating the appropriate terminal voltage reduced by the stator ohmic drop (the latter is obtained by multiplying the known value of the stator resistance by the appropriate stator current).

3.3.2.1.4 Reference stator flux-linkage generation

In Fig. 3.48(b), the stator flux-linkage error is generated by comparing the estimated stator flux-linkage modulus to its reference value ($|\bar{\psi}_{sref}|$). However, this reference value is obtained from the reference value of the electromagnetic torque. To obtain the function $|\bar{\psi}_{sref}|(|t_{eref}|)$, it is possible to utilize the concepts discussed in Section 3.2.2.1. Thus up to base speed, the function $|\bar{\psi}_{sref}|(|t_{eref}|)$ can be obtained by ensuring maximal torque per unit current, and above base speed, it can be obtained by ensuring maximal torque per unit flux. These two techniques are discussed below.

Application of the maximum torque/unit current criteria For the case of maximum torque per unit current the function $|\bar{\psi}_{sref}|(|t_{eref}|)$ can be obtained as follows by using the flux linkage and torque equations. The direct- and quadrature-axis stator flux linkages in the rotor reference frame are

$$\psi_{sd} = L_{sd} i_{sd} + \psi_F \tag{3.3-13}$$

$$\psi_{sq} = L_{sq} i_{sq}. \tag{3.3-14}$$

As emphasized above, it is also a goal to implement such a switching strategy which ensures minimal stator ohmic losses, i.e. maximal efficiency. This means that the desired electromagnetic torque should be achieved with minimum stator current. Thus i_{sd} is forced to zero (see also the discussion related to Fig. 3.2) by the appropriate switching vectors. If for simplicity a surface-mounted PM machine is assumed ($L_{sd} = L_{sq}$), then for maximum torque per unit current $i_{sd} = 0$ and the modulus of the stator flux-linkage space vector can be obtained from its components as

$$|\bar{\psi}_s| = (\psi_{sd}^2 + \psi_{sq}^2)^{1/2} = (\psi_F^2 + L_{sq}^2 i_{sq}^2)^{1/2}. \tag{3.3-15}$$

However, by considering eqn (3.1-3), the electromagnetic torque of the PM machine with surface-mounted magnets can be expressed as

$$t_e = \tfrac{3}{2} P \psi_F i_{sq}. \tag{3.3-16}$$

Thus the torque-producing stator current can be expressed as

$$i_{sq} = \frac{2t_e}{3P\psi_F}. \tag{3.3-17}$$

Substitution of eqn (3.3-17) into eqn (3.3-15) yields the required relationship between the flux-linkage modulus and the electromagnetic torque:

$$|\bar{\psi}_s| = \left\{\psi_F^2 + L_{sq}^2\left[\frac{2t_e}{3P\psi_F}\right]^2\right\}^{1/2}. \tag{3.3-18}$$

Thus the required function for the reference values is obtained from eqn (3.3-18) as

$$|\bar{\psi}_{sref}| = \left\{\psi_F^2 + L_{sq}^2\left[\frac{2t_{eref}}{3P\psi_F}\right]^2\right\}^{1/2}. \tag{3.3-19}$$

It should be noted that for a PM synchronous machine with interior magnets, $L_{sd} \neq L_{sq}$ and the electromagnetic torque also contains a reluctance torque component, and thus

$$t_e = \tfrac{3}{2} P[\psi_F i_{sq} + (L_{sd} - L_{sq}) i_{sd} i_{sq}], \tag{3.3-20}$$

and this expression should be used when $|\bar{\psi}_{sref}|$ is obtained from the torque reference.

Application of the maximum torque/unit flux criteria The appropriate forms of the function $|\bar{\psi}_{sref}|(t_{eref})$ call also be obtained when the maximum torque per unit flux criterion is used (in the high speed-range). In this case by using eqns (3.3-13) and (3.3-14), the stator flux-linkage modulus can be expressed as

$$|\bar{\psi}_s| = (\psi_{sd}^2 + \psi_{sq}^2)^{1/2} = [(L_{sd} i_{sd} + \psi_F)^2 + (L_{sq} i_{sq})^2]^{1/2}. \tag{3.3-21}$$

To obtain the expression of $|\bar{\psi}_{sref}|(t_{eref})$, eqn (3.3-21) has to be used together with the expression which can be obtained by using $dt_e/di_{sd} = 0$, where t_e is defined by eqn (3.3-20).

3.3.2.1.5 Optimum switching voltage vector table

The optimum switching voltage vector table shown in Fig. 3.48(b) can be obtained in various ways, e.g. by using physical considerations. This will only be discussed briefly, since a very detailed explanation is given in Section 4.6.2.2 for an induction motor, and similar considerations hold for the PMSM.

Similarly to that shown in Section 4.6.2.2, Table 4.3, for the induction motor, the optimum switching table also uses the information on the position of the stator flux-linkage space vector (in the stationary reference frame, whose axes are sD and sQ respectively). Since a six-step inverter is used, it is appropriate to define a minimum number of six sectors. Each of these sectors is 60 electrical degrees wide, and they cover the regions $\alpha(1), \alpha(2), \ldots, \alpha(6)$, where sector 1 spans from $-30°$ to $30°$, sector 2 spans from $30°$ to $90°, \ldots$, sector 6 spans from $270°$ to $330°$ (in the complex plane defined by the sD and sQ axes, where the sD-axis is at the zero degree position). These sectors are also shown in Fig. 3.48(a) together with the six active voltage switching vectors of the inverter ($\bar{u}_1, \bar{u}_2, \ldots, \bar{u}_6$).

For example, if the stator flux linkage is in the first sector and has to be increased, and the electromagnetic torque has to be positive (this corresponds to $d\psi = 1$, $dt_e = 1$

where $d\psi$ and dt_e are the discretized outputs of the flux-linkage and electromagnetic torque comparators shown in Fig. 3.48(b) respectively), then by considering the six active switching vectors also shown in Fig. 3.48(a), the switching voltage vector to be selected is $\bar{u}_2$. On the other hand if the stator flux linkage has to be increased, but the electromagnetic torque has to be negative (this corresponds to $d\psi=1$, $dt_e=-1$), then the switching vector $\bar{u}_6$ has to be selected. However, if the stator flux linkage has to be decreased, but the electromagnetic torque has to be positive ($d\psi=-1, dt_e=1$), then $\bar{u}_3$ has to be selected, and similarly if the stator flux linkage has to be increased, but the electromagnetic torque has to be negative ($d\psi=-1$, $dt_e=-1$) then this can be achieved by applying $\bar{u}_5$, etc.

For i_{sq} control ($i_{sd}=0$), the control of the stator flux linkages can be omitted. To produce the largest electromagnetic torque for a given stator current, in accordance with that shown above, it has to be ensured that the stator-current space vector in the rotor reference frame contains only a quadrature-axis current component. This is analogous to that in a separately excited d.c. motor, where it is achieved by the consecutive switching of the armature coils through the commutator. In this case the stator flux-linkage space vector increases with positively increasing torque and decreases with decreasing positive torque, etc. It follows that if the stator flux-linkage space vector is changed from a previous position described by $\bar{\psi}_{s1}=\psi_F+jL_{sq}i_{sq1}$ (note that $i_{sd}=0$), to a new position described by $\bar{\psi}_{s2}$, then since $i_{sd}=0$, the new position must be obtained by adding a quadrature-axis stator flux-linkage component $\Delta\bar{\psi}_s=jL_{sq}i_{sq2}$ to $\bar{\psi}_{s1}$. Thus

$$\bar{\psi}_{s2}=\bar{\psi}_{s1}+\Delta\bar{\psi}_s=\psi_F+jL_{sq}(i_{sq1}+i_{sq2}). \tag{3.3-22}$$

It follows that for a positive torque error (where $dt_e=1$), the optimum voltage switching vector is the one (of the six active switching vectors) which is closest to the rotor quadrature-axis. For a negative torque error (where the discretized output of the torque comparator is $dt_e=-1$), the optimum voltage switching vector is the one which is closest to the negative rotor quadrature-axis. For this purpose, the rotor position has also to be used. However, in this case, the calculation of $|\bar{\psi}_{sref}|(|t_{eref}|)$ and the estimation of $|\bar{\psi}_s|$ are not required.

3.3.2.2 Direct control of the electromagnetic torque and d-axis stator current using an optimal voltage switching look-up table (Scheme 2)

There are various possibilities for a DTC scheme of a PM synchronous machine. For example it is possible to control directly the d-axis stator current (i_{sd}) and the electromagnetic torque by the appropriate selection of the voltage switching vectors. This is shown in Fig. 3.49, where instead of using the stator flux-linkage control loop shown in Fig. 3.48(b), now there is an i_{sd} control loop.

In Fig. 3.49 the reference value of i_{sd} is compared with its actual value and this error is fed into a two-level hysteresis comparator, which outputs di_{sd} (discretized output), which can be 1 or -1. The inputs to the optimal voltage switching table are then di_{sd} and dt_e (dt_e is the digitized output of the two-level torque hysteresis

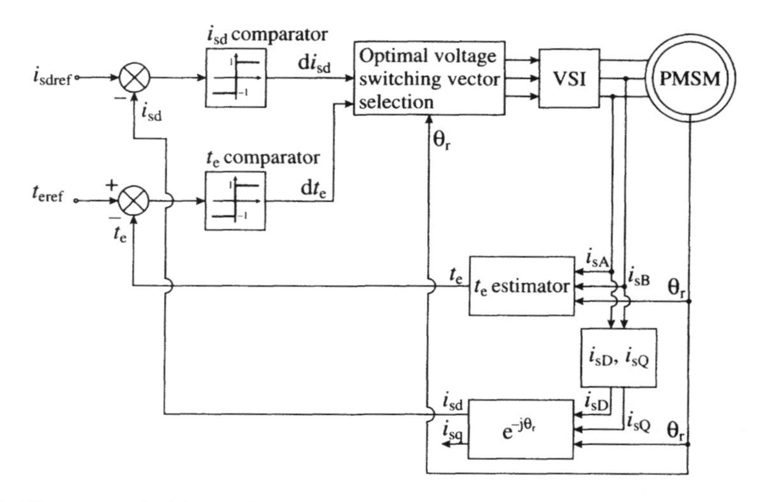

Fig. 3.49. Direct control of the d-axis stator current and electromagnetic torque of a PM synchronous motor supplied by VSI.

Table 3.2 Optimum switching voltage vector look-up table

di_{sd}	dt_e	$r(1)$	$r(2)$	$r(3)$	$r(4)$	$r(5)$	$r(6)$
1	1	$\bar{u}_2$	$\bar{u}_3$	$\bar{u}_4$	$\bar{u}_5$	$\bar{u}_6$	$\bar{u}_1$
	-1	$\bar{u}_6$	$\bar{u}_1$	$\bar{u}_2$	$\bar{u}_3$	$\bar{u}_4$	$\bar{u}_5$
-1	1	$\bar{u}_3$	$\bar{u}_4$	$\bar{u}_5$	$\bar{u}_6$	$\bar{u}_1$	$\bar{u}_2$
	-1	$\bar{u}_5$	$\bar{u}_6$	$\bar{u}_1$	$\bar{u}_2$	$\bar{u}_3$	$\bar{u}_4$

comparator and $dt_e=1$ or -1) and the rotor position. Table 3.2 shows the optimum switching voltage vector look-up table.

In this scheme, the actual value of i_{sd} is obtained from the measured stator currents (i_{s1}, i_{sQ}) as follows:

$$\bar{i}_s' = i_{sd} + ji_{sq} = \bar{i}_s \exp(-j\theta_r) = (i_{sD} + ji_{sQ}) \exp(-j\theta_r). \tag{3.3-23}$$

The electromagnetic torque can then be estimated by using i_{sd}, i_{sq} and the measured currents (and inductances); thus by using eqns (3.3-20) or (3.3-16) for the permanent-magnet synchronous machine with interior- or surface-mounted magnets respectively.

3.3.2.3 Direct control of the electromagnetic torque and reactive torque using an optimal switching voltage look-up table (Scheme 3)

It is possible to implement a direct-torque-controlled PM synchronous motor drive where the electromagnetic torque $t_e = (3/2)P\,\mathrm{Im}(\bar{\psi}_s^* \bar{i}_s)$ and also the reactive torque $t_r = (3/2)P\,\mathrm{Re}(\bar{\psi}_s^* \bar{i}_s) = (3/2)P\,\mathrm{Im}(j\bar{\psi}_s \bar{i}_s^*)$ are controlled directly. In this drive it is possible to achieve high dynamic performance, and the power factor can also be affected (by changing the reference value of the reactive torque). The control scheme is shown in Fig. 3.50(b).

In Fig. 3.50(b) the two inputs are the reference values of the electromagnetic torque (t_{eref}) and reactive torque (t_{rref}) respectively. The electromagnetic torque error (difference between the electromagnetic torque and its reference value) is fed into a three-level hysteresis comparator on the output of which dt_e is present. Similarly, the reactive torque error (difference between the reactive torque and its reference value) is fed into a two-level hysteresis comparator on the output of which dt_r is present. Both dt_e and dt_r are inputs to an optimal switching voltage

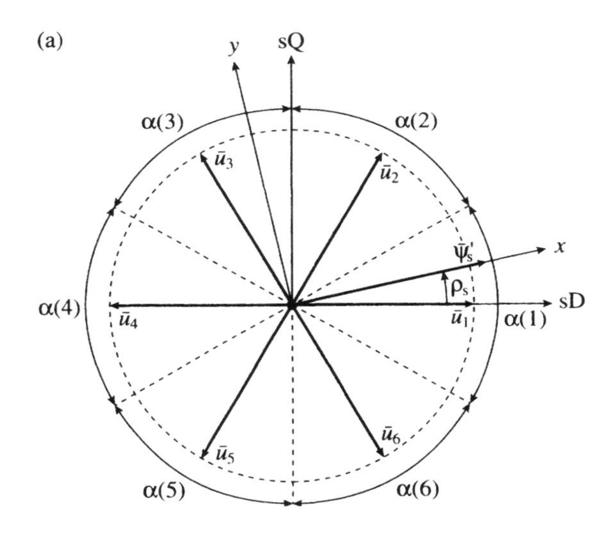

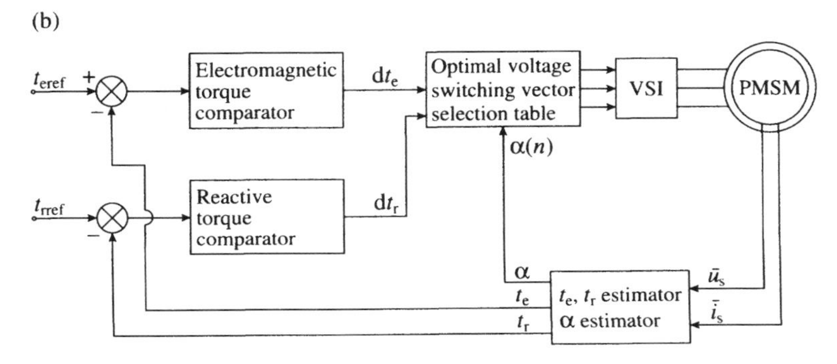

Fig. 3.50. Direct control of the electromagnetic torque and reactive torque of a PM synchronous motor supplied by VSI. (a) Six active switching vectors ($\bar{u}_1, \bar{u}_2, \ldots, \bar{u}_6$); (b) control scheme.

vector selection look-up table, together with the information on the sector where the stator flux-linkage space vector is located. There are six sectors: these are also shown in Fig. 3.50(a) (they are denoted by $\alpha(1), \alpha(2), \ldots, \alpha(6)$ respectively).

The drive scheme shown in Fig. 3.50(b) also contains an estimator, which also estimates two torque quantities. The first quantity is the instantaneous electromagnetic torque, and for this purpose

$$t_e = \frac{3}{2} P \operatorname{Im}(\bar{\psi}_s^* \bar{i}_s) = \frac{3}{2} P(\psi_{sD} i_{sQ} - \psi_{sQ} i_{sD}) \tag{3.3-24}$$

is used. The second quantity is the instantaneous reactive torque and is estimated by using

$$t_r = \frac{3}{2} P \operatorname{Im}(j\bar{\psi}_s \bar{i}_s^*) = \frac{3}{2} P(\psi_{sD} i_{sD} + \psi_{sQ} i_{sQ}). \tag{3.3-25}$$

Physically the electromagnetic torque is related to the corresponding mechanical power (active power), and the reactive torque is related to the reactive power, where the reactive power maintains the magnetizing flux of the machine. It should be noted that by using the stator voltage equation of the machine, it can be simply proved that $\operatorname{Im}(j\bar{\psi}_s \bar{i}_s^*)$ is related to the reactive power. The stator flux-linkage components can be estimated by using one of the techniques discussed above in Section 3.3.2.1.2. For example it is possible to have an implementation which uses the monitored stator currents (i_{sD} and i_{sQ}) and reconstructed stator voltages (u_{sD}, u_{sQ}). For this purpose the monitored measured d.c. link voltage and also the inverter switching states are used (see Section 3.1.3.2.1).

The optimum switching voltage vector look-up table can be obtained by using physical considerations. For this purpose Fig. 3.50(a) also shows the six active voltage switching vectors of the inverter ($\bar{u}_1, \bar{u}_2, \ldots, \bar{u}_6$) together with a possible stator flux-linkage space vector, which is aligned with the x-axis shown. To select the appropriate switching voltage vectors, the position of the stator flux-linkage space vector (with respect to the real-axis of the stationary reference frame, ρ_s) must also be detected. This position can be obtained e.g. by using $\rho_s = \tan^{-1}(\psi_{sQ}/\psi_{sD})$, or by using other techniques, where for example comparators are used, and the computation of trigonometric functions is completely avoided. There are six 60 electrical-degree-wide sectors to be considered, and the first sector spans angle $\alpha(1)$, which covers the region $-30°$ to $30°$, the second sector spans the angle $\alpha(2)$, etc. These angles are also shown in Fig. 3.50(a). For example, if the stator flux-linkage space vector is in sector 1, and $dt_e = 1$ and $dt_r = 1$ (discretized output values of the two hysteresis comparators are equal to one respectively), which means that both the electromagnetic torque and the reactive torque have to be increased, the switching vector $\bar{u}_2$ must be chosen. It can be seen that this is correct, since both the x-axis and y-axis components of $\bar{u}_2$ are positive. However $u_{2x} > 0$ is required to ensure that $t_{rref} - t_r > \Delta t_r$ ($dt_r = 1$), where $2\Delta t_r$ is the width of the hysteresis band of the reactive torque comparator, and similarly $u_{2y} > 0$ is required to ensure that $t_{eref} - t_e > \Delta t_e$ ($dt_e = 1$), where $2\Delta t_e$ is the width of the hysteresis band of the electromagnetic torque comparator. It is possible to obtain the required switching vectors by similar physical considerations for other values of the outputs of the hysteresis comparators and when the stator flux-linkage space vector is positioned in other sectors. In general, the following rules apply for the selection of the optimum switching vectors:

1. If $dt_r = 1$ then $u_x > 0$.
   Physically this means that in this case $t_{rref} - t_r > \Delta t_r$, and thus the x-axis component of the selected voltage switching vector ($\bar{u}_k$) must be positive ($u_x > 0$).
2. If $dt_r = 0$ then $u_x < 0$.
   Physically this means that in this case $t_{rref} - t_r < -\Delta t_r$ and thus the x-axis component of the selected voltage switching vector ($\bar{u}_k$) must be negative ($u_x < 0$).
3. If $dt_e = 1$ then $u_y > 0$.
   Physically this means that in this case $t_{eref} - t_e > \Delta t_e$ and thus the y-axis component of the selected voltage switching vector ($\bar{u}_k$) must be positive ($u_y > 0$).
4. If $dt_e = -1$ then $u_y < 0$.
   Physically this means that in this case $t_{eref} - t_e < -\Delta t_e$ and thus the y-axis component of the selected voltage switching vector ($\bar{u}_k$) must be positive ($u_y > 0$).

3.3.2.4 Direct torque control using a predictive algorithm for switching voltage vector selection (Scheme 4)

The use of powerful digital signal processors allow the on-line (real-time) computation of the switching vectors, leading to optimal solutions. Thus it is also possible to implement DTC schemes in which the required switching voltage vectors are obtained by using predictive algorithms. For this purpose two algorithms are briefly discussed below. In the first case a mathematical model of the permanent-magnet synchronous machine is used and the electromagnetic torque is estimated for each sampling period for all possible inverter modes. The predictive algorithm then selects the inverter switching states to give minimum deviation between the predicted electromagnetic torque and the reference torque. Since there are many possibilities for the mathematical model to be used, only the main concepts will be discussed below. This is followed by the description of a second predictive algorithm, which is obtained by modifying the first algorithm, and this technique ensures maximal torque/current ratio.

To obtain the predictive algorithm, first a suitable mathematical model of the permanent-magnet synchronous machine has to be derived. This can be based on the space-vector model of the PMSM. By using the direct- and quadrature-axis stator voltage differential equations expressed in the rotor reference frame, it is then possible to solve these in a sampling interval for the direct- and quadrature-axis stator current components (i_{sd}, i_{sq}). For this purpose it is assumed that within a sampling time interval, the rotor speed is constant (e.g. at the end of the time interval, the rotor speed is the same as at the beginning of the time interval). The obtained analytical expressions for the stator currents can then be substituted into the analytical expression for the electromagnetic torque. The thus-derived analytical expressions (for the stator currents and electromagnetic torque) depend on the machine parameters, voltage supplied by the inverter, rotor angle, and rotor speed. It follows that by using the initial values at the beginning of the sampling interval, it is possible to predict the electromagnetic torque and stator currents at the end of the sampling interval. The thus-obtained expression of the electromagnetic torque can then be used in the control system to generate the optimal voltage switching vectors applied to the inverter.

It is a principal task of this control algorithm to minimize in every sampling interval the difference between the reference and actual electromagnetic torque. This minimization can be performed in various ways. Since at the beginning of every sampling interval (t_0), the stator currents, speed, and position are known, it is possible to predict the electromagnetic torque of the permanent-magnet synchronous machine at the end of the sampling period ($t_0+\Delta T$) for all the inverter switching states, thus $t_e(t_0+\Delta T)$ is obtained. The control algorithm then selects that specific inverter switching state for which the error, $t_{eref}-t_e(t_0+\Delta T)$ is minimal. In this way minimal electromagnetic-torque error is ensured in every step. However, this algorithm has to be slightly changed, since it does not take account of the associated changes in the stator currents, and the modulus of the stator-current space vector may contain large oscillations, if some limiting actions are not provided by the control algorithm. This modification is now discussed.

If a modified algorithm is used which ensures maximum torque/current ratio, then by considering the expression of the elctromagnetic torque (for a given modulus of the stator flux-linkage space vector), it is possible to express the stator-current space vector in terms of the electromagnetic torque and the modulus of the stator flux-linkage space vector. It should be noted that for maximum torque/current, the angle of the stator-current space vector in the stationary reference frame is equal to the angle of the stator flux-linkage space vector plus 90°, and thus when this is considered in the expression for the electromagnetic torque, then a simple expression is obtained for the stator-current space vector (in terms of the electromagnetic torque and the stator flux-linkage modulus). If this new expression for the stator current vector is substituted into the analytical expression for the stator current vector (the derivation of which has been discussed above), then the obtained new expression can be used to select the optimal switching voltage vectors. In particular, it should be noted that the new expression for the stator-current space vector contains the stator-voltage space vector, which is the reference voltage space vector in the stationary reference frame ($\bar{u}_{sref}$). The appropriate switching state of the inverter is then determined by using space-vector modulation. Therefore $\bar{u}_{sref}$ is used to select the optimal switching voltage vectors in such a way that the two switching vectors ($\bar{u}_k$, $\bar{u}_{k+1}$) closest to $\bar{u}_{sref}$ are selected, and the amount of time during which these vectors are applied (t_a, t_b) is determined from [Vas 1992]

$$\bar{u}_{sref}\Delta T=\bar{u}_k t_a+\bar{u}_{k+1}t_b+\bar{u}_0 t_0. \qquad (3.3\text{-}26)$$

In eqn (3.3-26)

$$\Delta T = t_0 + t_a + t_b, \tag{3.3-27}$$

where $T_s = \Delta T$ is the sampling time. Equation (3.3-26) follows from the fact that the time average of the three switching states (two active states and one zero state) during the sampling interval is equal to the reference voltage space vector, and also the switching voltage vectors and the reference voltage vector are constant over one switching cycle. In eqn (3.3-26) the $\bar{u}_k$ are the switching vectors in the eight switching states of the voltage source inverter:

$$\bar{u}_k = \begin{cases} (2/3)U_d \exp[j(k-1)\pi/3] & k = 1, 2, \ldots, 6 \\ 0 & k = 7, 8, \end{cases} \tag{3.3-28}$$

where U_d is the d.c. link voltage, $k = 1, 2, \ldots, 6$ correspond to the active (non-zero) switching voltage vectors, and $k = 7, 8$ correspond to the two zero switching-voltage vectors. If eqn (3.3-28) is substituted into eqn (3.3-26), and the resulting equation is resolved into its real and imaginary parts, then t_a and t_b can be determined. Thus by using $\bar{u}_{sref} = u_{sDref} + ju_{sQref}$,

$$t_a = \frac{3\Delta T}{2U_d}\left[u_{sDref} - \frac{u_{sQref}}{\sqrt{3}}\right] \tag{3.3-29}$$

$$t_b = \left[\frac{\sqrt{3}}{U_d}\right]\Delta T u_{sQref}. \tag{3.3-30}$$

The time during which the appropriate zero switching vector is selected is

$$t_0 = \Delta T - t_a - t_b. \tag{3.3-31}$$

When this modified control algorithm is used, then the oscillations in the stator currents are reduced.

Finally it should be noted that it is also possible to have such predictive algorithms in direct-torque-controlled permanent-magnet synchronous motor drives, where the electromagnetic torque is directly controlled by using the electromagnetic torque versus rotor position versus current characteristics of the machine. The method of obtaining such characteristics has been discussed above in Section 3.3.2.1.3.

3.3.3 DTC OF THE VSI-FED SYNCHRONOUS RELUCTANCE MOTOR

3.3.3.1 General introduction

It has been discussed in Sections 1.2.1.3 and 3.2.2.1 that new types of synchronous reluctance motors have recently been introduced which have increased power factors and torque densities. For this purpose, special rotor structures are applied (e.g. axially laminated rotors, permanent-magnet-assisted axially laminated rotors, etc.). These drives offer the possibility of application in a wide range of areas, including robotic, spindle drive, and small traction applications.

In Section 3.2.2 the vector control of synchronous reluctance motors has been discussed. However, it is also possible to obtain high-dynamic performance by using direct torque control (DTC), and for this purpose first the direct electromagnetic torque and stator flux-linkage control of a synchronous reluctance motor supplied by a VSI is described in the next section (Scheme 1). In a 'convetional' VSI-fed direct-torque-controlled a.c. motor drive, those inverter switching states are selected which restrict the stator flux-linkage and electromagnetic torque errors within specified hysteresis bands and which result in fast electromagnetic torque response. The advantages and disadvantages of DTC have been discussed in Section 1.2.2 and the reader is referred to this, prior to studying the details of the drive schemes presented below. Furthermore, in general, the 'conventional' DTC of an a.c. machine supplied by a voltage-source inverter involves the direct control of the flux-linkage space vector (e.g. stator flux-linkage space vector) and electromagnetic torque by applying optimum voltage switching vectors of the inverter which supplies the motor. In addition to Scheme 1, a second scheme will also be discussed (in Section 3.3.3.3), where there is direct control of the electromagnetic torque and the angle of the stator-current space vector.

3.3.3.2 Direct control of the electromagnetic torque and stator flux linkage (Scheme 1)

3.3.3.2.1 General introduction

A direct control scheme is now discussed for the synchronous reluctance motor, where there is direct control of the electromagnetic torque and stator flux-linkage space vector. Fig. 3.51(b) shows the schematic of the DTC drive system, where the inverter is a six-pulse voltage source inverter.

In Fig. 3.51(b) the two inputs to the control scheme are the reference value of the electromagnetic torque (t_{eref}) and the reference value of the modulus of the stator flux-linkage space vector ($|\bar{\psi}_{sref}|$). The difference between the reference electromagnetic torque and the actual electromagnetic torque (t_e) is the input into a two-level torque hysteresis comparator, which outputs dt_e (dt_e is the discretized output). Similarly, the flux-linkage error, which is the difference between the reference stator flux-linkage and the modulus of the actual stator flux-linkage space vector, is fed into a two-level hysteresis comparator, which outputs $d\psi$. The outputs of the two comparators together with the information on the position of the stator flux-linkage space vector are fed into the optimal switching voltage vector look-up table.

In Fig. 3.51(b), there is also an estimator which is used to obtain the electromagnetic torque and also the stator flux-linkage space vector, $\bar{\psi}_s=|\bar{\psi}_s|\exp(j\rho_s)=\psi_{sD}+j\psi_{sQ}$ (its modulus and angle, or its two-axis components). The estimation of the stator flux linkages and electromagnetic torque is discussed in the next section.

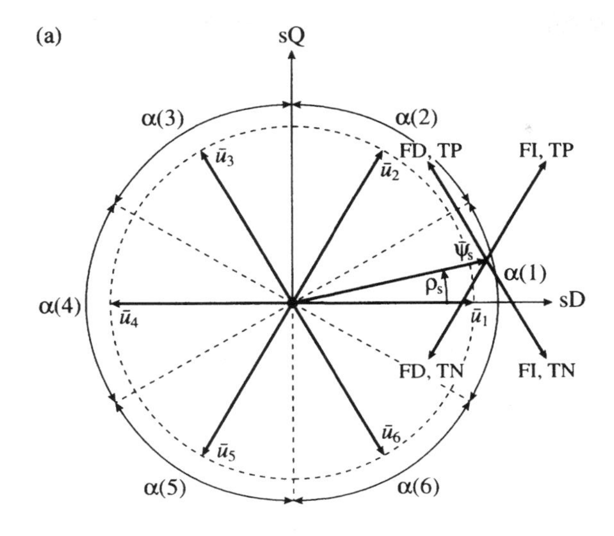

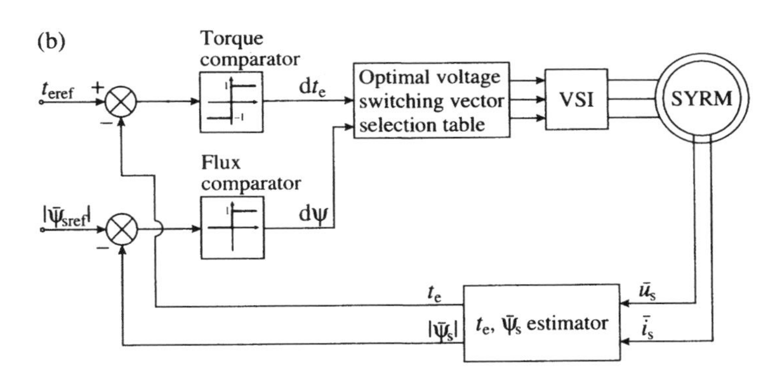

Fig. 3.51. Direct-torque-controlled synchronous reluctance motor drive. (a) Six switching vectors and stator-linkage space vector; FI: flux increase; FD: flux decrease; TP: torque positive; TN: torque negative; (b) drive scheme.

It follows from eqn (3.2-68) that the electromagnetic torque of the synchronous reluctance motor can also be expressed as follows:

$$t_e = \frac{\frac{3}{4}P(L_{sd}-L_{sq})|\bar{\psi}_s|^2\sin(2\delta)}{L_{sd}L_{sq}}, \tag{3.3-32}$$

where δ is the angle of the stator flux-linkage space vector with respect to the direct-axis of the rotor (see Fig. 3.45). In the DTC drive shown in Fig. 3.51(b) the stator voltages are controlled so that the modulus of the stator flux linkage is constant and the maximum electromagnetic torque is limited in such a way that the angle δ is not increased beyond 45°. The stator flux-linkage modulus is controlled by applying those voltage switching vectors (one of the six active inverter voltage vectors) which are directed towards the centre of the rotor to decrease the flux-linkage modulus and outwards to increase the modulus (see details in Section 3.3.3.2.3 below). Furthermore, the electromagnetic torque is controlled so that if an increase in the electromagnetic torque is required, then those switching voltage vectors are applied which advance the stator flux-linkage space vector in the positive direction of rotation. On the other hand, if the electromagnetic torque has to be decreased, switching voltage vectors which lag the stator flux-linkage space vector are applied. Since the stator flux-linkage space vector is the integral of the applied stator voltages (ohmic drop neglected), it moves in the direction of the applied switching voltage vectors, until the switching vectors are applied. Thus the voltage angle is indirectly controlled by the modulus of the stator flux-linkage space vector and the magnitude of the electromagnetic torque.

It should be noted that for the DTC synchronous reluctance motor drive, constant power operation above base speed can be conveniently implemented. When the motor reaches the base speed, then the inverter reaches its maximum voltage. Thus to increase the speed beyond the base speed, the voltage must be kept constant and since the voltage (modulus) is proportional to the product of the modulus of the stator flux-linkage space vector and the angular stator frequency, above base speed the modulus of the stator flux-linkage space vector has to be reduced (it is inversely proportional to the speed), $|\bar{\psi}_{\text{sref}}|=|\bar{\psi}_{\text{sbase}}|\omega_{\text{rbase}}/\omega_{\text{r}}$. Thus in Fig. 3.51(b) the reference value of the (modulus of the) stator flux-linkage space vector can be obtained on the output of the appropriate function, which is constant below base frequency, and above base frequency it is inversely proportional to the rotor speed; the input to this function is the rotor speed. For this purpose the monitored or estimated rotor speed can be used. For constant power operation (above base speed), the maximum torque can be obtained by considering that the mechanical power is equal to the product of the electromagnetic torque and speed, $p_{\text{m}}=t_{\text{e}}\omega_{\text{r}}=\text{constant}$. Thus $t_{\text{emaxref}}=t_{\text{emaxbase}}\omega_{\text{rbase}}/\omega_{\text{r}}$. It follows that above base speed, the maximum torque demand has to be reduced. If the maximum torque demand is not reduced, then the motor would lose synchronism (unstable operation would result), since the angle of the stator flux-linkage space vector with respect to the rotor direct-axis would exceed 45°.

It should be noted that efficiency could also be optimized at light loads by reducing the modulus of the stator flux-linkage space vector. However, this would result in a decreased dynamic performance.

3.3.3.2.2 Stator flux-linkage and electromagnetic torque estimation

Three different types of stator flux-linkage and electromagnetic torque estimators are discussed below. The first scheme can be used in a drive without a position sensor and the stator flux linkages and electromagnetic torque are obtained by monitoring the stator currents and stator voltages (or the stator voltages are reconstructed from the inverter switching functions by using the monitored value of the d.c. link voltage; see also Section 3.1.3.2.1). The second scheme uses the monitored stator currents and monitored rotor position, and also the saturation-dependent direct-axis inductance. However, in the third scheme, which also uses the monitored stator currents and rotor position, no saturation-dependent inductance is used.

Flux-linkage estimator 1: position-sensorless implementation In a drive without a position sensor, the electromagnetic torque can be estimated by considering eqn (2.1-157),

$$t_e = \tfrac{3}{2} P\bar{\psi}_s x \bar{i}_s = \tfrac{3}{2} P(\psi_{sD} i_{sQ} - \psi_{sQ} i_{sD}), \tag{3.3-33}$$

where the stator flux-linkage components (expressed in the stationary reference frame) can be estimated by the techniques described in Sections 3.1.3.2, 3.2.2.2.1, and 3.3.2.1.2. For example,

$$\psi_{sD} = \int (u_{sD} - R_s i_{sD})\,dt \tag{3.3-34}$$

$$\psi_{sQ} = \int (u_{sQ} - R_s i_{sQ})\,dt, \tag{3.3-35}$$

where u_{sD}, u_{sQ}, i_{sD}, and i_{sQ} are the direct- and quadrature-axis stator voltages and currents respectively in the stationary reference frame. Equations (3.3-34) and (3.3-35) require accurate knowledge of the stator resistance, especially at low stator frequencies, and incorrect values result in decreased performance. If the stator flux-linkage components are inaccurately estimated, then the modulus and the position of the stator flux-linkage space vector and also the estimated electromagnetic torque will be incorrect. However, errors in the stator flux linkages and electromagnetic torque cause torque oscillations and also steady-state torque errors. Furthermore, an error in the position of the stator flux-linkage space vector can cause incorrect switching vector selection (since the switching vector selection is also based on the position of the stator flux-linkage space vector, as discussed below in Section 3.3.3.2.3). However, it should be noted that more accurate methods are described in Sections 3.1.3 and 3.2.2.2, and it has been shown that the stator voltages can be reconstructed by using the inverter switching states and also the measured d.c. link voltage (see Section 3.1.3.2.1). In particular it should be noted that in Section 3.1.3.2 a very simple integrator drift compensation technique has been presented for a permanent-magnet synchronous machine. In that case, due to drift, the stator flux-linkage space vector was not a circle centred in the origin of the stator reference frame, but it was shifted from the origin. The drift compensation was performed by subtracting from the flux-linkage coordinates (ψ_{sD}, ψ_{sQ}) the offsets from the origin. This technique can give accurate results for the PM synchronous machine, since the main flux is produced by the magnets and the contribution of the stator currents is small. However, for the synchronous reluctance machine, the stator flux linkages are produced by the stator currents, and in the DTC drive, the stator flux-linkage vector is controlled to produce a perfect circle. Thus when there is d.c. offset, the actual magnitude of the stator flux-linkage space vector will vary as the machine is rotating. Therefore the d.c. offsets can be removed by recording the offset values when the motor is stopped, and then by subtracting these offsets from the measured values when the motor is running.

Flux-linkage estimator 2: estimation using a position sensor and (saturated) machine inductances In a drive where the rotor position is also monitored, the stator flux linkages and electromagnetic torque can be estimated as follows. The stator-current components in the rotor reference frame (i_{sd}, i_{sq}) can be obtained from the stator currents in the stator reference frame (i_{sD}, i_{sQ}) by using

$$\bar{i}_s' = i_{sd} + \mathrm{j}i_{sq} = \bar{i}_s \exp(-\mathrm{j}\theta_r) = (i_{sD} + \mathrm{j}i_{sQ}) \exp(-\mathrm{j}\theta_r). \tag{3.3-36}$$

Hence the stator flux-linkage components in the rotor reference frame can be estimated by using the stator currents i_{sd} and i_{sq} as follows:

$$\psi_{sd} = L_{sd} i_{sd} = L_{sd}(i_{sd}) i_{sd} \tag{3.3-37}$$

$$\psi_{sq} = L_{sq}(i_{sq}) i_{sq}, \tag{3.3-38}$$

where L_{sd} is the total direct-axis inductance (synchronous inductance) and L_{sq} is the quadrature-axis synchronous inductance. To obtain greater accuracy, the variations of the inductances with the currents should be considered, as shown in eqns (3.3-37) and (3.3-38), but L_{sq} can be considered to be constant. In eqns (3.3-37) and (3.3-38), the effects of cross-saturation couplings have been neglected. By using eqns (3.3-37) and (3.3-38), the stator flux-linkage space vector in the stator reference frame can be obtained as

$$\begin{aligned}\bar{\psi}_s &= \psi_{sD} + \mathrm{j}\psi_{sQ} = \bar{\psi}_s' \exp(\mathrm{j}\theta_r) = (\psi_{sd} + \mathrm{j}\psi_{sq}) \exp(\mathrm{j}\theta_r) \\ &= [L_{sd}(i_{sd}) i_{sd} + \mathrm{j}L_{sq}(i_{sq}) i_{sq}] \exp(\mathrm{j}\theta_r).\end{aligned} \tag{3.3-39}$$

The modulus of the stator flux-linkage space vector is then obtained as

$$|\bar{\psi}_s| = (\psi_{sD}^2 + \psi_{sQ}^2)^{1/2} = \{[L_{sd}(i_{sd}) i_{sd}]^2 + [L_{sq}(i_{sq}) i_{sq}]^2\}^{1/2}. \tag{3.3-40}$$

The angle of the stator flux-linkage space vector can be obtained by considering Fig. 3.52.

It follows from Fig. 3.52 that the position of the stator flux-linkage space vector (ρ_s) in the stationary reference frame is equal to its position (δ) in the rotor reference frame plus the rotor angle (θ_r). By also considering that $\delta = \tan^{-1}(\psi_{sq}/\psi_{sd})$, where ψ_{sd} and ψ_{sq} have been obtained by eqns (3.3.-37) and (3.3-38), it follows that

$$\rho_s = \theta_r + \delta = \theta_r + \tan^{-1}\left(\frac{\psi_{sq}}{\psi_{sd}}\right). \tag{3.3-41}$$

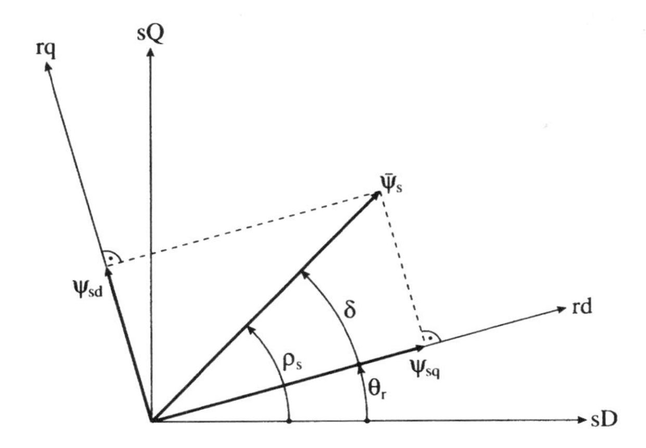

Fig. 3.52. Stator flux-linkage space vector in the stationary and rotor reference frames.

The electromagnetic torque can then be estimated by using the stator currents i_{sd}, i_{sq} and the inductances as follows:

$$t_e = \tfrac{3}{2} P \bar{\psi}_s x \bar{i}_s = \tfrac{3}{2} P(\psi_{sd} i_{sq} - \psi_{sq} i_{sd}) = \tfrac{3}{2} P[L_{sd}(i_{sd}) - L_{sq}] i_{sd} i_{sq}. \qquad (3.3\text{-}42)$$

However, this estimation technique requires a saturation-dependent inductance L_{sd} (it has been assumed that L_{sq} is constant). Nevertheless, it is possible to overcome this problem by using another estimator in which the stator currents and the rotor angle are also monitored, but the stator flux linkages and electromagnetic torque are obtained in a different way. This is discussed below.

Flux linkage estimator 3: estimation using a position sensor and no inductances It is also possible to estimate the electromagnetic torque by using a third technique, where the inductances L_{sd} and L_{sq} are not required. In this case the electromagnetic torque versus rotor position versus current characteristics of the machine are derived (see also Section 3.3.2.1 and Chapter 5). These characteristics can then be stored in a look-up table, and during the operation of the drive, by monitoring the rotor position and the stator current components, the electromagnetic torque can be estimated. For a given machine, the characteristics of the lookup table can be obtained in an initial self-commissioning stage by considering the fact that in general, the electromagnetic torque can be estimated from the rate of change of the coenergy:

$$t_e = \partial W_c / \partial \theta_r. \qquad (3.3\text{-}43)$$

However, the coenergy versus current (i) characteristic can be obtained from the total flux linkage (ψ) versus current versus rotor position characteristics by integration:

$$W_c = \int_0^i \psi \, dt \qquad (3.3\text{-}44)$$

The total flux linkage versus current versus rotor position characteristics can be obtained by utilizing the monitored stator voltages and currents at various rotor positions (see also Section 3.3.2.1).

3.3.3.2.3 Optimal switching voltage vector look-up table

The optimal switching voltage vector look-up table can be derived by using physical considerations as follows. For this purpose the six active (non-zero) switching vectors ($\bar{u}_1, \bar{u}_2, \ldots, \bar{u}_6$) of the inverter are also shown in Fig. 3.51(a).

The stator flux-linkage space vector, $\bar{\psi}_s = |\bar{\psi}_s| \exp(j\rho_s)$, can be positioned in one of the six sectors shown in Fig. 3.51(a) (due to the six-step inverter, a minimum of six sectors have to be considered). For example, the first sector spans the region $\pm 30°$, thus the angle $\alpha(1)$, which is also shown in Fig. 3.51(a), varies from $-30°$ to $+30°$, the angle $\alpha(2)$ covers the region $30°$ to $90°$, etc. It should be noted that the position of the stator flux-linkage space vector is described by the angle ρ_s, where $\rho_s = 0$ coincides with the sD-axis (real-axis of the stationary reference frame). If for example it is assumed that the stator flux-linkage space vector is located in sector 1, as shown in Fig. 3.51(a), then if it has to be increased (as dictated by the reference stator flux linkage) and the electromagnetic torque has to be increased as well (as dictated by the reference torque), thus the discretized comparator outputs in Fig. 3.51(b) are $d\psi = 1$ and $dt_e = 1$ respectively, then the switching voltage vector $\bar{u}_2$ has to be applied. It should be noted that, in general, if a switching voltage vector is applied, the flux-linkage space vector created (by this vector) will move in the direction of this switching vector (since it is the integral of the voltage vector), until the voltage vector is applied. It can be seen by considering Fig. 3.51(a) that $\bar{u}_2$ creates a stator flux-linkage component which indeed increases the original stator flux-linkage vector, and also the new stator flux-linkage vector will be in a position which is rapidly rotated in the positive direction (direction of rotation) with respect to the original stator flux-linkage space vector, and positive electromagnetic torque is developed. However, if for example, the stator flux linkage is again in the first sector, but it has to be decreased and the electromagnetic torque has to be positive (this corresponds to $d\psi = 0$, $dt_e = 1$), then it can be seen from Fig. 3.51(a) that switching voltage vector $\bar{u}_6$ has to be applied. In this case, due to $\bar{u}_6$, a stator flux-linkage component is created which will increase the original stator flux linkage, but the new stator flux-linkage space vector will be rapidly rotated in the negative direction with respect to the original stator flux-linkage space vector (and thus the electromagnetic torque is negative). If the stator flux linkage is in sector 1 and it has to be decreased and the electromagnetic torque has to be negative (thus $d\psi = 0$, $dt_e = -1$), then the switching vector $\bar{u}_5$ has to be applied (since this rapidly rotates the original flux-linkage space vector in the negative direction and also decreases it). Similar considerations hold when the stator flux-linkage space vector is in one of the other five sectors. Thus the optimum switching voltage selection look-up table shown in Table 3.3 can be constructed for the active voltage vectors.

Table 3.3 Optimum active voltage switching vector look-up table

$d\psi$	dt_e	$\alpha(1)$ sector 1	$\alpha(2)$ sector 2	$\alpha(3)$ sector 3	$\alpha(4)$ sector 4	$\alpha(5)$ sector 5	$\alpha(6)$ sector 6
1	1	$\bar{u}_2$	$\bar{u}_3$	$\bar{u}_4$	$\bar{u}_5$	$\bar{u}_6$	$\bar{u}_1$
	-1	$\bar{u}_6$	$\bar{u}_1$	$\bar{u}_2$	$\bar{u}_3$	$\bar{u}_4$	$\bar{u}_5$
0	1	$\bar{u}_3$	$\bar{u}_4$	$\bar{u}_5$	$\bar{u}_6$	$\bar{u}_1$	$\bar{u}_2$
	-1	$\bar{u}_5$	$\bar{u}_6$	$\bar{u}_1$	$\bar{u}_2$	$\bar{u}_3$	$\bar{u}_4$

Active switching vectors: $\bar{u}_1(1\ 0\ 0)$; $\bar{u}_2(1\ 1\ 0)$; $\bar{u}_3(0\ 1\ 0)$; $\bar{u}_4(0\ 1\ 1)$; $\bar{u}_5(0\ 0\ 1)$; $\bar{u}_6(1\ 0\ 1)$.

As expected, Table 3.3 is similar to the one obtained for the direct-torque- and flux-controlled induction motor supplied by a voltage-source inverter (see Section 4.6.2.2). To select the appropriate switching vectors, the position of the stator flux-linkage space vector (with respect to the real-axis of the stationary reference frame) must also be detected. This position can be obtained e.g. by using the estimated stator flux-linkage components and then by performing $\rho_s = \tan^{-1}(\psi_{sQ}/\psi_{sD})$. However, it is possible to avoid the computation of trigonometric functions by using comparators (see also Section 4.6.2.2). Since there are six active switching voltage vectors, for voltage vector selection, the complex plane must be divided at least into six sectors, each of which is 60 electrical degrees wide, when the position of the stator flux-linkage space vector is determined.

3.3.3.2.4 *Speed control loop; speed estimation*

For speed control purposes, a speed loop has to be added to the drive scheme shown in Fig. 3.51(b). For this purpose the reference speed input is subtracted from its actual value and the resulting error is fed into a speed controller (this can also be a fuzzy controller). The output of the speed controller gives the reference value of the electromagnetic torque.

In a speed-sensorless DTC synchronous reluctance motor drive, it is possible to obtain the speed signal without using a conventional speed sensor. Such implementations have been discussed in Section 3.2.2.2. For example, it is possible to obtain the speed signal by utilizing the speed of the stator flux-linkage space vector, which is $\omega_{ms} = d\rho_s/dt$, and for this purpose the estimated stator flux-linkage components can be used. Thus

$$\omega_{ms} = \frac{d\rho_s}{dt} = \frac{d[\tan^{-1}(\psi_{sQ}/\psi_{sD})]}{dt}. \tag{3.3-45}$$

By performing the differentiation, this can be put into the following form:

$$\omega_{ms} = \frac{\psi_{sD} d\psi_{sQ}/dt - \psi_{sQ} d\psi_{sD}/dt}{\psi_{sD}^2 + \psi_{sQ}^2}. \tag{3.3-46}$$

In the steady-state, the stator flux-linkage space vector is rotating exactly at the speed of the rotor. However, in the transient state, when there is a change in the electromagnetic torque reference, the stator flux-linkage space vector moves (relative to the rotor) to produce the required new electromagnetic torque. An increase in the reference electromagnetic torque causes the stator flux-linkage space vector to move in the negative direction of rotation. Thus if the rotor speed is estimated from ω_{ms}, it will be too low (underestimated), until the new steady-state is reached. Similarly, if there is a decrease of the reference electromagnetic torque, then the rotor speed estimated from ω_{ms} will be too large (overestimated). If the rate of change of the torque is limited, this effect is minimized, but if there are large changes in the electromagnetic torque, then this effect cannot be neglected.

3.3.3.3 Direct control of the electromagnetic torque and stator current (Scheme 2)

3.3.3.3.1 General introduction

It is possible to implement an instantaneous torque-controlled synchronous reluctance motor drive, such that

- the modulus of the stator flux linkage can be varied with the load (to obtain a maximum torque/control scheme);
- the switching voltage selection is based on minimizing the errors in the electromagnetic torque and stator-current position and by also considering the position of the stator-current space vector.

It can be seen that, with respect to the conventional DTC scheme (discussed in the previous section), this scheme is different. This is due to the fact that in the conventional scheme the switching vector selection is performed by minimizing the errors in the electromagnetic torque and stator flux linkage, and for this purpose the position (sector) of the stator flux-linkage space vector is also considered. Since, as shown below, the electromagnetic torque is proportional to the modulus of the stator-current space vector and also the sine of (2ε), where ε is the position of the stator-current space vector with respect to the real-axis of the rotor, the electromagnetic torque can be controlled by acting on the modulus of the stator-current space vector. Furthermore, the sign of the angle of the stator-current error determines the sign of the rotation of the stator-current space vector. Thus the electromagnetic torque error and the sign of the stator-current angle error determine the selected switching states of the inverter together with the information on the sector where the stator-current space vector is positioned.

By considering that $\bar{i}_s' = i_{sd} + ji_{sq} = |\bar{i}_s|\exp(\varepsilon)$, where ε is the angle of the stator-current space vector with respect to the direct-axis of the rotor, it follows from eqn (3.3-42) that the electromagnetic torque of the synchronous reluctance motor can also be expressed as

$$t_e = \tfrac{3}{2} P(L_{sd} - L_{sq}) i_{sd} i_{sq} = \tfrac{3}{4} P(L_{sd} - L_{sq}) |\bar{i}_s|^2 \sin(2\varepsilon). \tag{3.3-47}$$

It can be seen that, in addition to the difference between the direct- and quadrature-axis inductances, the electromagnetic torque depends only on the modulus ($|\bar{i}_s|$) and angle (ε) of the stator-current space vector. Thus the synchronous reluctance motor will react very fast to the changes in the stator currents and it is not necessary to keep the modulus of the stator flux constant to have very fast torque response. In fact, the torque/current ratio becomes maximal if $\varepsilon = 45°$, which is equivalent to $i_{sd} = i_{sq}$, and in this case $\bar{i}_s' = |\bar{i}_s| / \sqrt{2}$ and

$$t_{emax} = \tfrac{3}{8} P(L_{sd} - L_{sq}) |\bar{i}_s|^2. \tag{3.3-48}$$

Thus the electromagnetic torque can be quickly changed by changing the modulus of the stator-current space vector and if $i_{sd} = i_{sq} (\varepsilon = 45°)$ is ensured, then the torque/current ratio is maximal. This control strategy can be maintained until the stator flux linkage is below its rated value. At high loads, the stator flux linkage normally exceeds its rated value when the direct- and quadrature-axis stator currents are equal. It follows that the required control law is to keep $i_{sd} = i_{sq}$ at light loads, and at higher loads the i_{sd} current is kept constant but i_{sq} is increased (this results in increased stator-flux modulus). However, if $i_{sd} = i_{sq}$, then it follows from eqn (3.3-40) that

$$|\bar{\psi}_s| \big|_{\varepsilon = 45} = (\psi_{sD}^2 + \psi_{sQ}^2)^{1/2} = |\bar{i}_s| \frac{(L_{sd}^2 + L_{sq}^2)^{1/2}}{\sqrt{2}}, \tag{3.3-49}$$

and if $|\bar{i}_s|$ changes, then the stator flux modulus changes as well.

To summarize: in the synchronous reluctance machine, to have fast electromagnetic torque response, it is not necessary to keep the stator-flux modulus constant. Quick torque response can be obtained by changing the stator-current modulus and maximal torque/current can be ensured by imposing $i_{sd} = i_{sq}$ at light loads.

It follows from eqn (3.3-40), and by also considering that $i_{sd} + j i_{sq} = |\bar{i}_s| \exp(\varepsilon)$, that in general the modulus of the stator flux-linkage space vector is a function of the angle ε. By assuming a given stator-current modulus, then since $L_{sd} > L_{sq}$, if $\varepsilon = 90°$ then $|\bar{\psi}_s|$ is minimal and if $\varepsilon = 0°$ (or $180°$), then it is maximal, and $\varepsilon = 45°$ corresponds to the maximum torque/ampere condition. This is shown in Fig. 3.53.

As discussed above, ε is the angle of the stator-current space vector with respect to the direct-axis of the rotor. Since the control strategy tries to maintain the $\varepsilon = 45°$ position of the stator currents, it is important to know if there is any error in this angle, since if there is an error, it has to be corrected by applying the appropriate switching vector of the inverter. Obviously if a rotor position (θ_r) sensor is used, then by monitoring the stator current components in the stationary reference frame (i_{sD}, i_{sQ}) and then by transforming them into the current components in the rotor reference frame i_{sd}, i_{sq} [by using eqn (3.3-36), thus by using $i_{sd} + j i_{sq} = (i_{sD} + j i_{sQ}) \exp(-j\theta_r) = |\bar{i}_s| \exp(\varepsilon)$], then ε can be estimated and any deviation from the required value is known. However, it will now be shown that information on this deviation (angle error of the stator current $\Delta\varepsilon$) can also be

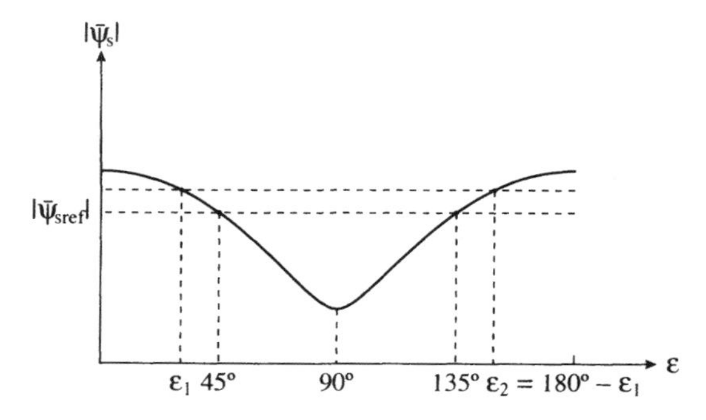

Fig. 3.53. Stator flux modulus as a function of the angle ε.

obtained without using a rotor position sensor. It should also be considered that it is not important to know the exact value of this angle error; it is only necessary to know if it is positive or negative, since this determines the selection of the appropriate switching vectors together with the error in the electromagnetic torque and also the position of the stator-current space vector (in the stationary reference frame). This position can be simply determined by monitoring the stator current components (i_{sD}, i_{sQ}) and e.g. $\alpha_s = \tan^{-1}(i_{sQ}/i_{sD})$ or $\alpha_s = \cos^{-1}(i_{sD}/|\bar{i}_s|)$ gives this angle, but on the other hand it is possible to avoid the use of any trigonometric functions by using comparators.

3.3.3.3.2 *Estimation of* $\Delta\varepsilon$

The reference value of the stator flux-linkage modulus corresponds to the value at $\varepsilon = 45°$, and this stator flux modulus is defined by eqn (3.3-49). It follows from Fig. 3.53 that if there is an error in the angle of the stator-current space vector, i.e. $\varepsilon \neq 45°$, e.g. $\varepsilon < 45°$, then the actual stator flux-linkage modulus will be less than this reference value, and a positive flux-linkage error occurs ($\Delta|\bar{\psi}_s| > 0$). Similarly, if $\varepsilon > 45°$, then the actual stator flux-linkage modulus will be larger than the reference value, and a negative flux-linkage error occurs ($\Delta|\bar{\psi}_s| < 0$). However, it can be seen from Fig. 3.53 that the same flux-linkage error can arise for two different values of ε (which are displaced by 90°). Thus it is not possible to determine in a unique way the corresponding error ($\Delta\varepsilon$) of the stator current angle from a flux-linkage error ($\Delta|\bar{\psi}_s|$). However, this ambiguity can be resolved by also considering that for the two stator current angles (ε_1 and $\varepsilon_2 = 180° - \varepsilon_1$) which yield the same flux-linkage error, the modulus of the electromagnetic torque is the same, but their signs are opposite [this follows directly from eqn (3.3-47)]. It follows that the sign of the stator-current angle error can be obtained from the sign of the electromagnetic torque and the sign of the flux-linkage error as $-\text{sign}(t_e)\text{sign}(\Delta|\bar{\psi}_s|)$. This sign determines the direction in which the stator-current space vector has to be rotated (to enforce the position $\varepsilon = 45°$). This information is then used in the switching vector selection look-up table, where those switching states are selected which ensure that the torque error and stator current errors are reduced to preset levels by considering also the actual position of the stator-current space vector (in the stationary reference frame).

3.3.3.3.3 Drive scheme

The modified direct-torque-control drive scheme of the synchronous reluctance motor is shown in Fig. 3.54.

In Fig. 3.54 the input is the reference value of the electromagnetic torque. This is then compared to its actual value and the resulting torque error is an input into the switching voltage vector selection table. As discussed above, there are also two other inputs to the switching vector selection table. These are the sign of the current angle error $[s=-\text{sign}(t_e)\text{sign}(\Delta|\bar{\psi}_s|)]$ and the position information of the stator-current space vector with respect to the stationary reference frame (α_s) [e.g. $\alpha_s=\tan^{-1}(i_{sQ}/i_{sD})$], but it is not the exact position which has to be known, but only the sector where the stator-current space vector is positioned. For this purpose it is adequate to use a minimal number of six sectors, similarly to that used above in the drive scheme shown in Fig. 3.51.

The electromagnetic torque can be estimated in various ways, e.g. by using eqns (3.3-33), (3.3-34), and (3.3-35) given above. The stator flux-linkage error can be estimated in several ways. For example, by also considering eqn (3.3-40), it can be obtained as

$$\Delta|\bar{\psi}_s|=|\bar{\psi}_{sref}|-|\bar{\psi}_s|=|\bar{\psi}_{sref}|-(\psi_{sD}^2+\psi_{sQ}^2)^{1/2}. \qquad (3.3\text{-}50)$$

If the machine inductances are known, then it is possible to generate the required $|\bar{\psi}_{sref}|$ characteristics as a function of the stator current angle for different values of the stator-current vector modulus. Such characteristics are shown in Fig. 3.55. By knowing the monitored values of the stator flux amplitude $|\bar{\psi}_s|=(\psi_{sD}^2+\psi_{sQ}^2)^{1/2}$

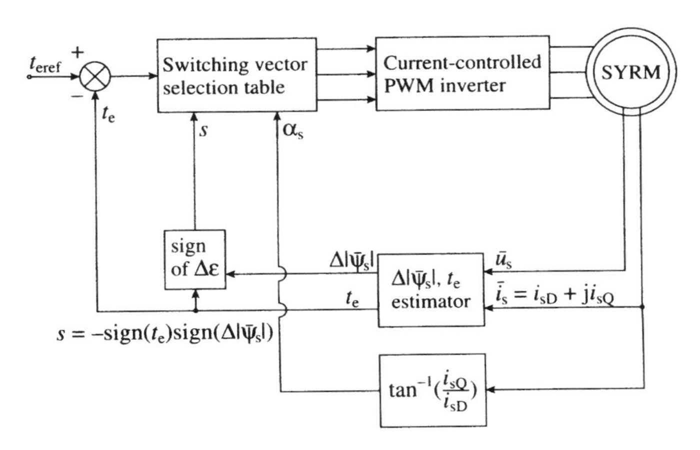

Fig. 3.54. Modified direct-torque-control scheme.

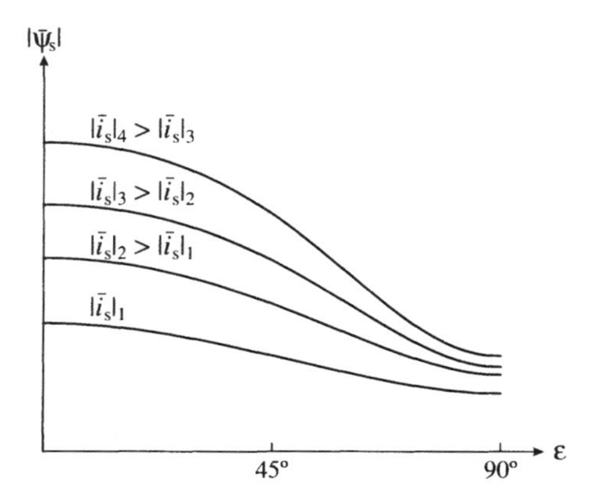

Fig. 3.55. Stator flux linkage versus stator current angle characteristics for different stator-current space-vector moduli.

and also the stator current modulus $|\bar{i}_s|$, the stator flux-linkage error $\Delta|\bar{\psi}_s|$ can be determined from the corresponding curve in Fig. 3.55 (as discussed above, if for a given $|\bar{\psi}_s|$ and $|\bar{i}_s|$, the stator-current angle determined from the curve is less than 45°, then the flux-linkage error is positive, and if it is larger then 45°, then it is negative). The sign of the stator flux-linkage error is then used, together with the sign of the electromagnetic torque error, to obtain the sign of the stator-current angle error, as discussed above:

$$s = -\text{sign}(t_e)\text{sign}(\Delta|\bar{\psi}_s|). \tag{3.3-51}$$

However, prior to starting up the drive, it is also possible to perform a simple test, which generates the reference stator flux-linkage modulus versus stator-current angle curves for different values of the stator-current vector modulus. It is an advantage of this measurement that the inductances L_{sd} and L_{sq} do not have to be known, but the stator flux modulus can still be estimated. The obtained data for the stator flux reference can then be stored in a look-up table, and this can be used in the $\Delta|\bar{\psi}_s|$ estimation block shown in Fig. 3.54.

If step changes are made to the electromagnetic torque reference, then the actual torque will follow these changes very quickly. At high values of the electromagnetic torque the stator flux-linkage modulus is quickly increased and at low values it is quickly decreased.

3.3.3.3.4 Switching vector selection

As discussed above, the electromagnetic torque is controlled by acting on the stator-current amplitude and the sign of the angle of the stator-current error determines the sign of the rotation of the stator-current space vector. The selection of the appropriate inverter switching states is now discussed by using simple physical considerations.

For example, if it is assumed that the electromagnetic torque error is positive, the current angle error is also positive, and the stator-current space vector lags the stator direct axis e.g. by 5°, then it follows that the switching vector $\bar{u}_2$ has to be selected. This is due to the fact that this ensures the required increase of the modulus of the stator-current space vector (which results in the desired increase of the electromagnetic torque) and also in this way the error in the stator-current angle is reduced. On the other hand if the stator-current space vector is in the same position as above, but the electromagnetic torque error is negative and the current angle error is positive, then the switching vector $\bar{u}_3$ has to be selected, since this reduces the stator current modulus (and thus reduces the electromagnetic torque) and also in this way the stator current angle is increased. In the same way it is possible to obtain the required switching vectors when the stator-current space vector is in the other five sectors.

3.3.4 DTC OF A VSI-FED ELECTRICALLY EXCITED SYNCHRONOUS MOTOR

It is also possible to achieve direct torque control of a VSI-fed electrically excited synchronous motor. For this purpose the fact can be utilized that the electromagnetic torque is produced by the interaction of the magnetizing flux-linkage space vector and the stator flux-linkage space vector, $t_e=(3/2)P(\bar{\psi}_m \times \bar{\psi}_s)/L_{s1}$, where L_{s1} is the stator leakage inductance. A quick change in the torque can be produced by rapidly rotating the stator flux-linkage space vector, so the angle between the stator flux-linkage space vector and the magnetizing flux-linkage space vector increases rapidly. This can be achieved by applying the appropriate voltage switching vectors to the inverter (see also Section 3.3.2.1).

Similarly to that shown in Fig. 3.48(b), there are three inputs to the optimum switching table. These are the outputs of the flux and torque hysteresis comparators and also the sector where the stator flux-linkage space vector is located. The DTC control scheme of the VSI-fed electrically excited synchronous motor also requires a stator flux estimator. For this purpose it is possible to use the various techniques discussed in the present book (improved flux estimators, observers, artificial-intelligence-based techniques, etc.). For example, it is also possible to employ a hybrid flux estimator, which uses the voltage model together with the current model of the machine (see also Section 4.6.2.6.3). At higher speeds the voltage model can be used ($\bar{\psi}_s=\int(\bar{u}_s-R_s\bar{\imath}_s)\,dt$), where $\bar{u}_s$ can be reconstructed from the monitored d.c. link voltage by using the inverter switching states and a model for the voltage drops across the inverter switches. For high accuracy a thermal model can be used to correct the stator resistance. However, at lower speeds, a current model has also to be used; such a model has been discussed in Section 3.2.1 for a salient-pole electrically excited synchronous machine. The two models are used to obtain a combined flux, which is obtained as a weighted mean of the individual fluxes, depending on the machine speed.

3.3.5 DTC OF A CSI-FED ELECTRONICALLY EXCITED SYNCHRONOUS MOTOR

3.3.5.1 General introduction

In the present section the direct torque control (DTC) of a load-commutated CSI-fed synchronous motor drive is briefly discussed. However, for better understanding of the concepts related to direct torque control, the reader is first referred to Sections 1.2.2 and 3.3.3. In the synchronous motor drive with a load-commutated current-source inverter a constant frequency mains power is rectified by a controlled rectifier, and a controlled d.c. voltage is obtained. A large d.c. link inductor filters the d.c. link current, and the machine side inverter (load-commutated CSI inverter) is thus supplied by a constant current. Load commutation is ensured by overexcitation of the synchronous motor to yield a leading power factor. The drive can inherently operate in all the four quadrants; the switching frequency of the inverter determines the motor speed. When the effects of commutation are neglected, the stator-current waveforms are analogous to those of the CSI-fed induction machine.

3.3.5.2 Drive scheme

It is shown below that the electromagnetic torque of a CSI-fed synchronous motor is produced by the interaction of the stator subtransient flux linkages and the stator currents. Direct torque control of a CSI-fed synchronous motor involves the direct control of the subtransient stator flux linkage (ψ_s'') and the electromagnetic torque (t_e) by applying the optimum current switching vectors. Furthermore, in a direct-torque-controlled (DTC) synchronous motor drive supplied by a current-source inverter, it is possible to control directly the modulus of the subtransient flux-linkage space vector ($|\bar{\psi}_s''|$) through the rectifier supplying the field current, and also the electromagnetic torque by the supply frequency of the load-commutated CSI. For this purpose the appropriate optimal inverter-current switching vectors are produced by using an optimal current switching vector table. This contains the six possible active current switching vectors ($\bar{i}_1, \bar{i}_2, \ldots, \bar{i}_6$) and also the non-active (zero) switching vectors ($\bar{i}_0$). The six active current switching vectors are also shown in Fig. 3.56(a). Optimum selection of the switching vectors is made to restrict the torque error within the torque hysteresis band. The inputs to the optimal current switching table are the discretized torque error (dt_e), which is the output of a 3-level hysteresis comparator. A 3-level comparator is used, since this corresponds to 0, 1, and -1 torque errors. The optimum switching look-up table also requires knowledge of the position of the subtransient stator flux-linkage space vector, since it must be known in which of the six sectors is the flux-linkage space vector. The basic scheme of the DTC load-commutated CSI-fed synchronous motor drive is shown in Fig. 3.56(b).

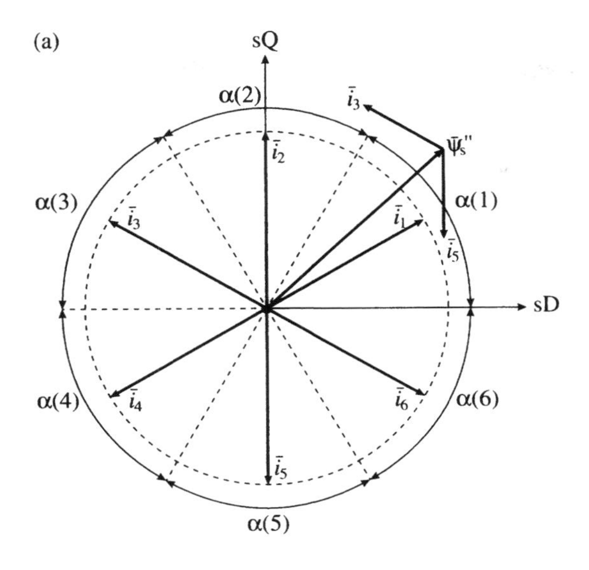

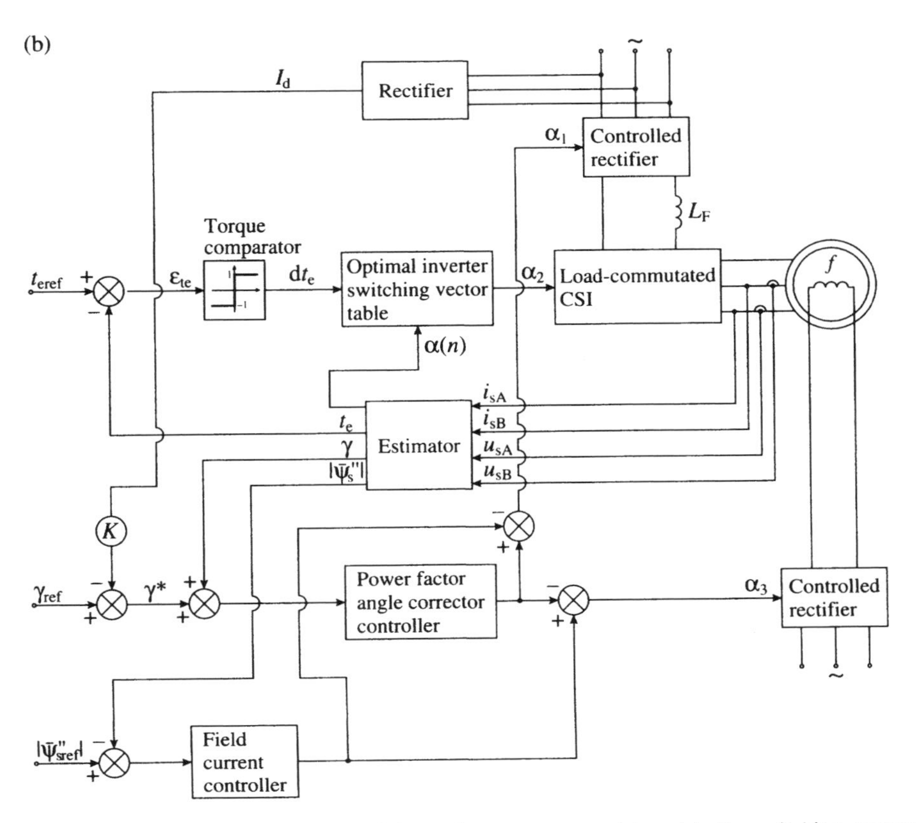

Fig. 3.56. DTC load-commutated CSI-fed synchronous motor drive. (a) Six switching vectors; (b) drive scheme.

In Fig. 3.56(b) the subtransient flux-linkage reference is compared with its actual value, which is estimated by using eqns (3.3-54) and (3.3-55) shown below (as $\bar{\psi}_s''=\psi_D+j\psi_Q$) or when the improved estimator is used, it can be obtained by considering the flux linkage obtained by eqn (3.3-64) given below. The flux-linkage error is the input to the field current controller. This flux error also acts on the rectifier in the field circuit to produce the required leading power factor (which ensures load commutation, as discussed above). However, to account for parameter detuning and overlap angle, a power factor angle-corrector controller is added as a feedforward signal to the field current controller. This feedforward signal also acts as a feedback signal to the rectifier supplying the CSI. The reference angle, γ_{ref}, is also an input to the drive system in Fig. 3.56(b), where in general, $90-\gamma$ is the angle between the space vectors $\bar{i}_s$ and $\bar{\psi}_s''$. It can be seen that γ is approximately equal to the power factor angle (ϕ), $\gamma=\phi+u/2$, where u is the overlap angle. In Fig. 3.56(b) the signal $-KI_d$ is added to this, to take account of the increased overlap angle (due to the increased d.c. link current I_d). The resulting $\gamma^*=\gamma_{ref}-KI_d$ angle is then added to the estimated angle γ, and the obtained signal is fed into the power factor angle-corrector controller. The details of the estimator shown in Fig. 3.56(b) which provides t_e, γ, and $|\bar{\psi}_s''|$ on its outputs are discussed below. It should be noted that at low frequencies forced commutation is required, since the generated e.m.f.s required for load commutation are very low, even if there is large overexcitation, since they also depend on the speed. Therefore it is also problematic to start the motor from rest.

For precise speed control a speed feedback loop has to be incorporated. In this case the torque reference is obtained on the output of a speed controller, which can be a PI controller or a fuzzy-logic controller, etc. The input to the speed controller is the difference between the reference speed and the actual speed. It is possible to have a speed-sensorless implementation, where the rotor speed is obtained by using one of the techniques described in earlier sections. The hysteresis band of the torque comparator may be increased with the rotor speed, since in this case PWM of the currents is avoided at high speeds.

3.3.5.3 Stator flux-linkage and electromagnetic torque estimation

The drive scheme contains an electromagnetic torque and stator subtransient flux-linkage estimator. The electromagnetic torque can be estimated from the terminal quantities by considering that, according to eqn (2.1-167), it is produced by the interaction of the stator flux linkages and stator current. By using space vectors in the stationary reference frame, and also considering that in the synchronous machine the stator flux-linkage space vector ($\bar{\psi}_s$) is equal to the subtransient stator flux-linkage space vector ($\bar{\psi}_s''$) plus $L_s''\bar{i}_s$ [Vas 1992], where L_s'' is the subtransient inductance,

$$t_e=\tfrac{3}{2}P\bar{\psi}_s\times\bar{i}_s=\tfrac{3}{2}P\bar{\psi}_s''\times\bar{i}_s=\tfrac{3}{2}\frac{L_m}{L_r}(\psi_D i_{sQ}-\psi_Q i_{sD}) \qquad (3.3\text{-}52)$$

is obtained, since $\bar{\psi}_s''=\psi_D+j\psi_Q$ and the vectorial product $L_s''\bar{i}_s\times\bar{i}_s=0$. In general, when the synchronous machine has damper windings, the subtransient direct- and quadrature-axis inductances, L_{sd}'' and L_{sq}'', are not equal; thus L_s'' is not equal to the transient inductance (L_s'), but an approximation can be made by assuming that $L_s''\approx(L_{sd}''+L_{sq}'')/2$. This is a good approximation if the subtransient saliency in the two axes is not much different. In eqn (3.3-52) $\bar{\psi}_s''=\psi_D+j\psi_Q$ is the subtransient flux-linkage space vector (stator flux linkage behind the subtransient inductance), and $\bar{i}_s=i_{sD}+ji_{sQ}$ is the stator-current space vector, and both vectors are expressed in the stationary reference frame. The stator currents and stator voltages are monitored and the stator subtransient flux-linkage space vector can be obtained by using

$$\frac{d\bar{\psi}_s''}{dt}=\bar{u}_s-R_s\bar{i}_s-L_s''\frac{d\bar{i}_s}{dt}, \tag{3.3-53}$$

where L_s'' is the subtransient inductance defined above. The derivative in eqn (3.3-53) is a back e.m.f. Thus the stator subtransient flux-linkage components are obtained as

$$\psi_D=\int\left(u_{sD}-R_si_{sD}-L_s''\frac{di_{sD}}{dt}\right)dt=\int(u_{sD}-R_si_{sD})\,dt-L_s''i_{sD} \tag{3.3-54}$$

$$\psi_Q=\int\left(u_{sQ}-R_si_{sQ}-L_s''\frac{di_{sQ}}{dt}\right)dt=\int(u_{sQ}-R_si_{sQ})\,dt-L_s''i_{sQ}. \tag{3.3-55}$$

The angle γ (which is the angle between the space vectors $j\bar{\psi}_s''$ and $\bar{i}_s$) can be estimated by using i_{sD}, i_{sQ}, ψ_D, and ψ_Q. The electromagnetic torque and the subtransient stator flux-linkage space vector can be obtained in other ways as well. The errors associated with open-loop integrators at low frequencies can also be avoided by using an improved estimator, which is now discussed. The estimator shown in Fig. 3.56(b) ouputs t_e, γ, and $|\bar{\psi}_s''|$.

If the rotor position (θ_r) is also measured, then the stator-current space vector in the rotor reference frame can be obtained from the stator-current space vector in the stationary reference frame as

$$\bar{i}_s'=i_{sd}+ji_{sq}=\bar{i}_s\exp(-j\theta_r)=(i_{sD}+ji_{sQ})\exp(-j\theta_r). \tag{3.3-56}$$

The stator flux linkages (not the subtransient stator flux linkages) in the rotor reference frame can be estimated by using the stator currents i_{sd} and i_{sq} as

$$\psi_{sd}=L_di_{sd}+L_{md}i_F=L_{s1}i_{sd}+L_{md}(i_{sd}+i_F) \tag{3.3-57}$$

$$\psi_{sq}=L_{sq}i_{sq}, \tag{3.3-58}$$

where i_F is the measured field current, L_{sd} is the total direct-axis inductance, $L_{sd}=L_{s1}+L_{md}$, L_{s1} is the stator leakage inductance, L_{md} is the d-axis magnetizing inductance, and L_{sq} is the q-axis inductance. To obtain greater accuracy, the variations of the inductances with the currents should also be considered. Thus if cross-saturation coupling is neglected then in general L_{md} varies with the d-axis magnetizing current: $i_{md}=i_{sd}+i_F$. Furthermore, $L_{sq}=L_{s1}+L_{mq}$ varies with the q-axis magnetizing current: $i_{mq}=i_{sq}$. Thus $L_{md}=L_{md}(i_{md})$, $L_{mq}=L_{mq}(i_{mq})$. Therefore the stator flux-linkage components in the rotor reference frame can be expressed as

$$\psi_{sd}=L_{s1}i_{sd}+L_{md}(i_{md})(i_{sd}+i_F) \tag{3.3-59}$$

$$\psi_{sq}=L_{s1}i_{sq}+L_{mq}(i_{mq})i_{sq}. \tag{3.3-60}$$

The stator flux-linkage space vector in the stator reference frame can be obtained from

$$\bar{\psi}_s=\bar{\psi}'_s\exp(j\theta_r)=(\psi_{sd}+j\psi_{sq})\exp(j\theta_r). \tag{3.3-61}$$

Finally the subtransient flux-linkage space vector can be obtained from

$$\bar{\psi}''_s=\bar{\psi}_D+j\psi_Q=\bar{\psi}_s-L''_s\bar{i}_s. \tag{3.3-62}$$

However, as shown in Section 3.1.3.2.1, enhanced performance at low frequencies can be achieved if the obtained $\bar{\psi}_s$ is added to $T(\bar{u}_s-R_s\bar{i}_s)$, where T is appropriately chosen, and then the resulting component is multiplied by $1/(1+pT)$, where $p=\mathrm{d}/\mathrm{d}t$. Thus

$$\hat{\bar{\psi}}_s=\frac{\bar{\psi}_s+T(\bar{u}_s-R_s\bar{i}_s)}{1+pT}. \tag{3.3-63}$$

It should be noted that in eqn (3.3-63) $\bar{u}_s-R_s\bar{i}_s$ is the rate of change of the stator flux-linkage space vector obtained from the measured stator voltages and currents. Thus it is obtained by using the stator voltage equation, which gives accurate estimation at higher stator frequencies. It follows that the stator subtransient flux-linkage space vector can be more accurately estimated by considering

$$\hat{\bar{\psi}}''_s=\hat{\bar{\psi}}_s-L''_s\bar{i}_s \tag{3.3-64}$$

and the thus obtained $\hat{\psi}_D$ and $\hat{\psi}_Q$ could be used for a more accurate estimation of the angle γ (which is the angle between $j\hat{\bar{\psi}}''_s$ and $\bar{i}_s$). The electromagnetic torque can then be obtained as

$$t_e=\tfrac{3}{2}P\hat{\bar{\psi}}''_s\times\bar{i}_s. \tag{3.3-65}$$

3.3.5.4 Optimal switching current vector selection

The optimal inverter current switching table can be obtained by considering the positions of the subtransient stator flux-linkage space vector in one of the six sectors (e.g. the first sector is in the region span by angle α_1 shown in Fig. 3.56(a); the second region spans the angle α_2, etc.).

It can be seen that, for example, if the subtransient stator flux-linkage space vector is in the first sector, then for positive electromagnetic torque, the switching

Table 3.4 Optimum active voltage switching vector look-up table

dt_e	$\alpha(1)$ sector 1	$\alpha(2)$ sector 2	$\alpha(3)$ sector 3	$\alpha(4)$ sector 4	$\alpha(5)$ sector 5	$\alpha(6)$ sector 6
1	$\bar{\imath}_3$	$\bar{\imath}_4$	$\bar{\imath}_5$	$\bar{\imath}_6$	$\bar{\imath}_1$	$\bar{\imath}_2$
0	$\bar{\imath}_0$	$\bar{\imath}_0$	$\bar{\imath}_0$	$\bar{\imath}_0$	$\bar{\imath}_0$	$\bar{\imath}_0$
−1	$\bar{\imath}_5$	$\bar{\imath}_6$	$\bar{\imath}_1$	$\bar{\imath}_2$	$\bar{\imath}_3$	$\bar{\imath}_4$

current vector $\bar{\imath}_3$ has to be applied. This is due to the fact that a stator current vector has to be selected which produces positive torque and is located at an angle greater than 90° in the positive direction from the subtransient flux-linkage space vector since, as discussed above, the synchronous machine is overexcited. However, for negative electromagnetic torque, $\bar{\imath}_5$ has to be applied, since the selected stator current has to produce negative torque and it must lag the subtransient flux-linkage space vector by an angle greater than 90°. For zero electromagnetic torque, one of the zero current switching vectors ($\bar{\imath}_0$) has to be selected. Similarly, if the subtransient flux-linkage space vector is in the second sector, then for positive torque $\bar{\imath}_4$, and for negative torque $\bar{\imath}_6$, must be selected, etc. The thus-obtained optimum current switching vector table is shown in Table 3.4.

This is in agreement with that shown in Table 4.4 in Section 4.6.3.4. It should be noted that it is possible to obtain improved switching tables by using fuzzy logics or non-artificial-intelligence-based techniques, e.g. in which the number of sectors is increased (see also Section 4.6.2.4).

Bibliography

Adkins, B. and Harley, R. G. (1978). *The general theory of alternating current machines.* Chapman and Hall, London.

Alaküla, M. and Vas, P. (1989). Influence of saturation on the dynamic performance of vector-controlled salient-pole synchronous machines. *4th International Conference on Electrical Machines and Drives*, IEE, London, pp. 298–302.

Arkadan, A. A., Demerdash, N. A., Vaidya, J. G., and Shah, M. J. (1988). Impact of load on winding inductances of permanent magnet generators with multiple damping circuits using energy perturbation. *IEEE Transactions on Energy Conversion* **3**, 880–9.

Bado, A., Bolognani, S., and Zigliotto, M. (1992). Effective estimation of speed and rotor position of PM synchronous motor drive by a Kalman filtering technique. *IEEE Power Electronics Specialists Conference*, pp. 951–7.

Bausch, H. (1987). Large power variable speed a.c. machines with permanent magnet excitation. *Electric Energy Conference, Adelaide*, pp. 265–71.

Bayer, K. H., Waldmann, H., and Weibelzahl, M. (1972). Field-oriented control of a synchronous machine with the new Transvektor control system. *Siemens Review* **39**, 220–3.

Becerra, R. C., Jahns, T. M., and Ehsani, M. (1991). Four-quadrant sensorless brushless ECM drive. *IEEE Applied Power Electronics Conference and Exposition*, pp. 202–9.

Betz, R. E. (1992). Theoretical aspects of control of synchronous reluctance machines. *Proceedings of the IEE* (Pt. B) **139**, 355–64.

Binns, K. J. and Chaaban, F. B. (1988). The relative merits of rare earth permanent magnet materials for use in the excitation of permanent magnet machines. In *Proceedings of the International Conference on Electrical Machines, Pisa,* 463–6.

Bojtor, L. and Schmidt, I. (1990). Microprocessor controlled converter-fed synchronous motor using subtransient flux model. *ICEM, Cambridge* (*USA*), pp. 427–33.

Boldea, I., Fu, Z., and Nasar, S. A. (1991). Digital simulation of a vector current controlled axially-laminated anisotropic (ALA) rotor synchronous motor servo-drive. *Electrical Machines and Power Systems* **19**, 419–24.

Boldea, I. and Trica, A. (1990). Torque vector controlled (TVC) voltage-fed induction motor drives: very slow speed performance via sliding mode control. *ICEM, Cambridge* (*USA*), pp. 1216–17.

Bolognani, S. and Buja, G. S. (1987). Dynamic characteristics of a P.W.M. voltage source inverter-fed P.M. brushless motor drive under field orientation. *Electric Energy Drive Conference, Adelaide*, pp. 91–5.

Bolognani, S. (1991). A torque angle calculator for sensorless reluctance motor drives. *EPE, Firenze*, 4.013–4.017.

Bolognani, S., Oboe, R., and Zigliotto, M. (1994). DSP-based extended Kalman filter estimation of speed and rotor position of a PM synchronous motor. *IECON, Bologna*, pp. 2097–102.

Bolognani, S. and Zigliotto, M. (1995). Parameter sensitivity of the Kalman filter applied to a sensorless synchronous motor drive. *EPE, Seville*, 3.375–3.380.

Bose B. K. (1986). *Power electronics and drives.* Prentice Hall, Englewood Cliffs, New Jersey.

Bose, B. K. (1987). A high performance inverter-fed system of an interior permanent magnet synchronous machine. *IEEE IAS Annual Meeting, Atlanta*, pp. 269–76.

Bose, B. K. and Szczesny, P. M. (1988). A microcomputer-based control and simulation of an advanced IPM synchronous machine drive system for electrical vehicle propulsion. *IEEE Transactions on Industrial Electronics* **35**, 547–59.

Chalmers, B. J. (1988). *Electric motor handbook.* Butterworths, London.

Chricozzi, E. *et al.* (1995). Fuzzy self-tuning PI control of PMS synchronous motor drives. *PEDS'95*, pp. 749–54.

Colby, R. S. and Novotny, D. W. (1986). Efficient operation of surface mounted P.M. synchronous motors. *IEEE IAS Annual Meeting, Denver*, pp. 806–13.

Consoli, A. and Abela, A. (1986). Transient performance of permanent magnet a.c. motor drives. *IEEE Transactions on Industry Applications* **IA-22**, 32–41.

Consoli, A., Musumeci, S., Raciti, A., and Testa, A. (1994). Sensorless vector and speed control of brushless motor drives. *IEEE Transactions on Ind. Electronics* **IE-41**, 91–6.

Consoli, A., Scarcella, G., and Testa, A. (1996). Self-synchronising torque control of reluctance motor drives. *IEEE PESC, Baveno*, pp. 344–9.

Corley, M. J. and Lorenz, R. D. (1996). Rotor position and velocity estimation for a permanent magnet synchronous machine at standstill and high speeds. *IEEE IAS Meeting*, 1/36–1/41.

Dhaouadi, R., Mohan, T., and Norum, L. (1991). Design and implementation of an extended Kalman filter for state estimation of a permanent magnet synchronous motor. *IEEE Transactions on Power Electronics* **6**, 491–7.

Dote, Y. and Kinoshita, S. (1990). *Brushless servomotors: fundamentals and applications.* Oxford University Press.

Endo, T., Tajima, F. *et al.* (1983). Microcomputer controlled brushless motor without a shaft mounted sensor. *International Power Electronics Conference, Tokyo*, pp. 1477–86.

Ertugrul, N. and Acarnley, P. (1994). A new algorithm for sensorless operation of permanent magnet motors. *IEEE Transactions on Ind. Applications* **IA-30**, 126–33.

Ferrais, P., Vagati, A., and Villata, F. (1980). P.M. brushless motor drives: a self commutation system without rotor position sensors. *9th Symposium on Incremental Motion Contro, Systems and Devices*, pp. 305–12.

Fletcher, J. E., Williams, B. W. and Green, T. C. (1995). *IEE Colloquium on Advances in Control Systems for Electrical Drives*, 11.1–11.5.

Franceschini, G., Fratta, A., Rosso, G., Troglia, G. P., and Vagati, A. (1993). Performance of synchronous reluctance motors in servo-drive applications. *PCIM, Nürnberg*, pp. 15–27.

Franceschini, G., Fratta, A., Perrache, C., and Vagati, A. (1994). Control of high performance synchronous reluctance motor drives. *PCIM, Nürnberg*, pp. 117–26.

Fratta, A. and Vagati, A. (1987). A reluctance motor drive for high dynamic performance applications. *IEEE IAS Annual Meeting, Atlanta*, pp. 295–302.

Fratta, A. *et al.* (1990). Design criteria of an IPM machine suitable for field weakening operation *ICEM, Cambridge (USA)*, pp. 1059–65.

Fratta, A. and Vagati, A. (1992). Synchronous reluctance versus induction motor: a comparison. *PCIM, Nürnberg*, pp. 179–86.

French, C. and Acarnley, P. (1996). Direct torque control of permanent magnet drives. *IEEE Transactions on Ind. Applications* **IA-32**, 1080–8.

Germano, A., Parasiliti, F., and Tursini, M. (1994). Sensorless speed control of a PM synchronous motor drive by Kalman filter. *ICEM, Paris*, pp. 540–4.

Grundman, S., Krause, M., and Muller, V. (1995). Application of fuzzy control for PWM voltage source inverter fed permanent magnet motor. *EPE, Seville*, 1.524–1.529.

Harnefors, L. and Nee, H. P. (1995). Robust control of a.c. machines using the internal model control method. *IEEE IAS Meeting*, pp. 303–9.

Harnefors, L. and Nee, H. P. (1996). On the dynamics of a.c. machines and sampling rate selection for discrete-time vector control. *ICEM, Vigo*, pp. 251–6.

Honsinger, V. B. (1982). The fields and parameters of interior type permanent magnet a.c. machines. *IEEE Transactions on PAS* **PAS-101**, 867–76.

Hussels, P. (1984). Flux and torque control of synchronous machine with compensation of inaccuracy due to variations of motor parameters. In IEE Conference Publication 234, *Power Electronics and Variable Speed Drives*, pp. 377–80. Institution of Electrical Engineers, London.

Iizuka, K., Uzuhashi, H., Kano, M., Endo, T., and Mohri, K. (1985). Microcomputer control for sensorless brushless motor. *IEEE Transactions on Ind. Applications* **IA-21**, 695–701.

Jahns, T. M. (1987). Flux-weakening regime operation of an interior permanent-magnet synchronous motor drive. *IEEE Transactions on Industry Applications* **IA-23**, 681–9.

Jahns, T. M., Kliman, G. B., and Neumann, T. W. (1986). Interior permanent-magnet synchronous motors for adjustable-speed drives. *IEEE Transactions on Industry Applications* **IA-22**, 738–47.

Jones, L. A. and Lang, J. H. (1989). A state observer for the permanent magnet synchronous motor. *IEEE Transactions on Ind. Electronics* **IE-36**, 374–82.

Jufer, M. (1985). Self-commutation of brushless d.c. motors without encoders. *EPE, Brussels*, 3.275–3.280.

Jufer, M. (1995). Indirect sensors for electric drives. *EPE, Seville*, 1.836–1.839.

Kondo, S., Takahashi, A., and Nishida, T. (1995). Armature current locus based estimation method of rotor position of permanent magnet synchronous motor without mechanical sensor. *IEEE IAS Meeting*, pp. 55–60.

Krause, P. C. (1986). *Analysis of electrical machinery*. McGraw-Hill, New York.

Kreidler, L., Testa, A., and Lipo, T. A. (1993). Position sensorless synchronous reluctance motor drive using stator phase voltage third harmonic. *IEEE IAS Meeting*, pp. 679–86.

Kulkarni, A. B. and Ehsani, M. (1992). A novel position sensor elimination technique for the interior permanent magnet synchronous motor drive. *IEEE Transactions on Ind. Applications* **IA-28**, 144–50.

Lagerquist, R., Betz, R. B., and Miller, T. J. E. (1992). DSP96002 based high performance digital vector controller for synchronous reluctance motors. *ICEM*, Vol. 3, pp. 903–7.

Lagerquist, R. and Boldea, I. (1994). Sensorless control of the synchronous reluctance motor. *IEEE Transactions on Ind. Applications* **IA-30**, 673–81.

Lawrenson, P. J. (1965). Two-speed operation of salient-pole reluctance machines. *Proceedings of the IEEE* (Pt. B) **117**, 545–51.

Lawrenson, P. J. and Agu, L. A. (1964). Theory and performance of polyphase reluctance machines. *Proceedings of the IEEE* **111**, 1435–45.

Lawrenson, P. J. and Gupta, S. K. (1967). Development in the performance and theory of segmental reluctance motors. *Proceedings of the IEE* **114**, 645–50.

Leonhard, W. (1985). *Control of electrical drives*. Springer-Verlag, Berlin.

Leonhard, W. (1986). Microcomputer control of high dynamic performance a.c. drives—a survey. *Automatica* **22**, 1–19.

Leonhard, W. (1988). Adjustable-speed a.c. drives. *Proceedings of the IEEE* **76**, 455–70.

Leonhard, W. (1988). Field-orientation for controlling a.c. machines—principle and application. *3rd International Conference on Power Electronics and Variable Speed Drives*, pp. 227–82. IEE, London.

Lessmeier, R., Schumacher, W., and Leonhard, W. (1986). Microprocessor-controlled a.c. servo drives with synchronous or induction motors: which is preferable? *IEEE Transactions on Industry Applications* **IA-22**, 812–19.

Letas, H. H. and Leonhard, W. (1983). Dual-axis servo-drive in cylindrical coordinates using permanent magnet synchronous motors with microprocessor control. *Conference Proceedings of Comunel 83*, pp. II.23–II.30, Toulouse.

Liu, S. and Stiebler, M. (1989). A continuous-discrete time state estimator for a synchronous motor fed by a PWM inverter. *EPE, Aachen*, pp. 865–9.

Low, T. S., Tseng, K. J., Lee, T. H., Lim, K. W., and Lock, K. S. (1990). Strategy for the instantaneous torque control of permanent magnet brushless d.c. drives. *Proceeding of the IEE* (Pt. B) **137**, 355–63.

Maes, J. and Melkebeek, J. A. A. (1996). Direct torque control, a sliding mode approach. *Speedam, Capri*, A6. 15–A6.20.

Matsui, N. and Shigyo, M. (1992). Brushless d.c. motor control without position and speed sensors. *IEEE Transactions on Ind. Applications* **IA-28**, 120–7.

Matsuo, T. and Lipo, T. A. (1995) Rotor position detection scheme for synchronous reluctance motor based on current measurements. *IEEE Transactions on Ind. Applications* **31**, 860–8.

Matuonto, M. and Monti, A. (1995). High performance field oriented control for cycloconverter-fed synchronous machine. *EPE, Seville*, 3.458–3.463.

Miller, T. J. E. (1981). Methods for testing permanent magnet polyphase motors. *IEEE IAS Annual Meeting*, pp. 494–9.

Miller, T. J. E. (1981). Transient performance of permanent magnet a.c. machines. *IEEE IAS Annual Meeting*, pp. 500–3.

Monti, A., Roda, A., and Vas, P. (1996). A new fuzzy approach to the control of the synchronous reluctance machine. *EPE, PEMC, Budapest*, 3/106–3/110.

Moynihan, J. F., Egan, M. G., and Murphy, J. M. D. (1994). The application of state observers in current regulated pm synchronous drives. *IEEE IECON*, pp. 20–5.

Moynihan, J. F., Bolognani, S., Kavangh, R. C., Egan, M. G., and Murphy, J. M. D. (1993). Single sensor current control of a.c. servodrives using digital signal processors. *EPE, Brighton*, pp. 415–21.

Murphy, J. M. D. and Turnbull, F. G. (1987). *Power electronic control of a.c. motors*. Pergamon Press, Oxford.

Novotny, D. W. and Lorenz, R. D. (1985). Introduction to field orientation and high performance a.c. drives. *Tutorial Course, IEEE IAS Annual Meeting, Toronto*.

Nakano, T., Ohsawa, H., and Endoh, K. (1984). A high-performance cycloconverter-fed synchronous machine drive system. *IEEE Transactions on Industry Applications* **IA-20**, 1278–84.

Ogasawara, S. and Akagi, H. (1991). An approach to position sensorless drive for brushless d.c. motors. *IEEE Transations on Ind. Applications* **27**, 920–33.

Ogasawara, S., Nishimura, M., Akagi, H., Nabae, A., and Nakanishi, Y. (1986). A high performance a.c. servo system with permanent magnet synchronous motors. *IEEE Transactions on Industrial Electronics* **IE-33**, 87–91.

Pillay, P. and Krishnan, R. (1987). Modelling, analysis and simulation of a high performance, vector controlled, permanent magnet synchronous motor drive. *IEEE IAS Annual Meeting, Atlanta*, pp. 254–61.

Rahman, M. A. and Little, T. A. (1984). Dynamic performance analysis of permanent magnet synchronous motors. *IEEE Transactions on PAS* **PAS-103**, 1277–82.

Richter, E. and Neumann, T. W. (1984). Saturation effects in salient-pole synchronous machines with permanent magnet excitation. In *Proceedings of International Conference on Electrical Machines, Lausanne*, pp. 603–6.

Sabanovic, A. and Bilalovic, F. (1989). Sliding mode control of a.c. drives. *IEEE IAS Meeting*, pp. 70–5.

Sattler, P. K. and Starker, K. (1989). Estimation of speed and position of an inverter-fed permanent excited synchronous machine. *EPE, Aachen*, pp. 1207–12.

Schiferl, R. and Lipo, T. A. (1986). Power capability of salient-pole permanent magnet synchronous motors in variable-speed drive applications. *IEEE IAS Annual Meeting*, pp. 23–31.

Schroedl, M. (1988). Detection of the rotor position of a permanent magnet synchronous machine at standstill. *ICEM, Pisa*, pp. 195–7.

Schroedl, M. (1991). Digital implementation of a sensorless control algorithm for permanent magnet synchronous motors. *Proc. Int. Conf. SM 100, ETH Zurich*, pp. 430–5.

Schroedl, M. (1993). Digital implementation of a sensorless algorithm for permanent magnet synchronous motors. *EPE, Brighton*, pp. 430–5.

Schroedl, M. (1994). Sensorless control of permanent magnet synchronous motors. *Electric Machines and Power Systems* **22**, 173–85.

Sebastian, T. and Slemon, G. R. (1987). Transient modelling and performance of variable speed permanent magnet motors. *IEEE IAS Annual Meeting*, pp. 35–9.

Shinkawa, O., Tabuta, K., *et al*. (1993). Wide operation of a sensorless brushless d.c. motor having an interior permanent magnet motor. *Power Conversion Conference, Yokohama*, pp. 364–70.

Sneyers, B., Novotny, D. W., and Lipo, T. A. (1985). Field-weakening in buried permanent magnet a.c. motor drives. *IEEE Transactions on Industry Applications* **IA-21**, 398–404.

Stronach, A. F., Vas, P., and Neuroth, M. (1997). Implementation of intelligent self-organising controllers in DSP-controlled electromechanical drives. *Proceedings of the IEE, Pt. D* **144**, 324–30.

Takeda, Y., Morimoto, S., Hirasa, T., and Fuchi, K. (1988). Most suitable control method for permanent magnet synchronous motors. In *Proceedings of the International Conference on Electrical Machines, Pisa*, pp. 53–8.

Tso, S. K. and Wai, L. K. (1987). A novel implementation of high-performance synchronous motor drives. *BICEM*, pp. 667–70.

Vagati, A., Fratta, A., Franceshini, G., and Rosso, P. (1996). AC motors for high-performance drives: a design-based comparison. *IEEE Transactions on Ind. Application*, **IA-32**, 1211–19.

Vagati, A. (1994). The synchronous reluctance solution: a new alternative in a.c. drives. *IECON, Bologna.*

Vas, P. (1992). *Electrical machines and drives: a space-vector theory approach.* Oxford University Press.

Vas, P. (1993). *Parameter estimation, condition monitoring, and diagnosis of electrical machines.* Oxford University Press.

Vas, P. and Drury, W. (1994). Vector-controlled drives. *PCIM, Nürnberg*, pp. 213–28.

Vas, P. and Drury, W. (1996). Future trends and developments in electric machines and drives (in Italian). *Energia Elettrica* **73**, 11–21.

Vas, P., Alaküla, M., Brown, J. E., and Hallenius, K. E. (1988). Field-oriented control of saturated a.c. machines. *3rd International Conference on Power Electronics and Variable-speed drives*, pp. 283–6. IEE, London.

Vauhkonen, V. and Mård, M. (1987). Compensated field-oriented control of cycloconverter-fed synchronous motors in rolling-mill drives. In *Proceedings of Beijing International Conference on Electrical Machines, Beijing*, pp. 679–82.

Wu, R. and Slemon, G. R. (1990). A permanent magnet motor drive without a shaft sensor. *IEEE IAS Meeting*, pp. 553–8.

4 Vector and direct torque control of induction machines

Induction machines have been used for over a hundred years. Because of their simplicity, ruggedness, reliability, efficiency, low cost, compactness, and economical and volume manufacturing advantages, induction machines with a squirrel-cage rotor are the most widely used machines at fixed speed. This is especially the case at lower power levels, where on the rotor there is a squirrel-cage winding, which is manufactured by die-casting. However, recent developments in the field of variable-speed drives have made possible the large-scale application of variable-speed induction motor drives.

Although the induction machine is superior to the d.c. machine with respect to size, weight, rotor inertia, efficiency, maximum speed, reliability, cost, etc., because of its highly non-linear dynamic structure with strong dynamic interactions, it requires more complex control schemes than, say, a separately excited d.c. machine. The general dynamic model of the induction machine can be represented by a sixth-order state-space equation, where the inputs to the stator are voltage and frequency and the outputs can be rotor speed, rotor position, electromagnetic torque, stator or rotor flux linkages, magnetizing flux linkage, stator or rotor currents, magnetizing current or a combination of these. Furthermore, the cost of a.c. power converters is higher for the variable-speed induction motor drive than for the converters which can supply d.c. machines. Until recently the cost of the introduction of the variable-speed induction motor has been prohibitive and the complexity of control has made its development difficult and acceptance reluctant. However, the rapid developments in the field of power electronics, whereby better and more powerful semiconductor devices are available (with higher switching speeds, high conducting currents and very high blocking voltages, which can be turned on and off, etc.) and where the power devices and circuits are packaged into modular form, and the existence of powerful and inexpensive microprocessors, which allow the complex control functions of the a.c. drive to be performed by utilizing software instead of expensive hardware, mean that a.c. drives employing induction machines can be considered as economical alternatives to adjustable-speed d.c. drives.

Some of the other functional advantages of the application of microprocessors or digital techniques are:

- Cost reduction in control electronics;
- Improved reliability due to the reduction of the number of components;
- Standard universal hardware is required and the only changes are to the software, which is very flexible and can be easily modified;
- Digital transmission requires a minimal amount of cabling and is very tolerant to noise; it eliminates drift and electromagnetic interference problems;

- Very high accuracy, excellent repeatability, linearity, and stability with different setting ranges;
- Centralized operator communications, monitoring, and diagnostics. The diagnostics programs monitor the operation of the system. Some parts of the programs monitor the various semiconductor devices (this is continuously performed), current/voltage transducers, supply voltages, speed/rotor angle detectors, serial data communication, etc.;
- Complex, high-speed arithmetic and capability of decision making.

Some of the service and diagnostic features of the application of microprocessors or digital circuits are:

- Only a few standard modules are required without any special adjustments;
- There is the possibility of plug-in memory modules for user programs, parameters and modifications to functions, etc.;
- Powerful system software for on-line measurements, control parameter setting (e.g. current-control parameter setting, speed-control parameter setting, etc.), and testing;
- Automatic location of hardware faults with the help of system and user software.

It is expected that the present trends in technology will result in cost reduction and performance improvement of a.c. drives and will thus lead to an even more widespread application of these drives. The MOS bipolar power devices, which include the Insulated Gate Bipolar Transistor (IGBT), and the MOS-Controlled Thyristor (MCT), can be turned on and off with a MOS gate and have such excellent characteristics that it is expected that in future they will revolutionize power converters.

The method of vector control described in earlier sections for synchronous machines can be extended to induction machines (see also Section 1.2.1.2). However, whilst in the electrically excited synchronous machine and the permanent-magnet synchronous machine, the space angle between the field winding and the direct axis of the stator and the space angle between the magnet flux and stator direct axis can be directly measured, for the induction machine the space angle of the rotor flux-linkage space phasor with respect to the direct axis of the stator is not a directly measurable quantity. Furthermore, in the case of the converter-fed induction machine, as well as supplying active power, the converter must supply reactive power for magnetization, since there is no external excitation as in the case of the synchronous machine. Thus both excitation (reactive) and torque-producing (active) currents must simultaneously exist in the stator windings of the induction machine.

The methods discussed in this chapter can be applied to induction machines with quadrature-phase stator windings ('two-phase' machine) or to induction machines with three-phase stator windings. The induction machine can have either a wound rotor (slip-ring machine) or a rotor with squirrel-cage winding, and it is possible to implement vector control on both types of induction machines. The squirrel cage can be of the double-cage or the single-cage type.

The mechanism of electromagnetic torque production in smooth-air-gap machines has been described in Section 2.1.6 and various forms of the expressions for the electromagnetic torque have been given in Section 2.1.8. It has been shown that in special reference frames fixed to the magnetizing flux-linkage, the stator flux-linkage, or the rotor flux-linkage space phasor, the expression for the electromagnetic torque of the smooth-air-gap machine is similar to the expression for the torque of the separately excited d.c. machine. This suggests that torque control of the induction machine can be performed by the decoupled control of the flux- and torque-producing components of the stator currents, which is similar to controlling the field and armature currents in the separately excited d.c. machine. However, it should be noted that in the squirrel-cage induction machine it is not possible to monitor the rotor currents directly. The stator currents of the induction machine can be separated into the flux- and torque-producing components by utilizing the transformations described in Section 2.1.8, whilst in the d.c. machine, as a result of the decoupled orthogonal field and armature axes, it is straightforward to achieve independent control of the flux and torque. It follows from Section 2.1.8 that the implementation of vector control (stator flux-oriented control, rotor flux-oriented control, or magnetizing flux-oriented control) requires information on the modulus and space angle (position) of the stator-flux, rotor-flux, or magnetizing-flux space phasors respectively. The control can be performed in a reference frame fixed to the stator-flux, rotor-flux, or magnetizing-flux space phasor respectively, and the direct- and quadrature-axis stator currents are obtained in the corresponding reference frame. These stator currents are similar to the field and armature currents of the separately excited d.c. machine.

In the case of induction machines rotor-flux-oriented control is usually employed, although it is possible to implement stator-flux- and also magnetizing-flux-oriented control, as described below. With rotor-flux-oriented control there are two main implementations to obtain the modulus and space angle of the rotor flux-linkage space phasor. When the so-called direct rotor-flux-oriented control is used (flux-feedback control), these quantities are directly measured (by using Hall-effect sensors, search coils, or tapped stator windings of the machine, etc.) or they are calculated from a so-called flux model. However, in one specific form of the so-called indirect rotor-flux-oriented control (flux-feedforward control), the modulus and space angle of the rotor flux-linkage space phasor are obtained by utilizing the monitored stator currents and the rotor speed. The space angle of the rotor flux-linkage space phasor is then obtained as the sum of the monitored rotor angle (θ_r) and the computed reference value of the slip angle (θ_{sl}), where the latter quantity gives the position of the rotor flux-linkage space phasor relative to the direct axis of the rotor, as shown below. However, it will also be shown in this chapter that the slip angle (θ_{sl}) can be calculated from the reference values of the torque- and flux-producing stator currents and strongly depends on the rotor parameters (rotor time constant, which is equal to the ratio of the rotor inductance to the rotor resistance) of the machine under consideration, and when inaccurate parameters are used it is not possible to achieve correct field orientation. The rotor time constant can change as a result of saturation, variation of the temperature, and effects of current displacement. Thus it is preferable to implement schemes which incorporate some form of on-line parameter adaption. In practice, despite its parameter sensitivity and the fact that it requires the use of shaft encoders, which are expensive, the indirect method has gained more widespread application, since it does not require flux sensors or a flux model.

It should be noted that with stator-flux-oriented control or magnetizing-flux-oriented control there can also be two main different implementations, direct and indirect, and they will be discussed later in this section.

Direct torque control (DTC) of induction machines is similar to that described for synchronous machines in Section 3.3 (see also Section 1.2). In the present chapter various direct-torque-control schemes will be described for voltage-source inverter-fed induction motors and also for a current-source inverter-fed induction motors.

Induction machines represent an alternative to synchronous machines in high-performance servo-drive applications. Although both the permanent-magnet synchronous machine and the induction machine are suitable for this type of application, for the selection of the appropriate machine, the following points have to be considered:

- Induction machines are more difficult to control.
- If field weakening is excluded, the drive containing the induction machine requires an inverter with higher rating (in the PM synchronous machine there are no rotor losses, and if there is field weakening, the machine is overexcited and draws large stator currents resulting in large stator losses).
- Because of the rotor losses, an induction machine normally requires forced cooling (a PM synchronous machine has no rotor losses and natural cooling can be sufficient).
- For the same torque produced, the efficiency (which has direct effects on the size of the machine) for the induction machine is lower (a PM synchronous machine has high efficiency because there are no rotor losses).
- An induction machine can be designed for higher flux densities (in a PM synchronous machine this is restricted by the magnets).
- An induction machine is cheaper (in a PM synchronous machine the price of the required magnets is still high).
- In the case of the drive containing the induction machine, field-weakening (constant-power mode) is easily achieved over a wide speed range (in a PM synchronous machine this is not the case).
- The drive with the induction machine requires the application of a more complicated microcomputer. (The microcomputer can be simpler for the permanent-magnet synchronous machine, since it does not require the application of a signal processor; or the machine can even be designed without any microprocessor.)

Vector control can be applied to an induction machine supplied by a voltage-source inverter or by a current-source inverter or by a cycloconverter. It is simpler to implement the control when controlled-current operation is achieved. The vector-controlled induction machine can achieve four-quadrant operation with high dynamic response. In the following sections, first the rotor-flux-oriented control of induction machines is discussed in detail and, as they are very similar, this will be followed by shorter discussions of the stator-flux-oriented and magnetizing-flux-oriented control techniques. Various 'sensorless' control schemes are also described. However, the direct torque control of VSI-fed induction motors will also be discussed in great detail. For this purpose different DTC schemes will be described. This will then be followed by the DTC of CSI-fed induction motors.

4.1 Rotor-flux-oriented control of induction machines

In this part of the chapter the rotor-flux-oriented control of induction machines is discussed for the case when the machine is supplied by impressed stator voltages, impressed stator currents, or impressed rotor currents. Both direct and indirect methods are discussed.

4.1.1 CONTROL OF AN INDUCTION MACHINE SUPPLIED BY A VOLTAGE-SOURCE INVERTER

4.1.1.1 General introduction

It is assumed that the induction machine is supplied by a pulse-width modulated (PWM) voltage-source thyristor inverter, whose switching frequency is low, usually in the range of 100 Hz–1 kHz. Such converters are used up to ratings of several 100 kW and one of the main applications is the high-dynamic-performance position-controlled servo drive. At lower power levels, inverters containing transistors are used where the switching frequency is high. At higher power levels, converters employing thyristors and gate turn-off thyristors are used.

Because of the low switching frequency, it is not possible to achieve fast closed-loop current control of the stator currents with, for example, sinusoidal stator current references in the steady state. In this drive the stator currents can contain time harmonics with large amplitudes. Since the stator currents cannot be assumed to be impressed by fast control loops, it is necessary to utilize the stator voltage equations as well, once the equation of the stator reference voltages are established. However, it will be shown in Section 4.1.2 that a significant simplification arises in the equations and thus in the drive control system if the induction machine can be considered to be supplied by current sources. This is the case for an induction machine supplied by a voltage-source transistor inverter with fast current control (used at lower power levels at a high switching frequency, usually above 15 kHz), for a cycloconverter-fed induction machine (used for high-power, low-speed applications), where the stator currents are approximately sinusoidal and where the stator currents can be controlled by individual current control loops, or for an induction machine supplied by, say, a current-source thyristor inverter.

4.1.1.2 Stator voltage equations in the rotor-flux-oriented reference frame

In this section the stator voltage equations are derived and formulated in the reference frame fixed to the rotor-flux linkage space phasor. There are many ways to obtain the stator voltage equations in this reference frame. However, a straightforward method is followed here, whereby the space-phasor forms of the voltage equations formulated in the general reference frame (see Section 2.1.7) are used directly. The rotor-current space phasor is deliberately expressed in terms of the so-called rotor magnetizing-current space phasor and thus the resulting voltage equations will contain the modulus and space angle of the rotor magnetizing-current space phasor (or rotor flux-linkage space phasor), which are necessary to implement vector control.

It has been shown in Section 2.1.8, eqn (2.1-195), that the so-called rotor magnetizing-current space phasor in the rotor-flux-oriented reference frame ($\bar{i}_{mr}$) is obtained by dividing the rotor flux-linkage space phasor established in this reference frame ($\bar{\psi}_{r\psi r}$) by the magnetizing inductance (L_m). Thus under linear magnetic conditions (where the magnetizing inductance is constant), $\bar{i}_{mr}$ and $\bar{\psi}_{r\psi r}$ are proportional. For convenience, eqn (2.1-195) is repeated here, and this also gives the relationship between the rotor magnetizing space phasor to the stator and rotor current space phasors:

$$\bar{i}_{mr}=\frac{\bar{\psi}_{r\psi r}}{L_m}=\bar{i}_{s\psi r}+(1+\sigma_r)\bar{i}_{r\psi r}, \tag{4.1-1}$$

where σ_r is the rotor leakage factor (it is the ratio of the rotor leakage inductance to the magnetizing inductance, as shown in Section 2.1.8).

By considering eqns (2.1-148) and (2.1-150), which give the stator space-phasor voltage and stator flux-linkage space phasor equations in the general reference frame respectively, and by assuming linear magnetic conditions (i.e. L_m=constant and the leakage inductances are also constant), the following stator voltage equation is obtained in the reference frame fixed to the rotor flux-linkage space phasor, which rotates at the speed ω_{mr} (this speed is defined in eqn (2.1-188) as the first time-derivative of the space angle ρ_r, which is the space angle of the rotor magnetizing-current space phasor with respect to the direct axis of the stationary reference frame):

$$\bar{u}_{s\psi r}=R_s\bar{i}_{s\psi r}+L_s\frac{d\bar{i}_{s\psi r}}{dt}+L_m\frac{d\bar{i}_{r\psi r}}{dt}+j\omega_{mr}L_s\bar{i}_{s\psi r}+j\omega_{mr}L_m\bar{i}_{r\psi r}. \tag{4.1-2}$$

It follows from eqn (2.1-189) that in the special rotor-flux-oriented (x,y) reference frame, the space phasor of the stator currents can be expressed in terms of the space phasor of the stator currents established in the stationary reference frame ($\bar{\imath}_s$) as

$$\bar{\imath}_{s\psi r}=i_{sx}+\mathrm{j}i_{sy}=\bar{\imath}_s\,\mathrm{e}^{-\mathrm{j}\rho_r}=(i_{sD}+\mathrm{j}i_{sQ})\,\mathrm{e}^{-\mathrm{j}\rho_r} \tag{4.1-3}$$

and a similar expression can be obtained for the stator-voltage space phasor, since the stator quantities in the same reference frame must be transformed by the same transformation,

$$\bar{u}_{s\psi r}=\bar{u}_s\,\mathrm{e}^{-\mathrm{j}\rho_r}=u_{sx}+\mathrm{j}u_{sy}=(u_{sD}+\mathrm{j}u_{sQ})\,\mathrm{e}^{-\mathrm{j}\rho_r}. \tag{4.1-4}$$

From eqn (4.1-1) and the fact that as a result of the special selection of the reference frame the rotor magnetizing-current space phasor is coaxial with the direct-axis and thus $\bar{\imath}_{mr}=|\bar{\imath}_{mr}|$, the rotor-current space phasor in the special reference frame is obtained in terms of the stator-current and rotor magnetizing-current space phasors as

$$\bar{\imath}_{r\psi r}=\frac{|\bar{\imath}_{mr}|-\bar{\imath}_{s\psi r}}{1+\sigma_r}. \tag{4.1-5}$$

Substitution of eqn (4.1-5) into eqn (4.1-2) yields the following differential equation for the stator currents, if both sides of the equation are divided by the stator resistance R_s and if it is expressed in the form required by a time-delay element:

$$T_s'\frac{\mathrm{d}\bar{\imath}_{s\psi r}}{\mathrm{d}t}+\bar{\imath}_{s\psi r}=\frac{\bar{u}_{s\psi r}}{R_s}-\mathrm{j}\omega_{mr}T_s'\bar{\imath}_{s\psi r}-(T_s-T_s')\left(\mathrm{j}\omega_{mr}|\bar{\imath}_{mr}|+\frac{\mathrm{d}|\bar{\imath}_{mr}|}{\mathrm{d}t}\right). \tag{4.1-6}$$

T_s' is the stator transient time constant of the machine, $T_s'=L_s'/R_s$, where L_s' is the stator transient inductance, $L_s'=(L_s-L_m^2/L_r)$, and can be expressed in terms of the total leakage factor σ and the stator inductance as $L_s'=\sigma L_s$. Thus it is possible to define σ as $1-L_m^2/(L_sL_r)$. T_s is the stator time constant, $T_s=L_s/R_s$, and thus it is also possible to express T_s' as σT_s. It follows that the term (T_s-T_s') can be expressed in terms of the total leakage factor and the stator time constant as $(1-\sigma)T_s$.

By resolving eqn (4.1-6) into its real (x) and imaginary axis (y) components, the following two-axis differential equations are obtained for the stator currents:

$$T_s'\frac{\mathrm{d}i_{sx}}{\mathrm{d}t}+i_{sx}=\frac{u_{sx}}{R_s}+\omega_{mr}T_s'i_{sy}-(T_s-T_s')\frac{\mathrm{d}|\bar{\imath}_{mr}|}{\mathrm{d}t} \tag{4.1-7}$$

$$T_s'\frac{\mathrm{d}i_{sy}}{\mathrm{d}t}+i_{sy}=\frac{u_{sy}}{R_s}-\omega_{mr}T_s'i_{sx}-(T_s-T_s')\omega_{mr}|\bar{\imath}_{mr}|. \tag{4.1-8}$$

The relationship between the stationary-axis voltage (u_{sD}, u_{sQ}) and stator current components (i_{sD}, i_{sQ}) and the corresponding voltage (u_{sx}, u_{sy}) and current (i_{sx}, i_{sy}) components can be obtained by utilizing the transformations defined by eqns (4.1-3) and (4.1-4) respectively. It follows from eqns (4.1-7) and (4.1-8) that with respect to the stator currents i_{sx} and i_{sy}, the induction machine behaves as a first-order time-delay element, whose time constant is equal to the stator transient time constant of the machine, and its gain is equal to the inverse of the stator resistance. However, it can be seen that there is an unwanted coupling between the stator circuits on the two axes. For the purposes of rotor-flux-oriented-control, it is the direct-axis stator current i_{sx} (rotor flux-producing component) and the quadrature-axis stator current i_{sy} (torque-producing component) which must be independently controlled. However, since the voltage equations are coupled, and the coupling term in u_{sx} also depends on i_{sy} and the coupling term in u_{sy} also depends on i_{sx}, u_{sx}, and u_{sy} cannot be considered as decoupled control variables for the rotor flux and electromagnetic torque. The stator currents i_{sx} and i_{sy} can only be independently controlled (decoupled control) if the stator voltage equations [eqns (4.1-7) and (4.1-8)] are decoupled and the stator current components i_{sx} and i_{sy} are indirectly controlled by controlling the terminal voltages of the induction machine. Various implementations of the necessary decoupling circuit will be described in the next section.

4.1.1.3 Decoupling circuits in the rotor-flux-oriented reference frame

In this section the decoupling circuits valid for the case of an ideal drive are derived first and then the derivation is given of the necessary decoupling circuits if there is an inverter time delay.

Decoupling circuits of an ideal drive: From eqns (4.1-7) and (4.1-8) there is in the direct-axis voltage equation a rotational voltage coupling term $\omega_{mr}L_s' i_{sy}$, and thus the quadrature-axis stator current i_{sy} affects the direct-axis stator voltage u_{sx}. Similarly, in the quadrature-axis voltage equation there is a rotational voltage coupling term $-\omega_{mr}L_s' i_{sx}-(L_s-L_s')\omega_{mr}|\bar{i}_{mr}|$ and thus the direct-axis stator current i_{sx} affects the quadrature-axis voltage u_{sy}. However, by assuming an ideal drive (which has no extra time delays apart from the natural ones, i.e. there is no inverter dead time, no delay due to signal processing, etc., and for which the parameters are those used in the equations) and by assuming constant rotor-flux operation ($|\bar{i}_{mr}|$=const.), it follows from eqns (4.1-7) and (4.1-8) that the stator current components can be independently controlled if the decoupling rotational voltage components,

$$u_{dx}=-\omega_{mr}L_s' i_{sy} \tag{4.1-9}$$

$$u_{dy}=\omega_{mr}L_s' i_{sx}+(L_s-L_s')\omega_{mr}|\bar{i}_{mr}|, \tag{4.1-10}$$

are added to the outputs ($\hat{u}_{sx}, \hat{u}_{sy}$) of the current controllers which control i_{sx} and i_{sy} respectively. This can be proved by considering that $\hat{u}_{sx}+\hat{u}_{dx}$ yields the direct-axis terminal voltage component, and the voltage on the output of the direct-axis current controller is

$$\hat{u}_{sx}=R_s i_{sx}+L_s'\frac{di_{sx}}{dt}. \tag{4.1-11}$$

Similarly, $\hat{u}_{sy}+u_{dy}$ gives the quadrature-axis stator voltage component and the voltage on the output of the quadrature-axis current controller is

$$\hat{u}_{sy}=R_s i_{sy}+L_s'\frac{di_{sy}}{dt}. \tag{4.1-12}$$

Thus $\hat{u}_{sx}$ and $\hat{u}_{sy}$ directly control the stator currents i_{sx} and i_{sy} through a simple time delay element with the stator transient time constant T_s'. The required decoupling circuit is shown in Fig. 4.1, where the decoupling voltages u_{dx} and u_{dy} are obtained from i_{sx}, i_{sy}, $|\bar{i}_{mr}|$, and ω_{mr} by using eqns (4.1-9) and (4.1-10) respectively.

In the decoupling circuit described above the inputs are the stator currents i_{sx} and i_{sy}, the rotor magnetizing current $|\bar{i}_{mr}|$, and the angular speed of the rotor flux-linkage space phasor, ω_{mr}. The current components i_{sx} and i_{sy} can be obtained from the measured three-phase stator currents by utilizing the transformations described by eqn (4.1-3), and $|\bar{i}_{mr}|$ and ω_{mr} can be obtained by utilizing a so-called flux-model, which can be obtained by considering the rotor voltage equations, as shown in the following section. However, it is also possible to have an implementation of the decoupling circuit where the stator currents are not utilized as input quantities, but instead, the output voltages of the stator current controllers ($\hat{u}_{sx}$, $\hat{u}_{sy}$) are used. For this purpose eqns (4.1-11) and (4.1-12) are utilized and the stator currents i_{sx}, i_{sy} are expressed in terms of the output voltages of the current controllers. When these are substituted into eqns (4.1-9) and (4.1-10), the following expressions are obtained for the decoupling voltage components:

$$u_{dx}=-\omega_{mr}\frac{T_s'}{1+T_s'p}\hat{u}_{sy} \tag{4.1-13}$$

$$u_{dy}=-\omega_{mr}\left[(L_s-L_s')|\bar{i}_{mr}|+\frac{T_s'}{1+T_s'p}\hat{u}_{sx}\right], \tag{4.1-14}$$

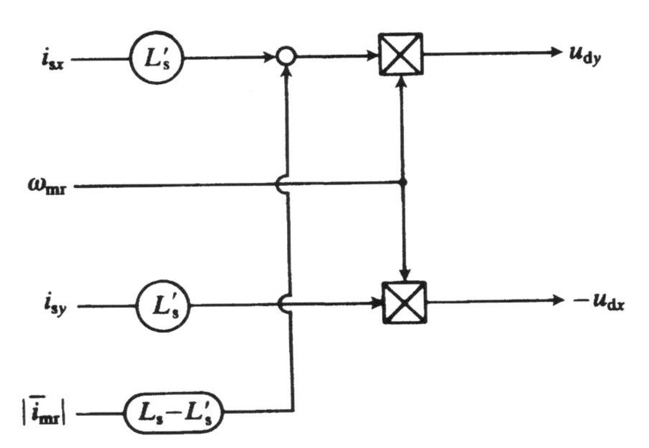

Fig. 4.1. Decoupling circuit to obtain decoupling voltages u_{dx}, u_{dy} (if $|\bar{i}_{mr}|=$constant). Inputs are i_{sx}, i_{sy}, ω_{mr}, $|\bar{i}_{mr}|$.

where $p = \mathrm{d}/\mathrm{d}t$ and it should be noted that these equations are valid if the rotor flux is constant. The decoupling circuit corresponding to eqns (4.1-13) and (4.1-14) is shown in Fig. 4.2, where the two blocks contain a lag element with time constant T_s' and gain T_s'.

It can be seen that the decoupling circuit shown in Fig. 4.2 also utilizes the stator resistance of the induction machine in addition to the machine parameters which have been used in the implementation shown in Fig. 4.1. The direct-axis reference value of the inverter voltage (in the rotor-flux-oriented reference frame) is again obtained as the sum of the direct-axis voltage obtained from the decoupling circuit and the direct-axis voltage which is present on the output of the direct-axis stator current controller. Similarly, the quadrature-axis reference value of the inverter voltage (in the rotor-flux-oriented reference frame) is obtained as the sum of the quadrature-axis voltage obtained from the decoupling circuit and the quadrature-axis voltage which is present on the output of the quadrature-axis stator current controller. To obtain the three-phase stator voltage references of the inverter, these have to be transformed into the two-axis stator voltage components by the application of the transformation $e^{j\rho_r}$ [see eqn (4.1-4)] and this transformation is followed by the two-phase to three-phase transformation.

Decoupling circuits considering the effect of the inverter time delay: It has been assumed above that the drive is ideal. However, the main difference between a real drive and an idealized drive is that in the real drive there are dead-time

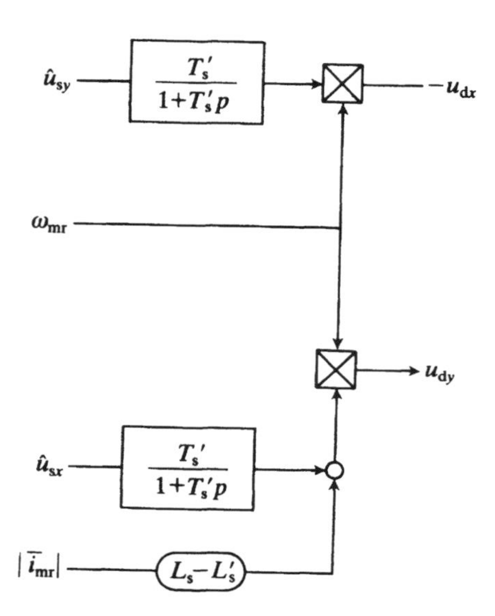

Fig. 4.2. Decoupling circuit to obtain decoupling voltages u_{dx}, u_{dy} (if $|\bar{i}_{mr}| = \text{constant}$). Inputs are $\hat{u}_{sx}$, $\hat{u}_{sy}$, ω_{mr}, $|\bar{i}_{mr}|$.

components due to the delay of the inverter, signal processing (if a microcomputer or microprocessor is employed), etc. These have to be taken account of, since they cause unwanted coupling terms and it is therefore not possible to perform the required decoupling of the stator circuits using the circuit described above. The unwanted coupling can cause instabilities, etc. and make the vector control scheme totally inoperative. It is now shown how the decoupling circuits have to be modified in order to take account of these effects.

For simplicity it is assumed that the total dead time T is concentrated at the end of the signal process, between the control circuits and the machine, i.e. in the inverter. The lag effect of the inverter is described by a first-order delay element, thus in the stationary reference frame

$$T\frac{\mathrm{d}\bar{u}_\mathrm{s}}{\mathrm{d}t}+\bar{u}_\mathrm{s}=\bar{u}_\mathrm{sref} \tag{4.1-15}$$

holds, where $\bar{u}_\mathrm{s}$ is the space phasor of the stator voltages in the reference frame fixed to the stator and $\bar{u}_\mathrm{sref}$ is its reference value. By using eqn (4.1-4), it is possible to transform eqn (4.1-15) into the rotor-flux-oriented reference frame. Thus the inverter can be described by the following first-order differential equation in the rotor-flux-oriented reference frame:

$$T\frac{\mathrm{d}\bar{u}_{\mathrm{s}\psi\mathrm{r}}}{\mathrm{d}t}+\mathrm{j}\omega_\mathrm{mr}T\bar{u}_{\mathrm{s}\psi\mathrm{r}}+\bar{u}_{\mathrm{s}\psi\mathrm{r}}=\bar{u}_{\mathrm{s}\psi\mathrm{rref}}, \tag{4.1-16}$$

resolution of which into direct- and quadrature-axis components yields

$$T\frac{\mathrm{d}u_{\mathrm{s}x}}{\mathrm{d}t}+u_{\mathrm{s}x}=\omega_\mathrm{mr}Tu_{\mathrm{s}y}+u_{\mathrm{s}x\mathrm{ref}} \tag{4.1-17}$$

$$T\frac{\mathrm{d}u_{\mathrm{s}y}}{\mathrm{d}t}+u_{\mathrm{s}y}=-\omega_\mathrm{mr}Tu_{\mathrm{s}x}+u_{\mathrm{s}y\mathrm{ref}}. \tag{4.1-18}$$

Thus it can be seen that the two equations are coupled; in the direct-axis equation there is an unwanted coupling $\omega_\mathrm{mr}Tu_{\mathrm{s}y}$ and in the quadrature-axis $-\omega_\mathrm{mr}Tu_{\mathrm{s}x}$. When the delay time is zero, the coupling disappears as expected. If the machine is operated in the field-weakening range (above base speed), it follows from eqns (4.1-17) and (4.1-18) that the undesirable decoupling terms will be the largest during field weakening. In the constant-flux range (below base speed), the effect of the delay time can be neglected. Equations (4.1-17) and (4.1-18) can be implemented by the circuit shown in Fig. 4.3.

Thus it is possible to obtain a combined decoupling circuit of the machine and inverter system, which is a series connection of a decoupling circuit of the machine (as described in Fig. 4.1 or Fig. 4.2) and the decoupling circuit of the inverter as shown in Fig. 4.3. If, for example, the decoupling circuit of the machine used is the one shown in Fig. 4.1, the combined decoupling circuit takes the form shown in Fig. 4.4.

However, it is not possible to decouple the machine and the drive system completely when this method is used, since the machine and the inverter represent

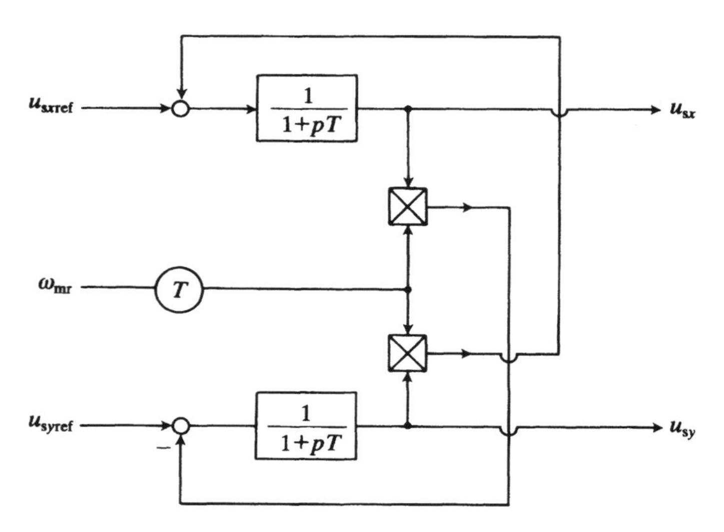

Fig. 4.3. Circuit for inverter in rotor-flux-oriented reference frame.

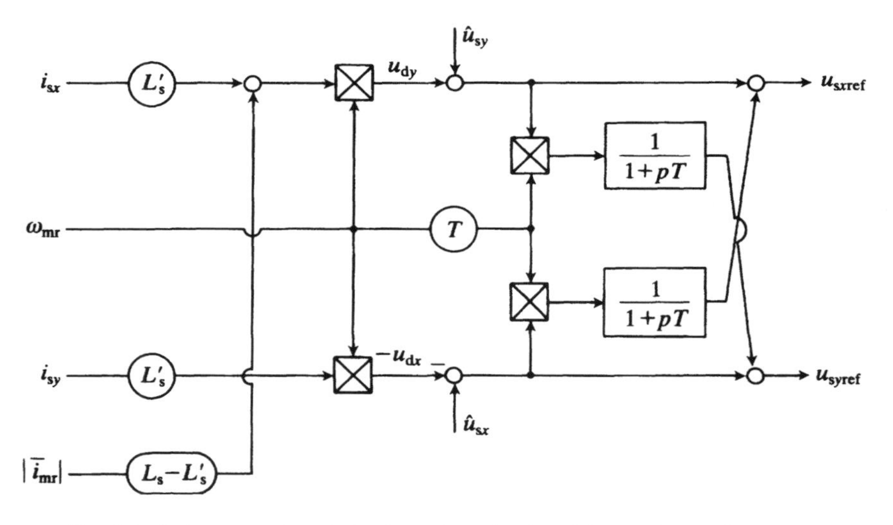

Fig. 4.4. Series connection of the decoupling circuits of the machine and inverter.

a total system, which cannot be decoupled by considering the machine and the inverter individually. However, by combining the equations of the machine with the equations of the inverter, it is possible to obtain a full decoupling circuit for the machine–inverter drive system as a whole. This will now be discussed.

For simplicity $|\bar{i}_{mr}|$ = constant is assumed, but similar considerations hold when this assumption is not made. Thus substitution of the expression for u_{sx}, which can be obtained by considering eqn (4.1-7), into eqn (4.1-17) yields the following expression for the direct-axis stator voltage reference:

$$\begin{aligned} u_{sxref} &= [R_s + (L_s' + TR_s)p + TL_s'p^2]i_{sx} - \omega_{mr}(L_s' + TR_s + 2TL_s'p)i_{sy} \\ &\quad - \omega_{mr}^2[TL_s'i_{sx} + T(L_s - L_s')|\bar{i}_{mr}|] - TL_s'i_{sy}(p\omega_{mr}) \\ &= \hat{u}_{sx} + \hat{u}_{dx}, \end{aligned} \tag{4.1-19}$$

where $p = d/dt$. Similarly, substitution for u_{sy} [given by eqn (4.1-8) into eqn (4.1-18)] gives

$$\begin{aligned} u_{syref} &= [R_s + (L_s' + TR_s)p + TL_s'p^2]i_{sy} + \omega_{mr}(L_s' + TR_s + 2TL_s'p)i_{sx} \\ &\quad + \omega_{mr}^2 TL_s'i_{sy} + [TL_s'i_{sx} + T(L_s - L_s')|\bar{i}_{mr}|](p\omega_{mr}) \\ &= \hat{u}_{sy} + \hat{u}_{dy} \end{aligned} \tag{4.1-20}$$

for the quadrature-axis stator voltage reference. If $T = 0$, u_{dx}, and u_{dy} in eqns (4.1-19) and (4.1-20) take the forms of eqns (4.1-9) and (4.1-10) respectively, which is an expected result. However, when T is not zero, the required full decoupling circuit corresponding to the full inverter–machine system is shown in Fig. 4.5 and this has been obtained by considering eqns (4.1-19) and (4.1-20).

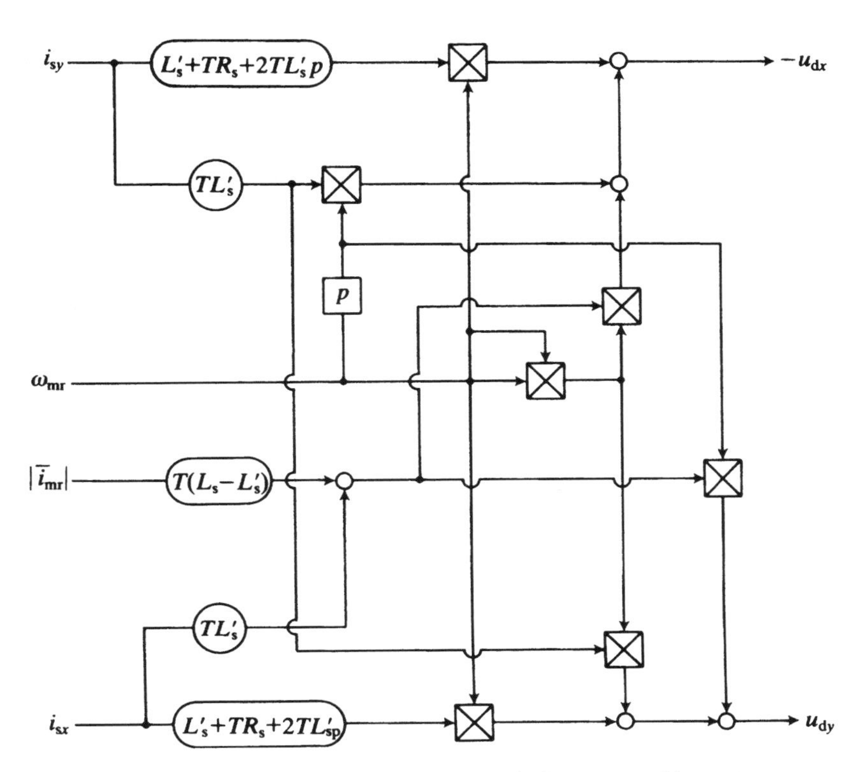

Fig. 4.5. Full decoupling circuit considering the inverter–machine system.

Although the circuit shown in Fig. 4.5 is much more complex than the one shown in Fig. 4.4, when this is used in the implementation of the vector-controlled drive, the drive will operate correctly for any value of the total dead time T.

In this and following sections various flux models are described, which yield the modulus and speed (or phase angle) of the rotor flux (or rotor magnetizing current) space phasor.

4.1.1.4 Flux models

4.1.1.4.1 Rotor voltage equations for the rotor flux model in the rotor-flux-oriented reference frame

The rotor voltage equations expressed in the rotor-flux-oriented reference frame can be used to obtain the modulus and phase angle of the rotor-flux space phasor or they can be used to obtain the modulus of the so-called rotor magnetizing current $|\bar{\imath}_{mr}|$ and its speed ω_{mr}. This flux model is derived in the present section.

By considering eqn (2.1-149), which gives the rotor voltage equation in the space-phasor form established in the general reference frame, the rotor voltage equation of the induction machine in the special rotor-flux-oriented reference frame will take the following form:

$$0=R_r\bar{\imath}_{r\psi r}+\frac{d\bar{\psi}_{r\psi r}}{dt}+j(\omega_{mr}-\omega_r)\bar{\psi}_{r\psi r}, \tag{4.1-21}$$

where $\bar{\psi}_{r\psi r}$ is the rotor flux-linkage space phasor in the rotor-flux-oriented reference frame and has been defined in eqns (2.1-190) and (2.1-193). Alternatively, it follows from eqn (4.1-1) that, because in the special reference frame $\bar{\imath}_{mr}=|\bar{\imath}_{mr}|$,

$$\bar{\psi}_{r\psi r}=L_m|\bar{\imath}_{mr}|, \tag{4.1-22}$$

which gives a linear relationship if the magnetizing inductance is assumed to be constant. Substitution of eqn (4.1-22) into eqn (4.1-21) yields the following rotor voltage differential equation if the effects of main flux saturation are neglected (L_m = constant):

$$0=R_r\bar{\imath}_{r\psi r}+L_m\frac{d|\bar{\imath}_{mr}|}{dt}+j(\omega_{mr}-\omega_r)L_m|\bar{\imath}_{mr}|. \tag{4.1-23}$$

By substituting eqn (4.1-5) into eqn (4.1-23) and dividing by R_r, which is the rotor resistance, finally the following equation is obtained, which is deliberately put into the form which is similar to the differential equation describing a time-delay element:

$$T_r\frac{d|\bar{\imath}_{mr}|}{dt}+|\bar{\imath}_{mr}|=\bar{\imath}_{s\psi r}-j(\omega_{mr}-\omega_r)T_r|\bar{\imath}_{mr}|. \tag{4.1-24}$$

By resolving into real- and imaginary-axis components, the following extremely simple equations are obtained which describe the flux model in the rotor-flux-oriented reference frame:

$$T_r \frac{d|\bar{i}_{mr}|}{dt} + |\bar{i}_{mr}| = i_{sx} \tag{4.1-25}$$

$$\omega_{mr} = \omega_r + \frac{i_{sy}}{T_r|\bar{i}_{mr}|}. \tag{4.1-26}$$

In eqn (4.1-26) the term $i_{sy}/(T_r|\bar{i}_{mr}|)$ represents the angular rotor frequency (angular slip frequency of the rotor flux) ω_{sl}, and it follows that the angular speed of the rotor flux is equal to the sum of the angular rotor speed and the angular slip frequency of the rotor flux. If $|\bar{i}_{mr}|$ is constant, it follows from eqn (4.1-25) that $|\bar{i}_{mr}| = i_{sx}$, which is in accordance with Section 2.1.8. The modulus of the rotor flux-linkage space phasor can be kept at a desired level by controlling the direct-axis stator current i_{sx}, as seen from eqn (4.1-25), but if there is no field weakening (below base speed) the electromagnetic torque is determined by the quadrature-axis stator current i_{sy}, in accordance with eqn (2.1-197).

Figure 4.6 shows the flux models of the induction machine in the rotor-oriented reference frame, based on eqns (4.1-25) and (4.1-26). The implementation shown in Fig. 4.6(a) utilizes the monitored values of the stator currents (i_{sA}, i_{sB}, i_{sC}), the monitored value of the rotor speed (ω_r), and the rotor time constant (T_r). The three stator currents are monitored and transformed into their two-axis components by the application of the three-phase to two-phase transformation. In the absence of zero-sequence currents, it is sufficient to monitor only two stator currents. The direct- and quadrature-axis stator currents, which are formulated in the stationary reference frame fixed to the stator (i_{sD}, i_{sQ}), are then transformed into the two-axis stator current components in the rotor-flux-oriented reference frame (i_{sx}, i_{sy}), by utilizing the transformation given in eqn (4.1-3). The current component i_{sx} serves as an input to a first-order time-delay element with gain 1 and time constant T_r, the output of which is the modulus of the rotor magnetizing current. This is proportional to the modulus of the rotor magnetizing flux-linkage space phasor and the proportionality factor is equal to the magnetizing inductance of the machine (which has been assumed to be constant). The quadrature-axis stator current i_{sy} is divided by $T_r|\bar{i}_{mr}|$, thus yielding the angular slip frequency of the rotor flux, and when the rotor speed is added to this, finally ω_{mr} is obtained. Integration of ω_{mr} yields the angle ρ_r, which defines the position of the rotor-flux space phasor with respect to the real axis of the stationary reference frame. This angle is used in the transformation block $e^{-j\rho_r}$.

However, it is possible to have a similar flux model where, instead of using the monitored values of the rotor speed, the monitored values of the rotor angle (θ_r) are utilized. This is shown in Fig. 4.6(b). It can be seen that the outputs of this flux-model are again $|\bar{i}_{mr}|$, ρ_r, and $|\bar{\psi}_r|$, but not the speed of the rotor flux-linkage space phasor ω_{mr}.

(a)

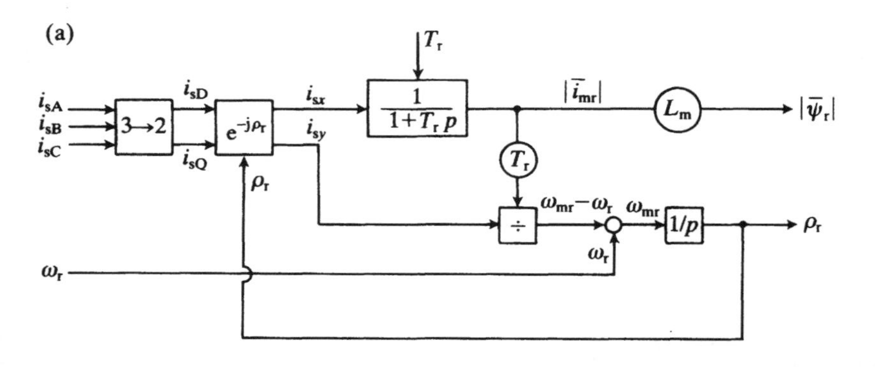

(b)

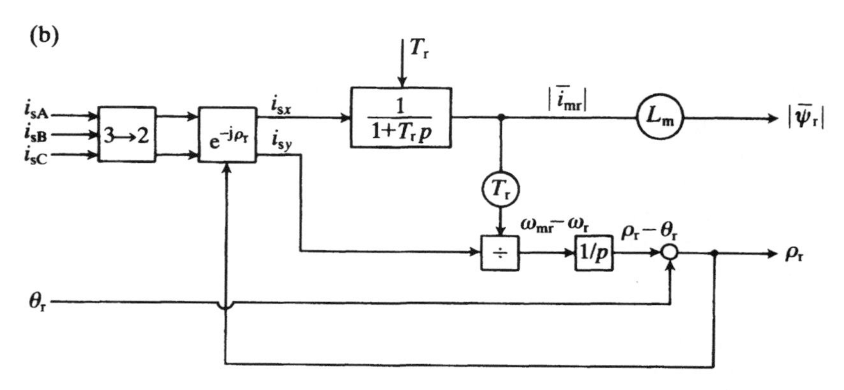

Fig. 4.6. Flux models in the rotor-flux-oriented reference frame. (a) Flux model with inputs i_{sA}, i_{sB}, i_{sC}, ω_r; (b) flux model with inputs i_{sA}, i_{sB}, i_{sC}, θ_r.

It is an important feature of eqns (4.1-25) and (4.1-26) that there is a strong dependency on the rotor time constant. If there is an inaccurate value of T_r in the flux models described above, it could lead to an unwanted coupling between the x and y axes, and therefore to a deteriorated dynamic performance of the drive with unwanted instabilities. This problem can be avoided by, for example, the application of on-line parameter-adaption schemes which yield accurate values of the rotor time constant. However, without parameter adaption more acceptable performance can be obtained at lower power levels than at higher power levels. The technique of Model Reference Adaptive Control (MRAC) can also be used to obtain an identification algorithm for the rotor resistance of a squirrel-cage induction machine. In Section 8.2 other models are described to obtain the rotor parameters of induction machines. It should be noted that in contrast to the large sensitivity to the rotor parameters, the implemented drive is less sensitive to variations in the stator parameters (which are, however, needed in the decoupling circuit).

4.1.1.4.2 Rotor voltage equations for flux models in the stationary reference frame

There are several ways to obtain the modulus and phase angle of the rotor-flux space phasor by utilizing certain machine parameters and various monitored quantities, such as the actual stator currents, the rotor speed and the stator voltages. Two different implementations will now be discussed. Since it is the intention to use actual stator voltages and/or currents, the equations will be formulated in the stationary reference frame fixed to the stator.

Flux model utilizing the monitored rotor speed and stator currents: It is possible to obtain the modulus and phase angle of the rotor-flux space phasor from a circuit, which in addition to the monitored rotor speed or rotor angle, also utilizes the monitored stator currents. However, in contrast to the implementation shown in Fig. 4.6, the stator currents do not have to be transformed into their values in the rotor-flux-oriented reference frame. For this purpose again the rotor voltage space-phasor equation is used, but in order to have an equation which directly contains the stator currents expressed in the reference frame fixed to the stator, the rotor voltage equation formulated in the stationary reference frame must be used. It should be noted that the effects of magnetic saturation are again neglected.

It follows from eqn (2.1-125) that in the stationary reference frame the rotor voltage equation takes the form,

$$0 = R_r \bar{i}_r' + \frac{d\bar{\psi}_r'}{dt} - j\omega_r \bar{\psi}_r', \tag{4.1-27}$$

where $\bar{i}_r'$ and $\bar{\psi}_r'$ are the rotor-current and rotor-flux space phasors respectively, but expressed in the stationary reference frame. Similarly to the definition used in eqn (4.1-1), the rotor magnetizing-current space phasor expressed in the stationary reference frame is obtained by dividing the rotor flux-linkage space phasor expressed in the stationary reference frame by the magnetizing inductance,

$$\bar{i}_{mr} = \frac{\bar{\psi}_r'}{L_m} = \bar{i}_s + (1+\sigma_r)\bar{i}_r', \tag{4.1-28}$$

where $\bar{i}_{mr}$ and $\bar{i}_s$ are the rotor magnetizing current and stator phasor currents respectively in the stationary reference frame. By considering eqn (4.1-28), the rotor-current space phasor can be expressed in terms of the rotor magnetizing space phasor, and when this expression is substituted into eqn (4.1-27), the following equation is obtained:

$$T_r \frac{d\bar{i}_{mr}}{dt} = \bar{i}_s - \bar{i}_{mr} + j\omega_r T_r \bar{i}_{mr}. \tag{4.1-29}$$

In the stationary reference frame, the rotor magnetizing-current space phasor and the stator-current space phasor can be expressed in terms of their direct- and quadrature-axis components as $\bar{i}_{mr} = i_{mrD} + ji_{mrQ}$ and $\bar{i}_s = i_{sD} + ji_{sQ}$, and therefore resolution of eqn (4.1-29) into real and imaginary axis components gives the following two differential equations:

$$T_r \frac{di_{mrD}}{dt} = i_{sD} - i_{mrD} - \omega_r T_r i_{mrQ} \tag{4.1-30}$$

$$T_r \frac{di_{mrQ}}{dt} = i_{sQ} - i_{mrQ} + \omega_r T_r i_{mrD}. \tag{4.1-31}$$

An implementation of eqns (4.1-30) and (4.1-31) is shown in Fig. 4.7, where first the three-phase stator currents are transformed into their two-axis components by the application of the three-phase to two-phase transformation. According to eqns (4.1-30) and (4.1-31), first the signals $i_{sD} - i_{mrD} - \omega_r T_r i_{mrQ}$ and $i_{sQ} - i_{mrQ} + \omega_r T_r i_{mrD}$ are obtained. These are then divided by the rotor time constant (T_r) and are integrated to yield the direct- and quadrature-axis rotor magnetizing current components (i_{mrD}, i_{mrQ}). A rectangular-to-polar converter is used to obtain the modulus ($|\bar{i}_{mr}|$) and the phase angle (ρ_r) of the rotor magnetizing flux-linkage space phasor. If required, $|\bar{i}_{mr}|$ can be multiplied by the magnetizing inductance (L_m) to yield the modulus of the rotor flux-linkage space phasor. This scheme is also dependent on the rotor time constant of the machine

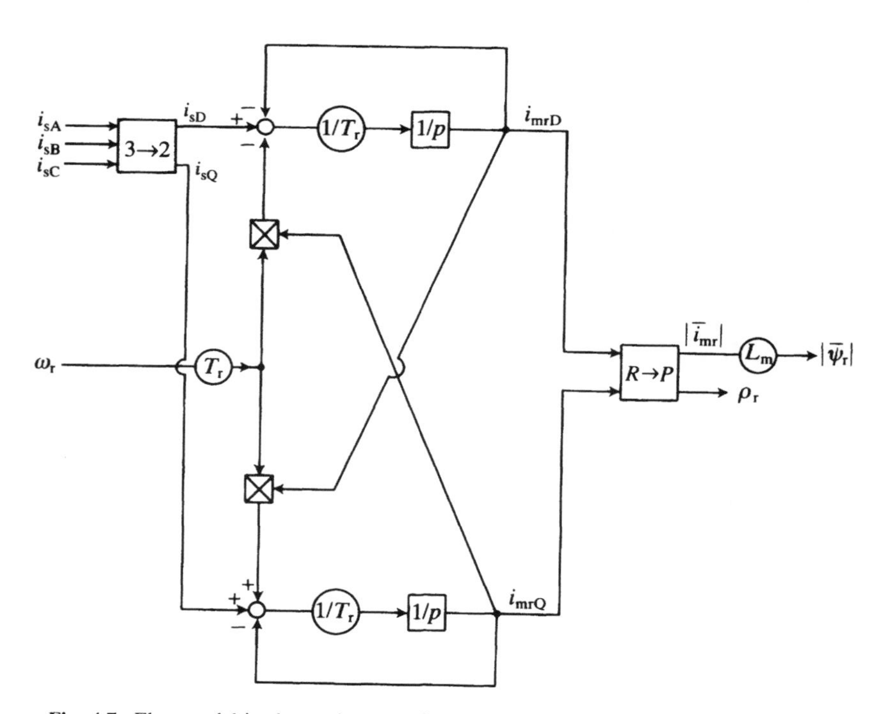

Fig. 4.7. Flux model in the stationary reference frame (the inputs are i_{sA}, i_{sB}, i_{sC}, ω_r).

and can be used over the entire speed range, including standstill. When this model is used, the angle ρ_r at the output of the model has to be differentiated to obtain ω_{mr} which is required in the decoupling circuits described in Figs 4.1, 4.2, 4.3, and 4.4. When compared with the flux models shown in Fig. 4.6, the flux model described in Fig. 4.7 yields less accurate values of the modulus and position of the rotor flux-linkage space phasor.

From eqn (4.1-29), it can be shown that in the steady state, where $p=\mathrm{d}/\mathrm{d}t=\mathrm{j}\omega_1$, the rotor magnetizing current can be expressed in terms of the stator current phasor as

$$\bar{I}_{mr}=\frac{\bar{I}_s}{1+\mathrm{j}(\omega_1-\omega_r)T_r}, \tag{4.1-32}$$

from which it follows that at high speeds, where the difference $(\omega_1-\omega_r)=s\omega_1$ (where s is the slip) is small, a small error in the monitored value of the rotor speed will result in a large error in the rotor magnetizing current (in its modulus and space angle), and this is especially pronounced in its phase angle. The other source of errors is the rotor time constant, which is also temperature dependent, although it follows from eqn (4.1-32) that only at no load ($s=0$) is the rotor magnetizing current (or the rotor flux) not influenced by the rotor time constant. This is of course a physically expected result, since at no load there are no currents in the rotor. However, it will be shown in the next section that it is possible to utilize a linear combination of the stator and rotor voltage equations, and these equations are not so sensitive to the rotor speed. Thus an implementation based on these equations requires less accurate rotor speed monitoring. Furthermore, it will also be shown that the system to be considered is affected by changes in the temperature only in the low speed region.

Flux model utilizing monitored rotor speed and stator voltages and currents: From eqns (2.1-148) and (2.1-149), which describe the stator voltage and flux-linkage equations respectively in the space-phasor form, it follows that in the stationary reference frame the stator voltage equations can be put into the following form, if again the effects of magnetic saturation are neglected:

$$\bar{u}_s=R_s\bar{\imath}_s+L_s\frac{\mathrm{d}\bar{\imath}_s}{\mathrm{d}t}+L_m\frac{\mathrm{d}\bar{\imath}_r'}{\mathrm{d}t}, \tag{4.1-33}$$

where R_s and L_s are the stator resistance and self-inductance L_m is the magnetizing inductance, and $\bar{\imath}_s$ and $\bar{\imath}_r'$ are the space phasors of the stator and rotor currents respectively in the stationary reference frame.

If in eqn (4.1-33) the rotor-current space phasor is expressed in terms of the rotor magnetizing current ($\bar{\imath}_{mr}$) defined by eqn (4.1-28), the following space-phasor voltage equation is obtained for the stator:

$$\bar{u}_s=R_s\bar{\imath}_s+\sigma L_s\frac{\mathrm{d}\bar{\imath}_s}{\mathrm{d}t}+(1-\sigma)L_s\frac{\mathrm{d}\bar{\imath}_{mr}}{\mathrm{d}t}, \tag{4.1-34}$$

where σ is the resultant leakage constant $\sigma=(1-L_m^2/L_sL_r)$. Thus it follows from eqn (4.1-34) that

$$(1-\sigma)T_s\frac{d\bar{i}_{mr}}{dt}=\frac{\bar{u}_s}{R_s}-\bar{i}_s-T_s'\frac{d\bar{i}_s}{dt}, \tag{4.1-35}$$

where $T_s=L_s/R_s$ are $T_s'=L_s'/R_s$ are the stator time constant and stator transient time constant respectively. When this equation is added to eqn (4.1-29), the following differential equation is obtained:

$$\frac{d\bar{i}_{mr}}{dt}[T_r+T_s(1-\sigma)]=\frac{\bar{u}_s}{R_s}+(j\omega_rT_r-1)\bar{i}_{mr}-T_s'\frac{d\bar{i}_s}{dt}. \tag{4.1-36}$$

It follows that if certain machine parameters and the stator voltages and currents (in the stationary reference frame) are known, it is possible to use eqn (4.1-36) directly to obtain the rotor magnetizing-current space phasor. However, for this purpose, eqn (4.1-36) must be resolved into its real- and imaginary-axis components. Since in the stationary reference frame $\bar{i}_{mr}=i_{mrD}+ji_{mrQ}$, $\bar{i}_s=i_{sD}+ji_{sQ}$, and $\bar{u}_s=u_{sD}+ju_{sQ}$, it follows from eqn (4.1-35) that its real- and imaginary-axis forms are

$$\frac{di_{mrD}}{dt}[T_r+T_s(1-\sigma)]+i_{mrD}=\frac{u_{sD}}{R_s}-\omega_rT_ri_{mrQ}-T_s'\frac{di_{sD}}{dt} \tag{4.1-37}$$

$$\frac{di_{mrQ}}{dt}[T_r+T_s(1-\sigma)]+i_{mrQ}=\frac{u_{sQ}}{R_s}+\omega_rT_ri_{mrD}-T_s'\frac{di_{sQ}}{dt}. \tag{4.1-38}$$

An implementation of eqns (4.1-37) and (4.1-38) is shown in Fig. 4.8, where the input quantities are the monitored value of the rotor speed ω_r, and the monitored values of the three-phase stator voltages and currents. These are then transformed into the two-axis components of the stator currents (i_{sD}, i_{sQ}) and stator voltages (u_{sD}, u_{sQ}) by the application of the three-phase to two-phase transformation. There are two integrators, labelled in the block '$1/p$' ($p=d/dt$). The two-axis components of the rotor magnetizing currents i_{mrD}, i_{mrQ} are then converted into the modulus $|\bar{i}_{mr}|$ and phase angle ρ_r of the rotor magnetizing-current space phasor with a rectangular-to-polar converter. It follows that the machine parameters to be used in this circuit are the stator resistance, the total leakage constant, and the stator and rotor time constants. When off-line parameter identification is used, these can be obtained by the application of conventional tests. Other techniques are described in Section 8.2.

It can be shown by considering eqn (4.1-36) that in the steady state the following expression is obtained for the rotor magnetizing current phasor (by utilizing $d/dt=j\omega_1$):

$$\bar{I}_{mr}=\frac{\bar{U}_s/R_s-j\omega_1T_s'\bar{I}_s}{1+j[(\omega_1-\omega_r)T_r+\omega_1T_s(1-\sigma)]}, \tag{4.1-39}$$

from which it follows that at high speeds, where $\omega_1-\omega_r$ is small, a small error in the measured value of the rotor speed will not influence $\bar{I}_{mr}$ as much as in

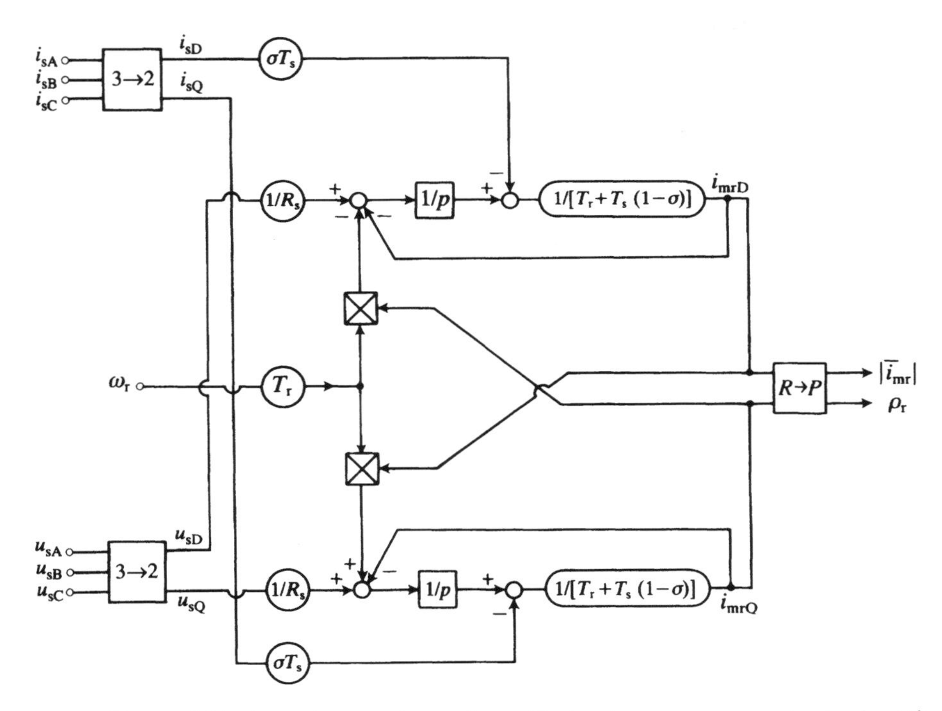

Fig. 4.8. Flux model in the stationary reference frame (the inputs are u_{sA}, u_{sB}, u_{sC}, i_{sA}, i_{sB}, i_{sC}, ω_r).

eqn (4.1-32), since at high values of the rotor speed in the imaginary part of the denominator of eqn (4.1-39), the term $\omega_1 T_s(1-\sigma)$ will dominate and this is independent of the speed.

To see the influence of the changes in temperature, eqn (4.1-39) is now rearranged into a form where the temperature-dependent parameters are only present in the denominator. For this purpose both the numerator and the denominator of eqn (4.1-39) are multiplied by $R_s/L_s = 1/T_s$ and, with $T_s' = \sigma T_s$,

$$\bar{I}_{mr} = \frac{\bar{U}_s/L_s - j\omega_1 \sigma \bar{I}_s}{1/T_s + j[(\omega_1-\omega_r)T_r/T_s + \omega_1(1-\sigma)]}. \qquad (4.1\text{-}40)$$

When $\omega_1 = \omega_r$, the denominator becomes $1/T_s + j\omega_1(1-\sigma)$, and the only parameter influenced by the change in the stator temperature is T_s, since it depends on R_s, but the effect is negligible. However, at low speeds or at standstill, large errors can arise in $\bar{I}_{mr}$, due to the change of R_s caused by the variation of the temperature. Higher precision can be obtained if instead of using the 'cold' value of the stator resistance a 'hot' value is used, if the temperature is sensed and a simple compensation circuit is implemented, or if a thermal model is used to correct the stator resistance.

It should be noted that instead of eqn (4.1-36), it is possible to use another equation, obtained in a similar way but by using different linear combinations of eqns (4.1-29) and (4.1-35). By increasing the effect of eqn (4.1-29), some improvement can be achieved in the sensitivity of $\bar{I}_{mr}$ in the low-speed region, but this would be obtained at the expense of greater sensitivity to the errors (in the monitored rotor speed and due to temperature variation) at higher speeds.

4.1.1.4.3 Flux model utilizing monitored stator voltages and currents; improved flux models

It is possible to establish a flux model which does not use the monitored rotor speed, but only the monitored values of the stator voltages and stator currents for the determination of the modulus and the space angle of the rotor magnetizing-current space phasor. For this purpose eqn (4.1-34) can be used. In the stationary reference frame $\bar{u}_s = u_{sD} + ju_{sQ}$ and $\bar{i}_s = i_{sD} + ji_{sQ}$, and therefore resolution of eqn (4.1-34) into its real- and imaginary-axis components yields

$$(1-\sigma)T_s \frac{di_{mrD}}{dt} = \frac{u_{sD}}{R_s} - i_{sD} - T_s' \frac{di_{sD}}{dt} \tag{4.1-41}$$

$$(1-\sigma)T_s \frac{di_{mrQ}}{dt} = \frac{u_{sQ}}{R_s} - i_{sQ} - T_s' \frac{di_{sQ}}{dt}. \tag{4.1-42}$$

In Fig. 4.9(a) there is shown an implementation of eqns (4.1-41) and (4.1-42), where the input quantities are the monitored values of the three-phase stator voltages and currents respectively ($u_{sA}, u_{sB}, u_{sC}, i_{sA}, i_{sB}, i_{sC}$) and the required parameters are the stator resistance, the total leakage constant, and the stator inductance. Similarly to Fig. 4.8, two integrators are used and the modulus ($|\bar{i}_{mr}|$) and the phase angle (ρ_r) of the rotor magnetizing-current space phasor are obtained from its two-axis components by the application of a rectangular-to-polar converter.

It follows from eqns (4.1-41) and (4.1-42) that when these equations are multiplied by the stator resistance, the direct- and quadrature-axis magnetizing-current components in the stator reference frame are essentially obtained by the integration of the direct- and the quadrature-axis magnetizing voltages respectively. These voltages are obtained by monitoring the two-axis components of the terminal voltages and by reducing them by the corresponding stator ohmic drops and by the corresponding voltage drops across the transient inductance of the stator (L_s'). When the magnetizing inductance is assumed to be very large (infinite), a voltage drop across the transient stator inductance is equal to the voltage drop across a fictitious resultant leakage inductance, which is obtained by connecting the leakage inductances of the stator and rotor in series. It should be noted that such an assumption leads to only a few per cent error in the determination of the stator transient inductance. However, at low stator frequencies, the stator ohmic drops will dominate and accurate ohmic voltage drop compensation must be performed prior to the integration. However, due to the temperature

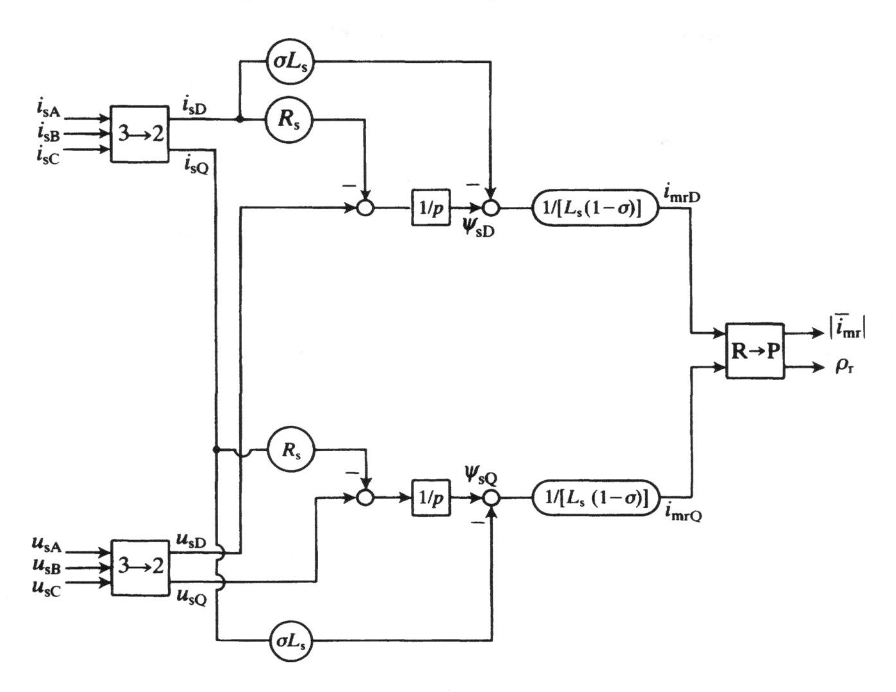

Fig. 4.9(a). Flux model in the stationary reference frame (the inputs are u_{sA}, u_{sB}, u_{sC}, i_{sA}, i_{sB}, i_{sC}).

dependency of the stator resistance, this is difficult to perform and with such an implementation a lower frequency limit for useful operation is approximately 3 Hz with a 50 Hz supply. It should also be noted that at low frequencies it is not possible to perform drift-free analogue integration.

It should be noted that to obtain the direct- and quadrature-axis stator voltages, in practice only two stator line voltages are required (e.g. u_{AC} and u_{BA}). This has been discussed in detail in Section 3.1.3.1 where it was shown that

$$u_{sD}=u_{sA}=\tfrac{1}{3}(u_{BA}-u_{AC})$$

$$u_{sQ}=\frac{1}{\sqrt{3}}(u_{sB}-u_{sC})=\frac{1}{\sqrt{3}}u_{CB}=-\frac{1}{\sqrt{3}}(u_{AC}+u_{BA}).$$

It has also been discussed in Section 3.1.3.1, that i_{sD} and i_{sQ} can be obtained by using only two stator line currents (e.g. i_{sA} and i_{sB}), since by assuming $i_{sA}+i_{sB}+i_{sC}=0$, finally

$$i_{sD}=i_{sA}$$

$$i_{sQ}=\frac{1}{\sqrt{3}}(i_{sB}-i_{sC})=\frac{1}{\sqrt{3}}(i_{sA}+2i_{sB})$$

are obtained.

In Fig. 4.9(a), the quantities which are integrated are the direct- and quadrature-axis stator flux linkages in the stationary reference frame (ψ_{sD}, ψ_{sQ}). Various aspects of this type of stator flux estimator have been discussed in detail in Section 3.1.3.1 and the reader is advised to study that section. Section 3.1.3.1 also discussed the effects of integrator drift and a technique to obtain more accurate estimations of the stator flux-linkage components by subtracting the corresponding drift components of the drift vector. Furthermore, it should also be noted that in addition to the scheme shown in Fig. 4.9(a), it is possible to construct another rotor flux model, where the integration drifts are reduced at low frequency. For this purpose, instead of open-loop integrators, closed-loop integrators are introduced and a rotor flux model is now presented where the stator flux linkages are obtained by using closed-loop integrators. Since

$$\bar{\psi}_s = |\bar{\psi}_s| \exp(j\rho_s) = \psi_{sD} + j\psi_{sQ},$$

in the rotor flux model shown in Fig. 4.9(b) the direct-axis stator flux can be obtained as $\psi_{sD} = |\bar{\psi}_s| \cos \rho_s$ and the quadrature-axis stator flux can be obtained as $\psi_{sD} = |\bar{\psi}_s| \sin \rho_s$ (this is performed by the polar-to-rectangular (P→R) converter shown). However, $|\bar{\psi}_s|$ and ρ_s are obtained by considering the stator voltage equation in the stator-flux-oriented reference frame, eqn (3.1-56). This is now repeated here for convenience:

$$\bar{u}_{s\psi s} = R_s \bar{i}_{s\psi s} + \frac{d|\bar{\psi}_s|}{dt} + j\omega_{ms}|\bar{\psi}_s|.$$

In this voltage equation $\omega_{ms} = d\rho_s/dt$ is the speed of the reference frame, which is equal to the speed of the stator flux-linkage space vector. In agreement with eqns (3.1-57) and (3.1-58), resolution of this into its real and imaginary components gives the rate of change of the stator flux modulus as $d|\bar{\psi}_s|/dt = u_{sx} - R_s i_{sx}$ and the speed of the stator flux space vector as $\omega_{ms} = (u_{sy} - R_s i_{sy})/|\bar{\psi}_s|$, where the subscripts x and y denote the direct- and quadrature-axis of the stator-flux-oriented reference frame respectively. These voltage equations are used in Fig. 4.9(b) to yield the first part of the circuit, which is the same as that shown

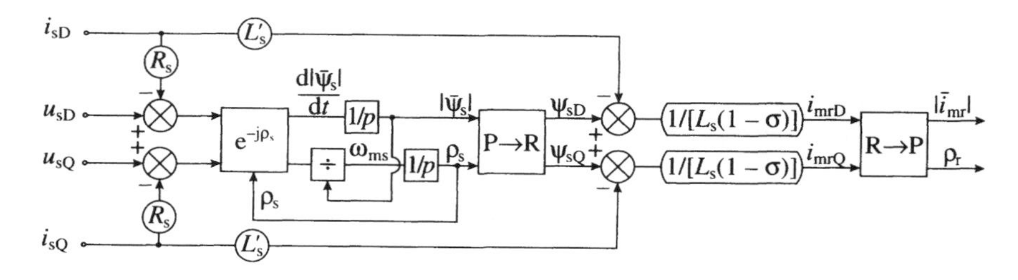

Fig. 4.9(b). Rotor flux model where stator flux linkages are obtained in the stator-flux-oriented reference frame.

in Fig. 3.20(b), and the extra parts of the circuit shown in Fig. 4.9(b) are the same as the corresponding parts shown in Fig. 49(a). It can be seen that these extra parts correspond physically to the fact that the rotor flux-linkage space vector (expressed in the stationary reference frame), $\bar{\psi}'_r = L_m|\bar{i}_{mr}|\exp(j\rho_r)$, is related to the stator flux-linkage space vector (expressed in the stationary reference frame) by

$$\bar{\psi}'_r = \frac{L_r}{L_m}(\bar{\psi}_s - \sigma L_s \bar{i}_s),$$

where $\sigma L_s = L'_s$ is the transient stator inductance.

It should also be noted that, similarly to the discussion in Section 3.1.3.1, it is also possible to use flux estimators such that the stator voltages are not monitored but are reconstructed. If the machine is supplied by a voltage-source inverter, then it is possible to reconstruct the stator voltages from the d.c. link voltage (U_d) by using the inverter switching states (S_A, S_B, S_C). The details are shown in Section 3.1.3.1. However, it follows from eqn (3.1-39) that

$$\bar{u}_s = \tfrac{2}{3}U_d(S_A + aS_B + a^2 S_C) = u_{sD} + ju_{sQ};$$

thus the direct- and quadrature-axis voltages can be reconstructed as

$$u_{sD} = \tfrac{2}{3}U_d\left(S_A - \frac{S_B}{2} - \frac{S_C}{2}\right)$$

$$u_{sQ} = \frac{1}{\sqrt{3}}U_d(S_B - S_C).$$

It is possible to construct a stator flux or rotor flux estimator, in which the drift problems associated with 'pure' open-loop integrators at low frequency are avoided by a band-limited integration of the high-frequency components and by replacing the inaccurate flux estimation at frequencies below $1/T$ by its reference value in a smooth transition. For this purpose a first-order delay element is used, as shown in Fig. 4.9(c).

The inputs to the stator flux estimator shown in Fig. 4.9(c) are the monitored values of the stator-voltage space vector ($\bar{u}_s$) and stator-current space vector ($\bar{i}_s$), expressed in the stationary reference frame. In addition, the third input is the modulus of the reference value of the stator flux-linkage space vector ($|\bar{\psi}_{sref}|$), which is also expressed in the stationary reference frame. It should be noted that since the stationary reference frame is used, the reference flux-linkage space vector $\bar{\psi}_{sref}$ contains two components, ψ_{sDref} and ψ_{sQref}. The space vector of the induced stator voltages is

$$\bar{u}_{si} = \bar{u}_s - R_s \bar{i}_s = \frac{d\bar{\psi}_s}{dt}$$

and in an open-loop stator flux estimator using a 'pure' integrator, its integrated value ($\int \bar{u}_{si}\,dt$) would yield the stator flux-linkage space vector $\bar{\psi}_s$. However, in

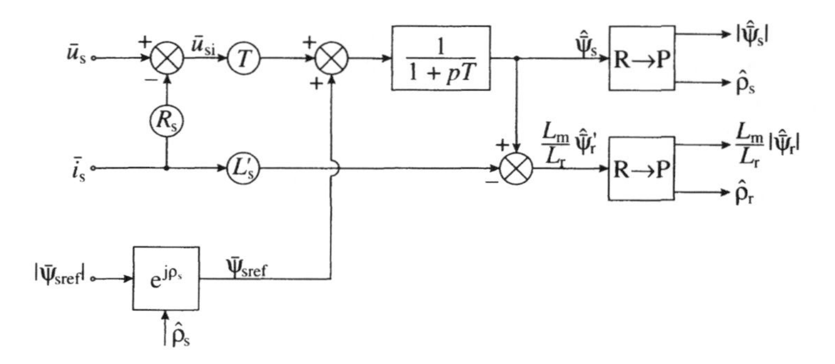

Fig. 4.9(c). Flux estimator using a time-delay element.

Fig. 4.9(c), $\bar{u}_{si}$ is multiplied by T. The reference stator flux-linkage space vector is added to $T\bar{u}_{si}$ yielding $T\bar{u}_{si}+\bar{\psi}_{sref}$. This is then the input to the first-order delay element, $1/(1+pT)$, on the output of which the estimated value of the stator flux-linkage space vector $\hat{\bar{\psi}}_s$ is obtained. It can be seen that if $\bar{\psi}_{sref}$ is equal to $\bar{\psi}_s$ then the output $(\hat{\bar{\psi}}_s)$ is exactly $\bar{\psi}_s$. In Fig. 4.9(c) an R→P converter is used which yields the modulus $|\hat{\bar{\psi}}_s|$ and also angle $\hat{\rho}_s$ of the stator flux-linkage space vector in the stationary reference frame. However, it is also possible to obtain an estimate of the rotor flux-linkage space vector expressed in the stationary reference frame, $\bar{\psi}'_r$, by considering that

$$\bar{\psi}_s = L'_s\bar{i}_s + \frac{L_m}{L_r}\bar{\psi}'_r,$$

where L'_s is the transient stator inductance, and this estimation is also shown in Fig. 4.9(c). Thus $(L_m/L_r)\hat{\bar{\psi}}'_r$ is obtained, and by using another R→P converter, $(L_m/L_r)|\hat{\bar{\psi}}'_r|$ and $\hat{\rho}_r$ are obtained. This scheme can be used in both vector drives and also in direct-torque-controlled drives. In a vector drive with stator-flux-oriented control, the space vector $\hat{\bar{\psi}}_{sref}$ can be obtained from the reference stator flux vector modulus $|\hat{\bar{\psi}}_{sref}|$, which is one of the inputs to a stator-flux-oriented control scheme, by considering that

$$\hat{\bar{\psi}}_{sref} = |\hat{\bar{\psi}}_{sref}|\exp(j\hat{\rho}_s),$$

where $\hat{\rho}_s$ is the angle of the stator flux-linkage space vector with respect to the real-axis of the stationary reference frame. It should be noted that for low-frequency operation, in the scheme shown in Fig. 4.9(c), the time-delay element has approximately a transfer function of unity and thus its output becomes almost equal to $\bar{\psi}_{sref}$ since $\bar{u}_{si}$ is small. This is in agreement to that emphasized above. Thus the integration is avoided at low stator frequencies. However, at high stator frequencies, the transfer function of the time-delay element is approximately equal to that of an integrator and it follows that the two actions are switched over smoothly around the stator frequency of $1/T$. The time constant must be selected so that it should minimize the estimation error when switching over takes place.

A suitable value for T is $T = T_r$, where T_r is the rotor time constant; this gives an estimator with minimum parameter sensitivity. The flux estimator scheme shown in Fig. 4.9(c) could be expanded to obtain the electromagnetic torque and also the rotor speed (see also Section 4.5.3.1 which discusses various techniques for the estimation of the rotor speed using flux-linkage estimates).

It is also possible to obtain a rotor flux-linkage estimator whose inputs are the stator voltage and current space vectors as above, but the third input is the reference value of the modulus of the rotor flux-linkage space vector, $|\bar{\psi}_{\text{rref}}|$, which, for example, is a known quantity in a vector drive employing rotor-flux-oriented control. Such a scheme is shown in Fig. 4.9(d).

It can be seen that in Fig. 4.9(d) the estimated value of the rotor flux-linkage space vector in the stationary reference frame is $\hat{\bar{\psi}}'_r$, and by using a R→P converter its modulus $|\hat{\bar{\psi}}'_r|$ and its angle $\hat{\rho}_r$ are obtained. Furthermore, by using the measured stator currents and also the estimated rotor flux linkages (in the stationary reference frame), the torque-producing stator current component i_{sy} is also obtained as

$$i_{sy} = \frac{(\hat{\bar{\psi}}'_r \times \bar{i}_s)}{|\hat{\bar{\psi}}'_r|^2} = \frac{(\hat{\psi}_{rd} i_{sQ} - \hat{\psi}_{rq} i_{sD})}{(\hat{\psi}^2_{rd} + \hat{\psi}^2_{rq})^{1/2}}.$$

This can be proved by using eqn (2.1-192) and $\cos \rho_r = \psi_{rd}/|\bar{\psi}_{rq}|$, $\sin \rho_r = \psi_{rq}/|\bar{\psi}_r|$. The angular slip frequency, ω_{sl}, can also be obtained, since by using the second term on the right-hand side of eqn (4.1-26),

$$\hat{\omega}_{sl} = \frac{(L_m/T_r) i_{sy}}{|\hat{\bar{\psi}}'_r|}.$$

However, by using eqn (4.1-26), the angular rotor speed can also be obtained as $\hat{\omega}_r = \hat{\omega}_{mr} - \hat{\omega}_{sl}$ where $\hat{\omega}_{mr} = \mathrm{d}\hat{\rho}_r/\mathrm{d}t$, where the estimation of $\hat{\rho}_r$ has been discussed above; thus $\hat{\rho}_r = \tan^{-1}(\hat{\psi}_{rq}/\hat{\psi}_{rd})$ or $\hat{\rho}_r = \cos^{-1}(\hat{\psi}_{rd}/|\hat{\bar{\psi}}'_r|)$, etc., where $|\hat{\bar{\psi}}'_r| = (\hat{\psi}^2_{rd} + \hat{\psi}^2_{rq})^{1/2}$. It is an advantage of this scheme that it can give accurate estimates of the flux linkages and thus the electromagnetic torque as well

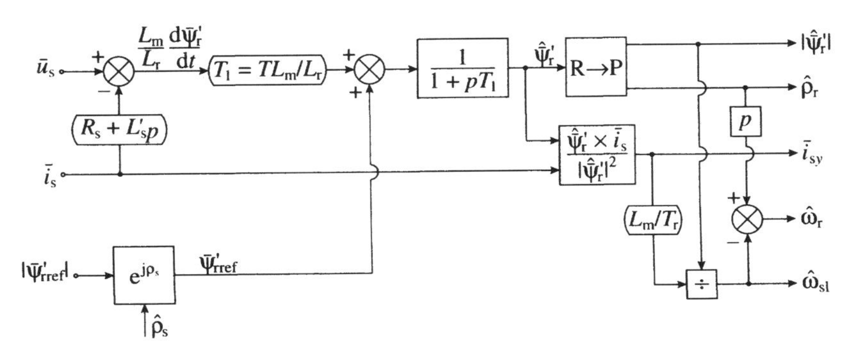

Fig. 4.9(d). Rotor flux estimator using a time-delay element.

$[\hat{t}_e=(3/2)P(L_m/L_r)(\hat{\psi}_{rd}i_{sQ}-\hat{\psi}_{rq}i_{sD})$ or the torque can be expressed directly in terms of the angular slip frequency], and also the rotor speed even at very low frequencies. Thus it can be used in speed-sensorless vector- and direct-torque-controlled induction motor drives (see also Section 4.5.3.1), but it is very important to select the appropriate T_1 time constant used in Fig. 4.9(d) and very large and very small values of T_1 should be avoided.

Another possibility to improve at low speed the stator flux-linkage and rotor flux-linkage estimates obtained by the application of the stator voltage model $[\bar{\psi}_s=\int(\bar{u}_s-R_s\bar{i}_s)\,dt;\ \bar{\psi}'_r=(L_r/L_m)(\bar{\psi}_s-L'_s\bar{i}_s)=|\bar{\psi}_r|\exp(j\rho_r)]$ is to use some form of stabilizing feedback in the voltage model. The correction term makes the modified voltage model insensitive to parameter deviation and measurement errors. The derivation of the modified voltage model is discussed briefly below.

As discussed earlier, due to feedforward integration, the voltage model is sensitive to parameter variations and measurement error and at low stator frequencies, these effects become more important. At zero stator frequency the induced voltage is zero and $\bar{u}_s=R_s\bar{i}_s$. It can be seen that at low frequency even a small deviation of the stator resistance will lead to an integration offset yielding incorrect estimates of $\bar{\psi}_s$ and $\bar{\psi}'_r$. However, it is possible to obtain the rotor flux-linkage space vector (and stator flux-linkage space vector) by using other models, thus $\bar{\psi}'^*_r=|\bar{\psi}^*_r|\exp(j\rho^*_r)$ is obtained. It is then possible to obtain the difference vector $\Delta\bar{\psi}'_r=\bar{\psi}'^*_r-\bar{\psi}'_r$ and its components can be used to obtain correction voltage components, which are fed back to the voltage model. The obtained modified voltage model then contains modified stator voltage components, where a modified component voltage is the sum of the original voltage and the feedback voltage.

It can be assumed that at low frequency and also at standstill the modulus of the rotor flux-linkage space vector is constant, thus $|\bar{\psi}^*_r|=$constant, since this will not degrade the dynamic performance of the drive (at low speed). However, its position (ρ^*_r) can also be estimated even at standstill, e.g. by using parameter-insensitive techniques, including the technique discussed in Section 3.2.2.2.3, where saturation effects are utilized. Thus $\bar{\psi}'^*_r=|\bar{\psi}^*_r|\exp(j\rho^*_r)$ is known, and then by using

$$\Delta\bar{\psi}'_r=\bar{\psi}'^*_r-\bar{\psi}'_r=\Delta\psi_{rx}+j\Delta\psi_{ry},$$

the component rotor flux-linkage deviations are obtained as

$$\Delta\psi_{rx}=|\bar{\psi}^*_r|\sin\rho$$

$$\Delta\psi_{ry}=|\bar{\psi}^*_r|(\cos\rho-1),$$

where $\rho=\rho^*_r-\rho_r$ is the angle between the two rotor flux-linkage space vectors $\bar{\psi}'_r$ and $\bar{\psi}'^*_r$ and x and y are the real and imaginary axes of the reference frame fixed to the rotor flux-linkage space vector $\bar{\psi}'_r$. These rotor flux-linkage deviation components are then used in the voltage feedback of the modified voltage model.

4.1.1.5 Expression for the electromagnetic torque; utilization of the steady-state equivalent circuit

By considering the expression for the electromagnetic torque shown in eqn (2.1-197), which is repeated here for convenience,

$$t_e = \frac{3}{2} P \frac{L_m^2}{L_r} |\bar{i}_{mr}| i_{sy}, \tag{4.1-43}$$

it follows that if the parameters of the machine are considered to be constant and $|\bar{i}_{mr}|$ is constant, then the electromagnetic torque is proportional to the quadrature-axis stator current expressed in the rotor-flux-oriented reference frame and thus the torque will respond instantaneously with i_{sy}. If there is a change in i_{sx}, it follows by considering eqns (4.1-25) and (4.1-26) that there will be a delayed response in the torque, and the delay is determined by the rotor time constant.

In the steady state, eqn (4.1-43) can also be derived by considering the steady-state equivalent circuit of the induction machine. This will now be discussed. The expression for the electromagnetic torque will be obtained by utilizing the relationship between the air-gap power and the torque, and the expression for the air-gap power will be obtained directly from the steady-state equivalent circuit.

In the steady state, if the induction machine is supplied by a sinusoidal symmetrical three-phase supply voltage system, the stator voltages are

$$u_{sA}(t) = \sqrt{2} U_s \cos(\omega_1 t) \qquad u_{sB} = \sqrt{2} U_s \cos(\omega_1 t - 2\pi/3)$$

$$u_{sC} = \sqrt{2} U_s \cos(\omega_1 t - 4\pi/3),$$

where U_s is the r.m.s. value of the line-to-neutral voltages. Thus it follows from the definition of the stator-voltage space phasor in the stationary reference frame given by eqn (2.1-61) that $\bar{u}_s = \sqrt{2} U_s e^{j\omega_1 t}$. The space phasor of the stator currents can be similarly defined. Thus by considering that in the steady state, the instantaneous values of the stator currents are

$$i_{sA}(t) = \sqrt{2} I_s \cos(\omega_1 t - \phi_s) \qquad i_{sB} = \sqrt{2} I_s \cos(\omega_1 t - \phi_s - 2\pi/3)$$

$$i_{sC} = \sqrt{2} I_s \cos(\omega_1 t - \phi_s - 4\pi/3),$$

it follows, by the application of eqn (2.1-4), that the space phasor of the stator currents in the stationary reference frame is

$$\bar{i}_s = \sqrt{2} I_s e^{j(\omega_1 t - \phi_s)} = \sqrt{2} \bar{I}_s e^{j(\omega_1 t)}.$$

Thus by considering eqns (4.1-27), (4.1-28), and (4.1-34) and the steady-state expressions for $\bar{u}_s$ and $\bar{i}_s$, the resulting equations can be represented by the equivalent circuit shown in Fig. 4.10(a).

It is a main feature of the equivalent circuit shown in Fig. 4.10(a) that, in contrast to the well-known steady-state equivalent circuit of the induction

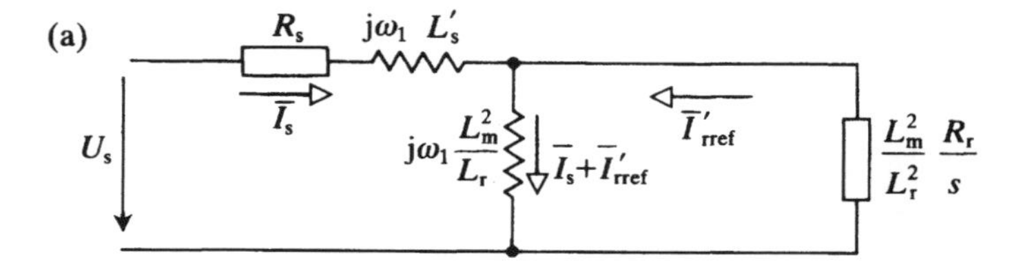

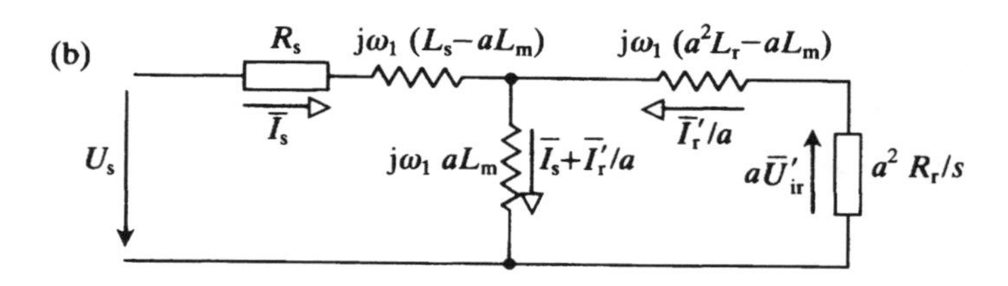

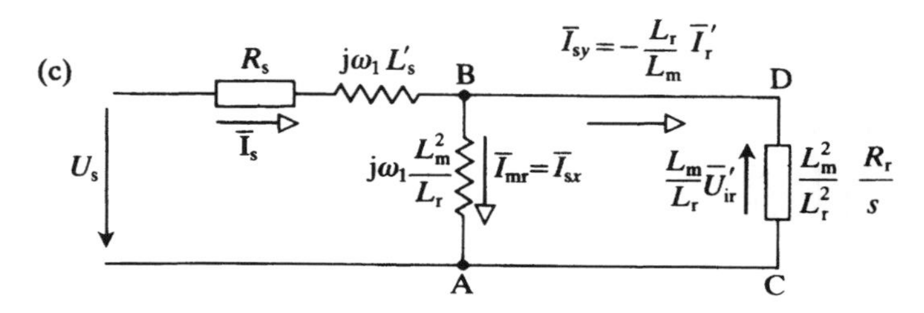

Fig. 4.10. Steady-state equivalent circuits of the induction machine. (a) Equivalent circuit where the rotor leakage inductance is not present in the rotor branch. (b) Equivalent circuit incorporating the effects of the general turns ratio. (c) Equivalent circuit which contains the torque- and flux-producing stator-current components.

machine, now the rotor leakage inductance is not present in the rotor branch. It should be noted that the rotor resistance (R_r) is divided by the slip (s) and R_r/s is multiplied by a^2, where a is a specially selected turns ratio $a=L_m/L_r$. In Fig. 4.10(a) the referred value of the rotor current phasor in the steady-state is present; this is $\bar{I}'_{rref}=\bar{I}'_r/a$ and $\bar{I}_{mr}=\bar{I}_s+\bar{I}'_{rref}$ is the rotor magnetizing-current phasor in the steady state in accordance with that stated in Section 2.1.3 and also in agreement with eqn (4.1-28). Furthermore, in the equivalent circuit shown, the referred value of the magnetizing inductance is present, which is equal to $aL_m=L_m^2/L_r$ and owing to the special referring factor, instead of the stator leakage inductance, the stator transient inductance ($L'_s=\sigma L_s$) is present.

The equivalent circuit shown in Fig. 4.10(a) could have also been obtained from the space-phasor stator and rotor voltage equations in the stationary reference frame, but by considering a general value of the turns ratio (a). For this purpose, first eqns (2.1-35) and (2.1-51) are considered, which define the stator and rotor flux-linkage space phasors respectively ($\bar{\psi}_s, \bar{\psi}'_r$) in the stationary reference frame. By considering the steady state and adding the term $aL_m\bar{I}_s - aL_m\bar{I}_s$ to the expression for $\bar{\Psi}_s$,

$$\bar{\Psi}_s = (L_s - aL_m)\bar{I}_s + aL_m(\bar{I}_s + \bar{I}'_r/a)$$

is obtained. The referred value of the rotor flux-linkage space phasor is obtained by multiplying $\bar{\Psi}'_r$ by the general turns ratio (a), and by adding the term $aL_m\bar{I}'_r/a - aL_m\bar{I}'_r/a$. Thus

$$a\bar{\Psi}'_r = (a^2L_r - aL_m)\bar{I}'_r/a + aL_m(\bar{I}_s + \bar{I}'_r/a)$$

is obtained. When these expressions are combined with the stator and rotor voltage equations, the equivalent circuit shown in Fig. 4.10(b) is obtained. It follows from Fig. 4.10(b) that if the general turns ratio is selected as $a = L_m/L_r$, in the rotor branch the inductive part will vanish and the equivalent circuit shown in Fig. 4.10(a) is obtained.

It is possible to redraw the equivalent circuit shown in Fig. 4.10(a) as the equivalent circuit shown in Fig. 4.10(c), which contains the torque and rotor flux-producing stator-current components ($\bar{I}_{sy}, \bar{I}_{sx}$). It follows from Fig. 4.10(c) that the stator current $\bar{I}_s$ is divided into $\bar{I}_{mr} = \bar{I}_{sx}$ and $\bar{I}_{sy}$ components. The current $\bar{I}_{mr} = \bar{I}_{sx}$ flows through the referred magnetizing reactance $\omega_1 L_m^2/L_r$ and the current $\bar{I}_{sy}$ flows through the referred rotor resistance $(L_m^2/L_r^2)R_r/s$. Furthermore, $(L_m/L_r)\bar{U}'_{ir}$ is the referred value of the voltage drop across the referred rotor resistance and thus $\bar{U}'_{ir} = -j\omega_1\bar{\Psi}'_r$. In Fig. 4.10(c) the current $\bar{I}_{sx}$ is the rotor flux-producing stator-current component and this can be proved by considering that the voltage across points A and B ($\bar{U}_{AB}$) is equal to the voltage across points C and D ($\bar{U}_{CD}$) and thus $\bar{I}_{sx}\, j\omega_1 L_m^2/L_r = j\omega_1(L_m/L_r)\bar{\Psi}'_r$ and it follows that $\bar{\Psi}'_r = L_m\bar{I}_{sx}$, so $\bar{I}_{sx}$ is indeed the rotor flux-producing stator current.

That $\bar{I}_{sy}$ is the torque-producing stator current can be proved by considering that in the steady state, the electromagnetic torque can be obtained as $T_e = P_{ag}/\omega_1$, where P_{ag} is the air-gap power, i.e. the power that crosses the air-gap. By considering the equivalent circuit shown in Fig. 4.10(c), P_{ag} can be expressed as $P_{ag} = (3P/2)|\bar{U}_{AB}||\bar{I}_{sy}|$, where $|\bar{U}_{AB}| = \omega_1(L_m^2/L_r)|\bar{I}_{sx}|$. Thus the electromagnetic torque can be expressed as

$$T_e = (3P/2)(L_m^2/L_r)|\bar{I}_{sx}||\bar{I}_{sy}| = (3P/2)(L_m/L_r)|\bar{\Psi}'_r||\bar{I}_{sy}|.$$

As expected, this is similar to eqn (4.1-43).

The equivalent circuit shown in Fig. 4.10(c) can also be used to obtain the expression for the angular slip frequency in the steady-state. If the modulus of the voltage across points A and B $|\bar{I}_{sx}|\,\omega_1 L_m^2/L_r$ is equal to the modulus of the voltage across points C and D, which in terms of the referred rotor resistance can be expressed as $|\bar{I}_{sy}|(L_m^2/L_r^2)R_r/s$, it follows that $s\omega_1 = |\bar{I}_{sy}|/(T_r|\bar{I}_{sx}|)$, where $s\omega_1 = \omega_{sl}$ is the angular slip frequency, and T_r is the rotor time constant ($T_r = L_r/R_r$). The same result also follows from eqn (4.1-26) if the steady state is considered, and $\omega_{mr} - \omega_r = s\omega_1$.

Since in the steady state the torque-producing stator current varies linearly with the angular slip frequency, and thus there is no pull-out slip or pull-out torque (see also Section 4.3.4), static instability does not arise in the induction machine subjected to rotor-flux-oriented control. However, in contrast to this, when, say, stator-flux-oriented control of the induction machine is performed, as shown in Section 4.2, the torque-producing stator current does not vary linearly with the angular slip frequency and theoretically static instability can arise. The details of this will be discussed in Section 4.3.4, but it should be noted that from the general equivalent circuit shown in Fig. 4.10(b), if the general turns ratio is chosen to be $a=L_s/L_m$, the inductive term in the stator branch is eliminated. In this case the current $\bar{I}_s+(L_m/L_s)\bar{I}_r'$ flows across the referred value of the magnetizing inductance and this current is equal to (L_m/L_s) times the so-called stator magnetizing current, which in the steady state takes the form $\bar{I}_{ms}=(L_s/L_m)\bar{I}_s+\bar{I}_r'$. A similar definition of this current will be used in Section 4.2. By choosing $a=L_s/L_m$, it is possible to use the equivalent circuit shown in Fig. 4.10(b) to obtain the expression for the electromagnetic torque in the steady state in terms of the stator flux-producing stator current component and the torque-producing stator current component, or to obtain the expression for the angular slip frequency in terms of the same two current components.

4.1.1.6 Implementation of the PWM VSI-fed vector-controlled induction machine

4.1.1.6.1 Implementation of the PWM VSI-fed induction machine drive utilizing a speed sensor and the actual values of i_{sx}, i_{sy}

This implementation of the PWM voltage-source inverter-fed induction machine drive uses the concepts discussed in previous sections. Figure 4.11 shows the schematic of the rotor-flux-oriented control of a voltage-source inverter-fed induction machine.

In this and the following sections two different implementations are described for the PWM voltage-source inverter-fed induction machine drive which have found widespread applications. The first one utilizes one of the decoupling circuits described above, which uses the actual values of the direct- and quadrature-axis stator currents in the rotor-flux-oriented reference frame (i_{sx}, i_{sy}) and contains a closed-loop control of the rotor position, rotor speed, electromagnetic torque, and rotor flux. In the second, simpler implementation, decoupling is achieved by using the reference values of the same stator current components (i_{sxref}, i_{syref}); there is closed-loop control of the rotor speed, which is obtained by the application of a sensor. It should be noted that it is possible to have closed-loop control of the flux- and torque-producing currents and of the rotor speed in a scheme where a decoupling circuit is utilized which is similar to the one used in the first implementation, and uses the actual values of i_{sx} and i_{sy}. However, the rotor speed is not obtained by a sensor, but is estimated by utilizing Model Reference Adaptive Control.

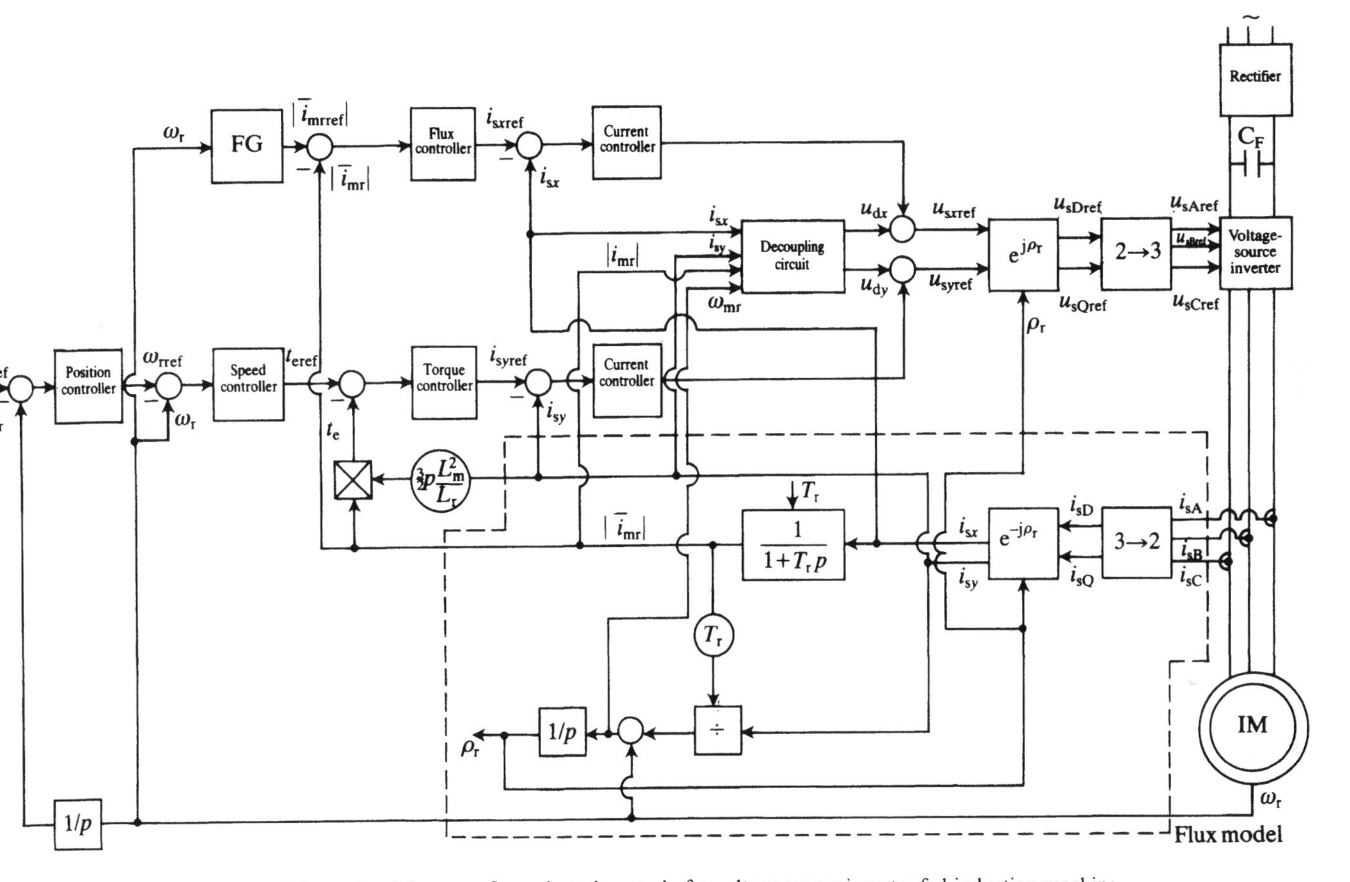

Fig. 4.11. Schematic of the rotor-flux-oriented control of a voltage-source inverter-fed induction machine.

In Fig. 4.11 the I.M. is supplied by a voltage-source PWM inverter, which is supplied by a diode bridge rectifier through a filter capacitor. The output voltages of the inverter are controlled by a pulse-width modulation technique. There are many types of modulation techniques which will not be discussed in detail here, but it should be noted that the so-called suboscillation technique has gained widespread application. When this technique is used, the output voltage of the inverter is generated by comparing a triangular carrier wave of frequency (switching frequency) f_s (in order to avoid beat effects, usually $f_s = Nf_1$, where f_1 is the fundamental frequency and N is the pulse number, an integer) with a sinusoidal modulating wave with fundamental frequency; the natural points of intersection determine the switching instants. The pulse-width modulated output voltages contain time harmonics and these lead to unwanted losses in the machine. There are a number of so-called harmonic elimination techniques, but when such a technique is used, although certain time harmonics (e.g. the fifth and the seventh) are almost eliminated, other harmonics will have increased magnitudes (e.g. the ninth and the eleventh). However, since the harmonic losses in the machine are determined by the r.m.s. value of the harmonic currents, it is these currents which have to be minimized in order to reduce these losses. This concept has lead to the development of the so-called minimum harmonic current technique. It is a common feature of these techniques that they are based on off-line computation of optimal pulse patterns for steady-state operation of the drive under consideration. Thus the optimum conditions can only be obtained in the steady state and not in the transient state. Furthermore, at low stator frequencies, off-line methods can be ineffective. To overcome this problem, on-line computation of the pulse patterns is required and by using on-line optimization algorithms it is possible to ensure, for example, minimum harmonic torque at minimum switching frequency or minimum torque pulsations. When the inverter switching frequency is reduced, the efficiency of the drive is increased as a result of the decrease in the switching losses, which can be especially important in thyristor inverters with forced commutation.

The flux model shown in Fig. 4.6 is used to obtain the angle ρ_r, which is used in the transformation blocks $e^{j\rho_r}$ and $e^{-j\rho_r}$. Furthermore, this flux model is used to obtain the angular speed of the rotor flux ω_{mr} and the modulus of the rotor magnetizing current $|\bar{i}_{mr}|$, since these are also used in the decoupling circuit, which is now assumed to be the one shown in Fig. 4.1. The modulus of the rotor magnetizing-current space phasor is also used to obtain the electromagnetic torque, in accordance with eqn (4.1-43).

A position controller, which can be a proportional controller (since the rotor speed ω_r is equal to the first time derivative of the rotor angle θ_r) provides as its output the reference value of the rotor speed (ω_{rref}). The speed controller, a PI controller, provides the reference torque (t_{eref}), and the torque controller, also a PI controller, gives the reference value of the quadrature-axis stator current in the rotor-flux-oriented reference frame (i_{syref}). Since in the steady state, the fundamental components of the stator voltages increase with stator frequency, but only a specific maximal output voltage of the inverter is available, it follows that with constant rotor magnetizing current ($|\bar{i}_{mr}|$), at a specific frequency, these voltages would exceed the maximally available value. Thus above a certain frequency, an increase of speed is only possible if the rotor magnetizing current is reduced. In Fig. 4.11 field weakening is achieved by the application of the function generator FG, the output of which is the reference value of the modulus of the rotor magnetizing-current space phasor ($|\bar{i}_{mrref}|$), which is speed dependent. Below base speed a constant (maximal) value is obtained (this value is only limited by main flux saturation), and above base speed this is reduced, inversely proportional to the rotor speed. The reference signal $|\bar{i}_{mrref}|$ is compared with the actual value of the rotor magnetizing current and the error serves as input to the flux controller, also a PI controller. Its output is the direct-axis stator current reference expressed in the rotor-flux-oriented reference frame (i_{sxref}). In Fig. 4.11 it is possible to leave out the quadrature-axis stator current controller, since the torque controller performs a similar role, and in this case there is no need to feed back the current i_{sy}; there is only the torque feedback, which is, however, proportional to the quadrature-axis stator current if the magnetizing inductance is constant.

The error signals $i_{sxref}-i_{sx}$ and $i_{syref}-i_{sy}$ serve as inputs to the respective current controllers and the outputs from these are added to the corresponding outputs of the decoupling circuit, as described above for the ideal drive. Thus the direct- and quadrature-axis reference stator voltages u_{sxref} and u_{syref} are obtained; these are established in the rotor-flux-oriented reference frame, and therefore they have to be transformed by $e^{j\rho_r}$ to obtain the two-axis stator voltages references in the stationary reference frame (u_{sDref}, u_{sQref}). This is followed by the application of the two-phase to three-phase transformation indicated by the '2→3' block and finally the reference values of the three-phase stator voltages are obtained. These signals are used to control the pulse-width modulator which transforms these reference signals into appropriate on–off switching signals to command the inverter phases.

With the given scheme it is possible to produce full torque even at standstill, since the stator currents flow even at standstill and supply the magnetizing currents. The implementation shown in Fig. 4.11 gives a high dynamic performance drive. In addition to the high dynamic response resulting from the decoupling control, there is no pull-out effect and if there is too quick a change of the speed reference or if the induction machine is overloaded, the electromagnetic torque cannot exceed the specified maximal level, since the speed error signal will saturate.

It is possible to obtain a fully digital implementation of the total control system, where all the control tasks and pulse-width modulation (which can be an on-line technique) are performed by a single 16-bit microprocessor, the currents are sensed in an analog manner (e.g. by the application of Hall sensors), and the rotor position or rotor speed is obtained by using a single optical encoder. Since the pulse-width modulation and all the control tasks are performed by the same microprocessor, this allows the application of optimization criteria to the modulation, since all the machine quantities (e.g. voltages and currents) are known. The possibilities for using various optimization schemes have been discussed above.

It should be noted that there are other ways to obtain the reference value of the rotor magnetizing current. For example, it is also possible to have an implementation in which the function generator in Fig. 4.11 is not present but the two-axis voltages u_{sxref}, u_{syref} are used to obtain the absolute value of the reference stator-voltage space phasor, $|\bar{u}_{sref}|=(u_{sxref}^2+u_{syref}^2)^{1/2}$. Field weakening can be automatically performed by the application of an extra control loop which contains a limiting stator voltage controller (a PI controller) the input of which is the error $u-|\bar{u}_{sref}|$, where u is a constant voltage reference (the ceiling voltage of the inverter), and the output is $|\bar{i}_{mref}|$. The voltage controller attempts to remove the error $(u-|\bar{u}_{sref}|)$ by adjusting the rotor magnetizing current $|\bar{i}_{mr}|$. When the motor operates below base speed, $|u_{sref}|<u$ and the voltage controller will be saturated and the maximal $|\bar{i}_{mr}|$ is produced, so the rotor flux is kept at the saturation limit of the induction machine. However, when the speed increases above base speed, the inverter will approach its maximal output voltage and when the maximal output voltage of the inverter is reached ($|\bar{u}_{sref}|$ is equal to u), the voltage controller will give a smaller value of $|\bar{i}_{mr}|$ in order to limit the modulus of the stator voltages. Thus it follows that field weakening is automatically performed with this implementation. In contrast to permanent-magnet synchronous machines, where there is only a limited field weakening range (because the large air-gap causes a small synchronous reactance), in the induction machine, field weakening can be achieved over a wide speed range with constant power. This feature is extremely useful in spindle-drive applications but it can be utilized in position-controlled feed drives as well.

4.1.1.6.2 Implementation of the PWM VSI-fed induction machine utilizing a speed sensor and the reference values of i_{sxref}, i_{syref}

Decoupling circuit: In this section an implementation of the rotor-flux-oriented control is presented for a voltage-source inverter-fed induction machine where the necessary decoupling circuit uses the reference values of the direct- and quadrature-axis components of the stator currents i_{sxref}, i_{syref}. Operation with constant rotor flux is assumed and the effects of magnetic saturation are neglected.

When the rotor flux ($\bar{\psi}_r$) is constant and since $\bar{\psi}_r=L_m\bar{i}_{mr}$ and thus under linear magnetic conditions the rotor-magnetizing current $|\bar{i}_{mr}|$ is also constant, it follows from eqn (4.1-25) that $|\bar{i}_{mr}|=i_{sx}$, where i_{sx} is the direct-axis component of the stator current in the rotor-flux-oriented reference frame. Under these conditions it follows from eqns (4.1-7) and (4.1-8) that if the term $L_s'\mathrm{d}i_{sy}/\mathrm{d}t$ is neglected, the stator voltage components in the rotor-flux-oriented reference frame are:

$$u_{sx}=R_si_{sx}-\omega_{mr}L_s'i_{sy} \tag{4.1-44}$$

$$u_{sy}=R_si_{sy}+\omega_{mr}L_s'i_{sx}. \tag{4.1-45}$$

It should be noted that in eqn (4.1-44), the stator transient inductance L_s' is present and in eqn (4.1-45) the stator inductance L_s is present in the direct- and quadrature-axis rotational voltage terms respectively. Furthermore, it follows from eqns (4.1-44) and (4.1-45) that in the direct-axis voltage equation, the rotational voltage is affected by the quadrature-axis stator current (i_{sy}) and in the quadrature-axis voltage equation, the rotational term is influenced by the direct-axis stator current (i_{sx}). Thus the rotor flux (or i_{sx}) is not solely determined by the direct-axis stator voltage expressed in the rotor-flux-oriented reference frame (u_{sx}), but is also influenced by the quadrature-axis stator current and similarly, the torque-producing stator-current component is not solely determined by the quadrature-axis stator voltage expressed in the rotor-flux-oriented reference frame u_{sy}, but is also dependent on the direct-axis stator current. However, it is possible to have an independent control of the flux-producing stator-current component (i_{sx}) and the torque-producing stator-current component (i_{sy}), when the unwanted coupling terms are cancelled and thus the rotor flux is controlled by u_{sx} and the electromagnetic torque is controlled by u_{sy}, which are independent of each other. It follows from eqns (4.1-44) and (4.1-45) that in the decoupling circuit the actual values of i_{sx} and i_{sy} have to be used. However, by assuming that the lag time between i_{sy} and u_{sy} is small and thus can be neglected, it is possible to utilize the reference currents in the decoupling circuit and in this case $\hat{u}_{sx}=R_s i_{sxref}$ and $\hat{u}_{sy}=R_s i_{syref}$. This concept is utilized in the implementation of the rotor-flux-oriented control of the voltage-source inverter-fed induction machine drive described in the following section.

Drive implementation: The schematic of the rotor-flux-oriented control of a voltage-source inverter-fed induction machine drive utilizing the concepts described above is shown in Fig. 4.12. This is simpler than the drive described in Fig. 4.11, but it can yield a satisfactory dynamic response.

In Fig. 4.12, the reference value of the rotor flux is $|\bar{\psi}_{rref}|$ and when it is divided by the magnetizing inductance, the rotor magnetizing current is obtained, which is equal to the direct-axis stator current reference i_{sxref}. The reference value of the rotor speed (ω_{rref}) is compared with its actual value (ω_r) and the error serves as input to the speed controller (a PI controller). The output of the speed controller is the torque reference, which is, however, proportional to the quadrature-axis stator current reference (i_{syref}).

The direct- and quadrature-axis stator current references are used in the decoupling circuit, which utilizes the principles described above. Thus i_{sxref} is multiplied by the stator resistance (R_s) and the value obtained is reduced by $-\omega_{mr}L_s' i_{syref}$. Thus the direct-axis stator reference voltage component expressed in the rotor-flux-oriented reference frame (u_{sxref}) is obtained. Similarly, the quadrature-axis stator current reference is multiplied by the stator resistance and the rotational voltage component $\omega_{mr}L_s i_{sxref}$ is added to this voltage, and thus the quadrature-axis stator voltage reference expressed in the rotor-flux-oriented reference frame (u_{syref}) is obtained. The voltage references u_{sxref} and u_{syref} are transformed into the two-axis voltage components of the stationary reference frame (u_{sDref}, u_{sQref}), by utilizing the transformation $e^{j\rho_r}$, where ρ_r is the space angle of the rotor flux-linkage space phasor with respect to the real axis of the stationary reference frame. For completeness the transformed values are also

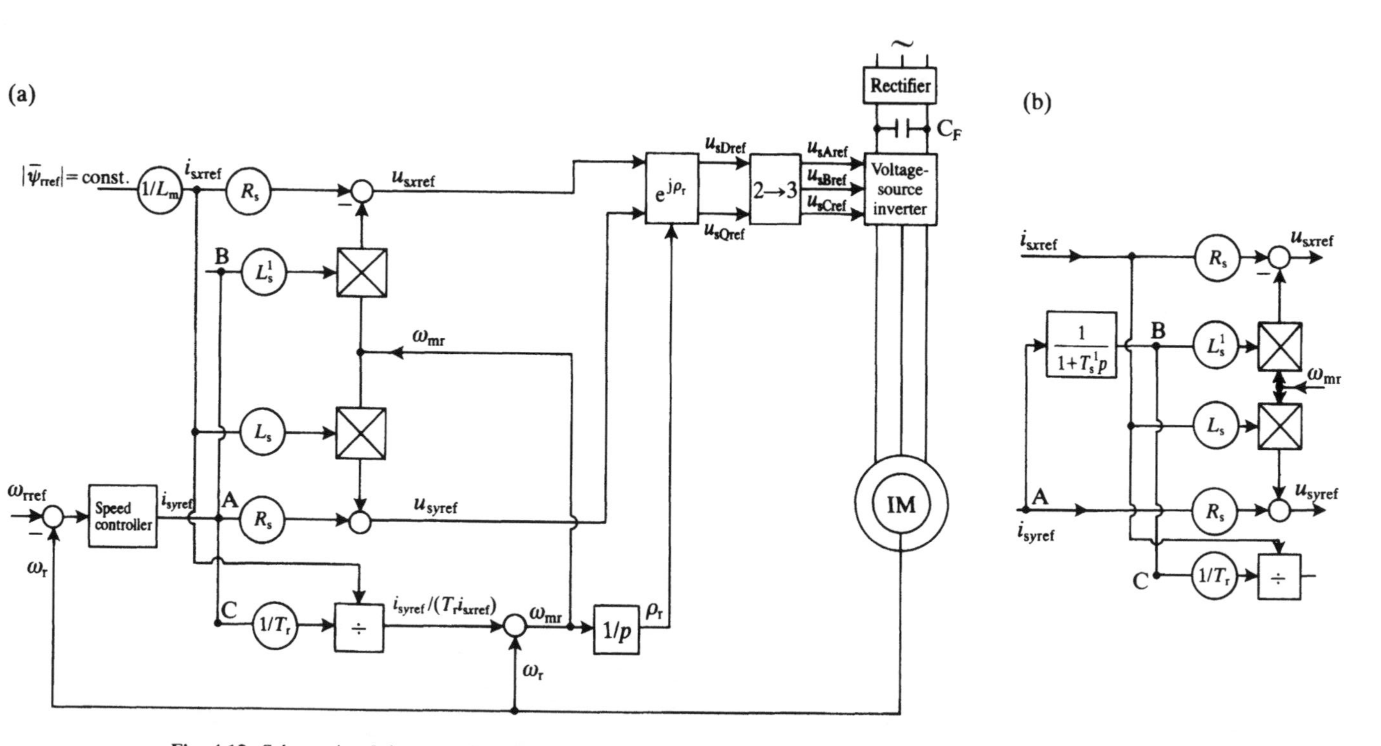

Fig. 4.12. Schematic of the rotor-flux-oriented control of a voltage-source induction machine drive. (a), (b) See text.

given in their expanded forms. Thus by considering eqn (4.1-4), but using reference values of the voltages,

$$u_{sDref} + ju_{sQref} = (u_{sxref} + ju_{syref})\, e^{j\rho_r} \tag{4.1-46}$$

or, from the resolution of eqn (4.1-46) into its real and imaginary components,

$$u_{sDref} = \cos \rho_r\, u_{sxref} - \sin \rho_r\, u_{syref} \tag{4.1-47}$$

$$u_{sQref} = \sin \rho_r\, u_{sxref} + \cos \rho_r\, u_{syref}. \tag{4.1-48}$$

These two-axis voltage references are then transformed into their three-phase reference values by the application of the two-phase to three-phase transformation indicated by the '2→3' block. It is possible to obtain the three-phase reference values by simply considering that in the absence of zero-sequence voltages, the projections of the voltage space phasor on the corresponding axes yield the instantaneous values of the phase voltages [see eqns (2.1-70), (2.1-71), and (2.1-72)]:

$$u_{sAref} = \mathrm{Re}(\bar{u}_{sref}) = \mathrm{Re}(u_{sDref} + ju_{sQref}) = u_{sDref}, \tag{4.1-49}$$

where $\bar{u}_{sref}$ is the reference value of the stator-voltage space phasor in the stationary reference frame,

$$u_{sBref} = \mathrm{Re}(a^2\bar{u}_{sref}) = \frac{-u_{sDref}}{2} + \sqrt{3}\,\frac{u_{sQref}}{2} \tag{4.1-50}$$

and

$$u_{sCref} = \mathrm{Re}(a^2\bar{u}_{sref}) = \frac{-u_{sDref}}{2} - \sqrt{3}\,\frac{u_{sQref}}{2}. \tag{4.1-51}$$

Of course it is also possible to use the Euler forms of the space-phasor equations. Thus the space phasor of the stator voltage references in the rotor-flux-oriented reference frame can be expressed as

$$\bar{u}'_{sref} = u_{sxref} + ju_{syref} = (u^2_{sxref} + u^2_{syref})^{1/2}\, e^{j\phi_u} \tag{4.1-52}$$

where $\phi_u = \tan^{-1}(u_{syref}/u_{sxref})$.

From eqn (4.1-46), in the stationary reference frame the space phasor of the stator reference voltages is

$$\bar{u}_{sref} = \bar{u}'_{sref}\, e^{j\rho_r} = (u^2_{sxref} + u^2_{syref})^{1/2}\, e^{j(\rho_r + \phi_u)}, \tag{4.1-53}$$

where $\bar{u}'_{sref}$ has been substituted by the expression given in eqn (4.1-52). Thus by considering eqns (4.1-49), (4.1-50), and (4.1-51), the three-phase stator voltage reference values can be put into the following form:

$$u_{sAref} = \mathrm{Re}(\bar{u}_{sref}) = (u^2_{sxref} + u^2_{syref})^{1/2}\, e^{j(\rho_r + \phi_u)} \tag{4.1-54}$$

$$u_{sBref} = \mathrm{Re}(a^2\bar{u}_{sref}) = (u^2_{sxref} + u^2_{syref})^{1/2}\, e^{j(\rho_r + \phi_u - 2\pi/3)} \tag{4.1-55}$$

$$u_{sCref} = \mathrm{Re}(a\bar{u}_{sref}) = (u^2_{sxref} + u^2_{syref})^{1/2}\, e^{j(\rho_r + \phi_u + 2\pi/3)} \tag{4.1-56}$$

which could be utilized directly to obtain the three-phase reference voltages.

It is also possible to have an implementation where the decoupling and the transformation of the stator reference voltage components from the rotor-flux-oriented reference frame into the stationary reference frame is performed in one step. For this purpose eqns (4.1-44) and (4.1-45) are substituted into eqns (4.1-47) and (4.1-48) respectively, and the reference values of the stator current components (i_{sxref}, i_{syref}) are used in the decoupling circuit instead of their actual values (i_{sx}, i_{sy}). The following two new voltage equations are then obtained:

$$u_{sDref} = (R_s + L_s p)\cos\rho_r\, i_{sxref} - (R_s + L_s' p)\sin\rho_r\, i_{syref} \tag{4.1-57}$$

$$u_{sQref} = (R_s + L_s p)\sin\rho_r\, i_{sxref} - (R_s + L_s' p)\cos\rho_r\, i_{syref}, \tag{4.1-58}$$

an implementation of which is shown in Fig. 4.13. The inputs to the circuit shown in Fig. 4.13 are the reference values of the flux- and torque-producing currents (i_{sxref}, i_{syref}) and the angle ρ_r. The angle ρ_r is obtained from i_{sxref}, i_{syref} and the monitored rotor speed, as described below. The outputs of the circuit shown in Fig. 4.13 are the direct- and quadrature-axis stator voltage references in the stationary reference frame (u_{sDref} and u_{sQref}).

The angular frequency of the stator voltages is ω_{mr} and is obtained from i_{syref} by utilizing eqns (4.1-25) and (4.1-26). It follows from eqn (4.1-25) that if $|\bar{i}_{mr}|$ is constant, $|\bar{i}_{mr}| = i_{sx}$ and $\omega_{mr} = \omega_r + i_{sy}/(T_r i_{sx})$. Thus in accordance with the discussion presented above, by using the reference values of the stator currents instead of the actual values, finally

$$\omega_{mr} = \omega_r + \frac{i_{syref}}{T_r i_{sxref}} \tag{4.1-59}$$

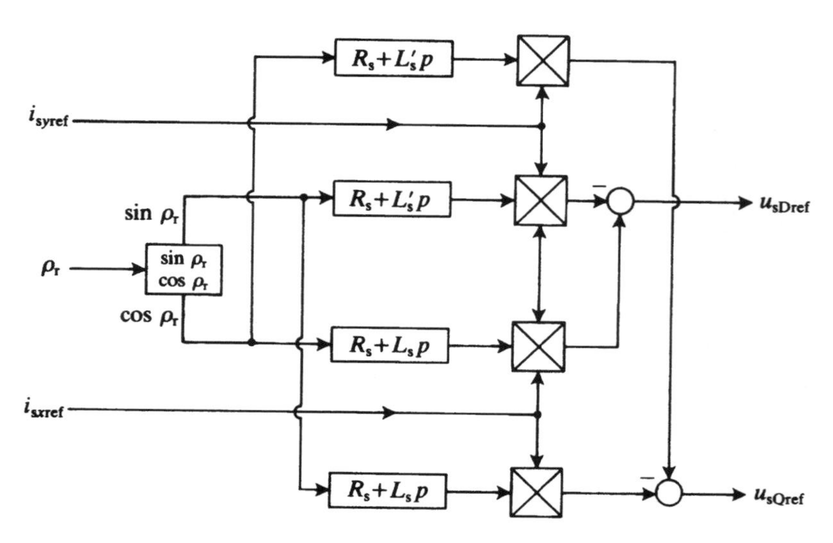

Fig. 4.13. Circuit to obtain reference voltages.

is obtained. In Fig. 4.12 eqn (4.1-59) is utilized to obtain ω_{mr}, where ω_r is the monitored value of the rotor speed. The integration of ω_{mr} gives the angle ρ_r which is required in the transformations described above. The integration is carried out by the block labelled '$1/p$'. It follows that accurate values for the rotor flux position are only obtained if the rotor time constant used is accurate. Under linear magnetic conditions this changes, mainly due to the variation of the rotor resistance. However, it is possible to implement relatively simple compensation of the rotor resistance variation because, according to experiments performed for the drive described in Fig. 4.12, the actual value of i_{sy} is approximately proportional to the inverse of the actual (correct) value of the rotor resistance.

The implementation shown in Fig. 4.12 uses eqns (4.1-44) and (4.1-45), and eqn (4.1-45) has been obtained by neglecting the term $L'_s \mathrm{d}i_{sy}/\mathrm{d}t$, where L'_s is the transient stator inductance. This term can be neglected in a standard cage induction machine, where the transient inductance of the stator is much smaller than the magnetizing inductance, which is large. In this case the transient inductance of the stator is approximately equal to the sum of the stator and rotor leakage inductances. However, in induction machines for servo-applications, the magnetizing inductance is smaller than for normal applications, and in this case the effect of L'_s should be incorporated in the controller. If this is not done and there is a step change in the torque reference, there can be an unwanted overshoot in the electromagnetic torque. A solution to this problem can be obtained in several ways, and two of these are now briefly discussed.

The first solution can be obtained by adding the missing term $L'_s \mathrm{d}i_{sy}/\mathrm{d}t$ to eqn (4.1-45). In this way a quick response will be obtained, but it is necessary to differentiate the torque-producing stator current component. A second solution can be obtained by incorporating a small delay term which corresponds to the stator transient inductance in the decoupling compensator of Fig. 4.12. In this case the actual torque producing stator current (i_{sy}) will be delayed by the stator transient time constant ($T'_s = L'_s/R_s$) and thus, for accurate decoupling, the torque-producing stator current reference (i_{syref}) should also be delayed by the same time constant. This can be achieved by inserting a new block in Fig. 4.12(a) between points A and B, which contains a first-order lag described by $1/(1+T'_s p)$, by connecting point C to point B and disconnecting points C and A as shown in Fig. 4.12(b). With this implementation, improved dynamic response is obtained and the unwanted overshoot in the electromagnetic torque is reduced. Compared to the first method, the second method results in a slightly slower response.

It is important to note that the implementation shown in Fig. 4.12 utilizes the fact that the rotor flux is constant, and thus is not suitable under field-weakening conditions. However, it can be extended to operate under field weakening. For this purpose eqns (4.1-7), (4.1-8), and (4.1-25) must be considered, but under the assumption that $|\bar{i}_{mr}|$ is not constant. In this case when the expression for $|\bar{i}_{mr}|$ obtained from eqn (4.1-25) is substituted into the direct-axis voltage equation defined by eqn (4.1-7), the resulting direct-axis voltage equation will contain three terms. These are the direct-axis voltage across the stator impedance, which is equal to $(R_s+L_s p)i_{sx}$, the direct-axis rotational voltage component $-\omega_{mr}L_s' i_{sy}$, and finally an extra term $-L_m^2 p^2 i_{sx}/(R_r+L_r p)$, where $p=d/dt$. This third component contains the second time derivative of the direct-axis stator current component i_{sx}. Thus to get a direct relationship between u_{sx} and i_{sx}, the rotational component must be eliminated (this was achieved previously when $|\bar{\imath}_{mr}|$ was assumed to be constant) and since the third component disturbs the simplified structure of the x-axis, it should also be cancelled. It should be noted that when $|\bar{\imath}_{mr}|=$constant, it follows by considering eqn (4.1-25) that this term is only present in the equations in the form of $i_{sx}=|\bar{\imath}_{mr}|$. The required cancellation of the unwanted third component can be performed by the application of such a direct-axis stator current (i_{sx}) controller, which ensures that $pi_{sx}=0$, i.e. the rate of change of the direct-axis stator current in the rotor-flux-oriented reference frame is zero. For this purpose it is also possible to use a model reference-frame proportional-output error-feedback controller and the actual control current at the output of this controller will replace the current i_{sxref} in Fig. 4.12, which is related to i_{sx} through the transfer function $1/(1+T_s p)$, where T_s is the stator time constant. It follows from eqn (4.1-25) that $|\bar{\imath}_{mr}|=i_{sx}/(1+T_r p)$ where T_r is the rotor time constant and this current has to replace i_{sxref} in the terms $L_s i_{sxref}$ and $i_{syref}/(T_r i_{sxref})$ shown in Fig. 4.12.

Under field-weakening operation, it is possible to obtain i_{sxref} similarly to the method shown in Fig. 4.11, or it is possible to determine it from an optimal value of the rotor flux reference $|\bar{\psi}_{rref}|=L_m|\bar{\imath}_{mref}|$, which has to be dependent on the rotor speed and which gives maximum efficiency of the induction machine subjected to rotor-flux-oriented control. The speed dependency of the rotor flux reference follows from the fact that two loss components, the hysteresis and eddy-current losses, are functions of the stator frequency, which, however, is the sum of the slip frequency and the rotor speed (see eqn (4.1-26)). However, when deciding which way to obtain i_{sxref}, it should be considered that in the transient state, the main goal is to obtain quick torque response, while efficiency optimization is more important in the steady state.

4.1.2 CONTROL OF AN INDUCTION MACHINE WITH IMPRESSED CURRENTS

4.1.2.1 General introduction

In contrast to inverters which function as voltage sources, inverters which function as current sources are becoming the main power sources for high-performance a.c. machine drives. In the lower-power region PWM inverters with fast current control are employed. In this case, because of the chopping mode of the inverter and the high gain of the current loops, the a.c. machine follows the current reference signals very quickly. At higher power levels the Current-Source Inverter (CSI) is used where a supply-side controlled rectifier is connected through a large d.c. link reactor to the motor-side inverter. Its major advantages are:

- its simplicity;
- its inherent ability for regeneration (by reversing the d.c. link voltage) and reversal (by electronic reversal of the phase sequence of the machine currents, which is carried out by changing the sequence of operation of the inverter switches);
- it does not require silicon-controlled rectifiers with high switching speeds—it can use converter-grade thyristors. (As shown later, this follows because commutation causes voltage spikes superimposed on the nearly sinusoidal stator voltages and these can be reduced by the application of large commutating capacitors or by reducing the transient inductance of the stator. Thus commutation is relatively slow and therefore it is not necessary to use thyristors with high switching speed);
- the inverter will recover from a short-circuit across any two of its output terminals and is undamaged by the misfire of the output thyristors (because there is a large d.c. link reactor which prevents quick changes of the link current).

However, the current response of the conventional CSI with auto-sequential commutation is much slower than that of the current-controlled PWM voltage inverter. Furthermore, at low speed there can exist unwanted torque pulsations (this will be discussed later).

By the application of the conventional CSI, rapid control of both the phase and amplitude of an a.c. current cannot be achieved. In this case, the amplitude of the a.c. currents is determined by the magnitude of the d.c. link current, and as a result of the large filter inductance, the current response to an input command is greatly influenced by the d.c. link parameters (rectifier phase-control delay time, rated value of the rectifier output voltage, gain of the d.c. link current loop, inductance of the d.c. link). Although the response can be minimized by establishing high gain in the current-regulator loop of the controlled rectifier, the output of which is the d.c. link current, the large gain can result in control problems because of the saturation of the output voltage of the controlled rectifier and ripple current instabilities at light load. Thus the current response of the conventional CSI is slower than for the current-controlled PWM inverter. It should be noted that when a so-called notched auto-sequentially commutated inverter is employed, where the d.c. link current is kept constant and thus the influence of the d.c. link parameters on the dynamics of the system response can be neglected, rapid control of both the phase and amplitude of the a.c. current can be achieved.

In this section the rotor-flux-oriented control of induction machines will be discussed for four cases. The induction machine can be supplied by: (i) a voltage-source PWM inverter with fast current control; (ii) a conventional current-source inverter with auto-sequential commutation (CSI); (iii) a cycloconverter with fast current control; or (iv) impressed rotor currents. The main assumptions are those used in Section 4.1.1, but unless stated otherwise, the time lag of the inverter is neglected although it could be considered by using the technique described in Section 3.1.1.

When a high-dynamic-performance induction machine drive with impressed currents is designed, ideally the following requirements should be satisfied:

- Smooth speed response without cogging or torque pulsations at low speed (it will be shown that sometimes it is difficult to satisfy this requirement);
- Smooth speed reversals under any torque condition;
- Capability of four-quadrant operation (this can be achieved by changing the phase sequence of the stator currents and the polarity of the d.c. link voltage);
- Operation of the drive with constant full torque below base speed and above base speed with reduced flux (field-weakening operation).

There are many control schemes known for induction machine drives with impressed currents, but rotor-flux-oriented control has emerged as one of the most frequently used techniques. The application of this technique yields fast dynamic response and most of the requirements mentioned above can also be satisfied. Similarly to that discussed in Section 4.1.1, in order to obtain high dynamic performance, the stator currents of the machine are transformed into flux- and torque-producing current components (i_{sx}, i_{sy}). These are defined in eqn (4.1-3), and Fig. 2.16 shows the relationship between various quantities in the reference frames fixed to the stator and rotor flux-linkage space phasors. Below base speed the modulus of the rotor magnetizing current ($|\bar{i}_{mr}|$), defined by eqn (4.1-1), is maintained at its maximum possible value but is limited by magnetic saturation. Above base speed, $|\bar{i}_{mr}|$ is reduced (field-weakening operation). The electromagnetic torque is controlled by the quadrature-axis stator current i_{sy}, (see, for example, eqn (2.1-197) and Section 2.1.8).

4.1.2.2 Control of the induction machine supplied by a current-controlled PWM inverter

In this section the rotor-flux-oriented control of an induction machine supplied by a voltage-source inverter with fast current control will be described. The inverter can be a transistorized inverter with high switching frequency, such as the one shown in Fig. 3.5(b). Two types of implementation will be given, which use the direct method and the indirect method. As discussed at the beginning of this chapter, when the direct method is used (flux-feedback control), the space angle of the rotor flux-linkage space phasor is obtained by direct measurements (e.g. by using Hall sensors) or by using a so-called flux model. However, when the indirect method is used (feedforward control), the space angle of the rotor flux-linkage space phasor is obtained as the sum of the monitored rotor angle (θ_r) and the computed reference value of the slip angle (θ_{sl}), where the slip angle gives the position of the rotor flux-linkage space phasor relative to the rotor (or more precisely relative to the direct axis of the reference frame fixed to the rotor).

4.1.2.2.1 Implementation using the direct method

Figure 4.14 shows the schematic of the direct implementation of the rotor-flux-oriented control of an induction machine supplied by a current-controlled PWM inverter. This is simpler than the implementation shown in Fig. 4.12 for a voltage-controlled PWM inverter, since now the stator currents are impressed by fast current control loops and thus the scheme does not utilize the stator voltage equations and there is no decoupling circuit in the implementation of Fig. 4.14.

In Fig. 4.14 the same cascade control structure is utilized as in Fig. 4.12. The monitored value of the rotor speed (ω_r) is integrated (the integration is denoted by $1/p$) to give the actual value of the rotor angle (θ_r). This is compared with its reference value (θ_{rref}) and the resulting error serves as the input to the position controller, which is a PI controller. The output of this is the reference value of the rotor speed (ω_{rref}). When this is compared with the monitored value of the rotor speed, the error signal is supplied to the input of the speed controller, also a PI controller, and the output of which is the reference value of the electromagnetic torque (t_{eref}). Comparison of t_{eref} and the actual value of the torque t_e gives an error which serves as input to the torque controller, again a PI controller, and its output is the reference value of the quadrature-axis stator current expressed in the rotor-flux-oriented reference frame (i_{syref}).

The direct-axis stator current reference (i_{sxref}), which is expressed in the rotor-flux-oriented reference frame, is obtained as the output of the flux controller (PI controller) the input of which is the difference between the reference value of the rotor magnetizing current $|\bar{\imath}_{mrref}|$ and the actual value of the rotor magnetizing current $|\bar{\imath}_{mr}|$. The reference current $|\bar{\imath}_{mrref}|$ is obtained as the output of the function generator FG, which allows field weakening to be implemented. Thus the input of FG is the monitored rotor speed and below base speed FG gives a constant value of $|\bar{\imath}_{mrref}|$, while above base speed $|\bar{\imath}_{mrref}|$ is inversely proportional to the rotor speed.

In accordance with eqn (4.1-3), the stator current references i_{sxref} and i_{syref} are first transformed into the two-axis stator current references of the stationary reference frame (i_{sDref}, i_{sQref}) by the application of the transformation $e^{j\rho_r}$ where ρ_r is the angle of the rotor magnetizing-current space phasor with respect to the direct-axis (sD) of the stationary reference frame. For convenience Fig. 4.15 shows the relationship of the space phasor of the stator currents in the stationary and in the special, rotor-flux-oriented reference frame. The space phasor of the rotor magnetizing current is also shown. Figure 4.15 is similar to Fig. 2.16.

Thus by resolving eqn (4.1-3) into its real and imaginary axes components, but using reference values, it follows that

$$i_{sDref} = \cos\rho_r\, i_{sxref} - \sin\rho_r\, i_{syref} \tag{4.1-60}$$

$$i_{sQref} = \sin\rho_r\, i_{sxref} + \cos\rho_r\, i_{syref}. \tag{4.1-61}$$

These two-axis current references are then transformed into their three-phase reference values ($i_{sAref}, i_{sBref}, i_{sCref}$) by the application of the two-phase to

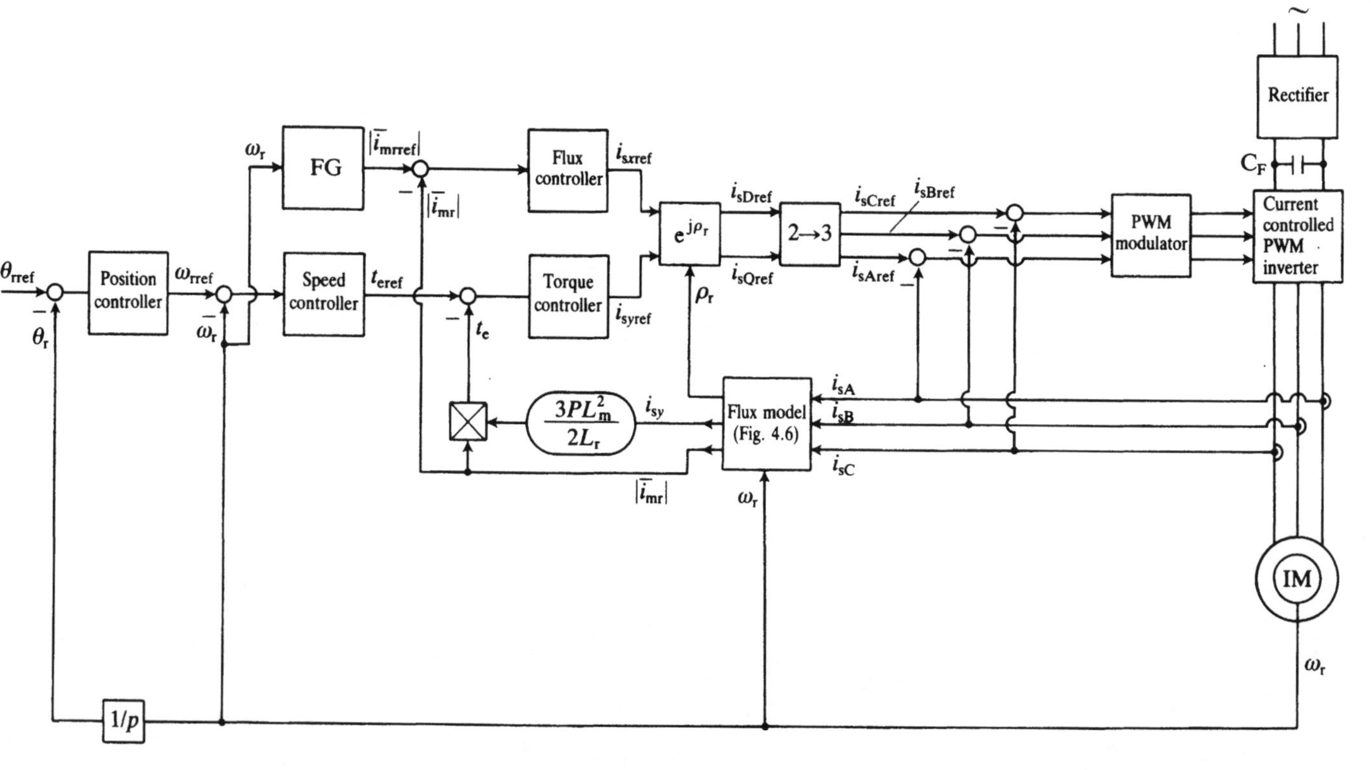

Fig. 4.14. Schematic of the direct implementation of the rotor-flux-oriented control of an induction machine supplied by a current-controlled PWM inverter.

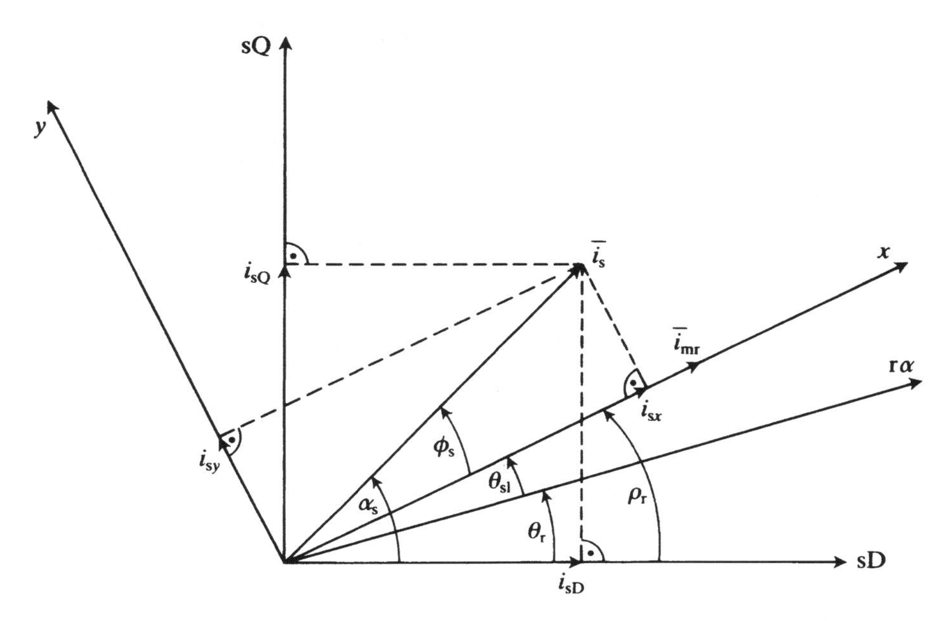

Fig. 4.15. Stator-current and rotor-current space phasors in the stationary and other reference frames.

three-phase transformation indicated by the block labelled '2→3'. These are used, together with the monitored three-phase currents (i_{sA}, i_{sB}, i_{sC}), to obtain the gate signals necessary for the inverter which supplies the induction machine.

As in eqns (4.1-49), (4.1-50), and (4.1-51) in the case of voltages, it is possible to obtain the three-phase reference currents by simply considering that in the absence of zero-sequence currents, the projections of the current space phasor on the corresponding axes yield the instantaneous values of the phase currents; thus

$$i_{sAref} = \mathrm{Re}(\bar{i}_{sref}) = \mathrm{Re}(i_{sDref} + \mathrm{j}i_{sQref}) = i_{sDref}, \tag{4.1-62}$$

where $\bar{i}_{sref}$ is the reference value of the stator-current space phasor in the stationary reference frame,

$$i_{sBref} = \mathrm{Re}(a^2\bar{i}_{sref}) = \frac{-i_{sDref}}{2} + \sqrt{3}\,\frac{i_{sQref}}{2} \tag{4.1-63}$$

and

$$i_{sCref} = \mathrm{Re}(a\bar{i}_{sref}) = \frac{-i_{sDref}}{2} - \sqrt{3}\,\frac{i_{sQref}}{2}. \tag{4.1-64}$$

Of course it is also possible to utilize the Euler forms of the space-phasor equations. Thus the space phasor of the stator current references in the rotor-flux-oriented reference frame can be expressed as

$$\bar{i}'_{sref} = i_{sxref} + \mathrm{j}i_{syref} = (i^2_{sxref} + i^2_{syref})^{1/2}\,\mathrm{e}^{\mathrm{j}\phi_s}, \tag{4.1-65}$$

where

$$\phi_s = \tan^{-1} \frac{i_{syref}}{i_{sxref}} \tag{4.1-66}$$

and ϕ_s is the space angle of the stator-current space phasor with respect to the real axis of the rotor-flux-oriented reference frame, as shown in Fig. 4.15.

From eqn (4.1-3) or Fig. 4.15 in the stationary reference frame the space phasor of the stator reference currents is obtained as

$$\bar{i}_{sref} = \bar{i}'_{sref}\, e^{j\rho_r} = (i^2_{sxref} + i^2_{syref})^{1/2}\, e^{j(\rho_r + \phi_s)}, \tag{4.1-67}$$

where $(\rho_r + \phi_s) = \alpha_s$ and α_s is the space angle of the stator-current space phasor shown in Fig. 4.15. In eqn (4.1-67) $\bar{i}'_{sref}$ has been substituted by the expression given in eqn (4.1-65). Thus from eqns (4.1-62), (4.1-63), (4.1-64), and (4.1-67), the three-phase current references can be put into the form,

$$i_{sAref} = \text{Re}(\bar{i}_{sref}) = (i^2_{sxref} + i^2_{syref})^{1/2}\, e^{j(\rho_r + \phi_s)} \tag{4.1-68}$$

$$i_{sBref} = \text{Re}(a^2 \bar{i}_{sref}) = (i^2_{sxref} + i^2_{syref})^{1/2}\, e^{j(\rho_r + \phi_s - 2\pi/3)} \tag{4.1-69}$$

$$i_{sCref} = \text{Re}(a \bar{i}_{sref}) = (i^2_{sxref} + i^2_{syref})^{1/2}\, e^{j(\rho_r + \phi_s + 2\pi/3)}. \tag{4.1-70}$$

Equations (4.1-68), (4.1-69) and (4.1-70) could be used directly to obtain the three-phase reference currents.

In Fig. 4.14 the monitored stator currents, together with the monitored rotor speed, are inputs to the flux model which has been described in Fig. 4.6. This contains the rotor time constant T_r. The outputs of the flux model are the rotor magnetizing current ($|\bar{i}_{mr}|$), the torque-producing stator current (i_{sy}), and the spatial position of the rotor flux-linkage space phasor (ρ_r). In accordance with eqn (4.1-43), the electromagnetic torque (t_e) is obtained by multiplying i_{sy} by the constant $3PL^2_m/(2L_r)$, where P is the number of pole pairs, L_m is the magnetizing inductance of the machine, and L_r is the self-inductance of the rotor.

The behaviour of the induction machine subjected to rotor-flux-oriented control shown in Fig. 4.14 is similar to that of the separately excited d.c. machine. The stator frequency is equal to the first time derivative of ρ_r. It should be noted that since the flux model uses the rotor time constant, the accuracy of the output signals of the flux model depends on the rotor time constant. Accurate values of T_r can be obtained, for example, by using Model Reference Adaptive Control to establish an on-line rotor time-constant estimator.

4.1.2.2.1 *Implementation using the indirect method*

The implementation of indirect rotor-flux-oriented control of an induction machine supplied by a current-controlled PWM inverter is similar to that of the direct method, except that the space angle of the rotor magnetizing-current space phasor (ρ_r) is obtained as the sum of the rotor angle (θ_r) and the reference value of the slip angle (θ_{sl}). These angles are shown in Fig. 4.15 and the required equations follow from eqn (4.1-59), according to which the speed of the rotor magnetizing current space phasor is

$$\omega_{\mathrm{mr}}=\omega_{\mathrm{r}}+\omega_{\mathrm{slref}} \tag{4.1-71}$$

where ω_{r} is the rotor speed,

$$\omega_{\mathrm{r}}=\frac{\mathrm{d}\theta_{\mathrm{r}}}{\mathrm{d}t}, \tag{4.1-72}$$

and ω_{slref} is the reference value of the slip frequency,

$$\omega_{\mathrm{slref}}=\frac{i_{\mathrm{syref}}}{T_{\mathrm{r}}i_{\mathrm{sxref}}} \tag{4.1-73}$$

where i_{sxref} and i_{syref} are the reference values of the direct- and quadrature-axis stator currents in the rotor-flux-oriented reference frame. Furthermore

$$\omega_{\mathrm{mr}}=\frac{\mathrm{d}\rho_{\mathrm{r}}}{\mathrm{d}t}. \tag{4.1-74}$$

Thus it follows from eqns (4.1-71)–(4.1-74) that

$$\rho_{\mathrm{r}}=\int\omega_{\mathrm{mr}}\,\mathrm{d}t=\int(\omega_{\mathrm{r}}+\omega_{\mathrm{slref}})\,\mathrm{d}t=\theta_{\mathrm{r}}+\theta_{\mathrm{slref}}. \tag{4.1-75}$$

As shown in Fig. 4.15, the slip angle θ_{slref} gives the position of the rotor magnetizing-current space phasor with respect to the direct axis (rα-axis) of the reference frame fixed to the rotor. Thus by considering eqns (4.1-72) and (4.1-73)

$$\rho_{\mathrm{r}}=\theta_{\mathrm{r}}+\int\omega_{\mathrm{slref}}\,\mathrm{d}t=\theta_{\mathrm{r}}+\int\frac{i_{\mathrm{syref}}}{T_{\mathrm{r}}i_{\mathrm{sxref}}}\,\mathrm{d}t, \tag{4.1-76}$$

which serves as the basis for obtaining the angle ρ_{r} when the indirect method is used in the implementation of the rotor-flux-oriented control of the induction machine. According to eqn (4.1-76) the division of the direct- and quadrature-axis stator currents ($i_{\mathrm{sxref}}, i_{\mathrm{syref}}$) is controlled by the slip frequency ω_{sl} and the two reference currents are used to determine the required slip frequency. When the rotor angle and the reference value of the slip frequency angle are added, the position of the rotor magnetizing-current space phasor is obtained. To obtain accurate values of ρ_{r}, the addition of the two angles must be performed very accurately; this can be done digitally, which also has the advantage of avoiding drift problems associated with analog implementation.

The schematic of the drive is shown in Fig. 4.16. In Fig. 4.16(a) the implementation uses Cartesian coordinates and in Fig. 4.16(b) polar coordinates are used. However, for simplicity, in Fig. 4.16(b) only that part of the scheme is shown which corresponds to the part shown in Fig. 4.16(a) within the dashed line.

In Fig. 4.16 an incremental rotor position sensor is used to obtain the rotor angle (θ_{r}) and the rotor speed (ω_{r}). The actual value of the rotor angle (θ_{r}) is compared with its reference value (θ_{rref}) and the resulting error serves as the input

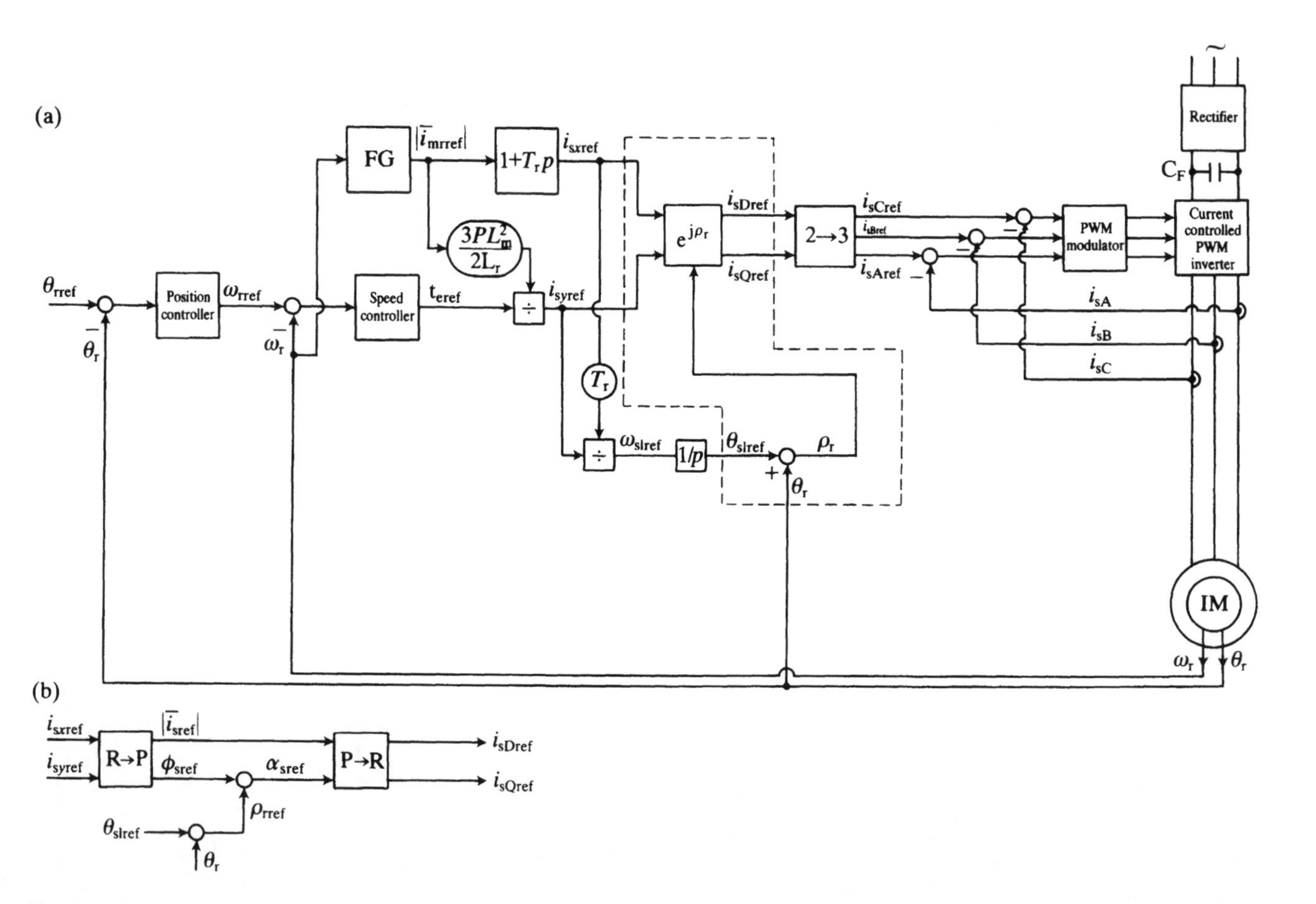

Fig. 4.16. Schematic of the indirect rotor-flux-oriented control of a current-controlled PWM inverter-fed induction machine. (a) In Cartesian coordinates; (b) in polar coordinates.

to the position controller, which is a PI controller. The output of the position controller is the reference value of the rotor speed (ω_{rref}). This is compared with the monitored value of the rotor speed and the error is supplied as the input to the speed controller, which is also a PI controller. Its output is the reference value of the electromagnetic torque (t_{eref}) and in accordance with eqn (4.1-43) this is divided by the constant $3PL_{\mathrm{m}}^2/(2L_{\mathrm{r}})$ to yield the reference value of the stator current expressed in the rotor-flux-oriented reference frame (i_{syref}).

To enable field weakening to be performed, the monitored rotor speed serves as the input to the function generator FG shown in Fig. 4.16, the output of which is the reference value of the rotor magnetizing current ($|\bar{i}_{\mathrm{mrref}}|$). Below base speed FG gives a constant value of $|\bar{i}_{\mathrm{mrref}}|$, while above base speed $|\bar{i}_{\mathrm{mrref}}|$ is inversely proportional to the rotor speed.

It follows from eqn (4.1-25) that, when the actual currents are replaced by their reference values,

$$i_{\mathrm{sxref}} = (1 + T_{\mathrm{r}}p)|\bar{i}_{\mathrm{mrref}}| \tag{4.1-77}$$

and eqn (4.1-77) is used in the implementation shown in Fig. 4.16 to derive the reference value of the direct-axis stator current component expressed in the rotor-flux-oriented reference frame (i_{sxref}).

As described in the previous section, the stator current references $i_{\mathrm{sxref}}, i_{\mathrm{syref}}$ are first transformed into the two-axis stator current references of the stationary reference frame ($i_{\mathrm{sDref}}, i_{\mathrm{sQref}}$) by the application of the transformation $e^{j\rho_{\mathrm{r}}}$, where ρ_{r} is the angle of the rotor magnetizing-current space phasor with respect to the direct-axis (sD) of the stationary reference frame. These two-axis current references are then transformed into their three-phase reference values ($i_{\mathrm{sAref}}, i_{\mathrm{sBref}}, i_{\mathrm{sCref}}$) by the application of the two-phase to three-phase transformation indicated by the block labelled '2→3'. The corresponding equations have been given in the previous section. The reference stator currents together with the monitored three-phase stator currents ($i_{\mathrm{sA}}, i_{\mathrm{sB}}, i_{\mathrm{sC}}$) are used to obtain the gate signals necessary for the inverter which supplies the induction machine.

Finally eqn (4.1-76) is utilized to obtain the space angle ρ_{r} from $i_{\mathrm{sxref}}, i_{\mathrm{syref}}$, and the monitored value of the rotor angle. The block labelled '$1/p$' performs the required integration of the slip frequency.

In Fig. 4.16(b) the reference currents i_{sxref} and i_{syref} are converted by a rectangular-to-polar converter into the modulus of the stator-current space phasor ($|\bar{i}_{\mathrm{sref}}|$) and also into the phase-angle of the stator-current space phasor with respect to the real axis of the rotor magnetizing-current reference frame (ϕ_{sref}), which according to Fig. 4.15 is equal to $\alpha_{\mathrm{sref}} - \rho_{\mathrm{rref}}$. Thus when ρ_{rref} is added to this angle, α_{sref} is obtained, which according to Fig. 4.15 is the phase-angle of the stator-current space phasor with respect to the real axis of the stationary reference frame. Thus $|\bar{i}_{\mathrm{sref}}|$ and α_{sref} are inputs to a polar-to-rectangular converter, the outputs of which are i_{sDref} and i_{sQref}. In accordance with Fig. 4.15, the angle ρ_{rref} is obtained by adding the rotor angle to the slip frequency angle and the latter is obtained by the integration of the angular slip frequency. The slip angle is obtained in the same way as in Fig. 4.16(a).

The indirect implementation given in Fig. 4.16 is similar to the implementation used in a CSI-fed induction machine drive with the conventional slip frequency control. In that case the inverter frequency command is obtained as the sum of the slip frequency command and the rotational frequency of the motor, and the slip frequency command is obtained from the measured d.c. link current by utilizing a function generator. This function generator ensures that below base speed the magnetizing flux of the machine should be constant. In contrast to maintaining constant magnetizing flux, in rotor-flux-oriented control the rotor flux is kept constant. Furthermore, since the traditional slip-frequency control method of the CSI-fed induction machine is based on steady-state equations, during transient operation (i.e. when the machine is accelerated or decelerated) the angle between the stator-current space phasor and the rotor magnetizing-current space phasor is not controlled correctly. In contrast to this, when rotor-flux-oriented control is performed, the correct relationship is achieved between these two space phasors even under transient conditions.

4.1.2.3 Control of the induction machine supplied by a conventional Current-Source Inverter (CSI)

4.1.2.3.1 General introduction

In this section the rotor-flux-oriented control of an induction machine supplied by a conventional CSI with auto-sequential commutation is described. Again two implementations are discussed which correspond to the direct and indirect methods of obtaining the space angle of the rotor flux-linkage space phasor.

Figure 4.17 shows the schematic of a conventional six-step CSI with auto-sequential commutation. In Fig. 4.17, a.c. power is first converted to d.c., which is then inverted by a square-wave six-step inverter to produce variable-frequency a.c. currents. Thus the supply-side mains-commutated thyristor converter provides, through a high-inductance filter, the smooth d.c. link current (i_D). This is supplied to the machine-side inverter. To maintain a constant direct link current specified by the reference value i_{Dref}, there is a current control loop, where i_{Dref} is compared with i_D and the difference serves as input to the current controller (which is a PI controller), whose output is used to control the firing angle of the supply-side converter. The phase-controlled rectifier can be replaced by an uncontrolled (diode) rectifier followed by a d.c. chopper to produce the required variable d.c. voltage source. Furthermore, it should also be noted that when the CSI supplies the stator windings of an induction machine, the d.c. link current is influenced by the back e.m.f. of the induction machine and under transient conditions the application of the PI current controller in the d.c. link current loop cannot give satisfactory current response. However, this problem can be overcome by the application of a function generator which compensates for the motor back e.m.f. This will be discussed in more detail below.

In Fig. 4.17, the machine-side converter contains six force-commutated thyristors. The function of the capacitors is to effect the successful commutation of

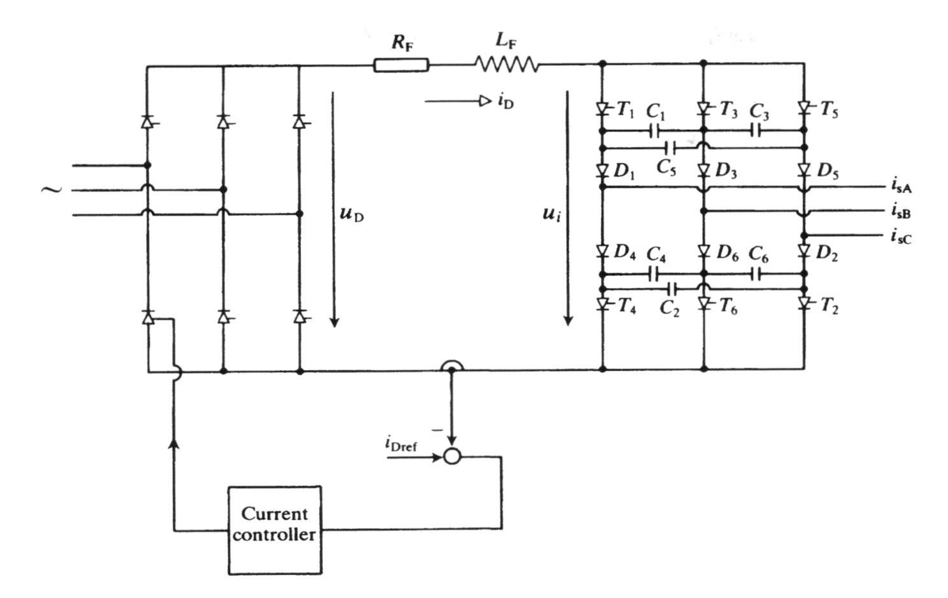

Fig. 4.17. Schematic of conventional six-step CSI with auto-sequential commutation.

current from each thyristor to the next thyristor in the same upper or lower row, i.e. thyristor T_1 is extinguished when T_3 is fired, T_2 is turned-off when T_4 is fired, etc. The six diodes decouple the commutating capacitors (C_1–C_6) from the input and output terminals of the machine-side converter, except during commutation. Thus they allow the capacitors to remain charged to relatively high voltages. During commutation, the capacitors become charged to peak voltages, which are mainly determined by the magnitude of the motor current to be commutated. Thus the inverter can commutate satisfactorily over a wide range of output voltage and frequency.

As a result of the commutation, there are voltage spikes superimposed on the almost sinusoidal stator voltages. The amplitudes of these are proportional to the link current and to the square root of the stator transient inductance (L_s') and are inversely proportional to the square root of the commutating capacitance. They can be reduced by increasing the capacitance of the commutation capacitors. Since the capacitor voltages appear directly on the semiconductor components, this increases their required voltage rating even when the fundamental component of the voltage waveform is relatively small. These voltage spikes are one of the limiting factors of the horsepower rating of CSI-fed induction machine drives. Another limiting factor in extending the horsepower rating of the machine is the rather limited speed range, which is again due to commutation problems. Although generally it is desirable to increase the output frequency of the inverter to the highest possible value, since this can result in a reduction of the size and cost of the machine, and there exist solid-state devices with high switching frequency, the commutation period for the CSI (which is affected by the commutating capacitor and the commutating inductance L_s') is significantly lower than for a voltage-source inverter of equal rating.

The d.c. link current is switched through the inverter thyristors to produce three-phase, six-stepped line currents. Each thyristor conducts for $2\pi/3$ and at any instant one upper and one lower thyristor is conducting. If it is assumed that only two phases conduct at any instant, six distinct modes of operation result. A more detailed description of the operation of the machine-side converter will now be given. For this purpose it is assumed that the diodes and the thyristors are ideal switches, that the d.c. link current is constant and does not contain ripples, and that commutation of the currents is completed in less than one-sixth of a period (T) of a stator current. The operation can be divided into three stages. If initially two thyristors T_1 and T_2 conduct, as shown in Fig. 4.18(a), the stator currents of the machine are $i_{sA}=i_D$, $i_{sB}=0$, $i_{sC}=-i_D$, and thus from the definition of the space phasor of the stator currents $\bar{i}_s$ (eqn (2.1-4)),

$$\bar{i}_s=\frac{2i_D(1-a^2)}{3}=\frac{2i_D\,e^{j\pi/6}}{\sqrt{3}}. \qquad (4.1\text{-}78)$$

Thus the space phasor of the stator currents is in position 1 in Fig. 4.18(e). The first stage begins at the instant when thyristor T_3 is triggered and simultaneously the gating signal is removed from T_1. Thus capacitor C_1 (which was charged at a previous commutation) begins to discharge and turns off T_1. As soon as T_1 turns off, the d.c. link currents flows through capacitor C_1 and the series connected capacitors C_3 and C_5 (Fig. 4.18(b)). The first stage continues until diode D_3 starts to conduct and at this instant the second stage begins (Fig. 4.18(c)). Thus there is three-phase conduction, where none of the stator currents are zero, so the space phasor of the stator currents must move from position 1 in the direction of position 2 shown in Fig. 4.18(e). During the second stage, the current gets diverted from the capacitor bank, diode D_1 and stator phase sA to diode D_3 and stator phase sB, and the second stage ends when the current in the capacitor bank is zero (D_1 is blocked, T_3, D_3 and T_2, D_2 conduct—see Fig. 4.18(d)). Since $i_{sA}=0$, $i_{sB}=i_D$ and $i_{sC}=-i_D$, it follows by considering eqn (2.1-4) that the space phasor of the stator currents is

$$\bar{i}_s=\frac{2a-a^2}{3}=\frac{2i_D j}{\sqrt{3}} \qquad (4.1\text{-}79)$$

and this corresponds to position 2 of the stator-current space phasor in Fig. 4.18(e). The third stage lasts until thyristor T_4 is triggered to turn off T_2 and the d.c. link current flows across T_3, D_3, the machine and T_2, D_2.

It follows that the locus of the stator currents over one period of the stator currents is a hexagon. During one-sixth of a cycle, the space phasor remains in a fixed position and during three-phase conduction it rotates by $2\pi/6$ along the hexagon. The locus is a closed curve (hexagon) and is symmetrical with respect to the origin of the stator reference frame as a result of the periodicity. It also

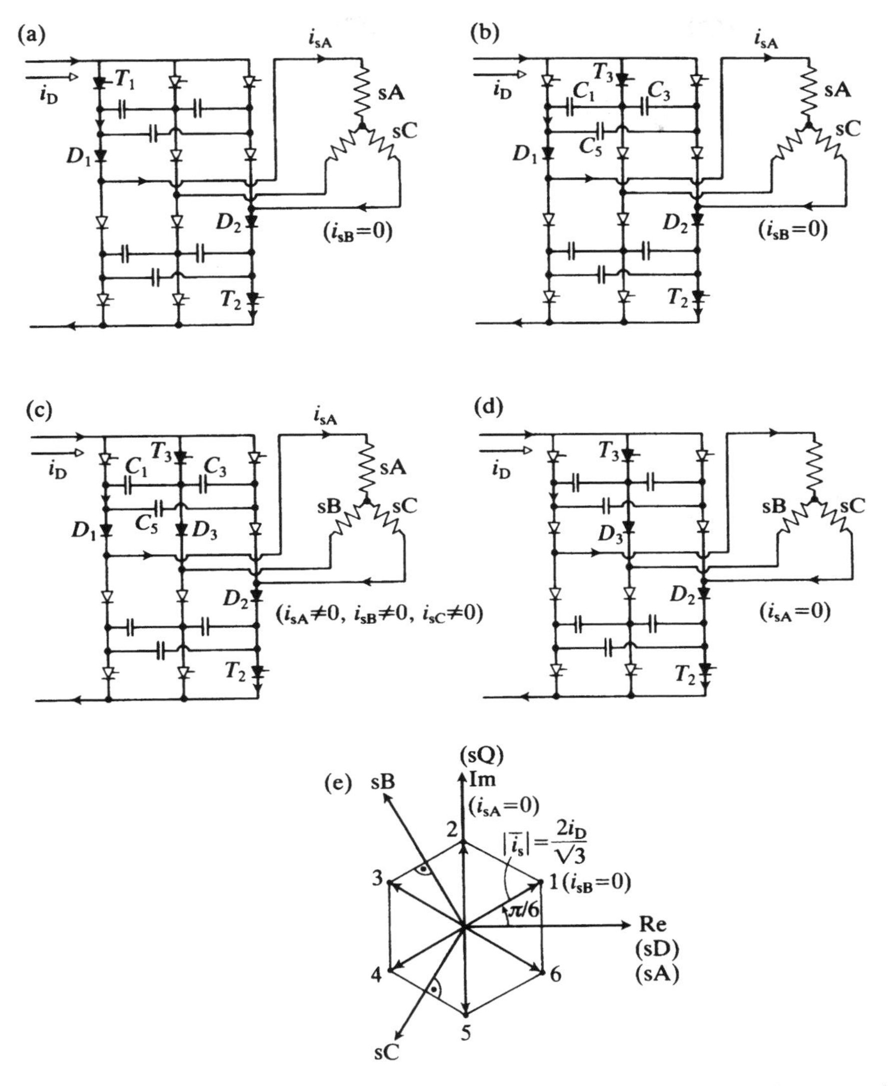

Fig. 4.18. Commutation process. (a) Initial stage of commutation; (b) first stage of commutation; (c) second stage commutation; (d) third stage of commutation; (e) locus of stator-current space phasor.

follows that the magnitude of the three-phase currents is controlled by the regulation of the d.c. link current. The frequency of the three-phase currents is determined by the rate of switching of the inverter thyristors.

When the CSI shown in Fig. 4.17 supplies the stator windings of an induction machine, it should be noted that at low speed there can be undesirable torque pulsations, which are mainly due to the non-sinusoidal stator currents produced by the inverter. The pulsations occur at six times the operating frequency and at very low frequency the frequency of the pulsations is reduced so much that the rotor moves in steps instead of rotating smoothly. This unsmooth rotation makes this drive unsuitable for servo-drive applications which require continuous position control.

In the case of the CSI-fed induction machine, one possibility for torque-ripple control is the pulse-width modulation of the stator currents by multiple commutations of the inverter in each half cycle. When this technique is used, the stator currents are notched at specific points to eliminate certain time harmonics. Each notch added to the current waveforms allows the elimination of a specific time harmonic, i.e. an appropriate double-notched waveform at each quarter cycle can result in the elimination of the fifth and seventh time harmonics in the stator currents and thus in the elimination of the sixth-harmonic torque. Pulse-width modulation can be achieved by the application of an angle loop, where the loop acts as a switching regulator which advances or reverses the sense of rotation of the space phasor of stator currents. A stator-current-angle control loop is utilized in the implementation of a rotor-flux-oriented control of a CSI-fed induction machine described in the next section.

4.1.2.3.2 Implementation using the direct method

In the present section the rotor-flux-oriented control of a CSI-fed induction machine is described, where the angle of the rotor magnetizing-current space phasor is obtained by using the so-called direct method.

As discussed above, in the CSI-fed induction machine the d.c. link current is determined by the supply-side converter. Since the modulus of the stator-current space phasor ($|\bar{\imath}_s|$) must be proportional to the d.c. link current (i_D), this is also determined by the supply-side converter. However, the space angle of the stator-current space phasor, α_s, which is also shown in Fig. 4.15, is determined by the machine-side inverter. Both $|\bar{\imath}_s(t)|$ and $\alpha_s(t)$ are continuous functions. By considering that one of the three stator currents is always zero outside the commutation intervals, the locus of the stator-current space phasor is a six-pointed star, whose radius is proportional to the d.c. link current, and according to eqns (4.1-78) and (4.1-79) is equal to $|\bar{\imath}_s|=(2i_D)/\sqrt{3}$. It has been shown that when currents flow in all the three stator phases, the locus is characterized by a hexagon which connects the end points of the six-pointed star.

Outside the commutation interval the switching state of the inverter is characterized by $e^{j\beta}$, where the angle β changes discontinuously in increments of $\pm\pi/3$. The firing signals for the inverter can be derived by the application of a six-step bidirectional ring counter, which is stepped clockwise and anticlockwise. This method is utilized in Fig. 4.19.

The outer control loops of the drive shown in Fig. 4.19 are similar to those presented in Fig. 4.14. The instantaneous value of the angular rotor speed (ω_r) is obtained together with the instantaneous value of the rotor angle (θ_r). This is compared with the reference value of the rotor angle (θ_{rref}) and the resulting error

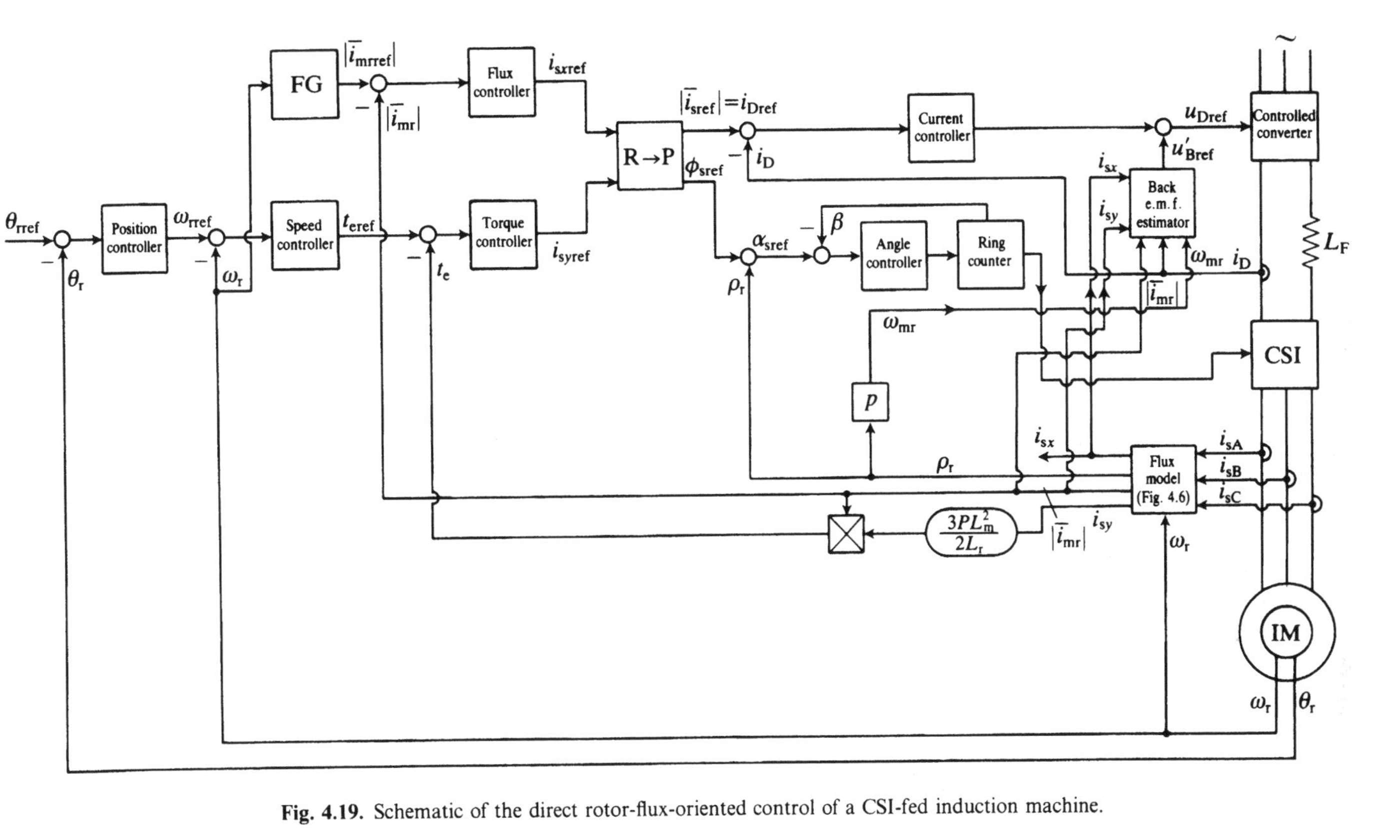

Fig. 4.19. Schematic of the direct rotor-flux-oriented control of a CSI-fed induction machine.

is the input to the position controller, which is a PI controller. The output of this is the reference value of the rotor speed (ω_{rref}). When this is compared with the monitored value of the rotor speed (ω_{r}), the resulting error is supplied to the input of the speed controller, also a PI controller, the output of which is the reference value of the electromagnetic torque (t_{eref}). Comparison of the torque reference with its actual value (t_{e}) gives an error, which serves as input to the torque controller (PI controller) of which the output is the reference value of the quadrature-axis stator current expressed in the rotor-flux-oriented reference frame (i_{syref}).

The direct-axis stator current reference (i_{sxref}), which is also expressed in the rotor-flux-oriented reference frame, is obtained as the output of the flux controller (PI controller), the input of which is the difference between the reference value of the rotor magnetizing current ($|\bar{i}_{\mathrm{mrref}}|$) and its actual value ($|\bar{i}_{\mathrm{mr}}|$). The reference current $|\bar{i}_{\mathrm{mrref}}|$ is obtained as the output of the function generator FG, which allows field weakening to be implemented. Therefore the input of FG is the monitored rotor speed and below base speed FG outputs a constant value of $|\bar{i}_{\mathrm{mrref}}|$, while above base speed, $|\bar{i}_{\mathrm{mrref}}|$ is inversely proportional to the rotor speed. The two-axis stator current references are then inputs to the R–P block which performs the rectangular-to-polar conversion and its outputs are the modulus ($|\bar{i}_{\mathrm{sref}}|$) and the space angle ($\phi_{\mathrm{sref}}$) of the reference value of the stator-current space phasor in the rotor-flux-oriented reference frame. The angle (ϕ_{s}) has also been shown in Fig. 4.15.

In Fig. 4.19 the monitored stator currents, together with the monitored rotor speed, are inputs to the flux model. This has been described in Fig. 4.6. The outputs of the flux model are the rotor magnetizing current ($|\bar{i}_{\mathrm{mr}}|$), the torque-producing stator current (i_{sy}), and the spatial position of the rotor flux-linkage space phasor (ρ_{r}). In accordance with eqn (4.1-43), the electromagnetic torque is obtained by multiplying the torque-producing current by the constant $3PL_{\mathrm{m}}^{2}/(2L_{\mathrm{r}})$, where P is the number of pole pairs, L_{m} is the magnetizing inductance of the machine, and L_{r} is the self-inductance of the rotor.

In Fig. 4.19 the angle ϕ_{sref} is first added to the angle ρ_{r} to yield the space angle of the reference value of the stator-current space phasor with respect to the real-axis of the stator reference frame (α_{sref}). This angle has also been shown in Fig. 4.15. The angle α_{sref} is then compared with the angle β described above, which corresponds to the state of the inverter, and the resulting error serves as the input to the angle controller, which is a PI controller. With the aid of the angle-control loop, a six-step ring counter is switched forward and backward between two adjacent states. Another advantage of using the angle-control loop is that, at low speed, the torque pulsations, which have been discussed in the previous section and are caused by the interaction of the approximately sinusoidal rotor magnetizing currents and the non-sinusoidal stator currents, will be reduced because of the reduced distortions in the stator currents. However, as discussed in the previous section, the CSI-fed induction machine is not suitable for servo-drive applications, where at very low speed controlled torque is required. In contrast to this, in the voltage-source PWM inverter-fed induction machine there is no low-speed cogging effect.

As discussed above, the amplitude of the stator currents is controlled through the d.c. link current (i_D) and conventionally a PI controller is used for this purpose. However, with this type of control the d.c. link current is influenced by the back e.m.f. of the motor. During transient conditions the conventional control cannot give satisfactory current response and, as a consequence, the performance of the drive deteriorates. The influence of the back e.m.f. can be reduced by increasing the gain of the PI current controller, which results in the narrowing of the bandwidth of the d.c. link current loop. However, this will not yield the desired current response since, as will be shown below, the back e.m.f. is a function of several variables whose dynamics are faster or as fast as that of the d.c. link current. Furthermore, the bandwidth of the d.c. link current loop is determined by a large time constant ($T_D = L/R$ where R and L will be defined in eqns (4.1-87) and (4.1-88) respectively) and also by the stability requirements on the current response. A solution to this problem, and thus improved current response, can be obtained if, in addition to the PI current controller, a function generator is incorporated in the control to compensate for the back e.m.f.s of the induction machine. The details of the design of the required back-e.m.f. estimator will now be discussed. For this purpose the voltage equation of the d.c. link, which is valid under transient conditions, has to be obtained, and it has to contain the back e.m.f. of the machine but referred to the inverter input.

First the space phasor of the back e.m.f. of the machine is obtained. It is assumed that there are no time harmonics in the d.c. link current and voltage, and none on the output of the inverter, apart from the fundamental. The effects of magnetic saturation are also neglected. Equation (4.1-6), the space-phasor form of the stator voltage equation in the rotor-flux-oriented reference frame, is

$$\bar{u}_{s\psi r} = (R_s + L_s' p)\bar{i}_{s\psi r} + j\omega_{mr} L_s' \bar{i}_{s\psi r} + (L_s - L_s')(p + j\omega_{mr})\bar{i}_{mr}, \qquad (4.1\text{-}80)$$

where $p = d/dt$, R_s and L_s are the resistance and self-inductance of a stator winding respectively, L_s' is the transient inductance of the stator, $\bar{u}_{s\psi r}$ and $\bar{i}_{s\psi r}$ are the stator voltage and current space phasors in the rotor-flux-oriented reference frame, and $\bar{i}_{mr}$ is the rotor magnetizing-current space phasor in the same reference frame ($\bar{i}_{mr} = |\bar{i}_{mr}|$). ω_{mr} is the speed of the reference frame and is equal to the sum of the rotor speed and the slip speed. The third term on the right-hand side of eqn (4.1-80) gives the space phasor of the back e.m.f.s of the induction machine,

$$\bar{u}_B = (L_s - L_s')(p + j\omega_{mr})|\bar{i}_{mr}|. \qquad (4.1\text{-}81)$$

This will be utilized in the back-e.m.f. estimator, but it has to be referred to the inverter input.

The dynamic voltage equation of the d.c. link will now be obtained. It follows from Fig. 4.17 that the output voltage of the controlled rectifier (u_D) and the input voltage of the inverter (u_i) are related by

$$u_D = (R_F + L_F p)i_D + u_i, \qquad (4.1\text{-}82)$$

where R_F and L_F are the resistance and inductance of the filter (smoothing reactor) and i_D is the d.c. link current. Furthermore, by assuming that there are no losses in the inverter (and thus the power into the inverter is equal to the output power of the inverter), it follows from eqn (2.1-87) that

$$u_i i_D = \tfrac{2}{3}\mathrm{Re}(\bar{u}_{s\psi r}\bar{i}^*_{s\psi r}), \tag{4.1-83}$$

where the asterisk denotes the complex conjugate and Re denotes the real part of a complex quantity.

If the expression for the voltage space phasor defined by eqn (4.1-80) is substituted into eqn (4.1-83), but its third term is replaced by u_B, the following equation is obtained after some rearrangement of the various terms:

$$\frac{\tfrac{2}{3}\mathrm{Re}(\bar{u}_B\bar{i}^*_{s\psi r})}{i_D} = u_i - \frac{2}{3}(R_s + L'_s p)\frac{|\bar{i}_s|^2}{i_D} = u'_B. \tag{4.1-84}$$

The second term after the first equal sign of eqn (4.1-84) contains the ohmic and leakage voltage drops referred to the inverter input and the term on the left-hand-side is the motor back e.m.f. referred to the inverter input (u'_B). When the expression for $\bar{u}_B$ given by eqn (4.1-81) is substituted into the left-hand side of eqn (4.1-84), and substituting $\bar{i}_{s\psi r} = i_{sx} + j i_{sy}$,

$$u'_B = \frac{\tfrac{2}{3}\mathrm{Re}(\bar{u}_B\bar{i}^*_{s\psi r})}{i_D} = \tfrac{2}{3}L_s(1-\sigma)(p|\bar{i}_{mr}|i_{sx} + \omega_{mr}|\bar{i}_{mr}|i_{sy})/i_D, \tag{4.1-85}$$

where σ is the resultant leakage constant, $\sigma = 1 - L_m^2/(L_s L_r)$. It can be seen that u'_B depends on the level of the rotor magnetizing current (rotor flux level), flux- and torque-producing stator currents, the d.c. link current and also on the rotor speed (this is contained in ω_{mr}). Thus by taking u_i from eqn (4.1-82) and substituting into eqn (4.1-84), the following equation is obtained, which describes the dynamics of the d.c. link in terms of u'_B:

$$u_D = (R + Lp)i_D + u'_B, \tag{4.1-86}$$

where u'_B is defined by eqn (4.1-85) and

$$R = R_F + \frac{2}{3}\left(R_s\frac{|\bar{i}_s|^2}{i_D^2}\right) \tag{4.1-87}$$

$$L = L_F + \frac{2}{3}\left(L'_s\frac{|\bar{i}_s|^2}{i_D^2}\right). \tag{4.1-88}$$

It follows from eqn (4.1-88) that u'_B acts as a feedback in the drive. The back e.m.f. can be compensated by the application of a back-e.m.f. estimator, which is also shown in Fig. 4.19, and which is based on eqn (4.1-85).

In general, it follows from eqn (4.1-85), that the inputs to the back-e.m.f. estimator shown in Fig. 4.19 are the d.c. link current (i_D), the flux-producing stator current component (i_{sx}), the torque-producing stator current component (i_{sy}), and the speed of the rotor flux (ω_{mr}). When the back-e.m.f. estimator is used together with the PI current controller in the d.c. link loop, the d.c. link current is no longer disturbed by the back e.m.f. of the machine and its transient behaviour is only determined by the PI current controller and a first-order lag, whose time constant is equal to L/R, where L and R are defined in eqns (4.1-87) and (4.1-88) respectively. When the rotor flux is constant, $|\bar{i}_{mr}|$=constant, eqn (4.1-85) takes a very simple form:

$$u'_B = \frac{2}{3} L_s(1-\sigma) \frac{\omega_{mr}|\bar{i}_{mr}|i_{sy}}{i_D}. \tag{4.1-89}$$

This can be arranged into various forms for various implementations and, for example, by considering eqn (4.1-43) it follows that the product $|\bar{i}_{mr}|i_{sy}$ is proportional to the electromagnetic torque (t_e) and eqn (4.1-89) takes the form

$$u'_B = \frac{t_e\omega_{mr}}{Pi_D}, \tag{4.1-90}$$

where P is the number of pole pairs. In eqn (4.1-90) ω_{mr} can be replaced by using eqn (4.1-26) and thus

$$u'_B = \left[t_e\left(\omega_r + \frac{i_{sy}}{T_r|\bar{i}_{mr}|}\right)\right] \Big/ Pi_D, \tag{4.1-91}$$

where ω_r is the rotor speed and T_r is the rotor time constant. It is also possible to use eqn (4.1-43) and to eliminate i_{sy} in eqn (4.1-91) to yield

$$u'_B = \left[t_e\left(\omega_r + \frac{t_e R_r}{c|\bar{i}_{mr}|^2}\right)\right] \Big/ Pi_D, \tag{4.1-92}$$

where $c = \frac{3}{2}PL_m^2$ and R_r is the rotor resistance (referred to the stator winding). A very simple realization of the u'_B compensator is obtained if t_e is replaced by its reference value.

It should be noted that when the motor back-e.m.f. estimator is incorporated into the drive control, the improvement in the d.c. link current response also results in improved torque and flux responses. Furthermore, back e.m.f.s are also present in the output current control loops of the current-controlled PWM voltage-source inverter-fed induction machine. In this case these e.m.f.s can be obtained by the transformation of u'_B into the actual back e.m.f.s, and current control with e.m.f. compensation can be established, where the compensator can be designed on the basis of the space-phasor equation of the back e.m.f.s.

4.1.2.3.3 Implementation using the indirect method

In this section the rotor-flux-oriented control of a CSI-fed induction machine is described, where the angle of the rotor magnetizing-current space phasor is obtained by using the so-called indirect method. Fig. 4.20 shows the schematic of the drive.

The implementation shown in Fig. 4.20 resembles the implementation shown in Fig. 4.16, which contains the control scheme of the indirect rotor-flux-oriented control of a current-controlled PWM inverter-fed induction machine. The outer control loops shown in Fig. 4.20 are the same as those shown in Fig. 4.16, and

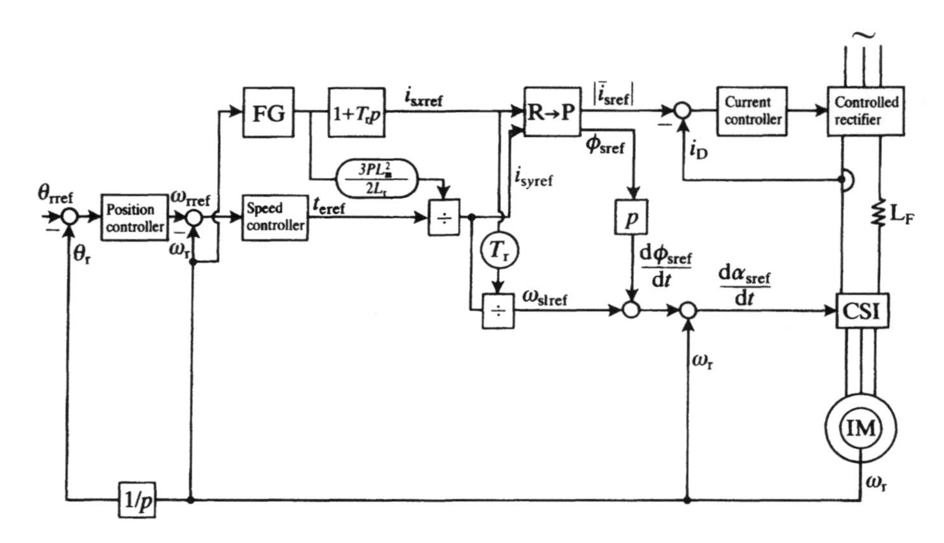

Fig. 4.20. Schematic of the indirect rotor-flux-oriented control of a CSI-fed induction machine.

therefore they are not described here. The reference values of the direct- and quadrature-axis stator currents expressed in the rotor-flux-oriented reference frame (i_{sxref}, i_{syref}) are obtained in the same way as shown in Fig. 4.16. Furthermore the reference value of the slip frequency (ω_{slref}) is obtained identically to that shown in Fig. 4.16, by utilizing i_{sxref} and i_{syref} and the known value of the rotor time constant (T_r), which can also be obtained by on-line identification.

In Fig. 4.20, as in Fig. 4.19, the reference stator currents i_{sxref}, i_{syref} are converted into polar coordinates by a rectangular-to-polar converter, and thus the outputs of the R→P converter are the modulus of the reference value of the stator-current space phasor ($|\bar{i}_{sref}|$) together with its space angle (ϕ_{sref}), which angle is shown in Fig. 4.15 and is the angle of the stator-current space phasor with respect to the real axis of the rotor-flux-oriented reference frame. It follows from Fig. 4.15 that

$$\alpha_{sref} = \theta_r + \theta_{slref} + \phi_{sref}, \tag{4.1-93}$$

where α_s is the space angle of the stator-current space phasor with respect to the real axis of the stationary reference frame, θ_r is the rotor angle, which gives the position of the real axis of the reference frame fixed to the rotor with respect to the real axis of the stationary reference frame, θ_{sl} is the slip angle, which gives the position of the rotor flux-linkage space phasor with respect to the real axis of the reference frame fixed to the rotor, and ϕ_s is the angle of the stator-current space phasor with respect to the real axis of the rotor-flux-oriented reference frame. The subscript 'ref' denotes the reference value. Thus differentiation of eqn (4.1-93) with respect to time yields

$$p\alpha_{sref} = p\theta_r + p\theta_{slref} + p\phi_{sref} = \omega_1 = \omega_r + \omega_{slref} + p\phi_{sref}, \tag{4.1-94}$$

where $p=\mathrm{d}/\mathrm{d}t$, ω_1 is the required stator frequency, ω_r is the rotor speed, and ω_{slref} is the reference value of the slip frequency, which is defined in eqn (4.1-73), repeated here for convenience:

$$\omega_{\mathrm{slref}}=\frac{i_{\mathrm{syref}}}{T_r i_{\mathrm{sxref}}}. \tag{4.1-95}$$

Thus in Fig. 4.20 the stator frequency ω_1 is obtained by considering eqns (4.1-94) and (4.1-95). The supply-side converter is controlled in the same way as in Fig. 4.17.

It should be noted that to obtain lower torque pulsations at low speeds, the technique described in Fig. 4.19 could be utilized. Furthermore, a more accurate value of the slip frequency can be obtained by utilizing the actual values of the flux- and torque-producing stator currents (i_{sx}, i_{sy}) instead of their reference values ($i_{\mathrm{sxref}}, i_{\mathrm{syref}}$), since the presence of the d.c. link inductor makes the stator current slow to respond to a change in the reference currents. During transient operation the slip frequency determined from i_{sxref} and i_{syref} differs from the slip frequency which is determined from the actual values of these currents and the difference is large at the beginning of the transient. As a consequence, an inappropriate value of the slip frequency results, and this affects adversely the torque response of the drive and leads to deteriorated performance. The actual values of the stator current components i_{sx} and i_{sy} can be obtained in several ways, including the utilization of the appropriate transformations of the monitored stator currents i_{sA}, i_{sB}, i_{sC}. In this case, by considering eqns (2.1-112) and (2.1-113), first the direct- and quadrature-axis stator currents in the stationary reference frame are obtained as follows, if it is assumed that there are no zero-sequence currents:

$$i_{sD}=i_{sA} \tag{4.1-96}$$

$$i_{sQ}=\frac{1}{\sqrt{3}}(i_{sB}-i_{sC}). \tag{4.1-97}$$

These are then transformed into the two-axis currents established in the rotor-flux-oriented reference frame by utilizing eqn (2.1-192):

$$i_{sx}=\cos\rho_r i_{sD}+\sin\rho_r i_{sQ} \tag{4.1-98}$$

$$i_{sy}=-\sin\rho_r i_{sD}+\cos\rho_r i_{sQ}. \tag{4.1-99}$$

However, this requires the angle ρ_r, which is the angle of the rotor magnetizing-current space phasor with respect to the real axis of the stationary reference frame. Furthermore, the two transformations required lead to large computation time in the case of a digital implementation or to complicated hardware in the case of analog implementation.

However, a much simpler implementation can be obtained if i_{sx} and i_{sy} are derived from the monitored d.c. link current (i_D) and the angle of the stator-current space phasor with respect to the real axis of the rotor-flux-oriented reference frame (ϕ_s). This angle is also shown in Fig. 4.15. Since the commutation period is short in the CSI inverter, it can be neglected and according to Fig. 4.18(e) the modulus of the stator-current space phasor $|\bar{i}_s|$ is proportional to the d.c. link current (i_D),

$$|\bar{i}_s| = \frac{2i_D}{\sqrt{3}}, \tag{4.1-100}$$

and the amplitude of the fundamental current is equal to $(2\sqrt{3}/\pi)i_D$. Thus, from Fig. 4.15, the space phasor of the stator currents in the rotor-flux-oriented reference frame can be expressed a

$$\bar{i}_{s\psi r} = i_{sx} + ji_{sy} = |\bar{i}_s|\, e^{j\phi_s} \tag{4.1-101}$$

and the following expressions are obtained for the flux- and torque-producing stator current components:

$$i_{sx} = ci_D \cos \phi_s \tag{4.1-102}$$

$$i_{sy} = ci_D \sin \phi_s, \tag{4.1-103}$$

where c is a constant. Thus i_{sx} and i_{sy} can be obtained in a simple manner from i_D and ϕ_s and the required transformation is not dependent on any machine parameter. A possible implementation is shown in Fig. 4.21, where ϕ_s is replaced by ϕ_{sref}, which is the space angle of the reference value of the stator-current space phasor.

In Fig. 4.21 i_{sxref} and i_{syref} are converted by a rectangular-to-polar converter into $|\bar{i}_{sref}|$ and ϕ_{sref}. As shown in Fig. 4.20, the slip frequency ω_{sl} is added to $d\phi_{sref}/dt$ and to this is added the monitored rotor speed to yield $\omega_1 = d\alpha_{sref}/dt$. The monitored d.c. link current is first multiplied by the constant c, and ci_D and ϕ_{sref} are inputs to a polar-to-rectangular converter, the outputs of which are i_{sy} and i_{sx}. By utilizing eqn (4.1-25), $|\bar{i}_{mr}|$ is obtained from i_{sx} in the block which contains $1/(1+T_r p)$, where $p = d/dt$ and T_r is the rotor time constant. The slip frequency

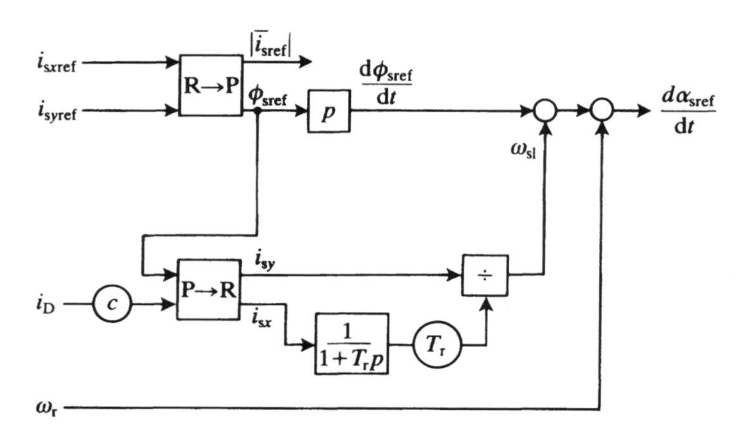

Fig. 4.21. Determination of i_{sx}, i_{sy}, and ω_{sl} by utilizing i_D and ϕ_{sref}.

(ω_{sl}) is obtained by utilizing eqn (4.1-26) and thus $\omega_{sl}=i_{sy}/(T_r|\bar{\imath}_{mr}|)$. The full rotor-flux-oriented control scheme can be obtained by adding the parts of the circuit shown in Fig. 4.20 to Fig. 4.21.

4.1.2.4 Control of the induction machine supplied by a current-controlled cycloconverter

A cycloconverter-fed induction machine drive is very suitable for high-power, low-speed applications. Some applications include rolling steel mills, temper mills, etc. and the characteristics of rolling mills have been described in Section 3.2.1. These characteristics can be satisfied by the vector-controlled cycloconverter-fed induction machine. Owing to the advantages of the induction machine over the d.c. machine (reliability, lower moment of inertia, ruggedness, absence of commutator and brushes, etc.) the cycloconverter-fed induction machine offers additional advantages over d.c. drives, and conventional d.c. drives can be replaced by high-performance vector-controlled cycloconverter-fed induction machines.

As a result of the application of line-commutated bridge converters, the output frequency of the cycloconverter varies between 0 and 25 Hz if the converter is supplied at a frequency of 50 Hz and between 0 and 20 Hz if the supply frequency is 60 Hz. This limits the speed range of the cycloconverter, but in low-speed drives this is of no importance. If necessary, the speed range can be increased by increasing the number of phases on the line-side of the cycloconverter, by employing a transformer with the required number of secondary windings between the line and the cycloconverter. However, it is a disadvantage of this technique that as a result of the increased number of phases, the number of thyristors has to be increased in the converter. The high input reactive power of the cycloconverter drive is also a disadvantage, but there are several well-known techniques to reduce this (asymmetrical voltage control, neutral potential bias control, etc.).

In this section the rotor-flux-oriented control of a current-controlled cycloconverter-fed induction machine is described. As described in Section 3.2.1 for the cycloconverter-fed salient-pole synchronous machine, each stator phase of the induction machine is supplied by two anti-parallel three-phase six-pulse bridge converters. These bridge converters have been shown in Fig. 3.20. The stator currents of the induction machine are separately controlled by fast current loops. Figure 4.22 shows an implementation of the drive.

The control structure is similar to that used for d.c. machines. The difference between the monitored rotor speed (ω_r) and the reference rotor speed (ω_{rref}) serves as the input to the speed controller, which is a PI controller, and this produces the required torque reference (t_{eref}). The function generator FG has the same role as, for example, in Fig. 4.14; this allows field weakening to be achieved and its output is the reference value of the rotor magnetizing current ($|\bar{\imath}_{mrref}|$). The output of the flux controller (which has the same role as in Fig. 4.14) is the reference value of the flux-producing stator current component (i_{sxref}) and its input is $|\bar{\imath}_{mrref}|-|\bar{\imath}_{mr}|$, where the actual value of the rotor magnetizing current ($|\bar{\imath}_{mr}|$) is obtained by utilizing the flux model shown in Fig. 4.6.

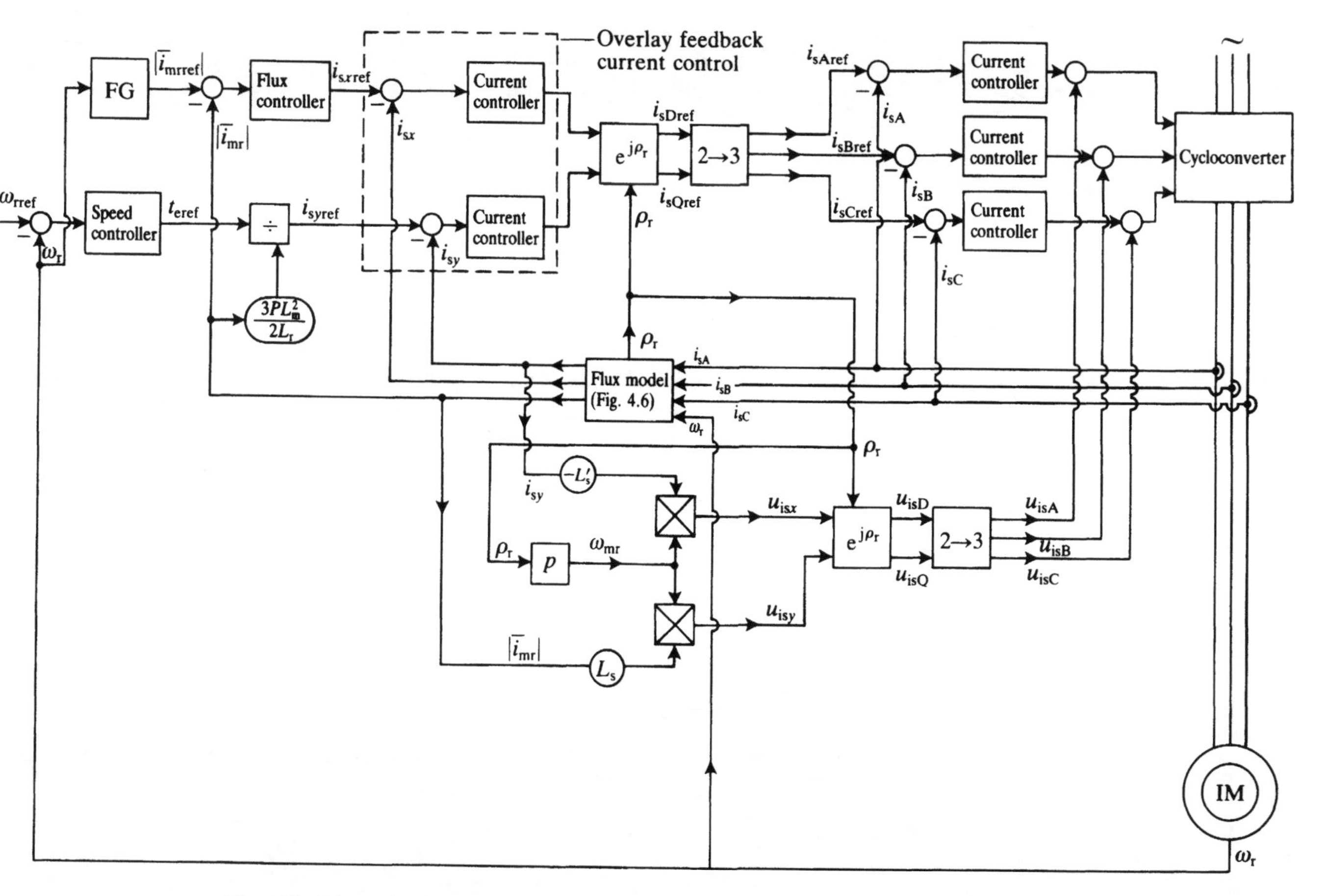

Fig. 4.22. Schematic of the rotor-flux-oriented control of a current-controlled cycloconverter.

The inputs to the flux model are the rotor speed (ω_r) and the stator currents (in the absence of zero-sequence currents only two line currents have to be monitored). The outputs of the flux model are the flux- and torque-producing stator currents (i_{sx}, i_{sy}), which are established in the rotor-flux-oriented reference frame, ρ_r, which is the angle of the rotor magnetizing-current space phasor with respect to the real axis of the stationary reference frame, and $|\bar{i}_{mr}|$, which is the modulus of the rotor magnetizing-current space phasor. These quantities are also shown in Fig. 4.15. In accordance with eqn (2.1-197), the reference value of the torque-producing stator current (i_{syref}) is obtained by dividing the torque reference by the factor $3P|\bar{i}_{mr}|L_m^2/(2L_r)$, where P is the number of pole pairs and L_m and L_r are the magnetizing inductance and rotor self-inductance respectively.

To obtain high accuracy, overlay current control loops are used to control the flux- and torque-producing stator currents. The outputs of the controllers in these loops are inputs to the transformation block $e^{j\rho_r}$, which is required to obtain the reference values of the direct- and quadrature-axis stator currents (i_{sDref}, i_{sQref}) established in the stationary reference frame, as discussed in connection with Fig. 4.14. These currents are then transformed into their three-phase components by the application of the three-phase to two-phase transformation, and are then compared with their actual values. The resulting errors are inputs to the appropriate current controllers which control the cycloconverter.

It has been noted in Section 3.2.2 that at higher output frequencies of the cycloconverter, induced a.c. voltages will disturb the current control system and there will result phase and amplitude errors, which even in the steady-state cannot be eliminated by the current controllers which control the cycloconverter. However, one possible solution to this is to add feedforward voltages to the outputs of these current controllers. Thus in Fig. 4.22 e.m.f. compensation is realized by feedforward of the induced voltages ($u_{isA}, u_{isB}, u_{isC}$). These are obtained from their two-axis values (u_{isD}, u_{isQ}) by the application of the two-phase to three-phase transformation. However, these are related to the induced voltages established in the rotor-flux-oriented reference frame (u_{isx}, u_{isy}) by the transformation $e^{j\rho_r}$. The required induced voltages u_{isx} and u_{isy} can be obtained from eqns (4.1-7) and (4.1-8) and by using the fact that in the steady state $|\bar{i}_{mr}|=$constant and considering eqn (4.1-25), which yields $|\bar{i}_{mr}|=i_{sx}$; so finally

$$u_{isx}=-\omega_{mr}L_s'i_{sy} \tag{4.1-104}$$

$$u_{isy}=\omega_{mr}L_s|\bar{i}_{mr}|. \tag{4.1-105}$$

ω_{mr} is the speed of the rotor-flux-oriented reference frame ($\omega_{mr}=d\rho_r/dt$), and L_s and L_s' are the self-inductance and transient inductance of the stator respectively.

4.1.2.5 Control of the slip-ring induction machine with impressed rotor currents

Rotor-flux-oriented control can also be used for the slip-ring induction machine with impressed rotor currents. For this purpose the converter in the rotor circuit can be a current-controlled cycloconverter (e.g. a six-pulse cycloconverter as described in the previous section), a pulse-width modulated inverter with current control (again described earlier) or, for example, a static converter cascade. In the last case there is dual converter in the rotor and this is shown in Fig. 4.23.

In the implementation shown in Fig. 4.23 a forced commutated thyristor converter is connected to the slip rings of the induction machine and this is connected via a d.c. reactor to a naturally commutated controlled-thyristor bridge converter (special form of static Scherbius cascade). Thus slip power can flow in both directions (to the rotor from the supply and from the rotor to the supply) and the speed of the induction machine can be controlled in the supersynchronous and subsynchronous ranges. Below the synchronous speed in the motoring mode and above the synchronous speed in the regeneration mode the rotor-side converter operates as a rectifier and the line-side converter as an inverter, and slip power is returned to the stator, usually through a transformer. Below the synchronous speed in the regeneration mode and above the synchronous speed in the motoring mode, the rotor-side converter operates as an inverter and the line-side converter operates as a rectifier, and slip power is supplied to the rotor. At the synchronous speed, slip power is taken from the supply to excite the rotor windings, and in this case the machine behaves as a synchronous machine. It should be noted that if the rotor-side converter was a line-commutated bridge converter, near to the synchronous speed, when the a.c. voltage available to commutate the converter is small, the commutating ability of the converter would be lost. It is for this reason that a force-commutated rotor-side converter is used in the special form of the static Scherbius cascade shown in Fig. 4.23, and thus full torque capability can be produced near standstill.

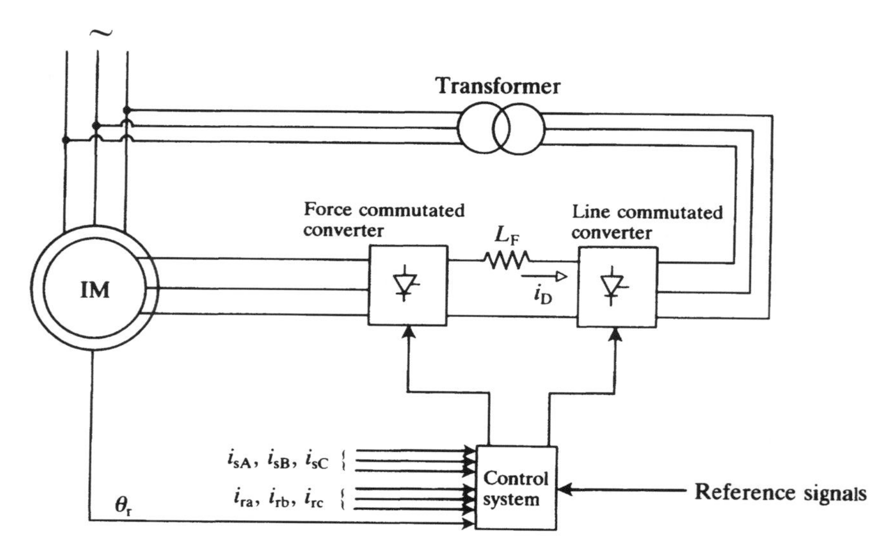

Fig. 4.23. Schematic of the control of a doubly-fed induction machine with converter cascade in the rotor.

Since impressed rotor currents are assumed, the rotor voltage equations can be disregarded from the point of view of the dynamics of the drive under consideration and this results in a simplified dynamic structure of the drive. Thus only the stator voltage equations have to be considered.

The space-phasor form of the stator voltage equation in the rotor-flux-oriented reference frame has been given as eqn (4.1-6) and its two-axis forms have been given as eqn (4.1-7) and eqn (4.1-8). These contain the rotor magnetizing current ($|\bar{i}_{mr}|$), which has been defined in eqn (4.1-1). These equations together with the equation of the electromagnetic torque, eqn (4.1-43), could be used to obtain the implementations of the rotor-flux-oriented control of a slip-ring induction machine with impressed rotor currents. However, in contrast to the implementations so far discussed for induction machines, where the rotor currents were not directly measurable quantities, in the slip-ring machine it is possible to monitor these and this results in simpler implementations than, say, in the case of a squirrel-cage machine. If the stator currents and the rotor currents are measured together with the rotor angle (θ_r) then $|\bar{i}_{mr}|$ and ρ_r, which is the space angle of the rotor magnetizing-current space phasor with respect to the direct-axis of the stationary reference frame (see Fig. 4.15), are known quantities. Mathematically this follows from eqn (4.1-1), which is now repeated here for convenience,

$$\bar{i}_{mr} = \bar{i}_{s\psi r} + (1+\sigma_r)\bar{i}_{r\psi r}, \tag{4.1-106}$$

where $\bar{i}_{s\psi r}$ and $\bar{i}_{r\psi r}$ are the space phasors of the stator currents and rotor currents respectively in the rotor-flux-oriented reference frame. By considering eqn (4.1-3) and eqn (2.1-194) they can be expressed in terms of the space phasor of the stator currents and rotor currents expressed in the stationary reference frame ($\bar{i}_s, \bar{i}_r'$) as

$$\bar{i}_{s\psi r} = \bar{i}_s \, e^{-j\rho_r} \tag{4.1-107}$$

$$\bar{i}_{r\psi r} = \bar{i}_r' \, e^{-j\rho_r}. \tag{4.1-108}$$

Substitution of eqns (4.1-107) and (4.1-108) into eqn (4.1-106) yields

$$\bar{i}_{mr} = \left[\bar{i}_s + \frac{L_r}{L_m}\bar{i}_r'\right] e^{-j\rho_r}. \tag{4.1-109}$$

Thus it follows from eqn (2.1-133) and eqn (4.1-109) that

$$\bar{i}_{mr} = |\bar{i}_{mr}| \, e^{j\rho_r} = \bar{i}_s + \frac{L_r}{L_m}\bar{i}_r' = \bar{i}_s + \frac{L_r}{L_m}\bar{i}_r \, e^{j\theta_r}, \tag{4.1-110}$$

where in terms of their two-axis components,

$$\bar{i}_s = i_{sD} + j i_{sQ} \tag{4.1-111}$$

$$\bar{i}_r = i_{r\alpha} + j i_{r\beta} \tag{4.1-112}$$

$$\bar{i}_r \, e^{j\theta_r} = i_{rd} + j i_{rq} \tag{4.1-113}$$

$$\bar{i}_{mr} = i_{mrD} + j i_{mrQ} \tag{4.1-114}$$

and θ_r is the rotor angle. Equations (4.1-110)–(4.1-113) can be used to obtain $|\bar{i}_{mr}|$ and ρ_r from the measured values of the stator and rotor currents and the rotor angle. Physically eqn (4.1-110) gives the rotor magnetizing-current space phasor in the stationary reference frame and, as expected, this contains two components; these are the stator-current space phasor and a component which would be equal to the rotor-current space phasor if there was no rotor leakage. In this case (when the rotor leakage inductance is zero) $L_r = L_m$ and the rotor magnetizing-current space phasor is identical to the conventional magnetizing-current space phasor, which is equal to the sum of the stator- and rotor-current space phasors. Equation (4.1-110) can also be obtained directly by dividing the expression for the rotor flux-linkage space phasor expressed in the rotor reference frame (see eqn (2.1-134)) by the magnetizing inductance and this technique has been used for the derivation of eqn (4.1-28), which gives the same definition for the rotor magnetizing-current space phasor as eqn (4.1-110).

It should be noted that when $|\bar{i}_{mr}|$ and ρ_r are determined from eqn (4.1-110), there is no dependency on the rotor resistance, in contrast to the derivation based on the flux models described in Figs 4.6 and 4.7. This is an advantage, since the rotor resistance varies as a result of changes in the temperature and the skin effect. By using eqn (4.1-110), Fig. 4.24 shows a possible implementation for obtaining $|\bar{i}_{mr}|$ and ρ_r.

It follows from Fig. 4.24 that the monitored three-phase stator currents (i_{sA}, i_{sB}, i_{sC}) are first transformed into the real- and imaginary-axis components of the stator-current space phasor expressed in the stationary reference frame ($\bar{i}_s$) by

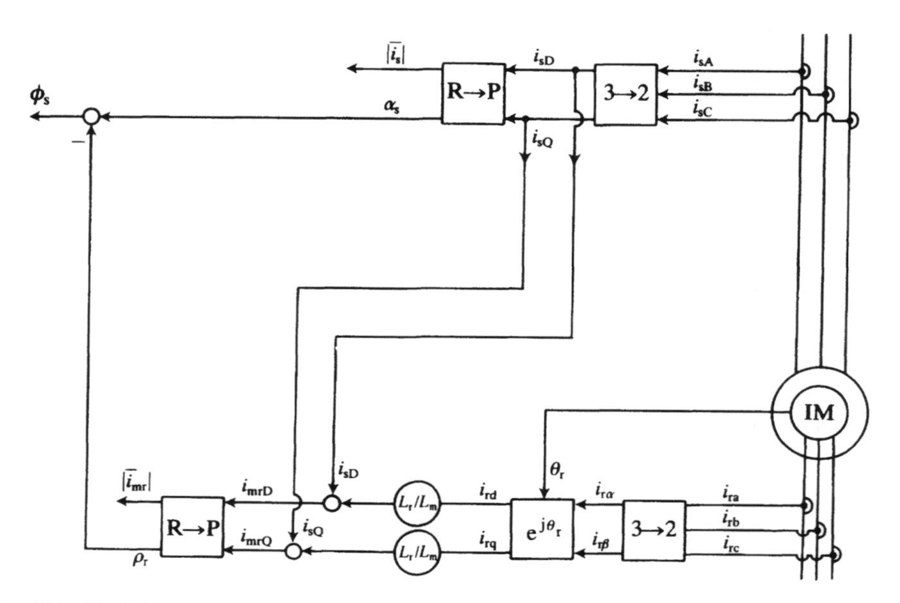

Fig. 4.24. Obtaining $|\bar{i}_{mr}|$, $|\bar{i}_s|$, ρ_r, and ϕ_s by monitoring the stator and rotor currents and the rotor angle.

the application of the three-phase to two-phase transformation. The monitored rotor currents (i_{ra}, i_{rb}, i_{rc}) are similarly transformed into the $i_{r\alpha}, i_{r\beta}$ components (which are the direct- and quadrature-axis rotor currents in the rotor reference frame) by the application of the three-phase to two-phase transformation. These are then transformed into i_{rd} and i_{rq}, which are the two-axis rotor current components in the stator reference frame, by the application of the transformation $e^{j\theta_r}$ (see eqn (4.1-113) and eqn (2.1-40)). Finally by considering eqn (4.1-110) i_{rd} and i_{rq} are multiplied by L_r/L_m and the newly obtained rotor currents are added to i_{sD} and i_{sQ} respectively to yield the two-axis components of the rotor magnetizing-current space phasor in the stationary reference frame (i_{mrD}, i_{mrQ}). These are inputs to a rectangular-to-polar converter the outputs of which are $|\bar{i}_{mr}|$ and ρ_r.

By utilizing the implementation shown in Fig. 4.24, several other quantities can also be obtained which could be required in various implementations of the rotor-flux-oriented control of slip-ring induction machines with impressed rotor currents. Thus when ρ_r is differentiated, ω_{mr}, the speed of the rotor flux-linkage space phasor, is obtained. From the monitored rotor angle (θ_r) and ρ_r, the slip angle θ_{sl} is obtained (this angle is shown in the phasor diagram of Fig. 4.15). It follows from eqn (4.1-107) that the space phasor of the stator currents in the rotor-flux-oriented reference frame is

$$(i_{sx}+ji_{sy})=(i_{sD}+ji_{sQ})e^{-j\rho_r},$$

and resolution into real and imaginary terms gives eqns (4.1-98) and (4.1-99) which could be used to obtain i_{sx} and i_{sy} from i_{sD} and i_{sQ}. The modulus ($|\bar{i}_s|$) and the space angle (α_s) of the stator-current space phasor expressed in the stationary reference frame are obtained in Fig. 4.24 by the application of a rectangular-to-polar converter by using i_{sD} and i_{sQ} as input quantities. The angle α_s is also shown in Fig. 4.15. Finally, in accordance with Fig. 4.15, in Fig. 4.24 the torque angle (ϕ_s), which is the angle of the stator-current space phasor with respect to the real axis of the rotor-flux-oriented reference frame, is obtained as $\phi_s=\alpha_s-\rho_r$.

For the implementation of the rotor-flux-oriented slip-ring machine with impressed rotor currents, when the rotor speed is monitored and when this is compared with its reference value, the error signal can serve as input to a speed controller (PI controller) the output of which is the torque reference.

For the rotor-flux-oriented control of the doubly-fed induction machine with the static converter cascade shown in Fig. 4.23, the torque reference can be used to obtain the reference value of the d.c. link current (i_D) which flows across the reactor connected between the two converters. Thus the line-side converter can be controlled by using the closed loop control of the d.c. link current, where the difference between the reference d.c. link current and the actual d.c. link current is fed into a PI controller. The rotor-side converter can be controlled by using the closed loop control of the torque angle. For this purpose a PI controller can be used, the input of which is the difference between the reference actual values of the torque angle (the latter can be obtained with the help of Fig. 4.24). A more detailed description will be given in Section 4.2.6, where stator-flux-oriented control of the same drive will be discussed.

The vector control of the slip-ring motor with impressed rotor currents using a cycloconverter will be discussed in detail in Section 4.2.6, where stator-flux-oriented control is discussed, since it will be shown that in the stator-flux-oriented system, the real- and imaginary-axis components of the rotor currents (i_{rx}, i_{ry}) correspond directly to the two-axis stator currents expressed in the stator-flux-oriented reference frame. Thus control of the i_{rx} current can be used to maintain the reactive stator current at a fixed value (or at a value dependent on the voltage) and the current i_{ry} can be used for torque control. Beside the speed-independent control of active and reactive powers, it is also an advantage of the application of this control scheme that it leads to the avoidance of the instability problems associated with doubly fed machines.

4.2 Stator-flux-oriented control of induction machines

In this part of the chapter stator-flux-oriented control will be applied to induction machines. The main assumptions are the same as those used in connection with rotor-flux-oriented control of induction machines and, in particular, the effects of magnetic saturation are neglected. Since there is a large similarity between stator-flux-oriented control and rotor-flux-oriented control (see also Section 1.2.1.2), emphasis will only be laid on the differences between these two control techniques.

4.2.1 EXPRESSIONS FOR THE ELECTROMAGNETIC TORQUE

It follows from Sections 1.2.1.2 and 2.1.8 that in the stator-flux-oriented reference frame the electromagnetic torque of an induction machine is proportional to the product of the modulus of the stator magnetizing-current space phasor ($|\bar{i}_{ms}|$) and the torque-producing stator current component (i_{sy}); the torque can be controlled by independent control of these two currents. Thus from eqn (2.1-209),

$$t_e = c|\bar{i}_{ms}|i_{sy}, \qquad (4.2\text{-}1)$$

where $c = 3PL_m/2$ is a constant, P is the number of pole pairs, and L_m is the magnetizing inductance of the machine. In eqn (4.2-1) the stator magnetizing-current space phasor can be put in terms of the stator- and rotor-current space phasors expressed in the stator-flux-oriented reference frame ($\bar{i}_{s\psi s}, \bar{i}_{r\psi s}$) as, if eqn (2.1-208) is considered,

$$|\bar{i}_{ms}| = \bar{i}_{r\psi s} + \frac{\bar{i}_{s\psi s} L_s}{L_m} = (i_{rx} + ji_{ry}) + (i_{sx} + ji_{sy})\frac{L_s}{L_m}, \qquad (4.2\text{-}2)$$

where L_s is the self-inductance of the stator and i_{rx}, i_{ry}, i_{sx}, i_{sy} are the two-axis components of the rotor- and stator-current space phasors respectively, expressed in the stator-flux-oriented reference frame. In eqn (2.1-201) the relation between the space phasors of the stator currents expressed in the stator (stationary) and stator-flux-oriented reference frames has been given, and it follows that

$$\bar{i}_{s\psi s} = i_{sx} + ji_{sy} = (i_{sD} + ji_{sQ})\,e^{-j\rho_s}, \qquad (4.2\text{-}3)$$

where i_{sD} and i_{sQ} are the direct- and quadrature-axis components of the stator currents in the stationary reference frame and the angle ρ_s gives the position of the stator magnetizing current with respect to the real-axis (sD) of the stationary reference frame (this is shown in Fig. 2.17). The rotor-current space phasor in the reference frame fixed to the rotor can be expressed in terms of its two-axis components as $\bar{i}_r = i_{r\alpha} + \mathrm{j}i_{r\beta} = |\bar{i}_r|\,\mathrm{e}^{\mathrm{j}\alpha_r}$, and thus it follows from eqn (2.1-206) that the rotor-current space phasor in the stator-flux-oriented reference frame can be expressed as

$$\bar{i}_{r\psi s} = i_{rx} + \mathrm{j}i_{ry} = (i_{r\alpha} + \mathrm{j}i_{r\beta})\,\mathrm{e}^{\mathrm{j}\theta_r}\,\mathrm{e}^{-\mathrm{j}\rho_s} = |\bar{i}_r|\,\mathrm{e}^{\mathrm{j}(\alpha_r + \theta_r - \rho_s)}. \tag{4.2-4}$$

It is evident that the stator and rotor current components i_{sx}, i_{sy}, i_{rx}, i_{ry} defined above are not equal to the similar current components defined by eqns (2.1-192) and (2.1-194), which are established in the rotor-flux-oriented reference frame. Only for simplicity is the real axis of the stator-flux-oriented reference frame denoted as x, and its imaginary axis as y, but these should not be confused with the real and imaginary axes of the rotor-flux-oriented reference frame. For better understanding, the stator magnetizing-current space phasor and the rotor-current space phasor are shown in Fig. 4.25, where the new reference frame (x,y) is also shown, together with the angles α_r, θ_r, and ρ_s.

It follows from Fig. 4.25 that the angle $\phi_r = (\alpha_r + \theta_r - \rho_s)$, which is the torque angle, is the angle of the rotor-current space phasor with respect to the stator magnetizing-current space phasor.

It should be noted that in the steady state, the definition of $\bar{i}_{ms}$ (expressed in the stationary reference frame) can also be obtained by considering Fig. 4.10(b) if the general turns ratio is selected as $a = L_s/L_m$.

A new expression will now be obtained for the electromagnetic torque which, instead of the quadrature-axis stator current (i_{sy}), contains the modulus of the rotor-current space phasor and the torque angle. This will be useful for the implementation of stator-flux-oriented control of doubly fed induction machines. It follows from eqn (4.2-2) that the torque-producing stator current can be expressed as

$$i_{sy} = -i_{ry}\frac{L_m}{L_s} \tag{4.2-5}$$

and i_{ry} can be obtained from eqn (4.2-4) as $i_{ry} = |\bar{i}_r|\sin(\alpha_r + \theta_r - \rho_s)$. By substitution of this expression into eqn (4.2-5) and considering eqn (4.2-1), the following expression is obtained for the electromagnetic torque:

$$t_e = -3|\bar{i}_{ms}||\bar{i}_r|\sin(\alpha_r + \theta_r - \rho_s)\frac{L_m^2}{2L_s}. \tag{4.2-6}$$

This expression will be used in Section 4.2.6, where stator-flux-oriented control of a doubly fed three-phase induction machine with impressed rotor currents will be discussed.

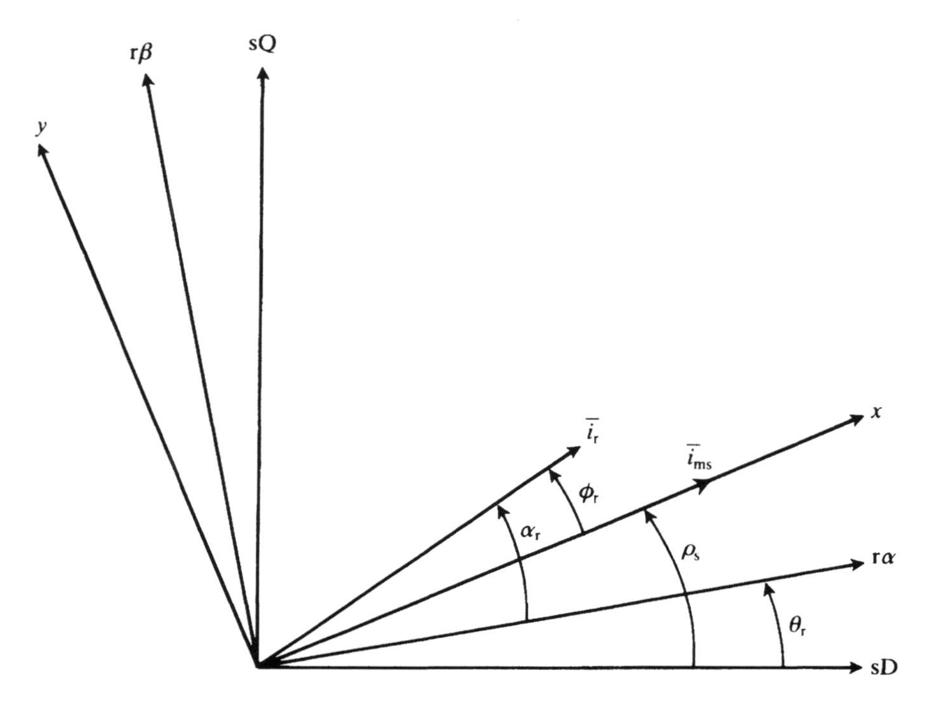

Fig. 4.25. Stator magnetizing-current space phasor and rotor-current space phasor.

4.2.2 STATOR VOLTAGE EQUATIONS FOR THE SQUIRREL-CAGE MACHINE

In this section, the space-phasor form of the stator voltage equation is first obtained in the reference frame fixed to the stator magnetizing flux-linkage space phasor. This will then be resolved into its two-axis components. For this purpose the voltage equations formulated in the general reference frame are directly utilized. These have been given in Section 2.1.7. The stator-current space phasor will be expressed in terms of the stator magnetizing-current space phasor, which has been defined in the previous section. Thus the resulting stator voltage equation will contain the modulus and space angle (or speed) of the stator magnetizing-current space phasor. These quantities are required in the implementation of stator-flux-oriented control. Two alternatives will then be considered, which correspond to a squirrel-cage machine and to a slip-ring induction machine where there is direct access to the rotor currents.

By considering eqns (2.1-148) and (2.1-150), which define the stator voltage and flux-linkage space-phasor equations in the general reference frame respectively, and by neglecting the effects of magnetic saturation, the following stator voltage equation is obtained in the special reference frame fixed to the stator-linkage space phasor, which rotates at a speed $\omega_{ms}=d\rho_s/dt$ (the angle ρ_s has been defined in the previous section):

$$\bar{u}_{s\psi s}=R_s\bar{i}_{s\psi s}+L_s\frac{d\bar{i}_{s\psi s}}{dt}+L_m\frac{d\bar{i}_{r\psi s}}{dt}+j\omega_{ms}L_s\bar{i}_{s\psi s}+j\omega_{ms}L_m\bar{i}_{r\psi s}, \tag{4.2-7}$$

where as in eqn (4.2-3), which defines the stator-current space phasor in the stator-flux-oriented reference frame, the stator-voltage space phasor in the same reference frame can be expressed as

$$\bar{u}_{s\psi s}=u_{sx}+ju_{sy}=(u_{sD}+ju_{sQ})e^{-j\rho_s}=\bar{u}_s e^{-j\rho_s}. \tag{4.2-8}$$

In eqn (4.2-7) $\bar{i}_{r\psi s}$ is the space phasor of the rotor currents in the stator-flux-oriented reference frame and has been defined in eqn (4.2-4).

It is assumed that the squirrel-cage induction machine is supplied by a voltage-source inverter, as described in Section 4.1.1, and the stator currents cannot be impressed by fast current control loops. Thus to obtain the required reference voltages of the inverter, it is necessary to use the stator voltage equations.

It follows from eqn (4.2-7) that if a stator-flux-oriented control scheme for a squirrel-cage induction machine has to be obtained, the rotor currents have to be eliminated from the equations since these cannot be measured directly. However, it follows from eqn (4.2-2) that

$$\bar{i}_{r\psi s}=|\bar{i}_{ms}|-\frac{L_s\bar{i}_{s\psi s}}{L_m} \tag{4.2-9}$$

and substitution of eqn (4.2-9) into eqn (4.2-7) yields

$$\bar{u}_{s\psi s}=R_s\bar{i}_{s\psi s}+L_m\frac{d|\bar{i}_{ms}|}{dt}+j\omega_{ms}L_m|\bar{i}_{ms}|. \tag{4.2-10}$$

This is much simpler than the corresponding voltage equation expressed in the rotor-flux-oriented reference frame (eqn (4.1-6)). It should be noted that eqn (4.2-10) can also be obtained directly from eqn (2.1-148) by considering that $\omega_{mg}=\omega_{ms}$ and $\bar{\psi}_{ms}=L_m\bar{i}_{ms}$. As a direct consequence of the selection of the stator-flux-oriented reference frame, there is no separate stator-leakage voltage component present in eqn (4.2-10). Furthermore, it does not contain the derivative of the stator-current space phasor, which is present when the rotor-flux-oriented reference frame is used. If the modulus of the stator magnetizing-current space phasor is constant, then eqn (4.2-10) takes an extremely simple form, which does not contain the derivative $d|\bar{i}_{ms}|/dt$. When eqn (4.2-10) is resolved into its two-axis components,

$$i_{sx}=\frac{u_{sx}}{R_s}-\frac{L_m}{R_s}\frac{d|\bar{i}_{ms}|}{dt} \tag{4.2-11}$$

$$i_{sy}=\frac{u_{sy}}{R_s}-\omega_{ms}L_m\frac{|\bar{i}_{ms}|}{R_s}. \tag{4.2-12}$$

These equations are much simpler than eqns (4.1-7) and (4.1-8). For stator-flux-oriented control, i_{sx} and i_{sy} have to be independently controlled. However, it follows from eqn (4.2-12) that there is an unwanted coupling term and thus u_{sx} and u_{sy} cannot be considered as decoupled variables for the stator flux and electromagnetic torque. The stator current components i_{sx} and i_{sy} can only be independently controlled if eqns (4.2-11) and (4.2-12) are decoupled and the stator currents are indirectly controlled by controlling the terminal voltages of the induction machine. As in Section 4.1.1, it follows from eqns (4.2-11) and (4.2-12) that the required decoupling voltages are

$$u_{dx}=0$$

$$u_{dy}=\omega_{ms}L_m|\bar{i}_{ms}|. \tag{4.2-13}$$

Thus a very simple decoupling circuit can be established, the inputs of which are ω_{ms} and $|\bar{i}_{ms}|$ and the output of which is u_{dy}. It follows from the simplicity of the stator voltage equations that when an induction machine is supplied by impressed stator voltages, the stator-flux-oriented control scheme will be simpler than the rotor-flux-oriented control scheme. When a squirrel-cage machine is supplied by a converter which functions as a current source, the stator equations can be omitted from the equations of the drive.

4.2.3 ROTOR VOLTAGE EQUATIONS FOR THE SQUIRREL-CAGE INDUCTION MACHINE

The rotor voltage equations expressed in the stator-flux-oriented reference frame can be used to obtain the modulus and speed (or phase angle) of the stator flux-linkage space phasor (or stator magnetizing-current space phasor). The required flux model can be obtained by using the rotor voltage equations formulated in the stator-flux-oriented reference frame.

For this purpose, eqn (2.1-153) is used, which gives the space-phasor form of the rotor voltage equation in the general reference frame, but it is assumed that all the inductance parameters are constant. Thus in the stator-flux-oriented reference frame ($\omega_g=\omega_{ms}$, $\bar{u}_{rg}=\bar{u}_{r\psi s}$, $\bar{i}_{sg}=\bar{i}_{s\psi s}$, $\bar{i}_{rg}=\bar{i}_{r\psi s}$), this takes the form,

$$\bar{u}_{r\psi s}=R_r\bar{i}_{r\psi s}+L_r\frac{d\bar{i}_{r\psi s}}{dt}+L_m\frac{d\bar{i}_{s\psi s}}{dt}+j(\omega_{ms}-\omega_r)(L_r\bar{i}_{r\psi s}+L_m\bar{i}_{s\psi s}), \tag{4.2-14}$$

where ω_r is the rotor speed and R_r and L_r are the rotor resistance and rotor self-inductance. Equation (4.2-14) contains the rotor-current space phasor, but since for a squirrel-cage machine this cannot be directly obtained, it is now eliminated. By the substitution of eqn (4.2-9) into eqn (4.2-14) and by considering that, since the rotor windings can be considered as short-circuited windings, $\bar{u}_{r\psi s}=0$,

$$0=R_r\left[|\bar{i}_{ms}|-\frac{L_s}{L_m}\bar{i}_{s\psi s}\right]+L_r\frac{d|\bar{i}_{ms}|}{dt}-\left(\frac{L_s'L_r}{L_m}\right)\frac{d\bar{i}_{s\psi s}}{dt}+j\omega_{sl}\left[L_r|\bar{i}_{ms}|-\left(\frac{L_s'L_r}{L_m}\right)\bar{i}_{s\psi s}\right] \tag{4.2-15}$$

is obtained, where

$$\omega_{sl}=\omega_{ms}-\omega_r \tag{4.2-16}$$

is the angular slip frequency. This is equal to the first time derivative of the slip angle θ_{sl}, which is equal to $\rho_s-\theta_r$, as shown in Fig. 4.25, and is the angle of the stator flux-linkage space phasor (or stator magnetizing-current space phasor) with respect to the real axis (rα) of the rotor reference frame. In eqn (4.2-15) L'_s is the transient stator inductance. It should be noted that eqn (4.2-15) is more complicated than the rotor voltage equation formulated in the rotor-flux-oriented reference frame (eqn (4.1-24)). This has the consequence that when the induction machine is supplied by impressed stator currents, the implementation of stator-flux-oriented control will be more complicated than the implementation of the rotor-flux-oriented control. Resolution of eqn (4.2-15) into its real- and imaginary-axis components yields

$$\frac{L_m}{L'_s}\frac{d|\bar{i}_{ms}|}{dt}+\frac{L_m}{L_sT'_r}|\bar{i}_{ms}|=\frac{di_{sx}}{dt}+\frac{i_{sx}}{T'_r}-\omega_{sl}i_{sy} \tag{4.2-17}$$

$$\omega_{sl}\left(\frac{L_m|\bar{i}_{ms}|}{L'_s}-i_{sx}\right)=\frac{di_{sy}}{dt}+\frac{i_{sy}}{T'_r}, \tag{4.2-18}$$

where T'_r is the transient rotor time constant.

For an induction machine with impressed stator currents eqns (4.2-17) and (4.2-18) have to be considered when stator-flux-oriented control is used. However, it follows from eqn (4.2-17) that there exists a coupling between the torque-producing stator current component i_{sy} and the stator magnetizing current. Therefore if there is a change in the torque-producing current, and if this is not followed by an appropriate change in i_{sx}, there will exist an unwanted transient in the stator magnetizing current. However, this undesirable coupling can be eliminated by the utilization of a decoupling circuit, as will now be described.

If the stator magnetizing current is regulated by a flux controller (as shown in Fig. 4.26), the input of which is the difference between the reference value ($|\bar{i}_{msref}|$) and the actual value ($|\bar{i}_{ms}|$) of the stator magnetizing current, and it is assumed that the output of this controller (which can be a PI controller) is the current $\hat{i}_{sx}$, then the required decoupling current i_{dx} has to be added to $\hat{i}_{sx}$ to yield the reference value of the stator current along the real axis of the stator-flux-oriented reference frame (i_{sxref}). It follows that

$$i_{sx}=\hat{i}_{sx}+i_{dx} \tag{4.2-19}$$

and the substitution of eqn (4.2-19) into eqn (4.2-17) yields

$$\frac{L_m}{L'_s}\frac{d|\bar{i}_{ms}|}{dt}+\frac{L_m}{L_sT'_r}|\bar{i}_{ms}|=\frac{d\hat{i}_{sx}}{dt}+\frac{\hat{i}_{sx}}{T'_r}+\frac{di_{dx}}{dt}+\frac{i_{dx}}{T'_r}-\omega_{sl}i_{sy}. \tag{4.2-20}$$

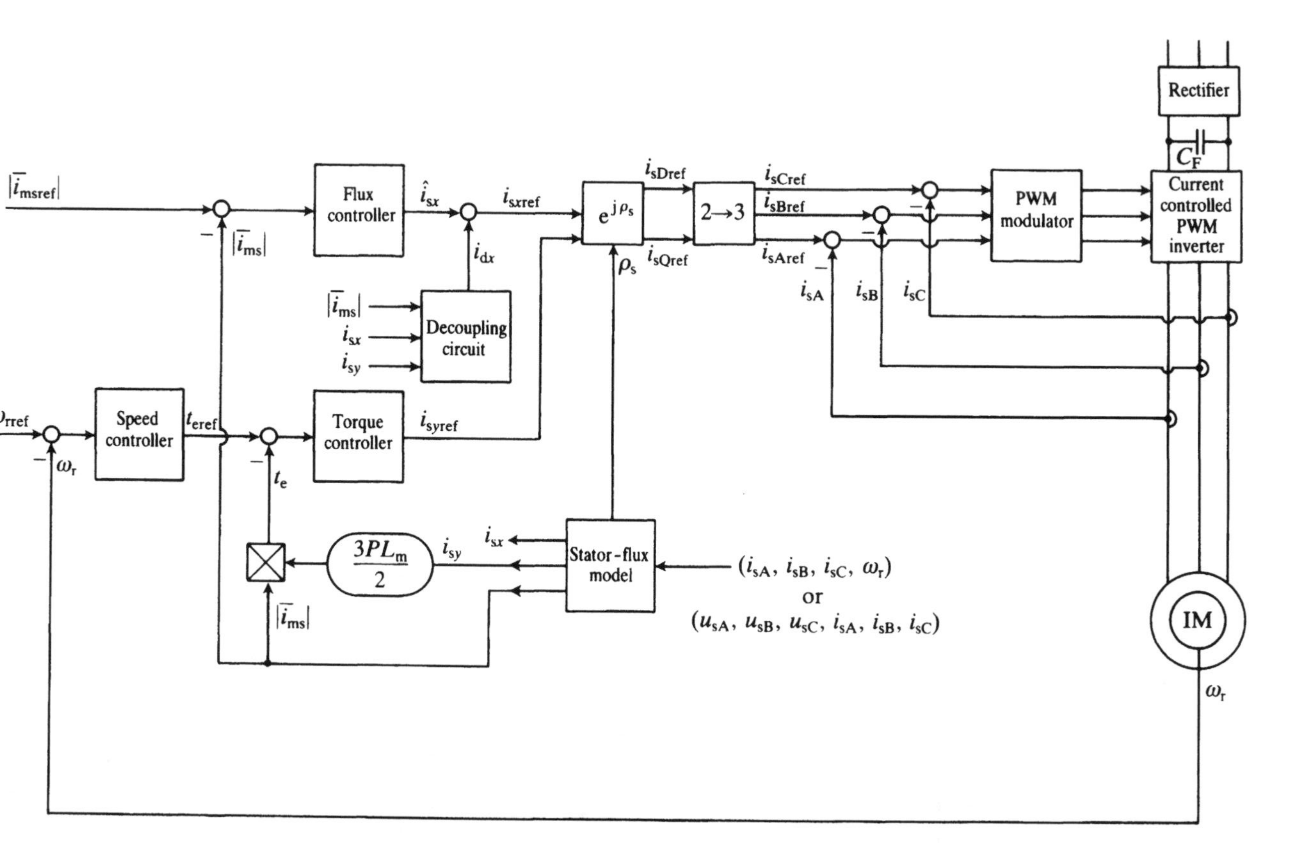

Fig. 4.26. Schematic of the stator-flux-oriented control of an induction machine with a current-controlled PWM inverter.

It follows from eqn (4.2-20) that $|\bar{i}_{ms}|$ can be decoupled from i_{sy} if the three last terms on the right-hand side of eqn (4.2-20) are zero, i.e.

$$\left(p+\frac{1}{T_r'}\right)i_{dx}=\omega_{sl}i_{sy}, \tag{4.2-21}$$

where $p=d/dt$. Thus it follows from eqn (4.2-21) that the required current i_{dx} can be expressed as

$$i_{dx}=\omega_{sl}\frac{T_r'}{1+T_r'p}i_{sy}. \tag{4.2-22}$$

In eqn (4.2-22) the angular slip frequency ω_{sl} is present. This can be obtained by considering eqn (4.2-18). It follows from eqn (4.2-18) that

$$\omega_{sl}=\frac{(1+T_r'p)L_s i_{sy}}{T_r(L_m|\bar{i}_{ms}|-L_s' i_{sx})}. \tag{4.2-23}$$

Equation (4.2-22), together with eqn (4.2-23), defines the required decoupling circuit. Thus when the correct machine parameters are used in these equations, the correct value of i_{sx} is obtained for a given value of i_{sy}, and thus $|\bar{i}_{ms}|$ is not altered by any changes in the torque reference even during transient operation. By using the decoupling circuit defined by eqns (4.2-22) and (4.2-23), Fig. 4.26 shows as an example the schematic of the stator-flux-oriented control of an induction machine supplied by a current-regulated PWM inverter (which has also been used in the implementation shown in Fig. 4.14).

Since this scheme is similar to the one shown in Fig. 4.14 only the differences will be pointed out. The direct-axis stator reference current is obtained as the sum of $\hat{i}_{sx}$ (which is the output of the flux controller) and i_{dx}, which is the output of the decoupling circuit described above. A flux model is used to obtain the modulus ($|\bar{i}_{ms}|$) and the space angle (ρ_s) of the stator magnetizing-current space phasor. The torque-producing stator current component (i_{sy}) is obtained from the two-axis stator currents of the stationary reference frame (i_{sD}, i_{sQ}) by considering eqn (4.2-3). The flux model can use at its inputs the actual stator currents (i_{sA}, i_{sB}, i_{sC}) and the monitored rotor speed (ω_r) and thus by considering eqns (4.2-16), (4.2-17), and (4.2-18) its outputs are i_{sy}, $|\bar{i}_{ms}|$, and ω_{ms}, and ρ_s is obtained by the integration of ω_{ms}. There are several other alternatives and if, for example, the stator voltages and currents are monitored, it follows from eqn (2.1-124) that the space phasor of the stator flux linkages ($\bar{\psi}_s$) can be obtained by the integration of $\bar{u}_s-R_s\bar{i}_s$, where $\bar{u}_s$ and $\bar{i}_s$ are the space phasors of the stator voltages and currents respectively in the stationary reference frame (in terms of their two-axis components $\bar{u}_s=\bar{u}_{sD}+ju_{sQ}$, and $\bar{i}_s=i_{sD}+ji_{sQ}$). When $\bar{\psi}_s$ is divided by L_m, the space phasor of the stator magnetizing current in the stationary reference frame is obtained, and its modulus is equal to $|\bar{i}_{ms}|$ and its phase angle is equal to ρ_s. However, when this technique is used, as pointed out previously, at lower stator frequencies the stator resistance will dominate and accurate ohmic voltage-drop compensation must be performed. Figure 4.27 shows this type of solution, but it should be noted that as shown in Section 3.1.3.1, in practice only two stator line voltages are required (e.g. u_{AC} and u_{BA}), and thus the direct- and quadrature-axis stator voltages can be obtained by using $u_{sD}=(1/3)(u_{BA}-u_{AC})$ and $u_{sQ}=-(1/\sqrt{3})(u_{AC}+u_{BA})$.

It has also been shown in Section 3.1.3.1 that the direct- and quadrature-axis stator currents i_{sD} and i_{sQ} can be obtained by using only two stator line currents (e.g i_{sA} and i_{sB}) if $i_{sA}+i_{sB}+i_{sC}=0$ holds, and in this case $i_{sD}=i_{sA}$ and $i_{sQ}=(1/\sqrt{3})(i_{sA}+2i_{sB})$ are obtained. Further details on other aspects of this scheme, including drift compensation, can be found in Sections 3.1.3.1 and 4.1.1.4, where improved schemes are also presented and where the reconstruction of the stator voltages from the monitored d.c. link voltage and inverter switching states is also discussed. For example, it should be noted that, as discussed in Section 4.1.1.4 in connection to Fig. 4.9(b), it is also possible to utilize directly the stator voltage equations in the stator-flux-oriented reference frame (eqns (4.2-11) and (4.2-12)). It follows from eqn (4.2-11) that the rate of change of the stator flux modulus is $u_{sx}-R_s i_{sx}$ and from eqn (4.2-12) the speed of the stator flux-linkage space vector (ω_{ms}) is obtained as $\omega_{ms}=(u_{sy}-R_s i_{sy})/|\bar{\psi}_s|$. These equations are also used in Fig. 4.28.

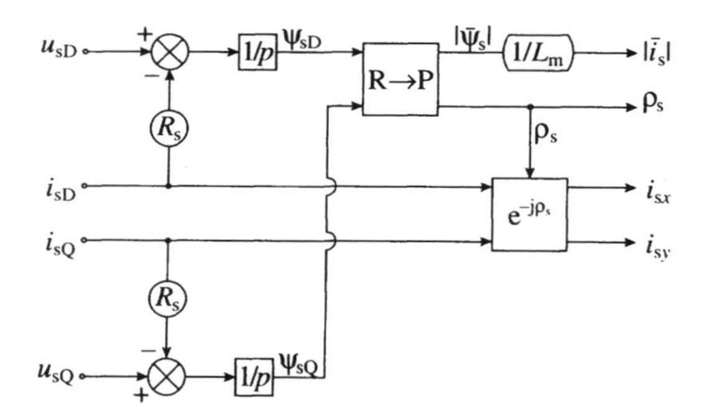

Fig. 4.27. Flux model (in stationary reference frame) and transformed stator currents.

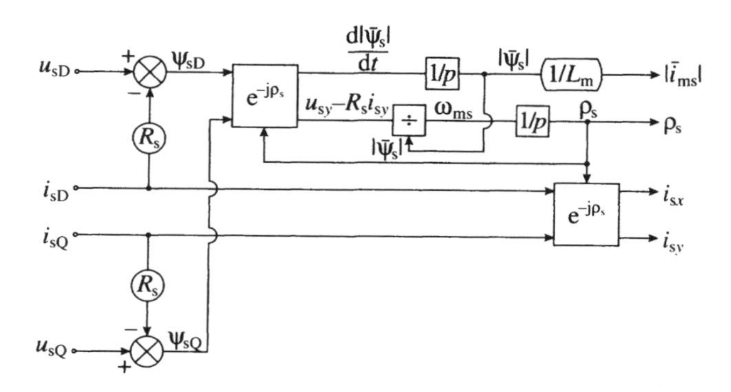

Fig. 4.28. Flux model (in stator-flux-oriented reference frame) and transformed stator currents.

In contrast to the scheme of Fig. 4.27, where open-loop integrations are present, in Fig. 4.28 closed-loop integrations are performed, and at low frequency this reduces the integrator drifts.

If the stator flux-linkage components are accurately estimated, then they can also be used for the estimation of the rotor speed in a speed-sensorless drive. For example, the scheme shown in Fig. 4.9(c) can also be used in a speed-sensorless vector-controlled drive even at low frequencies (see also Section 4.5.3.1).

4.2.4 STATOR VOLTAGE EQUATIONS FOR THE DOUBLY FED INDUCTION MACHINE

For a doubly fed machine, it is useful to express $\bar{i}_{s\psi s}$ in terms of $|\bar{i}_{ms}|$ and $\bar{i}_{r\psi s}$. Thus it follows from eqn (4.2-2) that

$$\bar{i}_{s\psi s}=\frac{L_m}{L_s}(|\bar{i}_{ms}|-\bar{i}_{r\psi s}) \tag{4.2-24}$$

and when this is substituted into eqn (4.2-7),

$$\bar{u}_{s\psi s}=\frac{R_s L_m}{L_s}(|\bar{i}_{ms}|-\bar{i}_{r\psi s})+\frac{L_m \mathrm{d}|\bar{i}_{ms}|}{\mathrm{d}t}+\mathrm{j}\omega_{ms}L_m|\bar{i}_{ms}|. \tag{4.2-25}$$

This can be put into the form

$$T_s\frac{\mathrm{d}|\bar{i}_{ms}|}{\mathrm{d}t}+|\bar{i}_{ms}|(1+\mathrm{j}\omega_{ms}T_s)=\bar{u}_{s\psi s}\frac{L_s}{R_s L_m}+\bar{i}_{r\psi s}, \tag{4.2-26}$$

where $T_s=L_s/R_s$ is the stator time constant. By considering the definitions of the stator-voltage, stator-current, and rotor-current space phasors in terms of their two-axis components established in the stator-flux-oriented reference frame (eqns (4.2-3), (4.2-4), and (4.2-8)), resolution of eqn (4.2-26) into its real- and imaginary-axis components yields

$$T_s\frac{\mathrm{d}|\bar{i}_{ms}|}{\mathrm{d}t}+|\bar{i}_{ms}|=\frac{L_s}{R_s L_m}u_{sx}+i_{rx} \tag{4.2-27}$$

$$\omega_{ms}T_s|\bar{i}_{ms}|=\frac{L_s}{R_s L_m}u_{sy}+i_{ry}. \tag{4.2-28}$$

For the doubly fed induction machine, the stator voltage components can be obtained as follows. By assuming a sinusoidal symmetrical three-phase supply voltage system, with frequency ω_1, the stator-voltage space phasor in the stationary reference frame can be expressed as

$$\bar{u}_s=\sqrt{2}U_s \mathrm{e}^{\mathrm{j}\omega_1 t}, \tag{4.2-29}$$

where U_s is the r.m.s. value of the line-to-neutral voltages and t denotes time. Thus substitution of eqn (4.2-29) into eqn (4.2-8) gives

$$\bar{u}_{s\psi s}=\sqrt{2}U_s \mathrm{e}^{\mathrm{j}(\omega_1 t-\rho_s)}=u_{sx}+\mathrm{j}u_{sy} \tag{4.2-30}$$

and thus

$$u_{sx}=\sqrt{2}U_s\cos(\omega_1 t-\rho_s) \tag{4.2-31}$$

$$u_{sy}=\sqrt{2}U_s\sin(\omega_1 t-\rho_s). \tag{4.2-32}$$

Compared with eqns (4.2-11) and (4.2-12), which contain the stator currents i_{sx}, i_{sy}, eqns (4.2-27) and (4.2-28) contain the rotor currents i_{rx}, i_{ry} because the rotor currents are subjected to stator-flux-oriented control. As mentioned earlier, two implementations will be discussed, where there is a current-controlled cycloconverter in the rotor of the slip-ring induction machine and where there is a static converter cascade in the rotor.

4.2.5 ROTOR VOLTAGE EQUATIONS FOR THE DOUBLY FED INDUCTION MACHINE

By substitution of eqn (4.2-24) into eqn (4.2-14), the space-phasor form of the rotor voltage equation in the stator-flux-oriented reference frame is

$$\bar{u}_{r\psi s}=(R_r+L_r'p)\bar{i}_{r\psi s}+\frac{L_m^2}{L_s}p|\bar{i}_{ms}|+j\omega_{sl}\left[\frac{L_m^2}{L_s}|\bar{i}_{ms}|+L_r'\bar{i}_{r\psi s}\right], \tag{4.2-33}$$

where it follows from eqn (4.2-16) that $\omega_{sl}=\omega_{ms}-\omega_r$ is the angular slip frequency, ω_{ms} is the speed of the reference frame, and ω_r is the rotor speed. Equation (4.2-33) can be resolved into its direct- (x) and quadrature-axis (y) components and thus

$$T_r'\frac{di_{rx}}{dt}+i_{rx}=\frac{u_{rx}}{R_r}+\omega_{sl}T_r'i_{ry}-(T_r-T_r')\frac{d|\bar{i}_{ms}|}{dt} \tag{4.2-34}$$

and

$$T_r'\frac{di_{ry}}{dt}+i_{ry}=\frac{u_{ry}}{R_r}-\omega_{sl}T_r'i_{rx}-\omega_{sl}(T_r-T_r')|\bar{i}_{ms}|. \tag{4.2-35}$$

These equations are very similar to eqns (4.1-7) and (4.1-8) and they can be used to obtain the implementation of stator-flux-oriented control if the rotor currents are not impressed. When these two equations are compared with eqns (4.1-7) and (4.1-8), it can be seen that instead of the stator current components, the rotor current components are now present and the rotor magnetizing-current space phasor is replaced by the stator magnetizing-current space phasor. Furthermore, the stator voltage components are replaced by the rotor voltage components, the stator resistance is replaced by the rotor resistance, the speed of the rotor magnetizing-current space phasor is replaced by the speed of the stator magnetizing-flux-space phasor, and finally the stator time constant and stator transient time constant are replaced by the rotor time constant and rotor transient time constant respectively. It also follows from eqns (4.2-34) and (4.2-35) that there are unwanted coupling terms between the rotor circuits along the

x and y axes. These can be eliminated by utilizing the following decoupling voltage components, which are obtained by considering the rotational voltages in eqns (4.2-34) and (4.2-35) and assuming constant-stator-flux operation ($|\bar{\imath}_{ms}|=$ constant):

$$u_{drx} = -\omega_{sl} L_r' i_{ry} \qquad (4.2\text{-}36)$$

$$u_{dry} = \omega_{sl} L_r' i_{rx} + \omega_{sl}(L_r - L_r')|\bar{\imath}_{ms}|. \qquad (4.2\text{-}37)$$

These have to be added to the outputs of the current controllers which control the rotor currents i_{rx} and i_{ry}.

With impressed rotor currents it is not necessary to consider the rotor voltage equations but, to achieve improved dynamic performance, appropriate terms must be added to the outputs of the rotor current controllers, which in the case of $|\bar{\imath}_{ms}|=$constant compensate for the last term in eqn (4.2-35), which acts as a disturbance for rotor current controllers. Thus it follows from eqn (4.2-35) that in the stator-flux-oriented reference frame the necessary feedforward signal takes the form

$$\frac{u_{iy}}{R_r} = \omega_{sl}(T_r - T_r')|\bar{\imath}_{ms}| \qquad (4.2\text{-}38)$$

or, in space-phasor form,

$$\frac{\bar{u}_{i\psi s}}{R_r} = \frac{\mathrm{j}u_{iy}}{R_r} = \mathrm{j}\omega_{sl}(T_r - T_r')|\bar{\imath}_{ms}|. \qquad (4.2\text{-}39)$$

By considering Fig. 4.25, according to which the angle between the real axis of the stator-flux-oriented reference frame and the real axis of the reference frame fixed to the rotor is $(\rho_s - \theta_r)$, in the reference frame fixed to the rotor, $\bar{u}_{i\psi s}/R_r$ can be expressed as

$$\frac{\bar{u}_i}{R_r} = \bar{u}_{i\psi s}\frac{e^{j(\rho_s - \theta_r)}}{R_r} = \mathrm{j}\omega_{sl}(T_r - T_r')|\bar{\imath}_{ms}|\, e^{j(\rho_s - \theta_r)} = i_\alpha + \mathrm{j}i_\beta, \qquad (4.2\text{-}40)$$

where $\bar{u}_i$ is the space phasor of the induced voltages in the rotor reference frame, and i_α and i_β are the required currents which have to be used as feedforward signals. Equation (4.2-40) also follows by considering the transformation of a space-phasor quantity from its natural reference frame into the general reference frame. Thus from eqn (2.1-143), the induced-voltage space phasor in the rotor reference frame ($\bar{u}_i$) can be expressed in terms of the induced-voltage space phasor in the general reference frame as $\bar{u}_i = \bar{u}_{ig} e^{j(\theta_g - \theta_r)}$. Thus when, instead of the general reference frame, the stator-flux-oriented reference frame is used, $\bar{u}_{ig} = \bar{u}_{i\psi s}$, $\theta_g = \rho_s$, and finally $\bar{u}_i = \bar{u}_{i\psi s} e^{j(\rho_s - \theta_r)}$ is obtained, which agrees with the induced voltage space phasor used in eqn (4.2-40). Equation (4.2-40) will be utilized in the implementations discussed in the following two sections. When the currents i_α and i_β are added to the rotor currents $i_{r\alpha ref}$, $i_{r\beta ref}$, which are the reference values of the rotor currents expressed in the rotor reference frame, then the effects of the rotational voltages in the rotor windings due to the stator flux are cancelled (the stator flux has been assumed to be constant) and improved dynamic performance is obtained.

4.2.6 CONTROL OF THE DOUBLY FED INDUCTION MACHINE WITH IMPRESSED ROTOR CURRENTS

First the stator-flux-oriented control of a doubly fed induction machine employing a cycloconverter in the rotor circuit is described. This is followed by a discussion on the stator-flux-oriented control of the doubly fed machine employing a converter cascade in the rotor.

4.2.6.1 Control of the induction machine with a current-controlled cycloconverter in the rotor circuit

In many countries, pumps, fans, and blowers can be responsible for approximately half the electricity consumed by industry. Pump drives are used, for example, in sewage treatment, water supply applications, circulating water pumps, etc., and fans are used for mine ventilation, wind-tunnel applications, etc. These can represent very large loads—over 10 MW being common—and there is an increasing trend to use variable-speed drives in an attempt to improve the efficiency. In these applications the load torque is proportional to the square of the speed and there is a restricted range of speed control. For such applications the doubly fed slip-ring induction machine with its stator connected to a constant-frequency power supply and its rotor connected to a static converter represents an ideal solution. It is one of the most important features of a slip recovery system that, if the application requires only limited range of speed control (in pump and fan drives only a speed variation of 10–30% is usually required), then the control system has to be rated only for slip power, which is only a fraction of the stator power. The resulting reduction in the converter costs makes the slip recovery systems very attractive in all applications where only a limited speed-range is required and where energy saving is an important factor.

In this section a current-controlled cycloconverter is used in the rotor. The cycloconverter used is similar to that described in Section 3.2.1 (also used in Section 4.1.2), and it consists of three sets of anti-parallel six-pulse thyristor bridges without circulating currents. The rotor currents of the slip-ring machine are separately controlled by fast current loops.

If sinusoidal stator currents and the decoupling of the active and reactive stator power is required, it is useful to employ stator-flux-oriented control which ensures the required decoupling by appropriate control of the amplitude and phase angle of the rotor-current space phasor. The application of stator-flux-oriented control also has the advantage of stabilizing the drive. It should be noted that another area of application for a doubly fed induction machine with stator-flux-oriented control can be in large wind power stations, where wind gusts can cause power and torque fluctuations which can result in unwanted mechanical stresses on the mechanical transmission system (shaft, gears) and can cause undesirable power fluctuations. When the doubly fed machine is used with a restricted speed range, the wind energy can be temporarily stored in the inertia of the rotor blades and the undesirable effects (power fluctuations, mechanical stresses) are reduced.

Figure 4.29 shows the schematic of the stator-flux-oriented control of an induction machine with a current-controlled cycloconverter in the rotor. The induction machine can operate as a motor or a generator in the supersynchronous and subsynchronous speed ranges with lagging or leading stator currents.

In Fig. 4.29 the input signals to the control system are the monitored stator and rotor currents, the monitored rotor angle (θ_r), and the active and reactive stator powers, respectively P_s and Q_s. The cycloconverter is connected to the grid via a three-phase transformer. The turns ratio of the transformer can be selected to minimize the voltage rating of the cycloconverter. The drive can be started in several ways. One possibility is to employ rotor resistances and to short-circuit them sequentially as the machine comes up to the operating speed when the cycloconverter is switched on. It is also possible to short-circuit the stator and to supply the rotor from the cycloconverter (which is connected to the grid) and which operates as a variable-frequency power supply. This arrangement corresponds to an inverted induction machine. The machine will accelerate and at approximately half speed the stator is connected to the grid. The rotor frequency will be approximately one half of the supply frequency and the motor can accelerate further.

It will now be shown that in the stator-flux-oriented system, the real- and imaginary-axis components of the rotor currents (i_{rx}, i_{ry}) correspond directly to

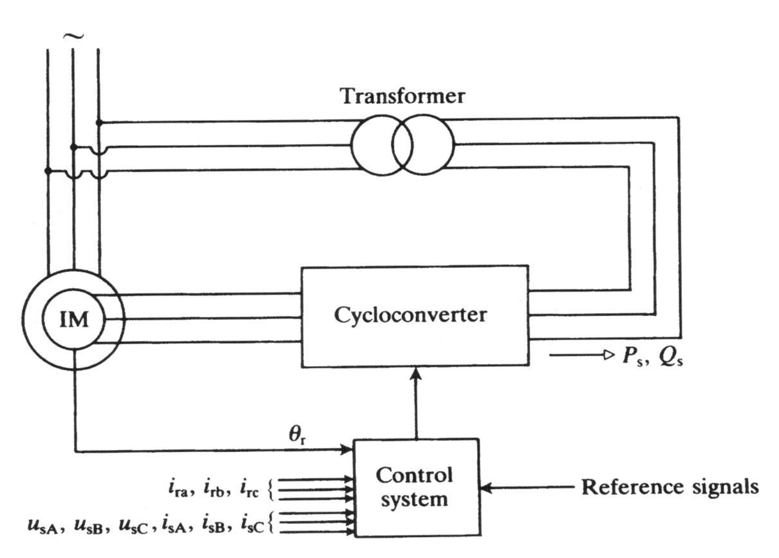

Fig. 4.29. Simplified schematic of the stator-flux-oriented control of an induction machine with a current-controlled cycloconverter in the rotor.

the two-axis stator currents (i_{sx}, i_{sy}) expressed in the stator-flux-oriented reference frame. Thus control of the i_{rx} current can be used to maintain the reactive stator current (i_{sx}) at a fixed value (or at a value dependent on the voltage) and the current i_{ry} can be used to control the electromagnetic torque.

By resolving eqn (4.2-24) into its two-axis components, it follows that the stator current components in the stator-flux-oriented reference frame (i_{sx}, i_{sy}) can be expressed in terms of the rotor current components established in the same reference frame (i_{rx}, i_{ry}) as follows:

$$i_{sy} = -i_{ry}\frac{L_m}{L_s}, \tag{4.2-41}$$

which is the same as eqn (4.2-5) but is repeated here for convenience, and

$$i_{sx} = (|\bar{i}_{ms}| - i_{ry})\frac{L_m}{L_s}. \tag{4.2-42}$$

It follows from eqn (4.2-41) that the quadrature-axis rotor current (i_{ry}) is proportional to the torque-producing (active) stator current component (i_{sy}). However, in eqn (4.2-42) the stator magnetizing-current space phasor is also present, and this depends on the stator voltage. This dependency will now be obtained. For this purpose, it is assumed that the stator resistance (R_s) of the induction machine can be neglected. This is usually justified in machines with a rating over 10 kW. Furthermore it is assumed that the frequency of the power supply on the stator is constant, $\omega_1 = \text{constant}$. In the steady state the modulus of the stator magnetizing current [which has been defined in eqn (4.2-2)] is constant ($|\bar{i}_{ms}| = \text{constant}$) and the speed of the stator magnetizing-current space phasor ($\omega_{ms} = d\rho_s/dt$, where ρ_s is the angle of the stator magnetizing-current space phasor with respect to the real axis of the stationary reference frame and is shown in Fig. 4.25) is also constant.

Thus it follows from eqn (4.2-25) that

$$|\bar{u}_s| = \omega_1 L_m |\bar{i}_{ms}|, \tag{4.2-43}$$

where $|\bar{u}_s|$ is the modulus of the stator-voltage space phasor and from eqn (4.2-30) $|\bar{u}_s| = \sqrt{2}U_s$ is obtained. It follows from eqn (4.2-43) that the modulus of the stator magnetizing current is

$$|\bar{i}_{ms}| = \frac{|\bar{u}_s|}{\omega_1 L_m}, \tag{4.2-44}$$

which is an expected result since a steady state has been assumed together with negligible stator resistance. It should be noted that the same result follows from eqns (4.2-27)–(4.2-32), which under the same assumptions yield

$$u_{sx} = 0 \qquad u_{sy} = |\bar{u}_s| \qquad |\bar{i}_{ms}| = \frac{u_{sy}}{L_m \omega_1}. \tag{4.2-45}$$

Thus from eqns (4.2-42) and (4.2-44),

$$i_{sx} = (|\bar{u}_s|/\omega_1 - L_m i_{rx})/L_s, \tag{4.2-46}$$

from which it follows that, since the machine parameters (L_s, L_m) and also $|\bar{u}_s|$ and ω_1 are constant, the direct-axis (reactive) stator currents (i_{sx}) can be controlled by the direct-axis rotor current (i_{rx}). These results will be used in the implementation of the stator-flux-oriented control of the doubly fed machine shown in Fig. 4.30.

As in eqn (2.1-87), the stator active power can be defined as

$$P_s = \tfrac{3}{2}\,\mathrm{Re}(\bar{u}_{s\psi s}\bar{i}^*_{s\psi s}) = \tfrac{3}{2}(u_{sx}i_{sx} + u_{sy}i_{sy}) \tag{4.2-47}$$

and the stator reactive power as

$$Q_s = \tfrac{3}{2}\,\mathrm{Im}(\bar{u}_{s\psi s}\bar{i}^*_{s\psi s}) = \tfrac{3}{2}(u_{sy}i_{sx} - u_{sx}i_{sy}). \tag{4.2-48}$$

From eqn (4.2-45) $u_{sx} = 0$ and $\bar{u}_{sy} = |\bar{u}_s|$, and it follows from eqns (4.2-47) and (4.2-48) that

$$P_s = \tfrac{3}{2}|\bar{u}_s| i_{sy} \tag{4.2-49}$$

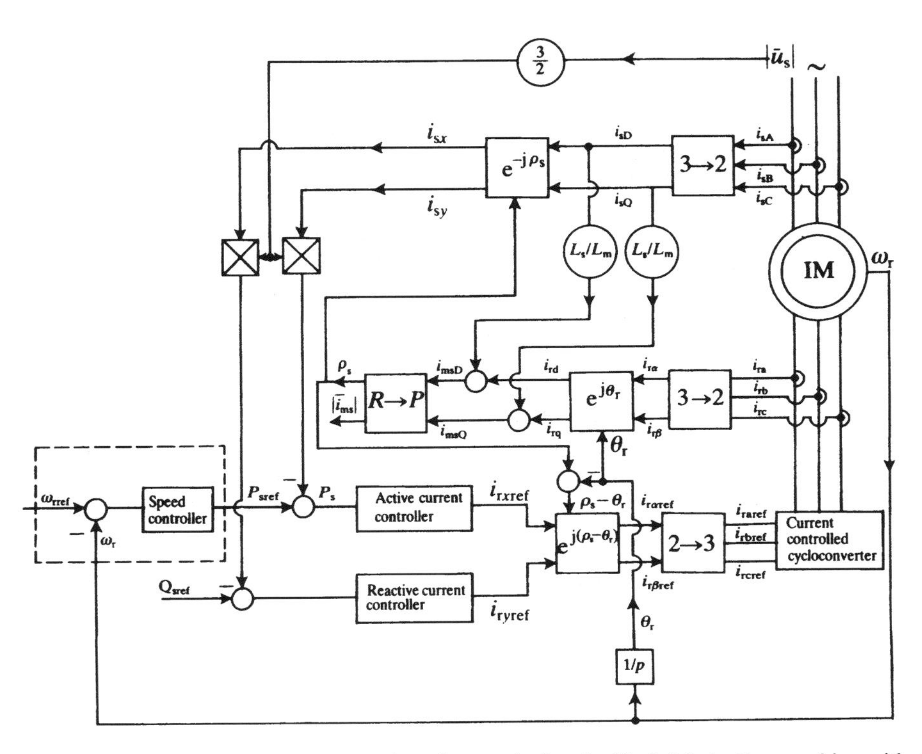

Fig. 4.30. Schematic of the stator-flux-oriented control of a doubly-fed induction machine with a current-controlled cycloconverter in the rotor.

and, since $|\bar{u}_s|$ is a constant, P_s is equal to a constant times the torque-producing current, thus it also follows that

$$Q_s = \tfrac{3}{2}|\bar{u}_s| i_{sx} \tag{4.2-50}$$

and this is equal to a constant times the stator reactive current. Equations (4.2-49) and (4.2-50) are utilized in the implementation shown in Fig. 4.30. The modulus of the stator-voltage space phasor can be obtained from the monitored values of the stator voltages.

In Fig. 4.30 the space angle of the stator magnetizing-current space phasor (ρ_s) is obtained from the direct-axis (i_{msD}) and quadrature-axis (i_{msQ}) components of the stator magnetizing-current space phasor expressed in the stationary reference frame, by utilizing a rectangular-to-polar converter. This can be obtained by considering eqns (4.2-2), (4.2-3), and (4.2-4) and thus

$$\begin{aligned} \bar{i}_{ms} = i_{msD} + j i_{msQ} &= \frac{L_s}{L_m}(i_{sD} + j i_{sQ}) + (i_{r\alpha} + j i_{r\beta})\, e^{j\theta_r} \\ &= \frac{L_s}{L_m}(i_{sD} + j i_{sQ}) + (i_{rd} + j i_{rq}) = |\bar{i}_{ms}|\, e^{j\rho_s}, \end{aligned} \tag{4.2-51}$$

where i_{sD}, i_{sQ} are the stator current components in the stationary reference frame and are obtained from the monitored three-phase stator currents (i_{sA}, i_{sB}, i_{sC}) by the application of the three-phase to two-phase transformation. The two-axis components of the rotor currents in the rotor reference frame ($i_{r\alpha}$, $i_{r\beta}$) are obtained from the monitored rotor currents (i_{ra}, i_{rb}, i_{rc}) by the application of the three-phase to two-phase transformation. The direct- and quadrature-axis rotor currents in the stator reference frame (i_{rd}, i_{rq}) are obtained from the rotor currents in the rotor reference frame ($i_{r\alpha}, i_{r\beta}$) by the application of the transformation $e^{j\theta_r}$ where θ_r is the rotor angle. It should be noted that eqn (4.2-51) can be obtained directly from eqn (2.1-128), which defines the stator flux-linkage space phasor in the stationary reference frame, if it is divided by the magnetizing inductance. It is also possible to obtain i_{msD}, i_{msQ} by using monitored stator voltages and currents. For this purpose $i_{msD} + j i_{msQ} = (\psi_{sD} + j\psi_{sQ})/L_m$ could be utilized, where $\psi_{sD} + j\psi_{sQ}$ can be obtained by the integration of $(u_{sD} + j u_{sQ}) - R_s(i_{sD} + j i_{sQ})$, where R_s is the stator resistance and u_{sD}, u_{sQ} are the two-axis components of the stator voltages in the stator reference frame.

In Fig. 4.30 the stator currents i_{sx}, i_{sy} are obtained from i_{sD}, i_{sQ} by utilizing eqn (4.2-3). The difference between the reference and actual values of the stator active power is the input to the active current controller, which is a PI controller and the output of which is the direct-axis reference rotor current component established in the stator-flux-oriented reference frame (i_{rxref}). Similarly, the reference value of the stator active power is compared with its actual value and their difference serves as the input to the reactive current controller, also a PI controller, and the output of which is i_{ryref}. The currents i_{rxref} and i_{ryref} are transformed into the two-axis rotor current references in the rotor reference frame $i_{r\alpha ref}$, $i_{r\beta ref}$ by using eqn (4.2-4). Finally by using the two-phase to three-phase transformation, these are transformed into the three-phase rotor current references ($i_{\text{raref}}, i_{\text{rbref}}, i_{\text{rcref}}$), which are required to control the cycloconverter.

It is also possible to implement a speed control loop, as shown in Fig. 4.30. In this case the output of the speed controller yields the reference value of the stator active power. To achieve improved dynamic performance, the currents i_α and i_β, defined in eqn (4.2-40), should be added to the currents $i_{\text{r}\alpha\text{ref}}$ and $i_{\text{r}\beta\text{ref}}$ respectively, since then the effects of the voltage induced in the y-axis rotor winding due to the stator flux are cancelled. However, the same effect can be achieved by adding the current defined in eqn (4.2-38) to the reference current i_{ryref}.

4.2.6.2 Control of the induction machine with a static converter cascade in the rotor

It is possible to implement stator-flux-oriented control of a doubly fed induction machine with a static converter cascade in the rotor. This is similar to the control scheme described in Fig. 4.23, but there rotor-flux-oriented control was employed. Figure 4.31 shows the schematic of the drive system.

In the implementation shown in Fig. 4.31 a force-commutated thyristor converter is connected to the slip rings of the induction machine and this is connected via a d.c. link reactor to a naturally commutated controlled-thyristor bridge converter (a special form of static Scherbius cascade). Thus slip power can flow in both directions and the speed of the induction machine can be controlled in the supersynchronous and subsynchronous speed ranges. If the rotor-side converter was a line-commutated bridge converter, near to synchronous speed, when the a.c. voltage available to commutate this converter is small, the commutating ability of this converter would be lost. It is for this reason that a force-commutated rotor-side converter is used in the implementation shown in Fig. 4.31.

It follows from eqn (4.2-6) that in the stator-flux-oriented reference frame, if the stator magnetizing current $|\bar{i}_{\text{ms}}|$ is constant, the electromagnetic torque of the doubly fed induction machine is proportional to the quadrature-axis rotor current component (i_{ry}) and the torque can be controlled in the same way as for a separately excited d.c. machine where there is independent control of the excitation flux and the field current. For convenience eqn (4.2-6) is repeated here in a slightly different form,

$$t_{\text{e}}=c|\bar{i}_{\text{ms}}|i_{\text{ry}}=c|\bar{i}_{\text{ms}}||\bar{i}_{\text{r}}|\sin(\alpha_{\text{r}}+\theta_{\text{r}}-\rho_{\text{s}}), \tag{4.2-52}$$

where $c=-3L_{\text{m}}^2/(2L_{\text{s}})$, $|\bar{i}_{\text{r}}|$ is the modulus of the rotor-current space phasor and the various angles are shown in Fig. 4.25. Thus the angle $\phi_{\text{r}}=(\alpha_{\text{r}}+\theta_{\text{r}}-\rho_{\text{s}})$ is the angle of the rotor-current space phasor with respect to the real axis of the stator-flux-oriented reference frame (torque angle). In Fig. 4.31 the modulus of the rotor-current space phasor is controlled by the commutated converter and the angle ϕ_{r} is controlled by the machine-side converter.

In Fig. 4.31 the line-side converter is controlled by utilizing the closed-loop control of the d.c. link current, where the difference between the reference d.c. link current and the actual d.c. link current is fed into a PI controller. The torque

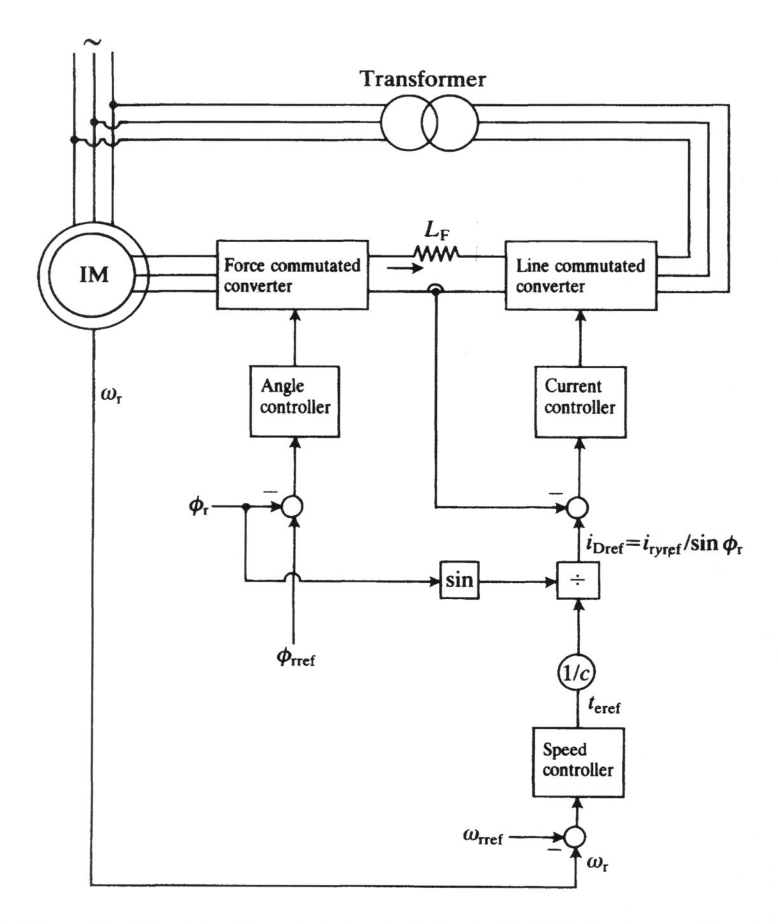

Fig. 4.31. Schematic of the stator-flux-oriented control of a doubly-fed induction machine with a static converter cascade in the rotor.

reference (t_{eref}) is used to obtain the reference value of the d.c. link current (i_{Dref}) and the torque reference appears as the output of the speed controller, also a PI controller, and the input of which is the difference between the reference rotor speed (ω_{rref}) and the monitored rotor speed (ω_r). The reference value i_{Dref} is obtained as $i_{ryref}/\sin\phi_r$.

The rotor-side converter is controlled by utilizing the closed-loop control of the torque angle. For this purpose an angle controller is used, also a PI controller, the input of which is the difference between the reference (ϕ_{rref}) and the actual (ϕ_r) value of the torque angle.

It is possible to obtain the torque angle in several ways, and a method which does not utilize the stator voltages is now described. It has been shown that the torque angle is a function of three angles, $\phi_r=\alpha_r+\theta_r-\rho_s$. These can be obtained as follows. If eqn (4.2-51) is utilized, then the angle ρ_s can be obtained by monitoring the stator and rotor currents and the rotor angle, as shown in Fig. 4.30, and during the process of the determination of the angle ρ_s the rotor current components established in the rotor reference frame ($i_{r\alpha}$ and $i_{r\beta}$) are also obtained. Thus the angle α_r, which according to Fig. 4.25 is the angle of the rotor-current space phasor with respect to the real axis of the rotor reference frame, can be obtained if these two currents are inputs to a rectangular-to-polar converter. Finally, the rotor angle (θ_r) can be obtained by direct measurement.

4.3 Magnetizing-flux-oriented control of induction machines

In this part of the chapter the magnetizing-flux-oriented control of induction machines will be discussed. The main assumptions are the same as those used in connection with rotor-flux-oriented control and stator-flux-oriented control of induction machines, and in particular the effects of magnetic saturation are neglected. Because of the similarity between magnetizing-flux-oriented control and stator- and rotor-flux-oriented control, (see Section 4.2.4.2), magnetizing-flux-oriented control will only be described briefly. However, it will be shown how this technique can be used in the case of a double-cage induction machine, whose space-phasor model will also be developed.

4.3.1 EXPRESSION FOR THE ELECTROMAGNETIC TORQUE

From the results presented in Section 2.1.8, it follows that the electromagnetic torque produced by an induction machine is proportional to the product of the modulus of the magnetizing-current space phasor ($|\bar{i}_{mm}|$) and the torque-producing stator current component (i_{sy}), and the torque can be controlled by the independent control of these two current components. It follows from Section 2.1.8 and also from Fig. 4.32 that in the magnetizing-flux-oriented reference frame $|\bar{i}_{mm}|=i_{mx}$, where i_{mx} is the direct-axis component of the magnetizing flux-linkage space phasor in the magnetizing-flux-oriented reference frame.

Thus from eqns (2.1-186) and the relationship $|\bar{i}_{mm}|=i_{mx}$, it follows that the electromagnetic torque can be expressed as

$$t_e=\tfrac{3}{2}PL_m i_{mx} i_{sy}=\tfrac{3}{2}PL_m|\bar{i}_{mm}|i_{sy}, \tag{4.3-1}$$

where P is the number of pole pairs and L_m is the magnetizing inductance of the machine. In eqn (4.3-1) the magnetizing-current space phasor can be expressed in terms of the stator- and rotor-current space phasors in the magnetizing-flux-oriented reference frame $\bar{i}_{s\psi m}$, $\bar{i}_{r\psi m}$, which for simplicity are denoted by $\bar{i}_{sm}$ and $\bar{i}_{rm}$, as follows, if eqns (2.1-181), (2.1-184), and (2.1-185) are considered,

$$|\bar{i}_{mm}|=\bar{i}_{sm}+\bar{i}_{rm}=i_{sx}+\mathrm{j}i_{sy}+i_{rx}+\mathrm{j}i_{ry}, \tag{4.3-2}$$

where i_{sx} and i_{sy} are the direct- and quadrature-axis components of the stator currents in the magnetizing-flux-oriented reference frame and they are also shown in Fig. 4.32. i_{rx} and i_{ry} are the direct- and quadrature-axis components of the rotor currents in the magnetizing-flux-oriented reference frame. It should be noted that the space phasor of the stator currents in the magnetizing-flux-oriented reference frame ($\bar{i}_{sm}$) can be expressed as follows, if eqn (2.1-181) is considered,

$$\bar{i}_{sm}=\bar{i}_s e^{-j\mu_m}=(i_{sD}+ji_{sQ})e^{-j\mu_m}=i_{sx}+ji_{sy}, \tag{4.3-3}$$

where $\bar{i}_s$ is the space phasor of the stator currents in the stationary reference frame, i_{sD} and i_{sQ} are its two-axis components in the same reference frame, and μ_m is the angle of the magnetizing-current (or magnetizing flux-linkage) space phasor with respect to the real axis of the stationary reference frame, as shown in Fig. 4.32. Similarly, the space phasor of the rotor currents in the magnetizing-flux-oriented reference frame can be expressed as follows, if eqn (2.1-185) is considered,

$$\bar{i}_{rm}=\bar{i}_r e^{-j(\mu_m-\theta_r)}=(i_{r\alpha}+ji_{r\beta})e^{-j(\mu_m-\theta_r)}=i_{rx}+ji_{ry}. \tag{4.3-4}$$

In eqn (4.3-4) $\bar{i}_r$ is the space phasor of the rotor currents in the reference frame fixed to the rotor, $i_{r\alpha}$ and $i_{r\beta}$ are its two-axis components in the rotor reference frame, and θ_r is the rotor angle, also shown in Fig. 4.32.

If the induction machine is supplied by impressed stator voltages or currents, eqn (4.3-1) is utilized when the magnetizing-flux-oriented control scheme is developed, but for a doubly fed induction machine, as in Section 4.2.1, the expression for the torque is put into a form which contains the rotor currents. Thus it follows from eqn (4.3-2) that

$$i_{sy}=-i_{ry} \tag{4.3-5}$$

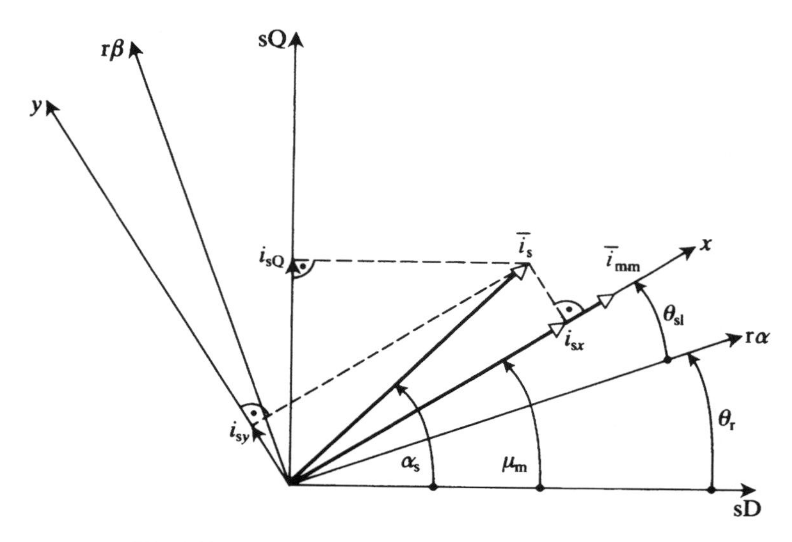

Fig. 4.32. Stator- and magnetizing-current space phasors.

and thus eqn (4.3-1) can be put into the form

$$t_e = -\tfrac{3}{2} P L_m |\bar{i}_{mm}| i_{ry} = -\tfrac{3}{2} P L_m |\bar{i}_{mm}| |\bar{i}_r| \sin(\alpha_r + \theta_r - \mu_m), \quad (4.3\text{-}6)$$

where i_{ry} has been obtained from eqn (4.3-4), and the Euler form of the space phasor of the rotor currents in the rotor reference frame, $\bar{i}_r = |\bar{i}_r| e^{j\alpha_r}$ (the angle α_r is the angle of $\bar{i}_r$ with respect to the rα-axis shown in Fig. 4.32) has been used. Equation (4.3-6) is similar to eqn (4.2-6), but as well as the differences in the constant parts, the modulus of the magnetizing-current space phasor is now present instead of the modulus of the stator magnetizing-current space phasor and the angle ρ_s is replaced by the angle μ_m. Equation (4.3-6) can be utilized for magnetizing-flux-oriented control of the doubly fed induction machine with impressed rotor currents.

4.3.2 STATOR VOLTAGE EQUATIONS

Here the space-phasor form of the stator voltage equation expressed in the magnetizing-flux-oriented reference frame will be obtained and this will be resolved into its two-axis components. For this purpose the voltage equations formulated in the general reference frame are utilized directly, these have been given in Section 2.1.7.

By considering eqns (2.1-148) and (2.1-150), which are the stator-voltage and flux-linkage space-phasor equations in the general reference frame, and by neglecting the effects of magnetic saturation, the space-phasor voltage equation of the stator will be as follows in the reference frame fixed to the magnetizing flux-linkage space phasor, which rotates at the speed $\omega_m = d\mu_m/dt$, where the angle μ_m has been defined above:

$$\bar{u}_{sm} = R_s \bar{i}_{sm} + L_s \frac{d\bar{i}_{sm}}{dt} + L_m \frac{d\bar{i}_{rm}}{dt} + j\omega_m (L_s \bar{i}_{sm} + L_m \bar{i}_{rm}). \quad (4.3\text{-}7)$$

R_s and L_s are the stator resistance and self-inductance respectively and L_m is the magnetizing inductance and, as for the transformation of the stator currents given by eqn (4.1-27), $\bar{u}_{sm}$ is related to the space phasor of the stator voltages in the stationary reference frame ($\bar{u}_s$) by $\bar{u}_{sm} = \bar{u}_s e^{-j\mu_m} = (u_{sD} + ju_{sQ}) e^{-j\mu_m} = u_{sx} + ju_{sy}$. By considering eqn (4.3-2), the rotor-current space phasor can be eliminated from eqn (4.3-7), since it follows that

$$\bar{i}_{rm} = |\bar{i}_{mm}| - \bar{i}_{sm} \quad (4.3\text{-}8)$$

and when eqn (4.3-8) is substituted into eqn (4.3-7),

$$\bar{u}_{sm} = R_s \bar{i}_{sm} + L_{sl} \frac{d\bar{i}_{sm}}{dt} + L_m \frac{d|\bar{i}_{mm}|}{dt} + j\omega_m (L_{sl} \bar{i}_{sm} + L_m |\bar{i}_{mm}|), \quad (4.3\text{-}9)$$

where L_{sl} is the leakage inductance of a stator winding, and $L_{sl} = L_s - L_m$ has been used. Physically it follows from eqn (4.3-9) that the stator-voltage space phasor is balanced by the stator ohmic drop ($R_s \bar{i}_{sm}$), a voltage drop across the stator leakage inductance ($L_{sl}d\bar{i}_{sm}/dt$), the magnetizing voltage drop ($L_m d|\bar{i}_{mm}|/dt$), and a rotational voltage component due to the rotation of the chosen reference frame: that last component is equal to $j\omega_m(L_{sl}\bar{i}_{sm}+L_m|\bar{i}_{mm}|)$, where $L_{sl}\bar{i}_{sm}$ is the stator leakage flux-linkage space phasor and $L_m|\bar{i}_{mm}|$ is the magnetizing flux-linkage space phasor, and of course both are expressed in the magnetizing-flux-oriented reference frame.

Equation (4.3-9) is more complicated than eqn (4.2-10), which was expressed in the stator-flux-oriented reference frame and does not contain the derivative of the stator currents. Furthermore, in contrast to eqn (4.2-10), which does not contain a separate stator leakage voltage drop, in eqn (4.3-9) the stator leakage voltage drop is also present. Thus if the induction machine is supplied by impressed stator voltages, it is simpler to use the stator-flux-oriented control scheme than the magnetizing-flux-oriented control scheme. When compared with the corresponding stator voltage equation expressed in the rotor-flux-oriented reference frame (eqn 4.1-6), whilst in eqn (4.3-9) L_{sl} is present in the two corresponding terms, in eqn (4.1-6) the stator transient inductance (L_s') is present. Furthermore, in eqn (4.3-9) in two respective terms the modulus ($|\bar{\psi}_{mm}|=L_m|\bar{i}_{mm}|$) and the derivative of the magnetizing flux-linkage space phasor are present, whereas in the corresponding two terms of eqn (4.1-6) $|\bar{i}_{mr}|(L_s-L_s')=(L_m/L_r)|\bar{\psi}_{r\psi r}|$ and its derivative are present, where L_r is the self-inductance of the rotor and $|\bar{\psi}_{r\psi r}|$ is the modulus of the rotor flux-linkage space phasor.

Resolution of eqn (4.3-9) into its real- and imaginary-axis components yields

$$i_{sx}=\frac{u_{sx}}{R_s}-T_{sl}\frac{di_{sx}}{dt}-\frac{L_m}{R_s}\frac{d|\bar{i}_{mm}|}{dt}+\omega_m T_{sl}i_{sy} \tag{4.3-10}$$

$$i_{sy}=\frac{u_{sy}}{R_s}-T_{sl}\frac{di_{sy}}{dt}-\frac{\omega_m(L_{sl}i_{sx}+L_m|\bar{i}_{mm}|)}{R_s}, \tag{4.3-11}$$

where $T_{sl}=L_{sl}/R_s$ is the stator leakage time-constant. Equations (4.3-10) and (4.3-11) must be considered for the magnetizing-flux-oriented control scheme of an induction machine supplied by impressed stator voltages. However, for this purpose, a decoupling circuit must be utilized, as discussed in the next section.

Decoupling circuit

It follows from eqns (4.3-10) and (4.3-11) that there are unwanted coupling terms, and independent control of the flux- and torque-producing stator currents i_{sx} and i_{sy} is only possible if eqns (4.3-10) and (4.3-11) are decoupled and i_{sx} and i_{sy} are indirectly controlled by controlling the terminal voltages of the induction machine. For this purpose, as described previously, the following decoupling voltages are introduced, if it is assumed that the magnetizing flux is constant,

$$u_{dx}=-\omega_m L_{sl}i_{sy} \tag{4.3-12}$$

$$u_{dy}=\omega_m(L_{sl}i_{sx}+L_m|\bar{i}_{mm}|). \tag{4.3-13}$$

These equations are similar to eqns (4.1-9) and (4.1-10), which define the decoupling voltages in the case of rotor-flux-oriented control and the required decoupling circuit shown in Fig. 4.33 is similar to the one shown in Fig. 4.1.

Thus the flux- and torque-producing stator current components can be independently controlled if u_{dx} and u_{dy} are added to the outputs of the current controllers ($\hat{u}_{sx}$, $\hat{u}_{sy}$) which control i_{sx} and i_{sy} respectively, since $\hat{u}_{sx}+u_{dx}$ gives the direct-axis terminal voltage component in the magnetizing-flux-oriented reference frame and if the magnetizing flux is constant, it follows from eqn (4.3-10) that the voltage at the output of the direct-axis current controller is

$$\hat{u}_{sx}=R_s i_{sx}+L_{sl}\frac{di_{sx}}{dt}. \tag{4.3-14}$$

Thus $\hat{u}_{sx}$ directly controls i_{sx} through a simple delay element with the stator leakage time constant ($T_{sl}=L_{sl}/R_s$). Similarly, $\hat{u}_{sy}+u_{dy}$ gives the quadrature-axis terminal voltage in the same reference frame and it follows from eqn (4.3-11) that the voltage on the output of the quadrature-axis current controller is

$$\hat{u}_{sy}=R_s i_{sy}+L_{sl}\frac{di_{sy}}{dt} \tag{4.3-15}$$

and $\hat{u}_{sy}$ directly controls i_{sy} through a simple delay element with the time constant T_{sl}.

In the decoupling circuit shown in Fig. 4.33, the stator currents i_{sx} and i_{sy} can be obtained from the monitored three-phase stator currents by using eqn (4.3-3) and $|\bar{i}_{mm}|$ and ω_m (and μ_m) can be obtained either from a flux model which utilizes the rotor equations of the machine or by using direct measurements (e.g. Hall-sensors, special search coils, etc.). If the effect of inverter time delay is to be considered in the decoupling circuit, the method developed in Section 4.1.1 can be used and a decoupling circuit is obtained which is similar to that shown in Fig. 4.5.

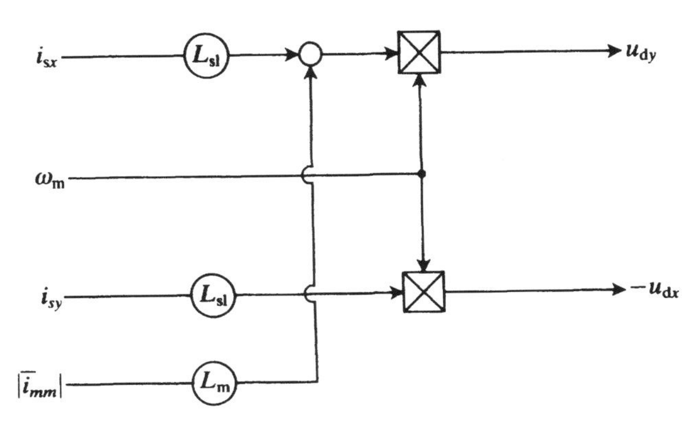

Fig. 4.33. Decoupling circuit to obtain the decoupling voltages u_{dx}, u_{dy}.

When the machine is supplied by impressed stator currents, the stator voltage equations can be omitted from the equations of the drive.

4.3.3 ROTOR VOLTAGE EQUATIONS

The rotor voltage equations expressed in the magnetizing-flux-oriented reference frame can be used to obtain the modulus of the magnetizing flux-linkage space phasor ($|\bar{\psi}_{mm}|=L_m|\bar{i}_{mm}|$) and its speed ($\omega_m$) or its phase angle ($\mu_m$). For this purpose eqn (2.1-153) is used which gives the rotor voltage equation in a general reference frame. Thus by neglecting the effects of saturation, it follows from eqn (2.1-153) that in the magnetizing-flux-oriented reference frame ($\omega_g=\omega_m$, $\bar{i}_{rg}=\bar{i}_{rm}$, $\bar{i}_{sg}=\bar{i}_{sm}$) the rotor voltage equation is as follows if short-circuited rotor windings are assumed:

$$0=R_r\bar{i}_{rm}+L_r\frac{d\bar{i}_{rm}}{dt}+L_m\frac{d\bar{i}_{sm}}{dt}+j(\omega_m-\omega_r)(L_r\bar{i}_{rm}+L_m\bar{i}_{sm}), \tag{4.3-16}$$

where the stator- and rotor-current space phasors in the magnetizing-flux-oriented reference frame ($\bar{i}_{sm}$ and $\bar{i}_{rm}$) have been defined in eqns (4.3-3) and (4.3-4) respectively and ω_r is the rotor speed. Since eqn (4.3-16) contains the rotor-current space phasor, and the rotor currents cannot be directly measured in the case of a squirrel-cage induction machine, $\bar{i}_{rm}$ is eliminated by considering eqn (4.3-8) and

$$0=R_r(|\bar{i}_{mm}|-\bar{i}_{sm})+L_r\frac{d|\bar{i}_{mm}|}{dt}-L_{rl}\frac{d\bar{i}_{sm}}{dt}+j(\omega_m-\omega_r)(L_r|\bar{i}_{mm}|-L_{rl}\bar{i}_{sm}). \tag{4.3-17}$$

This equation is much more complicated than eqn (4.1-24), which defines the rotor voltage equation in the rotor-flux-oriented reference frame. Thus if the machine is supplied by impressed stator currents, magnetizing-flux-oriented control leads to more complicated implementation than the implementation based on rotor-flux-oriented control. When eqn (4.3-17) is compared with eqn (4.2-15), which gives the rotor voltage equation in the stator-flux-oriented reference frame, it can be seen that the two equations are similar.

Resolution of eqn (4.3-17) into its direct- and quadrature-axis components in the magnetizing-flux-oriented reference frame yields

$$\frac{|\bar{i}_{mm}|+T_r\,d|\bar{i}_{mm}|/dt}{T_{rl}}=\frac{di_{sx}}{dt}+\frac{i_{sx}}{T_{rl}}-\omega_{sl}i_{sy} \tag{4.3-18}$$

$$\omega_{sl}\left(|\bar{i}_{mm}|\frac{T_r}{T_{rl}}-i_{sx}\right)=\frac{di_{sy}}{dt}+\frac{i_{sy}}{T_{rl}}, \tag{4.3-19}$$

where $\omega_{sl}=\omega_m-\omega_r$ is the angular slip frequency and T_r and T_{rl} are the rotor time constant and rotor leakage time constant respectively ($T_r=L_r/R_r, T_{rl}=L_{rl}/R_r$). For an induction machine with impressed stator currents eqns (4.3-18) and (4.3-19) have to be considered. However, there is unwanted coupling between these two equations and as a consequence $|\bar{i}_{mm}|$ will be altered by a change in the torque-producing stator current component. This coupling can be removed by the application of a decoupling circuit, as obtained in the following section, where the magnetizing-flux-oriented control scheme of an induction machine supplied by a current-controlled PWM inverter is also described.

Decoupling circuit; control scheme of a current-controlled PWM inverter-fed induction machine

If the magnetizing current is controlled by a flux controller (which can be a PI controller) and the output of this controller is $\hat{i}_{sx}$, then the required decoupling current along the direct axis of the magnetizing-flux-oriented reference frame i_{dx} has to be added to $\hat{i}_{sx}$ to give the reference value i_{sxref}. Thus

$$i_{sx}=\hat{i}_{sx}+i_{dx} \tag{4.3-20}$$

and, when eqn (4.3-20) is substituted into eqn (4.3-18),

$$\frac{|\bar{i}_{mm}|+T_r\,\mathrm{d}|\bar{i}_{mm}|/\mathrm{d}t}{T_{rl}}=\frac{\mathrm{d}\hat{i}_{sx}}{\mathrm{d}t}+\frac{\hat{i}_{sx}}{T_{rl}}+\frac{\mathrm{d}i_{dx}}{\mathrm{d}t}+\frac{i_{dx}}{T_{rl}}-\omega_{sl}i_{sy}. \tag{4.3-21}$$

Thus $|\bar{i}_{mm}|$ can be decoupled from the torque-producing stator current if i_{dx} is selected in such a way that the term $\omega_{sl}i_{sy}$ should disappear in eqn (4.3-21). Thus it follows from eqn (4.3-20) that this can be satisfied if

$$i_{dx}=\frac{\omega_{sl}i_{sy}T_{rl}}{1+T_{rl}p} \tag{4.3-22}$$

where $p=\mathrm{d}/\mathrm{d}t$. The angular slip frequency in eqn (4.3-22) can be obtained from eqn (4.3-19) and it follows that it takes the form,

$$\omega_{sl}=\frac{(1+T_{rl}p)i_{sy}}{T_r|\bar{i}_{mm}|-T_{rl}i_{sx}}. \tag{4.3-23}$$

Equations (4.3-22) and (4.3-23) can be used to construct the required decoupling circuit. When this is used, the magnetizing-flux-oriented control scheme of an induction machine supplied by a current-controlled PWM inverter will be similar to that shown in Fig. 4.26, which uses stator-flux-oriented control. Naturally for the magnetizing-flux-oriented control the transformation $e^{j\rho_s}$ in Fig. 4.26 has to be replaced by the transformation $e^{j\mu_m}$ and $|\bar{i}_{ms}|$ and $|\bar{i}_{msref}|$ have to be replaced by $|\bar{i}_{mm}|$ and $|\bar{i}_{mmref}|$ respectively.

The modulus and space angle of the magnetizing flux-linkage space phasor can also be obtained from the stator voltages and currents. For this purpose the stator voltage equation is formulated in the stationary reference frame. Thus by considering eqns (2.1-124) and (2.1-128), and also $L_s=L_{sl}+L_m$ and by neglecting the effects of magnetic saturation,

$$\bar{u}_s=R_s\bar{i}_s+L_{sl}\frac{\mathrm{d}\bar{i}_s}{\mathrm{d}t}+\frac{\mathrm{d}\bar{\psi}_m}{\mathrm{d}t}=u_{sD}+\mathrm{j}u_{sQ}, \tag{4.3-24}$$

where $\bar{i}_s = i_{sD} + ji_{sQ}$ and $\bar{\psi}_m$ is the magnetizing flux-linkage space phasor in the stationary reference frame which has also been used in eqn (2.1-178),

$$\bar{\psi}_m = L_m(\bar{i}_s + \bar{i}_r e^{j\theta_r}) = L_m \bar{i}_m = L_m(i_{mD} + ji_{mQ}) = L_m |\bar{i}_{mm}| e^{j\mu_m}. \tag{4.3-25}$$

In eqn (4.3-25) $\bar{i}_m$ is the magnetizing flux-linkage space phasor in the stationary reference frame, which in polar form is $|\bar{i}_{mm}|e^{j\mu_m}$ (for simplicity, instead of $|\bar{i}_{mm}|$ the notation $|\bar{i}_m|$ can be used, since the modulus of the magnetizing-current space phasor in the stationary reference, $|\bar{i}_m|$, is equal to the modulus of the same quantity in the magnetizing-flux-oriented reference frame $|\bar{i}_{mm}|$). Thus it follows from eqns (4.3-24) and (4.3-25) that

$$\frac{di_{mD}}{dt} = u_{sD} - R_s i_{sD} - L_{sl} \frac{di_{sD}}{dt} \tag{4.3-26}$$

$$\frac{di_{mQ}}{dt} = u_{sQ} - R_s i_{sQ} - L_{sl} \frac{di_{sQ}}{dt} \tag{4.3-27}$$

and i_{mD} and i_{mQ} can be obtained by considering eqns (4.3-26) and (4.3-27) and by monitoring the terminal voltages and currents of the machine. Thus $i_{mD} + ji_{mQ} = |\bar{i}_{mm}| e^{j\mu_m}$ can be used to obtain the modulus and the space angle of the magnetizing-current space phasor. In this case a flux model is obtained which is similar to that for an induction machine supplied by a voltage inverter. It is a disadvantage of the application of eqns (4.3-26) and (4.3-27) that they depend on the stator resistance and stator leakage inductance respectively and at low stator frequencies the stator ohmic drops will dominate. Thus prior to the integration of eqns (4.3-26) and (4.3-27), accurate stator voltage-drop compensation must be performed, which is a difficult task.

Equations (4.3-18) and (4.3-19) can be used to obtain the indirect magnetizing-flux-oriented control scheme of an induction machine with impressed stator currents. For this purpose the following expression for i_{sx} is obtained from eqn (4.3-18),

$$i_{sx} = \frac{(1 + T_r p)|\bar{i}_{mm}| + i_{sy} T_{rl} \omega_{sl}}{1 + T_{rl} p}, \tag{4.3-28}$$

where ω_{sl} has been defined in eqn (4.3-23).

Equations (4.3-28) and (4.3-23) are used in the indirect scheme shown in Fig. 4.34, where the induction machine is supplied by a current-controlled PWM inverter.

In Fig. 4.34(a) one of the inputs is the reference value of the modulus of the magnetizing current $|\bar{i}_{mmref}|$. The quadrature-axis stator current reference i_{syref} is obtained from the torque reference (t_{eref}) by utilizing eqn (4.3-1) and the torque reference is obtained as the output of the speed controller. In Fig. 4.34(a), within the dashed lines, the two-axis stator current references expressed in the stationary reference frame i_{sDref} and i_{sQref} are obtained by the application of the transformation $e^{j\mu_m}$ in accordance with eqn (4.3-3). These are then transformed into the three-phase reference values by the application of the three-phase to two-phase

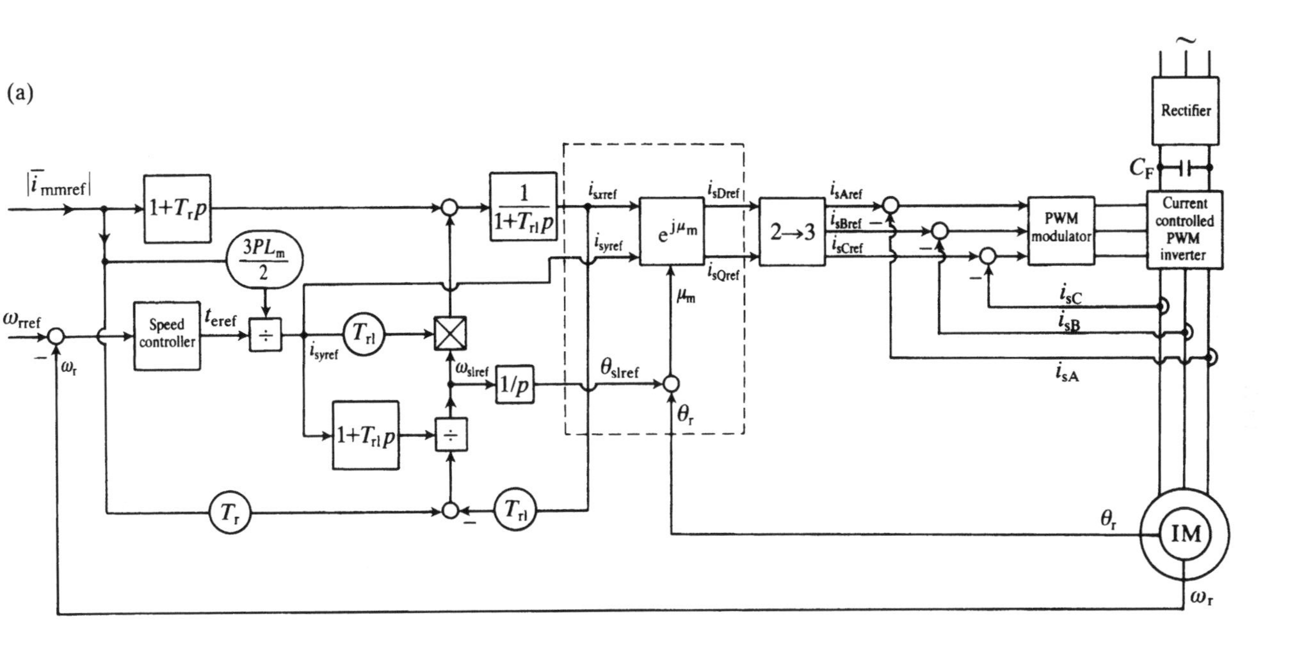

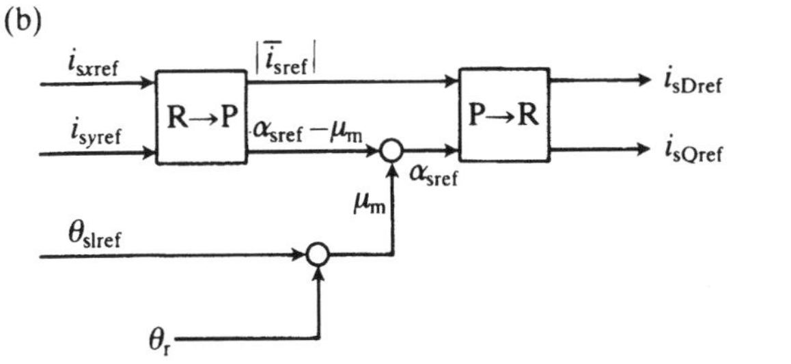

Fig. 4.34. Indirect magnetizing-flux-oriented control of a current-controlled PWM inverter-fed induction machine. (a) In Cartesian coordinates; (b) in polar coordinates.

transformation. The angle μ_m is obtained in two steps; first the angular slip frequency is integrated and this gives the slip angle θ_{sl}, which is also shown in Fig. 4.32 and is the angle of the magnetizing-current space phasor with respect to the real axis of the rotor reference frame. During the second step, the monitored value of the rotor angle (θ_r) is added to the slip angle, and thus the angle of the magnetizing-current space phasor with respect to the direct-axis of the stationary reference frame (μ_m) is obtained, as shown in Fig. 4.32.

In Fig. 4.34(a) the part shown within the dashed lines can be replaced by the part shown in Fig. 4.34(b). In this case, i_{sxref} and i_{syref} are inputs to a rectangular-to-polar converter, the inputs of which are the reference value of the modulus of the stator-current space phasor ($|\bar{i}_{sref}|$) and the space angle of the stator-current space phasor with respect to the direct axis of the magnetizing-flux-oriented reference frame ($\alpha_{sref}-\mu_m$) (see Fig. 4.32). From Fig. 4.32 the sum of the slip angle and the rotor angle give μ_m, and when this is added to ($\alpha_{sref}-\mu_m$), α_{sref} is obtained, which is the angle of the stator-current space phasor with respect to the direct axis of the stationary reference frame. Finally, a polar-to-rectangular converter is used to obtain i_{sDref} and i_{sQref}.

When the implementation given in Fig. 4.34(a) is compared with the implementation shown in Fig. 4.16(a), which corresponds to indirect rotor-flux-oriented control of the induction machine supplied by a current-controlled PWM inverter, it can be seen that it is much simpler to implement the rotor-flux-oriented control.

Furthermore, in the steady state, with constant-magnetizing-flux control, the electromagnetic torque has a pull-out value (t_{emax}) and this corresponds to the pull-out value of the angular slip frequency (ω_{slmax}). Beyond the pull-out value of the slip frequency, the induction machine can have a static stability limit. However, in contrast to this, in the steady state, with constant-rotor-flux control, there is no pull-out torque (if there is no limitation on the stator currents) and there is no static stability limit. With constant-stator-flux control, in the steady state there is also a pull-out torque and thus a static stability limit. These conclusions can be proved by considering the equations for the torque in the steady state. This will be discussed in the next section and the stability limits will be derived.

4.3.4 STEADY-STATE STABILITY LIMIT

For magnetizing-flux-oriented control it follows from eqn (4.3-17) that in the steady state (when all the derivative terms are zero), the stator-current space phasor in the magnetizing-flux-oriented reference frame can be expressed as

$$\bar{i}_{sm}=|\bar{i}_{mm}|\frac{R_r+j\omega_{sl}L_r}{R_r+j\omega_{sl}L_{rl}}=i_{sx}+ji_{sy}. \tag{4.3-29}$$

The torque producing stator current (i_{sy}) is equal to the imaginary part of eqn (4.3-29) and thus

$$i_{sy}=\frac{\omega_{sl}R_rL_m|\bar{i}_{mm}|}{R_r^2+\omega_{sl}^2L_{rl}^2}=\frac{\omega_{sl}R_r|\bar{\psi}_{mm}|}{R_r^2+\omega_{sl}^2L_{rl}^2}, \tag{4.3-30}$$

where $|\bar{\psi}_{mm}|$ is the modulus of the magnetizing flux-linkage space phasor. Thus by the substitution of eqn (4.3-30) into eqn (4.3-1), the following expression is obtained for the electromagnetic torque:

$$t_e = \frac{\frac{3}{2}PR_r|\bar{\psi}_{mm}|^2\omega_{sl}}{R_r^2+\omega_{sl}^2L_{rl}^2}. \tag{4.3-31}$$

This (and the torque-producing stator current) will have their maxima at the following value of the angular slip frequency,

$$\omega_{slmax} = \pm\frac{R_r}{L_{rl}} = \pm\frac{1}{T_{rl}}, \tag{4.3-32}$$

where T_{rl} is the leakage time constant of the rotor. Equation (4.3-32) can also be proved by differentiation of the electromagnetic torque (or the torque-producing stator current) with respect to the angular slip frequency, and setting this first derivative to zero. Substitution of eqn (4.3-32) into eqn (4.3-31) yields

$$t_{emax} = \pm\frac{\frac{3}{4}P|\bar{\psi}_{mm}|^2}{L_{rl}} = \pm\frac{\frac{3}{4}PL_m^2|\bar{i}_{mm}|^2}{L_{rl}} = \tfrac{3}{2}P|\bar{\psi}_{mm}|i_{symax}, \tag{4.3-33}$$

where i_{symax} is the maximal value of the torque-producing stator current. When the angular slip frequency is higher than ω_{slmax}, static instability arises. It follows from eqns (4.3-32) and (4.3-33) that the pull-out angular slip frequency depends only on the rotor leakage time constant and does not depend on the magnetizing flux. However, the pull-out torque is proportional to the square of the modulus of the magnetizing flux-linkage space phasor and thus a small increase of the magnetizing flux will cause a significant increase in the electromagnetic torque.

It is also possible to obtain the limit of static stability in terms of the torque-producing stator current. From eqn (4.3-33)

$$i_{symax} = \frac{\pm L_m|\bar{i}_{mm}|}{2L_{rl}} \tag{4.3-34}$$

and it follows that unstable operation arises if the torque-producing stator current is larger than $L_m|\bar{i}_{mm}|/(2L_{rl})$. If $L_{rl}/L_m = 1/30$ which is a typical value, instability only arises if $i_{sy} > 15|\bar{i}_{mm}|$, which is an unusually high value under steady-state conditions.

The static-stability limit in the case of stator-flux-oriented control with constant stator flux can be similarly determined by utilizing eqns (4.2-1) and (4.2-15). It follows from eqn (4.2-15) that the stator-current space phasor in the steady state is

$$\bar{i}_{s\psi s} = \frac{L_m}{L_s}\frac{|\bar{i}_{ms}|(R_r + j\omega_{sl}L_r)}{R_r + j\omega_{sl}L_r'} \tag{4.3-35}$$

and the torque-producing stator current (i_{sy}) is equal to its imaginary-axis component. The maximum of this is at

$$\omega_{slmax} = \pm\frac{1}{T_r'}, \tag{4.3-36}$$

where T_r' is the rotor transient time constant and thus i_{symax} can also be obtained. By the substitution of i_{symax} into eqn (4.2-1), the pull-out torque is obtained as

$$t_{emax} = \pm\frac{3}{4}P\left(\frac{L_m}{L_s}\right)^2 L_m^2|\bar{i}_{ms}|^2/L_r' = \pm\frac{3}{4}P\left(\frac{L_m}{L_s}\right)^2|\bar{\psi}_s|^2/L_r'$$

$$= \tfrac{3}{2}PL_m|\bar{i}_{ms}|i_{symax}, \qquad (4.3\text{-}37)$$

where $|\bar{\psi}_s|$ is the modulus of the stator flux-linkage space phasor. If the angular slip frequency is larger than ω_{slmax}, static instability arises. It follows from eqn (4.3-36) that this critical value of the angular slip frequency depends only on the rotor transient time constant and is not determined by the flux level. However, the pull-out torque depends on the square of the modulus of the stator linkage space phasor. It follows from eqn (4.3-37) that

$$i_{symax} = \pm\frac{L_m^3|\bar{i}_{ms}|}{2L_s^2L_r'} \qquad (4.3\text{-}38)$$

and static instability arises if $i_{sy} > L_m^3|\bar{i}_{ms}|/(2L_s^2L_r')$. By assuming equal stator and rotor leakages ($L_{sl} = L_{rl}$), the following inequality is obtained

$$i_{sy} > \frac{(1-\sigma)|\bar{i}_{ms}|}{2\sigma(1+L_{rl}/L_m)},$$

where σ is the resultant leakage constant, $\sigma = 1 - L_m^2/(L_sL_r)$. Thus when typical values of σ and L_{rl}/L_m are chosen, e.g. $\sigma = 1/15$ and $L_{rl}/L_m = 1/30$, it follows that the instability only arises if the torque-producing stator current is greater than $6.78|\bar{i}_{ms}|$. However, in the steady state it is unusual to have such a high torque-producing current and thus this instability does not cause a serious practical problem. It is also possible to obtain eqn (4.3-37) or eqn (4.3-38) by considering the general steady-state equivalent circuit of the induction machine shown in Fig. 4.10(b) and by utilizing the fact that the general turns ratio is equal to $a = L_s/L_m$.

Finally for the case of rotor-flux-oriented control it follows from eqn (4.1-26), that in the steady state the torque-producing stator current can be expressed as

$$i_{sy} = \omega_{sl}T_r|\bar{i}_{mr}|. \qquad (4.3\text{-}39)$$

This varies linearly with ω_{sl} (when $|\bar{i}_{mr}|$ = constant and T_r = constant), and thus the torque will also vary linearly with the angular slip frequency and there is no pull-out slip and no pull-out torque. This has also been discussed in Section 4.1.1.

4.3.5 MAGNETIZING-FLUX-ORIENTED CONTROL OF DOUBLE-CAGE INDUCTION MACHINES

4.3.5.1 General introduction

In all the previous sections, the squirrel-cage induction machine has been represented by its single-cage model and the effects of deep bars have not been considered. However, during transient operation, in an induction machine with deep bars, the deep-bar effect can significantly influence the rotor time constants of the machine, which are important parameters during vector control. A similar situation exists in a squirrel-cage machine with double-cage rotor. If the rotor time constants are inaccurate, the angular slip frequency (ω_{sl}) will be inaccurate, and this will result in an inaccurate slip angle (θ_{sl}). Thus the angle of the magnetizing flux-linkage space phasor (μ_m), which is equal to the sum of the slip angle and the monitored rotor angle (θ_r) (see Fig. 4.32), will also be inaccurate. Therefore the desired decoupling of flux- and torque-producing stator current components (i_{sx}, i_{sy}) cannot be achieved and this can lead to unwanted oscillations and a degradation of the performance of the induction machine. Compensation for the deep bar effects can improve the performance of the drive. For this purpose the angular slip frequency has to be calculated in such a way that it contains the effects of the deep bars (or double cage). In this section, the space-phasor model of the double-cage induction machine is developed and magnetizing-flux-oriented control of the double-cage machine with impressed stator currents is discussed in detail.

In practice, squirrel-cage rotors can have deep-bar rotors (with narrow and deep bars) or double-cage rotors with an outer and inner cage. In the case of the deep-bar rotor, it is possible to consider a bar with a large number of layers, and the leakage inductance of the bottom layer is greater than that of the top layer since the bottom layer links a greater amount of leakage flux. Thus the leakage inductance of the outer layer will be low and the leakage inductance of the inner layer will be high. Since all the layers can be considered to be electrically connected in parallel, when a.c. current flows in the cage (e.g. when the machine is started and the frequency of the rotor currents is high and equal to the stator frequency), the current in the upper layer (with low leakage reactance) will be larger than the current in a lower layer (with high leakage reactance). Thus the current will mainly flow in the top layers and the uneven rotor current distribution will result in an increase of effective rotor resistance and thus high starting torque is produced. When the machine accelerates, the rotor frequency decreases, so that at nominal speed the rotor frequency is very low and thus the effective rotor resistance is almost equal to the d.c. value. Thus in the induction machine with deep-bar rotor, high starting torque (resulting from high effective rotor resistance) and good running performance (resulting from low effective rotor resistance) are automatically ensured. As mentioned above, a rotor bar can be divided into several layers. In the dynamic state, to achieve high accuracy, several layers must be considered, but even the application of a model where a rotor bar is divided into two equal sections can give a substantial improvement in the computed characteristics, when compared with the application of a model where the rotor bar is not divided into layers. When a rotor bar is divided into two equal layers, the resistances of the two layers will be the same.

Similar conditions can be achieved by the application of a double-cage winding. In this case, the outer bars have smaller cross section, and thus they have a larger resistance than that of the inner bars. Furthermore, the leakage inductance of the inner bars is larger than that of the outer bars, since the inner bars link more leakage flux. During starting, most of the rotor current will flow in the outer bars which have high resistance and small leakage reactance. Thus high starting torque is produced. At nominal speed, the rotor frequency is so low that the reactance of the inner cage is considerably lower than its resistance and the effective rotor resistance will be low and will be approximately equal to the d.c. resistance of the two cages in parallel.

In the following sections the dynamic model of the double-cage induction machine is established by the utilization of space-phasor theory. Although the goal is to obtain the dynamic model in the magnetizing-flux-oriented reference frame, which could be directly obtained by considering the voltage equations in the general reference frame, for better understanding of the various quantities (space phasors of various flux linkages, space phasors of rotor currents in the upper and lower cages, etc.) first the space-phasor voltage equations of the stator and rotor are established in the stationary reference frame. This will be followed by transformation of these into the magnetizing-flux-oriented reference frame.

4.3.5.2 Stator- and rotor-voltage space-phasor equations in the stationary reference frame

The main assumptions are those used previously, in particular, the effects of magnetic saturation are neglected and the rotor is assumed to consist of two cages. Figure 4.35 shows the construction of the rotor. It is assumed that there is no mutual leakage between the stator and the upper cage (outer cage). Furthermore, in accordance with the discussion presented above, the leakage inductance of the upper cage is neglected. It is assumed that an upper bar and a lower bar occupy the same slot.

As in eqn (2.1-124), the space-phasor form of the stator voltage equation in the stationary reference frame is

$$\bar{u}_s = R_s \bar{i}_s + \frac{d\bar{\psi}_s}{dt}, \tag{4.3-40}$$

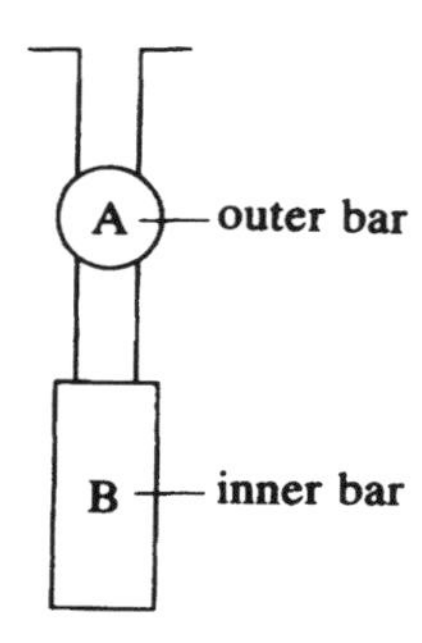

Fig. 4.35. Schematic of a double-cage.

where $\bar{u}_s$ and $\bar{\imath}_s$ are the space phasors of the stator voltages and currents respectively in the stationary reference frame, R_s is the stator resistance and $\bar{\psi}_s$ is the space phasor of the stator flux linkages in the stationary reference frame. This contains three components. The first component is a flux linkage due to the stator currents, $L_s\bar{\imath}_s$, where L_s is the self-inductance of the stator, which is the sum of the stator leakage inductance L_{sl} and the magnetizing inductance $L_m (L_s = L_{sl} + L_m)$. The second component is a mutual flux linkage due to the rotor currents in the upper bars (denoted by A in Fig. 4.35), $L_{mA}\bar{\imath}'_{rA}$, where L_{mA} is the mutual inductance between the stator and the upper cage and $\bar{\imath}'_{rA}$ is the space phasor of the rotor currents in the upper bars, but expressed in the stationary reference frame. Since the mutual leakage inductance between the stator and the upper cage has been neglected, it follows that $L_{mA} = L_m$. Finally the third component is a mutual flux-linkage component which is due to the rotor currents in the lower bars (denoted by B in Fig. 4.35), $L_{mB}\bar{\imath}'_{rB}$, where L_{mB} is the mutual inductance between the stator and the bottom cage and $\bar{\imath}'_{rB}$ is the space phasor of the rotor currents in the bottom bars, but expressed in the stationary reference frame. From the assumptions made above $L_{mA} = L_{mB} = L_m$. Thus $\bar{\psi}_s$ can be expressed as

$$\bar{\psi}_s = L_{sl}\bar{\imath}_s + L_m(\bar{\imath}_s + \bar{\imath}'_{rA} + \bar{\imath}'_{rB}) = L_{sl}\bar{\imath}_s + \bar{\psi}'_m, \tag{4.3-41}$$

where $\bar{\psi}'_m$ is the magnetizing flux-linkage space phasor in the stationary reference frame,

$$\bar{\psi}'_m = L_m(\bar{\imath}_s + \bar{\imath}'_{rA} + \bar{\imath}'_{rB}) = L_m\bar{\imath}'_m = L_m|\bar{\imath}_{mm}|\,e^{j\mu_m} \tag{4.3-42}$$

and in eqn (4.3-42) $\bar{\imath}'_m$ is the magnetizing current in the stationary reference frame, $|\bar{\imath}_{mm}|$ is its modulus and μ_m is its angle with respect to the real axis of the stationary reference frame, as shown in Fig. 4.32.

The space-phasor forms of the rotor voltage equations corresponding to the upper and lower cages, when expressed in the stationary reference frame, are similar to eqn (2.1-125),

$$0 = R_{rA}\bar{\imath}'_{rA} + \frac{d\bar{\psi}'_{rA}}{dt} - j\omega_r\bar{\psi}'_{rA} \tag{4.3-43}$$

and

$$0 = R_{rB}\bar{\imath}'_{rB} + \frac{d\bar{\psi}'_{rB}}{dt} - j\omega_r\bar{\psi}'_{rB}. \tag{4.3-44}$$

In eqn (4.3-43), $\bar{\psi}'_{rA}$ is the space phasor of the rotor flux linkages which link the upper cage and it is expressed in the stationary reference frame. $\bar{\psi}'_{rB}$ is the space phasor of the rotor flux linkages which link the bottom cage and it is also expressed in the stationary reference frame. From the assumptions made above, $\bar{\psi}'_{rA}$ will contain three components, two of which are leakage components, $L_{AB}\bar{\imath}'_{rA}$ and $L_{AB}\bar{\imath}'_{rB}$, where L_{AB} is the mutual leakage inductance between the upper and lower cages. The third component is equal to the magnetizing flux-linkage space phasor $\bar{\psi}'_m$ defined by eqn (4.3-42). Thus

$$\bar{\psi}'_{rA} = L_{AB}(\bar{i}'_{rA} + \bar{i}'_{rB}) + \bar{\psi}'_m. \tag{4.3-45}$$

In eqn (4.3-45) it has also been assumed that the leakage inductance of the upper cage is neglected. However, the leakage inductance of the lower cage ($L_{rBl} = L_{rl}$) is not neglected and thus $\bar{\psi}'_{rB}$ in eqn (4.3-44) can be expressed in a similar way to eqn (4.3-45), but it also has to include the leakage flux component $L_{rl}\bar{i}'_{rB}$. Thus

$$\bar{\psi}'_{rB} = L_{rl}\bar{i}'_{rB} + L_{AB}(\bar{i}'_{rA} + \bar{i}'_{rB}) + \bar{\psi}'_m. \tag{4.3-46}$$

Equations (4.3-40)–(4.3-46) define the voltage equations of the double-cage machine under dynamic conditions in the stationary reference frame. In the following section these equations will be expressed in the magnetizing-flux-oriented reference frame. It should be noted that, since there are two rotor flux-linkage space phasors (corresponding to the upper and lower cages respectively), in the case of the double-cage machine, it is not possible to establish a rotor-flux-oriented reference frame, where the stator currents can be decoupled into flux- and torque-producing components. This is in contrast to the single-cage induction machine, where it has been possible to use the rotor-flux-oriented reference frame and the stator currents have been decoupled into flux- and torque-producing components.

4.3.5.3 Stator- and rotor-voltage space-phasor equations in the magnetizing-flux-oriented reference frame

The voltage equations of the double-cage machine, expressed in the magnetizing-flux-oriented reference frame, can be obtained directly from the corresponding voltage equations expressed in the stationary reference frame by utilizing the appropriate transformations between these two reference frames. These will be described in the present section. The stator voltage equations expressed in the magnetizing-flux-oriented reference frame are required for the magnetizing-flux-oriented control scheme of the double-cage machine supplied by impressed stator voltages, but are not required when the same machine is supplied by impressed stator currents. In the latter case only the rotor voltage equations are required.

4.3.5.3.1 Stator voltage equations

By considering eqns (4.3-40) and (4.3-41), together with the required transformations of the stator voltage, current and flux-linkage space phasors [see eqns (4.3-48), (4.3-49), and (4.3-50)], it follows that in the special reference frame fixed to the magnetizing flux-linkage space phasor, which rotates at the speed of ω_m, the stator-voltage space-phasor equation will be

$$\bar{u}_{sm} = R_s\bar{i}_{sm} + \frac{d\bar{\psi}_{sm}}{dt} + j\omega_m\bar{\psi}_{sm}. \tag{4.3-47}$$

This is similar to eqn (2.1-148), which gives the stator-voltage space-phasor equation in the general reference frame. The required transformations used for the derivation of eqn (4.3-47) are as follows. The space phasor of the stator currents in the magnetizing-flux-oriented reference frame ($\bar{i}_{sm}$) has been defined in eqn (2.1-181), but for convenience it is repeated here,

$$\bar{i}_{sm}=\bar{i}_s e^{-j\mu_m}=(i_{sD}+ji_{sQ})e^{-j\mu_m}=i_{sx}+ji_{sy}, \tag{4.3-48}$$

and the space phasor of the stator voltages in the same reference frame can similarly be defined as

$$\bar{u}_{sm}=\bar{u}_s e^{-j\mu_m}=(u_{sD}+ji_{sQ})e^{-j\mu_m}=u_{sx}+ju_{sy}. \tag{4.3-49}$$

In these equations the components with the subscripts D and Q are direct- and quadrature-axis components in the stationary reference frame and the components with subscripts x and y are the direct- and quadrature-axis components in the magnetizing-flux-oriented reference frame (see Fig. 4.32). In eqn (4.3-47), $\bar{\psi}_{sm}$ is the space phasor of the stator flux linkages in the magnetizing-flux-oriented reference frame and the following transformation has been utilized during the derivation of eqn (4.3-47):

$$\bar{\psi}_{sm}=\bar{\psi}_s e^{-j\mu_m}=(\psi_{sD}+j\psi_{sQ})e^{-j\mu_m}=\psi_{sx}+j\psi_{sy}, \tag{4.3-50}$$

which is similar to the transformation defined by eqns (4.3-48) and (4.3-49). By substitution of eqn (4.3-41) into eqn (4.3-50) and by utilizing the definition of $\bar{i}_{sm}$ given in eqn (4.3-48) and the definition of $\bar{\psi}'_m$ given in eqn (4.3-42),

$$\begin{aligned}\bar{\psi}_{sm}&=\bar{\psi}_s e^{-j\mu_m}=(L_{sl}\bar{i}_s+\bar{\psi}'_m)e^{-j\mu_m}=L_{sl}\bar{i}_s e^{-j\mu_m}+\bar{\psi}'_m e^{-j\mu_m}\\&=L_{sl}\bar{i}_{sm}+L_m|\bar{i}_{mm}|=\psi_{sx}+j\psi_{sy},\end{aligned} \tag{4.3-51}$$

where ψ_{sx} and ψ_{sy} are the direct- and quadrature-axis components of the stator flux-linkage space phasor in the magnetizing-flux-oriented reference frame, and it follows from eqn (4.3-51) that they can be expressed as

$$\psi_{sx}=L_{sl}i_{sx}+L_m|\bar{i}_{mm}|=L_{sl}i_{sx}+|\bar{\psi}_{mm}| \tag{4.3-52}$$

$$\psi_{sy}=L_{sl}i_{sy}, \tag{4.3-53}$$

where $|\bar{\psi}_{mm}|$ is the modulus of the magnetizing flux-linkage space phasor and is equal to the space phasor of the magnetizing flux linkages in the magnetizing-flux-oriented reference frame. Thus the two-axis forms of the stator voltage equation in the magnetizing-flux-oriented reference frame are obtained by resolving eqn (4.3-47) into its real- and imaginary-axis components,

$$u_{sx}=R_s i_{sx}+\frac{d\psi_{sx}}{dt}-\omega_m\psi_{sy} \tag{4.3-54}$$

$$u_{sy}=R_s i_{sy}+\frac{d\psi_{sy}}{dt}+\omega_m\psi_{sx} \tag{4.3-55}$$

where ψ_{sx} and ψ_{sy} have been defined in eqns (4.3-52) and (4.3-53) respectively.

By multiplying eqn (4.3-42) by $e^{-j\mu_m}$, it follows that the space phasor of the magnetizing flux linkages in the magnetizing-flux-oriented reference frame is

$$\bar{\psi}'_m e^{-j\mu_m} = L_m(\bar{i}_s e^{-j\mu_m} + \bar{i}'_{rA} e^{-j\mu_m} + \bar{i}'_{rB} e^{-j\mu_m})$$

$$= L_m|\bar{i}_{mm}| = |\bar{\psi}_{mm}| = \psi_{mx} + j\psi_{my}, \quad (4.3\text{-}56)$$

where according to eqn (4.3-48) $\bar{i}_s e^{-j\mu_m} = \bar{i}_{sm}$ is the stator-current space phasor in the magnetizing-flux-oriented reference frame and the rotor-current phasors $\bar{i}'_{rA} e^{-j\mu_m}$ and $\bar{i}'_{rB} e^{-j\mu_m}$ are the space phasors of the rotor currents in the upper and lower cages respectively, but expressed in the magnetizing-flux-oriented reference frame,

$$\bar{i}_{rAm} = \bar{i}'_{rA} e^{-j\mu_m} = i_{rAx} + j i_{rAy} \quad (4.3\text{-}57)$$

$$\bar{i}_{rBm} = \bar{i}'_{rB} e^{-j\mu_m} = i_{rBx} + j i_{rBy}. \quad (4.3\text{-}58)$$

It should be noted that $\bar{i}'_{rA}$ and $\bar{i}'_{rB}$ are the space phasors of the rotor currents in the upper and lower cages respectively, but expressed in the stationary reference frame. Equations (4.3-57) and (4.3-58) are similar to eqn (4.3-4), which defines the space phasor of the rotor currents in the magnetizing-flux-oriented reference frame for a single-cage machine. In eqns (4.3-57) and (4.3-58) i_{rAx}, i_{rAy}, and i_{rBx}, i_{rBy} are the two-axis rotor currents of the upper and lower cages respectively in the magnetizing-flux-oriented reference frame. Thus by resolving eqn (4.3-56) into its real- and imaginary-axis components,

$$\psi_{mx} = |\bar{\psi}_{mm}| = L_m(i_{sx} + i_{rAx} + i_{rBx}) \quad (4.3\text{-}59)$$

$$\psi_{my} = 0 = L_m(i_{sy} + i_{rAy} + i_{rBy}). \quad (4.3\text{-}60)$$

These two equations will be used below, when the state-variable equations of the double-cage induction machine with impressed stator currents are obtained, to express $i_{rAx} + i_{rBx}$ in terms of $|\psi_{mm}|$ and i_{sx}, and to express $(i_{rAy} + i_{rBy})$ in terms of i_{sy}.

Equations (4.3-54), (4.3-55), (4.3-52), and (4.3-53) have to be considered if the double-cage induction machine is supplied by impressed stator voltages. They can be put into a form which is similar to eqns (4.3-10) and (4.3-11) for the single-cage machine. Thus the required decoupling circuit can be constructed in a similar way to that shown in Fig. 4.33 and it should be noted that $|\bar{\psi}_{mm}| = L_m|\bar{i}_{mm}|$. In such a decoupling circuit, the stator currents i_{sx} and i_{sy} can be obtained from the monitored three-phase stator currents by utilizing eqn (4.3-48), and $|\bar{i}_{mm}|$ and ω_m (and μ_m) can be obtained either from a flux-model which uses the rotor equations of the machine expressed in the magnetizing-flux-oriented reference frame or by using direct measurements (e.g. Hall-sensors, special search coils, etc.). The rotor voltage equations will be obtained in the following section. They will also be used to obtain the magnetizing-flux-oriented control scheme of the double-cage machine with impressed stator currents.

4.3.5.3.2 *Rotor voltage equations*

By utilizing the required transformations [see eqns (4.3-57), (4.3-58), (4.3-63), and (4.3-64)], it follows from eqns (4.3-43) and (4.3-44) that the rotor-voltage space-phasor equations in the magnetizing-flux-oriented reference frame will take the form,

$$0=R_{rA}\bar{i}_{rAm}+\frac{d\bar{\psi}_{rAm}}{dt}+j(\omega_m-\omega_r)\bar{\psi}_{rAm} \tag{4.3-61}$$

$$0=R_{rB}\bar{i}_{rBm}+\frac{d\bar{\psi}_{rBm}}{dt}+j(\omega_m-\omega_r)\bar{\psi}_{rBm}. \tag{4.3-62}$$

These are similar to the rotor voltage equations of a single-cage machine expressed in the same reference frame, which can be obtained from eqn (2.1-149). For the derivation of eqns (4.3-61) and (4.3-62), eqns (4.3-57) and (4.3-58) have also been utilized; they give the space-phasor forms of the rotor currents of the upper and lower cages respectively in the magnetizing-flux-oriented reference frame ($\bar{i}_{rAm}, \bar{i}_{rBm}$). A similar definition has been used for the space phasors $\bar{\psi}_{rAm}$ and $\bar{\psi}_{rBm}$, which are the space phasors of the rotor flux linkages for the upper and lower cages, expressed in the magnetizing-flux-oriented reference frame,

$$\bar{\psi}_{rAm}=\bar{\psi}'_{rA}\,e^{-j\mu_m}=\psi_{rAx}+j\psi_{rAy} \tag{4.3-63}$$

$$\bar{\psi}_{rBm}=\bar{\psi}'_{rB}\,e^{-j\mu_m}=\psi_{rBx}+j\psi_{rBy}, \tag{4.3-64}$$

where $\bar{\psi}'_{rA}$ and $\bar{\psi}'_{rB}$ have been defined in eqns (4.3-45) and (4.3-46) respectively.

By substituting eqn (4.3-45) into eqn (4.3-63) and considering eqns (4.3-57), (4.3-58), and (4.3-56),

$$\bar{\psi}_{rAm}=[L_{AB}(\bar{i}'_{rA}+\bar{i}'_{rB})+\bar{\psi}'_m]\,e^{-j\mu_m}=L_{AB}(\bar{i}_{rAm}+\bar{i}_{rBm})+|\bar{\psi}_{mm}|. \tag{4.3-65}$$

Similarly, substituting eqn (4.3-46) into eqn (4.3-64) and considering eqns (4.3-57), (4.3-58), and (4.3-56) gives

$$\begin{aligned}\bar{\psi}_{rBm}&=[L_{rl}\bar{i}'_{rB}+L_{AB}(\bar{i}'_{rA}+\bar{i}'_{rB})+\bar{\psi}'_m]\,e^{-j\mu_m}\\&=L_{rl}\bar{i}_{rBm}+L_{AB}(\bar{i}_{rAm}+\bar{i}_{rBm})+|\bar{\psi}_{mm}|.\end{aligned} \tag{4.3-66}$$

Resolution of eqns (4.3-61), (4.3-62), (4.3-65), and (4.3-66) into their real- and imaginary-axis components give

$$0=R_{rA}i_{rAx}+\frac{d\psi_{rAx}}{dt}-(\omega_m-\omega_r)\psi_{rAy} \tag{4.3-67}$$

$$0=R_{rA}i_{rAy}+\frac{d\psi_{rAy}}{dt}+(\omega_m-\omega_r)\psi_{rAx} \tag{4.3-68}$$

$$0=R_{rB}i_{rBx}+\frac{d\psi_{rBx}}{dt}-(\omega_m-\omega_r)\psi_{rBy} \tag{4.3-69}$$

$$0=R_{rB}i_{rBy}+\frac{d\psi_{rBy}}{dt}+(\omega_m-\omega_r)\psi_{rBx}, \tag{4.3-70}$$

where

$$\psi_{rAx}=L_{AB}(i_{rAx}+i_{rBx})+|\bar{\psi}_{mm}| \qquad (4.3\text{-}71)$$

$$\psi_{rAy}=L_{AB}(i_{rAy}+i_{rBy}) \qquad (4.3\text{-}72)$$

$$\psi_{rBx}=L_{rl}i_{rBx}+L_{AB}(i_{rAx}+i_{rBx})+|\bar{\psi}_{mm}| \qquad (4.3\text{-}73)$$

$$\psi_{rBy}=L_{rl}i_{rBy}+L_{AB}(i_{rAy}+i_{rBy}), \qquad (4.3\text{-}74)$$

where $|\bar{\psi}_{mm}|$ has been defined in eqn (4.3-59). The stator voltage equations, eqns (4.3-54) and (4.3-55), which use the expressions for the stator flux linkages, eqns (4.3-52) and (4.3-53), together with the rotor voltage equations, eqns (4.3-67)–(4.3-70), which use the rotor flux-linkage equations, eqns (4.3-71)–(4.3-74), describe the dynamic behaviour of the double-cage induction machine if the equation of motion is also considered. The expression for the electromagnetic torque will be obtained in the next section.

4.3.5.3.3 The electromagnetic torque

The electromagnetic torque can be obtained in a similar way to that shown in eqn (4.3-1) for the single-cage machine. Physically this must be the case since it has been shown earlier that the electromagnetic torque can be considered to be produced by the interaction of the stator flux linkages and the stator currents and from this point of view, it is irrelevant how many cages there are on the rotor. Mathematically, it is possible to prove in several ways that the expression for the electromagnetic torque in the double-cage machine is similar to that given by eqn (4.3-1), but, of course, the modulus of the magnetizing flux-linkage space phasor ($|\bar{\psi}_{mm}|$) is different for the double-cage machine than for the single-cage machine, since as well as the stator currents, the rotor currents of both rotor cages contribute to the magnetization current. Furthermore the torque-producing stator current (i_{sy}) is different in the double-cage machine from that of the single-cage machine. It is very convenient to utilize a form of the torque equation which contains the stator flux-linkage space phasor, rather than other forms where two rotor flux-linkage space phasors have to be used.

In eqn (2.1-167) the electromagnetic torque has been defined by utilizing the space phasors of the stator flux linkages and the stator current in the general reference frame. Thus when the magnetizing-flux-oriented reference frame is used, it follows from eqn (2.1-167) that

$$t_e=\tfrac{3}{2}P\bar{\psi}_{sm}\times\bar{i}_{sm}, \qquad (4.3\text{-}75)$$

where $\bar{\psi}_{sm}$ and $\bar{i}_{sm}$ have been defined in eqns (4.3-48) and (4.3-51) respectively. By performing the vector product, the following expression is obtained for the electromagnetic torque, if the stator flux-linkage components are replaced by the expressions given in eqns (4.3-52) and (4.3-53)

$$\begin{aligned} t_e&=\tfrac{3}{2}P(\psi_{sx}i_{sy}-\psi_{sy}i_{sx})\\ &=\tfrac{3}{2}P[(L_{sl}i_{sx}+|\bar{\psi}_{mm}|)i_{sy}-L_{sl}i_{sy}i_{sx}]=\tfrac{3}{2}P|\bar{\psi}_{mm}|i_{sy}, \end{aligned} \qquad (4.3\text{-}76)$$

which as expected agrees with the form of eqn (4.3-1). In eqn (4.3-76) the modulus of the magnetizing flux-linkage space phasor, ($|\bar{\psi}_{\text{mm}}|$), is defined in eqn (4.3-59) and the torque-producing stator current (i_{sy}) is related to the direct-axis and quadrature-axis stator currents in the stationary reference frame by eqn (4.3-48).

4.3.5.4 Magnetizing-flux-oriented control of the double-cage induction machine with impressed stator currents

In this chapter the indirect magnetizing-flux-oriented control scheme of the double-cage induction machine with impressed stator currents is developed. For this purpose the rotor equations derived in the stationary reference frame are utilized together with the expression for the electromagnetic torque given in eqn (4.3-76). However, the rotor voltage equations are reformulated in such a way that the stator currents i_{sx}, i_{sy} are retained. $|\bar{\psi}_{\text{mm}}|$ is an input command and μ_{m} is controlled indirectly. Since there are four rotor voltage equations, two other state variables also have to be selected. These are chosen to be the real- and imaginary-axis components ($\psi_{\text{rBl}x}$, $\psi_{\text{rBl}y}$) of the space phasor of the rotor leakage flux linkages in the lower cage. This space phasor has been defined in eqn (4.3-66) as $\bar{\psi}_{\text{rBl}} = L_{\text{rl}}\bar{i}_{\text{rBm}}$ and thus its two-axis components in the magnetizing-flux-oriented reference frame are

$$\psi_{\text{rBl}x} = L_{\text{rl}} i_{\text{rB}x} \tag{4.3-77}$$

$$\psi_{\text{rBl}y} = L_{\text{rl}} i_{\text{rB}y}. \tag{4.3-78}$$

These are selected as state variables since they are essential for the description of the double-cage (deep-bar) effect. The direct-axis stator current (i_{sx}) and the angular slip frequency (ω_{sl}) are strongly dependent on these. It should be noted that L_{rl} is the leakage inductance of the bottom cage, and in the transient state, the non-uniform current distribution between the upper and lower cages depends on the energy stored in this inductance.

The four rotor voltage equations, eqns (4.3-67)–(4.3-70), are now rearranged to contain the required four state variables. For this purpose, first the rotor flux-linkage components given by eqns (4.3-71)–(4.3-74) are substituted into eqns (4.3-67)–(4.3-70). The resulting equations will thus contain the four rotor current components ($i_{\text{rA}x}, i_{\text{rA}y}, i_{\text{rB}x}, i_{\text{rB}y}$):

$$0 = R_{\text{rA}} i_{\text{rA}x} + \frac{\mathrm{d}|\bar{\psi}_{\text{mm}}|}{\mathrm{d}t} + L_{\text{AB}} \frac{\mathrm{d}(i_{\text{rA}x} + i_{\text{rB}x})}{\mathrm{d}t} - \omega_{\text{sl}} L_{\text{AB}}(i_{\text{rA}y} + i_{\text{rB}y}) \tag{4.3-79}$$

$$0 = R_{\text{rA}} i_{\text{rA}y} + L_{\text{AB}} \frac{\mathrm{d}(i_{\text{rA}y} + i_{\text{rB}y})}{\mathrm{d}t} + \omega_{\text{sl}}[|\bar{\psi}_{\text{mm}}| + L_{\text{AB}}(i_{\text{rA}x} + i_{\text{rB}x})] \tag{4.3-80}$$

$$0 = R_{\text{rB}} i_{\text{rB}x} + \frac{\mathrm{d}|\bar{\psi}_{\text{mm}}|}{\mathrm{d}t} + \frac{\mathrm{d}(L_{\text{rl}} i_{\text{rB}x})}{\mathrm{d}t} + L_{\text{AB}} \frac{\mathrm{d}(i_{\text{rA}x} + i_{\text{rB}x})}{\mathrm{d}t}$$

$$- \omega_{\text{sl}}[L_{\text{rl}} i_{\text{rB}y} + L_{\text{AB}}(i_{\text{rA}y} + i_{\text{rB}y})] \tag{4.3-81}$$

$$0 = R_{\mathrm{rB}} i_{\mathrm{rB}y} + \frac{\mathrm{d}(L_{\mathrm{rl}} i_{\mathrm{rB}y})}{\mathrm{d}t} + L_{\mathrm{AB}} \frac{\mathrm{d}(i_{\mathrm{rA}y} + i_{\mathrm{rB}y})}{\mathrm{d}t}$$

$$+ \omega_{\mathrm{sl}}[L_{\mathrm{rl}} i_{\mathrm{rB}x} + L_{\mathrm{AB}}(i_{\mathrm{rA}x} + i_{\mathrm{rB}x}) + |\bar{\psi}_{\mathrm{mm}}|], \tag{4.3-82}$$

where $\omega_{\mathrm{sl}} = (\omega_{\mathrm{m}} - \omega_{\mathrm{r}})$ is the angular slip frequency.

However, the rotor current components can be eliminated by expressing them in terms of the four state-variables, since it follows from eqns (4.3-77) and (4.3-78) that

$$i_{\mathrm{rB}x} = \frac{\psi_{\mathrm{rBl}x}}{L_{\mathrm{rl}}} \tag{4.3-83}$$

$$i_{\mathrm{rB}y} = \frac{\psi_{\mathrm{rBl}y}}{L_{\mathrm{rl}}}, \tag{4.3-84}$$

and from eqns (4.3-59) and (4.3-60) that

$$i_{\mathrm{rA}x} = \frac{|\bar{\psi}_{\mathrm{mm}}|}{L_{\mathrm{m}}} - i_{\mathrm{s}x} - i_{\mathrm{rB}x} = \frac{\bar{\psi}_{\mathrm{mm}}}{L_{\mathrm{m}}} - i_{\mathrm{s}x} - \frac{\psi_{\mathrm{rBl}x}}{L_{\mathrm{rl}}} \tag{4.3-85}$$

$$i_{\mathrm{rA}y} = -i_{\mathrm{s}y} - i_{\mathrm{rB}y} = -i_{\mathrm{s}y} - \frac{\psi_{\mathrm{rBl}y}}{L_{\mathrm{rl}}}, \tag{4.3-86}$$

where $i_{\mathrm{rB}x}$ and $i_{\mathrm{rB}y}$ have been substituted by the expressions given by eqns (4.3-83) and (4.3-84) respectively. Thus when eqns (4.3-83)–(4.3-86) are substituted into the rotor voltage equations, eqns (4.3-79)–(4.3-82), the following four equations are obtained. The direct-axis rotor voltage equation of the upper cage yields

$$\frac{\mathrm{d}|\bar{\psi}_{\mathrm{mm}}|}{\mathrm{d}t} = -\frac{R_{\mathrm{rA}}|\bar{\psi}_{\mathrm{mm}}|}{L_{\mathrm{m}} + L_{\mathrm{AB}}} + \frac{L_{\mathrm{m}}}{L_{\mathrm{m}} + L_{\mathrm{AB}}}\left(R_{\mathrm{rA}} i_{\mathrm{s}x} + L_{\mathrm{AB}} \frac{\mathrm{d}i_{\mathrm{s}x}}{\mathrm{d}t} + \frac{R_{\mathrm{rA}}\psi_{\mathrm{rBl}x}}{L_{\mathrm{rl}}} - L_{\mathrm{AB}}\omega_{\mathrm{sl}} i_{\mathrm{s}y}\right). \tag{4.3-87}$$

The quadrature-axis rotor voltage equation of the upper cage yields

$$\omega_{\mathrm{sl}} = \left[R_{\mathrm{rA}}\left(i_{\mathrm{s}y} + \frac{\psi_{\mathrm{rBl}y}}{L_{\mathrm{rl}}}\right) + L_{\mathrm{AB}} \frac{\mathrm{d}i_{\mathrm{s}y}}{\mathrm{d}t}\right] \Big/ \psi_{\mathrm{rA}x}, \tag{4.3-88}$$

where in the denominator,

$$\psi_{\mathrm{rA}x} = |\bar{\psi}_{\mathrm{mm}}| \frac{(L_{\mathrm{m}} + L_{\mathrm{AB}})}{L_{\mathrm{m}}} - L_{\mathrm{AB}} i_{\mathrm{s}x} \tag{4.3-89}$$

is the direct-axis flux-linkage component of the upper cage. It follows from the direct-axis rotor voltage equation of the lower cage that

$$\frac{\mathrm{d}\psi_{\mathrm{rBl}x}}{\mathrm{d}t} = -\frac{R_{\mathrm{rB}}\psi_{\mathrm{rBl}x}}{L_{\mathrm{rl}}} - \frac{L_{\mathrm{AB}} + L_{\mathrm{m}}}{L_{\mathrm{m}}} \frac{\mathrm{d}|\bar{\psi}_{\mathrm{mm}}|}{\mathrm{d}t} + L_{\mathrm{AB}} \frac{\mathrm{d}i_{\mathrm{s}x}}{\mathrm{d}t} - \omega_{\mathrm{sl}} L_{\mathrm{AB}} i_{\mathrm{s}y} + \omega_{\mathrm{sl}}\psi_{\mathrm{rBl}y}. \tag{4.3-90}$$

Finally, the following equation is obtained from the quadrature-axis rotor voltage equation of the lower cage:

$$\frac{d\psi_{rBly}}{dt} = -\frac{R_{rB}\psi_{rBly}}{L_{rl}} - \omega_{sl}\psi_{rBlx} + L_{AB}\frac{di_{sy}}{dt} - \omega_{sl}\psi_{rAx}, \tag{4.3-91}$$

where ψ_{rAx} has been defined in eqn (4.3-89). However, eqn (4.3-91) contains the first derivative di_{sy}/dt and this can be eliminated by using eqn (4.3-88). Thus from eqn (4.3-88) ψ_{rAx} is expressed in terms of ω_{sl}, i_{sy}, and ψ_{rBly} and, when this is substituted into eqn (4.3-91),

$$\frac{d\psi_{rBly}}{dt} = -\frac{R_{rA} + R_{rB}}{L_{rl}}\psi_{rBly} - R_{rA}i_{sy} - \omega_{sl}\psi_{rBlx}. \tag{4.3-92}$$

Thus eqns (4.3-87), (4.3-88), (4.3-89), (4.3-90), and (4.3-92), together with the expression of the electromagnetic torque, eqn (4.3-76), can be used to obtain the indirect magnetizing-flux-oriented control scheme of the double-cage induction machine supplied by impressed stator currents. For this purpose it should be considered that if the reference value of the modulus of the magnetizing flux-linkage space phasor ($|\bar{\psi}_{mmref}|$) and the reference value of the torque-producing stator current (i_{syref}) are known, in the indirect control scheme i_{sxref} and ω_{slref} have to be determined. Thus these quantities have to be expressed in terms of ($|\bar{\psi}_{mmref}|$) and (i_{syref}), but in accordance with eqns (4.3-87)–(4.3-90) and eqn (4.3-92), they will depend on the leakage flux components ψ_{rBlx} and ψ_{rBly}. Therefore to obtain the control scheme, the equations given above have to be arranged for i_{sx} and ω_{sl}, and the leakage fluxes of the bottom cages (ψ_{rBlx} and ψ_{rBly}) have to be expressed in terms of $|\bar{i}_{mm}|$, i_{sx}, and i_{sy}. Thus when eqn (4.3-87) is rearranged, the following equation is obtained for the direct-axis stator current:

$$(1 + T_{AB}p)i_{sx} = (1 + T_{mAB}p)|\bar{i}_{mm}| + T_{AB}\omega_{sl}i_{sy} - \frac{\psi_{rBlx}}{L_{rl}}, \tag{4.3-93}$$

where $p = d/dt$ and the mutual time constants associated with the upper cage are

$$T_{AB} = \frac{L_{AB}}{R_{rA}}, \qquad T_{mAB} = T_m + T_{AB}, \quad \text{and} \quad T_m = \frac{L_m}{R_{rA}}.$$

Equation (4.3-88) yields the following expression for the angular slip frequency:

$$\omega_{sl} = \frac{i_{sy}(1 + T_{AB}p) + \psi_{rBly}/L_{rl})}{T_{mAB}|\bar{i}_{mm}| - T_{AB}i_{sx}}. \tag{4.3-94}$$

In eqns (4.3-93) and (4.3-94) the rotor leakage-flux linkages are also present. These are obtained from eqns (4.3-89) and (4.3-90) as

$$(p + T_{Bl})\psi_{rBlx} = -pR_{rA}(T_{mAB}|\bar{i}_{mm}| - T_{AB}i_{sx}) - \omega_{sl}L_{AB}i_{sy} + \omega_{sl}\psi_{rBly}, \tag{4.3-95}$$

where $T_{Bl} = R_{rB}/L_{rl}$ is the leakage time constant of the lower cage,

$$(p + T_{ABl})\psi_{rBly} = -R_{rA}i_{sy} - \omega_{sl}\psi_{rBlx} \tag{4.3-96}$$

and T_{ABl} is the total leakage time constant of the upper and lower cages,

$$T_{ABl} = T_{Bl} + T_{Al} \quad \text{where} \quad T_{Al} = \frac{R_{rA}}{L_{rl}}.$$

Equations (4.3-93)–(4.3-96) are used in Fig. 4.36, which shows the magnetizing-flux-oriented indirect control scheme of the double-cage induction machine with impressed stator currents. The converter which supplies the induction machine can be either a CSI or a current-controlled PWM inverter.

In Fig. 4.36 the required reference values of the stator currents are obtained in a way similar to that shown in Fig. 4.34 for the single-cage machine. Thus by using the reference values of the direct- and quadrature-axis stator current components expressed in the magnetizing-flux-oriented reference frame (i_{sxref}, i_{syref}), it is possible to obtain the reference value of the space phasor of the stator currents in the magnetizing-flux-oriented reference frame,

$$(i_{sxref} + \mathrm{j}i_{syref}) = |\bar{i}_{sref}|\, \mathrm{e}^{\mathrm{j}(\alpha_{sref} - \mu_m)} \tag{4.3-97}$$

(see Fig. 4.32). Equation (4.3-97) can be used to obtain the reference values of the direct- and quadrature-axis stator currents in the stationary reference frame (i_{sDref}, i_{sQref}), by either of the methods shown in Fig. 4.34(a) or Fig. 4.34(b). If the method of Fig. 4.34(b) is used, $|\bar{i}_{sref}|$ and $(\alpha_{sref} - \mu_m)$ are obtained by polar-to-rectangular conversion from i_{sxref} and i_{syref}, since eqn (4.3-97) is utilized. When the angular slip frequency (ω_{sl}) obtained from eqn (4.3-94) is integrated, the slip angle (θ_{sl}) is obtained and when the monitored rotor angle (θ_r) is added to this, the angle of the magnetizing flux-linkage space phasor with respect to the direct-axis of the magnetizing-oriented reference frame (μ_m) is obtained (this also follows from Fig. 4.32) as

$$\mu_m = \theta_{slref} + \theta_r. \tag{4.3-98}$$

Thus when μ_m is added to $(\alpha_{sref} - \mu_m)$, α_{sref} is obtained, which is the space angle of the stator-current space phasor with respect to the direct axis of the stationary reference frame (see Fig. 4.32).

In Fig. 4.36, the circuit which produces the required leakage-flux components of the bottom cage (ψ_{rBlx}, ψ_{rBly}), contains a differentiator (indicated by the block containing the operator p) and furthermore, it is asymmetrical. However, it is possible to construct a circuit which produces the required rotor leakage fluxes, where differentiation is not required and which is symmetrical. For this purpose, eqn (4.3-95) is rearranged to resemble the simple structure of eqn (4.3-96), but eqn (4.3-96) itself will not be changed.

Thus when the expression for the voltage component $R_{rA}|\bar{\psi}_{mm}|/L_{rl}$ is taken from eqn (4.3-93) [or it may be easier to use eqn (4.3-87) directly] and the expression is added to the expression for the rotational voltage, $\omega_{sl}\psi_{rBly}$, which can be obtained from eqn (4.3-95) [or eqn (4.3-90)], the following voltage equation is obtained:

$$(p + T_{ABl})\psi_{rBlx} = R_{rA}(|\bar{i}_{mm}| - i_{sx}) + \omega_{sl}\psi_{rBly}. \tag{4.3-99}$$

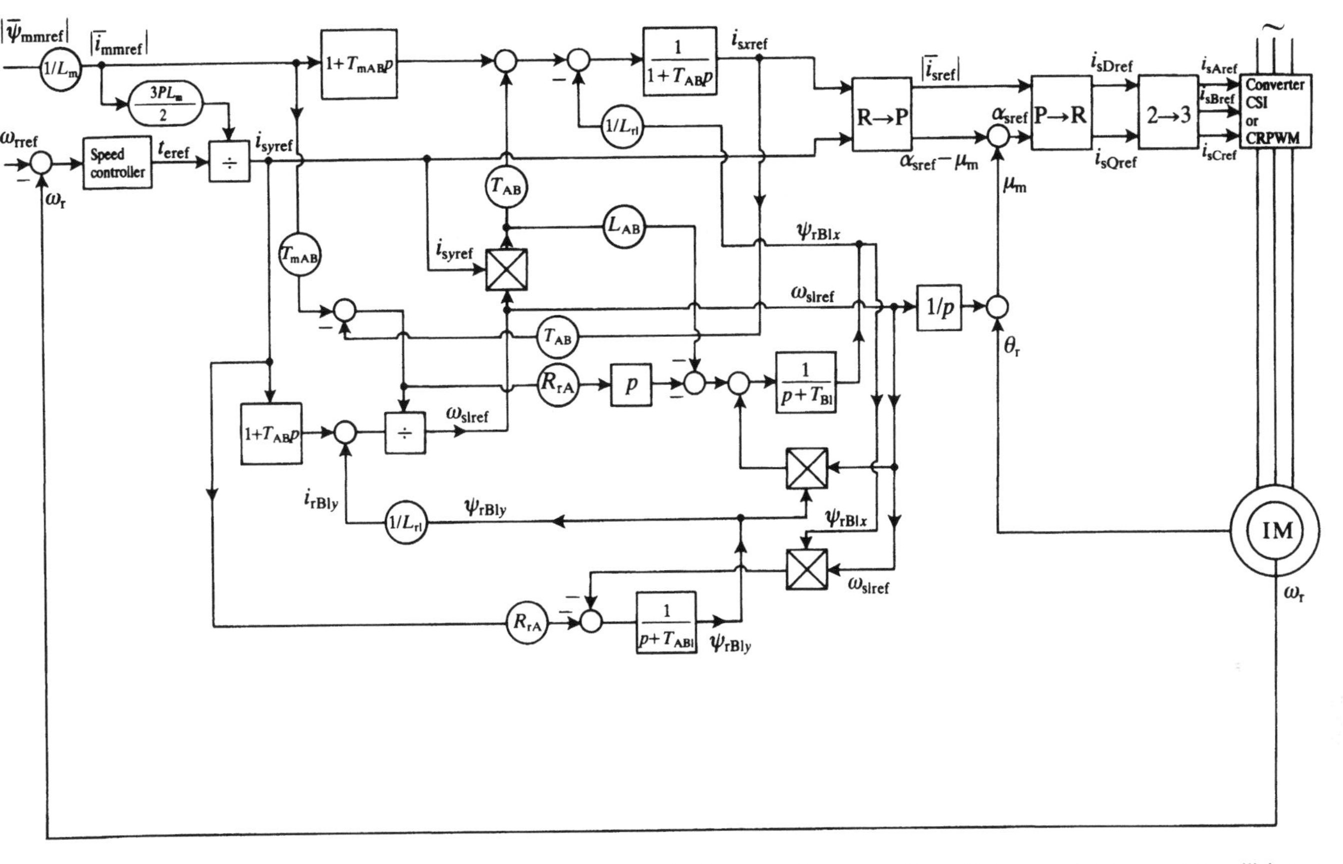

Fig. 4.36. Schematic of the indirect magnetizing-flux-oriented control of a double-cage induction machine with impressed stator currents utilizing an asymmetrical rotor flux-linkage estimator.

The structure of eqn (4.3-99) resembles that of eqn (4.3-96) and thus eqns (4.3-96) and (4.3-99) describe a rotor flux-linkage estimator circuit which is symmetrical and does not contain a differentiator. Figure 4.37 shows the schematic of the indirect magnetizing-flux-oriented control of the double-cage induction machine with impressed stator currents where the rotor flux-linkage estimator circuit is symmetrical.

In the implementations shown in Fig. 4.36 and Fig. 4.37, the quadrature-axis stator current reference (i_{syref}) is obtained from the torque reference (t_{eref}), by using eqn (4.3-76). The torque reference is obtained as the output of the speed controller, which can be a PI controller. When the vector-control schemes shown in Fig. 4.36 and Fig. 4.37 are compared with the scheme shown in Fig. 4.34, which corresponds to the single-cage machine, there is a resemblance, although as a result of the double-cage, extra parts are present in Fig. 4.36 and Fig. 4.37. The main difference between the magnetizing-flux-oriented control schemes corresponding to the single-cage and double-cage machines is that in the scheme for the double-cage machine, an estimation circuit is present, which derives the direct- and quadrature-axis leakage flux-linkage components of the bottom cage. However, the required flux linkages can be obtained in real time at very high speed by using digital signal processors.

For indirect vector control of both the single-cage machine and the double-cage machine, it has been necessary to calculate the angular slip frequency by, for example, the circuits shown in Fig. 4.34(a) and Fig. 4.37 respectively. It follows from eqn (4.3-94) which defines ω_{sl}, by considering eqns (4.3-93), (4.3-95), and (4.3-96), and by setting all the derivative terms to be zero, that the expression for ω_{sl} is obtained in the steady state. This gives a fourth-order equation for ω_{sl}. It follows from this equation that if $i_{sx}/|\bar{\imath}_{mm}|$ and i_{sy}/i_{sx} are small and ω_{sl} is also small, the angular slip frequency of the double-cage machine takes the same form as the angular slip frequency of the single-cage machine. However, while for the single-cage machine the angular slip frequency is proportional to the rotor resistance (R_r) for the double-cage machine it is proportional to a resultant rotor resistance, which is obtained by connecting in parallel the resistors of the upper and lower cages, giving $R_{rA}R_{rB}/(R_{rA}+R_{rB})$. Usually the rotor resistance of the single-cage machine R_r and the resultant rotor resistance of the double-cage machine $R_{rA}R_{rB}/(R_{rA}+R_{rB})$ are different and when a double-cage machine is subjected to vector control and ω_{sl} is obtained by utilizing the equations of the double-cage machine, since the resistances of the upper and lower cages are independent of the rotor slip frequency, more accurate values of ω_{sl} are obtained than by utilizing the single-cage equations. This also holds at low values of the torque-producing current. It can also be shown, by considering the steady-state expression for ω_{sl}, that when ω_{sl} is calculated as a function of $i_{sx}/|\bar{\imath}_{mm}|$ or of i_{sy}/i_{sx}, it will have a maximum value. In the steady state this limits the stability region of the double-cage induction machine subjected to magnetizing-flux-oriented control.

When eqn (4.3-94) is used to obtain the expression for the angular slip frequency in the transient state, by also considering eqns (4.3-93), (4.3-95), and (4.3-96) it is

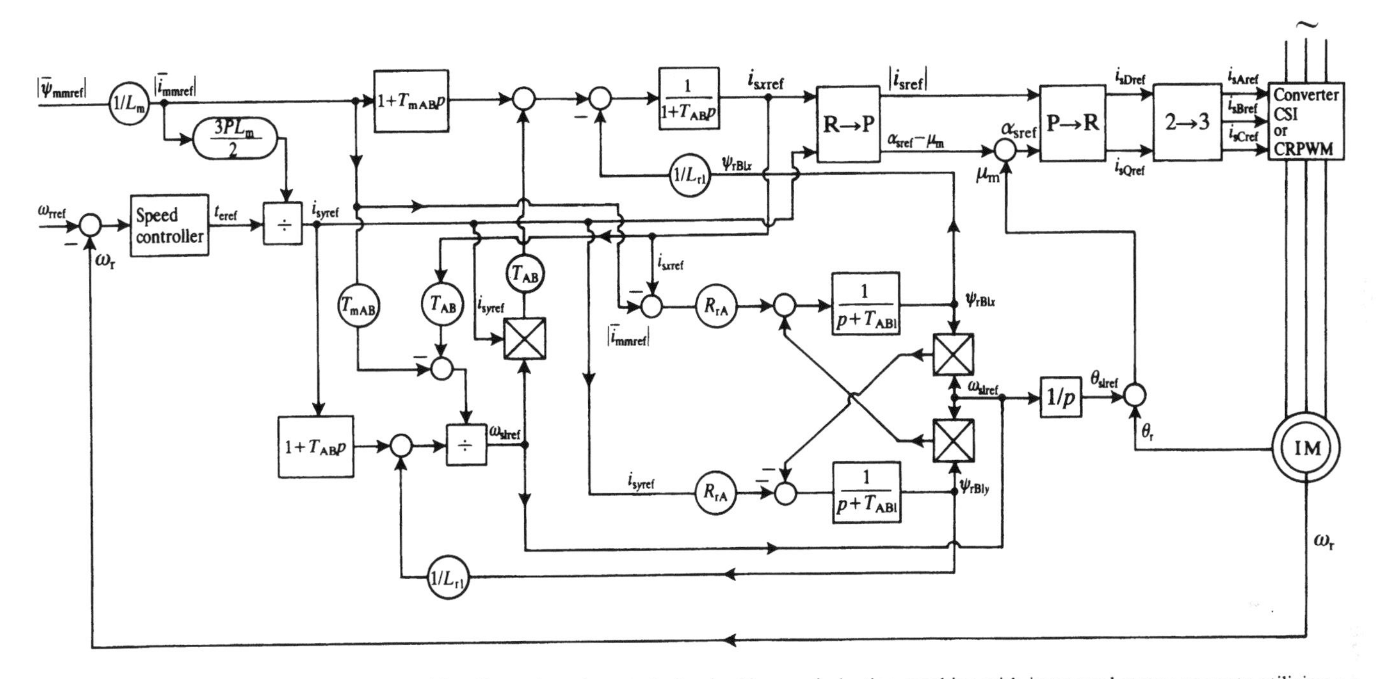

Fig. 4.37. Schematic of the indirect magnetizing-flux-oriented control of a double-cage induction machine with impressed stator currents utilizing a symmetrical rotor flux-linkage estimator.

possible to show that in the transient state, ω_{sl} can be expressed in terms of i_{sy}, i_{sx} and the various inductance and resistance parameters of the machine. If ω_{sl} is small, the rotor time constant for slow and fast variations of i_{sx} decreases from $T_{r1}=L_m/[R_{rA}R_{rB}/(R_{rA}+R_{rB})]$ to $T_{r2}=L_{rl}/(R_{rA}+R_{rB})$. If, say, a machine with deep rotor bars is considered and it is assumed that a rotor bar is divided into two equal sections, it follows that $R_{rA}=R_{rB}=R_r$. If $L_m/L_{rl}=30$, the ratio of T_{r1} to T_{r2} is equal to 120, while if $L_m/L_{rl}=10$, it is equal to 40. This shows that there can be a significant change in the rotor time constant. Thus to obtain fast torque response for high-performance a.c. servo applications, it is very important to incorporate the effects of the double-cage in the vector control scheme.

4.4 Artificial-intelligence-based vector-controlled induction motor drives

In the present section vector-controlled drives using fuzzy-logic controllers and also fuzzy-neural controllers will be discussed. The main purpose of using artificial-intelligence-based controllers is to reduce the tuning efforts associated with the controllers and also to obtain improved responses. By using minimum configuration artificial-intelligence-based controllers, it is possible to have DSP implementations which do not have excessive memory and computation requirements.

4.4.1 VECTOR DRIVES WITH FUZZY CONTROLLERS

4.4.1.1 General introduction

In the literature there are many papers which discuss various vector drives with fuzzy-logic controllers. Most of these contain a single fuzzy-logic controller, which is a speed controller, and only a few papers discuss implementations and mainly concentrate on simulations. However, in the present section, a fully digital vector-controlled induction motor drive is discussed where all the controllers are fuzzy, i.e. the drive contains four fuzzy-logic controllers and the experimental results are also shown. The DSP used is the Texas Instruments TMS320C30. The drive considered is a vector drive using rotor-flux-oriented control and the machine is supplied by a voltage-source inverter and is similar to that discussed in Section 4.1.1.6.

Recent developments in the application of control theory are such that conventional techniques for the design of controllers are being replaced by alternatives that adopt radically different design strategies by making extensive use of artificial-intelligence-based (neural, fuzzy, fuzzy-neuro and genetic) concepts (see also Chapter 7). These methods are characterized by the different amount and type of the necessary *a priori* knowledge describing the system and the required performance. There is a strong industrial need for the development and exploitation of systems incorporating controllers based on these novel methods because of the numerous advantages offered (see Section 7.1). Recently, fuzzy-logic control has emerged as an attractive area for research in the control application of fuzzy set theory. The main feature is the construction of fuzzy-logic controllers (FLCs) which utilize the linguistic knowledge of human experts.

4.4.1.2 General structure of a fuzzy-logic controller

There are many types of fuzzy-logic controllers (FLCs), but now the Mamdani-type of fuzzy-logic controller [Mamdani 1974] is used. As shown in Fig. 4.38(a), in general this type of fuzzy-logic controller contains four main parts, two of which perform transformations; these are:

- fuzzifier (transformation 1);
- knowledge base;
- inference engine;
- defuzzifier (transformation 2).

The *fuzzifier* performs measurement of the input variables (input signals, real variables), scale mapping, and fuzzification (transformation 1). Thus all the monitored input signals are scaled and fuzzification means that the measured signals (crisp input quantities which have numerical values) are transformed into fuzzy quantities (which are also referred to as linguistic variables in the literature). This transformation is performed by using membership functions. For example, if an input signal is small, e.g. it is a speed error and has a crisp value of 0.001, then it belongs to the 'POSITIVE SMALL' fuzzy set, or -0.001 would belong to the 'NEGATIVE SMALL' fuzzy set; other speed errors may belong to other fuzzy

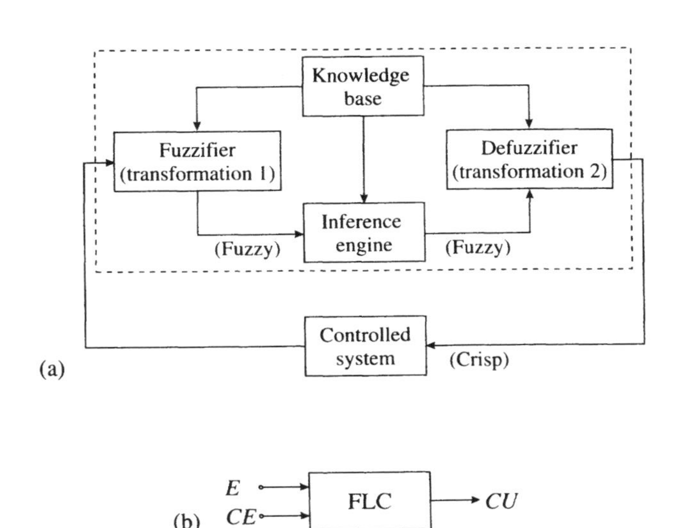

Fig. 4.38. Mamdani type of fuzzy-logic controller (FLC). (a) Schematic block diagram of a control system containing a Mamdani-type of FLC; (b) FLC with two inputs.

sets (e.g. POSITIVE LARGE, POSITIVE MEDIUM, etc.). In a 'conventional' fuzzy-logic controller (not a fuzzy-neural controller), the number of membership functions and the shapes of these are initially determined by the user. A membership function has a value between 0 and 1, and it indicates the degree of belongingness of a quantity to a fuzzy set. If it is absolutely certain that the quantity belongs to the fuzzy set, then its value is 1 (it is 100% certain that this quantity belongs to this set), but if it is absolutely certain that it does not belong to this set then its value is 0. Similarly if, for example, the quantity belongs to the fuzzy set to an extent of 50%, then the membership function is 0.5. The membership functions can take many forms including triangular, Gaussian, bell-shaped, trapezoidal, etc. For example, if the rotor speed error is 0.001, this is the input to the membership function of the positive small (PS) rotor speed errors, $\mu_{\omega r}^{PS}(E)$, and e.g. on the output of this 0.99 is obtained, $\mu_{\omega r}^{PS}(E=0.001)=0.99$, which means that it is 99% certain that the 0.001 rotor speed error is positive small. The initial forms of the membership functions can be obtained by using expert considerations or by clustering the input data. Final tuning of the membership functions can be performed on the DSP-controlled drive. Practical aspects of choosing the appropriate scaling factors and membership functions are also discussed below.

The *knowledge base* consists of the data base and the linguistic-control rule base. The data base provides the information which is used to define the linguistic control rules and the fuzzy data manipulation in the fuzzy-logic controller. The rule base (expert rules) specify the control goal actions by means of a set of linguistic control rules. In other words, the rule base contains rules which are those which would be provided by an expert. The FLC looks at the input signals and by using the expert rules determines the appropriate output signals (control actions). The rule base contains a set of if–then rules (see below). The main methods of developing the rule base are:

- using the experience and knowledge of an expert for the application and the control goals;
- modelling the control action of the operator;
- modelling the process;
- using a self-organized fuzzy controller;
- using artificial neural networks.

When the initial rules are obtained by using expert physical considerations, these can be formed by considering that the three main objectives to be achieved by the fuzzy-logic controller are:

- removal of any significant errors in the process output by suitable adjustment of the control output;
- ensuring a smooth control action near the reference value (small oscillations in the process output are not transmitted to the control input);
- preventing the process output exceeding user specified values.

The *inference engine* is the kernel of a fuzzy-logic controller and has the capability both of simulating human decision-making based on fuzzy concepts and of inferring fuzzy control actions by using fuzzy implication and fuzzy-logic rules of inference. In other words, once all the monitored input variables are transformed into their respective linguistic variables (by transformation 1 discussed above), the inference engine evaluates the set of if–then rules (given in the rule base) and thus a result is obtained which is again a linguistic value for the linguistic variable. This linguistic result has then to be transformed into a real output value of the FLC and this is why there is a second transformation in the FLC.

The second transformation is performed by the *defuzzifier* which performs scale mapping as well as defuzzification. The defuzzifier yields a non-fuzzy, real control action from the inferred fuzzy control action by using membership functions. There are many defuzzification techniques, but due to the simplicity of its implementation and simpler training algorithms, the centre of gravity method is adopted here. Physically this corresponds to taking a weighted average of the control action contributions from each of the various fuzzy rules. Since each of the rules (in the full rule base) can be considered to be rules provided by subexperts, this type of defuzzification simply means that the final decision is being made by taking the weighted average of all the recommendations of the subexperts.

When a classical controller (e.g. PI or PID) is used, then the input to the controller is the error signal. For example, for a PI speed controller, the input is the speed error, which is the difference between the reference speed and the actual speed, $E(k)=\omega_{\mathrm{rref}}(k)-\omega_{\mathrm{r}}(k)$. However, when a fuzzy-logic controller is used, there is more than one input to the controller. In the most frequently used fuzzy-logic controller, there are two inputs, these are the error (E) and the change of the error (CE) as shown in Fig. 4.38(b), and

$$E(k)=\omega_{\mathrm{rref}}(k)-\omega_{\mathrm{r}}(k) \qquad CE(k)=E(k)-E(k-1).$$

This type of fuzzy-logic controller is used in the vector drive described in the present section, but it should be noted that there are other types of fuzzy-logic controllers as well, where the number of inputs is higher. It is a goal of the fuzzy-logic controller to obtain on its output a signal which is based on E and CE, e.g. this signal can be CU, which is the change of the output signal. If this is known, then it is possible to obtain the output signal from the change of the output by using $u(k)=u(k-1)+CU(k)$. It should also be noted that it is also possible to use another fuzzy-logic controller, where instead of obtaining the change of the output quantity, the output quantity is directly obtained.

As discussed above, in the heart of the fuzzy-logic controller there is a rule base and this contains the individual rules (subrules). In general, these linguistic rules are in the form of IF–THEN rules and take the form:

IF (E is A **AND** CE is B) **THEN** (CU is C),

where A, B, C are fuzzy subsets for the universe of discourse of the error, change of the error and change of the output respectively. For example, A can denote the subset NEGATIVE LARGE of the error, etc. To be more precise, for example, for a fuzzy speed controller, the following rules can be obtained by physical considerations:

Rule 1: **IF** (E is ZE **AND** CE is ZE) **THEN** (CU is ZE)
Rule 2: **IF** (E is ZE **AND** CE is NS) **THEN** (CU is NS)
Rule 3: **IF** (E is PS **AND** CE is NS) **THEN** (CU is ZE)

and so on, where ZE, NS, PS denote the fuzzy sets ZERO, NEGATIVE SMALL, and POSITIVE SMALL respectively.

In general, for given numerical values of E and CE (which are the measured scaled inputs to the FLC), several rules can be activated simultaneously. This follows from the fact that a given numerical value can be a member of more than one fuzzy set (this is determined by the membership functions). Example 4.1 shows a numerical example of obtaining the output signal of a Mamdani-type of FLC, if two rules are assumed only for simplicity, and the two numerical (crisp) inputs to the FLC are $x_1=4$ and $y_1=8$. It can be seen that $x_1=4$ belongs to both the fuzzy set $A1$ and fuzzy set $A2$; the belongingness is described by the membership functions $\mu^{A1}(x)$ and $\mu^{A2}(x)$ respectively, and for the given x_1 value they are $\mu^{A1}(x_1)=2/3$ (x_1 belongs to fuzzy set $A1$ to a degree of 66.6%) and $\mu^{A2}(x_1)=1/3$ (x_1 also belongs to the fuzzy set $A2$ to a degree of 33%). Although in general, several rules apply at the same time (several rules are 'fired' simultaneously), only a single control action must be performed, so the main task is to combine in an appropriate manner the contribution of each rule to obtain the control action. The inference engine determines the controller output on the basis of the contribution of each rule. In fuzzy-logic terms, the composition operation is used to perform this task. There are many types of composition operations, but the most commonly used is the sup–min (supremum–minimum) composition. In practice, the sup–min principle is applied to one rule at a time, which means that first, for each rule, an output membership function is computed, which gives the strength of the rule under consideration. This is obtained by using the 'min' operator (minimum operator) which corresponds to the fuzzy AND operation in the rules; thus the minimums of the appropriate membership functions are computed and then the combined fuzzy output is obtained by the supremum operator, or in other words, the outputs of each rule are combined into a single fuzzy set. When the supremum operator is applied, it means that after obtaining for each rule a truncated membership function (due to the min operator), then due to the supremum operator, a combined membership function is obtained, which is the outer envelope of the individual membership functions. For better understanding, this is illustrated in Example 4.1 which is related to a Mamdani-type of FLC, where for simplicity and there are only two rules. There are two inputs (x,y) and an output (z) and the input membership functions and output membership functions are given (they are triangular membership functions). It can be seen that for rule 1 the trapesoidal membership function $\mu^{C1}(z)$ is obtained (its height is equal to the firing strength of the rule, $w_1=2/3$, and for rule 2 the membership function $\mu^{C2}(z)$ is obtained (with height $w_2=1/3$), and the combined membership function, $\mu(z)$, is obtained from the outer envelopes of the two output membership functions. This combined output membership function is then used together with an appropriate defuzzification technique to obtain the crisp output value of the controller. In Example 4.1 the centre of gravity (COG) defuzzification method has been used, and the crisp output value of the controller is $z^*=4.7$.

Example 4.1 Mamdani-type of FLC with two rules

Inputs $x_1=4$; $y_1=8$
Determine (output): z^*

Rule 1: IF x is $A1$ AND y is $B1$ THEN z is $C1$
Rule 2: IF x is $A2$ AND y is $B2$ THEN z is $C2$

Given membership functions: $\mu^{A1}(x)$, $\mu^{A2}(x)$, $\mu^{B1}(x)$, $\mu^{B2}(x)$, $\mu^{C1}(z)$, $\mu^{C2}(z)$

Input membership functions (for given inputs x_1, y_1)
In rule 1: $\mu^{A1}(x_1)=2/3$; $\mu^{B1}(y_1)=1$
In rule 2: $\mu^{A2}(x_1)=1/3$; $\mu^{B2}(y_1)=2/3$

Firing strength of rule $i(w_i)$; Fuzzy AND$\Rightarrow$min operator
Firing strength of rule 1: $w_1=\min[\mu^{A1}(x_1), \mu^{B1}(y_1)]=\min(2/3, 1)=2/3$
Firing strength of rule 2: $w_2=\min[\mu^{A2}(x_1), \mu^{B2}(y_1)]=\min(1/3, 2/3)=1/3$

Defuzzification Given: combined output membership function $\mu(z)$; determine: z^*

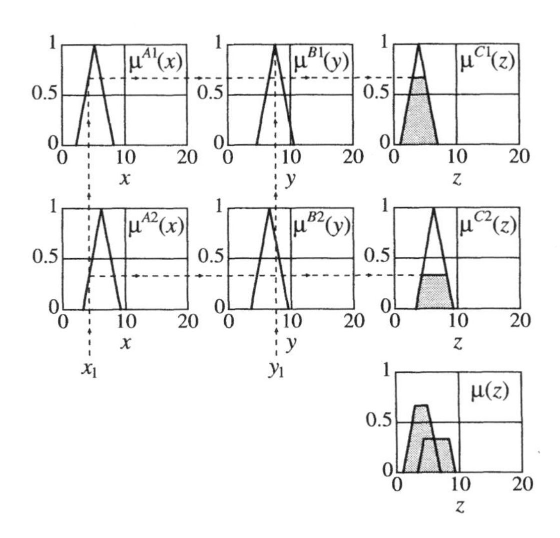

Centre of gravity defuzzification (COG)

$$z^{*COG} = \sum z_k \mu(z_k) \Big/ \sum \mu(z_k) \qquad (k = 1, 2, \ldots, 9)$$

$$z^{*COG} = \frac{2^*(1/3) + 3^*(2/3) + 4^*(2/3) + 5^*(2/3) + 6^*(1/3) + 7^*(1/3) + 8^*(1/3) + 9^*0}{1/3 + 2/3 + 2/3 + 2/3 + 1/3 + 1/3 + 1/3 + 0}$$

$$= 4.7$$

One possible initial rule base that can be used in drive systems consists of 64 linguistic rules as shown in Table 4.1. The 64 rules can be considered to be rules provided by 64 subexperts. It should be noted that a classical fixed-term controller (e.g. a PID controller) cannot achieve the three main control objectives listed above, unless extra logic is added. This extra heuristic logic is contained implicitly within the rules shown in Table 4.1.

In Table 4.1, the following fuzzy sets are used: *NM* = NEGATIVE MEDIUM, *NS* = NEGATIVE SMALL, *NZ* = NEGATIVE ZERO, *PZ* = POSITIVE ZERO, *PS* = POSITIVE SMALL, *PM* = POSITIVE MEDIUM, *PL* = POSITIVE LARGE. For example, it follows from Table 4.1 that the first 'expert rule' is: **IF** (*E* is *NL* **AND** *CE* is *NL*) **THEN** (*CU* is *NL*), where *CU* denotes the change of the output. It should be emphasized that although this initial rule base has given satisfactory responses in the DSP-controlled vector drive under consideration, it has been possible to significantly reduce the number of rules without degrading the drive performance. The reduced rule-base contains only 16 rules. The reduced rule-base has great influence on the overall complexity of the fuzzy system; this includes computational complexity and memory requirements. This is an important factor for industrial users, most of whom at present are reluctant to employ fuzzy techniques in variable-speed drives, since they associate fuzzy controllers with large computational and memory requirements. Fuzzy-controlled drives with minimal configuration fuzzy controllers is an important research topic [Vas 1996].

Table 4.1 Fuzzy rule base with 64 rules

		CE							
		NL	*NM*	*NS*	*NZ*	*PZ*	*PS*	*PM*	*PL*
	NL	*NL*	*NL*	*NL*	*NL*	*NL*	*NM*	*NS*	*NZ*
	NM	*NL*	*NL*	*NL*	*NM*	*NM*	*NS*	*NZ*	*PS*
	NS	*NL*	*NL*	*NM*	*NS*	*NS*	*NZ*	*PS*	*PM*
E	*NZ*	*NL*	*NM*	*NS*	*NZ*	*NZ*	*PS*	*PM*	*PL*
	PZ	*NL*	*NM*	*NS*	*PZ*	*PZ*	*PS*	*PM*	*PL*
	PS	*NM*	*NS*	*PZ*	*PS*	*PS*	*PM*	*PL*	*PL*
	PM	*NS*	*PZ*	*PS*	*PM*	*PM*	*PL*	*PL*	*PL*
	PL	*PZ*	*PS*	*PM*	*PL*	*PL*	*PL*	*PL*	*PL*

4.4.1.3 Vector drive with four fuzzy controllers; design and tuning of fuzzy controllers

The overall structure of the system is shown in Fig. 4.39. The system contains a voltage-source inverter, a 3 kW squirrel-cage induction motor, analog circuits for the voltage and current transformations and a TMS320C30 DSP system board and interface board (DMB). The two boards are installed in a host computer. There are four A/D channels in the DMB which enable the input of four signals. The DSP software contains the algorithms for the fuzzy controllers, vector-control, PWM generation, machine soft-starting, and A/D calibration. During operation, up to fifteen drive quantities can be simultaneously shown on the host computer display in real-time. This greatly facilitates the tuning of the drive system. The PWM scheme uses an asynchronous symmetrical space-vector modulation technique. The digital signal conversion block is used for inverter fault detection, inverter protection, and the generation of the six firing (gating) signals.

As shown in Fig. 4.40, in the induction motor (IM) drive there are four fuzzy controllers; controller 1 is the speed controller, controller 2 is the torque-producing stator-current controller, controller 3 is the rotor-magnetizing-current (rotor-flux) controller, and controller 4 is the flux-producing stator-current controller. The four digital controllers implemented have been initially tuned by using a simulation of the drive system and fine tuning has been achieved on-line in the DSP implementation. The decoupling circuit has been described in Section 4.1.1.3 and is required in a VSI-fed drive (see also Fig. 4.11), to produce the required stator voltages u_{sxref}, u_{syref}, which are also shown in Fig. 4.40. These are voltages in the rotor-flux-oriented reference frame, so they have to be transformed into

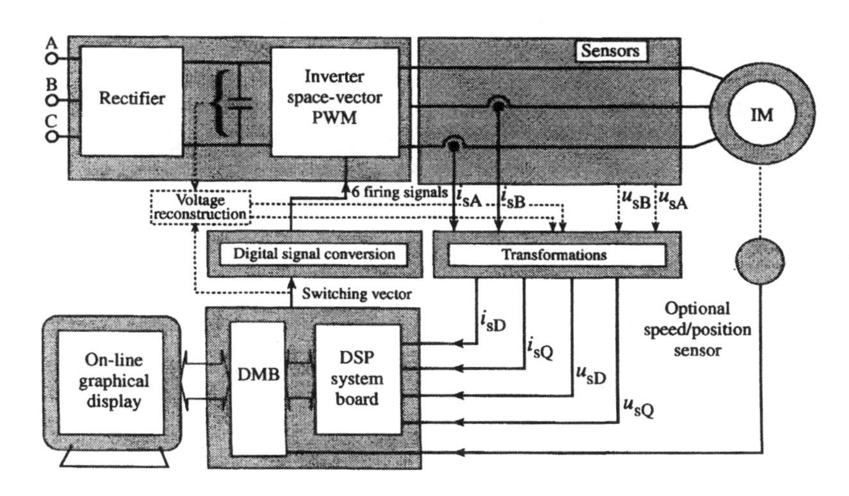

Fig. 4.39. Schematic diagram of vector-controlled PWM voltage-source inverter-fed induction motor drive.

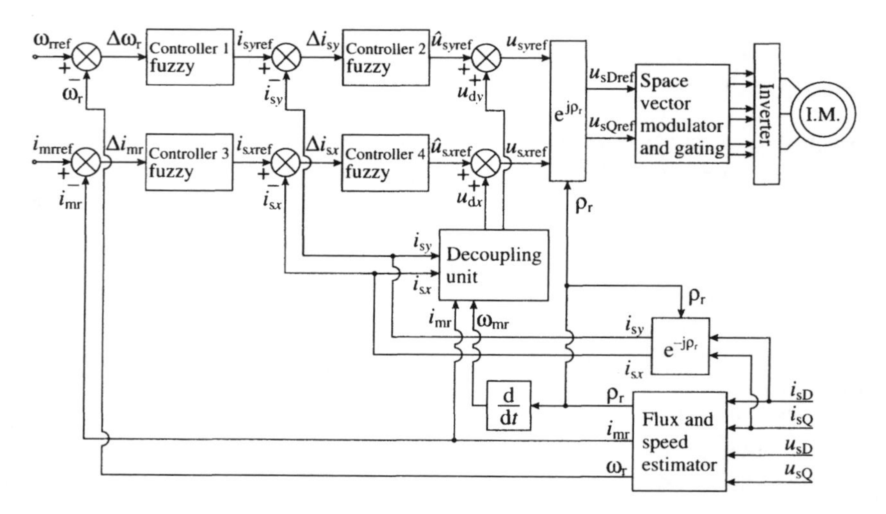

Fig. 4.40. Vector-controlled PWM voltage-source inverter-fed induction motor drive with four fuzzy controllers.

their stationary reference frame values (u_{sDref}, u_{sQref}) by using the complex transformation $\exp(j\rho_r)$, where ρ_r is the angle of the rotor flux-linkage space vector. The inverse transformation $\exp(-j\rho_r)$ is used in Fig. 4.40 to transform the stator current components expressed in the stationary reference frame (i_{sD}, i_{sQ}) into the stator currents in the rotor-flux-oriented reference frame (i_{sx}, i_{sy}).

It should be noted that classical PI and PID controllers which are used in conventional converter-fed a.c. drive systems are mainly tuned using *ad hoc* methods. Several techniques exist which provide initial values of the controller parameters, the most commonly used being based on the Ziegler–Nichols methods. However, these techniques can be time-consuming and fixed controllers cannot necessarily provide acceptable dynamic performance over the complete operating range of the drive. Performance will degrade mainly because of machine non-linearities, parameter variation, etc. Adaptive controllers can be used to overcome these problems and also to eliminate the need to perform detailed *a priori* controller design. Alternatively, performance-index-based optimal control techniques can be adopted, but these may suffer from convergence-related problems. The fundamental aim of drive system optimization is to obtain the smallest overshoot with the shortest rise and settling times. It is not generally possible to fulfil all these criteria simultaneously, but a solution can be obtained by using some type of an integral criteria (performance index). When an integral criteria is used, the integrands are products of the control error and the time, each raised to some power. When the integral square error (ISE) criterion is used, the performance index is the integral of the square of the error. To obtain better damping of the controlled variable, the ITSE criterion is used. In this case the integral of the error squared multiplied by time is used. Further improvements of the settling time can be obtained by using the ISTSE (integral of time squared multiplied by the error squared) criterion, but the computation time is increased. The ITAE criteria is commonly used in variable-speed drives, where the integrand is the absolute value of the error multiplied by time. However, the determination of the optimum controller parameters based on performance indexes may be very time consuming and, as mentioned above, may suffer from convergence-related problems.

For a practical DSP implementation of the four fuzzy controllers the following aspects should be considered:

1. *Discretization of membership functions* For ease of real-time implementation, and for the purposes of fuzzification and defuzzification, triangular membership functions have also been considered, and they can be conveniently described by using discrete values in a look-up table.
2. *Rule base look-up table* It should be noted that when a microprocessor or DSP-based fuzzy controller is designed, a look-up table for the rule base can also be used. This contains the discrete values of the change of the output of the fuzzy controller. However, care must be taken when a look up table is selected, otherwise errors could arise due to improper selection.

The absence of widespread industrial applications of fuzzy controllers in drives is also related to the fact that there are no straightforward procedures for the tuning of fuzzy controllers. However, it is possible to give guidelines for the developer of these controllers. The main tuning steps are:

1. *Tuning of input and output scaling factors* The output scaling factor in a fuzzy-logic controller has great influence on the stability of the system. The input scaling factor has great influence on the sensitivity of the fuzzy-logic controller with respect to the optimal choice of the operating ranges of the input signals. Both scaling factors are thus set during the initial tuning stage.
2. *Tuning of input and output membership functions* The number and shape of the input and output membership functions have a considerable influence on the controller behaviour. The second stage of the tuning process is concerned with the tuning of the membership functions. A variety of membership-function forms has been considered, including triangular and Gaussian membership functions. Finally, triangular membership functions have been selected.
3. *Tuning of the rules* It is possible to achieve optimal tuning by the appropriate adjustment of the base.

An extensive range of digital simulations has been performed to obtain the appropriate values of the scaling factors, membership functions, and rule base. This was followed by the real-time implementation of the fuzzy-logic controllers in various drive systems.

It should be noted that in addition to fuzzy controllers, fuzzified controllers (FPIC) [Vas 1995] may also be used. Sometimes industry prefers such a solution, since it only requires small changes to be made to an existing system. FPICs can improve the performance of the system incorporating conventional PI controllers by using fuzzy logic. Small changes of the values of the controller coefficients may lead to considerable improvement in performance. In a FPIC, the rule base contains the following type of rules:

$$\text{if } E \text{ is } A \text{ and } CE \text{ is } B \text{ then } CP \text{ is } C$$

$$\text{if } E \text{ is } A \text{ and } CE \text{ is } B \text{ then } CI \text{ is } C,$$

where E and CE describe the error and change in error respectively, CP and CI represent the changes of the proportional term and the integral term of the PI part in the FPIC separately. For example, there can be in total 72 rules. In these rules the linguistic variables A, B, and C may take the values PM, NS, etc. The parameters of the PI part of the fuzzified controller can be adjusted according to

$$P(k)=P(k-1)+K_{\text{p}}CP \tag{4.4-1}$$

for the proportional term and

$$I(k)=I(k-1)+K_{\text{i}}CI \tag{4.4-2}$$

for the integral term. In these equations K_{p} and K_{i} are proportional and integral coefficients on which the degree of dynamic parameter adjustment depends, $P(k)$ and $P(k-1)$ describe the proportional term of the PI part at the kth and $(k-1)$th sampling times respectively, and $I(k)$ and $I(k-1)$ correspond to the integral term of the PI part at these sampling instants. It should be emphasized that the performances of the FLC are far better than those of both the FPIC and optimal PI controller. In contrast to the computational times for the fuzzy-logic-based system, those for the optimized controller are excessive.

4.4.1.4 Experimental results

Figure 4.41 shows some experimental characteristics obtained for the DSP-controlled induction motor drive system with four fuzzy controllers. It can be seen from Fig. 4.41(a) that the drive starts from rest. The reference rotor speed (ω_{rref}) is first set to zero then to 220 rad s^{-1} and then to -220 rad s^{-1} and finally to zero again. As shown in Fig. 4.41(c), the torque-producing stator current (i_{sy}) is maintained at its maximum value during both acceleration and deceleration. It follows well its reference value (i_{syref}). It can be seen from Fig. 4.41(b) that the rotor magnetizing current ($|\bar{i}_{\text{mr}}|$) is effectively constant. As expected, the angular slip frequency (ω_{sl}) has a similar shape as i_{sy}, since if $|\bar{i}_{\text{mr}}|=$constant, and the rotor time constant (T_{r}) is also considered to be constant, then from eqn (4.1-26) $\omega_{\text{sl}}=i_{sy}/(T_{\text{r}}|\bar{i}_{\text{mr}}|)=ci_{sy}$, where c is a constant.

It is very important to note that the tuning effort required with the fuzzy-controlled drive is significantly less than that required for the drive using four PI controllers. Similar conclusions for other fuzzy drives have also been obtained by Texas Instruments [Beierke 1995]. Furthermore, there is also some improvement in the responses, e.g. the stator flux-producing current (i_{sx}) for the fuzzy-controlled

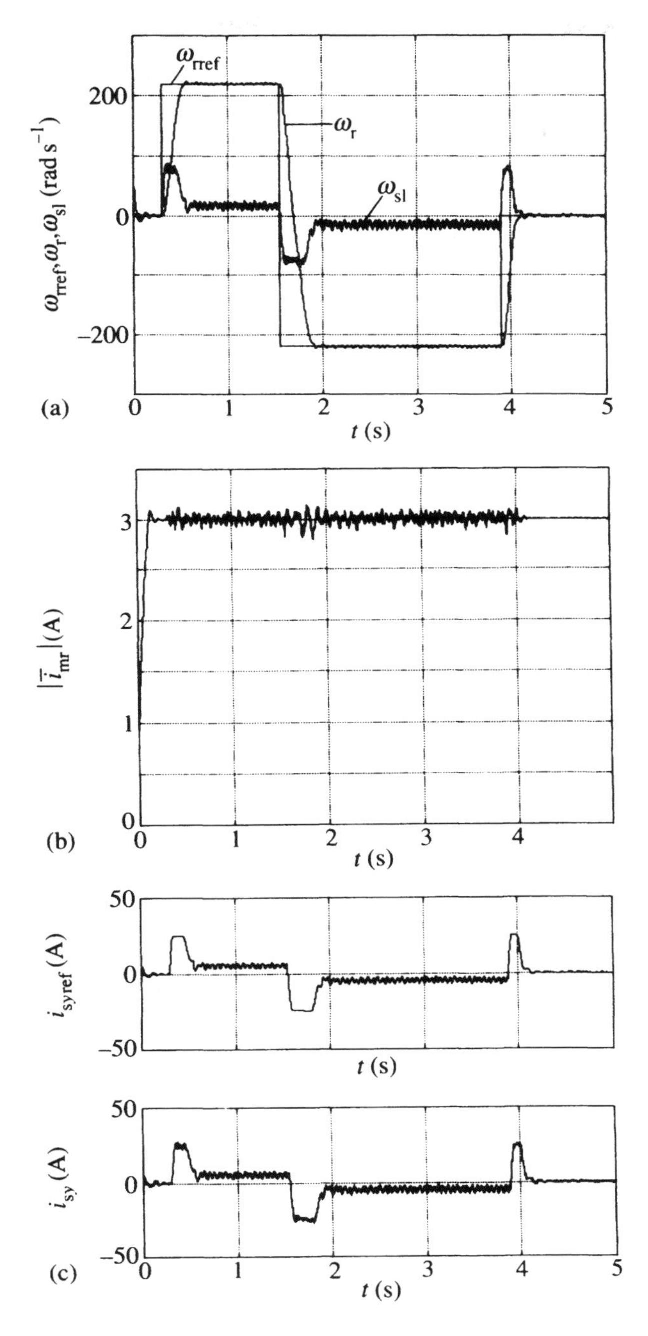

Fig. 4.41. Vector-controlled induction motor drive responses using four fuzzy controllers (experimental results). (a) Rotor speed reference, rotor speed, angular slip frequency ($\omega_{\mathrm{rref}}, \omega_{\mathrm{r}}, \omega_{\mathrm{sl}}$); (b) rotor magnetizing current ($|\bar{i}_{\mathrm{mr}}|$); (c) torque-producing stator-current reference and its actual value ($i_{\mathrm{syref}}, i_{\mathrm{sy}}$).

drive shows an improved transient profile. An extensive simulation of the complete drive system has also been performed, which has also incorporated the effects of the space vector pulse-width modulator. For this purpose the simulation software described in [Vas 1993] has been extended to incorporate the four fuzzy-logic controllers.

Figures 4.42(a) and (b) show the highly non-linear 3-D controller profile of the implemented speed and flux fuzzy-logic controllers. For illustration purposes Figs 4.42(c) and 4.42(d) show the controller profiles of the discretized speed and flux PI controllers used above. Although the control profiles of the PI controllers are simpler, which also correspond to simpler digital implementations, it is still extremely useful to implement the more complicated fuzzy controllers resulting in more complex control surfaces. This is due to the fact that the fuzzy controllers are adaptive controllers. Therefore, by appropriate tuning it is possible to obtain better dynamic characteristics under all operating conditions. It should be noted that even the very first laboratory implementation of the discretized fuzzy controllers has been successful. This also proves the existence of the reduction in tuning effort associated with fuzzy controllers.

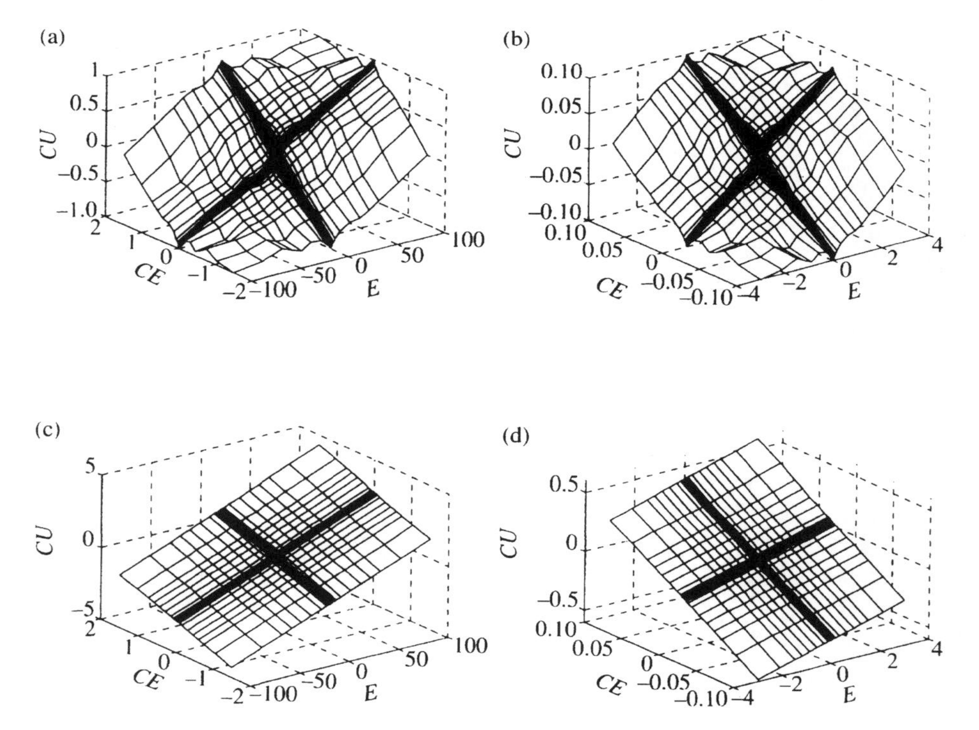

Fig. 4.42. PI and fuzzy-controller profiles. (a) Fuzzy speed-controller profile; (b) Fuzzy flux-controller profile; (c) PI speed-controller profile; (d) PI flux-controller profile.

4.4.2 VECTOR DRIVE WITH FUZZY-NEURAL CONTROLLERS

4.4.2.1 General introduction

In the previous section, a fuzzy-controlled vector drive has been described for a voltage-source inverter-fed induction motor. This contained four fuzzy-logic controllers and all the fuzzy-logic controllers required the use of their own rule base and membership functions (for fuzzification and defuzzification). However, it is possible to use such controllers where these rules and membership functions do not have to be known *a priori*. For this purpose fuzzy-neural controllers can be used (see also Chapter 7).

Fuzzy-neural control emerged as a powerful technique more than a decade ago. In fuzzy-neural control the ideas of a fuzzy-logic controller and an artificial neural network (ANN) structure are combined. The fuzzy-neural network is automatically constructed by a learning process. In a connectionist fuzzy-neural controller, the input and output nodes of the ANN represent the input and output signals, and in the hidden layer nodes take the roles of membership functions and rules. The learning algorithm for this network can be hybrid, combining unsupervised and supervised methods. The unsupervised learning produces the number of fuzzy sets for each input and output variable, the number of fuzzy rules, the rules themselves, and the centres and widths of the membership sets. This information is used to establish a fuzzy-neural controller which is then trained using a back-propagation algorithm to further tune the centres and widths of the membership functions. The structure of the controller is fixed. Hybrid learning outperforms purely supervised learning by reducing training times.

There are many architectures which can be used for fuzzy-neural controllers. The fuzzy-neural controller used here contains five layers: an input layer, a layer for the fuzzy membership sets, a fuzzy AND layer, a fuzzy OR layer, and an output layer (see Fig. 4.43 below). The input layer contains the input nodes, which represent linguistic variables. These distribute each input variable to its membership functions. There are three hidden layers: layer 2 generates the appropriate membership values, layer 3 defines the preconditions of the rule nodes, and the nodes in the fourth layer connect the output of the fuzzy AND nodes to the consequences in the rules. The last layer is the output layer and it performs defuzzification. Tuning can conveniently be performed using a back-propagation-type of algorithm.

The direct implementation of conventional fuzzy-logic controllers suffers from the disadvantage that there is no formal procedure for the direct incorporation of the expert knowledge during the development of the controller. The structure and detail of the fuzzy controller (number of rules, the rules themselves, number and shape of membership functions, etc.) is achieved through a time-consuming tuning process which is essentially *ad hoc*. The ability to automatically 'learn' characteristics and structure which may be obscure to the human observer is, however, an inherent feature of neural networks. However, in applying artificial neural networks to control problems, there is no general systematic approach to choosing the network type or structure and it is also difficult to relate the final trained network (in terms of the activation functions, weights, etc.) to the original physical problem (see also Section 7.1). The combination of a fuzzy-logic controller with the structure of a neural network does, however, offer the control system designer the opportunity to make use of the advantages of both—the ability of the fuzzy logic to take account of expert human knowledge and the learning ability of the neural network—to overcome their respective disadvantages—the lack of a formal learning procedure for the fuzzy controller and the lack of a clear correlation with the physical problem when using artificial neural networks. This approach therefore provides a means of combining the use of imprecise, linguistic information but which has a clear physical significance with formalized mathematical structures and training algorithms. A fuzzy-neural controller offers a structure which enables 'automated' design, requiring a minimum of human intervention for the tuning.

4.4.2.2 Fuzzy-neural controller design

This section provides a detailed description of both general and application-specific features of the development of a fuzzy-neural controller implementation. This is based on the Mamdani type of fuzzy logic system shown in Fig. 4.38(a), the basic elements of which are a fuzzy rule base, with associated fuzzy sets, and a fuzzy inference engine, with associated operators. The artificial neural network used is multi-layer feedforward ANN. The training algorithms used are of the competitive learning and the back-propagation types. A variety of activation functions associated with the nodes are used and are described in detail.

4.4.2.2.1 Structure of fuzzy-neural controllers

The fuzzy-neural controller structure used is represented by a neural network consisting of five layers. Figure 4.43 shows an example of the network structure for a controller with two inputs and a single output.

In Fig. 4.43 the first layer is an input layer with one node for each controller input variable. The nodes in this layer act as single-input, multi-output, 'fan-out' nodes distributing each input variable to each of its associated membership function nodes in the second layer. The weights (strength of connection) between a jth node in the kth layer and an ith node in the previous layer are denoted by w_{ji}^k. The interconnection weights between the first and second layers are all unity and constant. The second layer is made up of nodes representing Gaussian membership functions. The total number of nodes in this layer is equal to the total number of fuzzy sets associated with the input variables. The input function, f, the activation function, g, and the output function, h, are such that the overall the node output, y, is given by

$$y = \exp[-(x^2 - m)^2/\sigma^2], \tag{4.4-3}$$

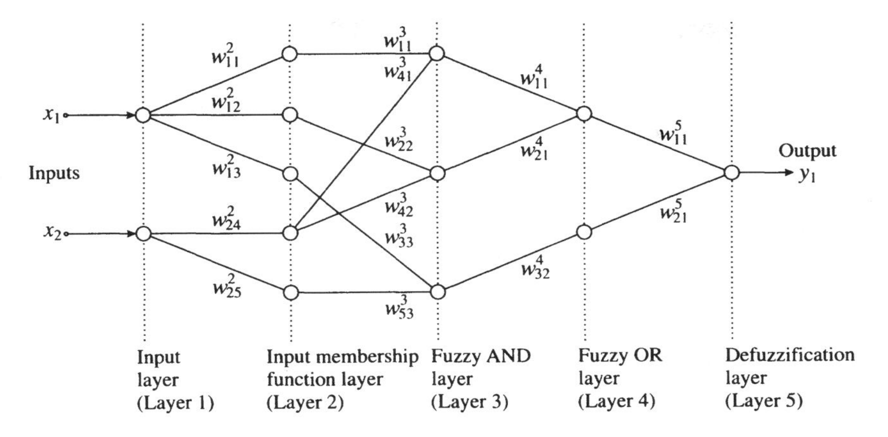

Fig. 4.43. Fuzzy-neural network.

where m and σ denote the centre and width of a fuzzy set respectively. The interconnection weights between the second and third layers are all unity and constant. The third layer is made up of nodes implementing the fuzzy intersection form of the fuzzy AND operator. Each node represents a fuzzy rule and the node output, y, is given by

$$y=\min(x_1, x_2 \ldots, x_k). \tag{4.4-4}$$

The interconnection weights between the third and fourth layers are also all unity and constant. The fourth layer is made up of nodes implementing the bounded sum form of the fuzzy OR operator. The number of nodes is equal to the total number of fuzzy sets associated with the controller output variables. The node output, y, is given by

$$y=\min\left(1, \sum_{i=1}^{k} x_i\right). \tag{4.4-5}$$

The fifth and final layer comprises nodes implementing the centre-of-area defuzzification algorithm, with one node for each output variable. The weights of the interconnections between the nodes in the fourth and fifth layers are the products of the centre and width of the membership function associated with the fuzzy set for each layer-four node output variable. The node output, y, is given by

$$y=\frac{\sum_{i=1}^{k} m_i \sigma_i x_i}{\sum_{i=1}^{k} \sigma_i x_i}. \tag{4.4-6}$$

In order to establish the fuzzy-neural controller structure for a particular application the following must be determined:

- The number of membership sets associated with each input variable, i.e. the number of nodes in the second layer and the connections between layers 1 and 2.

- The number of membership sets associated with each output variable, i.e. the number of nodes in the fourth layer.
- What the fuzzy rules are, i.e. the connections between layers 2 and 3 and between layers 3 and 4.
- The initial estimates for the centre and width of each input variable fuzzy-set-membership function, i.e. the parameters of the activation functions of the layer-2 nodes.
- The initial estimates for the centre and width of each output variable fuzzy-set-membership function, i.e. the connection weights between layers 4 and 5 and the parameters of the activation function of the layer-5 nodes.

The determination of the required structure and the initial values of the membership function parameters is done in an initial tuning stage. Subsequent to this, a further tuning, using a back-propagation type of algorithm, adjusts the membership function parameters and the network connection weights to produce the final trained fuzzy-neural controller. The details of these two stages are given in the following sections.

4.4.2.2.2 Determination of the fuzzy-neural network structure and parameters

This section describes in detail the first stage of the fuzzy-neural controller development discussed above. This stage uses a separate, competitive-learning type of artificial neural network to determine the number of fuzzy rules, the fuzzy rules themselves, and the number and centre of the membership sets. The results from this give both the structure and the initial parameter and weight settings for the controller network.

The network used in this first stage consists of a layer of input nodes followed by a single layer of competitive nodes trained using a supervised instar learning algorithm. The number of nodes in the input layer is equal to the number of controller input and output variables—a total of three in the present application, two variables representing the controller inputs—error and change in error, the third representing the change in controller output. The number of nodes in the competitive layer is equal to the maximum number of fuzzy rules possible for the problem under consideration. For example, if eight membership sets are assumed for each of the three controller variables then the number of output nodes is 512.

The network is fully connected, i.e. all first layer nodes are connected to all nodes in the second layer. The initial setting of the weights associated with the network are assigned as uniformly distributed random values. The weights connecting each of the input nodes to an output node are regarded as a weight vector; all such vectors are normalized to a magnitude of unity. The input to the network consists of a sequence of training vectors. Each training vector comprises the values of the controller inputs and the corresponding controller output.

The network is then trained by establishing which weight vector is 'closest' to the input (training) vector. This is done by forming the inner product of the training vector with each of the weight vectors and finding the maximum of the resulting products. This identifies the 'winning' output node which is assigned an output of unity; all other network outputs are set to zero. The weight vector associated with the 'winning' output node is adjusted to become closer to the training vector; all other remain unchanged. A count is kept of the number of times each output node 'wins'. At the end of training, the training vectors have been 'clustered' by association with the 'winning' output nodes.

The clusters associated with output nodes which 'win' represent the fuzzy rules implied by the training data. A straightforward application of this procedure may, however, result in an unnecessarily high number of rules. The outcome of this stage of the training is therefore modified. This procedure eliminates 'weak' rules, of both a similar and conflicting nature, thereby enabling a minimal rule base to be established from the training data. Any available expert linguistic information can be included, as required, at this stage of the controller development.

Initial estimates of the centres of the membership functions are established from the mean of each variable associated with the remaining vector clusters. The widths of the membership functions are established using the 'nearest neighbour' and 'overlap' concepts [Stronach *et al.* 1996]. As an alternative, the standard deviation of the vector clusters can be used but experience has shown that problems arise in cases where standard deviation values are insufficiently large and overlap between adjacent membership sets is 'inadequate'.

4.4.2.2.3 Fuzzy-neural controller tuning

This section describes the second-phase tuning of the fuzzy-neural controller development outlined above. The results of the first stage training, the number of fuzzy sets for each variable and the fuzzy rules, are used to establish the structure of the fuzzy-neural controller network, as described in Section 4.4.2.2.1 and typified by the network shown in Fig. 4.43. The network is initialized using the values for the membership function centres and widths obtained from the first-stage tuning. Once established, the network structure remains fixed during the second stage of tuning. Therefore the network interconnection weights between layers one and two, between layers two and three, and between layers three and four do not change in the second stage of tuning. Only the interconnection weights between layers four and five and the parameters (centres and widths) of the membership functions are altered. The second stage of tuning to improve these values uses a back-propagation-type algorithm with the same set of training data as was used for the first-stage competitive network. The final, trained form of this network gives the implementation form of the controller.

4.4.2.2.4 Alternative fuzzy-neural controller designs

In addition to the approach described in detail above, a number of alternative methods may be adopted for establishing a fuzzy-neural controller.

A back-propagation-based method which does not use clustering allows all fuzzy logic system parameters to be updated in a single optimization procedure. This approach can also incorporate linguistic information but the high dependence on a non-linear optimization search results in a possible slow convergence to a local minimum. The numbers of fuzzy sets for each variable must be specified *a priori*. It has been found that the proposed clustering-based approach results in fewer such sets. An 'order of magnitude' reduction is typically achieved. Investigations using a much simpler 'nearest neighbour' type of clustering algorithm do not show such a reduction in the number of fuzzy sets.

An approach based on an orthogonal least-squares solution establishes fuzzy basis functions and uses these to set up a form of fuzzy-logic controller which is linear in the unknown parameters. This approach, however, is computationally expensive and again requires the *a priori* specification of the initial number of basis functions to be used from which a specified number of 'best' functions are selected.

Look-up-table methods have the advantage of simplicity and the resulting controller can be established in a one-pass operation. There is, however, no optimization and *a priori* specification of the numbers of fuzzy sets for each variable is again required. Equivalent and conflicting rules are eliminated by calculating a weight for each rule; rules with the highest weights are retained. Since there is no prior clustering of the training data, the elimination of such rules involves an exhaustive, and hence computationally expensive, evaluation of rule weights leading to an increased development time for the controller.

The method adopted and described above has been found to result in a considerably simplified controller structure when compared to those obtained by any of the alternative methods, and the proposed procedure also has the advantage that the number of fuzzy sets and fuzzy rules can be determined by an automatic procedure.

4.4.2.3 Vector drive with a self-organizing fuzzy-neural controller; minimal configuration

The vector drive considered is similar to that shown in Fig. 4.40, where controller 1 is the speed controller, controller 2 is the torque-producing current controller, controller 3 is the magnetizing current controller, and controller 4 is the flux-producing current controller. Conventionally these are implemented as PI-type controllers, but any of these may be replaced by a fuzzy-neural controller, and now the speed controller (controller 1) is a fuzzy-neural controller.

There are various possibilities for obtaining the training data, e.g. it can be obtained from the PI-controlled drive. For this purpose, approximately 1500 data sets for the error, change of error, and controller output were recorded with the drive responding to a sequence of changes in reference speed going from standstill to 75% rated speed, followed by a speed reversal to -75% rated speed, followed by a reference speed change bringing the drive to rest. Data from this test was used to establish and train the fuzzy-neural speed controller by the two-stage, off-line procedure described in Sections 4.4.2.2.2 and 4.4.2.2.3. The first-stage training was carried out for approximately 60 epochs and the second-stage tuning was carried out for a further 30 epochs.

It is a feature of the proposed method that it results in a significant reduction (up to 50%) in the number of fuzzy sets required for each variable when compared to classical non-self-organizing fuzzy speed controllers used in drives. A corresponding reduction in the number of fuzzy rules is also obtained, in this case up to an order of magnitude. These reductions afford a significant advantage over conventional approaches to controller design, particularly regarding the DSP requirements of a practical implementation.

The results for the real-time performance obtained for the drive using the TMS320C30 DSP with the fuzzy-neural controller replacing the PI speed controller are shown in Figs 4.44(a) to 4.44(c). The sequence of speed reference changes is as indicated above. It follows from Fig. 4.44(a) that a satisfactory speed response is achieved. The torque-producing stator current component (i_{sy}) shown in Fig. 4.44(b) has the expected form and follows the reference value (i_{syref}) generated by the fuzzy-neural controller. As expected the rotor magnetizing current ($|\bar{i}_{mr}|$) shown in Fig. 4.44(c) is effectively constant.

It should be noted that similar experimental results have been obtained for the induction motor drive when various speed controllers have been used: for the case when a fuzzy speed controller was used, for the drive with a PI speed controller, and also for the drive with an artificial-neural speed controller. However, in the present scheme the fuzzy rules and the membership functions have been obtained by an automatic, systematic procedure. In addition, a reduced number of membership functions and fuzzy rules has been established.

Figure 4.45 shows the extremely non-linear 3-D controller characteristic of the fuzzy-neural speed controller. This results from a considerably reduced rule base established from the first phase of the network development followed by a second-stage tuning process with a duration of around 100 epochs.

In summary it can be concluded that the combination of artificial neural network and fuzzy-logic concepts allows the advantages of each approach to be utilized whilst at the same time overcoming their respective disadvantages. The design procedure adopted allows for two-stage tuning of the controller. The first stage involves a method for the determination of the detailed structure of the fuzzy-neural controller, the determination of a minimal rule-base together with initial estimates for the centres and widths of the membership functions. The first stage allows for the inclusion of expert linguistic information. The second stage involves further tuning of the membership functions of the fuzzy-neural controller using a back-propagation-type training algorithm. The fuzzy-neural-controller form of a vector-controlled induction motor drive has been compared to that using conventional PI-type and non-self-tuning fuzzy controllers. Although the responses obtained in these cases are similar, it is important to note that the two-stage approach results in a reduced overall development time compared to the conventional approaches. The proposed technique produces minimal fuzzy set and rule base configurations for the fuzzy-neural controller. This leads to simpler real-time implementation. It is believed that in the future fuzzy-neural controllers will find widespread applications in various variable-speed drives.

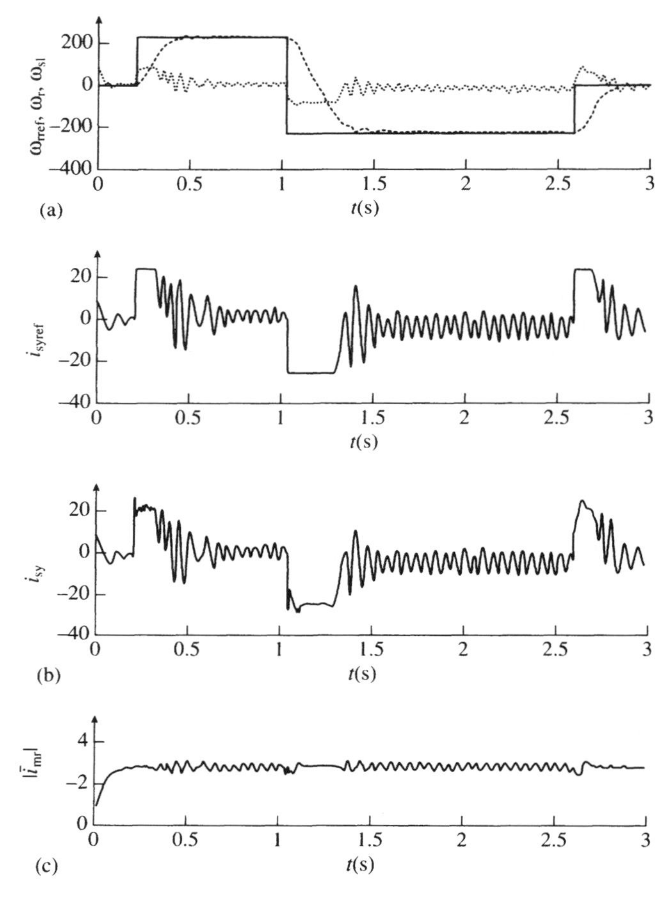

Fig. 4.44. Fuzzy-neural-controlled vector drive experimental responses. (a) Reference speed, rotor speed, slip frequency ($\omega_{rref}, \omega_r, \omega_{sl}$); (b) torque-producing stator current reference, torque-producing stator current (i_{syref}, i_{sy}); (c) rotor magnetizing current ($|\bar{i}_{mr}|$).

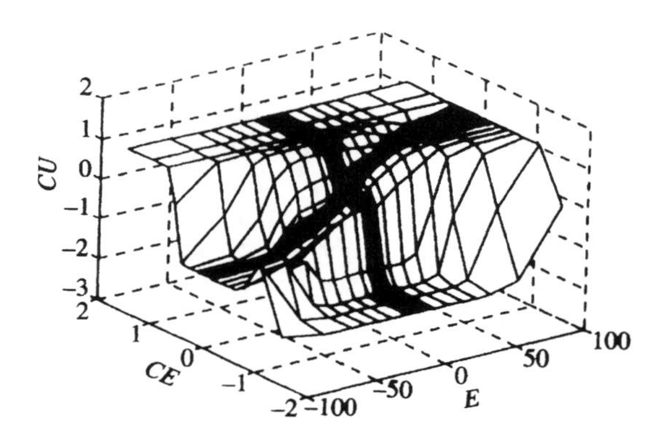

Fig. 4.45. Non-linear fuzzy-neural speed controller profile.

4.5 Main techniques of sensorless control of induction motors

4.5.1 GENERAL INTRODUCTION

Many attempts have been made in the past to extract the speed or position signal of an induction machine; Chapter 5 of a recent book [Vas 1993] describes in detail various solutions. However, the first attempts have been restricted to techniques which are only valid in the steady state, and as an example, such techniques will first be briefly described. These can be used in low-cost drive applications, not requiring high dynamic performance. However, other techniques will also be described which are applicable for high-performance applications in vector- and direct-torque-controlled drives.

It is a common feature of many of the 'sensorless' techniques that they depend on machine parameters: they may depend on the temperature, saturation levels, frequency, etc. To compensate for the parameter variations, various parameter-adaptation schemes have also been proposed in the literature. In an ideal sensorless drive, speed information and control is provided with an accuracy of 0.5% or better, from zero speed to the highest speed, for all operating conditions and independent of saturation levels and parameter variations. In the only industrial sensorless implementation of a direct-torque-controlled drive, an improved mathematical model is used to estimate the speed, and it is claimed that the drive can work even at zero speed. However, it is believed that in the not-distant future, artificial-intelligence-based techniques (fuzzy, neural, fuzzy-neural, etc.) will have a more dominant role in sensorless drives and in particular, complicated mathematical models will not be required for speed or position estimation.

Conventional techniques are not suitable to achieve stable, very low speed operation in a speed or position sensorless high-performance induction motor drive. In this case the estimation of the speed is problematic, due to parameter mismatch and noise. If the induction motor is represented by its fundamental model, then it becomes unobservable at zero frequency. However, as will be discussed in Section 4.5.3, it is possible to utilize various effects (rotor slot harmonics, saliency, etc.) for the speed estimation, but e.g. the rotor slot harmonic signal is insufficient in bandwidth as a speed feedback signal in a high-performance drive, and specially introduced saliencies require non-standard rotors. On the other hand, it is possible to estimate the rotor position at very low rotor speed and even at zero frequency with high bandwidth in a standard squirrel cage induction motor by utilizing the rotor angle variation of the machine inductances $u_{sA}/(di_{sA}/dt)$, $u_{sB}/(di_{sB}/dt)$, and $u_{sC}/(di_{sC}/dt)$.

In Chapter 7 artificial-intelligence-based speed-sensorless drives are also discussed.

4.5.2 SLIP AND SPEED ESTIMATORS FOR LOW-PERFORMANCE APPLICATIONS

It is possible to construct a low-cost slip-sensing device, which uses the stator voltages and currents of the induction motor. This can be obtained by simply considering the steady-state equivalent circuit of the induction motor, and then it can be seen that for small values of the slip, the slip (s) can be expressed as

$$s=\frac{(R_r'/3)\omega_1 T_e}{|\bar{U}_m|^2}, \tag{4.5-1}$$

where R_r' is the referred rotor resistance, ω_1 is the stator angular frequency, T_e is the steady-state electromagnetic torque, and $|\bar{U}_m|$ is the absolute value of the magnetizing (air-gap) voltage. Thus the slip monitor uses the monitored electromagnetic torque and also the magnetizing voltage, which can be simply obtained by subtracting the ohmic and stator leakage voltage drops from the terminal voltage. The electromagnetic torque can be obtained from monitored stator voltages and currents (by using the fact that the torque can be expressed as the cross-vectorial product of the stator flux-linkage and current space vectors, and the stator flux-linkage vector can be obtained by integrating the stator-voltage space vector reduced by the stator-voltage space vector of the ohmic drops).

As an alternative, the electromagnetic torque can be obtained from the air-gap power,

$$T_e=\frac{P_{gap}}{\omega_1}, \tag{4.5-2}$$

which can be determined as the difference of the d.c. link power and the power losses for the inverter, stator, and choke:

$$P_{gap}=P_{dc}-P_{loss}=U_{dc}I_{dc}-(P_s+P_{inv}+P_{choke}). \tag{4.5-3}$$

In this scheme the d.c. link voltage and d.c. link current are monitored. The inaccuracy in the determination of the various losses has a significant effect on the accuracy of the estimated slip at low values of the slip, since at low slip values, these losses constitute a large percentage of the input power. The steady-state angular rotor speed can then be obtained as

$$\omega_r=\omega_1(1-s). \tag{4.5-4}$$

It is important to note that the monitoring schemes described above cannot be used under dynamic conditions and the first scheme can only be used for the estimation of the slip near to its rated value, i.e. for small values of the slip. Thus the speed range is limited. Furthermore, in the first scheme, when obtaining the magnetizing voltage, the stator leakage voltage contains the derivative of the stator currents, and this can cause significant problems due to the noise content in the stator currents. The techniques described can only be used in low-performance applications and not in high-performance drives.

It is also possible to utilize the Kloss formula for the determination of the steady-state slip, if the pull-out slip (s_{max}) and pull-out torque (T_{emax}) values are known:

$$\frac{T_e}{T_{emax}} \approx \frac{2}{(s/s_{max} + s_{max}/s)}, \tag{4.5-5}$$

where the electromagnetic torque can be estimated by using one of the techniques described above. Furthermore, it follows from eqn (4.5-5) that for low values of the slip, the slip estimation can be based on

$$\frac{T_e}{T_{emax}} \approx \frac{2s}{(s_{max})} \tag{4.5-6}$$

and at large values of the slip it can be obtained from

$$\frac{T_e}{T_{emax}} \approx \frac{2s_{max}}{s}. \tag{4.5-7}$$

In a conventional speed-controlled voltage-source inverter-fed induction motor drive with open-loop flux control using the steady-state machine model (V/f control), the speed is monitored and the reference value of the slip frequency is obtained on the output of a speed controller as shown in Fig. 4.46.

In this scheme the reference angular stator frequency is obtained by using

$$\omega_{1ref} = \omega_{slref} + \omega_r \tag{4.5-8}$$

and ω_{1ref} is an input to the function generator $f(\omega_1)$, which outputs the modulus of the reference stator voltages, $|\bar{u}_{sref}|$. If the effects of the stator resistance are neglected, then to ensure constant stator flux, $|\bar{u}_{sref}|$ is varied linearly with the stator frequency (thus the function f is a linear function of ω_1). However, at low stator frequencies, it is important to consider the effects of the stator resistance, and then the stator voltage has to be boosted (thus the function f is not linear). Such an induction motor drive has good steady-state performance but low dynamic performance, since it is based on the steady-state equivalent circuit. Similarly to all types of open-loop drives, such an open-loop drive is sensitive to various secondary effects (parameter variations, e.g. due to temperature variation).

In a low-cost implementation of the V/f drive, no slip compensation is used, thus there is no need to monitor the speed. Such a low-cost, low-dynamic-performance

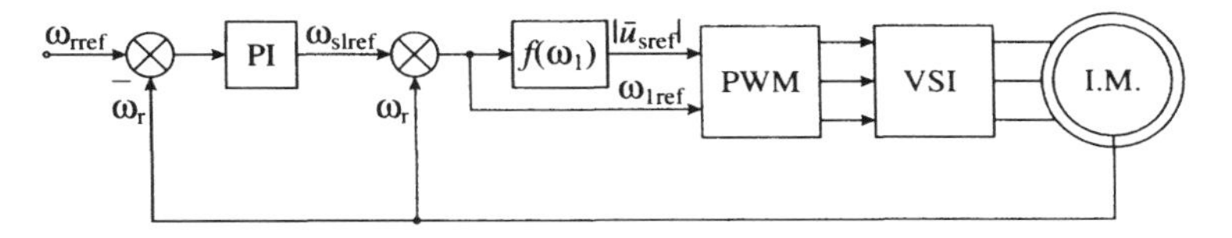

Fig. 4.46. V/f control scheme using speed sensor.

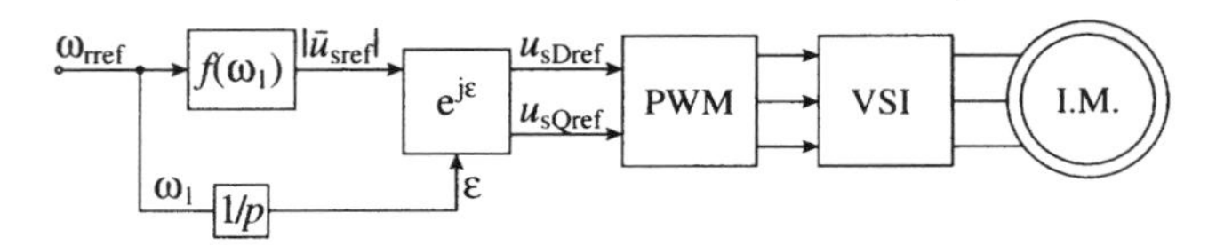

Fig. 4.47. Low-cost V/f control scheme without speed sensor.

VSI-fed induction motor drive scheme is shown in Fig. 4.47. In this scheme, the slip frequency is not considered, thus $\omega_1 = \omega_{\text{rref}}$ is assumed.

When the stator frequency is integrated, it gives the position (ε) of the stator-voltage space vector (with respect to the direct-axis of the stationary reference frame). Thus

$$\varepsilon(t) = \int \omega_1(t)\,\mathrm{d}t. \tag{4.5-9}$$

On the other hand, the output of the $f(\omega_1)$ function generator gives the stator-voltage reference modulus ($|\bar{u}_{\text{sref}}|$). This function generator implements the V/f characteristic as discussed above. In the stator-voltage-oriented reference frame, the reference stator-voltage space vector has only a direct-axis component, $u_{\text{sxref}} = |\bar{u}_{\text{sref}}|$, and its quadrature-axis components (u_{syref}) is zero. It follows that the two-axis components of the reference stator-voltage space vector in the stationary reference frame ($u_{\text{sDref}}, u_{\text{sQref}}$) can be obtained by using the transformation

$$u_{\text{sDref}} + \mathrm{j}u_{\text{sQref}} = \exp(\mathrm{j}\varepsilon)(u_{\text{sxref}} + \mathrm{j}u_{\text{syref}}). \tag{4.5-10}$$

Thus $u_{\text{sDref}} = |\bar{u}_{\text{sref}}| \cos \varepsilon$ and $u_{\text{sQref}} = |\bar{u}_{\text{sref}}| \sin \varepsilon$, which are the inputs to the modulator, as shown in Fig. 4.47. If the three-phase reference stator voltages are required, these can also be obtained as

$$u_{\text{sAref}} = |\bar{u}_{\text{sref}}| \cos \varepsilon \qquad u_{\text{sBref}} = |\bar{u}_{\text{sref}}| \cos(\varepsilon - 2\pi/3) \qquad u_{\text{sCref}} = |\bar{u}_{\text{sref}}| \cos(\varepsilon - 4\pi/3). \tag{4.5-11}$$

However, the scheme shown in Fig. 4.47 will result in speed errors caused by a load (for large loads the error is large), since it has been assumed above that the stator frequency is equal to the reference speed, which is only correct if the slip is zero (there is no load). It follows that to ensure good steady-state speed regulation, slip compensation must be employed.

In various implementations of the V/f control scheme with slip compensation, it is possible to obtain the slip frequency using the concepts discussed above for slip estimation. However, a simple scheme shown in Fig. 4.48 can also be obtained, which contains the slip frequency compensation, but where the slip frequency is obtained in another easier way. For this purpose it is utilized that the slip frequency is proportional to the torque, which is however proportional to the quadrature-axis stator current (i_{sy}) in the stator-voltage-oriented reference frame (this reference frame was introduced in connection with the scheme

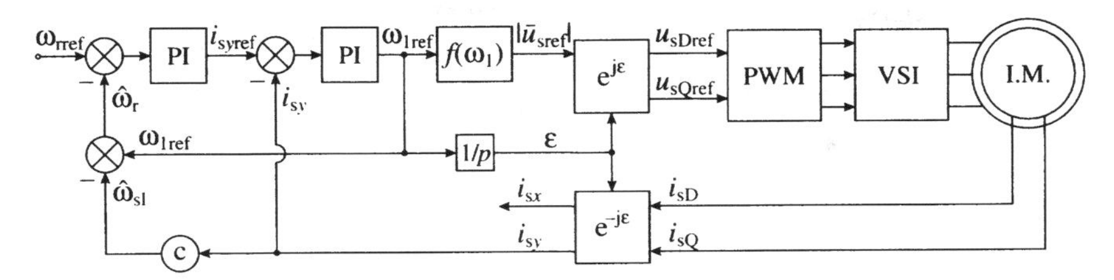

Fig. 4.48. Speed-sensorless V/f scheme with slip frequency compensation and using the position of the stator-voltage space vector.

described in Fig. 4.47). However, for this purpose the active stator current (i_{sy}) component must be monitored. This can be obtained from the monitored stator currents, since the stator-current space vector in the stator-voltage-oriented reference frame can be expressed as

$$i_{sx}+ji_{sy}=\exp(-j\varepsilon)(i_{sD}+ji_{sQ}), \tag{4.5-12}$$

where $i_{sD}+ji_{sQ}$ is the stator-current space vector in the stator (stationary) reference frame. Thus $i_{sy}=-i_{sD}\sin\varepsilon+i_{sQ}\cos\varepsilon$. This is the reason why the $\exp(-j\varepsilon)$ transformation is also present in the control scheme shown in Fig. 4.48. The voltage transformation is the same as in the scheme of Fig. 4.47. The slip frequency can be estimated by considering that it is proportional to the active stator current, thus $\hat{\omega}_{sl}=ci_{sy}$, where c is a constant. The speed controller in Fig. 4.47 outputs the reference value of the active stator current (i_{syref}) and when this is compared with the actual value of the active current, the error is an input to the active stator-current controller. This outputs the reference value of the angular stator frequency (ω_{1ref}), and when integrated gives the angle ε as discussed above. Furthermore, the difference of ω_{1ref} and the estimated slip frequency ($\hat{\omega}_{sl}$) gives the estimated speed ($\hat{\omega}_r$). By considering that $i_{sA}+i_{sB}+i_{sC}=0$, the two-axis stator currents i_{sD}, i_{sQ} can be obtained by only measuring two stator currents (e.g. i_{sA}, i_{sB}), where i_{sD} is proportional to i_{sA} and i_{sQ} is proportional to $i_{sB}-i_{sC}=i_{sB}+2i_{sA}$.

It follows that this scheme contains closed-loop control of both the speed and the active stator current. It can be used for the accurate steady-state speed control of the induction machine for any value of the load. However, the dynamic performance of the drive is low, since the scheme is based on the steady-state equivalent circuit of the machine. Speed estimation schemes which allow high dynamic performance are discussed in the next section.

4.5.3 SLIP, SPEED, ROTOR ANGLE, AND FLUX-LINKAGE ESTIMATORS FOR HIGH-PERFORMANCE APPLICATIONS

In the present section, various techniques are described which can be used in high-performance drives for the estimation of the slip, rotor speed, rotor angle, and various machine flux linkages. Thus the following techniques are described:

1. Open-loop estimators using monitored stator voltages/currents and improved schemes;
2. Estimators using the spatial-saturation stator phase third-harmonic voltage;
3. Estimators using saliency (geometrical, saturation) effects;
4. Model reference adaptive systems (MRAS);
5. Observers (Kalman, Luenberger);
6. Estimators using artificial intelligence (neural network, fuzzy-logic-based systems, fuzzy-neural networks, etc.).

4.5.3.1 Open-loop speed estimators using monitored stator voltages and currents; improved schemes

4.5.3.1.1 General introduction

In the present section various rotor speed and slip frequency estimators are obtained by considering the voltage equations of the induction machine. The schemes described below use the monitored stator voltages and currents or the monitored stator currents and reconstructed stator voltages. Some of these estimation schemes are used in recently introduced commercially available speed-sensorless induction motor drives. However, it is important to note that in general, the accuracy of open-loop estimators depends greatly on the accuracy of the machine parameters used. At low rotor speed, the accuracy of the open-loop estimators is reduced, and in particular, parameter deviations from their actual values have great influence on the steady-state and transient performance of the drive system which uses an open-loop estimator. Furthermore, higher accuracy is achieved if the stator flux is obtained by a scheme which avoids the use of pure integrators (see also Sections 3.1.3.2, 3.2.2.2, and 4.1.1.4).

In general, open-loop speed estimators depend on various parameters of the induction machine. The stator resistance (R_s) has important effects on the stator flux linkages, especially at low speeds, and if the rotor flux linkage is obtained from the stator flux linkage, then the rotor flux-linkage accuracy is also influenced by the stator resistance. However, it is possible to have a rather accurate estimate of the appropriate 'hot' stator resistance by using a thermal model of the induction machine.

In some schemes, the rotor flux-linkage estimation requires the rotor time constant, which can also vary, since it is the ratio of the rotor self-inductance and the rotor resistance, and the rotor resistance can vary due to temperature effects and skin effect (current-displacement effect), and the rotor self-inductance can vary due to skin effect and saturation effects. The changes of the rotor resistance due to temperature changes are usually slow changes. Due to main flux saturation, the magnetizing inductance (L_m) can change and thus the stator self-inductance

($L_s = L_{sl} + L_m$) and rotor self-inductance ($L_r = L_{rl} + L_m$) can also change even if the leakage inductances (L_{sl}, L_{rl}) are constant. The changes of the rotor self-inductance due to saturation can be fast. Due to leakage flux saturation, L_{sl}, L_{rl} and the stator transient inductance (L'_s) can also change. In a vector-controlled drive, where the rotor flux amplitude is constant, the variations of L_m are small. In a machine with closed rotor slots, the stator transient inductance varies due to leakage flux saturation, and is a function of the stator currents, due to the saturable closing bridge around the rotor slots. In a torque-controlled drive, the effects of incorrect parameters result in incorrect torque, flux, degradation of system performance, etc.

In Section 8.1 and also in a recent book [Vas 1993] on-line parameter identification techniques are discussed, where the most important four induction motor parameters (stator resistance, rotor time constant, stator transient inductance, stator self-inductance) are also identified during a self-commissioning stage of a torque-controlled induction motor drive system (vector drive or direct-torque-controlled drive). Various commercial drives use parameter estimation techniques which are identical or similar to that described in Section 8.1. It should be noted that in this self-commissioning stage, the machine is at standstill during all measurements. The stator resistance is identified by injecting a d.c. current in the stator winding and measuring the corresponding d.c. voltage in the steady state. The rotor time constant is identified from the exponential voltage waveform when a step change in the stator current is applied. Finally the stator transient time constant is identified from the slope of the stator currents when a step voltage is applied.

Five rotor-speed estimator schemes are described below, but where possible, small modifications of a scheme are also discussed within the scheme. In every subsection, first the expression for the rotor speed used by the scheme is given and this is then followed by a mathematical proof and the description of various aspects of implementation details.

4.5.3.1.2 Rotor speed estimation scheme 1

$$\boxed{\omega_r = \left[-\frac{d\psi_{rd}}{dt} - \frac{\psi_{rd}}{T_r} + \frac{L_m}{T_r} i_{sD} \right] \Big/ \psi_{rq}}$$

It is possible to obtain an expression for the rotor speed directly by using the rotor-voltage space-vector equation expressed in the stationary reference frame ($\omega_g = 0$). It follows from eqn (2.1-149) that for the induction motor, the direct-axis rotor-voltage equation becomes

$$0 = R_r i_{rd} + \frac{d\psi_{rd}}{dt} + \omega_r \psi_{rq}, \tag{4.5-13}$$

where

$$\psi_{rd} = L_r i_{rd} + L_m i_{sD}. \tag{4.5-14}$$

Equation (4.5-13) can be rearranged for the rotor speed, but it contains the direct-axis rotor current which cannot be measured directly. However, by considering eqn (4.5-14) it can be expressed in terms of the direct-axis stator current and direct-axis rotor flux and when this is substituted into eqn (4.5-13),

$$\omega_r = \left[-\frac{d\psi_{rd}}{dt} - \frac{\psi_{rd}}{T_r} + \frac{L_m}{T_r} i_{sD} \right] \Big/ \psi_{rq} \tag{4.5-15}$$

is obtained. This equation can be used for the estimation of the rotor speed. However, for this purpose, the direct- and quadrature-axis rotor flux linkages have to be estimated. Various techniques have been discussed in Section 4.1.1 for the estimation of the stator flux and rotor magnetization current from terminal quantities. For example, the rotor flux-linkage space vector can be obtained from the monitored terminal voltages and currents by using the circuit of Fig. 4.9(a), and by utilizing $\psi_{rd} = L_m i_{mrD}$, $\psi_{rq} = L_m i_{mrQ}$, or $\bar{\psi}_r' = L_m |\bar{i}_{mr}| \exp(j\rho_r)$. However, other circuits have been shown in Figs 4.9(b), 4.9(c), and 4.9(d) which result in smaller integrator drift at low frequency (two of the circuits use closed-loop integrators). However, as discussed in Section 3.1.3.1 and also in Section 4.1.1, it is also possible to reconstruct the stator voltages by using the monitored d.c. link voltage together with the inverter switching functions. For clarity, some of the equations will now be summarized.

In eqn (4.5-15) T_r is the rotor time constant ($T_r = L_r / R_r$) and it can be seen that this equation contains the derivative of the direct-axis rotor flux-linkage component. By using eqns (2.1-150) and (2.1-151), the rotor flux-linkage space vector can be expressed in terms of the stator flux-linkage space vector, and in the stationary reference frame $\bar{\psi}_r' = (L_r / L_m)(\bar{\psi}_s - L_s' \bar{i}_s)$ is obtained, where L_s' is the stator transient inductance. Resolution of this into its real and imaginary parts gives

$$\psi_{rd} = \frac{L_r}{L_m} (\psi_{sD} - L_s' i_{sD}) \tag{4.5-16}$$

$$\psi_{rq} = \frac{L_r}{L_m} (\psi_{sQ} - L_s' i_{sQ}). \tag{4.5-17}$$

The derivative of ψ_{rq} is used in eqn (4.5-15), and if magnetic saturation is neglected, it follows from eqn (4.5-16) that

$$\frac{d\psi_{rd}}{dt} = \frac{L_r}{L_m} \left(\frac{d\psi_{sD}}{dt} - L_s' \frac{di_{sD}}{dt} \right). \tag{4.5-18}$$

Eqns (4.5-16)–(4.5-18) contain the stator flux-linkage components which can be obtained from eqn (2.1-148) in the stationary reference frame by using

$$\frac{d\psi_{sD}}{dt} = u_{sD} - R_s i_{sD} \tag{4.5-19}$$

$$\frac{d\psi_{sQ}}{dt} = u_{sQ} - R_s i_{sQ}. \tag{4.5-20}$$

However, greater accuracy can be obtained by using the flux estimation techniques discussed in Section 4.1.1.4.3. The rotor speed can finally be obtained from eqn (4.5-15), by considering eqns (4.5-16)–(4.5-20). To summarize:

$$\omega_r = \left[-\frac{d\psi_{rd}}{dt} - \frac{\psi_{rd}}{T_r} + \frac{L_m}{T_r} i_{sD} \right] \Big/ \psi_{rq}, \tag{4.5-21}$$

where

$$\frac{d\psi_{rd}}{dt} = \frac{L_r}{L_m}\left(u_{sD} - R_s i_{sD} - L_s' \frac{di_{sD}}{dt} \right) \tag{4.5-22}$$

and ψ_{rd} is obtained from this by integration. Similarly, ψ_{rq} is obtained from

$$\frac{d\psi_{rq}}{dt} = \frac{L_r}{L_m}\left(u_{sQ} - R_s i_{sQ} - L_s' \frac{di_{sQ}}{dt} \right) \tag{4.5-23}$$

by integration. A possible implementation is shown in Fig. 4.49. It can be seen that this requires several machine parameters, some of which vary with temperature, skin effect, and saturation. Thus the speed can only be obtained accurately if these parameters are accurately known. It is possible to have a rather accurate estimate of the stator resistance by utilizing the 'cold' value of the stator resistance together with a thermal model of the induction machine. At low speeds the accuracy of this scheme is limited, but improvements can be obtained by using the flux estimation techniques discussed in Section 4.1.1.4.3. It should be noted that a similar scheme could be obtained if the speed is estimated by using the quadrature-axis rotor equation in the stationary reference frame. The stator voltages can be monitored or reconstructed from the d.c. link voltage and the inverter switching states, as discussed in Section 3.1.3.1 and Section 4.1.1.

As mentioned above, it is possible to improve the speed estimation by employing improved flux estimators (e.g. those shown in Figs 4.9(b), 4.9(c), or 4.9(d)). For illustration purposes one of the improved schemes which is obtained by utilizing the technique described in connection with Fig. 4.9(b) is shown in Fig. 4.50.

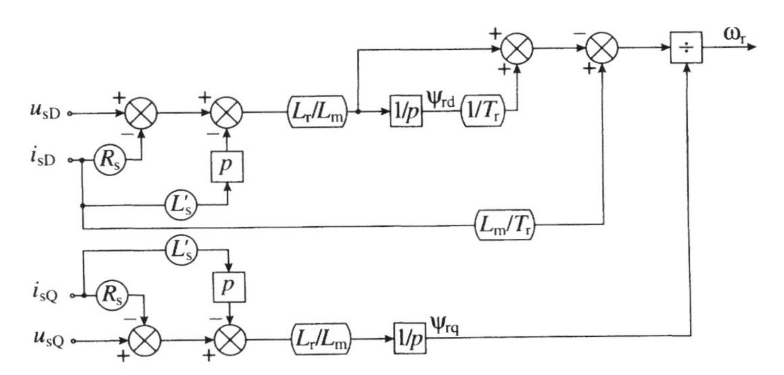

Fig. 4.49. Rotor speed estimator using two integrators, two differentiators, and 5 machine parameters.

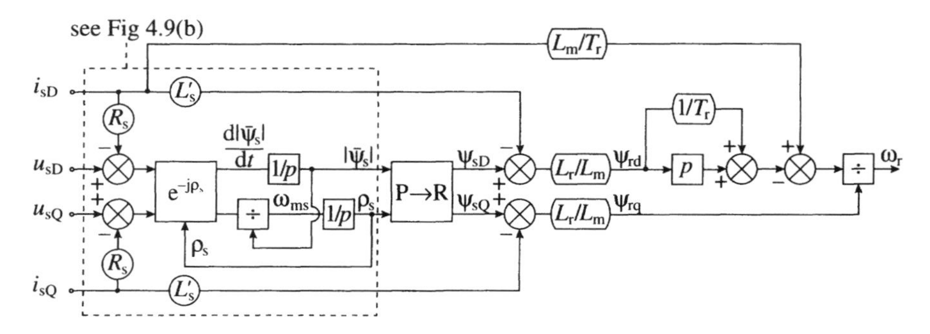

Fig. 4.50. Rotor speed estimator using 2 integrators, 1 differentiator, and 4 machine parameters.

In this case, first, the modulus ($|\bar{\psi}_s|$) and angle (ρ_s) of the stator flux-linkage space vector are obtained (by the circuit shown in Fig. 4.9(b)) and then the rotor flux linkages are obtained by using eqns (4.5-16) and (4.5-17). These are then used in eqn (4.5-21).

The estimator shown in Fig. 4.50 contains a differentiator to obtain $d\psi_{rd}/dt$. When a suitable numerical algorithm is used, this can be performed accurately. However, it is also possible to use another combined scheme, where $d\psi_{rd}/dt$ is obtained by using eqn (4.5-22) and ψ_{rq} is obtained by using the technique described in Fig. 4.50.

4.5.3.1.3 Rotor speed estimation scheme 2

$$\omega_r = -\left[u_{sD} - \left(R_s + \frac{L_s}{T_r}\right)i_{sD} - L_s'\frac{di_{sD}}{dt} + \frac{\psi_{sD}}{T_r}\right] \Big/ (\psi_{sQ} - L_s' i_{sQ})$$

or

$$\omega_r = \left[u_{sQ} - \left(R_s + \frac{L_s}{T_r}\right)i_{sQ} - L_s'\frac{di_{sQ}}{dt} + \frac{\psi_{sQ}}{T_r}\right] \Big/ (\psi_{sD} - L_s' i_{sD})$$

It is possible to obtain another rotor speed estimation scheme, which can be obtained from eqn (4.5-15) by the substitution of eqns (4.5-16), (4.5-17), and (4.5-22). Thus

$$\omega_r = -\left[u_{sD} - \left(R_s + \frac{L_s}{T_r}\right)i_{sD} - L_s'\frac{di_{sD}}{dt} + \frac{\psi_{sD}}{T_r}\right] \Big/ (\psi_{sQ} - L_s' i_{sQ}). \qquad (4.5\text{-}24)$$

It should be noted that this expression can also be directly obtained by considering the quadrature-axis rotor voltage equation, eqn (4.5-13), and by substitution into this eqns (4.5-17), (4.5-22), and $i_{rd} = (\psi_{sD} - L_s i_{sD})/L_m$ (which follows from the

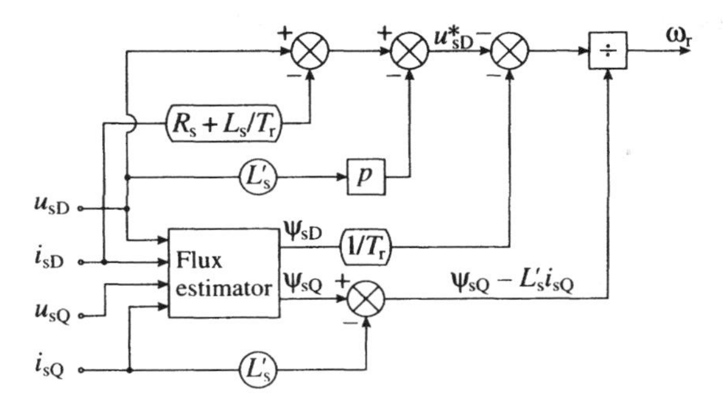

Fig. 4.51. Rotor speed estimator using 2 integrators, 1 differentiator, and 4 machine parameters.

definition of the direct-axis stator flux linkage, $\psi_{sD}=L_s i_{sD}+L_m i_{rd}$). A scheme using eqn (4.5-24) is shown in Fig. 4.51. The flux estimator block can be implemented by using any of the improved techniques discussed earlier.

The speed estimator shown in Fig. 4.51 requires four machine parameters and accuracy of the speed estimator greatly depends on these. However, it is possible to obtain satisfactory results down to 1–2 Hz.

Equation (4.5-24) has been derived by using the quadrature-axis rotor equation, but if the direct-axis rotor voltage equation is used, then a similar expression can be obtained:

$$\omega_r=\left[u_{sQ}-\left(R_s+\frac{L_s}{T_r}\right)i_{sQ}-L'_s\frac{di_{sQ}}{dt}+\frac{\psi_{sQ}}{T_r}\right]\Big/(\psi_{sD}-L'_s i_{sD}). \qquad (4.5\text{-}25)$$

An implementation based on eqn (4.5-25) requires the same complexity as that shown in eqn Fig. 4.51. However, it can be seen from (4.5-25) that if another speed estimator is used where the quadrature-axis stator flux is forced to be zero, then the scheme is simplified and the numerator will not contain the quadrature-axis stator flux-linkage space vector. This is indeed the case if the rotor voltage equation is expressed in a reference frame rotating with the speed of the stator flux. This is now discussed below.

4.5.3.1.4 Rotor speed estimation scheme 3

$$\boxed{\omega_r=u^*_{sy}/(|\bar{\psi}_s|-L'_s i_{sx})}$$

There are many possibilities for obtaining the expression for the rotor speed for this scheme, but it is very straightforward to use first the rotor voltage equation in the stationary reference frame, and the resulting equation is then transformed into the stator-flux-oriented reference frame. By following this approach, of course eqns (4.5-24) and (4.5-25) will also result. For this purpose the rotor-voltage space vector equation is considered in the stationary reference frame. This can be obtained from eqn (2.1-149) by assuming $\omega_g=0$ and thus $0=R_r\bar{i}_r'+d\bar{\psi}_r'/dt-j\omega_r\bar{\psi}_r'$ is obtained. However, the rotor-current space vector can be eliminated by considering $\bar{i}_r'=(\bar{\psi}_s-L_s\bar{i}_s)/L_m$ and the rotor flux-linkage space vector can be substituted by $\bar{\psi}_r'=(L_r/L_m)\,(\bar{\psi}_s-L_s'\bar{i}_s)$. Hence

$$\frac{d\bar{\psi}_r'}{dt}=\frac{L_r}{L_m}\left(\bar{u}_s-R_s\bar{i}_s-L_s'\frac{d\bar{i}_s}{dt}\right)$$

and substitution of the expressions for $\bar{i}_r'$, $\bar{\psi}_r'$, and $d\bar{\psi}_r'/dt$ into the rotor voltage equation gives

$$\bar{u}_s-\left(R_s+\frac{L_s}{T_r}\right)\bar{i}_s-L_s'\frac{d\bar{i}_s}{dt}=-\frac{\bar{\psi}_s}{T_r}+j\omega_r(\bar{\psi}_s-L_s'\bar{i}_s). \tag{4.3-26}$$

It should be noted that, as expected, the real and imaginary parts of eqn (4.5-26) yield eqns (4.5-24) and (4.5-25) respectively. Equation (4.5-26) is a specific form of the rotor voltage equation in the stationary reference frame. When this is transformed into the reference frame rotating with the stator flux-linkage space vector, whose speed is $\omega_{ms}=d\rho_s/dt$ (stator-flux-oriented reference frame, where ρ_s is the angle of the stator flux-linkage space vector with respect to the real-axis of the stator reference frame as shown in Fig. 4.25), then

$$\left[\bar{u}_s-\left(R_s+\frac{L_s}{T_r}\right)\bar{i}_s-L_s'\frac{d\bar{i}_s}{dt}\right]\exp(-j\rho_s)$$
$$=-\bar{\psi}_s\exp(-j\rho_s)/T_r+j\omega_r[\bar{\psi}_s\exp(-j\rho_s)-L_s'\bar{i}_s\exp(-j\rho_s)] \tag{4.5-27}$$

is obtained. However, on the right-hand side of eqn (4.5-27),

$$\bar{\psi}_s\exp(-j\rho_s)=\bar{\psi}_s'=\psi_{sx}+j\psi_{sy}=|\bar{\psi}_s|$$

and

$$\bar{i}_s\exp(-j\rho_s)=\bar{i}_s'=i_{sx}+ji_{sy}$$

are the stator-flux and stator current space vectors respectively in the stator-flux-oriented reference frame (whose real and imaginary axes are denoted by x and y respectively). Thus it follows that

$$\left[\bar{u}_s-\left(R_s+\frac{L_s}{T_r}\right)\bar{i}_s-L_s'\frac{d\bar{i}_s}{dt}\right]\exp(-j\rho_s)=-\frac{|\bar{\psi}_s|}{T_r}+j\omega_r(|\bar{\psi}_s|-L_s'\bar{i}_s'). \tag{4.5-28}$$

The real and imaginary parts of eqn (4.5-28) give

$$u_{sx}^*=-\frac{|\bar{\psi}_s|}{T_r}+\omega_r L_s' i_{sy} \tag{4.5-29}$$

$$u_{sy}^*=\omega_r(|\bar{\psi}_s|-L_s' i_{sx}), \tag{4.5-30}$$

where

$$u^*_{sx}+ju^*_{sy}=\left[\bar{u}_s-\left(R_s+\frac{L_s}{T_r}\right)\bar{i}_s-L'_s\frac{d\bar{i}_s}{dt}\right]\exp(-j\rho_s) \tag{4.5-31}$$

and $\bar{u}_s=u_{sD}+ju_{sQ}$, $\bar{i}_s=i_{sD}+ji_{sQ}$. It follows from eqn (4.5-30) that

$$\omega_r=u^*_{sy}/(|\bar{\psi}_s|-L'_s i_{sx}). \tag{4.5-32}$$

As expected, the numerator of eqn (4.5-32) contains only u^*_{sy} and not the quadrature-axis stator flux. This is physically due to the fact that in the stator-flux-oriented reference frame the quadrature-axis stator flux is zero. This should be contrasted with eqn (4.5-25), whose numerator contains the quadrature-axis stator flux (expressed in the stationary reference frame). Equation (4.5-32) has been implemented in Fig. 4.52. This scheme again depends on four machine parameters and can give satisfactory results even at very low speeds. The voltages can either be monitored or they can be reconstructed from the measured d.c. link voltage and inverter switching states. A speed estimator based on eqn (4.5-32) can be effectively used in a stator-flux-oriented vector control scheme even at relatively low stator frequency. The flux estimator can use any of the improved techniques discussed earlier, including the estimator shown in Fig. 4.9(b) or (c).

It should be noted that if the rotor speed is determined by eqn (4.5-32), then by using this in eqn (4.5-29), it is possible to obtain an expression for the rotor time constant, and this can be used to monitor the changes of T_r.

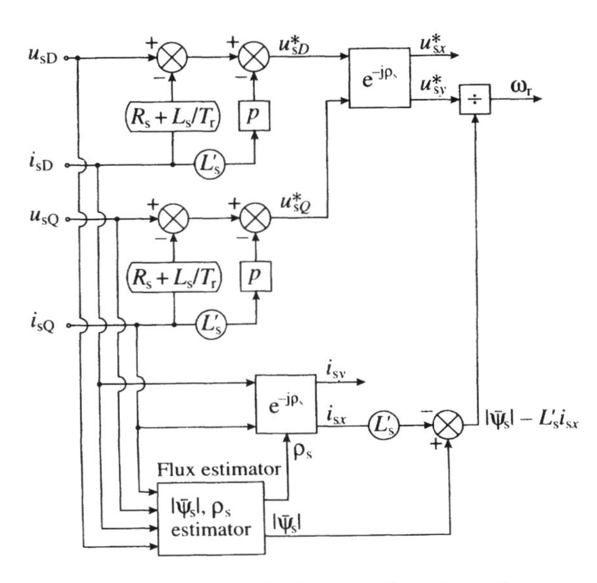

Fig. 4.52. Rotor speed estimator using scheme 3.

4.5.3.1.5 Rotor speed estimation scheme 4

$$\omega_r = \omega_{mr} - \omega_{sl} = \left(\psi_{rd}\frac{d\psi_{rq}}{dt} - \psi_{rq}\frac{d\psi_{rd}}{dt}\right) \Big/ |\bar{\psi}_r|^2 - \frac{L_m}{T_r|\bar{\psi}_r|^2}(-\psi_{rq}i_{sD} + \psi_{rd}i_{sQ})$$

or

$$\omega_r = \omega_{mr} - \omega_{sl} = \left(\psi_{rd}\frac{d\psi_{rq}}{dt} - \psi_{rq}\frac{d\psi_{rd}}{dt}\right) \Big/ |\bar{\psi}_r|^2 - \frac{2t_e R_r}{3P|\bar{\psi}_r|^2}$$

or

$$\omega_r = \left(\psi_{sD}\frac{d\psi_{sQ}}{dt} - \psi_{sQ}\frac{d\psi_{sD}}{dt}\right) \Big/ (\psi_{sD}^2 + \psi_{sQ}^2) - \frac{d}{dt}[\sin^{-1}(t_e/(c|\bar{\psi}_s||\bar{\psi}_r|))] - \frac{2t_e R_r}{3P|\bar{\psi}_r|^2}$$

Various rotor speed estimators are now described, which use the speed of the rotor flux-linkage space vector or the speed of the stator flux-linkage space vector. These require the estimate of the appropriate slip frequency. This type of scheme is one of those schemes which are used in recently introduced speed-sensorless, commercially available high-performance induction motor drives. In these commercial implementations the stator voltages are not monitored but are reconstructed from the d.c. link voltage and switching states of the inverter (see Sections 3.1.1 and 4.1.1). It should be noted that a similar scheme has been described in Section 3.1.3 for a permanent-magnet synchronous machine [e.g. see eqn (3.1-48)], but in the PM synchronous machine the speed estimation is simpler, since the rotor speed is equal to the speed of the stator flux (slip speed is zero).

It is possible to implement an angular rotor-slip-frequency estimator (ω_{sl}) by considering the rotor voltage equation of the induction motor. It has been shown in Section 4.1.1 that if the rotor voltage equation is expressed in the rotor-flux-oriented reference frame, then eqn (4.1-26) is obtained, which contains the angular slip frequency, $\omega_{sl} = i_{sy}/(T_r|\bar{i}_{mr}|)$. If the expression of i_{sy} given by (2.1-192) is substituted into this, then

$$\omega_{sl} = \frac{i_{sy}}{T_r|\bar{i}_{mr}|} = \frac{-i_{sD}\sin\rho_r + i_{sQ}\cos\rho_r}{T_r|\bar{i}_{mr}|} \tag{4.5-33}$$

is obtained, where ρ_r is the angle of the rotor flux with respect to the real-axis of the stationary reference frame. By considering that the rotor flux modulus can be expressed as $|\bar{\psi}_r| = L_m|\bar{i}_{mr}|$, where $|\bar{i}_{mr}|$ is the modulus of the rotor magnetizing-current space vector, and also using $\sin\rho_r = \psi_{rq}/|\bar{\psi}_r|$ and $\cos\rho_r = \psi_{rd}/|\bar{\psi}_r|$, which expressions follow from $\bar{\psi}_r' = |\bar{\psi}_r|\exp(j\rho_r)$, eqn (4.5-33) can be rewritten as

$$\omega_{sl} = \frac{L_m}{T_r|\bar{\psi}_r|^2}(-\psi_{rq}i_{sD} + \psi_{rd}i_{sQ}). \tag{4.5-34}$$

Equation (4.5-34) can be used to monitor the angular slip frequency by monitoring the stator currents, and also by using the rotor flux linkages. The rotor flux linkages can be estimated in various ways, e.g. by using eqns (4.5-16) and (4.5-17). These expressions include the stator flux linkages, which can be estimated from the stator voltages and currents as described above (see also Section 4.1.1.4). It should be noted that it is possible to reformulate eqn (4.5-34) by considering the expression for the electromagnetic torque given by eqn (2.1-191), where $\psi_{rx}=|\bar{\psi}_r|$ is the modulus of the rotor flux. Thus

$$\omega_{sl}=\frac{2t_e R_r}{3P|\bar{\psi}_r|^2} \tag{4.5-35}$$

is obtained, where t_e is the electromagnetic torque, P is the number of pole pairs, and R_r is the rotor resistance. This expression contains the modulus of the rotor flux and also the electromagnetic torque. The electromagnetic torque can also be substituted by $t_e=(3P/2)(\psi_{sD}i_{sQ}-\psi_{sQ}i_{sD})$, where ψ_{sD} and ψ_{sQ} are the stator flux linkages in the stationary reference frame, and their monitoring has also been discussed above.

It is also possible to obtain an expression for the rotor speed by considering that from eqn (4.1-26) it follows that

$$\omega_r=\omega_{mr}-\omega_{sl}, \tag{4.5-36}$$

where ω_{mr} is the speed of the rotor flux (relative to the stator) $\omega_{mr}=\mathrm{d}\rho_r/\mathrm{d}t$, and ω_{sl} the angular slip frequency [e.g. given by eqns (4.5-34) and (4.5-35)]. In other words, ω_{mr} is the speed of the rotor flux-linkage space vector with respect to the rotor. It is possible to obtain an expression for ω_{mr} in terms of the rotor flux-linkage components by expanding the expression for the derivative $\mathrm{d}\rho_r/\mathrm{d}t$. Since the rotor flux-linkage space vector expressed in the stationary reference frame is

$$\bar{\psi}_r'=\psi_{rd}+j\psi_{rq}=|\bar{\psi}_r|\exp(j\rho_r),$$

thus $\rho_r=\tan^{-1}(\psi_{rq}/\psi_{rd})$ and it follows that

$$\omega_{mr}=\frac{\mathrm{d}\rho_r}{\mathrm{d}t}=\frac{\mathrm{d}}{\mathrm{d}t}[\tan^{-1}(\psi_{rq}/\psi_{rd})]=\frac{\psi_{rd}\,\mathrm{d}\psi_{rq}/\mathrm{d}t-\psi_{rq}\,\mathrm{d}\psi_{rd}/\mathrm{d}t}{\psi_{rd}^2+\psi_{rq}^2}. \tag{4.5-37}$$

The numerator contains $|\bar{\psi}_r|^2$. By substituting eqn (4.5-37) into eqn (4.5-36) and by also considering (4.5-34) or (4.5-35), finally we obtain

$$\omega_r=\omega_{mr}-\omega_{sl}=\frac{\psi_{rd}\,\mathrm{d}\psi_{rq}/\mathrm{d}t-\psi_{rq}\,\mathrm{d}\psi_{rd}/\mathrm{d}t}{|\bar{\psi}_r|^2}-\frac{L_m}{T_r|\bar{\psi}_r|^2}(-\psi_{rq}i_{sD}+\psi_{rd}i_{sQ}) \tag{4.5-38}$$

and

$$\omega_r=\omega_{mr}-\omega_{sl}=\frac{\psi_{rd}\mathrm{d}\psi_{rq}/\mathrm{d}t-\psi_{rq}\mathrm{d}\psi_{rd}/\mathrm{d}t}{|\bar{\psi}_r|^2}-\frac{2t_e R_r}{3P|\bar{\psi}_r|^2}. \tag{4.5-39}$$

A rotor speed estimator can then be constructed which uses the monitored stator currents and the rotor flux components, which, however, can be obtained from the stator flux linkages as discussed above. The stator flux linkages can be obtained by using monitored stator currents and monitored or reconstructed stator voltages as discussed above. The accuracy of a speed estimator using eqn (4.5-38) or (4.5-39) depends greatly on the machine parameters used, and also on the model used for the estimation of the rotor flux-linkage components. A possible implementation is shown in Fig. 4.53. This contains the following three machine parameters: R_s, L_s', and $k=L_m/L_r$. However, an improved scheme is obtained if improved flux-linkage estimation is used. For this purpose it is also possible to use any of the techniques discussed earlier (see Section 4.1.1.4) and then, for example, the speed estimator shown in Fig. 4.9(d) is obtained.

For digital implementation it is possible to use various forms, including the following discrete form:

$$\omega_r(k)=[\psi_{rd}(k-1)\psi_{rq}(k)-\psi_{rq}(k-1)\psi_{rd}(k)]/|\bar{\psi}_r(k)|^2$$
$$-[L_m/(T_r|\bar{\psi}_r(k)|^2)][-\psi_{rq}(k)i_{sD}(k)+\psi_{rd}(k)i_{sQ}(k)], \qquad (4.5\text{-}40)$$

where $|\bar{\psi}_r(k)|^2=[\psi_{rd}(k-1)]^2+[\psi_{rq}(k-1)]^2$. Since this equation contains a modelling error, which results in an error of the estimated rotor speed, in practice a low-pass filter can be used to remove this error.

It should be noted that in a vector-controlled drive, rather accurate estimation of the rotor speed may be obtained if the slip frequency term in eqn (4.5-39) is replaced by its reference value. For example, in a vector-controlled drive with rotor-flux-oriented control where i_{syref} and $|\bar{i}_{mrref}|$ are used, $\omega_{slref}=i_{syref}/(T_r|\bar{i}_{mrref}|)$ or $\omega_{slref}=(2t_{eref}R_r)/(3P|\bar{\psi}_{ref}|^2)$.

It is also possible to estimate the rotor speed in another way, which is similar to that described by eqn (4.5-39), but which instead of using the speed of the rotor flux (ω_{mr}), uses the speed of the stator flux, ω_{ms}. If the direct and quadrature-axis

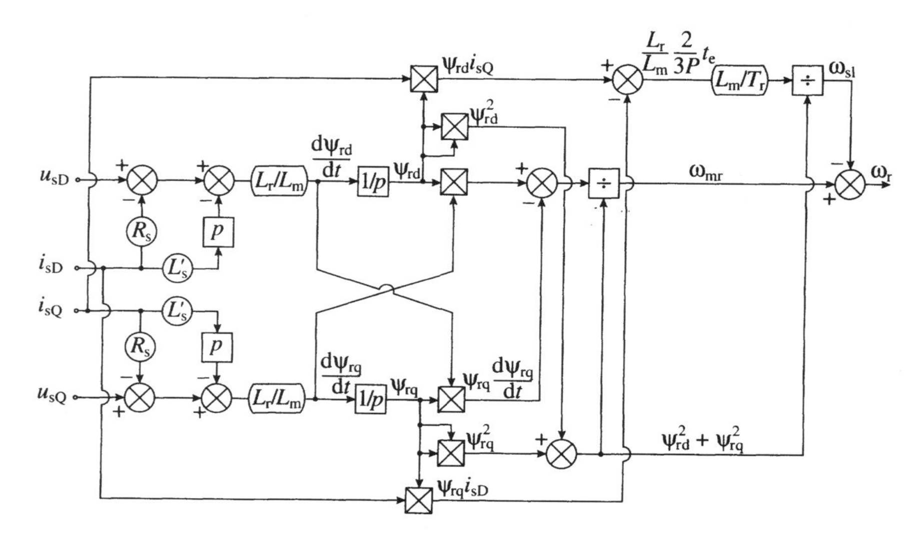

Fig. 4.53. Rotor speed estimator using 2 integrators, 2 differentiators, and 3 parameters.

stator flux linkages (ψ_{sD}, ψ_{sQ}) in the stationary reference frame are known (they are estimated by one of the techniques described above, which uses the monitored stator voltages and currents or the reconstructed stator voltages and currents), then since $\bar{\psi}_s = \psi_{sD} + j\psi_{sQ} = |\bar{\psi}_s| \exp(j\rho_s)$, where ρ_s is the angle of the stator flux-linkage space vector with respect to the real-axis of the stationary reference frame, it follows that

$$\omega_{ms} = \frac{d\rho_s}{dt} = \frac{d}{dt}[\tan^{-1}(\psi_{sQ}/\psi_{sD})]. \tag{4.5-41}$$

In eqn (4.5-41) ω_{ms} is the speed of the stator flux-linkage space vector with respect to the stator. By performing the differentiation, this can be put into the following form:

$$\omega_{ms} = \frac{\psi_{sD} d\psi_{sQ}/dt - \psi_{sQ} d\psi_{sD}/dt}{\psi_{sD}^2 + \psi_{sQ}^2}, \tag{4.5-42}$$

where the numerator contains $|\bar{\psi}_s|^2$. This is similar to eqn (4.5-37). By using eqns (4.5-19) and (4.5-20), it is possible to express ω_{ms} in terms of the stator voltages and stator currents. The upper part of the circuit shown in Fig. 4.54 shows a possible implementation of eqn (4.5-42).

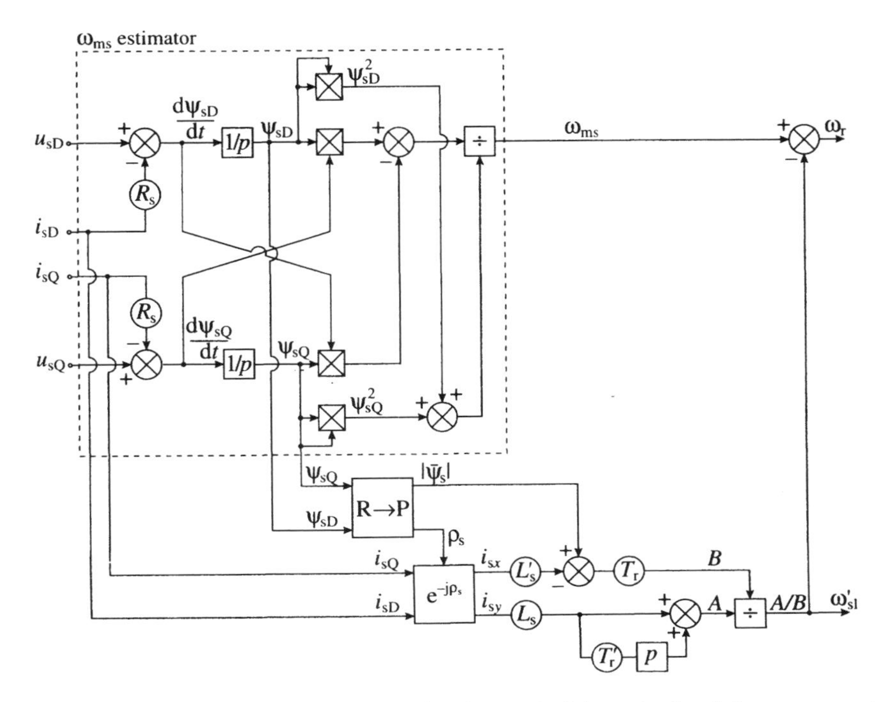

Fig. 4.54. Estimation of the rotor speed by using the speed of the stator flux-linkage space vector.

This requires the use of two integrators and only one machine parameter: the stator resistance. It should be noted that by using the concepts discussed in Sections 3.1.3.1 and 4.1.1, it is possible to obtain other schemes where the problems associated with pure integrator initial values and drift are reduced. For example, the pure integrator can be replaced by a low-pass filter.

To obtain a rotor speed estimator, which uses ω_{ms} [defined by eqn (4.5-42)], it is possible to proceed in various ways. A specific solution can be obtained by considering eqns (4.2-16) and (4.2-23), from which $\omega_r = \omega_{ms} - \omega'_{sl}$, where now ω'_{sl} is the speed of the stator flux-linkage space vector relative to the rotor (not the speed of the rotor flux relative to the rotor, which is denoted by ω_{sl}):

$$\omega'_{sl} = \frac{L_s(i_{sy} + T'_r \, di_{sy}/dt)}{T_r(|\bar{\psi}_s| - L'_s i_{sx})}. \tag{4.5-43}$$

It should be noted that eqn (4.5-43) has been directly obtained from the rotor voltage equation of the induction machine, but expressed in the stator-flux-oriented reference frame (which rotates at the speed of ω_{ms}). Thus the stator currents i_{sx} and i_{sy} are the stator currents in the stator-flux-oriented reference frame, and they can be obtained from the stator currents i_{sD}, i_{sQ} (expressed in the stationary reference frame) by considering $i_{sx} + ji_{sy} = (i_{sD} + ji_{sQ})\exp(-j\rho_s)$. Hence by using eqn (4.5-43), the scheme shown in Fig. 4.54 can be extended to obtain ω'_{sl}. For this purpose $\cos\rho_s$ and $\sin\rho_s$ can be obtained from the stator flux-linkage components shown in Fig. 4.54, by using $\cos\rho_s = \psi_{sD}/|\bar{\psi}_s|$ and $\sin\rho_s = \psi_{sQ}/|\bar{\psi}_s|$ to obtain i_{sx} and i_{sy}, or ρ_s can be obtained by using a rectangular-to-polar converter (where the inputs are ψ_{sD} and ψ_{sQ} and the outputs are $|\bar{\psi}_s|$ and ρ_s). By using the obtained i_{sx}, i_{sy}, and $|\bar{\psi}_s|$, the slip frequency ω'_{sl} can be obtained by the application of eqn (4.5-43), and finally when this is subtracted from ω_{ms}, the rotor speed, ω_r is obtained. This is also shown in Fig. 4.54. However, it can be seen that this speed estimator is more complicated than the one shown in Fig. 4.53.

It should be noted that it is possible to obtain other solutions as well, which use the speed of the stator and rotor flux-linkage space vectors. For example, such a scheme can be derived by considering that the rotor speed is equal to the sum of the speed of the stator flux-linkage space vector, $\omega_{ms} = d\rho_s/dt$, minus the speed difference between the stator and rotor flux-linkage space vectors, $\omega_d = d\rho/dt$, minus the speed of the rotor flux-linkage space vector (relative to the rotor), $\omega_{sl} = d\theta_{sl}/dt$. This last term is now defined as in eqn (4.5-35), and can be expressed as in eqns (4.5-34) or (4.5-35). Thus

$$\omega_r = \omega_{ms} - \omega_d - \omega_{sl}. \tag{4.5-44}$$

Equation (4.5-44) can be simply proved by considering Fig. 4.55 which shows the stator and rotor flux-linkage space vectors and their angles with respect to different reference frames. As discussed above, ρ_s is the angle of the stator flux-linkage space vector with respect to the real axis of the stator reference frame, ρ_r is the angle of the rotor flux-linkage space vector with respect to the real axis

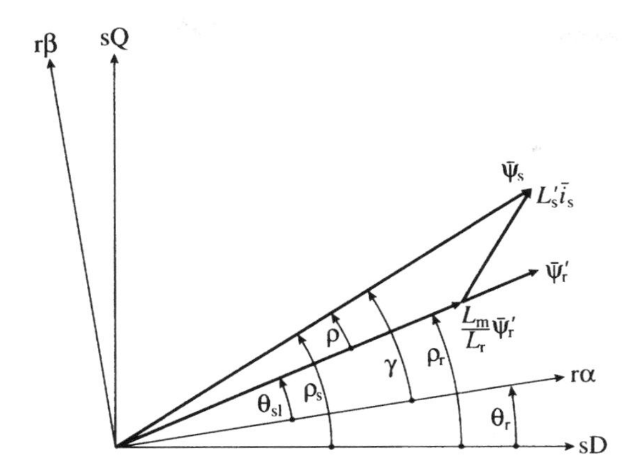

Fig. 4.55. Stator flux and rotor flux-linkage space vectors, and their positions in various reference frames.

of the stator reference frame, and ρ is the angle between the stator and rotor flux-linkage space vectors, $\rho=\rho_s-\rho_r$. It follows that

$$\omega_d=\frac{d\rho}{dt}=\frac{d\rho_s}{dt}-\frac{d\rho_r}{dt}=\omega_{ms}-\omega_{mr}$$

is indeed the difference between the speed of the stator flux-linkage space vector and the speed of the rotor flux-linkage space vector. Furthermore, as shown in Fig. 4.55, θ_r is the rotor angle, and $\theta_r=\rho_s-\rho-\theta_{sl}$. Thus

$$\omega_r=\frac{d\theta_r}{dt}=\frac{d\rho_s}{dt}-\frac{d\rho}{dt}-\frac{d\theta_{sl}}{dt}=\omega_{ms}-\omega_d-\omega_{sl},$$

in agreement with eqn (4.5-44). In eqn (4.5-44), ω_{ms} can be obtained as given by eqn (4.5-42), and ω_{sl} can be obtained as shown by eqn (4.5-35). However, ω_d can be obtained by considering that

$$t_e=\tfrac{3}{2}PL_m/(L'_s/L_r)|\bar{\psi}_s||\bar{\psi}_r|\sin\rho=c|\bar{\psi}_s||\bar{\psi}_r|\sin\rho,$$

where $c=(3/2)PL_m/(L'_s/L_r)$, from which

$$\omega_d=\frac{d\rho}{dt}=\frac{d}{dt}[\sin^{-1}(t_e/(c|\bar{\psi}_s||\bar{\psi}_r|))].$$

It should be noted that $t_e=c|\bar{\psi}_s||\bar{\psi}_r|\sin\rho$ follows from $t_e=(3/2)P\bar{\psi}_s\times\bar{i}_s$, by considering that $\bar{i}_s=[\bar{\psi}_s-(L_m/L_r)\bar{\psi}'_r]/L'_s$ [this last equation follows from eqns (2.1-150) and (2.1-151)]. Finally eqn (4.5-44) can be put into the form

$$\omega_r=\frac{\psi_{sD}d\psi_{sQ}/dt-\psi_{sQ}d\psi_{sD}/dt}{\psi_{sD}^2+\psi_{sQ}^2}-\frac{d}{dt}[\sin^{-1}(t_e/(c|\bar{\psi}_s||\bar{\psi}_r|))]-\frac{2t_eR_r}{3P|\bar{\psi}_r|^2},\qquad(4.5\text{-}45)$$

where it is possible to use different expressions for the electromagnetic torque [e.g. $t_e=(3/2)P(\psi_{sD}i_{sQ}-\psi_{sQ}i_{sD})$ or $t_e=(3/2)P(L_m/L_r)(\psi_{rd}i_{sQ}-\psi_{rq}i_{sD})$]. Equation (4.5-45) can also be used for the estimation of the rotor speed of an induction machine, but this is not simpler than that described by eqn (4.5-38) or (4.5-39).

In Fig. 4.55, the angle $\rho_s-\rho$ is equal to the angle ρ_r and thus

$$\omega_r=\frac{d\theta_r}{dt}=\frac{d}{dt}(\rho_s-\rho)-\frac{d\theta_{sl}}{dt}=\omega_{mr}-\omega_{sl},$$

which, as expected, agrees with eqn (4.5-36). It follows that when the rotor speed is determined by using the speed (ω_{mr}) of the rotor flux-linkage space vector and the speed of the rotor flux with respect to the rotor (ω_{sl}), then $\omega_r=\omega_{mr}-\omega_{sl}$ holds. However, when the rotor speed is determined by using the speed of the stator flux-linkage space vector (ω_{ms}) and also the speed of the rotor flux-linkage space vector with respect to the rotor (ω_{sl}), then an extra speed term, ω_d, has also to be considered, and $\omega_r=\omega_{ms}-\omega_d-\omega_{sl}$. On the other hand, it also follows that the rotor speed can be determined by considering that $\theta_r=\rho_s-(\rho+\theta_{sl})=\rho_s-\gamma$, where as shown in Fig. 4.55, $\gamma=\rho+\theta_{sl}$ is the angle of the stator flux-linkage space vector with respect to the real axis of the rotor, since it follows that

$$\omega_r=\frac{d\theta_r}{dt}=\frac{d\rho_s}{dt}-\frac{d\gamma}{dt}=\omega_{ms}-\omega_{sl}'.$$

This means that the rotor speed is equal to the speed of the stator flux-linkage space vector (with respect to the stator) minus the speed of the stator flux-linkage space vector with respect to the rotor, ω_{sl}' (which has been given by eqn (4.5-43). This last approach has been used in the scheme of Fig. 4.54.

Various other simplified rotor-speed estimation schemes can be obtained directly by considering eqns (4.5-44) and (4.5-45), if it is assumed that the speeds of the stator and rotor flux-linkage space vectors are equal. In this case $\omega_d=0$, and

$$\omega_r\approx\frac{\psi_{sD}d\psi_{sQ}/dt-\psi_{sQ}d\psi_{sD}/dt}{\psi_{sD}^2+\psi_{sQ}^2}-\frac{2t_eR_r}{3P|\bar{\psi}_r|^2}. \qquad (4.5\text{-}46)$$

This is similar in form to eqn (4.5-39), but eqn (4.5-39) is more accurate, since eqn (4.5-46) assumes that $\omega_d=0$.

In summary it should be noted that among the various speed estimators discussed above in connection with the Scheme shown in Fig. 4.53, those using eqns (4.5-38) or (4.5-39) are the simplest and most accurate. These rely heavily on the accuracy of the estimated flux-linkage components. If the stator voltages and currents are used to obtain the flux-linkage estimates, then by considering the thermal variations of the stator resistance (e.g. by using a thermal model), and also by using appropriate saturated inductances, the estimation accuracy can be greatly improved. However, a speed-sensorless high-performance torque-controlled drive (vector- or direct-torque-controlled drive) using this type of speed estimator will only work successfully at extremely low speeds (including zero speed) if the flux-linkage estimator is a closed-loop observer.

4.5.3.1.6 Rotor speed estimation scheme 5

$$\omega_r = \left[\frac{d\psi_{rq}}{dt}\psi'_{sD} - \frac{d\psi_{rd}}{dt}\psi'_{sQ}\right] \Big/ [\psi_{rd}\psi'_{sD} + \psi_{rq}\psi'_{sQ}]$$

$$\psi'_{sD} = \psi_{sD} - L_s i_{sD} \qquad \psi'_{sQ} = \psi_Q - L_s i_{sQ}$$

It is also possible to obtain directly an expression for the rotor speed from the rotor space-vector voltage equation of the induction machine in the stationary reference frame, by eliminating from this the unmeasurable rotor-current space vector. However, due to its limitations, this estimation technique has been deliberately considered as the last scheme, since it will be shown that this direct approach cannot be used in two cases: under sinusoidal steady-state conditions and also when the rotor flux is constant. The rotor voltage equation can also be used for the estimation of the rotor resistance, but again, the estimation cannot be used in the sinusoidal steady-state and when the rotor flux is constant. The details will be considered below, and these are mainly useful for educational purposes.

It follows from eqn (2.1-149) that

$$0 = R_r \bar{i}'_r + \frac{d\bar{\psi}'_r}{dt} - j\omega_r \bar{\psi}'_r, \tag{4.5-47}$$

where the primed rotor quantities denote quantities in the stationary reference frame. However, by considering that the stator flux-linkage space vector can be expressed as $\bar{\psi}_s = L_s \bar{i}_s + L_m \bar{i}'_r$, it follows that the rotor-current space vector expressed in the stationary reference frame is

$$\bar{i}'_r = \frac{\bar{\psi}_s - L_s \bar{i}_s}{L_m}. \tag{4.5-48}$$

Substitution of eqn (4.5-48) into (4.5-47) gives the rotor voltage equation in terms of the stator flux-linkage, rotor flux-linkage, and stator-current space vector respectively, and when this voltage equation is resolved into its real and imaginary parts,

$$R_r \frac{(\psi_{sD} - L_s i_{sD})}{L_m} + \omega_r \psi_{rq} = -\frac{d\psi_{rd}}{dt} \tag{4.5-49}$$

$$R_r \frac{(\psi_{sQ} - L_s i_{sQ})}{L_m} - \omega_r \psi_{rd} = -\frac{d\psi_{rq}}{dt} \tag{4.5-50}$$

are obtained. Equations (4.5-49) and (4.5-50) can be put into the form

$$a_1 x_1 + b_1 x_1 = c_1 \tag{4.5-51}$$

$$a_2 x_2 + b_2 x_2 = c_2, \tag{4.5-52}$$

where

$$a_1=\frac{\psi_{sD}-L_s i_{sD}}{L_m} \qquad b_1=\psi_{rq} \qquad a_2=\frac{\psi_{sQ}-L_s i_{sQ}}{L_m} \qquad b_2=-\psi_{rd}$$
$$c_1=-\frac{d\psi_{rd}}{dt} \qquad c_2=-\frac{d\psi_{rq}}{dt} \qquad x_1=R_r \quad \text{and} \quad x_2=\omega_r. \tag{4.5-53}$$

Equations (4.5-51) and (4.5-52) can be solved for the 'unknowns' x_1 and x_2 when the resulting determinant $D=a_1b_2-a_2b_1$ is not zero, and finally

$$x_1=R_r=-\frac{L_m[\psi_{rd}(d\psi_{rd}/dt)+\psi_{rq}(d\psi_{rq}/dt)]}{\psi_{rd}(\psi_{sD}-L_s i_{sD})+\psi_{rq}(\psi_{sQ}-L_s i_{sQ})} \tag{4.5-54}$$

$$x_2=\omega_r=\frac{(d\psi_{rq}/dt)(\psi_{sD}-L_s i_{sD})-(d\psi_{rd}/dt)(\psi_{sQ}-L_s i_{sQ})}{\psi_{rd}(\psi_{sD}-L_s i_{sD})+\psi_{rq}(\psi_{sQ}-L_s i_{sQ})} \tag{4.5-55}$$

are obtained. These equations depend on the stator flux linkages, rotor flux linkages and stator currents, where the various flux-linkage components can be obtained from the stator voltages and currents by considering eqns (4.5-16), (4.5-17), (4.5-19), and (4.5-20) as follows, or by using the improved flux estimators described earlier:

$$\psi_{sD}=\int(u_{sD}-R_s i_{sD})\,dt \tag{4.5-56}$$

$$\psi_{sQ}=\int(u_{sQ}-R_s i_{sQ})\,dt \tag{4.5-57}$$

$$\psi_{rd}=\frac{L_r}{L_m}(\psi_{sD}-L_s' i_{sD}) \tag{4.5-58}$$

$$\psi_{rq}=\frac{L_r}{L_m}(\psi_{sQ}-L_s' i_{sQ}). \tag{4.5-59}$$

In general the rotor speed can be determined from eqn (4.5-55), which can also be put into the following form by considering eqns (4.5-48), (4.5-58), and (4.5-59):

$$\omega_r=\frac{\bar{i}_r'\times(d\bar{\psi}_r'/dt)}{\bar{i}_r'\bullet\bar{\psi}_r'}=\frac{|\bar{i}_r'||d\bar{\psi}_r'/dt|\sin\alpha}{|\bar{i}_r'||\bar{\psi}_r'|\cos\beta}$$
$$=\frac{|d\bar{\psi}_r'/dt|\sin\alpha}{|\bar{\psi}_r'|\cos\beta} \tag{4.5-60}$$

where $\times$ denotes the vectorial product and $\bullet$ denotes the scalar (dot) product, α is the phase angle between the rotor-current space vector and the space vector $d\bar{\psi}_r'/dt$, and β is the angle between the rotor-current space vector and the rotor flux-linkage space vector expressed in the stationary reference frame ($\bar{\psi}_r'$). However, in the sinusoidal steady-state of an induction machine, where the stator voltages and currents are sinusoidal, eqn (4.5-55) cannot be used, since its numerator and denominator become zero at every instant of time. This can be simply proved by considering that it follows from eqn (4.5-47) [by replacing the derivative by $j\omega_1$] that in the sinusoidal steady-state (where the supply frequency is ω_1) the rotor-current space vector ($\bar{i}_r'$) is in phase with $d\bar{\psi}_r'/dt$ (thus $\alpha=0$) and $\bar{i}_r'$ is in quadrature to the rotor flux vector $\bar{\psi}_r'$ (thus $\beta=90°$). It follows that $|d\bar{\psi}_r'/dt|\sin\alpha=0$, and $|\bar{\psi}_r'|\cos\beta=0$, and it can be seen that both the numerator and denominator of eqn (4.5-60) are zero in the sinusoidal steady-state. Furthermore, if steady-state is not assumed, but $|\bar{\psi}_r'|=$constant is assumed, then $d\bar{\psi}_r'/dt=j\omega_{mr}\bar{\psi}_r'$, where $\omega_{mr}=d\rho_r/dt$ is the speed of the rotor flux-linkage space vector, since

$$\frac{d\bar{\psi}_r'}{dt}=\frac{d}{dt}[|\bar{\psi}_r'|\exp(j\rho_r)]=\exp(j\rho_r)\frac{d}{dt}(|\bar{\psi}_r'|)+j\frac{d\rho_r}{dt}|\bar{\psi}_r'|\exp(j\rho_r)=j\omega_{mr}\bar{\psi}_r',$$

and it can be seen that the phase shift between $d\bar{\psi}_r'/dt$ and $\bar{\psi}_r'$ is 90° (similarly to that in the steady-state). Thus it follows from eqn (4.5-60) that if the angle of $\bar{i}_r'$ is δ_i (with respect to the real-axis of the stationary reference frame) and ρ_r is the angle of $\bar{\psi}_r'$ (with respect to the real-axis of the stationary reference frame), then

$$\omega_r=\frac{|d\bar{\psi}_r'/dt|\sin\alpha}{|\bar{\psi}_r'|\cos\beta}=\frac{\omega_{mr}|\bar{\psi}_r'|\sin(\rho_r+\pi/2-\delta_i)}{|\bar{\psi}_r'|\cos(\rho_r-\delta_i)}=\omega_{mr},$$

and hence the rotor speed is equal to the speed of the rotor flux-linkage space vector. However, in general, this is not correct; it is only correct when the slip frequency is zero ($\omega_{sl}=0$) [e.g. it follows from eqn (4.5-39) that when $\omega_{sl}=0$, $\omega_r=\omega_{mr}-\omega_{sl}=\omega_{mr}$]. The slip frequency is only zero now because the rotor resistance, which can be obtained from eqn (4.5-54), is zero when the rotor flux-linkage space vector is assumed to be constant. This follows since under the special condition of $|\bar{\psi}_r'|=$constant, the flux-linkage space vector and its derivative are displaced by 90° and therefore their dot product ($\bar{\psi}_r'\bullet d\bar{\psi}_r'/dt$) is zero, which leads to $R_r=0$ in eqn (4.5-54), which can also be expressed in general as

$$R_r=-\frac{\bar{\psi}_r'\bullet(d\bar{\psi}_r'/dt)}{\bar{i}_r'\bullet\bar{\psi}_r'}.$$

Although eqn (4.5-55) cannot be used in the sinusoidal steady-state of an induction machine, it can be used for an induction machine supplied by non-sinusoidal voltages and currents, e.g. the case of a PWM-fed induction machine. This follows from the fact that in this case the special phase relationships described above ($\alpha=0, \beta=90°$) do not hold in every instant of time. However, due to the flux ripples, resulting from the non-sinusoidal stator voltages, both the denominator and numerator of the speed expression will have periodic waveforms, which contain many zero-crossings, and at the time instants where the denominator zero-crossings occur, it is not possible to perform the required division. It is also important to note that, since the amplitudes of the harmonic components (ripples) present in the waveforms of the numerator and denominator are small, high accuracy of the estimated rotor speed can only be obtained if high accuracy voltage and current sensors are used together with high accuracy A/D converters.

In the steady-state of a PWM-fed machine, the numerator and denominator have similar waveforms, but the amplitudes of these two are different, and thus the ratio of the numerator and denominator gives a constant value.

It has been stated above that when the denominator of eqn (4.5-60) becomes zero (at the zero-crossing instants), it is impossible to use eqn (4.5-60). However, this problem can be avoided by considering that the numerator and denominator waveforms have different amplitudes, and the same zero-crossings. Therefore the correct speed estimate will result if the absolute value of the numerator is divided by the absolute value of the denominator, but since the estimated speed can be positive or negative, this ratio has to be multiplied by the sign of the expression given by eqn (4.5-60).

Thus

$$\omega_r = \text{sign}\left\{\frac{\bar{i}_r' \times (\mathrm{d}\bar{\psi}_r'/\mathrm{d}t)}{\bar{i}_r' \bullet \bar{\psi}_r'}\right\} \frac{|[\bar{i}_r' \times (\mathrm{d}\bar{\psi}_r'/\mathrm{d}t)]|}{|(\bar{i}_r' \bullet \bar{\psi}_r')|}. \tag{4.5-61}$$

Furthermore, when the absolute values in eqn (4.5-61) are small, the estimated rotor speed may become inaccurate. This problem can be avoided by low-pass filtering the absolute values of the numerator and denominator waveforms respectively. To obtain high accuracy, the two low-pass filters must be identical. The speed estimator described above depends on various machine parameters, R_s, L_s', L_s, L_r/L_m, but does not depend on the rotor resistance. The stator resistance has important effects on the stator flux linkages, especially at low speeds. Due to main flux saturation L_m can change and $L_s = L_{sl} + L_m$, $L_r = L_{rl} + L_m$ can also change even if the leakage inductances are constant. However, due to leakage flux saturation, L_{sl}, L_{rl}, and L_s' can also change. In a vector-controlled drive, where the rotor flux amplitude is constant, the variations of L_m are small. It follows that in this drive, accurate speed estimation based on eqn (4.5-61) depends mostly on R_s and L_s'. In an induction machine with closed rotor slots, L_s' varies due to leakage flux saturation, and L_s' is a function of the stator currents, due to the saturable closing bridge around the rotor slots. Furthermore, higher accuracy is obtained if the stator flux is obtained by a scheme which avoids the use of pure integrators (some solutions for this are discussed elsewhere in the book, e.g. see Sections 3.1.3, 4.1.1).

It should be noted that, in general, eqn (4.5-55) can be used for the identification of the rotor resistance. As mentioned above, this can be put into the following form:

$$R_r = -\frac{\bar{\psi}_r' \bullet (\mathrm{d}\bar{\psi}_r'/\mathrm{d}t)}{\bar{i}_r' \bullet \bar{\psi}_r'}. \tag{4.5-62}$$

In the sinusoidal steady-state, incorrect rotor resistance is obtained, since in the steady state the rotor flux-linkage space vector and the derivative of the rotor flux-linkage space vector are displaced by 90° (see above), and thus the numerator in eqn (4.5-62) is zero. However, in the sinusoidal steady-state, the denominator of eqn (4.5-62) is also zero, since the rotor-current space vector and the rotor flux-linkage space vector are also displaced by 90° (see above). Furthermore, it has been shown above that when $|\bar{\psi}'_r|$=constant, this expression gives the incorrect $R_r=0$, since in this case the rotor flux-linkage space vector and its derivative are displaced by 90°.

It should be noted that since $|\bar{\psi}'_r|$=constant results in $d\bar{\psi}'_r/dt=j\omega_{mr}\bar{\psi}'_r$ (see above), and thus $d\psi_{rd}/dt=-\omega_{mr}\psi_{rq}$ and $d\psi_{rq}/dt=\omega_{mr}\psi_{rd}$, it follows from (4.5-49) and (4.5-50) that $\omega_r=\omega_{mr}$, which is in agreement with that shown above, using a different derivation.

It should also be noted that, as shown above, the simultaneous estimation of the rotor resistance and rotor speed is impossible if the rotor flux-linkage space vector does not change (e.g. this occurs in the sinusoidal steady-state, or in a drive where the voltages and currents are not sinusoidal but constant rotor flux linkage is imposed). This impossibility can also be simply proved in a slightly different way by directly considering eqn (4.5-47). It follows that when $|\bar{\psi}'_r|$=constant, and thus $d\bar{\psi}'_r/dt=j\omega_{mr}\bar{\psi}'_r$, then

$$0=R_r\bar{i}'_r+j(\omega_{mr}-\omega_r)\bar{\psi}'_r=R_r\bar{i}'_r+j\omega_{sl}\bar{\psi}'_r$$

(ω_{sl} is the slip frequency) and hence $0=(R_r/\omega_{sl})\bar{i}'_r+j\bar{\psi}'_r$. However, since for given $\bar{u}_s$ and $\bar{i}_s$ (and machine parameters) the rotor flux-linkage space vector and rotor-current space vector ($\bar{\psi}'_r$ and $\bar{i}'_r$) are uniquely determined, e.g. $\bar{\psi}_s=\int(\bar{u}_s-R_s\bar{i}_s)\,dt$, $\bar{\psi}'_r=(L_r/L_m)(\bar{\psi}_s-L'_s\bar{i}_s)$, and $\bar{i}'_r=(\bar{\psi}_s-L_s\bar{i}_s)/L_m$, it follows that it is only possible to estimate the combined parameter R_r/ω_{sl} and it is not possible to have a separate, simultaneous estimation of R_r and ω_{sl} (or R_r and ω_r). In the sinusoidal steady-state, the same result can also be obtained directly by considering the equivalent circuit of the induction machine (Fig. 4.10), which shows that when the rotor flux is constant ($L_m|\bar{i}_{mr}|$=const.), only the ratio of the rotor resistance and slip can be estimated.

However, when the special conditions (sinusoidal steady-state or rotor flux constant) do not apply, and when the induction machine is supplied by a voltage-source PWM inverter, thus there are time harmonics in the stator voltages and currents (and therefore in the flux linkages as well), these special phase relationships do not apply at every instant of time and R_r can be estimated by using eqn (4.5-62). To avoid the problems associated with the zero-crossing of the numerator and denominator waveforms of this expression, a similar technique is used as before. However, since the rotor resistance is always positive, there is no need to take the sign of the ratio given by eqn (4.5-62) and the rotor resistance is estimated from

$$R_r=\frac{|-[\bar{\psi}'_r\bullet(d\bar{\psi}'_r/dt)]|}{|(\bar{i}'_r\bullet\bar{\psi}'_r)|}. \tag{4.5-63}$$

In practice, the rotor resistance can be estimated by dividing the low-pass filtered absolute value of $|-[\bar{\psi}'_r\bullet(d\bar{\psi}'_r/dt)]|$ by the low-pass filtered absolute value $|(\bar{i}'_r\bullet\bar{\psi}'_r)|$. As before, the two low-pass filters used in the rotor resistance estimator must have identical characteristics in order to obtain high accuracy. This scheme can be used in a torque-controlled drive. When this rotor resistance estimator is used in a speed-sensorless torque-controlled drive which requires the rotor resistance, and e.g. the rotor speed estimator is described by eqn (4.5-61), further improvements in the speed response are expected.

Finally it should be noted that the accuracy of open-loop estimators depends greatly on the accuracy of the machine parameters used. In general, at low rotor speed the accuracy of the open-loop estimators is reduced, and in particular, parameter deviations from their actual values have great influence on the steady-state and transient performance of the drive system. The robustness against parameter mismatch and noise (in the measured signals) can be greatly reduced by using closed-loop observers for the estimation of the state-variables (e.g. flux linkages, speed, etc.). Various closed-loop observers are discussed in Section 4.5.3.5.

4.5.3.2 Estimators using spatial saturation third-harmonic voltage

4.5.3.2.1 General introduction

In the present section saturation phenomena are utilized to obtain the rotor speed and also the magnitude and position of the magnetizing flux-linkage space vector. This information can then also be used to obtain the rotor flux-linkage space vector and the stator flux-linkage space vector, and their applications in vector-controlled drives is discussed. The same saturation phenomena can also be used to obtain a signal proportional to the electromagnetic torque.

For better understanding, first the main features of this technique will now be summarized. In a symmetrical three-phase induction motor with stator windings without a neutral point, the sum of the stator voltages is monitored. This is a spatial saturation third-harmonic voltage component and is due to stator and rotor teeth saturation, which saturation condition is normal in a standard motor. The third-harmonic voltage is then integrated to give the third-harmonic flux, and the fundamental component of the magnetizing flux (ψ_{m1}) is then determined by using a saturation function that can be obtained experimentally by performing the conventional no-load test (a look-up table stored in the memory of a microprocessor or DSP can be used which gives the relationship of the fundamental magnetizing flux and third-harmonic flux). The obtained magnetizing-flux modulus is large and is practically free of noise.

It is also possible to obtain the angle of the magnetizing-flux space vector by utilizing the monitored third-harmonic voltage. For this purpose the current in stator phase sA is also monitored and the displacement angle (γ) between this stator current maximum and the fundamental magnetizing-flux maximum is determined. It should be noted that the fundamental magnetizing-flux component does not have to be monitored, since only the location of its maximum point is required and this can be obtained from the third-harmonic stator voltage (the suitable zero-crossing point of this voltage). By knowing the angle of the fundamental magnetizing-current space vector with respect to the stator-current space vector (γ), and also by measuring the angle of the stator-current space vector with respect to the direct-axis of the stationary reference frame (α_s), it is possible to obtain the angle of the magnetizing-flux space vector with respect to the direct-axis of the stationary reference frame (μ_m). The details will be described below.

By using the modulus of the fundamental magnetizing vector and also its phase angle with respect to the stationary reference frame, the magnetizing flux-linkage space vector $\bar{\psi}_m = |\bar{\psi}_m| \exp(j\mu_m)$ is known, and this can be used to construct the rotor flux-linkage space vector or the stator flux-linkage space vector. The magnetizing flux-linkage, rotor flux-linkage, and stator flux-linkage space vectors can be used in vector control schemes using magnetizing-flux-oriented, rotor-flux-oriented, or stator-flux-oriented control. The rotor speed can be obtained, for example, by also considering the first derivative of the angle of the stator-flux space vector with respect to the direct-axis of the stationary reference frame (see also Section 4.5.3.1).

4.5.3.2.2 Physical picture due to saturation; teeth and core saturation

For better utilization, a modern induction machine is designed to operate in the saturated region of the magnetizing characteristic. If the stator windings of a symmetrical three-phase induction machine are assumed to be sinusoidally distributed, the stator currents will create a sinusoidally distributed m.m.f in an unsaturated machine and this establishes a sinusoidal air-gap flux density distribution. However, if the machine is saturated, the sinusoidal distribution becomes distorted. Two cases must be considered, saturation of the stator/rotor teeth and saturation of the stator/rotor core.

Saturation in the teeth is more common in a practical machine, since the volume of the iron is generally greater in the core where a much higher flux density exists. Therefore the techniques described below will only utilize teeth saturation phenomena, although some physical aspects of core saturation will also be briefly discussed for clarity. Considering the saturation of the teeth, the teeth with the highest flux density will saturate first, thus the sinusoidal flux density distribution will become flattened. The flattened flux density distribution, $B_{sat}^{teeth}(\theta)$ is shown in Fig. 4.56 together with the fundamental flux density distribution, $B_{gap1}(\theta)$. It can be seen that its peak value is smaller than the peak of the fundamental distribution. The flattening of the curve is mainly caused by the presence of a third-harmonic air-gap flux density component $B_{gap3}^{teeth}(\theta)$, which is also shown in Fig. 4.56. By using Fourier analysis, it can be shown that the flattened curve contains other odd harmonics as well, but the dominant component is the third harmonic for all saturation levels. The speed of the third-harmonic component is identical to that of the fundamental component and both rotate in the same direction. It also follows that if a model of the induction machine with saturated teeth is required, it can be simply obtained by superposition of the third harmonic on the fundamental air-gap flux component. It can also be seen that teeth saturation decreases the permeability of the iron parts and this increases the reluctances of the teeth for the region around the resultant air-gap flux density.

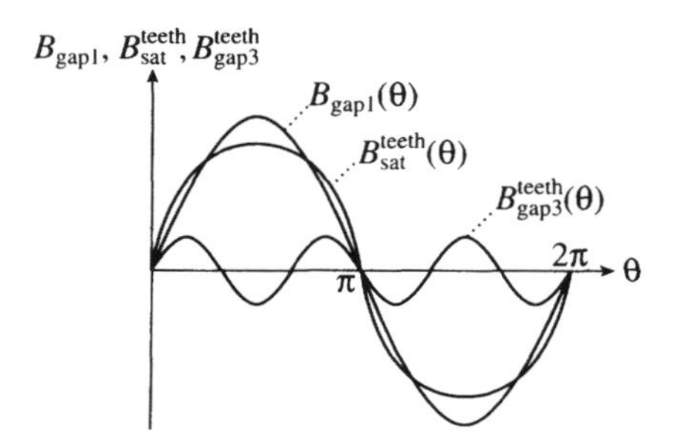

Fig. 4.56. Fundamental, flattened, and third-harmonic flux density distributions for teeth saturation.

By keeping the necessary assumption of constant and infinite iron permeability, the reluctance variation of the teeth can also be viewed as a variation of the air-gap reluctance. However, the air-gap has a constant permeability, thus the air-gap length has to be made variable (to ensure variation of the air-gap reluctance). It is therefore also possible to model an induction motor with saturated teeth by incorporating a variable air-gap length into its linear model, where the air-gap length is a function of the saturation level and spatial position. This also explains the resemblance between a saturated smooth-air-gap two-axis model and a non-saturated two-axis salient-pole model, which has been discussed in detail in [Vas 1992], where in the latter case, there exists real (physical) air-gap length variation.

When saturation occurs in the stator or rotor core, it also distorts the air-gap flux density distribution. However, in this case, since the core with the highest flux density will saturate first, the core flux density distribution will have a flattened sinusoidal form, and the air-gap flux density distribution will be peaked. This can also be explained by the fact that the air-gap flux-density distribution is proportional to the spatial derivative of the core flux-density distribution. This can be understood by considering the fluxes in the two-pole induction machine shown in Fig. 4.57. It can be seen that in the magnetic axis of the coil (which corresponds to the position $\theta=\pi/2$), the core flux is zero, but the air-gap flux is maximal, but 90° away (which corresponds to $\theta=0$), the core flux is maximal and the air-gap flux is zero. It clearly follows that the core flux-density distribution is the spatial integral of the teeth flux-density distribution.

As in the case of teeth saturation, again the third harmonic is the dominant component $B_3^{core}(\theta)$, but this third harmonic is in phase opposition to that produced by teeth saturation (see Fig. 4.56) thus enforcing the fundamental component, and leading to the peaked resultant core density distribution, $B_{sat}^{core}(\theta)$. This is shown in Fig. 4.58.

It is important to note that that the presence of a third-harmonic component in the air-gap flux density distribution is only a sufficient but not necessary condition for the occurrence of saturation. It has been shown that the mechanisms of teeth and core saturation cause opposite effects; the two types of third-harmonic

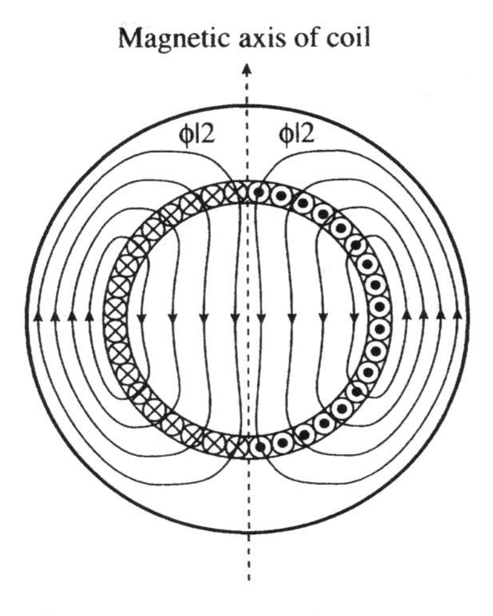

Fig. 4.57. Fluxes in the air-gap and core of a two-pole induction machine.

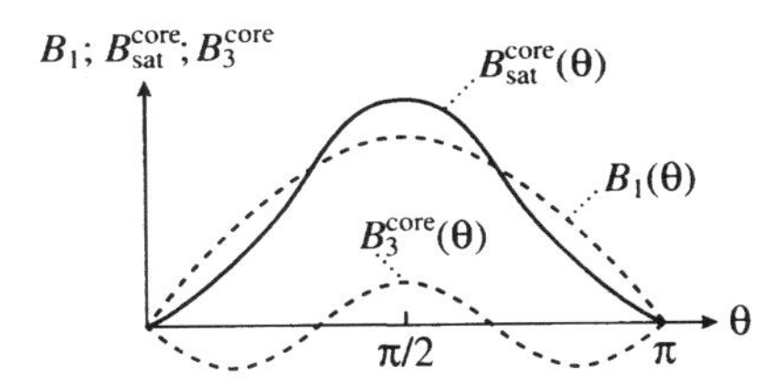

Fig. 4.58. Fundamental, third-harmonic, and peaked flux-density distributions for core saturation.

waves are in direct phase opposition. Thus a highly saturated machine does not necessarily have a high third-harmonic air-gap flux component.

4.5.3.2.3 General aspects of magnetizing flux-linkage monitoring using saturation phenomena

For an induction machine with wye-connected stator windings without a neutral point, it is possible to have a simple direct-vector-control scheme utilizing saturation effects in the stator/rotor teeth. For this purpose, first the fundamental component of the magnetizing-flux space vector (e.g. its modulus $|\bar{\psi}_{m1}|$ and its phase angle γ, with respect to the stator current vector) can be obtained by processing of the monitored (spatial saturation) zero-sequence third-harmonic voltage (u_{s3}) together with the monitored stator line current (e.g. i_{sA}), and also by utilizing a saturation function (f) obtained by a conventional no-load test. The saturation function f gives the relationship between the rms value of the third-harmonic magnetizing voltage and the rms value of the fundamental magnetizing voltage under no-load condition. The zero-sequence third-harmonic voltage component (u_{s3}) is due to the stator and rotor teeth saturation, and this saturation condition is normal in a modern induction motor.

The zero-sequence voltage can be obtained by adding the three stator phase voltages (e.g. by using potential transformers, or by using operational amplifiers as discussed below) and by eliminating the high frequency slot harmonic voltage component, u_{sh}, (by using an analog or digital filter). Of course for this purpose access is required to the neutral point of the stator windings. Figure 4.59 shows the fundamental stator voltage, u_{s1}, and the third-harmonic zero-sequence stator voltage u_{s0}, which is modulated by the high frequency slot harmonics, $u_{s0}=u_{s3}+u_{sh}$.

For rotor-flux-oriented control, the rotor flux-linkage space vector (its modulus and phase angle) can be obtained from $|\bar{\psi}_{m1}|$ and γ, as discussed below. Figures 4.60 and 4.61 show two implementations respectively of obtaining the zero-sequence stator voltage, u_{s0}. In Fig. 4.60 three potential transformers are used and the secondary windings of the transformers are connected in series.

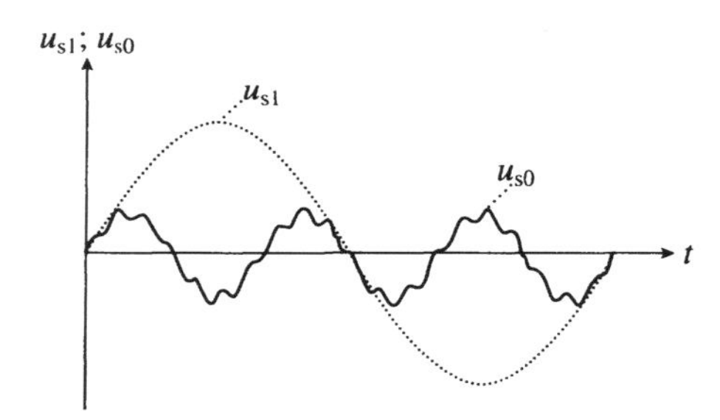

Fig. 4.59. Fundamental stator voltage and zero-sequence stator voltage with slot ripples.

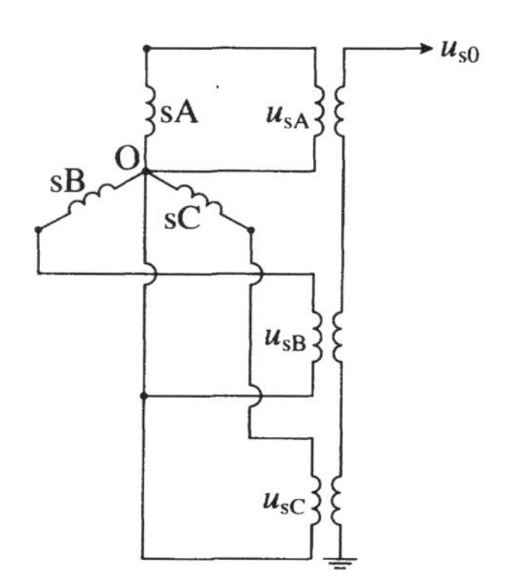

Fig. 4.60. Monitoring of the zero-sequence stator voltage using three potential transformers.

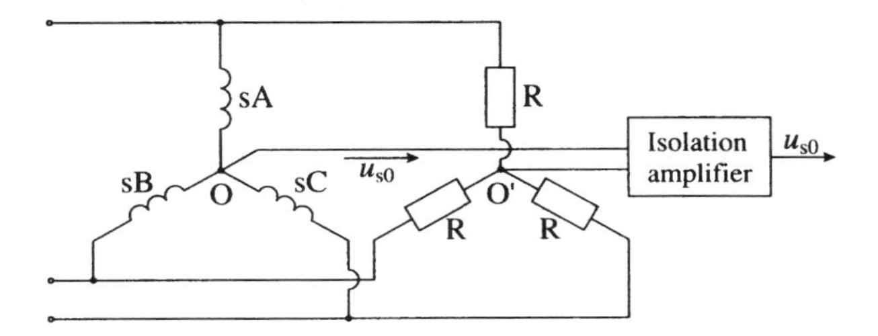

Fig. 4.61. Monitoring of the zero-sequence stator voltage using three identical external resistors.

However, in Fig. 4.61, three identical resistors are connected in star, and this arrangement is connected in parallel with the stator windings of the induction motor.

In Fig. 4.61 the zero-sequence stator voltage is obtained across the neutral point of the stator winding (point 0) and the pseudo-neutral point of the three external resistors (0′). This must then be isolated. If the induction motor is directly supplied from a PWM inverter, high transient intercoil voltages are also created at the line end of the windings. The magnitude of these transient voltages varies due to reflections at the end points and therefore the switching transients, which are present at both the pseudo-star (0′) and motor-star (0) point, may not exactly cancel when the motor signal is formed. Thus additional filtering of this feedback signal is necessary.

It has been discussed above that due to teeth saturation a flattened magnetizing flux-density distribution will arise, which contains a dominant third-harmonic flux component. This links the stator windings and induces a third-harmonic voltage component in each stator phase (u_{sA3}, u_{sB3}, u_{sC3}), which are all in phase, forming a zero-sequence set of stator voltages. However, if the stator windings of the saturated induction machine are connected in star with no neutral connection, no third-harmonic currents (zero-sequence current components) will circulate in the stator and therefore there will be no third-harmonic ohmic and leakage voltage drops across the stator impedance. Hence the third harmonic stator voltage (u_{s0}), whose monitoring has been discussed above, can be used to obtain directly the third-harmonic magnetizing voltage ($u_{m3}=u_{s0}$). The third-harmonic magnetizing flux (ψ_{m3}) can be obtained from this by integration:

$$\int u_{s0}\,dt=\int (u_{s3}+u_{sh})\,dt=\int u_{s3}\,dt=\psi_{m3}$$

(u_{sh} is filtered out, e.g. by a low-pass filter with high cut-off frequency).

It should be noted that if the stator winding distribution contains a third-harmonic spatial component, then the zero-sequence currents can flow in the stator. Furthermore, induction of a third-harmonic voltage in the stator windings due to stator and rotor currents is also possible. However, for a star-connected machine, without neutral return, these components can only exist if positive-sequence or negative-sequence third-harmonic current components flow in the stator windings. The summation of these voltages will cancel out these harmonics. If the machine is delta-connected, zero-sequence third-harmonic currents can flow around the delta path, provided that the stator winding distribution contains the third-harmonic spatial components. In this case, with practical windings, the current is only very small (a few per cent of the rated current).

The third-harmonic stator voltage (which has been shown to be equal to the third-harmonic magnetizing voltage) is always in phase with the fundamental magnetizing voltage. Thus the fundamental magnetizing flux (ψ_{m1}) is in phase with the third-harmonic magnetizing flux (ψ_{m3}) for all load conditions. This is shown in Fig. 4.62, where the temporal variations of ψ_{m1} and ψ_{m3} are shown for no-load and rated conditions respectively. However, since ψ_{m3} is small, for clarity, $40\psi_{m3}$ is shown in both diagrams.

The fact that the fundamental magnetizing flux is in phase with the third-harmonic magnetizing flux for all load conditions is a direct consequence of the physical phenomenon that when saturation of the teeth occurs, the magnetizing-flux density distribution keeps its flattened sinusoidal form, independent of the load. This can also be seen by considering that the third-harmonic rotor currents

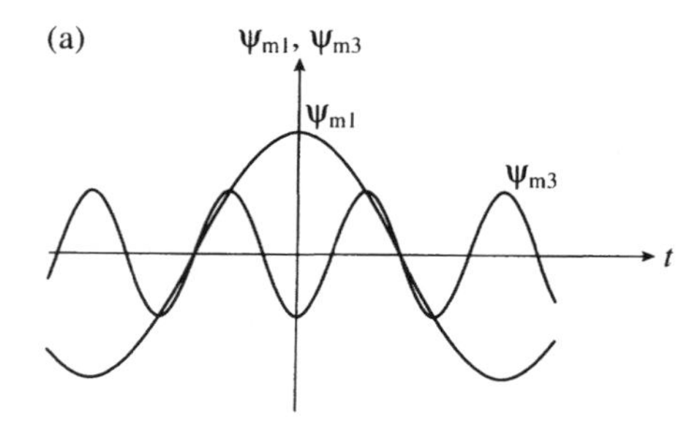

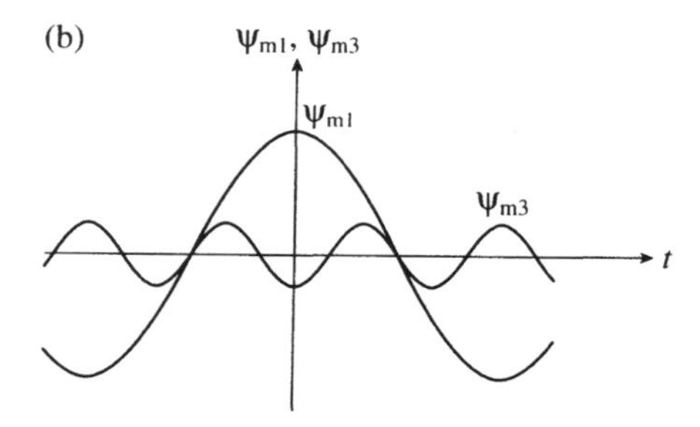

Fig. 4.62. Variation of ψ_{m1} and ψ_{m3} at no-load and rated load. (a) No-load condition; (b) rated load condition.

are not large enough to produce a rotor m.m.f. that would change the position of the third-harmonic magnetizing flux with respect to the fundamental magnetizing flux. It follows that the third-harmonic magnetizing flux always maintains a constant position with respect to the fundamental magnetizing flux. This is why the position of the third-harmonic magnetizing flux coincides with the position of the fundamental magnetizing flux, and thus the third-harmonic magnetizing voltage can be used to obtain the position (angle) of the fundamental magnetizing flux. This principle is utilized below in the next section.

To conclude: when the three stator voltages are added, the resultant zero-sequence voltage (u_{s0}) contains a third-harmonic component (u_{s3}) and the high frequency slot harmonic component (u_{sh}), $u_{s0}=u_{s3}+u_{sh}$. However, the third-harmonic component is dominant, since the fundamental and other, non-triple, so-called characteristic harmonics (e.g. 5th, 7th, 11th, etc.) which could be present in the currents and thus in the air-gap m.m.f. and magnetizing flux, will cancel. The slot harmonic component can be filtered out, as discussed above.

4.5.3.2.4 Estimation of the modulus of the magnetizing flux-linkage space-vector and the angle between the stator current vector and the magnetizing flux-linkage space-vectors

As shown above, by assuming a star-connected induction machine, where saturation of the teeth is present, the third-harmonic stator voltage u_{s3} will be a component of the stator zero-sequence voltage u_{s0} (which is obtained by the addition of the individual stator phase voltages, $u_{s0}=u_{sA}+u_{sB}+u_{sC}$), and $u_{s0}=u_{s3}+u_{sh}$, where u_{sh} is the high-frequency slot harmonic voltage component. The voltage component, u_{s3}, can be obtained from u_{s0}, by filtering out the high frequency slot harmonic voltage component (by using a bandpass filter). The monitored third-harmonic stator voltage, u_{s3}, and a stator line current (i_{sA}) are used for the determination of the phase angle (γ) of the fundamental component of the magnetizing flux.

As shown in Fig. 4.63, γ is equal to the displacement angle between the maximum of the stator current i_{sA} and maximum of the fundamental magnetizing

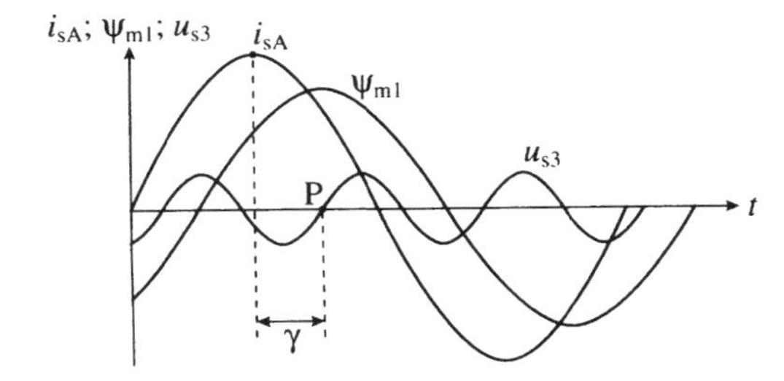

Fig. 4.63. Time variation of stator current i_{sA}, third-harmonic stator voltage u_{s3}, and fundamental magnetizing flux ψ_{m1}.

flux ψ_{m1}. This would imply that first ψ_{m1} has to be determined. However, the fundamental magnetizing flux does not have to be monitored, since for the determination of γ, only the location of the maximum point of ψ_{m1} is required, and this can be obtained from the suitable zero-crossing point of the third-harmonic voltage u_{s3} (since the zero-crossing of the voltage u_{s3} corresponds to a maximum value of the flux ψ_{m1}). The zero-crossing point is P in Fig. 4.63.

The displacement angle, γ, is very small for no-load condition, since a small mechanical output power is developed only to overcome windage and friction losses. However, when the induction machine is loaded, this angle will increase in response to the torque required by the load. It should be noted that this angle is positive for motoring and negative for generating.

The determination of the modulus of ψ_{m1} will next be discussed. However, it should be noted that for this purpose it does not have to be monitored; it can be obtained from the modulus of the third-harmonic magnetizing flux, ψ_{m3}, which is obtained from the monitored voltage u_{s3}.

The integration of the third-harmonic stator voltage u_{s3} gives the third-harmonic magnetizing flux linkage ψ_{m3},

$$\psi_{m3} = \int u_{s3}(t)\,dt, \tag{4.5-64}$$

and from this the amplitude of the third-harmonic magnetizing flux, $|\psi_{m3}|$, can be directly obtained. However, the integration of the offsets and noise present in the third-harmonic voltage can lead to inaccurate third-harmonic magnetizing-flux-linkage estimates. It should be noted that at low speeds a highly distorted third-harmonic voltage signal is obtained since the signal/noise ratio for the third-harmonic voltage increases considerably, which makes difficult a correct flux estimation. However, if proper filtering and averaging is used, useful information can be obtained. This information can be derived from the zero-crossings of u_{s3} to estimate the position (θ_{s3}) of the third-harmonic voltage (the estimated value

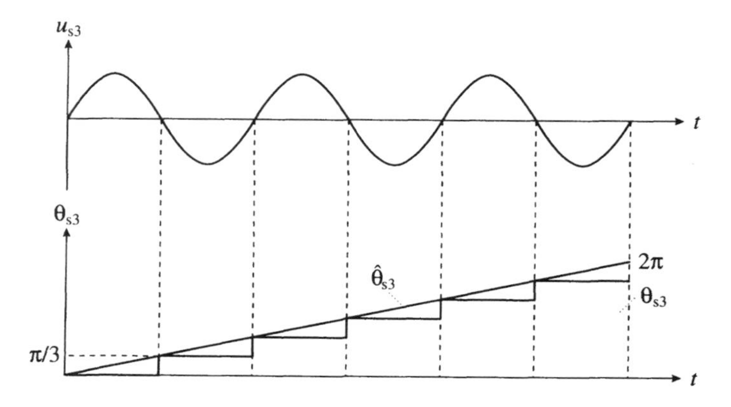

Fig. 4.64. Absolute position detection using u_{s3} and the phase voltages.

is $\hat{\theta}_{s3}$), and the position of the magnetizing flux linkage with an accuracy of 60 electrical degrees. As shown in Fig. 4.64 at every zero crossing of u_{s3}, a 60 electrical degree increment in the third-harmonic voltage position (and with the corresponding shift of the third-harmonic magnetizing flux) is obtained. It follows that an exact magnetizing flux-linkage position is available in a convenient manner.

However, for absolute position sensing further measurements must be made and θ_{s3} must also be estimated during positions between the zero crossings of u_{s3}. This problem (absolute position sensing) can be solved by using the relative positions of the stator phase voltages to detect the absolute positions of u_{s3}. If the voltage drops across the stator impedances are neglected, then at each zero crossing of u_{s3}, one of the phase voltages is zero (others have opposite signs), and this can be used to determine which of the six zero crossings has taken place, in order to obtain the absolute position. However, in a realistic approach the voltage drops must be compensated for each stator phase. It is possible to obtain an estimate for the stator impedance using the measured value of the phase voltage (at a zero crossing of u_{s3}) and also of the corresponding stator current. Thus the absolute position of the fundamental magnetizing flux can be determined at each zero crossing of u_{s3}. Problems arise at extremely low speeds, where high values of harmonics in the stator voltages may lead to incorrect flux estimation. The problem of absolute position sensing could also be solved by using a relative incremental scheme. In this case the induction machine is initially positioned in a known position and any further zero-crossings of u_{s3} are used to increment or decrement the absolute position information depending on the sign of the speed. This method is easier to implement than the first one, since it does not require the measurement of the phase voltages and currents or the stator impedance voltage-drop corrections. However, in this case, any false zero-crossings of u_{s3} will lead to an unacceptable position error.

As mentioned above, the second problem to be solved is the accurate estimation of the position between the zero-crossings of u_{s3}, since only at the zero-crossings of u_{s3} is the exact position available (only six steps have been used above with 60 degree accuracy). Between the six instants, a special estimation technique is required. A solution is obtained by using the position information of the stator-voltage space vector expressed in the stator reference frame (which requires the monitoring of the stator voltages) between the zero crossings of u_{s3}. However, mainly at low speeds, several factors influence the monitored voltages. These are: measurement noise, impedance voltage drop, harmonics, etc. Therefore it is better to use the variation of the position of the stator-voltage space vector for the determination of the position between the zero-crossings. In an alternative scheme, the position between the zero-crossings can be determined by utilizing the average speed of the magnetizing flux, ω_{av}, then the position is incremented by $\pi/3$ radians between two switchings which occur at consecutive zero-crossing instants t_1 and t_2, thus $\omega_{av}=(\pi/3)(t_2-t_1)$. Such an estimator will work well in the steady state, but in the transient state correction terms are required to take account of the motor and load models.

A similar signal/noise problem to that discussed above occurs when the motor operates at a flux level which is lower than the rated level, since the amplitude of

u_{s3} reduces due to the lower saturation level. Thus an accurate estimation of the flux under these conditions requires special considerations.

The amplitude of the third-harmonic stator voltage is a function of the saturation level in the induction machine, which is dictated by the amplitude of the fundamental magnetizing flux. Thus there exists a function f between the third-harmonic stator voltage (amplitude) and the fundamental magnetizing voltage (amplitude). This function is the so-called saturation function. It can be obtained from a conventional no-load test (where the rms value of the fundamental stator voltage, u_{s1}, is measured together with the rms value of the third harmonic voltage, u_{s3}), yielding the required relationship for the amplitudes: $|u_{s1}|=f(|u_{s3}|)$. Since $|u_{s3}|=|u_{m3}|$ for a star-connected machine, where u_{m3} is the third-harmonic magnetizing voltage, thus $|\psi_{m1}|=f(|\psi_{m3}|)$ also holds. Therefore the amplitude of the fundamental magnetizing flux-linkage space vector can be obtained as $|\psi_{m1}|=f(|\psi_{m3}|)$. It follows that for the microprocessor or DSP implementation of the saturation function, a look-up table stored in the memory of the microprocessor or DSP can be used, which gives the relationship between the fundamental magnetizing flux and the third-harmonic flux (relationship of rms value of fundamental stator line voltage and rms value of third-harmonic stator voltage under no-load condition). In general, the thus obtained magnetizing-flux modulus is large, it is practically free of noise, and problems arise only at low frequency.

4.5.3.2.5 Estimation of the magnetizing-, stator-, and rotor flux-linkage space vectors; utilization in torque control schemes; estimation of the electromagnetic torque

By using the technique described above, the modulus of the magnetizing-flux space vector $|\bar{\psi}_m|=|\psi_{m1}|$ (the subscript 1 is usually omitted in the literature) is obtained together with the angle γ, which is the angle between the magnetizing flux-linkage space vector and the stator-current space vector. These quantities, together with the angle of the stator-current space vector, α_s, can be used to obtain the magnetizing flux-linkage space vector and also the stator and rotor flux-linkage space vectors, which are required in various vector control implementations. For clarity Fig. 4.65 shows the magnetizing-flux space vector, stator-current space vector, and the relationship of various angles.

In Fig. 4.65 μ_m is the angle of the magnetizing flux-linkage space vector with respect to the real axis of the stationary reference frame. By using the angle α_s (stator-current vector angle with respect to the real-axis stationary reference frame), which can be obtained as

$$\alpha_s=\tan^{-1}(i_{sQ}/i_{sD}), \tag{4.5-65}$$

the angle μ_m can be obtained by using

$$\mu_m=\alpha_s-\gamma. \tag{4.5-66}$$

Thus the magnetizing flux-linkage space vector in the stationary reference frame is

$$\bar{\psi}_m=\psi_{mD}+j\psi_{mQ}=|\bar{\psi}_m|\exp(j\mu_m)=|\bar{\psi}_m|\exp[j(\alpha_s-\gamma)], \tag{4.5-67}$$

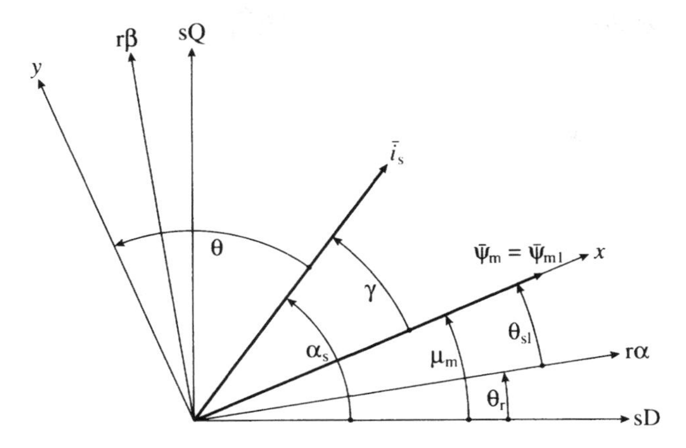

Fig. 4.65. Magnetizing flux-linkage and stator-current space vectors and the various angles.

or alternatively its components in the stationary reference frame are

$$\psi_{mD} = |\bar{\psi}_m| \cos(\alpha_s - \gamma) \tag{4.5-68}$$

$$\psi_{mQ} = |\bar{\psi}_m| \sin(\alpha_s - \gamma). \tag{4.5-69}$$

Of course in the magnetizing-flux-oriented reference frame (whose real axis is x, as shown in Fig. 4.65), $\bar{\psi}'_m = \bar{\psi}_m \exp(-j\mu_m) = \psi_{mx} + j\psi_{my}$, thus

$$\psi_{mx} = |\bar{\psi}_m| \tag{4.5-70}$$

$$\psi_{my} = 0. \tag{4.5-71}$$

The magnetizing flux-linkage space vector can then be used to obtain the stator or rotor flux space vectors. Thus the space vector of the stator flux linkages in the stationary reference frame can be obtained by considering

$$\bar{\psi}_s = L_s \bar{i}_s + L_m \bar{i}'_r = L_{sl} \bar{i}_s + L_m(\bar{i}_s + \bar{i}'_r) = L_{sl} \bar{i}_s + L_m \bar{i}_m,$$

but since $\bar{\psi}_m = L_m \bar{i}_m$, thus

$$\bar{\psi}_s = L_{sl} \bar{i}_s + \bar{\psi}_m = L_{sl} \bar{i}_s + |\bar{\psi}_m| \exp[j(\alpha_s - \gamma)] = \psi_{sD} + j\psi_{sQ}, \tag{4.5-72}$$

where L_{sl} is the stator leakage inductance. The rotor flux-linkage space vector can be obtained as follows in terms of the magnetizing-flux space vector by considering that

$$\bar{\psi}'_r = L_r \bar{i}'_r + L_m \bar{i}_s = L_r \bar{i}'_r + (L_r - L_{rl})(\bar{i}_s + \bar{i}'_r) = -L_{rl} \bar{i}_s + L_r \bar{i}_m$$

$$\bar{\psi}_m = L_m \bar{i}_m,$$

thus

$$\bar{\psi}'_r = -L_{rl} \bar{i}_s + \frac{L_r}{L_m} \bar{\psi}_m = -L_{rl} \bar{i}_s + \frac{L_r}{L_m} |\bar{\psi}_m| \exp[j(\alpha_s - \gamma)] = \psi_{rd} + j\psi_{rq}, \tag{4.5-73}$$

or in terms of the stator flux-linkage space vector, similarly to that shown in eqns (4.5-16) and (4.5-17), as

$$\bar{\psi}'_r = \frac{L_r}{L_m}(\bar{\psi}_s - L'_s \bar{i}_s) = \psi_{rd} + j\psi_{rq}, \tag{4.5-74}$$

where L_{rl} and L_r are the rotor leakage and rotor self-inductance respectively and L_m and L'_s are the magnetizing inductance and stator transient inductance respectively. In general these inductances are saturation dependent, and for correct flux-linkage estimation this has to be considered. It follows from eqn (4.5-73) that if it is used in a rotor-flux-oriented control scheme, then even if the magnetizing flux linkage can be obtained with high accuracy, the estimated rotor flux linkage will be sensitive to the rotor leakage inductance.

The flux space vectors obtained above can be used in various torque-controlled drives. In sensorless schemes, it is possible to obtain from these the rotor speed (or rotor position), similarly to that described in Section 4.5.3.1, where the rotor speed is obtained by utilizing the speed of the flux vector under consideration.

For better understanding, Fig. 4.66 shows the schematic of the flux estimators using the spatial saturation third-harmonic stator voltage. In Fig. 4.66(a), first the zero-sequence voltage u_{s0} is obtained by the technique shown in Figs 4.60 or 4.61, e.g. by a circuit which sums the three line-to-neutral stator voltages of the induction machine. This is followed by an isolation circuit on the output of which the isolated zero-sequence voltage is present. This could be an isolation amplifier. By using a filter (low-pass filter), which removes the high-frequency slot harmonics, the third-harmonic stator voltage, u_{s3}, is obtained. The third harmonic magnetizing-flux modulus, $|\bar{\psi}_{m3}|$, is then obtained by integration and taking the absolute value of the integrated signal. For this purpose a low-pass digital filter with a low cut-off frequency (e.g. 0.1 Hz) can be used to minimize the integration error at low frequencies. This signal is then applied to a function generator, where the function f is the saturation function described above and which can be obtained by using the no-load magnetizing curve. In a digital implementation this is a look-up table. On the output of this, the modulus of the fundamental magnetizing flux-linkage space vector amplitude, $|\bar{\psi}_m|$, is obtained. The angle of the magnetizing flux-linkage space vector, μ_m, is obtained as follows. The monitored direct- and quadrature-axis stator currents (i_{sD}, i_{sQ}) are inputs to a rectangular-to-polar (R→P) converter, on the outputs of which the angle α_s and the modulus $|\bar{i}_s|$ of the stator-current space vector is present. This angle is then added to the angle γ, which is the angle between the magnetizing-flux space vector and the stator-current space vector (as shown in Fig. 4.65). The angle γ is obtained on the output of the γ-estimator block, which obtains γ by using u_{s3} and i_{sA} as described above. To be more specific, this angle is obtained by considering that it is the angle between the maximum value of the line current in stator phase sA and the zero-crossing of u_{s3} (since the zero-crossing occurs at the maximum value of the fundamental magnetizing flux). As discussed above, the angle γ for no-load conditions is very small, since the mechanical output power is developed only to overcome the windage and friction losses. However, when the machine is

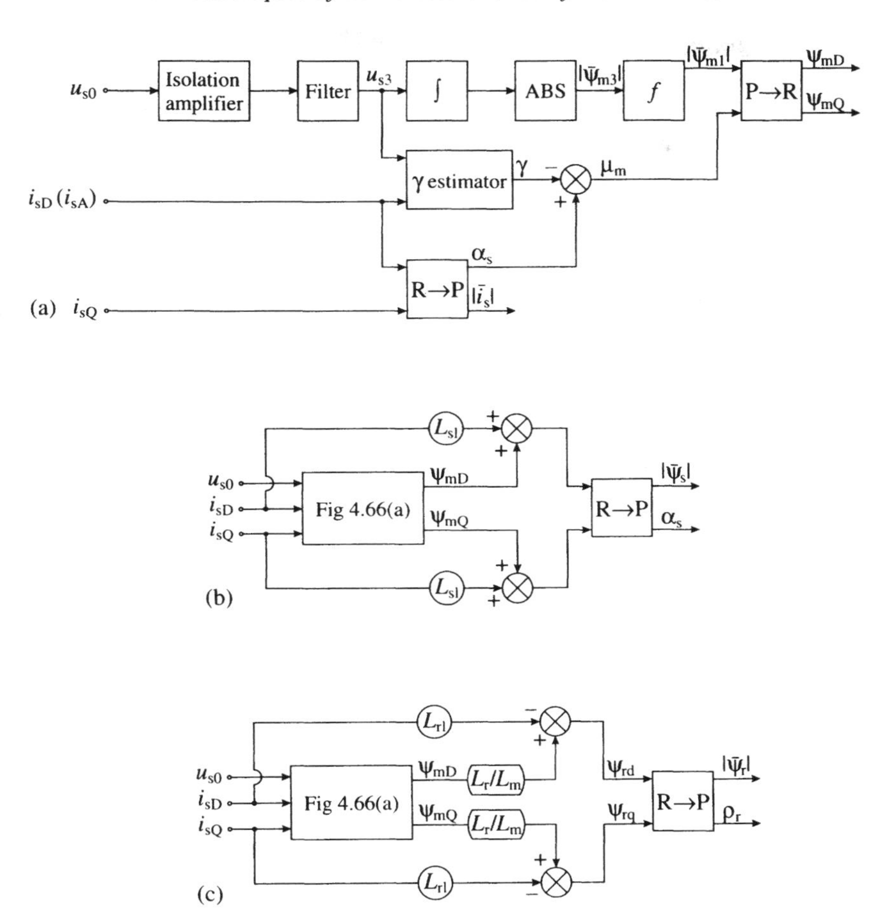

Fig. 4.66. Schematic of flux-linkage estimators using the saturation third-harmonic stator voltage. (a) Estimation of magnetizing flux linkages; (b) estimation of stator flux linkages; (c) estimation of rotor flux linkages.

loaded, this angle increases to respond to the torque required by the load. Finally, in the last part of the circuit shown in Fig. 4.66(a), a polar-to-rectangular (P→R) converter is used, on the output of which the direct-and quadrature-axis components of the magnetizing-flux space vector are obtained in the stationary reference frame, ψ_{mD}, ψ_{mQ}.

For completeness, in Fig. 4.66(b) the schematic of a circuit is shown, which is based on eqn (4.5-72) and can be used to obtain the direct- and quadrature-axis components of the stator flux-linkage components in the stationary reference frame, ψ_{sD}, ψ_{sQ}. A rectangular-to-polar converter is then used to obtain the modulus and phase angle of the stator flux-linkage space vector. On the output of the circle with

L_{sl} inside, the appropriate leakage flux component is present and when this is added to the corresponding magnetizing-flux component, the corresponding total stator flux-linkage component is obtained. It should be noted that in this circuit the only extra machine parameter to be used is the stator leakage inductance.

Similarly, in Fig. 4.66(c) a circuit is shown which is based on eqn (4.5-73), on the outputs of which the modulus and phase angle of the rotor-flux space vector in the stationary reference frame are obtained. It should be noted that $L_r/L_m = 1 + L_{rl}/L_m$, so this circuit only requires two machine parameters, the rotor leakage inductance and the magnetizing inductance (L_{rl}, L_m). For normal operating conditions the constancy of the rotor leakage inductance is a very good approximation. If the effects of leakage flux saturation cannot be neglected, the entire scheme may become problematic, since due to leakage flux saturation, the zero-sequence stator voltage u_{s0} will contain an extra voltage component (in addition to the high-frequency component described above). If the magnetizing inductance changes (with the flux or load), a correction can be applied to L_m via a non-linear function, which relates the estimated value of the magnetizing inductance to the magnitude of the third-harmonic magnetizing flux component, $|\psi_{m3}|$. This function can be obtained from the saturation function described above and can be realized by a look-up table in a digital implementation.

Finally it should be noted that the instantaneous value of the electromagnetic torque can also be determined by using the fundamental component of the magnetizing flux-linkage space vector (which has been determined from the third-harmonic stator voltage component and also the phase angle of the fundamental magnetizing flux, with respect to the stator current in phase sA). For this purpose it is utilized that the electromagnetic torque is produced by the interaction of the stator flux-linkage space vector and stator-current space vector, and the stator flux-linkage space vector can be expressed in terms of the magnetizing flux-linkage space vector as $\bar{\psi}_s = L_{sl}\bar{i}_s + \bar{\psi}_m$, where L_{sl} is the stator leakage inductance. Thus

$$t_e = \tfrac{3}{2}P\bar{\psi}_s \times \bar{i}_s = \tfrac{3}{2}P\bar{\psi}_{m1} \times \bar{i}_s = \tfrac{3}{2}P\bar{\psi}_m|\bar{i}_s|\sin\gamma = \tfrac{3}{2}P|i_s|f(|\psi_{m3}|)\sin\gamma, \qquad (4.5\text{-}75)$$

where P is the number of pole-pairs, f is the saturation function described above, and γ is the angle of the magnetizing flux with respect to the stator current i_{sA}. All the other expressions derived earlier for the electromagnetic torque can also be used, e.g. if they contain the magnetizing flux-linkage components, then these can be substituted by their values shown in eqns (4.5-68) and (4.5-69). However, if they contain the stator or rotor flux-linkage components, then these components can be substituted by their values shown in eqns (4.5-72), (4.5-73), and (4.5-74).

It should be noted that there are many vector-controlled induction motor implementations in the literature (e.g. Kreindler *et al.* 1994) which utilize the saturation third-harmonic stator voltage for the estimation of the modulus of the fundamental magnetizing flux linkages and which contain a speed or position sensor. The main advantage compared to other schemes is that this type of solution is less dependent on machine parameters. The sensorless torque-controlled drive implementations of an induction machine using the saturation third-harmonic voltage will not be discussed in further detail, since they can be readily derived by using the information provided on sensorless drives presented in the present and other sections.

4.5.3.3 Estimators using saliency (geometrical, saturation) effects

4.5.3.3.1 General introduction and various schemes

In high-performance induction motor drives it is possible to estimate the rotor speed, rotor position, and various flux linkages of a squirrel-cage induction motor by utilizing different types of geometrical effects (normal slotting, inherent air-gap asymmetry, intentional rotor magnetic saliency created by spatial modulation of rotor-slot leakage inductance) and saliency effects created by saturation. Thus the rotor speed can also be estimated by utilizing slot harmonics or eccentricity harmonics. Furthermore it is also possible to estimate fiux-linkage space vector position (and using this information, the rotor position) by using flux-linkage detection based upon saturation-induced inductance variations (see also Sections 3.1.3.6 and 3.2.2.2.3), where test voltages are applied to the stator terminals of the machine and the flux-linkage space-vector position is detected from measured current responses. However, in the presence of injected high frequency stator voltages, for the estimation of flux-linkage space-vector position and the rotor position, it is also possible to use saliency effects, which are due to magnetic saturation (main flux or leakage flux saturation) or saliency effects intentionally created by using special rotor construction, where spatial modulation of the rotor leakage inductance is created (e.g. by periodically varying the rotor slot-opening widths or by varying the depths of the rotor slot openings, etc.).

In the following sections the various techniques listed above will be discussed in some detail. The first technique to be discussed uses slot-harmonic effects and for this purpose two alternatives will be examined: the first one uses monitored stator voltages and the second uses monitored stator currents.

4.5.3.3.2 Estimators using rotor slot harmonics

The slip frequency and rotor speed of a three-phase induction machine can also be estimated by utilizing the rotor slot harmonics. However, in a speed-sensorless high-performance drive, up to now, due to the measurement bandwidth limitation, such an estimator has not been directly used for rotor speed estimation, but it has only been used indirectly to help the tuning of MRAS speed estimators (see also Section 4.5.3.4). It is a great advantage of this technique that variations in the motor parameters do not influence the accuracy of the estimation and the technique can also be used for all loads. The rotor slot harmonics can be detected by using various techniques:

- Utilizing monitored stator voltages;
- utilizing monitored stator currents.

The details of these two techniques are presented below, but in a speed-sensorless high-dynamic performance drive, the second technique is preferred since the monitoring of the stator currents is always required, but the voltage monitoring can be avoided.

Slip frequency and speed estimation using monitored stator voltages If a symmetrical three-phase induction motor is supplied by a system of symmetrical three-phase voltages, the air-gap flux will contain space harmonics. Neglecting the effects of magnetic saturation, some of these space harmonics are due to the non-sinusoidal distribution of the stator windings; these are usually referred to as m.m.f. space harmonics. However, due to slotting, there are also slot harmonics produced by the variation of the reluctance due to stator and rotor slots; these are the so-called stator and rotor slot harmonics respectively. The rotor slot harmonics can be utilized for the detection of the angular slip frequency and the angular rotor speed of an induction machine. When the air-gap m.m.f. contains slot harmonics, slot-harmonic voltages are induced in the stator windings when the rotor rotates. Both the amplitude and frequency of these depend on the rotor speed. However, it is difficult to extract information on the rotor speed from the magnitude, because it depends not only on the rotor speed but also on the magnitude of the flux level and the loading conditions. The rotor speed and the slip frequency are obtained from the frequency of these slot-harmonic voltages. It should be noted that if the induction machine does not have skewed rotor slots, stronger slot harmonics will be produced, but usually skewed rotor slots are employed, in order to reduce the audible noise and to eliminate asynchronous crawling (created by these harmonics) during line starts.

In a stator phase of an induction machine the magnitude of the induced slot-harmonic voltages is small and thus to separate these from the dominating fundamental voltage, the stator phase voltages are added. In general, if the monitored stator voltages of the induction machine are added, the resulting voltage ($u_s = u_{sA} + u_{sB} + u_{sC}$) will contain a slot-harmonic component u_{sh}, due to the fundamental m.m.f. wave, but due to main flux saturation it will also contain a third-harmonic component u_{s3}. Furthermore, when an inverter supplies the induction machine, the stator voltage (u_s), will also contain extra time harmonic voltages, u_{shk}, where k is the time-harmonic order. Thus in general

$$u_s = u_{sh} + u_{s3} + u_{shk}. \tag{4.5-76}$$

The frequency of the slot harmonic voltage components (u_{sh}, u_{shk}) is given below and it will be shown how it is related to the stator frequency and rotor speed (and the number of rotor slots, which is assumed to be known). It can be proved (e.g. see [Vas 1993]) that if the stator voltages of the induction machine (u_{sA}, u_{sB}, u_{sC}) are added, and if the m.m.f. distribution is assumed to be sinusoidal, then the resulting stator voltage $u_s = u_{sA} + u_{sB} + u_{sC}$ will contain the rotor slot harmonic voltages (u_{sh}) and the frequency of their dominant component (fundamental slot-harmonic frequency) is

$$f_{sh} = N_r f_r \pm f_1 = 3Nf_1 - N_r f_{sl} \qquad N_r = 3N \mp 1. \tag{4.5-77}$$

In eqn (4.5-77) f_r is the rotational frequency of the rotor, $f_r=\omega_r/2\pi$, where ω_r is the angular rotor speed. Furthermore, f_1 is the stator frequency ($f_1=\omega_1/2\pi$, where ω_1 is the angular stator frequency) and f_{sl} is the slip frequency, $f_{sl}=f_1-f_r=(\omega_1-\omega_r)/2\pi=\omega_{sl}/2\pi$, where ω_{sl} is the angular slip frequency. Finally N_r is the number of rotor slots per pole-pair ($N_r=Z_r/P$), where Z_r is the number of rotor slots and P is the number of pole-pairs. It follows that the rotor slot-harmonic frequency only depends on f_1, f_r, and N_r. By considering that $\omega_r=\omega_1(1-s)$, where s is the slip, it is also possible to express f_{sh} as

$$f_{sh}=\left[\frac{Z_2(1-s)}{P}\pm 1\right]f_1. \qquad (4.5\text{-}78)$$

When the induction machine is supplied by a three-phase inverter, in general, the output currents and voltages of the inverter contain time harmonics (u_{sk}, i_{sk}). Since in the output voltage of the three-phase inverter, there are no harmonic voltages with harmonic orders $3k$, where $k=1,2,\ldots$ the voltage (u_s) which is obtained by adding the three stator phase voltages (u_{sA}, u_{sB}, u_{sC}) does not contain time harmonics if the induction motor is symmetrical. However, slot-harmonic voltages are present in the stator winding due to the time-harmonic fluxes produced by the time-harmonic currents. The slot-harmonic frequency due to the kth time harmonic can be expressed as

$$f_{shk}=\begin{cases}N_r f_r \pm kf_1=3Nf_1\pm 6mf_1-N_r f_{sl} & k=6m-1\\ N_r f_r \mp kf_1=3Nf_1\mp 6mf_1-N_r f_{sl} & k=6m+1\end{cases} \qquad (4.5\text{-}79)$$

where $N_r=3N\pm 1$, $m=1,2,\ldots$

If saturation of the main flux paths occurs, a third-harmonic voltage (u_3) is produced in each stator phase voltage. These third-harmonic voltages are in phase with each other and therefore are present in the sum of the phase voltages. It should be noted that in general the magnitude of the third-harmonic voltage in a stator phase is smaller than the magnitude of the fundamental voltage in the corresponding stator phase, but the slot-harmonic voltage is also small and thus u_3 cannot be ignored even if the motor operates at rated voltage. The third-harmonic voltage is approximately one fifth of the slot-harmonic component.

If an induction motor with star-connected stator windings is assumed with the neutral point accessible, then the summation of the stator voltages can be performed by using three potential transformers, whose secondary windings are connected in series (e.g. see Fig. 4.60). It is also possible to use operational amplifiers to add the three stator voltages, but in a DSP-controlled drive the addition can be simply done numerically. Thus u_s is obtained. It follows from eqn (4.5-76) that u_{sh} can be obtained by removing the voltage components u_{s3} and u_{shk}. This can be achieved by using various circuits (e.g. see [Vas 1993]) and then the frequency of the rotor slot harmonic (f_{sh}) can be obtained from u_{sh}. By subtracting the stator frequency, $f_{sh}-f_1=N_r f_r$ is obtained and multiplication of this by $2\pi/N_r$ gives the angular rotor speed $\omega_r=2\pi(f_{sh}-f_1)/N_r$. The slip frequency can be obtained by using $f_{sl}=f_1-f_r$, where $f_r=\omega_r/2\pi$. However, special considerations are required in the low speed range, because at low speeds the amplitude of the slot-harmonic voltages decrease.

Slip frequency and speed estimation using monitored stator currents The rotor speed can also be obtained by using the monitored stator line currents and performing harmonic spectral estimation. This is the preferred technique, since in a speed-sensorless high-dynamic-performance torque-controlled induction motor drive there is always the need for current monitoring, and it is useful to reduce the number of sensors required (by eliminating the voltage sensors). The stator-current slot harmonics can be similarly obtained as given above by eqn (4.5-79), thus if a stator line current of a PWM inverter-fed induction motor is monitored, then $f_{shk}=N_r f_r \pm k f_1$ holds, where $f_r=f_1(1-s)$. For example, Fig. 4.67 shows the measured frequency spectrum of a loaded inverter-fed induction motor (the slip is 0.01133) which has $Z_2=28$ rotor slots, $2p=4$, $f_1=50$ Hz.

As expected, it follows from Fig. 4.67 that due to the inverter, in the window shown, the time harmonics $k \times 50$ Hz are present, these are $3\times 50=150$ Hz; $5\times 50=250$ Hz; $7\times 50=350$ Hz; $9\times 50=450$ Hz. Furthermore, in agreement with

$$f_{shk}=N_r f_r(1-s)\pm k f_1=14\times 50\times(1-s)\pm k\times 50,$$

the slot-harmonic frequencies $14\times 50\times 0.988-5\times 50=442$ Hz; $14\times 50\times 0.988-7\times 50=342$ Hz, and $14\times 50\times 0.988-9\times 50=242$ Hz are also present.

For the purpose of speed estimation, in a PWM inverter-fed induction motor drive, a line current is monitored, scaled, and low-pass filtered (to eliminate high frequency PWM harmonics) and e.g. digital FFT can be used to detect the

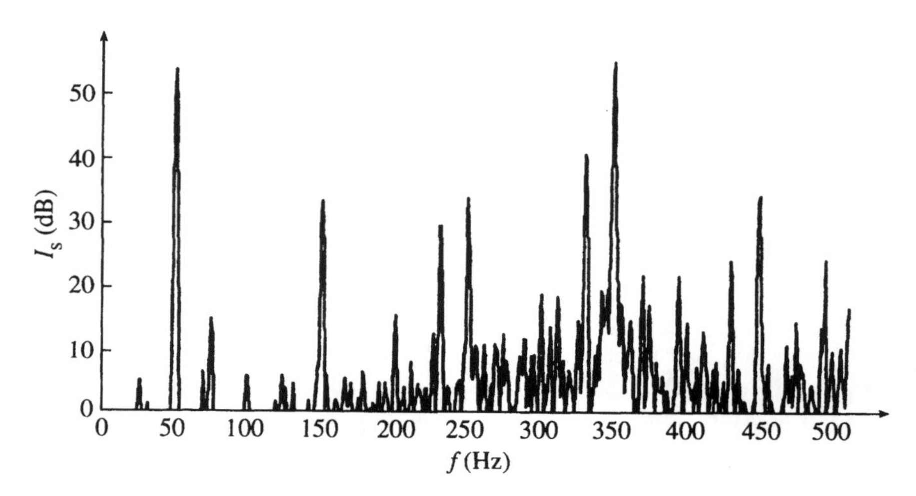

Fig. 4.67. Measured stator-current frequency spectrum of an induction motor ($Z_2=28$ rotor slots, $2p=4$, $f_1=50$ Hz).

speed-dependent rotor slot harmonic (f_{sh}). When f_{sh} is known, by using eqn (4.5-77) the rotor speed can be obtained as

$$\omega_r = 2\pi f_r = 2\pi \frac{(f_{sh} \mp f_1)}{N_r}, \tag{4.5-80}$$

where f_1 is the fundamental stator frequency and N_r is the number of rotor slots per pole-pair, $N_r = Z_r/P$, where Z_r is the number of slots and P is the number of pole-pairs. It follows from eqn (4.5-80) that the accuracy of the estimated rotor speed depends on the accuracy of the measurement of f_{sh} and f_1. In a speed-sensorless vector-controlled drive, with rotor-flux-oriented control, the derivative of the angle of the rotor flux-linkage space vector can be used to obtain f_1. The accuracy of obtaining f_{sh} by using the Fast Fourier Transform (FFT) is also discussed below. It is important to note that the rotor-speed detection scheme based on eqn (4.5-80) requires the knowledge of the number of rotor slots per pole-pair.

The five main steps of the f_{sh} estimation are as follows:

1. Identification of f_1 (see also discussion above).

2. Determination of the no-load slot harmonic (f_{sh0}) around a specific stator harmonic. For example by considering $k = \pm 1$, the no-load slot harmonic frequency is $f_{sh0} = N_r f_1 \pm k f_1 = N_r f_1 + f_1$ (in general it follows from eqn (4.5-77) that $f_{sh} = f_{sh0} - N_r s f_1$, where s is the slip).

3. Defining the width of the slot-harmonic tracking window; $\Delta f_{sh} = N_r f_{slmax}$, where f_{slmax} is the maximum value of the slip frequency (e.g. rated slip frequency). For motoring operation, the window is placed at $[f_{sh0} - \Delta f_{sh}, f_{sh0}]$, while for motoring and generating operations it is placed at $[f_{sh0} - \Delta f_{sh}, f_{sh0} + \Delta f_{sh}]$.

4. Searching for the highest-amplitude harmonic (highest spectrum line) in the window which is a non-triple harmonic of f_1.

5. Increasing the accuracy of the f_{sh} estimation (e.g. by interpolation, see below).

It is possible to isolate the rotor-slot harmonic from its neighbouring harmonics at any load over a wide speed-range if an adequate frequency resolution is used. The separation of the slot harmonics from their no-load values increases with load (this follows from $f_{sh} = f_{sh0} - N_r s f_1$). A digital FFT can give satisfactory results above a few Hz, but below this, noise and other factors prevent a correct estimation. For digital implementation, a real-value FFT based on the split-radix algorithm (Soresen *et al.* 1987) is ideal, since it requires a minimal number of multiplications and additions. For the purposes of the algorithm, 2^N samples are used over an acquisition time T_a with sampling frequency f_s. This gives a spectrum of base resolution $1/T_a$ over the frequency range 0 to $f_s/2$. In a practical application, the determination of the spectral lines is terminated at the maximum frequency of interest (which also depends on N_r and the maximum f_1). It is important to note that the resulting spectrum also includes the PWM harmonics of the excitation frequency, which will interfere with the detection of the slot harmonics. Thus when using the FFT approach, in a PWM-fed induction motor drive, care should be taken since the slot harmonics can cross over a succession of PWM harmonics as the rotor speed decreases. The inverter harmonics interfere with the detection of the slot harmonics and this interference at low speed contributes to a low speed limit of a few Hertz for speed estimation. It should also be considered that the FFT will provide inaccurate frequency estimates when the frequencies to be detected are not integer multiples of $1/T_a$. This has two immediate effects. The first one is that for an estimated frequency, a spectrum quantization error (Δf) is present (with respect to the real frequency), which results in a speed error. The second effect is the spectral leakage effect (spreading of the energy distribution of each harmonic; also see below) and this effect is important when small-amplitude harmonics are close to large-amplitude harmonics. At low speeds, the slot harmonics are close to the PWM harmonics, and this effect must be considered. Machines with a large number of rotor slots have rotor-slot harmonics at higher frequencies, and the frequency at which the rotor-slot harmonics become close to large PWM harmonics is lower, and thus the rotor speed estimation can be performed at lower frequencies.

The accuracy of the speed estimation can be increased by using appropriate windowing (e.g. Hanning data window, Hamming window, rectangular window, etc.) and interpolation techniques. Windowing means that the FFT analyser sees the signal to be analysed through a window (it only sees a short length of the signal). However, in general, major discontinuities exist at the window edges. When these are transformed into the frequency domain, false results are obtained. The effect of windowing causes sidelobes to appear either side of the peaks. Some of the energy in the signal is leaking away into these sidelobes (hence the terminology: spectral leakage). This effect can be prevented by avoiding the discontinuities by arranging for the window length to be an exact multiple of the signal period. However, the problem is that most signals contain more than one fundamental frequency and also on most analysers the window length is not adjustable. This means that these discontinuities have to be accommodated some way. In practice, the effect is suppressed by using 'weighting'. Weighting is a function applied to the samples of the signal prior to processing by the FFT algorithm. For the case of rectangular weighting, all the samples have a weighting of one (from this point of view all the samples are considered to be of equal importance). However, there exist other weightings, which reduce the importance of the samples at the edges of the window, and correspondingly increase the importance, or weight, of the samples towards the middle. For example, the Hamming weighting is such a technique. The main effects of these weightings are to

- reduce the discontinuity to zero;
- modulate the signal by the 'shape' of the window;
- reduce the sidelobe height in the frequency domain;
- increase the effective bandwidth.

The increased bandwidth is an important factor for practical applications in torque-controlled speed-sensorless drives. The bandwidth considered is that of each point in the frequency domain. In a perfect system, each point would represent a perfect band-pass filter of very small width and virtually 'brick-wall' characteristics either side. However, in practice, the filter has a finite width and finite cut-off slope, which determines the selectivity of the system. The time and frequency domains have an inverse relationship. If the window shapes are considered in the time domain, it is clear that the maximum effective window width is obtained with a rectangular window. All other windows can intuitively be seen to have a reduced effective width (e.g. consider the 3 dB points). Reduced width in the time domain is equivalent to increased width in the frequency domain, thus the effect of windows other than rectangular is to increase the bandwidth of each frequency point. Therefore this effectively reduces the ability of the FFT analyser to resolve the close components. It also reduces the so-called 'picket fence' effect.

The accuracy of the FFT can be improved by using the method of interpolation. It has been emphasized above that spectral leakage causes the spreading of the energy distribution of each harmonic. This results in a number of spectral lines for each harmonic. If the largest peak exists at the ith spectral line with amplitude A_i, then the speed estimation accuracy can be increased by using interpolation, which means that the frequency resolution is enhanced by using an appropriate frequency correction factor, δ. For a rectangular window using interpolation, $\delta = 1/(1+x)$, where x is obtained from $x = A_i/\max(A_{i+1}, A_{i-1})$, where A_{i+1}, A_{i-1} are known magnitudes of the surrounding desired frequency. For the Hanning window, when interpolation is used, the expression of the correction factor is simple,

$$\delta = \frac{2-x}{1+x},$$

and for the Hamming window it is more complicated,

$$\delta = \frac{(x-2)(0.16x^2-0.18)}{(x+1)(0.16x^2-0.32x-0.92)}.$$

However, in practice, the application of the Hanning window can give accurate results in many cases. The interpolation technique is analogous to the use of a phase-locked-loop frequency multiplier acting on an encoder pulse train [Vas 1993].

As an example of the speed estimation of a $2p=4$, 50 Hz, 1.8 kW three-phase induction machine with 28 rotor slots Table 4.2 shows the speeds which have been estimated with interpolation (I) and without interpolation (NI) and using

Table 4.2 Estimation of the rotor speed

Δf(Hz)	ω_r(rpm)(I, Rectangular)	ω_r(rpm)(I, Hanning)	ω_r(rpm)(NI, Rectangular)
0.5	1430.74	1430.83	1429.96
1	1430.15	1430.71	1428.03
2	1429.38	1430.14	1426.65

rectangular and Hamming windows, for various spectrum quantisation (Δf) errors. It should be noted that the correct speed is 1431 rpm.

As expected, larger errors are present when no interpolation is used, and the errors are larger when the rectangular window is used compared to the errors for the case of using the Hanning window. This is due to the fact that the rectangular window exhibits larger spectral leakage. The Hanning window can give relatively accurate results even for large values of the spectrum quantization error. For the example discussed, with the Hamming window, the speed errors for the three quantization errors are 0.17 rpm, 0.29 rpm, and 0.86 rpm respectively. The example given only illustrates some possible improvements, but it is possible to obtain larger accuracies.

It has been shown above that the described speed-detection scheme requires the knowledge of the number of rotor slots per pole-pair. However, if information is not available from the manufacturer, then it is possible to obtain information on this parameter by conducting additional (experiments see discussion following eqn (4.5-82) below).

In practice, the air-gap m.m.f. also contains space harmonics due to various types of asymmetries, e.g. eccentricity ([Vas 1993], [Hurst *et al.* 1994]). By neglecting the effects of magnetic saturation, and assuming sinusoidally distributed stator windings, the stator current contain harmonics with the frequencies

$$f=[(cZ_2 \pm n_d)(1-s)/P \pm k]f_1, \tag{4.5-81}$$

where f_1 is the fundamental stator frequency, c is any integer, Z_2 is the number of rotor slots, and n_d is the eccentricity order number, which for static eccentricity is zero and for dynamic eccentricity is 1. Furthermore, in eqn (4.5-81) s is the slip, P is the number of pole-pairs, and k is the order of the time harmonics ($k=1$, 3, 5, 7, etc.). It should be noted that if $c=1$ and $k=1$ and $n_d=0$, eqn (4.5-81) reduces to eqn (4.5-77). It follows from eqn (4.5-81) that for 'pure' dynamic eccentricity ($c=0$, $n_d=1$, $k=1$), the current harmonics due to eccentricity are

$$f=[1 \pm (1-s)/P]f_1. \tag{4.5-82}$$

This expression is independent of the Z_2 and only requires knowledge of the pole-pair number, which can however be obtained by simple measurement. The most commonly used pole-pair number is four. It follows from eqn (4.5-82) that if the pole-pair number is known, and in an initial test the harmonics associated with 'pure' dynamic eccentricity are first detected by FFT from the measured stator current, then it is possible to estimate the slip (and also the speed). It should be noted that although the harmonics caused by pure eccentricity enable Z_2-independent speed and slip estimation, they provide much lower slip resolution than the slot harmonics for a given sampling time. Thus the slot harmonics can be used to provide the accurate speed estimation and the eccentricity harmonics are only used to give extra information for initialization of the slot-harmonic estimator. By using the estimated slip and also eqn (4.5-81), it is then possible to use a search algorithm to obtain Z_2, n_d, and k, from a table containing possible values of these parameters. When these are known, these parameters can be used in a speed-sensorless induction motor drive using current-harmonic spectral estimation. The number of rotor slots influences crawling, cogging, acoustic noise, stray losses, etc. In a squirrel-cage induction motor with die-casted rotor, the number of rotor slots is generally lower than the number of stator slots. For induction machines in the 4 kW–100 kW range, the most common number of poles is four, and the stator slot numbers usually employed are 36, 48, 60, 72. The rotors of the 4-pole induction machine usually have eight slots less than the stator. This information can be used to obtain an initial estimate of the number of rotor slots.

4.5.3.3.3 Estimation utilizing saturation-induced saliency

In Sections 3.1.3.6.1 and 3.2.2.2.3 a simple technique has been discussed for the rotor position estimation of PM synchronous machines and also synchronous reluctance machines. This technique [Schroedl 1988] can also be used for induction machines even at standstill. This is based on the fact that in the induction motor, due to saturation of the stator and rotor teeth, the stator inductances depend not only on the level of saturation but also on the position of the main flux. It follows from the stator voltage equation of the saturated induction machine that at standstill the rate of change of the stator currents can be expressed as $d\bar{i}_s/dt = \bar{u}_s/L$. In this equation L is the complex stator transient inductance of the induction machine, whose magnitude and angle depend on the magnetic operating point and the direction of the magnetizing flux-linkage space vector. By applying appropriate stator-voltage test vectors ($\bar{u}_s$), the rate of change of the stator-current space vector ($d\bar{i}_s/dt$) can be measured. The angle of the magnetizing flux-linkage space vector can then be obtained, since the locus of the modulus of the complex transient inductance is an ellipse and the minimum of this ellipse is in the direction of the magnetizing flux-linkage space vector.

It is also possible to estimate flux-linkage space-vector position in an induction machine by the tracking of high-frequency magnetic saliency created by magnetic saturation (main flux or leakage flux saturation) at zero or low rotor speeds [Jansen and Lorenz 1995*b*]. For this purpose high-frequency voltages are injected in the stator. The injected voltages can also be produced by the PWM VSI used in the drive scheme. It will be shown below that this technique is only possible in a saturated smooth-air-gap machine, because of the physically existing cross-saturation effect (this effect is due to the saliency caused by saturation).

It is well known (see e.g. [Vas 1992] and also Section 6.1 of the present book) that as a consequence of magnetic saturation, saliency is created and the stator direct- and quadrature-axis inductances become asymmetrical ($L_{sD} \neq L_{sQ}$), and also coupling will exist between the two axes (cross-saturation effect, $L_{DQ} \neq 0$). It has been shown [Vas 1992] that due to saturation, all the inductances are functions of the saliency position (μ), and

$$L_{sD} = L_1 + \Delta L \cos(2\mu) \qquad L_{sQ} = L_1 - \Delta L \cos(2\mu) \qquad L_{DQ} = L_{QD} = \Delta L \sin(2\mu).$$

It then follows that due to saturation, the stator transient inductances in the direct- and quadrature-axis (of the stationary reference frame) are also asymmetrical. It can be concluded that due to saturation, the stator transient inductances are spatially modulated, i.e. in the stationary reference frame the stator transient inductances also depend on the position of the saturation-induced saliency. Thus the direct- and quadrature-axis stator transient inductances (L'_{sD}, L'_{sQ}) and also the cross-coupling transient inductances ($L'_{DQ}=L'_{QD}$) are functions of the angle μ, and similarly to that shown above,

$$L'_{sD}=L'_1+\Delta L'\cos(2\mu) \qquad L'_{sQ}=L'_1-\Delta L'\cos(2\mu) \qquad L'_{DQ}=L'_{QD}=\Delta L'\sin(2\mu),$$

where $L'_1=(L'_{sd}+L'_{sq})/2$ and $\Delta L'=(L'_{sd}-L'_{sq})/2$. In these expressions L'_{sd} and L'_{sq} are the direct- and quadrature-axis transient stator inductances in the direct- and quadrature-axes of the existing saliency respectively. As expected the cross-saturation coupling and inductance asymmetry disappears when $L'_{sd}=L'_{sq}$.

The stator current responses will now be obtained, when high-frequency stator voltages are injected into the stator windings. These currents can then be used to obtain information on the position of the saliency. For high stator frequencies, the stator equations of the saturated induction machine in the stationary reference frame can be well approximated by the voltages across the appropriate stator transient inductances. Thus when a symmetrical three-phase high-frequency stator voltage system with amplitude U_{si} and angular frequency ω_i (ω_i is high) is injected into the stator, in the steady state the stator currents will be displaced from the stator voltages by 90° (since the stator transient inductances are dominant), and it follows from the stator voltage equations (in the stationary reference frame) that the space vector of the stator currents in the stationary reference frame can be expressed as

$$\bar{i}_{si}=I_{sio}\exp(j\omega_i t)+I_{si1}\exp[j(2\mu-\omega_i t)].$$

This current response can be measured.

It can be seen that the first term in the current response is independent of the angle μ, thus in practice the second term can be used to estimate μ. For completeness it should be noted that the amplitudes of the two current components are

$$I_{sio}=(U_{si}/\omega_i)L'_1/(L'^2_1-\Delta L'^2)$$

$$I_{si1}=(U_{si}/\omega_i)\Delta L'/(L'^2_1-\Delta L'^2).$$

It follows that in the absence of cross-saturation (no saliency) $\Delta L'=0$ and it is not possible to estimate the second term in the expression of the currents (since $I_{si1}=0$). Furthermore, the first term is a direct measure of the saliency present; it characterizes the average stator transient inductance.

By resolving the stator-current space vector ($\bar{i}_{si}$) into its real- and imaginary-axis components, i_{sDi} and i_{sQi} are obtained. The angle μ, which is present in the stator-current components, can then be extracted in a number of ways. One possibility is to use a demodulation scheme involving heterodyning [Jansen and Lorenz 1995*a*], where the direct-axis stator current (i_{sDi}) is multiplied by

$\sin(2\hat{\mu}-\omega_i t)$, and the quadrature-axis stator current (i_{sQi}) is multiplied by $\cos(2\hat{\mu}-\omega_i t)$, where $\hat{\mu}$ is the estimated saliency angle (in electrical radians), and then the difference of these two signals is obtained:

$$\varepsilon = i_{sQi}\cos(2\hat{\mu}-\omega_i t) - i_{sDi}\sin(2\hat{\mu}-\omega_i t).$$

When the expressions for the direct- and quadrature-axis stator currents are substituted into this expression, then the newly obtained expression is $\varepsilon = I_{si0}\sin[2(\omega_i t-\hat{\mu})]+I_{si1}\sin[2(\mu-\hat{\mu})]$. The second term of this contains the spatial position information, and approaches zero as $\hat{\mu}\rightarrow\mu$ (where $\hat{\mu}$ is the estimated and μ is the actual saliency position). The first term can be removed by a low-pass filter. The remaining second part (heterodyned and filtered signal, ε_f) is in the form of a linear position error (as $\hat{\mu}\rightarrow\mu$),

$$\varepsilon_f = I_{si1}\sin[2(\mu-\hat{\mu})] \approx 2I_{si1}(\mu-\hat{\mu}).$$

This signal can then be used to drive a controller, described by K_1+K_2/p, where K_1 and K_2 are gains of the controller. It follows that the saliency position is obtained as $\hat{\mu}=\int\hat{\omega}\,dt$, where $\hat{\omega}=(K_1+K_2/p)\varepsilon_f$. By using the saliency position, it is also possible to obtain estimates of the rotor flux linkages. For this purpose a flux model can be used; thus the measured stator-current components of the induction machine are first transformed into the reference frame rotating with the saliency [by using the transformation $\exp(-j\hat{\mu})$], and the obtained transformed stator currents are then used as inputs into the flux model of the machine [e.g. eqns (4.1-25) and (4.1-26)]. The obtained flux-linkage components can then be transformed into their stationary axis components ($\hat{\psi}_{rd}, \hat{\psi}_{rq}$), by using the transformation $\exp(j\hat{\mu})$.

The best machine configurations for saturation-induced saliency tracking appear to be those with open or semi-closed rotor slots and are designed so that main flux saturation has a much greater impact on the stator transient inductance than localized leakage saturation. It is important to note that robust tracking of saturation-induced saliency may require operation at flux levels which are considerably higher than normal or rated. The maximum operational speed is then limited by core loss and/or stator voltage. Field-weakening greatly beyond base speed may then be not possible. The generation of the high-frequency injected stator voltages by the inverter requires the inverter to remain in a PWM or pulse-density modulation mode. It follows that to obtain a wide range operation, including field weakening, the scheme discussed above must be combined with another scheme which is suitable for high-frequency operation.

Another technique has been described recently [Blaschke *et al.* 1996], which allows sensorless vector control at zero flux frequency. With this scheme it is possible to hold a load stationary for a long time even at zero frequency. For this purpose, saturation effects are again utilized to obtain the position of the rotor flux-linkage space vector (namely that the dynamic inductance is not equal to the static magnetizing inductance, i.e. there is magnetic saliency in the induction machine due to saturation). The quantities to be measured are the stator voltages and currents. As shown in Fig. 4.68(c), a special test-current space vector ($\Delta\bar{i}_s$) is

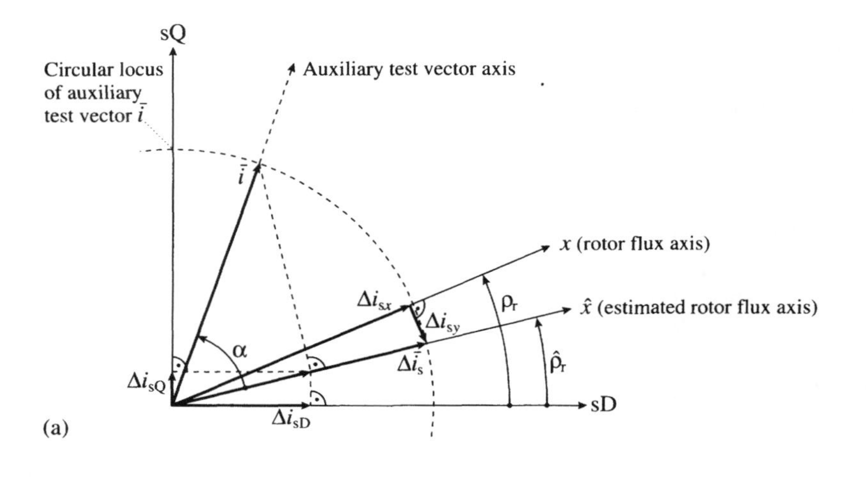

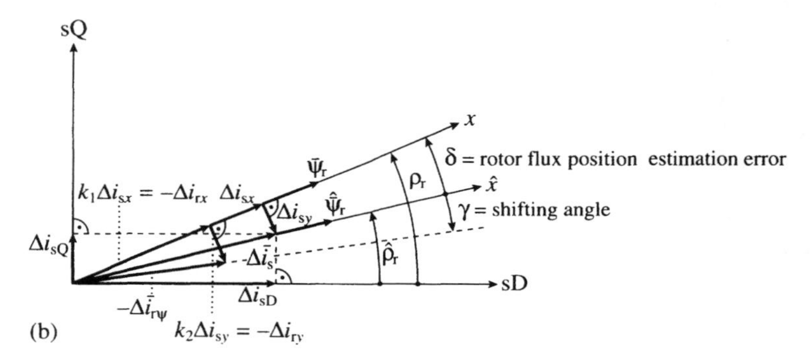

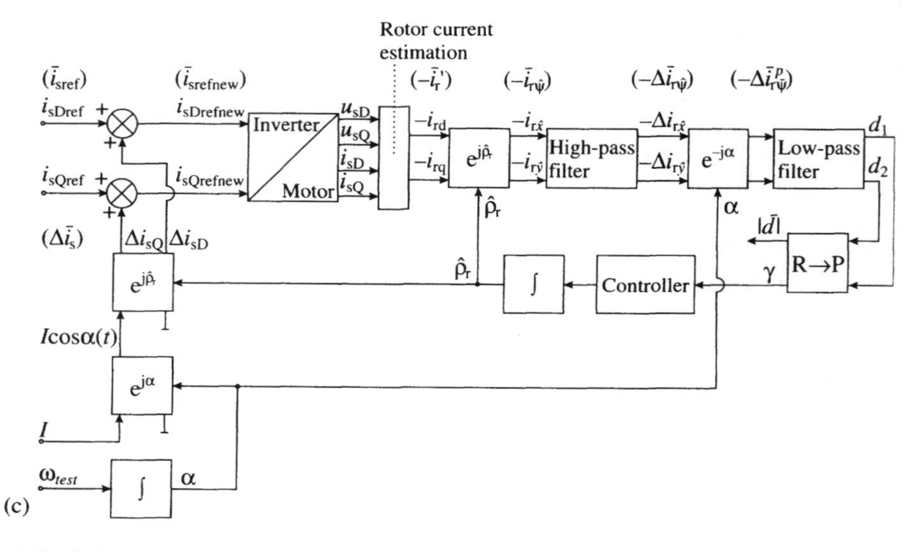

Fig. 4.68. Estimation of rotor flux angle at zero and low frequencies. (a) Vector diagram, production of test vector; (b) vector diagram, effect of saturation on rotor currents; (c) basic estimation scheme.

added to the reference stator-current space vector ($\bar{i}_{sref}$), where the test-current space vector is pulsating in parallel with the direction of the estimated rotor-flux vector axis. It also follows that the angle between the test-current space vector and the direct-axis of the stationary reference frame is equal to the position of the estimated rotor flux-linkage space vector ($\hat{\rho}_r$). Due to the special selection of the test-current space vector, it has no influence on the produced electromagnetic torque. As shown in Fig. 4.68(c), due to the test currents, modified stator reference currents are obtained; these can be described by $\bar{i}_{srefnew} = \bar{i}_{sref} + \Delta\bar{i}_s$, where $\Delta\bar{i}_s = [I\cos\alpha(t)]\exp(j\hat{\rho}_r)$, and $\hat{\rho}_r$ is the estimated position of the rotor flux-linkage space vector (with respect to the real axis of the stationary reference frame). The transformation $\exp(j\hat{\rho}_r)$ is required to ensure that the test-current space vector in the stationary reference frame is coaxial with the real axis of the estimated rotor flux-linkage space vector (see also Fig. 4.68(b)). In the expression above, I is the amplitude of the test current (it is also shown in Fig. 4.68(c), since it is one of the inputs) and the angle $\alpha(t)$ is obtained by integration of a constant test frequency, ω_{test}, since $\omega_{test} = d\alpha/dt$ holds (see also Fig. 4.68(c)). The stator-current test signal, $I\cos\alpha(t)$, can be considered as a projection of an auxiliary test-current vector $\bar{i}$ (which rotates at the speed of ω_{test}) on the real axis of the estimated rotor flux-linkage space vector, and this projection moves along a straight curve when the auxiliary vector rotates. This is shown in Fig. 4.68(a). In this way, the test vector $\Delta\bar{i}_s$ is indeed pulsating in the direction ($\hat{x}$) of the estimated rotor flux-linkage space vector ($\hat{\bar{\psi}}_r$).

It is a very important feature of the scheme that due to magnetic saturation, the rotor-current space vector produced by the test stator-current space vector ($-\Delta\bar{i}_{r\psi}$) is displaced by an angle (γ) with respect to the test-current space vector (see also Fig. 4.68(b)); this is now discussed. This follows from the fact that by assuming fast and small variations of the stator currents, it can be shown by using eqns (4.1-5) and (6.1-18) that

$$\Delta\bar{i}_{r\psi} = \Delta i_{rx} + j\Delta i_{ry},$$

where

$$\Delta i_{rx} = -[1/(1 + L_{rl}/L)]\Delta i_{sx} = k_1\Delta i_{sx}$$

$$\Delta i_{ry} = -[1/(1 + L_{rl}/L_m)]\Delta i_{sy} = k_2\Delta i_{sy},$$

where k_1 and k_2 are saturation-dependent gains, L_{rl} is the rotor leakage inductance, and L and L_m are the dynamic and static inductances respectively (L is the tangent slope and L_m is the static (chord) slope of the magnetizing curve as discussed in Chapter 6). Under saturated conditions $L < L_m$, and thus $k_2 > k_1$ and therefore the test motion of the stator-current space vector ($\Delta\bar{i}_s$) is transferred into the motion of the rotor-current space vector without any delay, and the current transfer in the x-axis (axis coaxial with the rotor flux-linkage space vector) takes place with smaller gain than current transfer in the y-axis (axis in space-quadrature to the rotor flux-linkage space vector). In other words, under saturated conditions the space vector $-\Delta\bar{i}_{r\psi}$ is displaced from the space vector $\Delta\bar{i}_s$

by the shifting angle γ as shown in Fig. 4.68(b). In a saturated machine, it is this effect which causes the shifting angle γ to be different from zero when the estimated rotor flux position ($\hat{\rho}_r$) is different from the actual rotor flux position (ρ_r), i.e. when the error angle is $\delta = \hat{\rho}_r - \hat{\rho}_r \neq 0$. It is the angle γ which is used to obtain the position of the rotor flux-linkage space vector. It is important to note that if $\rho_r = \hat{\rho}_r$ ($\delta = 0$) then $\gamma = 0$ and when the shifting angle is not zero, in general, it has an opposite direction to the error angle. It follows from the above that, to obtain the shifting angle, the negative rotor-current components $-\Delta i_{r\hat{x}}$, $-\Delta i_{r\hat{y}}$ due to the test stator-current space vector have to be known in the estimated-rotor-flux-oriented reference frame. The estimation of the required rotor currents is now discussed briefly.

The current components $-\Delta i_{r\hat{x}}$ and $-\Delta i_{r\hat{y}}$, which are also shown in Fig. 4.68(c), can be obtained in a few steps. As shown in Fig. 4.68(c), first the negative rotor-current space vector expressed in the stationary reference frame, $-\bar{i}_r' = -(i_{rd} + j i_{rq})$, is monitored and this is then transformed into the negative rotor-current space vector in the estimated-rotor-flux-oriented reference frame $-\bar{i}_{r\hat{\psi}} = -(i_{r\hat{x}} + j i_{r\hat{y}}) = -\bar{i}_r' \exp(-j\hat{\rho}_r)$. This transformation is then followed by high-pass filtering, which yields the pulsating negative rotor-current space vector (in the estimated-rotor-flux-oriented reference frame) due to the test stator-current space vector $-\Delta\bar{i}_{r\hat{\psi}} = -(\Delta i_{r\hat{x}} + j\Delta i_{r\hat{y}})$. This pulsating current vector can be described by two counter-rotating vectors, which have the same amplitude. The component which rotates at the constant speed of ω_{test} is the positive-sequence system and the other, which rotates at the $-\omega_{test}$ speed, corresponds to a negative-sequence system. As shown in Fig. 4.68(c), the positive-sequence current vector $-\Delta\bar{i}_{r\hat{\psi}}^{p}$ is obtained by performing the transformation $-\Delta\bar{i}_{r\hat{\psi}} \exp(-j\alpha)$ and then by low-pass filtering of this signal. The resulting vector is a stationary vector, $\bar{d} = d_1 + jd_2$, which is displaced from the axis of the auxiliary test-current vector ($\bar{i}$) by angle γ, and thus γ can be obtained, e.g. by considering $\gamma = \tan^{-1}(d_2/d_1)$. This shifting angle is then input into a controller on the output of which the derivative of the estimated rotor flux-linkage position is obtained, this is then integrated to yield $\hat{\rho}_r$. The estimated $\hat{\rho}_r$ is changed until $\gamma = 0$ is obtained (and in this case $\hat{\rho}_r = \rho_r$).

It has been discussed above that the scheme requires the monitored values of the rotor currents in the stationary reference frame. However, these can be obtained from the measured stator voltages and currents by considering the stator voltage equation in the stationary reference frame. The stator voltage equation is simplified by considering a high test frequency and also that the speed of the rotor flux-linkage space vector, $\omega_{mr} = d\rho_r/dt$, is low. The details of the rotor-current estimation circuit shown in Fig. 4.77(c) can be found in the literature, where the technique has been verified in a vector-controlled induction machine, fed by a current-controlled PWM inverter [Blaschke *et al.* 1996]. The test results reported prove that the scheme can work at exactly zero flux frequency, i.e. a loaded machine was running in a stable manner at exactly zero frequency for a long period.

Finally it should be noted that instead of utilizing saturation effects, it is also possible to estimate the rotor position of the induction machine by tracking of high-frequency saliency created by spatial modulation of the rotor leakage inductance. This is briefly discussed in the next section.

4.5.3.3.4 Estimation utilizing saliency introduced by special rotor construction

For the estimation of the rotor position and rotor speed even at zero and low speeds, it is also possible to use saliency effects, which are intentionally created by using special rotor construction, where spatial modulation of the rotor leakage inductance is created. This can be achieved, for example, by periodically varying the rotor-slot opening widths or by varying the depths of the rotor-slot openings, etc. [Jansen and Lorenz 1995b]. The technique resembles the one described in the previous section, but the main difference is that in the present case there exists a deliberately introduced saliency due to physical asymmetry, whilst in the previous case magnetic saliency existed due to saturation. It should, however, be noted that the introduced rotor saliency is symmetric about each pole, i.e. it is 90 mechanical degrees for a four-pole machine. In order to be able to track the saliency, similarly to that discussed in the previous section, symmetrical three-phase high-frequency voltages are injected into the stator (the amplitude and angular frequency of these are U_{si} and ω_i respectively). First, it is shown below that when special (asymmetrical) rotor constructions are used, the stator transient inductances due to asymmetry are position dependent. It is this position dependency of the inductances which causes position-dependent current responses, when the stator is supplied by high-frequency voltages. By measuring these stator currents, it is then possible to extract the information on the rotor position.

Due to the special rotor construction, the rotor leakage inductances (L_{rld}, L_{rlq}) in the direct and quadrature axes of the rotor reference frame are different. However, when these are transformed into the stator reference frame, rotor-position (θ_r)-dependent inductances,

$$L_{rlD}=L_1+\Delta L\cos(2\theta_r) \qquad L_{rlQ}=L_1-\Delta L\cos(2\theta_r) \qquad L_{rlDQ}=L_{rlQD}=\Delta L\sin(2\theta_r)$$

are obtained, where $L_1=(L_{rld}+L_{rlq})/2$ and $\Delta L=(L_{rld}-L_{rlq})/2$. A simple proof of this can be obtained e.g. by performing the transformation of the rotor leakage-flux space vector expressed in the rotor reference frame, $\bar{\psi}_{rl}=L_{rld}i_{rd}+jL_{rlq}i_{rq}$, into the rotor flux-linkage space vector expressed in the stationary reference frame, $\bar{\psi}'_{rl}=\bar{\psi}_{rl}\exp(j\theta_r)$, and then by resolving the obtained flux linkages into their real- and imaginary-axis components. For this purpose the expressions of the rotor-current space vector in the rotor reference frame, $\bar{i}_r=i_{rd}+ji_{rq}$, and also in the stationary reference frame, $\bar{i}'_r=\bar{i}_r\exp(j\theta_r)$, are introduced. Thus

$$\begin{aligned}\bar{\psi}'_{rl}&=\bar{\psi}_{rl}\exp(j\theta_r)=(L_{rld}i_{rd}+jL_{rlq}i_{rq})\exp(j\theta_r)\\&=[(L_{rld}+L_{rlq})/2]\bar{i}_r\exp(j\theta_r)+[(L_{rld}-L_{rlq})/2]\bar{i}_r^*\exp(j\theta_r)\\&=L_1\bar{i}'_r+\Delta L\bar{i}_r'^*\exp(2j\theta_r)\end{aligned}$$

is obtained, and resolution into its components yields the inductances defined above. As expected, the cross-coupling inductances ($L_{rlDQ}=L_{rlQD}$) are zero, if $L_{\mathrm{rld}}=L_{\mathrm{rlq}}$. As a consequence of this rotor leakage asymmetry, four stator transient inductances can be defined: the direct-axis transient inductance, $L'_{\mathrm{sD}}=L_{\mathrm{sl}}+L_{\mathrm{rlD}}=L'_1+\Delta L'\cos(2\theta_{\mathrm{r}})$, the quadrature-axis transient inductance $L'_{\mathrm{sQ}}=L_{\mathrm{sl}}+L_{\mathrm{rlQ}}=L'_1-\Delta L'\cos(2\theta_{\mathrm{r}})$, and also the cross-coupling transient inductances $L'_{\mathrm{DQ}}=L'_{\mathrm{QD}}=\Delta L'\sin(2\mu)$. In these expressions $L'_1=L_1$ and $\Delta L=\Delta L'$, where as shown above $L_1=(L_{\mathrm{rld}}+L_{\mathrm{rlq}})/2$ and $\Delta L=(L_{\mathrm{rld}}-L_{\mathrm{rlq}})/2$.

When high-frequency stator voltages ($\bar{u}_{\mathrm{si}}$) are injected into the stator of the induction machine, it follows from the stator voltage equation of the induction machine in the stationary reference frame that $\bar{u}_{\mathrm{si}}\approx \mathrm{d}\bar{\psi}_{\mathrm{s}}/\mathrm{d}t$, where the stator flux-linkage component is mainly a leakage flux component. This contains the appropriate stator transient inductances (sum of stator and rotor leakage inductances). It follows that the stator currents due to the injected stator voltages can be formally expressed in the same way as in the previous section. Thus the current response to the injected stator voltages is obtained as

$$\bar{i}_{\mathrm{si}}=I_{\mathrm{si0}}\exp(\mathrm{j}\omega_{\mathrm{i}}t)+I_{\mathrm{si1}}\exp[\mathrm{j}(2\theta_{\mathrm{r}}-\omega_{\mathrm{i}}t)],$$

where

$$I_{\mathrm{si0}}=(U_{\mathrm{si}}/\omega_{\mathrm{i}})(L_{\mathrm{sl}}+L'_1)/[(L_{\mathrm{sl}}+L'_1)^2-\Delta L'^2],$$

$$I_{\mathrm{si1}}=(U_{\mathrm{si}}/\omega_{\mathrm{i}})\Delta L'/[(L_{\mathrm{sl}}+L'_1)^2-\Delta L'^2].$$

The second component of the stator current contains the rotor angle, and this can be used to estimate the rotor position, by using the same technique as discussed in the previous section.

It follows that the signal

$$\varepsilon=(i_{\mathrm{sQi}})\cos(2\hat{\theta}_{\mathrm{r}}-\omega_{\mathrm{i}}t)-(i_{\mathrm{sDi}})\sin(2\hat{\theta}_{\mathrm{r}}-\omega_{\mathrm{i}}t)$$

is first obtained, which can be expressed as

$$\varepsilon=I_{\mathrm{si0}}\sin[2(\omega_{\mathrm{i}}t-\hat{\theta}_{\mathrm{r}})]+I_{\mathrm{si1}}\sin[2(\theta_{\mathrm{r}}-\hat{\theta}_{\mathrm{r}})].$$

The second term contains the rotor position information, and approaches zero as $\hat{\theta}_{\mathrm{r}}\rightarrow\theta_{\mathrm{r}}$ (where $\hat{\theta}_{\mathrm{r}}$ is the estimated and θ_{r} is the actual rotor position). The first term can be removed by a low-pass filter. The remaining second part (heterodyned and filtered signal) is in the form of a linear position error (as $\hat{\theta}_{\mathrm{r}}\rightarrow\theta_{\mathrm{r}}$), thus $\varepsilon_{\mathrm{f}}=I_{\mathrm{si1}}\sin[2(\theta_{\mathrm{r}}-\hat{\theta}_{\mathrm{r}})]\approx 2I_{\mathrm{si1}}(\theta_{\mathrm{r}}-\hat{\theta}_{\mathrm{r}})$. This can then be used to drive a controller, K_1+K_2/p, where K_1 and K_2 are gains of the controller, thus the rotor position is obtained as $\hat{\theta}_{\mathrm{r}}=\int\hat{\omega}_{\mathrm{r}}\,\mathrm{d}t$, where $\hat{\omega}_{\mathrm{r}}=(K_1+K_2/p)\varepsilon_{\mathrm{f}}$ is the estimated rotor speed.

As discussed in the previous section, saturation causes magnetic saliencies and as a consequence, an extra term will be added to $I_{\mathrm{si1}}\sin[2(\theta_{\mathrm{r}}-\hat{\theta}_{\mathrm{r}})]$. However, this extra term can be avoided by appropriate machine design. For example, the effects of leakage saturation can be avoided in the rotor by selecting minimal slot-opening widths.

Finally it should be noted that variation of the rotor slot-opening width may also influence adversely the magnetizing inductance and may result in torque pulsations. If the depths of the rotor slot openings are varied in order to create a spatial modulation of the rotor leakage inductance, then the magnetizing inductance is almost unchanged, but the rotor lamination asymmetry can be undesirable for a manufacturer. The rotor leakage inductance modulation can also be achieved by a variation in rotor slot opening fill, but this can also have some adverse effects.

4.5.3.4 Model reference adaptive systems (MRAS)

4.5.3.4.1 General introduction

In Section 4.5.3.1 various open-loop speed and flux-linkage estimators have been discussed. These have utilized the stator and rotor voltage equations of the induction machine. However, the accuracy of these open-loop observers depends strongly on the machine parameters. In closed-loop estimators the accuracy can be increased. Five rotor speed observers using the Model Reference Adaptive System (MRAS) are described below. In a MRAS system, some state variables, x_d, x_q (e.g. rotor flux-linkage components, ψ_{rd}, ψ_{rq}, or back e.m.f. components, e_d, e_q, etc.) of the induction machine (which are obtained by using measured quantities, e.g. stator voltages and currents) are estimated in a reference model and are then compared with state variables $\hat{x}_d$, $\hat{x}_q$ estimated by using an adaptive model. The difference between these state variables is then used in an adaptation mechanism, which outputs the estimated value of the rotor speed ($\hat{\omega}_r$) and adjusts the adaptive model until satisfactory performance is obtained. Such a scheme is shown in Fig. 4.69(a), where the compact space-vector notation is used. However, Fig. 4.69(b) corresponds to an actual implementation, and here the components of the space vectors are shown.

The appropriate adaptation mechanism can be derived by using Popov's criterion of hyperstability. This results in a stable and quick response system, where the differences between the state-variables of the reference model and adaptive model (state errors) are manipulated into a speed tuning signal (ε), which is then an input into a PI-type of controller (shown in Fig. 4.69(c)), which outputs the estimated rotor speed. Four schemes will be discussed in the following sections; these use the speed tuning signals $\varepsilon_\omega=\mathrm{Im}(\bar{\psi}'_r\hat{\bar{\psi}}'^*_r)$, $\varepsilon_e=\mathrm{Im}(\bar{e}\hat{\bar{e}}^*)$, $\varepsilon_{\Delta e}=\mathrm{Im}(\Delta\bar{e}\bar{i}_s^*)$ and $\varepsilon_{\Delta e'}=\mathrm{Im}(\Delta\bar{e}p\bar{i}_s^*)$ respectively, and are indicated in Fig. 4.69(c). In these expressions $\bar{\psi}'_r$ and $\bar{i}_s$ denote the rotor flux-linkage and stator-current space vectors respectively in the stationary reference frame, $\bar{e}$ denotes the back-e.m.f. space vector also in the stationary reference frame $\bar{e}=(L_m/L_r)\,\mathrm{d}\bar{\psi}'_r/\mathrm{d}t$, and finally the error back-e.m.f. space vector is defined as $\Delta\bar{e}=\bar{e}-\hat{\bar{e}}$. The symbol ^ denotes the quantities estimated by the adaptive model. In addition to these four schemes, artificial-intelligence-assisted MRAS speed estimators are also discussed below in Section 4.5.3.4.6 and also in Section 4.5.3.6, estimators which do not contain any mathematical adaptive model, and the adaptation mechanism is incorporated into the tuning of the appropriate artificial-intelligence-based network (which can be a neural network, a fuzzy-neural network, etc.).

To improve the performance of the observers described, various practical techniques are also discussed which avoid the use of pure integrators. One of the schemes described is robust to stator resistance and rotor resistance variations,

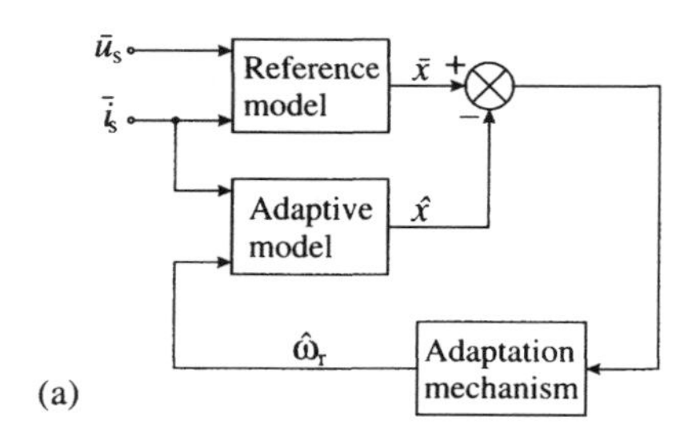

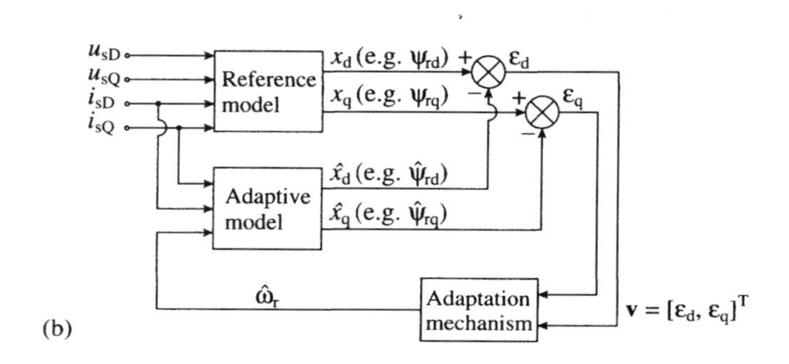

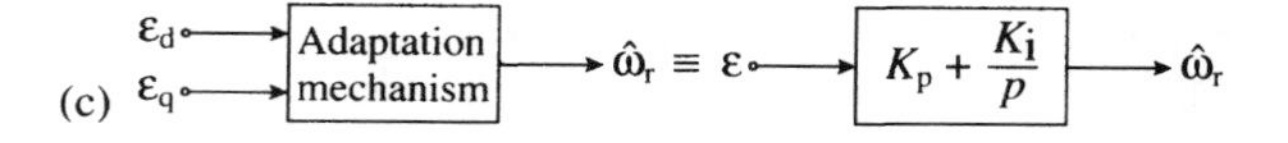

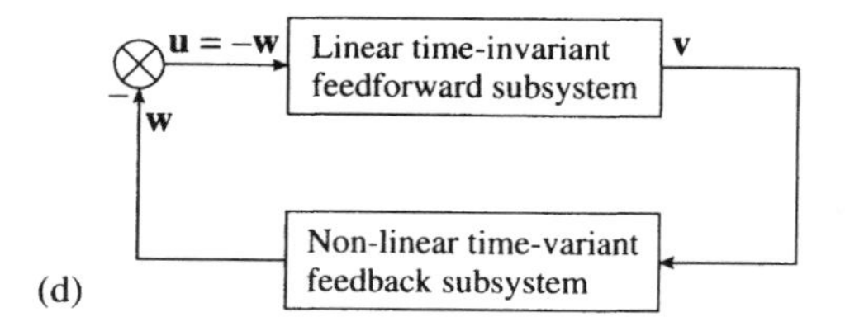

Fig. 4.69. MRAS-based speed estimator scheme. (a) Basic scheme (using space-vector notation); (b) basic scheme (using space-vector components); (c) adaptation mechanism; speed tuning signal

$$\varepsilon = \begin{cases} \varepsilon_\omega = \mathrm{Im}(\bar{\psi}'_r \hat{\bar{\psi}}'^*_r) & \text{where } \bar{\psi}'_r = \psi_{rd} + j\psi_{rq} \\ \varepsilon_e = \mathrm{Im}(\bar{e}\hat{\bar{e}}^*) & \text{where } \bar{e} = e_d + je_q = \dfrac{L_m}{L_r}\psi_{rd} + j\dfrac{L_m}{L_r}\psi_{rq} \\ \varepsilon_{\Delta e} = \mathrm{Im}(\Delta\bar{e}\bar{i}_s^*) & \text{where } \Delta\bar{e} = \bar{e} - \hat{\bar{e}} = (e_d - \hat{e}_d) + j(e_q - \hat{e}_q); \\ \varepsilon_{\Delta e'} = \mathrm{Im}(\Delta\bar{e}p\bar{i}_s^*) & \end{cases}$$

(d) equivalent non-linear feedback system.

and can even be used at very low speeds, e.g. 0.3 Hz (but not zero speed). All the observers described below use monitored stator currents and stator voltages, but in a voltage-source inverter-fed drive, it is not necessary to monitor the stator voltages, since it is possible to reconstruct them by using the inverter switching states and also the monitored value of the d.c. link voltage (see also Section 3.1.3.2.1). An artificial-intelligence-based MRAS speed estimator seems to offer the most satisfactory performance even at very low speeds.

4.5.3.4.2 Application of Popov's hyperstability theorem and integral inequality

This section is only given here for completeness and contains a short description of the selection of the appropriate adaptation mechanism, proves why there is a PI controller in the schemes described below, and also shows the form of the speed tuning signal to be used. However, this section can be skipped by those who do not require a proof of the fact that the application of Popov's theorem yields the various MRAS systems described, e.g. that shown in Fig. 4.69, which contains a PI controller and the appropriate speed tuning signal.

In general, a model reference adaptive speed estimator system can be represented by an equivalent non-linear feedback system which comprises a feedforward time-invariant linear subsystem and a feedback non-linear time-varying subsystem. This is shown in Fig. 4.69(d).

In Fig. 4.69(d) the input to the linear time-variant system is $\mathbf{u}$ (which contains the stator voltages and currents), its output is $\mathbf{v}$, which is the speed tuning signal (generalized error), $\mathbf{v}=[\varepsilon_d, \varepsilon_q]^T$. The output of the non-linear time-variant system is $\mathbf{w}$, and $\mathbf{u}=-\mathbf{w}$. The rotor speed estimation algorithm (adaptation mechanism) is chosen according to Popov's hyperstability theory, whereby the transfer function matrix of the linear time-invariant subsystem must be strictly positive real and the non-linear time-varying feedback subsystem satisfies Popov's integral inequality, according to which $\int \mathbf{v}^T\mathbf{w}\,\mathrm{d}t \geqslant 0$ in the time interval $[0, t_1]$ for all $t_1 \geqslant 0$. Thus to obtain the adaptation mechanism, first the transfer function $\mathbf{F}(s)$ of the linear time-invariant feedforward subsystem has to be obtained. It can be shown by lengthy calculations that in all the schemes described below this is strictly positive real.

A possible proof uses the state-variable form of the error equation, $\mathrm{d}\mathbf{v}/\mathrm{d}t=\mathbf{A}\mathbf{v}-\mathbf{w}$, which is obtained by subtracting the state-variable equations of the adjustable model from the state-variable equations of the reference model. The equations of the adjustable model (rotor model) are the rotor voltage equations in the stationary reference frame, which can be obtained from eqn (4.1-27) by eliminating the rotor-current space vector. Thus

$$\hat{\bar{\psi}}'_r=\int\left\{\left[-\frac{1}{T_r}+\mathrm{j}\omega_r\right]\hat{\bar{\psi}}'_r+\frac{L_m}{T_r}\bar{i}_s\right\}\mathrm{d}t$$

is obtained, which is then resolved into its two-axis components. The reference model (stator model) equations are the two-axis stator voltage equations in the stationary reference frame, which have been defined by eqns (4.5-22) and (4.5-23), and the space-vector form of this is

$$\bar{\psi}_r' = \frac{L_r}{L_m}\left[\int(\bar{u}_s - R_s\bar{i}_s)dt - L_s'\bar{i}_s\right].$$

In the resulting state-variable form of the error equations, the state matrix is **A**, where

$$\mathbf{A} = \begin{bmatrix} -1/T_r & -\omega_r \\ \omega_r & -1/T_r \end{bmatrix}.$$

The feedforward path transfer matrix of the linear time-invariant subsystem shown in Fig. 4.69(d) is $\mathbf{F}(s) = [s\mathbf{I} - \mathbf{A}]^{-1}$, where $\mathbf{I}$ is an identity matrix. It follows from the derivation of the error state equation that $\mathbf{w} = [\hat{\omega}_r - \omega_r][-\hat{x}_q x_d]^T$ (where $\hat{x}_d$ and $\hat{x}_q$ are the states estimated by the adaptive model), thus when $\mathbf{w}$ is substituted into Popov's integral inequality, $\int \mathbf{v}^T\mathbf{w}\,dt \geq 0$, it can be shown that this inequality can be satisfied by letting $\hat{\omega}_r = (K_p + K_i/p)\varepsilon$. In this equation $1/p$ represents an integrator and ε is the appropriate speed tuning signal. In general, the state variables in the reference and adaptive models are x_d, x_q and $\hat{x}_d$, $\hat{x}_q$ respectively and when these are the rotor flux linkages ψ_{rd}, ψ_{rq}, $\hat{\psi}_{rd}$, $\hat{\psi}_{rq}$, then the speed tuning signal is $\text{Im}(\bar{\psi}_r'\hat{\bar{\psi}}_r'^*)$, where the asterisk denotes the complex conjugate. This signal is used in the first scheme discussed below, thus $\varepsilon_\omega = \text{Im}(\bar{\psi}_r'\hat{\bar{\psi}}_r'^*)$. In the second scheme to be discussed it has a similar form, $\varepsilon_\varepsilon = \text{Im}(\bar{e}\hat{\bar{e}}^*)$, etc. It can be seen that when a specific state variable is used (on the outputs of the reference and adaptive models), then a corresponding speed tuning signal of a specific form is obtained by using Popov's integral inequality. It has been discussed above that when the rotor speed to be estimated is changed in the adaptive model in such a way that the difference between the output of the reference model and the adaptive model is zero, then the estimated rotor speed is equal to the actual rotor speed. The error signal actuates the rotor-speed identification algorithm, which makes this error converge asymptotically to zero. The physical reason for the integrator (in the PI controller) is that this ensures that the error converges asymptotically to zero.

4.5.3.4.3 *Scheme 1: speed tuning signal is* $\varepsilon_\omega = \text{Im}(\bar{\psi}_r'\hat{\bar{\psi}}_r'^*)$

As shown in the previous section, it is possible to estimate the rotor speed by using two estimators (a reference-model-based estimator and an adaptive-model-based one), which independently estimate the rotor flux-linkage components in the stator reference frame (ψ_{rd}, ψ_{rq}), and by using the difference between these flux-linkage estimates to drive the speed of the adaptive model to that of the actual speed. The expressions for the rotor flux linkages in the stationary reference frame can be obtained by using the stator voltage equations of the induction machine (in the stationary reference frame). These give eqns (4.5-22) and (4.5-23), which are now rearranged for the rotor flux linkages:

$$\psi_{rd} = \frac{L_r}{L_m}\left[\int(u_{sD} - R_s i_{sD})\,dt - L_s' i_{sD}\right] \qquad (4.5\text{-}83)$$

$$\psi_{rq} = \frac{L_r}{L_m}\left[\int (u_{sQ} - R_s i_{sQ})\,dt - L'_s i_{sQ}\right], \tag{4.5-84}$$

where L'_s is the stator transient inductance. These two equations represent a so-called stator *voltage* model, which does not contain the rotor speed and is therefore a reference model. However, when the rotor voltage equations of the induction machine are expressed in the stationary reference frame, they contain the rotor fluxes and the speed as well. They can be obtained from eqn (4.1-27) and, by eliminating the rotor-current space vector by using $\bar{i}'_r = (\bar{\psi}'_r - L_m \bar{i}_s)/L_r$ or from eqns (4.1-30) and (4.1-31) by considering that $\psi_{rd} = L_m i_{mrd}$, $\psi_{rq} = L_m i_{mrq}$, thus

$$\hat{\psi}_{rd} = \frac{1}{T_r}\int (L_m i_{sD} - \hat{\psi}_{rd} - \omega_r T_r \hat{\psi}_{rq})\,dt \tag{4.5-85}$$

$$\hat{\psi}_{rq} = \frac{1}{T_r}\int (L_m i_{sQ} - \hat{\psi}_{rq} + \omega_r T_r \hat{\psi}_{rd})\,dt, \tag{4.5-86}$$

where T_r is the rotor time constant. These two equations correspond to a *current* model, which contains the rotor speed, and therefore represent the adjustable (adaptive) model. The reference and adaptive models are used to estimate the rotor flux linkages and the angular difference of the outputs of the two estimators $\varepsilon_\omega = \mathrm{Im}(\bar{\psi}'_r \hat{\bar{\psi}}'^*_r) = \psi_{rq}\hat{\psi}_{rd} - \psi_{rd}\hat{\psi}_{rq}$ is used as the speed tuning signal. This tuning signal is the input to a linear controller (PI controller) which outputs the estimated rotor speed as shown in Fig. 4.70.

As discussed in the previous section, the reason for using a PI controller and $\varepsilon_\omega = \mathrm{Im}(\bar{\psi}'_r \hat{\bar{\psi}}'^*_r) = \psi_{rq}\hat{\psi}_{rd} - \psi_{rd}\hat{\psi}_{rq}$ is that this will give a stable non-linear feedback system, and a rigorous proof uses Popov's hyperstability criterion. The PI controller tunes the rotor speed value when the error between the two rotor flux-linkage space vectors is not zero ($\bar{\psi}'_r \neq \hat{\bar{\psi}}'_r$). In other words, when the rotor speed to be estimated ($\hat{\omega}_r$) is changed in the adjustable model in such a way that the difference between the output of the reference model and the output of the

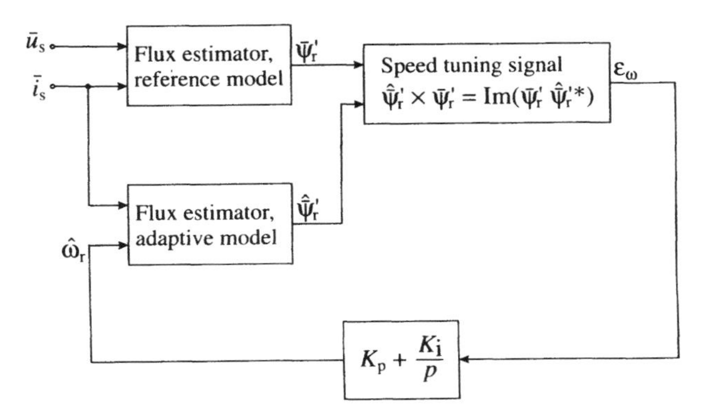

Fig. 4.70. MRAS-based rotor speed observer using rotor flux linkages for the speed tuning signal and employing pure integrators.

adjustable model becomes zero, then the estimated rotor speed is equal to the actual rotor speed (ω_r). The error signal actuates the rotor-speed identification algorithm, which makes this error converge to zero. The algorithm is chosen to give quick and stable response. It should be noted that it is also possible to construct similar MRAS schemes which, however, use the back-e.m.f. components instead of the rotor flux-linkage components, or they can use the components of the power vector ($\bar{u}_s \bar{i}_s^*$), etc. Some aspects of these schemes will be discussed at the end of the present section and it will be shown that they have some advantages (e.g. absence of pure integrator, reduced noise sensitivity, etc.) over the scheme using the rotor flux estimates in the speed tuning signal.

It follows from Fig. 4.70 that the estimated speed can be expressed as

$$\hat{\omega}_r = K_p \varepsilon_\omega + K_i \int \varepsilon_\omega \, dt, \tag{4.5-87}$$

and it can be shown by using the linearized state-variable equations that arbitrary K_p and K_i cannot be used to obtain satisfactory performance. The complete

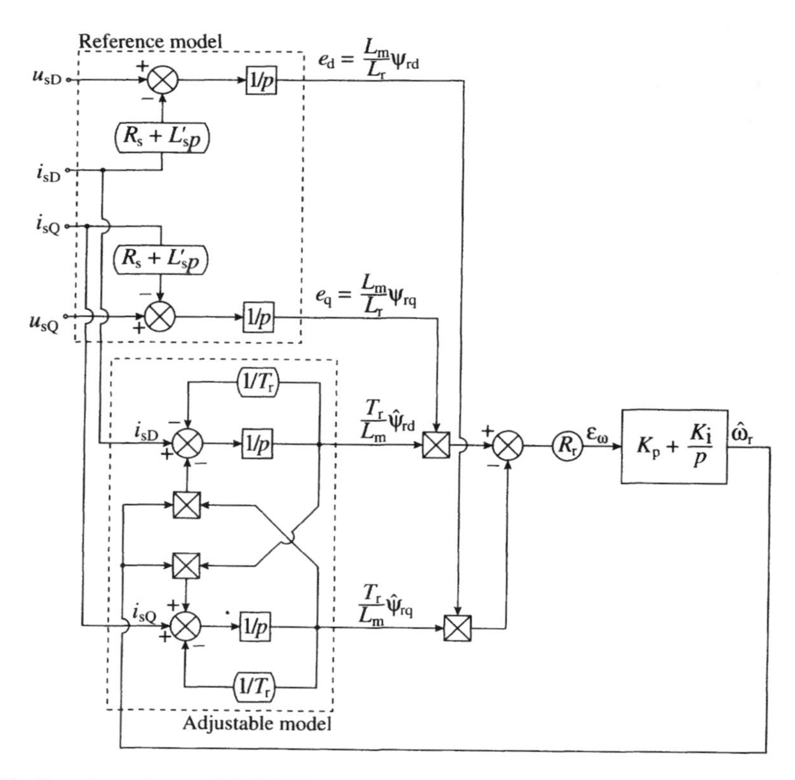

Fig. 4.71. Complete scheme of MRAS-based speed observer using rotor flux linkages for the speed tuning signal.

scheme of the MRAS-based rotor speed observer is shown in Fig. 4.71. It should be noted that in Fig. 4.71 the speed tuning signal ε_ω is multiplied by $1/R_r$, and thus to obtain ε_ω, it is multiplied by R_r. However, this separate multiplication is not required, since R_r could be incorporated in the proportional and integral gain constants of the PI controller. The stator voltages can be obtained from the monitored terminal voltages or they can be obtained in an inverter-fed drive by reconstructing the stator voltages from the inverter switching states and by using the monitored value of the d.c. link voltage (see Sections 3.1.1 and 4.1.1 for further detail).

In practice, the reference model is difficult to implement due to the pure integrators required by eqns (4.5-83) and (4.5-84), which have initial value and drift problems. However, to avoid these problems, in a practical implementation a low-pass filter, with the transfer function $1/(p+1/T)$, can be used instead of a pure integrator. However, since $1/(p+1/T)=(1/p)[p/(p+1/T)]$, thus in the practical system, the reference model (which contains $1/p$) is followed by a high-pass filter $[p/(p+1/T)]$, as shown in Fig. 4.72.

In this case the modified rotor flux linkages of the reference model (ψ'_{rd}, ψ'_{rq}) can be obtained from eqns (4.5-83) and (4.5-84) and thus

$$\left(p+\frac{1}{T}\right)\psi'_{rd}=\frac{L_r}{L_m}(u_{sD}-R_s i_{sD}-L'_s p i_{sD}) \qquad (4.5\text{-}88)$$

$$\left(p+\frac{1}{T}\right)\psi'_{rq}=\frac{L_r}{L_m}(u_{sQ}-R_s i_{sQ}-L'_s p i_{sQ}). \qquad (4.5\text{-}89)$$

Since the output of the modified reference model gives the modified rotor flux linkages ψ'_{rd} and ψ'_{rq}, the adaptive model must also be adjusted to give the corresponding modified values, and therefore the high-pass filter block $[p/(p+1/T)]$ is placed in front of the original adaptive model in Fig. 4.72. In practice the cut-off frequency of the high-pass filter is a few Hertz (e.g. $T=0.05$ s gives $f=1/(2\pi/T)=3.2$ Hz). Below the cut-off frequency, the rotor speed estimation

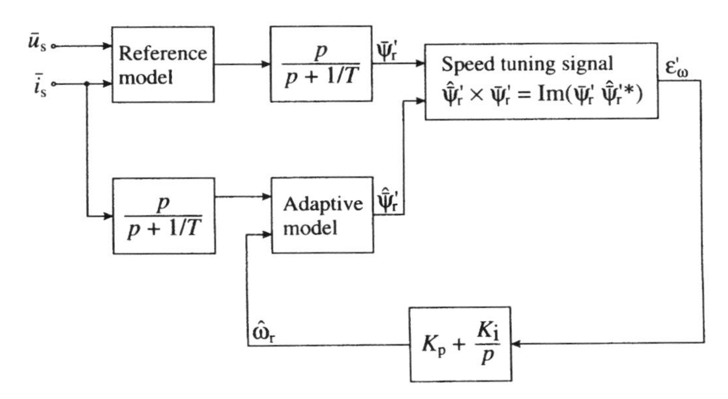

Fig. 4.72. Practical MRAS-based rotor speed observer using rotor flux linkages for the speed tuning signal and avoiding pure integrators.

becomes inaccurate. Furthermore, at low speeds, the stator voltages are small and therefore an accurate value of the stator resistance is required to have a satisfactory response. However, when the MRAS scheme described above is used in a vector-controlled induction motor drive, speed reversal through zero during a fast transient process is possible, but if the drive is operated at zero frequency for more than a few seconds, then speed control is lost due to the incorrect flux-linkage estimation. It should be noted that the problems associated with pure integrators in the reference model can be also avoided by using another flux estimator in the reference model (e.g. see Sections 3.1.3.2 and 4.1.1.4, where such stator flux estimators are also described, which avoid the use of pure integrators, use the monitored stator voltages and currents, and contain a feedback of the flux angle). However, it is also possible to avoid the use of pure integrators by using another speed tuning signal in the MRAS system, which does not require pure integration. Since in the above scheme the integrator was present only because the rotor flux linkage is estimated from the stator flux linkage, $\bar{\psi}'_r=(L_r/L_m)(\bar{\psi}_s-L'_s\bar{i}_s)$, and the stator flux-linkage estimation from the stator voltages and currents requires an integration, it follows that if a signal (e.g. back e.m.f.) proportional to $d\bar{\psi}'_r/dt$ is used (the space vector of the back e.m.f. is $\bar{e}=(L_m/L_r)\,d\bar{\psi}'_r/dt$), then no integration is required. Such a scheme is discussed in the next section.

4.5.3.4.4 Scheme 2: speed tuning signal is $\varepsilon_e=Im(\bar{e}\hat{\bar{e}}^*)$

It has been mentioned above that it is also possible to construct other MRAS schemes which, instead of using the rotor flux-linkage estimates in the speed tuning signal, use the back e.m.f.s or some other quantities, e.g. the components of the power vector ($\bar{u}_s\bar{i}_s^*$), etc. When the back e.m.f.s are used, then the problems associated with the pure integrators in the reference model disappear, since in this case the reference model does not contain any integrator. The equations for the direct- and quadrature-axis back e.m.f.s follow from eqns (4.5-83) and (4.5-84):

$$e_d=\frac{L_m}{L_r}\frac{d\psi_{rd}}{dt}=u_{sD}-R_s i_{sD}-L'_s\frac{di_{sD}}{dt} \tag{4.5-90}$$

$$e_q=\frac{L_m}{L_r}\frac{d\psi_{rq}}{dt}=u_{sQ}-R_s i_{sQ}-L'_s\frac{di_{sQ}}{dt}, \tag{4.5-91}$$

and these are used in a new reference model, shown in Fig. 4.73.

As expected, the component back e.m.f.s e_d and e_q can be obtained without integration. Similarly to scheme described in the previous section (Scheme 1), the direct- and quadrature-axis stator voltages can be obtained from monitored terminal voltages, or in an inverter-fed drive, they can be reconstructed by using the inverter switching states and the monitored d.c. link voltage.

The corresponding back e.m.f. equations for the adaptive model are obtained from eqns (4.5-85) and (4.5-86):

$$\hat{e}_d=\frac{L_m}{L_r}\frac{d\hat{\psi}_{rd}}{dt}=\frac{L_m}{L_r}\frac{(L_m i_{sD}-\psi_{rd}-\omega_r T_r\psi_{rq})}{T_r} \tag{4.5-92}$$

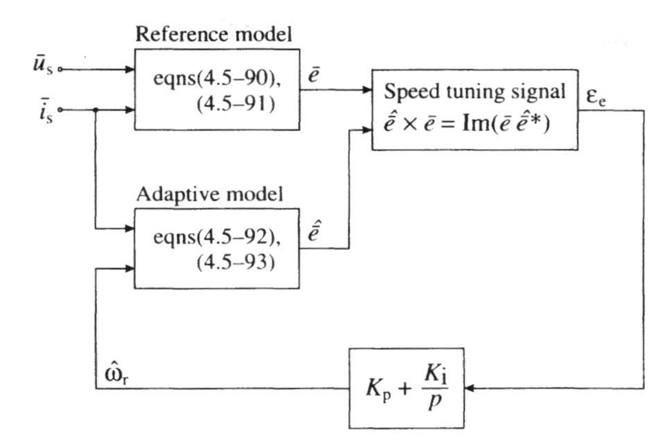

Fig. 4.73. MRAS-based speed observer using back e.m.f.s for the speed tuning signal.

$$\hat{e}_q = \frac{L_m}{L_r}\frac{d\hat{\psi}_{rq}}{dt} = \frac{L_m}{L_r}\frac{(L_m i_{sQ} - \psi_{rq} + \omega_r T_r \psi_{rd})}{T_r}. \tag{4.5-93}$$

Equations (4.5-92) and (4.5-93) are used in the adaptive model shown in Fig. 4.73. In Fig. 4.73 the speed tuning signal is $\varepsilon_e = \mathrm{Im}(\bar{e}\hat{\bar{e}}^*) = \hat{\bar{e}} \times \bar{e} = e_q\hat{e}_d - e_d\hat{e}_q$, which is proportional to the angular difference between the back e.m.f. vectors of the reference and adjustable models respectively, $\bar{e}$, $\hat{\bar{e}}$ ($\times$ denotes the cross-vectorial product). The full scheme is shown in Fig. 4.74.

It should be noted that in Fig. 4.74 the speed tuning signal ε_e is multiplied by L'_m/T_r, which has been assumed to be constant, where $L'_m = L_m^2/L_r$ is the referred value of the magnetizing inductance (this inductance was used in Fig. 4.10(a)). Thus multiplication by the constant L'_m/T_r is used to obtain ε_e, which is then the input to the PI controller. However, this multiplication can be omitted since L'_m/T_r can be incorporated in the proportional and integral gain constants of the PI controller (which become $K'_p = K_p L'_m/T_r$ and $K'_i = K_i L'_m/T_r$).

When the scheme shown in Fig. 4.74 is employed in a speed-sensorless vector-controlled drive, satisfactory performance can be obtained even at low speeds if an accurate value of the stator resistance is used, since the reference model does not contain pure integration. However, the stator resistance varies with temperature, and this effects the stability and performance of the speed observer, especially at very low speeds. A MRAS scheme which is insensitive to stator resistance variation can be obtained by using such a speed tuning signal, which is obtained from a quantity which does not contain the stator resistance. This is discussed in the next section.

4.5.3.4.5 Scheme 3: speed tuning signal is $\varepsilon_{\Delta e} = Im(\Delta\bar{e}\bar{i}_s^*)$

In Schemes 2 and 3 described above, the speed tuning signal is obtained from $\mathrm{Im}(\bar{\psi}'_r\hat{\bar{\psi}}'^*_r)$ and from $\mathrm{Im}(\bar{e}\hat{\bar{e}}^*)$ respectively. In the new scheme (Scheme 3), the speed

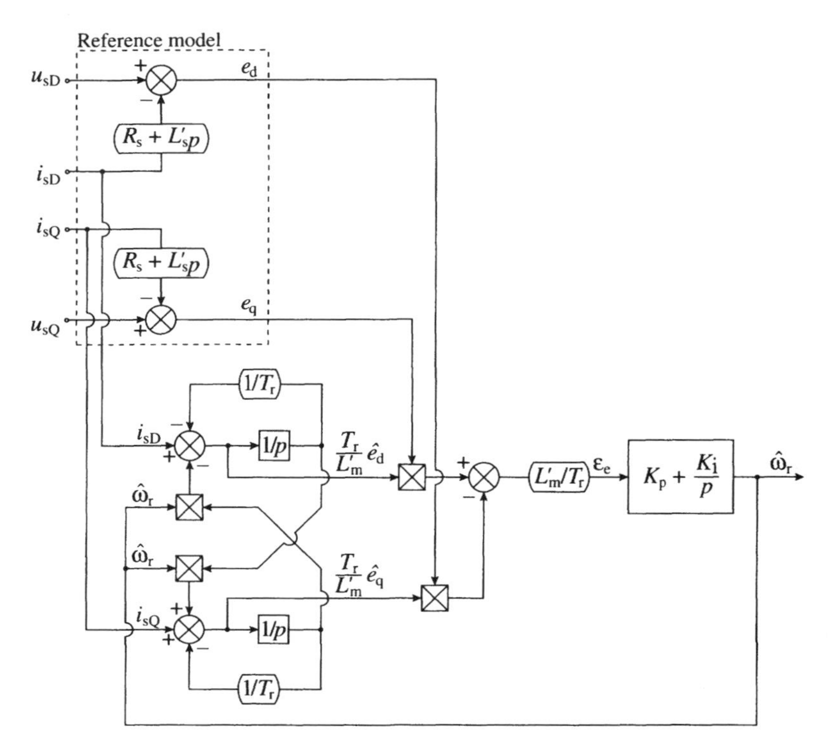

Fig. 4.74. Complete scheme of MRAS-based speed observer using back e.m.f.s for the speed tuning signal.

tuning signal is deliberately chosen to be $\mathrm{Im}(\Delta\bar{e}\bar{i}_s^*)$, where $\Delta\bar{e}=\bar{e}-\hat{\bar{e}}$ and $\bar{e}$, $\hat{\bar{e}}$ are the space vectors of the back e.m.f.s in the reference model and adaptive model respectively. It follows that $\mathrm{Im}(\Delta\bar{e}\bar{i}_s^*)=\bar{i}_s\times\Delta\bar{e}=\bar{i}_s\times\bar{e}-\bar{i}_s\times\hat{\bar{e}}$, and by considering eqns (4.5-90), (4.5-91), and $\bar{e}=e_d+je_q$, $\bar{u}_s=u_{sD}+ju_{sQ}$,$\bar{i}_s=i_{sD}+ji_{sQ}$,

$$y=\bar{i}_s\times\bar{e}=\bar{i}_s\times\left(\bar{u}_s-L'_s\frac{d\bar{i}_s}{dt}\right) \qquad (4.5\text{-}94)$$

is obtained, which is the output of the reference model. It can be seen that this does not contain the stator resistance, and this is why y has been chosen to be a component of the speed tuning signal. In other words, since the stator-voltage space vector ($\bar{u}_s$) is equal to the sum of the stator ohmic voltage drop ($R_s\bar{i}_s$) plus $L'_s d\bar{i}_s/dt$, plus the back e.m.f. $\bar{e}=(L_m/L_r)\,d\bar{\psi}_r/dt$, therefore the vectorial product $\bar{i}_s\times\bar{u}_s$ does not contain the stator resistance and takes the form $\bar{i}_s\times\bar{u}_s=\bar{i}_s\times L'_s d\bar{i}_s/dt+\bar{i}_s\times\bar{e}$, and this gives eqn (4.5-94) as expected. The first term on the right-hand side of eqn (4.5-94) is ($\bar{i}_s\times\bar{u}_s$), the reactive input power. Similarly to the other two schemes discussed in the previous two sections, the stator voltage components u_{sD}, u_{sQ} can be obtained from the monitored line voltages, or in an inverter-fed induction motor drive, they can be reconstructed from the inverter switching states and the monitored value of the d.c. link voltage.

The output of the adaptive model is obtained by considering eqns (4.5-92), (4.5-93) and $\hat{\bar{e}}=e_d+je_q$, as follows:

$$\hat{y}=\bar{i}_s\times\hat{\bar{e}}=\bar{i}_s\times\frac{L_m}{L_r}\left[\frac{L_m}{T_r}\bar{i}_s+\frac{\bar{\psi}_r'(j\omega_r-1)}{T_r}\right]=\frac{L_m}{L_r}\left[\frac{1}{T_r}\bar{\psi}_r'\times\bar{i}_s+\omega_r(\bar{i}_s\times j\bar{\psi}_r')\right]. \quad (4.5\text{-}95)$$

Figure 4.75 shows the schematic of the rotor speed observer using the speed tuning signal $\varepsilon_{\Delta e}=\mathrm{Im}(\Delta\bar{e}\bar{i}_s^*)$. The reference model is represented by eqn (4.5-94) and the adaptive model by eqn (4.5-95).

The component forms of eqns (4.5-94) and (4.5-95) yield

$$y=\bar{i}_s\times\bar{e}=u_{sQ}i_{sD}-u_{sD}i_{sQ}-L_s'\left(i_{sD}\frac{di_{sQ}}{dt}-i_{sQ}\frac{di_{sD}}{dt}\right) \quad (4.5\text{-}96)$$

and

$$\hat{y}=\bar{i}_s\times\hat{\bar{e}}=i_{sD}\hat{e}_q-i_{sQ}\hat{e}_d$$
$$=\frac{L_m}{L_r}\left[\frac{1}{T_r}(\psi_{rd}i_{sQ}-\psi_{rq}i_{sD})+\omega_r(\psi_{rd}i_{sD}+\psi_{rq}i_{sQ})\right] \quad (4.5\text{-}97)$$

respectively. Equations (4.5-96) and (4.5-97) can then be used in a final implementation of the rotor speed observer. When this observer is used in a vector-controlled drive, it is possible to obtain satisfactory performance even at very low speeds. The observer can track the actual rotor speed with a bandwidth that is only limited by noise, so the PI controller gains should be as large as possible. The scheme is insensitive to stator resistance variations and it can be shown that if in a rotor-flux-oriented vector drive, the same T_r is used as in the MRAS-based speed observer, then the drive will be robust to the variation of T_r as well [Peng *et al.* 1994]. Thus rotor-flux alignment is maintained, despite the fact that

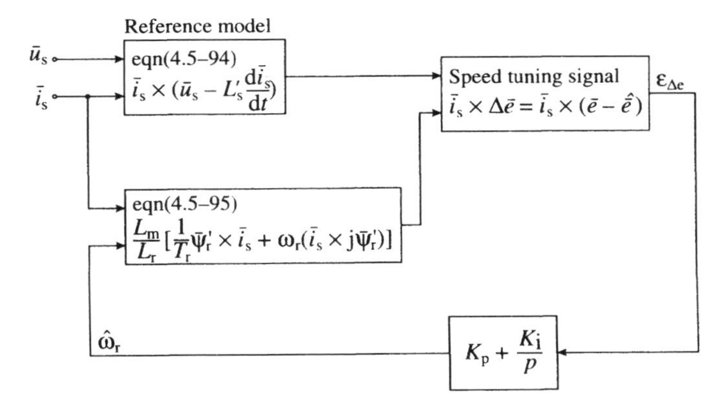

Fig. 4.75. Rotor speed observer using the tuning signal $\mathrm{Im}(\Delta\bar{e}\bar{i}_s^*)$.

an incorrect value of the rotor time constant is used. This is due to the accumulative effects of the errors. This also holds for the other MRAS-based speed observers described above. However, the accuracy of the speed estimation system discussed above depends on the transient stator inductance (L_s') and also on the referred magnetizing inductance (L_m^2/L_r). The latter quantity is not too problematic, since it does not change with the temperature. Furthermore, deviation of T_r from its correct value produces a steady-state error in the estimated speed and this error can become significant at low speed. In general, the scheme can be used at very low speeds as well, e.g. at 0.3 Hz, but not at zero stator frequency. Special care must be taken for the design of this scheme for applications where there are negative step changes in the torque-producing stator current, otherwise it is difficult to keep the speed stable.

Finally it should be noted that by using a simple space-vector model of the induction machine, which takes into account the effects of iron losses, it is possible to estimate the effects of iron losses on the estimated rotor speed. It can then be shown that in the field-weakening range the iron losses may have an important effect on the accuracy of the speed estimation. In a simple space-vector model of the induction machine, the iron losses can be considered in a similar way to that used in the conventional steady-state equivalent circuit of the induction machine. Thus in the model expressed in the stationary reference frame, the resistor taking account of iron losses is connected in parallel with the magnetizing inductance. In a model expressed in the general reference frame rotating at the general speed ω_g, this resistance is connected in parallel with a circuit which contains the sum of the transformer voltage $L_m p \bar{i}_{mg}$ ($p = \mathrm{d}/\mathrm{d}t$ and $\bar{i}_{mg}$ is the magnetizing-current space vector in the general reference frame) and the rotational voltage $\mathrm{j}\omega_g \bar{\psi}_{mg}$ ($\bar{\psi}_{mg}$ is the magnetizing flux-linkage space vector in the general reference frame). For simplicity the magnetizing inductance has been assumed to be constant, but it is possible to use a similar dynamic model of the induction machine which takes account of both the iron losses and saturation effects. This could be used for the analysis and compensation of the speed estimation error caused by iron losses and main flux saturation in a MRAC-based speed-sensorless induction motor drive.

4.5.3.4.6 *Scheme 4: speed tuning signal is* $\varepsilon_{\Delta e'} = \mathrm{Im}(\Delta \bar{e} p \bar{i}_s^*)$

As discussed in the MRAS scheme of the previous section, the presence of the stator transient inductance (L_s') is undesirable, since this affects the accuracy of the estimated rotor speed in the entire speed range. For the purposes of the present scheme it is a goal to eliminate the need for using the stator transient inductance. Furthermore, it is another goal that the MRAS scheme should not require pure integration (similarly to the schemes described previously). It will be shown that it is possible to fulfil these conditions by using a suitable speed-tuning signal, but accurate speed estimation can only be achieved if the stator resistance and rotor time constant are accurately known (e.g. they are adapted on-line).

By examining the stator and rotor voltage space-vector equations of the induction machine, it can be physically deduced that both of these requirements can be fulfilled if the speed tuning signal contains the derivative of the stator currents and is chosen as $\varepsilon_{\Delta e'} = \mathrm{Im}(\Delta \bar{e} p \bar{i}_s^*)$. Similarly to the notation used in the earlier schemes, in this expression $\bar{e}$ is the space vector of the back e.m.f.s, $\bar{e} = (L_m/L_r)\mathrm{d}\bar{\psi}_r'/\mathrm{d}t$, p is the differential operator ($p = \mathrm{d}/\mathrm{d}t$) and $\Delta \bar{e} = \bar{e} - \hat{\bar{e}}$ is the error back-e.m.f. space vector (difference between the back-e.m.f. space vector of the reference model and that of the adaptive model of the MRAS speed estimator scheme). Alternatively, this tuning signal can also be expressed as $\varepsilon_{\Delta e'} = p\bar{i}_s \times \Delta \bar{e}$. It should be noted that by using the expressions for $\bar{e}$ and $\hat{\bar{e}}$ which can be obtained from the stator and rotor voltage equations respectively (both expressed in the stationary reference frame), it is possible to obtain the tuning signal in terms of the direct- and quadrature-axis components of the stator currents, stator voltages, and rotor flux linkages (all expressed in the stationary reference frame). Thus the following expression is obtained

$$\varepsilon_{\Delta e'} = [u_{sQ} p i_{sD} - u_{sD} p i_{sQ} - R_s(i_{sQ} p i_{sD} - i_{sD} p i_{sQ})] - \frac{L_m}{L_r}[\hat{\omega}_r(\psi_{rd} p i_{sD} + \psi_{rq} p i_{sQ}) + (\psi_{rd} p i_{sQ} - \psi_{rq} p i_{sD} + i_{sQ} p i_{sD} - i_{sD} p i_{sQ})/T_r],$$

where $\hat{\omega}_r$ is the estimated speed, which is used in the adjustable model of the MRAS speed-estimation scheme. As expected, this expression does not contain the direct- and quadrature-axis voltage drops across the stator transient inductance. The term within the first square brackets is implemented by the use of the reference model, which uses at its inputs the actual stator voltage and stator current components (u_{sD}, u_{sQ}, i_{sD}, i_{sQ}), but it can be seen that, in contrast to the scheme described in the previous section, it also uses R_s. Thus the success of an accurate speed estimation at low speeds depends on accurate knowledge of the stator resistance. It can also be seen that in space-vector terms this term is equal to $[(p\bar{i}_s) \times \bar{e}]$ or alternatively $[(p\bar{i}_s) \times (p\bar{\psi}_s)] = [(p\bar{i}_s) \times (\bar{u}_s - R_s\bar{i}_s)]$ and obviously $(p\bar{i}_s) \times \bar{i}_s \neq 0$. Therefore it is an advantage that the derivative of the stator flux-linkage space vector is present here and not the stator flux-linkage space vector itself, and thus there is no pure integration involved. Furthermore, it can also be seen that the term within the second square brackets contains the rotor flux-linkage components, which are obtained in the adjustable model, which uses the rotor voltage equations of the induction machine [eqns (4.5-85) and (4.5-86)] together with the monitored stator currents (i_{sD}, i_{sQ}) and the estimated rotor speed ($\hat{\omega}_r$). However, this term also depends on the rotor time constant. In space-vector terms, when this term is multiplied by (L_m/L_r), then it can be expressed as $[(p\bar{i}_s) \times \hat{\bar{e}}]$.

Similarly to all the other MRAS estimation schemes described in the previous sections, by using a rigorous mathematical proof, which e.g. involves Popov's hyperstability theory (discussed in Section 4.5.3.4.2), or Lyapunov's stability theory (discussed in Section 4.5.3.5.1), it can be shown that the adaptation mechanism is again $K_p + K_i/p$, where K_p and K_i are gain constants. Thus the estimated speed used by the adjustable model of the MRAS system is obtained from $\hat{\omega}_r = (K_p + K_i/p)\varepsilon_{\Delta e'}$.

4.5.3.4.7 Scheme 5: MRAS-based system with AI-based adaptive model and various possibilities for speed tuning signal

All the MRAS-based schemes described in Sections 4.5.3.4.3–4.5.3.4.6 contain a reference model and an adaptive model. An input to the adaptive model has been the estimated rotor speed, which was the output of a suitable adaptation mechanism, which utilized at its inputs the difference of the estimated state variables of the reference and adaptive models. Furthermore, in all the four schemes discussed above, the adaptation mechanism is based on using Popov's hyperstability criterion. This has eventually resulted in an adaptation mechanism in which the estimated state variables of the reference and adaptive model were manipulated into a speed tuning signal, which was then input into a PI controller, containing the proportional and integrator gains of the adaptation mechanism (K_p and K_i). This approach has also required the use of a mathematical model for the adaptive model. A digital implementation of such a scheme is relatively simple and, in terms of processor requirements, Scheme 2 described in Section 4.5.3.4.4 is the simplest. However, greater accuracy and robustness can be achieved if this mathematical model is not used at all and instead, an artificial-intelligence-based non-linear adaptive model is employed. It is then also possible to eliminate the need for the separate PI controller, since this can be integrated into the tuning mechanism of the artificial-intelligence-based model.

The artificial-intelligence-based model can take various forms: it can be an artificial neural network (ANN), or a fuzzy-neural network, etc. (see also Chapter 7 and Section 4.4), and there is also the possibility of using different types of speed tuning signals. Furthermore, the adaptation-mechanism input signals can take various forms, as in the non-artificial-intelligence-based, conventional type of schemes discussed above. Thus there are various possibilities for the speed tuning signal. It follows that if only one ANN configuration is considered (e.g. a back-propagation feedforward multi-layer ANN), and only one specific fuzzy-neural network is used (see also Chapter 7 and Section 4.4), then it is possible to have eight different implementations by considering the four different types of speed tuning signals discussed in the previous sections. However, the possibilities are even greater, since there are many types of ANNs and fuzzy-neural networks. It is believed that some of these solutions can give high accuracy and are robust to parameter variations even at extremely low stator frequency. For illustration purposes of this technique, one specific solution will be described in Section 4.5.3.6, where the ANN contains adjustable and constant weights, and the adjustable weights are proportional to the rotor speed.

4.5.3.5 Observers

4.5.3.5.1 General, Luenberger, and Kalman observers

In general an estimator is defined as a dynamic system whose state variables are estimates of some other system (e.g. electrical machine). There are basically two forms of the implementation of an estimator: open-loop and closed-loop, the distinction between the two being whether or not a correction term, involving the estimation error, is used to adjust the response of the estimator. A closed-loop estimator is referred to as an observer (see also Section 3.1.3.5).

In Section 4.5.3.1 different types of open-loop speed estimators have been discussed. In open-loop estimators, especially at low speeds, parameter deviations have a significant influence on the performance of the drive both in the steady-state and transient state. However, it is possible to improve the robustness against parameter mismatch and also signal noise by using closed-loop observers.

An observer can be classified according to the type of representation used for the plant to be observed. If the plant is considered to be deterministic, then the observer is a deterministic observer; otherwise it is a stochastic observer. The most commonly used observers are Luenberger and Kalman types (Du *et al.* 1994). The Luenberger observer (LO) is of the deterministic type and the Kalman filter (KF) is of the stochastic type. The basic Kalman filter is only applicable to linear stochastic systems, and for non-linear systems the extended Kalman filter (EKF) can be used, which can provide estimates of the states of a system or of both the states and parameters (joint state and parameter estimation). The EKF is a recursive filter (based on the knowledge of the statistics of both the state and noise created by measurement and system modelling), which can be applied to a non-linear time-varying stochastic system. The basic Luenberger observer is applicable to a linear, time-invariant deterministic system. The extended Luenberger observer (ELO) is applicable to a non-linear time-varying deterministic system. In summary it can be seen that both the EKF and ELO are non-linear estimators and the EKF is applicable to stochastic systems and the ELO to deterministic systems. The deterministic extended Luenberger observer (ELO) is an alternative solution for real-time implementations in industrial drive systems. The simple algorithm and the ease of tuning of the ELO may give some advantages over the conventional extended Kalman filter.

Various types of speed observers are discussed in the present section, which can be used in high-performance induction machine drives. These include a full-order (fourth-order) adaptive state observer (Luenberger observer) which is constructed by using the equations of the induction machine in the stationary reference frame by adding an error compensator. Furthermore, the extended Kalman filter (EKF) and the extended Luenberger observer (ELO) are also discussed. In the full-order adaptive state observer the rotor speed is considered as a parameter, but in the EKF and ELO the rotor speed is considered as a state variable. Furthermore, as discussed above, whilst the ELO is a deterministic observer, the EKF is a stochastic observer which also uses the noise properties of measurement and system noise. It is shown that when the appropriate observers are used in high-performance speed-sensorless torque-controlled induction motor drives (vector-controlled drives, direct-torque-controlled drives), stable operation can be obtained over a wide speed-range, including very low speeds. Various industrial a.c. drives already incorporate observers and it is expected that, in the future, observers will have an increased role in industrial high-performance vector- and direct-torque-controlled drives.

4.5.3.5.2 Application of a full-order adaptive state observer

In the present subsection, an adaptive speed and stator-resistance observer is discussed for the induction machine. However, first, a state estimator is described which can be used to estimate the rotor flux linkages of an induction machine. This estimator is then modified so it can also yield the speed estimate, and thus an adaptive speed estimator is derived (to be precise, a speed-adaptive flux observer is obtained). To obtain a stable system, the adaptation mechanism is derived by using the state-error dynamic equations together with Lyapunov's stability theorem. The design of the observer is also discussed. In an inverter-fed drive system, the observer uses the monitored stator currents together with the monitored (or reconstructed) stator voltages, or reference stator voltages. However, when the reconstructed stator voltages or reference voltages are used, some error compensation schemes must also be used and some aspects of this are also discussed below.

A state observer is a model-based state estimator which can be used for the state (and/or parameter) estimation of a non-linear dynamic system in real time. In the calculations, the states are predicted by using a mathematical model (the estimated states are denoted by $\hat{\mathbf{x}}$), but the predicted states are continuously corrected by using a feedback correction scheme. This scheme makes use of actual measured states ($\mathbf{x}$) by adding a correction term to the predicted states. This correction term contains the weighted difference of some of the measured and estimated output signals (the difference is multiplied by the observer feedback gain, $\mathbf{G}$) Based on the deviation from the estimated value, the state observer provides an optimum estimated output value ($\hat{\mathbf{x}}$) at the next input instant. In an induction motor drive a state observer can also be used for the real-time estimation of the rotor speed and some of the machine parameters, e.g. stator resistance. This is possible since a mathematical dynamic model of the induction machine is sufficiently well known. For this purpose the stator voltages and currents are monitored on-line and for example, the speed and stator resistance of the induction machine can be obtained by the observer quickly and precisely. The accuracy of the state observer also depends on the model parameters used. By using a DSP, it is possible to implement the adaptive state observer conveniently in real time. The state observer is simpler than the Kalman observer (e.g. see Section 3.1.3.5), since no attempt is made to minimize a stochastic cost criterion.

To obtain the full-order non-linear speed observer, first the model of the induction machine is considered in the stationary reference frame and then an error compensator term is added to this. The simplest derivation uses the stator and rotor space-vector equations of the induction machine given by eqns (2.1-148)–(2.1-151). These yield the following state variable equations in the stationary reference frame ($\omega_g = 0$) if the space vectors of the stator currents ($\bar{i}_s$)

and rotor flux linkages ($\bar{\psi}'_r$) are selected as state variables:

$$\frac{d}{dt}\begin{bmatrix}\bar{i}_s\\ \bar{\psi}'_r\end{bmatrix}=\begin{bmatrix}-[1/T'_s+(1-\sigma)/T'_r] & -[L_m/(L'_sL_r)](-1/T_r+j\omega_r)\\ L_m/T_r & -1/T_r+j\omega_r\end{bmatrix}\begin{bmatrix}\bar{i}_s\\ \bar{\psi}'_r\end{bmatrix}+\begin{bmatrix}\bar{u}_s/L'_s\\ 0\end{bmatrix}, \tag{4.5-98}$$

where $\bar{u}_s$ is the space vector of stator voltages, L_m and L_r are the magnetizing inductance and rotor self-inductance respectively, L'_s is the stator transient inductance, $T'_s=L'_s/R_s$ and $T'_r=L'_r/R_r$ are the stator and rotor transient time constants respectively, and $\sigma=1-L_m^2/(L_sL_r)$ is the leakage factor. Since $\bar{i}_s=i_{sD}+ji_{sQ}$, $\bar{\psi}'_r=\psi_{rd}+j\psi_{rq}$, eqn (4.5-98) can also be put into the following component form:

$$\frac{d\dot{\mathbf{x}}}{dt}=\mathbf{Ax}+\mathbf{Bu} \tag{4.5-99}$$

$$\mathbf{A}=\begin{bmatrix}-[1/T'_s+(1-\sigma)/T'_r]\mathbf{I}_2 & [L_m/(L'_sL_r)][\mathbf{I}_2/T_r-\omega_r\mathbf{J}]\\ L_m\mathbf{I}_2/T_r & -\mathbf{I}_2/T_r+\omega_r\mathbf{J}\end{bmatrix}$$

$$\mathbf{B}=[\mathbf{I}_2/L'_s,\mathbf{O}_2]^T$$

$$\mathbf{J}=\begin{bmatrix}0 & -1\\ 1 & 0\end{bmatrix}$$

The output equation is defined as

$$\mathbf{i}_s=\mathbf{Cx}. \tag{4.5-100}$$

In eqn (4.5-99), $\mathbf{x}=[\mathbf{i}_s,\boldsymbol{\psi}'_r]$ is the state vector, which contains the stator-current column vector, $\mathbf{i}_s=[i_{sD},i_{sQ}]^T$ and also the rotor flux-linkage column vector, $\boldsymbol{\psi}'_r=[\psi_{rd},\psi_{rq}]^T$.

Furthermore, $\mathbf{u}$ is the input column vector, which contains the direct- and quadrature-axis stator voltages $\mathbf{u}=\mathbf{u}_s=[u_{sD},u_{sQ}]^T$, $\mathbf{A}$ is the state matrix, which is a four-by-four matrix and is dependent on the speed (ω_r). Furthermore, $\mathbf{B}$ is the input matrix, $\mathbf{I}_2$ is a second-order identity matrix, $\mathbf{I}_2=\text{diag}(1,1)$, and $\mathbf{O}_2$ is a two-by-two zero matrix. In eqn (4.5-100) $\mathbf{C}$ is the output matrix $\mathbf{C}=[\mathbf{I}_2,\mathbf{O}_2]^T$. It can be seen that the space-vector form of the state equations, eqn (4.5-98), and its component form, eqn (4.5-99), are very similar. However, in the component equation matrix $\mathbf{J}$ is used in contrast to j, which is used in the space-vector equations, and also the identity matrix $\mathbf{I}_2$ is present in the component equations instead of the 1 in the space vector equations. Both of these equations can be used to design the observer.

By using the derived mathematical model of the induction machine, e.g. if the component form of the equations, eqn (4.5-99), is used, since this is required in an actual implementation, and adding the correction term described above, which contains the difference of actual and estimated states, a full-order state-observer, which estimates the stator currents and rotor flux linkages, can be described as follows:

$$\frac{\mathrm{d}\hat{\mathbf{x}}}{\mathrm{d}t}=\hat{\mathbf{A}}\mathbf{x}+\mathbf{B}\mathbf{u}+\mathbf{G}(\mathbf{i}_s-\hat{\mathbf{i}}_s) \tag{4.5-101}$$

$$\hat{\mathbf{A}}=\begin{bmatrix} -[1/T_s'+(1-\sigma)/T_r']\mathbf{I}_2 & [L_m/(L_s'L_r)][\mathbf{I}_2/T_r-\hat{\omega}_r\mathbf{J}] \\ L_m\mathbf{I}_2/T_r & -\mathbf{I}_2/T_r+\hat{\omega}_r\mathbf{J} \end{bmatrix}$$

and the output vector is

$$\hat{\mathbf{i}}_s=\mathbf{C}\hat{\mathbf{x}}, \tag{4.5-102}$$

where ^ denotes estimated values. It can be seen that the state matrix of the observer ($\hat{\mathbf{A}}$) is a function of the rotor speed, and in a speed-sensorless drive, the rotor speed must also be estimated. The estimated rotor speed is denoted by $\hat{\omega}_r$, and in general $\hat{\mathbf{A}}$ is a function of $\hat{\omega}_r$, It is important to note that the estimated speed is considered as a parameter in $\hat{\mathbf{A}}$; however, in some other types of observers (e.g. extended Kalman filter), the estimated speed is not considered as a parameter, but it is a state variable. In eqns (4.5-101) and (4.5-102) the estimated state variables are $\hat{\mathbf{x}}=[\hat{\mathbf{i}}_s, \hat{\psi}_r']^T$ and $\mathbf{G}$ is the observer gain matrix, which is selected so that the system will be stable. In eqn (4.5-101) the gain matrix is multiplied by the error vector $\mathbf{e}=(\mathbf{i}_s-\hat{\mathbf{i}}_s)$, where $\mathbf{i}_s$ and $\hat{\mathbf{i}}_s$ are the actual and estimated stator-current column vectors respectively, $\mathbf{i}_s=[i_{sD}, i_{sQ}]^T$, $\hat{\mathbf{i}}_s=[\hat{i}_{sD}, \hat{i}_{sQ}]^T$. By using eqns (4.5-101) and (4.5-102) it is possible to implement a speed estimator which estimates the rotor speed of an induction machine by using the adaptive state observer shown in Fig. 4.76.

In Fig. 4.76 the estimated rotor flux-linkage components and the stator-current error components are used to obtain the error speed tuning signal, which can be

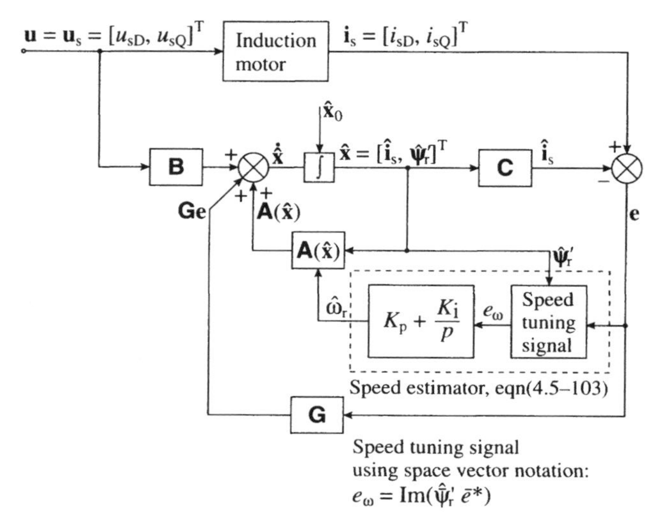

Fig. 4.76. Adaptive speed observer (speed-adaptive flux observer).

put into a very compact, simple form when the space-vector notation is used: $e_\omega = \mathrm{Im}(\hat{\bar{\psi}}_r' \bar{e}^*)$, where $\hat{\bar{\psi}}_r = \psi_{rd} + j\psi_{rq}$ and $\bar{e} = e_{sD} + je_{sQ}$. The estimated speed is obtained from the speed tuning signal by using a PI controller thus,

$$\hat{\omega}_r = K_p(\hat{\psi}_{rq} e_{sD} - \hat{\psi}_{rd} e_{sQ}) + K_i \int (\hat{\psi}_{rq} e_{sD} - \hat{\psi}_{rd} e_{sQ})\, dt, \qquad (4.5\text{-}103)$$

where K_p and K_i are proportional and integral gain constants respectively, $e_{sD} = i_{sD} - \hat{i}_{sD}$ and $e_{sQ} = i_{sQ} - \hat{i}_{sQ}$ are the direct- and quadrature-axis stator current errors respectively. The adaptation mechanism is similar to that discussed in the MRAS-based speed estimators in Section 4.5.3.4 where the speed adaptation has been obtained by using the state-error equations of the system considered. Since the goal is to obtain a stable observer, a rigorous proof of eqn (4.5-103) also uses the state-error equation of the system, together with Lyapunov's stability theorem. It should be noted that, instead of determining the stability of the observer itself, it is advantageous to determine the stability of its error dynamics. By using the error dynamics, the objective of a stability analysis is very clear: the states in the error dynamics should decay to the origin. To obtain the error dynamics, eqn (4.5-101) is subtracted from eqn (4.5-99), yielding the following observer-error equation:

$$\frac{d\mathbf{e}}{dt} = \frac{d}{dt}(\mathbf{x} - \hat{\mathbf{x}}) = (\mathbf{A} - \mathbf{GC})(\mathbf{x} - \hat{\mathbf{x}}) + (\mathbf{A} - \hat{\mathbf{A}})\hat{\mathbf{x}} = (\mathbf{A} - \mathbf{GC})\mathbf{e} - \Delta\mathbf{A}\hat{\mathbf{x}}. \qquad (4.5\text{-}104)$$

In eqn (4.5-104) $\mathbf{e} = \mathbf{x} - \hat{\mathbf{x}}$ is the estimation-error column vector (of the stator currents and rotor flux linkages) and the error state matrix is

$$\Delta\mathbf{A} = \hat{\mathbf{A}} - \mathbf{A} = \begin{bmatrix} \mathbf{O}_2 & -(\hat{\omega}_r - \omega_r)\mathbf{J}(L_m/L_r)L_s' \\ \mathbf{O}_2 & (\hat{\omega}_r - \omega_r)\mathbf{J} \end{bmatrix}, \qquad (4.5\text{-}105)$$

where matrix $\mathbf{J}$ has been defined in eqn (4.5-99). It can be seen that the error dynamics are described by the eigenvalues of $\mathbf{A} - \mathbf{GC}$ and these could also be used to design a stable observer (gain matrix). However, to determine the stability of the error dynamics of the observer, it is also possible to use Popov's hyperstability theorem (which has been used in Section 4.5.3.4 for MRAS-based systems) or Lyapunov's stability theorem, which gives a sufficient condition for the uniform asymptotic stability of a non-linear system by using a Lyapunov function V. This function has to satisfy various conditions, e.g. it must be continuous, differentiable, positive definite, etc. Such a function exists and the following Lyapunov function is introduced:

$$V = \mathbf{e}^T\mathbf{e} + (\hat{\omega}_r - \omega_r)/c, \qquad (4.5\text{-}106)$$

where c is a positive constant. This function is zero when the error ($\mathbf{e}$) is zero and when the estimated speed ($\hat{\omega}_r$) is equal to the actual speed (ω_r). Since a sufficient condition for uniform asymptotic stability is that the derivative Lyapunov function, dV/dt, is negative definite, the derivative of V is now obtained. By using the chain differentiation rule, it follows from eqn (4.5-106) that the time derivative of V becomes as follows:

$$\frac{dV}{dt}=\mathbf{e}\left[\frac{d(\mathbf{e}^{T})}{dt}\right]+\mathbf{e}^{T}\left[\frac{d\mathbf{e}}{dt}\right]+2\frac{d\hat{\omega}_r}{dt}\frac{(\hat{\omega}_r-\omega_r)}{c} \tag{4.5-107}$$

By substitution of $d\mathbf{e}/dt$ by its expression given by eqn (4.5-104), eqn (4.5-107) becomes

$$\frac{dV}{dt}=\mathbf{e}^{T}[(\mathbf{A}-\mathbf{GC})^{T}+(\mathbf{A}-\mathbf{GC})]\mathbf{e}-(\hat{\mathbf{x}}^{T}\Delta\mathbf{A}^{T}\mathbf{e}+\mathbf{e}^{T}\Delta\mathbf{A}\hat{\mathbf{x}})+2\frac{d\hat{\omega}_r}{dt}\frac{(\hat{\omega}_r-\omega_r)}{c}. \tag{4.5-108}$$

When $\mathbf{e}=\mathbf{x}-\hat{\mathbf{x}}$, $\mathbf{x}=[\mathbf{i}_s, \boldsymbol{\psi}_r']^{T}$ and $\hat{\mathbf{x}}=[\hat{\mathbf{i}}_s, \hat{\boldsymbol{\psi}}_r']^{T}$ are substituted into eqn (4.5-108), where $\mathbf{i}_s=[i_{sD}, i_{sQ}]^{T}$ and $\boldsymbol{\psi}_r'=[\psi_{rd}, \psi_{rq}]^{T}$, finally the derivative of the Lyapunov function can be expressed as follows:

$$\frac{dV}{dt}=\mathbf{e}^{T}[(\mathbf{A}-\mathbf{GC})^{T}+(\mathbf{A}-\mathbf{GC})]\mathbf{e}-2\frac{L_m}{L_r}(\hat{\omega}_r-\omega_r)\frac{(e_{sD}\hat{\psi}_{rq}-e_{sQ}\hat{\psi}_{rd})}{L_s'}+\frac{2}{c}(\hat{\omega}_r-\omega_r)\frac{d\hat{\omega}_r}{dt}, \tag{4.5-109}$$

where $e_{sD}=i_{sD}-\hat{i}_{sD}$ and $e_{sQ}=i_{sQ}-\hat{i}_{sQ}$. Since a sufficient condition for uniform asymptotic stability is that the derivative Lyapunov function, dV/dt, is negative definite, e.g. V has to decrease when the error is not zero, this can be satisfied by ensuring that the sum of the last two terms in eqn (4.5-109) is zero (the other terms on the right-hand side of eqn (4.5-109) are always negative). Thus it follows that the adaptive scheme (adjustment law) for the speed estimation is obtained as

$$\frac{d\hat{\omega}_r}{dt}=K_i(e_{sD}\hat{\psi}_{rq}-e_{sQ}\hat{\psi}_{rd}), \tag{4.5-110}$$

where $K_i=cL_m/(L_s'L_r)$ (c is a positive constant introduced above). From eqn (4.5-110) the speed is estimated as follows:

$$\hat{\omega}_r=K_i\int(e_{sD}\hat{\psi}_{rq}-e_{sQ}\hat{\psi}_{rd})\,dt. \tag{4.5-111}$$

However, to improve the response of the speed observer, this is modified to

$$\hat{\omega}_r=K_p(e_{sD}\hat{\psi}_{rq}-e_{sQ}\hat{\psi}_{rd})+K_i\int(e_{sD}\hat{\psi}_{rq}-e_{sQ}\hat{\psi}_{rd})\,dt \tag{4.5-112}$$

and eqn (4.5-112) is used for speed estimation, which agrees with eqn (4.5-103). In summary it can be seen that an adaptive observer can be used to obtain the rotor flux estimates ($\hat{\psi}_{rq}, \hat{\psi}_{rd}$) and the rotor speed is estimated by using the estimated rotor flux linkages and using the stator current errors (e_{sD}, e_{sQ}). This is why the precise name of this speed observer is 'speed-adaptive flux observer'. The same result can also be obtained by applying Popov's hyperstability theorem.

If the chosen PI constants K_p and K_i are large, then the convergence of the rotor speed estimation will be fast. However, in a PWM inverter-fed induction machine, the estimated speed will be rich in higher harmonics, due to the PWM inverter. Thus the PI gains must be limited when the stator voltages and currents are obtained asynchronously with the PWM pattern. It should be noted that eqn (4.5-108) can also be used to obtain the gain matrix. If the observer gain matrix **G** is chosen so the first term of eqn (4.5-108) is negative semi-definite, then the speed observer will be stable. To ensure stability (at all speeds), the conventional procedure is to select observer poles which are proportional to the motor poles [Kubota *et al.* 1990] (the proportionality constant is k, and $k \geqslant 1$. This makes the observer dynamically faster than the induction machine. However, to make the sensitivity to noise small, the proportionality constant is usually small. Thus by using this conventional pole-placement technique, the gain matrix is obtained as

$$\mathbf{G} = -\begin{bmatrix} g_1\mathbf{I}_2 + g_2\mathbf{J} \\ g_3\mathbf{I}_2 + g_4\mathbf{J} \end{bmatrix} \tag{4.5-113}$$

which yields a two-by-four matrix. The four gains in **G** can be obtained from the eigenvalues of the induction motor as follows:

$$\begin{aligned}
g_1 &= -(k-1)\left(\frac{1}{T_s'} + \frac{1}{T_r'}\right) \\
g_2 &= (k-1)\hat{\omega}_r \\
g_3 &= (k^2-1)\left\{-\left[\frac{1}{T_s'} + \frac{(1-\sigma)}{T_r'}\right]\frac{L_s' L_m}{L_r} + \frac{L_m}{T_r}\right\} + L_s'\frac{L_m}{L_r}(k-1)\left(\frac{1}{T_s'} + \frac{1}{T_r'}\right) \\
g_4 &= -(k-1)\hat{\omega}_r\frac{L_s' L_m}{L_r}
\end{aligned} \tag{4.5-114}$$

It follows that the four gains depend on the estimated speed, $\hat{\omega}_r$. With this selection, the estimated states converge to the actual states in all the speed range. The observer can be implemented by using a DSP (e.g. Texas Instruments TMS320C30). However, in a discrete implementation of the speed observer, for small sampling time and low speed, accurate computation is required, otherwise (due to computational errors) stability problems can occur (roots are close to stability limit). For this purpose, another pole-placement procedure could be used, which ensures that the low-speed roots are moved away from the stability limit. Some of the issues related to DSP implementation are now discussed.

For DSP implementation the discretized form of the observer, eqn (4.5-101), and the adaptation mechanism, eqn (4.5-112), has to be used. Thus the discretized observer is described by

$$\hat{\mathbf{x}}(k+1) = \mathbf{A}_d\hat{\mathbf{x}}(k) + \mathbf{B}_d\mathbf{u}(k) + \mathbf{G}_d[\mathbf{i}_s(k) - \hat{\mathbf{i}}_s(k)],$$

where $\mathbf{G}_d$ is the discretized observer gain matrix,

$$\mathbf{A}_d = \exp(\mathbf{A}T) \approx \mathbf{I}_4 + \mathbf{A}T + \frac{(\mathbf{A}T)^2}{2}$$

is the discretized system matrix, where T is the sampling time, and

$$\mathbf{B}_\mathrm{d}=\int_0^\tau [\exp(\mathbf{A}T)]\mathbf{B}\,\mathrm{d}\tau \approx \mathbf{B}T+\frac{\mathbf{AB}T^2}{2}.$$

Since matrices $\mathbf{A}$ and $\mathbf{A}_\mathrm{d}$ depend on the rotor speed, the gain matrix has to be computed at each time step and, as discussed above, the observer poles are chosen to be proportional to the poles of the induction machine. To make the scheme insensitive to measurement noise, the proportionality constant (k) is selected to be low. In a real-time implementation, due to the complexity of the induction motor model, first the gain matrix $\mathbf{G}$ is updated directly and then the discretized gain matrix $\mathbf{G}_\mathrm{d}$ is computed. However, this pole-placement technique may have some disadvantages and may not ensure good observer dynamics. It is a disadvantage that it requires extensive computation time, due to the updating of matrix $\mathbf{G}$ and the discretization procedure. The observer dynamics can be adversely affected by the fact that, for small sampling time and low rotor speed, the discrete-root locus is very close to the stability limit and if there are computational errors, then instability may arise. However, it is possible to overcome some of these difficulties by using a simpler pole-placement technique. For this purpose, the symmetrical structure of $\mathbf{G}$ is preserved, but the elements of $\mathbf{G}$ are determined so that at low speed the poles are further displaced from the stability limit. Since these gain matrix elements can lead to a higher sensitivity on noise, beyond a specific rotor-speed value they are decreased. In this way two different, constant gain matrices ($\mathbf{G}$, $\mathbf{G}'$) are predetermined and used according to the rotor speed (one for speed values less than a specified value and the other for speed values higher than this specific value). In the first 2-by-4 discretized gain matrix, $\mathbf{G}_\mathrm{d}$, the elements are $g_{11\mathrm{d}}=g_{12\mathrm{d}}=-g_{21\mathrm{d}}=g_{22\mathrm{d}}=g_{31\mathrm{d}}=g_{32\mathrm{d}}=-g_{41\mathrm{d}}=g_{42\mathrm{d}}=c$, where c is a constant. The second discretized observer gain matrix $\mathbf{G}'_\mathrm{d}$ can also be approximated by a 2-by-4 matrix, which has four zero elements, $g_{12\mathrm{d}'}=g_{21\mathrm{d}'}=g_{31\mathrm{d}'}=g_{34\mathrm{d}'}=0$ but four constant elements, $g_{11\mathrm{d}'}=g_{22\mathrm{d}'}=c_1$ and $g_{32\mathrm{d}'}=-g_{42\mathrm{d}'}=c_2$, where c_1 and c_2 are constants. This approach leads to a reduced computation time.

The speed estimator discussed above will only give correct speed estimates if correct machine parameters are used in the system matrix and in the input matrix. However, these also contain the stator and rotor transient time constants (T'_s and T'_r), which also vary with the temperature (since they depend on the temperature-dependent stator and rotor resistances respectively). The variation of the stator resistance has significant influence on the estimated speed, especially at low speeds. On the other hand, in a high-dynamic-performance induction motor drive, where the rotor flux is constant, the influence of the rotor resistance variation is constant, independent of the speed, since the speed estimation error and the rotor resistance error cannot be separated from the stator variables (see also the discussion in Section 4.5.3.1, Scheme 5). The influence of the stator resistance variation on the speed estimation can be removed by using an adaptive stator-resistance estimation scheme. A rigorous mathematical derivation of the stator resistance estimator is similar to that shown above for the speed estimator, and in this case again the state error equations of the observer have first to be obtained. These are obtained by subtracting from the original machine equations the new observer equation, which now contains a new state matrix, but this contains the stator resistance to be estimated $\hat{R}_s$ (where $\hat{R}_s = R_s + \Delta R_s$, and ΔR_s is the stator resistance error). It then follows that the following matrix form of the state error equation contains the error state matrix due to the stator resistance mismatch:

$$\Delta \mathbf{A}_{Rs} = \begin{bmatrix} -\mathbf{I}_2/L_s' & 0_2 \\ 0_2 & 0_2 \end{bmatrix}, \tag{4.5-115}$$

where $\mathbf{I}_2$ and 0_2 are second-order identity and zero matrices respectively. By using the state error equations and applying Popov's hyperstability theorem, or Lyapunov's stability theorem, the estimate of the stator resistance can be obtained. Both of these approaches yield the following stator resistance estimate:

$$\hat{R}_s = -K_i' \int (e_{sD}\hat{i}_{sD} + e_{sQ}\hat{i}_{sQ})\,dt - K_p'(e_{sD}\hat{i}_{sD} + e_{sQ}\hat{i}_{sQ}), \tag{4.5-116}$$

and e.g. this can be rigorously checked by using a similar procedure to that described above in conjunction with Lyapunov's stability theorem. When this stator resistance estimation is used on-line in a speed-sensorless high-performance induction motor drive employing torque control (vector control or direct torque control) which also uses the on-line estimation of the rotor speed based on eqn (4.5-112), the drive can be operated in a stable manner in a very wide speed-range, including extremely low speeds. However, it should be noted that the speed observer discussed above uses the monitored stator voltages. In a PWM inverter-fed induction machine the stator voltages contain harmonics due to the inverter and also the degree of voltage measurement deteriorates at low speeds. These problems can be overcome in various ways, e.g. by reconstructing the stator voltages from the inverter switching states by using the monitored d.c. link voltage (see also Sections 3.1.3.1, 4.1.1). Alternatively, in a drive system, it is also possible to use the reference voltages, e.g. these are the inputs to a space-vector PWM modulator (see also Section 3.1.3.5.1), where the reference voltages have been used in a PMSM drive. If this technique is used for the induction machine, to obtain high accuracy, it is necessary to compensate the error between the reference and actual stator voltages by the estimation of the voltage error. For this purpose the adaptive observer shown in Fig. 4.76 must be complemented by a voltage-error estimator block. This voltage error contains both a d.c. and an a.c. component, corresponding to a constant bias error between the real and reference voltage and also to an amplitude error. These errors are present due to the dead time which is required to prevent the short circuits of the inverter arms, errors caused by A/D quantization, voltage drops of the switching devices, etc. The modified observer with stator-voltage error compensator is simple to implement and significantly improves the system behaviour. The extra parts of this observer scheme contain a stator-voltage error estimator. It can be shown by simple considerations that this voltage error estimator outputs the integral of $\mathbf{G}(\mathbf{i}_s - \hat{\mathbf{i}}_s)$, and this signal is manipulated into other signals which are then added to the reference voltages to obtain the correct stator voltages.

4.5.3.5.3 Application of the extended Kalman filter (EKF)

In the present section the extended Kalman filter (EKF) is used for the estimation of the rotor speed of an induction machine. The details of the EKF have been discussed in Section 3.1.3.5, where it has been used in PMSM drives. Two induction-motor models will be derived which can be used by the EKF; one of them contains 5 induction machine parameters and the other one uses only 4 parameters. The details and practical aspects of the EKF algorithm are also discussed. The EKF is suitable for use in high-performance induction motor drives, and it can provide accurate speed estimates in a wide speed-range, including very low speeds. It can also be used for joint state and parameter estimation. However, it is computationally more intensive than the full-order state observer described in the previous section.

The EKF is a recursive optimum stochastic state estimator which can be used for the joint state and parameter estimation of a non-linear dynamic system in real-time by using noisy monitored signals that are disturbed by random noise. This assumes that the measurement noise and disturbance noise are uncorrelated. The noise sources take account of measurement and modelling inaccuracies. The EKF is a variant of the Kalman filter, but the extended version can deal with a non-linear system. It should be noted that in the full-order state-observer (speed-adaptive flux observer) discussed in Section 4.5.3.5.2, the noise has not been considered (it is a deterministic observer, in contrast to the EKF which is a stochastic observer). Furthermore, in the speed-adaptive flux observer, the speed was considered as a parameter, but in the EKF it is considered as a state. Similarly to the speed-adaptive flux observer, where the state variables are adapted by the gain matrix (**G**), in the EKF the state variables are adapted by the Kalman gain matrix (**K**).

In a first stage of the calculations of the EKF, the states are predicted by using a mathematical model of the induction machine (which contains previous estimates) and in the second stage, the predicted states are continuously corrected by using a feedback correction scheme. This scheme makes use of actual measured states by adding a term to the predicted states (which are obtained in the first stage). The additional term contains the weighted difference of the measured and estimated output signals. Based on the deviation from the estimated value, the EKF provides an optimum output value at the next input instant. In an induction motor drive the EKF can be used for the real-time estimation of the rotor speed, but it can also be used for joint state and parameter estimation. For this purpose the stator voltages and currents are measured (or the stator voltages are reconstructed from the d.c. link voltage and the inverter switching signals) and, for example, the speed of the machine can be obtained by the EKF quickly and precisely.

The main design steps for a speed-sensorless induction motor drive implementation using the discretized EKF algorithm are as follows (see also Section 3.1.3.5):

1. Selection of the time-domain induction machine model;
2. Discretization of the induction machine model;
3. Determination of the noise and state covariance matrices **Q**, **R**, **P**;
4. Implementation of the discretized EKF algorithm; tuning.

These steps are now discussed.

1. Time-domain, augmented induction-machine model For the purpose of using an EKF for the estimation of the rotor speed of an induction machine, it is possible to use various machine models. For example, it is possible to use the equations expressed in the rotor-flux-oriented reference frame ($\omega_g=\omega_{mr}$), or in the stator-flux-oriented reference frame ($\omega_g=0$). When the model expressed in the rotor-flux-oriented reference frame is used, the stator current components in the rotor-flux-oriented reference frame (i_{sx}, i_{sy}) are also state variables and the input and output matrices contain the sine and cosine of ρ_r, which is the angle of the rotor flux-linkage space vector (with respect to the direct-axis of the stationary reference frame). This is due to the fact that in the state-variable equations the actual stator voltages must be transformed into the rotor-flux-oriented reference frame (to give the required transformed input voltages, u_{sx}, u_{sy}) and also since the output matrix must contain the actual stator currents (i_{sD}, i_{sQ}) which are obtained from i_{sx}, i_{sy}. These transformations introduce extra non-linearities, but these transformations are not present in the model established in the stationary reference frame. The main advantages of using the model in the stationary reference frame are:

- reduced computation time (e.g. due to reduced non-linearities);
- smaller sampling times;
- higher accuracy;
- more stable behaviour.

It follows from eqn (4.5-98) or (4.5-99) that the two-axis state-space equations of the induction machine in the stationary reference are as follows, when the stator currents and rotor flux linkages are the state variables and these are augmented with the estimated quantity, which in this application is the rotor speed (ω_r):

$$\frac{d}{dt}\begin{bmatrix} i_{sD} \\ i_{sQ} \\ \psi_{rd} \\ \psi_{rq} \\ \omega_r \end{bmatrix} = \begin{bmatrix} -1/T_s'^* & 0 & L_m/(L_s' L_r T_r) & \omega_r L_m/(L_s' L_r) & 0 \\ 0 & -1/T_s'^* & -\omega_r L_m/(L_s' L_r) & L_m/(L_s' L_r T_r) & 0 \\ L_m/T_r & 0 & -1/T_r & -\omega_r & 0 \\ 0 & L_m/T_r & \omega_r & -1/T_r & 0 \\ 0 & 0 & 0 & 0 & 0 \end{bmatrix} \begin{bmatrix} i_{sD} \\ i_{sQ} \\ \psi_{rd} \\ \psi_{rq} \\ \omega_r \end{bmatrix} + \begin{bmatrix} 1/L_s' & 0 \\ 0 & 1/L_s' \\ 0 & 0 \\ 0 & 0 \\ 0 & 0 \end{bmatrix} \begin{bmatrix} u_{sD} \\ u_{sQ} \end{bmatrix} \tag{4.5-117}$$

and

$$\begin{bmatrix} i_{sD} \\ i_{sQ} \end{bmatrix} = \begin{bmatrix} 1 & 0 & 0 & 0 & 0 \\ 0 & 1 & 0 & 0 & 0 \end{bmatrix} [i_{sD} i_{sQ} \psi_{rd} \psi_{rq} \omega_r]^T = \mathbf{C}[i_{sD} i_{sQ} \psi_{rd} \psi_{rq} \omega_r]^T, \quad (4.5\text{-}118)$$

where $T_s'^*$ is defined as

$$\frac{1}{T_s'^*} = \frac{1}{T_s'} + \frac{(1-\sigma)}{T_r'} = \frac{R_s + R_r(L_m/L_r)^2}{L_s'} = \frac{R_s + R_{rref}}{L_s'}. \quad (4.5\text{-}119)$$

It should be noted that in eqn (4.5-117) it has been assumed that the rotor speed derivative is negligible, $d\omega_r/dt = 0$. Although this last equation corresponds to infinite inertia, however, in reality this is not true, but the required correction is performed by the Kalman filter (by the system noise, which also takes account of the computational inaccuracies). Furthermore, it should be noted that the effects of saturation of the magnetic paths of the machine have been neglected. This assumption is justified, since it can be shown that the EKF is not sensitive to changes in the inductances, since any changes in the stator parameters are compensated by the current loop inherent in the EKF. The application of eqn (4.5-117) in the EKF will give not only the rotor speed but also the rotor flux-linkage components (and as a consequence the angle and modulus of the rotor flux-linkage space vector will also be known). This is useful for high-performance drive implementations. It is important to emphasize that the rotor speed has been considered as a state variable and the system matrix **A** is non-linear—it contains the speed, $\mathbf{A} = \mathbf{A}(\mathbf{x})$.

The compact forms of eqns (4.5-117) and (4.5-118) are

$$\frac{d\mathbf{x}}{dt} = \mathbf{A}\mathbf{x} + \mathbf{B}\mathbf{u} \quad (4.5\text{-}120)$$

$$\mathbf{y} = \mathbf{C}\mathbf{x}, \quad (4.5\text{-}121)$$

where

$$\mathbf{A} = \begin{bmatrix} -1/T_s'^* & 0 & L_m/(L_s' L_r T_r) & \omega_r L_m/(L_s' L_r) & 0 \\ 0 & -1/T_s'^* & -\omega_r L_m/(L_s' L_r) & L_m/(L_s' L_r T_r) & 0 \\ L_m/T_r & 0 & -1/T_r & -\omega_r & 0 \\ 0 & L_m/T_r & \omega_r & -1/T_r & 0 \\ 0 & 0 & 0 & 0 & 0 \end{bmatrix} \quad \mathbf{B} = \begin{bmatrix} 1/L_s' & 0 \\ 0 & 1/L_s' \\ 0 & 0 \\ 0 & 0 \\ 0 & 0 \end{bmatrix}$$

$$\mathbf{C} = \begin{bmatrix} 1 & 0 & 0 & 0 & 0 \\ 0 & 1 & 0 & 0 & 0 \end{bmatrix}, \quad (4.5\text{-}122)$$

and $\mathbf{x} = [i_{sD} i_{sQ} \psi_{rd} \psi_{rq} \omega_r]^T$ is the state vector, **u** is the input vector, $\mathbf{u} = [u_{sD} u_{sQ}]^T$, **A** is the system matrix, and **C** is the output matrix.

Equation (4.5-117) contains five machine parameters, these are $R_s + R_{rref}$, L_s', L_m, L_r, T_r, where $R_s + R_{rref}$ is a combined machine parameter: it is the sum of the stator resistance and the referred value of the rotor resistance, $R_r^{\text{rref}}=R_r(L_m/L_r)^2$. This referred rotor resistance (R_{rref}) is also present in the equivalent circuit shown in Fig. 4.10(a) and 4.10(c). However, it is possible to obtain another state-space model of the induction machine which contains only four machine parameters; these are: ($R_s R_{\text{rref}}$, L_s', T_r) or (R_s, L_s', L_M, T_r), where L_M is the referred value of the magnetizing inductance, $L_M=L_m^2/L_r$ (and $R_{\text{rref}}=L_M/T_r$ also holds). These four parameters are the ones which are present in the equivalent circuit of Fig. 4.10(a) (or Fig. 4.10(c)), and this is a consequence of the fact that the special referring factor $a=L_m/L_r$ was used. To obtain the state-variable equations with these four machine parameters, instead of using the rotor flux linkages as state variables, the referred values of the rotor flux linkages are used as state variables. This also follows directly from eqn (4.5-117), since in the stator equations all the parameters which are multiplied by the rotor flux linkages contain L_m/L_r. For convenience, the equivalent circuit of the induction machine using these four parameters is shown in Fig. 4.77. These four parameters can be conveniently determined in the self-commissioning stage of an induction motor drive by standstill tests, as discussed in Chapter 8. Such a self-commissioning stage is used in a wide range of commercial drives.

The space vector of the referred rotor flux linkages in the stationary reference frame ($\bar{\psi}'_{\text{rref}}$) is obtained from the non-referred rotor flux-linkage space vector ($\bar{\psi}'_r$) as follows:

$$\bar{\psi}'_{\text{rref}}=\psi_{\text{rdref}}+\text{j}\psi_{\text{rqref}}=\frac{L_m}{L_r}\bar{\psi}'_r=\frac{L_m}{L_r}(\psi_{\text{rd}}+\text{j}\psi_{\text{rq}}). \tag{4.5-123}$$

Thus when the state variables are i_{sD}, i_{sQ}, ψ_{rdref}, ψ_{rqref} and the augmented state-variable is ω_r, then the following state-variable equation is obtained from eqn (4.5-117):

$$\frac{\text{d}}{\text{d}t}\begin{bmatrix} i_{sD} \\ i_{sQ} \\ \psi_{\text{rdref}} \\ \psi_{\text{rqref}} \\ \omega_r \end{bmatrix}=\begin{bmatrix} -1/T_s'^{*} & 0 & 1/(L_s'T_r) & \omega_r/L_s' & 0 & i_{sD} \\ 0 & -1/T_s'^{*} & -\omega_r/L_s' & 1/(L_s'T_r) & 0 & i_{sQ} \\ L_M/T_r & 0 & -1/T_r & -\omega_r & 0 & \psi_{\text{rdref}} \\ 0 & L_M/T_r & \omega_r & -1/T_r & 0 & \psi_{\text{rqref}} \\ 0 & 0 & 0 & 0 & 0 & \omega_r \end{bmatrix}\begin{bmatrix} 1/L_s' & 0 \\ 0 & 1/L_s' \\ 0 & 0 \\ 0 & 0 \\ 0 & 0 \end{bmatrix}\begin{bmatrix} u_{sD} \\ u_{sQ} \end{bmatrix} \tag{4.5-124}$$

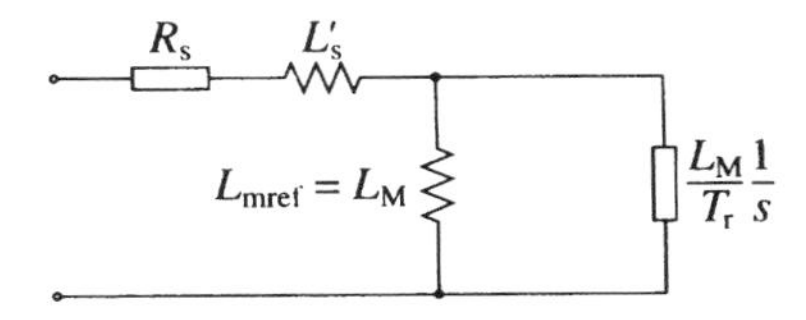

Fig. 4.77. Equivalent circuit of an induction machine using special referred parameters ($a=L_m/L_r$).

and

$$\begin{bmatrix} i_{sD} \\ i_{sQ} \end{bmatrix} = \begin{bmatrix} 1 & 0 & 0 & 0 & 0 \\ 0 & 1 & 0 & 0 & 0 \end{bmatrix} [i_{sD} i_{sQ} \psi_{rdref} \psi_{rqref} \omega_r]^T. \tag{4.5-125}$$

It can be seen that the output equation [eqn (4.5-124)] contains the same output matrix $\mathbf{C}$ as before.

Of course when eqns (4.5-124) and (4.5-125) are used in the EKF, then in addition to the rotor speed, the referred rotor flux-linkage components are obtained, and not the unreferred values. However, this is not a disadvantage, since the angle of $\bar{\psi}'_{rref}$ is the same (ρ_r) as for $\bar{\psi}'_r$ and the rotor magnetizing-current modulus can be obtained as $|\bar{i}_{mr}| = |\bar{\psi}_{rref}|/L_M$ (where L_M is the referred magnetizing inductance), in contrast to the usual $|\bar{i}_{mr}| = |\bar{\psi}'_r|/L_m$. As discussed in Section 4.1.1, in vector drives with rotor-flux-oriented control these quantities ($|\bar{i}_{mr}|$ and ρ_r) are the usual outputs of the rotor flux model. Furthermore, by considering eqn (4.1-43), the electromagnetic torque can also be expressed in terms of the referred rotor flux-linkage space vector as

$$t_e = \frac{3}{2} P \frac{L_m}{L_r} |\bar{\psi}_r| i_{sy} = \frac{3}{2} P |\bar{\psi}_{rref}| i_{sy} \tag{4.5-126}$$

or as

$$\begin{aligned} t_e &= \frac{3}{2} P \frac{L_m}{L_r} \bar{\psi}'_r \times \bar{i}_s = \frac{3}{2} P \frac{L_m}{L_r} (\psi_{rd} i_{sQ} - \psi_{rq} i_{sD}) = \frac{3}{2} P \bar{\psi}'_{rref} \times \bar{i}_s \\ &= \frac{3}{2} P (\psi_{rdref} i_{sQ} - \psi_{rqref} i_{sD}). \end{aligned} \tag{4.5-127}$$

Thus in a torque-controlled high-performance drive, if the referred rotor flux linkages are determined by the EKF, then it is possible to obtain the electromagnetic torque in terms of the referred rotor flux linkages and stator currents, and L_m/L_r is not present in the expression for the electromagnetic torque.

2. *Discretized augmented induction-machine model* For digital implementation of the EKF, the discretized machine equations are required. These can be obtained from eqns (4.5-120), (4.5-121) or (4.5-124), (4.5-125) as follows:

$$\mathbf{x}(k+1) = \mathbf{A}_d \mathbf{x}(k) + \mathbf{B}_d(k) \mathbf{u}(k) \tag{4.5-128}$$

$$\mathbf{y}(k) = \mathbf{C} \mathbf{x}(k). \tag{4.5-129}$$

It should be noted that for the sampled value of $\mathbf{x}$ in the t_k instant, the notation $\mathbf{x}(t_k)$ could be used, or the corresponding more simple $\mathbf{x}(k)$, which strictly means $\mathbf{x}(kT)$, which corresponds to sampling at the kth instant, and the sampling time is T ($T = t_{k+1} - t_k$). In eqns (4.5-128) and (4.5-129) $\mathbf{A}_d$ and $\mathbf{B}_d$ are the discretized system and input matrices respectively,

$$\mathbf{A}_d = \exp[\mathbf{A}T] \approx \mathbf{I} + \mathbf{A}T + \frac{(\mathbf{A}T)^2}{2} \tag{4.5-130}$$

$$\mathbf{B}_d = \approx \mathbf{B}T + \frac{\mathbf{AB}T^2}{2}, \tag{4.5-131}$$

and it should be noted that the discrete output matrix is $\mathbf{C}_d = \mathbf{C}$, where $\mathbf{C}$ has been defined in eqns (4.5-121) and (4.5-125). It should be noted that when the last terms are ignored in eqns (4.5-130) and (4.5-131), then they require very short sampling times to give a stable and accurate discretized model. However, a better approximation is obtained with the second-order series expansion, when these last terms are also considered. In general, to achieve adequate accuracy, the sampling time should be appreciably smaller than the characteristic time-constants of the machine. The final choice for this should be based on obtaining adequate execution time of the full EKF algorithm and also satisfactory accuracy and stability. The second-order technique obviously increases the computational time.

As an example of the discretized equations, the discrete form of eqn (4.5-120) is now first given, if the second-order terms are neglected in eqns (4.5-130) and (4.5-131):

$$\mathbf{x}(k+1) = \mathbf{A}_d\mathbf{x}(k) + \mathbf{B}_d\mathbf{u}(k) \tag{4.5-132}$$

$$\mathbf{y}(k) = \mathbf{C}\mathbf{x}(k). \tag{4.5-133}$$

$$\mathbf{A}_d = \begin{bmatrix} 1-T/T_s'^* & 0 & TL_m/(L_s'L_rT_r) & \omega_r TL_m/(L_s'L_r) & 0 \\ 0 & 1-T/T_s'^* & -\omega_r TL_m/(L_s'L_r) & TL_m/(L_s'L_rT_r) & 0 \\ TL_m/T_r & 0 & 1-T/T_r & -T\omega_r & 0 \\ 0 & TL_m/T_r & T\omega_r & 1-T/T_r & 0 \\ 0 & 0 & 0 & 0 & 1 \end{bmatrix}$$

$$\mathbf{B}_d = \begin{bmatrix} T/L_s' & 0 \\ 0 & T/L_s' \\ 0 & 0 \\ 0 & 0 \\ 0 & 0 \end{bmatrix} \quad \mathbf{C}_d = \begin{bmatrix} 1 & 0 & 0 & 0 & 0 \\ 0 & 1 & 0 & 0 & 0 \end{bmatrix}, \tag{4.5-134}$$

where

$$\mathbf{x}(k) = [i_{sD}(k) i_{sQ}(k) \psi_{rd}(k) \psi_{rq}(k) \omega_r(k)]^T$$

$$\mathbf{u}(k) = [u_{sD}(k) u_{sQ}(k)]^T.$$

By considering the system noise $\mathbf{v}(k)$ ($\mathbf{v}$ is the noise vector of the states), which is assumed to be zero-mean, white Gaussian, which is independent of $\mathbf{x}(k)$, and which has covariance matrix $\mathbf{Q}$, the system model becomes

$$\mathbf{x}(k+1) = \mathbf{A}_d\mathbf{x}(k) + \mathbf{B}_d\mathbf{u}(k) + \mathbf{v}(k). \tag{4.5-135}$$

By considering a zero-mean, white Gaussian measurement noise, $\mathbf{w}(k)$ (noise in the measured stator currents), which is independent of $\mathbf{y}(k)$ and $\mathbf{v}(k)$ and whose covariance matrix is $\mathbf{R}$, the output equation becomes

$$\mathbf{y}(k)=\mathbf{C}\mathbf{x}(k)+\mathbf{w}(k). \tag{4.5-136}$$

3. Determination of the noise and state covariance matrices **Q**, **R**, **P** The goal of the Kalman filter is to obtain unmeasurable states (e.g. rotor speed) by using measured states, and also statistics of the noise and measurements [i.e. covariance matrices **Q**, **R**, **P** of the system noise vector, measurement noise vector, and system state vector ($\mathbf{x}$) respectively]. In general, by means of the noise inputs, it is possible to take account of computational inaccuracies, modelling errors, and errors in the measurements. The filter estimation ($\hat{\mathbf{x}}$) is obtained from the predicted values of the states ($\mathbf{x}$) and this is corrected recursively by using a correction term, which is the product of the Kalman gain (**K**) and the deviation of the estimated measurement output vector and the actual output vector ($\mathbf{y}-\hat{\mathbf{y}}$). The Kalman gain is chosen to result in the best possible estimated states. Thus the filter algorithm contains basically two main stages, a prediction stage and a filtering stage. During the prediction stage, the next predicted values of the states $\mathbf{x}(k+1)$ are obtained by using a mathematical model (state-variable equations) and also the previous values of the estimated states. Furthermore, the predicted state covariance matrix (**P**) is also obtained before the new measurements are made, and for this purpose the mathematical model and also the covariance matrix of the system (**Q**) are used. In the second stage, which is the filtering stage, the next estimated states, $\hat{\mathbf{x}}(k+1)$ are obtained from the predicted estimates $\mathbf{x}(k+1)$ by adding a correction term $\mathbf{K}(\mathbf{y}-\hat{\mathbf{y}})$ to the predicted value. This correction term is a weighted difference between the actual output vector ($\mathbf{y}$) and the predicted output vector ($\hat{\mathbf{y}}$), where **K** is the Kalman gain. Thus the predicted state estimate (and also its covariance matrix) is corrected through a feedback correction scheme that makes use of the actual measured quantities. The Kalman gain is chosen to minimize the estimation-error variances of the states to be estimated. The computations are realized by using recursive relations. The algorithm is computationally intensive, and the accuracy also depends on the model parameters used. A critical part of the design is to use correct initial values for the various covariance matrixes. These can be obtained by considering the stochastic properties of the corresponding noises. Since these are usually not known, in most cases they are used as weight matrices, but it should be noted that sometimes simple qualitative rules can be set up for obtaining the covariances in the noise vectors (see also Section 3.1.3.5.1 and the last part of the present section, Section 4.5.3.5.2). With advances in DSP technology, it is possible to conveniently implement an EKF in real time.

The system noise matrix **Q** is a five-by-five matrix, the measurement noise matrix **R** is a two-by-two matrix, so in general this would require the knowledge of 29 elements. However, by assuming that the noise signals are not correlated, both **Q** and **R** are diagonal, and only 5 elements must be known in **Q** and 2 elements in **R**. However, the parameters in the direct and quadrature axes are the same, which means that the two first elements in the diagonal of $\mathbf{Q}$ are equal ($q_{11}=q_{22}$), the third and fourth elements in the diagonal of $\mathbf{Q}$ are equal ($q_{33}=q_{44}$), so $\mathbf{Q}=\text{diag}(q_{11}, q_{11}, q_{33}, q_{33}, q_{55})$ contains only 3 elements which have to be known. Similarly, the two diagonal elements in $\mathbf{R}$ are equal ($r_{11}=r_{22}=r$), thus $\mathbf{R}=\text{diag}(r, r)$. It follows that in total only 4 noise covariance elements must be known.

4. Implementation of the discretized EKF algorithm; tuning As discussed above, the EKF is an optimal, recursive state estimator, and the EKF algorithm contains basically two main stages, a prediction stage and a filtering stage. During the prediction stage, the next predicted values of the states $\mathbf{x}(k+1)$ [which will be denoted by $\mathbf{x}^*(k+1)$] and the predicted state covariance matrix ($\mathbf{P}$) [which will be denoted by $\mathbf{P}^*$] are also obtained. For this purpose the state-variable equations of the machine are used, and also the system covariance matrix ($\mathbf{Q}$). During the filtering stage, the filtered states ($\hat{\mathbf{x}}$) are obtained from the predicted estimates by adding a correction term to the predicted value ($\mathbf{x}^*$); this correction term is $\mathbf{Ke}=\mathbf{K}(\mathbf{y}-\hat{\mathbf{y}})$, where $\mathbf{e}=(\mathbf{y}-\hat{\mathbf{y}})$ is an error term, and it uses measured stator currents, $\mathbf{y}=\mathbf{i}_s$, $\hat{\mathbf{y}}=\hat{\mathbf{i}}_s$. This error is minimized in the EKF. The EKF equation is

$$\frac{d\hat{\mathbf{x}}}{dt}=\mathbf{A}(\hat{\mathbf{x}})\hat{\mathbf{x}}+\mathbf{Bu}+\mathbf{K}(\mathbf{i}_s-\hat{\mathbf{i}}_s). \qquad (4.5\text{-}137)$$

The structure of the EKF is shown in Fig. 4.78. The state estimates are obtained by the EKF algorithm in the following seven steps:

Step 1: *Initialization of the state vector and covariance matrices*

Starting values of the state vector $\mathbf{x}_0=\mathbf{x}(t_0)$ and the starting values of the noise covariance matrixes $\mathbf{Q}_0$ (diagonal 5×5 matrix) and $\mathbf{R}_0$ (diagonal 2×2 matrix) are set, together with the starting value of the state covariance matrix $\mathbf{P}_0$ (which is a 5×5 matrix), where $\mathbf{P}$ is the covariance matrix of the state vector (the terminology

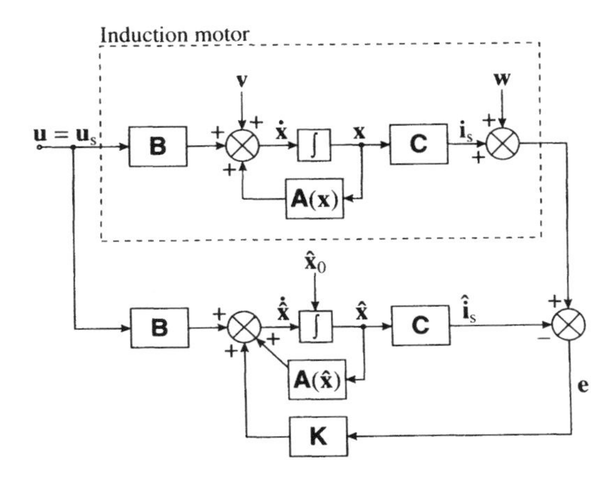

Fig. 4.78. Structure of the EKF.

'covariance matrix of prediction', 'error covariance matrix' are also used in the literature). The starting-state covariance matrix can be considered as a diagonal matrix, where all the elements are equal. The initial values of the covariance matrices reflect on the degree of knowledge of the initial states: the higher their values, the less accurate is any available information on the initial states. Thus the new measurement data will be more heavily weighted and the convergence speed of the estimation process will increase. However, divergence problems or large oscillations of state estimates around a true value may occur when too high initial covariance values are chosen. A suitable selection allows us to obtain satisfactory speed convergence, and avoids divergence problems or unwanted large oscillations.

The accuracy of the state estimation is affected by the amount of information that the stochastic filter can extract from its mathematical model and the measurement data processing. Some of the estimated variables, especially unmeasured ones, may indirectly and weakly be linked to the measurement data, so only poor information is available to the EKF. Finally it should be noted that another important factor which influences the estimation accuracy is due to the different sizes of the state variables, since the minimization of the trace of the estimation covariance ($\hat{\mathbf{P}}$) may lead to high-percentage estimation errors for the variables with small size. This problem can be overcome by choosing normalized state-variables.

Step 2: *Prediction of the state vector*

Prediction of the state vector at sampling time $(k+1)$ from the input $\mathbf{u}(k)$, state vector at previous sampling time, $\hat{\mathbf{x}}(k)$, by using $\mathbf{A}_d$ and $\mathbf{B}_d$ is obtained by performing

$$\mathbf{x}^*(k+1|k)=\mathbf{x}^*(\mathbf{k}+1)=\mathbf{A}_d\hat{\mathbf{x}}(k)+\mathbf{B}_d\mathbf{u}(k). \qquad (4.5\text{-}138)$$

The notation $\mathbf{x}^*(k+1|k)$ means that it is a predicted value at the $(k+1)$-th instant, and it is based on measurements up to the kth instant. However, to simplify the notation, this has been replaced by $\mathbf{x}^*(k+1)$. Similarly, $\hat{\mathbf{x}}(k|k)$ has been replaced by $\hat{\mathbf{x}}(k)$.

Thus from eqn (4.5-135)

$$\mathbf{x}^*(k+1)=[a\ b\ c\ d\ e]^{\mathrm{T}},$$

where

$$\begin{aligned}
a&=(1-T/T_s'^*)\hat{\imath}_{sD}(k)+[TL_m/(L_s'L_rT_r)\hat{\psi}_{rd}(k)]\\
&\quad+[\hat{\omega}_r(k)TL_m/(L_s'L_r)]\hat{\psi}_{rq}(k)+(T/L_s')u_{sD}(k)\\
b&=(1-T/T_s'^*)\hat{\imath}_{sQ}(k)-[\hat{\omega}_r(k)TL_m/(L_s'L_r)\hat{\psi}_{rd}(k)]\\
&\quad+[TL_m/(L_s'L_rT_r)]\hat{\psi}_{rq}(k)+(T/L_s')u_{sQ}(k)\\
c&=(TL_m/T_r)\hat{\imath}_{sD}(k)+(1-T/T_r)\hat{\psi}_{rd}(k)-T\hat{\omega}_r(k)\hat{\psi}_{rq}(k) \qquad (4.5\text{-}139)\\
d&=(TL_m/T_r)\hat{\imath}_{sQ}(k)+(1-T/T_r)\hat{\psi}_{rq}(k)+T\hat{\omega}_r(k)\hat{\psi}_{rd}(k)\\
e&=\hat{\omega}_r(k)
\end{aligned}$$

Step 3: *Covariance estimation of prediction*

The covariance matrix of prediction is estimated as

$$\mathbf{P}^*(k+1)=\mathbf{f}(k+1)\hat{\mathbf{P}}(k)\mathbf{f}^{\mathrm{T}}(k+1)+\mathbf{Q}, \tag{4.5-140}$$

where, to have simpler notation, $\mathbf{P}^*(k+1|k)$ has been replaced by $\mathbf{P}^*(k+1)$, $\hat{\mathbf{P}}(k|k)$has been replaced by $\hat{\mathbf{P}}(k)$ [where $k|k$ denotes prediction at time k based on data up to time k] and $\mathbf{f}(k+1|k)$ has been replaced by $\mathbf{f}(k+1)$, where $\mathbf{f}$ is the following gradient matrix:

$$\mathbf{f}(k+1)=\frac{\partial}{\partial \mathbf{x}}\left(\mathbf{A}_{\mathrm{d}}\mathbf{x}+\mathbf{B}_{\mathrm{d}}\mathbf{u}\right)\Bigg|_{\mathbf{x}=\hat{\mathbf{x}}(k+1)}. \tag{4.5-141}$$

By using eqn (4.5-134),

$$\mathbf{f}(k+1)=\begin{bmatrix} 1-T/T_s'^* & 0 & TL_m/(L_s'L_rT_r) & \omega_r TL_m/(L_s'L_r) & [TL_m/(L_s'L_r)]\psi_{rq} \\ 0 & 1-T/T_s'^* & -\omega_r TL_m/(L_s'L_r) & TL_m/(L_s'L_rT_r) & -[TL_m/(L_s'L_r)]\psi_{rd} \\ TL_m/T_r & 0 & 1-T/T_r & -T\omega_r & T\psi_{rq} \\ 0 & TL_m/T_r & T\omega_r & 1-T/T_r & T\psi_{rd} \\ 0 & 0 & 0 & 0 & 1 \end{bmatrix} \tag{4.5-142}$$

where $\omega_r=\hat{\omega}_r(k+1)$, $\psi_{rd}=\hat{\psi}_{rd}(k+1)$, $\psi_{rq}=\hat{\psi}_{rq}(k+1)$. There are 17 elements in $\mathbf{f}$ which are constant and 8 elements ($f_{14}, f_{15}, f_{23}, f_{25}, f_{34}, f_{35}, f_{43}, f_{45}$) which are variable. In a practical DSP application it is useful to compute first the gradient matrix elements, since this contains the constants required in $\mathbf{x}^*(k+1)$. This leads to reduced memory requirements and reduced computational time, since space is only reserved once for the constant elements and the products involving the speed and the flux linkages have to be computed only once in one recursive step.

Step 4: *Kalman filter gain computation*

The Kalman filter gain (correction matrix) is computed as

$$\mathbf{K}(k+1)=\mathbf{P}^*(k+1)\mathbf{h}^{\mathrm{T}}(k+1)[\mathbf{h}(k+1)\mathbf{P}^*(k+1)\mathbf{h}^{\mathrm{T}}(k+1)+\mathbf{R}]^{-1}. \tag{4.5-143}$$

For the induction machine application, the Kalman gain matrix contains two columns and five rows. For simplicity of the notation, $\mathbf{P}^*(k+1)$ has been replaced by $\mathbf{P}^*(k+1)$, and $\mathbf{h}(k+1)$ is a gradient matrix, defined as

$$\mathbf{h}(k+1)=\frac{\partial}{\partial \mathbf{x}}\left[\mathbf{C}_{\mathrm{d}}\mathbf{x}\right]\Bigg|_{\mathbf{x}=\mathbf{x}^*(k+1)} \tag{4.5-144}$$

$$\mathbf{h}(k+1)=\begin{bmatrix} 1 & 0 & 0 & 0 & 0 \\ 0 & 1 & 0 & 0 & 0 \end{bmatrix}. \tag{4.5-145}$$

Step 5: *State-vector estimation*

The state-vector estimation (corrected state-vector estimation, filtering) at time $(k+1)$ is performed as:

$$\hat{\mathbf{x}}(k+1)=\mathbf{x}^*(k+1)+\mathbf{K}(k+1)[\mathbf{y}(k+1)-\hat{\mathbf{y}}(k+1)], \tag{4.5-146}$$

where for simplicity of the notation, $\hat{\mathbf{x}}(k+1|k+1)$ has been replaced by $\hat{\mathbf{x}}(k+1)$, $\mathbf{x}^*(k+1|k)$ has been replaced by $\mathbf{x}^*(k+1)$, and

$$\hat{\mathbf{y}}(k+1)=\mathbf{C}_d\mathbf{x}^*(k+1), \tag{4.5-147}$$

where

$$\hat{\mathbf{y}}(k+1)=[i^*_{sD}(k+1),\ i^*_{sQ}(k+1)]^T. \tag{4.5-148}$$

Step 6: *Covariance matrix of estimation error*

The error covariance matrix can be obtained from

$$\hat{\mathbf{P}}(k+1)=\mathbf{P}^*(k+1)-\mathbf{K}(k+1)\mathbf{h}(k+1)\mathbf{P}^*(k+1), \tag{4.5-149}$$

where for simplicity of notation $\hat{\mathbf{P}}(k+1|k+1)$ has been replaced by $\hat{\mathbf{P}}(k+1)$, and $\mathbf{P}^*(k+1|k)$ has been replaced by $\mathbf{P}^*(k+1)$.

Step 7: *Put* $k=k+1$, $\mathbf{x}(k)=\mathbf{x}(k-1)$, $\mathbf{P}(k)=\mathbf{P}(k-1)$ *and goto Step 1.*

For the realization of the EKF algorithm it is very convenient to use a signal processor (e.g. Texas Instruments TMS320C30, TMS320C40, TMS320C50, etc.), due to the large number of multiplications required and also due to the fact that all of the computations have to be performed fast: within one sampling interval. The calculation time of the EKF is 130 μs on the TMS320C40. However, the calculation times can be reduced by using optimized machine models.

The EKF described above can be used under both steady-state and transient conditions of the induction machine for the estimation of the rotor speed. By using the EKF in the drive system, it is possible to implement a PWM inverter-fed induction motor drive without the need of an extra speed sensor. It should be noted that accurate speed sensing is obtained in a very wide speed-range, down to very low values of speed (but not zero speed). However, care must be taken in the selection of the machine parameters and covariance values used. The speed estimation scheme requires the monitored stator voltages and stator currents. Instead of using the monitored stator line voltages, the stator voltages can also be reconstructed by using the d.c. link voltage and inverter switching states, but especially at low speeds it is necessary to have an appropriate dead-time compensation, and also the voltage drops across the inverter switches (e.g. IGBTs) must be considered (see also Section 3.4.3.2.1). Furthermore in a VSI inverter-fed vector-controlled induction motor drive, where a space-vector modulator is used, it is also possible to use the reference voltages (which are the inputs to the modulator) instead of the actual voltages, but this requires an appropriate error compensation (see also Section 4.5.3.5.1 on the full-order state observer).

The tuning of the EKF involves an iterative modification of the machine parameters and covariances in order to yield the best estimates of the states. Changing the covariance matrices **Q** and **R** affects both the transient duration and steady-state operation of the filter. Increasing **Q** corresponds to stronger system noises, or larger uncertainty in the machine model used. The filter gain matrix elements will also increase and thus the measurements will be more heavily weighted and the filter transient performance will be faster. If the covariance **R** is increased, this corresponds to the fact that the measurements of the currents are subjected to a stronger noise, and should be weighted less by the filter. Thus the filter gain matrix elements will decrease and this results in slower transient performance. Finally it should be noted that in general, the following qualitative tuning rules can be obtained: Rule 1: If **R** is large then **K** is small (and the transient performance is faster). Rule 2: If **Q** is large then **K** is large (and the transient performance is slower). However, if **Q** is too large or if **R** is too small, instability can arise. It is possible to derive similar rules to these rules, and to implement a fuzzy-logic-assisted system for the selection of the appropriate covariance elements.

In summary it can be stated that the EKF algorithm is computationally more intensive than the algorithm for the full-order state observer described in the previous section. The EKF can also be used for joint state and parameter estimation. It should be noted that to reduce the computational effort and any steady state error, it is possible to use various EKFs, which utilize reduced-order machine models and different reference frames (e.g. a reference frame fixed to the stator current vector).

4.5.3.5.4 Application of the extended Luenberger observer (ELO)

To develop a speed-sensorless high-dynamic-performance a.c. drive system, both the rotor speed and flux-linkage signals have to be estimated. Conventionally these have been obtained by using open-loop estimators. A voltage model or a current model of the induction machine, or a combination of these (hybrid estimators), are often used to estimate the speed and some other machine quantities. The problem with this kind of approach is that the estimation error introduced in the speed estimation cannot be overcome and can lead to an increasing deviation of the result from the actual value. If the error is significant, the detuning of the drive system may result. Therefore, for a speed-sensorless control system, it may be advisable that the estimation of the speed and the estimation of the flux should not be treated separately because there always exist mutual interactions between the two estimation systems. Otherwise, any error introduced by the flux-linkage estimator may be amplified by the action of the speed estimator, and vice versa. Thus it is a main objective to describe below an implementation of a real-time joint-flux-and-speed ELO estimation scheme.

It has been discussed in Section 4.5.3.5.1 that the basic Luenberger observer is applicable to a linear, time-invariant deterministic system. The extended Luenberger observer (ELO) is applicable to a non-linear time-varying deterministic system. However, in the past, extended Kalman filters (EKF) have been used almost uniquely for the joint state and parameter estimation problem in a.c. drive systems (see also Section 4.5.3.5.3). However, this stochastic approach appears to have some inherent disadvantages. Difficulties may arise in situations where the noise content of the system and associated measurements are too low. The tuning of the EKF is *ad hoc*. In contrast to the standard linear Kalman filter, the EKF is not optimal (this important point is not well recognized) [Du *et al.* 1995]. There may be a bias problem when the assumed characteristics of stochastic noises do not match those of the real ones [Ljung 1979, Du *et al.* 1995]. Perhaps the most adverse drawback is that there is no means in the EKF design and implementation which can be utilized to tune its dynamic performance without affecting its steady-state accuracy. Therefore this traditional approach may sometimes not be efficient due to its computation burden, and sometimes even be unacceptable due to its bias.

Basic Luenberger observer A basic Luenberger observer (LO) can only be applied to the estimation of states of a linear-time-invariant system described by

$$\dot{\mathbf{x}}(t)=\mathbf{A}\mathbf{x}(t)+\mathbf{B}\mathbf{u}(t) \tag{4.5-150}$$

$$\mathbf{y}(t)=\mathbf{C}\mathbf{x}(t), \tag{4.5-151}$$

and the LO is described by

$$\dot{\hat{\mathbf{x}}}(t)=\mathbf{A}\hat{\mathbf{x}}(t)+\mathbf{B}\mathbf{u}(t)+\mathbf{G}[\mathbf{y}(t)-\mathbf{C}\hat{\mathbf{x}}(t)], \tag{4.5-152}$$

where $\hat{\mathbf{x}}(t)$ is the state vector of the estimates and $\mathbf{G}$ is the observer gain matrix.

Extended Luenberger observer In contrast to the LO, the ELO can be applied to the estimation of the states of a non-linear time-invariant system, described by

$$\dot{\mathbf{x}}(t)=\mathbf{f}[\mathbf{x}(t)]+\mathbf{B}\mathbf{u}(t) \tag{4.5-153}$$

$$\mathbf{y}(t)=\mathbf{C}\mathbf{x}(t). \tag{4.5-154}$$

Equations (4.5-153) and (4.5-154) represent an extended induction-motor model. The model of a symmetrical three-phase single-cage induction machine can be described by five first-order differential equations, where four equations correspond to the stator and rotor voltage equations. The simplest derivation uses the stator and rotor space-vector equations of the induction machine given by eqns (2.1-148)–(2.1-151). These yield the following state-variable equations in the reference frame rotating with the rotor flux-linkage space vector ($\omega_g=\omega_{mr}$) if the space vectors of the stator currents ($\bar{i}_s'$) and rotor flux linkages ($\bar{\psi}_r'$) are selected as state variables:

$$\frac{d}{dt}\begin{bmatrix}\bar{i}_s'\\ \bar{\psi}_r'\end{bmatrix}=\begin{bmatrix}-1/T_s'^{*}+j\omega_{mr} & -[L_m/(L_s'L_r)](-1/T_r+j\omega_r)\\ L_m/T_r & -1/T_r+j(\omega_r-\omega_{mr})\end{bmatrix}\begin{bmatrix}\bar{i}_s'\\ \bar{\psi}_r'\end{bmatrix}+\begin{bmatrix}\bar{u}_s'/L_s'\\ 0\end{bmatrix} \tag{4.5-155}$$

where $\bar{u}_s'$ is the space vector of stator voltages (in the rotor-flux-oriented reference frame), L_m and L_r are the magnetizing inductance and rotor self-inductance respectively, and L_s' is the stator transient inductance. Furthermore, in eqn (4.5-155) $T_s'^{*}$ has been defined by eqn (4.5-119), which is now repeated here for convenience:

$$\frac{1}{T_s'^{*}}=\frac{1}{T_s'}+\frac{(1-\sigma)}{T_r'}=\frac{R_s+R_r(L_m/L_r)^2}{L_s'}=\frac{R_s+R_{rref}}{L_s'}, \tag{4.5-156}$$

where $T_s' = L_s'/R_s$ and $T_r = L_r'/R_r$ are the stator and rotor transient time-constants respectively and $\sigma = 1 - L_m^2/(L_s L_r)$ is the leakage factor. Resolving eqn (4.5-155) into its real- and imaginary-axis components gives the following state-variable form of the voltage equations:

$$\dot{\mathbf{x}}_n(t) = \mathbf{f}_n[\mathbf{x}_n(t), \omega_r] + \mathbf{B}_n \mathbf{u}(t), \tag{4.5-157}$$

where $\mathbf{x}_n = [i_{sx}, i_{sy}, \psi_{rx}, \psi_{ry},]^T$ is the state vector of the states, $\mathbf{u} = [u_{sx}, u_{sy}]^T$ contains the stator-voltage components in the rotor-flux-oriented reference frame, and the input matrix is

$$\mathbf{B}_n = \begin{bmatrix} 1/L_s' & 0 & 0 & 0 \\ 0 & 1/L_s' & 0 & 0 \end{bmatrix}^T. \tag{4.5-158}$$

Furthermore, $\mathbf{f}_n(\mathbf{x}_n, \omega_r) = \mathbf{A}_n(\omega_r)\mathbf{x}_n$, where the state matrix of the induction motor is defined as $\mathbf{A}_n(\omega_r)$, where

$$\mathbf{A}_n(\omega_r) = \begin{bmatrix} -1/T_s'^* & \omega_{mr} & L_m/(L_r L_s T_r') & \omega_r L_m/(L_r L_s') \\ -\omega_{mr} & -1/T_s'^* & -\omega_r L_m/(L_r L_s') & L_m/(L_r L_s T_r') \\ L_m/T_r & 0 & -1/T_r & \omega_{mr} - \omega_r \\ 0 & L_m/T_r & \omega_r - \omega_{mr} & -1/T_r \end{bmatrix} \tag{4.5-159}$$

It should be noted that it is possible to obtain a somewhat simpler state matrix if, instead of the rotor flux-linkage components, the referred values of these are used as state variables [see also derivation of eqn (4.5-124)]. In agreement with the definition of the referred rotor flux-linkage space vector [eqn (4.5-123)], in this case the L_m/L_r coefficients are not present in the stator voltage equation, and as a direct consequence, in the rotor voltage equation, L_m is replaced by the referred magnetizing inductance $L_M = L_m^2/L_r$.

The fifth equation of the induction machine corresponds to the equation of motion,

$$\dot{\omega}_r = c_1 (i_{sy}\psi_{rx} - i_{sx}\psi_{ry}) - c_2 t_L. \tag{4.5-160}$$

In eqn (4.1-160) the first term is the electromagnetic torque given by eqn (2.1-177), and e.g. for a two-pole machine c_1 and c_2 are defined as $c_1 = 3L_m/(2JL_r)$, $c_2 = 1/J$, where J is the inertia. It should be noted that when the referred flux linkages are used as state variables, then c_1 does not contain L_m/L_r. The damping torque caused by windage and friction has been ignored in the equation of motion. However, the equation of motion contains the load torque, which in general is unknown. This makes the direct estimation of the rotor speed difficult. However, there are three possible approaches to this problem. The first approach is to ignore the load torque and the second approach is to assume that the rate of change of the rotor speed is zero,

$$\dot{\omega}_r = 0. \tag{4.5-161}$$

In this case satisfactory rotor-speed estimates can be obtained if the mechanical time constant is much larger than the electromagnetic time constants. The third method uses the equation of motion but assumes a constant load torque, so that both the speed and the load torque can be estimated. In the present book the second approach is discussed, since it is believed that in general this gives the most accurate estimates.

The first step for the joint rotor speed and rotor flux-linkage estimation is to construct the extended motor model. For this purpose an extended state vector is defined as $\mathbf{x}=[\mathbf{x}_n, \mathbf{x}_p]^T$, where in general $\mathbf{x}_p$ denotes the parameter vector to be estimated, and for the present application $\mathbf{x}_p=\omega_r$. Thus it follows from eqns (4.5-157) and (4.5-161), and also considering that the equation defining the measurement variables $\mathbf{y}=[i_{sx}, i_{sy}]^T$ is

$$\mathbf{y}(t)=\mathbf{C}_n\mathbf{x}_n(t) \tag{4.5-162}$$

where

$$\mathbf{C}_n=\begin{bmatrix}1 & 0 & 0 & 0\\ 0 & 1 & 0 & 0\end{bmatrix}, \tag{4.5-163}$$

and the extended induction motor model can be described by

$$\dot{\mathbf{x}}(t)=\mathbf{f}[\mathbf{x}(t)]+\mathbf{Bu}(t)=[\mathbf{f}_n(\mathbf{x}_n, \mathbf{x}_p), \mathbf{f}_p(\mathbf{x}_n, \mathbf{x}_p)]^T+[\mathbf{B}_n, 0]^T\mathbf{u}(t) \tag{4.5-164}$$

$$\mathbf{y}(t)=\mathbf{Cx}(t)=[\mathbf{C}_n, 0][\mathbf{x}_n, \mathbf{x}_p]^T. \tag{4.5-165}$$

In eqn (4.5-165) $\mathbf{y}$ is the extended measurement vector.

In contrast to the design of an LO, which results in a time-invariant observer, the ELO is fundamentally time-varying and requires continuous updating of the observer coefficients. Therefore the design of the ELO is based on the linearized form of eqn (4.5-164). Thus the linearized extended induction motor model (around the estimate of the augmented state $\mathbf{x}$ in the last step, denoted by $\tilde{\mathbf{x}}$) is obtained as

$$\dot{\mathbf{x}}(t)=\mathbf{f}'_{\tilde{\mathbf{x}}}[\tilde{\mathbf{x}}(t)]+\mathbf{Bu}(t)+\mathbf{g}[\tilde{\mathbf{x}}(t)]=\mathbf{A}(\tilde{\mathbf{x}})\mathbf{x}+\mathbf{Bu}(t)+\mathbf{g}(\tilde{\mathbf{x}}), \tag{4.5-166}$$

where $\tilde{\mathbf{x}}(t)$ is the reference trajectory, which is usually chosen as the estimate in the last step, $\hat{\mathbf{x}}(t-\tau)$. In eqn (4.5-166) $\mathbf{f}'_{\tilde{\mathbf{x}}}=\mathrm{d}\mathbf{f}/\mathrm{d}\tilde{\mathbf{x}}$ is the system Jacobian matrix (gradient matrix), which is a 5-by-5 matrix,

$$\mathbf{A}=\frac{\mathrm{d}\mathbf{f}}{\mathrm{d}\tilde{\mathbf{x}}}=\begin{bmatrix}\mathbf{A}_n(\tilde{\mathbf{x}}_p) & \mathbf{A}_2(\tilde{\mathbf{x}}_n)\\ 0 & 0\end{bmatrix}, \tag{4.5-167}$$

where

$$\mathbf{A}_2(\tilde{\mathbf{x}}_n)=\frac{\partial(\mathbf{A}_n\mathbf{x}_n)}{\partial\tilde{\omega}_r}=[c\tilde{\psi}_{ry}, -c\tilde{\psi}_{rx}, -c\psi_{ry}, c\tilde{\psi}_{rx}]^T \tag{4.5-168}$$

$(c=L_m/L'_sL_r)$.

The term $\mathbf{g}[\tilde{\mathbf{x}}(t)]$ is defined as $\mathbf{g}[\tilde{\mathbf{x}}(t)]=\mathbf{f}(\tilde{\mathbf{x}})-\mathbf{f}_{\tilde{\mathbf{x}}}'(\tilde{\mathbf{x}})\tilde{\mathbf{x}}$ and thus

$$\mathbf{g}(\tilde{\mathbf{x}})=[\mathbf{A}(\tilde{\mathbf{x}})-\mathbf{A}(\tilde{\mathbf{x}})]\tilde{\mathbf{x}}. \tag{4.5-169}$$

The resulting full-order extended Luenberger observer can be written as

$$\dot{\hat{\mathbf{x}}}(t)=\mathbf{A}[\hat{\mathbf{x}}(t-\tau)]\hat{\mathbf{x}}(t)+\mathbf{B}\mathbf{u}(t)+\mathbf{G}[\hat{\mathbf{x}}(t-\tau)][\mathbf{y}(t)-\mathbf{C}\hat{\mathbf{x}}(t-\tau)]+\mathbf{g}[\hat{\mathbf{x}}(t-\tau)], \tag{4.5-170}$$

where

$$\mathbf{g}[\hat{\mathbf{x}}(t-\tau)]=\mathbf{f}[\hat{\mathbf{x}}(t-\tau)]-\mathbf{f}_{\hat{\mathbf{x}}(t-\tau)}'[\hat{\mathbf{x}}(t-\tau)]\hat{\mathbf{x}}(t-\tau) \tag{4.5-171}$$

and

$$\mathbf{f}_{\hat{\mathbf{x}}(t-\tau)}'[\hat{\mathbf{x}}(t-\tau)]=\mathbf{A}[\hat{\mathbf{x}}(t-\tau)]. \tag{4.5-172}$$

It should be noted that in eqn (4.5-170), the gain matrix $\mathbf{G}$ is not constant, but depends on the past estimates of the system state vector.

In applications where all of the states are not required, a reduced ELO may be used (Du *et al.* 1995). This is described by

$$\dot{\mathbf{z}}(t)=\mathbf{F}[\hat{\mathbf{x}}(t-\tau)]\mathbf{z}(t)+\mathbf{G}[\hat{\mathbf{x}}(t-\tau)]\mathbf{y}(t)+\mathbf{H}[\hat{\mathbf{x}}(t-\tau)]\mathbf{u}+\mathbf{D}[\hat{\mathbf{x}}(t-\tau)], \tag{4.5-173}$$

where the coefficient matrices are chosen to satisfy the constraints

$$\begin{aligned}\mathbf{T}f_{\hat{\mathbf{x}}}'(t-\tau)[\hat{\mathbf{x}}(t-\tau)]-\mathbf{F}[\hat{\mathbf{x}}(t-\tau)]\mathbf{T}&=\mathbf{GC}\\ \mathbf{TB}&=\mathbf{H}\\ \mathbf{D}&=\mathbf{T}\mathbf{g}[\hat{\mathbf{x}}(t-\tau)],\end{aligned} \tag{4.5-174}$$

where $\mathbf{T}$ is a transformation matrix. If $\mathbf{T}$ is chosen as a unity matrix, with dimension equal to the number of components in $\mathbf{x}(t)$, then a full-order ELO results. Equation (4.5-174) is used to determine the gain matrix $\mathbf{G}$.

For convenience, the ELO algorithm is now summarized for a continuous-time ELO.

Step 1 Specify the system function f and derive the system Jacobian $f_{\hat{\mathbf{x}}}'$.

Step 2 Input data: input the input coefficient matrix $\mathbf{B}$ and output coefficient matrix $\mathbf{C}$.

Step 3 Initialize all observer states.

Step 4 Specify required observer poles. Construct suitable observer matrix $\mathbf{F}$ with specified poles.

Step 5 Evaluate $\mathbf{g}$ using eqn (4.5-171).

Step 6 Compute gain matrix $\mathbf{G}$:
(a) for a full-order ELO: use $\mathbf{F}=\mathbf{A}\text{-}\mathbf{GC}$, with $\mathbf{A}=\mathbf{f}_{\hat{\mathbf{x}}}'$ evaluated at the present time instant;
(b) for a reduced-order ELO, use eqn (4.5-174) to obtain the gain matrix $\mathbf{G}$, the transformation matrix $\mathbf{T}$, the observer input coefficient matrix $\mathbf{H}$, and the matrix $\mathbf{D}$.

Step 7 Using a suitable numerical integration routine, compute the state vector:
(a) for the full-order ELO compute the state vector $\hat{\mathbf{x}}$ at the next time instant by solving eqn (4.5-170);
(b) for the reduced-order ELO compute the state vector $\mathbf{z}$ at the next time instant by solving eqn (4.5-173) and calculate $\hat{\mathbf{x}}$ using the relationship $\mathbf{z}=\mathbf{T}\hat{\mathbf{x}}$. In practice, direct evaluation of this equation is avoided [Orlowska-Kowalska 1989; Du *et al.* 1995].

Step 8 Go to Step 4

Finally, to indicate clearly the types of problems to which the ELO is applicable, to emphasize the main design features, and also to allow comparisons with the EKF, the important features of the ELO are now summarized:

- it is applicable to a majority of industrial systems since they can be regarded as deterministic;
- its performance can be altered by adjusting the gain matrix so that rapid convergence of the estimates and a robust design may be obtained;
- its computational requirements may be no more demanding than that for other estimation algorithms such as the EKF, as matrix inversion is required in the implementation of the EKF;
- its design incorporates a considerable amount of flexibility owing to the fact that redundancy exists in the constraints imposed by eqn (4.5-174) when specifying the gain matrix $\mathbf{G}$. This flexibility can be used to meet any prescribed criteria such as speed of response, speed of convergence, robustness against parameter drift, etc.

For real-time implementation a discrete form of the ELO is required, where the estimate at the previous sampling instant is often taken as the reference trajectory, thus

$$\hat{\mathbf{x}}(t-\tau)=\hat{\mathbf{x}}(t-T_s), \qquad (4.5\text{-}175)$$

where T_s is the sampling time. It should be noted that at time t, $\hat{\mathbf{x}}(t-T_s)$ is constant, so that for a full-order ELO, eqn (4.5-170) at the kth sampling time becomes

$$\dot{\hat{\mathbf{x}}}(t)=\mathbf{F}[\hat{\mathbf{x}}(t-T_s)]\hat{\mathbf{x}}(t)+\mathbf{B}\mathbf{u}(t)+\mathbf{G}[\hat{\mathbf{x}}(t-T_s)]\mathbf{y}(t)+\mathbf{g}[\hat{\mathbf{x}}(t-T_s)]$$

$$\mathbf{F}[\hat{\mathbf{x}}(t-T_s)]=\mathbf{A}[\hat{\mathbf{x}}(t-T_s)]-\mathbf{GC} \qquad (4.5\text{-}176)$$

$(kT_s<t\leqslant(k+1)T_s)$.

It is important to note that this represents a linear time-invariant system over the current sampling interval. Thus by using

$$\mathbf{\Phi}_k=\exp(\mathbf{F}_k T_s) \qquad (4.5\text{-}177)$$

$$\mathbf{\Gamma}_k=\int_0^{T_s}\exp(\mathbf{F}_k t)\,\mathrm{d}t \qquad (4.5\text{-}178)$$

then the estimation is obtained as

$$\hat{\mathbf{x}}_{k+1|k} = \mathbf{\Phi}_k \hat{\mathbf{x}}_{k|k-1} + \mathbf{\Gamma}_k(\mathbf{G}_k \mathbf{y}_k + \mathbf{B}\mathbf{u}_k + \mathbf{g}_k). \quad (4.5\text{-}179)$$

If $\mathbf{F}_k$ is chosen as a constant matrix, $\mathbf{\Phi}_k$ and $\mathbf{\Gamma}_k$ will also be constant and their values can be computed in advance. Thus the computation time of eqn (4.5-179) is significantly reduced.

To allow comparison with the EKF, the main features of the EKF are summarized here:

- it is based on an extension of a readily implementable algorithm, the Kalman filter;
- it is effective for applications in industrial systems which can be regarded as stochastic in nature;
- its performance can be tuned by adjusting the covariance matrices;
- its design incorporates no flexibility owing to the fact that a constraint implying optimality must be satisfied. Thus no additional prescribed performance criteria such as speed of response, speed convergence, robustness against parameter variation, etc. can be accommodated directly into the design procedure;
- in contrast to the basic Kalman filter, *ad hoc* covariance matrix tuning adjustments, and the non-linear nature of the plant, may result in a non-optimal estimator;
- *ad hoc* covariance matrix adjustments may also result in a bias problem [Ljung 1979].

It appears from a comprehensive study of the ELO in high-performance torque-controlled induction motor drives that the ELO is always capable of producing unbiased estimates. It is important to note that the transient behaviour of the ELO can be conveniently tuned by the common control-system design procedure of adjusting the pole positions. This tuning does not degrade the steady-state performance. When the ELO is used for joint rotor speed and rotor flux-linkage estimation in a torque-controlled induction motor drive, then accurate speed estimates can be obtained even for fast speed transients. However, when the speed is very low, its non-linear observability becomes weak for joint flux-linkage and speed estimation. However, this weak observability does not constitute a severe problem if the drive does not operate constantly around zero speed.

4.5.3.6 Estimators using artificial intelligence

In Section 3.1.3.7 and Chapter 7 the possibilities of using different types of artificial-intelligence-based speed and position estimators are discussed. The two main solutions considered use artificial neural networks (ANN) and fuzzy-neural networks, and the advantages of the fuzzy-neural implementations are also highlighted in Section 3.1.3.7 and Chapter 7. It is believed that this type of approach will find increasing application in the future. This is mainly due to the fact that the development time of such an estimator is short and the estimator can be made robust to parameter variations and noise. Furthermore, in contrast to all conventional schemes, it can avoid the direct use of a speed-dependent mathematical model of the machine. Two types of ANN-based speed estimators are discussed below for an induction machine. The first one uses a simple two-layer ANN, where the rotor speed is proportional to the appropriate weights, and this ANN is part of a Model Reference Adaptive System (MRAS). In this system, the ANN takes the role of the adaptive model. However, in the more general, second implementation, which is not related to a MRAS, a multi-layer feedforward ANN with hidden layers is used, and the rotor speed is directly present on the output of the ANN.

4.5.3.6.1 MRAS containing two-layer ANN: speed proportional to weights

In Section 4.5.3.4 various MRAS-based speed estimation schemes have been described. All of them have contained a reference model and also an adaptive model. The inputs to the adaptive model were the estimated rotor speed and the stator currents. The estimated rotor speed was the output of a suitable adaptation mechanism, which utilized at its inputs the difference of the estimated state variables of the reference and adaptive models. The main differences between the various MRAS-based speed estimator schemes lie basically in the type of speed tuning signal used, but the adaptation mechanisms in all cases have used Popov's hyperstability criterion. This has eventually resulted in an adaptation mechanism in which the estimated state-variables of the reference and adaptive models were manipulated into a speed tuning signal, which was then input into a PI controller containing the proportional and integrator gains (K_p and K_i) of the adaptation mechanism (e.g. see Fig. 4.68). This approach has also required the use of a mathematical model for the adaptive model. However, greater accuracy and robustness can be achieved if this mathematical model is not used at all and instead, an artificial-intelligence-based no-linear adaptive model is employed. It is then also possible to eliminate the need for the separate PI controller, since this can be integrated into the tuning mechanism of the appropriate artificial-intelligence-based model.

An AI-based model can take various forms: it can be an artificial neural network (ANN) or a fuzzy-neural network, etc. Furthermore, different types of adaptation mechanism input signals (speed tuning signal) can be used, similarly to that used in the conventional schemes discussed in Section 4.5.3.4. Thus there are various possibilities for the speed tuning signal. It follows that if only one ANN configuration is considered (e.g. a back-propagation multi-layer feedforward ANN), and only one specific fuzzy-neural network is used (e.g. ANFIS [Jang 1993]), then it is possible to have eight different implementations by considering four different types of speed tuning signals. However, in practice the possibilities are even greater, since there are many types of ANNs and fuzzy-neural networks. It is believed that some of these solutions can give high accuracy and are robust to parameter variations even at extremely low stator frequency.

For illustration purposes, one specific implementation of this technique will now be discussed where the ANN contains adjustable and constant weights, and the adjustable weights are proportional to the rotor speed. The adjustable weights are changed by using the error between the outputs of the reference model and the adjustable model, since any mismatch between the actual speed and the estimated speed results in an error between the outputs of the reference and adaptive estimators.

Figure 4.79 shows the MRAS-based speed estimation scheme, which contains an ANN. It follows from Fig. 4.79 that the inputs to the reference model are the monitored stator voltages and currents of the induction machine. The outputs of the reference model are the rotor flux-linkage components in the stationary reference frame (ψ_{rd}, ψ_{rq}). These are obtained by considering eqns (4.5-83) and (4.5-84) in Section 4.5.3.4, which are now repeated here for convenience:

$$\psi_{rd} = \frac{L_r}{L_m}\left[\int (u_{sD} - R_s i_{sD})\,dt - L_s' i_{sD}\right] \tag{4.5-180}$$

$$\psi_{rq} = \frac{L_r}{L_m}\left[\int (u_{sQ} - R_s i_{sQ})\,dt - L_s' i_{sQ}\right]. \tag{4.5-181}$$

These two equations do not contain the rotor speed and describe the reference model. However, when the rotor voltage equations of the induction machine are expressed in the stationary reference frame, they contain the rotor flux linkages and the rotor speed as well. These are the equations of the adaptive model and have been given by eqns (4.5-85) and (4.5-86) respectively, which are also given here for convenience:

$$\hat{\psi}_{rd} = \frac{1}{T_r}\int (L_m i_{sD} - \hat{\psi}_{rd} - \omega_r T_r \hat{\psi}_{rq})\,dt \tag{4.5-182}$$

$$\hat{\psi}_{rq} = \frac{1}{T_r}\int (L_m i_{sQ} - \hat{\psi}_{rq} + \omega_r T_r \hat{\psi}_{rd})\,dt. \tag{4.5-183}$$

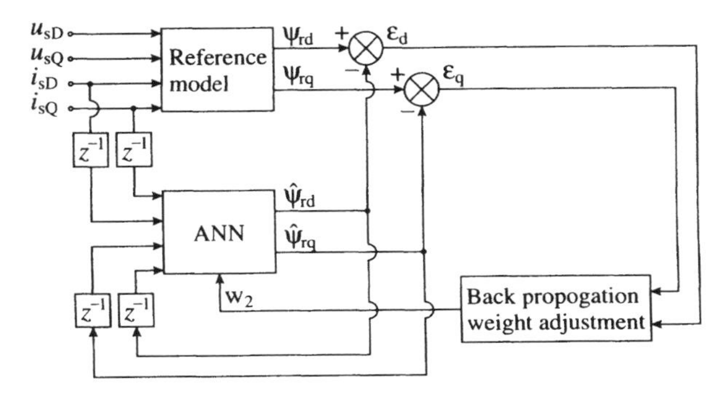

Fig. 4.79. MRAS-based rotor-speed estimator containing an ANN.

In these equations $\hat{\psi}_{rd}$ and $\hat{\psi}_{rq}$ are the rotor flux linkages estimated by the adaptive model, and they are also shown in Fig. 4.79. Equations (4.5-182) and (4.5-183) contain the rotor speed, which in general is changing, and it is our purpose to estimate this speed by using an ANN. For this purpose, eqns (4.5-182) and (4.5-183) can be implemented by a two-layer ANN which contains variable weights, and the variable weights are proportional to the rotor speed (a proof of this will also be given below).

For given stator voltages and currents and induction machine parameters, the actual rotor speed (ω_r) must be the same as the speed estimated by the ANN ($\hat{\omega}_r$), when the outputs of the reference model and the adaptive model are equal. In this case the errors $\varepsilon_d=\psi_{rd}-\hat{\psi}_{rd}$ and $\varepsilon_q=\psi_{rq}-\hat{\psi}_{rq}$ are zero (these errors are also shown in Fig. 4.79).

When there is any mismatch between the rotor speed estimated by the ANN and the actual rotor speed, then these errors are not zero, and they are used to adjust the weights of the ANN (or in other words the estimated speed). The weight adjustment is performed in such a way that the error should converge fast to zero.

To obtain the required weight adjustments in the ANN, the sampled data forms of eqns (4.5-182) and (4.5-183) are considered. By using the backward difference method, e.g. by considering that the rate of change of an estimated rotor flux linkage can be expressed as

$$\frac{d\hat{\psi}_{rd}(t)}{dt}=\frac{\hat{\psi}_{rd}(k)-\hat{\psi}_{rd}(k-1)}{T}, \qquad (4.5\text{-}184)$$

where T is the sampling time, the sampled data forms of the equations for the rotor flux linkages can be written as

$$\frac{\hat{\psi}_{rd}(k)-\hat{\psi}_{rd}(k-1)}{T}=-\frac{\hat{\psi}_{rd}(k-1)}{T_r}-\frac{\omega_r\hat{\psi}_{rq}(k-1)}{T}+\frac{L_m}{T_r}i_{sD}(k-1) \qquad (4.5\text{-}185)$$

$$\frac{\hat{\psi}_{rq}(k)-\hat{\psi}_{rq}(k-1)}{T}=-\frac{\hat{\psi}_{rq}(k-1)}{T_r}+\frac{\omega_r\hat{\psi}_{rd}(k-1)}{T}+\frac{L_m}{T_r}i_{sQ}(k-1). \qquad (4.5\text{-}186)$$

Thus the rotor flux linkages at the kth sampling instant can be obtained from the previous $(k-1)$-th values as

$$\hat{\psi}_{rd}(k)=\hat{\psi}_{rd}(k-1)\left(1-\frac{T}{T_r}\right)-\omega_rT\hat{\psi}_{rq}(k-1)+\frac{L_mT}{T_r}i_{sD}(k-1) \qquad (4.5\text{-}187)$$

$$\hat{\psi}_{rq}(k)=\hat{\psi}_{rq}(k-1)\left(1-\frac{T}{T_r}\right)+\omega_rT\hat{\psi}_{rd}(k-1)+\frac{L_mT}{T_r}i_{sQ}(k-1). \qquad (4.5\text{-}188)$$

By introducing $c=T/T_r$ and assuming that the rotor time constant (T_r) is constant, the following weights are introduced:

$$\begin{aligned} w_1&=1-c\\ w_2&=\omega_rcT_r=\omega_rT\\ w_3&=cL_m. \end{aligned} \qquad (4.5\text{-}189)$$

It can be seen that w_1 and w_3 are constant weights, but w_2 is a variable weight, and is proportional to the speed. Thus equations (4.5-187) and (4.5-188) take the following forms:

$$\hat{\psi}_{rd}(k)=w_1\hat{\psi}_{rd}(k-1)-w_2\hat{\psi}_{rq}(k-1)+w_3 i_{sD}(k-1) \qquad (4.5\text{-}190)$$

$$\hat{\psi}_{rq}(k)=w_1\hat{\psi}_{rq}(k-1)+w_2\hat{\psi}_{rd}(k-1)+w_3 i_{sQ}(k-1). \qquad (4.5\text{-}191)$$

These equations can be visualized by the very simple two-layer ANN shown in Fig. 4.80. This contains four input nodes. The input signals to these input nodes are the past values of the estimated rotor flux-linkage components expressed in the stationary reference frame $[\hat{\psi}_{rd}(k-1), \hat{\psi}_{rq}(k-1)]$, and also the past values of the stator current components expressed in the stationary reference frame $[i_{sD}(k-1), i_{sQ}(k-1)]$. There are two output nodes which output the present values of the estimated rotor flux-linkage components $[\hat{\psi}_{rd}(k), \hat{\psi}_{rq}(k)]$. Thus all the nodes are well defined. The connections between the nodes are represented by weights (synapses), and a weight shows the strength of the connection considered. In general a weight can be positive or negative, corresponding to excitatory and inhibitory weights.

In the ANN shown in Fig. 4.80, the adaptive w_2 weights are shown with thick solid lines and, as noted above, these are proportional to the speed ($w_2=\omega_r cT_r=\omega_r T$), where the proportionality factor is the sampling time. The adaptive weights are adjusted so that $E=(1/2)\boldsymbol{\varepsilon}^2(k)$ should be a minimum, where $\boldsymbol{\varepsilon}(k)=\boldsymbol{\psi}_r(k)-\hat{\boldsymbol{\psi}}_r(k)$, $\boldsymbol{\psi}_r(k)=[\psi_{rd}(k), \psi_{rq}(k)]^T$, $\hat{\boldsymbol{\psi}}_r(k)=[\hat{\psi}_{rd}(k), \hat{\psi}_{rq}(k)]^T$. Thus the weight adjustments to give minimum squared error have to be proportional to the negative of the gradient of the error with respect to the weight, $-\partial E/\partial w_2$, since in this way it is possible to move progressively towards the optimum solution, where the squared error is minimal. The proportionality factor is the so-called learning rate, η, which is a positive constant, and larger learning rates yield larger changes in the weights. In practice as large a value is chosen for the learning rate as possible, since this gives the fastest learning, but a large learning rate can yield

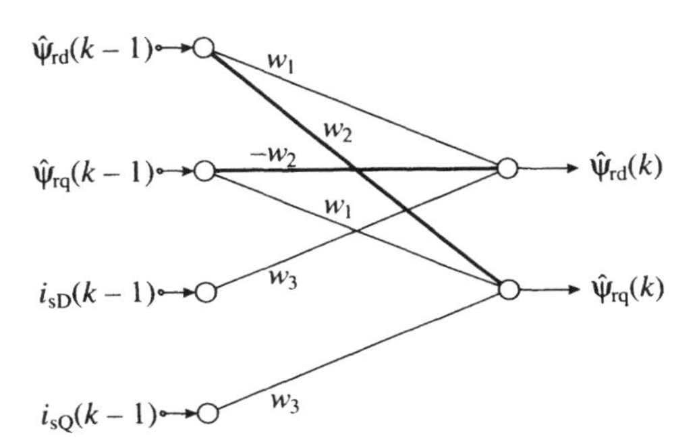

Fig. 4.80. ANN representation of estimated rotor flux linkages.

oscillations in the output of the ANN. It follows from the above that the mathematical expression for the weight adjustment has to be

$$\Delta w_2(k) = -\eta \frac{\partial E}{\partial w_2} \qquad (4.5\text{-}192)$$

(see also Chapter 7). However, eqn (4.5-192) can be rewritten as follows

$$\Delta w_2(k) = -\eta \frac{\partial E}{\partial w_2} = -\eta \frac{\partial E}{\partial \hat{\boldsymbol{\psi}}_r(k)} \frac{\partial \hat{\boldsymbol{\psi}}_r(k)}{\partial w_2}. \qquad (4.5\text{-}193)$$

Equation (4.5-193) is now expanded. By using the definition of E, the term $\partial E/\partial \hat{\boldsymbol{\psi}}_r(k)$, which is present on the right-hand side of eqn (4.5-193), can be expressed as

$$\frac{\partial E}{\partial \hat{\boldsymbol{\psi}}_r(k)} = \frac{1}{2} \frac{\partial[\varepsilon^2(k)]}{\partial \hat{\boldsymbol{\psi}}_r(k)} = \frac{1}{2} \frac{\partial\{[\boldsymbol{\psi}_r(k) - \hat{\boldsymbol{\psi}}_r(k)]^2\}}{\partial \hat{\boldsymbol{\psi}}_r(k)} = -\boldsymbol{\varepsilon}^{\mathrm{T}}(k). \qquad (4.5\text{-}194)$$

Furthermore, the other term $\partial \hat{\boldsymbol{\psi}}_r(k)/\partial w_2$ which is present in eqn (4.5-193) can be expanded as follows:

$$\frac{\partial \hat{\boldsymbol{\psi}}_r(k)}{\partial w_2} = [-\hat{\psi}_{rq}(k-1), \hat{\psi}_{rd}(k-1)]^{\mathrm{T}}. \qquad (4.5\text{-}195)$$

This follows directly from Fig. 4.80 or by considering eqns (4.5-190) and (4.5-191). Thus the substitution of eqns (4.5-194) and (4.5-195) into eqn (4.5-193) gives the following expression, if $\boldsymbol{\varepsilon}(k) = [\varepsilon_d(k), \varepsilon_q(k)]^{\mathrm{T}}$ is also considered:

$$\begin{aligned}\Delta w_2(k) &= -\eta \frac{\partial E}{\partial w_2} = \eta \boldsymbol{\varepsilon}^{\mathrm{T}}(k)[-\hat{\psi}_{rq}(k-1), \hat{\psi}_{rd}(k-1)]^{\mathrm{T}} \\ &= \eta[-\varepsilon_d(k)\hat{\psi}_{rq}(k-1) + \varepsilon_q(k)\hat{\psi}_{rd}(k-1)],\end{aligned} \qquad (4.5\text{-}196)$$

where $\varepsilon_d(k) = \psi_{rd}(k) - \hat{\psi}_{rd}(k)$ and $\varepsilon_q(k) = \psi_{rq}(k) - \hat{\psi}_{rq}(k)$. Equation (4.5-196) is a well-known type of expression in neural networks using the method of steepest gradient for weight adjustment (see also Chapter 7), and it can be seen that the appropriate errors are multiplied by the appropriate inputs of the neural network shown in Fig. 4.80. When the expressions for the errors $(\varepsilon_d, \varepsilon_q)$ are substituted into eqn (4.5-196), it follows that

$$\Delta w_2(k) = \eta\{-[\psi_{rd}(k) - \hat{\psi}_{rd}(k)]\hat{\psi}_{rq}(k-1) + [\psi_{rq}(k) - \hat{\psi}_{rq}(k)]\hat{\psi}_{rd}(k-1)\} \qquad (4.5\text{-}197)$$

is obtained. Thus in Fig. 4.80 the weight adjustments can be obtained from

$$\begin{aligned}w_2(k) &= w_2(k-1) + \Delta w_2(k) \\ &= w_2(k-1) + \eta\{-[\psi_{rd}(k) - \hat{\psi}_{rd}(k)]\hat{\psi}_{rq}(k-1) \\ &\quad + [\psi_{rq}(k) - \hat{\psi}_{rq}(k)]\hat{\psi}_{rd}(k-1)\}.\end{aligned} \qquad (4.5\text{-}198)$$

It has been discussed above that for rapid learning, the learning rate (η) has to be selected to be large, but this can lead to oscillations in the outputs of the ANN. However, to overcome this difficulty, a so-called momentum term is added to eqn (4.5-198), which takes into account the past [$(k-1)$-th] weight changes on the present [kth] weight. This ensures accelerated convergence of the algorithm (see also Chapter 7). Thus the current weight adjustment $\Delta w_2(k)$ described by eqn (4.5-198) is supplemented by a fraction of the most recent weight adjustment, $\Delta w_2(k-1)$:

$$\begin{aligned} w_2(k) &= w_2(k-1)+\Delta w_2(k)+\alpha\Delta w_2(k-1) \\ &= w_2(k-1)+\eta\{-[\psi_{rd}(k)-\hat{\psi}_{rd}(k)]\hat{\psi}_{rq}(k-1) \\ &\quad +[\psi_{rq}(k)-\hat{\psi}_{rq}(k)]\hat{\psi}_{rd}(k-1)\}+\alpha\Delta w_2(k-1), \end{aligned} \qquad (4.5\text{-}199)$$

where α is a positive constant called the momentum constant. The term $\alpha\Delta w_2(k-1)$ is called the momentum term, and is a scaled value of the most recent weight adjustment. Usually α is in the range between 0.1 and 0.8. The inclusion of the momentum term into the weight adjustment mechanism can significantly increase the convergence, which is extremely useful when the ANN shown in Fig. 4.80 is used to estimate in real-time the speed of the induction machine. Since it follows from eqn (4.5-189) that the weight w_2 is proportional to the speed, $w_2=\omega_r T$, finally the estimated rotor speed can be obtained from

$$\begin{aligned} \hat{\omega}_r(k) &= \hat{\omega}_r(k-1)+\frac{\Delta w_2(k)}{T}+\frac{\alpha}{T}\Delta w_2(k-1) \\ &= \hat{\omega}_r(k-1)+\frac{\eta}{T}\{-[\psi_{rd}(k)-\hat{\psi}_{rd}(k)]\hat{\psi}_{rq}(k-1) \\ &\quad +[\psi_{rq}(k)-\hat{\psi}_{rq}(k)]\hat{\psi}_{rd}(k-1)\}+\frac{\alpha}{T}\Delta w_2(k-1). \end{aligned} \qquad (4.5\text{-}200)$$

The simple structure of the ANN shown in Fig. 4.80 has various advantages. In a multi-layer feedforward neural network, which contains one or several hidden layers, and which uses the back-propagation algorithm (see Section 4.5.3.6.2), a supervised (off-line) training stage is required before the ANN can be used. This is usually a slow process. In contrast to this, the simple two-layer ANN shown above does not require a separate learning stage, since the learning takes place during the on-line speed estimation process. However, the multi-layer feedforward ANN with hidden layers can give more accurate speed estimates, especially at low speeds. It should be noted that it has been assumed above that T_r=constant, but in practice this is not a valid assumption and as a result incorrect speed can be estimated. Furthermore, the reference model has used open-loop integration, but this can be avoided by using other techniques developed earlier (see Sections 3.1.3.2, 4.1.1.4, 4.5.3.4).

It is also possible to implement other algorithms, where the error is not obtained from the respective rotor flux linkages, but it is an error between other quantities which can also be obtained from the terminal voltages and currents of the induction machine, and e.g. which do not depend on the stator resistance (see various schemes in Section 4.5.3.4). The application of such algorithms can lead to improved speed estimation in the low-speed region, but in general, more accurate estimation can be achieved by using the technique discussed in the following section.

4.5.3.6.2 Multi-layer feedforward ANN: speed is an output quantity

In addition to the speed estimation scheme shown in Fig. 4.79, which resembles a classical MRAS estimator (but the adaptive model is an ANN), it is also possible to use other ANN-based speed estimators for an induction machine, which are not related to a MRAS estimator. Such a system can use various types of learning techniques, some of which can be fuzzy-assisted. However, it is also possible to use fuzzy-neural networks for this purpose and an appropriate design can be made to be robust to noise and parameter variations, even at extremely low speeds. It is expected that such solutions will also appear in commercial drive applications in the future.

For example, as shown in Section 7.3.2.2, it is possible to use a multi-layer feedforward ANN speed estimator for an induction machine, which has four layers: an input layer, two hidden layers, and an output layer. The inputs to the ANN are the stator voltages and stator currents: $u_{sD}(k)$, $u_{sD}(k-1)$, $u_{sQ}(k)$, $u_{sQ}(k-1)$, $i_{sD}(k)$, $i_{sD}(k-1)$, $i_{sQ}(k)$, $i_{sQ}(k-1)$ and the output is the rotor speed, $\omega_r(k)$. The activation functions used in the hidden layers can be, for example, tansigmoid functions. In Section 7.3.2.2.1 two ANNs are considered for the speed estimation of a specific induction machine; they have 8-9-7-1 structure (there are eight input nodes, nine hidden nodes in the first hidden layer, seven hidden nodes in the second hidden layer, and a single output node) and 8-8-6-1 structure. It should be noted that other induction machines may require other ANN structures. In Section 7.3.2.2.2 a much more complex ANN with the structure 8-12-10-4 is applied, since it was the goal to obtain on its outputs four quantities: the rotor speed, the electromagnetic torque, and the direct- and quadrature-axis stator flux linkages in the stationary reference frame (ψ_{sD}, ψ_{sQ}). In Section 7.3.2.2.3 the details of an ANN speed estimator used in a voltage-source inverter-fed induction motor drive employing rotor-flux-oriented control are given, and for this purpose various ANN structures (8-15-13-1), (9-9-8-1) and (5-12-11-1) are considered. It is shown that the rotor speed estimation can be obtained when, instead of using an ANN with 8 inputs, there are 9 inputs, where a past rotor speed value, $\omega_r(k-1)$, is also present at the inputs.

It is believed that the ANNs discussed in the present section can be used in various torque-controlled induction motor drives. Different artificial-intelligence-based approaches for fully digital DSP-controlled speed-sensorless drives are under investigation by the research group (Intelligent Motion Control Group) of the author at Aberdeen University (see also Chapter 7).

4.6 Direct torque control (DTC) of induction machines

4.6.1 GENERAL INTRODUCTION

High-dynamic-performance instantaneous electromagnetic-torque-controlled induction motor (and other a.c. motor) drives have been used for more than 20 years. Based on the pioneering works of Blaschke, Hasse, and Leonhard, vector-controlled drives have become increasingly popular, and have become the standard in the drives industry. The most significant industrial contributions in this field have been made by Siemens. Direct-torque-controlled induction motor drives were developed more than 10 years ago by Depenbrock [Depenbrock 1985] and Takahashi [Takahashi and Noguchi 1985]. However, at present, ABB is the only industrial company who have introduced (in 1995) a commercially available direct-torque-controlled induction motor drive. This is a significant industrial contribution, and in numerous papers by ABB it has been claimed that 'direct torque control' is the latest a.c. motor control method developed by ABB (e.g. see [Tiitinen 1996]).

In the present section, the general and fundamental aspects of the direct torque control of induction machines is examined in great detail by using both mathematical and physical analyses of the processes involved. Torque and flux estimation, optimum switching-vector selection, reduction of torque and flux ripples, field weakening, speed-sensorless implementations, and predictive schemes are also discussed in great detail. Some aspects of direct torque control have also been discussed in Sections 1.2.2 and 3.3.

4.6.2 DTC OF A VSI-FED INDUCTION MOTOR

4.6.2.1 General, mathematical, and physical fundamentals of producing fast torque response

In a direct-torque-controlled (DTC) induction motor drive, supplied by a voltage source inverter, it is possible to control directly the stator flux linkage (or rotor flux linkage, or magnetizing flux linkage) and the electromagnetic torque by the selection of optimum inverter switching modes. The selection is made to restrict the flux and torque errors within respective flux and torque hysteresis bands, to obtain fast torque response, low inverter switching frequency, and low harmonic losses. In the present section such a DTC drive will be described in which, in addition to controlling the electromagnetic torque, the controlled flux linkage is the stator flux linkage. However, it should be noted that it is possible to have other implementations in which the rotor flux linkage or the magnetizing flux linkage is controlled. DTC allows very fast torque responses and flexible control of an induction machine. Due to the importance of DTC drives, and also due to the great interest in these types of drives, and also since it is expected that various DTC drives will emerge in the future, for better understanding a very detailed description will be given below.

In general, in a symmetrical three-phase induction machine, the instantaneous electromagetic torque is proportional to the cross-vectorial product of the stator flux-linkage space vector and the stator-current space vector,

$$t_e = \tfrac{3}{2} P \bar{\psi}_s \times \bar{i}_s, \tag{4.6-1}$$

where $\bar{\psi}_s$ is the stator flux-linkage space vector and $\bar{i}_s$ is the stator-current space vector. In eqn (4.6-1), both space vectors are expressed in the stationary reference frames. By considering that $\bar{\psi}_s = |\bar{\psi}_s| \exp(j\rho_s)$, where ρ_s is the angle of the stator flux-linkage space vector with respect to the direct-axis of the stator reference frame (see Fig. 4.81), and $\bar{i}_s = |\bar{i}_s| \exp(j\alpha_s)$ (see Fig. 4.81), it is possible to put eqn (4.6-1) into the following form:

$$t_e = \tfrac{3}{2} P |\bar{\psi}_s||\bar{i}_s| \sin(\alpha_s - \rho_s) = \tfrac{3}{2} P |\bar{\psi}_s||\bar{i}_s| \sin\alpha, \tag{4.6-2}$$

where $\alpha = \alpha_s - \rho_s$ is the angle between the stator flux-linkage and stator-current space vector.

It can be shown by using the voltage equations of the induction machine that, for a given value of the rotor speed, if the modulus of the stator flux-linkage space vector is kept constant and the angle ρ_s is changed quickly, then the electromagnetic torque can be rapidly changed. The mathematical proof is now briefly discussed, but it should be noted that a much simpler proof will also be presented below, which gives a very clear physical description of the processes involved.

For the purposes of the mathematical proof, the electromagnetic torque response of the machine for a step change in ρ_s at $t=0$ is now derived. Therefore the time variation of the electromagnetic torque must be determined. For this purpose, first the rotor-current space vector (formulated in the stationary reference frame) is expressed in terms of the stator flux-linkage space vector, $\bar{i}_r' = (\bar{\psi}_s - L_s\bar{i}_s)/L_m$, and also the rotor flux-linkage space vector is expressed in terms of the stator flux-linkage space vector by using $\bar{\psi}_r' = L_r\bar{i}_r' + L_m\bar{i}_s$, where $\bar{i}_r' = (\bar{\psi}_s - L_s\bar{i}_s)/L_m$, thus $\bar{\psi}_r' = (L_r/L_m)(\bar{\psi}_s - L_s'\bar{i}_s)$. The thus-obtained expressions for

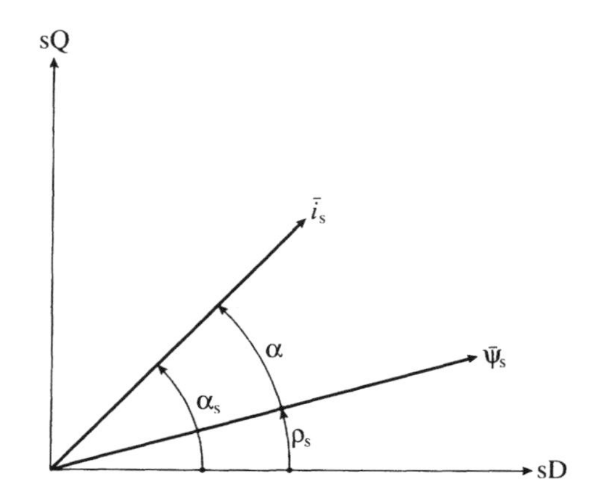

Fig. 4.81. Stator flux-linkage and stator current space vectors.

$\bar{i}'_r$ and $\bar{\psi}'_r$ are substituted into the rotor voltage equation expressed in the stationary reference frame (which is obtained from eqn (2.1-125) as $0 = R_r\bar{i}'_r + d\bar{\psi}'_r/dt - j\omega_r\bar{\psi}'_r$). The derived rotor voltage equation contains $\bar{i}_s$ and $\bar{\psi}_s$, and this can be used to express the stator-current space vector in terms of the stator flux-linkage space vector. This expression for the stator-current space vector is then substituted into eqn (4.6-1). However, by also utilizing in this expression the fact that the stator flux-linkage space vector modulus is constant ($|\bar{\psi}_s| = c_1$), thus $\bar{\psi}_s = |\bar{\psi}_s|\exp(j\rho_s) = c_1\exp(j\rho_s)$ and therefore $d\bar{\psi}_s/dt = j|\bar{\psi}_s|d\rho_s/dt$, finally it is possible to obtain an equation for the electromagnetic torque whose inverse Laplace transform gives the required temporal variation of the electromagnetic torque. An examination of this expression shows that for constant $|\bar{\psi}_s|$, the rate of change of the increasing electromagnetic torque is almost proportional to the rate of change of ρ_s. Thus by forcing the largest $d\rho_s/dt$ under the condition of constant stator flux-linkage modulus, the fastest (minimum) electromagnetic torque response time is obtained.

In other words, if such stator voltages are imposed on the motor, which keep the stator flux constant (at the demanded value), but which quickly rotate the stator flux-linkage space vector into the position required (by the torque demand), then fast torque control is performed. It follows that if in the DTC induction motor drive, the developed actual electromagnetic torque of the machine is smaller than its reference value, the electromagnetic torque should be increased as fast as possible by using the fastest $d\rho_s/dt$. However, when the electromagnetic torque is equal to its reference value, the rotation is stopped. If the stator flux-linkage space vector is accelerated in the forward direction, then positive electromagnetic torque is produced, and when it is decelerated backwards, negative electromagnetic torque is produced. However, the stator flux-linkage vector can be adjusted by using the appropriate stator-voltage space vector, which is generated by the VSI inverter which supplies the induction machine (see details of voltage generation below). To summarize: the electromagnetic torque can be quickly changed by controlling the stator flux-linkage space vector, which, however, can be changed by using the appropriate stator voltages (generated by the inverter which supplies the induction motor). It can be seen that there is direct stator-flux and electromagnetic torque control achieved by using the appropriate stator voltages. This is why this type of control is usually referred to as direct torque control.

It is very useful to consider another form of the expression for the instantaneous electromagnetic torque, which gives an extremely clear physical picture of the processes involved but leads to the same results as shown above. By considering that $\bar{\psi}_s = L_s\bar{i}_s + L_m\bar{i}'_r$ and $\bar{\psi}'_r = L_r\bar{i}'_r + L_m\bar{i}_s$, where again the primed rotor quantities are expressed in the stationary reference frame, it follows that $\bar{i}_s = \bar{\psi}_s/L'_s - [L_m/(L_rL'_s)]\bar{\psi}'_r$; thus eqn (4.6-1) takes the following form:

$$
\begin{aligned}
t_e &= \tfrac{3}{2}P\frac{L_m}{L'_sL_r}\bar{\psi}'_r \times \bar{\psi}_s = \tfrac{3}{2}P\frac{L_m}{L'_sL_r}|\bar{\psi}'_r||\bar{\psi}_s|\sin(\rho_s - \rho_r) \\
&= \tfrac{3}{2}P\frac{L_m}{L'_sL_r}|\bar{\psi}'_r||\bar{\psi}_s|\sin\gamma.
\end{aligned}
\tag{4.6-3}
$$

In eqn (4.6-3) γ is the angle between the stator and rotor flux-linkage space vectors, $\gamma=\rho_s-\rho_r$, where ρ_r is the angle of the rotor flux-linkage space vector with respect to the real-axis of the stationary reference frame, as shown in Fig. 4.82.

The rotor time constant of a standard squirrel-cage induction machine is large (e.g. a typical value is greater than 0.1 s, but it should be noted that for larger machines this is much larger); thus the rotor flux linkage changes only slowly compared to the stator flux linkage. It can be assumed to be constant. This also follows from the rotor voltage equation of the induction machine if the stator flux linkage is assumed to be constant. However, if the stator and rotor flux linkages are assumed to be constant, it follows from eqn (4.6-3) that the electromagnetic torque can be rapidly changed by changing γ in the required direction (which is determined by the torque command). This is the essence of direct torque control. However, as discussed below, the angle γ can be easily changed by switching on the appropriate stator-voltage space vector (produced by the appropriate inverter voltage). If the modulus of the stator flux-linkage space vector is not constant (e.g. in the field-weakening range), then it still possible to control both the angle γ and $|\bar{\psi}_s|$ by switching on the appropriate inverter voltage.

In contrast to a vector-controlled induction motor drive, where the stator currents are used as control quantities, in the direct-torque-controlled drive, the stator flux linkages are controlled. It should be noted that if $|\bar{\psi}_r|$=constant, then it follows from eqn (2.1-191) that

$$t_e=\tfrac{3}{2}P\frac{L_r}{L_m}|\bar{\psi}_r|i_{sy},$$

where i_{sy}, is the torque-producing stator current (quadrature-axis stator current in the rotor-flux-oriented reference frame whose axes are denoted by x and y respectively in Fig. 4.82), and the electromagnetic torque can be quickly changed

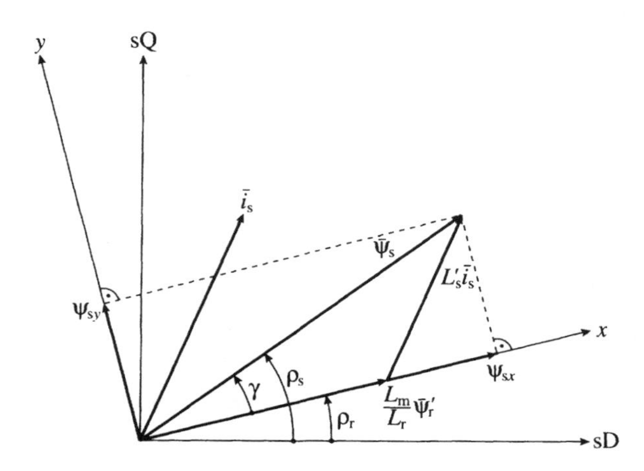

Fig. 4.82. Stator flux-linkage, rotor flux-linkage, and stator-current space vectors.

by quickly changing i_{sy}. In a vector-controlled drive, the stator currents are the controlled quantities (i_{sy} controls the torque and i_{sx} controls the rotor flux). This is one of the reasons why, in a vector-controlled drive employing rotor-flux-oriented control, the stator currents expressed in the stationary reference frame must be transformed into the stator currents in the rotor-flux-oriented reference frame. However, it follows from Fig. 4.82 [or eqn (4.6-3)] that $|\bar{\psi}_s|\sin\gamma = \psi_{sy}$ is the torque-producing stator flux-linkage component and ψ_{sx} is collinear with the rotor flux. Thus the flux is controlled by the direct-axis stator flux and the torque is controlled by the quadrature-axis stator flux and again it can be seen that, in contrast to vector control, now the flux-linkage components are the control quantities. Equation (4.6-3) is similar to that of a synchronous machine, where the electromagnetic torque is controlled by the load angle between the stator and rotor flux linkages. During a short transient, the rotor flux is almost unchanged, thus rapid changes of the electromagnetic torque can be produced by rotating the stator flux in the forward direction (phase advancing) or by rotating it in the negative direction (retarding), or by stopping it, according to the demanded torque. In summary: in the direct-torque-controlled drive rapid instantaneous torque control can be achieved by quickly changing the position of the stator flux-linkage space vector (relative to the rotor flux-linkage space vector), or in other words, by quickly changing its speed (speed of the stator flux-linkage space vector). However, the stator flux-linkage space vector (both its modulus and its angle) can be changed by the stator voltages.

If for simplicity it is assumed that the stator ohmic drops can be neglected, then $d\bar{\psi}_s/dt = \bar{u}_s$, and it can be seen that the inverter voltage ($\bar{u}_s = \bar{u}_i$) directly impresses the stator flux, and thus the required stator-flux locus will be obtained by using the appropriate inverter voltages (obtained by using the appropriate inverter switching states). For better understanding this is now discussed in detail. It follows from $d\bar{\psi}_s/dt = \bar{u}_s$ that in a short Δt time, when the voltage vector is applied, $\Delta\bar{\psi}_s = \bar{u}_s\Delta t$. Thus the stator flux-linkage space vector moves by $\Delta\bar{\psi}_s$ in the direction of the stator-voltage space vector at a speed which is proportional to the magnitude of the stator-voltage space vector (which is proportional to the d.c. link voltage). By selecting step-by-step the appropriate stator voltage vector, it is then possible to change the stator flux in the required way. Decoupled control of the torque and stator flux is achieved by acting on the radial and tangential components of the stator flux-linkage space vector in the locus. These two components are directly proportional (stator ohmic drop was neglected) to the components of the stator-voltage space vector in the same directions, and thus they can be controlled by the appropriate inverter switchings. It should be emphasized that for torque production, the angle γ plays a vital role, or in other words, the relative position of the stator and rotor flux-linkage space vectors determines the electromagnetic torque. By assuming a slow motion of the rotor flux-linkage space vector, if a stator-voltage space vector is applied, which causes a quick movement of the stator flux-linkage space vector away from the rotor flux linkage, then the electromagnetic torque will increase since the angle γ is increased. However, if a voltage space vector is applied (a zero-voltage space vector, see below) which almost stops the rotation of the stator flux-linkage space vector, then the electromagnetic torque will decrease, since the rotor flux-linkage space vector is still moving and the angle γ decreases. If the duration of the zero-voltage space vector is sufficiently long, then since the stator flux-linkage space vector will almost not move (in practice it will move slightly due to the stator ohmic voltage drop), then the rotor flux-linkage space vector will overtake the stator flux-linkage space vector, the angle γ will change its sign, the electromagnetic torque will change its direction.

By considering the six-pulse VSI shown in Fig. 4.83(a), there are six non-zero active voltage-switching space vectors ($\bar{u}_1, \bar{u}_2, \ldots \bar{u}_6$) and two zero space vectors ($\bar{u}_7, \bar{u}_8$). These are shown in Fig. 4.83(c), and Fig. 4.83(b) shows the corresponding eight switching states. The six active inverter-switching vectors can be expressed as

$$\bar{u}_s = \bar{u}_k = \tfrac{2}{3} U_d \exp[j(k-1)\pi/3] \qquad k=1,2,\ldots,6, \tag{4.6-4}$$

where U_d is the d.c. link voltage (see also the book on space-vector theory [Vas 1992]). However, for $k=7,8$, $\bar{u}_k=0$ holds for the two zero switching states where the stator windings are short-circuited, $\bar{u}_s=\bar{u}_k=0$. It follows from the definition of the switching vectors given above that since $\bar{u}_s=u_{sD}+ju_{sQ}$, $\bar{u}_1$ is aligned with the real axis (sD) of the stationary reference frame, and this definition is used in Europe. It should be noted that in the USA, usually the quadrature-axis of the stator reference frame is aligned with $\bar{u}_1$, which means that all switching vectors would be displaced by 90° in the positive direction with respect to the switching vectors defined above.

Since $\Delta\bar{\psi}_s=\bar{u}_s\Delta t$, it can be seen that the stator flux-linkage space vector will move fast if non-zero (active) switching vectors are applied, for a zero switching vector it will almost stop (it will move very slowly due to the small ohmic voltage drop). For a six-pulse VSI, the stator flux linkage moves along a hexagonal path with constant linear speed, due to the six switching vectors. For a sinusoidal PWM (where the inverter switching states are chosen to give stator flux-linkage variations which are almost sinusoidal), a suitable sequence of the zero and active (non-zero) switching vectors is applied to obtain the required flux-linkage locus. In the DTC drive, at every sampling period, the switching vectors are selected on the basis of keeping the stator flux-linkage errors in a required tolerance band (hysteresis band), and keeping the torque error in the hysteresis band. It is assumed that the widths of these hysteresis bands are $2\Delta\psi_s$ and $2\Delta t_e$ respectively. (The factor 2 appears in this definition since it is assumed that e.g. for the stator flux linkage, the upper limit value is above the reference value by $\Delta\psi_s$, and the lower limit value is below the reference value by $\Delta\psi_s$, thus the width of the hysteresis band is indeed $2\Delta\psi_s$). If the stator flux-linkage space vector lies in the kth sector, where $k=1,2,\ldots,6$, its magnitude can be increased by using the switching vectors $\bar{u}_k$, $\bar{u}_{k+1}$, $\bar{u}_{k-1}$; however, its magnitude can be decreased by selecting $\bar{u}_{k+2}$, $\bar{u}_{k-2}$ and $\bar{u}_{k+3}$. Obviously the selected voltage switching vectors affect the electromagnetic torque as well. The speed of the stator flux-linkage

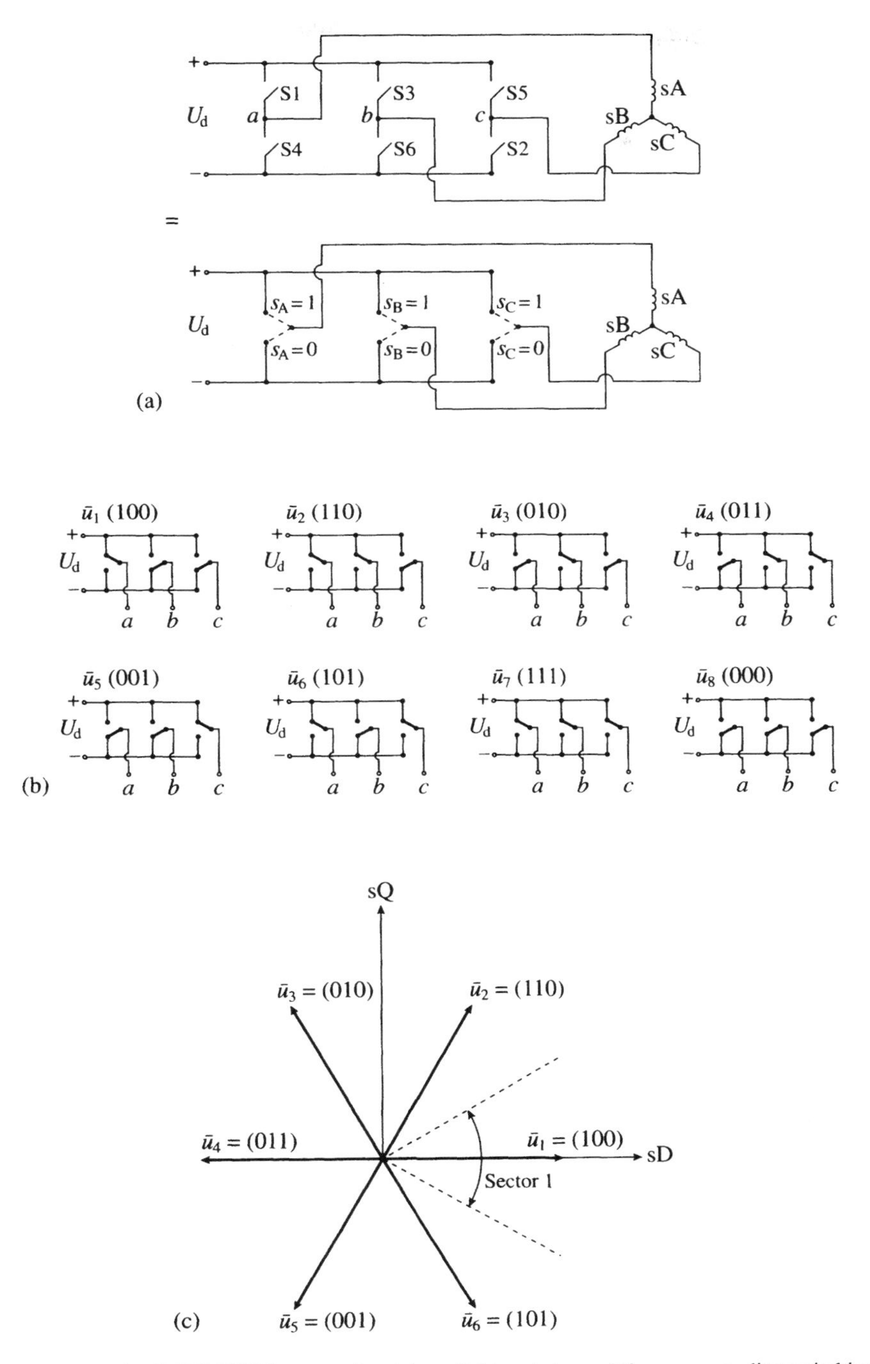

Fig. 4.83. Schematic of PWM VSI inverter, the eight switching states and the corresponding switching space vectors. (a) PWM VSI; (b) switching stages; (c) switching-voltage space vectors.

space vector is zero if a zero switching vector is selected, and it is possible to change this speed by changing the output ratio between the zero and non-zero voltage vectors. It is important to note that the duration of the zero states has a direct effect on the electromagnetic torque oscillations.

As shown above, the stator flux-linkage space vector is basically the integral of the stator-voltage space vector and it will move in the direction of the stator-voltage space vector for as long as the voltage space vector is applied. Thus if a reduced stator flux-linkage space-vector modulus is required, the stator flux-linkage space-vector modulus can be controlled by applying switching voltage vectors which are directed towards the centre of the rotor, and if an increased stator flux-linkage space-vector modulus is required, it is controlled by applying voltage vectors which are directed out from the centre of the rotor. This is illustrated in the example of Fig. 4.84, where $|\bar{\psi}_{sref}|$ is the reference value of the stator flux-linkage space vector.

It is our goal to keep the modulus of the stator flux-linkage space vector ($|\bar{\psi}_s|$) within the hysteresis band (denoted by the two circles), whose width is $2\Delta\psi_s$ as shown in Fig. 4.84. The locus of the flux-linkage space vector is divided into several sectors, and due to the six-step inverter, the minimum number of sectors required is six. The six sectors are also shown in Fig. 4.84. It is assumed that initially the stator flux-linkage space vector is at position P_0, thus is in sector 1. Assuming that the stator flux-linkage space vector is rotating anticlockwise, it follows that since at position P_0 the stator flux-linkage space-vector flux is at the upper limit ($|\bar{\psi}_{sref}|+\Delta\psi_s$), it must be reduced. This can be achieved by applying the suitable switching vector, which is the switching vector $\bar{u}_3$, as shown in

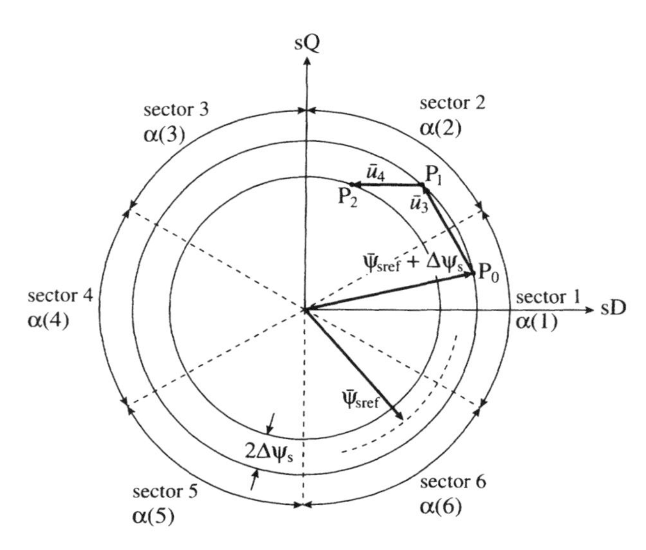

Fig. 4.84. Control of the stator flux-linkage space vector: stator flux-linkage space-vector locus [stator flux variations ($\Delta\psi_s$)], and inverter switching vectors.

Fig. 4.84. Thus the stator flux-linkage space vector will move rapidly from point P_0 to point P_1, and it can be seen that point P_1 is in sector 2. As mentioned above, altogether there are six 60°-wide sectors (due to the six-step inverter). It can also be seen that at point P_1 the stator flux-linkage space vector is again at its upper limit. On the other hand, it should be noted that if the stator flux space vector moves in the clockwise direction from point P_0, then the switching vector $\bar{u}_5$ would have to be selected, since this would ensure the required rotation and also the required flux decrease. Since at point P_1, the stator flux-linkage space vector again reaches the upper limit, it has again to be reduced when it is rotated anticlockwise, hence for this purpose the switching vector $\bar{u}_4$ has to be selected, and then $\bar{\psi}_s$ moves from point P_1 to point P_2 as shown in Fig. 4.84, which is also in sector 2. It should be noted that if, for example, at point P_1 a quick anticlockwise rotation is required, then it can be seen that the quickest rotation could be achieved by applying the switching vector $\bar{u}_6$. On the other hand, if at point P_1 the rotation of the stator flux-linkage space vector has to be stopped, then a zero switching vector would have to be applied, so either $\bar{u}_7$ or $\bar{u}_8$ can be applied. However, since prior to this the last switching was performed by the application of the switching vector $\bar{u}_3=\bar{u}_3(010)$, which means that the first switch is connected to the (lower) negative d.c. rail, the second switch is connected to the (upper) positive d.c. rail, and the third switch is connected to the negative d.c. rail as shown in Fig. 4 83(b), to minimize the number of switchings, the state $\bar{u}_8(000)$ is selected, since this requires only switching of the second switch (from 1 to 0), in contrast to selecting $\bar{u}_7(111)$, which would require two switchings (of the first switch from 0 to 1 and of the third switch from 0 to 1). If the stator flux-linkage space vector is at point P_2, then the lower limit ($|\bar{\psi}_{sref}|-\Delta\psi_s$) is reached and the stator flux-linkage space vector can be rotated in the anticlockwise direction to point P_3 by increasing it, and for this purpose the switching vector $\bar{u}_3$ gives the fastest rotation. It can be seen that point P_3 is still in sector 2. If on the other hand, the flux-linkage space vector has to be rotated from P_2 in the opposite direction (clockwise), then by selecting the switching vector which rotates $\bar{\psi}_s$ from P_2 (where the flux-linkage space vector is at the lower limit, thus flux increase is required) in the fastest way in the clockwise direction gives the switching vector $\bar{u}_1$, etc.

As discussed above, stopping the rotation of the stator flux-linkage space vector corresponds to the case when the electromagnetic torque does not have to be changed (reference value of the electromagnetic torque is equal to its actual value). However, when the electromagnetic torque has to be changed (in the clockwise or anticlockwise direction) than the stator flux-linkage space vector has to be rotated in the appropriate direction. For example, when the stator flux rotates anticlockwise, and if an increase in the electromagnetic torque is required, then e.g. if the stator flux-linkage space vector is in the second sector at point P_1 where the flux linkage has to be decreased, then the electromagnetic torque increase can be achieved by applying switching vector $\bar{u}_4$. On the other hand, if the stator flux-linkage space vector is in the second sector, but a torque decrease is required, but the flux-linkage has to be increased, then this can be achieved by applying the switching vector $\bar{u}_1$, since this moves the stator flux-linkage space vector in the clockwise direction (which is the direction for the negative torque), and also increases the stator flux linkage. If the stator flux-linkage space vector is in the second sector and a torque decrease is required, but the stator flux linkage has to be decreased, then switching vector $\bar{u}_1$ has to be applied, etc. Figure 4.85 shows the position of the various stator flux-linkage vectors if the stator flux-linkage space vector is in one of the six sectors. It is also shown which switching vector has to be selected to obtain the required increase or decrease of the stator flux linkage and the required increase or decrease of the electromagnetic torque (by creating positive or negative torques).

It can be seen that, in general, if an increase of the torque is required, then the torque is controlled by applying voltage vectors that advance the flux-linkage space vector in the direction of rotation and if a decrease is required, voltage vectors are applied which oppose the direction of the torque. If zero torque is required then that zero switching vector is applied ($\bar{u}_7$ or $\bar{u}_8$), which minimizes

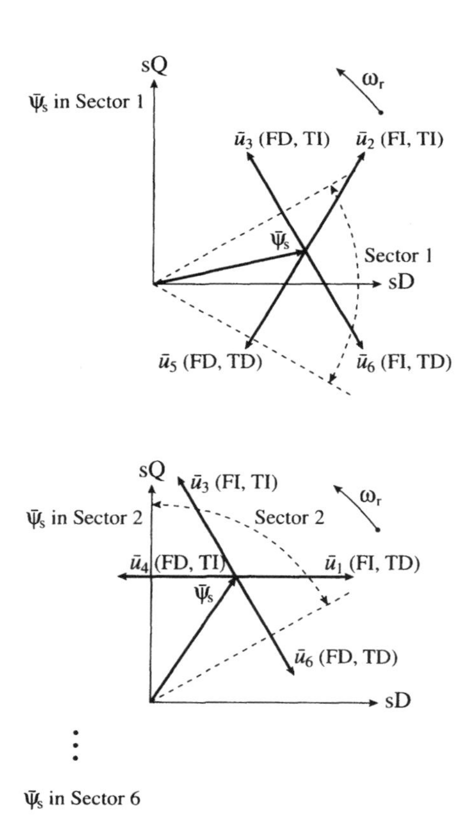

Fig. 4.85. Position of various stator flux-linkage space vectors, and selection of the optimum switching voltage vectors. FI: flux increase; FD: flux decrease; TI: torque increase; TD: torque decrease.

the inverter switching. It follows that the angle of the stator-voltage space vector is indirectly controlled through the flux vector modulus and torque, and increasing torque causes an increased angle. The torque demand is reduced to a choice of increase (positive torque), decrease (negative torque), or zero. Similarly, the stator flux-linkage vector modulus is limited to a choice of increase (flux increase) or decrease.

4.6.2.2 Optimum switching vector selection (drive with non-predictive switching vector selection)

The results obtained in the previous section can be tabulated in the so-called optimum switching vector selection table shown in Table 4.3. This gives the optimum selection of the switching vectors for all the possible stator flux-linkage space-vector positions [six positions, corresponding to the six sectors shown in Fig. 4.84, where sector 1 is in the range of $\alpha(1)$, sector 2 is in the range of $\alpha(2), \ldots$, sector 6 is in the range of $\alpha(6)$] and the desired control inputs (which are the reference values of the stator flux-linkage modulus and the electromagnetic torque respectively). If a stator flux increase is required then $d\psi = 1$; if a stator flux-linkage decrease is required $d\psi = 0$. The notation corresponds to the fact that the digital output signals of a two-level flux hysteresis comparator are $d\psi$, where

$$d\psi = 1 \text{ if } |\bar{\psi}_s| \leqslant |\bar{\psi}_{sref}| - |\Delta\psi_s|$$

$$d\psi = 0 \text{ if } |\bar{\psi}_s| \geqslant |\bar{\psi}_{sref}| + |\Delta\psi_s|.$$

If a torque increase is required then $dt_e = 1$, if a torque decrease is required then $dt_e = -1$, and if no change in the torque is required then $dt_e = 0$. The notation corresponds to the fact that the digital output signals of a three-level hysteresis comparator are dt_e, where for anticlockwise rotation (forward rotation)

$$dt_e = 1 \quad \text{if} \quad |t_e| \leqslant |t_{eref}| - |\Delta t_e|$$

$$dt_e = 0 \quad \text{if} \quad t_e \geqslant t_{eref},$$

Table 4.3 Optimum voltage switching vector look-up table

$d\psi$	dt_e	$\alpha(1)$ sector 1	$\alpha(2)$ sector 2	$\alpha(3)$ sector 3	$\alpha(4)$ sector 4	$\alpha(5)$ sector 5	$\alpha(6)$ sector 6
	1	$\bar{u}_2$	$\bar{u}_3$	$\bar{u}_4$	$\bar{u}_5$	$\bar{u}_6$	$\bar{u}_1$
1	0	$\bar{u}_7$	$\bar{u}_8$	$\bar{u}_7$	$\bar{u}_8$	$\bar{u}_7$	$\bar{u}_8$
	−1	$\bar{u}_6$	$\bar{u}_1$	$\bar{u}_2$	$\bar{u}_3$	$\bar{u}_4$	$\bar{u}_5$
	1	$\bar{u}_3$	$\bar{u}_4$	$\bar{u}_5$	$\bar{u}_6$	$\bar{u}_1$	$\bar{u}_2$
0	0	$\bar{u}_8$	$\bar{u}_7$	$\bar{u}_8$	$\bar{u}_7$	$\bar{u}_8$	$\bar{u}_7$
	−1	$\bar{u}_5$	$\bar{u}_6$	$\bar{u}_1$	$\bar{u}_2$	$\bar{u}_3$	$\bar{u}_4$

Active switching vectors: $\bar{u}_1(100)$; $\bar{u}_2(110)$; $\bar{u}_3(010)$; $\bar{u}_4(011)$; $\bar{u}_5(001)$; $\bar{u}_6(101)$
Zero switching vectors: $\bar{u}_7(111)$; $\bar{u}_8(000)$.

and for clockwise rotation (backward rotation)

$$dt_e = -1 \quad \text{if} \quad |t_e| \geq |t_{eref}| + |\Delta t_e|$$

$$dt_e = 0 \quad \text{if} \quad t_e \leq t_{eref}.$$

The selection of the width of the hysteresis bands has important effects, a too-small value may have the effect of losing the control, e.g. the stator flux linkage may exceed the values required by the tolerance band (the width of which is $2|\Delta\psi_s|$). The duration of the zero states directly influences the torque oscillations.

The optimum switching look-up table requires knowledge of the position of the stator flux-linkage space vector, since it must be known in which sector is the stator flux-linkage space vector. For this purpose the angles $\alpha(1), \alpha(2), \ldots, \alpha(6)$ shown in Fig. 4.84 are required. Since $\bar{\psi}_s = |\bar{\psi}_s| \exp(j\rho_s) = \psi_{sD} + j\psi_{sQ}$, the stator flux angle (ρ_s) can be determined by using the estimated values of the direct- and quadrature-axis stator flux linkages in the stationary reference frame (ψ_{sD}, ψ_{sQ}); thus

$$\rho_s = \tan^{-1}(\psi_{sQ}/\psi_{sD}). \tag{4.6-5}$$

Alternatively

$$\rho_s = \cos^{-1}(\psi_{sD}/|\bar{\psi}_s|) \tag{4.6-6}$$

can also be used, where $|\bar{\psi}_s| = (\psi_{sD}^2 + \psi_{sQ}^2)^{1/2}$ or the expression

$$\rho_s = \sin^{-1}(\psi_{sQ}/|\bar{\psi}_s|) \tag{4.6-7}$$

can be used. The angle ρ_s can then be used to obtain the angles $\alpha(1)$, $\alpha(2)$, etc. However, it is possible to eliminate the need for using trigonometric functions (e.g. the inverse tangent, or inverse sine or inverse cosine), since it is not the accurate position of the stator flux-linkage space vector which has to be known, but only the sector (number) in which the stator flux-linkage space vector is positioned. This information can be simply obtained by considering only the signs of the various stator flux-linkage components, and this allows a simple implementation which requires only the use of comparators. For this purpose it should be considered that e.g. in sector 1, $\psi_{sD} > 0$, but since in sector 1, ψ_{sQ} can be both positive and negative, the sign of ψ_{sQ} will not give any useful information on the position of the stator flux-linkage space vector in sector 1. However, instead of ψ_{sQ} it is possible to use the stator flux linkage in stator phase sB (ψ_{sB}), and it follows from Fig. 4.86 that $\psi_{sB} < 0$ if $\bar{\psi}_s$ is in the first sector (at point P_1 in Fig. 4.86, where it has the value $\bar{\psi}_{s1}$). Similarly if $\bar{\psi}_s$ is in sector 2, then $\psi_{sD} > 0$, $\psi_{sQ} > 0$, and $\psi_{sB} > 0$, etc. These results are summarized in Table 4.4.

It can be seen that the sign of ψ_{sQ} does not give useful information in sectors 1 and 4, since in both of these sectors this sign can be both positive (+) and negative (−). In accordance with the very simple physical picture, Table 4.4 shows correctly that for ψ_{sD} three plus signs (in sectors 6, 1, 2) are followed by three minus signs (in sectors 3, 4, 5) and similarly for ψ_{sB} three plus signs (in sectors 2, 3, 4) are followed by three minus signs (in sectors 5, 6, 1).

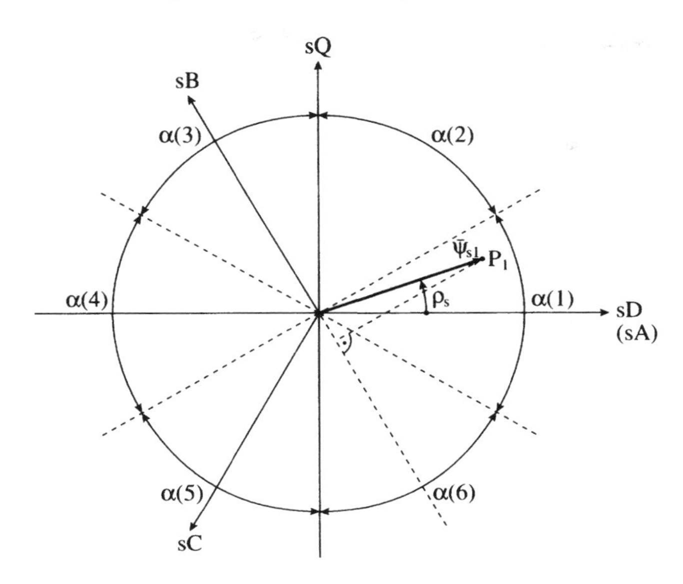

Fig. 4.86. Relationship of the space vector $\bar{\psi}_s$ to the stator flux-linkage components ψ_{sD}, ψ_{sQ}, and ψ_{sB}.

Table 4.4 Selection of the stator flux-linkage space-vector sector

Sectors / Signs of flux linkages	α(1) sector 1	α(2) sector 2	α(3) sector 3	α(4) sector 4	α(5) sector 5	α(6) sector 6
Sign of ψ_{sD}	+	+	−	−	−	+
Sign of ψ_{sQ}	(nu; − +)	+	+	(nu; + −)	−	−
Sign of ψ_{sB} = sign of $[\sqrt{3}\psi_{sD} - \psi_{sQ}]$	−	+	+	+	−	−

nu = not useful

It should also be noted for the determination of the appropriate sector that the computation of ρ_s by using eqn (4.6-5), eqn (4.6-6), or eqn (4.6-7) can also be avoided by using another technique, where first the signs of ψ_{sD} and ψ_{sQ} are determined. These give information on the quadrant where the space vector of the stator flux linkages is located (there are four quadrants, and every quadrant spans 90 °: the first quadrant starts at the sD-axis and spans until the sQ-axis, the second quadrant starts at the sQ-axis and spans until the negative sD-axis, etc.). Since every quadrant contains only one full sector and half of another sector, thus there are two possible sectors (in a quadrant), but the specific sector where $\bar{\psi}_s$ is located can be obtained by also using the ratio of ψ_{sQ}/ψ_{sD}. However, it is also possible to use other techniques as well, which minimize the computation burden.

It should also be noted that the application of the switching vectors shown in Table 4.3 yields excellent results when the speed of the machine is not too low. However, at very low speeds flux control can be lost. For example, when the machine is started, problems can occur. In this case although at $t=0$ the constant reference stator flux-linkage is applied and at $t>t_1$ a step electromagnetic torque reference is applied, the modulus of the actual stator flux-linkage space vector will be zero until $t=t_1$, and even after t_1 it will not reach its reference value and will vary. Thus instead of having the circular path corresponding to the constant reference flux, the space-vector locus of the stator flux-linkage space vector will not be circular, but will be a six-sided symmetrical locus, (not a symmetrical hexagon), where during 1/6 of the cycle the modulus of the stator flux-linkage space vector changes (at a 'corner' point it is maximum, then it decreases, and then it reaches the maximum value again at the next 'corner' point). The problems are related to the inappropriate use of the switching voltage vectors in the low speed region.

Improved DTC schemes, including improved switching vector selection schemes and predictive switching vector selection schemes, will be discussed in Sections 4.6.2.4 and 4.6.2.9.

4.6.2.3 Fundamentals of stator flux-linkage estimation; estimation problems

In the DTC induction motor drive the stator flux-linkage components have to be estimated due to two reasons. First, these components are required in the optimum switching vector selection table discussed in the previous section. Secondly, they are also required for the estimation of the electromagnetic torque. It should be noted that, in general, it follows directly from the stator voltage equation in the stator reference frame that

$$\psi_{sD}=\int(u_{sD}-R_s i_{sD})\,\mathrm{d}t \tag{4.6-8}$$

$$\psi_{sQ}=\int(u_{sQ}-R_s i_{sQ})\,\mathrm{d}t, \tag{4.6-9}$$

and as shown below

$$\psi_{sB}=\frac{\sqrt{3}\psi_{sD}-\psi_{sQ}}{2}. \tag{4.6-10}$$

If the non-power-invariant forms of the space vectors are used, then

$$\bar{\psi}_s=\tfrac{2}{3}(\psi_{sA}+a\psi_{sB}+a\psi_{sC})=\psi_{sD}+\mathrm{j}\psi_{sQ}$$

$$\psi_{sD}=\psi_{sA}=\int(u_{sD}-R_s i_{sD})\,\mathrm{d}t,$$

where $u_{sD}=u_{sA}$ and $i_{sD}=i_{sA}$. Furthermore

$$\psi_{sQ}=\frac{\psi_{sB}-\psi_{sC}}{\sqrt{3}}=\int(u_{sQ}-R_s i_{sQ})\,\mathrm{d}t,$$

where $u_{sQ}=(u_{sB}-u_{sC})/\sqrt{3}$ and $i_{sQ}=(i_{sB}-i_{sC})/\sqrt{3}$. However, since $\psi_{sC}=-(\psi_{sA}+\psi_{sB})$, thus $\psi_{sB}=[\sqrt{3}\psi_{sQ}-\psi_{sD}]/2$ is obtained in agreement with eqn (4.6-10).

As discussed in Section 3.1.3, it is not necessary to use three stator-voltage sensors and three stator-current sensors since it is possible to show, by considering $u_{sA}+u_{sB}+u_{sA}=0$ and $i_{sA}+i_{sB}+i_{sC}=0$, that u_{sD} and u_{sQ} can be obtained by monitoring only two stator line voltages (e.g. u_{BA}, u_{AC}), and i_{sD}, i_{sQ} can be obtained by monitoring only two stator currents (e.g. i_{sA}, i_{sB}). Thus

$$u_{sD}=\tfrac{1}{3}(u_{BA}-u_{AC}) \qquad u_{sQ}=-\frac{(u_{AC}+u_{BA})}{\sqrt{3}} \qquad i_{sD}=i_{sA} \qquad i_{sQ}=\frac{i_{sA}+2i_{sB}}{\sqrt{3}}.$$

It follows from eqn (4.6-10) that the sign of ψ_{sB} can be obtained by examining the sign of the flux linkage $[\sqrt{3}\psi_{sD}-\psi_{sQ}]$ (physically this corresponds to twice the stator flux linking stator phase sB).

It is very important to note that the performance of the DTC drive using eqns (4.6-8) and (4.6-9) will depend greatly on the accuracy of the estimated stator flux-linkage components, and these depend on the accuracy of the monitored voltages and currents, and also on an accurate integration technique. However, errors may occur in the monitored stator voltages and stator currents due to the following factors: phase shift in the measured values (due to the sensors used), magnitude errors due to conversion factors and gain, offsets in the measurement system, quantization errors in the digital system, etc. Furthermore, an accurate value has to be used for the stator resistance. For accurate flux estimation, the stator resistance must be adapted to temperature changes. The integration can become problematic at low frequencies, where the stator voltages become very small and are dominated by the ohmic voltage drop. At low frequencies the voltage drop of the inverter must also be considered. This is a typical problem associated with open-loop flux estimators used in other a.c. drives as well, which use measured terminal voltages and currents.

Drift compensation is also an important factor in a practical implementation of the integration, since drift can cause large errors of the flux position. In an analog implementation the source of drift is the thermal drift of analog integrators. However, a transient offset also arises from the d.c. components which result after a transient change. If an open-loop speed estimator is used in the DTC induction motor drive (see Section 4.6.2.10), which utilizes the estimated stator flux-linkage components, the speed is determined by also using the flux-linkage space-vector position, thus a drift in the flux-linkage space vector will cause incorrect and oscillatory speed values. An open-loop flux linkage estimator can work well down to 1–2 Hz, but not below this unless special techniques are used (see also Sections 3.1.3.2, 4.1.1.4, 4.5.3.1).

4.6.2.4 Stator-flux-based DTC induction motor drives

4.6.2.4.1 Basic DTC schemes

Figure 4.87 shows the schematic of one simple form of the DTC induction motor drive, employing a VSI inverter. In this scheme the stator flux is the

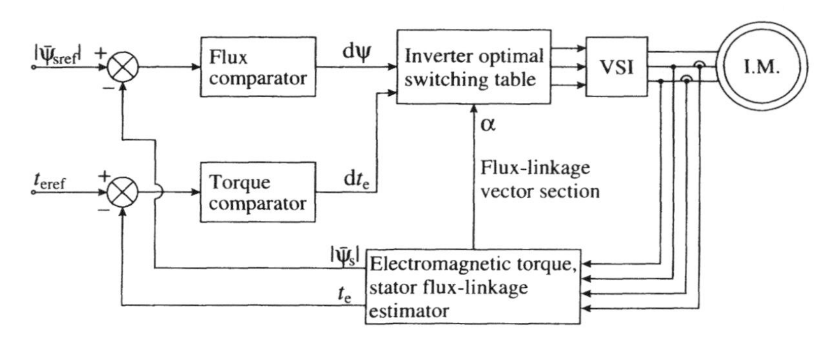

Fig. 4.87. Schematic of stator-flux-based DTC induction motor drive with VSI.

controlled flux, thus it will be referred to as a stator-flux-based DTC induction motor drive.

In Fig. 4.87 a voltage-source (VSI) six-pulse inverter-fed stator-flux-based DTC induction motor drive is shown. As discussed above, direct torque control involves the separate control of the stator flux and torque through the selection of optimum inverter switching modes. The optimum switching table has been shown in Section 4.6.2.2, Table 4.3. In Fig. 4.87 the reference value of the stator flux-linkage space vector modulus, $|\bar{\psi}_{\mathrm{sref}}|$, is compared with the actual modulus of the stator flux-linkage space vector, $|\bar{\psi}_{\mathrm{s}}|$, and the resulting error is fed into the two-level stator flux hysteresis comparator. Similarly, the reference value of the electromagnetic torque (t_{eref}) is compared with its actual value (t_{e}) and the electromagnetic torque error signal is fed into the three-level torque hysteresis comparator. The outputs of the flux and torque comparators ($\mathrm{d}\psi$, $\mathrm{d}t_{\mathrm{e}}$) are used in the inverter optimal switching table (look-up table), which also uses the information on the position of the stator flux-linkage space vector.

In Fig. 4.87, the flux-linkage and electromagnetic torque errors are restricted within their respective hysteresis bands, which are $2\Delta\psi_{\mathrm{s}}$ and $2\Delta t_{\mathrm{e}}$ wide respectively. The flux hysteresis band mainly affects the stator-current distortion in terms of low-order harmonics, and the torque hysteresis band affects the switching frequency and thus the switching losses. The DTC scheme requires flux-linkage and electromagnetic torque estimators. As discussed in Section 4.6.2.3, the stator flux-linkage components can be obtained by integrating appropriate monitored terminal voltages reduced by the ohmic losses, as shown by eqns (4.6-8) and (4.6-9), but at low frequencies large errors can occur due to the variation of the stator resistance, integrator drift, and noise. However, it is not necessary to monitor the stator voltages since they can be reconstructed by using the inverter switching modes and the monitored d.c. link voltage (see also Section 3.1.3.2.1). Improved stator flux estimators are also discussed in Section 4.5.3.1. For completeness, some simple schemes are briefly discussed below in Section 4.6.2.6. The electromagnetic torque can be estimated by using eqn (4.6-1); thus

$$t_{\mathrm{e}}=\tfrac{3}{2}P(\psi_{\mathrm{sD}}i_{\mathrm{sQ}}-\psi_{\mathrm{sQ}}i_{\mathrm{sD}}). \qquad (4.6\text{-}11)$$

Closed-loop speed control can be obtained by using a speed controller (e.g. a PI controller or a fuzzy-logic controller, etc.), whose output gives the torque reference, and the input to the speed controller is the difference between the reference speed and the actual speed.

4.6.2.4.2 Reduction of stator flux and torque ripples

In the DTC induction motor drive there are torque and flux ripples, since none of the inverter switching vectors is able to generate the exact stator voltage required to produce the desired changes in the electromagnetic torque and stator flux linkage in most of the switching instances. However, the ripples in the electromagnetic torque and stator flux linkages can be reduced by using various techniques, some of which involve the use of high switching frequencies or changed inverter topology, but it is also possible to use schemes which do not involve high switching frequency and change of inverter topology (e.g. duty ratio control).

In a DTC induction motor drive, increased switching frequency is desirable since it reduces the harmonic content of the stator currents, and also leads to reduced torque harmonics. However, if high switching frequency is used, this will result in significantly increased switching losses (leading to reduced efficiency) and increased stress on the semiconductor devices of the inverter. This is the reason why inverters of higher power-rating (e.g. in diesel–electric traction) are operated at low switching frequency (to reduce the switching losses), e.g. they are a few hundred Hertz. Furthermore, in the case of high switching frequency, a fast processor is required since the control processing time becomes small. This increases the costs. However, it should be noted that in the ABB DTC drive a high-speed, 40 MHz digital signal processor is used [together with application-specific integrated circuit (ASIC) hardware] to determine the switching frequency of the inverter. The inverter switches in the ABB drive are supplied with the optimum switching pattern every 25 microseconds.

When changed inverter topology is used, it is possible to use an increased number of switches, but this will also increase the costs. For a delta-connected induction machine the switching state number can be increased by the use of two GTO inverters connected in parallel [Takahashi and Ohmori 1989]. Although in this system the number of non-zero switching states is increased to 18, zero-sequence stator currents are produced which have to be reduced by using special techniques, which also require the monitored zero-sequence stator current. However, it is also possible to use schemes, e.g. duty ratio control, which do not involve using inverters with a higher number of switches.

In the conventional DTC induction motor drive (discussed in detail above) a voltage vector is applied for the entire switching period, and this causes the stator current and electromagnetic torque to increase over the whole switching period. Thus for small errors, the electromagnetic torque exceeds its reference value early during the cycle, and continues to increase, causing a high torque ripple. This is then followed by switching cycles in which the zero switching vectors are applied in order to reduce the electromagnetic torque to its reference value. A solution can be obtained where the ripples in the torque and flux can be reduced by employing a selected inverter switching vector not for the entire switching period, as in the conventional DTC induction motor drive, but only for a part of the switching period (which is defined as the duty ratio, δ) and by using the zero switching vector for the rest of the period. The time for which a non-zero voltage vector has to be applied is chosen so as to increase the electromagnetic torque to its reference value. When the electromagnetic torque reaches its reference value, a zero switching vector is applied. During the application of the zero switching vector, zero voltage is forced on the machine, and thus the electromagnetic torque is almost constant; it only decreases slightly (as also discussed earlier). The average input voltage to the induction motor during the application of each switching vector is δU_d (U_d is the inverter d.c. link voltage). By varying the duty ratio between 0 and 1, it is possible to apply any voltage between 0 and U_d during each switching period. This increases the choice of the voltage vector, which is limited by the number of switching vectors in the conventional DTC induction motor drive. As stated above, the duty ratio is selected to give a voltage vector whose average over the switching cycle gives the desired torque change, thus resulting in reduced torque ripples.

The duty ratio of each switching state is a non-linear function of the electromagnetic torque error and stator flux-linkage error, and it is also a function of the position of the stator flux-linkage space vector. Thus it is difficult to model this non-linear function. However, by using a fuzzy-logic-based DTC system, it is possible to perform fuzzy-logic-based duty-ratio control, where the duty ratio is determined during every switching cycle. In such a fuzzy-logic system, there are two inputs, the electromagnetic torque error $e_{te}=t_{eref}-t_e$, and the stator flux-linkage position ρ_s. The output of the fuzzy-logic controller (FLC) is the duty ratio (δ). The fuzzy-logic duty-ratio estimator is shown in Fig. 4.88(a).

The general aspects of fuzzy-logic controllers (FLC) have been discussed in Section 4.4.1. A Mamdani-type of FLC contains a rule base, a fuzzifier, and also a defuzzifier. The fuzzy logic controller shown in Fig. 4.88(a) is a Mamdani-type of controller and contains a rule base, but this comprises two groups of rules, each of which contains nine rules (these are expert statements). The first group is used when the stator flux linkage is smaller than its reference value and the second group of rules is used when it is greater than its reference value. The rules can also be generated by simulating a non-fuzzy DTC induction motor drive using different switching states. As shown in Table 4.5, altogether there are 18 simple rules and there are only a minimal number (three) fuzzy sets used for the two input variables and also for the three output variables; these are: small, medium, and large.

It can be seen that there are five rules in group 2 which are the same as in group 1; the four different rules are shown in bold.

The nine membership functions are selected as follows. The membership function 'small' for the smallest stator-flux position $\rho_s=0$ is $\mu_{\rho s}^{\text{small}}(\rho_s=0\,°)=1$, since it is absolutely certain that $\rho_s=0$ belongs to the fuzzy set of small position angles (the absolute certainty gives the membership value 1, e.g. 50% certainty

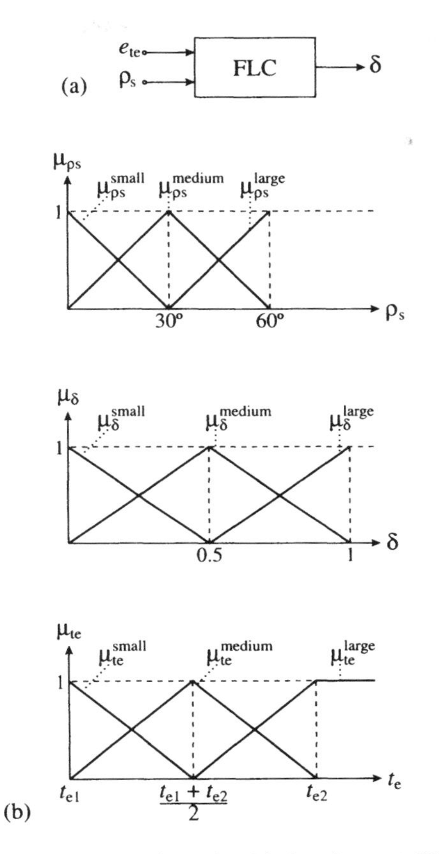

Fig. 4.88. Fuzzy-logic duty-ratio estimator, and membership functions. (a) FLC duty-ratio estimator; (b) membership functions.

would give a membership value of 0.5). Similarly, the membership function 'large' for the largest stator-flux position $\rho_s = 60°$ is $\mu_{\rho s}^{large}(\rho_s = 60°) = 1$, since it is absolutely certain that $\rho_s = 60°$ belongs to the large position angle. The membership function 'medium' for the medium stator-flux position $\rho_s = 30°$ is $\mu_{\rho s}^{medium}(\rho_s = 30°) = 1$, since it is absolutely certain that $\rho_s = 30°$ belongs to the medium position angle. Thus these three points are the extreme points of the $\mu_{\rho s}(\rho_s)$ membership functions shown in Fig. 4.88(b). The three membership functions are obtained by connecting the extreme points with straight lines, and are shown in the top part of Fig. 4.88(b). The three membership functions for the duty ratio can be similarly obtained and are also shown in the middle part of Fig. 4.88(b); these are $\mu_\delta^{small}(\delta)$, $\mu_\delta^{medium}(\delta)$, $\mu_\delta^{large}(\delta)$. The three membership functions for the electromagnetic torque error $[\mu_{te}^{small}(t_e), \mu_{te}^{medium}(t_e), \mu_{te}^{large}(t_e)]$ can also be constructed in a similar way, but of course these vary with the electromagnetic torque, and the constant values shown for the torque (t_{e1}, t_{e2}) depend on the specific machine used. These three membership functions are shown in the bottom part of Fig. 4.88(b).

Table 4.5 Fuzzy rule-base

Group 1 rules					
If e_{te} is small	and	ρ_s is	small	then	δ is medium
If e_{te} is small	and	ρ_s is	medium	then	δ is small
If e_{te} is small	and	ρ_s is	large	then	δ is small
If e_{te} is medium	and	ρ_s is	small	then	δ is medium
If e_{te} is medium	and	ρ_s is	medium	then	δ is medium
If e_{te} is medium	and	ρ_s is	large	then	δ is medium
If e_{te} is large	and	ρ_s is	small	then	δ is large
If e_{te} is large	and	ρ_s is	medium	then	δ is large
If e_{te} is large	and	ρ_s is	large	then	δ is large
Group 2 rules					
If e_{te} is small	**and**	**ρ_s is**	**small**	**then**	**δ is small**
If e_{te} is small	and	ρ_s is	medium	then	δ is small
If e_{te} is small	**and**	**ρ_s is**	**large**	**then**	**δ is medium**
If e_{te} is medium	and	ρ_s is	small	then	δ is medium
If e_{te} is medium	and	ρ_s is	medium	then	δ is medium
If e_{te} is medium	**and**	**ρ_s is**	**large**	**then**	**δ is large**
If e_{te} is large	**and**	**ρ_s is**	**small**	**then**	**δ is medium**
If e_{te} is large	and	ρ_s is	medium	then	δ is large
If e_{te} is large	and	ρ_s is	large	then	δ is large

The ripples in electromagnetic torque, stator flux, stator currents, and speed are reduced by the duty-ratio-controlled DTC. However, to obtain minimal torque ripples, the decrease in the electromagnetic torque during the application of the zero switching vector has to be minimized. Since this decrease of the electromagnetic torque also depends on the modulus of the reference stator flux-linkage space vector, an optimized stator flux-linkage reference value has to be used. For this purpose such a reference stator flux linkage is selected which is just large enough to generate the reference electromagnetic torque. This implies that the maximum electromagnetic torque reference has to be found and the optimum stator flux reference corresponds to this. The maximum torque reference can be obtained by using eqn (4.3-37), and equating the torque reference to this gives

$$t_{eref}=\frac{3}{4}P\left(\frac{L_m}{L_s}\right)^2\frac{|\bar{\psi}_s|^2}{L_r'}, \tag{4.6-12}$$

where L_r' is the rotor transient inductance. Thus it follows by using $|\bar{\psi}_{sref}|=|\bar{\psi}_s|$ that the optimized reference flux linkage for the given electromagnetic torque reference is

$$|\bar{\psi}_{sref}|=\left[\frac{4L_s^2L_r'}{3PL_m^2}\right]^{1/2}. \tag{4.6-13}$$

When the optimized reference stator flux linkage given by eqn (4.6-13) is used in the DTC induction motor drive with duty-ratio control, then the torque ripples are reduced.

When the conventional DTC induction motor drive is operated in the zero-speed region, problems occur as discussed at the end of Section 4.6.2.2. This is due to the fact that, during magnetization, the stator flux comparator selects only non-zero switching vectors (see optimum switching table, Table 4.3 shown in Section 4.6.2.2), and at this time the output of the torque comparator takes one state and the inverter cannot apply zero switching vectors to the motor. However, one possible solution to this problem can be obtained by the application of an additional carrier signal to the input of the electromagnetic torque comparator [Kazmierkowski and Sulkovski 1991]. The injected carrier signal is e.g. a 500 Hz square wave and is only applied in the zero speed region. This forces the zero switching vectors and improves both stator flux-linkage and stator-current waveforms. Furthermore, it ensures robust start and operation in the zero-speed region. However, many other types of solutions can also be obtained which solve the problems in the low speed region, and it is possible to get a satisfactory solution even without using the zero switching vectors (at low speed). However, it is useful to implement another technique where the zero vectors are applied, since in this case the switching frequency is reduced. Such a technique can also be developed by using simple physical considerations and when this is employed, high-dynamic performance is obtained in addition to the reduced switching frequency in the steady state.

A new switching vector selection scheme is proposed in [Damiano *et al.* 1997], which can be used at low speed as well.

4.6.2.5 Main features, advantages, and disadvantages of DTC

The main features of the DTC are:

- direct control of flux and torque (by the selection of optimum inverter switching vectors);
- indirect control of stator currents and voltages;
- approximately sinusoidal stator fluxes and stator currents;
- possibility for reduced torque oscillations; torque oscillations depend on duration of zero-switching vectors;
- high dynamic performance;
- inverter switching frequency depends on widths of flux and torque hysteresis bands.

The main advantages of the DTC are:

- absence of coordinate transformations (which are required in most of the vector-controlled drive implementations);
- absence of separate voltage modulation block (required in vector drives);
- absence of voltage decoupling circuits (required in voltage-source vector drives);
- absence of several controllers (e.g. in a VSI PWM-fed induction motor drive employing rotor-flux-oriented control, there are minimally four controllers);

- only the sector where the flux-linkage space vector is located, and not the actual flux-linkage space-vector position, has to be determined (and the minimum accuracy required is 60 electrical degrees, in contrast to approx. minimum 1.4 degrees in vector drives);
- minimal torque response time.

However, the main disadvantages of a conventional DTC are:

- possible problems during starting and low speed operation and during changes in torque command;
- requirement for flux and torque estimators (same problem exists for vector drives);
- changing switching frequency;
- high ripple torque.

However, it is possible to overcome some of these difficulties, e.g. the difficulties during starting, low speed operation, high ripple torque, etc. (see later sections). In the only known industrially available DTC drive (ABB, 1996), torque response times typically better than 2 ms have been claimed [Tiitinen 1996] together with high torque-control linearity even down to low frequencies including zero speed. It has also been claimed by ABB that the new a.c. drive technology rests chiefly on a new motor model which enables the computation of the motor states without using a speed or position sensor. The motor model used by ABB is a mathematical model (using various machine parameters i.e. the stator resistance, mutual inductance, etc.) and not an artificial-intelligence-based model (e.g. it does not use a neural network or a fuzzy-neural network, etc.). Further details of the ABB DTC induction motor drive are discussed in Section 4.6.2.11.

4.6.2.6 Improved stator flux-linkage estimation

4.6.2.6.1 Application of low-pass filters

As also shown in the first book totally devoted to vector controlled drives [Vas 1990], due to the initial value and drift problems associated with pure integrators, it is useful to replace the pure integrators by a low-pass filter; thus the stator flux linkages can be obtained from

$$\bar{u}_s = R_s \bar{i}_s + \frac{(1+pT)}{T} \bar{\psi}_s. \tag{4.6-14}$$

In eqn (4.6-14) $p = \mathrm{d}/\mathrm{d}t$, and T is a suitably chosen time constant which gives a low cut-off frequency and thus allows eqn (4.6-14) to approximate a pure integration in the widest speed range [e.g. $T = 0.2$ gives a cut-off frequency of $f = 1/(2\pi T) = 0.795$ Hz]. It follows from eqn (4.6-14) that

$$\psi_{sD} = \frac{u_{sD} - R_s i_{sD}}{p + 1/T} \tag{4.6-15}$$

$$\psi_{sQ}=\frac{u_{sQ}-R_s i_{sQ}}{p+1/T}. \tag{4.6-16}$$

4.6.2.6.2 *Application of first-order delay elements*

It is also possible to use a stator flux estimator in which the drift problems associated with 'pure' open-loop integrators at low frequency are avoided by a band-limited integration of the high-frequency components, and by replacing the inaccurate flux estimation at frequencies below $1/T$ by its reference value in a smooth transition. For this purpose a first-order delay element $1/(1+pT)$ is used, thus the stator flux-linkage space vector is obtained from

$$\bar{\psi}_s=\frac{T(\bar{u}_s-R_s\bar{i}_s)+\bar{\psi}_{sref}}{1+pT}, \tag{4.6-17}$$

where $\bar{\psi}_{sref}$ is the stator flux-linkage space vector in the stationary reference frame, $\psi_{sref}=|\bar{\psi}_{sref}|\exp(j\rho_s)$. As discussed in Section 4.1.1.4, the inputs to this stator flux estimator are the measured values of the stator-voltage space vector ($\bar{u}_s$) and stator-current space vector ($\bar{i}_s$), expressed in the stationary reference frame. However, there is also a third input, which is the modulus of the reference value of the stator flux-linkage space vector ($|\bar{\psi}_{sref}|$). It should be noted that since the stationary reference frame is used, $\bar{\psi}_{sref}$ contains two components, ψ_{sDref} and ψ_{sQref}. In eqn (4.6-17) the space vector of the induced stator voltages is $\bar{u}_{si}=\bar{u}_s-R_s\bar{i}_s$ and in an open-loop stator flux-linkage estimator using a 'pure' integrator, its integrated value ($\int\bar{u}_{si}\,dt$) would yield the stator flux-linkage space vector $\bar{\psi}_s$. However, in eqn (4.6-17) $\bar{u}_{si}$ is multiplied by T and the reference stator flux-linkage space vector is added to $T\bar{u}_{si}$, yielding $T\bar{u}_{si}+\bar{\psi}_{sref}$. This is then the input to the first-order delay element, $1/(1+pT)$, on the output of which the estimated value of the stator flux-linkage space vector is obtained.

4.6.2.6.3 *Application of hybrid flux estimators*

It is also possible to use hybrid stator flux estimators, where the rotor voltage equation is also utilized (see also Section 3.3.4). Thus the hybrid flux estimator uses two models; a stator-voltage-equation-based model and also a rotor-voltage-equation-based model. Such an estimator utilizes the fact that at high speeds, accurate stator flux estimation can be obtained by using the stator voltage equation, but at low speeds accurate stator flux estimation can be obtained by using the rotor voltage equation. Several simple solutions can be obtained, but it is important to have smooth transition from the stator-voltage-equation-based estimation to the rotor-voltage-based flux estimation. If a speed sensor can be used, then it is relatively simple to construct the hybrid model by the direct use of the stator and rotor voltage equations, and the inputs are then the measured stator voltages and currents and also the rotor speed.

It is also possible to improve the estimation of the stator flux linkages by using the stabilizing feedback in the voltage model discussed in Section 4.1.1.4.

4.6.2.6.4 Application of observers

To obtain greater accuracy it is possible to use observers (Luenberger, Kalman). In particular, greater robustness to parameter variations can be obtained by using a full-order Luenberger observer (see also Section 4.5.3.3.1), yielding estimations of the stator flux-linkage components which are less sensitive to parameter variations. For this purpose, the observer contains a correction term, which contains the product of the observer gain matrix and the difference between the measured and estimated stator-current components. If the rotor speed is monitored, then it is very simple to use a full-order Luenberger observer which in general is described by

$$\frac{d\hat{\mathbf{x}}}{dt} = \mathbf{A}\hat{\mathbf{x}} + \mathbf{Bu} + \mathbf{K}(\mathbf{y}_s - \hat{\mathbf{y}}_s), \qquad (4.6\text{-}18)$$

where $\hat{\mathbf{x}}$ denotes the estimated states, which are the stator currents and flux-linkage components in the stationary reference frame, $\hat{\mathbf{x}} = [\hat{i}_{sD}, \hat{i}_{sQ}, \hat{\psi}_{sD}, \hat{\psi}_{sQ}]^T$, and $\mathbf{u}$ is the input vector, $\mathbf{u} = [u_{sD}, u_{sQ}]^T$. Furthermore, matrices $\mathbf{A}$ and $\mathbf{B}$ can be obtained from the stator and rotor voltage equations of the induction machine in the stationary reference frame (the state-variable form of these equations is $d\mathbf{x}/dt = \mathbf{Ax} + \mathbf{Bu}$). For example, since $\bar{i}_s$ and $\bar{\psi}_s$ are state variables and the stator voltage equation is $\bar{u}_s = R_s\bar{i}_s + d\bar{\psi}_s/dt$, it immediately follows that the state-variable form of the component stator voltage equations are

$$\frac{d\psi_{sD}}{dt} = -R_s i_{sD} + u_{sD}$$

$$\frac{d\psi_{sQ}}{dt} = -R_s i_{sQ} + u_{sQ}.$$

Thus the system matrix $\mathbf{A}$, which is a 4-by-4 matrix, will contain the following elements: $a_{31} = -R_s$, $a_{32} = a_{33} = a_{34} = 0$, and $a_{42} = -R_s$, $a_{41} = a_{43} = a_{44} = 0$. The other 8 elements of $\mathbf{A}$ can be obtained by considering the rotor voltage equation ($0 = R_r\bar{i}_r' + d\bar{\psi}_r'/dt - j\omega_r\bar{\psi}_r'$) and by eliminating $\bar{i}_r'$ and $\bar{\psi}_r'$ (for this purpose $\bar{i}_r' = (\bar{\psi}_r' - L_m\bar{i}_s)/L_r$ and $\bar{\psi}_r' = (L_r/L_m)(\bar{\psi}_s - L_s'\bar{i}_s)$ are used). Thus the elements a_{12}, a_{14} and a_{21} and a_{23} will contain the rotor speed, and there are four other non-zero elements as well. Thus the system matrix $\mathbf{A}$ contains the rotor speed (ω_r), which is obtained by using a speed sensor.

In eqn (4.6-18) $\mathbf{K}$ is the observer gain matrix, $\mathbf{y}$ is the actual output vector, $\mathbf{y} = \mathbf{Cx}$, $\hat{\mathbf{y}}$ is the estimated output vector, $\hat{\mathbf{y}} = \mathbf{C}\hat{\mathbf{x}}$. The output matrix $\mathbf{C}$ is a two-by-four matrix: $\mathbf{C} = [\mathbf{I}_2, \mathbf{0}_2]$, where $\mathbf{I}_2$ is a second-order identity matrix and $\mathbf{0}_2$ is a second-order null matrix. The estimation error of the stator currents and stator flux linkages (error dynamics) can be obtained by defining the error as $\mathbf{e} = \mathbf{x} - \hat{\mathbf{x}}$. Thus if eqn (4.6-18) is subtracted from $d\mathbf{x}/dt = \mathbf{Ax} + \mathbf{Bu}$, then

$$\frac{d\mathbf{e}}{dt} = (\mathbf{A} - \mathbf{KC})\mathbf{e} \qquad (4.6\text{-}19)$$

is obtained. This is simpler than the observer error equation given by eqn (4.5-104), since eqn (4.5-104) also contains a term with $\Delta\hat{\mathbf{A}}=\hat{\mathbf{A}}-\mathbf{A}$, but now this term is not present since now $\mathbf{A}=\hat{\mathbf{A}}$. It follows from eqn (4.6-19) that the speed of error convergence is determined by the poles of the matrix $\mathbf{A}-\mathbf{KC}$. The gain matrix $\mathbf{K}$ contains four rows and two columns, so it contains 8 elements:

$$\mathbf{K}^{\mathrm{T}}=\begin{bmatrix} k_{11} & k_{12} & k_{13} & k_{14} \\ k_{21} & k_{22} & k_{23} & k_{24} \end{bmatrix}$$

For fast convergence, the elements of the gain matrix $\mathbf{K}$ are selected in such a way that the observer poles should be more negative (faster) then the motor poles. However, care should be taken, since if the observer poles are made too negative, the observer gain may become too large, leading to stability problems resulting from amplified measurement noise. In addition, the motor poles change with the rotor speed, and thus the gain matrix can become large at low speeds if the observer poles are positioned in a fixed location. However, this problem can be overcome by selecting observer poles which are proportional to those of the induction motor (see also Section 4.5.3.5.1). In this way fast convergence and improved robustness to measurement noise is achieved. The poles of the induction motor are described by two sets of complex conjugate pairs, $\Lambda_1=a_1\pm jb_1$, $\Lambda_2=a_2\pm jb_2$ (see for example the book on space vector theory [Vas 1992]); these can be computed for the entire speed range. It is then possible to obtain values of $\mathbf{K}$ such that the observer gains are proportional to the motor gains (it can be shown that in $\mathbf{K}$, the gains k_{12} and k_{21} are proportional to the rotor speed). When such an estimator is used, improved robustness to parameter variations is obtained (compared to the open-loop flux-linkage estimator using pure integrators), and in particular more accurate stator flux-linkage estimates can be obtained even if the stator resistance is not known accurately. However, when inaccurate stator resistance is used, this will also result in inaccurate stator flux-linkage estimates. A substantial improvement can be obtained for stator flux-linkage estimation if a thermal model of the machine is used and this is utilized for accurate stator resistance calculation. However, the thermal model can also be implemented by an observer.

It is also possible to use a Kalman filter for stator flux estimation. (The details of the Kalman filter are described in Section 4.5.3.5.2). When an extended Kalman filter is used, it is possible to estimate the rotor speed in addition to the stator flux linkages and some machine parameters (joint state and parameter estimation).

The stator flux-linkage components can also be estimated by using a model reference adaptive control (MRAS) system. The details of the MRAS systems are described in Section 4.5.3.4, so such a system will not be discussed here.

Finally it is very important to note that it is possible to obtain accurate stator flux-linkage estimates by using an artificial neural network, or a fuzzy-neural estimator. It is also possible to combine fuzzy-logic techniques with conventional observer-based techniques (see also Chapter 7).

4.6.2.7 Field weakening

In the speed range above rated speed, field weakening has to be performed. In a vector-controlled induction motor drive with stator-flux-oriented control, the conventional technique is to decrease the stator flux-linkage reference in inverse proportion to the rotor speed. Thus the electromagnetic torque will decrease inversely with speed. It has been discussed above that in a direct-torque-controlled induction motor, fast torque control is achieved by fast changes of the angle (torque angle) between the stator and rotor flux linkages, and this can be achieved by phase advancing (increase of the torque) or by retarding (decrease of torque) the stator flux-linkage space vector.

It is assumed that the induction machine is supplied by the voltage-source inverter and the inverter is working in the six-step square-wave operation mode. This allows us to exploit the full voltage of the inverter. For the six-step square-wave operation, in the steady state the rotor flux-linkage space vector rotates on an almost circular locus, with a synchronous speed. However, the locus of the stator flux-linkage space vector is a hexagon and the angle between the stator flux and rotor flux-linkage space vectors is constant. This angle (torque angle) at rated torque is around 0.2 rad for a usual induction machine at rated flux. However, in the field-weakening range the rotor speed is in the order of few hundred, rad s^{-1}, and it follows that fast torque control can be achieved by very fast retardation of the stator flux linkage which, however, can be performed by stopping it only for a moment, by applying a short zero stator voltage vector (pulse), as shown in Fig. 4.89(a).

In Fig. 4.89(a), point P_1 is the momentary stop point. This short zero stator-vector pulse is also called a stop pulse. For example, at 50 Hz, a load angle reduction of 0.2 rad can be achieved in 0.6 ms. It should be noted that if the inverter has a restricted minimum time between commutations, the stop pulse cannot be applied

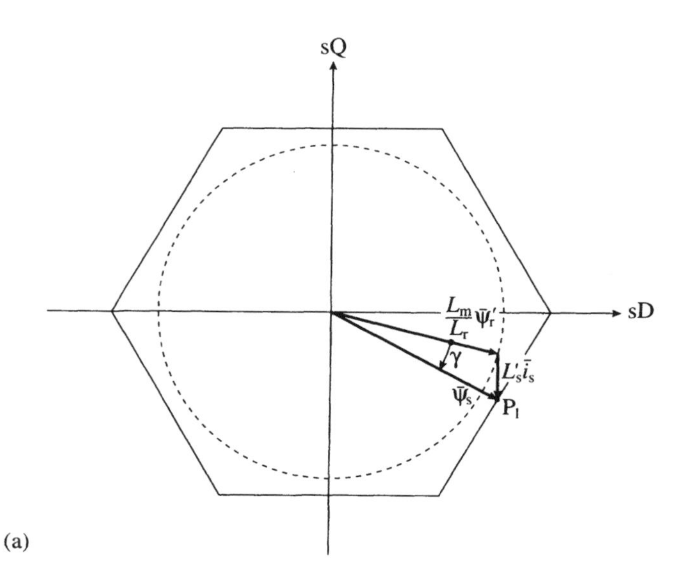

(a)

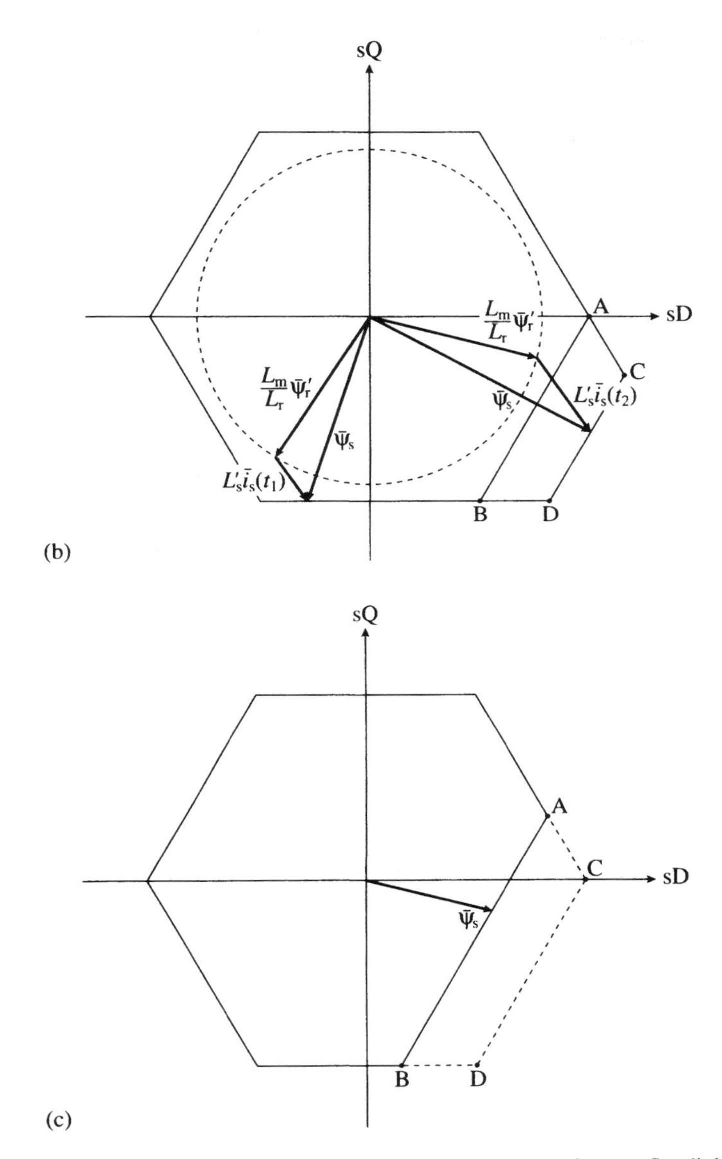

Fig. 4.89. Field-weakening operation: phase retard and phase advance of stator flux-linkage space vector. (a) Phase retard (torque reduction) by applying stop pulse; (b) phase retard (torque reduction) by increasing peripheral length of hexagon around a corner; (c) phase advance (torque increase) by decreasing peripheral length of a hexagon side.

and in this case small retardation of the stator flux linkage is possible by using another technique, which is the deformation technique, shown in Fig. 4.89(b). In this case, the peripheral length of the stator flux linkage hexagon is increased from AB to AC + CD + BD. This causes stator flux-linkage retardation (torque reduction) in the field-weakening range, since the peripheral length along the corner of the stator flux-linkage locus is increased from the original AB length to the length $l = AC + CD + BD$. Thus if the stator flux-linkage space vector is rotating from point B to point A, it will reach point B later than in the original locus and therefore the stator flux-linkage vector is phase retarded. It can also be seen from Fig. 4.89(b) that there is a momentary stator flux increase accompanied by a momentary increase of the stator current (since $\bar{i}_s = [\bar{\psi}_s - (L_m/L_r)\bar{\psi}'_r/L'_s]$, and the rotor flux is constant). Thus torque reduction (phase retardation of the stator flux-linkage space vector) in the field-weakening range can be performed in two ways; by employing a stop pulse or by performing the hexagon deformation shown in Fig. 4.89(b) [Angquist 1986]. However, if the inverter commutation time does not impose any restrictions (see above), then the stop-pulse technique should be used, since this does not result in momentary high values of the stator currents and it also gives faster torque reduction.

If on the other hand a torque increase must be performed, then the stator flux-linkage space vector must be phase advanced. However, since the stator flux-linkage space vector is already running at its maximum speed, phase advance can only be achieved by shortening the peripheral length of the stator flux-linkage locus (hexagon) along one of its sides. This produces a momentary flux weakening as shown in Fig. 4.89(c). It follows from Fig. 4.89(c) that in the original flux hexagon, the peripheral length along the hexagon side is $l = AC + CD + BD$. However, in the new flux locus, this length has been shortened to AB ($AB < l$), and therefore when the stator flux-linkage space vector moves along the locus in the direction from point A to point B, it will reach point B sooner. Since the time of movement through this new locus side is very short (typically between 1 to 5 ms), the rotor flux-linkage space vector will mainly follow the mechanical motion of the rotor and thus the stator flux-linkage space vector is phase advanced.

4.6.2.8 Optimal control of efficiency

In a torque-controlled induction motor drive, in addition to ensuring high dynamic performance, it is also possible to obtain optimal efficiency. This can be very important, for example in a battery-driven electrical vehicle application of the DTC induction motor drive. There are many possibilities to maximize the efficiency, and for this purpose the motor, the inverter, and the controller must be considered.

In general, some of the possible methods of optimal efficiency control are:

Conventional techniques

Constant flux control In this case constant flux is used until the voltage limit of the inverter is reached, and for higher supply frequencies (when the machine speed is larger than the base speed), the flux is reduced in inverse proportion to the supply frequency (conventional method of field weakening). However, when this technique is used, the drive may not achieve maximum torque capability over the full speed region, since the curve which gives the optimal flux level as a function of the rotor speed may differ from the inverse of the rotor speed.

Stator current-ratio control In this case the ratio of the direct- and quadrature-axis stator currents is controlled. This can involve constant, modified, or optimized current-ratio control. For example, for constant stator-current-ratio control, when the induction motor is operating with rotor-flux-oriented quantities, the ratio of the direct- and quadrature-axis stator currents is constant. This ratio can be unity to give maximum torque-to-stator-current ratio. When the modified stator current ratio is used, saturation effects are also considered. For optimal stator-current-ratio control, the stator current ratio depends on the stator frequency as well [Nilsen and Kasteenpohja 1995].

Optimal flux control In this case the control strategy can be found by measuring the efficiency in many points of the speed torque plane, and e.g. by using a look-up table for the flux reference as a function of the torque reference. These measurements also contain information on the saturation effects, but temperature variations are uncompensated.

Artificial-intelligence-based control

Due to the non-linearities involved, it is possible to use very effectively a simple fuzzy-logic-based efficiency controller, which can optimize the efficiency on-line by minimizing the d.c. link power. For this purpose the d.c. link voltage and d.c. link current are also measured and optimum stator-current ratio is obtained, which gives maximum efficiency.

When a conventional, non-artificial-intelligence-based efficiency control technique is used, the control strategy should be a combination of constant current-ratio control and optimal current-ratio control.

4.6.2.9 Improved switching-vector selection schemes; predictive schemes

4.6.2.9.1 General introduction; various predictive schemes

In a DTC induction motor drive using the switching table shown in Section 4.6.2.2, sluggish response can be obtained during start-up and during a change in the reference stator flux linkage and reference electromagnetic torque. Although the switching voltage vectors are determined by the stator flux-linkage position and the errors in the electromagnetic torque and modulus of the stator flux-linkage space vector, large and small errors are not distinguished. Thus the switching vectors chosen for large errors (e.g. during start-up or during a step change in the torque) are the same as the switching vectors chosen for normal operation, when these errors are small. However, it is possible to choose switching vectors which are in accordance with the range of the errors (and stator flux-linkage space-vector position). In this case the responses at start-up and change in reference stator flux and reference electromagnetic torque can be increased. There are several possible solutions, e.g. it is possible to use a fuzzy-logic-based system. In this system those switching states are used during start-up which give faster increase of the stator flux (during this time the change in torque is small). However, when the stator flux-linkage error becomes small, those switching states are chosen which give faster increase of the electromagnetic torque.

It should also be noted that in addition to using a fuzzy-logic-based optimal switching-vector selection strategy, it is also possible to have improvements by using a non-artificial-intelligence-based scheme, where the stator flux-linkage and electromagnetic torque errors are not quantized to two and three levels, and where more than six stator flux-linkage sectors are used. Although it is possible to have an implementation of this system which also uses a look-up table, the computational requirements will be increased.

It is also possible to implement DTC drive schemes in which the required switching voltage vectors are obtained by using predictive algorithms. For this purpose a suitable mathematical model of the induction machine is used and the electromagnetic torque is estimated for each sampling period for all possible inverter modes. The predictive algorithm then selects the inverter switching states to give minimum deviation between the predicted electromagnetic torque and the reference torque. This approach is discussed in some detail below; however, since there are many possibilities for the mathematical model to be used, only the main concepts will be discussed.

For the purposes of the predictive switching-vector estimation scheme, a suitable mathematical model of the induction machine can be obtained by considering the stator and rotor voltage equations (4.1-6), (4.1-24), and the expression of the electromagnetic torque, eqn (4.1-43). Thus the voltage equations in the rotor-flux-oriented reference frame (rotating at the speed of ω_{mr}) are as follows, if for simplicity the effects of magnetic saturation are neglected:

$$\bar{u}_s' = R_s \bar{i}_s' + \frac{L_s' d\bar{i}_s'}{dt} + j\omega_{mr} L_s' \bar{i}_s' + \frac{L_m}{L_r}\left(j\omega_{mr}|\bar{\psi}_r| + \frac{d|\bar{\psi}_r|}{dt}\right) \tag{4.6-20}$$

$$0 = \frac{|\bar{\psi}_r|}{T_r'} + \frac{d|\bar{\psi}_r|}{dt} - \frac{L_m \bar{i}_s'}{T_r'} + j(\omega_{mr} - \omega_r)|\bar{\psi}_r| \tag{4.6-21}$$

$$t_e = \frac{3}{2} P \frac{L_m}{L_r} |\bar{\psi}_r| i_{sy}. \tag{4.6-22}$$

In these equations, the primed quantities are expressed in the rotor-flux-oriented reference frame, L_s' is the stator transient inductance, T_r' is the rotor transient time constant. It is possible to combine these three equations into a single (vector) equation for the stator-voltage space vector expressed in the stationary reference frame, which contains the electromagnetic torque, modulus of the rotor flux-linkage space vector, the rotor speed, and the machine parameters. By assuming that within a sampling time interval, the rotor speed is constant, and also by assuming that the reference electromagnetic torque is a step function, a simple and compact (vector) expression for $\bar{u}_s$ is obtained in terms of the reference values of the electromagnetic torque and reference value of the rotor flux-linkage modulus. This voltage space vector is the reference voltage space vector in the stationary reference frame ($\bar{u}_{sref}$). Thus the appropriate switching state of the inverter can then be determined by using space-vector modulation.

It is also possible to estimate the required stator voltage switching vectors by using a predictive scheme in which there is deadbeat control of the electromagnetic torque and stator flux linkage over a constant switching cycle. For this purpose the stator-voltage space vector is calculated, which is required to control the electromagnetic torque and stator flux-linkage space vector on a cycle-by-cycle basis, by using the electromagnetic torque error and stator flux-linkage errors of the previous cycle, and also the estimated back e.m.f. of the induction machine. It is one of the advantages of this algorithm that it gives constant switching frequency. However, it is important to note that under overmodulation (when $a>1$, see eqn (4.6-43) below) and general transient conditions (when there is a transient in both the flux and torque references), deadbeat control is not possible. However, when there is only a torque transient (which is the most important practical case), it is possible to have deadbeat control of the flux, as discussed in Section 4.6.2.9.2, and in this case two switching voltage vectors are selected in a sampling interval. Similarly, if there is a transient in the stator flux linkage, then it is possible to have deadbeat control of the torque; this will also be discussed in Section 4.6.2.9.2 and again, two switching voltage vectors are applied in a sampling interval. If there is a simultaneous transient in the torque and flux then a single switching vector has to be selected for the entire sampling period. However, if there is no transient in the flux and torque (steady-state) then it is possible to implement deadbeat control of the flux and torque by using the technique described in the next section.

4.6.2.9.2 A predictive control algorithm in the steady state

The eight steps of a predictive algorithm (in an arbitrary nth sampling period T_s) in the steady state are now described.

Step 1 ***Estimation of stator flux linkages from monitored voltages and currents***

If t_n is the time at the beginning of an arbitrary T_s sampling period, then by using the monitored values of stator currents and voltages ($u_{sD}, u_{sQ}, i_{sD}, i_{sQ}$) the stator flux linkages, ψ_{sD}, ψ_{sQ}, (at the beginning of period T_s) can be obtained by using a suitable technique, e.g.

$$\psi_{sD}(t_n)=\int(u_{sD}-R_s i_{sD})\,dt$$

$$\psi_{sQ}(t_n)=\int(u_{sQ}-R_s i_{sQ})\,dt.$$

However, it should be noted that the stator voltages do not have to be monitored, since it is possible to reconstruct the stator voltages from the monitored d.c. link voltage and inverter switching states (e.g. see Section 3.1.3.2.1). Furthermore, it is also possible to utilize any of the simple improved flux-estimation schemes described earlier (e.g. see Sections 4.1.1.4, 4.6.2.6).

Step 2 ***Estimation of back-e.m.f. components from flux linkages and currents***

The back-e.m.f. components (e_{sD}, e_{sQ}) are estimated by using eqn (4.6-24) given below, thus

$$\bar{e}=e_{sD}+je_{sQ}=\frac{d\bar{\psi}_s}{dt}-L_s'\frac{d\bar{i}_s}{dt} \qquad (\bar{\psi}_s=\psi_{sD}+j\psi_{sQ}; \quad \bar{i}_s=i_{sD}+ji_{sQ}).$$

Step 3 ***Estimation of the electromagnetic torque from flux linkages and currents***

The 'present' value of the electromagnetic torque is estimated, e.g. by using

$$t_e(t_n)=\tfrac{3}{2}P(\psi_{sD}i_{sQ}-\psi_{sQ}i_{sD}).$$

Step 4 ***Estimation of change of electromagnetic torque***

$$\Delta t_e=t_e(t_n)-t_{eref}$$

Step 5 ***Estimation of required direct-axis stator reference voltage v_{sDref} (by assuming R_s=0)***

Estimate v_{sDref} by solving eqn (4.6-34) given below, thus solve

$$\left[T_s^2+\left(\frac{\psi_{sQ}T_s}{\psi_{sD}}\right)^2\right]v_{sDref}^2+\left[2a\psi_{sQ}\left(\frac{T_s}{\psi_{sD}}\right)^2+2T_s\psi_{sD}+\frac{2\psi_{sQ}^2T_s}{\psi_{sD}}\right]v_{sDref}$$

$$+2aT_s\frac{\psi_{sQ}}{\psi_{sD}}+\left(\frac{aT_s}{\psi_{sD}}\right)^2+\psi_{sD}^2+\psi_{sQ}^2-\psi_{sref}^2=0,$$

where

$$a=\frac{2\Delta t_e L_s'}{3PT_s}+(\psi_{sD}e_{sQ}-\psi_{sQ}e_{sD})$$

(see eqn (4.6-30) below).

Step 6 ***Estimation of required direct-axis stator reference voltage v_{sQref} (by assuming R_s=0)***

Estimate v_{sQref} by using eqn (4.6-29) given below, thus

$$v_{sQref}=\frac{\psi_{sD}v_{sDref}+a}{\psi_{sD}}.$$

Step 7 ***Estimation of the required reference voltage space vector $\bar{u}_{sref}$ (adding correction term due to $R_s\neq 0$)***

Estimate the reference voltage space vector by using eqn (4.6-35) given below, thus

$$\bar{u}_{sref}=u_{sDref}+ju_{sQref}=\bar{v}_{sref}+R_s\bar{i}_s(t_n)=v_{sDref}+jv_{sQref}+R_s[i_{sD}(t_n)+ji_{sQ}(t_n)].$$

Step 8 ***Estimation of time duration of adjacent and zero switching-voltage vectors (t_a, t_b, and t_0)***

Estimate the time duration of the switching voltage vectors which are adjacent to the reference stator voltage vector (t_a, t_b) by using eqns (4.6-39) and (4.60-40) or eqns (4.6-41)–(4.6-43). Estimate the time duration of the zero switching vector (t_0) by using eqn (4.6-44). Thus e.g.

$$t_a = \frac{3T_s}{2U_d}\left[u_{sDref} - \frac{u_{sQref}}{\sqrt{3}}\right]$$

$$t_b = \frac{\sqrt{3}}{U_d} T_s u_{sQref}$$

$$t_0 = T_s - t_a - t_b.$$

If a solution exists, then output switching states and switching intervals to base drives of the inverter. If solution does not exist (since it is not possible to have deadbeat control of stator flux linkage and electromagnetic torque), then proceed according to the method described below. However, first a mathematical and physical proof of the equations used will be presented.

Mathematical and physical proof of predictive steady-state algorithm

1. Estimation of change of torque over a sampling period

In this computation step, the change of the electromagnetic torque is obtained over a period which is half of the switching period, in terms of the stator flux-linkage components, the reference values of the stator voltage components, and the back-e.m.f. components. For this purpose the stator voltage equation of the induction machine is used together with the expression for the electromagnetic torque. In contrast to the general predictive scheme described above, the rotor voltage equation is not used for the determination of the required switching voltage vectors, since the stator voltage equation is formulated in terms of the back e.m.f. (which implicitly contains the rotor flux) and the change of the electromagnetic torque is obtained in terms of the back e.m.f.

It follows from eqn (4.1-34) that the stator voltage equation of the induction machine in the stationary reference frame can be expressed as

$$\bar{u}_s = R_s\bar{i}_s + \frac{d\bar{\psi}_s}{dt} = R_s\bar{i}_s + L_s'\frac{d\bar{i}_s}{dt} + \bar{e} = R_s\bar{i}_s + \bar{v}, \tag{4.6-23}$$

where $\bar{e}$ is the space vector of the back e.m.f. and $\bar{v} = L_s' d\bar{i}_s/dt + \bar{e}$ is the voltage behind the stator transient inductance (L_s'). Thus if the stator ohmic voltage drop is neglected, it is possible to estimate the direct- and quadrature-axis back e.m.f. components as

$$\bar{e} = e_{sD} + je_{sQ} = \frac{d\bar{\psi}_s}{dt} - L_s'\frac{d\bar{i}_s}{dt}. \tag{4.6-24}$$

However, it should be noted that if it is assumed that $\bar{\psi}_s$ and $\bar{e}$ are sinusoidal, then $\bar{e} = j\omega_1(\bar{\psi}_s - L_s'\bar{i}_s)$, where ω_1 is the excitation frequency, and it can be simply estimated by using $\omega_1=[\bar{\psi}_s \times (\bar{u}_s - R_s\bar{i}_s]/|\bar{\psi}_s|^2$. This can be proved by considering eqn (2.1-124), and by substitution of d/dt by jω_1.

It also follows from eqn (4.6-23) that if the stator ohmic voltage drop is neglected, then the rate of change of the stator-current space vector is simply

$$\frac{d\bar{i}_s}{dt}=(\bar{v}-\bar{e})/L_s'. \tag{4.6-25}$$

If T_s is the time, which is equal to half the switching period, and it is assumed that this is sufficiently short, then $d\bar{i}_s/dt \approx \Delta\bar{i}_s/T_s$ and the change of the stator current can be obtained from eqn (4.6-25) as

$$\Delta\bar{i}_s=\frac{T_s(\bar{v}-\bar{e})}{L_s'}. \tag{4.6-26}$$

By assuming that the stator electrical time constant is much longer than T_s (the change of the stator current over the T_s period is linear), the change of the electromagnetic torque over the time T_s can be obtained as $(3/2)P\bar{\psi}_s \times \Delta\bar{i}_s$, and thus by considering eqn (4.6-26),

$$\Delta t_e=t_{ref}-t_e\approx\frac{3}{2}P(\bar{\psi}_s\times\Delta\bar{i}_s)\approx\frac{3}{2}PT_s\bar{\psi}_s\times\frac{(\bar{v}_{sref}-\bar{e}\,)}{L_s'} \tag{4.6-27}$$

$$=\frac{3}{2}PT_s\frac{(\bar{\psi}_s\times\bar{v}_{sref}-\bar{\psi}_s\times\bar{e}\,)}{L_s'}$$

is obtained. It can be seen that the change of the electromagnetic torque over a period T_s can be obtained from the stator voltage reference (which is now $\bar{v}_{sref}=\bar{u}_{sref}$ since the stator ohmic drop has been neglected) and also the voltage vector ($\bar{e}$) behind the stator transient inductance. By using the two-axis form of eqn (4.6-27), and considering that $\bar{e}=e_{sD}+je_{sQ}$, $\bar{\psi}_s=\psi_{sD}+j\psi_{sQ}$, $\bar{u}_{sref}=u_{sDref}+ju_{sQref}$, we obtain

$$\Delta t_e=\frac{3PT_s}{2L_s'}[(\psi_{sD}v_{sQref}-\psi_{sQ}v_{sDref})+(\psi_{sQ}e_{sD}-\psi_{sD}e_{sQ})]. \tag{4.6-28}$$

It follows from eqn (4.6-28) that the reference value of the quadrature-axis stator voltage is

$$v_{sQref}=\frac{\psi_{sD}v_{sDref}+a}{\psi_{sD}}, \tag{4.6-29}$$

where

$$a=\frac{2\Delta t_e L_s'}{3PT_s}+(\psi_{sD}e_{sQ}-\psi_{sQ}e_{sD}). \tag{4.6-30}$$

2. Stator voltages required for deadbeat control

The change of the stator flux can be obtained from eqn (4.6-23) and, by neglecting the ohmic drop (which is a valid assumption if the stator frequency is above a few Hz) and considering $d\bar{\psi}_s/dt \approx \Delta\bar{\psi}_s/T_s$,

$$\Delta\bar{\psi}_s=T_s\bar{v} \tag{4.6-31}$$

is obtained. However, since $|\bar{\psi}_{\text{sref}}|-|\bar{\psi}_{\text{s}}|=\Delta|\bar{\psi}_{\text{s}}|=\Delta\psi_{\text{s}}$, it follows by considering eqn (4.6-31) that

$$\psi_{\text{sref}}=|\bar{\psi}_{\text{sref}}|=|\bar{\psi}_{\text{s}}|+\Delta|\bar{\psi}_{\text{s}}|=|\bar{\psi}_{\text{s}}(t_n)+T_{\text{s}}\bar{v}_{\text{ref}}|, \qquad (4.6\text{-}32)$$

where t_n is the beginning of an nth T_{s} period and $\bar{\psi}_{\text{s}}(t_n)=\psi_{\text{sD}}+\text{j}\psi_{\text{sQ}}$ is the space vector of the stator flux linkages at the beginning of the sampling period (the flux-linkage components are known, since they are determined by integrating the appropriate stator voltage). Equation (4.6-32) can be used to obtain the stator voltage required for deadbeat control of the stator flux linkages. By using the two-axis form of eqn (4.6-32),

$$\psi_{\text{sref}}^2=(\psi_{\text{sD}}+v_{\text{sDref}}T_{\text{s}})^2+(\psi_{\text{sQ}}+v_{\text{sQref}}T_{\text{s}})^2, \qquad (4.6\text{-}33)$$

where ψ_{sD} and ψ_{sQ} are the direct- and quadrature-axis stator flux linkages at the beginning of the nth sampling period. Thus the reference stator-voltage components v_{sDref}, v_{sQref} can be obtained from eqns (4.6-29) and (4.6-33). Substitution of eqn (4.6-29) into (4.6-33) gives a quadratic equation, where the unknown is v_{sDref},

$$\left[T_{\text{s}}^2+\left(\frac{\psi_{\text{sQ}}T_{\text{s}}}{\psi_{\text{sD}}}\right)^2\right]v_{\text{sDref}}^2+\left[2a\psi_{\text{sQ}}\left(\frac{T_{\text{s}}}{\psi_{\text{sD}}}\right)^2+2T_{\text{s}}\psi_{\text{sD}}+\frac{2\psi_{\text{sQ}}^2T_{\text{s}}}{\psi_{\text{sD}}}\right]v_{\text{sDref}}$$
$$+2aT_{\text{s}}\frac{\psi_{\text{sQ}}}{\psi_{\text{sD}}}+\left(\frac{aT_{\text{s}}}{\psi_{\text{sD}}}\right)^2+\psi_{\text{sD}}^2+\psi_{\text{sQ}}^2-\psi_{\text{sref}}^2=0.$$

However, eqn (4.6-34) gives two solutions for v_{sDref}, but the solution with the smallest absolute value is chosen, since this corresponds to the smallest direct-axis stator voltage which is required to drive the stator flux linkage and electromagnetic torque to their reference values. When the obtained v_{sDref} is substituted into eqn (4.6-29), finally v_{sQref} is obtained. Thus $\bar{v}_{\text{sref}}=v_{\text{sDref}}+\text{j}v_{\text{sQref}}$ is obtained (which corresponds to the reference-voltage space vector if the stator ohmic voltage drop is neglected). Therefore the corrected reference stator-voltage space vector (which incorporates the effects of the stator ohmic voltage drop) is obtained as

$$\bar{u}_{\text{sref}}=\bar{v}_{\text{sref}}+R_{\text{s}}\bar{i}_{\text{s}}(t_n). \qquad (4.6\text{-}35)$$

In eqn (4.6-35) the stator ohmic drop from the previous cycle $R_{\text{s}}\bar{i}_{\text{s}}(t_n)$ is added to $\bar{v}_{\text{sref}}$. However, this is justified, since as discussed above the change of the stator current is assumed to be linear over the period T_{s} (since the ohmic drop is small compared to the voltage drop across the stator transient inductance).

3. Inverter switching-state determination in the steady state (space-vector PWM)

The appropriate switching state of the inverter is determined by using space-vector pulse-width modulation (PWM). Therefore $\bar{u}_{\text{sref}}$ [defined by eqn (4.6-35)] is used to select the optimal switching voltage vectors in such a way that the two switching vectors $(\bar{u}_k, \bar{u}_{k+1})$ closest to $\bar{u}_{\text{sref}}$ are selected, and the amount of time during which these vectors are applied $(t_{\text{a}}, t_{\text{b}})$ is determined from

$$\bar{u}_{\text{sref}}T_{\text{s}}=\bar{u}_kt_{\text{a}}+\bar{u}_{k+1}t_{\text{b}}+\bar{u}_0t_0. \qquad (4.6\text{-}36)$$

In eqn (4.6-36)

$$T_s = t_0 + t_a + t_b, \tag{4.6-37}$$

where T_s is the sampling time. Equation (4.6-36) follows from the fact that the time average of the three switching states (two active states and one zero state) during the sampling interval is equal to the reference voltage space vector, and also the switching voltage vectors and the reference voltage vector are constant over one switching cycle. In eqn (4.6-36), $\bar{u}_k$ are the switching vectors in the 8 switching states of the voltage source inverter [see also eqn (4.6-4)]:

$$\bar{u}_k = \begin{cases} \frac{2}{3}U_d \exp[j(k-1)\pi/3] & k=1,2,\ldots,6 \\ 0 & k=7,8 \end{cases} \tag{4.6-38}$$

where U_d is the d.c. link voltage, $k=1,2,\ldots,6$ corresponds to the active (non-zero) switching voltage vectors, and $k=7,8$ correspond to the two zero switching voltage vectors. The different switching states have been shown in Fig. 4.83(b). If eqn (4.6-38) is substituted into eqn (4.6-36) and the resulting equation is resolved into its real and imaginary parts, then t_a and t_b can be determined. Thus by using $\bar{u}_{sref} = u_{sDref} + ju_{sQref}$,

$$t_a = \frac{3T_s}{2U_d}\left[u_{sDref} - \frac{u_{sQref}}{\sqrt{3}}\right] \tag{4.6-39}$$

$$t_b = \frac{\sqrt{3}}{U_d} T_s u_{sQref}, \tag{4.6-40}$$

or by using polar coordinates, $\bar{u}_{sref} = |\bar{u}_{sref}| \exp(j\alpha_{sref})$:

$$t_a = a\left[T_s \frac{(\cos\alpha_{sref} - \sin\alpha_{sref})}{\sqrt{3}}\right] \tag{4.6-41}$$

$$t_b = 2aT_s \frac{\sin\alpha_{sref}}{\sqrt{3}}, \tag{4.6-42}$$

where 'a' is the modulation index,

$$a = \frac{3|\bar{u}_{sref}|}{2U_d}. \tag{4.6-43}$$

The time during which the appropriate zero switching vector is selected is

$$t_0 = T_s - t_a - t_b. \tag{4.6-44}$$

Switching from the zero state to two adjacent states involves commutation of each inverter leg exactly once, thus T_s is a half-period of the switching frequency. Hence the stator flux linkage and the electromagnetic torque are controlled twice per switching frequency. Furthermore, when this control algorithm is used, then the oscillations in the stator currents are reduced.

Under transient conditions, it is not possible to have the deadbeat control. This is due to the fact that, in this case, there is no sufficient d.c. link voltage to cause an adequate change in the electromagnetic torque and/or in the stator flux linkage to achieve deadbeat control. It follows from eqns (4.6-37), (4.6-39), and (4.6-40) that when there is an electromagnetic torque transient and/or stator flux-linkage transient (e.g. step changes), then the electromagnetic torque and stator flux-linkage errors are large in one switching period and thus $t_a+t_b>T_s$ is obtained. This means that $\bar{u}_{sref}$ is too large to be synthesized in a single switching period and thus an alternative control technique must be used. To summarize: the deadbeat control scheme presented above is suitable for the determination of the switching vectors in the steady state of the drive, and it is one advantage of the technique that the switching frequency is constant.

4.6.2.9.3 Predictive control in the transient state

In the following the predictive control strategy (switching vector selection) to be used under transient conditions is discussed. If there is a transient in the electromagnetic torque reference (e.g. step change), then the controller must drive the electromagnetic torque in the required direction (to reduce the torque error over the period T_s) while still maintaining deadbeat stator flux-linkage control. If there is a transient in the reference stator flux linkage, then the stator flux linkage has to be driven in the direction of its reference, while maintaining deadbeat control of the electromagnetic torque. In this way the selection of the appropriate switching vectors can be performed by first considering the position of the stator flux-linkage space vector (with respect to the sD-axis of the stationary reference frame), ρ_s, and then the sign of the electromagnetic torque or stator flux-linkage error is also considered. This is now discussed in some detail.

First a transient in the torque reference is assumed, where the electromagnetic torque cannot be driven to its reference value in a single T_s period (i.e. it is not possible to have deadbeat torque control). This is the most important case which arises most frequently, and in general a step change in the stator flux linkages is not required. In most of the cases the flux linkage is changed only during field weakening. However, in this operating mode, when the flux starts to be weakened, there is an almost linear change and then it varies continuously. Thus to obtain a simplified predictive switching-vector selection scheme, the simultaneous torque and flux transient can be neglected.

For the case when there is only torque transient, the switching vectors are chosen *a priori* in such a way that they should drive the torque in the desired direction, but allowing deadbeat stator flux-linkage control. If, for example, the stator flux-linkage space vector is in sector 1 shown in Fig. 4.84, which spans the region $-30°$ to $30°$ [angle $\alpha(1)$], then the switching vectors $\bar{u}_2$ and $\bar{u}_3$ create flux linkages which will increase the original stator flux linkage and will also cause an increase of the electromagnetic torque. Similarly, switching vectors $\bar{u}_5$, $\bar{u}_6$ result in a decreased stator flux-linkage space vector and in a decreased electromagnetic torque. Thus $\bar{u}_2$ and $\bar{u}_3$ (which correspond to switching states 2 and 3) can be used to control the flux linkage to its reference value over the T_s period, while continuously increasing the torque over the entire interval. Thus it is possible to construct a switching vector selection table for the transient state, where two switching vectors (kth and $(k+1)$-th vectors) are selected by using the torque error and the sector number $n=1, 2, \ldots, 6$ (where the stator flux linkage is located). This is shown in Table 4.6. The sector where a stator flux-linkage space vector is located can be determined from the position of the stator flux-linkage space vector (see discussion above in Section 4.6.2.2). In Table 4.6, k and $k+1$ denote the selected kth and $(k+1)$-th switching states respectively.

Thus by using the selected switching vectors $\bar{u}_k$, $\bar{u}_{k+1}$, the stator flux linkage is controlled in deadbeat fashion similarly to that given by eqn (4.6-32); thus

$$\psi_{\text{sref}} = |\bar{\psi}_s| + \Delta|\bar{\psi}_s| = |\bar{\psi}_s(t_n) + \bar{u}_k t_k + \bar{u}_{k+1} t_{k+1}|, \tag{4.6-45}$$

where for duration t_k the switching vector $\bar{u}_k$ is applied and for duration t_{k+1} the switching vector $\bar{u}_{k+1}$ is applied. In the transient state the zero switching vectors are not used, since it is desired to drive the electromagnetic torque in one direction in the fastest possible way during the switching period. Thus $t_0=0$, and it follows from eqn (4.6-44) that

$$T_s = t_k + t_{k+1}. \tag{4.6-46}$$

It should be noted that skipping the zero states while still switching between adjacent states is equivalent to pulse dropping in a conventional sinusoidal PWM. For a given value of ψ_{sref}, eqns (4.6-45) and (4.6-46) can then be solved for t_k and t_{k+1}. Thus by applying the appropriate switching vectors for the appropriate duration, the stator flux linkage is controlled to its reference value, and the electromagnetic torque is driven continuously to its desired value in the proper direction, with maximum voltage applied to the inverter (due to $t_0=0$, the voltage reference is maintained on the boundary of the hexagonal locus; thus the inverter output voltage is maximal).

If the second case is considered, when there is a transient of the stator flux linkage (i.e. the stator flux linkage cannot be driven to its reference value in a single T_s period), then again the voltage vector selection is made *a priori*. In this case, if for example the stator flux-linkage space vector is in sector 1, then $\bar{u}_1$ and $\bar{u}_2$ increase the flux and $\bar{u}_3$ and $\bar{u}_4$ decrease the flux, etc. The appropriate switching vector selection table is shown in Table 4.7.

Thus by using the selected switching vectors $\bar{u}_k$, $\bar{u}_{k+1}$, the electromagnetic torque is controlled in deadbeat fashion similarly to that given by eqn (4.6-27); thus

$$\Delta t_e = \frac{3P}{2L_s'}[\bar{\psi}_s \times (t_k \bar{u}_k + t_{k+1}\bar{u}_{k+1} - T_s \bar{e}]. \tag{4.6-47}$$

Table 4.6 Switching-voltage vector selection if there is an electromagnetic torque transient

$\text{sgn}(t_e - t_{\text{eref}})$	k						$k+1$					
	$\alpha(1)$ sect 1	$\alpha(2)$ sect 2	$\alpha(3)$ sect 3	$\alpha(4)$ sect 4	$\alpha(5)$ sect 5	$\alpha(6)$ sect 6	$\alpha(1)$ sect 1	$\alpha(2)$ sect 2	$\alpha(3)$ sect 3	$\alpha(4)$ sect 4	$\alpha(5)$ sect 5	$\alpha(6)$ sect 6
0	$\bar{u}_2$	$\bar{u}_3$	$\bar{u}_4$	$\bar{u}_5$	$\bar{u}_6$	$\bar{u}_1$	$\bar{u}_3$	$\bar{u}_4$	$\bar{u}_5$	$\bar{u}_6$	$\bar{u}_1$	$\bar{u}_2$
1	$\bar{u}_5$	$\bar{u}_6$	$\bar{u}_1$	$\bar{u}_2$	$\bar{u}_3$	$\bar{u}_4$	$\bar{u}_6$	$\bar{u}_1$	$\bar{u}_2$	$\bar{u}_3$	$\bar{u}_4$	$\bar{u}_5$

Table 4.7 Switching voltage vector selection if there is a stator flux-linkage transient

$\text{sgn}(\lvert\bar{\psi}_{\text{sref}}\rvert-\psi_{\text{sref}})$	k						$k+1$					
	$\alpha(1)$ sect 1	$\alpha(2)$ sect 2	$\alpha(3)$ sect 3	$\alpha(4)$ sect 4	$\alpha(5)$ sect 5	$\alpha(6)$ sect 6	$\alpha(1)$ sect 1	$\alpha(2)$ sect 2	$\alpha(3)$ sect 3	$\alpha(4)$ sect 4	$\alpha(5)$ sect 5	$\alpha(6)$ sect 6
0	$\bar{u}_1$	$\bar{u}_2$	$\bar{u}_3$	$\bar{u}_4$	$\bar{u}_5$	$\bar{u}_6$	$\bar{u}_2$	$\bar{u}_3$	$\bar{u}_4$	$\bar{u}_5$	$\bar{u}_6$	$\bar{u}_1$
1	$\bar{u}_3$	$\bar{u}_4$	$\bar{u}_5$	$\bar{u}_6$	$\bar{u}_1$	$\bar{u}_2$	$\bar{u}_4$	$\bar{u}_5$	$\bar{u}_6$	$\bar{u}_1$	$\bar{u}_2$	$\bar{u}_3$

Table 4.8 Switching-voltage vector selection if there are both stator flux-linkage and electromagnetic torque transients

$\text{sgn}(t_{\text{e}}-t_{\text{eref}})$	$\text{sgn}(\lvert\bar{\psi}_{\text{s}}\rvert-\psi_{\text{sref}})$	$\alpha(1)$ sector 1	$\alpha(2)$ sector 2	$\alpha(3)$ sector 3	$\alpha(4)$ sector 4	$\alpha(5)$ sector 5	$\alpha(6)$ sector 6
0	0	$\bar{u}_2$	$\bar{u}_3$	$\bar{u}_4$	$\bar{u}_5$	$\bar{u}_6$	$\bar{u}_1$
0	1	$\bar{u}_3$	$\bar{u}_4$	$\bar{u}_5$	$\bar{u}_6$	$\bar{u}_1$	$\bar{u}_2$
1	1	$\bar{u}_5$	$\bar{u}_6$	$\bar{u}_1$	$\bar{u}_2$	$\bar{u}_3$	$\bar{u}_4$
1	0	$\bar{u}_6$	$\bar{u}_1$	$\bar{u}_2$	$\bar{u}_3$	$\bar{u}_4$	$\bar{u}_5$

Equations (4.6-46) and (4.6-47) yield the required t_k and t_{k+1}. Thus if the selected switching vector $\bar{u}_k$ is applied for duration t_k and $\bar{u}_{k+1}$ is applied for duration t_{k+1}, then deadbeat control of the electromagnetic torque is performed, while driving the stator flux-linkage space vector in the desired direction. It should be noted that, as before, the zero switching vectors are not used. Disregarding the zero switching states while still switching between adjacent states is equivalent to pulse-dropping in a conventional sinusoidal PWM.

Finally the third condition is considered: this is the case when there is a transient in both the stator flux linkage and electromagnetic torque. In this case a single switching state is selected for the entire switching period, which drives both the electromagnetic torque and stator flux linkage in the desired directions as quickly as possible. For example, if the stator flux-linkage space vector is in the first sector, and both the flux and the torque have to be increased, then by considering the six switching vectors, $\bar{u}_2$ has to be selected. If the flux has to be increased but the torque has to be decreased then $\bar{u}_6$ has to be selected, etc. The switching vectors are given in Table 4.8.

Obviously this agrees with the corresponding parts in Table 4.3 shown in Section 4.6.2.2.

Thus, in general, the algorithm for the switching voltage vector selection is such that first the computational steps (Steps 1–8) given above for the steady-state estimation are performed. However, if in Step 8 it is found that there is no positive solution (for t_{a} and t_{b}), e.g. $t_k+t_{k+1}\leqslant T_{\text{s}}$ cannot be satisfied, then it is first assumed that there is a transient in the electromagnetic torque reference. Thus Table 4.6 is used for the selection of the appropriate two inverter switching states and then eqns (4.6-45) and (4.6-46) are solved to yield t_k and t_{k+1}. If this gives positive solutions, then $\bar{u}_k$ and $\bar{u}_{k+1}$ are applied for the duration of t_k and t_{k+1} respectively. However, if a positive solution still does not exist, then it is assumed that a transient in the stator flux linkage is present and therefore eqns (4.6-46) and (4.6-47) are solved. If positive solutions exist then the appropriate switching vectors shown in Table 4.7 are applied for the durations t_k and t_{k+1} respectively. However, if a solution does not exist then the switching states shown in Table 4.8 are applied (corresponding switching states and switching times are outputs to the inverter base drives). The predictive algorithm can be implemented by a DSP, but it should be considered that this is a computationally very intensive scheme and at higher switching frequencies very fast computations are required.

It should be noted that, similarly to other types of predictive control schemes, when the present predictive scheme is used, a single T_s-period steady-state error occurs. This is due to the fact that the estimation of the electromagnetic torque and stator flux linkage is based on inputs from the preceding period. A single-period delay is required to allow time to estimate the switching signals. An ideal deadbeat controller would require the estimations to be performed in zero time. Furthermore, the predictive controller is based only on the electromagnetic torque and stator flux-linkage errors. The combination of this fact with the single-period delay results in a steady-state error of the electromagnetic torque. However, this does not present any problem if a speed controller is also used.

An alternative predictive switching-vector selection technique is now briefly discussed, which is not as computationally intensive as the one described above. It is assumed that there is only a torque transient. Although the alternative technique does not result in deadbeat control of the stator flux linkage, only minimal performance degradation will result. This is achieved by limiting the magnitude of the stator voltage reference to the maximum instantaneous value allowable with space-vector modulation. This is shown in Fig. 4.90.

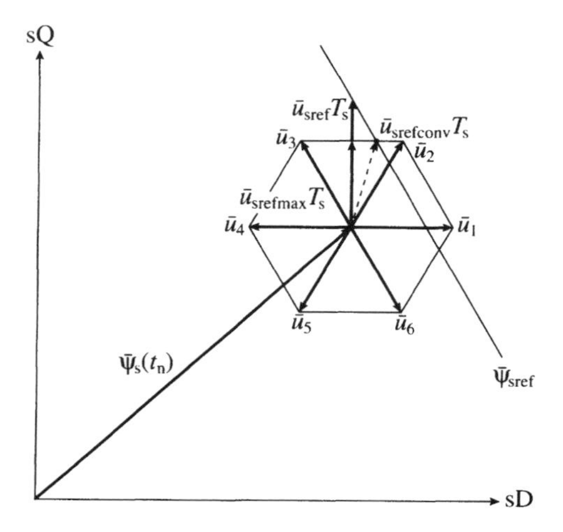

Fig. 4.90. Alternative predictive switching-vector selection.

It is shown in Fig. 4.90 that every time the voltage reference vector ($\bar{u}_{sref}$) resulting from a deadbeat control algorithm is outside the voltage hexagon, the magnitude of the voltage input to the space-vector modulator is limited to the maximum inverter voltage ($\bar{u}_{srefmax}$), which is also shown in Fig. 4.90. The voltage $\bar{u}_{srefmax}$ has the same angle as the original voltage reference (α_{sref}). Thus the magnitude of the voltage reference vector is given as

$$|\bar{u}_{smaxref}| = (1/3)U_d\sqrt{3}/\sin(\alpha_{sref} + \pi/3) \tag{4.6-48}$$

and the angle is unchanged from that computed by using the algorithm given above. In Fig. 4.90 the average value of the stator voltage over one sampling period (T_s) computed by the original (not the alternative) predictive scheme is $\bar{u}_{srefconv}$; this is shown by a vector with broken lines. The tip of this vector is at the intersection of the output voltage hexagonal boundary and the reference stator flux linkage. However, when the alternative predictive scheme is used, the average voltage ($\bar{u}_{srefmax}$) over the same period (T_s) lies on the voltage boundary at the same angle as $\bar{u}_{sref}$. It should be noted that the flux-linkage error increases as the magnitude of $\bar{u}_{sref}$ increases. It is interesting to note that $\bar{u}_{sref}$ is leading $\bar{u}_{srefconv}$ for the transient condition when the electromagnetic torque has to be increased, and it lags $\bar{u}_{srefconv}$ for the transient condition when the electromagnetic torque has to be decreased. It follows that the torque-producing current is larger with the alternative scheme then with the original predictive scheme. This leads to improved dynamic performance, despite the fact that the improved scheme does not provide deadbeat control of the flux magnitude. This alternative predictive scheme can also be used when overmodulation occurs (i.e. in the transition region between continuous PWM and six-step operation). This operating region is a particular case of the torque/flux transients. Similarly to that discussed above, in the transition region the voltage reference vector $\bar{u}_{sref}$ lies outside the hexagon boundary. In the limit, as $\bar{u}_{sref}$ becomes large in magnitude, the inverter voltage $\bar{u}_{srefmax}$ jumps from one corner of the hexagon to the next, which is equivalent to six-step operation.

4.6.2.10 Speed-sensorless DTC drive implementations

For a speed-sensorless DTC drive it is possible to use any of the techniques described in Section 4.5. Thus the following techniques can be used is a speed-sensorless DTC drive (see Section 4.5.1 for a brief discussion):

1. Open-loop and improved estimators using monitored stator voltages/currents;
2. Estimators using the spatial saturation stator-phase third-harmonic voltage;
3. Estimators using saliency (geometrical, saturation) effects;
4. Model reference adaptive systems (MRAS);
5. Observers (Kalman, Luenberger);
6. Estimators using artificial intelligence (neural network, fuzzy-logic-based systems, fuzzy-neural networks, etc.).

Such techniques have been discussed in detail in Sections 4.5.3.1–4.5.3.6 and thus these will not be repeated here. However, it should be noted that so far only one specific solution using the first technique has been applied for a DTC induction motor drive (in the ABB DTC induction motor drive), but it is expected that the other techniques shown above will also be used in the future. In particular it is expected that estimators using artificial intelligence will play an increased role. Such estimators are also robust to parameter variations.

The speed signal can be required in a DTC induction motor drive for two reasons. One reason is that it can be used for stator flux-linkage estimation, if the stator flux-linkage estimator requires the rotor speed signal (e.g. when a hybrid stator flux-linkage estimator is used, which uses both the stator voltage equation and also the rotor voltage equation, as discussed above in Section 4.6.2.6.3, and the rotor voltage equation contains the rotor speed). The other reason is that a rotor speed signal is required if the drive contains a speed-control loop (in this case the speed controller outputs the torque reference, and the input to the speed controller is the difference between the reference speed and the estimated speed). In many variable-speed drive applications torque control is required, but speed control is not necessary. An example of an application where the electromagnetic torque is controlled without precise speed control is traction drives. In traction applications (diesel-electric locomotives, electrical cars, etc.) the electromagnetic torque is directly controlled, i.e. the electromagnetic torque is the commanded signal, and it is not the result of a speed error signal.

The key to the success of simple open-loop speed estimation schemes is the accurate estimation of the stator (or rotor) flux-linkage components. If the flux linkages are accurately known, then it is possible to estimate the rotor speed by simple means, which utilize the speed of the estimated flux-linkage space vector. It is this technique which is used in the DTC induction motor drive manufactured by ABB, which is the only known industrial DTC drive at present. However, the same technique is also used in some other commercially available drives (vector drives), thus this will now be briefly discussed. In these commercial implementations the stator voltages are not monitored but are reconstructed from the d.c. link voltage and switching states of the inverter (see Section 3.1.3.2.1).

It follows from eqn (4.5-36) that the rotor speed can be expressed as

$$\omega_r = \omega_{mr} - \omega_{sl}, \tag{4.6-49}$$

where ω_{mr} is the speed of the rotor flux (relative to the stator) $\omega_{mr} = d\rho_r/dt$, and ω_{sl} is the angular slip frequency, given by eqns (4.5-34) and (4.5-35), which are now repeated here for convenience:

$$\omega_{sl} = \frac{L_m}{T_r|\bar{\psi}_r|^2}(-\psi_{rq}i_{sD} + \psi_{rd}i_{sQ}) \tag{4.6-50}$$

$$\omega_{sl} = \frac{2t_e R_r}{3P|\bar{\psi}_r|^2}. \tag{4.6-51}$$

Physically, ω_{sl} is the speed of the rotor flux-linkage space vector with respect to the rotor. It is possible to obtain an expression for ω_{mr} in terms of the rotor flux-linkage components by expanding the expression for the derivative $d\rho_r/dt$. Since the rotor flux-linkage space vector expressed in the stationary reference frame is $\bar{\psi}'_r = \psi_{rd} + j\psi_{rq} = |\bar{\psi}_r| \exp(j\rho_r)$, thus $\rho_r = \tan^{-1}(\psi_{rq}/\psi_{rd})$, it follows that the derivative can be expanded to give

$$\omega_{mr} = \frac{d\rho_r}{dt} = \frac{d}{dt}[\tan^{-1}(\psi_{rq}/\psi_{rd})]$$

$$= \left(\psi_{rd}\frac{d\psi_{rq}}{dt} - \psi_{rq}\frac{d\psi_{rd}}{dt}\right) \Big/ (\psi_{rd}^2 + \psi_{rq}^2). \qquad (4.6\text{-}52)$$

The numerator of eqn (4.6-52) contains $|\bar{\psi}_r|^2 = (\psi_{rd}^2 + \psi_{rq}^2)$. Substitution of eqn (4.6-52) into eqn (4.6-49), and by also considering eqns (4.6-50) or (4.6-51), we obtain

$$\omega_r = \left(\psi_{rd}\frac{d\psi_{rq}}{dt} - \psi_{rq}\frac{d\psi_{rd}}{dt}\right) \Big/ |\bar{\psi}_r|^2 - \frac{L_m}{T_r|\bar{\psi}_r|^2}(\psi_{rd}i_{sQ} - \psi_{rq}i_{sD}) \qquad (4.6\text{-}53)$$

and

$$\omega_r = \left(\psi_{rd}\frac{d\psi_{rq}}{dt} - \psi_{rq}\frac{d\psi_{rd}}{dt}\right) \Big/ |\bar{\psi}_r|^2 - \frac{2t_e R_r}{3P|\bar{\psi}_r|^2}. \qquad (4.6\text{-}54)$$

Such a speed estimator is used in the ABB DTC induction motor drive. A rotor speed estimator based on eqn (4.6-53) or (4.6-54) can then be constructed, which uses the monitored stator currents and the rotor flux components which, however, can be obtained from the stator flux linkages by considering eqns (4.5-16) and (4.5-17):

$$\psi_{rd} = \frac{L_r}{L_m}(\psi_{sD} - L'_s i_{sD}) \qquad (4.6\text{-}55)$$

$$\psi_{rq} = \frac{L_r}{L_m}(\psi_{sQ} - L'_s i_{sQ}). \qquad (4.6\text{-}56)$$

In eqns (4.6-55) and (4.6-56) the stator flux linkages can be obtained by using monitored stator currents and monitored or reconstructed stator voltages, as discussed earlier in Section 3.1.3.2. Thus by using the inverter switching functions S_A, S_B, S_C (see also Fig. 4.83 in Section 4.6.2.1) the stator-voltage space vector (expressed in the stationary reference frame) can be obtained, by using the switching states and the d.c. link voltage U_d, as

$$\bar{u}_s = \frac{2}{3}U_d(S_A + aS_B + a^2 S_C) = u_{sD} + ju_{sQ}, \qquad (4.6\text{-}57)$$

where

$S_A = 1$ when upper switch in phase sA of inverter (S1) is ON and lower switch (S4) is OFF

$S_A = 0$ when upper switch in phase sA of inverter (S1) is OFF and lower switch (S4) is ON

$S_B=1$ when upper switch in phase sB of inverter (S3) is ON and lower switch (S6) is OFF
$S_B=0$ when upper switch in phase sB of inverter (S3) is OFF and lower switch (S6) is ON
$S_C=1$ when upper switch in phase sC of inverter (S5) is ON and lower switch (S2) is OFF
$S_C=0$ when upper switch in phase sC of inverter (S5) is OFF and lower switch (S2) is ON.

It follows from eqn (4.6-57) that

$$u_{sD}=\frac{2}{3}U_d\left(S_A-\frac{S_B}{2}-\frac{S_C}{2}\right)=U_d\left[S_A-\frac{(S_A+S_B+S_C)}{3}\right] \tag{4.6-58}$$

$$u_{sQ}=U_d\frac{(S_B-S_C)}{\sqrt{3}}. \tag{4.6-59}$$

Thus the kth sampled value of the estimated stator flux-linkage space vector can be obtained as

$$\bar{\psi}_s(k)=\bar{\psi}_s(k-1)+\tfrac{2}{3}U_dT_s[S_A(k-1)+aS_B(k-1)+a^2S_C(k-1)]-R_sT_s\bar{i}_s(k-1), \tag{4.6-60}$$

where T_s is the sampling time (flux control period) and $\bar{i}_s$ is the stator-current space vector. Resolution of eqn (4.6-60) into its real- and imaginary-axis components gives ψ_{sD} and ψ_{sQ}. It is important to note that eqn (4.6-60) is sensitive to voltage errors caused by:

- dead-time effects (e.g. at low speeds, the pulse-widths become very small and the dead time of the inverter switches must be considered);
- the voltage drop in the power electronic devices;
- the fluctuation of the d.c. link voltage (but due to this, the d.c. link voltage must be monitored);
- the variation of the stator resistance (this resistance variation sensitivity is also a feature of the method using the monitored stator voltages).

However, it is possible to have a speed-sensorless DTC drive implementation in which the dead-time effects are also considered (see also Sections 3.1.3.2 and 8.2) and the thermal variation of the stator resistance is also incorporated into the control scheme (e.g. by using a thermal model of the induction machine).

The accuracy of a speed estimator using eqn (4.6-53) or (4.6-54) depends greatly on the machine parameters used, and also on the model used for the estimation of the rotor flux-linkage components. A possible implementation has been shown in Section 4.5.2.1, Fig. 4.53. The speed estimator requires the following machine parameters: R_s, L'_s, $k_r = L_m/L_r$, and L_m/T_r. For digital implementation it is possible to use various forms, including the following discrete form:

$$\omega_r(k) = \frac{\psi_{rd}(k-1)\psi_{rq}(k) - \psi_{rq}(k-1)\psi_{rd}(k)}{|\bar{\psi}_r(k)|^2} - \frac{L_m}{T_r|\bar{\psi}_r(k)|^2}[-\psi_{rq}(k)i_{sD}(k) + \psi_{rd}(k)i_{sQ}(k)], \tag{4.6-61}$$

where $|\bar{\psi}_r(k)|^2 = [\psi_{rd}(k-1)]^2 + [\psi_{rq}(k-1)]^2$. Since this equation contains a modelling error, which results in an error of the estimated rotor speed, in practice a low-pass filter can be used to remove this error.

It is also possible to estimate the rotor speed in another way, which is similar to that described by eqn (4.6-49), but which instead of using the speed of the rotor flux (ω_{mr}) can use the speed of the stator flux ω_{ms}. It follows from eqns (4.2-16) and (4.2-23) that in this case the rotor speed can be expressed as

$$\omega_r = \omega_{ms} - \omega'_{sl}, \tag{4.6-62}$$

where

$$\omega'_{sl} = \frac{L_s(i_{sy} + T'_r \mathrm{d}i_{sy}/\mathrm{d}t)}{T_r|\bar{\psi}_s| - L'_s i_{sx}} \tag{4.6-63}$$

is the speed of the stator flux-linkage space vector relative to the rotor (not the speed of the rotor flux relative to the rotor, which has been denoted by ω_{sl}). If the direct- and quadrature-axis stator flux linkages (ψ_{sD}, ψ_{sQ}) in the stationary reference frame are known (they are estimated by one of the techniques described in Section 4.6.2.6, which uses the monitored stator voltages and currents or the reconstructed stator voltages and currents), then since $\bar{\psi}_s = \psi_{sD} + j\psi_{sQ} = |\bar{\psi}_s|\exp(j\rho_s)$, where ρ_s is the angle of the stator flux-linkage space vector with respect to the real axis of the stationary reference frame, it follows that

$$\omega_{ms} = \frac{\mathrm{d}\rho_s}{\mathrm{d}t} = \frac{\mathrm{d}}{\mathrm{d}t}[\tan^{-1}(\psi_{sQ}/\psi_{sD})]. \tag{4.6-64}$$

By performing the differentiation, this can be put into the following form:

$$\omega_{ms} = \frac{\psi_{sD}\mathrm{d}\psi_{sQ}/\mathrm{d}t - \psi_{sQ}\mathrm{d}\psi_{sD}/\mathrm{d}t}{\psi_{sD}^2 + \psi_{sQ}^2}, \tag{4.6-65}$$

where the numerator contains $|\bar{\psi}_s|^2$. By using eqns (4.6-15) and (4.6-16), it is possible to express ω_{ms} in terms of the stator voltages and stator currents. To obtain a rotor speed estimator, which uses ω_{ms} [defined by eqn (4.6-64)], it is possible to proceed in various ways.

It should be noted that eqn (4.6-62) has been directly obtained from the rotor voltage equation of the induction machine, but this expression was expressed in the stator-flux-oriented reference frame (which rotates at the speed of ω_{ms}). Thus in eqn (4.6-63) the stator currents i_{sx} and i_{sy} are the stator currents in the stator-flux-oriented reference frame, and they can be obtained from the stator currents i_{sD}, i_{sQ} (expressed in the stationary reference frame) by considering

$$i_{sx}+ji_{sy}=(i_{sD}+ji_{sQ})\exp(-j\rho_s)=(i_{sD}+ji_{sQ})(\cos\rho_s-j\sin\rho_s).$$

However, $\cos\rho_s$ and $\sin\rho_s$ can be obtained by using $\cos\rho_s=\psi_{sD}/|\bar{\psi}_s|$ and $\sin\rho_s=\psi_{sQ}/|\bar{\psi}_s|$, or ρ_s can be obtained by using a rectangular-to-polar converter (where the inputs are ψ_{sD} and ψ_{sQ} and the outputs are $|\bar{\psi}_s|$ and ρ_s.

It is possible to obtain other solutions as well which use the speed of both the stator and rotor flux-linkage space vectors. For example, such a scheme can be derived by considering that the rotor speed is equal to the sum of the speed of the stator flux-linkage space vector, $\omega_{ms}=d\rho_s/dt$, minus the speed difference between the stator and rotor flux-linkage space vectors, $\omega_d=d\rho/dt$, minus the speed of the rotor flux-linkage space vector (relative to the rotor), $\omega_{sl}=d\theta_{sl}/dt$. Thus it follows from eqn (4.5-44) that

$$\omega_r=\omega_{ms}-\omega_d-\omega_{sl}. \tag{4.6-66}$$

In Section 4.5.3.1 a simple proof of eqn (4.6-66) has been given by considering Fig. 4.55. As stated above, ρ_s is the angle of the stator flux-linkage space vector with respect to the real axis of the stator reference frame, ρ_r is the angle of the rotor flux-linkage space vector with respect to the real axis of the stator reference frame, and ρ is the angle between the stator and rotor flux-linkage space vectors, $\rho=\rho_s-\rho_r$. It follows that $\omega_d=d\rho/dt=d\rho_s/dt-d\rho_r/dt=\omega_{ms}-\omega_{mr}$ is indeed the difference between the speed of the stator flux-linkage space vector and the speed of the rotor flux-linkage space vector. Furthermore, θ_r is the rotor angle, and $\theta_r=\rho_s-\rho-\theta_{sl}$. Thus $\omega_r=d\theta_r/dt=d\rho_s/dt-d\rho/dt-d\theta_{sl}/dt=\omega_{ms}-\omega_d-\omega_{sl}$ is obtained, in agreement with by eqn (4.6-66). This can be put into the following form (see the derivation of eqn (4.5-45)):

$$\omega_r=\frac{\psi_{sD}d\psi_{sQ}/dt-\psi_{sQ}d\psi_{sD}/dt}{\psi_{sD}^2+\psi_{sQ}^2}-\frac{d}{dt}[\sin^{-1}(t_e/(c|\bar{\psi}_s||\bar{\psi}_r|))]-(2t_eR_r)/(3P|\bar{\psi}_r|^2), \tag{4.6-67}$$

where it is possible to use different expressions for the electromagnetic torque, e.g.

$$t_e=\tfrac{3}{2}P(\psi_{sD}i_{sQ}-\psi_{sQ}i_{sD}) \tag{4.6-68}$$

or

$$t_e=\frac{3}{2}P\frac{L_m}{L_r}(\psi_{rd}i_{sQ}-\psi_{rq}i_{sD}). \tag{4.6-69}$$

Equation (4.6-67) can also be used for the estimation of the rotor speed of an induction machine, but this is not simpler than eqns (4.6-53) or (4.6-54).

In summary it should be noted that the speed estimators discussed above depend heavily on the accuracy of the used flux-linkage components. If the stator voltages and currents are used to obtain the flux estimates, then by considering the thermal variations of the stator resistance (e.g. by using a thermal model), and also by using appropriate saturated inductances (see also Sections 4.6.2.11, 8.2.6), the estimation accuracy can be greatly improved. However, a speed-sensorless high-performance direct-torque-controlled drive using this type of speed estimator will only work successfully at ultra-low speeds (including the zero speed) if the flux estimator is some type of closed-loop observer.

It is expected that in the future universal induction motor drives will emerge which will also incorporate direct torque control. Control techniques Plc (UK) was the first in the world to introduce a universal drive, but at present it does not incorporate direct torque control.

4.6.2.11 The ABB DTC induction motor drive

ABB has introduced in 1995 the first industrial, speed-sensorless DTC induction motor drive in the world. This contains the ACS 600 frequency converter (inverter), which uses power-plate IGBT modules and is shown in Fig. 4.91. The inverter switching directly controls the motor flux linkages and electromagnetic torque.

The ACS 600 product family (ACS 601, 603, 604, 607) suits many applications and operating environments, with a large selection of a.c. voltages (380 V–690 V), power (2.2 kW–630 kW), and enclosure (IP 00, IP 20, IP 21, IP 22, IP 54) ratings, combined with highly flexible communication capabilities. The drive is suitable for 95% of all industrial applications (including pumps, fans, mixers, conveyors, lifts, elevators, cranes, hoists, winders, centrifuges, extruders, etc.).

The ACS 600 can accurately control the rotor speed and electromagnetic torque (and stator flux linkages) without encoder or tachogenerator feedback. The schematic block diagram of the drive is shown in Fig. 4.92. This is similar to that shown in Fig. 4.87 and thus the reader is referred to the many details discussed earlier (in Sections 4.6.1, 4.6.2.1–4.6.2.10) and only some of the main features will be described below. The primary control variables are the electromagnetic torque and stator flux linkage.

In Fig. 4.92 two stator currents are measured together with the d.c. link voltage. The two stator currents can be used to obtain the direct- and quadrature-axis stator currents in the stationary reference frame, $i_{sD}=i_{sA}$, $i_{sQ}=(i_{sA}+2i_{sB})/\sqrt{3}$. By using the measured d.c. link voltage (U_d) and the switching signals of the inverter, the stator voltages are reconstructed (see eqn (4.6-57) in Section 4.6.2.10). However, as discussed in Section 4.6.2.10, for an accurate reconstruction of the stator voltages, the interlock delay (dead time, which is programmed in the DSP switching logic to prevent short-circuits of the d.c. link), and voltage drops across the semiconductor switches of the inverter (IGBTs), must also be considered, especially at low rotor speed.

The stator currents together with the stator voltages are inputs to an adaptive induction motor model also shown in Fig. 4.92 (which contains an observer), which estimates in real time (using a DSP) the modulus of the stator flux-linkage space vector, $|\bar{\psi}_s|$, its position (with respect to the real-axis of the stationary reference frame), ρ_s, the electromagnetic torque (t_e), and rotor speed (ω_r) in every

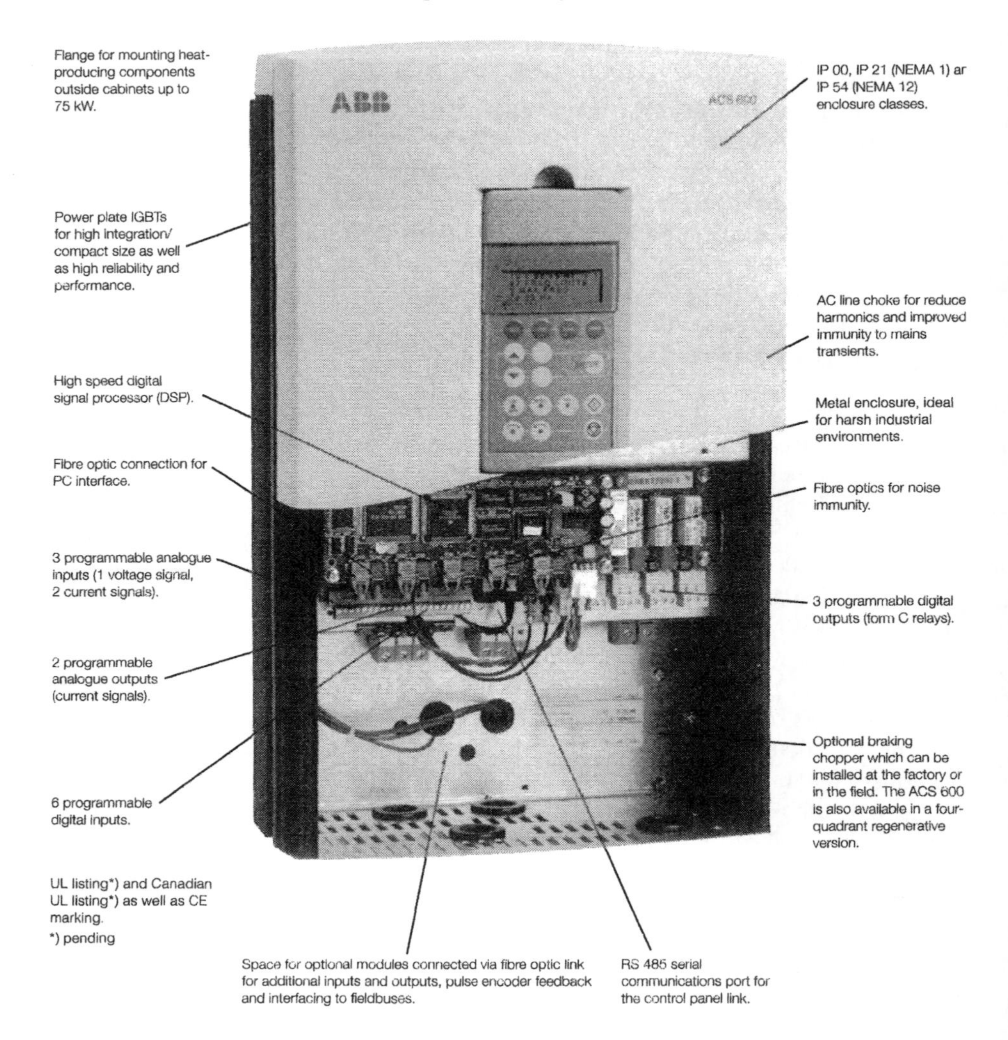

Fig. 4.91. ACS 600 frequency converter (Courtesy of ABB Industry Oy, Helsinki).

25 microseconds. As discussed in Sections 4.6.2.3 and 4.6.2.6, it is very important to consider that the main role of the motor model is to estimate accurately the stator flux-linkage components, since the modulus and position of the stator flux-linkage space vector, the electromagnetic torque, and also the rotor speed are estimated from the stator flux-linkage components. The reader is referred to the previous sections which discuss different techniques of stator flux linkage, electromagnetic torque, and rotor speed estimation. In particular, Sections 4.6.2.3 and 4.6.2.6 discuss the estimation of the stator flux linkages together with the application of various drift compensation techniques. Furthermore, eqns (4.6-11)

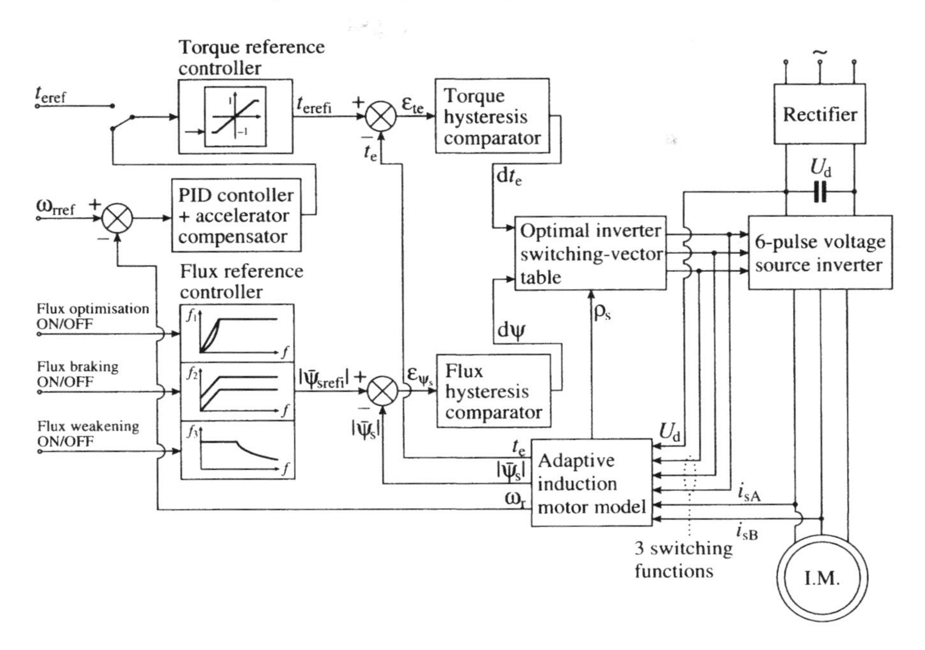

Fig. 4.92. Schematic block diagram of ABB DTC induction motor drive.

and (4.6-54) give the expressions for the electromagnetic torque and rotor speed in terms of flux-linkage components. The rotor flux-linkage components present in eqn (4.6-54) can be obtained from the stator flux linkages by using eqns (4.6-55) and (4.6-56).

To create one possible motor model, the only data which have to be entered by the user into the initialization software of the ABB DTC drive are the nameplate data (e.g. rated motor speed, rated stator current, rated stator voltage, rated power, rated frequency). However, to obtain a more accurate motor model, some of the parameters used in the motor model are initialized during a motor identification run stage (self-commissioning stage, which is also sometimes referred to as the auto-calibration stage). During the start-up of the drive system, the ACS 600 operates the motor for approximately one minute. The control circuitry monitors the response of the induction motor to the applied power and determines various machine parameters (e.g. the stator resistance, stator inductance, magnetizing inductance, saturation coefficients for these two inductances, inertia of the motor, etc.) and enhances the mathematical model of the motor. For example, the stator resistance is required, since in principle, at high speed, the stator flux linkages are obtained by integrating the appropriate stator voltage reduced by the corresponding ohmic drop, $\psi_{sD}=\int(u_{sD}-R_s i_{sD})\,dt$, $\psi_{sQ}=\int(u_{sQ}-R_s i_{sQ})\,dt$. Since the rotor speed is determined by using the speed of the rotor flux-linkage space vector [see eqn (4.6-54)], this also requires the use of various inductance parameters of the induction machine.

There are two identification (ID) run alternatives: the standard ID run, and the reduced ID run. During the standard ID run, the motor is uncoupled from the load. However, the reduced ID run has to be used if the load cannot be disengaged from the induction motor or if stator flux reduction is not allowed. For example, stator flux reduction is not allowed with a braking motor, where the brake is switched on if the motor voltage or flux reduces significantly. To obtain the most accurate motor model and the best possible control performance, the standard ID run has to be selected. If the identification run is not selected, a rapid motor identification is automatically performed (when the START command is issued). During the first start-up procedure, the motor is run at zero speed for several seconds to allow the estimation of the required parameters in the advanced motor model. It should be noted that various techniques of estimating on-line and in real-time the different parameters of an induction motor at standstill have been also discussed in detail in a recent book [Vas 1993], but some aspects are also described in Chapter 8 of the present book.

There are several parameters (e.g. stator resistance) which are continuously updated in the adaptive induction motor model during the operation of the drive. For example, the stator resistance is updated by using a thermal model of the induction motor. In the ABB DTC drive, the temperature of the motor is estimated by assuming that the motor ambient temperature is 30 °C. The motor temperature can be estimated by using two curves: the variation of the motor load with time and the variation of the motor thermal-time with time.

During the motor identification run, the speed controller shown in Fig. 4.92 is automatically tuned. This contains a classical PID controller (and e.g. not a fuzzy controller). However, it is also possible to adjust the PID gains manually. In the auto-tune run mode, the automatic tuning of the PID gains is achieved. This also uses the combined load and machine inertia. With auto-tuning, it is possible to achieve better dynamic performance (faster speed responses) than with manual tuning. According to ABB the static speed controller error is typically between $\pm 0.1\%$ and $\pm 0.5\%$ of motor rated speed. Higher speed regulation can only be achieved by using a pulse encoder, and in this case the static speed error is typically in the $\pm 0.01\%$ range (if an encoder with 1024 pulses/revolution is used). The dynamic speed control error is typically $\pm 0.4\%$s at 100% load torque step when an encoder or tachogenerator is not used ('sensorless' drive), and it is typically $\pm 0.1\%$s when an incremental pulse encoder is used. However, it should be noted that the dynamic speed error strongly depends on the speed controller tuning.

For many applications, speed control is the most important function of an inverter-fed drive. In a DTC-fed drive, speed control is not a part of the inner inverter control as in traditional inverter-fed drives. In Fig. 4.92 the input to the speed controller is the speed error $(\omega_{\mathrm{rref}} - \omega_{\mathrm{r}})$ and on the output of the speed controller, the reference electromagnetic torque is produced. The speed controller consists of a PID controller and also an acceleration compensator. The acceleration compensator (feedforward coupling from the speed reference derivative) is extremely useful for minimizing control deviation during starting, acceleration, and deceleration. The PID controller can be tuned to be more as a load compensator. The PID controller and acceleration compensator are tuned by an automatic tuning method, which is based on the identification of the mechanical time constant of the drive. By identifying the mechanical time constant in an initial self-tuning stage (see below), or during normal operation, it is possible to tune the PID controller for maximum effectiveness.

In Fig. 4.92 the input to the torque-control loop is either the external torque reference (t_{eref}) or the torque reference on the output of the speed controller. The electromagnetic torque reference controller outputs the internal torque reference signal (t_{erefi}). Within the electromagnetic torque reference controller, the speed control output is limited by the torque limits and the d.c. bus voltage. In this way the motor torque is prevented from exceeding the pull-out torque and the inverter is protected from overload. The torque limit calculation also uses the maximum inverter current and maximum motor current. In Fig. 4.92, the stator flux-linkage reference controller produces the internal stator flux-linkage modulus reference ($|\bar{\psi}_{srefi}|$). The ability to control and modify this modulus provides a convenient way to implement the three functions (f_1, f_2, f_3) shown in Fig. 4.92, which are flux optimization, flux braking, and field weakening. Flux optimizing provides automatic flux adaptation to load variation, yielding higher efficiency. Flux braking provides the highest possible braking torque without additional hardware. Field weakening enables higher than rated rotor speed. These functions are now briefly discussed.

Flux optimization By using the motor model, the optimal magnetizing level (stator flux space-vector modulus) can also be estimated as a function of the load (see also Section 4.6.2.8). In the flux optimization mode, flux optimization reduces the total energy consumption of the motor (thus improves efficiency) and noise level, when the drive operates below nominal load. The total efficiency of the motor and the drive can be improved by 1% to 10%, depending on the operating point (load torque and rotor speed).

Flux braking Flux braking is a technique where the mechanical energy of the load is converted into heat inside the motor by increasing the flux. The inverter supplying the motor can provide greater deceleration by raising the level of magnetization in the motor, without any extra hardware. By increasing the stator flux, the energy generated by the motor during braking can be converted into thermal energy. The inverter monitors the status of the motor continuously, and also during flux braking. Thus flux braking can be used for stopping the motor and also for changing from one speed to another speed. It should be noted that the latter is not possible with d.c. (injection) braking, which is a widely used technique (see also the various braking techniques discussed in [Vas 1992]). Some of the other benefits of flux braking compared to d.c. injection braking are as follows:

- The braking action starts immediately after the STOP command is given. With the conventional d.c. (injection) braking, there is typically a 500 ms delay after the STOP command, before the braking can be started. This delay is essential, since d.c. injection is only possible after the motor flux is sufficiently reduced.

- There is more efficient motor cooling. During flux braking the stator currents increase and with d.c. braking the rotor currents increase. However, the stator cools more efficiently than the rotor.

It is also possible to have effective braking by using resistor braking. However, in this case the inverter must be equipped with extra hardware: a braking chopper and a braking resistor.

Field weakening Below base speed, a constant value of the stator flux modulus is used (the stator voltage increases) and above base speed (where the inverter ceiling voltage is reached), this modulus is reduced inversely with the speed (field-weakening). Some aspects of field weakening are also discussed in Section 4.6.2.7.

In Fig. 4.92, the actual values (estimated values) of the electromagnetic torque (t_e) and stator space-vector flux-linkage modulus ($|\bar{\psi}_s|$) are compared to their internal reference values (in every 25 microseconds), which are present on the outputs of the electromagnetic torque and stator flux-linkage reference controllers respectively. The resulting errors (ε_{te}, $\varepsilon_{\psi s}$) are inputs to the electromagnetic torque and stator flux-linkage comparators respectively, which according to ABB are two-level hysteresis comparators. Depending on the outputs of the two comparators and also on the position of the stator flux-linkage space vector, an optimum switching vector selector, which uses an optimal voltage switching-vector look-up table, determines the optimum inverter switchings. For this purpose a very fast, 40 MHz digital signal processor together with ASIC hardware is used. All the control signals are transmitted via optical links for high-speed data transmission. Thus it can be seen that every switching is determined separately based on the values of the electromagnetic torque and stator flux linkages (modulus and angle of stator flux-linkage space vector), and not in a predetermined pattern as in some other a.c. drives. In the DTC drive there is no need for a separate voltage- and frequency-controlled pulse-width-modulator. Due to the extremely fast torque response (e.g. less than 2 ms), the drive can instantly react to dynamic changes such as sudden load changes, power loss, overvoltage conditions, etc. Since the switching vector selection determines the motor voltages and currents, which in turn influence the electromagnetic torque and stator flux linkages, the control loops are closed.

In the DTC induction motor drive the stator flux-linkage modulus and electromagnetic torque are kept within their respective preset hysteresis bands. The inverter switchings are altered if the values of the actual torque and stator flux linkage differ from their reference values more than that allowed by their respective hysteresis bands. When the rotating stator flux-linkage space vector reaches the upper or lower hysteresis limit, a suitable voltage switching vector is selected, which changes the direction of the stator flux-linkage space vector and thus forces it to be in the required hysteresis band. The physical and other aspects of the selection of the optimal switching table have also been discussed earlier in

Sections 4.6.2.1, 4.6.2.2, and 4.6.2.9. It should be noted that at low frequencies the switching voltage vectors in the tangential direction (e.g. voltage vectors $\bar{u}_3$, $\bar{u}_6$, etc.) have a very strong influence on the electromagnetic torque. In the DTC drive there is also the possibility of controlling the switching frequency of the inverter by modifying the hysteresis parameters as a function of the electrical frequency.

In the ABB DTC drive, fast and precise torque control can be achieved without using a speed sensor. For example it is possible to have an electromagnetic torque rise time less than 5 ms with 100% reference electromagnetic torque. By applying a torque reference instead of a speed reference, a specific motor torque can be maintained, and the speed adjusts automatically to maintain the reference torque. By using direct torque control, it is also possible to have maximized starting torque which is controllable and is also smooth.

According to ABB, it is possible to operate the DTC controlled induction motor even at zero speed (not zero stator frequency) and the motor can develop rated torque at this speed, without using any pulse encoder or tachogenerator. This is an important feature for various applications, e.g. elevators, lifts, hoists, extruders, etc. However, if long-term operation at zero-speed is required, a pulse encoder may be required. By activating the so-called motor d.c. hold feature, it is possible to lock the rotor at zero speed. When both the reference and actual (estimated) rotor speed drop below a preset d.c. hold value, the inverter stops the drive and starts to inject d.c. into the motor. When the reference speed rises again above the d.c. hold speed, the normal operation of the inverter is resumed.

Before start-up, the motor is automatically magnetized (by the inverter), and the highest possible breakaway torque (even twice the rated torque) is guaranteed. By adjusting the premagnetizing time, it is possible to fix the motor start with a mechanical brake release. The automatic start feature and d.c. magnetizing cannot be activated at the same time. The automatic start feature outperforms the flying start and ramp start features normally found in other frequency converters. Since the ACS 600 can detect the state of the induction motor rapidly (within a few milliseconds), starting is immediate under all conditions. For example, there is no restarting delay with DTC.

Figure 4.93 shows some experimental results obtained with the speed sensorless ABB DTC induction motor drive using a voltage-source inverter (ACS 600) [Tiitinen 1996]. For the purposes of the experiments, two standard squirrel-cage induction motors were coupled using a torque measurement shaft. The inverters which supply the two motors have a common d.c. link circuit. The first motor did not have an encoder fitted, but the second motor (load motor) was equipped with a tachometer, to enable comparisons to be made between measured and estimated rotor speeds. The induction motor ratings are: 15 kW, 30 A, 380 V, number of poles 4; rated stator frequency 50 Hz, rated rotor speed 1460 r.p.m. The rating of the prototype ACS 600 inverter used is 25 kVA, 400 V.

Figure 4.93(a) shows the measured temporal variation of the electromagnetic torque for a 70% torque reference step at 25 Hz. The electromagnetic torque has been estimated by using the measured stator currents and estimated stator flux linkages [by using $c(\psi_{sD}i_{sQ}-\psi_{sQ}i_{sD})$]. The very fast torque response (less than 2 ms)

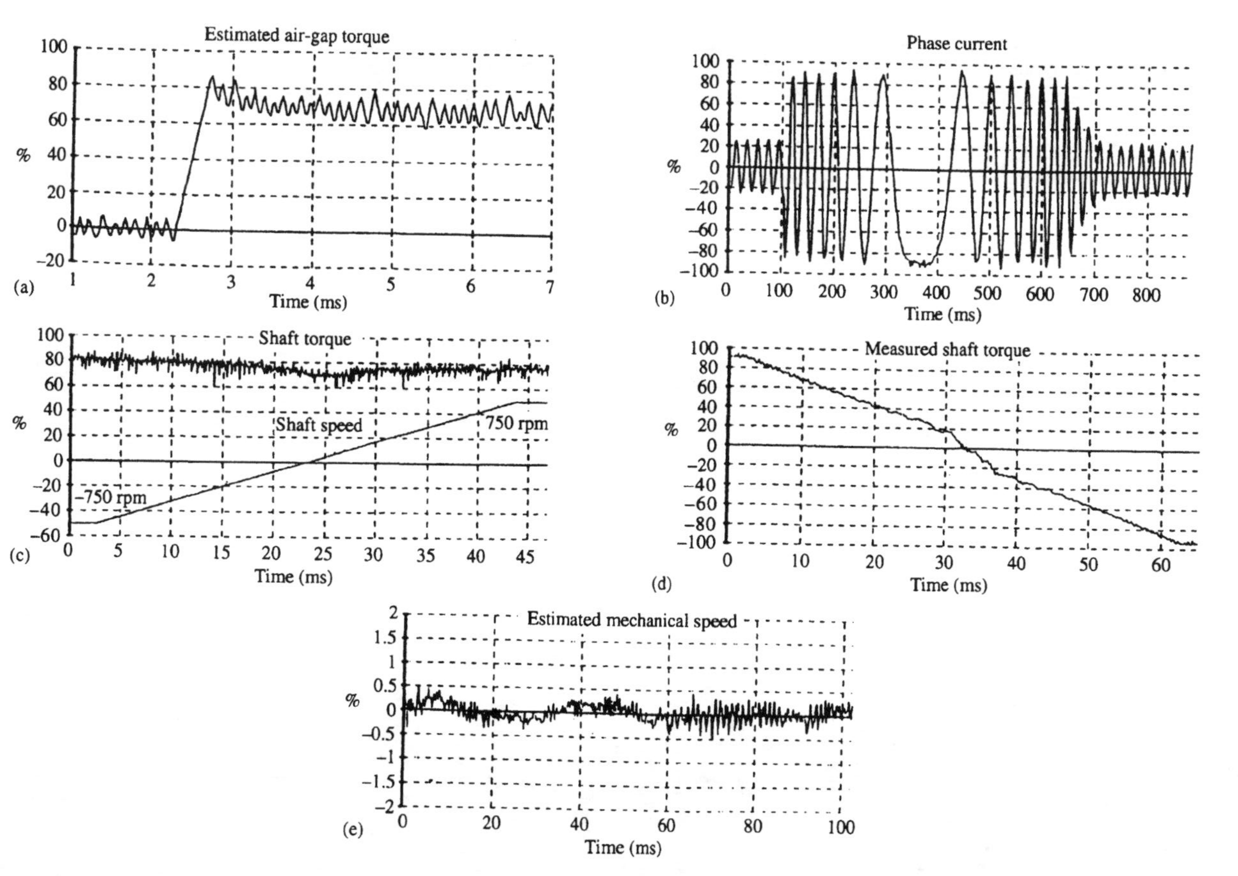

Fig. 4.93. Experimental results for speed-sensorless ABB DTC induction motor drive (Courtesy of ABB Industry Oy, Helsinki). (a) Estimated electromagnetic torque, for 70% reference torque step at 25 Hz; (b) measured stator phase current for fast reversal with constant 20% load; (c) measured shaft torque and measured rotor speed for slow reversal with constant (80%) torque reference; (d) measured shaft torque at zero speed; (e) estimated rotor speed (torque ramp from positive-rated value to negative-rated value at zero speed).

can be seen from this figure. Figure 4.93(b) shows the variation of a stator phase current during fast reversal with constant 20% load torque and it can be seen that the expected curve has been obtained. Figure 4.93(c) shows the measured shaft torque and measured rotor speed for slow reversal with constant (80%) torque reference. Furthermore, Fig. 4.93(d) shows the variation of the measured shaft torque ramp (from positive-rated torque to negative-rated torque) at zero speed. The linearity error is below 10%. Finally Fig. 4.93(e) shows the estimated rotor speed for the case shown in Fig. 4.93(d); the estimation has been performed by using the motor model, which utilizes the speed of the rotor flux linkages to obtain the rotor speed, as discussed in connection with eqn (4.6-54). It should be noted that the accuracy of the rotor-speed estimates has been found to be very satisfactory in the entire speed range.

4.6.3 DTC OF A CSI-FED INDUCTION MACHINE

4.6.3.1 General introduction

In the present section the direct torque control (DTC) of a CSI-fed induction motor drive is discussed briefly. However, for better understanding of the concepts related to direct torque control, the reader is first referred to Section 4.6.2, which discusses in great detail various aspects of the DTC of a VSI-fed induction motor drive. As discussed in Section 4.1.2, in a CSI-fed induction motor drive, a smooth d.c. link current is supplied to the machine-side inverter. The d.c. link current is obtained by using a phase-controlled rectifier and a high-inductance filter. The machine-side inverter contains six force-commutated thyristors (see Fig. 4.17). The d.c. link current is switched through the inverter thyristors to produce the required a.c. currents. Figure 4.18 has also shown the locus of the stator-current space vector (in the stationary reference frame), which corresponds to the six-stepped line currents. It has been shown that during one sixth of a cycle, the stator-current space vector remains in a fixed position (e.g. in the first cycle, it is in position 1 shown in Fig. 4.18, but all the six fixed positions have been shown). The six current vectors corresponding to these fixed positions are the six active (non-zero) switching current vectors used below.

4.6.3.2 Drive scheme

Direct torque control of a CSI-fed induction motor involves the direct control of the rotor flux linkage (or stator flux linkage) and the electromagnetic torque by applying the optimum current switching vectors. Furthermore, in a direct-torque-controlled (DTC) induction motor drive supplied by a current-source inverter, it is possible to control directly the modulus of the rotor flux-linkage space vector $|\bar{\psi}_r|$ through the rectifier voltage, and the electromagnetic torque (t_e) by the supply frequency of the CSI. For this purpose the appropriate optimal inverter current-switching vectors are produced by using an optimal current switching-vector table. This contains the six possible active current switching vectors ($\bar{i}_1, \bar{i}_2, \ldots, \bar{i}_6$)

and also the non-active (zero) switching vectors ($\bar{i}_0$). The six active current switching vectors are also shown in Fig. 4.94(a), together with the locus of the stator-current space vector, which is a hexagon. Optimum selection of the switching vectors is made to restrict the electromagnetic torque error within the torque hysteresis band. An input to the optimal current switching table is the discretized torque error (dt_e), which is the output of a 3-level hysteresis comparator. A three-level comparator is used, since this corresponds to 0, 1, and -1 torque

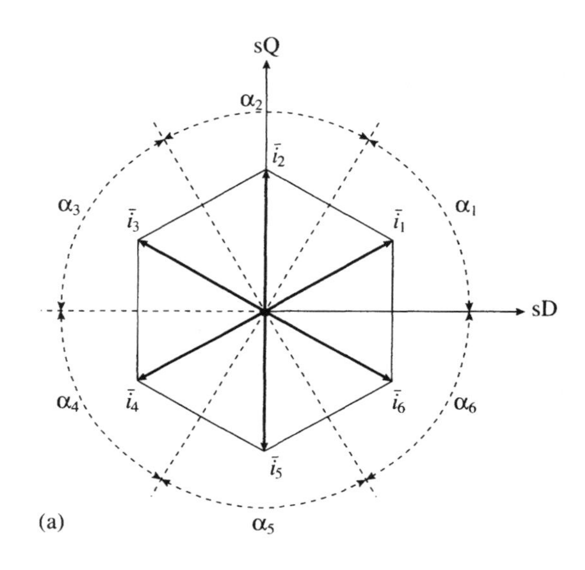

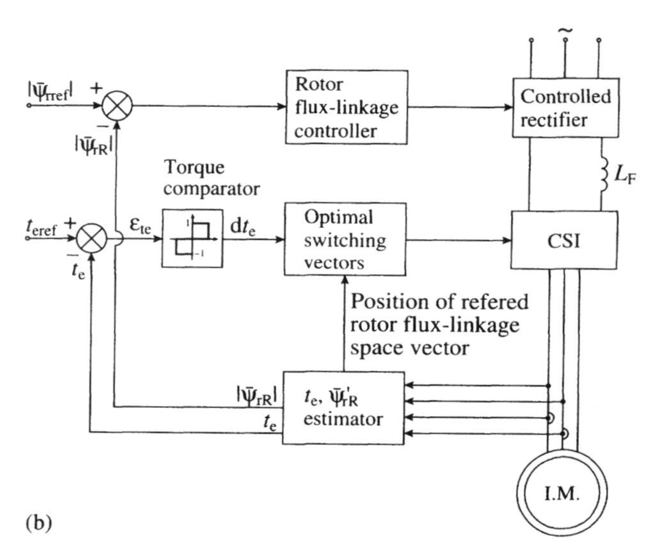

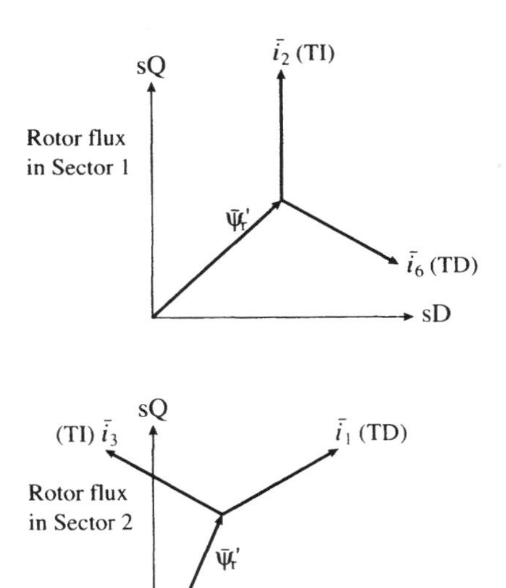

Fig. 4.94. Basic scheme of the DTC CSI-fed induction motor drive and the active current switching vectors. (a) Active current switching vectors; (b) drive scheme; (c) selection of switching vectors; TI: torque increase; TD: torque decrease.

errors. The optimum switching look-up table also requires knowledge of the position of the rotor flux-linkage space vector, since it must be known in which of the six sectors is the rotor flux-linkage space vector. The basic scheme of the DTC CSI-fed induction motor drive is shown in Fig. 4.94(b). The drive scheme contains an electromagnetic torque and rotor flux-linkage estimator.

4.6.3.3 Stator flux-linkage and electromagnetic torque estimation

The electromagnetic torque can be estimated from the terminal quantities by considering eqn (2.1.176), thus

$$t_e = \frac{3}{2} P \frac{L_m}{L_r} \bar{\psi}_r' \times \bar{i}_s = \frac{3}{2} P \frac{L_m}{L_r} (\psi_{rd} i_{sQ} - \psi_{rq} i_{sD}), \tag{4.6-70}$$

where P is the number of pole pairs, $\bar{\psi}_r' = \psi_{rd} + j\psi_{rq}$ is the rotor flux-linkage space vector, and $\bar{i}_s = i_{sD} + j i_{sQ}$ is the stator-current space vector, and both vectors are expressed in the stationary reference frame. The stator currents are monitored and the rotor flux-linkage components can be obtained by considering the stator voltage equation in the stationary reference frame, which gives

$$\frac{d\bar{\psi}_r'}{dt} = \left(u_s - R_s i_s - L_s' \frac{d\bar{i}_s}{dt} \right) \frac{L_r}{L_m}, \tag{4.6-71}$$

where L_s' is the stator transient inductance. Thus the rotor flux-linkage components are obtained as

$$\psi_{rd} = \frac{L_r}{L_m}\int\left(u_{sD} - R_s i_{sD} - L_s'\frac{di_{sD}}{dt}\right)dt \tag{4.6-72}$$

$$\psi_{rq} = \frac{L_r}{L_m}\int\left(u_{sQ} - R_s i_{sQ} - L_s'\frac{di_{sQ}}{dt}\right)dt. \tag{4.6-73}$$

It can be seen that if eqns (4.6-72) and (4.6-73) are used to estimate the rotor flux-linkage components, then in addition to the stator resistance, the stator transient inductance must also be used together with the inductance ratio L_r/L_m. However, it is also possible to have an implementation where L_r/L_m is not required. In this case, instead of estimating the rotor flux linkages, their referred values are estimated. By using eqns (4.5-123) and (4.6-71), it follows that the direct- and quadrature-axis referred rotor flux-linkage components (in the stationary reference frame) can be obtained from $\bar{\psi}_{rR}' = \int(\bar{u}_s - R_s\bar{i}_s)\,dt - L_s'\bar{i}_s$ (where for simplicity the subscript R denotes the referred value). Thus

$$\psi_{rdR} = \int(u_{sD} - R_s i_{sD})\,dt - L_s' i_{sD} \tag{4.6-74}$$

$$\psi_{rqR} = \int(u_{sQ} - R_s i_{sQ})\,dt - L_s' i_{sQ}. \tag{4.6-75}$$

In this case, the electromagnetic torque can be obtained by using eqn (4.5-127); thus

$$t_e = \frac{3}{2}P\bar{\psi}_{rR}' \times \bar{i}_s = \frac{3}{2}P(\psi_{rdR} i_{sQ} - \psi_{rqR} i_{sD}). \tag{4.6-76}$$

Since in the CSI-fed drive, the currents are limited by the forced commutation circuit, saturation of the leakage flux paths can be neglected and thus L_s' can be assumed to be constant. However, in addition to this open-loop flux estimator it is possible to use more accurate schemes described earlier (e.g. see Sections 3.1.3.2.1 and 4.6.2.6). The angle of the referred rotor flux-linkage space vector (which is the same as the angle of the non-referred rotor flux-linkage space vector) can be estimated by using $\rho_r = \tan^{-1}(\psi_{rqR}/\psi_{rdR})$, or $\rho_r = \cos^{-1}(\psi_{rdR}/|\bar{\psi}_{rR}|)$, or $\rho_r = \sin^{-1}(\psi_{rqR}/|\bar{\psi}_{rR}|)$, where $|\bar{\psi}_{rR}| = (\psi_{rdR}^2 + \psi_{rqR}^2)^{1/2}$, but similarly to the case discussed in Section 4.6.2.2, it is possible to avoid the use of trigonometric functions.

In Fig. 4.94(b), the rotor flux-linkage controller acts directly on the rectifier voltage. The physical reason for the presence of this control loop is the fact that the modulus of the rotor flux-linkage space vector can be changed by changing the amplitude of the stator voltage. This also follows from the stator voltage equation [eqn (4.6-71)] given above. However, the stator voltages can be changed by changing the rectifier voltage. On the other hand it can also be shown by simple physical considerations that the electromagnetic torque can be controlled by changing the stator frequency, but this also follows by considering the rotor voltage equation, eqn (4.1-24), together with $|\bar{\psi}_r| = L_m|\bar{i}_{mr}|$. Thus

$$0 = -\frac{L_m}{T_r}\bar{i}_s + \left(\frac{1}{T_r} + p - j\omega_r\right)\bar{\psi}'_r \tag{4.6-77}$$

is obtained, where $p = j\omega_1$ if it is assumed that the rotor flux linkages are sinusoidal, in which case $\bar{\psi}'_r = |\bar{\psi}_r|\exp(j\omega_1 t)$ and $T_r = L_r/R_r$ is the rotor time constant. By considering eqn (4.6-74) and also eqn (4.6-70), it follows that when the position of the stator-current space vector is quickly changed, a quick change in the electromagnetic torque and also in the angular slip frequency $(\omega_{sl} = \omega_1 - \omega_r)$ is obtained. This is similar to that in the voltage-source inverter-fed DTC induction motor drive, but in the VSI-fed drive rapid change in the electromagnetic torque is obtained by rapidly changing the appropriate voltage switching vector.

The torque reference can be obtained on the output of a speed controller, which can be PI controller. The input to the speed controller is the difference between the reference speed and the actual speed. It is possible to have a speed-sensorless implementation, where the rotor speed is obtained by using one of the techniques described in Section 4.6.2.10. The hysteresis band of the torque comparator may be increased with rotor speed, since in this case PWM of the currents is avoided at high speeds.

4.6.3.4 Optimal current switching vector selection

It follows from the similarity of the switching states of the current-source inverter and that of the voltage-source inverter that the required optimum inverter current switching-vector selection table must resemble the switching table obtained for the voltage switching vectors in Section 4.6.2.2, but now current vectors replace the voltage vectors. However, as discussed above, for the system with the CSI, the flux error is not an input to the optimum switching table (since the rotor flux is directly controlled through the rectifier voltage, and not through the inverter switching states). Thus it is formally possible to use the first part of the switching table given for the VSI-fed drive, but all the voltages have to be changed to currents. However, for better understanding, the required switching table is also derived below in another way.

The optimal inverter current switching table can be obtained by considering the positions of the rotor flux-linkage space vector in one of the six sectors (e.g. the first sector is in the region spanned by angle α_1 shown in Fig. 4.94(a), the second region spans the angle α_2, etc.). It can be seen that e.g. if the rotor flux-linkage space vector is in the first sector, then for positive electromagnetic torque the switching current vector $\bar{i}_2$ has to be applied. This is due to the fact that a stator current vector has to be selected which produces positive electromagnetic torque and is located at an angle less than 90° in the positive direction from the rotor flux-linkage space vector, since the machine behaves like an RL-circuit. However,

Table 4.9 Optimum current switching-vector look-up table for a CSI-fed induction machine

dt_e	$\alpha(1)$ sector 1	$\alpha(2)$ sector 2	$\alpha(3)$ sector 3	$\alpha(4)$ sector 4	$\alpha(5)$ sector 5	$\alpha(6)$ sector 6
1	$\bar{i}_2$	$\bar{i}_3$	$\bar{i}_4$	$\bar{i}_5$	$\bar{i}_6$	$\bar{i}_1$
0	$\bar{i}_0$	$\bar{i}_0$	$\bar{i}_0$	$\bar{i}_0$	$\bar{i}_0$	$\bar{i}_0$
−1	$\bar{i}_6$	$\bar{i}_1$	$\bar{i}_2$	$\bar{i}_3$	$\bar{i}_4$	$\bar{i}_5$

Table 4.10 Optimum current switching vectors for a CSI-fed synchronous motor

dt_e	$\alpha(1)$ sector 1	$\alpha(2)$ sector 2	$\alpha(3)$ sector 3	$\alpha(4)$ sector 4	$\alpha(5)$ sector 5	$\alpha(6)$ sector 6
1	$\bar{i}_3$	$\bar{i}_4$	$\bar{i}_5$	$\bar{i}_6$	$\bar{i}_1$	$\bar{i}_2$
0	$\bar{i}_0$	$\bar{i}_0$	$\bar{i}_0$	$\bar{i}_0$	$\bar{i}_0$	$\bar{i}_0$
−1	$\bar{i}_5$	$\bar{i}_6$	$\bar{i}_1$	$\bar{i}_2$	$\bar{i}_3$	$\bar{i}_4$

for negative electromagnetic torque, $\bar{i}_6$ has to be applied, since the selected stator current has to produce negative torque and it must lag the rotor flux-linkage space vector by an angle less than 90°. For zero electromagnetic torque one of the zero-current switching vectors ($\bar{i}_0$) has to be applied. Similarly, if the rotor flux-linkage space vector is in the second sector, then for positive torque $\bar{i}_3$, and for negative torque $\bar{i}_1$, must be selected, etc. The rotor flux-linkage space vector and the appropriate current switching vectors are shown in Fig. 4.94(c) for the two cases discussed above, when $\bar{\psi}'_r$ is in the first and second sector respectively. The thus-obtained optimum current switching-vector table is shown in Table 4.9.

It should be noted that, as expected, this switching table formally resembles the top part of the optimum switching-vector table (Table 4.3) used in Section 4.6.2.2 for a VSI-fed DTC induction motor drive. Furthermore, it should also be noted that in a synchronous motor drive with a load-commutated current-source inverter, if direct torque control is employed, the control scheme is similar to that described above. However, the optimum switching table will be different, since due to the fact that the machine is overexcited, the stator current vector has to be located at a position which is greater than 90° with respect to the controlled flux (which is a subtransient flux). Thus in sector 1, for positive torque (instead of $\bar{i}_2$) $\bar{i}_3$ is present, and for negative torque (instead of $\bar{i}_6$) $\bar{i}_5$ is present, etc., thus Table 4.10 is used in the CSI-fed synchronous motor drive (see also Section 3.3.4).

Bibliography

Acarnley, P. P. and Finch, J. W. (1987). Review of control techniques for field-orientation in a.c. drives. 22 Universities Power Engineering Conference, Sunderland, paper 9.08.

Akamatsu, M., Ikeda, K., Tomei, H., and Yano, S. (1982). High performance IM drive by coordinate control using a controlled current inverter. *IEEE Transactions on Industry Applications* **IA-18**, 382–92.

Albrecht, P. and Vollstedt, W. (1982). Microcomputer control of a variable speed double-fed induction generator operating on the fixed frequency grid. *Conference on Microelectronics in Power Electronics and Electrical Drives, Darmstadt*, pp. 363–8.

Angquist, L. (1986). Stator flux control of asynchronous motor in the field-weakening region. *Int. Conference on the evolution and modern aspects of induction machines, Turin*, **82**, pp. 458–64.

Atkinson, D. J., Acarnley, P. P., and Finch, J. W. (1991). Observers for induction motor state and parameter estimation. *IEEE Transactions on Ind. Applications* **IA-27**, 1119–27.

Baader, U., Depenbrock, M., and Gierse, G. (1992). Direct self control (DSC) of inverter-fed induction machine: a basis for speed control without speed measurement. *IEEE Transactions on Ind. Applications* **IA-28**, 581–8.

Bassi, E., Benzi, F., Bolognani, S., and Buja, G. S. (1987). Torque-angle control scheme for field-oriented CSIM drives. *Beijing International Conference on Electrical Machines, Beijing*, pp. 368–70.

Bauman, T. (1996). A new adaptation method used for the rotor time constant of the field oriented induction machine. *PEMC, Budapest*, 2/386–2/390.

Bausch, H. and Hontheim H. (1985). Transient performance of induction machines with field-oriented control. *Second International Conference on Electrical Machines—Design and Applications*, IEE, London, pp. 199–203.

Bausch, H., Hontheim, H., and Kolletschke, D. (1986). The influence of decoupling methods on the dynamic behaviour of a field-oriented controlled induction machine. In *Proceedings of International Conference on Electrical Machines, München*, pp. 648–51.

Bayer, K. H. and Blaschke, F. (1977). Stability problems with the control of induction machines using the method of field orientation. *IFAC Symposium on Control in Power Electronics and Drives, Düsseldorf*, pp. 483–92.

Bedford, B. D. and Hoft, R. G. (1964). *Principles of inverter circuits*. Wiley, New York.

Beguenane, R. and Capolino, G. A. (1995). Design of slip frequency detector with improved accuracy for induction motor rotor parameters updating. *EPE, Seville*, pp. 417–22.

Beierke, S. (1995). Enhanced fuzzy controlled a.c. motor using DSP. Embedded Systems Conference, San Jose, USA, 545–53.

Beierke, S., Vas, P., Simor, B., and Stronach, A. F. (1997). DSP-controlled sensorless a.c. vector drives using the extended Kalman filter. *PCIM, Nurnberg*, pp. 31– 42.

Beierke, S., Vas, P., and Stronach, A. F. (1997). TMS320 DSP implementation of fuzzy-neural controlled induction motor drives. EPE, Trondheim, pp. 2.449–2.453.

Bellini, A., Figalli, G., and Ulivi, G. (1986). A microcomputer based direct field oriented control of induction motors. In *Proceedings of International Conference on Electrical Machines, München*, pp. 652–65.

Ben-Brahim L. and Kawamura, A. (1991). A fully digitised field oriented controlled induction motor drive using only current sensors. *IEEE Transactions on Ind. Electronics* **IE-39**, 241–9.

Ben-Brahim, L. and Kurosawa, R. (1993). Identification of induction motor speed using neural networks. *IEEE PCC, Yokohama*, pp. 689–94.

Blaschke, F. (1972). The principle of field-orientation as applied to the new Transvektor closed-loop control system for rotating-field machines. *Siemens Review* **34**, 217–20.

Blaschke, F., Van der Burgl, I., and Vandenput, A. (1996). Sensorless direct field orientation at zero flux frequency, *IEEE IAS Meeting*, 1/189–1/196.

Bodson, M., Chiasson, J. N., and Novotnak, R. T. (1995). A systematic approach to selecting the flux references for torque minimization in induction motors. *IEEE Transactions on Control Systems Tech.* **3**, 388–97.

Boldea, I. and Nasar, S. A. (1986). *Electric machine dynamics.* Macmillan, New York.

Boldea, L. *et al.* (1987). Modified sliding mode (MSM) versus PI control of a current-controlled field-oriented induction motor drive. *Electrical Machines and Power Systems* **16**, 209–23.

Boldea, L. and Nasar, S. A. (1988). Torque vector control (TVC)—a class of fast and robust torque-speed and position digital controllers for electric drives. *Electric Machines and Power Systems* **15**, 135–47.

Bolognani, S. and Buja, G. S. (1986). DC link current control for high performance CSIM drive systems. *IEEE IAS Annual Meeting*, pp. 112–16.

Bolognani, S. and Buja, G. S. (1988). Parameter variation and computation error effects in indirect field-oriented induction motor drives. In *Proceedings of International Conference on Electrical Machines, Pisa*, pp. 545–9.

Borsting, H. and Vadstrup, P. (1995). Robust speed and position estimation in induction motors. *EPE, Seville*, 1.089–1.093.

Bose, B. K. (1986). *Power electronics and drives.* Prentice Hall, Englewood Cliffs, New Jersey.

Bose, B. K. and Simoes, M. G. (1995). Speed sensorless hybrid vector controlled induction motor drive. *IEEE IAS Meeting*, pp. 137–43.

Bousak, M., Capolino, G. A., and Phuoc, V. T. N. (1991). Speed measurement in vector-controlled induction machines. *EPE, Firenze*, 3.653–3.658.

Bowes, S. R. (1984). Steady-state performance of P.W.M. inverter drives. *Institute of Electrical Engineers Proc.* (*Pt. B*) **130**, 229–44.

Brahim, L. B. and Kurosawa, R. (1993). Identification of induction motor speed using neural networks. *IEEE PCC, Yokohama*, pp. 689–94.

Breshnahan, K., Zelaya, H. and Evans, P. D. (1995). Analysis of rotor flux reference frame by partition of its control structure. *EPE, Seville*, 3.387–3.392.

Bunte, A. and Grotstollen, H. (1993). Parameter identification of an inverter-fed induction motor at standstill with the correlation method. *EPE, Brighton*, pp. 97–102.

Casadei, D., Grandi, G., and Serra, G. (1993). Study and implementation of a simplified and efficient digital vector controller for induction motors. *IEE EMD, Oxford*, pp. 196–201.

Casadei, D., Grandi, G., and Serra, G. (1993). Rotor flux oriented torque control of induction machines based on stator flux vector control. *EPE, Brighton*, pp. 67–72.

Cirrincione, M., Rizzo, R., Vitale, G., and Vas, P. (1995). Neural networks in motion control: theory, an experimental approach and verification on a test bed. *PCIM, Nurnberg*, pp. 131–41.

Crowder, R. M. (1995). Electric drives and their controls. Clarendon Press, Oxford.

Damiano, A., Vas, P., Marongiu, I., and Stronach, A. F. (1997). Comparison of speed sensorless DTC drives. *PCIM, Nurnberg*, pp. 1–12.

Deng, D. and Lipo, T. A. (1986). A modified control method for fast response current source inverter drives. *IEEE Transactions on Industry Applications* **IA-22**, 653–65.

De Doncker, R., Vandenput, A., and Geysen, W. (1986). A digital field-oriented controller using the double-cage induction motor model. *IEEE Power Electronics Specialists Conference, Vancouver*, pp. 502–9.

De Doncker, R. (1987). Field oriented controllers with rotor deep bar compensation circuits. *IEEE IAS Annual Meeting, Atlanta*, pp. 142–9.

De Doncker, R. and Novotny, D. W. (1988). The universal field-oriented controller. *IEEE IAS Annual Meeting, Pittsburgh*, pp. 450–6.

Depenbrock, M. (1985). Direkte Selbstregelung (DSR) für hochdynamische Drehfeld-antriebe mit Stromrichterschaltung. *ETZA* **7**, 211–18.

Depenbrock, M. (1988). Direct self-control (DSC) of inverter-fed induction machine. *IEEE Transactions on Power Electronics* **3**, 420–9.

Dewan, S. B., Slemon, G. R., and Straughen, A. (1984). *Power semiconductor drives*. Wiley, New York.

Doki, S., Sangwongwanich, S., and Okuma, S. (1992). Implementation of speed sensorless field oriented vector control using adaptive sliding observer. *IEEE, IECON*, pp. 453–8.

Drury, W. (1994). Variable-speed drives for sensorless vector control of induction machines. *EPE Journal* **4**, 5–6.

Du, T. and Brdys, M. A. (1993). Shaft, speed, load torque and rotor flux estimation of induction motor drive using an extended Luenberger observer. *IEE EMD*, pp. 179–84.

Du, T., Vas, P., Stronach, A. F., and Brdys, M. A. (1994). Applications of Kalman filters and extended Luenberger observers in induction motor drives. *PCIM, Nurnberg*, pp. 369–86.

Du, T., Vas, P., and Stronach, F. (1995). Design and application of extended observers for joint state and parameter estimation in high-performance a.c. drives. *IEE Proc. (Pt. B)* **142**, 71–8.

Feller, P. (1983). Speed control of an a.c. motor by state variables feedback with decoupling. *IFAC Symposium on Control in Power Electronics and Drives, Lausanne, Switzerland*, pp. 87–93.

Ferrah, A., Bradley, K. J., and Asher, G. M. (1992). Sensorless speed detection of inverter fed induction motors using rotor slot harmonics and fast Fourier transforms. *IEEE PES Meeting*, pp. 279–86.

Ferraris, P., Fratta, A., Vagati, A., and Villata, F. (1986). About the vector control of induction motors for special applications without speed sensor. *Int. conference on the evolution and modern aspects of induction machines, Torino*, pp. 444–50.

Finney, D. (1988). *Variable-frequency a.c. motor drive systems*. Peter Peregrinus Ltd., London.

Fornel, B., Pietrzak-David, M., and Roboam, X. (1996). State observers for the control of the a.c. variable speed drives. *PEMC, Budapest*, 2/1–2/8.

Franceschini, G., Pastorelli, M., Profumo, F., Tassoni, C., and Vagati, A. (1990). About the choice of flux observer in induction servomotors. *IEEE IAS Meeting*, pp. 1235–42.

Fratta, A., Vagati, A., and Villata, F. (1988). Vector control of induction motors without transducers. *IEEE Power Electronics Specialists Conference, Kyoto*, pp. 839–46.

Garces, L. J. (1980). Parameter adaptation of the speed-controlled static a.c. drive with a squirrel-cage induction motor. *IEEE Transactions on Industry Applications* **IA-16**, 173–8.

Gastli, A., Tomita, M., Takeshita, T., and Matsui, N. (1993). Improvement of a stator-flux-oriented speed-sensorless control of an induction motor. *IEEE PCC, Yokohama*, pp. 415–20.

Gheysens, R., Cherif, H., and Poloujadoff, M. (1990). Speed determination of a squirrel cage induction motor by indirect method. *IPEC, Tokyo*, pp. 1137–43.

Griva, G., Habetler, T. G., Profumo, F., and Pastorelli, M. (1995). Performance evaluation of a direct torque controlled drive in the continuous PWM-square wave transition range. *IEEE Transactions on Power Electronics* **10**, 464–71.

Grotstollen, H. and Wiesing, J. (1995). Torque capability and control of a saturated induction motor over a wide range of flux weakening. *IEEE Transaction on Ind. Electronics* **42**, 374–81.

Habetler, T. G. and Divan, D. M. (1991). Control strategies for direct torque control using discrete pulse modulation. *IEEE Transactions on Ind. Applications* **27**, 893–901.

Habetler, T., Profumo, F., Pastorelli, M., and Tolbert, L. M. (1992). Direct torque control of induction machines using space vector modulation. *IEEE Transactions on Ind. Applications* **IA-28**, 1045–52.

Haemmerli, B., Tanner, R., and Zwicky, R. (1987). A rotor speed detector for induction machines utilising rotor slot harmonics and active three-phase injection. *EPE, Grenoble*, pp. 599–603.

Harashima, F., Kondo, S., Ohnishi, K., Kajita, M., and Susono, M. (1985). Multimicroprocessor-based control system for quick response induction motor drive. *IEEE Transactions on Industry Applications* **IA-21**, 602–9.

Hasse, K. (1972). Drehzahlregelverfahren für schnelle Umkehrantriebe mit stromrichtergespeisten Asynchron-Kurzschlusslaufermotoren. *Regelungstechnik* **20**, 60–6.

Hasse, K. (1977). Control of cycloconverters for feeding asynchronous machines. *IFAC Conference on Control in Power Electronics and Drives, Düsseldorf*, pp. 537–46.

Healey, R. C. (1995). The implementation of a vector control scheme based on an advanced motor model. *EPE, Seville*, 3.017–3.022.

Heinemann, G. and Leonhard, W. (1990). Self-tuning field-oriented control of an induction motor drive, *IPEC, Tokyo*, pp. 465–72.

Henneberger, G., Brunsbach, B. J., and Klepsch, T. H. (1991). Field oriented control of synchronous and asynchronous drives without mechanical sensors using a Kalman filter. *EPE, Firenze*, 3.664–3.671.

Ho, E. Y. Y. and Sen, P. C. S. Decoupling control of induction motor drives. *IEEE Transactions on Industrial Electronics* **35**, 253–62.

Hofman, H., Sanders, S. R., and Sullivan, C. (1995). Stator-flux-based vector control of induction machines in magnetic saturation. *IEEE IAS Meeting, Orlando*, pp. 152–8.

Hofman, W. and Krause, M. (1993). Fuzzy control of a.c. drives fed by PWM inverters. *PCIM, Nurnberg*, pp. 123–32.

Holtz, J. and Stadtfeld, S. (1983). Field-oriented control by forced motor currents in a voltage-fed induction motor drive. *IFAC symposium on Control in Power Electronics and Electrical Drives, Lausanne*, pp. 103–10.

Holtz, J. and Bube, E. (1991). Field-oriented asynchronous pulse-width modulation for high-performance a.c. drives operating at low switching frequency. *IEEE Transactions on Ind. Applications* **27**, 574–9.

Holtz, J. (1994). Speed estimation and sensorless control of a.c. drives. *IEEE IECON, Maui*, pp. 649–54.

Hori, Y. and Umeno, T. (1990). Flux observer based field orientation controllers for high performance torque control. *Proc. IPEC, Tokyo*, pp. 1219–26.

Huang, L., Tadakoro, Y., and Matsuse, K. (1994). Deadbeat flux level control of direct field oriented high horsepower motor using adaptive rotor flux observer. *IEEE Transactions on Ind. Applications* **30**, 954–62.

Hurst, K. D., Habetler, T. G., Griva, G., and Profumo, F. (1994). Speed sensorless field oriented control of induction machines using current harmonics spectral estimation. *IEEE IAS Meeting*, pp. 601–7.

Ilas, C., Bettini, A., Ferraris, L., Griva, G., and Profumo, F. (1994). Comparison of different schemes without shaft sensors for field orientated control drives. *IEEE IECON, Maui*, pp. 1579–88.

Ilas, C. and Magureanu, R. (1996). DSP-based sensorless direct field oriented control of induction motor drives. *PEMC, Budapest*, 2/309–2/313.

Irisa, T., Takata, S., Ueda, R., and Sonoda, T. (1984). Effect of machine structure identification of spatial position and its magnitude of rotor flux in induction motor vector control. *IEEE Conference on Power Electronics and Variable Speed Drives, London*, pp. 352–6.

Ishida, M. and Iwata, K. (1984). A new slip frequency detector of an induction motor utilising rotor slot harmonics. *IEEE Transactions on Ind. Applications* **IA-20**, 575–82.

Ishida, M. and Iwata, K. (1987). Steady-state characteristics of a torque and speed control system of an induction motor utilising rotor slot harmonics for a slip frequency sensing. *IEEE Transactions on Power Electronics* **2**, 257–63.

Islam, S. M. and Somouah, C. B. (1989). An efficient high performance voltage decoupled induction motor drive with excitation control. *IEEE Transactions on Energy Conversion* **4**, 109–17.

Iwasaki, T. and Kataoka, T. (1989). Application of an extended Kalman filter to parameter identification of an induction motor. *IEEE IAS Meeting, San Diego*, pp. 251–3.

Jang, J. G. (1993). ANFIS: adaptive-network-based fuzzy inference system. *IEEE Transactions on System, Man and Cybernetics* **23**, 665–84.

Jansen, P. L. and Lorenz, R. D. (1993). Accuracy limitations of velocity and flux estimation in direct field oriented induction machines. *EPE, Brighton*, pp. 312–18.

Jansen, P. L. and Lorenz, R. D. (1995a). Transducerless position and velocity estimation in induction and salient a.c. machines. *IEEE Transactions on Ind. Applications* **IA-31**, 240–7.

Jansen, P. L. and Lorenz, R. D. (1995b). Transducerless field orientation concepts employing saturation induced saliency in induction machines. *IEEE IAS Meeting*, pp. 174–81.

Joetten, R. and Maeder, G. (1982). Control methods for good dynamic performance induction motor drives based on current and voltage as measured quantities. *International Semiconductor Power Converter Conference, IEEE/IAS, Orlando*, pp. 397–407.

Jonsson, R. and Leonhard, W. (1995). Control of an induction motor without a mechanical sensor, based on the principle of natural field orientation. *IPEC, Yokohama*, pp. 298–303.

Kanmachi, T. and Takahashi, I. (1993). Sensorless speed control of an induction motor with no influence on secondary resistance variation. *IEEE IAS Meeting*, pp. 408–13.

Kawamura, A. and Hoft, R. (1983). An analysis of induction motor for field oriented or vector control. *IEEE Power Electronics Specialists Conference, Albuquerque, New Mexico*, pp. 91–100.

Kazmierkowski, M. P. and Köpcke, H. J. (1983). Comparison of dynamic behaviour of frequency converter fed induction machine drives. *IFAC Symposium on Control in Power Electronics and Drives, Lausanne*, pp. 313–20.

Kazmierkowski, M. P. and Köpcke, H. J. (1985). A simple control system for current source inverter-fed induction motor drives. *IEEE Transactions on Ind. Applications* **IA-21**, 617–23.

Kazmierkowski, M. P. and Sulkovski, W. (1991). A novel vector control scheme for transistor PWM inverter-fed induction motor drives. *IEEE Transactions on Ind. Electronics* **38**, 41–7.

Kim, S. H. and Sul, S. K. (1995). Maximum torque control of an induction machine in the field-weakening region. *IEEE Transactions on Ind. Applications* **IA-31**, 787–94.

Kim, Y. R., Sul, S. K., and Park, M. H. (1992). Speed sensorless vector control of an induction motor using an extended Kalman filter. *IEEE IAS Meeting*, pp. 594–9.

Kreindler, L., Moreira, J. C., Testa, A., and Lipo, T. A. (1994). Direct field orientation controller using the stator phase voltage third harmonic. *IEEE Transactions on Ind. Applications* **IA-30**, 441–7.

Krishnan, R., Doran, F. C., and Latos, T. S. (1987). Identification of thermally safe load cycles for an induction motor position servo. *IEEE Transactions on Industry Applications* **23**, 636–43.

Krishnan, R. and Doran, F. C. (1987). Study of parameter sensitivity in high-performance inverter-fed induction motor drive systems. *IEEE Transactions on Industry Applications* **IA-23**, 623–35.

Kubota, H. and Matsuse, K. (1994). Speed sensorless field oriented control of induction machines. *IEEE IECON*, pp. 1611–15.

Kubota, H., Matsuse, K., and Nakano, T. (1990). New adaptive flux observer for induction motor drives. *IEEE IECON*, pp. 921–6.

Landau, Y. D. (1986). *Adaptive control, the model reference approach*. Marcel Dekker, New York.

Lee, T. K., Cho, S. B., and Hyun, D. S. (1992). Sensorless vector control of induction motor compensating the variation of rotor resistance. *IEEE IECON*, pp. 72–6.

Leonhard, W. (1985). *Control of electrical drives*. Springer-Verlag, Berlin.

Leonhard, W. (1986). Microcomputer control of high dynamic performance a.c. drives—a survey. *Automatica* **22**, 1–19.

Leonhard, W. (1988). Adjustable-speed a.c. drives. *Proceedings of the IEEE* **76**, 455–70.

Leonhard, W. (1988). Field-orientation for controlling a.c. machines—Principle and application. *3rd International Conference on Power Electronics and Variable Speed Drives, London*, pp. 277–82. IEE, London.

Levi, E. (1995). Improvements in operation of vector-controlled induction machines by application of modified machine models. *IEE Colloquium on Advances in Control Systems for Electric Drives*, 3.1–3.8.

Li, Y. D., Shao, J. W., Cao, J. T., and Ji., Z. Y. (1993). Direct torque control of induction motors using DSP. *ICEM, Paris*, pp. 18–23.

Ljung, L. (1979). Asymptotic behaviour of the extended Kalman filter as a parameter estimator for linear system. *IEEE Transactions on Automatic Control* **AC-24**, 36–50.

Lorenz, R. D., Lucas, M. O., and Lawson, D. B. (1986). Synthesis of a state-variable motion controller for high-performance field oriented induction machine drives. *IEEE IAS Annual Meeting, Denver*, pp. 80–5.

Lorenz, R. D. and Yang, S. M. (1992). Efficiency-optimised flux trajectories for closed-cycle operation of field-orientation induction machine drives. *IEEE Transactions on Ind. Applications* **28**, 574–9.

Luenberger, D. G. (1971). An introduction to observers. *IEEE Transactions on Automatic Control* **16**, 596–602.

Mamdani, E. H. (1974). Application of fuzzy algorithms for simple dynamic plant. *Proc. IEE* **427**, 4585–4588.

Manninen, V. (1995). Application of direct torque control modulation technology to a line converter. *EPE, Seville*, 1.292–1.296.

Matsuse, K. and Tadokoro, Y. (1993). Deadbeat flux level control of direct field oriented induction servomotor using adaptive rotor flux observer. *EPE, Brighton*, pp. 336–41.

Miyashita, I., Imayanagida, A., and Koga, T. (1994). Recent industrial application of speed sensorless vector control in Japan. *IEEE IECON*, pp. 1573–8.

Murphy, J. M. D. and Turnbull, F. G. (1987). *Power electronic control of a.c. motors*. Pergamon Press, Oxford.

Nilsen, R. and Kasteenpohja, T. (1995). Direct torque controlled induction motor drive utilised in an electrical vehicle. *EPE, Seville*, pp. 2.877–2.882.

Nippert, T. and Schollmeyer, H. (1987). Sensorless speed measuring device for asynchronous machines. *Electric Energy Conference, Adeleide*, pp. 433–6.

Noguchi, T. and Takahashi, I. (1984). Quick torque response control of an induction motor using a new concept. *IEEJ Tech. Meeting on Rotating Machines*, paper RM84-76, pp. 61–70.

Novotny, D. W. and Lorenz, R. D. (1985). Introduction to field orientation and high performance a.c. drives. Tutorial Course. *IAS Annual Meeting, IEEE*, 6–7 October. Toronto.

Ohnishi, K. and Miyachi, K. (1982). Principles of constant magnitude regulation of secondary flux based on slip frequency control in induction motor drive. In *Proceedings of International Conference on Electrical Machines, Budapest*, pp. 203–6.

Ohnishi, K., Suzuki, H., Miyachi, K., and Terashima, M. (1985). Decoupling control of secondary flux and secondary current in induction motor drive with controlled voltage source and its comparison with volts/hertz control. *IEEE Transactions on Industry Applications* **IA-21**, 241–7.

Ohnishi, K., Ueda, Y., and Miyachi, K. (1986). Model reference adaptive system against rotor resistance variation in induction motor drive. *IEEE Transactions on Industrial Electronics* **IA-22**, 217–23.

Ohtani, T., Takada, N., and Tanaka, K. (1992). Vector control of induction motor without shaft encoder. *IEEE Transactions on Ind. Applications* **IA-28**, 157–64.

Okuyama, T., Fujimoto, N., Matsui, T., and Kubota, Y. (1986). A high performance speed control scheme of induction motor without speed and voltage sensors. *IEEE IAS Annual Meeting, Denver*, pp. 106–10.

Orlowska-Kowalska, T. (1989). Application of extended Luenberger observer for the flux and rotor time constant estimation in induction motor drives. *IEE Proc.* (*Pt. D*) **136**, 324–30.

Peng, F. Z., Fukao, T., and Lai, J. S. (1994). Robust speed identification for speed-sensorless vector control of induction motors. *IEEE Transactions on Ind. Applications* **IA-30**, 1234–40.

Plunkett, A. B., Kliman, G. B., and Boyle, M. J. (1985). Digital techniques in evaluation of high-efficiency induction motors for inverter drives. *IEEE Transactions on Industry Applications* **IA-21**, 456–63.

Polmann, A. (1986). Software pulsewidth modulation for microprocessor control of a.c. drives. *IEEE Transactions on Industry Applications* **IA-22**, 691–6.

Popov, V. M. (1973). Hyperstability of control systems. Springer-Verlag, New York.

Rácz, I. and Vas, P. (1985). Oscillatory behaviour of induction machines employed in railway traction. *IEE EMD, London*, pp. 168–72.

Rowan, T. M., Kerkman, R. J., and Lipo, T. A. (1987). Operation of naturally sampled current regulators in transition mode. *IEEE Transactions on Industry Applications* **IA-23**, 586–96.

Rubin, N. P., Harley, R. G., and Diana, G. (1992). Evaluation of various slip estimation techniques for an induction machine operating under field-oriented control conditions. *IEEE Transactions on Ind. Applications* **IA-28**, 1367–76.

Saito, K., Kamiyama, K., and Sukegawa, T. (1986). A multi-processor based, fully digital, a.c. drive system for rolling mills. *IEEE IAS Annual Meeting, Denver*, pp. 36–41.

Sangwongwanich, S. (1995). Speed sensorless vector control of induction motor stability analysis and realisation. *IPEC, Yokohama*, Vol. 1, pp. 310–15.

Sathikumar, S. and Vithayathil, J. (1984). Digital simulation of field-oriented, control of induction motor. *IEEE Transactions on Industrial Electronics* **IE-31**, 141–8.

Schauder, C. (1992). Adaptive speed identification for vector control of induction motors without rotational transducers. *IEEE Transactions on Ind. Applications* **IA-28**, 1054–61.

Schauder, C. D., Choo, F. H., and Roberts, M. T. (1983). High-performance torque-controlled induction motor drives. *IEEE Transactions on Industry Applications* **IA-19**, 349–55.

Schroedl, M. (1988). Detection of rotor position of permanent magnet synchronous machine at standstill. ICEM, Pisa, 195–7.

Schroedl, M. (1992). Sensorless control of induction motors at low speed and standstill. *ICEM, Manchester*, pp. 863–7.

Simoes, G. and Bose, B. K. (1995). Neural network based estimation of feedback signals for a vector controlled induction motor drive. *IEEE Transactions on Ind. Applications* **IA-31**, 620–9.

Sobczyk, T. J., Vas, P., and Tassoni, C. (1996). Models for induction motors with air-gap asymmetry for diagnostic purposes. *ICEM, Vigo*, pp. 79–84.

Soresen, H. V., Jones, D. L., Heideman, M. T., and Burrus, C. S. (1987). Real-valued fast Fourier transform algorithms. *IEEE Transactions on ASSP* **35**, 849–63.

Stepina, J. (1979). Non-transformational matrix analysis of electrical machinery. *Electrical Machines and Electromechanics* **4**, 255–68.

Stronach, A. F., Vas, P., and Lees, P. (1994). The application of self-tuning digital controllers in variable-speed drive. *IEE, PEVD94, London*, pp. 119–24.

Stronach, F. and Vas, P. (1995). Variable-speed drives incorporating multi-loop adaptive controllers. *Proc. IEE* (*Pt. D*) **142**, 411–20.

Stronach, A. F. and Vas, P. (1995). Fuzzy-neural control of variable-speed drives. *PCIM, Nurnberg*, pp. 117–29.

Stronach, A. F., Vas, P., and Neuroth, M. (1997). Implementation of intelligent self-organising controllers in DSP-controlled electromechanical drives. *IEE Proc.* (*Pt. D*) **144**, pp. 324–330.

Sugimoto, H. and Tamai, S. (1987). Secondary resistance identification of an induction motor applied model reference adaptive system and its characteristics. *IEEE Transactions on Industry Applications* **IA-23**, 296–303.

Svensson, T. (1983). Current reference estimator for inverter-fed a.c. machines. *IFAC Symposium on Control in Power Electronics and Electrical Drives, Lausanne*, pp. 539–46.

Tajima, H. and Hori, Y. (1993). Speed sensorless field-orientation control of the induction machine. *IEEE Transactions on Ind. Applications* **IA-29**, 175–80.

Takahashi, I. and Noguchi, T. (1985). A new quick response and high efficiency control strategy of an induction motor. *IEEE IAS Meeting*, pp. 496–502.

Takahashi, I. and Ohmori, Y. (1989). High performance direct torque control of an induction machine. *IEEE Transactions on Ind. Applications* **IA-25**, 257–64.

Tamai, S., Sugimoto, H., and Yano, M. (1987). Speed sensorless vector control of induction motor with model reference adaptive system. *IEEE IAS Annual Meeting, Atlanta*, pp. 189–95.

Tiitinen, P. (1996). The next generation motor control method, DTC, direct torque control. *PEDES*, pp. 37–43.

Tsuji, M., Yamada, E., Izumi, K., and Oyama, J. (1984). Stability analysis of a current-source inverter-fed induction motor under vector control. In *Conference Proceedings of International Conference on Electrical Machines, Lausanne*, pp. 867–74.

Vagati, A. and Villata, F. (1984). AC servo system for position control. In *Conference Proceedings of International Conference on Electrical Machines, Lausanne*, pp. 871–4.

Vagati, A. and Villata, F. (1985). Field-oriented control, a particular implementation. In *Conference Record of Motor-Con, Hannover*, pp. 160–72.

Vagati, A., Fratta, A., Franceshini, G., and Rosso, P. (1996). AC motors for high-performance drives: a design-based comparison. *IEEE Transactions on Ind. Applications* **IA-32**, 1211–19.

Vas, P. (1990). *Vector control of a.c. machines*. Oxford University Press.

Vas, P. (1992). *Electrical machines and drives: a space-vector theory approach*. Oxford University Press.

Vas, P. (1993). *Parameter estimation, condition monitoring, and diagnosis of electrical machines*. Oxford University Press.

Vas, P. (1995). Application of space-vector techniques to electrical machines and variable-speed drives. *EPE tutorial, Seville*, pp. 1–130.

Vas, P. (1995). Artificial intelligence applied to variable-speed drives. *PCIM Seminar, Nurnberg*, pp. 1–149.

Vas, P. (1996). Recent trends and development in the field of machines and drives, application of fuzzy, neural and other intelligent techniques. *Workshop a.c. motor drives technology, IEEE/IAS/PELS, Vicenza*, pp. 55–74.

Vas, P. and Alaküla, M. (1989). Field-oriented control of saturated induction machines. *IEEE Transactions on Energy Conversion*, IEEE Power Engineering Society Summer Meeting, Long Beach, California, paper 89SM 738–6EC.

Vas, P. and Alaküla, M. (1990). Field-oriented control of saturated induction machines. *IEEE Transactions on Energy Conversion* **5**, 218–24.

Vas, P. and Drury, W. (1994). Vector-controlled drive. *PCIM, Nurnberg*, pp. 213-28.

Vas, P. and Drury, W. (1995). Future trends and developments of electrical machines and drives. *PCIM, Nurnberg*, pp. 1–28.

Vas, P. and Stronach, A. F. (1995). DSP-controlled intelligent high performance a.c. drives: present and future. Colloquium on *Vector control and direct torque control of induction motors*, IEE, London, 7/1–7/8.

Vas, P. and Drury, W. (1996). Future developments and trends in electrical machines and variable-speed drives. *Energia Elettrica* **73**, 11–21.

Vas, P. and Stronach, A. F. (1996). Design and application of multiple fuzzy controllers for servo drives. *Archiv für Elektrotechnik* **79**, 1–12.

Vas, P. and Stronach, A. F. (1996). Minimum configuration soft-computing-based DSP controlled drives. *PCIM, Nurnberg*, pp. 299–315.

Vas, P. and Stronach, A. F. (1996). Adaptive fuzzy-neural control of high-performance drives. *IEE PEVD*, Nottingham, pp. 424–9.

Vas, P., Brown, J. E., and Hallenius, K. E. (1987). Small-signal analysis of smooth-air-gap machines using space vector theory. *Electric Energy Conference, Adelaide*, pp. 210–14.

Vas, P., Li, J., and Stronach, A. F. (1993). Simulation of various vector-controlled drives. *ICEMA, Adelaide*, Vol. 2, pp. 297–302.

Vas, P., Li, J., Stronach, A. F., and Lees, P. (1994). Artificial neural network-based control of electromechanical system. *4th European Conference on Control, IEE, Coventry*, pp. 1065–70.

Vas, P., Stronach, A. F., Neuroth, M., and Du, T. (1995). Fuzzy, pole-placement and PI controller design for high-performance drives. *Stockholm Power Tech.*, 1–6.

Vas, P., Stronach, F., Neuroth, M., and Du, T. (1995). A fuzzy controlled speed-sensorless induction motor drive with flux estimators. *IEE EMD, Durham*, pp. 315–19.

Vas, P., Drury, W., and Stronach, A. F. (1996). Recent developments in artificial intelligence-based drives: a review. *PCIM, Nurnberg*, pp. 59–71.

Vas, P., Stronach, A. F., and Drury, W. (1996). Artificial intelligence in drives. *PCIM, Europe*, Vol. 8, pp. 152–5.

Vas, P., Stronach, A. F., Neuroth, and Neuroth, M. (1998). Application of conventional and AI-based techniques in sensorless high-performance torque-controlled induction motor drives. *IEE Colloquium*, **23** February.

Verghese, G. C. and Sanders, S. R. (1985). Observers for faster flux estimation in induction machines. *IEEE Power Electronics Specialists Conference.*

Wade, S., Dunningan, M. W., and Williams, B. W. (1995). Improvements for induction machine vector control. *EPE, Seville*, 1.542–1.546.

Wheeler, W. P., Clare, J. C., and Summer, M. (1995). A bi-directional rectifier for use with a voltage fed inverter in high performance variable speed a.c. motor drive system. *EPE, Seville*, 1.095–1.100.

Wieser, R. and Schroedl, M. (1996). High dynamic emf based rotor flux detection in induction motors using measurements only. *Speedam, Capri*, B1.1–B1.6.

Williams, B. W., Goodfellow, J. K., and Green, T. C. (1990). Sensorless speed measurement of inverter driven squirrel cage induction motors without rotational transducers. *IEE PEVD*, pp. 297–300.

Wolfbank, T. M. (1996). High dynamic low frequency filtering of the flux angle in field oriented control of induction motor without speed and position sensor. *ICEM, Vigo*, I/7–I/11.

Wu, Z. K. and Strangas, E. G. (1988). Feedforward field orientation control of an induction motor using a PWM voltage source inverter and standardized single-board computers. *IEEE Transactions on Industrial Electronics* **35**, 75–9.

Xu, X. and Novotny, D. W. (1991). Implementation of direct stator flux orientation control on a versatile DSP based system. *IEEE Transactions on Ind. Applications* **27**, 694–700.

Xu, X., De Doncker, R., and Novotny, D. W. (1988). Stator flux-orientation control of induction machines in the field-weakening region. *IEEE IAS Annual Meeting, Pittsburgh*, pp. 437–43.

Xu, X., De Doncker, R., and Novotny, D. W. (1988). A stator flux oriented induction machine drive. *IEEE Power Electronics Specialists Conference, Kyoto*, pp. 870–6.

Yamamura, S. (1986). AC *motors for high-performance applications.* Marcel Dekker, New York.

Yong, S. I., Choi, J. W., and Sul, S. K. (1994). Sensorless vector control of the induction machine using high frequency current injection. *IEEE IAS Meeting*, pp. 503–8.

Zhang, J., Thiagaranjan, V., Grant, T., and Barton, T. H. (1988). New approach to field orientation control of a CSI induction motor drive. *IEEE Proc. B.* **135**, 1–7.

Zhang, J. and Barton, T. H. (1987). Microprocessor based primary current control for cage induction motor drive. *IEEE Power Electronics Specialists Conference, Blacksburg*, pp. 616–24.

5 Torque control of switched reluctance motors

In the present chapter, first the fundamentals of switched reluctance motor drives are discussed. This is then followed by a description of various position estimators which can be used in position-sensorless switched reluctance motor drives. Finally the two main possibilities for instantaneous torque-controlled switched reluctance motor drives are discussed. The first solution uses profiled stator currents, and thus the drive contains a current controller, but the second solution uses profiled stator flux linkages, and the drive scheme contains a stator flux-linkage controller. It is shown that the second scheme is simpler to implement. Both schemes can operate in a wide speed range and offer the potential of using switched reluctance motors in many applications, where so far only other types of motors have been used.

5.1 Switched reluctance motor drive fundamentals; main techniques of position-sensorless implementations

5.1.1 GENERAL INTRODUCTION

Pioneering work on switched reluctance motors has been performed in the U.K. by Professor P. J. Lawrenson and Professor T. J. E. Miller. Recently switched reluctance motors (SRM) have gained increased interest, mainly due to the simple technology involved. Furthermore, they have high efficiency over a wide speed range and they require a reduced number of switching devices. They also have superior fault tolerance characteristics. The converter supplying the SRM has totally independent circuits for each stator winding and direct control is achieved through the stator currents. In a conventional SRM drive, a position sensor is used to switch the stator currents at the appropriate instants, but in more recent implementations, the position sensor is eliminated by estimating the rotor position. However, most of the difficulty in understanding the operation and design of switched reluctance motor drives originates from the fact that the SRM is a doubly salient, highly non-linear machine. In the SRM only the stator windings are excited, so the SRM is a singly excited doubly salient machine. In contrast to a variable reluctance stepper motor, the SRM produces continuous motion.

5.1.2 SRM STRUCTURE; STATOR EXCITATION

The switched reluctance machine consists of a salient-pole stator (made of laminated steel) and a salient-pole rotor which is usually made from normal electrical steel laminations. However, in high-efficiency applications the lamination is thinner than normally used for a.c. motors and is, for example, silicon steel, to reduce the eddy current losses. The reason is that in an SRM the switching frequency is higher than for an a.c. motor with comparable rating and speed.

The rotor does not require any rotor conductors or permanent magnets. The number of stator and rotor poles are generally different ($N_s \neq N_r$), e.g. some possible combinations are: $N_s=6$, $N_r=4$; $N_s=8$, $N_r=6$; $N_s=12$, $N_r=10$, etc. [Miller 1993]. This ensures that the rotor is never in a position where the electromagnetic torque due to a stator current in any stator phase is zero. Switched reluctance motors with a higher number of stator and rotor poles have less torque ripple. By choosing a combination where there are two more stator poles than rotor poles (one more stator pole-pair than rotor pole-pair), high average torque and low switching frequency of the converter (which supplies the machine) is ensured. It is also possible to choose the pole numbers so that instead of $N_s-N_r=2$, the difference is negative: $N_s-N_r=-2$, e.g. $N_s=6$, $N_r=8$. For the same N_s, the advantage of a larger N_r is a smaller step angle and, possibly, smaller torque ripples. In such a design the stator pole width is reduced, leading to the decrease of the inductance ratio of a stator phase winding in the aligned and misaligned positions, which is a disadvantage (see also below). An odd number of rotor poles is to be avoided, since this creates an imbalance in the forces acting on the structure. It should also be noted that it is also possible to have configurations where there are two teeth per stator pole, e.g. a machine with 6 stator poles and two teeth per stator pole, the rotor having 10 poles [Finch *et al.* 1984]. This will lead to an increase of the torque/ampere but the core losses will also increase. Therefore such a configuration is better suited to low-speed applications.

The stator windings of the SRM are concentrated windings (which are cheaper than distributed windings). Stator windings on diametrically opposite poles (teeth) are connected in series to form a stator phase. Thus it can be seen that the machine with six stator poles and 4 rotor poles is a three-phase machine (generic three-phase machine). The machine with $N_s=12$, $N_r=8$ is also a three-phase machine, but there are four stator poles per stator phase. A three-phase machine has starting torque capabilities in either forward or reverse directions. If $N_s=2$, $N_r=2$, a single-phase machine is obtained, but this machine can only be started by using extra starting arrangements (e.g. by using stepped air-gap). If $N_s=4$ and $N_r=2$, then the generic two-phase machine is obtained, and this can also be started by using some form of extra starting technique (e.g. stepped air-gap). If $N_s=8$, $N_r=6$, the machine is a four-phase machine, which can start from any rotor position, and gives a smoother torque than its three-phase counterpart.

In the SRM a sequence of current pulses is applied to each stator phase by using the appropriate power converter. The operation of the SRM is simple: a pair of stator poles is excited when a pair of opposite rotor poles is approaching and these rotor poles align themselves to these. The excitation is turned off before the rotor and stator poles come into alignment. The physical principle is that when a stator pole-pair is energized, the corresponding rotor pole-pair is attracted toward the energized stator pole pair to minimize the reluctance of the magnetic path. Thus by energizing the consecutive stator phases in succession, it is possible to develop constant torque in either direction of rotation.

Table 5.1 SRM and VR stepper motor main features

	SRM	**VR stepper motor**
Stator	salient pole	salient pole
Rotor	salient pole	salient pole
Stator winding	concentrated	concentrated
Rotor winding	—	—
Stator currents	unidirectional	unidirectional
Control	position feedback	open-loop
Motion	continuous	in steps

For better understanding, Table 5.1 compares the main features of the SRM and the variable reluctance (VR) stepper motor.

5.1.3 ELECTROMAGNETIC TORQUE PRODUCTION; MAIN TECHNIQUES OF SENSORLESS OPERATION

It follows from the discussion above that in the SRM, the electromagnetic torque is developed by the tendency of the magnetic circuit to adopt a position with minimum reluctance of the magnetic paths and is independent of the direction of the current flow. Thus only unidirectional currents are required, as shown in Table 5.1. This is the reason why a simple converter can be used in a SRM drive. Figure 5.1 shows an SRM with 6 stator poles and 4 rotor poles at three positions.

The first position is an unaligned position, where the rotor is not aligned with stator phase sA. In this position stator phase sA is energized with unidirectional current i_{sA} supplied by a converter and torque is produced which moves the rotor into the second, intermediate position shown. In this case the rotor poles have partial overlap with the stator sA-phase poles. The rotor then moves to the third position where the rotor poles are aligned with the stator sA poles. In the unaligned position the stator inductance is minimum (reluctance is maximum). The rotor then moves to reduce this; the stator inductance is thus larger in the intermediate position (reluctance is reduced) and in the aligned position the stator inductance is maximum (reluctance minimum). If it is assumed that the effects of magnetic saturation are neglected, then the variation of the stator inductances with the rotor position (for a given stator current) is shown in Fig. 5.2. A stator winding inductance varies cyclically with the rotor position and has an angular period given by $2\pi/N_r$. The excitation current is applied during each cyclic variation of the inductance. It can be seen from Fig. 5.2 that in an SRM with three phase stator, $L_{sA}=L_{sA}(\theta_r)$; $L_{sB}(\theta_r)=L_{sA}(\theta_r-2\pi/3)$; $L_{sC}(\theta_r)=L_{sA}(\theta_r-4\pi/3)$. If the effects of magnetic saturation are also considered then the curves shown in Fig. 5.2 become flatter at the aligned position.

In general, the equations of the SRM can be described as follows. For one stator phase (e.g. sA) the voltage equation in the stator reference frame is

$$u_{sA}=R_s i_{sA}+\frac{d\psi_{sA}}{dt}, \tag{5.1-1}$$

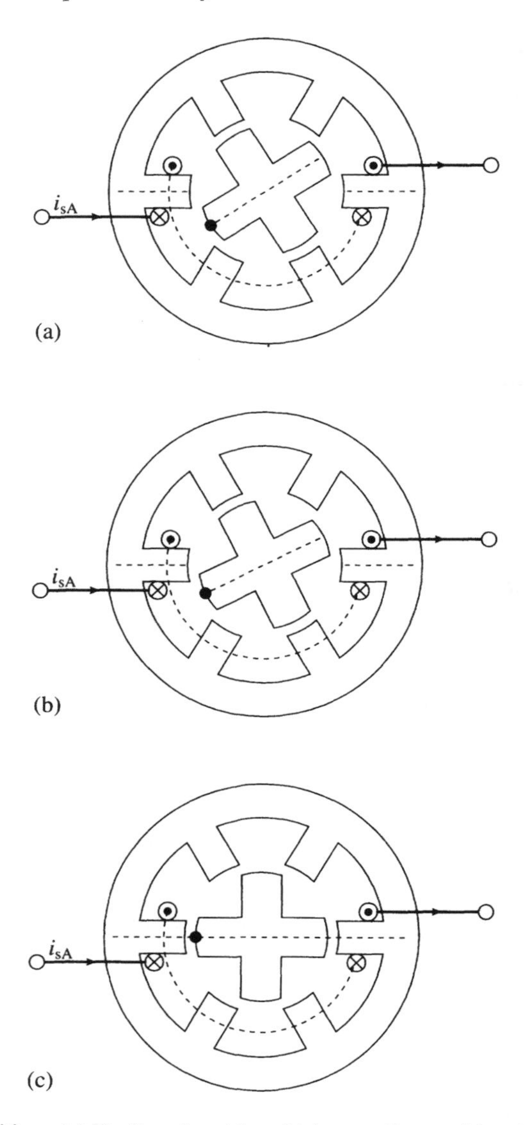

Fig. 5.1. SRM positions. (a) Unaligned position; (b) intermediate position; (c) aligned position.

where R_s is the stator resistance and ψ_{sA} is the flux which links stator winding sA. If there is no mutual coupling between the stator phases, then in general $\psi_{sA}=L_{sA}i_{sA}$, where L_{sA} is the stator self-inductance (shown in Fig. 5.2). In general, this flux linkage depends on the current and also the rotor position, as shown in Fig. 5.3.

The physical picture is as follows. At rotor position $\theta_r=\theta_{r1}$, when the rotor poles of the SRM are aligned with the stator poles, the stator flux linkage for a

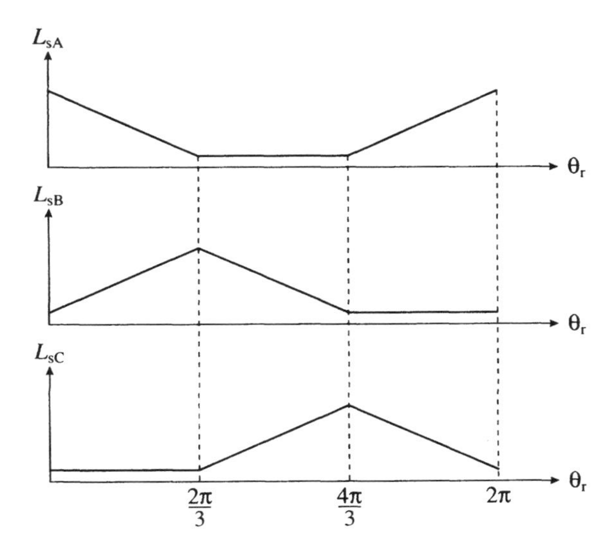

Fig. 5.2. Stator self-inductance dependency on the rotor position, if magnetic saturation is neglected.

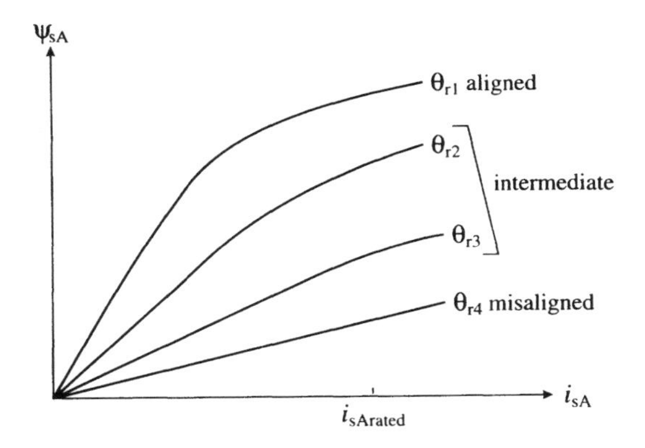

Fig. 5.3. Flux linkage versus stator current for various rotor positions.

given stator phase current is maximal. However, at this aligned position, the relationship between the flux and the current is extremely non-linear in a well-designed machine, since at rated stator current (see $i_{sArated}$ in Fig. 5.3), the tooth iron is magnetically saturated. At another extreme rotor position, e.g. $\theta_r = \theta_{r4}$, where the rotor poles are misaligned with the stator poles, there is large air-gap between the stator and rotor poles, and despite the local saturation of the pole corners, the stator flux is linearly dependent on the stator phase current. It follows that the slopes of the curves in Fig. 5.3, which are also the values of L_{sA} at different rotor positions, increase from the value corresponding to the misaligned position to the value corresponding to the aligned position. It should be noted that if the effects of magnetic saturation are neglected, then all the curves shown in Fig. 5.3 would be linear.

Assuming that there is no mutual coupling between the stator phases, eqn (5.1-1) can be put into the following form:

$$u_{sA} = R_s i_{sA} + \frac{\partial \psi_{sA}}{\partial i_{sA}} \frac{di_{sA}}{dt} + \frac{\partial \psi_{sA}}{\partial \theta_r} \omega_r, \tag{5.1-2}$$

where ω_r is the rotor speed, $\omega_r = d\theta_r/dt$, and $\partial\psi_{sA}/\partial i_{sA}$ is a dynamic (incremental) inductance,

$$L_d = \frac{\partial \psi_{sA}}{\partial i_{sA}}. \tag{5.1-3}$$

If the flux varies linearly with the currents, then the dynamic inductance is equal to the static inductance (ψ_{sA}/i_{sA}). The third component in eqn (5.1-2) is a rotational voltage (motional e.m.f., which is also sometimes called back e.m.f.). The dynamic inductance changes its value due to rotor position and also saturation (current level). It follows from Fig. 5.3 that, for example, at rotor position θ_{r1} (aligned position) the gradient of the stator flux-linkage characteristic ($\partial\psi_{sa}/\partial i_{sA}$), which is equal to the dynamic inductance, is large for small values of the stator current, but it decreases rapidly as the stator current increases to $i_{sArated}$. However, at rotor position e.g. $\theta_r = \theta_{r4}$ (misaligned position), the dynamic inductance is independent of the value of the stator current. It follows that, in general, the dynamic inductance is dependent on both the rotor position and also on the stator current, and this dependency can also be used for the estimation of the rotor position in a position-sensorless SRM drive (as discussed below in Section 5.2.1).

The electromagnetic torque produced by one stator phase of the SRM, e.g. stator phase sA, can be expressed as follows:

$$t_e = \left. \frac{\partial W_c(\theta_r, i_{sA})}{\partial \theta_r} \right|_{i_{sA} = \text{const}}, \tag{5.1-4}$$

where W_c is the coenergy (area under magnetization curve, e.g. area under the appropriate curve in Fig. 5.3 in the region to $i_{sArated}$). Thus the electromagnetic torque, for constant stator current, is equal to the rate of change of the coenergy of the magnetic field with respect to the rotor angle. When the effects of magnetic saturation are neglected, $W_c = (1/2)L_{sA} i_{sA}^2$, and eqn (5.1-4) becomes

$$t_e = \frac{1}{2} i_{sA}^2 \frac{dL_{sA}}{d\theta_r}, \tag{5.1-5}$$

where L_{sA} is the stator self-inductance (shown in Fig. 5.2). It follows from eqn (5.1-5) that the electromagnetic torque is zero when the stator inductance is maximum (thus $dL_{sA}/d\theta_r = 0$). This corresponds to the aligned position. However, when the rotor is in a non-aligned position, the electromagnetic torque is not zero. These conclusions are in agreement with those made above. Since the torque is not zero in the unaligned positions, this torque causes the rotor to align with stator phase sA. The direction of the torque is always towards the nearest aligned position. Thus positive torque can only be produced if the rotor is between misaligned and aligned positions in the forward direction.

Thus it follows from physical considerations and also from eqn (5.1-5) that to obtain positive (motoring) torque, the stator phase current has to be switched on during the rising inductance region of L_{sA}. To obtain negative (braking) torque, it has to be switched on during the decreasing part of the corresponding stator inductance region. However, it should be noted that to obtain maximal motoring torque, the current in a stator phase should be switched on during the constant inductance region so it could build up before the region of increasing inductance starts. In a conventional SRM drive, the detection of these regions is made by using a position sensor. In 'sensorless' drives the position information is obtained without using a position sensor (e.g. by using an observer). It should also be noted that the current in a stator phase has to be switched off before the end of the increasing inductance region, since in this case the current can decay to zero, and no negative torque is produced. In the literature, the rotor position where a current is switched on is called the turn-on angle (θ_0), the rotor position where the current is turned off is called the turn-off angle (θ_c). The conduction angle (dwell angle) is thus defined as $\theta_d = \theta_c - \theta_0$ and the position where the current becomes zero is called the extinction angle, θ_q. It also follows from eqn (5.1-5) that the electromagnetic torque is independent of the direction of the stator-current flow, and thus a unipolar converter is sufficient to drive the SRM, in contrast to the bidirectional converters used with non-reluctance type of a.c. machines. This is in agreement with the statement above, where it was shown that only unipolar (unidirectional) stator currents are required in the SRM. However, it should be considered that eqn (5.1-5) has only been introduced so that a simple physical explanation could be given for the torque production, but the SRM is operated under highly non-linear saturated magnetic conditions and not in the unsaturated mode. If the SRM is operated in the magnetic linear range, then its torque (per unit volume) would be very low.

It should also be noted that when the rotor reaches the appropriate position, other stator phases will be excited. With $i_{sA}=0$, i_{sB} is switched on; this will cause the rotor to become aligned with stator phase sB, etc. The direction of rotation is controlled by the stator phase excitation sequence, e.g. sA, sB, sC, sA ... sequence gives clockwise rotation and sA, sC, sB, sA ... sequence yields counter-clockwise rotation. The speed of the SRM drive can be changed by varying the stator frequency. If the fundamental switching frequency is f, then $f = \omega_r N_r$, where ω_r is the angular rotor speed and N_r is the number of rotor poles. The non-uniform torque production causes torque ripples (and noise), but the torque ripples could be reduced by increasing N_r, but this leads to higher core losses. In a SRM drive, the torque ripples can be reduced by appropriate stator current or stator flux-linkage profiling (see Section 5.3) or by optimizing the machine geometry.

It can be concluded that, in contrast to other types of a.c. and d.c. motors, the SRM cannot run directly from a.c. or d.c. (the stator flux is not constant), but the stator flux must be established from zero at every step and the converter must supply unipolar current pulses, which are timed accurately by using information on the rotor position. In a conventional SRM, this information is obtained using a position sensor, but in position-sensorless SRM drives, it is possible to use different techniques to extract this information. The main techniques are:

- detection of the rotor position from the monitored stator currents;
- position estimation using an observer (Kalman filter, Luenberger observer, etc.);
- stator flux- and stator current-based position estimation;
- artificial-intelligence-based position estimation.

5.1.4 CONVERTER TOPOLOGIES; SRM WAVEFORMS

There are many types of converter configurations used with SRM drives. One simple power converter is shown in Fig. 5.4; this is an asymmetrical half-bridge converter. It can be seen that in Fig. 5.4 the two-quadrant converter is connected to the stator phase windings. This configuration allows the motor to be rated close to the maximum switch voltage, which is an important factor when the d.c. supply voltage or the available switch voltage is limited. The converter allows flexible modes of current control, and it can be seen that there are two switches (e.g. IGBT transistors) and two free-wheeling diodes in each phase.

The input voltage to the converter is the d.c. link voltage U_d (d.c. bus voltage) which can be obtained by using a rectifier. The free-wheeling diodes are used to obtain the reversed d.c. link voltage ($-U_d$) when the two switches are turned off. A single pulse operation is first discussed. At instant t_0, which corresponds to position θ_0 (as discussed above), two switches in a phase (e.g. stator phase sA) are turned on by using the position information which detects a rotor-pole edge (which is moving in the direction of the phase to be excited). Due to the switch-on, the current in this stator phase sA (i_{sA}) will increase and the stator flux linkage ψ_{sA} also increases. It follows from eqn (5.1-1) that if the stator ohmic voltage drop is neglected, and $+U_d$ is applied, then ψ_{sA} has to increase linearly. This is also shown in the third curve of Fig. 5.5, but the first curve shows the variation of the idealized stator inductance L_{sA} versus the rotor position, which is the same as shown in the top curve of Fig. 5.3 (idealized means that the effects of magnetic saturation and field fringing are neglected).

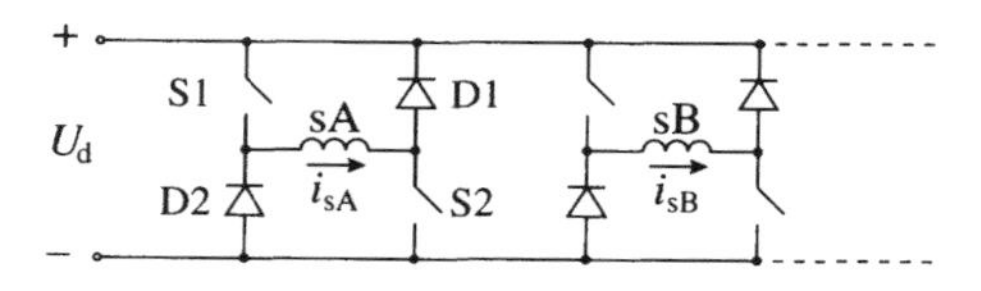

Fig. 5.4. Asymmetrical half-bridge converter.

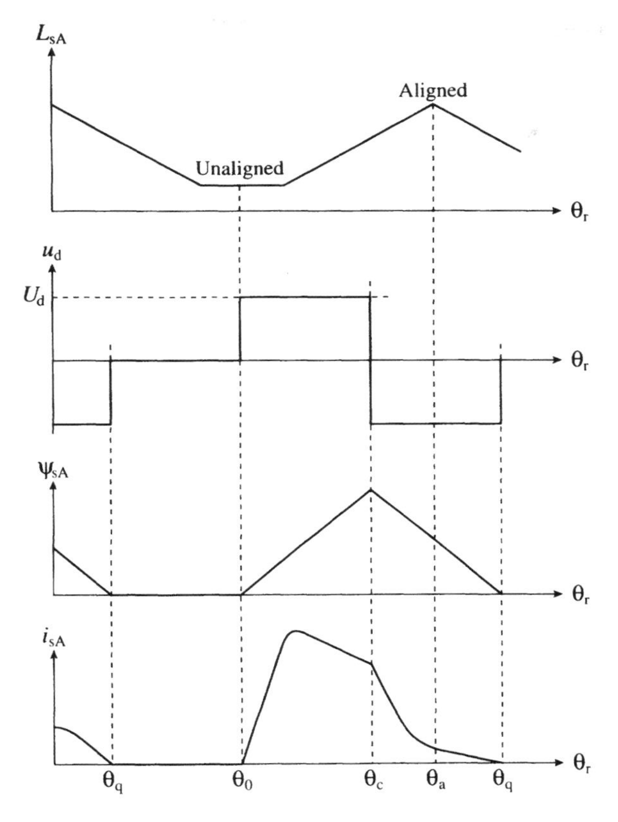

Fig. 5.5. SRM single-pulse waveforms (adapted from [Miller 1993]).

In the unaligned region, the increase of i_{sA} is linear, since the d.c. voltage U_d is applied, and the stator inductance is constant and the stator ohmic voltage drop is neglected. However, when the poles begin to overlap (beginning of intermediate stage), the stator inductance increases and current i_{sA} further increases, but this increase is non-linear and the current reaches its maximum value. It then decreases to a desired value when it is switched off at time t_c corresponding to rotor position θ_c. At the maximum value of the current, where $di_{sA}/dt=0$, the back e.m.f. is approximately equal to U_d (it is equal to U_d if the ohmic voltage drop is neglected), but then it increases, since the stator flux linkage is still increasing and the rotor speed is constant. This increase of the back e.m.f. then causes i_{sA} to decrease. At position θ_c the switches are turned off, and the stator i_{sA} current flows across the corresponding two free-wheeling diodes and the reversed voltage $-U_d$ is applied, as also shown in Fig. 5.5. Thus the stator flux linkage must decrease linearly if the ohmic voltage drop is neglected; see eqn (5.1-1). However, this position (θ_c) must be carefully selected so that maximum torque should be obtained. Thus at position θ_c winding sA is connected to

$-U_d$, and i_{sA} will further decrease at a higher rate. Physically this is due to the fact that, in this case, the stator flux linkage (ψ_{sA}) starts to decrease. It can be seen that in this case the back e.m.f. augments the negative d.c. voltage. At the aligned position θ_a, the rate of the decrease of i_{sA} is reduced, and i_{sA} eventually reaches zero at position θ_q as shown in Fig. 5.5. At θ_a the rate of change of i_{sA} is reduced since L_{sA} starts to decrease, $dL_{sA}/d\theta_r$ becomes negative, and thus the back e.m.f. reverses. It follows that the back e.m.f. acts against the negative d.c. link voltage, causing the rate of decrease of i_{sA} to be reduced. In this region the back e.m.f. may exceed the supply voltage and this could cause an increasing current. To avoid this, in the single-phase operation mode, θ_c is selected to be slightly smaller than the angle at the aligned position. When i_{sA} reaches zero, the corresponding free-wheeling diodes open, and the current remains zero until the switch pair closes again. It is important to note that the current is never negative and is driven to zero for a specific duration once per electrical cycle to avoid counter-productive torque.

It should be noted that at low speed, chopping operation is required to control the currents. Chopping is required because, at low speeds, the SRM does not have sufficient back e.m.f. to limit the currents. For this purpose it is also possible to use a hysteresis controller, which controls the switch ON and OFF states of the switches shown in Fig. 5.4. In this case the turn-off angle (θ_c) means that after θ_c the negative d.c. link voltage ($-U_d$) is applied to reduce the stator current and positive voltage (U_d) is not applied again during this cycle. If a hysteresis current controller is used to control the current (to limit it to a small hysteresis band around the reference value), then a typical current waveform is shown in Fig. 5.6.

In Fig. 5.6 from θ_0 (unaligned position) to θ_c ($\theta_c<\theta_a$, where θ_a is the aligned position), the supply voltage is switched between $+U_d$ and $-U_d$. The stator flux-linkage waveform can then be obtained by considering eqn (5.1-1) and this will give a rising, but oscillatory, stator flux linkage. As a consequence, between θ_0 and θ_c, the stator current i_{sA} first rises and then it oscillates, but the oscillations are limited to the required band, due to the use of the hysteresis current controller. At θ_c the supply voltage is reversed, thus the current i_{sA} starts to decrease slowly (at a large time constant) because this is near to the aligned position, where the stator inductance is maximum. Since the d.c. link voltage is fixed, the switching frequency decreases as the dynamic inductance of the stator winding increases (for the idealized machine, the dynamic inductance is equal to the stator idealized self-inductance, which increases from θ_0 to θ_a). The current oscillations will also result in torque ripples. The tail end of i_{sA} (from θ_c to θ_q) can produce a reversal of the torque as the rotor pulls away from the stator pole, carrying the stator flux.

5.1.5 PRESENT AND FUTURE RESEARCH WORK; SRM APPLICATIONS

At present, extensive research work is under progress by various manufacturers to obtain SRMs with improved noise and torque ripple characteristics. Work is also in progress on new power converters.

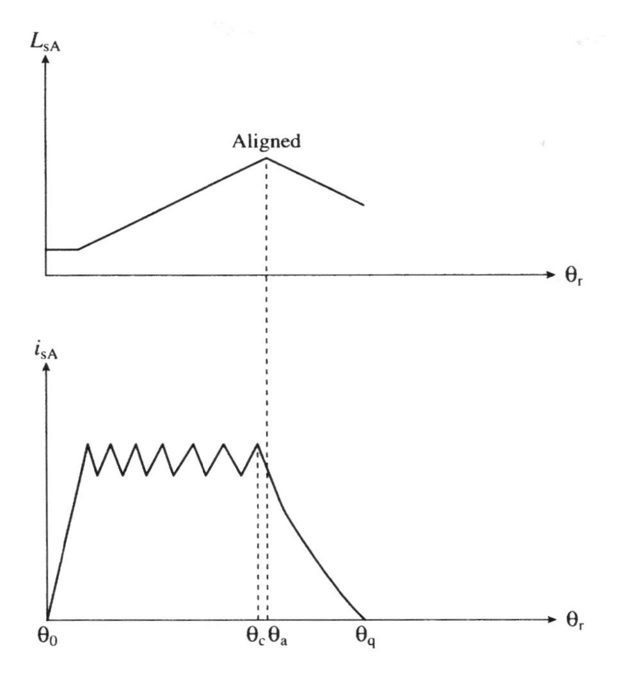

Fig. 5.6. SRM chopping waveforms.

Recently a new design of a three-phase SRM with $N_s = 12$, $N_r = 8$ has been reported [Mecrow 1992], where fully pitched stator windings are used. Such an arrangement allows a greater volume of the copper in the motor to be excited at any one time and this results in higher motor output.

Further work is also expected in the field of position-sensorless SRM drives and in the field of high-performance applications. At present no standard SRM drive has emerged (with a specific single motor and converter combination), although the greatest attention has been given to three-phase SRMs. This is mainly due to the fact that the three-phase SRM contains the lowest number of stator phase windings which allow the motor to start in either direction from any position without any extra starting measures taken.

Possible applications of SRMs include (when appropriate techniques are used to reduce the torque ripples):

- general-purpose industrial drives;
- application-specific drives: compressors, fans, pumps, centrifuges;
- domestic drives: food processors, washing machines, vacuum cleaners;
- electric vehicle applications;
- aircraft applications;
- servo-drives.

5.1.6 SRM DRIVE DESIGN FACTORS

Some important factors which influence the design of a switched reluctance motor drive for a specific application are: the running speed, torque requirements, supply voltage, cost, etc. These have a strong influence on the selection of the appropriate number of stator phase windings and these aspects are now briefly discussed.

Running speed The rotor speed determines directly the switching frequency of the currents, as shown above. The switching frequency is also related to the number of steps per revolution. Thus low-speed machines (e.g. where speed is less than 1000 r.p.m.) can be designed with a high number of poles but without high iron losses. However, a high number of poles does not necessarily mean a high number of stator phases, and e.g. the SRM with $N_s=8$ and $N_r=4$ is a two-phase machine, and the SRM with $N_s=12$ and $N_r=8$ is a three-phase machine. On the other hand, a high-speed machine (e.g. where the speed is higher than 5000 r.p.m.) should have a low number of poles and hence a low number of stator phase windings, since this ensures minimal iron losses.

Torque requirements The torque requirements determine the rating of the motor and also the number of stator phase windings. If very smooth torque is required, the required stator-phase number should be three or higher, since in this case appropriate current or flux-linkage control will yield relatively constant torque. However, in an application where only small starting torque is required, and where there can be a higher torque ripple, it is possible to use a single-phase SRM.

Starting requirements An SRM with three or a higher number of stator phases can start from any rotor position, in both the forward and reverse directions. However, an appropriate design of the rotor and/or the stator of a two-phase motor will ensure that it can start from any position (but it will have a preferred direction of rotation). If the SRM has a single-phase stator winding, the motor can only start if the rotor can be positioned in the correct position (relative to the stator) before the correct stator current is applied.

Supply voltage The available supply voltage has a strong influence on the power converter topology to be selected and also determines the number of turns in each stator winding (to obtain the desired torque–speed characteristics).

Costs An SRM with fewer stator phase windings requires a simpler power converter. For example, a single-phase SRM requires only a converter with one switching device (e.g. IGBT transistor).

5.2 Position-sensorless SRM drives

It has been discussed in Section 5.1.3 that the main techniques for obtaining information on the rotor position of an SRM are as follows:

1. Detection of the rotor position from the monitored stator currents;
2. Position using an observer (Kalman filter, Luenberger observer, etc.);
3. Stator flux- and stator current-based position estimation;
4. Artificial-intelligence-based position estimation (e.g. using ANN, fuzzy-neural network, etc.).

These techniques are now briefly discussed.

5.2.1 DETECTION OF ROTOR POSITION FROM THE MONITORED STATOR CURRENTS

It has been shown in Section 5.1.3 that the dynamic inductance is a function of both the rotor position and the stator current. Thus it is also possible to estimate the rotor position in the SRM (or other doubly-salient motors) by using the fact that the stator dynamic inductance varies with the rotor position and the stator currents. Hence for the purposes of the rotor position estimation using this technique, a stator current is monitored.

It follows from eqns (5.1-2) and (5.1-3) that, in general, the rate of change of a stator phase current can be expressed as

$$\frac{di_s}{dt}=\frac{u_s-R_si_s-\omega_r(\partial\psi_s/\partial\theta_r)}{L_d}, \tag{5.2-1}$$

where ω_r is the rotor speed, $\omega_r=d\theta_r/dt$, θ_r is the rotor angle (rotor position), and $L_d=\partial\psi_s/\partial i_s$ is a dynamic (incremental) inductance, which is a function of the rotor angle and a stator current, and therefore it is possible to estimate the rotor position from the monitored value of a stator current (its rate of change). It also follows from eqn (5.2-1) that this technique can also be used at standstill ($\omega_r=0$), where

$$\frac{di_s}{dt}=\frac{u_s-R_si_s}{L_d}. \tag{5.2-2}$$

In an SRM with chopper drives (see also Section 5.1), the variation of the dynamic inductance with the stator current can be neglected and thus it can be assumed that the dynamic inductance is only a function of the rotor position. This allows for a simple estimation of the rotor position from the chopped stator-current characteristics.

During chopping, a stator current oscillates around the required level at a rate determined by the dynamic inductance L_d of the stator phase winding at that particular current level. Since L_d is only dependent on θ_r, the rotor position can be obtained from the chopped stator current. It should be noted that for normal operation of an SRM, the phase current is either 'on' (and carries the rated current) or 'off' (where the current is zero). In both cases, the current oscillation during chopping occurs around a constant current level and thus the variation of the dynamic inductance with the current level is negligible.

The estimation of the rotor angle is now discussed for the first case, whereby the chopping characteristics (oscillating stator current i_s) are monitored in a stator phase winding which carries rated current. In this operation the stator current varies between a constant mean current I_s; thus it varies between $I_s+\Delta I_s/2$ and $I_s-\Delta I_s/2$. Thus during rise time t_{rise}, i_s increases by the value ΔI_s, and during the decay time, t_{decay}, it decreases by the same value. By assuming that the current oscillation is small relative to the mean current level I_s, thus $\Delta I_s \ll I_s$, the rise or decay of the current can be considered to be linear, and it follows from eqn (5.2-1) that

$$t_{rise}=\frac{L_d \Delta I_s}{u_s-R_s i_s-\omega_r(\partial\psi_s/\partial\theta_r)}. \tag{5.2-3}$$

A similar expression can be obtained for the decay time. It can be seen that the rise and decay times of the phase current are proportional to the dynamic inductance and also depend on the motional e.m.f. $[\omega_r(\partial\psi_s/\partial\theta_r]$. However continuous rotor-angle estimation is not possible during chopping at rated current, since it follows from the discussion presented in Section 5.1.3, in connection to Fig. 5.3, that there are several positions (e.g. four in Fig. 5.3) corresponding to each value of the dynamic inductance. Thus there is an ambiguity in θ_r corresponding to a given L_d, and hence rise time and decay time, and therefore it is not possible to have continuous rotor-angle estimation. However, it is possible to estimate a specific rotor angle. In particular, at aligned and misaligned positions, the rate of change of the flux linkage with the rotor angle is zero, and thus the motional e.m.f. is zero and hence

$$t_{rise}=\frac{L_d \Delta I_s}{u_s-R_s i_s}. \tag{5.2-4}$$

By monitoring the stator current i_s, I_s and ΔI_s can also be determined, and t_{rise} can also be determined. Thus by knowing the stator resistance and also the d.c. link voltage ($u_s=U_d$), the dynamic inductance can be determined by using eqn (5.2-4). However, by using the variation of L_d with θ_r, finally θ_r can be estimated. The variation of L_d with θ_r can be determined from the flux linkage–stator current characteristics (shown in Fig 5.3) for various rotor angles, by using eqn (5.1-3). This method can be used in a wide speed range.

5.2.2 POSITION ESTIMATION USING AN OBSERVER (EKF, ELO)

By utilizing measured values of the stator voltages and currents, it is possible to estimate the rotor position of the SRM by using an extended Kalman filter or an extended Luenberger observer. Such an approach also requires a suitable model of the SRM. For this purpose it is possible to establish a model of the SRM in which the inputs are the stator phase voltages ($u_{sA}, u_{sB}, \ldots$), the outputs are the stator phase currents ($i_{sA}, i_{sB}, \ldots$) and the state-variables are stator flux linkages ($\psi_{sA}, \psi_{sB}, \ldots$), the speed ($\omega_r$), and the rotor position (θ_r).

The voltage equation for one of the stator phases has been given by eqn (5.1-1), but when the flux linkages are the state variables, by using $i_{sA} = \psi_{sA}/L_{sA}$, this takes the following form:

$$u_{sA} = \frac{R_s \psi_{sA}}{L_{sA}} + \frac{d\psi_{sA}}{dt}, \tag{5.2-5}$$

where L_{sA} also depends on the rotor position. The other stator equations have a similar form, e.g for the nth stator phase of an SRM with N stator phases,

$$u_{sn} = \frac{R_s \psi_{sn}}{L_{sn}} + \frac{d\psi_{sn}}{dt} \qquad (n = 1, 2, \ldots, N) \tag{5.2-6}$$

where the stator inductance of stator phase sn is L_{sn} and it is a function of θ_r, and the flux linking stator phase sn is ψ_{sn}. It can also be seen from Fig 5.2 (which corresponds to a three-phase SRM) that the variation of the stator inductance in the second stator phase is similar to that in the first stator phase, but it is displaced by $2\pi/3$, e.g. $L_{sB}(\theta_r) = L_{sA}(\theta_r - 2\pi/3)$; and in general, for the nth stator phase, $L_{sn}(\theta_r) = L_{sA}[\theta_r - 2\pi(n-1)/N]$. Since the rotor speed is also an unknown, the equation of motion is also utilized in the model of the SRM; thus if the load torque is zero,

$$\frac{d\omega_r}{dt} = \frac{D}{J}\omega_r + \frac{t_e}{J}, \tag{5.2-7}$$

where J is the rotor inertia, D is the viscous damping, and t_e is the total instantaneous electromagnetic torque contributed by all the stator phases. The expression for the total electromagnetic torque in terms of ψ_{sn} and L_{sn} will now be derived.

The contribution of one stator phase (nth stator phase) to the total electromagnetic torque can be expressed as shown by eqn (5.1-5) as follows (this assumes a non-saturable machine):

$$t_{en} = \frac{1}{2} i_{sn}^2 \frac{dL_{sn}}{d\theta_r} \tag{5.2-8}$$

but by considering that $i_{sn} = \psi_{sn}/L_{sn}$ and $L_{sn} = L_{sn}(\theta_r)$,

$$t_{en} = \frac{1}{2}\psi_{sn}^2 \frac{d}{d\theta_r}[L_{sn}(\theta_r)]^{-1}. \tag{5.2-9}$$

Thus the total electromagnetic torque can be expressed in terms of the flux linkages and the position-dependent stator inductances as

$$t_e = \sum_n t_{en} = \frac{1}{2}\sum_n \psi_{sn}^2 \frac{d}{d\theta_r}[L_{sn}(\theta_r)]^{-1}. \tag{5.2-10}$$

When eqn (5.2-10) is substituted into eqn (5.2-7), the final form of the equation of motion is obtained as

$$\frac{\mathrm{d}\omega_r}{\mathrm{d}t} = \frac{D}{J}\omega_r + \frac{1}{2J}\sum_n \psi_{sn}^2 \frac{\mathrm{d}}{\mathrm{d}\theta_r}[L_{sn}(\theta_r)]^{-1}. \qquad (5.2\text{-}11)$$

Finally, the rotor speed can be obtained from the rate of change of the rotor position:

$$\omega_r = \frac{\mathrm{d}\theta_r}{\mathrm{d}t}. \qquad (5.2\text{-}12)$$

Equations (5.2-6), (5.2-11), and (5.2-12) correspond to a model of the SRM with the stator flux linkages, rotor speed, and rotor position as state variables and they are then utilized by a Kalman filter or Luenberger observer for the estimation of the states.

The algorithm for the extended Kalman filter has been described in great detail in Section 4.5.3.5.3, thus it will not be repeated here. Similarly, the algorithm for the extended Luenberger observer has been discussed in Section 4.5.3.5.4. The states (stator flux linkages, rotor speed, rotor position) can then be estimated by an observer which uses the monitored values of the stator voltages and currents. This technique can yield accurate results over a wide speed range, but is computationally intensive. However, it is expected that in the future various observer-based techniques will gain wider acceptance and these solutions will use high-speed digital signal processors. The cost of such DSPs is expected to fall sharply in the future.

5.2.3 STATOR FLUX AND STATOR CURRENT-BASED POSITION ESTIMATION

It is possible to estimate the rotor position from the estimated stator inductance L_{sn} (which is dependent on the rotor position). For example Fig. 5.3 shows the flux linkage versus stator current curves for various rotor positions. However, the stator inductance can be obtained by considering that it is the ratio of the appropriate stator flux linkage and stator current,

$$L_{sn}(\theta_r) = \frac{\psi_{sn}}{i_{sn}}. \qquad (5.2\text{-}13)$$

Thus it follows from Fig. 5.3 that the slopes, which are equal to the stator inductance at various rotor positions, increase from the unaligned position to the aligned position. The stator flux linkage can be obtained by integrating the stator voltage minus the ohmic drop (see eqn (5.2-6)); thus

$$\psi_{sn} = \int (u_{sn} - R_s i_{sn})\,\mathrm{d}t. \qquad (5.2\text{-}14)$$

This type of stator flux estimation has also been used for the induction machine in Section 4.5.3.1. However, this approach is particularly suitable for the SRM, since the flux returns to zero at each electrical cycle, and thus allows the integrator to be reset, preventing large error accumulation. This is a simple estimation technique. which can also be used over a very wide speed-range, including zero speed. However, it should be noted that it has been assumed that there is only current flow in one stator phase at a given time or, in other words, the mutual coupling between the stator phases has been neglected. In an SRM, where these mutual effects cannot be neglected, all the stator currents must be measured: these are $i_{s1}, i_{s2}, \ldots, i_{sn}$ and the stator flux linking the nth stator winding can be expressed as

$$\psi_{sn} = M_{n1} i_1 + M_{n2} i_2 + \cdots + L_{sn} i_n, \qquad (n = 1, 2, \ldots, N) \qquad (5.2\text{-}15)$$

where M_{n1}, M_{n2}, ... are the mutual inductances between stator phase n and stator phase 1; stator phase n and stator phase 2, etc. In this case ψ_{sn} can still be obtained by the integration given by eqn (5.2-14), but L_{sn} cannot be obtained in the simple manner described by eqn (5.2-13), since now eqn (5.2-15) holds. To obtain the rotor position, it is necessary to know accurately the values of the mutual inductances as well.

5.2.4 REDUCED ON-LINE INDUCTANCE ESTIMATORS

It is possible to obtain the rotor position in an SRM drive by using a simple inductance estimation scheme which utilizes the fact that the stator phase self-inductances are functions of the rotor position, and the mutual inductances are small between the stator phases. The simple technique discussed below is aimed at providing accurate rotor position at zero and low rotor speeds. For the purposes of the estimation of a stator self-inductance, appropriate test stator voltages are applied (by using the inverter which supplies the SRM) and the rate of change of the stator currents is measured. It is shown below that if the stator voltage change in the test stator phase s1 is measured (during the time $\Delta t = t_2 - t_1$), and it is $\Delta U_{s1} = U_{s1}(t_1) - U_{s2}(t_2)$, and the change of the rate of change of the stator current in the same stator phase is $\mathrm{d}\Delta i_{s1}/\mathrm{d}t$, then an approximate value of the stator inductance is obtained simply by evaluating $\Delta U_{s1}/(\mathrm{d}\Delta i_{s1}/\mathrm{d}t)$. It is also shown that when this approximate inductance is used, the relative error due to the simplifying assumptions is small and it is equal to the leakage constant (σ) of the motor. The accurate inductance value can be obtained from $[\Delta U_{s1}/(\mathrm{d}\Delta i_{s1}/\mathrm{d}t)](1+\sigma)$, but for small mutual inductance values σ is small; thus the approximation $\Delta U_{s1}/(\mathrm{d}\Delta i_{s1}/\mathrm{d}t)$ is valid.

For simplicity a two-phase stator winding is assumed with stator windings s1 and s2. Thus e.g. the stator voltage equation for stator phase s1 is

$$U_{s1}(t) = R_{s1} i_{s1}(t) + L_{s1} \frac{\mathrm{d}i_{s1}(t)}{\mathrm{d}t} + M \frac{\mathrm{d}i_{s2}(t)}{\mathrm{d}t} + \omega_r(t) i_{s1}(t) \frac{\mathrm{d}L_{s1}}{\mathrm{d}\theta_r} + \omega_r(t) i_{s2}(t) \frac{\mathrm{d}M}{\mathrm{d}\theta_r}, \qquad (5.2\text{-}16)$$

where R_{s1} and L_{s1} are the stator resistance and self-inductance respectively of stator winding s1, M is the mutual inductance between stator phase s1 and s2, $\omega_r = d\theta_r/dt$ is the rotor speed, and i_{s1} and i_{s2} are the stator currents in stator phases s1 and s2 respectively. Equation (5.2-16) can be used to obtain the voltage equations for stator phase s1 at instants t_1 and t_2 respectively (which are the instants when measurements are performed). If for the purposes of inductance estimation two different test voltage pulses $U_{s1}(t_1)$, $U_{s1}(t_2)$ are applied to stator phase s1 (which is the test stator phase), then by subtracting the stator voltage equation of stator winding s1 at instant t_2 from the stator voltage equation at instant t_1,

$$U_{s1}(t_1) - U_{s1}(t_2) = L_{s1}\left[\frac{di_{s1}(t_1)}{dt} - \frac{di_{s1}(t_2)}{dt}\right] + \left[R_{s1} + \frac{\omega_r dL_{s1}}{d\theta_r}\right][i_{s1}(t_1) - i_{s2}(t_2)]$$
$$+ M\left[\frac{di_{s2}(t_1)}{dt} - \frac{di_{s2}(t_2)}{dt}\right] + \omega_r \frac{dM}{d\theta_r}[i_{s2}(t_1) - i_{s2}(t_2)] \quad (5.2\text{-}17)$$

is obtained. Since the purpose is to derive a simple expression for L_{s1}, various assumptions are now considered. Thus by assuming

$$i_{s1}(t_1) = i_{s1}(t_2) \quad (5.2\text{-}18)$$

and

$$i_{s2}(t_1) = i_{s2}(t_2) \qquad \frac{di_{s2}(t_1)}{dt} = \frac{di_{s2}(t_2)}{dt}, \quad (5.2\text{-}19)$$

it follows from eqn (5.2-17) that a very simple expression can be obtained for the stator self-inductance:

$$L_{s1\,simplified} = \frac{U_{s1}(t_1) - U_{s2}(t_2)}{di_{s1}(t_1)/dt - di_{s1}(t_2)/dt} = \frac{\Delta U_{s1}}{d\Delta i_{s1}/dt}. \quad (5.2\text{-}20)$$

This expression agrees with that given above in the introduction to the present subsection. The conditions needed to fulfil the constraints given by eqns (5.2-18) and (5.2-19) will be discussed below. It should be noted that eqn (5.2-18) has resulted in the disappearance of the resistive voltage and back-e.m.f. terms in the expression for the inductance. To satisfy eqn (5.2-18), two different voltage pulses $U_{s1}(t_1)$ and $U_{s1}(t_2)$ must be applied to the test stator phase s1. If the inverter voltage drops are neglected, there are three possible voltage levels: U_d, $-U_d$, and 0 (where U_d is the d.c. link voltage). The choice depends on the required position accuracy, resolution, and speed of the A/D conversions, etc. It can be seen from eqn (5.2-20) that it does not contain the current of stator phase s2. However, if only eqn (5.2-18) is satisfied but eqn (5.2-19) is not satisfied, then it follows from eqn (5.2-17) and by also considering eqn (5.2-20) that the stator self-inductance in stator phase s1 can be obtained from a more complicated expression:

$$L_{s1} = L_{s1\,simplified} - \frac{M[di_{s2}(t_1)/dt - di_{s2}(t_2)/dt] + \omega_r(dM/d\theta_r)[i_{s2}(t_1) - i_{s2}(t_2)]}{di_{s1}(t_1)/dt - di_{s1}(t_2)/dt}.$$
$$(5.2\text{-}21)$$

This equation contains the derivatives of the stator currents in stator phase s2 $di_{s2}(t_1)/dt$, $di_{s2}(t_2)/dt$. Obviously if the mutual inductance M is very small then the term in eqn (5.2-21) which contains the derivatives di_{s2}/dt can be neglected and at zero speed, the last term becomes zero. It follows that in this case $L_{s1}=L_{s1\,\text{simplified}}$.

In general, eqn (5.2-21) can be simplified and the derivatives of the stator currents in stator phase s2 can be eliminated. For this purpose, the derivatives of the currents in stator phase s2 are expressed in terms of the derivatives of the stator currents in stator phase sl by considering the stator voltage equations of stator phase s2 at time instants t_1 and t_2 respectively. The stator voltage equation of stator phase s2 is similar to eqn (5.2-16), thus

$$U_{s2}(t)=R_{s2}i_{s2}(t)+\frac{L_{s2}di_{s2}(t)}{dt}+M\frac{di_{s1}(t)}{dt}+\omega_r(t)i_{s2}(t)\frac{dL_{s2}}{d\theta_r}+\omega_r(t)i_{s1}(t)\frac{dM}{d\theta_r}. \tag{5.2-22}$$

Equation (5.2-22) can be used to obtain the voltage equations for stator phase s2 at instants t_1 and t_2 respectively. Then first $U_{s2}(t_1)=U_{s2}(t_2)$ is assumed, which means the application of constant voltage to stator phase s2 (while in stator phase sl two different voltage pulses are applied at instants t_1 and t_2 respectively, as discussed above). Secondly, $i_{s2}(t_1)=i_{s2}(t_2)$ is also assumed, which is the first condition implied by eqn (5.2-19). This is satisfied by short-circuiting stator phase s2 during the tests in stator phase sl, hence $U_{s2}(t_1)=U_{s2}(t_2)=0$. Thus it follows by subtracting the voltage equation of stator phase s2 at instant t_2 from the voltage equation of stator phase s2 at instant t_1, and also considering the constraint imposed by eqn (5.2-18), that

$$\frac{d\Delta i_{s2}}{dt}=\frac{di_{s2}(t_1)}{dt}-\frac{di_{s2}(t_2)}{dt}=-\frac{M}{L_{s2}}\frac{d\Delta i_{s1}}{dt}=-\frac{M}{L_{s2}}\left[\frac{di_{s1}(t_1)}{dt}-\frac{di_{s1}(t_2)}{dt}\right], \tag{5.2-23}$$

where L_{s2} is the self inductance of stator phase s2. When eqn (5.2-23) is substituted into eqn (5.2-21), and by also considering that $i_{s2}(t_1)=i_{s2}(t_2)$,

$$L_{s1}=L_{s1\,\text{simplified}}+\frac{M^2}{L_{s2}}=L_{s1\,\text{simplified}}(1+\sigma) \tag{5.2-24}$$

is obtained. In eqn (5.2-24) σ is the leakage factor,

$$\sigma=\frac{M^2}{L_{s1\,\text{simplified}}L_{s2}}, \tag{5.2-25}$$

which is small, and for small values of M the leakage factor σ is negligible. Thus it can be seen that the expression for L_{s1} can be well approximated by the expression for $L_{s1\,\text{simplified}}$.

When eqn (5.2-20) is used for the estimation of the stator inductance (and this inductance is then used to obtain the rotor position), then it can be seen that accurate position estimation can be made at low and zero rotor speed, because at higher speeds the denominator of eqn (5.2-20) may become small. Experimental results have been reported in the literature (Kokornaczyk and Stiebler 1996), proving the viability of the approach discussed above.

5.2.5 ARTIFICIAL-INTELLIGENCE-BASED POSITION ESTIMATORS

It is expected that artificial-intelligence-based position estimators will also be developed in the future for SRM drives. These could use artificial neural networks (ANN), fuzzy estimators, or fuzzy-neural estimators. Such estimators are known to be universal function estimators and, due to the highly non-linear nature of the SRM, they are ideal candidates in position-sensorless drives (see also Chapter 7).

For example, when a neural network is used for position estimation, the following procedures could be followed. If a simplified technique is used, then it is assumed that the static flux linkage versus current characteristics of the switched reluctance machine are known (see Fig. 5.3), and the ANN is used as a non-linear function approximator to approximate these curves. The position estimator is shown in Fig. 5.7. within the broken lines. The ANN can be a multi-layer feedforward neural network and the training can be based on back-propagation (see also Chapter 7). The training uses the static characteristic data, which are the input data pairs of the flux linkages $\psi_s(k)$ and stator currents $i_s(k)$, and the corresponding outputs, which are the rotor positions $\theta_r(k)$. The trained ANN can then be used in the position-sensorless drive and, as shown in Fig. 5.7 in this case u_s and i_s are measured, and the stator flux linkages are obtained from the stator voltages and currents by integration [see eqn (5.1-1)]; thus by using $\psi_s = \int(u_s - R_s i_s)\,dt$. However, due to errors in measured stator voltages, stator currents, offsets, variation of stator resistance due to temperature, etc., flux estimation errors may arise which directly influence the accuracy of the estimated rotor position. These errors may be eliminated by using various techniques (see also below).

It is also possible to have another, more accurate ANN-based position estimator, where the static characteristics are not utilized, so the role of the neural postion estimator shown within the broken lines in Fig. 5.7 is not to approximate the static characteristics, but its role is to approximate the relationship between the output quantity (rotor position) and the inputs (stator current, stator flux)

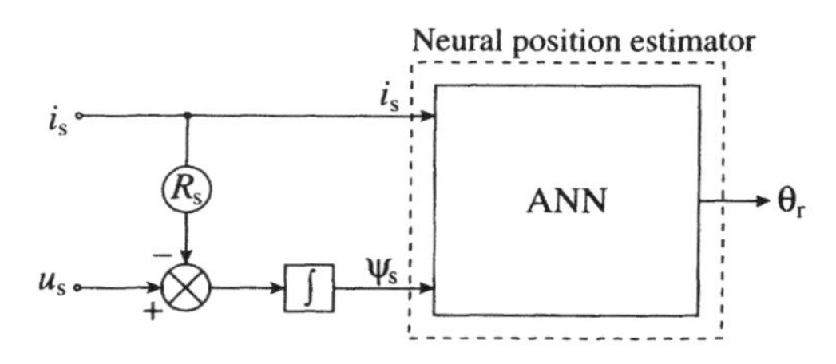

Fig. 5.7. ANN-based position estimator.

under real operating conditions. For this purpose in an initial test stage, the motor is run under its true operating conditions and the input and output data are measured in real time, and these are used to train the ANN (only during this training data-collection stage is a position sensor used).

Both schemes described above use an ANN where the flux linkage is also an input, and this has been obtained by using integration. However, it is possible to have another ANN-based position estimator scheme where the ANN directly estimates the position by also using the stator voltages and stator currents. The training of this ANN can be used on data which is obtained in an initial test phase (where there is also a position sensor).

When a multi-layer feedforward ANN is used, in the training stage the number of hidden layers and hidden neurons is obtained by trial and error (see also Chapter 7). However, it is also possible to replace the ANNs discussed above by fuzzy-neural estimators. By choosing a specific type of fuzzy-neural estimator, e.g. it can be Mamdani-type or Sugeno-type (see also Section 4.4.2 and Chapter 7.4), and also by choosing the number of fuzzy sets, the network structure is fixed. Thus the problem of finding the number of hidden layers and hidden neurons is eliminated. The training of the fuzzy-neural estimator can be based on the same data as used for the training of the ANN.

Finally it should be noted that it is also possible to use fuzzy position estimators. For example, instead of using the scheme shown in Fig. 5.7, the block surrounded by broken lines is replaced by the fuzzy estimator (which e.g. can be a Mamdani-type of fuzzy system) which uses linguistic rules which describe the relationship between the two inputs and the output:

$$\text{If } i_s(1) \text{ is } A \text{ and } \psi_s(2) \text{ is } B \text{ then } \theta_r(1) \text{ is } C$$

$$\text{If } i_s(2) \text{ is } D \text{ and } \psi_s(2) \text{ is } E, \text{ then } \theta_r(2) \text{ is } F,$$

where A, B, C, ... are fuzzy sets (see also Section 4.4.1). The fuzzy rules are then used to obtain the position, e.g. by the technique shown in Example 4.1 in Section 4.4. There are many possibilities for obtaining the rules, e.g. this can be achieved by clustering the data, using simulations, etc. However, it is also possible to have a fuzzy position estimator, which utilizes the monitored stator voltages and currents. When this second type of fuzzy estimator is used, problems related to integration, stator resistance variation, etc. can be eliminated.

5.3 Torque-controlled SRM drives

5.3.1 GENERAL INTRODUCTION; TORQUE RIPPLES

The instantaneous torque control of induction machines and permanent-magnet synchronous machines (PMSM) is now a standard feature; offered by many manufacturers. For this purpose both vector control and direct torque control implementations exist. In these drives it is possible to have separate control of the flux- and torque-producing stator currents. However, as discussed in

Section 5.1, a SRM is highly non-linear and it is doubly-salient, and this causes difficulties in both the analysis and control of the SRM. Such a doubly-salient, non-linear machine cannot be described by conventional space-vector theory which, however, has been extensively used for the development of various instantaneous control schemes in induction motor and permanent-magnet synchronous drives.

It follows from Section 5.1 that in a SRM the electromagnetic torque characteristics are highly non-linear and depend on both the stator currents and also on the rotor position. A simple SRM with constant stator-current excitation produces a torque which has high ripple content, and the torque is not linearly proportional to the current demand nor is it constant with respect to the rotor position. These are the main problems which so far have restricted the application of SRMs to low-performance drives, but it is possible to achieve high performance, as is discussed below. Furthermore, in the SRM, the stator flux linkage ψ_s, which also varies during the operation of the SRM drive, also depends on the stator current (i_s) and rotor angle (θ_s). Thus it follows that it is necessary to determine the $\psi_s(i_s, \theta_r)$ characteristics for the calculation of the static torque and dynamic behaviour of the SRM drive. It should be noted that it is also possible to use an artificial neural network (ANN) to predict the torque for given stator currents and rotor position or vice versa and it is possible to use an ANN to estimate the required stator currents for a given torque reference and position (or speed).

In an SRM, the torque ripples are also dependent on the speed of the motor. An SRM generally has three types of modes of operation: low-speed operation, where the stator currents are assumed to have square shapes, medium-speed operation, where the stator currents are chopped, and high speeds, where the stator current waveforms depend on many factors, including the back e.m.f. To be able to predict the torque ripples, both the current waveforms and static torque characteristics of the motor have to be considered. However, torque ripples have the greatest effects in low-speed operation: at higher speeds the inertia helps to reduce the speed ripples. In an SRM there are basically two techniques which can be used to reduce the torque ripples. In theory it is possible to optimize the geometry which leads to minimal torque ripples, but this is difficult since it also requires the prediction of the SRM performance for a specific operating condition. However, it is also possible to reduce the torque ripples by the appropriate shaping of the stator currents or stator flux linkages. A combination of the two techniques can also be used, but in the implementations discussed below, the optimization of the geometry will not be considered. Such an optimization could be performed by using finite-element analysis.

5.3.2 INSTANTANEOUS TORQUE CONTROL USING CURRENT OR FLUX CONTROLLERS

In an SRM, there are basically two possibilities for instantaneous torque control: it is possible directly to control the currents or it is possible directly to control the flux linkages. In both cases, by using the machine characteristics to determine the required excitations, it is possible to produce a high-performance SRM drive which gives constant torque with respect to rotor position and has linear torque/reference characteristics. In the two approaches to be discussed, the current and the flux linkage are chosen as reference values respectively. A simple control scheme will be described for these two cases: where torque control is achieved by using current controllers, or by using flux controllers.

It follows from the discussion presented in Section 5.1 that in an idealized SRM where saturation and field fringing effects are neglected, for constant stator phase current, in the non-aligned rotor positions the electromagnetic torque is constant [since i_s=const. and $dL_s/d\theta_r$ is linear, thus $t_e=(1/2)i_s^2 dL_s/d\theta_r$=const., and e.g. for motoring operation this constant is positive]. However, in a real machine, due to saturation effects and field fringing, the electromagnetic torque produced for a constant-phase current is a non-linear function of saturation. Thus to produce constant torque, profiled stator currents are required. The required current profiles are non-linear functions of the position and also the torque. As the aligned and misaligned positions are approached, the required stator current is greatly increased, as the torque/ampere falls towards zero. In the intermediate overlapping positions, the currents are almost constant. By considering a three-phase SRM (e g. with $N_s=12$ and $N_r=8$), during the positive-torque half cycle, only a single phase can contribute to positive torque. At low speed, the phase currents can be controlled so as not to produce any negative torque. As the torque/ampere falls rapidly at the aligned and unaligned positions, two phases must contribute to the torque production. It is the different positions over which torque production commutates from one phase to the next, and the torque profiles during this period, which distinguish the different modes of constant-torque production.

When designing a current reference waveform for constant torque operation, one of the main difficulties is to maintain constant torque over a wide speed-range by also considering the variation in the d.c. link voltage. This is due to the fact that the non-linear SRM characteristics make it difficult to take into account the rate of change of the flux linkage which is required for a given current reference. It follows from eqn (5.1-2) that in a stator phase, supplied by the d.c. voltage U_d,

$$U_d=R_s i_s+\frac{\partial\psi_s}{\partial i_s}\frac{di_s}{dt}+\frac{\partial\psi_s}{\partial\theta_r}\omega_r. \tag{5.3-1}$$

Thus the rate of change of the stator current can be obtained as

$$\frac{di_s}{dt}=\frac{U_d-R_s i_s-\omega_r(\partial\psi_s/\partial\theta_r)}{\partial\psi_s/\partial i_s} \tag{5.3-2}$$

or, since $\omega_r=d\theta_r/dt$ and hence $dt=d\theta_r/\omega_r$, it follows that

$$\frac{di_s}{d\theta_r}=\frac{U_d-R_s i_s-\omega_r(\partial\psi_s/\partial\theta_r)}{\omega_r(\partial\psi_s/\partial i_s)}. \tag{5.3-3}$$

If the ohmic voltage drop is neglected in eqn (5.3-3), then

$$\frac{di_s}{d\theta_r} = \frac{U_d - \omega_r(\partial\psi_s/\partial\theta_r)}{\omega_r(\partial\psi_s/\partial i_s)}. \tag{5.3-4}$$

Thus the maximum rate of change of the stator current is a non-linear function of the stator current and rotor position. Calculation of the phase current references for constant torque which the current controllers are able to track is made difficult by the non-linear variation of the current slew rate. However, if instead of current controllers, flux controllers are used, then it follows from eqn (5.1-1) that the maximum rate of change of the flux is

$$\frac{d\psi_s}{dt} = U_d - i_s R_s. \tag{5.3-5}$$

If the stator ohmic voltage drop is negligible (in comparison to the supply voltage, U_d), then it follows from eqn (5.3-5) by considering $dt = d\theta_r/\omega_r$, that

$$\frac{d\psi_s}{d\theta_r} = \frac{U_d}{\omega_r}. \tag{5.3-6}$$

Thus the maximum rate of the change of the stator flux linkage with respect to the rotor position can be simply determined from the maximum available supply voltage and speed. Equation (5.3-6) should be contrasted with eqn (5.3-4). It follows that by using a stator flux-linkage reference waveform (instead of a stator current reference waveform), it is much easier to take into account practical limitations (rotor speed and d.c. link voltage limitations) because if the stator ohmic drop is neglected, the rate of change of the flux linkage is an extremely simple expression: $d\psi_s/dt = U_d$. Thus the rate of change of the stator flux linkage is equal to the supply voltage. Furthermore, when a large change of the flux linkage is required, the converter output will tend to saturate, and the flux linkage will ramp at a linear rate (stator ohmic drop is neglected). These considerations lead to an implementation using stator flux-linkage control for constant torque operation. When such a scheme is used, the required stator flux-linkage reference waveform (as a function of the rotor position) will contain a series of linear ramps, each with a gradient not higher than the greatest actually achievable for a given speed and supply voltage. Neglecting the stator ohmic drop, each linear flux ramp can be realized by the application of a constant voltage to the stator phase. The gradient of each flux ramp determines the maximum speed at which it is possible to track the constant torque flux characteristic for a given d.c. link voltage. If the constant torque operation has to be extended to high-speed operation, the stator flux reference waveform has to be planned over the cycle as a whole, and cannot be performed on a point-by-point basis. By using linear flux-linkage ramps, it is possible to predict the operation later in the cycle, as the time taken to ramp the new stator flux-linkage value can be simply determined. The reference flux linkage versus rotor positions profile can be approximated by the curve shown in Fig. 5.8, and it can be seen that it contains several constant flux ramps. The curve is

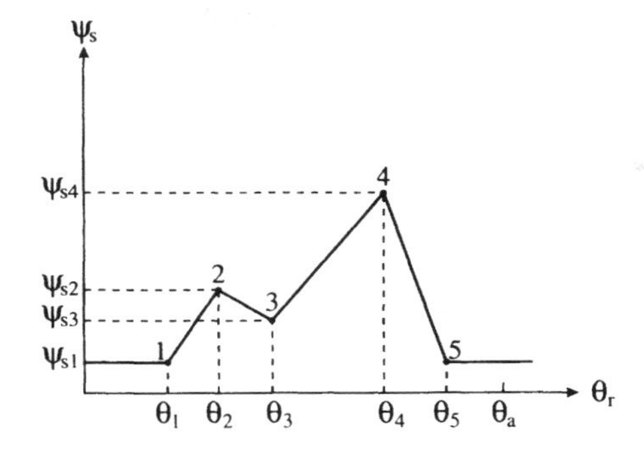

Fig. 5.8. Stator flux reference as a function of rotor position.

defined by five points, which have to be optimized (e.g. by simulations) to obtain minimum torque ripples. These points are connected by linear flux-linkage ramps.

In Fig. 5.8 the period from θ_1 to θ_5 can be considered as the phase conduction period, since ψ_{s1} (which is also shown in Fig. 5.8) is usually zero. The main torque-producing period is from θ_2 to θ_4, during which period the stator flux-linkage reference follows the machine constant-torque flux-linkage characteristic. Commutation of torque production between the stator phases occurs mainly over the periods θ_1 to θ_2 and also θ_4 to θ_5, but as the rotor speed increases, commutation extends beyond these intervals. When negative torque is produced for braking (or generating), the flux-linkage profile is reflected about the aligned position (corresponding to angle θ_a in Fig. 5.8). For constant torque operation, the flux-linkage points must be chosen as follows:

- the flux linkage at any point must not correspond to a phase current greater than the maximum value;
- the rate of change of the stator flux linkage between successive points must not exceed the maximum value possible for a given speed and d.c. link voltage (see eqn (5.3-4)).
- The sum of the torque produced by all the stator phases should minimize the rms torque ripple whilst maintaining the required average torque.

Figure 5.9 shows the block diagram for two types of instantaneous torque-controlled SRM drives. Figure 5.9(a) shows the SRM drive scheme when a current controller is used, and Fig. 5.9(b) corresponds to the SRM drive with a flux controller. In Fig. 5.9 (a) the input is the torque reference. This is used together with the rotor position and d.c. link voltage for the determination of the stator current references. For accurate instantaneous torque control, it is very important to have a very accurate estimation of the current references. The d.c. link voltage is obtained on the output of a three-phase uncontrolled rectifier. The position sensor can be an optical encoder. For practical implementation look-up tables can be used, which contain predetermined values obtained by using suitable

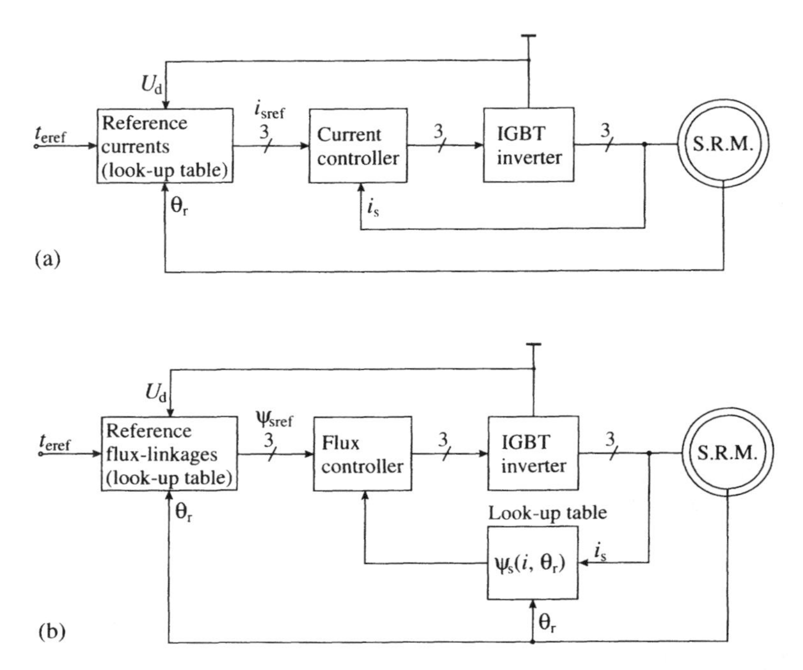

Fig. 5.9. Torque-controlled SRM drive. (a) SRM drive with a current controller; (b) SRM drive with a flux controller.

models. This is a drawback, since it would be more desirable to have a scheme where on-line calculation could be performed for any type of SRM, not just the SRM used in the specific application. However, this is problematic, due to the highly non-linear property of the SRM. The current references are then inputs to the current controller (e.g. hysteresis controller), which also requires the measured values of the stator currents. The currents can be measured by using Hall-effect current sensors. The current controller outputs the required switching signals for an IGBT converter. The SRM can be for example a three-phase machine, with $N_s = 12$ and $N_r = 8$. For the three-phase machine, the converter supplying the machine contains 6 legs, each of which contains an IGBT switch and a free-wheeling diode, and the converter then contains three asymmetrical half-bridges (as discussed earlier in Section 5.1.4, Fig 5.4).

In Fig. 5.9(b) a torque-controlled SRM drive scheme is shown, but instead of a current controller there is a flux linkage controller. This is a dead-beat controller, where the desired flux is reached in one sampling time without overshoot. The flux-linkage controller also requires the actual stator flux linkages, since the actual values are compared with the reference flux linkage values. The actual flux linkages can be obtained from the stator currents and the rotor position, but the infomation required can be stored in a look-up table. By using the flux-linkage error ($\Delta\psi_s$), the d.c. link voltage, and also the stator ohmic voltage drop correction (since the stator ohmic drop is neglected in the stator-flux reference calculation stage), the inverter switching times can be simply determined [by using $\Delta\psi_s = (U_d - R_s i_s)\Delta t$]. It should be noted that in both implementations, it is possible to use an artificial neural network or a fuzzy-neural network to obtain the required current or flux references but, as emphasized above, for precise torque control, very accurate estimates are required. It is a great advantage of such an approach that it does not require explicit use of any mathematical model of the SRM, which is highly non-linear. An ANN or fuzzy-neural network is ideal for approximating a highly non-linear function (see also Chapter 7).

Figure 5.10 shows a more detailed block diagram corresponding to the scheme of Fig. 5.9(a). It can be seen that the stator current estimation contains the switching angle estimation, where the angles θ_0 (turn-on angle) and θ_c (turn-off angle) have been defined in Section 5.1.3, and it can be seen that they also depend on the d.c. link voltage. Accurate torque control also requires accurate values of the turn-on and turn-off angles. In the motoring mode the torque is very sensitive to the turn-on angle and in the generating mode it is very sensitive to the turn-off angle. However, at higher speeds, the turn-on angle has less influence on the torque. In the chopping mode of operation, hysteresis current control is used. The torque-controlled SRM can operate in a wide speed-range, where the torque is constant (low-speed operation, chopping mode) and also where the torque is reduced (high-speed operation, constant power region).

The torque-controlled SRM is an ideal candidate for traction purposes or electric vehicles. Most of the real-time control tasks required in the two drive schemes described above can be conveniently performed by using a DSP, e.g. the Texas Instruments TMS320C30, TMS320C31, etc. It is possible to modify the two torque-control schemes presented above to position sensorless forms by utilizing concepts discussed in Section 5.2. It is expected that such control schemes will emerge in the future. Furthermore, it is also expected that integrated torque-controlled SRM drives will also be manufactured, where all the required controllers and power converter will be integrated in the machine.

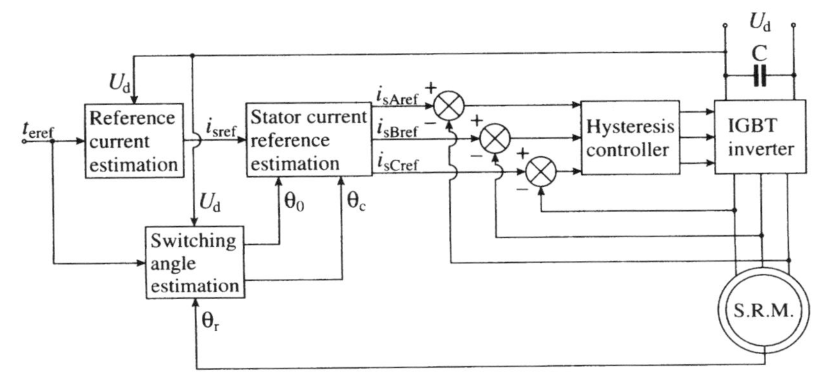

Fig. 5.10. Torque control of the SRM using current control.

Bibliography

Acarnley, P. P., French, C. D., and Al-Bahadly, I. H. (1995). Position estimation in switched reluctance drives. *EPE, Seville*, 3.765–3.779.

Acarnley, P. P., Hill, R. J., and Hooper, C. W. (1985). Detection of rotor position in stepping and switched motors by monitoring of current waveforms. *IEEE Transactions on Industrial Electronics* **IE-32**, 215–22.

Backhaus, K., Link, L., and Reinert, J. (1995). Investigations on a high-speed switched reluctance drive incorporating amorphous iron. *EPE, Seville*, 1.460–1.464.

Barrass, P. and Mecrow, B. C. (1996). Torque control of switched reluctance motor drives. *ICEM, Vigo*, pp. 254–9.

Bass, J. T., Ehsani, M., and Miller, T. J. E. (1987). Simplified electronics for torque control of sensorless switched reluctance motor drive. *IEEE Transactions on Ind. Electronics* **IE-34**, 234–9.

Bausch, H., Greif, K., Kanelis, K., and Nickel, A. (1996). Torque control of battery-supplied switched reluctance drives for electric vehicles. *ICEM, Vigo*, pp. 229–34.

Bolognani, S. and Zigliotto, M. (1996). Fuzzy logic control of a switched reluctance motor drive. *IEEE Transactions on Ind. Applications* **32**, 1063–8.

Elliott, C. R., Stephenson, J. M., and McCleland, M. L. (1995). Advances in switched reluctance drive system dynamic simulation. *EPE, Seville*, 3.622–3.626.

Finch, J. W., Harris, M. R., Musoke, A., and Metwally, H. M. B. (1984). Variable speed drives using multi tooth per pole switched reluctance motors. In *13th Incremental Motion Control Systems Society Symposium*, pp. 293–302.

Franceschini, G., Pirani, S., Rinaldi, M., and Tassoni, C. (1991). Spice assisted simulation of controlled electric drives: an application to switched reluctance drives. *IEEE Transactions on Ind. Applications* **27**, 1103–10.

Harris, M. R., Finch, J. W., Mallick, J. A., and Miller, T. J. (1986). A review of the integral horsepower switched reluctance drive. *IEEE Transactions on Ind. Applications* **IA-22**, 716–21.

Hopper, E. (1995). The development of switched reluctance motor applications. *PCIM Europe*, Vol. 7, pp. 236–41.

Husain, I. and Ehsani, M. (1994). Rotor position sensing in switched reluctance motor drives by measuring mutually induced voltages. *IEEE Transactions on Ind. Applications* **IA-30**, 665–72.

Jufer, M. and Crivii, M. (1995). Effects of phase current waveforms on the characteristics and acoustic noise of switched reluctance motors. *EPE, Seville*, 3.1003–3.1007.

Kjaer, P. C., Blaabjerg, F., Cossar, C., and Miller, T. J. E. (1995). Efficiency optimisation in current controlled variable-speed switched reluctance motor drives. *EPE, Seville*, 3.741–3.747.

Kokornaczyk, E. and Stiebler, M. (1996). A method of on-line inductance estimation applied to sensorless SR motors. *PEMC, Budapest*, 2/501–2/504.

Laurent, P., Gabsi, M., and Multon, B. (1993). Sensorless rotor position analysis using resonant method for switched reluctance motor. *IEEE IAS Meeting*, pp. 687–94.

Laurent, P., Multon, B., Hoang, E., and Gabsi, M. (1995). Sensorless position measurement based on PWM eddy current variation for switched reluctance machine. *EPE, Seville*, 3.787–3.792.

Lawrenson, P. J. (1965). Two-speed operation of salient-pole reluctance machines. *Proc. IEE. Pt. B* **117**, 545–51.

Lawrenson, P. J. and Gupta, S. K. (1967). Developments in the performance and theory of segmented-rotor reluctance machines. *Proc. IEE*, 645–53.

Lawrenson, P. J., Stephenson, J. M., Blenkinshop, P. T., Corda, J., and Fulton, N. N. (1980). Variable speed reluctance motors. *Proc. IEE* (*Pt. B*) **127**, 253–65.

Lawrenson, P. J. and Vamaruju, S. R. (1978). New 4/6 pole reluctance motor. *Electric Machines and Electromechanics*, 311–23.

Lumsdaine, A. and Lang. J. H. (1990). State observers for variable-reluctance motors. *IEEE Transactions on Industrial Electronics* **IE-37**, 133–42.

Lyons, J. P., MacMinn, S. R., and Preston, M. A. (1991) Flux/current methods for SRM rotor position estimation. *IEEE IAS Meeting*, pp. 482–7.

Mecrow, B. C. (1992). New winding configurations for doubly salient reluctance machines. *IEEE IAS Meeting*, 249–56.

Mecrow, B. C. (1993). Fully pitched switched reluctance and stepping motor arrangements. *Proc. IEE Pt. B* **140**, 61–70.

Miller, T. J. E. (1985). Converter volt-ampere requirements of the switched reluctance motor drive. *IEEE Transaction on Ind. Applics.* **IA-21**, 1136–44.

Miller, T. J. E. (1989). *Brushless permanent magnet and reluctance motor drives.* Clarendon Press, Oxford.

Miller, T. (1993). *Switched reluctance motors and their control.* Magna Physics Publishing and Clarendon Press, Oxford.

Miller, T. J. E. and McGilp, M. (1990). Nonlinear theory of the switched reluctance motor for rapid computer-aided design. *Proc. IEE Pt. B* **137**, 337–47.

Moallem, M. and Ong, C. M. (1991). Predicting the steady-state performance of a switched reluctance machine. *IEEE Transactions on Ind. Applications* **IA-27**, 1087–97.

Pollock, C. and Williams, B. W. (1990). Power convertor circuits for switched reluctance motors with minimum number of switches. *Proc. IEE Pt. B* **137**, 373–83.

Ray, W. F. and Davis, R. M. (1979). Inverter drive for doubly salient reluctance motor: its fundamental behaviour, linear analysis, and cost implications. *Electric Power Applications* **2**, 185–93.

Ray, W. F., Lawrenson, P. J., Davis, R. M., and Stephenson, N. N. (1985). High performance switched reluctance brushless drives. *IEEE Transactions on Ind. Applications* **IA-21**, 1769–76.

Stephenson, J. M. and Corda, J. (1979). Computation of torque and current in doubly-salient reluctance motors. *Proc. IEE Pt. B* **126**, 393–6.

Torrey, D. A. and Lang, J. H. (1990). Modeling a nonlinear variable reluctance drive. *Proc. IEE Pt. B* **137**, 314–26.

Vas, P. (1993). *Parameter estimation, condition monitoring and diagnosis of electrical machines.* Oxford University Press.

6 Effects of magnetic saturation

In this chapter the effects of main flux saturation the equations required for the various implementations of vector control (e.g. rotor-flux-oriented control, magnetizing-flux-oriented control, stator-flux-oriented control) are discussed for both smooth-air-gap and salient-pole machines. However, only the derivation and a brief description of the required equations is given, since these can be applied similarly to the corresponding equations derived in Chapters 3 and 4, which, however, are only valid under linear magnetic conditions. It will be shown that as a result of saturation, the linear equations are modified and to obtain the required decoupling between the flux- and torque-producing currents of the machine under consideration, it is these equations which have to be considered when the machine operates under saturated conditions.

As a result of saturation of the main flux paths, the magnetizing inductance and thus also the stator and rotor inductances (and the stator and rotor transient time constants) are not constant, but vary with saturation. The variation of the magnetizing inductance will be incorporated into the voltage equations. However, it will be shown that the voltage equations which are valid under saturated conditions differ from the equations which can be derived from their linear forms given in Chapters 3 and 4, where the currents have been used as state variables. The new voltage equations are not identical to the voltage equations which could be obtained by simply substituting the variable magnetizing inductance into the voltage equations which are valid under linear magnetic conditions, but they also contain new and/or modified terms.

The equations to be derived will be based on two-axis (or space-phasor) theory. Saturation of the main flux paths distorts the flux density distributions, which in the case of the linear theory (which neglects saturation) are sinusoidal in space (see Chapter 2) if sinusoidal m.m.f. distributions are assumed. The resulting space harmonics can, however, be neglected if sinusoidally distributed windings are assumed since only the fundamental sinusoidal component of a flux wave can produce flux linkages with sinusoidally distributed windings, and thus it is possible to use two-axis or space-phasor theories.

6.1 Vector control of smooth-air-gap machines by considering the effects of main flux saturation

The effects of main flux saturation are briefly discussed for vector-controlled induction machines with single-cage rotor. The various expressions for the electromagnetic torque will not be derived for the saturated machine, since formally they are the same as in the case of the unsaturated machine, since saturation does not introduce new terms into the expression for the electromagnetic torque, and the assumptions used in the previous section hold. Of course, as a result of saturation, the saturation-dependent machine parameters which are present in the different expressions for the torque, e.g. the magnetizing inductance, or the rotor self-inductance or the stator self-inductance, will be different to their unsaturated values and are variables which depend on the machine currents.

6.1.1 ROTOR-FLUX-ORIENTED CONTROL

In this section the stator and rotor voltage equations for an induction machine are described in the rotor-flux-oriented reference frame. However, because of saturation of the main flux paths, the magnetizing inductance and thus the stator and rotor self-inductances are not constant but vary with saturation. There are several ways to consider this variation with saturation and two approaches will be discussed. First the magnetizing inductance will be expressed as a function of the modulus of the rotor magnetizing-current space phasor and secondly it will be expressed in terms of the modulus of the magnetizing-current space phasor. It will be shown that when the second method is used, more complicated equations are obtained than in the first case, since in contrast to the rotor magnetizing-current space phasor, which is coaxial with the direct axis of the rotor-flux-oriented reference frame, the magnetizing flux-linkage space phasor is not coaxial with the direct axis of the rotor-flux-oriented reference frame and under saturated conditions, owing to the effects of cross-magnetization (cross-saturation), extra terms will be obtained in the equations. Furthermore, it will be shown that the equations derived in this section differ from the equations which can be derived from their linear forms given in Section 4.1 by simply replacing the constant magnetizing inductance by the variable magnetizing inductance.

In Section 4.1.1 the stator and rotor voltage equations in the rotor-flux-oriented reference frame were given for an induction machine, and the effects of magnetic saturation were neglected. In this section it is assumed that the effects of leakage flux saturation can be neglected but the effects of main flux saturation are incorporated in the analysis. All other assumptions are the same as those used in Section 4.1.

6.1.1.1 Rotor-flux-oriented control, expressing the magnetizing inductance as a function of the rotor magnetizing-current space phasor

In this section the stator and rotor voltage equations are obtained which contain the effects of main flux saturation and they are formulated in the reference frame fixed to the rotor-flux linkage space phasor. The method followed here is similar to that used in Section 4.1.1, where the space-phasor forms of the voltage equations formulated in the general reference frame were utilized directly for the derivation of the required forms of the voltage equations. The rotor-current space phasor is expressed in terms of the rotor magnetizing-current space phasor and therefore the resulting voltage equations will contain the modulus and space angle of the rotor magnetizing-current space phasor. These quantities are necessary to implement rotor-flux-oriented control.

Stator voltage equations: By considering eqns (2.1-148) and (2.1-150), which are the stator space-phasor voltage and stator flux-linkage space-phasor equations in the general reference frame, the following stator-voltage space-phasor equation is obtained in the reference frame fixed to the rotor flux-linkage space phasor, which rotates at the speed ω_{mr}, where it follows from eqn (2.1-188) that this speed is equal to the first time derivative of the space angle ρ_r of the rotor magnetizing-current space phasor with respect to the direct axis of the stationary reference frame,

$$\bar{u}_{s\psi r}=R_s\bar{\imath}_{s\psi r}+\frac{\mathrm{d}(L_s\bar{\imath}_{r\psi r})}{\mathrm{d}t}+\frac{\mathrm{d}(L_m\bar{\imath}_{r\psi r})}{\mathrm{d}t}+\mathrm{j}\omega_{mr}(L_s\bar{\imath}_{s\psi r}+L_m\bar{\imath}_{r\psi r}). \qquad (6.1\text{-}1)$$

L_s is the self-inductance of a stator winding and it is not constant since $L_s=L_{sl}+L_m$, where L_{sl} is the leakage inductance of a stator winding, which is assumed to be constant, but L_m varies with saturation. It is assumed that the magnetizing inductance is a non-linear function of the rotor magnetizing current $\bar{\imath}_{mr}(t)$, which in general varies with time, thus $L_m=L_m(|\bar{\imath}_{mr}|)$. It follows from eqn (4.1-28) that the magnetizing inductance can be defined as $L_m=\bar{\psi}_r'/\bar{\imath}_{mr}=|\bar{\psi}_r|/|\bar{\imath}_{mr}|$. However, it should be noted that in the theory of electrical machines it is usually defined as $L_m=|\bar{\psi}_m|/|\bar{\imath}_{mm}|$, which follows from eqn (2.1-184), where $|\bar{\psi}_r|$ and $|\bar{\imath}_{mm}|$ are the modulus of the magnetizing flux-linkage space phasor and magnetizing-current space phasor respectively ($|\bar{\imath}_{mm}|=|\bar{\imath}_m|$). It follows from eqn (2.1-184) that the magnetizing-current space phasor is equal to the sum of the stator-current and rotor-current space phasors.

Substitution of eqn (4.1-5) into eqn (6.1.1) yields the following stator voltage equation:

$$\bar{u}_{s\psi r}=R_s\bar{\imath}_{s\psi r}+\frac{\mathrm{d}(L_s'\bar{\imath}_{r\psi r})}{\mathrm{d}t}+\frac{\mathrm{d}(L_m^2|\bar{\imath}_{mr}|/L_r)}{\mathrm{d}t}+\mathrm{j}\omega_{mr}\left(L_s'\bar{\imath}_{s\psi r}+\frac{L_m^2|\bar{\imath}_{mr}|}{L_r}\right). \qquad (6.1\text{-}2)$$

L_s' is the stator transient inductance, $L_s'=(L_s-L_m^2/L_r)$, L_r is the self-inductance of the rotor and $L_r=L_{rl}+L_m$, where L_{rl} is the leakage inductance of the rotor, which is assumed to be constant. There are two derivatives in eqn (6.1-2) which contain L_m, which varies with saturation, these are

$$\frac{\mathrm{d}(L_s'\bar{\imath}_{s\psi r})}{\mathrm{d}t}=\frac{\mathrm{d}[(L_s-L_m^2/L_r)\bar{\imath}_{s\psi r}]}{\mathrm{d}t}$$

and

$$\frac{\mathrm{d}(L_m^2|\bar{\imath}_{mr}|/L_r)}{\mathrm{d}t}=\frac{\mathrm{d}}{\mathrm{d}t}\left[\frac{L_m^2\,|\bar{\imath}_{mr}|}{L_{rl}+L_m}\right].$$

By utilizing the chain differentiation rule, these can be expanded as

$$\frac{\mathrm{d}(L_s'\bar{\imath}_{s\psi r})}{\mathrm{d}t}=\frac{\mathrm{d}[(L_s-L_m^2/L_r)\bar{\imath}_{s\psi r}]}{\mathrm{d}t}=L_s\frac{\mathrm{d}\bar{\imath}_{s\psi r}}{\mathrm{d}t}-\frac{L_m^2}{L_r}\frac{\mathrm{d}\bar{\imath}_{s\psi r}}{\mathrm{d}t}$$
$$+\bar{\imath}_{s\psi r}\frac{\mathrm{d}(L_{sl}+L_m)}{\mathrm{d}t}-\bar{\imath}_{s\psi r}\frac{\mathrm{d}[L_m^2/(L_{rl}+L_m)]}{\mathrm{d}t}$$

$$=L_s'\frac{\mathrm{d}\bar{\imath}_{s\psi r}}{\mathrm{d}t}+\bar{\imath}_{s\psi r}\frac{\mathrm{d}L_m}{\mathrm{d}t}\left[1-\frac{L_m}{L_r^2}(2L_r-L_m)\right]$$

$$=L_s'\frac{\mathrm{d}\bar{\imath}_{s\psi r}}{\mathrm{d}t}+\bar{\imath}_{s\psi r}\left(\frac{L_{rl}}{L_r}\right)^2\frac{\mathrm{d}L_m}{\mathrm{d}t} \tag{6.1-3}$$

and

$$\frac{\mathrm{d}(L_m^2|\bar{\imath}_{mr}|/L_r)}{\mathrm{d}t}=\frac{L_m^2}{L_r}\frac{\mathrm{d}|\bar{\imath}_{mr}|}{\mathrm{d}t}+|\bar{\imath}_{mr}|\frac{\mathrm{d}[L_m^2/(L_{rl}+L_m)]}{\mathrm{d}t}$$

$$=\frac{L_m^2}{L_r}\frac{\mathrm{d}|\bar{\imath}_{mr}|}{\mathrm{d}t}+|\bar{\imath}_{mr}|\left(\frac{\mathrm{d}L_m}{\mathrm{d}t}\right)\left(1-\frac{L_{rl}^2}{L_r^2}\right). \tag{6.1-4}$$

In eqns (6.1-3) and (6.1-4) the term $\mathrm{d}L_m/\mathrm{d}t$ is also present. However, this can be expanded as

$$\frac{\mathrm{d}L_m}{\mathrm{d}t}=\frac{\mathrm{d}L_m}{\mathrm{d}|\bar{\imath}_{mr}|}\frac{\mathrm{d}|\bar{\imath}_{mr}|}{\mathrm{d}t}=\frac{L-L_m}{|\bar{\imath}_{mr}|}\frac{\mathrm{d}|\bar{\imath}_{mr}|}{\mathrm{d}t}, \tag{6.1-5}$$

where

$$L=\frac{\mathrm{d}\bar{\psi}_r'}{\mathrm{d}\bar{\imath}_{mr}}=\frac{\mathrm{d}|\bar{\psi}_r|}{\mathrm{d}|\bar{\imath}_{mr}|}. \tag{6.1-6}$$

In eqn (6.1-6) L is a dynamic (tangent-slope or incremental) inductance, and it is equal to the derivative of the modulus of the rotor flux-linkage space phasor with respect to the modulus of the rotor magnetizing-current space phasor. It should be noted that in contrast to this, L_m is the static (chord slope) inductance, and it follows from the considerations given above that it can also be expressed as

$$L_m=\frac{|\bar{\psi}_r|}{|\bar{\imath}_{mr}|}. \tag{6.1-7}$$

Figure 6.1 shows the variation of L and it should be noted that according to eqn (5.1-5) L and L_m are related by $L=L_m+|\bar{\imath}_{mr}|\mathrm{d}L_m/\mathrm{d}|\bar{\imath}_{mr}|$ and in Fig. 6.1 the variation $\mathrm{d}L_m/\mathrm{d}|\bar{\imath}_{mr}|$ is also shown.

It follows from eqns (6.1-5), (6.1-6), and (6.1-7) that under linear magnetic conditions $L=L_m$ and $\mathrm{d}L_m/\mathrm{d}t=0$ as expected, since in this case L_m=constant. Thus substitution of eqns (6.1-3) and (6.1-4) into eqn (6.1-2) gives the following equation, after the substitution of eqn (6.1-5),

$$\bar{u}_{s\psi r}=R_s\bar{\imath}_{s\psi r}+L_s'\frac{\mathrm{d}\bar{\imath}_{s\psi r}}{\mathrm{d}t}+\bar{\imath}_{s\psi r}\left(\frac{L_{rl}}{L_r}\right)^2\frac{L-L_m}{|\bar{\imath}_{mr}|}\frac{\mathrm{d}|\bar{\imath}_{mr}|}{\mathrm{d}t}$$

$$+\left[\frac{L_m^2}{L_r}+(L-L_m)\left(\frac{1-L_{rl}^2}{L_r^2}\right)\right]\frac{\mathrm{d}|\bar{\imath}_{mr}|}{\mathrm{d}t}+\mathrm{j}\omega_{mr}\left(\frac{L_s'\bar{\imath}_{s\psi r}+L_m^2|\bar{\imath}_{mr}|}{L_r}\right). \tag{6.1-8}$$

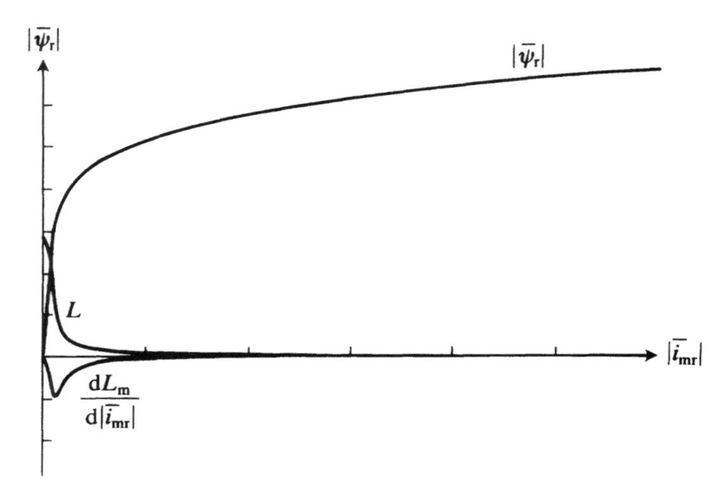

Fig. 6.1. Rotor magnetization curve. Variations of the dynamic inductance and the derivative of the magnetizing inductance.

When eqn (6.1-8) is divided by the stator resistance and is deliberately arranged into a form which is similar to eqn (4.1-6), finally the following space-phasor equation is obtained:

$$T'_s \frac{d\bar{i}_{s\psi r}}{dt} + \bar{i}_{s\psi r} = \frac{\bar{u}_{s\psi r}}{R_s} - \left\{ \frac{L - L_m}{R_s} \left[1 - \frac{L_{rl}^2}{L_r^2(1 - \bar{i}_{s\psi r}/|\bar{i}_{mr}|)} \right] + \frac{L_m^2}{L_r R_s} \right\} \frac{d|\bar{i}_{mr}|}{dt}$$

$$- j\omega_{mr} \left[T'_s \bar{i}_{s\psi r} + \frac{L_m^2}{L_r R_s |\bar{i}_{mr}|} \right]. \qquad (6.1\text{-}9)$$

It follows that as a result of the saturation of the main flux paths, the extra term

$$\Delta\bar{i}_s = \left\{ (L_m - L) \left[1 - \frac{L_{rl}^2}{L_r^2(1 - \bar{i}_{s\psi r}/|\bar{i}_{mr}|)} \right] \Big/ R_s \right\} \frac{d|\bar{i}_{mr}|}{dt} \qquad (6.1\text{-}10)$$

is also present in the voltage equation, but under linear magnetic conditions, when $L = L_m$, this term is zero. Furthermore, it follows from eqn (6.1-9) that the effects of changing saturation are also present in T'_s, since it depends on the changing L_m. By considering that $T_s - T'_s = (L_s - L'_s)/R_s = L_m^2/(L_r R_s)$, it can be seen that under linear magnetic conditions eqn (6.1-9) yields eqn (4.1-6).

By resolving eqn (6.1-9) into its real (x-axis) and imaginary (y-axis) components, the following two-axis voltage equations are obtained:

$$T'_s \frac{di_{sx}}{dt} + i_{sx} = \frac{u_{sx}}{R_s} + \omega_{mr} T'_s i_{sy} - (T_s - T'_s) \frac{d|\bar{i}_{mr}|}{dt} + \Delta i_{sx} \qquad (6.1\text{-}11)$$

$$T'_s \frac{di_{sy}}{dt} + i_{sy} = \frac{u_{sy}}{R_s} - \omega_{mr} T'_s i_{sx} - (T_s - T'_s)\omega_{mr}|\bar{i}_{mr}| + \Delta i_{sy} \qquad (6.1\text{-}12)$$

where

$$\Delta i_{sx} = \left\{ (L_m - L) \frac{[1 - (L_{rl}^2/L_r^2)(1 - i_{sx}/|\bar{i}_{mr}|)]}{R_s} \right\} \frac{d|\bar{i}_{mr}|}{dt} \tag{6.1-13}$$

$$\Delta i_{sy} = \frac{(L_m - L)(L_{rl}^2 i_{sy})}{L_r^2 R_s |\bar{i}_{mr}|} \frac{d|\bar{i}_{mr}|}{dt}. \tag{6.1-14}$$

Comparison with eqns (4.1-7) and (4.1-8) shows that under linear magnetic conditions eqns (6.1-11) and (6.1-12) agree with eqns (4.1-7) and (4.1-8). However, as a result of saturation, extra transformer terms arise along both the direct and quadrature axes and these result in the extra current terms Δi_{sx} and Δi_{sy} respectively. Both of these extra current components are proportional to the rate of change of the rotor magnetizing current. It should be noted that it is possible to obtain many other similar forms of these extra terms. For example, it follows from eqn (6.1-5) that it is possible to replace the factor $(L_m - L)$ by

$$-|\bar{i}_{mr}| \, dL_m/d|\bar{i}_{mr}| \quad \text{and} \quad \Delta i_{sy} = -\frac{L_{rl}^2}{R_s L_r^2} \frac{dL_m}{d|\bar{i}_{mr}|} \frac{d|\bar{i}_{mr}|}{dt} i_{sy}$$

$$= -\frac{L_{rl}^2}{R_s L_r^2} \frac{dL_m}{dt} i_{sy}.$$

If the rotor leakage inductance is assumed to be zero, then $\Delta i_{sy} = 0$, or in other words, there is no quadrature-axis extra induced e.m.f. due to saturation. This is an expected result, since in this case the rotor flux-linkage space phasor, which has only a direct-axis component in the rotor-flux-oriented reference frame, is equal to the magnetizing flux-linkage space phasor, which will also have a direct-axis component only, and therefore there cannot result a quadrature-axis transformer e.m.f. due to the change of this zero quadrature-axis magnetizing flux-linkage space phasor. Furthermore, if $L_{rl} = 0$, $\Delta i_{sx} \neq 0$, but $\Delta i_{sx} = [(L_m - L)/R_s] \, d|\bar{i}_{mr}|/dt$ or, in other words, in the direct-axis stator winding (x) there will be an extra direct-axis induced transformer e.m.f. due to saturation, which is equal to $-[(L_m - L)] \, d|\bar{i}_{mr}|/dt$. This is also an expected result, since when $L_{rl} = 0$, the direct-axis rotor flux-linkage component ($\psi_{rx} = |\bar{\psi}_r|$) is equal to the direct-axis magnetizing flux-linkage component $\psi_{mx} = |\bar{\psi}_r| = L_m |\bar{i}_{mr}|$ and its rate of change under linear magnetic conditions is equal to $d|\bar{\psi}_r|/dt = L_m d|\bar{i}_{mr}|/dt$. This latter voltage component plus the extra component voltage due to saturation gives the total voltage $L_m d|\bar{i}_{mr}|/dt - (L_m - L) d|\bar{i}_{mr}|/dt = L d|\bar{i}_{mr}|/dt$ and this is the correct result since when $L_{rl} = 0$, the total transformer e.m.f. in the x-axis stator winding must be equal to the rate of change of $|\bar{\psi}_r|$ which is indeed equal to $d|\bar{\psi}_r|/dt = (d|\psi_r|/d|\bar{i}_{mr}|) d|\bar{i}_{mr}|/dt = L d|\bar{i}_{mr}|/dt$.

By analogy it follows that if the stator voltage equations are formulated in the magnetizing-flux-oriented reference frame, where the direct-axis component of the magnetizing flux-linkage space phasor is equal to its modulus and its quadrature-axis component is zero, then under saturated conditions, in the direct-axis stator voltage equation there must be a transformer e.m.f., which is equal to the rate of change of the modulus of the magnetizing flux-linkage space phasor and in the quadrature-axis stator winding there will be no induced e.m.f. due to the rate of change of the quadrature-axis magnetizing flux-linkage component (the details of this will be discussed in Section 6.1.2). This is an advantage over the formulation of the voltage equations of a saturated machine in the stationary reference frame (fixed to the stator), where the magnetizing flux-linkage space phasor must contain both the direct-axis and also the quadrature-axis component, and therefore their rate of change is not zero and in general there will exist cross-magnetization coupling due to saturation (cross-saturation) between the windings in space-quadrature. The phenomenon of cross-saturation due to cross-magnetization in a smooth-air-gap machine is similar to the phenomenon of the demagnetizing effect of the cross-magnetizing armature reaction in a d.c. machine. Its physical existence can be proved experimentally and also a simple explanation of its existence can be given by utilizing space-phasor theory. This will be described below.

The relationship between the stationary-axis voltage components (u_{sD}, u_{sQ}), the stator current components (i_{sD}, i_{sQ}) and the corresponding voltage (u_{sx}, u_{sy}) and current (i_{sx}, i_{sy}) components can be obtained by utilizing the transformations defined in eqns (4.1-3) and (4.1-4) respectively. It follows from eqns (6.1-11)–(6.1-14) that with respect to the stator currents i_{sx} and i_{sy}, the induction machine behaves as a first-order time delay element, whose time constant is equal to the stator transient time constant of the machine and whose gain is equal to the inverse of the stator resistance. However, because of saturation, the stator transient time constant is not a real constant, since it depends on the magnetizing inductance, which changes with saturation. Furthermore, it can be seen that there is an unwanted coupling between the stator circuits in the two axes. For the purposes of rotor-flux-oriented control, it is the direct-axis stator current i_{sx} (rotor-flux-producing component) and the quadrature-axis stator current i_{sy} (torque-producing component) which must be independently controlled. It will now be assumed that the induction machine is supplied by impressed stator voltages. Since the voltage equations are coupled and the coupling term in u_{sx} also depends on i_{sy} and the coupling term in u_{sy} also depends on i_{sx}, u_{sx} and u_{sy} cannot be considered as decoupled control variables for the rotor flux and electromagnetic torque and the stator currents i_{sx} and i_{sy} can only be independently controlled (decoupled control) if the stator voltage equations are decoupled, and the stator current components i_{sx} and i_{sy} are indirectly controlled by controlling the terminal voltages of the induction machine. The required decoupling circuits can be obtained from the equations given above in the same way as described in Chapter 4.

If an induction machine with impressed stator currents is assumed, then the stator voltage equations can be omitted from the dynamic model of the drive and only the rotor equations have to be considered. These are obtained in the next section, where the effects of main-flux saturation are incorporated in the rotor voltage equations.

Rotor voltage equations: In this section the rotor voltage equations of the induction machine are obtained in the rotor-flux-oriented reference frame if the effects of main-flux saturation are incorporated in the equations. It follows from eqns (4.1-5), (4.1-22), and (4.1-23) that when the magnetizing inductance is changing due to saturation, the space-phasor form of the rotor voltage equations is

$$0=R_r L_m \frac{|\bar{i}_{mr}|-\bar{i}_{s\psi r}}{L_r}+\frac{d(L_m|\bar{i}_{mr}|)}{dt}+j(\omega_{mr}-\omega_r)L_m|\bar{i}_{mr}|, \qquad (6.1\text{-}15)$$

where $L_r=L_{rl}+L_m$ is the self-inductance of the rotor and changes with the saturation. In eqn (6.1-15) the derivative $d(L_m|\bar{i}_{mr}|)/dt$ is also present and this contains the non-linear element L_m. This can be expressed as

$$\frac{d(L_m|\bar{i}_{mr}|)}{dt}=\frac{d|\bar{\psi}_r|}{dt}=\frac{d|\bar{\psi}_r|}{d|\bar{i}_{mr}|}\frac{d|\bar{i}_{mr}|}{dt}=L\frac{d|\bar{i}_{mr}|}{dt}, \qquad (6.1\text{-}16)$$

where $|\bar{\psi}_r|$ is the modulus of the rotor flux-linkage space phasor, and it should be noted that of course the same result follows from the application of the chain differentiation rule

$$\begin{aligned}\frac{d(L_m|\bar{i}_{mr}|)}{dt}&=L_m\frac{d|\bar{i}_{mr}|}{dt}+|\bar{i}_{mr}|\frac{dL_m}{dt}\\&=L_m\frac{d|\bar{i}_{mr}|}{dt}+|\bar{i}_{mr}|\frac{dL_m}{d|\bar{i}_{mr}|}\frac{d|\bar{i}_{mr}|}{dt}\\&=\left(L_m+\frac{dL_m}{d|\bar{i}_{mr}|}\right)\frac{d|\bar{i}_{mr}|}{dt}=L\frac{d|\bar{i}_{mr}|}{dt}.\end{aligned}$$

This result is in agreement with that stated in the previous section. Substitution of eqn (6.1-16) into eqn (6.1-15) yields

$$T_r\left(\frac{L}{L_m}\right)\frac{d|\bar{i}_{mr}|}{dt}+|\bar{i}_{mr}|=\bar{i}_{s\psi r}-j(\omega_{mr}-\omega_r)T_r|\bar{i}_{mr}|, \qquad (6.1\text{-}17)$$

where $T_r= L_r/R_r=(L_{rl}+L_m)/R_r$ is the rotor time constant, which because of saturation is not constant. In this expression L_{rl} is the leakage inductance of the rotor, which has been assumed to be constant. Under linear magnetic conditions $L=L_m$, and eqn (6.1-17) yields eqn (4.1-24) as expected. On the left-hand side of eqn (6.1-17) the modified rotor time constant appears, and this can be put into the form

$$T_r^*=T_r(L/L_m)=(L_{rl}+L_m)(L/L_m)/R_r=T_{rl}L/L_m+L/R_r$$

where T_{rl} is the rotor leakage time constant, $T_{rl}=L_{rl}/R_r$. Thus the modified rotor time constant is equal to the sum of two rotor time constants. The first component is equal to the reduced value of the rotor leakage time constant and the reduction factor is equal to the ratio of the dynamic and static inductances, which ratio is equal to 1 under linear conditions. The second component is equal to the ratio of the dynamic inductance and the rotor resistance.

It follows from eqn (6.1-17) that it can be expressed in terms of T_r^* and T_r and therefore when eqn (6.1-17) is resolved into its real- (x-axis) and imaginary-axis (y-axis) components, the direct-axis rotor voltage equation will contain the modified rotor time constant T_r^* and the quadrature-axis rotor voltage equation will contain T_r. When these equations are deliberately put into a similar form to eqns (4.1-25) and (4.1-26), the following equations are obtained:

$$T_r^* \frac{d|\bar{\imath}_{mr}|}{dt} + |\bar{\imath}_{mr}| = i_{sx} \qquad (6.1\text{-}18)$$

$$\omega_{mr} = \omega_r + \frac{i_{sy}}{T_r|\bar{\imath}_{mr}|}. \qquad (6.1\text{-}19)$$

Under linear magnetic conditions these agree with eqns (4.1-24) and (4.1-25). However, it is important to note that owing to saturation both T_r and T_r^* are changing parameters and depend on the magnetizing inductance, which changes with $|\bar{\imath}_{mr}|$. By utilizing eqn (6.1-5), it is possible to put T_r^* into the form

$$T_r^* = L\left(\frac{T_{rl}}{L_m} + \frac{1}{R_r}\right) = \left(L_m + |\bar{\imath}_{mr}| \frac{dL_m}{d|\bar{\imath}_{mr}|}\right)\left(\frac{T_{rl}}{L_m} + \frac{1}{R_r}\right),$$

and the dependence on the magnetizing inductance is clearer.

On the basis of eqns (6.1-18) and (6.1-19) it is possible to obtain the flux models of the saturated induction machine in the rotor-oriented reference frame, which will be similar to those shown in Fig. 4.6(a) and Fig. 4.6(b), but will now contain both T_r^* and T_r.

6.1.1.2 Rotor-flux oriented control, expressing the magnetizing inductance as a function of the magnetizing-current space phasor

In this section the voltage equations required for the implementation of the various forms of rotor-flux-oriented control of induction machines are derived, but it is assumed that the magnetizing inductance is a non-linear function of the modulus of the magnetizing-current space phasor.

Rotor magnetizing current: It follows from eqn (4.1-5), by considering that the rotor self-inductance is the sum of the leakage and the magnetizing inductances ($L_r = L_{rl} + L_m$), that the modulus of the rotor magnetizing-current space phasor can be expressed as

$$|\bar{\imath}_{mr}| = \frac{L_r}{L_m}\bar{\imath}_{r\psi r} + \bar{\imath}_{s\psi r} = \frac{L_{rl}}{L_m}\bar{\imath}_{r\psi r} + (\bar{\imath}_{r\psi r} + \bar{\imath}_{s\psi r}) = \frac{L_{rl}}{L_m}\bar{\imath}_{r\psi r} + \bar{\imath}_{m\psi r}, \qquad (6.1\text{-}20)$$

where

$$|\bar{\imath}_{m\psi r}| = \bar{\imath}_{s\psi r} + \bar{\imath}_{r\psi r} = \bar{\imath}_s e^{-j\rho_r} + \bar{\imath}_r' e^{-j\rho_r} = (\bar{\imath}_s + \bar{\imath}_r')e^{-j\rho_r} = \bar{\imath}_m e^{-j\rho_r} = |\bar{\imath}_m| e^{j(\mu_m - \rho_r)} = |\bar{\imath}_m| e^{j\mu} \qquad (6.1\text{-}21)$$

is the rotor magnetizing-current space phasor in the rotor-flux-oriented reference frame. Furthermore, $\bar{\imath}_m$ is the space phasor of the magnetizing currents in the stationary reference frame, $|\bar{\imath}_m|$ is its modulus, and μ_m is the space angle of $\bar{\imath}_m$ with respect to the real axis of the stationary reference frame. However, the angle $\mu=\mu_m-\rho_r$ is the angle of $\bar{\imath}_m$ with respect to the real axis of the rotor-flux-oriented reference frame. In Fig. 6.2 $\bar{\imath}_m$ and $\bar{\imath}_{mr}$ are shown, together with the angles μ_m and μ.

Since in general the leakage inductance is not zero, $L_{rl}\neq 0$, the magnetizing-current space phasor is not coaxial with the rotor flux-linkage space phasor and thus $\bar{\imath}_m$ is not coaxial with $\bar{\imath}_{mr}$.

An alternative method to that used previously is now presented, where instead of expressing the stator and rotor voltage equations in terms of $|\bar{\imath}_{mr}|$, they are expressed in terms of $|\bar{\imath}_m|$. Thus by considering eqns (6.1-20) and (6.1-21), the relationship of $|\bar{\imath}_{mr}|$ and $|\bar{\imath}_m|$ is

$$|\bar{\imath}_{mr}|=(L_{rl}/L_m)\bar{\imath}_{r\psi r}+|\bar{\imath}_m|e^{j\mu}. \tag{6.1-22}$$

When this is differentiated with respect to time, under saturated conditions cross-coupling terms must arise in the direct- and quadrature-axis stator and rotor equations expressed in the rotor-flux-oriented reference frame, and the resulting equations will be more complicated than eqns (6.1-11), (6.1-12), (6.1-18) and (6.1-19). This will now be proved and the resulting equations will be derived. To obtain these equations, eqn (6.1-2) or (6.1-8) and eqn (6.1-15) or (6.1-17) can be used directly, together with eqn (6.1-22), but it is more convenient to use a different approach, as shown in the following section.

Stator voltage equations: It follows from eqns (6.1-1) and (6.1-21) that since the stator self-inductance (L_s) is equal to the sum of the stator leakage inductance

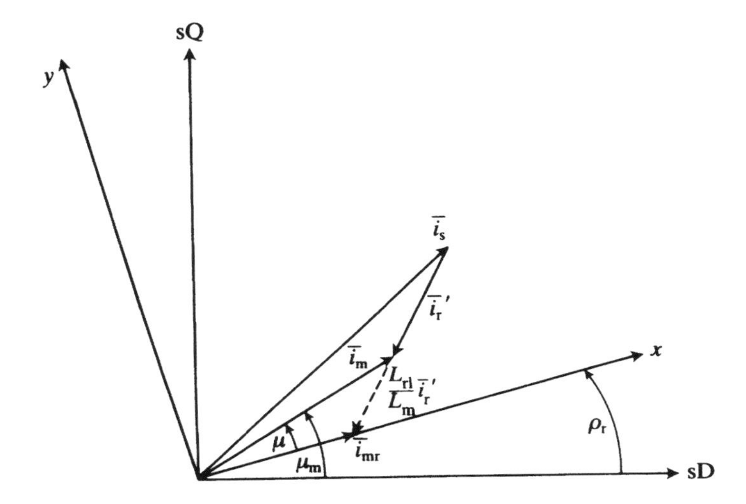

Fig. 6.2. Space phasor diagram of the rotor magnetizing current and the magnetizing current.

(L_{sl}) and the magnetizing inductance (L_m), and it is assumed that L_{sl} = constant and only the magnetizing inductance varies with saturation,

$$\bar{u}_{s\psi r} = R_s \bar{i}_{s\psi r} + L_{sl}\frac{d\bar{i}_{s\psi r}}{dt} + \frac{d(L_m|\bar{i}_m|e^{j\mu})}{dt} + j\omega_{mr}(L_{sl}\bar{i}_{s\psi r} + L_m|\bar{i}_m|e^{j\mu}). \quad (6.1\text{-}23)$$

In eqn (6.1-23) L_m can be obtained from the no-load curve of the induction machine, since $L_m = |\bar{\psi}_m|/|\bar{i}_m|$, where $|\bar{\psi}_m|$ and $|\bar{i}_m|$ are the moduli of the space phasors of the magnetizing flux-linkage space phasor and magnetizing-current space phasor respectively. It should be noted that $L_m = |\bar{\psi}_r|/|\bar{i}_{mr}|$ also holds [see eqn (6.1-7)], where $|\bar{\psi}_r|$ and $|\bar{i}_{mr}|$ are the moduli of the space phasors of the rotor flux linkage and rotor magnetizing current respectively. In eqn (6.1-23)

$$L_m|\bar{i}_m|e^{j\mu} = \bar{\psi}_{m\psi r} = L_m(\bar{i}_s + \bar{i}_r')e^{-j\rho_r} = \psi_{mx} + j\psi_{my} \quad (6.1\text{-}24)$$

is the magnetizing flux-linkage space phasor expressed in the rotor-flux-oriented reference frame. Since in general this is not coaxial with the rotor flux (it is only coaxial when $\mu = 0$, i.e. when $\mu_m = \rho_r$) in eqn (6.1-23), under saturated conditions, the derivative $d\bar{\psi}_{m\psi r}/dt$ will lead to cross-coupling between the direct- and quadrature-axis stator voltage equations. This will now be discussed in detail.

Resolution of eqn (6.1-23) into its real and imaginary axes components yields

$$u_{sx} = R_s i_{sx} + L_{sl}\frac{di_{sx}}{dt} + \frac{d\psi_{mx}}{dt} - \omega_{mr}(L_{sl}i_{sy} + \psi_{my}) \quad (6.1\text{-}25)$$

$$u_{sy} = R_s i_{sy} + L_{sl}\frac{di_{sy}}{dt} + \frac{d\psi_{my}}{dt} + \omega_{mr}(L_{sl}i_{sy} + \psi_{mx}), \quad (6.1\text{-}26)$$

where the direct- and quadrature-axis components of the magnetizing flux-linkage space phasor in the rotor-flux-oriented reference frame are

$$\psi_{mx} = L_m(i_{sx} + i_{rx}) = L_m i_{mx} \quad (6.1\text{-}27)$$

$$\psi_{my} = L_m(i_{sy} + i_{ry}) = L_m i_{my}. \quad (6.1\text{-}28)$$

In eqns (6.1-27) and (6.1-28) i_{mx} and i_{my} are respectively the direct-axis and imaginary-axis components of the magnetizing-current space phasor in the rotor-flux-oriented reference frame and in eqns (6.1-25) and (6.1-26) their first time derivatives ($d\psi_{mx}/dt$ and $d\psi_{my}/dt$) are present, which can be expanded as follows by utilizing the chain differentiation rule:

$$\frac{d\psi_{mx}}{dt} = \frac{d(L_m i_{mx})}{dt} = L_m\frac{di_{mx}}{dt} + i_{mx}\frac{dL_m}{dt} \quad (6.1\text{-}29)$$

$$\frac{d\psi_{my}}{dt} = \frac{d(L_m i_{my})}{dt} = L_m\frac{di_{my}}{dt} + i_{my}\frac{dL_m}{dt}. \quad (6.1\text{-}30)$$

In these equations the first time derivative of the magnetizing inductance is present, which owing to saturation is not zero. By the application of the chain differentiation rule and also by considering that $L_m = |\bar{\psi}_m|/|\bar{i}_{mr}|$,

$$\frac{dL_m}{dt} = \frac{dL_m}{d|\bar{i}_m|}\frac{d|\bar{i}_m|}{dt} = \frac{d}{d|\bar{i}_m|}(|\bar{\psi}_m|/|\bar{i}_m|)\frac{d|\bar{i}_m|}{dt} = \frac{L_d - L_m}{|\bar{i}_m|}\frac{d|\bar{i}_m|}{dt}, \qquad (6.1\text{-}31)$$

where $L_d = d|\bar{\psi}_m|/d|\bar{i}_m|$ is a dynamic (tangent slope) inductance, which under saturated conditions is not zero, but under linear magnetic conditions is equal to the (chord slope) magnetizing inductance L_m. In eqn (6.1-31) the first time-derivative of the modulus of the magnetizing-current space phasor is present, and since this depends on both the direct- and quadrature-axis magnetizing currents, which however contain the direct- and quadrature-axis stator and rotor currents respectively [see eqns (6.1-27) and (6.1-28)], it follows that dL_m/dt will contribute to the cross-saturation coupling terms. Mathematically, this can be proved by considering that

$$\frac{d|\bar{i}_m|}{dt} = \frac{d(i_{mx}^2 + i_{my}^2)^{1/2}}{dt} = \frac{i_{mx}}{|\bar{i}_m|}\frac{di_{mx}}{dt} + \frac{i_{my}}{|\bar{i}_m|}\frac{di_{my}}{dt} = \cos\mu\frac{di_{mx}}{dt} + \sin\mu\frac{di_{my}}{dt}, \qquad (6.1\text{-}32)$$

where μ has been defined above (see Fig. 6.2, $\mu = \mu_m - \rho_r$), i_{mx} and i_{my} have been defined in eqns (6.1-27) and (6.1-28) respectively and from eqns (6.1-24) and (6.1-27), (6.1-28) (or from Fig. 6.2)

$$i_{mx} = |\bar{i}_m|\cos\mu$$

$$i_{my} = |\bar{i}_m|\sin\mu. \qquad (6.1\text{-}33)$$

Thus by considering eqns (6.1-29), (6.1-30), (6.1-31), (6.1-32), and (6.1-33), finally

$$\frac{d\psi_{mx}}{dt} = L_{mx}\frac{di_{mx}}{dt} + L_{xy}\frac{di_{my}}{dt} \qquad (6.1\text{-}34)$$

$$\frac{d\psi_{my}}{dt} = L_{my}\frac{di_{my}}{dt} + L_{xy}\frac{di_{mx}}{dt}, \qquad (6.1\text{-}35)$$

where L_{mx} and L_{my} are the magnetizing inductances along the direct (x) and quadrature (y) axes respectively,

$$L_{mx} = L_d\cos^2\mu + L_m\sin^2\mu = L_d\cos^2(\mu_m - \rho_r) + L_m\sin^2(\mu_m - \rho_r) \qquad (6.1\text{-}36)$$

$$L_{my} = L_d\sin^2\mu + L_m\cos^2\mu = L_d\sin^2(\mu_m - \rho_r) + L_m\cos^2(\mu_m - \rho_r), \qquad (6.1\text{-}37)$$

and L_{xy} is the cross-coupling inductance between the x and y axes,

$$L_{xy} = \tfrac{1}{2}[(L_d - L_m)\sin(2\mu)] = \tfrac{1}{2}(L_d - L_m)\sin[2(\mu_n - \rho_r)]. \qquad (6.1\text{-}38)$$

Under linear magnetic conditions $L_d = L_m$ and $L_{mx} = L_{my} = L_m$ and $L_{xy} = 0$. However, owing to saturation of the main flux paths, in general the magnetizing inductances in the two axes are not equal, $L_{mx} \neq L_{my}$ and the cross-coupling inductance is not zero, $L_{xy}=0$. For the special selection of $\mu=\mu_m-\rho_r=0$, it follows that $L_{mx}=L_d$, $L_{my}=0$ and $L_{xy}=0$, which are expected results since in this case it follows from eqn (6.1-33) or Fig. 6.2 that $i_{mx}=|\bar{\imath}_m|$, $i_{my}=0$ and thus $\psi_{mx}=L_m i_{mx}=L_m|\bar{\imath}_m|$, and

$$\frac{d\psi_{mx}}{dt}=\frac{d\psi_{mx}}{d|\bar{\imath}_m|}\frac{d|\bar{\imath}_m|}{dt}=L_d\frac{d\bar{\imath}_{mx}}{dt}$$

and $\psi_{my}=0$; thus $d\psi_{my}/dt=0$. The same result also follows from eqns (6.1-33)–(6.1-38). Substitution of eqns (6.1-34), (6.1-35), (6.1-27), and (6.1-28) into eqns (6.1-25) and (6.1-26) yields the following stator voltage equations:

$$u_{sx}=R_s i_{sx}+L_{sx}\frac{di_{sx}}{dt}+L_{mx}\frac{di_{rx}}{dt}+L_{xy}\left(\frac{di_{sy}}{dt}+\frac{di_{ry}}{dt}\right)$$
$$-\omega_{mr}[L_{sl}i_{sy}+L_m(i_{sy}+i_{ry})] \qquad (6.1\text{-}39)$$

$$u_{sy}=R_s i_{sy}+L_{sy}\frac{di_{sy}}{dt}+L_{my}\frac{di_{ry}}{dt}+L_{xy}\left(\frac{di_{sx}}{dt}+\frac{di_{rx}}{dt}\right)$$
$$+\omega_{mr}[L_{sl}i_{sx}+L_m(i_{sx}+i_{rx})], \qquad (6.1\text{-}40)$$

where L_{mx}, L_{my}, and L_{xy} have been defined in eqns (6.1-36), (6.1-37), and (6.1-38) respectively, and

$$L_{sx}=L_{sl}+L_{mx}$$
$$L_{sy}=L_{sl}+L_{my} \qquad (6.1\text{-}41)$$

are the self-inductances of the stator winding along the direct- and quadrature-axes of the rotor-flux-oriented reference frame and, as a result of saturation, anisotropy (asymmetry) exists, $L_{sx}\neq L_{sy}$. Under linear magnetic conditions, however, $L_{sx}=L_{sy}=L_{sl}+L_m$, as expected. It follows from eqns (6.1-39) and (6.1-40) that even when the speed of the rotor-flux-oriented reference frame is zero ($\omega_{mr}=0$), the two stator voltage equations are coupled as a result of saturation.

Under linear magnetic conditions, $L_{sx}=L_{sy}=L_{sl}+L_m$ and it follows from the resolution of eqn (6.1-20) into its direct- and quadrature-axis components that

$$i_{rx}=(|\bar{\imath}_{mr}|-i_{sx})\frac{L_m}{L_r} \qquad (6.1\text{-}42)$$

$$i_{ry}=-i_{sy}\frac{L_m}{L_r}, \qquad (6.1\text{-}43)$$

and when eqns (6.1-42) and (6.1-43) are substituted into eqns (6.1-39) and (6.1-40), the stator equations derived in Section 4.1, eqns (4.1-7) and (4.1-8), are obtained. Comparison of the equations which are valid under linear and saturated conditions show that with saturation extra and modified terms are present. If $\rho_r=0$, the rotor-flux-oriented reference frame coincides with the stationary reference frame (see Fig. 6.2) and by considering $\omega_{mr}=d\rho_r/dt=0$, the voltage equations, in the stationary reference frame are obtained. Thus it follows from eqns (6.1-39), (6.1-40) and (6.1-36), (6.1-37), (6.1-38) that

$$u_{sD}=R_s i_{sD}+L_{sl}\frac{di_{sD}}{dt}+\frac{d\psi_{mD}}{dt} \tag{6.1-44}$$

$$u_{sQ}=R_s i_{sQ}+L_{sl}\frac{di_{sQ}}{dt}+\frac{d\psi_{mQ}}{dt}, \tag{6.1-45}$$

where ψ_{mD} and ψ_{mQ} are respectively the direct- and quadrature-axis components of the magnetizing flux-linkage space phasor in the stationary reference frame,

$$\psi_{mD}=L_m i_{mD}=L_m(i_{sD}+i_{rd}) \tag{6.1-46}$$

$$\psi_{mQ}=L_m i_{mQ}=L_m(i_{sQ}+i_{rq}) \tag{6.1-47}$$

$$\frac{d\psi_{mD}}{dt}=L_{mD}\frac{di_{mD}}{dt}+L_{DQ}\frac{di_{mQ}}{dt} \tag{6.1-48}$$

$$\frac{d\psi_{mQ}}{dt}=L_{mQ}\frac{di_{mQ}}{dt}+L_{DQ}\frac{di_{mD}}{dt}, \tag{6.1-49}$$

where i_{mD}, i_{sD}, i_{rd} and i_{mQ}, i_{sQ}, i_{rq} are the direct- and quadrature-axis components of the magnetizing current, stator-current, and rotor-current space phasors respectively in the stationary reference frame. L_{mD} and L_{mQ} are the magnetizing inductances along the direct and quadrature axes of the stationary reference frame, and L_{DQ} is the cross-coupling inductance between the stator windings sD and sQ, which are in space quadrature, but the same cross-coupling exists between all the windings of the stationary-axis model which are in space quadrature, e.g. between windings sD and rq, sQ and rd, etc.:

$$L_{mD}=L_d\cos^2\mu_m+L_m\sin^2\mu_m \tag{6.1-50}$$

$$L_{mQ}=L_d\sin^2\mu_m+L_m\cos^2\mu_m \tag{6.1-51}$$

$$L_{DQ}=(L_d-L_m)\sin\mu_m\cos\mu_m. \tag{6.1-52}$$

Eqns (6.1-44)–(6.1-52) can be put into a more compact form:

$$u_{sD}=R_s i_{sD}+L_{sD}\frac{di_{sD}}{dt}+L_{mD}\frac{di_{rd}}{dt}+L_{DQ}\left(\frac{di_{sQ}}{dt}+\frac{di_{rq}}{dt}\right)$$
$$-\omega_r[L_{sl}i_{sQ}+L_m(i_{sQ}+i_{rq})] \tag{6.1-53}$$

$$u_{sQ}=R_s i_{sQ}+L_{sQ}\frac{di_{sQ}}{dt}+L_{mQ}\frac{di_{rQ}}{dt}+L_{DQ}\left(\frac{di_{sD}}{dt}+\frac{di_{rd}}{dt}\right)$$
$$+\omega_r[L_{sl}i_{sD}+L_m(i_{sD}+i_{rd})], \tag{6.1-54}$$

where $L_{sD}=L_{sl}+L_{mD}$ and $L_{sQ}=L_{sl}+L_{mQ}$ are the self-inductances of the stator winding along the direct and quadrature axes of the stationary reference frame and, in general, as a result of saturation $L_{sD}\neq L_{sQ}$.

A very simple physical explanation can be given for the existence of the phenomenon of cross-saturation (cross-magnetization coupling due to saturation of the main flux paths). This is now described. For simplicity a symmetrical quadrature-phase machine is assumed and it is also assumed that the direct- and quadrature-axis m.m.f. distributions around the periphery of the machine (spatial distributions) are sinusoidal, in agreement with the discussion given above (see introduction to Chapter 6).

In Fig. 6.3(a) $\bar{f}_D = F_D$ is the m.m.f. space phasor corresponding to the case where there is only direct-axis excitation. By using the magnetization curve of the machine, which is assumed to be non-linear, the space phasor of the flux-linkage space phasor $\bar{\psi}_m$ is obtained. However, when both the direct-axis and quadrature-axis windings are excited, the quadrature-axis m.m.f. component F_Q is also present and the resultant m.m.f. space phasor will take the form $\bar{f} = F_D + jF_Q = |\bar{f}|e^{j\mu_m}$, where μ_m is the space angle of the m.m.f. space phasor with respect to the direct axis of the stationary reference frame as shown in Fig. 6.3(a). By using the same magnetization curve as before, the corresponding space phasor of the magnetizing flux linkages $\bar{\psi}'_m$ is obtained, $\bar{\psi}'_m = |\bar{\psi}'_m|e^{j\mu_m}$, and this is also shown in Fig. 6.3(a). It follows from Fig. 6.3(a) that the flux linkage along the direct axis has changed from $|\bar{\psi}_m|$ to $|\bar{\psi}'_m|\cos\mu_m$ and the reduction of the magnetizing flux linkage is $\Delta\bar{\psi}_m = |\bar{\psi}_m| - |\bar{\psi}'_m|\cos\mu_m$.

However, under linear magnetic conditions (by using the linear part of the previously used magnetizing curve), the space-phasor diagram shown in Fig. 6.3(b) is obtained. If the slope of the linear section of the magnetization curve is $\tan\alpha$ and only the direct axis is excited, the flux linkage along the direct axis is $\psi_{mD} = |\bar{f}|\cos\mu_m \tan\alpha$. However, if the quadrature-axis winding is also excited, the space phasor of the resultant flux linkages will be $\bar{\psi}'_m$, and its direct-axis component is equal to $\psi_{mD} = |\bar{\psi}'_m|\cos\mu_m$. As a result of linear magnetic conditions $|\bar{\psi}'_m| = |\bar{f}|\tan\alpha$, and thus $\psi'_{mD} = |\bar{f}|\cos\mu_m \tan\alpha$ and it follows that ψ'_{mD} is equal to ψ_{mD}, which is an expected result. Thus $\Delta\psi_m = \psi'_{mD} - \psi_{mD} = 0$. It is important to note that under linear magnetic conditions the magnetizing flux-linkage components correspond exactly (in per-unit terms) to the m.m.f. components, and

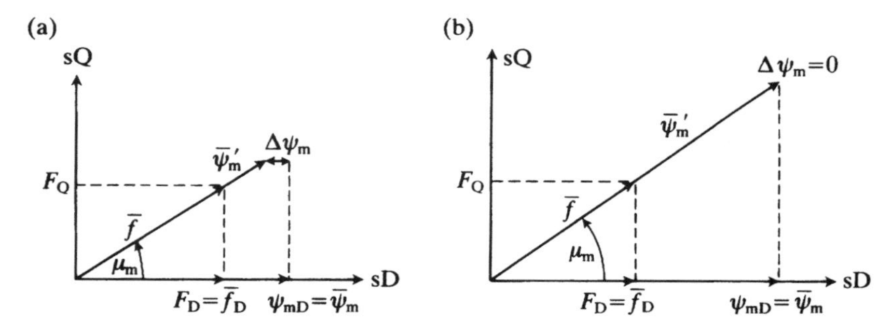

Fig. 6.3. Space-phasor diagram to prove the existence of cross-saturation. (a) Non-linear case, (b) linear case.

angular displacement and a change of the resultant magnetizing flux-linkage space phasor does not lead to an induced e.m.f. in the direct-axis winding, since the direct-axis component of the magnetizing flux-linkage space phasor is unchanged. However, under saturated magnetic conditions, the change in the magnitude of the resultant m.m.f. space phasor leads to a non-proportional change in the direct-axis magnetizing flux-linkage component and an induced e.m.f. in the direct-axis winding. It should be emphasized that an angular displacement of the magnetizing flux-linkage space phasor (relative to a stationary winding) is not in itself sufficient to induce voltage, nor is angular displacement with change in magnitude. The change of the quadrature-axis magnetizing flux linkage $\Delta\psi_m$ depends on the change of the quadrature-axis magnetizing current (Δi_{mQ}) and thus $\Delta\psi_m = L_{DQ}\Delta i_{mQ}$, where L_{DQ} is a cross-coupling inductance between the direct and quadrature axes which is zero under linear magnetic conditions.

It follows from Fig. 6.3 that if $\mu_m = 0$, there cannot be any cross-saturation coupling ($L_{DQ} = 0$) between the windings in space quadrature, since this corresponds to the case when only the direct-axis winding is excited and therefore there is no change in the resultant flux-density distribution. Furthermore, if $\mu_m = \pi/2$, this corresponds to the case when there is only quadrature-axis excitation, and therefore the resultant flux-density distribution is equal to the quadrature-axis flux-density distribution, and thus the cross-saturation coupling will be zero since there is no direct-axis excitation and the quadrature-axis flux density distribution does not change. Thus when $\mu_m = 0$ or $\mu_m = \pi/2$, $L_{DQ} = 0$, but when $\mu_m \neq 0$ it has to be proportional to $\cos\mu_m \sin\mu_m$. However, it has to be zero under linear magnetic conditions and has to be non-zero under saturated conditions. Thus L_{DQ} has to contain another term as well, and this is equal to $L_d - L_m$. In this term L_d is a dynamic inductance and it is equal to the slope of the magnetizing curve ($L_d = d|\bar{\psi}_m|/d|\bar{i}_m|$), where $|\bar{\psi}_m|$ and $|\bar{i}_m|$ are the moduli of the magnetizing flux-linkage and magnetizing-current space phasors respectively and L_m is the magnetizing inductance, $L_m = |\bar{\psi}_m|/|\bar{i}_m|$. Thus $L_{DQ} = (L_d - L_m)\sin\mu_m \cos\mu_m$ is obtained and a rigorous mathematical analysis yields the same expression for L_{DQ} [see the derivation of eqn (6.1-52)]. The change of the magnetizing flux linkage $\Delta\psi_m = L_{DQ}\Delta i_{mQ}$ yields the induced e.m.f. $L_{DQ}\,di_{mQ}/dt$, which is the transformer voltage induced in the direct-axis winding as a result of cross-saturation. It follows that under saturated conditions, solely due to saturation, even sinusoidally distributed windings in space quadrature can become coupled. If the windings are not sinusoidally distributed, the cross-coupling described above also exists, but in addition, there are some other extra voltage terms which have to be considered.

The existence of the phenomenon of cross-saturation can also be proved by simple experiments, or by the application of numerical analysis techniques, e.g. the finite-element method. For this purpose it is very useful to employ the so-called method of frozen permeabilities. When this method is used, it can be shown that in a saturated induction machine, the obtained cross-saturation coupling inductance L_{DQ} agrees with the L_{DQ} obtained by the application of the simple expression $L_{DQ} = (L_d - L_m)\sin\mu_m \cos\mu_m$. Furthermore, it can also be shown that in a saturated salient-pole machine, in addition to the effects of the physically existing saliency, the effects of cross-saturation are also present. This will be discussed in Section 6.2.

The stator voltage equations, eqns (6.1-39) and (6.1-40) contain the quadrature-axis rotor current component (i_{ry}) and its derivative ($\mathrm{d}i_{ry}/\mathrm{d}t$), but from eqn (6.1-43) i_{ry} can be substituted by the torque-producing stator current component (i_{sy}) and $\mathrm{d}i_{ry}/\mathrm{d}t$ can be expressed in terms of i_{sy}. The derivative $\mathrm{d}i_{ry}/\mathrm{d}t$ can be obtained directly by differentiation of eqn (6.1-43). Under linear magnetic conditions this yields $-(L_m/L_r)\mathrm{d}i_{sy}/\mathrm{d}t$, but under saturated conditions, owing to cross-saturation, it must also depend on the direct-axis magnetizing current (i_{mx}). The required derivative can be most simply obtained by utilizing eqn (2.1-151) and $L_r = L_{rl} + L_m$, where L_{rl} is the rotor leakage inductance, which is assumed to be constant, and thus the rotor flux-linkage space phasor in the rotor-flux oriented reference frame is obtained as

$$\bar{\psi}_{r\psi r} = |\bar{\psi}_r| = L_{rl}\bar{i}_{r\psi r} + \bar{\psi}_{m\psi r}, \tag{6.1-55}$$

where $\bar{i}_{r\psi r}$ has been defined in eqn (6.1-21) and $\bar{\psi}_{r\psi r}$ has been defined in eqn (6.1-24). The quadrature-axis component of eqn (6.1-55) yields

$$\psi_{ry} = 0 = L_{rl}i_{ry} + \psi_{my}, \tag{6.1-56}$$

where ψ_{my} has been defined by eqn (6.1-28). Equation (6.1-56) together with eqn (6.1-35) can be used to obtain $\mathrm{d}i_{ry}/\mathrm{d}t$ in terms of i_{sx}, i_{sy}, and i_{rx}. For this purpose eqn (6.1-56) is differentiated, and thus

$$0 = L_{rl}\frac{\mathrm{d}i_{ry}}{\mathrm{d}t} + \frac{\mathrm{d}\psi_{my}}{\mathrm{d}t}, \tag{6.1-57}$$

where $\mathrm{d}\psi_{my}/\mathrm{d}t$ has been given by eqn (6.1-35). Thus by substitution of eqn (6.1-35) into eqn (6.1-57),

$$\frac{\mathrm{d}i_{ry}}{\mathrm{d}t} = -\left(L_{my}\frac{\mathrm{d}i_{sy}}{\mathrm{d}t} + L_{xy}\frac{\mathrm{d}i_{sx}}{\mathrm{d}t} + L_{xy}\frac{\mathrm{d}i_{rx}}{\mathrm{d}t}\right)\Big/ L_{ry}, \tag{6.1-58}$$

where $L_{ry} = L_{rl} + L_{my}$ is the self-inductance of the rotor along the quadrature axis of the rotor-flux-oriented reference frame and L_{my} and L_{xy} have been defined by eqns (6.1-36) and (6.1-38) respectively. It should be noted that because of saturation, the self-inductance of the rotor winding along the direct axis (L_{rx}) differs from L_{ry}, and $L_{rx} = L_{rl} + L_{mx}$, where L_{mx} has been defined in eqn (6.1-37).

When eqns (6.1-58) and (6.1-43) are substituted into eqns (6.1-39) and (6.1-40) and eqn (6.1-41) is taken into account,

$$u_{sx} = R_s i_{sx} + L_{sl}\frac{\mathrm{d}i_{sx}}{\mathrm{d}t} + \left(L_{mx} - \frac{L_{xy}^2}{L_{ry}}\right)\left(\frac{\mathrm{d}i_{sx}}{\mathrm{d}t} + \frac{\mathrm{d}i_{rx}}{\mathrm{d}t}\right)$$
$$+ L_{xy}\left(1 - \frac{L_{my}}{L_{ry}}\right)\frac{\mathrm{d}i_{sy}}{\mathrm{d}t} - \omega_{mr}L_s' i_{sy} \tag{6.1-59}$$

and

$$u_{sy}=R_s i_{sy}+L_{sl}\frac{\mathrm{d}i_{sy}}{\mathrm{d}t}+L_{my}\left(1-\frac{L_{my}}{L_{ry}}\right)\frac{\mathrm{d}i_{sy}}{\mathrm{d}t}$$

$$+L_{xy}\left(1-\frac{L_{my}}{L_r}\right)\left(\frac{\mathrm{d}i_{sx}}{\mathrm{d}t}+\frac{\mathrm{d}i_{rx}}{\mathrm{d}t}\right)+\omega_{mr}[L_{sl}i_{sx}+L_m(i_{sx}+i_{rx})], \qquad (6.1\text{-}60)$$

where L_s' is the stator transient inductance, which in general varies with saturation, $L_s'=L_s-L_m^2/L_r$, and $L_r=L_{rl}+L_m$, $L_{ry}=L_{rl}+L_{my}$. Thus the stator voltage equations are coupled, and extra coupling (cross-saturation coupling) exists as a result of saturation. Under linear magnetic conditions ($L_{mx}=L_{my}=L_m$ and $L_{xy}=0$) the stator voltage equations are coupled owing to the existence of the rotational voltages $\omega_{mr}L_s'i_{sy}$ and $\omega_{mr}[L_{sl}i_{sx}+L_m(i_{sx}+i_{rx})]$.

The stator equations can be decoupled in the same way as that used in Section 4.1. Thus if an ideal drive is assumed, independent control of the stator currents i_{sx} and i_{sy} can be achieved by defining new voltages which directly control the stator current components. These new voltages can be defined as $\hat{u}_{sx}$, and $\hat{u}_{sy}$ and they are directly related to the stator currents i_{sx} and i_{sy} respectively,

$$\hat{u}_{sx}=\left[R_s+\left(L_{sx}-\frac{L_{xy}^2}{L_{ry}}\right)p\right]i_{sx}$$

$$=u_{sx}-\left(L_{mx}-\frac{L_{xy}^2}{L_{ry}}\right)pi_{rx}-L_{xy}\left(1-\frac{L_{my}}{L_{ry}}\right)pi_{sy}+\omega_{mr}L_s'i_{sy} \qquad (6.1\text{-}61)$$

$$\hat{u}_{sy}=\left[R_s+\left(L_{sy}-\frac{L_{my}^2}{L_{ry}}\right)p\right]i_{sy}$$

$$=u_{sy}-L_{xy}\left(1-\frac{L_{my}}{L_{ry}}\right)pi_{sx}-L_{xy}\left(1-\frac{L_{my}}{L_{ry}}\right)pi_{rx}-\omega_{mr}(L_s'i_{sx}+L_m i_{rx}), \qquad (6.1\text{-}62)$$

where according to eqn (6.1-41) $L_{sx}=L_{sl}+L_{mx}$ and $p=\mathrm{d}/\mathrm{d}t$. Thus the required decoupling circuit will be relatively complicated.

Rotor voltage equations: As for the stator voltage equations derived in the previous section, the rotor voltage equations can also be expressed in terms of L_m, which varies with the modulus of the magnetizing-current space phasor $|\bar{i}_m|$. For this purpose eqn (2.1-153) is used and it follows that in the rotor-flux-oriented reference frame ($\omega_g=\omega_{mr}$), by also considering eqn (6.1-21) and $L_r=L_{rl}+L_m$, where L_{rl} is the rotor leakage inductance which is assumed to be constant and L_m is the magnetizing inductance which varies due to saturation,

$$\bar{u}_{r\psi r}=0=R_r\bar{i}_{r\psi r}+L_{rl}\frac{\mathrm{d}\bar{i}_{r\psi r}}{\mathrm{d}t}+\frac{\mathrm{d}\bar{\psi}_{m\psi r}}{\mathrm{d}t}+\mathrm{j}(\omega_{mr}-\omega_r)(L_{rl}\bar{i}_{r\psi r}+\bar{\psi}_{m\psi r}), \qquad (6.1\text{-}63)$$

where $\bar{\psi}_{m\psi r}$ has been defined in eqn (6.1-24). Resolution of eqn (6.1-63) into its real- and imaginary-axis components gives the following voltage equations, if it is considered that in the rotor-flux-oriented reference frame the quadrature-axis rotor flux-linkage component is zero (and according to eqn (6.1-56) $\psi_{ry}=L_{rl}i_{ry}+\psi_{my}=0$),

$$0=R_r i_{rx}+L_{rl}\frac{di_{rx}}{dt}+\frac{d\psi_{mx}}{dt} \tag{6.1-64}$$

$$0=R_r i_{ry}+L_{rl}\frac{di_{ry}}{dt}+\frac{d\psi_{my}}{dt}+(\omega_{mr}-\omega_r)(L_{rl}i_{rx}+\psi_{mx}), \tag{6.1-65}$$

where $d\psi_{mx}/dt$ and $d\psi_{my}/dt$ are given in eqns (6.1-29) and (6.1-30) respectively and ψ_{mx} and ψ_{my} have been defined in eqns (6.1-27) and (6.1-28) respectively. In eqn (6.1-64) there is no term multiplied by $(\omega_{mr}-\omega_r)$, since the quadrature-axis rotor flux-linkage component is zero. In eqn (6.1-65) the derivative term di_{ry}/dt, which has been given by eqn (6.1-58), is present. Thus by substitution of eqns (6.1-29), (6.1-30), (6.1-27), (6.1-28), and (6.1-43) into eqns (6.1-64) and (6.1-65):

$$0=R_r i_{rx}+L_{rl}\frac{di_{rx}}{dt}+\left(L_{mx}-\frac{L_{xy}^2}{L_{ry}}\right)\frac{di_{rx}}{dt}+\left(L_{mx}-\frac{L_{xy}^2}{L_{ry}}\right)\frac{di_{sx}}{dt}+\left(L_{xy}-\frac{L_{xy}L_{my}}{L_{ry}}\right)\frac{di_{sy}}{dt} \tag{6.1-66}$$

$$0=-R_r\frac{L_m}{L_r}i_{sy}+(\omega_{mr}-\omega_r)(L_r i_{rx}+L_m i_{sx}), \tag{6.1-67}$$

where $L_r=L_{rl}+L_m$, $L_{ry}=L_{rl}+L_{my}$. In eqn (6.1-67) $\psi_{rx}=(L_r i_{rx}+L_m i_{sx})$ and under linear magnetic conditions ($L_{mx}=L_{my}=L_m$, $L_{xy}=0$), eqns (6.1-66) and (6.1-67) yield eqns (4.1-25) and (4.1-26) respectively. This can be proved by substitution of $\psi_{rx}=|\bar{\psi}_r|=L_m|\bar{i}_{mr}|$ into eqn (6.1-67), where $|\bar{i}_{mr}|$ is the modulus of the rotor magnetizing-current space phasor, and by substitution of the expression for i_{rx} given by eqn (6.1-42) into eqn (6.1-66). Whilst the rotor voltage equations under linear magnetic conditions imply the decoupled control of the rotor flux and torque, by the independent control of i_{sx} and i_{sy} respectively, it follows from eqns (6.1-66) and (6.1-67) that under saturated conditions this is not the case. It follows from eqn (6.1-66) that if i_{sx} is kept constant, as a result of the cross-saturation coupling a change in the torque-producing stator current (i_{sy}) will cause a change in the direct-axis rotor current i_{rx} and thus the rotor flux level (which depends on the rotor currents also) will change. It should be noted that eqn (6.1-66) is much more complicated than eqn (6.1-18). However, whilst eqn (6.1-18) has been derived on the basis that the magnetizing inductance is a function of the modulus of the rotor magnetizing-current space phasor, which has only a direct-axis component in the rotor-flux-oriented reference frame, eqn (6.1-66) has been obtained on the basis that the magnetizing inductance is a function of the modulus of the magnetizing-current space phasor, which does have direct-axis and quadrature-axis components in the same reference frame. Furthermore, it should also be noted that eqn (6.1-66) cannot be obtained from eqn (4.1-25) by simply replacing the constant magnetizing inductance by the variable magnetizing inductance which is a function of the modulus of the magnetizing-current space phasor.

For completeness, the rotor voltage equations are also given in the stationary reference frame. These can be most simply obtained by considering the real- and imaginary-axis components of eqn (6.1-63) and by utilizing the fact that $\omega_{mr}=0$. Furthermore, in the stationary reference frame the rotor-current space phasor $\bar{i}_r'$ can be expressed in terms of its two-axis components as $\bar{i}_r'=i_{rd}+ji_{rq}$ and the two-axis components of the magnetizing flux-linkage space phasor are ψ_{mD} and ψ_{mQ}, which have been defined in eqns (6.1-46) and (6.1-47) respectively. Thus

$$u_{rd}=0=R_r i_{rd}+L_{rl}\frac{di_{rd}}{dt}+\frac{d\psi_{mD}}{dt}+\omega_r(L_r i_{rq}+L_m i_{sQ}) \tag{6.1-68}$$

$$u_{rq}=0=R_r i_{rq}+L_{sl}\frac{di_{rq}}{dt}+\frac{d\psi_{mQ}}{dt}-\omega_r(L_r i_{rd}+L_m i_{sD}), \tag{6.1-69}$$

where $d\psi_{mD}/dt$ and $d\psi_{mQ}/dt$ are defined in eqns (6.1-48) and (6.1-49) respectively. Thus by the substitution of eqns (6.1-48) and (6.1-49) into eqns (5.1-68) and (6.1-69), the following rotor voltage equations are obtained for an induction machine when saturation of the main flux paths is present:

$$u_{rd}=0=R_r i_{rd}+L_{rd}\frac{di_{rd}}{dt}+L_{mD}\frac{di_{sD}}{dt}$$
$$+L_{DQ}\left(\frac{di_{sQ}}{dt}+\frac{di_{rq}}{dt}\right)+\omega_r(L_r i_{rq}+L_m i_{sQ}) \tag{6.1-70}$$

$$u_{rq}=0=R_r i_{rq}+L_{rq}\frac{di_{rq}}{dt}+L_{mQ}\frac{di_{sQ}}{dt}$$
$$+L_{DQ}\left(\frac{di_{sD}}{dt}+\frac{di_{rd}}{dt}\right)-\omega_r(L_r i_{rd}+L_m i_{sD}), \tag{6.1-71}$$

where $L_{rd}=L_{rl}+L_{mD}$ and $L_{rq}=L_{rl}+L_{mQ}$ are the self-inductances of the rotor in the direct and quadrature axes of the stationary reference frame and $L_r=L_{rl}+L_m$. As a result of saturation these are not equal, but under linear magnetic conditions $L_{rd}=L_{rq}=L_r$. In eqns (6.1-70) and (6.1-71) the effects of cross-saturation are also present and under saturated conditions in, say, the direct-axis rotor winding, a change in the quadrature-axis stator current will induce a transformer e.m.f. ($L_{DQ}di_{sQ}/dt$) in the direct-axis rotor winding.

If eqns (6.1-70) and (6.1-71) are combined with eqns (6.1-53) and (6.1-54), the stationary-axis model of a saturated smooth-air-gap machine is obtained as:

$$\begin{bmatrix} u_{sD} \\ u_{sQ} \\ u_{rd} \\ u_{rq} \end{bmatrix}=\begin{bmatrix} R_s+L_{sD}p & L_{DQ}p & L_{mD}p & L_{DQ}p \\ L_{DQ}p & R_s+L_{sQ}p & L_{DQ}p & L_{mQ}p \\ L_{mD}p & \omega_r L_m+L_{DQ}p & R_r+L_{rd}p & \omega_r L_r+L_{DQ}p \\ -\omega_r L_m+L_{DQ}p & L_{mQ}p & -\omega_r L_r+L_{DQ}p & R_r+L_{rq}p \end{bmatrix}\begin{bmatrix} i_{sD} \\ i_{sQ} \\ i_{rd} \\ i_{rq} \end{bmatrix}. \tag{6.1-72}$$

When eqn (6.1-72) is compared with eqn (2.1-123), which describes the linear model, it follows that as a result of saturation, all the inductance terms are modified and there will be 16 inductance terms in the impedance matrix of the saturated commutator model. Furthermore, in the saturated model the self-inductances of both the stator windings and the rotor windings are unequal, the magnetizing inductances along the two axes are unequal, and all the windings in space quadrature are coupled owing to cross-saturation. The cross-saturation coupling terms are present in the transformer e.m.f.s. It is important to note that in eqn (6.1-72) only $\omega_r L_m i_{sQ} + \omega_r L_r i_{rq}$ and $-(\omega_r L_m i_{sD} + \omega_r L_r i_{rd})$ are rotational voltage components which correspond to $\omega_r \psi_{rd}$ and $-\omega_r \psi_{rd}$ respectively. Thus under saturated conditions there will be no cross-saturation or other extra terms present in the expression for the electromagnetic torque as compared with the expression valid under linear magnetic conditions.

6.1.2 MAGNETIZING FLUX-ORIENTED CONTROL

In this section the effects of main-flux saturation are discussed for the equations of the saturated single-cage induction machine which have to be used if magnetizing-flux-oriented control is employed. For this purpose, the stator and rotor voltage equations are formulated in the magnetizing-flux-oriented reference frame. It will be shown that relatively simple equations arise, since in this reference frame, because the magnetizing flux-linkage space phasor has only one component, the effects of cross-saturation will be absent.

6.1.2.1 Stator voltage equations

It follows from eqns (2.1-148) and (2.1-150) that in the magnetizing-flux-oriented reference frame, which rotates at the speed $\omega_m = d\mu_m/dt$, where μ_m is the angle of the magnetizing-current space phasor with respect to the real axis of the stationary reference frame (see Fig. 6.2 or Section 4.3.1), by considering that the self-inductance of the stator (L_s) is equal to the sum of the stator leakage inductance (L_{sl}) and the magnetizing inductance (L_m), which varies due to saturation, the space-phasor voltage equation of the stator can be put into the form:

$$\bar{u}_{sm} = R_s \bar{\imath}_{sm} + L_{sl} \frac{d\bar{\imath}_{sm}}{dt} + \frac{d(L_m|\bar{\imath}_{mm}|)}{dt} + j\omega_m (L_{sl}\bar{\imath}_{sm} + L_m|\bar{\imath}_{mm}|), \qquad (6.1\text{-}73)$$

where L_{sl} is assumed to be constant and $|\bar{\imath}_{mm}|$ is the magnetizing-current space phasor in the rotor-flux-oriented reference frame,

$$|\bar{\imath}_{mm}| = |\bar{\imath}_m| = \bar{\imath}_{sm} + \bar{\imath}_{rm} = i_{sx} + ji_{sy} + (i_{rx} + ji_{ry}), \qquad (6.1\text{-}74)$$

where $|\bar{\imath}_m|$ is the modulus of the magnetizing-current space phasor in the stationary reference frame. In eqn (6.1-73) in general L_m is a non-linear function of the modulus of the magnetizing-current space phasor and in eqn (6.1-74) $\bar{\imath}_{sm}$ and $\bar{\imath}_{rm}$ are the space phasors of the stator and rotor current respectively in the magnetizing-flux-oriented reference frame [they have been defined in eqns (4.3-3) and (4.3-4)]. $\bar{u}_{sm}=\bar{u}_s e^{-j\mu_m}$ is the space phasor of the stator voltages in the same reference frame, where $\bar{u}_s$ is the space phasor of the stator voltages in the stationary reference frame. Resolution of eqn (6.1-73) into its direct- and quadrature-axis components yields

$$u_{sx}=R_s i_{sx}+L_{sl}\frac{di_{sx}}{dt}+\frac{d(L_m|\bar{i}_m|)}{dt}-\omega_m L_{sl} i_{sy} \tag{6.1-75}$$

$$u_{sy}=R_s i_{sy}+L_{sl}\frac{di_{sy}}{dt}+\omega_m(L_{sl} i_{sx}+L_m|\bar{i}_m|). \tag{6.1-76}$$

Under linear magnetic conditions, the magnetizing inductance is constant and eqns (6.1-75) and (6.1-76) yield eqns (4.3-10) and (4.3-11) respectively. However, under saturated conditions when L_m is a function of $|\bar{i}_m|$, the derivative term, $d/dt(L_m|\bar{i}_m|)\neq L_m d|\bar{i}_m|/dt$, which is the magnetizing voltage and is equal to the rate of change of the magnetizing flux-linkage space phasor, can be expressed as

$$\frac{d(L_m|\bar{i}_m|)}{dt}=\frac{d|\bar{\psi}_m|}{dt}=\frac{d|\bar{\psi}_m|}{d|\bar{i}_m|}\frac{d|\bar{i}_m|}{dt}=L_d\frac{d|\bar{i}_m|}{dt}, \tag{6.1-77}$$

where L_d is a dynamic inductance (also used in the previous section), $L_d=d|\bar{\psi}_m|/d|\bar{i}_m|$, where $|\bar{\psi}_m|$ is the modulus of the magnetizing flux-linkage space phasor in the stationary reference frame (and is equal to $|\bar{\psi}_{mm}|$, which is the modulus of the magnetizing flux-linkage space phasor in the magnetizing-flux-oriented reference frame). Substitution of eqn (6.1-77) into eqn (6.1-75) and rearrangement of the resulting equation, together with the rearranged form of eqn (6.1-76), yields

$$T_{sl}\frac{di_{sx}}{dt}+i_{sx}=\frac{u_{sx}}{R_s}+\omega_m T_{sl} i_{sy}-\frac{L_d}{R_s}\frac{d|\bar{i}_m|}{dt} \tag{6.1-78}$$

$$T_{sl}\frac{di_{sy}}{dt}+i_{sy}=\frac{u_{sy}}{R_s}-\omega_m\left(T_{sl} i_{sx}+\frac{|\bar{i}_m|L_m}{R_s}\right), \tag{6.1-79}$$

where $T_{sl}=L_{sl}/R_s$ is the stator leakage time constant. Under linear magnetic conditions L_m=constant, $L_d=L_m$ and eqns (6.1-78) and (6.1-79) agree with eqns (4.3-9) and (4.3-10) respectively. However, it follows from eqns (6.1-78) and (6.1-79) that when the effects of main-flux saturation are also considered, the stator voltage equations are very similar to those obtained under linear magnetic conditions, but in the direct-axis voltage equation the magnetizing inductance is replaced by the dynamic inductance and of course in the quadrature-axis voltage equation, instead of the constant magnetizing inductance, the variable magnetizing inductance has to be used. The simple forms of the equations are due to the fact that the magnetizing-current space phasor has only one component in the magnetizing-flux-oriented reference frame and therefore there cannot be any terms due to cross-saturation. However, the stator equations are still coupled and the required decoupling circuits can be obtained in a way similar to that described in Section 4.3.2. When the machine is supplied by impressed stator currents, only the rotor voltage equations have to be considered, and these are described in the next section.

6.1.2.2 Rotor voltage equations

From eqn (2.1-153), which is the rotor-voltage space phasor in the general reference frame, it follows that in the magnetizing-flux-oriented reference frame, which rotates at the speed ω_m, when the effects of main-flux saturation are considered with $L_r = L_{rl} + L_m$, where L_{rl} is the rotor leakage inductance and L_m is the magnetizing inductance which changes with saturation, $L_m = L_m(|\bar{\imath}_m|)$, the following rotor voltage equation for the induction machine is obtained:

$$0 = R_r \bar{\imath}_{rm} + L_{rl}\frac{d\bar{\imath}_{rm}}{dt} + \frac{d(L_m|\bar{\imath}_m|)}{dt} + j(\omega_m - \omega_r)(L_{rl}\bar{\imath}_{rm} + L_m|\bar{\imath}_m|), \qquad (6.1\text{-}80)$$

where $|\bar{\imath}_m|$ is defined in eqn (6.1-74). Owing to saturation of the main flux paths, in eqn (6.1-80) $d/dt(L_m|\bar{\imath}_m|) \neq L_m d|\bar{\imath}_m|/dt$ but is given by eqn (6.1-77). Thus by the substitution of eqn (6.1-77) into eqn (6.1-80),

$$0 = R_r \bar{\imath}_{rm} + L_{rl}\frac{d\bar{\imath}_{rm}}{dt} + L_d\frac{d(|\bar{\imath}_m|)}{dt} + j(\omega_m - \omega_r)(L_{rl}\bar{\imath}_{rm} + L_m|\bar{\imath}_m|). \qquad (6.1\text{-}81)$$

However, the space phasor of the rotor currents can be written in terms of the space phasors of the stator currents and the magnetizing current and it follows from eqn (6.1-74) that $\bar{\imath}_{rm} = |\bar{\imath}_m| - \bar{\imath}_{sm}$ and, when this expression is substituted into eqn (6.1-81),

$$0 = R_r(|\bar{\imath}_m| - \bar{\imath}_{sm}) + (L_{rl} + L_d)\frac{d|\bar{\imath}_m|}{dt} + L_{rl}\frac{d\bar{\imath}_{sm}}{dt} + j(\omega_m - \omega_r)(L_r|\bar{\imath}_m| - L_{rl}\bar{\imath}_{sm}). \qquad (6.1\text{-}82)$$

This equation is very similar to eqn (4.3-17) (which holds under linear magnetic conditions and it should be noted that $|\bar{\imath}_m| = |\bar{\imath}_{mm}|$), but whilst in eqn (4.3-17) the derivative of the modulus of the magnetizing-current space phasor is multiplied by the rotor self-inductance ($L_r = L_{rl} + L_m$), in eqn (6.1-82) $d|\bar{\imath}_m|/dt$ is multiplied by the inductance $(L_{rl} + L_d)$ and under saturated conditions this varies and is smaller than $L_{rl} + L_m$. Thus the resolution of eqn (6.1-82) into its direct- and quadrature-axis components in the magnetizing-flux-oriented reference frame gives

$$\left(|\bar{\imath}_m| + T\frac{d|\bar{\imath}_m|}{dt}\right)\Big/T_{rl} = \frac{di_{sx}}{dt} + \frac{i_{sx}}{T_{rl}} - \omega_{sl} i_{sy} \qquad (6.1\text{-}83)$$

$$\omega_{sl}\left(|\bar{\imath}_m|\frac{T_r}{T_{rl}} - i_{sx}\right) = \frac{di_{sy}}{dt} + \frac{i_{sy}}{T_{rl}}, \qquad (6.1\text{-}84)$$

where the various rotor time constants are defined as $T = (L_{rl} + L_d)/R_r$, $T_r = (L_{rl} + L_m)/R_r$, and $T_{rl} = L_{rl}/R_r$, and ω_{sl} is the angular slip frequency, $\omega_{sl} = \omega_m - \omega_r$.

Under linear magnetic conditions $L_d = L_m$=constant and eqns (6.1-83) and (6.1-84) yield eqns (4.3-18) and (4.3-19) respectively. It follows from eqns (6.1-83) and (6.1-84) that there is unwanted coupling between these equations and $|\bar{i}_m|$ is also a function of i_{sy} (it depends on all the currents of the machine). The required decoupling circuit can be obtained by utilizing a method which is similar to the one described in Section 4.3.3.

6.1.3 STATOR-FLUX-ORIENTED CONTROL

In this section the stator and rotor voltage equations of the saturated single-cage induction machine are formulated in the stator-flux-oriented reference frame, which rotates at the speed ω_{ms}, and the effects of main-flux saturation are considered.

6.1.3.1 Stator voltage equations

It follows from eqns (2.1-148) and (2.1-150) that, by considering $\omega_g = \omega_{ms}$

$$\bar{u}_{s\psi s} = R_s \bar{i}_{s\psi s} + \frac{d\bar{\psi}_{s\psi s}}{dt} + j\omega_{ms}\bar{\psi}_{s\psi s}, \tag{6.1-85}$$

where $\bar{u}_{s\psi s}$ and $\bar{i}_{s\psi s}$ are the space phasors of the stator voltages and stator currents respectively and have been defined in eqns (4.1-180) and (4.1-175) respectively, and

$$\bar{\psi}_{s\psi s} = |\bar{\psi}_s| = L_m|\bar{i}_{ms}| = \psi_{sx} + j\psi_{sy} \tag{6.1-86}$$

is the space phasor of the stator flux linkages in the stator-flux-oriented reference frame ($\psi_{sy}=0$). In eqn (6.1-86) $|\bar{i}_m|$ is the modulus of the stator magnetizing-current space phasor and it is related to the space phasor of the stator and rotor currents in the stator-flux-oriented reference frame as given by eqn (4.2-9). Substitution of eqn (6.1-86) into eqn (6.1-85) yields

$$\bar{u}_{s\psi s} = R_s \bar{i}_{s\psi s} + \frac{d(L_m|\bar{i}_{ms}|)}{dt} + j\omega_{ms}L_m|\bar{i}_{ms}|. \tag{6.1-87}$$

Under linear magnetic conditions L_m=constant and eqn (6.1-87) agrees with eqn (4.2-10). However, under saturated conditions, when the magnetizing inductance is changing and is a non-linear function of $|\bar{i}_{ms}|$,

$$\frac{d(L_m|\bar{i}_{ms}|)}{dt} = \frac{d|\bar{\psi}_s|}{d|\bar{i}_{ms}|}\frac{d|\bar{i}_{ms}|}{dt} = L^{ds}\frac{d|\bar{i}_{ms}|}{dt}. \tag{6.1-88}$$

In eqn (6.1-88) L^{ds} is a dynamic inductance, $L^{ds} = d|\bar{\psi}_s|/d|\bar{i}_{ms}|$, and substitution of eqn (6.1-88) into eqn (6.1-87) yields

$$\bar{u}_{s\psi s} = R_s \bar{i}_{s\psi s} + L^{ds}\frac{d(|\bar{i}_{ms}|)}{dt} + j\omega_{ms}L_m|\bar{i}_{ms}|. \tag{6.1-89}$$

This equation is very similar to eqn (4.2-10), which holds under linear magnetic conditions, but instead of the constant L_m now L^{ds} is present in the stator voltage equation. Thus the resolution of eqn (6.1-89) into its direct- and quadrature-axis components yields

$$i_{sx}=\frac{u_{sx}}{R_s}-\frac{L^{ds}}{R_s}\frac{d|\bar{i}_{ms}|}{dt} \qquad (6.1\text{-}90)$$

$$i_{sy}=\frac{u_{sy}}{R_s}-\frac{\omega_{ms}L_m|\bar{i}_{ms}|}{R_s}, \qquad (6.1\text{-}91)$$

which are similar to eqns (4.2-11) and (4.2-12), which hold under linear magnetic conditions. Equations (6.1-90) and (6.1-91) contain unwanted coupling terms and these can be removed by the application of the suitable decoupling circuit. It should be noted that cross-saturation coupling terms are not present in eqns (6.1-90) and (6.1-91) since in the stator-flux-oriented reference frame the stator magnetizing-current space phasor contains only a direct-axis component (the quadrature-axis component is zero) and the magnetizing inductance has been expressed in terms of the stator magnetizing-current space phasor. However, it is also possible to formulate the equations in the stator-flux-oriented reference frame, by expressing the magnetizing inductance in terms of the modulus of the magnetizing-current space phasor. In this case, under saturated conditions, the resulting equations will contain the cross-saturation coupling terms as well, similarly to the equations described in Section 6.1.1, since these will contain the rate of change of the magnetizing flux-linkage space phasor expressed in the stator-flux-oriented reference frame ($\bar{\psi}_{m\psi s}$), and $\bar{\psi}_{m\psi s}$ is not coaxial with the stator-flux-oriented reference frame. Full details of this will not be given here, since the approach is very similar to that used in Section 6.1.1, but a cursory derivation of the equations will be given.

The magnetizing flux-linkage space phasor expressed in the stator-flux-oriented reference frame can be defined in a similar way to the flux-linkage space phsor given by eqn (6.1-24) and thus

$$\bar{\psi}_{m\psi s}=L_m\bar{i}_{m\psi s}=L_m|\bar{i}_m|e^{j(\mu_m-\rho_s)}=\psi_{mx}+j\psi_{my},$$

where μ_m is the angle of the magnetizing flux-linkage space phasor with respect to the direct axis of the stationary reference frame and ρ_s is the angle of the stator magnetizing-current space phasor with respect to the real-axis of the stationary reference frame (see, for example, Fig. 2.17). It follows from the resolution of eqns (6.1-85) and (6.1-86) into their real- and imaginary-axis components that by utilizing $\psi_{sy}=0$ and $\psi_{sx}=L_{sl}i_{sx}+\psi_{mx}$,

$$u_{sx}=R_si_{sx}+L_{sl}\frac{di_{sx}}{dt}+\frac{d\psi_{mx}}{dt}, \qquad (6.1\text{-}92)$$

where, as in eqn (6.1-34), the rate of change of the direct-axis component of the magnetizing flux-linkage space phasor in the stator-flux-oriented reference frame can be expressed as

$$\frac{d\psi_{mx}}{dt}=L_{mx}\frac{di_{mx}}{dt}+L_{xy}\frac{di_{my}}{dt}, \qquad (6.1\text{-}93)$$

where i_{mx} and i_{my} are the magnetizing current components along the direct and quadrature axes of the stator-flux-oriented reference frame, $i_{mx}=i_{sx}+i_{rx}$, $i_{my}=i_{sy}+i_{ry}$, L_{mx} is the magnetizing inductance in the direct-axis,

$$L_{mx}=L_{d}\cos^2(\mu_m-\rho_s)+L_m\sin^2(\mu_m-\rho_s), \tag{6.1-94}$$

and L_{xy} is the cross-coupling inductance between the direct and quadrature axes of the stator-flux-oriented reference frame,

$$L_{xy}=\tfrac{1}{2}(L_d-L_m)\sin[2(\mu_m-\rho_s)]. \tag{6.1-95}$$

In eqns (6.1-94) and (6.1-95) L_d is a dynamic inductance, $L_d=\mathrm{d}|\bar{\psi}_m|/\mathrm{d}|\bar{\imath}_m|$, where $|\bar{\psi}_m|$ and $|\bar{\imath}_m|$ are the moduli of the magnetizing flux-linkage and magnetizing-current space phasors respectively. Substitution of eqn (6.1-93) into eqn (6.1-92) gives the direct-axis voltage equation and it follows that under saturated conditions this contains cross-saturation coupling terms.

A similar derivation could be performed for the quadrature-axis stator voltage equation. However, it should be noted that both the direct- and quadrature-axis stator equations contain derivatives of the rotor currents ($\mathrm{d}i_{rx}/\mathrm{d}t, \mathrm{d}i_{ry}/\mathrm{d}t$) and, according to eqn (4.2-2), the rotor currents are related to the stator currents as

$$i_{rx}=|\bar{\imath}_{ms}|-i_{sx}\frac{L_s}{L_m} \tag{6.1-96}$$

$$i_{ry}=-i_{sy}\frac{L_s}{L_m}. \tag{6.1-97}$$

6.1.3.2 Rotor voltage equations

The rotor voltage equations of the saturated induction machine in the stator-flux-oriented reference frame can be obtained by considering the general form of the rotor voltage equation, eqn (2.1-186). These equations can be used for the various implementations of stator-flux-oriented control of the saturated induction machine with impressed stator currents. The same method is followed as in Section 4.2.3, but it is assumed that the magnetizing inductance varies with saturation. Thus by using the same method as for the derivation of eqn (4.2-15), the following rotor-voltage space-phasor equation is obtained in the stator-flux-oriented reference frame:

$$0=R_r[|\bar{\imath}_{ms}|-(L_s/L_m)\bar{\imath}_{s\psi s}]+\frac{\mathrm{d}(L_r|\bar{\imath}_{ms}|)}{\mathrm{d}t}-\frac{\mathrm{d}[(L_s'L_r/L_m)\bar{\imath}_{s\psi s}]}{\mathrm{d}t}$$
$$+\mathrm{j}\omega_{sl}[L_r|\bar{\imath}_{ms}|-(L_s'L_r/L_m)\bar{\imath}_{s\psi s}], \tag{6.1-98}$$

where

$$\omega_{sl}=\omega_{ms}-\omega_r \tag{6.1-99}$$

is the angular slip frequency. In eqn (6.1-98) there are two derivative terms which contain the variable magnetizing inductance (which is a function of the modulus of the stator magnetizing-current space phasor, $|\bar{\imath}_{ms}|$) and these can be expanded by the application of the chain differentiation rule. Thus the first term can be expanded as

$$\frac{d(L_r|\bar{\imath}_{ms}|)}{dt}=L_{rl}\frac{d|\bar{\imath}_{ms}|}{dt}+\frac{d(L_m|\bar{\imath}_{ms}|)}{dt}=(L_{rl}+L^{ds})\frac{d|\bar{\imath}_{ms}|}{dt}, \tag{6.1-100}$$

where $L^{ds}=d|\bar{\psi}_s|/d|\bar{\imath}_{ms}|$ is a dynamic inductance, $|\bar{\psi}_s|$ and $|\bar{\imath}_{ms}|$ are the moduli of the stator flux-linkage space phasor and stator magnetizing-current space phasor respectively and under linear magnetic conditions $L^{ds}=L_m$. Also L_{rl} is the leakage inductance of the rotor. The second term of eqn (6.1-98) can be expanded as follows, if it is considered that the stator transient inductance can be expressed as $L_s'=L_s-L_m^2/L_r$:

$$\frac{d[(L_s'L_r/L_m)\bar{\imath}_{s\psi s}]}{dt}=\frac{L_s'L_r}{L_m}\frac{d\bar{\imath}_{s\psi s}}{dt}+\bar{\imath}_{s\psi s}\frac{d(L_sL_r/L_m-L_m)}{dt}. \tag{6.1-101}$$

The chain differentiation rule can be applied to the second term on the right-hand side of eqn (6.1-101), since it is a function of L_m which varies with saturation, and finally eqn (6.1-101) will take the following form:

$$\frac{d[(L_s'L_r/L_m)\bar{\imath}_{s\psi s}]}{dt}=\frac{L_s'L_r}{L_m}\frac{d\bar{\imath}_{s\psi s}}{dt}-\bar{\imath}_{s\psi s}\frac{L_{sl}L_{rl}}{L_m^2}\frac{dL_m}{dt}, \tag{6.1-102}$$

where L_{sl} is the leakage inductance of the stator. In eqn (6.1.102) the derivative dL_m/dt is present; under linear magnetic conditions this is zero, but when the magnetizing inductance changes with saturation, then as in eqn (6.1-5)

$$\frac{dL_m}{dt}=\frac{L^{ds}-L_m}{|\bar{\imath}_{ms}|}\frac{d|\bar{\imath}_{ms}|}{dt}. \tag{6.1-103}$$

Thus by substitution of eqn (6.1-103) into eqn (6.1-102),

$$\frac{d[(L_s'L_r/L_m)\bar{\imath}_{s\psi s}]}{dt}=\frac{L_s'L_r}{L_m}\frac{d\bar{\imath}_{s\psi s}}{dt}-\frac{\bar{\imath}_{s\psi s}}{|\bar{\imath}_{ms}|}(L^{ds}-L_m)\frac{L_{sl}L_{rl}}{L_m^2}\frac{d|\bar{\imath}_{ms}|}{dt} \tag{6.1-104}$$

is obtained, and substitution of eqns (6.1-104) and (6.1-100) into eqn (6.1-98) yields the final form of the space-phasor form of the rotor voltage equation:

$$0=R_r\left[|\bar{\imath}_{ms}|-\frac{L_s}{L_m}\bar{\imath}_{s\psi s}\right]+(L_{rl}+L^{ds})\frac{d|\bar{\imath}_{ms}|}{dt}-\frac{L_s'L_r}{L_m}\frac{d\bar{\imath}_{s\psi s}}{dt}$$
$$+\frac{\bar{\imath}_{s\psi s}}{|\bar{\imath}_{ms}|}(L^{ds}-L_m)\frac{L_{sl}L_{rl}}{L_m^2}\frac{d|\bar{\imath}_{ms}|}{dt}+j\omega_{sl}\left[L_r|\bar{\imath}_{ms}|-\frac{L_s'L_r}{L_m}\bar{\imath}_{s\psi s}\right], \tag{6.1-105}$$

and resolution into real- and imaginary-axis components yields

$$\frac{di_{sx}}{dt}+\frac{i_{sx}}{T_r'}-\omega_{sl}i_{sy}=\left[\frac{L_m}{L_s'}A+\frac{i_{sx}}{|\bar{\imath}_{ms}|}B(L^{ds}-L_m)\right]\frac{d|\bar{\imath}_{ms}|}{dt}+\frac{L_m}{L_sT_r'}|\bar{\imath}_{ms}| \tag{6.1-106}$$

and

$$\omega_{\text{sl}}\left(\frac{L_{\text{m}}|\bar{i}_{\text{ms}}|}{L'_{\text{s}}}-i_{\text{s}x}\right)=i_{\text{s}y}\left[\frac{1}{T'_{\text{r}}}-\frac{(L^{\text{ds}}-L_{\text{m}})B}{|\bar{i}_{\text{ms}}|}\right]\frac{\text{d}|\bar{i}_{\text{ms}}|}{\text{d}t}+\frac{\text{d}i_{\text{s}y}}{\text{d}t}, \qquad (6.1\text{-}107)$$

where

$$A=\frac{L_{\text{rl}}+L^{\text{ds}}}{L_{\text{r}}}$$

$$B=\frac{L_{\text{sl}}L_{\text{rl}}}{L_{\text{m}}L'_{\text{s}}L_{\text{r}}} \qquad (6.1\text{-}108)$$

and $L'_{\text{s}}=L_{\text{s}}-L_{\text{m}}^2/L_{\text{r}}$, $L_{\text{r}}=L_{\text{rl}}+L_{\text{m}}$ depend on the magnetizing inductance, which varies with saturation. Under linear magnetic conditions, eqns (5.1-106) and (6.1-107) yield eqns (4.2-17) and (4.2-18) respectively, since $L^{\text{ds}}=L_{\text{m}}=$constant. For stator-flux-oriented control of the saturated induction machine with impressed stator currents eqns (6.1-106) and (6.1-107) have to be considered. However, there is unwanted coupling between the torque-producing stator current component ($i_{\text{s}y}$) and the stator magnetizing-current component ($i_{\text{s}x}$). This coupling can be removed by the application of a suitable decoupling circuit. This can be derived in a similar way to the decoupling circuit described in Section 4.2.3.

In the equations derived above the magnetizing inductance has been assumed to be a non-linear function of the modulus of the stator magnetizing-current space phasor and the rotor voltage equations have been obtained in the stator-flux-oriented reference frame. Since the stator magnetizing space phasor has only a direct-axis component in this reference frame, no cross-saturation terms appear in the equations. However, if the magnetizing inductance is expressed in terms of the modulus of the magnetizing-current space phasor and the rotor voltage equations are expressed in the stator-flux-oriented reference frame, under saturated conditions cross-saturation terms will be present in the equations, since the magnetizing-current space phasor is not coaxial with the direct axis of the reference frame, and thus there is magnetization of both axes which will result in cross-saturation transformer e.m.f.s under saturated conditions. The derivation of the resulting rotor voltage equations is similar to the derivation of eqns (6.1-66) and (6.1-67) and thus it will not be discussed in detail. However, like eqn (6.1-63), the space-phasor voltage equation in the stator-flux-oriented reference frame can be expressed as

$$\bar{u}_{\text{r}\psi\text{s}}=0=R_{\text{r}}\bar{i}_{\text{r}\psi\text{s}}+L_{\text{rl}}\frac{\text{d}\bar{i}_{\text{r}\psi\text{s}}}{\text{d}t}+\frac{\text{d}\bar{\psi}_{\text{m}\psi\text{s}}}{\text{d}t}+\text{j}(\omega_{\text{ms}}-\omega_{\text{r}})(L_{\text{rl}}\bar{i}_{\text{r}\psi\text{s}}+\bar{\psi}_{\text{m}\psi\text{s}}), \qquad (6.1\text{-}109)$$

where the magnetizing flux-linkage space phasor in the stator-flux-oriented reference frame has been defined in the previous section as

$$\bar{\psi}_{\text{m}\psi\text{s}}=L_{\text{m}}\bar{i}_{\text{m}\psi\text{s}}=L_{\text{m}}|\bar{i}_{\text{m}}|\text{e}^{\text{j}(\mu_{\text{m}}-\rho_{\text{s}})}=\psi_{\text{m}x}+\text{j}\psi_{\text{m}y}.$$

Resolution of eqn (6.1-109) into its real- and imaginary-axis components gives the following equations, if it is considered that in the stator-flux-oriented reference frame the quadrature-axis stator flux-linkage component is zero

$(\psi_{sy}=L_{sl}i_{sy}+\psi_{my}=0)$:

$$0=R_r i_{rx}+L_{rl}\frac{di_{rx}}{dt}+\frac{d\psi_{mx}}{dt}-(\omega_{ms}-\omega_r)(L_{rl}i_{ry}-L_{sl}i_{sy}) \qquad (6.1\text{-}110)$$

$$0=R_r i_{ry}+L_{rl}\frac{di_{ry}}{dt}+\frac{d\psi_{my}}{dt}+(\omega_{ms}-\omega_r)(L_{rl}i_{rx}+\psi_{mx}), \qquad (6.1\text{-}111)$$

where, by considering eqn (6.1-93), which is repeated here for convenience,

$$\frac{d\psi_{mx}}{dt}=L_{mx}\frac{di_{mx}}{dt}+L_{xy}\frac{di_{my}}{dt}, \qquad (6.1\text{-}112)$$

and, as for eqn (6.1-35), the rate of change of the quadrature-axis magnetizing flux linkage component can be expressed as

$$\frac{d\psi_{my}}{dt}=L_{my}\frac{di_{my}}{dt}+L_{xy}\frac{di_{mx}}{dt}. \qquad (6.1\text{-}113)$$

In these equations L_{mx} and L_{my} are the magnetizing inductances along the direct (x) and quadrature (y) axes respectively in the stator-flux-oriented reference frame. The inductance L_{mx} has been defined by eqn (5.1-94) and L_{my} can be defined in a similar way to eqn (5.1-37) and therefore

$$L_{my}=L_d\sin^2(\mu_m-\rho_s)+L_m\cos^2(\mu_m-\rho_s). \qquad (6.1\text{-}114)$$

In eqns (6.1-112) and (6.1-113) L_{xy} is the cross-coupling inductance between the x and y axes, and it has been given in eqn (6.1-95). Substitution of eqns (6.1-112) and (6.1-113) into eqns (6.1-110) and (6.1-111) yields the rotor voltage equations in the stator-flux-oriented reference frame, and the effects of main-flux saturation are also incorporated in these. They will contain the cross-saturation terms as well, in agreement with the discussion presented above. The rotor voltage equations thus obtained can be used for the implementation of stator-flux-oriented control of an induction machine supplied by impressed stator currents, in a similar way to that described in Section 4.2.3. It follows from eqns (6.1-94), (6.1-114), and (6.1-95) that under linear magnetic conditions when $L_d=L_m$, $L_{mx}=L_{my}=L_m$ and $L_{xy}=0$ and thus by also considering eqns (6.1-96) and (6.1-97), it can be shown that the rotor voltage equations yield eqns (4.2-17) and (4.2-18).

6.1.4 MAXIMUM TORQUE CONTROL IN THE FIELD-WEAKENING RANGE

In a torque-controlled induction motor drive, the maximum output torque and output power also depend on the inverter current rating and the maximum voltage which the inverter can supply to the machine. Thus considering the limited voltage and current capabilities, it is useful to consider those control schemes which give maximum torque/ampere in the whole speed range. For this purpose, it is possible to implement various control strategies.

In a vector-controlled induction motor drive with rotor-flux-oriented control, above base speed in the field-weakening range, the conventional technique is to vary the modulus of the reference value of the rotor flux-linkage space vector in inverse proportion to the rotor speed. However, in this case maximum torque is not achieved in the field-weakening range. The conventional scheme is the first one to be described below. This is followed by discussing an improved technique. However this does not provide maximum torque in the field-weakening region, since it assumes that the rotor flux linkage is $\psi_{rx}=L_m i_{sx}$ and thus it follows by using eqn (2.1-191) that the electromagnetic torque is $t_e=(3/2)P(L_m^2/L_r)i_{sx}i_{sy}$, and by ignoring the effects of magnetic saturation, maximum torque is obtained by maximizing the product $i_{sx}i_{sy}$.

A third technique is also described below, where the effects of magnetic saturation are ignored but it is considered that the direct-axis rotor flux linkage during field weakening obeys eqn (4.1-25), thus by also considering eqn (4.1-22), $\psi_{rx}=L_m i_{sx}/(1+pT_r)$ is used in the field-weakening technique. This yields greatly improved torque capability. This follows from the fact that in this way an increased direct-axis rotor flux linkage is obtained, $\psi_{rx}=L_m i_{sx}+\Delta\psi_{rx}$ and thus the torque is increased to

$$t_e=(3/2)P(L_m/L_r)(\psi_{rx}+\Delta\psi_{rx})i_{sy},$$

where $\psi_{rx}=L_m i_{sx}$.

In the last part of the present section, the effects of magnetic saturation on the torque capability of an induction motor in the field-weakening range are considered. A detailed analysis is presented on the basis of combining the voltage and current limit curves and a modified rotor-flux-oriented control scheme is discussed which can be used in a wide speed range.

6.1.4.1 Conventional field-weakening technique

The inverter imposes limits on the maximum stator voltage and stator current. The maximum stator voltage U_{smax} is determined by the available d.c. link voltage and the PWM strategy, and the following expression must hold:

$$U_{sx}^2+U_{sy}^2\leqslant U_{smax}^2. \qquad (6.1\text{-}115)$$

In eqn (6.1-115) the direct- and quadrature-axis stator voltages U_{sx}, U_{sy} can be expressed in terms of the stator currents by considering eqns (4.1-7), (4.1-8), and also the fact that in the steady-state all the derivatives are zero:

$$U_{sx}=R_s I_{sx}-\omega_{mr}L_s' I_{sy} \qquad (6.1\text{-}116)$$

$$U_{sy}=R_s I_{sy}+\omega_{mr}\left[\frac{L_m}{L_r}|\bar{\psi}_r|+L_s' I_{sx}\right]. \qquad (6.1\text{-}117)$$

For high speeds, the ohmic voltage drops can be neglected in eqns (6.1-116) and (6.1-117), thus

$$U_{sx} = -\omega_{mr} L_s' I_{sy} \tag{6.1-118}$$

$$U_{sy} = \omega_{mr}\left[\frac{L_m}{L_r}|\bar{\psi}_r| + L_s' I_{sx}\right]. \tag{6.1-119}$$

With the conventional approach, $|\bar{\psi}_r| = L_m I_{sx}$ and eqn (6.1-119) is simplified to $U_{sy} = \omega_{mr} L_s L_{sx}$. Thus it follows by substitution of the expressions for U_{sx} and U_{sy} into eqn (6.1-115) that the limit for the steady-state stator currents is obtained as

$$(\omega_{mr} L_s' I_{sy})^2 + (\omega_{mr} L_s I_{sx})^2 \leqslant U_{smax}^2. \tag{6.1-120}$$

It follows from eqn (6.1-120) that the locus of the stator-current space vector is an ellipse, which can also be expressed as follows:

$$(I_{sx}/a)^2 + (I_{sy}/b)^2 \leqslant U_{smax}^2, \tag{6.1-121}$$

where a and b are the minor and major axes of the ellipse respectively, $a = U_{smax}/(\omega_{mr} L_s)$, $b = U_{smax}/(\omega_{mr} L_s')$ and ω_{mr} is the stator angular frequency. It follows that for different speeds, different ellipses are obtained and for larger speeds (for a given stator voltage), smaller ellipses are obtained (since a and b become smaller). However, the maximum stator current, I_{smax}, is also limited by the inverter current rating and the thermal rating of the induction motor. Thus the modulus of the stator-current space vector must also satisfy

$$I_{sx}^2 + I_{sy}^2 \leqslant I_{smax}^2, \tag{6.1-122}$$

which is the expression of a circle, and it follows that the stator-current space vector must stay inside the circle whose radius is I_{smax}. Figure 6.4 shows the circle (current-limited circle) and also various ellipses (voltage-limited ellipses), corresponding to different stator frequencies (rotor speeds).

When the conventional technique of field weakening is used, it is assumed that the direct-axis rotor flux linkage in the rotor-flux-oriented reference frame ($|\bar{\psi}_r| = \psi_{rx}$) is expressed as $\psi_{rx} = L_m i_{sx}$ and i_{sxref} is reduced in inverse proportion to the rotor speed; thus

$$I_{sxref} = \frac{I_{sxrated}}{\omega_r} \tag{6.1-123}$$

(where $I_{sxrated}$ is the direct-axis flux producing stator current, which produces rated rotor flux and ω_r is the p.u. rotor speed, which at the transition from the constant torque region to the constant power region is unity and above base speed it is larger than 1). In this case the reference value of the quadrature-axis stator current is

$$I_{syref} = (I_{smax}^2 - I_{sxref}^2)^{1/2}. \tag{6.1-124}$$

It follows that when the rotor flux is reduced in inverse proportion to the rotor speed, the reference stator-current space vector rotates along a circle in the

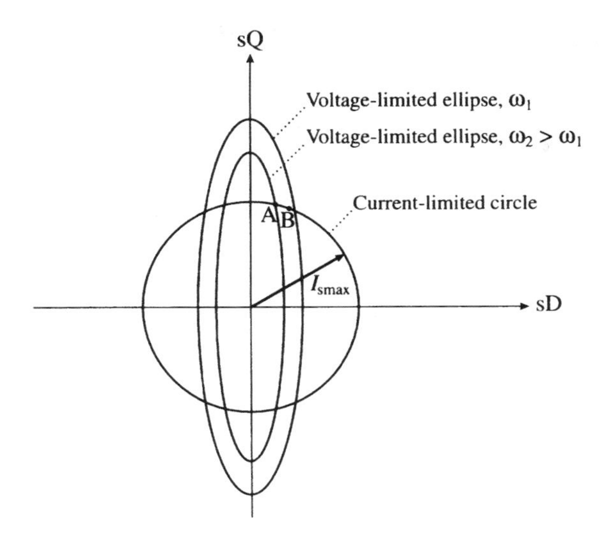

Fig. 6.4. Voltage and current limits in the field-weakening region.

anti-clockwise direction (e.g. from point B to point A in Fig. 6.4). If the stator-current space vector for a given frequency (speed) is inside the ellipse corresponding to that frequency (e.g. inside the ellipse corresponding to ω_1 at point B), field-weakening operation can be maintained. However, as discussed above, at larger frequencies the ellipses become smaller (e.g. this is shown in Fig. 6.4 as the ellipse for angular frequency ω_2), and a point is reached (point A in Fig. 6.4) where the stator-current space vector is at a position where the corresponding ellipse and circle intersect. At point A the inverter is saturated completely (maximum inverter voltage is reached) and there is no voltage margin to regulate the stator current. Operation at higher speeds requires that the stator-current space vector remains on the ellipse.

6.1.4.2 Field-weakening techniques yielding improved and maximal torque capability

Two techniques are described below, but the effects of magnetic saturation are neglected. Technique A gives improved torque capability, in contrast to the conventional field-weakening approach, discussed in the previous section, but the application of technique B gives higher torque in the field-weakening range.

Technique A: Field weakening yielding improved torque capability

When the maximum values of the stator voltage and current are specified, the frequency at which field weakening is started is not an independent variable. It can be obtained by considering eqns (6.1-121), (6.1-123), and (6.1-124). It follows that it is a function of the maximum stator voltage, maximum stator current, direct-axis rated current, and the inductances L_s and L'_s:

$$\omega_b = U_{smax}/[(L'_s I_{sxrated})^2 + L'^2_s(I^2_{smax} - I^2_{sxrated})]^{1/2}. \qquad (6.1\text{-}125)$$

In the constant output-power region part of the field weakening, the maximum torque is obtained [Kim *et al.* 1993] if the stator currents are chosen by combining the equations describing the voltage and current limits, eqn (6.1-120) and (6.1-122), and using the equality sign:

$$I_{smax} = \left[\frac{(U_{smax}/\omega_{mr})^2 - (L'_s I_{smax})^2}{L^2_s - L'^2_s}\right]^{1/2} \qquad (6.1\text{-}126)$$

$$I_{syref} = (I^2_{smax} - I^2_{sxref})^{1/2}. \qquad (6.1\text{-}127)$$

It should be noted that the physical reasons for using both voltage- and current-limiting conditions are also discussed in Section 6.1.4.3.

The upper part of the field-weakening range (constant speed × power range) starts at ω^*_b,

$$\omega^*_b = \frac{U_{smax}}{I_{smax}}\left[\frac{L^2_s + L'^2_s}{2L'^2_s L^2_s}\right]^{1/2} \qquad (6.1\text{-}128)$$

and optimum torque is obtained by considering the voltage limit only (see physical reasons for this in Section 6.1.4.3). Thus the optimal stator currents are selected by considering eqn (6.1-121) and $t_e = (3/2)P(L^2_m/L_r)i_{sx}i_{sy}$. Thus

$$I_{sxref} = \frac{U_{smax}}{\omega_{mr}\sqrt{2}L_s} \qquad (6.1\text{-}129)$$

$$I_{syref} = \frac{U_{smax}}{\omega_{mr}\sqrt{2}L'_s}. \qquad (6.1\text{-}130)$$

Technique B: Maximum torque capability

It has been discussed in the introduction to the present section that it is possible to obtain higher torque in the field-weakening range if it is considered that the direct-axis rotor flux linkage during field weakening obeys eqn (4.1-25); thus, by also considering eqn (4.1-22),

$$\psi_{rx} = \frac{L_m i_{sx}}{1 + T_r p} \qquad (6.1\text{-}131)$$

is used in the field-weakening technique, where $p = d/dt$ and T_r is the rotor time constant. This yields greatly improved torque capability. This follows from the fact that by solving eqn (6.1-131), an increased direct-axis rotor flux linkage is obtained,

$$\psi_{rx} = L_m i_{sx} + \Delta\psi_{rx}, \qquad (6.1\text{-}132)$$

where

$$\Delta\psi_{rx} = [\psi_{rx}(t_0) - L_m i_{sx}(t_0)]\exp[(t_0 - t)/T_r] \qquad (6.1\text{-}133)$$

and t_0 is the initial time. Thus the electromagnetic torque is increased to

$$t_e = \frac{3}{2} P \frac{L_m}{L_r} (\psi_{rx} + \Delta\psi_{rx}) i_{sy} = \frac{3}{2} P \frac{L_m}{L_r} (L_m i_{sx} + \Delta\psi_{rx}) i_{sy}, \qquad (6.1\text{-}134)$$

where $\omega_{rx} = L_m i_{sx}$.

The stator currents required to produce maximum torque in the first part of the field-weakening range (constant output-power range) are obtained similarly to the method described above with technique A. Thus first eqns (6.1-116), and (6.1-117) are used, but eqn (6.1-132) is substituted into eqn (6.1-117) and then the resulting equations are used together with the equations describing the voltage and current limits, eqns (6.1-115) and (6.1-122). It follows that the resulting expression for I_{sxref} is also a function of $(L_m/L_r)\Delta\psi_{rx}$ during a sampling interval. This can be approximated by $c = (L_m/L_r)[\psi_{rx}(t_0) - L_m i_{sx}(t_0)]$, where t_0 is the time at the beginning of each sampling period. Thus finally

$$I_{sxref} = [-cL_s + \{(cL_s)^2 + (L_s^2 - L_s'^2)[(U_{smax}/\omega_{mr})^2 - L_s'^2 I_{smax}^2 - c^2]\}^{1/2}]/(L_s^2 - L_s'^2) \qquad (6.1\text{-}135)$$

$$I_{syref} = (I_{smax}^2 - I_{sxref}^2)^{1/2} \qquad (6.1\text{-}136)$$

are obtained. It can be seen that if the extra rotor flux linkage component ($\Delta\psi_{rx}$) is neglected, then eqn (6.1-135) simplifies to eqn (6.1-126).

Similarly to that described above for technique A, in the second part (high-speed part) of the flux-weakening region, where the speed × power is constant, the stator currents which are required to produce maximal torque are obtained by considering the voltage-limit condition only, eqn (6.1-115), but not the current-limit condition. By using eqns (6.1-115) and (6.1-133), the electromagnetic torque can be expressed as

$$t_e = \frac{3}{2} P \frac{L_m}{L_s' L_r} [(U_{smax}/\omega_{mr})^2 - (L_s I_{sxref} + c)^2 (L_m i_{sxref} + \Delta\psi_{rx})]^{1/2}, \qquad (6.1\text{-}137)$$

and the stator currents required to produce maximum torque can be obtained by performing the differentiation of $dt_e/di_{sxref} = 0$.

6.1.4.3 Control of a saturated motor over a wide range of field weakening

In contrast to the previous two sections, in the present section the effects of magnetic saturation are also considered in a drive employing rotor-flux-oriented control. It has been shown (Grotstollen and Wiesing 1995) that when maximum torque and power are required over a wide range of field weakening, satisfactory results can only be achieved by considering magnetic saturation. For this purpose the saturated magnetizing inductance has to be utilized [Grotstollen and Wiesing 1995]:

$$L_m(|\bar{i}_m|) = L_\infty + L_\alpha \exp(-c_1|\bar{i}_m|) - L_\beta \exp(-c_2|\bar{i}_m|), \qquad (6.1\text{-}138)$$

where the various constants are obtained by curve fitting using the measured magnetizing curve of the machine. It is interesting to note that this contains a second exponential term to model the increasing slope of the magnetizing curve at low currents, which causes the first, almost linear, part of the magnetizing curve to be offset from the origin. It has been found [Grotstollen and Wiesing 1995] that this is important for modelling saturation at very high speed (where the rotor flux becomes small).

By considering eqn (2.1-191) the electromagnetic torque can be expressed as

$$t_e = \frac{3}{2} P \frac{L_m}{L_r} |\bar{\psi}_r| i_{sy}, \tag{6.1-139}$$

where $|\bar{\psi}_r| = L_m i_{sx}$ and $L_m = L_m(|\bar{i}_{ms}|)$. The modulus of the magnetizing-current space vector can be expressed as

$$|\bar{i}_m| = \{i_{sx}^2 + [i_{sy} L_{rl}/(L_{rl} + L_m)]^2\}^{1/2}. \tag{6.1-140}$$

This last expression follows from the fact that in the rotor-flux-oriented reference frame the magnetizing-current space vector can be expressed as $\bar{i}_m = \bar{i}_s' + \bar{i}_r' = i_{mx} + j i_{my}$, where $\bar{i}_s' = i_{sx} + j i_{sy}$ and $\bar{i}_r' = i_{rx} + j i_{ry}$, thus

$$|\bar{i}_m|^2 = i_{mx}^2 + i_{my}^2 = i_{sx}^2 + (i_{sy} + i_{ry})^2,$$

if $i_{mx} = i_{sx}$ is assumed. However, since $\psi_{ry} = 0 = L_r i_{ry}$, thus $i_{ry} = -L_m i_{sy}/L_r$ and therefore

$$|\bar{i}_m|^2 = i_{sx}^2 + (i_{sy} + i_{ry})^2 = i_{sx}^2 + i_{sy}^2(1 - L_m/L_r)^2 = i_{sx}^2 + [i_{sy} L_{rl}/(L_{rl} + L_m)]^2,$$

which gives eqn (6.1-140).

The effects of combining the voltage- and current-limiting constraints will now be discussed, and the variation of the electromagnetic torque as a function of the modulus of the rotor flux-linkage space vector will be obtained. The variation of electromagnetic torque with the rotor flux-linkage modulus is considered since, in the drive scheme employing rotor-flux-oriented control, the rotor flux-linkage modulus is a reference value used, and accurate setting of this is required to achieve optimum torque.

First only the current limit is considered, $i_{sx}^2 + i_{sy}^2 = i_{smax} = \text{const.}$ (second part of the flux-weakening region discussed in the previous section), and if $|\bar{\psi}_r| = L_m i_{sx}$ is used (the magnetizing inductance varies), then the electromagnetic torque versus rotor flux-linkage vector-modulus curve (current limit curve) can be determined by using eqn (6.1-139) together with eqns (6.1-138) and (6.1-140). This curve is also shown in Fig. 6.5, and this curve does not depend on the speed. It is very important to note that this curve is totally different from the one which can be obtained by neglecting magnetic saturation. The area under the corresponding curve which could be obtained by using an unsaturated machine model is much larger than the area under the curve shown in Fig. 6.5, which uses the saturated machine model. Furthermore, the maximum torque obtained from the 'unsaturated curve' is also much larger than that obtained from the 'saturated

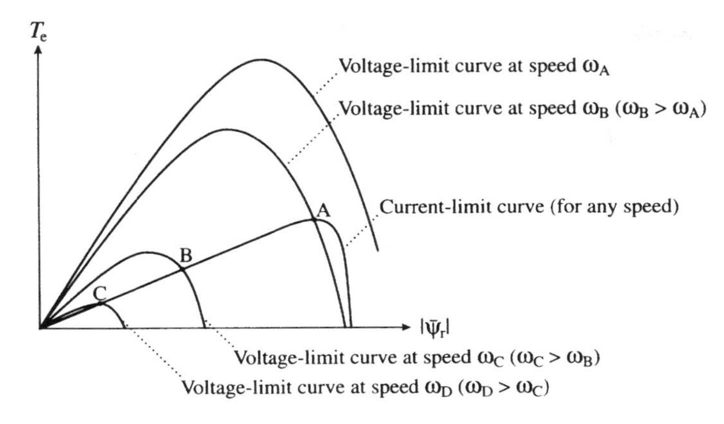

Fig. 6.5. Current and voltage limit curves.

curve' shown in Fig. 6.5. The error which would result by neglecting magnetic saturation also depends on the current limit. The area under the current limit curve in Fig. 6.5 (in which the permitted current is not exceeded) is called the permitted operating area.

As a second step, the torque–rotor flux curve is obtained if only the voltage limit is considered, e.g. when $u_{sx}^2 + u_{sy}^2 = u_{smax} = \text{const}$. By considering the direct- and quadrature-axis stator voltage equations in the steady-state, eqns (6.1-116) and (6.1-117), the rotor flux-linkage space-vector modulus and electromagnetic torque can be computed when the maximum stator voltage is applied. Thus it is possible to plot the torque–rotor flux characteristics for different (constant) values of the rotor speed. These curves are also shown in Fig. 6.5, and they are called the voltage-limit curves. It can be seen that in Fig. 6.5 four voltage-limit curves are shown, corresponding to four different speed values, where the speed values range from the small speed value (ω_A) to the very high speed value (ω_D). The peak of each voltage limit curve gives the maximum torque for the given speed without considering the current limitation. As expected, this torque is almost identical to the conventional maximum torque of a line-fed induction motor (computed for constant stator frequency). The voltage-limit curves mark the upper border of the possible operating region, which the drive cannot exceed due to limitation of the inverter voltage. The possible operating region is further decreased when the speed is increasing.

The actual torque capability of the induction motor can be determined by considering the voltage and current limits simultaneously. For this purpose the voltage and current-limit curves are combined (and these curves have been obtained by considering a saturated machine model). This is why the voltage and current curves have been plotted in the same diagram in Fig. 6.5. For each speed value, the area in which operation is permitted and also the point in this area where the torque is maximum can be identified. Thus the maximum torque (T_{emax}) and the corresponding rotor flux-linkage modulus (ψ_{ropt}) can be determined.

It follows from Fig. 6.5 that at low speeds, e.g. at speed ω_A (up to base speed), the current-limit curve (or at least its peak value) is located below the voltage-limit curve. Since the permitted operating area cannot be exceeded, the peak of the current-limit curve determines the maximum torque. It follows that the maximum torque does not depend on the actual rotor speed and is achieved when the machine is operated with constant rotor flux (ψ_{ropt}=const.), which is the rated flux. At speed ω_B, the corresponding voltage-limit curve intersects the peak of the current-limit curve at point A; this is the border of the basic (low speed) region, and until this point the rotor flux is constant. This is where the lower flux-weakening region starts (medium-speed region), and lasts until ω_B^*, and above this is the high-speed region.

At medium speeds, e.g. at ω_C, the intersection of the voltage-limit curve and the current-limit curve is at point B. However, now the peak of the current-limit curve is located outside the possible operating region, and cannot be attained. Thus the maximum torque which is permitted and is possible at speed ω_C is achieved when the drive is operated at the intersection of both limiting curves, where the voltage and also the current are maximal and where as a consequence, maximal apparent power is applied to the motor. When the speed varies, the point of maximum electromagnetic torque shifts on the current-limit curve and flux weakening has to be applied when the speed is increased ($\psi_{ropt} < \psi_{rated}$). A simple control technique to reach maximum torque regardless of rotor speed is to apply maximum stator currents to the motor, with as much flux-generating direct-axis component as allowed by the limited voltage. It is an important advantage of this control technique that it does not depend on any machine parameter or the actual value of the maximum inverter voltage or the rotor flux.

The upper border of the lower (first) flux-weakening region is reached at speed ω_D, where the peak of the voltage-limit curve reaches the current-limit curve (point C in Fig. 6.5). At higher speeds ($\omega_r > \omega_D$), the peak of this voltage-limit curve (or even the entire voltage-limit curve) is located below the current-limit curve. Thus maximum electromagnetic torque is determined by the voltage limit only and it appears at the peak of the voltage-limit curve. It should be noted that this physical property has also been used in Section 6.1.4.2 for the derivation of eqns (6.1-129) and (6.1-130). It follows that the control strategy must be changed, otherwise torque break-off will appear at high speed, due to the fact that the intersection of the current and voltage-limit curves, being the set-point on the lower flux-weakening region, is shifted to very low torque values and disappears with increasing sneed. As a likely method, the flux reference can be obtained by using a flux-speed characteristic which, however, should not be a hyperbola (the flux is decreased in inverse proportion to the speed).

A modified rotor-flux-oriented control strategy can then be implemented to achieve full utilization of the torque capability over the entire speed region. For this purpose the reference value of the flux-producing stator current (i_{sxref}) is obtained on the output of a flux controller and the torque-producing stator current is obtained (i_{syref}) on the output of the speed controller as shown in Fig. 6.6.

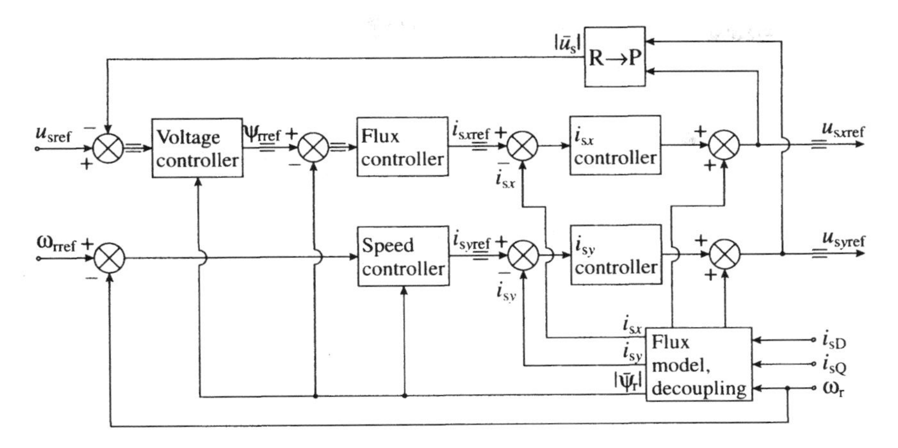

Fig. 6.6. Schematic of drive scheme with modified rotor-flux-oriented control.

The speed controller is adapted to the variations of the rotor flux. The input to the flux controller is $\psi_{rref}-\psi_r$ and ψ_{rref} is obtained on the output of a voltage controller [Grotstollen and Wiesing 1995]. The input to the voltage controller is $u_{sref}-|\bar{u}_s|$, where $|\bar{u}_s|=(u_{sx}^2+u_{sy}^2)^{1/2}$. The voltage controller has two main features. It increases the rotor flux when the voltage required by the motor does not exceed the value set by the voltage reference (u_{sref}). In this way, it is aiming to operate the motor at the voltage limit. Furthermore, flux is reduced automatically when the voltage required by the induction motor is too high. Thus the voltage requirement of the motor is automatically adjusted to the voltage capability of the inverter by varying the flux in the flux-weakening region.

6.2 Vector control of salient-pole machines by considering the effects of main-flux saturation

In this section the effects of main-flux saturation are incorporated into the five-winding model of the salient-pole synchronous machine discussed in Section 3.2. All the other assumptions are those used for the saturated smooth-air-gap machine. Only a brief discussion is given since the various forms of vector control technique can be developed in ways similar to those described in Sections 3.1 and 3.2.

6.2.1 STATOR VOLTAGE EQUATIONS

It follows from eqns (3.2-14) and (3.2-15) that in the reference frame fixed to the rotor the stator equations take the form:

$$u_{sd}=R_s i_{sd}+\frac{d\psi_{sd}}{dt}-\omega_r\psi_{sq} \tag{6.2-1}$$

$$u_{sq}=R_s i_{sq}+\frac{d\psi_{sq}}{dt}+\omega_r\psi_{sd}, \tag{6.2-2}$$

where the two-axis components of the stator flux linkages are

$$\psi_{sd}=L_{sl}i_{sd}+\psi_{md} \tag{6.2-3}$$

and

$$\psi_{sq}=L_{sl}i_{sq}+\psi_{mq}. \tag{6.2-4}$$

ψ_{md} and ψ_{mq} are the direct- and quadrature-axis components of the magnetizing flux linkages in the reference frame fixed to the rotor, and from eqns (3.2-1) and (3.2-2) they can be defined as

$$\psi_{md}=L_{md}i_{md}=L_{md}(i_{sd}+i_{r\alpha}+i_{rF}) \tag{6.2-5}$$

$$\psi_{mq}=L_{mq}i_{mq}=L_{md}(i_{sq}+i_{r\beta}). \tag{6.2-6}$$

i_{md} and i_{mq} are the magnetizing current components along the real and imaginary axes of the reference frame fixed to the rotor and L_{md} and L_{mq} are the magnetizing inductances in the direct and quadrature axes respectively. As a result of physical saliency these are different even under linear conditions. However, under saturated conditions they vary with the currents. This will now be discussed.

In the salient-pole machine, because of the physical saliency, in general, the magnetizing flux-linkage space phasor is not coaxial with the magnetizing-current space phasor (the resultant flux wave is not collinear with the resultant m.m.f. wave). In special cases, when the m.m.f. lies completely in one axis (the magnetizing, current space phasor has only a direct-axis or a quadrature-axis component) then the magnetizing flux-linkage space phasor is coaxial with the magnetizing-current space phasor. Thus it is more problematic to characterize the saturation level in the saturated salient-pole machine than in the saturated smooth-air-gap machine, where the magnetizing flux-linkage space phasor is coaxial with the magnetizing-current phasor and where the amplitude of the magnetizing flux-linkage space phasor is a non-linear function of the magnetizing current, as discussed in Section 6.1. In the salient-pole machine, because of symmetry, the magnetizing flux linkage along the direct axis (ψ_{md}) must be zero if the direct-axis magnetizing current is zero; thus under saturated conditions $\psi_{md}=L_{md}(i_{md}, i_{mq})i_{md}$, where L_{md} is a non-linear function of the magnetizing current components, and similarly $\psi_{mq}=L_{mq}(i_{md}, i_{mq})i_{mq}$, where L_{mq} is another non-linear function of the magnetizing currents. However, it is assumed that the pole tips and/or the teeth are the most saturated parts of the magnetic circuit, and therefore, as for the smooth-air-gap machine, it is assumed that in each axis the saturation level is determined by the amplitude of the magnetizing-current space phasor $|\bar{i}_m|=(i_{md}^2+i_{mq}^2)^{1/2}$. Thus ψ_{md}/i_{md} and ψ_{mq}/i_{mq} depend only on $|\bar{i}_m|$ and

$$\psi_{md}=L_{md}(|\bar{i}_m|)i_{md} \tag{6.2-7}$$

$$\psi_{mq}=L_{mq}(|\bar{i}_m|)i_{mq}. \tag{6.2-8}$$

The magnetizing inductances L_{md} and L_{mq} are chord-slope, static inductances and can be measured or computed by considering the direct- and quadrature-axis magnetizing curves respectively, with excitation in the corresponding axis only (when $i_{md}=|\bar{\imath}_m|$ and $i_{mq}=|\bar{\imath}_m|$ respectively). If the flux linkages are the state variables of the model, then it is only the chord-slope inductances which are present in the equations (together with other parameters), but if such a model is used, where the currents are the state variables, the 'tangent-slope' dynamic inductances in the two axes are also present. The direct-axis dynamic inductance can be defined as $L^d=\mathrm{d}\psi_{md}/\mathrm{d}i_{md}$, but again there is only excitation along the direct axis and thus $|i_{md}|=|\bar{\imath}_m|$. Similarly, the quadrature-axis dynamic inductance can be defined as $L^q=\mathrm{d}\psi_{mq}/\mathrm{d}i_{mq}$, and there is only excitation along the quadrature axis, thus $|i_{mq}|=|\bar{\imath}_m|$. The inductances L_{md}, L_{mq}, L_d and L_q will now be utilized in the expanded forms of the voltage equations.

Equations (6.2-1)–(6.2-6) describe the stator voltage equations in the presence of saturation of the main flux paths and can be used for various implementations of vector control. It is also possible to expand these equations, and by the substitution of eqns (6.2-3) and (6.2-4) into eqns (6.2-1) and (6.2-2),

$$u_{sd}=R_s i_{sd}+L_{sl}\frac{\mathrm{d}i_{sd}}{\mathrm{d}t}+\frac{\mathrm{d}\psi_{md}}{\mathrm{d}t}-\omega_r(L_{sl}i_{sq}+\psi_{mq}) \tag{6.2-9}$$

$$u_{sq}=R_s i_{sq}+L_{sl}\frac{\mathrm{d}i_{sq}}{\mathrm{d}t}+\frac{\mathrm{d}\psi_{mq}}{\mathrm{d}t}+\omega_r(L_{sl}i_{sd}+\psi_{md}). \tag{6.2-10}$$

R_s and L_{sl} are the resistance and leakage inductance of a stator winding respectively and L_{sl} is assumed to be constant. Equations (6.2-9) and (6.2-10) contain the magnetizing voltages along the direct and quadrature axes respectively and, by the application of the chain differentiation rule, they can be expanded in a similar way to that shown in eqns (6.1-34) and (6.1-35), so that

$$\frac{\mathrm{d}\psi_{md}}{\mathrm{d}t}=\frac{\mathrm{d}(L_{md}\,\mathrm{d}i_{md})}{\mathrm{d}t}=L_{mD}\frac{\mathrm{d}i_{md}}{\mathrm{d}t}+L_{dq}\frac{\mathrm{d}i_{mq}}{\mathrm{d}t} \tag{6.2-11}$$

$$\frac{\mathrm{d}\psi_{mq}}{\mathrm{d}t}=\frac{\mathrm{d}(L_{md}\,\mathrm{d}i_{mq})}{\mathrm{d}t}=L_{mQ}\frac{\mathrm{d}i_{mq}}{\mathrm{d}t}+L_{qd}\frac{\mathrm{d}i_{md}}{\mathrm{d}t}, \tag{6.2-12}$$

$$L_{mD}=L^d\cos^2\mu'+L_{md}\sin^2\mu' \tag{6.2-13}$$

$$L_{mQ}=L^q\sin^2\mu'+L_{mq}\cos^2\mu'. \tag{6.2-14}$$

In eqns (6.2-13) and (6.2-14) the angle $\mu'=\mu_m-\theta_r$ is the angle of the space phasor of the magnetizing currents with respect to the direct axis of the reference frame fixed to the rotor, μ_m is the angle of the same space phasor with respect to the direct axis of the stationary reference frame, and θ_r is the rotor angle. In eqns (6.2-11) and (6.2-12)

$$L_{dq}=(L^d-L_{md})\sin\mu'\cos\mu' \tag{6.2-15}$$

$$L_{qd}=(L^q-L_{mq})\sin\mu'\cos\mu' \tag{6.2-16}$$

and L_{dq} and L_{qd} are the cross-coupling (mutual) inductances between the d and q, and the q and d axes respectively. In the smooth-air-gap machine $L_{dq}=L_{qd}$ due to symmetry, but in the salient-pole machine this reciprocity only holds if certain criteria are satisfied. Namely, it can be shown that as for the reciprocity conditions of two stationary, non-linear inductors (denoted by a, b respectively), where the reciprocity condition for the mutual inductances ($M_{ab}=M_{ba}$) holds only if the inductances are assumed to be lossless or if the losses are modelled separately, in the salient-pole machine the reciprocity condition $L_{dq}=L_{qd}$ is only satisfied if the inductances of the machine are considered to be lossless (or if the losses are modelled separately).

Substitution of eqns (6.2-11) and (6.2-12) into eqns (6.2-9) and (6.2-10) respectively yield the following stator voltage equations, if eqns (6.2-5) and (6.2-6) are also substituted into eqns (6.2-9) and (6.2-10):

$$u_{sd}=R_s i_{sd}+L_{sl}\frac{di_{sd}}{dt}+L_{mD}\frac{d(i_{sd}+i_{r\alpha}+i_{rF})}{dt}+L_{dq}\frac{d(i_{sq}+i_{r\beta})}{dt}$$

$$-\omega_r[L_{sl}i_{sq}+L_{mq}(i_{sq}+i_{r\beta})] \qquad (6.2\text{-}17)$$

$$u_{sq}=R_s i_{sq}+L_{sl}\frac{di_{sq}}{dt}+L_{mQ}\frac{d(i_{sq}+i_{r\beta})}{dt}+L_{qd}\frac{d(i_{sd}+i_{r\alpha}+i_{rF})}{dt}$$

$$+\omega_r[L_{sl}i_{sd}+L_{md}(i_{sd}+i_{r\alpha}+i_{rF})]. \qquad (6.2\text{-}18)$$

It follows that in general, due to saturation, in the stator voltage equations four inductances, L_{md}, L_{mq}, L^d, and L^q have to be considered together with the angle of the magnetizing-current space phasor μ'. The equations also contain the effects of cross-saturation and thus there exists cross-saturation coupling between all the windings which are in space quadrature, resulting in cross-saturation transformer e.m.f.s. Under linear magnetic conditions $L_{mD}=L^d=L_{md}$ and $L_{mQ}=L^q=L_{mq}$, $L_{dq}=L_{qd}=0$ and the well-known equations are obtained.

If $i_{rF}=0$ and a smooth air-gap is considered, the static inductances are equal in the two axes and $L_{md}=L_{mq}=L_m$. Furthermore, the dynamic inductances are also equal and $L^d=L^q=L_d$. It follows from eqns (6.2-13), (6.2-14), and (6.2-15) that under these conditions, in the stationary reference frame the expressions for L_{mD}, L_{mQ}, and L_{dq} agree with eqns (6.1-50), (6.1-51), and (6.1-52) respectively and eqns (6.2-17) and (6.2-18) yield the stator voltage equations given by eqns (6.1-53) and (6.1-54) respectively.

6.2.2 ROTOR VOLTAGE EQUATIONS

The rotor voltage equations of the saturated five-winding models of the salient-pole machine can be obtained in a similar way to the equations derived in the previous section. It follows from eqns (3.2-18) and (3.2-19) that the voltage equation for the field winding can be put into the following form in the reference frame fixed to the rotor:

$$u_{rF}=R_{rF}i_{rF}+L_{Fl}\frac{di_{rF}}{dt}+\frac{d\psi_{md}}{dt}, \tag{6.2-19}$$

where R_{rF} is the resistance of the field winding and L_{Fl} is the leakage inductance of the field winding. Similarly, it follows from eqns (3.2-20) and (3.2-21) that the voltage equations of the direct- and quadrature-axis damper windings are

$$0=R_{r\alpha}i_{r\alpha}+L_{r\alpha l}\frac{di_{r\alpha}}{dt}+\frac{d\psi_{md}}{dt} \tag{6.2-20}$$

$$0=R_{r\beta}i_{r\beta}+L_{r\beta l}\frac{di_{r\beta}}{dt}+\frac{d\psi_{mq}}{dt}, \tag{6.2-21}$$

where $R_{r\alpha}$ and $R_{r\beta}$ are the resistances of the damper windings along the direct and quadrature axes respectively and $i_{r\alpha}$ and $i_{r\beta}$ are the currents in the two damper windings. Equations (6.2-19)–(6.2-21) contain the derivatives of the direct- and quadrature-axis magnetizing flux-linkage components, and they have been given by eqns (6.2-11) and (6.2-12) respectively. Thus when eqns (6.2-11) and (6.2-12) are substituted into eqns (6.2-19), (6.2-20), and (6.2-21), the following rotor voltage equations are obtained, if eqns (6.2-5) and (6.2-6) are also considered:

$$u_{rF}=R_{rF}i_{rF}+L_{Fl}\frac{di_{rF}}{dt}+L_{mD}\frac{d(i_{sd}+i_{r\alpha}+i_{rF})}{dt}+L_{dq}\frac{d(i_{sq}+i_{r\beta})}{dt} \tag{6.2-22}$$

$$0=R_{r\alpha}i_{r\alpha}+L_{r\alpha l}\frac{di_{r\alpha}}{dt}+L_{mD}\frac{d(i_{sd}+i_{r\alpha}+i_{rF})}{dt}+L_{dq}\frac{d(i_{sq}+i_{r\beta})}{dt} \tag{6.2-23}$$

$$0=R_{r\beta}i_{r\beta}+L_{r\beta l}\frac{di_{r\beta}}{dt}+L_{mQ}\frac{d(i_{sq}+i_{r\beta})}{dt}+L_{qd}\frac{d(i_{sd}+i_{r\alpha}+i_{rF})}{dt}. \tag{6.2-24}$$

In these equations the inductances L_{mD}, L_{mQ} and L_{dq}, L_{qd} are present, and these have been defined in eqns (6.2-13)–(6.2-16) respectively. Because of saturation of the main flux paths, these equations contain the effects of cross-saturation as well. Under linear magnetic conditions, the well-known forms of these equations are obtained. In general, the cross-saturation coupling terms (L_{dq}, L_{qd}) only disappear under linear magnetic conditions or if the magnetizing-current space phasor lies along the direct axis or the quadrature axis (when $\mu_m=0$ or $\mu_m=\pi/2$).

Bibliography

Anvari, H. A. and Faucher, J. (1988). Simulation considering the cross-magnetization effect. In *Conference Proceedings of International Conference on Electrical Machines, Pisa*, pp. 519–24.

Anvari, H. A., Faucher, J., and Trennoy, B. (1986). On the cross-coupling saturation effect and its computation by the finite difference solution. In *Conference Proceedings of International Conference on Electrical Machines, München*, pp. 451–4.
Boldea, I. and Nasar, S. A. (1987). Unified treatment of core losses and saturation in the orthogonal-axis model of electrical machines. *IEE Proc.* **B 134**, 355–63.
Brown, J. E., Kovács, K. P., and Vas, P. (1982). A method of including the effects of main flux saturation in the generalized equations of a.c. machines. *IEEE Transactions on PAS* **PAS-102**, 96–103.
Bunte, A. and Grotstollen, H. (1993). Parameter identification of an inverter-fed induction motor at standstill with a correlation method. *EPE, Brighton*, pp. 97–102.
Bunte, A. and Grotstollen, H. (1996). Off-line parameter identification of an inverter-fed induction motor at standstill. *EPE, Seville*, 3.492–3.496.
Delfino, B., Denegri, G. B., Pinceti, P., and Schiappacasse, A. (1985). Dynamic analysis of turbogenerator performance based on saturated equivalent circuit models. In *Universities Power Engineering Conference, Huddersfield*, pp. 234–7.
Grotstollen, H. and Wiesing, J. (1995). Torque capability and control of a saturated induction motor over a wide range of flux weakening. *IEEE Transaction on Ind. Electronics* **42**, 374–81.
Hallenius, K. E., Vas, P., and Brown, J. E. (1988). The analysis of a saturated self-excited asynchronous generator. *IEEE Power Engineering Society Winter Meeting, New York*, Paper 88 WM 017–6.
He, Y.-K. and Lipo, T. A. (1985). Computer simulation of an induction machine with spatially dependent saturation. *IEEE Transactions on Industry Applications* **IA-21**, 226–34.
Healey, R. C. (1995). The implementation of a vector control scheme based on an advanced motor model. *EPE, Seville*, 3.017–3.106.
Hofman, H., Sanders, S. R., and Sullivan, C. (1995). Stator-flux-based vector control of induction machines in magnetic saturation. *IEEE IAS Meeting, Orlando*, pp. 152–8.
Jack, A. G., Vas, P., and Brown, J. E. (1980). A finite element study of the phenomenon of cross-magnetization. In *Conference Proceedings of International Conference on Electrical Machines, München*, pp. 462–5.
Jing-Qiu, Q. and Yun-Qiu, T. (1988). Cross-magnetization effect and d,q axis coupling reactances. In *Conference Proceedings of International Conference on Electrical Machines, Pisa*, pp. 109–12.
Jotten, R. and Schierling, H. (1983). Control of the induction machine in the field weakening region. IFAC, 297–304.
Khater, F. M. H., Lorenz, R. D., Novotny, D. W., and Tang, K. (1987). Selection of flux level in field-oriented induction machine controllers with consideration of magnetic saturation effects. *IEEE Transactions on Industry Applications* **IA-23**, 276–82.
Kim, S. H., Sul, S. K., and Park, M. H. (1993). Maximum torque control of an induction machine in the field weakening region. *IEEE IAS Meeting*, pp. 401–7.
Leplat, P. M., Tounzi, A., Clenet, S., and Piriou, F. (1996). Study of induction machine using Park's model and finite element method. *ICEM, Vigo*, pp. 24–9.
Levi, E. (1995). Improvements in operation of vector-controlled induction machines by application of modified machine models. *IEE Colloquium on Advances in Control Systems for Electric Drives*, 3.1–3.8.
Macdonald, A. C. and Turner, P. J. (1983). Turbine generator reactances in the steady-state. IEE Colloquium on 'Finite element methods applied to electrical machines'. IEE, London, March, also 18 Universities Power Engineering Conference, Surrey.

Melkebeek, J. A. (1983). Magnetizing field saturation and dynamic behaviour of induction machines. Part I: An improved calculation method for induction machine dynamics. *IEE Proc. B.* **130**, 1–9.

Melkebeek, J. E., Vas, P., and Brown, J. E. (1985). Dynamic analysis of synchronous machines incorporating the effects of main flux saturation. *20 Universities Power Engineering Conference, Huddersfield*, pp. 238–41.

Melkebeek, J. A. and Willems, J. L. (1988). Reciprocity relations for the mutual inductances of orthogonal axis windings in saturated electrical machines. *IEEE IAS Annual Meeting*, pp. 104–10.

Minnich, S. H., Schulz, R. P., Baker, D. H., Sharma, D. K., Farmer, R. G., and Fish, J. H. (1987). Saturation function for synchronous generators from the finite elements. *IEEE Transactions on Energy Conversion* **2**, 680–2.

Nehl, T. W., Fouad, F. A., and Demerdash, N. A. (1982). Determination of saturated values of rotating machinery incremental and apparent inductances by energy perturbation method. *IEEE Transactions on PAS* **PAS-101**, 4441–52.

Robert, J. (1987). A simplified method for the study of saturation in a.c. machines. *International Association for Mathematics and Computers. In Simulation, Symposium on Modelling and Simulation of Electrical Machines, Converters and Power Systems, Quebec*, pp. 119–26.

Rácz, I. and Vas, P. (1985). Oscillatory behaviour of induction machines employed in railway traction. *2nd International Conference on Electrical Machines—Design and Applications, London*, pp. 168–75. IEE, London.

Ramshaw, R. S. and Xie, G. (1984). Nonlinear model of nonsalient synchronous machines. *IEEE Transactions on PAS* **PAS-103**, 1809–15.

Richter, E. and Neumann, T. W. (1984). Saturation effects in salient-pole synchronous machines with permanent magnet excitation. In *Proceedings of International Conference on Electrical Machines, Lausanne*, pp. 603–6.

Ruff, M. and Grotstollen, H. (1993). Identification of the saturated mutual inductance of an asynchronous motor at standstill by recursive squares algorithm. *EPE, Brighton*, pp. 103–8.

Shackshaft, G. (1979). Model of generator saturation for use in power system studies. *Proc. IEE* **126** 759–63.

Sobczyk, T. J. (1996). An energy-based approach to modelling the magnetic nonlinearity in a.c. machines. *ICEM, Vigo*, pp. 68–73.

Stiebler, M. H. and Ritter, C. (1985). Synchronous machine modelling including d/q flux saturation. *IEE Second International Conference on Electrical Machines and Drives, London*, pp. 108–12.

Sullivan, C. R., Brian, C. K., and Sanders, S. R. (1996). Control systems for induction machines with magnetic saturation. *IEEE Transactions on Ind. Electronics* **43**, 142–51.

Vas, P. (1981). Generalized analysis of saturated a.c. machines. *Archiv für Elektrotechnik* **63**, 57–62.

Vas, P. (1992). *Electrical machines and drives: A space-vector theory approach.* Oxford University Press.

Vas, P. (1987). Various models for saturated electrical machines. *International Conference on Saturated Electrical Machines, Paper 1, Liege, 2 December 1987.*

Vas, P. and Hallenius, K. E. (1989). A three-phase and a quadrature-phase slip-ring model for saturated smooth-air-gap electrical machines. *Archiv für Elektrotechnik* **72**, 59–68.

Vas, P. and Alaküla, M. (1989). Field-oriented control of saturated induction machines. *IEEE Transactions on Energy Conversion* **5**, 218–24.

Vas, P., Brown, J. E., and Hallenius, K. E. (1984). Cross-saturation effect in smooth-air-gap electrical machines. In *Proceedings of International Conference on Electrical Machines, Lausanne*, pp. 261–4.

Vas, P., Deleroi, W., and Brown, J. E. (1984). Transient analysis of smooth-air-gap machines incorporating the effects of main and leakage flux saturation. In *Proceedings of International Conference on Electrical Machines, Lausanne*, pp. 269–72.

Vas, P., Hallenius, K. E., and Brown, J. E. (1985). Transient analysis of saturated smooth-air-gap synchronous machines with and without damper windings. *11 International Association for Mathematics and Computers in Simulation World Congress, Oslo, Vol. 3,* pp. 289–92.

Vas, P., Hallenius, K. E., and Brown, J. E. (1985). Computer simulation of saturated cylindrical rotor synchronous machines. *11 International Association for Mathematics and Computers in Simulation World Congress, Oslo*, Vol. 3, pp. 303–6.

Vas, P., Hallenius, K. E., and Brown, J. E. (1986). Cross-saturation in smooth-air-gap electrical machines. *IEEE Transactions on Energy Conversion* **EC-1**, 103–12.

Xu, X., De Doncker, R., and Novotny, D. W. (1988). Stator flux orientation control of induction machines in the field weakening region. *IEEE IAS Meeting*, pp. 437–43.

7 Artificial-intelligence-based steady-state and transient analysis of electrical machines and drives; AI-based estimators

7.1 General introduction; advantages of AI-based systems

Traditionally mathematical-model-based analysis techniques have always been used for the steady-state and transient analysis of electrical machines. There are many techniques, but they mainly differ in their complexity and the many assumptions used. One of the most popular techniques relies on space-vector theory and is extensively used worldwide, and it is this technique which has also been extensively used in the present book for the development of various high-performance, instantaneous torque-controlled drives.

In general, when space-vector theory is used for the simulation of a variable-speed drive system, the stator and rotor voltage space-vector equations (in the appropriate reference frame) of the machine considered are solved together with the equation of motion and the equations governing the controllers and converter. This results in a system of first-order non-linear differential equations, plus several algebraic equations. In general, the solutions are obtained by using a numerical technique (e.g. Runge–Kutta technique). This way it is possible to compute various machine quantities, e.g. flux linkages, rotor speed, electromagnetic torque, etc. For implementation purposes (e.g. to implement speed, torque, flux-linkage, etc. estimators), conventionally again the mathematical-model-based techniques are used, e.g. space-vector theory.

However, it is possible to perform the simulations and also implementations of estimators, controllers, etc. (in real time) by using artificial-intelligence (AI) -based techniques (e.g. artificial neural networks (ANN), fuzzy-logic systems, fuzzy-neural networks, etc.), which do not require a mathematical model of the machine and drive system. Such a system is not restricted by the many assumptions used in conventional electrical machine and linear control theories. It can also yield the results more quickly than by using the conventional approach. When an AI-based system is implemented, in general, some of the main advantages are:

- The design does not require a mathematical model of the plant (e.g. the design can be based exclusively on using target system data; see also below).
- In a fuzzy-logic system, which is an expert system, the design can be based exclusively on linguistic information available from experts, or by using clustering (of data) or other techniques (e.g. when expert information is not available).
- In a fuzzy-neural system, which is an 'expert-guided' neural network, the design can be performed by using linguistic rules available from the experts or

clustering or other techniques (e.g. when expert information is not available). Thus a model (network) can also be constructed by using target system data only and there is no need to have prior knowledge of the fuzzy rules and membership functions (at the beginning of training). However, expert knowledge of the target system helps to initialize the network structure, and in this way the training of the network starts with reduced errors and the training time is also reduced.

- In a system using a conventional (non-fuzzy) neural network, if the ANN uses a supervised training technique, then the design is based on information available for the training and this information can originate from various sources, including measurements (response data). However, when the ANN does not use supervised training, e.g. it is a self-organizing neural network, the ANN uses available data (e.g. measured data, or data from other sources), and the ANN classifies the data according to its features.
- The tuning effort of an AI-based system can be less than that of a conventional system.
- The system may generalize extremely well (it can give good estimates when some new unknown input data are used) and thus be independent of particular characteristics of the drive.
- A solution may be obtained to problems which are intractable by conventional methods.
- Such a system exhibits good noise rejection properties.
- Such a system can be fault tolerant (e.g. if a neuron is destroyed or deleted in an ANN or a rule is eliminated in a fuzzy-neural network, then the AI-based system will continue to operate due to its parallel architecture, but the performance will deteriorate).
- Such a system can be easily extended and modified.
- Such a system can be robust to parameter variations.
- Such a system can be computationally less intensive (when a minimum configuration system is used [Vas 1996]).
- Such a system leads to reduced development times, etc. [Vas 1996; Beierke 1995].

There are many excellent publications which use artificial-intelligence-based techniques for various purposes in electrical machines and drives. These include parameter and state estimators (e.g. in many publications different types of electrical machine parameters are estimated by using ANNs), condition monitoring (e.g. various faults are detected in electrical machines by using ANNs or fuzzy logics), controller applications (e.g. in most of the cases a conventional speed controller is replaced by a fuzzy-logic speed controller), etc. Many publications are referenced in a recent review paper [Vas 1996]. Furthermore, various applications of artificial intelligence have also been presented in this book in different sections (e.g. for induction motors see Sections 4.4, 4.5.3.6, 4.6.2.4.2, etc.). However, it is believed, that this is the first time when an attempt is made to perform, in general, the steady-state and transient analysis of various electrical machines by using artificial intelligence. It should be noted that this has very important practical consequences and the results can be directly used for implementation purposes (e.g. to implement instantaneous speed, torque, position, flux-linkage estimators in various drives), as discussed below.

The similarity of the control schemes of d.c. and a.c. motors has been discussed in Section 1.2.1. Thus a d.c. drive can also be considered as a development platform for high-performance drives, and therefore the simulation of d.c. machines using AI-based techniques will also be presented. This approach also gives the possibility of comparing the complexity of an AI-based d.c. motor simulation to that of an AI-based a.c. motor. Furthermore, the AI-based simulation of induction machines will also be discussed.

7.2 Neural-network-based steady-state and transient analysis of d.c. machines, estimators

7.2.1 CONVENTIONAL ANALYSIS

For the purposes of analysis a separately excited d.c. motor is considered. For conventional transient analysis, in general, the voltage equations for the field winding and armature winding are required together with the equation of motion. Thus in general, this represents a system of three first-order, simultaneous differential equations. Assuming that the machine parameters, field and armature voltage, and load torque are known, it is then possible to solve the three differential equations for the three unknowns, which are the field current (i_f), armature current (i_a), and rotor speed (ω_r), and by knowing the machine currents, the electromagnetic torque (t_e) can also be determined. In a drive system, e.g. where the speed is varied by varying the armature voltage using a convertor, it is possible to perform the simulations by also considering the equations of the controllers and converter.

For illustration purposes Fig. 7.1 shows the speed versus time characteristic of a separately excited d.c. machine, with a constant field and armature voltage and zero load torque, after switching the machine on the supply. This curve has been obtained by numerical solution of the three differential equations discussed above, and this will be referred to as the conventional simulation technique.

However, it is also possible to perform the transient analysis of the d.c. machine by using an artificial neural network (ANN), since it is known that, in general, an appropriate ANN is a general non-linear function estimator. There are many types of ANNs, but it is beyond the scope of the present book to discuss these and in the next section only a multi-layer feedforward ANN with supervised training [Kosko 1992] will be discussed briefly.

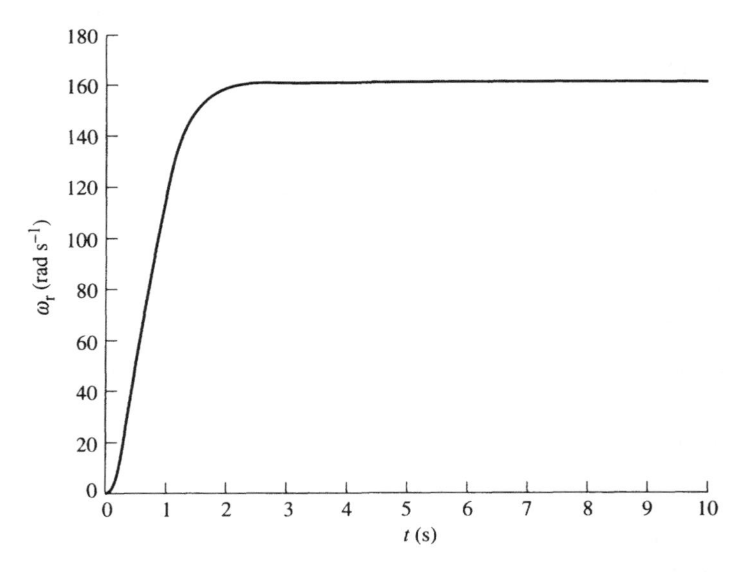

Fig. 7.1. Run-up speed of a separately excited d.c. machine (armature voltage 240 V).

7.2.2 MULTI-LAYER FEEDFORWARD NEURAL NETWORK; TRAINING

It is the goal to obtain an ANN, on the inputs of which there are some known signals (e.g. these could be measured in the d.c. drive system) and on the outputs of which there are the signals which have to be estimated (e.g. rotor speed, electromagnetic torque, etc.). For simplicity, in Fig. 7.2(a) there is shown the schematic of a d.c. machine with four input signals and one output signal and in Fig. 7.2(b) there is shown the equivalent neural network. Since the machine is non-linear, the goal is to obtain the ANN which is a non-linear function estimator.

In general, in a multi-layer feedforward artificial neural network there is an input layer, an output layer, and between the input and output layers there are so-called hidden layers. Such an ANN is shown in Fig. 7.2(c), but for simplicity only two hidden layers are shown.

In all the layers there are nodes, e.g. in the input layer shown in Fig. 7.2(c) there are four nodes due to the fact that four input signals have been assumed. For simplicity, in the first hidden layer there are two nodes and in the second hidden layer there is one hidden node; however, in general, there could be any number of hidden nodes. Finally in the output layer there is only one output node, since it has been assumed that there is only a single output signal. The input signals are direct inputs to the input nodes. All the nodes of a specific layer are connected by weights to all the nodes in the next layer, these weights are denoted by w, and denote the strength of the connection (e.g. zero weight would mean that there is no connection). Thus for example, input node 1 is connected to node 5 in the first hidden layer via weight w_{51}, node 5 is connected to node 7 via weight

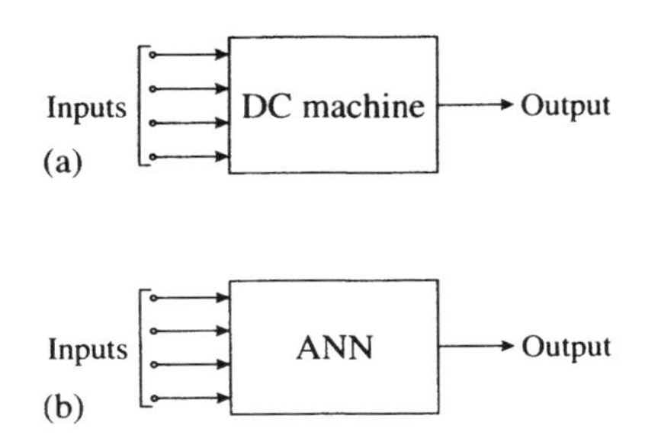

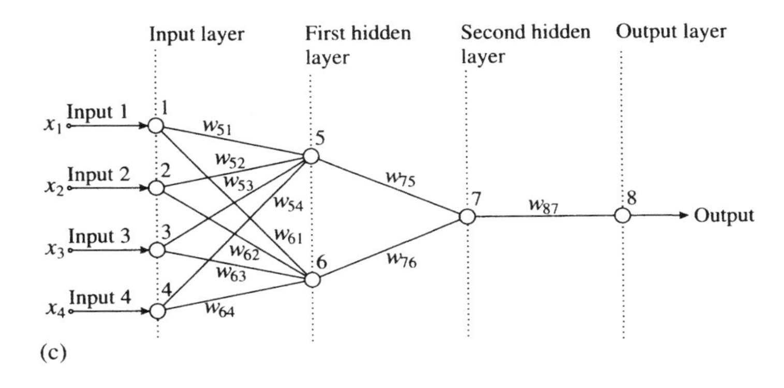

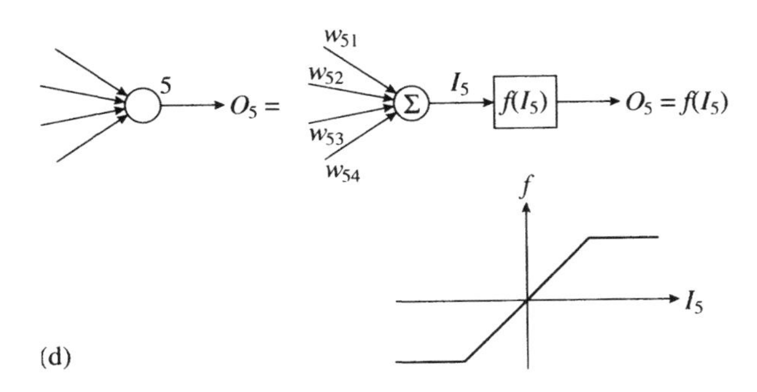

Fig. 7.2. Multi-layer feedforward artificial neural network with two hidden layers. (a) D.C. machine; (b) equivalent neural network; (c) multi-layer feedforward neural network structure; (d) computational node, activation function.

w_{75}, etc. The input nodes transmit the input signals directly to the nodes in the first hidden layer as shown in Fig. 7.2(c) and the input nodes do not perform any processing. In Fig. 7.2(c), the direction of signal transmission is also indicated, this is feed-forward (hence the name, multi-layer, feedforward ANN). However, all the other nodes (hidden nodes and output nodes) are computational (processing) units, which process the signal entering the particular node. In general, a processing node outputs a signal, which is a non-linear function of the weighted sum of the signals entering the node. Thus for example, node 5 receives four signals, which are transmitted from input node 1, input node 2, input node 3, and input node 4 respectively. Hence the signal received by node 5 (input signal to node 5) is

$$I_5 = x_1 w_{51} + x_2 w_{52} + x_3 w_{53} + x_4 w_{54} + b_5 = \sum_{i=1}^{4} w_{5i} x_i(t) + b_5, \tag{7.2-1}$$

where b_5 is a constant (which is often called a bias or threshold of the activation function of node 5), i.e. I_5 is equal to the sum of the weighted inputs to node 5, plus the bias b_5, but the output signal is a non-linear function of this,

$$O_5 = f(I_5). \tag{7.2-2}$$

This non-linear function f is the activation function. An ANN with biases can represent input-to-output mappings (required non-linear functions) more easily than one without biases. Since some training techniques (e.g. the back-propagation technique described below) requires the first derivative of the activation function, the activation function must be differentiable. Thus the non-linear activation function can also be a sigmoid, a hyperbolic tangent function, etc. It should be noted that the activation function of an output node is usually linear. The non-linearity for the hidden nodes is required since it is our goal to represent the d.c. machine by an equivalent non-linear system. In Fig. 7.2(d) there is also shown the structure of computational node 5 and the non-linear activation function. The structure of all the other nodes is similar. Thus it can be seen that, in general, if a kth neuron in layer L outputs the signal x_k^L, and the weight connecting this neuron to an ith neuron in the previous $(L-1)$ layer is w_{ki}^L, and the input to that ith neuron is x_i^{L-1}, then

$$x_k^L = f_k^L\left(\sum_i w_{ki}^L x_i^{L-1} + b_k^L\right), \tag{7.2-3}$$

where f_k^L is the activation function of node k in layer L. For example when a sigmoid function is chosen, then the output of the kth neuron is

$$x_k^L = 1 \bigg/ \left\{1 + \exp\left[-g\left(\sum_i w_{ki}^L x_i^{L-1} + b_k^L\right)\right]\right\}, \tag{7.2-4}$$

where g is the gain of the activation function and usually $g=1$ is chosen. Considering the neural network shown in Fig. 7.2(c), it can be visualized that, if there are known input signals and the goal is to obtain the correct output signal, then if the nodes are connected by weights which have random values, then even if a correct activation function is used in every node, it is almost certain that the output signal of the ANN will be incorrect (unless accidentally the correct weights and biases are chosen). However, if the correct weights could be obtained, it would be possible to obtain the correct output. From this point of view, for visualization purposes only, it could be assumed that if the biases are zero and the weights are varying resistors, the correct outputs could only be obtained by adjusting all resistors to their right values. But the problem is how to adjust these resistors to a value which will give the correct output signal for all possible input values.

In general, if the weight and bias adjustment is not performed systematically, it is impossible to get the correct outputs. However, there exists a technique, the so-called back-propagation technique, which is one of the most widely used neural network training techniques, which can give a solution to this problem. This is described below, but it should be noted that when a supervised ANN is used then the ANN is first trained by using known inputs and corresponding (known) outputs, and after this training stage there is an application stage, when the ANN is presented with inputs it has not seen before and then it outputs the required output signal.

The most frequently used training technique in a multi-layer feedforward ANN is the error back-propagation technique [Kosko 1992]. The multi-layer ANN can only implement the required mapping if the network contains a sufficient number of hidden layers and hidden nodes and adequate activation functions. There is no straightforward technique to select the optimal hidden layer and node numbers and the optimal activation functions to be used, and the solution is usually obtained by trial and error. The back-propagation training algorithm is basically a steepest descent method that searches for the minimum in the multi-dimensional error space; the errors of the output nodes are fed back in the network, and these are then used for weight and bias adjustments. The iterative weight adjustments for the output nodes are different to those for the hidden nodes. By using the back-propagation technique, the required non-linear function approximation can be obtained. The algorithm contains the learning rate parameter, and this has significant influence on the behavior of the network. To obtain increased learning and to avoid instability, a momentum term can also be used in the weight adaptation process. Sometimes the convergence of the network can be slow. It is possible to improve the convergence during the training of a neural network, by using genetic algorithms [Goldberg 1989]. Genetic algorithms are stochastic search and optimization algorithms based on the mechanics of evolution and natural genetics. They simulate the survival of the fittest among individuals over consecutive generations for solving a problem. Each generation consists of a population of individuals represented by chromosomes (a set of character strings that are analogous to the chromosomes in a DNA). Each individual represents a point in a search space and a possible solution. The individuals in the population are then made to go through a process of evolution.

In the training stage, input values are applied to the neural network and there are known outputs applied (which correspond to the input values). If the inputs are time-varying signals, then sampled values of these at the first sampling instant are applied to the ANN and the weights and biases are randomly initialized. The output signal of the ANN is then computed by using the technique described above (thus the input signals are propagated through the network to obtain the output signal). This output is then compared to the known output and the error is determined. If the error is zero, then obviously the correct weights and biases have been chosen (but this is most unlikely at this stage, as discussed above). The error is then back-propagated from the output layer to the input layer and the weights and biases are modified in such a way that the sum of the squares of the errors (global error) is minimized. It should be noted that the back-propagation is required, since otherwise the errors at the hidden layer nodes would not be known (since only the errors at the outputs of the ANN are known), and thus it would not be possible to adjust the weights and biases in the hidden nodes which would give minimal errors at the hidden nodes. The details of the back-propagation technique are described in the next section, but the reader can skip that section and can still solve problems using the back-propagation algorithm, since there are many softwares (e.g. MATLAB neural toolbox) in which this algorithm is incorporated into the software package and is easy to use.

7.2.3 BACK-PROPAGATION ALGORITHM

When the error back-propagation technique is used, the main task is to configure the weights and biases is such a way that the squared output error (between the desired and actual output) of the ANN should be minimum. It should be noted that, in general, if there are n external inputs $(x_1, x_2, \ldots, x_n)$ to a jth neuron, and there is also a bias input (internal input) to this neuron, then there are in total $n+1$ inputs to the neuron. The bias can also be incorporated by employing an additional input $x_0=1$ to the jth neuron, and using a corresponding weight of b_j, and in this case the training means the adjustment of all the weights (which includes b_j).

Initially the weights in the neural network are randomly selected, and therefore the output signals of the neural network will not be equal to the desired outputs. However, during training, the actual output signals are compared to the desired output signals (output pattern), and the weights are adjusted (iteratively by the back-propagation training algorithm), until the output error becomes smaller than a preset threshold value. During training, when the input patterns are applied, the total output square error (E) is the sum of the squared output error for all output layer neurons (E_p), thus

$$E = \sum_{p=1}^{P} E_p,$$

where $E_p = (1/2)(d_p - y_p)^2$; thus

$$E = \sum_{p=1}^{P} E_p = \frac{1}{2} \sum_{p=1}^{P} \sum_{k=1}^{K} (d_{pk} - y_{pk})^2. \tag{7.2-5}$$

In eqn (7.2-5) d_{pk} is the desired output vector of the kth output neuron (in the output layer) for the input pattern p, ($p = 1, 2, \ldots, P$, where P is the number of training patterns); y_{pk} is the actual output of the kth output neuron (in the output layer) for input pattern p, and K is the number of output nodes (in the output layer). The goal of the training is to reduce the total error (E) to a minimum e.g. by using the gradient descent technique. For this purpose, during training, the weights of the neural network are changed iteratively for all the P training patterns. A new value of a weight $w_{ji}(k+1)$ is obtained by adding a Δw_{ji} value to the old weight $w_{ji}(k)$, where this added value is $-\eta \partial E_p / \partial w_{ji}(k)$, where η is the learning factor (learning rate), which is a positive number. The training is considered to be satisfactory when the total error (summed over the P patterns) becomes smaller than a preset threshold value. The iterative process propagates the error backwards in the ANN. To ensure that the error converges to a global minimum, and does not stay in a local minimum (of the error surface), a momentum term $m[w_{ji}(k)-w_{ji}(k-1)]$ is also added to the weight adjustment mechanism, where m is the momentum.

With a successful learning process for the ANN, after a large number of exposures to the input and output data (epochs), the ANN can find the required relationship (which is usually non-linear) between the input and output data sets and configures the weights (and biases). This training process is then followed by supplying the trained ANN with the new input data, and the ANN then outputs the required output data. Thus there is a separate training and a recall mode of operation of the ANN, and the two modes should be distinguished.

To help the reader, the steps of the back-propagation algorithm are now given. This is mainly for better understanding, since many excellent softwares exist which contain this algorithm (e.g. MATLAB artificial-neural-network toolbox) and which are very convenient to use.

Back-propagation algorithm

Step 1 Initialization

Select learning rate (this is a positive numbers $\eta > 0$), choose maximum global error E_{max}, initialize weights and biases at small random values.

Step 2 Training starts

(a) Presentation phase: present random input (pattern) vector (**x**) to input layer, specify desired output vector (**d**) [each input–output pair can be presented for one or more cycles of the back-propagation training procedure, for multiple training samples of the input–output pairs can be presented cyclically until the weights become stabilized (will not change)].

(b) Compute layer outputs (y) by using appropriate weighted sums and activation functions.

Step 3 Training continues: check phase

Check if each output vector is equal to desired vector (associated with each input vector); stop when all input vectors generate correct output vectors (for specific set of weights and biases) or after maximum epochs.

Check phase 1: Total (cumulative) output error (E) determination

Compute total error (sum of all output errors in the entire training set):

$$E = \frac{1}{2} \sum_{p=1}^{P} \sum_{k=1}^{K} (d_{pk} - y_{pk}),$$

where d_{pk} is desired output of kth neuron, y_{pk} is actual output of kth neuron for pattern p, P is number of input patterns, and K is number of outputs in the network.

Check phase 2: Equivalent output error (delta) determination

Substep 1: Determine equivalent output error in the output layer (for simplicity sigmoid activation function with unity gain ($g=1$) is assumed, see eqn (7.2-4) above):

$$\delta_j = (d_j - y_j) y_j (1 - y_j) \qquad j = 1, \ldots, K \quad (K \text{ outputs}).$$

Substep 2: Determine the equivalent output error in hidden layer utilizing δ_j (for simplicity sigmoid with unity gain is assumed):

$$\delta_i = y_i (1 - y_i) \sum_{j=1}^{L} (\delta_j w_{ki}) \qquad j = 1, \ldots, L \quad (L \text{ hidden layers}).$$

Step 4 Training last phase—learning phase: weight adjustment at sampling time (k+1)

Use a recursive algorithm starting from the output layer and then move backwards to weights in the first hidden layer.

Substep 1: Output layer weight adjustment

Adjust output layer weights to

$$w_{ji}(k+1) = w_{ji}(k) + \Delta w_{ji} = w_{ji} + \eta \delta_j y_i \qquad (j = 1, \ldots, K, \quad i = 1, \ldots, L)$$

where $w_{ji}(k)$ is the output weight at time k and δ_j is the equivalent error obtained in Step 3, Substep 1.

Substep 2: Hidden layer weight adjustment

Adjust hidden weights to

$$w_{ih}(k+1) = w_{ih}(k) + \Delta w_{ih} = w_{ih} + \eta \delta_i y_h \qquad i = 1, \ldots, L, \quad h = 1, \ldots, M (M \text{ inputs})$$

where $w_{ih}(k)$ is the hidden weight at time k and δ_j is the equivalent error obtained in Step 3, Substep 2.

Step 5 Further pattern application

Present further pattern in training set and goto Step 2.

Step 6 Termination of training cycle

If $E < E_{\max}$ terminate training cycle, else start new training cycle and go to Step 2.

It has been mentioned above that to avoid local minima, a momentum term is also added to the weight adjustments; thus the total weight adjustment is

$$\Delta w_{ji}(k) + m[w_{ji}(k) - w_{ji}(k-1)],$$

where m is smaller than 1. It can be seen that the current weight adjustment is supplemented by a fraction of the most recent weight adjustments.

7.2.4 ANN-BASED ANALYSIS, ESTIMATION

In the following subsections, various multi-layer ANN configurations for a separately excited d.c. motor are described, which are suitable for the

- estimation of the rotor speed;
- estimation of the electromagnetic torque;
- simultaneous estimation of the rotor speed and torque;
- simultaneous estimation of the rotor speed, field current, and armature current.

To show the effects of the ANN structure on the estimation, different ANN structures (with different numbers of hidden layers and hidden nodes) will also be considered. However, it will be shown that for the d.c. motor under consideration, it is possible to use extremely simple ANNs. For example, for the estimation of the speed it was possible to obtain accurate results even during the transients by using a multi-layer feedforward ANN containing a single hidden layer with a single hidden node. Such an ANN is easy to implement in a DSP-controlled d.c. motor drive and the extra memory requirements are minimal.

7.2.4.1 Speed estimation: 3-4-2-1 ANN, 4-4-2-1 ANN, 4-1-1 ANN, 4-2-1 ANN, d.c. drive with closed-loop speed control

As mentioned above, various ANN structures have been examined for a separately excited d.c. motor to be able to compare their accuracy. Thus, first, ANNs with more complex structures are deliberately introduced, but these are followed by the simpler ANN-based virtual-speed estimators. It is shown below that the ANNs with simpler structures can also provide satisfactory responses. Results for a drive with closed-loop speed control are also given.

3-4-2-1 ANN

An ANN similar to that shown in Fig. 7.2(c) is first used to estimate the speed of a separately excited d.c. motor during the starting transients; thus the ANN contains four layers, an input layer, two hidden layers, and an output layer. However, now the input layer contains three nodes (which directly transmit the input signals to the nodes in the next layer); the first hidden layer contains 4 nodes, the second hidden layer contains 2 nodes, and the output layer contains a single node. Hence this neural network will be referred to as the 3-4-2-1 artificial neural network. Readers using other d.c. motors may find other ANN structures for their motors, but it should be noted that in general two hidden layers should always provide satisfactory results and an increase of the number of hidden layers and hidden nodes increases substantially the training effort. However, there are no general rules for the selection of the number of hidden neurons—these have been obtained by trial and error. This is an obvious disadvantage of using a conventional, supervised, multi-layer back-propagation neural network, but this problem can be eliminated e.g. by using a fuzzy-neural network, where the structure of the network is exactly determined (see also Section 4.4.2).

At every step of the training (training phase), three inputs are presented to the 3-4-2-1 ANN; these are the sampled values of the armature voltage, $u_a(k)$, and the armature current at two instants, $i_a(k)$ and $i_a(k-1)$. Although the armature voltage in this example is deliberately assumed to be constant, in general (due to speed-control purposes) the armature voltage varies, and this change will be considered in the ANNs discussed later in the present section. In the 3-4-2-1 ANN, there is only one output signal; this is the speed of the d.c. motor, and the sampled value of this has been used to train the ANN, this is $\omega_r(k)$.

By assuming a constant armature voltage, the training data set for the armature currents has been obtained by using the conventional simulation technique described above in Section 7.2.1 (for simplicity, a constant field voltage is assumed for all the cases to be discussed). However, it should be noted that the data set could also be obtained by measurements. Furthermore, for training purposes not all the data have been used, but only 40% of the entire data set (since the remaining data have been left for the checking (application) phase. The activation functions of the hidden layers were logsigmoid functions and the output activation function is purelin (linear activation function).

The training technique used is the back-propagation technique described in the previous section. Both gradient descent and Levenberg–Marquardt optimization techniques have been used. the latter has converged much faster, but this technique requires more memory than the gradient-descent technique. It should be noted that when the Levenberg–Marquardt technique is used, the weight adjustment is performed by $\Delta\mathbf{W}=(\mathbf{J}^T\mathbf{J}+\mu\mathbf{I})^{-1}\mathbf{J}^T\mathbf{E}$, where $\mathbf{J}$ is the Jacobian matrix of the derivatives of each error to each weight, $\mathbf{I}$ is an identity matrix, μ is a scalar, and $\mathbf{E}$ is an error vector. If the scalar μ is large, this weight adjustment is similar to that when the gradient descent method is used, but for small values of μ, it is similar to the Gauss–Newton method. Since the Gauss–Newton method is faster and more accurate near an error minimum, thus the goal is to transfer to the Gauss–Newton method as quickly as possible and therefore μ is decreased after each step (which decreases the error) and increased only when a step increases the error.

For checking purposes (application phase), the trained 3-4-2-1 ANN was subjected to the entire data set, and the results are shown in Fig. 7.3(a) and 7.3(b). It can be seen from Fig. 7.3(a) that the speed estimate is quite satisfactory in both the transient and steady states, and there is a very small spike around 2 s. The speed error, which is the difference between the actual speed shown in Fig. 7.1 and

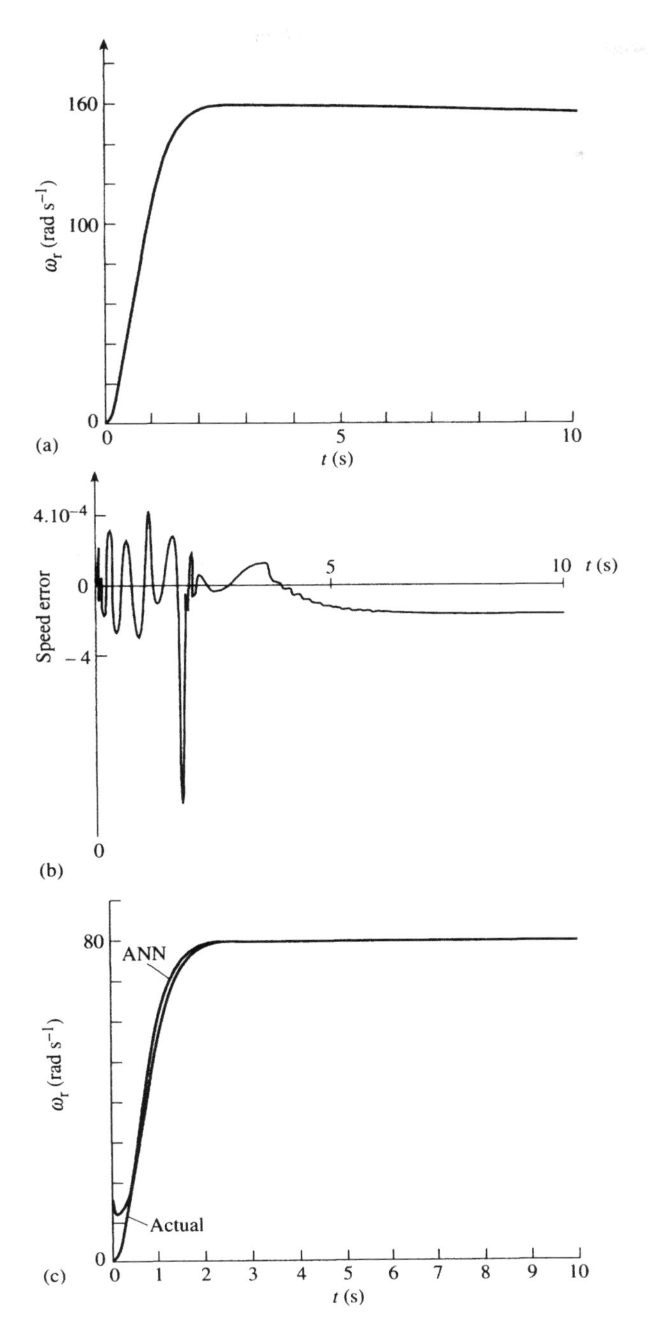

Fig. 7.3. Speed of d.c. machine. (a) Speed obtained by 3-4-2-1 ANN (armature voltage 240 V); (b) speed error; (c) actual speed and speed obtained by ANN (training with sinusoidal armature voltage, checking with constant armature voltage, 120 V).

that shown in Fig. 7.3(a), is also plotted in Fig. 7.3(b) and it can be seen that this error is quite small; there is a very small transient error and the error in the steady-state is almost zero. However, it is important to note that, as expected, when the trained ANN is subjected to new inputs corresponding to armature voltages different to the constant value used for the training above, significant errors will arise, even if the new armature voltage differs from the one used for training only by a small amount (say 1%).

To enable the ANN to estimate the speed correctly for different armature voltages, i.e. to be able to use it for simulation purposes instead of the conventional mathematical model, or to be able to use it as a virtual-speed sensor in a speed-sensorless d.c. motor drive, it was also trained with a varying armature voltage. Thus for training purposes different armature voltage waveforms were used, e.g. a triangular or a sinusoidal armature voltage. This also allows speed reversals.

For illustration purposes Fig. 7.3(c) shows the actual speed and also the speed obtained by the trained ANN, where the training has been performed by using a sinusoidal armature voltage of 240 V amplitude. However, in the checking phase, the trained ANN was used to estimate the speed if the armature voltage is a constant 120 V. It can be seen that in the initial part of the transient there is a large difference between the two curves. However, this can also be reduced by using a different network configuration, different initialization, etc. The trained ANN was also used to estimate the speed during the run-up of the machine for different cases corresponding to different armature voltages, and satisfactory results have been obtained. The trained ANN was also checked for different cases involving different loads, and again satisfactory responses have been obtained.

4-4-2-1 ANN

To enable to use the ANN in a speed-sensorless drive, it was trained by using a variable armature voltage. The results obtained by using the conventional mathematical model and that by using a 4-4-2-1 ANN are shown in Fig. 7.4(a). The 4 inputs to the 4-4-2-1 ANN correspond to $u_a(k)$, $u_a(k-1)$, $i_a(k)$, $i_a(k-1)$. The ANN was trained by using a triangular armature voltage waveform with an amplitude of 240 V. However, in the application phase, the trained 4-4-2-1 ANN was subjected to a triangular armature voltage whose maximum value was 300 V. It can be seen from Fig. 7.4(a) that there was only a small speed error.

Similarly to that discussed in connection with the 3-4-2-1 ANN, in another application phase, the trained ANN (using triangular armature voltage for training) was also tested for the run-up of the d.c. machine with different (but constant) armature voltages, and apart from the very initial parts of the transients, correct speed estimates were obtained. Although there was an initial transient error, the sign of the initial speed was always correct, and when the ANN-based virtual-speed estimator was employed in a 'speed-sensorless' drive, satisfactory responses were obtained.

Instead of using a triangular armature voltage waveform, the learning stage has also been performed by using a sinusoidal armature voltage (with an amplitude of 240 V) and the trained network has again given satisfactory results. For example, Fig. 7.4(b) shows the results obtained when the trained 4-4-2-1 ANN was used, but the amplitude of the applied armature voltage sine wave was 180 V. The trained ANN was also tested for the run-up of the d.c. machine with different (but constant) armature voltages, and the correct results have been obtained apart from a short part of the initial transients (see discussion above).

It can be concluded that a well-trained ANN is suitable for the speed estimation of the d.c. machine. Suitable speed estimates can be obtained when the armature voltage is changing and also when there are different loads on the machine. Hence it can also be used as a virtual speed sensor in a variable-speed d.c. motor drive (instead of a conventional speed sensor). However, it should be noted that it is also possible to use ANNs with less complexity, e.g. ANNs which contain only a single hidden layer, as shown below (4-1-1 ANN, 4-2-1 ANN).

4-1-1 ANN

For illustration purposes of the application of an ANN with a simple structure, Fig. 7.4(c) shows the results obtained by using a 4-1-1 ANN speed estimator and a conventional estimator, thus the ANN contains only one hidden layer with a single hidden neuron. The activation function used for the single hidden layer neuron is a logsigmoid function. For training purposes a sinusoidal armature voltage was used with an amplitude of 240 V. The training required only 50 epochs, and the sum of the error squares was 2.97×10^{-4}. In the application phase, the trained 4-1-1 ANN was subjected to a sinusoidal armature voltage with an amplitude of 180 V. It can be seen that the results are even better than the corresponding ones shown in Fig. 7.4(b) (which used a 4-4-2-1 ANN).

The trained 4-1-1 ANN was also tested for the run-up of the d.c. machine with different (but constant) armature voltages and the correct results were obtained, but as before, there was an error in the initial part of the transients.

4-2-1 ANN

To illustrate the application of another simple ANN with a single hidden layer, Fig. 7.4(d) shows results which are similar to those shown in Fig. 7.4(b), but these correspond to a 4-2-1 ANN speed estimator (one hidden layer with two hidden neurons) and a conventional speed estimator.

Although in general multi-layer ANNs with one or two hidden layers can be used for the estimation of the speed of a d.c. motor, it is also possible to use different types of ANNs, which lead to a further reduction of the computational effort.

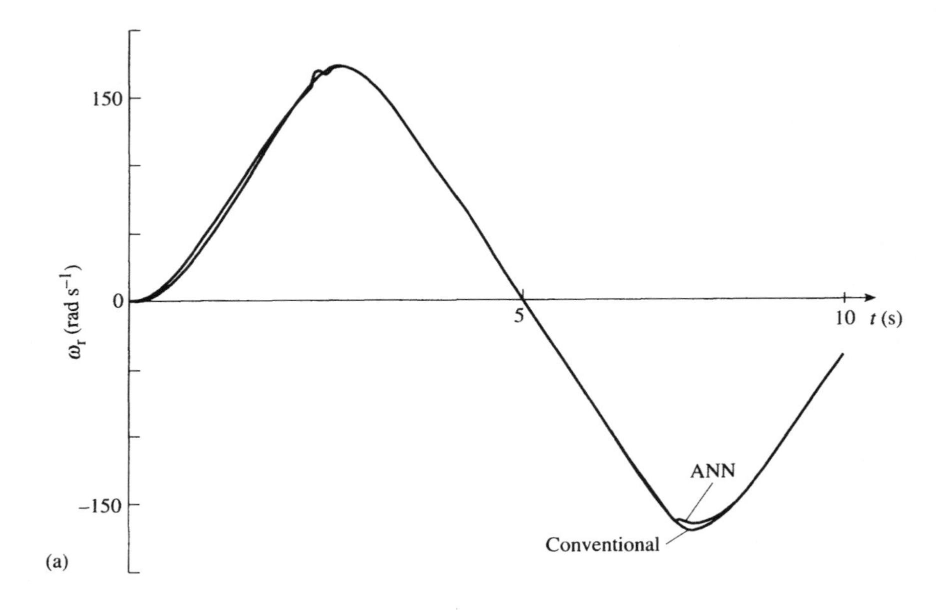
150
0
−150
ω_r (rad s^{-1})
5
10
t (s)
ANN
Conventional
(a)

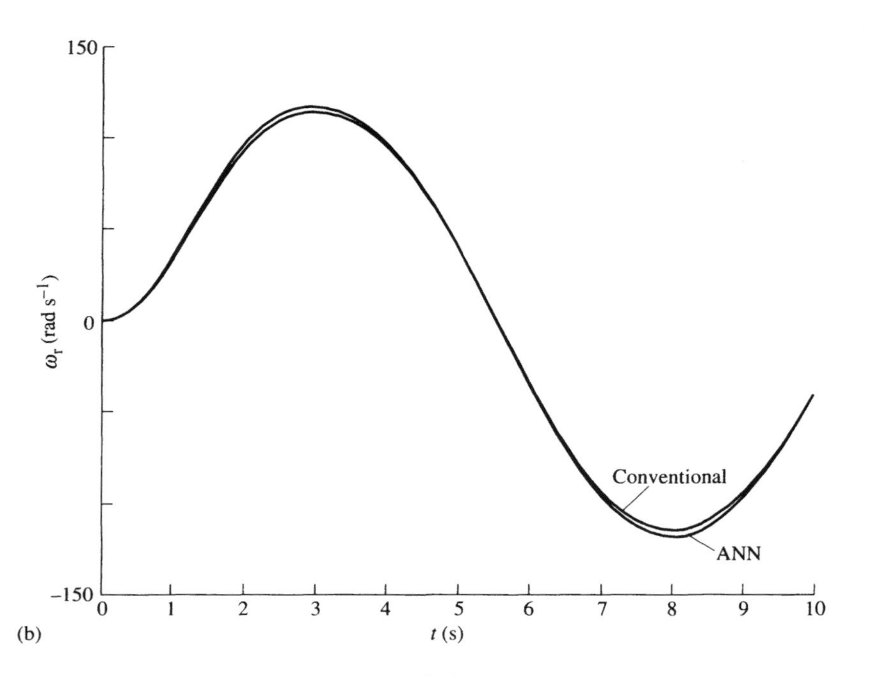
150
0
−150
ω_r (rad s^{-1})
0
1
2
3
4
5
6
7
8
9
10
t (s)
Conventional
ANN
(b)

Fig 7.4.

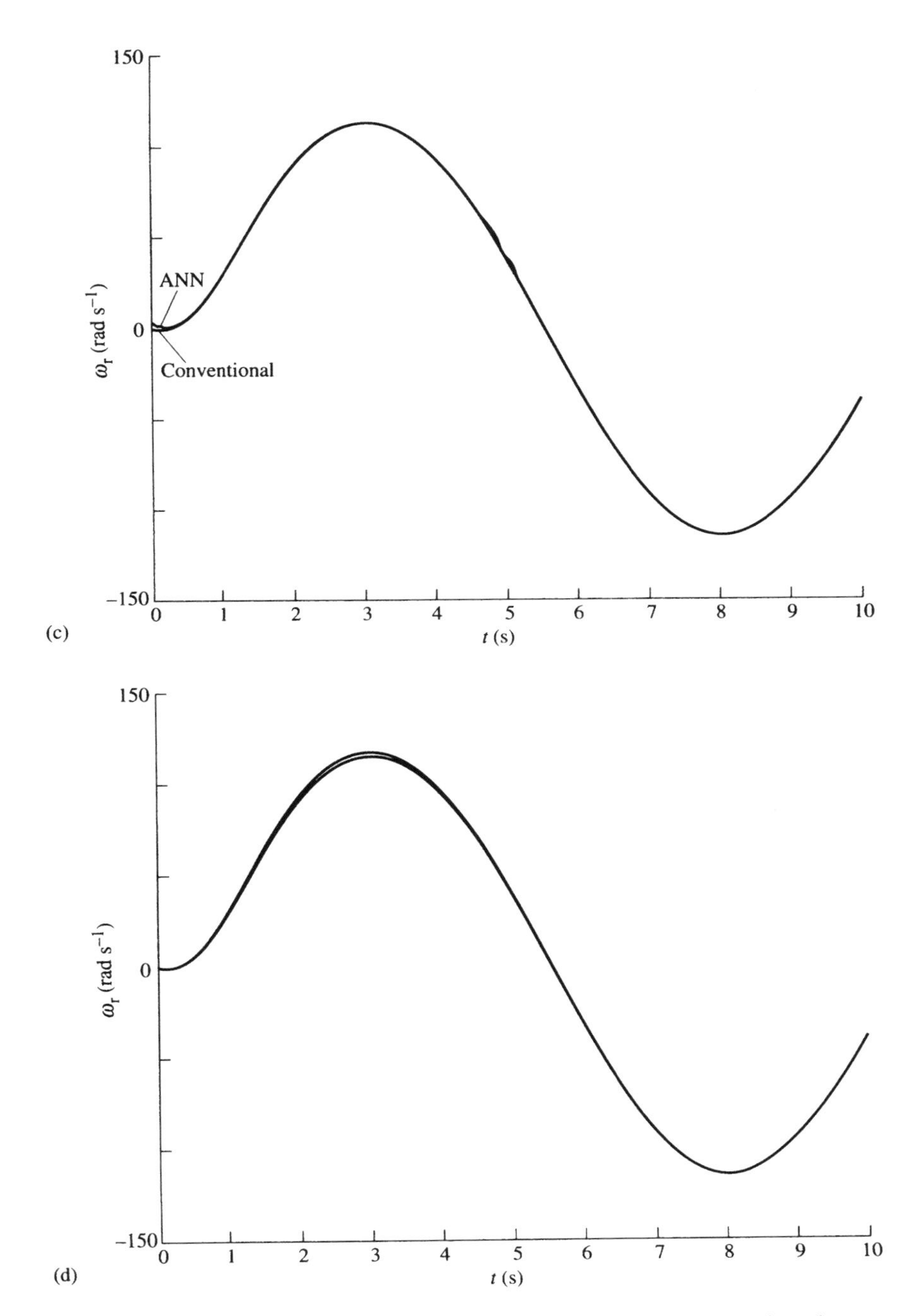

Fig. 7.4. Speed of d.c. machine. (a) Speed obtained by 4-4-2-1 ANN (training with triangular armature voltage, amplitude = 240 V, checking with triangular armature voltage, amplitude = 300 V). (b) Speed obtained by 4-4-2-1 ANN (training with sinusoidal armature voltage, amplitude = 240 V, checking with sinusoidal armature voltage, amplitude = 180 V). (c) Speed obtained by 4-1-1 ANN (training with sinusoidal armature voltage, amplitude = 240 V, checking with sinusoidal armature voltage, amplitude = 180V). (d) Speed obtained by 4-2-1 ANN.

D.c.-drive with closed-loop speed control

Various types of ANNs (multi-layer feedforward, multi-layer recursive, radial basis function ANN etc.) were considered as speed estimators in a speed-controlled d.c. drive. In all these cases it was possible to obtain well-trained ANNs (where the estimated and target data were almost totally identical). However, to investigate the generalization properties of an ANN, in the so-called checking stage (which follows the training stage), it was also subjected to input data which was not present during the training stage. This checking phase was also performed in the closed-loop speed-controlled drive, where a trained ANN provided the speed feedback signal. Different step reference speed signals were applied, and satisfactory results were obtained even when the drive operated outside the training region. However, to further investigate the generalisation capabilities of the trained ANN, in addition to the step response tests, random speed reference inputs signals were also applied in the speed-controlled d.c. drive, and satisfactory responses were obtained. Furthermore, speed reversals were also investigated, and Fig. 7.5 shows the corresponding results for three different cases. For this purpose a multi-layer feedforward ANN was used. It can be seen that in all the cases, excellent responses were obtained, and in particular it follows from the last diagram, that the drive with the ANN speed estimator works satisfactorily even in the very low speed range.

Finally it should be noted that various ANN-based DSP-controlled speed sensorless d.c. drives have also been successfully implemented: the details will be published elsewhere [Stronach and Vas 1997].

7.2.4.2 Electromagnetic torque estimation: 3-4-2-1 ANN, 4-4-2-1 ANN, 4-5-1 ANN

The ANN-based estimation of the instantaneous electromagnetic torque is discussed in the present section. For this purpose simple, multi-layer feedforward ANNs are investigated, where in the output layer there is only one node which outputs the electromagnetic torque. The trained ANNs described in the previous section cannot provide the electromagnetic torque on their output, since they have only been trained to provide the speed output (at every step of the training it was the speed and not the electromagnetic torque which was used on the output of the neural network).

First a 3-4-2-1 ANN was trained to estimate the electromagnetic torque during the run-up transient of an unloaded d.c. machine. The armature voltage was constant and the electromagnetic torque was determined by using the conventional mathematical model and also by using the 3-4-2-1 ANN. A part of the conventional simulation results provided the training data. The results are shown in Fig. 7.6, and it can be seen in Fig. 7.6(a) that the ANN predicts accurately the electromagnetic torque and there is only a small error in the region of 3–4 s. It should be noted that logsigmoid activation functions have been used in the

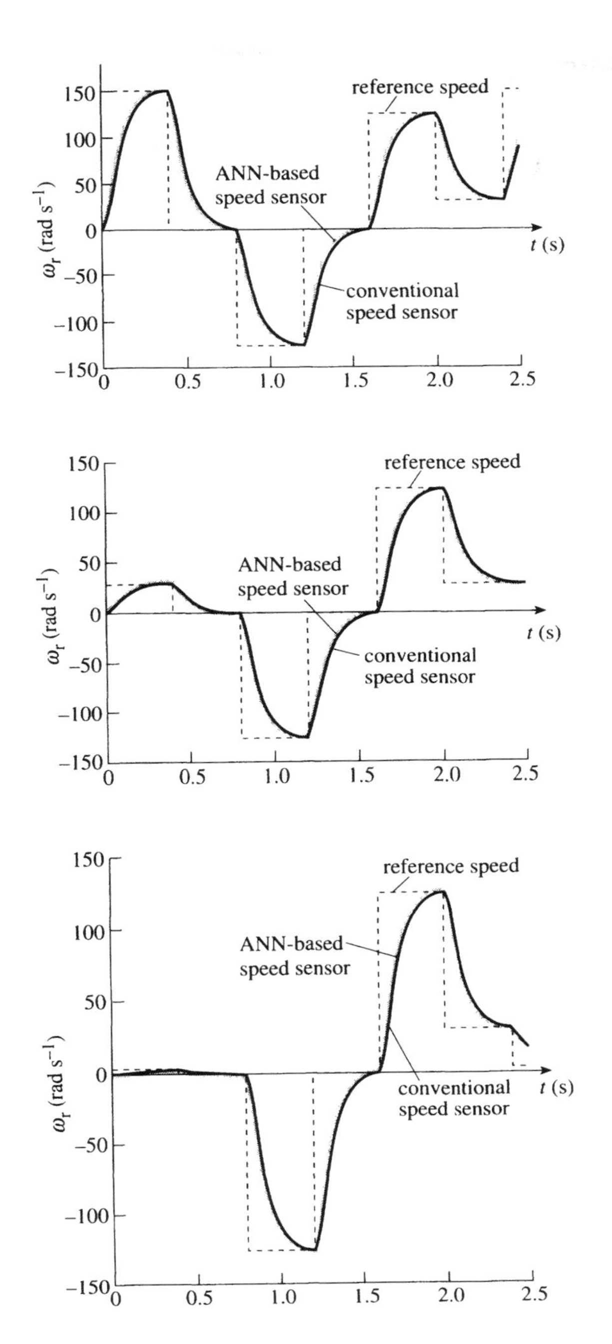

Fig. 7.5. Reversals in a speed-controlled d.c. drive using a multi-layer feedforward neural speed estimator.

hidden layer nodes. However, when tansigmoid (hyperbolic tangent) activation functions were used, the error completely disappeared, as shown in Fig. 7.6(b). Furthermore, it was also possible to reduce the complexity of the ANN and to obtain satisfactory results. It was found that an ANN with a single hidden layer, with the 4-5-1 structure, also gives the correct results, when the tansigmoid activation function is used in the hidden layer. However, it was also confirmed that the correct value of the electromagnetic torque can be obtained by an appropriately trained ANN (e.g. a 4-4-2-1 or 4-5-1 ANN) when the armature voltage is changed. This has also proved that this type of estimator can also be used in a variable-speed d.c. drive.

7.2.4.3 Simultaneous speed and torque estimation: 4-6-5-2 ANN, 4-5-3-2 ANN, 4-3-2 ANN

A more general ANN has also been investigated for the purposes of transient analysis. This contained two outputs, since it was a goal to estimate both the rotor speed and electromagnetic torque simultaneously by using a single ANN. However, since there are two outputs, the structure of the ANN has become more complicated; a satisfactory solution was obtained by using a 4-6-5-2 ANN.

The results are shown in Fig. 7.7 for the unloaded d.c. machine. In Fig. 7.7(a) the speed of the machine is shown if the armature voltage is a triangular waveform. The ANN was trained by using a triangular armature voltage waveform with a peak of 240 V; however, to check the generalization capabilities of the ANN, in Fig. 7.7(a) the results are shown when the peak of the triangular armature voltage is 120 V. Both the 'actual' speed and the one obtained by the ANN are shown, and it can be seen that there is a small error. In Fig. 7.7(b) the variation of electromagnetic torque is shown, and again it can be seen that the 'actual' torque and the one estimated by the ANN are in good agreement from 3–10 s, but around 1–3 s there is some error. In Fig. 7.7(c) the speed variation during run-up is shown, but the peak of the armature voltage is 300 V. It can be seen that there is very good agreement between the actual speed and the one estimated by the ANN. It is not surprising that the ANN-generated results shown in Fig. 7.7(c) are better than those in Fig. 7.7(a), since the training was performed with a triangular armature voltage wave with a peak of 240 V, and the amplitude of 300 V used to obtain the results of Fig. 7.7(c) is closer to this than 120 V (which was used to obtain the results in Fig. 7.7(a)). In Fig.7.7(d) the torque variations are shown, corresponding to the triangular armature voltage with a peak of 300 V. It can be seen that, for the same reasons as before, there is a better agreement between the actual and ANN-generated torque than in Fig. 7.7(b). It can be concluded that an appropriately trained ANN can provide an accurate simultaneous estimation of both the electromagnetic torque and rotor speed.

Another ANN was also trained with a configuration of 4-5-3-2, but instead of using a triangular armature voltage training waveform, a sinusoidal waveform was used. The activation functions used were the same as those used for the ANN in Fig. 7.7. The results are shown in Fig. 7.8 for two cases where the armature

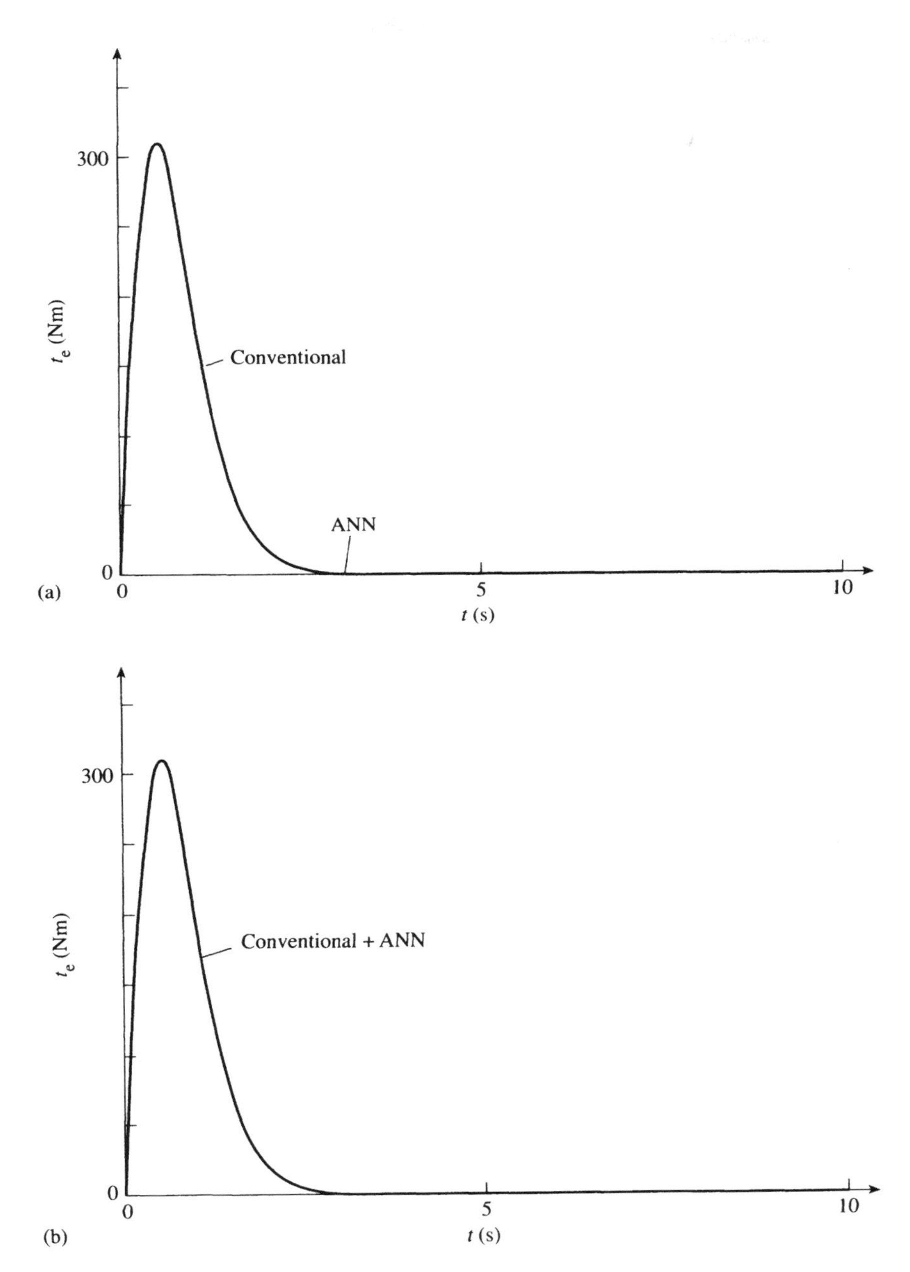

Fig. 7.6. Estimated electromagnetic torque of a d.c. motor during run-up using conventional mathematical model and 3-4-2-1 ANN. (a) Electromagnetic torque obtained by conventional mathematical model and ANN (using logsigmoid activation functions); (b) electromagnetic torque obtained by conventional mathematical model and ANN (using tansigmoid activation functions).

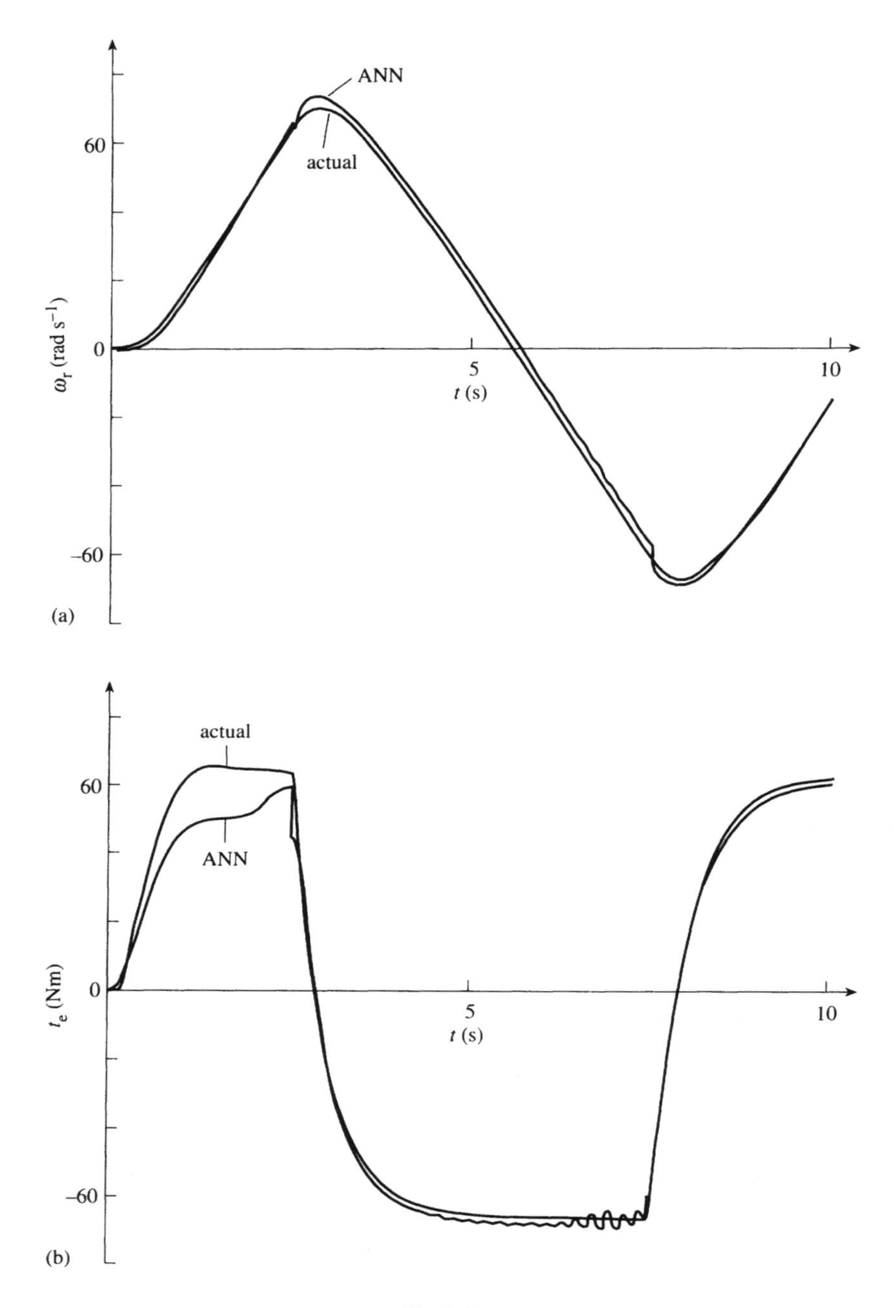

ANN
60
actual
0
ω_r (rad s^{-1})
5
10
t (s)
–60
(a)
actual
60
ANN
0
t_e (Nm)
5
10
t (s)
–60
(b)

Fig. 7.7.

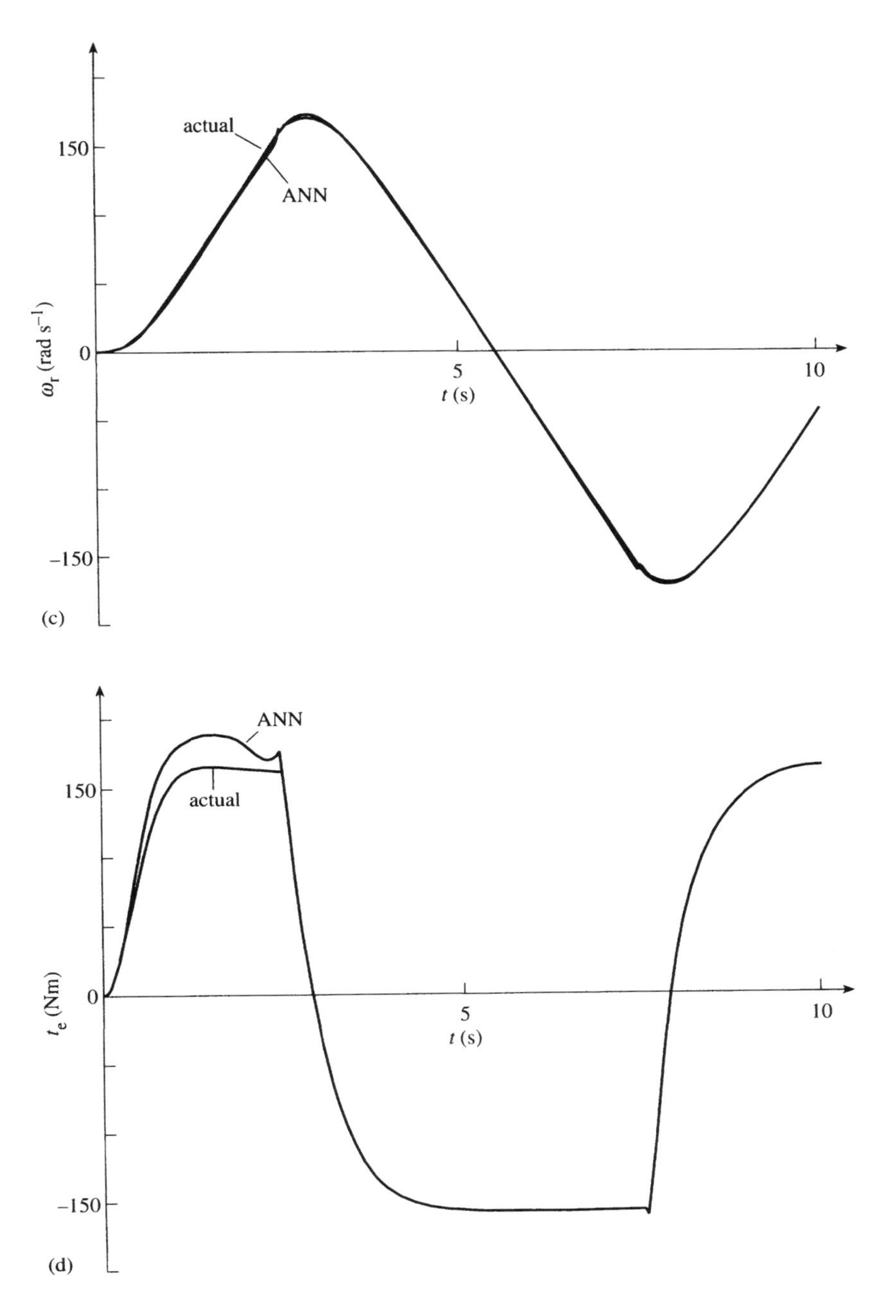

Fig. 7.7. Electromagnetic torque and speed of d.c. motor. (a) Speed waveforms using conventional mathematical model and 4-6-5-2 ANN (triangular wave peak 120 V); (b) torque waveforms using conventional mathematical model and 4-6-5-2 ANN (triangular wave peak 120 V); (c) speed waveforms using conventional mathematical model and 4-6-5-2 ANN (triangular wave peak 300 V); (d) torque waveforms using conventional mathematical model and 4-6-5-2 ANN (triangular wave peak 300 V).

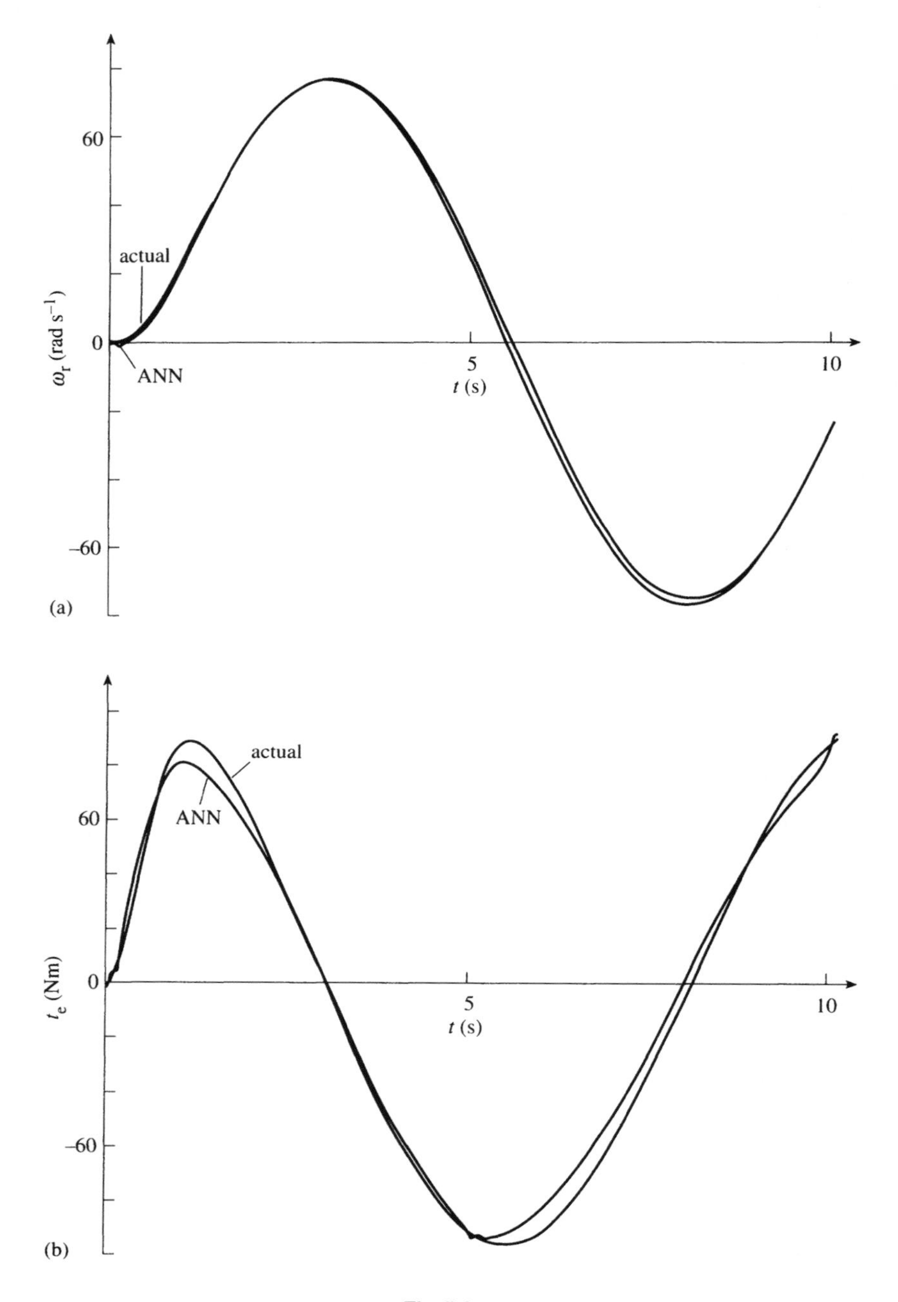
60
0
-60
actual
ANN
ω_r (rad s^{-1})
5
10
t (s)
(a)
60
0
-60
actual
ANN
t_e (Nm)
5
10
t (s)
(b)

Fig. 7.8

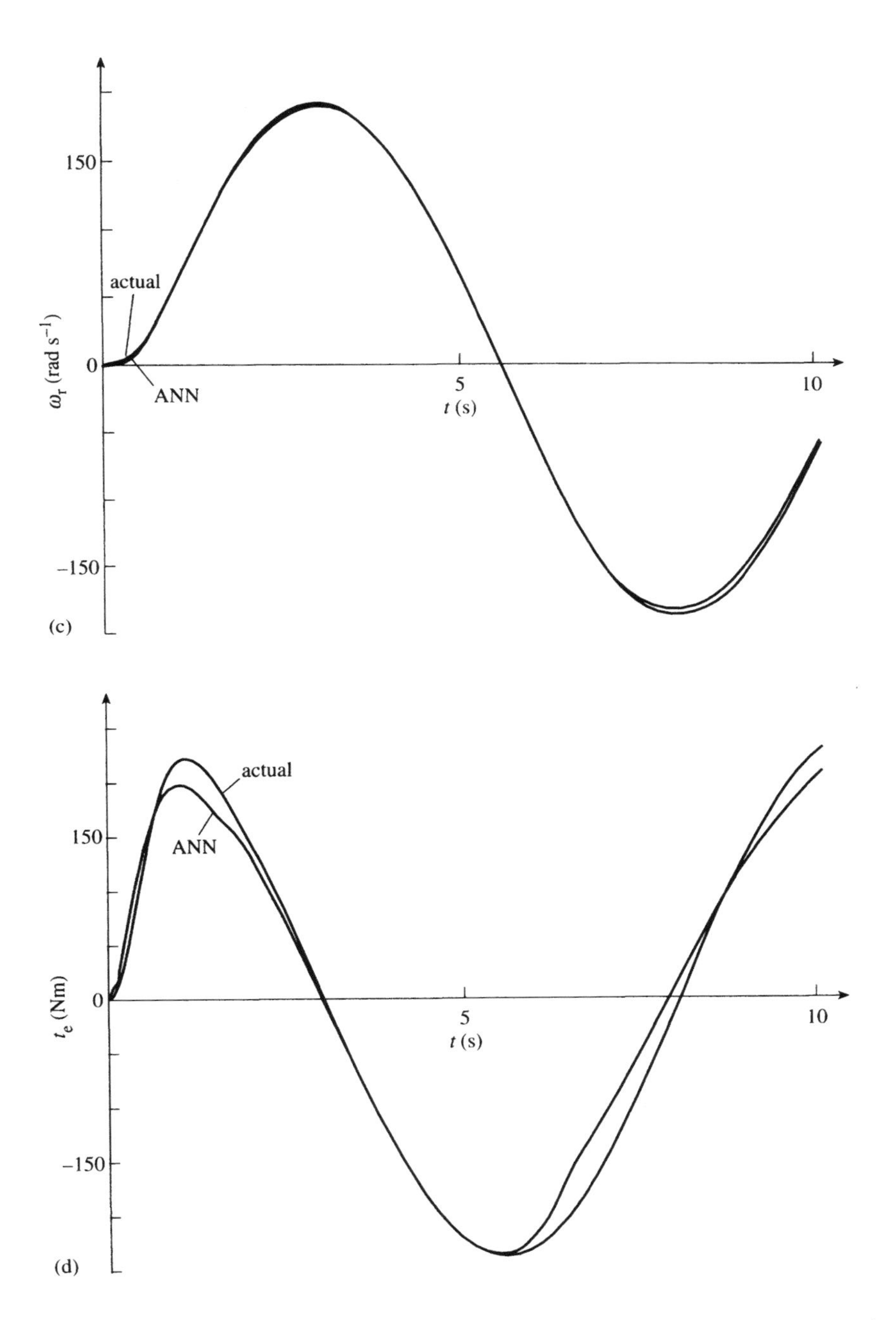

Fig. 7.8. Electromagnetic torque and speed of d.c. motor. (a) Speed waveforms using conventional mathematical model and 4-5-3-2 ANN (sinusoidal wave peak 120 V); (b) torque waveforms using conventional mathematical model and 4-5-3-2 ANN (sinusoidal wave peak 120 V); (c) speed waveforms using conventional mathematical model and 4-5-3-2 ANN (triangular wave peak 300 V); (d) Electromagnetic torque and speed of d.c. motor. Torque waveforms using conventional mathematical model and 4-5-3-2 ANN (sinusoidal wave peak 300 V).

voltage peak was 120 V and 300 V respectively. It should be noted that the training waveform had a peak of 240 V, and it can be seen that the ANN has learnt to generalize, and can obtain accurate estimates of both the speed and electromagnetic torque.

A simpler ANN (minimal configuration ANN) with the structure 4-3-2 was also successfully trained, which can be used for the simultaneous estimation of the speed and the torque. In 98 epochs the sum of the squared errors was 0.15×10^{-4}.

It should be noted that in contrast to the ANNs discussed above, it is also possible to obtain simultaneous estimates of the speed and torque by a modified ANN in which not all the nodes in the second hidden layer are connected to both output nodes. In this case some of the hidden-layer nodes are connected only to the first output node (say, speed node) and others only to the second output node (electromagnetic torque node). In this case changing the weights to reduce the electromagnetic torque error will not affect the speed error so much, but the number of second-layer hidden nodes will be increased.

7.2.4.4 Simultaneous speed, field, and armature current estimation: 6-4-4-3 ANN, 10-4-5-3 ANN

Instead of using the conventional mathematical model (differential equations) of the separately excited d.c. machine to obtain a simultaneous estimate of the rotor speed, armature current, and field current, it is also possible to use an ANN for this purpose. Thus various ANNs have been tested and e.g. a recursive ANN with 6-4-4-3 structure has been found to provide results which are acceptable, apart from the initial part of the starting transients.

The three outputs of the 6-4-4-3 ANN are $\omega_r(k)$, $i_a(k)$, and $i_f(k)$. The electromagnetic torque is not estimated by the ANN since it can be obtained by using the armature current and field current. Since the field voltage is constant, only $u_f(k)$ is an input (and other values of u_f at previous instants are not required) and further inputs to the ANN are the present and past value of the armature voltage [$u_a(k)$ and $u_a(k-1)$], the past value of the armature current, $i_a(k-1)$, the past value of the field current, $i_f(k-1)$, and the past value of the rotor speed, $\omega_r(k-1)$. Obviously with a conventional mathematical-model-based simulation, the numerical solution of the system of differential equations also requires the past values $i_a(k-1)$, $i_f(k-1)$, $\omega_r(k-1)$ at every time step. The first hidden layer contains 4 neurons and the second hidden layer also contains 4 neurons. An adequate number of hidden neurons has been obtained by trial and error following extensive simulations. The armature voltage waveform used for training was a sinusoidal wave, with an amplitude of 240 V (see also Section 4.2.2.1). The results are shown in Fig. 7.9. The activation functions are tansigmoids in the hidden layers. The training lasted for 240 epochs. During training the sum of the squared errors was 8.9×10^{-7}.

As expected, it follows from Fig. 7.9(a) that the speed waveform obtained by the 6-4-4-3 ANN is not as good as that obtained earlier by the 4-4-2-1 ANN,

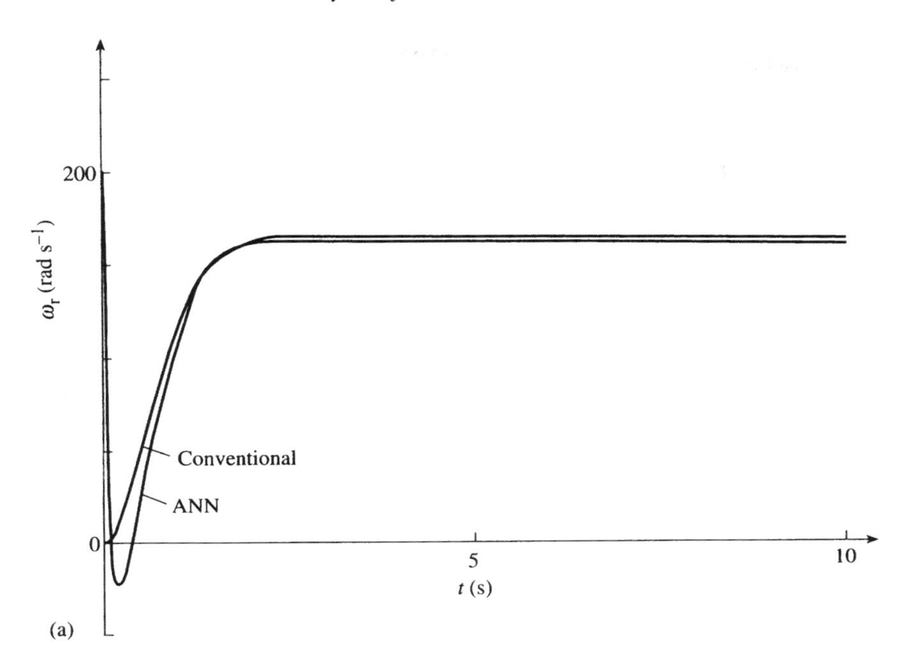
200
ω_r (rad s^{-1})
Conventional
ANN
0
5
10
t (s)
(a)

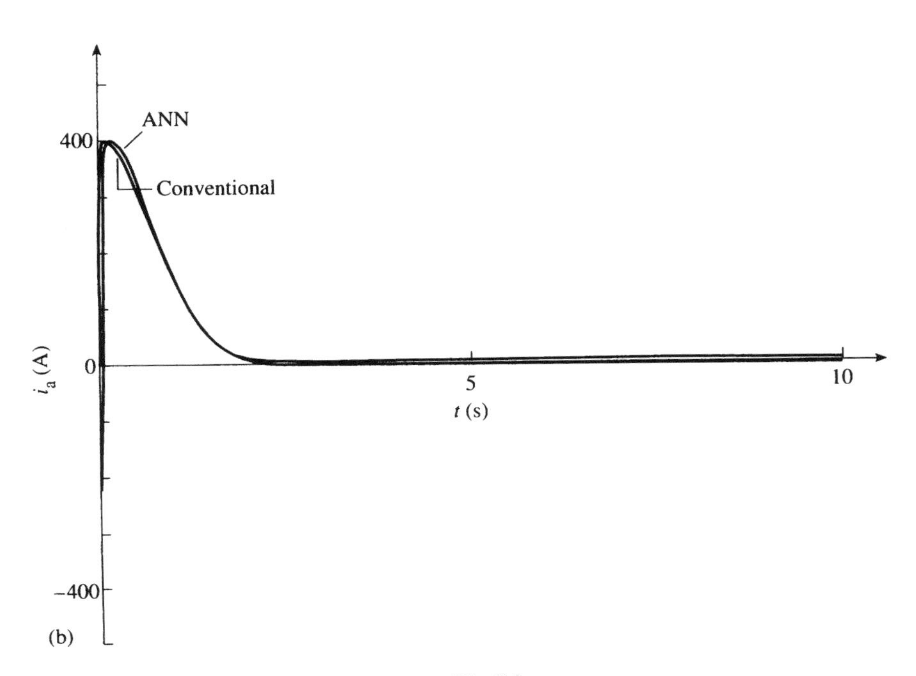
ANN
400
Conventional
i_a (A)
0
5
10
t (s)
−400
(b)

Fig. 7.9.

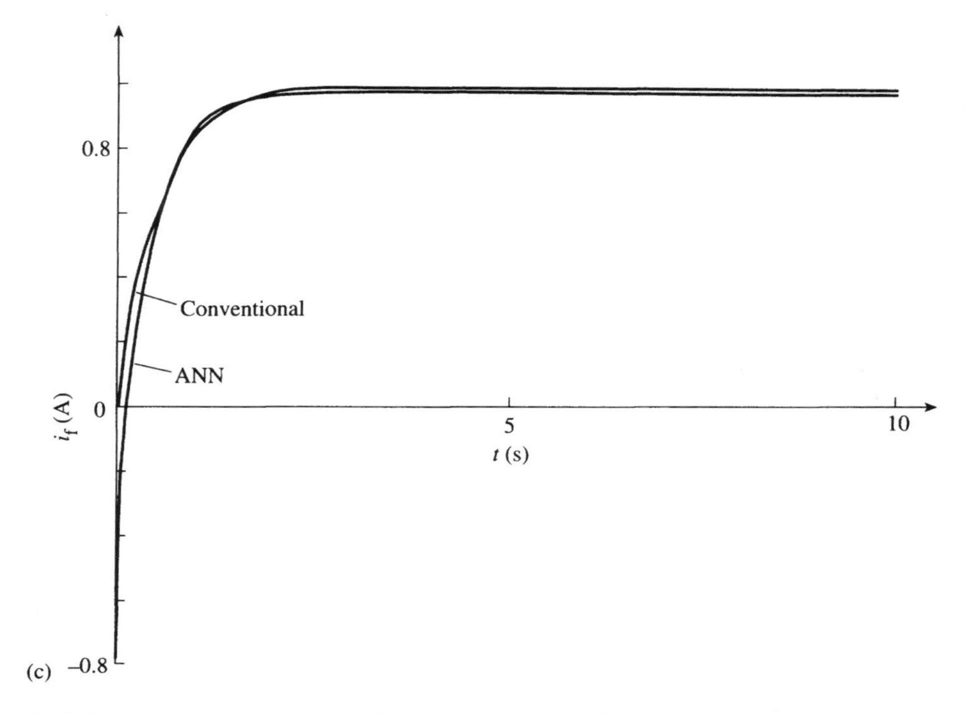

Fig. 7.9. Speed, armature current, and field current of d.c. machine. (a) Speed waveforms using 6-4-4-3 ANN and conventional mathematical model; (b) armature current waveforms using 6-4-4-3 ANN and conventional mathematical model; (c) field current waveform using 6-4-4-3 ANN and conventional mathematical model.

which had to estimate only a single quantity, which was the rotor speed, and not three quantities. It can be seen that during the initial transient up to 1 s, there is a speed error, but beyond this and in the steady state the speed obtained by the ANN is quite accurate. It can be seen from Fig. 7.9(b) that, apart from the initial transient, the armature current estimated by the ANN is satisfactory. The same holds for the field current.

Another ANN with 10-4-5-3 configuration has also been tested successfully. In this ANN the number of inputs has been deliberately increased to check the effects of the extra inputs. Thus $u_a(k-2)$, $i_a(k-2)$, $i_f(k-2)$, $\omega_r(k-2)$ were also inputs to the ANN in addition to those used previously in the 6-4-4-3 ANN. The 10-4-5-3 ANN has given a better speed estimate than the 6-4-4-3 ANN, i.e. the steady-state error has been reduced to zero and the initial negative speed shown in Fig. 7.9(a) has disappeared. The results are shown in Fig. 7.10.

It should be noted that an ANN with 10-4-4-3 configuration did not provide satisfactory results, and this is why the number of hidden neurons in the second hidden layer had to be increased from 3 to 4.

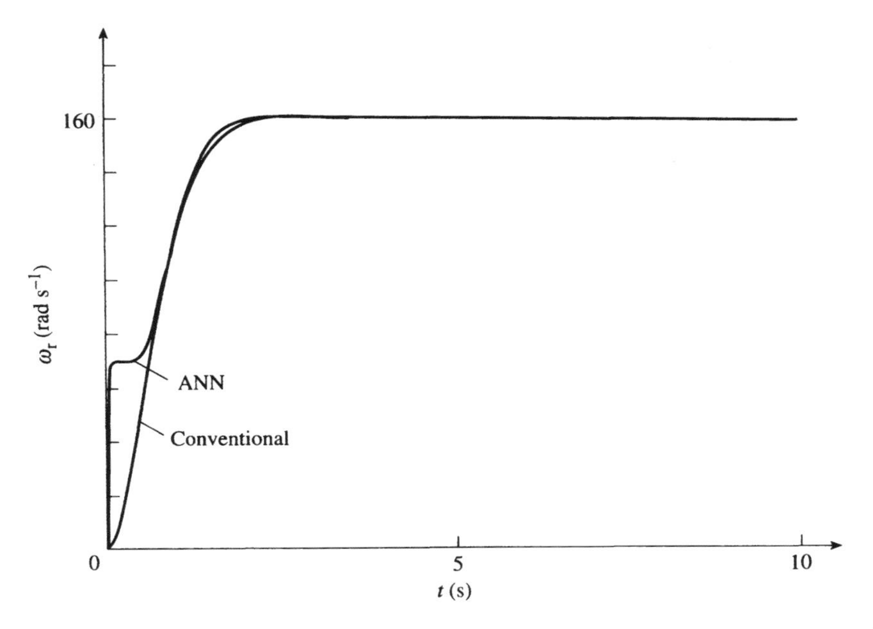

Fig. 7.10. Speed during run-up of d.c. machine (10-4-5-3 ANN and conventional mathematical model).

7.3 Neural-network-based steady-state and transient analysis of induction machines, estimators

7.3.1 CONVENTIONAL STEADY-STATE AND TRANSIENT ANALYSIS

It has been stated above that by using a well-trained multi-layer ANN, it is possible to estimate correctly the rotor speed and the electromagnetic torque of a d.c. machine. In the present section the estimation of the rotor speed and electromagnetic torque and other quantities (e.g. flux linkages) of induction machines (slip-ring and squirrel-cage) is described by using various ANNs.

Conventionally the transient analysis of a symmetrical three-phase induction machine is performed by using the real and imaginary parts of the stator and rotor space-vector voltage differential equations, together with the equation of motion. This results in a system of first-order non-linear differential equations, with five equations. In a variable-speed drive system these have to be complemented by the equations for the controllers and converter. Normally, the machine equations contain six machine parameters (R_s, R_r', L_{sl}, L_{rl}', L_M, J), but it is also possible to have such a transient model where only 5 parameters are used (stator resistance, stator transient inductance, referred magnetizing inductance, referred rotor resistance, inertia; see also Section 4.5.3.5.2 where the second machine model used for the EKF contains fewer parameters). However, it is also possible to estimate the various machine quantities in both the steady-state and transient state by using an ANN, and in this case these parameters are not required and the ANN can accurate describe the non-linear behaviour of the machine (no assumptions have to be made about any type of non-linearity). This is described in the next section. Such an estimator can also be used in 'sensorless' induction motor drives.

7.3.2 ANN-BASED STEADY-STATE AND TRANSIENT ANALYSIS, ESTIMATORS

For the ANN-based transient analysis of an induction machine, a multi-layer ANN has been considered with back-propagation training (see also Sections 7.2.2, 7.2.3). The run-up speed of an unloaded induction machine using a conventional mathematical model is shown in Fig. 7.11. For the purposes of obtaining the speed and other signals by using an ANN, several approaches have been considered.

7.3.2.1 ANNs for a slip-ring induction machine

It is assumed that the machine is a slip-ring induction machine and thus the rotor currents can be measured, and for training purposes (and only for training

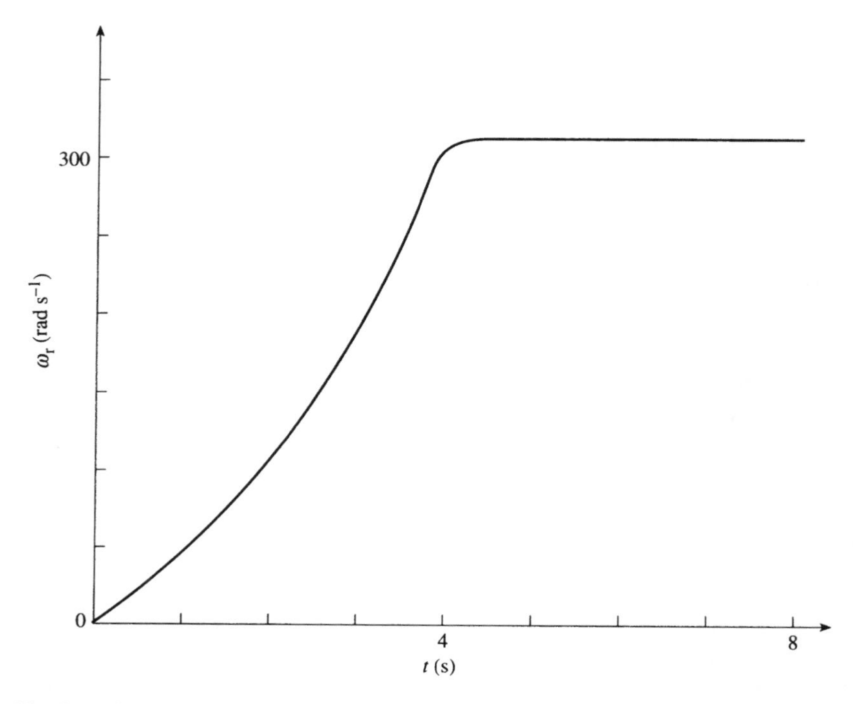

Fig. 7.11. Run-up speed of an induction motor using a conventional mathematical model.

purposes) there is a rotor position sensor, so the rotor currents can be transformed into their values in the stationary reference frame.

A multi-layer feedforward ANN with the structure 8-9-12-1 (8 input nodes, nine hidden nodes in the first hidden layer, 12 hidden nodes in the second hidden layer, and one output node) has been used, since this has given correct results. As mentioned in the previous sections, the structure of this type of ANN can only be determined by trial and error. It can be seen that this ANN is much more complex than those used for the d.c. machine, since the induction motor is a much more complicated non-linear plant than the d.c. machine. The inputs to the ANN are the sampled values of the stator and rotor currents, all expressed in the stationary reference frame, thus the 8 inputs are: $i_{sD}(k)$, $i_{sD}(k-1)$, $i_{sQ}(k)$, $i_{sQ}(k-1)$, $i_{rd}(k)$, $i_{rd}(k-1)$, $i_{rq}(k)$, and $i_{rq}(k-1)$. For training purposes only 60% of the training data set was used, but when the trained ANN was applied in the checking phase (when the data not seen previously by the network was also presented to the network), it gave correct results for those data as well. The activation functions of the hidden layers are tansigmoid functions.

Figure 7.12(a) shows the speed of the unloaded induction machine during the starting process (run-up), as obtained by the 8-9-12-1 ANN, and Fig. 7.12(b) shows the speed error (speed determined by computation minus speed determined by ANN). It can be seen that there is good agreement, and the error is small, typically less than 10^{-3}.

The ANN was then also tested by using input data when the machine was started with a load torque of 25 Nm. The results are shown in Fig. 7.13 and it can be seen that in general there is good agreement between the conventional and ANN based results. The error is typically less than 2×10^{-3}.

It is interesting to note that the same 8 input data have also been used with another ANN, which, however, contains only two layers: the input layer with 8 nodes and the output layer with 2 nodes, and there were no hidden layers (8-2 ANN). Physically this corresponds to the fact that under linear magnetic conditions the flux linkages are determined by the currents and there is a linear relationship between the flux linkages and currents, and thus a linear ANN can be used. This type of ANN is called an Adaline. The two output nodes correspond to the direct- and quadrature-axis stator flux linkages (ψ_{sD}, ψ_{sQ}). The trained neural network was then able to estimate correctly the instantaneous values of the stator flux linkages during run-up (without explicit knowledge of the machine inductances). The training was achieved very quickly, only after four epochs. The final weights were then checked, and as expected, they agreed with the corresponding inductances (L_s, L_m, L_r) of the induction machine. The stator flux-linkage error was typically less than 10^{-4} when the Adeline was used.

Another ANN was also trained to estimate the electromagnetic torque transients. Traditionally the electromagnetic torque can be computed by using the vectorial cross product of the stator flux-linkage space vector and stator-current space vector $(3/2)P(\bar{\psi}_s \times \bar{i}_s)$, which, however, yields $(3/2)PL_m\bar{i}'_r \times \bar{i}_s$ (where $\bar{i}'_r$ is the rotor-current space vector in the stationary reference frame), or in two-axis form $(3/2)PL_m$ $(i_{rd}i_{sQ}-i_{rq}i_{sD})$. Since the currents are inputs to the ANN, the ANN has to only

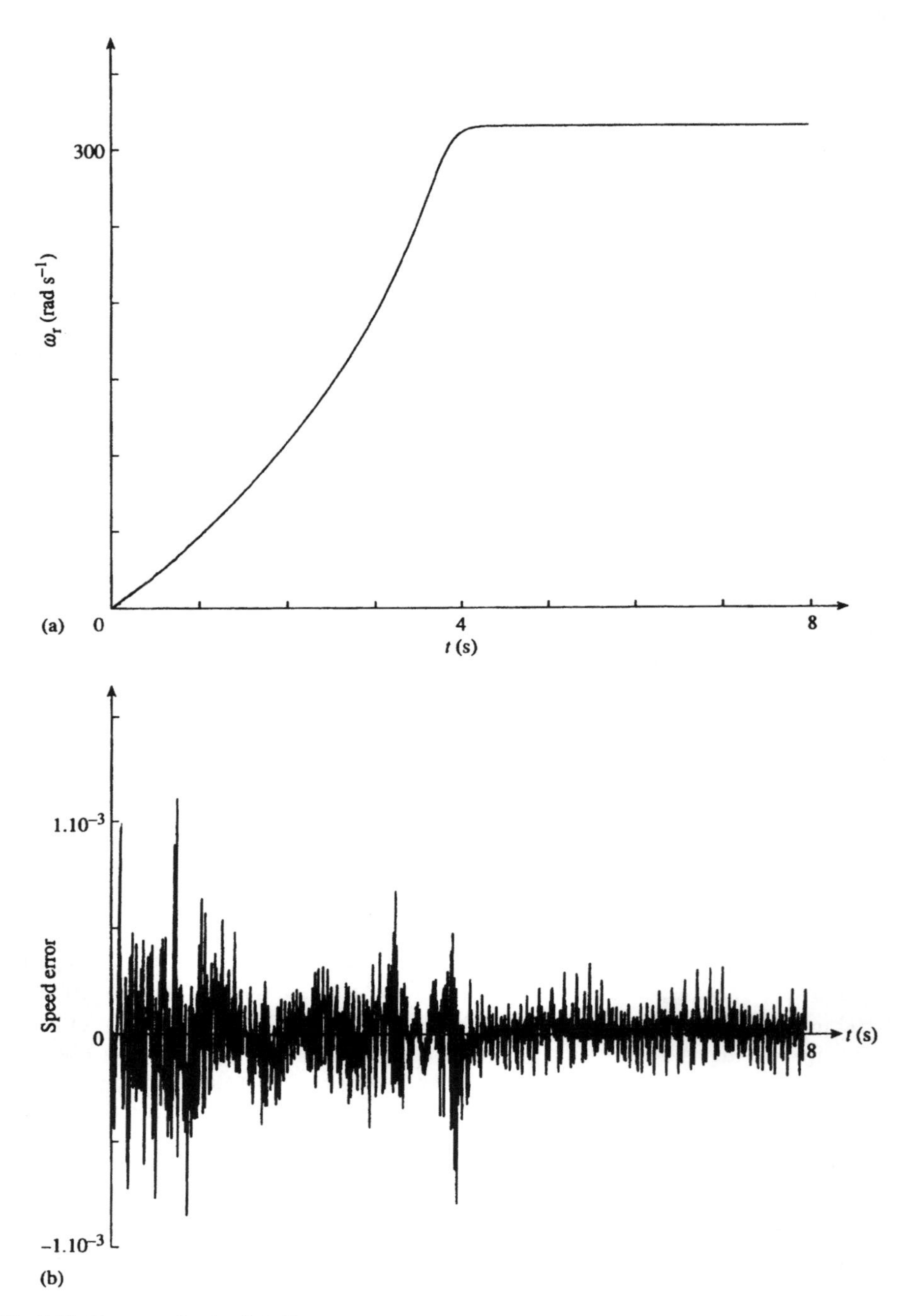

Fig. 7.12. Run-up of unloaded slip-ring induction machine. (a) 8-9-12-1 ANN-based speed estimation; (b) speed error.

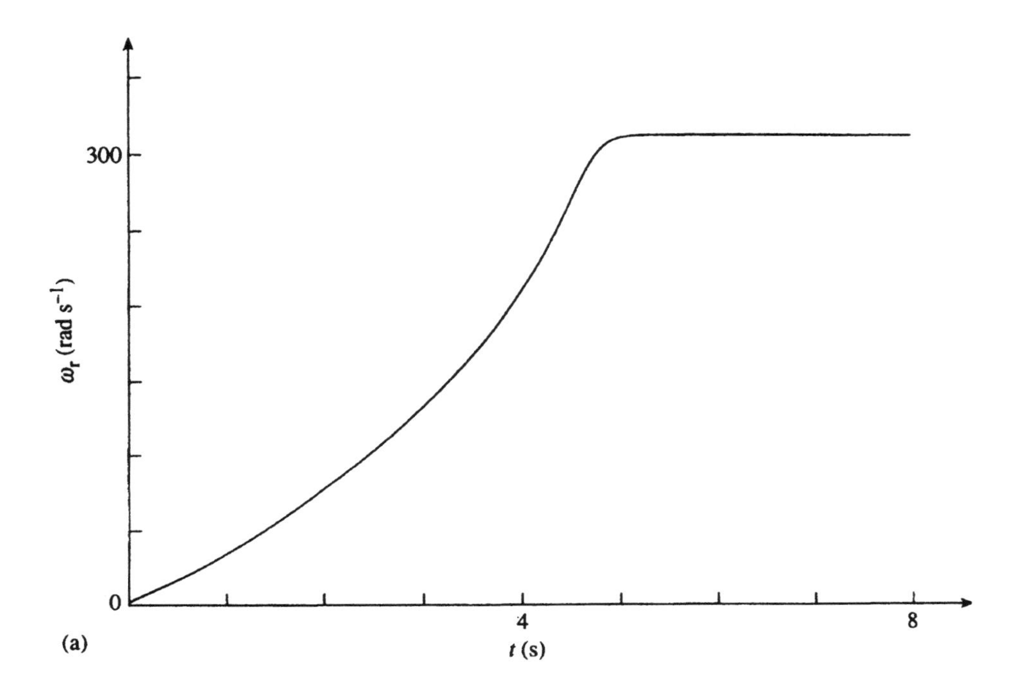

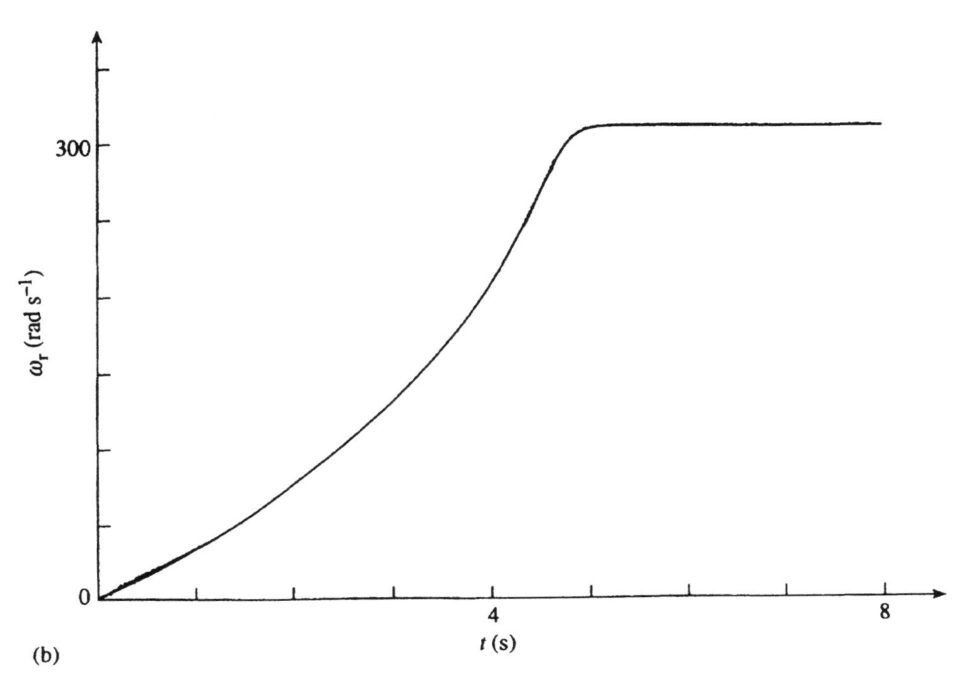

Fig. 7.13.

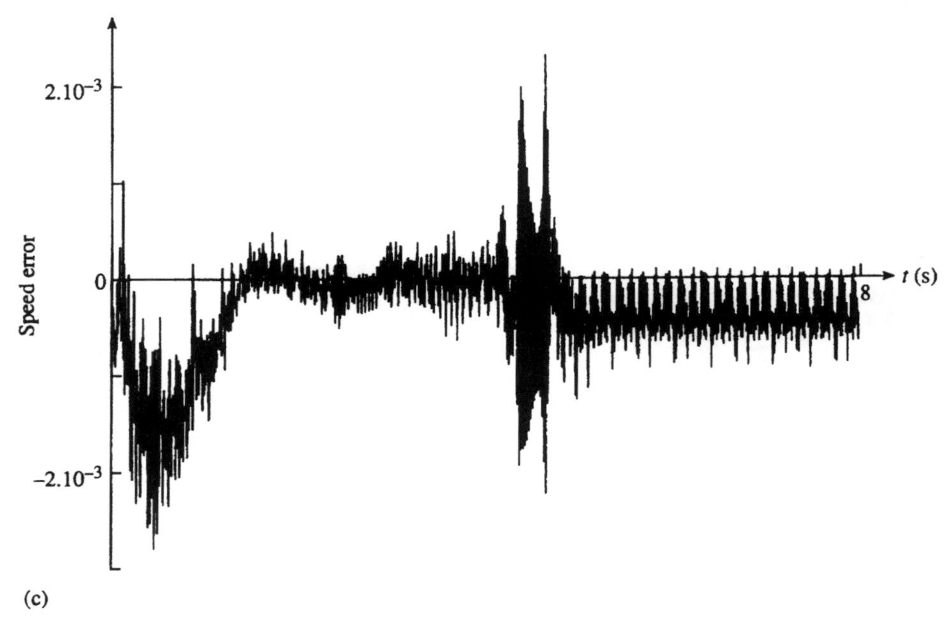

Fig. 7.13. Run-up speed of loaded induction machine. (a) Application of conventional mathematical model; (b) 8-9-12-1 ANN-based speed estimation; (c) speed error.

perform a simple operation ($i_{rd}i_{sQ} - i_{rq}i_{sD}$). Therefore, an ANN with a relatively simple structure was selected, and finally an 8-10-1 structure was obtained (where the inputs were the same as above, the output was the electromagnetic torque, and there were 10 hidden nodes). The activation functions of the hidden layer were tansigmoid functions. Figure 7.14 shows the variation of the electromagnetic torque of the unloaded induction machine during run-up. Figure 7.14(a) shows the results obtained by using the conventional mathematical model, but Fig. 7.14(b) shows the results obtained by the ANN. In Fig. 7.14(c) the variation of the torque error is shown (error = electromagnetic torque obtained by using conventional technique minus torque obtained by ANN). It follows that the ANN is capable of providing correct estimates. The error is typically less than 10^{-3}.

An ANN with increased complexity was also trained which contained two outputs, since it was required to have a simultaneous estimate of both the electromagnetic torque and speed. The ANN was then applied to a previously unseen data set, and the correct results were obtained.

The ANN described above is of limited use since it requires rotor currents. Therefore another scheme was also investigated which uses the stator voltages and stator currents as inputs, and such an estimator is suitable for a squirrel-cage induction machine.

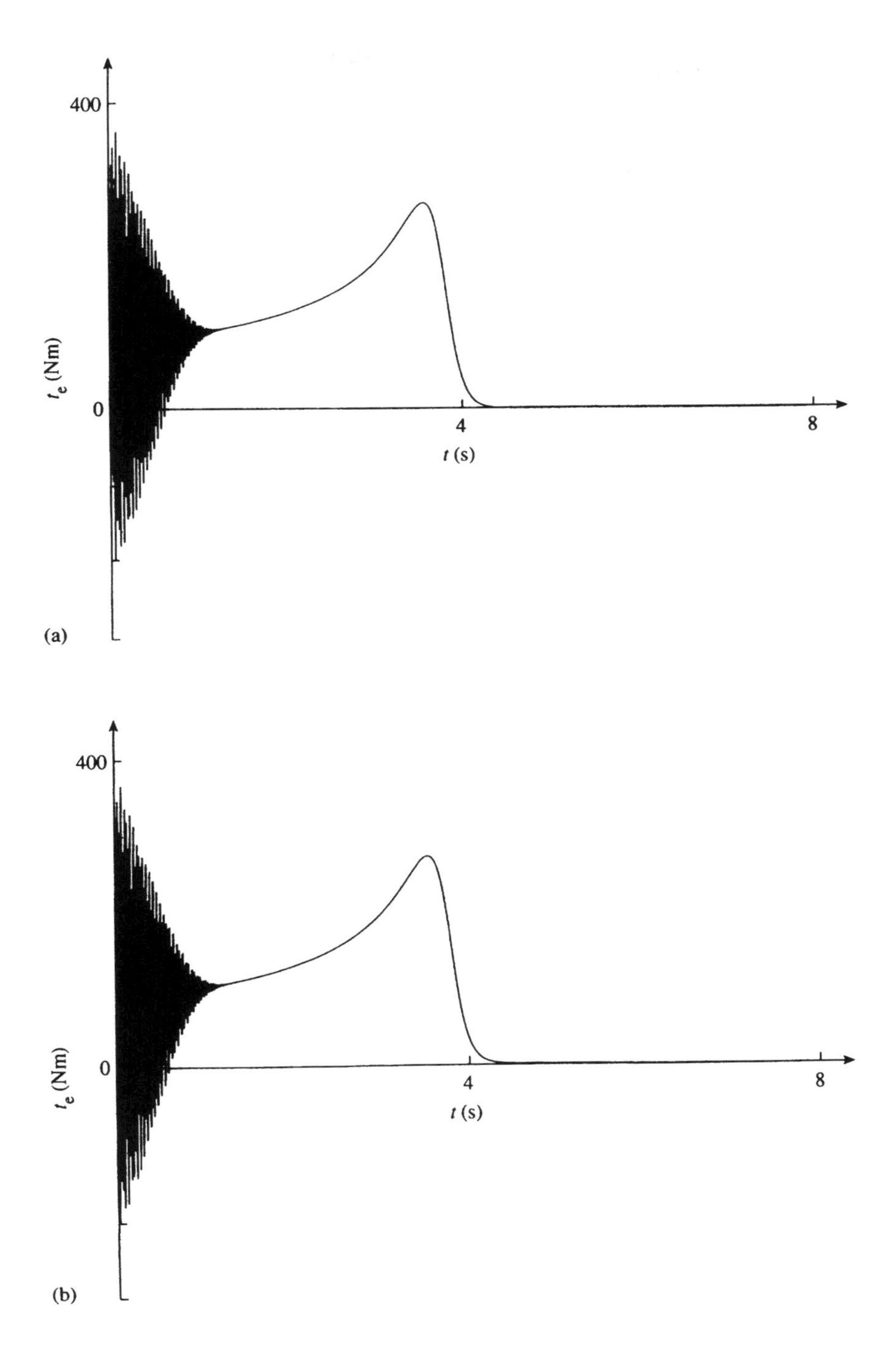

Fig. 7.14.

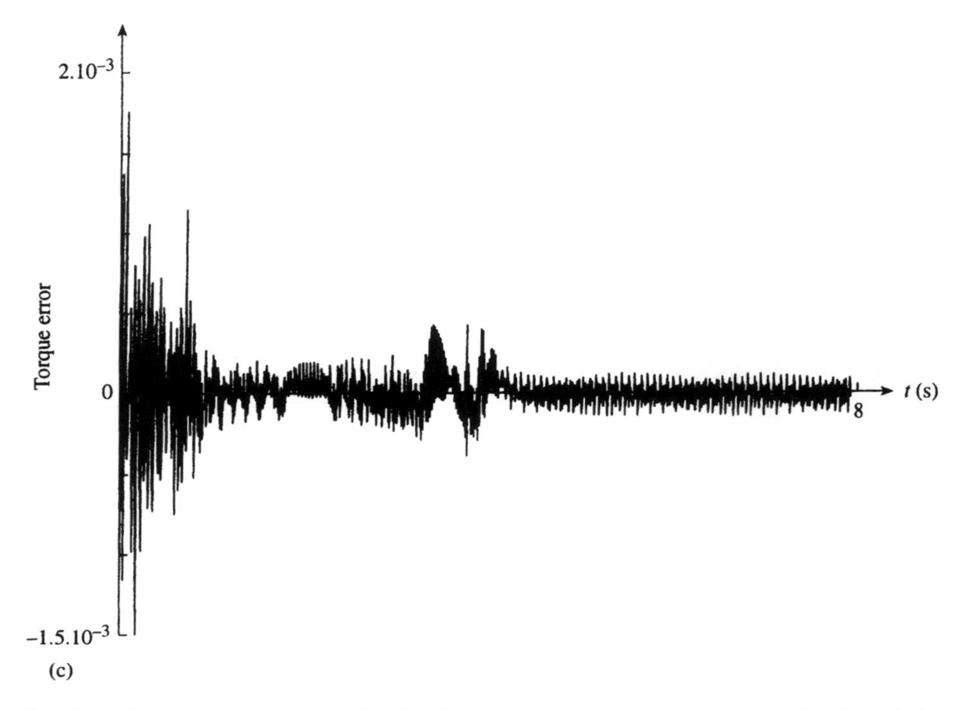

Fig. 7.14. Electromagnetic torque of induction machine during run-up. (a) Application of the conventional technique; (b) application of the 8-10-1 ANN; (c) electromagnetic torque error.

7.3.2.2 ANNs for a squirrel-cage induction machine

7.3.2.2.1 Speed estimator: 8-9-7-1 ANN, 8-8-6-1 ANN

An ANN-based speed estimator was also considered for a squirrel-cage induction machine. The multi-layer feedforward ANN had an 8-9-7-1 structure. The inputs to the ANN were the stator voltages and stator currents: $u_{sD}(k)$, $u_{sD}(k-1)$, $u_{sQ}(k)$, $u_{sQ}(k-1)$, $i_{sD}(k)$, $i_{sD}(k-1)$, $i_{sQ}(k)$, $i_{sQ}(k-1)$ [since for the squirrel-cage machine the rotor currents cannot be measured] and the output was the rotor speed, $\omega_r(k)$. The results obtained by the ANN for the transient speed of the unloaded machine after switch-on to the supply are shown in Fig. 7.15(a), and Fig. 7.15(b) shows the speed error (difference between 'actual' speed and speed estimated by ANN). It can be seen that the error is typically less than 3×10^{-3}. The activation functions used in the hidden layers were tansigmoid functions and for training purposes only half of the data were used. The sum of the squared errors during training was 5.8×10^{-4}.

Another ANN was also considered with a structure of 8-8-6-1 (thus the number of hidden neurons was reduced by one in the first and second hidden layers respectively). In this case the corresponding results are shown in Figs 7.16(a) and (b),

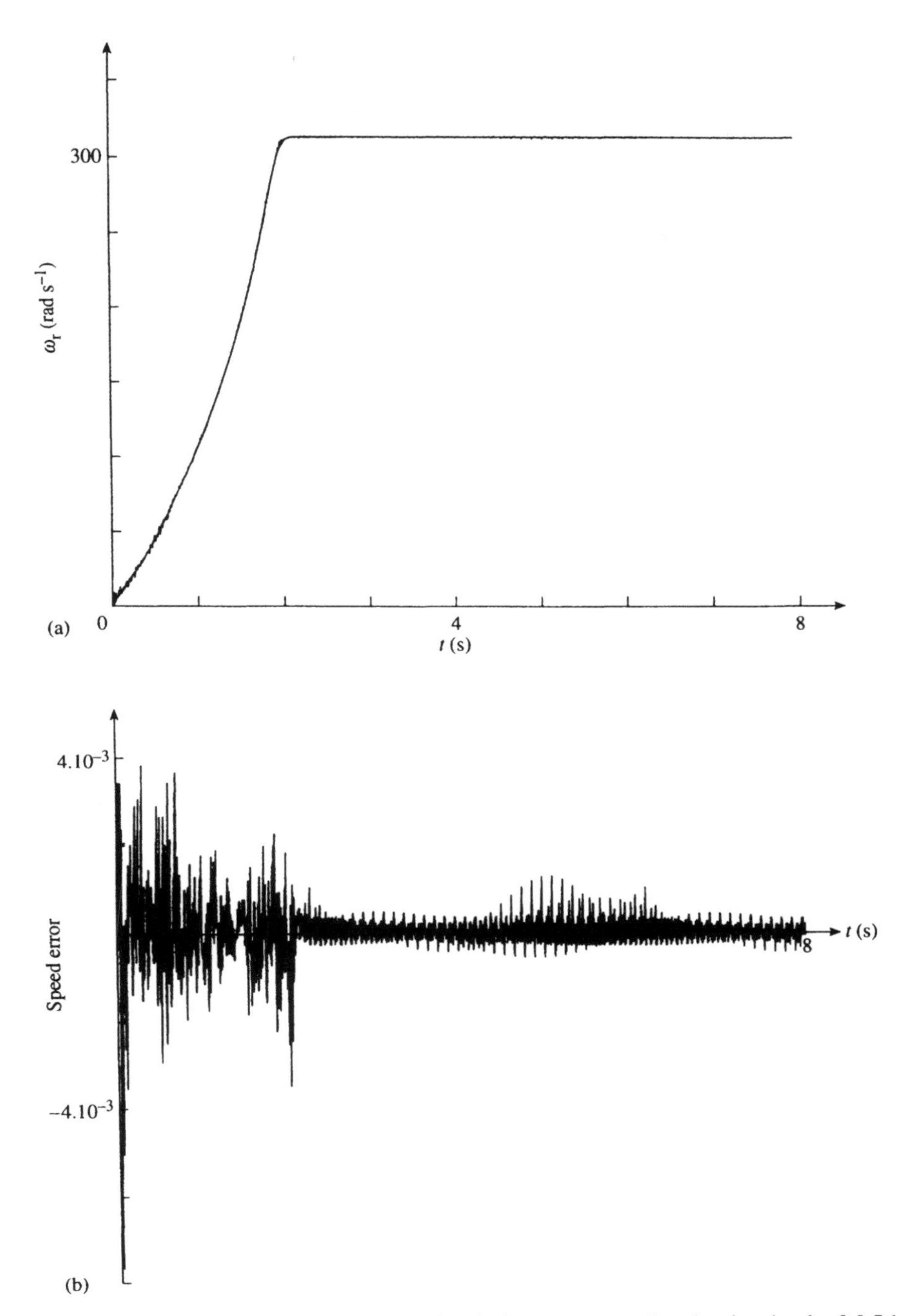

Fig. 7.15. Speed of unloaded squirrel-cage machine during run-up. (a) Speed estimation by 8-9-7-1 ANN; (b) speed error.

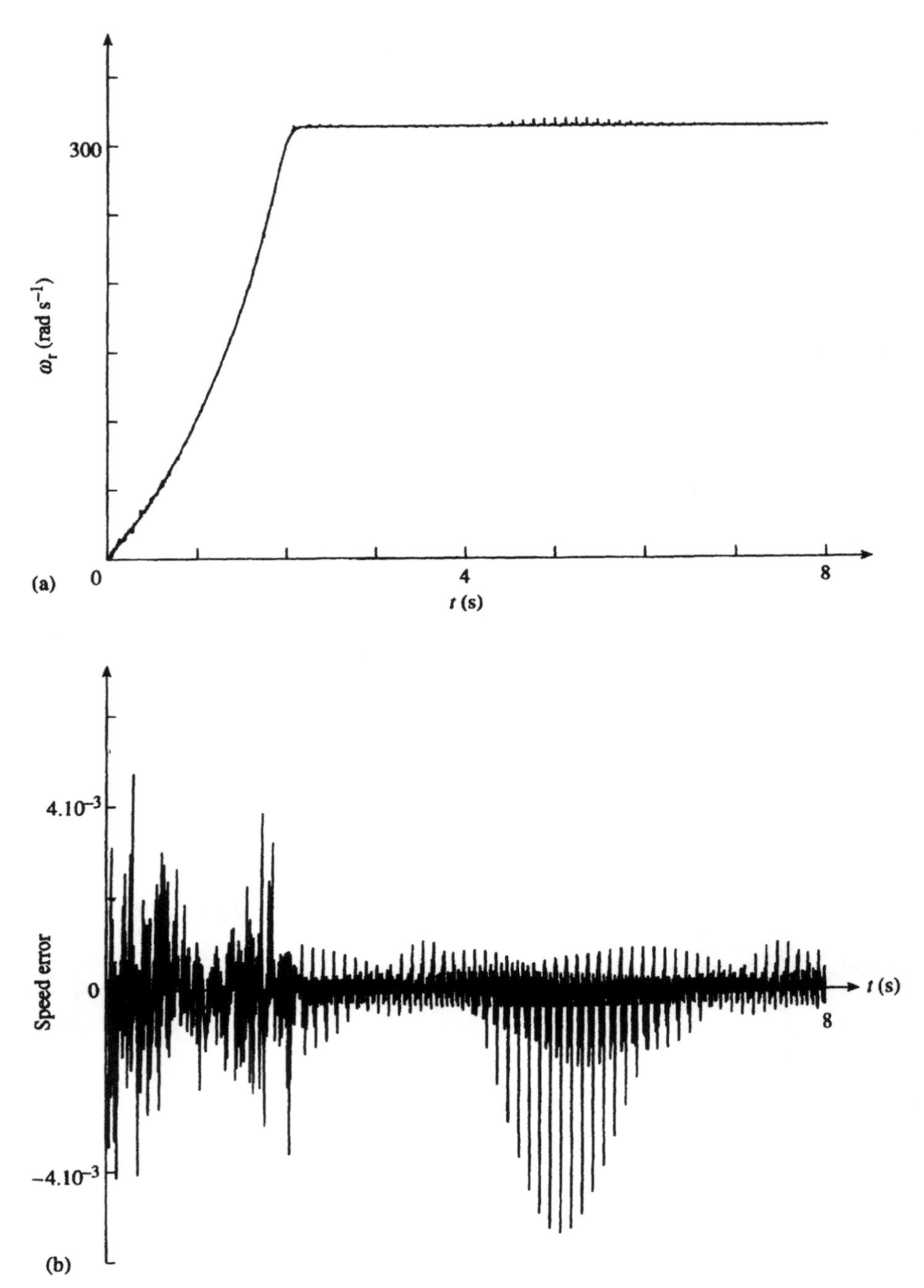

Fig. 7.16. Speed of unloaded squirrel-cage machine during run-up. (a) Speed obtained by conventional mathematical model and by 8-8-6-1 ANN; (b) speed error.

and it can be seen that there are some differences between the curves shown in Figs 7.16 and 7.15.

7.3.2.2.2 Speed, torque, and flux estimator: 8-12-10-4 ANN

A much more complex ANN has also been applied to the same machine as above. It was the aim to obtain on its outputs four quantities: the rotor speed, the electromagnetic torque, and the direct- and quadrature-axis stator flux linkages in the stationary reference frame (ψ_{sD}, ψ_{sQ}). The ANN structure was again found by trial and error and finally the 8-12-10-4 ANN gave correct results. Thus the inputs to the ANN are $u_{sD}(k)$, $u_{sD}(k-1)$, $u_{sQ}(k)$, $u_{sQ}(k-1)$, $i_{sD}(k)$, $i_{sD}(k-1)$, $i_{sQ}(k)$, $i_{sQ}(k-1)$; the outputs are $\omega_r(k)$, $t_e(k)$, $\psi_{sD}(k)$ and, $\psi_{sQ}(k)$, and there were two hidden layers with 12 and 10 nodes respectively. Figure 7.17(a) shows the 'actual' speed, Fig. 7.17(b) shows the speed estimated by the ANN, and Fig. 7.17(c) shows the speed error (actual speed minus speed estimated by ANN) during the run-up of the unloaded machine. Figure 7.16(d) shows the 'actual' electromagnetic torque, Fig. 7.17(e) shows the electromagnetic torque estimated by the ANN, and Fig. 7.17(f) shows the torque error. The activation functions used were tansigmoid, and only half of the data was used for training purposes. The sum of the squared errors during training was 0.06 (much higher than in the previous cases discussed above).

7.3.2.2.3 Speed estimator in a vector-controlled drive: 8-15-13-1 ANN, other ANNs

An ANN-based virtual-speed estimator in a vector-controlled voltage-source induction motor drive employing rotor-flux-oriented control was also investigated. The goal was to achieve speed-sensorless operation of the drive. Various ANN structures were considered.

The first multi-layer ANN to be used has a structure of 8-15-13-1. The eight inputs are the measured values of the stator voltages and stator currents, $u_{sD}(k)$, $u_{sD}(k-1)$, $u_{sQ}(k)$, $u_{sQ}(k-1)$, $i_{sD}(k)$, $i_{sD}(k-1)$, $i_{sQ}(k)$, $i_{sQ}(k-1)$ during run-up and speed reversal. For illustration purposes the measured stator voltage u_{sD} and stator current i_{sD} are shown in Figs 7.18(a) and (b) respectively.

During training the sum of the squared errors was 0.07, which is quite high. The speed obtained by a speed sensor is shown in Fig. 7.18(c), and the speed estimated by the 8-15-13-1 ANN is shown in Fig. 7.18(d). The activation functions used in the hidden layers are tansigmoids. It can be seen that the speed estimated by the ANN contains many oscillations, but it is possible to obtain better results using other neural networks. For example, in one version, the sum of the squared errors of the trained ANN is very small, 7.3×10^{-4}. Figure 7.19 shows the estimated rotor speed obtained by the ANN.

It is important to note that there are great improvements in the speed estimation. The number of epochs used during the training is 1300.

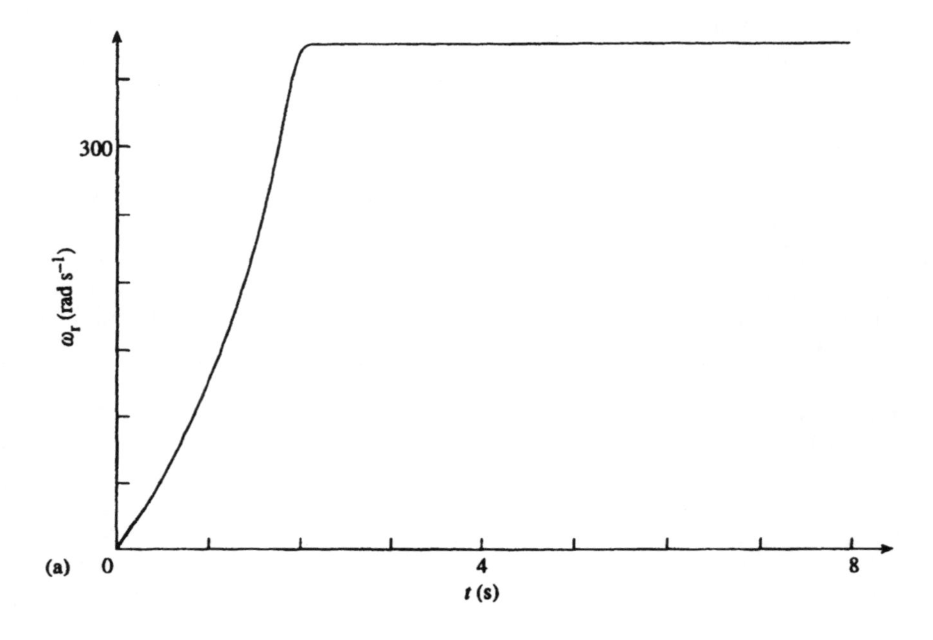

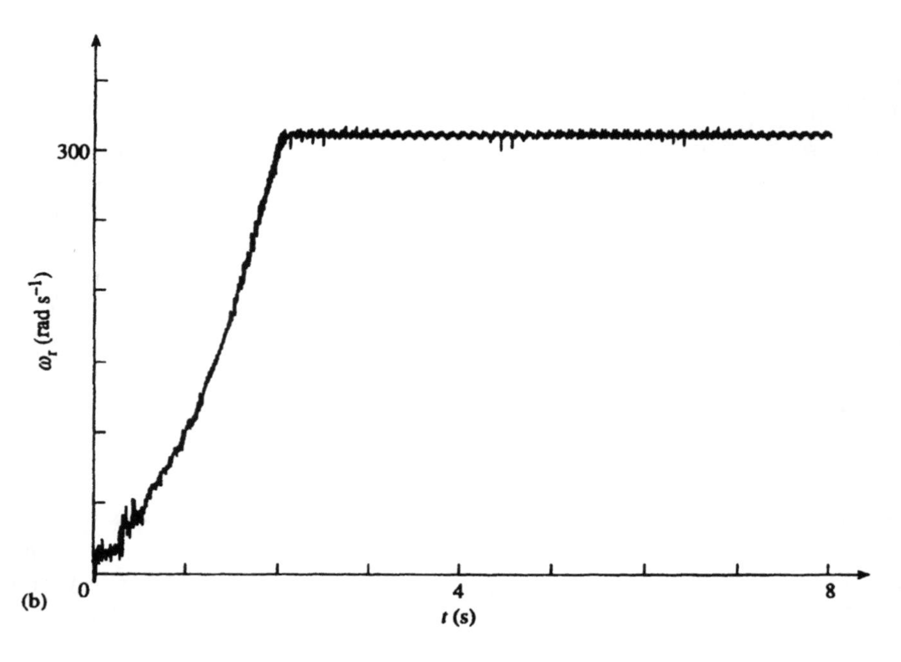

Fig. 7.17.

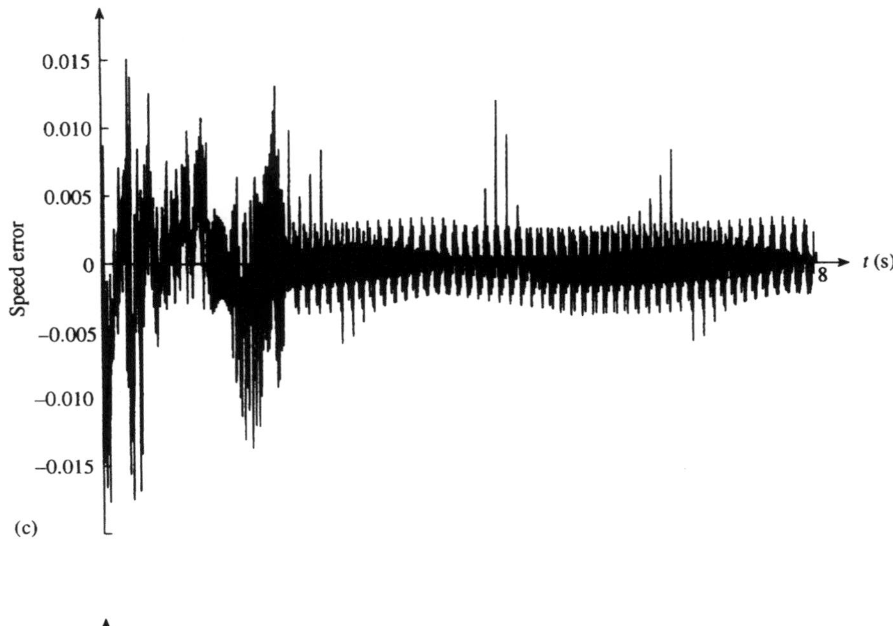

0.015
0.010
0.005
0
–0.005
–0.010
–0.015
Speed error
8
t (s)
(c)

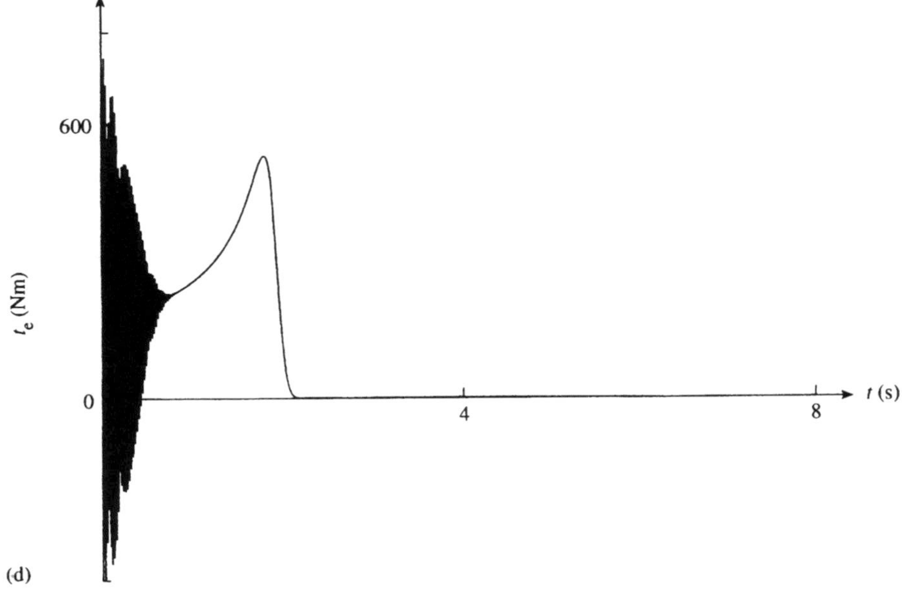

600
0
t_e (Nm)
4
8
t (s)
(d)

Fig. 7.17.

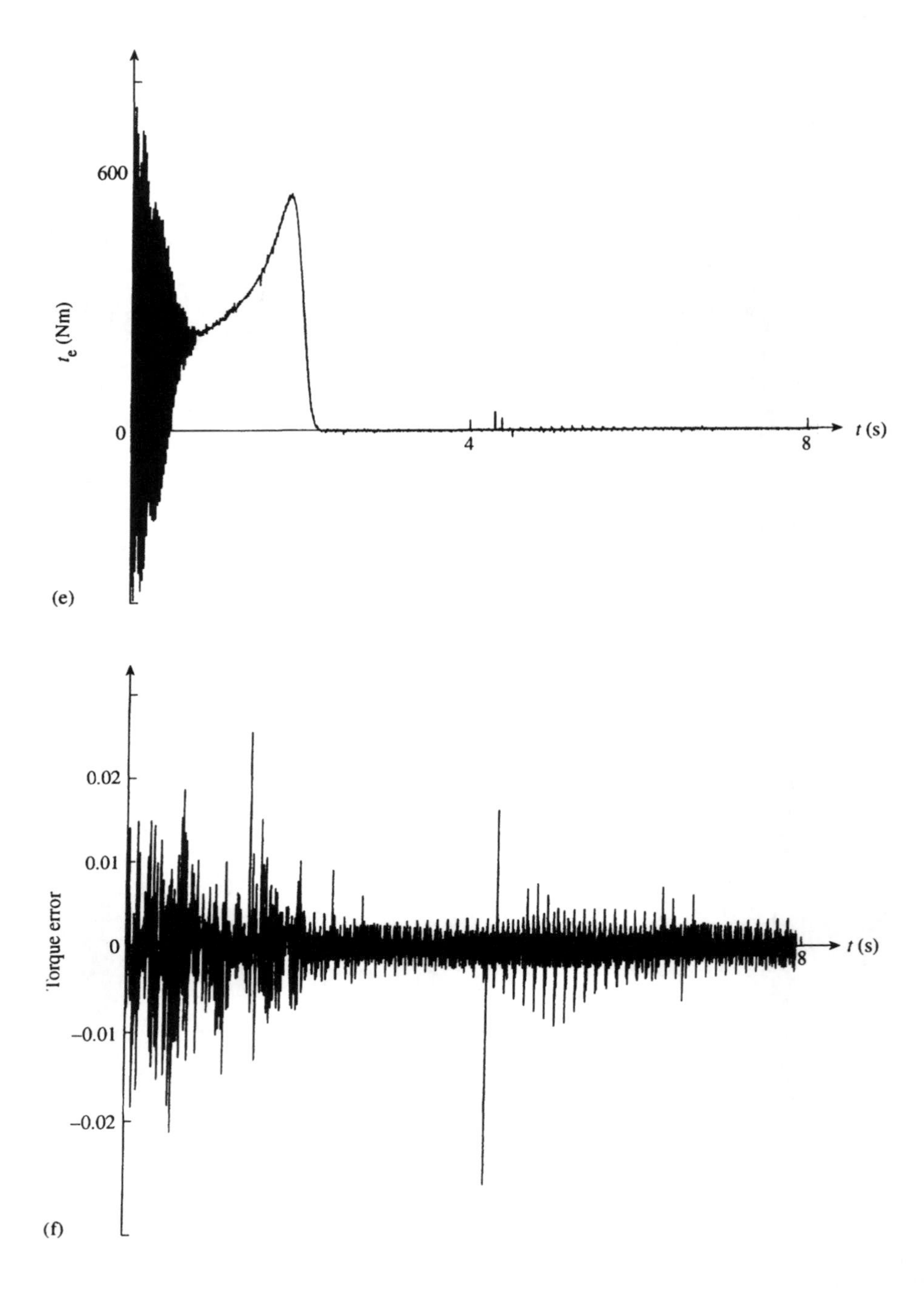

Fig. 7.17. Run-up speed and torque of unloaded squirrel-cage machine. (a) Actual speed; (b) speed estimated by 8-12-10-4 ANN; (c) speed error; (d) actual electromagnetic torque; (e) electromagnetic torque estimated by 8-12-10-4 ANN; (f) electromagnetic torque error.

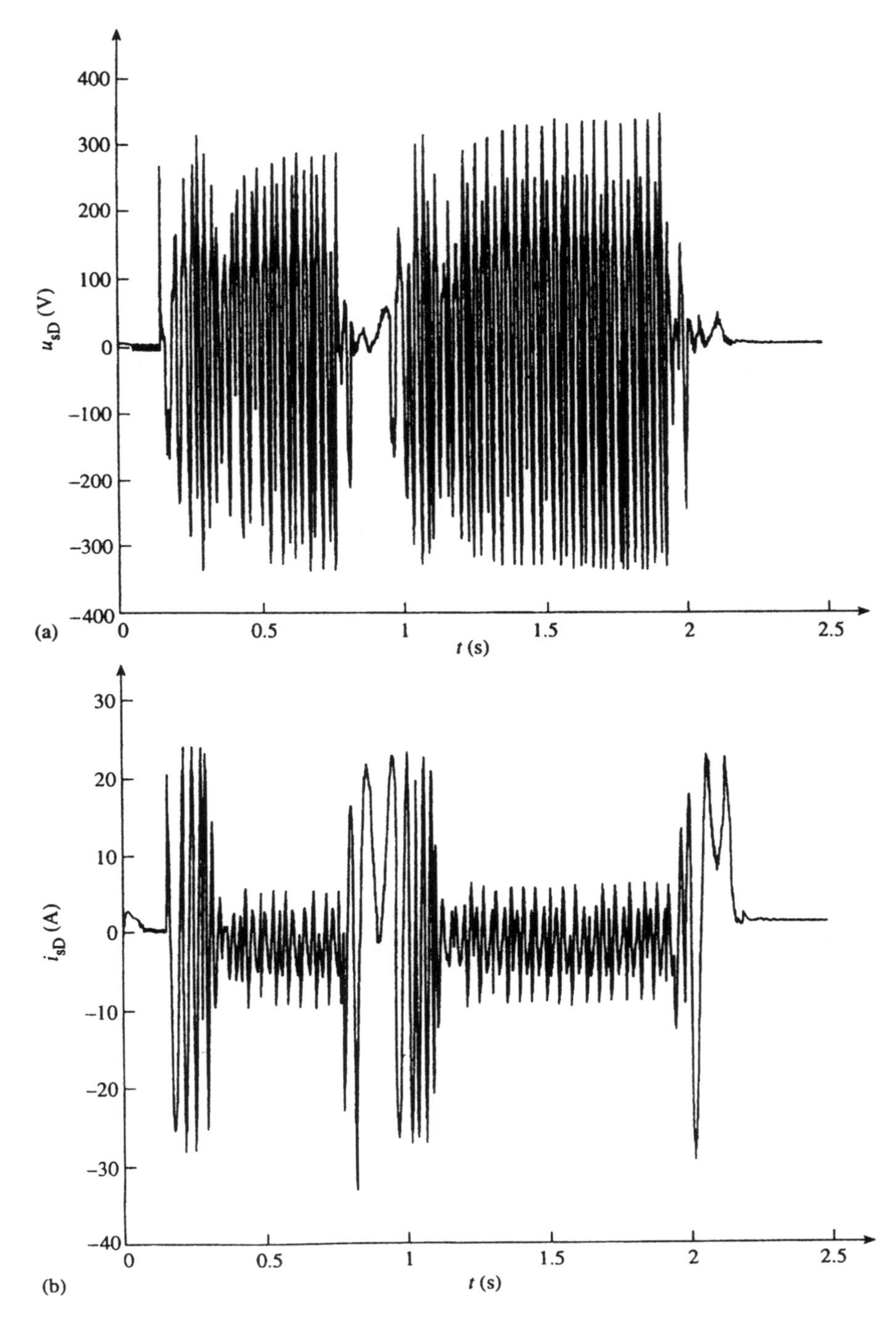

400
300
200
100
0
−100
−200
−300
−400
u_{sD} (V)
(a)
0
0.5
1
1.5
2
2.5
t (s)
30
20
10
0
−10
−20
−30
−40
i_{sD} (A)
(b)
t (s)

Fig. 7.18.

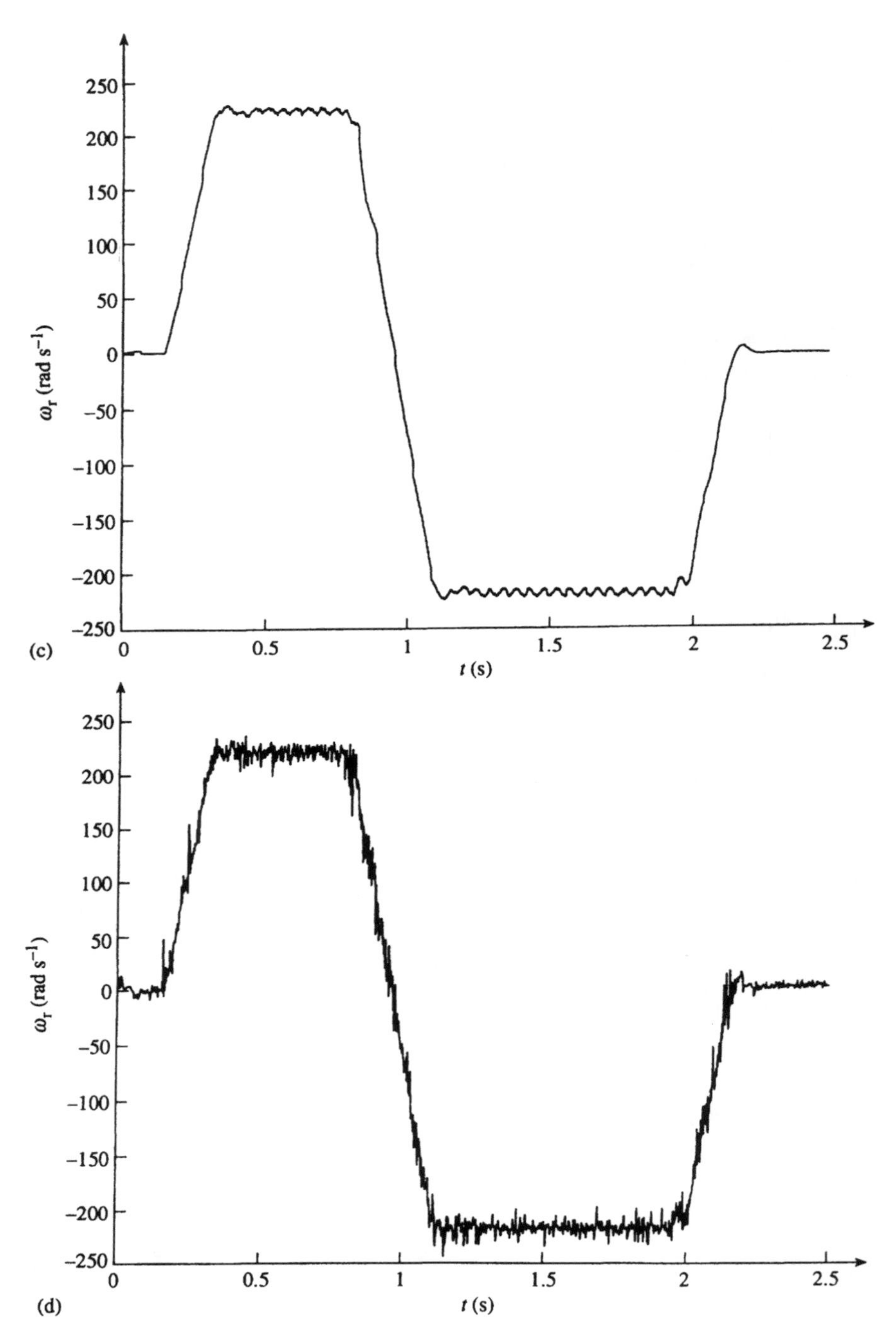

Fig. 7.18. Speed estimation in vector-controlled induction motor drive. (a) Measured u_{sD}; (b) measured i_{sD}; (c) measured speed using speed sensor; (d) speed estimated by 8-15-13-1 ANN.

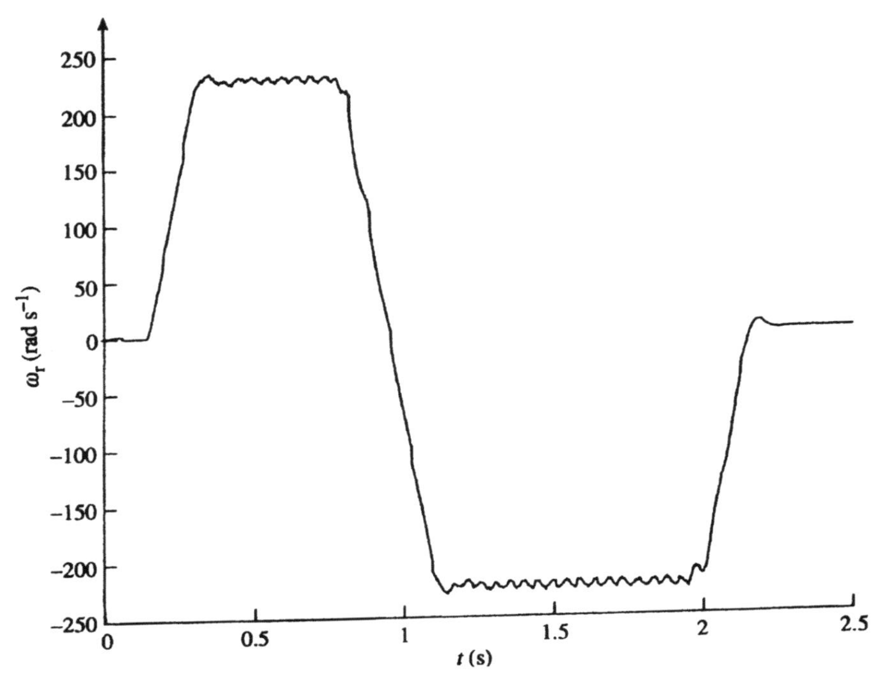

Fig. 7.19. Rotor speed estimated by ANN in vector-controlled induction motor drive.

The possibility of using an ANN-based speed estimator with a reduced number of inputs was also examined. For example, for an ANN which provided correct results, the sum of the squared errors is 9×10^{-4}. The number of epochs was 1500. The results are shown in Fig. 7.20. The obtained speed estimates agree well with those obtained by the application of a speed sensor (see curve shown in Fig. 7.18(c)). Various ANNs were implemented in a DSP-based speed-sensorless vector drive.

It is believed that ANN-based speed estimators will be used in other vector- and direct-torque-controlled drives. However, to overcome the difficulty of finding the appropriate ANN structure (appropriate number of hidden layers and hidden neurons) by trial and error, a fuzzy-neural network can also be used (see also Sections 4.4.2, 7.4). For this purpose various fuzzy-neural networks have also been successfully trained.

7.3.3 ANN-BASED HARMONIC ANALYSIS

In converter-fed a.c. drives, due to the switching of the converter there are always time harmonics present in the current and/or voltage waveforms. In contrast to conventional harmonic-estimation techniques, it is also possible to estimate the generated time harmonics by using ANN-based feature extractors. For this

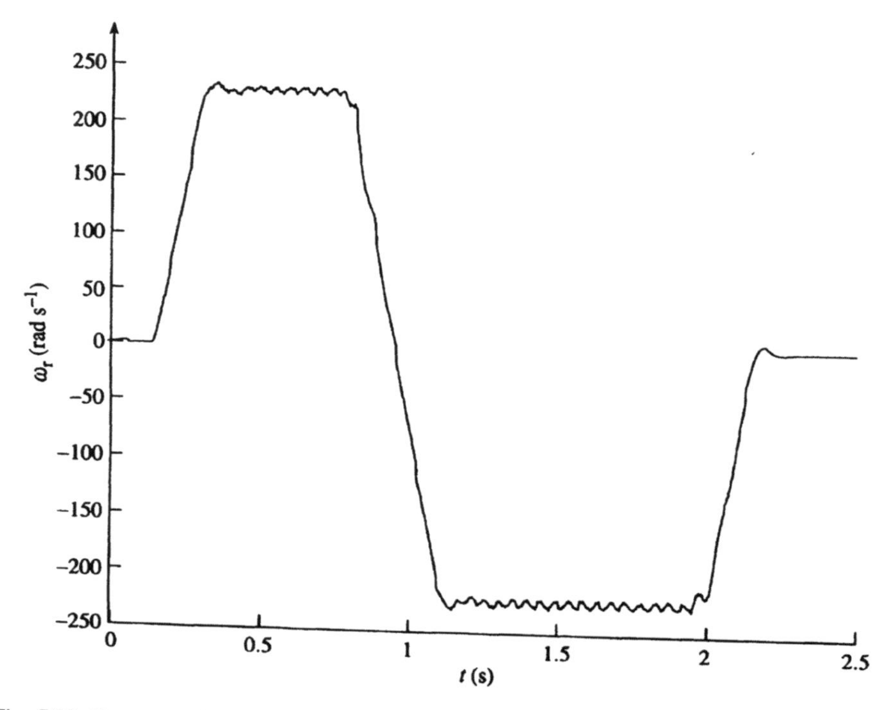

Fig. 7.20. Rotor speed estimated by ANN with reduced number of inputs in vector-controlled induction motor drive.

purpose the self-organizing Kohonen feature map can be used, which is an unsupervised ANN [Kohonen 1990].

In its most basic form the Kohonen network, which is a competitive neural network (often referred to as the Kohonen feature map), consists of a two-dimensional (rectangular) array of artificial neuron nodes, which form a topologically ordered map of data to which it is exposed. It has three main features:

- it speeds up convergence when the number of input data is high;
- it is suitable for clustering data with noise;
- it is topologically preserving (i.e. it provides both qualitative and quantitative information on the data).

It is one of the main advantages of this technique that unlabelled input data can be quickly organized into a map which can show characteristic features in the data structure, which otherwise are 'hidden'.

In the Kohonen feature map, the inputs are transmitted to the competitive layer (feature layer), which in the present application is a two-dimensional rectangular layer containing the output nodes. The input signals can be the sampled values of any kind of signal, e.g. they can be the sampled values of the stator currents of an inverter-fed induction machine. The weights are organized so that topologically close nodes are sensitive to inputs that are similar. Thus the output nodes are ordered in a natural manner (where the topology is preserved). In the competitive layer of the Kohonen feature map, several output nodes in the neighbourhood of the winner node can be activated at the same time, and all weights of neurons which belong to this neighbourhood are adapted so that their weights become closer to that of the winning neuron.

It should be noted that the feature array could have other forms as well, e.g. triangular, etc. It is interesting to note that there are no physical lateral interconnections between the output nodes, even though lateral interconnections play an important role in a biological neural network. The lateral interconnections in the Kohonen self-organizing map are modelled implicitly by the appropriate learning algorithm.

In the literature of electrical machines, harmonic analysis is almost exclusively performed on the temporal variation of the quantities (e.g. stator currents) considered. However, it has been shown in [Vas 1993] that there are advantages in performing harmonic analysis not on the temporal variations of the waveforms under consideration, but on their respective space-vector loci. The harmonic analysis described in [Vas 1993] is not based on artificial intelligence. However, space-vector loci contain specific features which can be easily recognized by human experts, e.g. when there are different time harmonics in the stator currents of an inverter-fed induction motor. This has resulted in the development of the Kohonen-neural-network-based harmonic analyser. Furthermore, since the space-vector current loci have different and well-distinguishable shapes in various reference frames, the harmonic analysis can be performed on the measured values of the currents expressed in different reference frames. For this purpose the currents have been considered in two reference frames: the stationary reference frame and also in a reference frame which rotates synchronously with the speed of the fundamental harmonic.

By considering data which are sampled values of the stator-current components of an induction motor in the stationary reference frame, Fig. 7.21 shows results obtained by using the Kohonen feature map. It can be seen from the space-vector locus obtained, that the stator current fifth-harmonic has been successfully identified. The three subplots correspond to the different epochs, and it can be seen that at epoch 300 the feature map has already started to unfold, and at epoch 2400 it has almost learnt the topology of the input space. In epoch 6000 it has learnt the main features of the input space well and the fifth harmonic has been identified, as seen from the six protrusions of the feature map. A large number of results have been obtained which have identified other harmonic components as well, establishing the validity of the proposed new approach.

When sampled data values of the currents expressed in the synchronously rotating reference frame described above have been used, the Kohonen feature-map has successfully identified the same harmonics as before, but of course now the locus has become an ellipse. The new technique can be used for rapid determination of stator-current harmonics. This can be used for condition monitoring

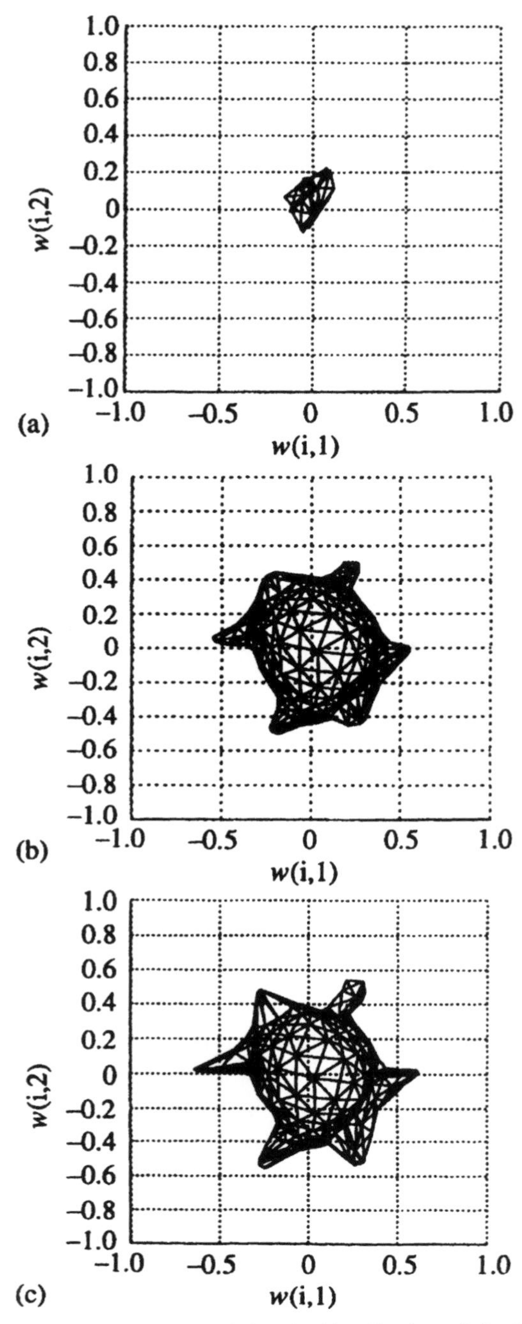

Fig. 7.21. Application of self-organizing ANN for the identification of the 5th harmonic by using sampled values of the currents in the stationary reference frame. (a) 300 epochs; (b) 2400 epochs; (c) 6000 epochs.

and diagnostic purposes and also for the development of new types of adaptive PWM modulators.

7.4 Fuzzy-neural-network-based steady-state and transient analysis of d.c. and induction machines, estimators

Recently, fuzzy-logic control has emerged as an attractive area for research in the control application of fuzzy set theory. The main feature is the construction of fuzzy-logic controllers (FLCs) which utilize the linguistic, imprecise knowledge of human experts. However, the implementation of conventional fuzzy-logic controllers suffers from the disadvantage that no *formal* procedures exist for the direct incorporation of the expert knowledge during the development of the controller. The structure of the fuzzy controller (number of rules, the rules themselves, number and shape of membership functions, etc.) is achieved through a time-consuming tuning process which is essentially manual in nature. The ability to 'learn' automatically the characteristics and structure which may be obscure to the human observer is, however, inherent in neural networks. A fuzzy-logic-type controller having a neural network structure offers the advantages of both—the ability of fuzzy logic to use expert human knowledge and the learning ability of the neural network—and overcomes their disadvantages—the lack of a formal learning procedure for the fuzzy controller and the lack of a clear correlation with the physical problem when using neural networks.

In a connectionist fuzzy-neural controller, the input and output nodes of the ANN represent the input and output signals and in the hidden layer, the nodes take the roles of membership functions and rules. The learning algorithm for this network can be hybrid, combining unsupervised and supervised methods. The unsupervised learning produces the number of fuzzy sets for each input and output variable, the number of fuzzy rules, the rules themselves, and the centres and widths of the membership sets. This information is used to establish a fuzzy-neural controller which is then trained using a back-propagation algorithm to tune the centres and widths of the membership functions further. The structure of the controller is fixed. Hybrid learning outperforms purely supervised learning by reducing training times.

There are many architectures which can be used for fuzzy-neural controllers. For example, a Mamdani-type of fuzzy-neural controller may contains five layers: an input layer, a layer for the fuzzy membership sets, a fuzzy AND layer, a fuzzy OR layer, and an output layer. The input layer contains the input nodes, which represent linguistic variables. These distribute each input variable to its membership functions. The second layer generates the appropriate membership values. The third layer defines the preconditions of the rule nodes. The nodes in the fourth layer connect the output of the fuzzy AND nodes to the consequences in the rules. The last layer (output layer) performs defuzzification. The details of such a network have been described in Section 4.4.2 and will not be repeated here. However, it is also possible to use a Sugeno-type of fuzzy-neural controller

[Sugeno 1985; Jang 1993], where the fact is utilized that the output of each rule is a linear combination of the input variables plus a constant term. In such a fuzzy-neural controller, the first layer is the input layer and the nodes in the second layer perform the membership functions (via the membership nodes). The nodes in the third layer perform the fuzzy AND operation (e.g. minimum operation); thus the outputs of the nodes give the firing strength of the corresponding rule. The nodes in the fourth layer compute the normalized firing strength of each rule. The outputs of the nodes of the fifth layer give the weighted consequent part of the rules, and finally in the last layer there is a single node which sums all the signals entering the node.

By using a fuzzy-neural network, it is possible to obtain various quantities of a d.c. or an induction machine. Thus it can be used to yield the instantaneous variations of the rotor speed, electromagnetic torque, etc. This is discussed below.

7.4.1 FUZZY-NEURAL SPEED ESTIMATOR FOR A D.C. MOTOR

In Section 7.2.4 speed estimation using ANNs for a d.c. machine has been discussed. The same d.c. machine is considered now, but the speed is estimated by using a fuzzy-neural network. In contrast to a speed estimator using a conventional multi-layer ANN, where the number of hidden layers and hidden nodes had to be obtained by trial and error, when a fuzzy-neural network is used the network structure can be obtained automatically.

First a Sugeno-type of fuzzy-neural network is considered (see the introduction to Section 7.4.) with four inputs (as in Section 7.2.4.1). Thus the inputs are $u_a(k)$, $u_a(k-1)$, $i_a(k)$, and $i_a(k-1)$, and it is the goal to estimate $\omega_r(k)$. To have a simple network, only three fuzzy sets are considered for both the armature voltage and armature current. Thus altogether six membership functions have been used, the initial values of which are shown in Figs 7.22(a) and (b). In total there were 81 rules.

The fuzzy-neural network was trained with a sinusoidally varying armature voltage (see also Section 7.2.4.1). The training lasted for four epochs only. Figure 7.21(c) shows the results obtained when a constant armature voltage $U_a = 240$ V was applied to the trained fuzzy-neural network. It can be seen that, despite the very limited training, the fuzzy-neural speed estimator gives correct results in the steady state. However, it was also possible to obtain improved estimation in the transient state after only 10 epochs of training by increasing the number of inputs from four to five, where the fifth input is the past value of the rotor speed. In this case, again three membership functions have been used for the armature voltage, another three membership functions for the armature currents, and finally three membership functions for the rotor speed. As a consequence, the number of rules has increased from the previous 81 to 243, but this is still less than the number of rules which would be obtained by using the original four inputs and increasing the number of membership functions from three to four (or even a higher number) for both the armature voltage and armature current. It

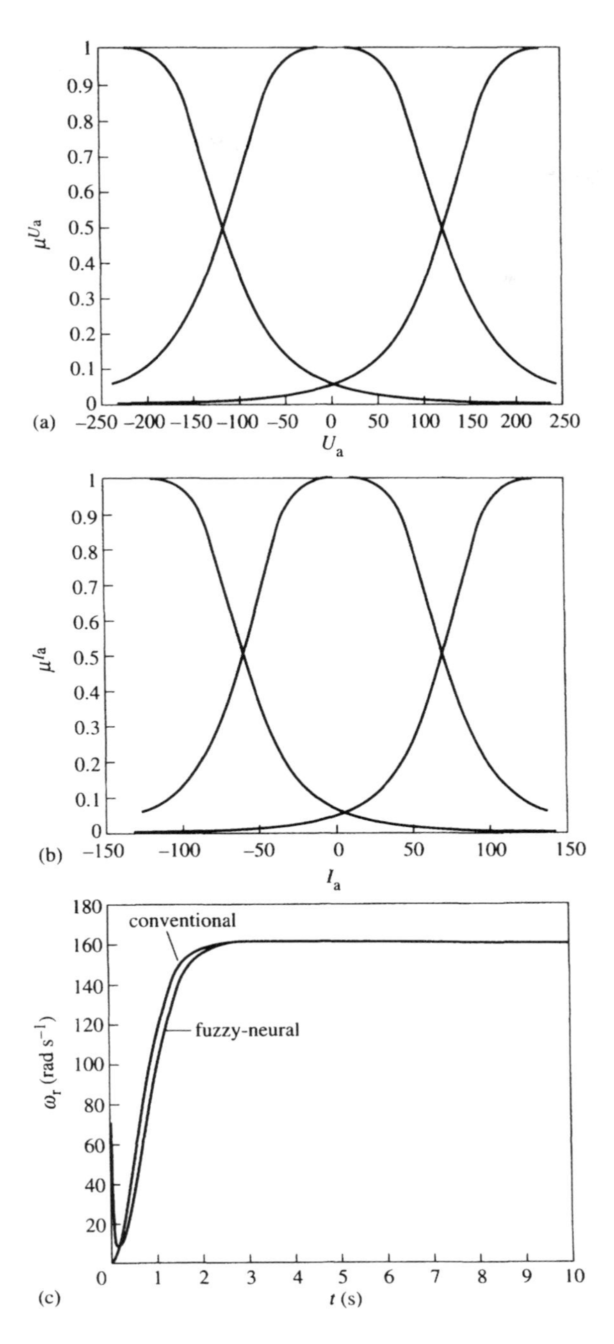

1
0.9
0.8
0.7
0.6
0.5
0.4
0.3
0.2
0.1
0
μ^{U_a}
(a)
−250 −200 −150 −100 −50 0 50 100 150 200 250
U_a
1
0.9
0.8
0.7
0.6
0.5
0.4
0.3
0.2
0.1
0
μ^{I_a}
(b)
−150 −100 −50 0 50 100 150
I_a
180
160
140
120
100
80
60
40
20
ω_r (rad s^{-1})
conventional
fuzzy-neural
(c)
0 1 2 3 4 5 6 7 8 9 10
t (s)

Fig. 7.22.

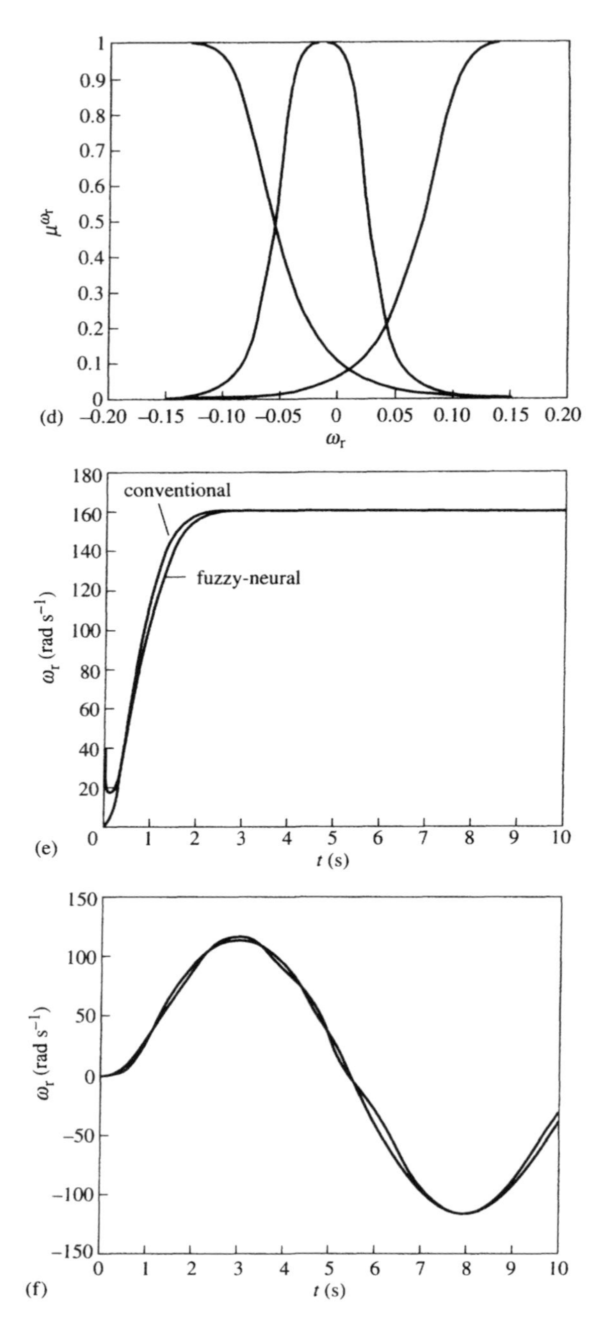

Fig. 7.22. Fuzzy-neural speed estimation. (a) Initial armature-voltage membership functions; (b) initial armature-current membership functions; (c) estimated speed obtained by using conventional and fuzzy-neural technique (network has 4 inputs and $U_a = 240$ V; (d) trained speed membership functions; (e) estimated speed obtained by using conventional and fuzzy-neural technique (network has 5 inputs and $U_a = 240$ V); (f) estimated speed obtained by using conventional and fuzzy-neural technique (network has 5 inputs and armature voltage is sinusoidal).

follows that this approach results in a simpler network structure. The three membership functions obtained after the training (for 10 epochs) are shown in Fig. 7.21(d) (the curves contain scaled values of the speed). These trained membership functions are slightly different from their initial values. Figure 4.21(e) shows the speed obtained by the trained fuzzy-neural estimator if the input is a constant armature voltage, $U_a = 240$ V. As expected, in comparison with the speed response obtained by the fuzzy-neural estimator shown in Fig. 4.21(c), an improved speed response is obtained. Finally Fig. 4.21(f) shows the speed estimated by the fuzzy-neural estimator when a sinusoidal armature voltage with an amplitude of 180 V was applied to the d.c. machine (the training was achieved by using a sinusoidal armature voltage with an amplitude of 240 V). It can be seen that satisfactory speed estimates are obtained.

Other fuzzy-neural networks have also been successfully trained to yield estimates of other machine quantities (e.g. electromagnetic torque) as well.

7.4.2 FUZZY-NEURAL SPEED ESTIMATORS FOR AN INDUCTION MOTOR

By using a fuzzy-neural network, it is possible to estimate various quantities of an induction machine. Thus it can be used to yield the instantaneous variations of the rotor speed, electromagnetic torque, stator flux linkages, etc. For example, by using a Mamdani-type and also a Sugeno-type of fuzzy-neural controller, the instantaneous rotor speed in a vector-controlled and also in a direct-torque-controlled induction motor drive has been obtained. For this purpose, similarly to the case discussed in Section 7.3.2.2, in one configuration the inputs to the fuzzy-neural network were the monitored values of the present and past values of the direct- and quadrature-axis stator voltages and stator currents. However, in contrast to the discussion for the d.c. motor in the previous section in which only three membership functions were used for both the armature voltage and armature current, the results were not acceptable for the induction motor when the same number of membership functions was used for the stator voltages and stator currents.

When the number of fuzzy sets for the stator voltages and stator currents was increased, improved results were obtained, but the complexity of the network has considerably increased, resulting in a higher training time. Finally it has been possible to find a compromise which resulted in an optimum solution yielding adequate responses and acceptable network structure. This has been possible by using in addition some expert knowledge of the speed estimation process.

It is believed that speed, torque, flux, etc. estimation using artificial-intelligence-based adaptive networks will have an increased role in the future. However, for this purpose, all expert information should be utilized, since this can also help to establish the initial network structure and can speed up the convergence of the estimation. There are many types of adaptive artificial-intelligence-based networks, but the key to successful industrial applications are minimum configuration networks.

Bibliography

Allesandri, A., Dagnino, S., Marchesoni, M., Mazzuccheli, M., and Saccani, P. (1995). A neural rotor flux observer for induction motor control. *EPE, Seville*, 3.011–3.016.

Anderson, J. and Rosenfield, E. (1988). *Neurocomputing*. MIT Press, Cambridge, M.A.

Beierke, S. (1995). Enhanced furry controlled a.c. motor using DSP. Embedded Systems Conference, San Jose, USA, 545–553.

Ben-Brahim, L. and Kurosawa, R. (1993). Identification of induction motor speed using neural networks. *IEEE PCC, Yokohama*, pp. 689–94.

Bianchi, N., Bolognani, S., and Zigliotto, M. (1995). Fuzzy logic automatic design of induction machines. *IEEE, Stockholm Powertek*, pp. 245–50.

Blumel, R., Schurak, E., and Schurig, J. (1990). PC 486 based fuzzy logic controller for VSI-fed induction motor drive. *PCIM, Nurnberg*, pp. 105–14.

Bolognani, S., Buja, G. S., and Zigliotto, M. (1993). Fuzzy logic control of a d.c. servo drive. *IEEE, Workshop on Neuro-Fuzzy Control, Muroran, Japan*, pp. 370–3.

Bolognani, M. and Zigliotto, M. (1992). Development of fuzzy controller for a d.c. drive. *IFAC*, pp. 235–40.

Brdys, M. A. and Sim, W. L. (1995). Fuzzy logic supervision of adaptive generalised predictive control of induction motor. *EPE, Seville*, 3.586–3.591.

Cerruto, E., Consoli, A., Raciti, A., and Testa, A. (1992). Adaptive fuzzy control of high performance motion systems. *IECON*, pp. 88–94.

Chen, S. and Billings, S. A. (1990). Non-linear system identification using neural networks. *Int. Journal Control* **51**, 1191–214.

Chricozzi, E. *et al.* (1995). Fuzzy self-tuning PI control of PMS synchronous motor drives. *PEDES'95*, pp. 749–54.

Cirrincione, M., Millemaci, S., Rizzo, R., and Vitale, G. (1993). A neural network-based system for driving a stepper motor. *ISMCR'93*, BM.I-43–BM.I-48.

Cirrincione, M., Rizzo, R., Vitale, G., and Vas, P. (1995). Neural networks in motion control: theory, an experimental approach and verification on a test bed. *PCIM, Nurnberg*, pp. 131–41.

Damiano, A., Vas, P., Marongiu, I., and Stronach, A. F. (1997). Comparison of speed sensorless DTC drives. *PCIM, Nurnberg*, pp. 1–12.

Demut, H. and Beale, M. (1994). *Neural network toolbox*. The Mathworks, Inc.

Fillippetti, F., Franceshini, G., and Tassoni, C. (1993). Neural network aided diagnostics of induction motor rotor faults. *IEEE IAS*, pp. 316–26.

Filippetti, F., Franceshini, G., and Tassoni, C. (1994). Synthesis of aritificial intelligence and neural network technologies in power electric systems diagnosis. *ICEM*, pp. 353–62.

Filippetti, F., Franceschini, G., Tassoni, C., and Vas, P. (1995). A fuzzy logic approach to on-line induction motor diagnostics based on stator current monitoring. *Stockholm Power Tech* **1**, 150–61.

Flaccomio, A., Cirrincione, G., Miceli, R., and Vas, P. (1995). Neural modelling of a system: theory and experimentation on a test bed for a d.c. machine. *IMACS, Berlin*, June 1995.

Galvan, E., Barrero, F. *et al.* (1993). A robust speed control of a.c. motor drives based on fuzzy reasoning. *IEEE IAS*, pp. 2055–8.

Goldberg, D. E. (1989). *Genetic algorithms in search, optimisation, machine learning*. Addison Wesley.

Grundman, S., Krause, M., and Muller, V. (1995). Application of fuzzy control for PWM voltage source inverter fed permanent magnet motor. *EPE, Seville*, 1.524–1.529.

Hecht-Nielsen, R. (1989). Theory of the backpropagation neural network. In *Proc. IEEE Int. Conf. on Neural Networks*, **Vol. I**, pp. 593–605.

Henneberger, G. and Otto, B. (1995). Neural network application to the control of electrical drives. *PCIM, Nurnberg*, pp. 103–16.

Henneberger, G., Otto, B., and Brosse, A. (1995). Investigation and comparison of neural network applications to the control of electrical machines. *EPE, Seville*, 1.303–1.307.

Hofman, W. and Krause, M. (1993). Fuzzy control of a.c. drives fed by PWM inverters. *PCIM, Nurnberg*, pp. 123–32.

Jang, J. G. (1993). ANFIS: adaptive-network-based fuzzy inference system. *IEEE Transactions on System, Man and Cybernetics* **23**, 665–84.

Jiang, H. and Penman, J. (1993). Using Kohonen feature maps to monitor the condition of synchronous generators. *Workshops on computer science, neural networks: techniques and applications, Liverpool*, paper 11.

Kohonen, T. (1990). The self-organising map. *Proc. IEEE* **78**, 1464–80.

Kosko, B. (1992). *Neural networks and fuzzy systems.* Prentice Hall.

Krzeminski, Z. (1995). Estimation of induction motor speed in fuzzy logic system. *EPE, Seville*, 3.563–3.568.

Lee, C. (1990). Fuzzy logic in control systems: fuzzy logic controller—Pts 1 and 2. *IEEE Trans. Syst. Man Cybern.* **20**, 404–35.

Lee, S. and Lee, E. (1974). Fuzzy sets and neural networks. *Journal of Cybernetics* **4**, 83–103.

Mamdani, E. H. (1974). Application of fuzzy algorithms for simple dynamic plant. *Proc. IEE* **121**, 1585–8.

Miki, I., Hagai, N., Nishiyama, S., and Yamada, T. (1991). Vector control of induction motor with fuzzy PI controller. *IEEE IAS*, pp. 341–6.

Monti, A., Roda, A., and Vas, P. (1996). A new fuzzy approach to the control of the synchronous reluctance machine. *EPE, PEMC, Budapest*, 3/106-3/110.

Narendra, K. S. and Parthasarathy, K. (1990). Identification and control of dynamical systems using neural networks. *IEEE Transactions on Neural Networks* **1**, 4-27.

Penman, J. and Yin, C. M. (1994). Feasibility of using unsupervised learning, artificial neural networks for the condition monitoring of electrical machines. *IEE Proc. Elect. Power Appl.* **141**, 317–22.

Penman, J., Stavrou, A., *et al.* (1994). Machine diagnostics with ANNs: possibilities and problems. *ICEM*, pp. 363–8.

Profumo, F., Kosc, P., and Fedak, V. (1994). AC drives for high performance applications using fuzzy logic controllers. *Trans. IEE, Japan*, 734–40.

Simoes, G. M. and Bose, K. L. (1994). Neural network based estimation of feedback signals for a vector controlled induction motion drive. *IEEE IAS*, pp. 471–9.

Stronach, A. F. and Vas, P. (1995). Fuzzy-neural control of variable-speed drives. *PCIM, Nurnberg*, pp. 117–29.

Stronach, A. F. and Vas, P. (1997). Application of artificial-intelligence-based speed estimators in high-performance electromechanical drives. *PCIM, Nurnberg*, pp. 265–78.

Stronach, A. F., Vas, P., and Neuroth, M. (1997). Implementation of intelligent self-organising controllers in DSP-controlled electromechanical drives. *IEE Proc. (Pt. D)* **144**, 325–30.

Sugeno, M. (1985). An introductory survey of fuzzy control. *Inform. Sci.* **36**, 59–83.

Takagi, T. and Sugeno, M. (1983). Derivation of fuzzy control rules from human operator's control action. *IFAC Symp. Fuzzy Inform., Knowledge Representation and Decision Analysis*, pp. 55–60.

Takagaki, T. and Sugeno, M. (1985). Fuzzy identification of systems and its applications to modeling and control. *IEEE Transactions on Systems, Man and Cybernetics* **15**, 116–32.

Takahashi, Y. (1990). Adaptive control via neural networks. *Journal of SICE* **29**, 729–33.

Truong, T. T. and Hofman, W. (1995). Fuzzy tuning for servo control: a double ratio approach. *EPE, Seville*, 1.379–1.383.

Vas, P. (1992). *Electrical machines and drives; a space-vector theory approach.* Oxford University Press.

Vas, P. (1993). *Parameter estimation, condition monitoring and diagnosis of electrical machines.* Oxford University Press.

Vas, P. (1995). Application of space-vector techniques to electrical machines and variable-speed drives. *EPE tutorial, Seville*, pp. 1–130.

Vas, P. (1995). Artificial intelligence applied to variable-speed drives. *PCIM Seminar, Nurnberg*, pp. 1–149.

Vas, P. (1996). Recent trends and development in the field of machines and drives, application of fuzzy, neural and other intelligent techniques. *Workshop on a.c. motor drives technology, IEEE/IAS/PELS, Vicenza*, pp. 55–74.

Vas, P. and Drury, W. (1994). Vector-controlled drives. *PCIM, Nurnberg*, pp. 213–28.

Vas, P. and Drury, W. (1995). Future trends and developments of electrical machines and drives. *PCIM, Nurnberg*, pp. 1–28.

Vas, P., Drury, W., and Stronach, A. F. (1996). Recent developments in artificial intelligence-based drives. *A review, PCIM, Nurnberg*, pp. 59–71.

Vas, P., Li, J., Stronach, A. F., and Lees, P. (1994). Artificial neural network-based control of electromechanical systems. *4th European Conference on Control, IEE, Coventry*, pp. 1065–70.

Vas, P. and Stronach, A. F. (1996). Design and application of multiple fuzzy controllers for servo drives. *Archiv für Elektrotechnik* **79**, 1–12.

Vas, P. and Stronach, A. F. (1996). Adaptive fuzzy-neural control of high-performance drives. *IEE PEVD, Nottingham*, pp. 424–9.

Vas, P. and Stronach, A. F. (1996). Minimum configuration soft-computing-based DSP controlled drives. *PCIM, Nurnberg*, pp. 299–315.

Vas, P., Stronach, A. F., and Drury, W. (1996). Artificial intelligence in drives. *PCIM Europe*, **Vol. 8**, pp. 152–5.

Vas, P., Stronach, A. F., Neuroth, M., and Du, T. (1995). Fuzzy pole-placement and PI controller design for high-performance drives. *Stockholm Power Tech*, 1–6.

Vas, P., Stronach, F., Neuroth, M., and Du, T. (1995). A fuzzy controlled speed-sensorless induction motor drive with flux estimators. *IEE EMD, Durham*, pp. 315–19.

Warwick, K., Irwin, G. W., and Hunt, K. J. (1992). *Neural networks for control and systems.* Peter Peregrinus Ltd., UK.

Woshrat, T. M. and Harley, R. G. (1993). Identification and control of induction machines using artificial neural networks, *IEEE, IAS*, pp. 703–9.

Zadeh, L. A. (1965). Fuzzy sets. *Information control* **8**, 338–53.

Zadeh, L. A. (1971). Similarity relations and fuzzy ordering. *Information Sciences* **3**, 177–206.

Zadeh, L. A. (1973). Outline of a new approach to the analysis of complex systems and decision processes. *IEEE Transactions on Systems, Man, and Cybernetics* **3**, 28–44.

8 Self-commissioning

All the instantaneous torque-controlled drives (vector drives, direct-torque-controlled drives) described in previous chapters use various machine parameters. On-line estimation of some of the machine parameters has been described in various sections, but in the present chapter different techniques are briefly described, which are suitable for the estimation of machine parameters by using simple tests prior to starting the drive system. It should be noted that a recent book (Vas 1993) discusses many types of parameter estimation and identification schemes. Furthermore, it is important to note that various modern industrial drives now incorporate self-commissioning techniques. In general, two main methods exist for self-commissioning:

1. Self-commissioning without identification. In this case, no signals are measured and there are two main approaches. It is a great advantage that the computational effort is minimal, but it is a disadvantage that the parameters cannot be absolutely correct.

   Approach 1: utilization of stored parameters. When this first approach is used, the motor is selected from a menu and a look-up table contains required precomputed parameters. This is a very simple system, which allows fast self-commissioning, but results in detuning effects if real parameters are different from the values stored in the look-up table. Such a system can be effective in the most recent types of integral motor drives, where the inverter and motor are integrated into a single unit.

   Approach 2: utilization of calculated parameters. When this second approach is used, the motor can be unknown and the parameters are obtained on the basis of the name-plate and other data. However, for many applications this can only provide a crude estimation of the parameters. This method is also problematic since the name-plate data do not give any account of any machine imperfectness. In some advanced industrial drives, this technique can also be selected from a menu (e.g. ABB DTC drive), but these drives also contain other self-commissioning techniques, where there is identification (see also below).

2. Self-commissioning with identification. In this case, various test signals (step signals, ramp signals, etc.) are applied to the machine and the responses are measured. It is also possible to use some name-plate and other data to ensure that the machine will not be destroyed during the tests (e.g. maximum current). It is also a goal to use the minimum number of sensors. For example, currents are always measured in every drive (e.g. to ensure safe operation of semiconductor switches). However, if it is possible to avoid the use of voltage and speed sensors, then they should not be used.

There are basically two types of approaches.

Approach 1: off-line identification. In this case suitable test signals are applied to the motor and the response data is measured. This is used to identify the parameters. However, it is a disadvantage that parameter variations are not considered, since these can only appear during the normal operation of the drive. When this approach is used it is also a goal not to produce torque (e.g. for an induction motor this can be ensured by applying single-phase stator voltage).

Approach 2: on-line identification. When this approach is used, initial, precomputed controller parameter values are required. Measurements are made during the operation of the drive and motor and controller parameters are continuously estimated, and controllers are continuously adapted. This approach imposes high computation burdens.

It is a common feature that those industrial drives which use self-commissioning with identification rely on mathematical-model-based techniques, but it is expected that in the future artificial-intelligence (AI)-based techniques will also be used (see also Chapter 7).

In many industrial applications, the a.c. machine and converter are sold by separate manufacturers and the parameters of the a.c. machine are not known. However, prior to starting the variable-speed drive system, some machine parameters have to be known. Although the name-plate data can also be used for this purpose, in many applications this does not yield accurate information on the parameters required. Thus in general extensive testing of the machine and tuning of the controllers is required, but this can be expensive and time-consuming and requires specially trained staff (which also increases the costs). The self-commissioning of high-performance a.c. drives is more important than that of d.c. drives, since a.c. drives are more complicated and the control of a.c. machines is more difficult. Thus the tuning of an a.c. drive is more time-consuming and requires better-qualified staff. However, it is possible to obtain various machine parameters and to tune the controllers by an automated process during a self-commissioning stage. For example, for a squirrel-cage induction machine, the required electrical parameters (stator and rotor resistances, rotor time constant, stator transient inductance, etc.) can be obtained from on-line measurements of stator currents and/or stator voltages, when the machine is at standstill and where the inverter in the drive is utilized to generate the signals required for the parameter estimation. However, identification techniques are preferred in which the machine does not produce a torque at standstill, since locking of the rotor is undesirable (e.g. for a three-phase induction motor this can be achieved by using a single-phase supply). Some of these techniques will be discussed below. When the machine parameters are known, then during the running of the drive, only a few parameter-related identification schemes have to be employed, e.g. those utilizing thermal models (which are used to update the initial resistance values obtained during the self-commissioning stage, etc.).

The three main techniques of on-line and off-line parameter estimation are:

1. Estimation in the time domain. For this purpose various techniques can be used, e.g. MRAS (described in earlier sections), Recursive Least Squares (RLS), Maximum Likelihood, etc.
2. Estimation in the frequency domain. In this case the test signals applied depend on the system and evaluation technique (sinusoidal signals with cross-correlation technique for a non-linear system). This can be a computationally intensive technique. Parameters are obtained from frequency characteristics. It is an advantage that this technique is also suitable for non-linear systems, but the input signal in this case must be perfectly sinusoidal. For a linear system, the test signal can be any other, periodic signal.
3. Estimation using artificial intelligence (AI). In this case artificial neural networks (e.g. multi-layer back-propagation ANN, self-organizing feature maps), fuzzy-neural networks, etc. can be used. Since neural networks and fuzzy-neural netwoks are general function estimators, they are very suitable in the case of non-linearities. By the application of minimum configuration networks, it is possible to obtain cost-effective, simple solutions.

8.1 Determination of the rotor time constant of an induction machine with impressed stator currents and subjected to indirect rotor-flux-oriented control

It follows from Section 4.1.1 that when indirect rotor-flux-oriented control of an induction machine is used, it is very important to have an accurate representation of the rotor time constant of the machine. This is a function of the rotor resistance and the rotor self-inductance. The former can increase by up to 100% of its nominal value for an increase of 170–180° in temperature. In systems which employ indirect rotor-flux-oriented control of an induction machine, where an encoder or a resolver is used, the position of the rotor flux-linkage space phasor with respect to the direct axis of the stationary reference frame (ρ_r, which is shown in Fig. 4.15), is obtained by the addition of the slip angle (θ_{sl}) and the monitored rotor angle (θ_r). The slip angle is obtained with a slip angle estimator which contains the effect of the open-circuit rotor time constant T_r (see eqn (4.1-26)). When an incorrect value of the rotor time constant is used for setting the gain of the slip controller, the performance of the machine will degrade under both steady-state and transient conditions. It is therefore important to utilize accurate values of the rotor time constant in the slip-angle estimator.

In Section 4.5.3.4 Model Reference Adaptive Control was used to obtain several circuits which continuously monitor the rotor speed. A similar technique could be used for the estimation of the rotor resistance. However, in this section, an alternative technique is described for the determination of the rotor time constant.

8.1.1 OBTAINING THE ROTOR TIME CONSTANT PRIOR TO STARTING THE MACHINE

Various methods are discussed for obtaining the rotor time constant of the induction machine; these utilize measurements made prior to starting the machine. Traditionally, the rotor time constant can be obtained by performing the locked-rotor and no-load tests. However, this also requires mechanical blocking of the rotor and therefore it is more difficult to automatize these tests than tests which avoid mechanical blocking of the rotor. In the next section, such a rotor time-constant measurement system is described, which can be used for the automatic determination of the rotor time constant prior to starting the induction machine. The blocking of the rotor is avoided by applying single-phase excitation to the stator. However, the no-load test is still used.

Measurement which utilizes single-phase excitation of the stator

It is possible to obtain the rotor time constant from measurements made on the induction machine prior to start-up and by utilizing single-phase excitation of the stator windings, by measuring one of the line-to-line stator voltages of the machine. This voltage is then processed to yield the rotor time constant of the machine. The test utilizes the inverter which supplies the induction machine.

It is assumed that the induction machine is supplied by a current-controlled PWM inverter (described in Section 3.1.1) and at instant t_0, single-phase sinusoidal current $i_s(t)=I_{sref}\cos(\omega_1 t)$ is supplied to two of the stator phases of the induction machine with the third stator current controlled to zero. The time variation of the applied stator current of the induction machine is shown in Fig. 8.1.

As a result of the single-phase excitation, the rotor will remain at standstill and therefore it does not have to be locked during the tests. After switching on the

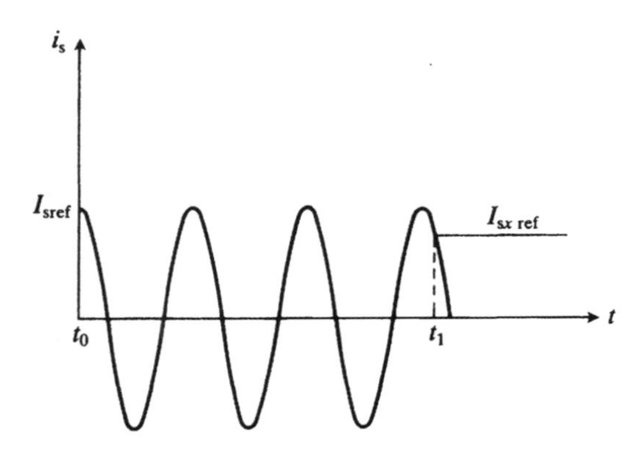

Fig. 8.1. Switching of the applied stator current.

a.c. current, there is a transient state followed by the steady state. In the steady state, at an appropriate instant t_1, where t_1 is much greater than the rotor time constant, the stator current is switched to a d.c. current, $i_s(t \geqslant t_1) = I_{sxref}$. This is also shown in Fig. 8.1. The instant t_1 is selected so that at this instant the instantaneous value of the stator current is equal to the reference value of the rotor flux-producing current component, which is a magnetizing current. The value of the magnetizing current can be obtained from a no-load test and it is dependent on the flux level in the machine.

Since the rotor speed is zero, the applied inverter angular frequency (ω_1) is equal to the angular slip frequency ($\omega_{sl} = \omega_1$). As a result of switching from a.c. current to d.c. current, a transient stator voltage will arise when the reference value of the angular slip frequency is not equal to the correct value. A physical analysis will be presented below. This transient voltage can be processed to yield the correct value of the rotor time constant. The inverter frequency is adjusted until this voltage transient is a minimum and when this is accomplished, from eqn (4.1-26), and $\omega_{sl} = \omega_1$, the rotor time constant can be obtained as

$$T_r = \frac{1}{\omega_1}\frac{I_{syref}}{I_{sxref}} = \frac{1}{\omega_1}\frac{(I_{sref}^2 - I_{sxref}^2)^{1/2}}{I_{sxref}}. \tag{8.1-1}$$

I_{syref} is the reference value of the torque-producing stator current, and the amplitude of the a.c. stator current can be expressed as $I_{sref} = (I_{sxref}^2 + I_{syref}^2)^{1/2}$.

It should be noted that, at standstill, the equivalent circuit shown in Fig. 4.10(c) is also valid under transient conditions, since for zero rotor speed all the speed voltages are zero and the induction machine can be described in a similar way to a transformer. This equivalent circuit can be used to give some physical considerations of the stator terminal voltages for two cases, when the reference value of the angular slip frequency is larger or smaller than the correct value of the angular slip frequency.

If the reference value of the angular slip frequency is larger than the correct value, both the actual magnetizing current and the rotor flux will be smaller than their reference values. Therefore the magnetizing current and the rotor flux are too small before the switching of the stator current. After the switching of the stator current to the d.c. value, the instantaneous value of the flux-producing stator current (i_{sx}) will be smaller than the instantaneous value of the stator current i_s. Since a current which flows across an inductor cannot be changed instantaneously, some trapped rotor current must flow across the rotor branch which contains the rotor resistance and its direction will be from point *B* to *D* in Fig. 4.10(c). The stator terminal voltage is the sum of three component voltages, the stator ohmic drop, the voltage across the transient inductance of the stator, and the voltage across the magnetizing branch, which contains the inductance (L_m^2/L_r). The second component can be neglected, since it corresponds to a leakage drop, which is small compared with the other two components. It follows from Fig. 4.10(c) that the voltage across the magnetizing branch is determined by the current across the rotor resistance $R_r L_m^2/L_r^2$. Eventually, the trapped rotor current will flow across the magnetizing branch since it represents a short circuit at zero frequency, and in the steady state, the rotor flux will be equal to the reference value. Initially, the stator terminal voltage will be higher than its steady-state value and this will decay.

If the reference value of the angular slip frequency is smaller than the correct value, then both the actual magnetizing current and the rotor flux are larger than their reference values before the switching of the stator current. After the switching of the stator current to the d.c. value, the instantaneous value of the rotor flux-producing stator current will be higher than the instantaneous value of the stator current and thus the trapped rotor current will be reversed compared with the previous case, and eventually this current will again decay to zero as the rotor flux decays to the reference value. Initially, the stator terminal voltage will be smaller than its steady-state value, since the transient voltage across the inductance L_m^2/L_r shown in Fig. 4.10(c) has a different polarity to the polarity of the ohmic stator drop.

It follows that the trapped rotor current will only be zero when the reference value of the angular slip frequency is equal to its correct value. Under this condition, both before and after the switching of the stator current, the current which flows across the magnetizing branch is equal to the reference value of the rotor flux-producing stator current. Thus no voltage will arise across the inductance L_m^2/L_r and the stator terminal voltage will be equal to the sum of the voltage across the stator resistance and the voltage across the stator transient inductance; the latter component can be neglected.

It follows from eqn (8.1-1) that for the determination of the rotor time constant it is necessary to utilize the two reference currents I_{sxref} and I_{syref}. However, to obtain high torque response, it is necessary to keep the flux-producing component constant. Thus the flux-producing current component must be held constant to yield the required flux in the machine when the reference value of the angular slip frequency is correct. It is preferable to perform the test at a constant value of I_{sxref}, which will produce the required flux in the machine. It follows from eqn (8.1-1) that the rotor time constant can be obtained with greater accuracy if the ratio I_{syref}/I_{sxref} is large. However, with a large ratio there will be large ohmic losses in the machine. If I_{syref} or I_{syref}/I_{sxref} are chosen to result in a small rotor-flux error, then by calculating the rotor flux error which is due to the presence of the incorrect value of the rotor time constant, it follows from the rotor voltage equations presented in Section 4.1.1 that an acceptable level of the rotor flux error can be obtained even if the current ratio (I_{sxref}/I_{syref}) is not too large.

To summarize the measurement procedure, it follows that by the application of a current-controlled PWM inverter, a single-phase a.c. current is supplied to the stator windings and this is switched to an appropriate d.c. value in the steady state. This d.c. value is equal to the reference value of the rotor flux-producing component of the stator current. Its value is obtained from a no-load test and is dependent on the required flux level. One of the line-to-line voltages is monitored and the frequency of the inverter is varied until a minimum of this voltage is reached and, under this condition, eqn (8.1-1) is used to obtain the rotor time constant. It follows that when this technique is used together with the indirect rotor-flux-oriented control system described in Section 4.1.2, the only extra signal to be monitored is a line-to-line stator voltage. The experiment is performed with a rated value of the flux reference, but the actual flux is equal to its rated value only when the correct value of the angular slip-frequency reference is used.

In the case of an induction machine supplied by a current-controlled PWM inverter, the measured terminal voltage transients must be low-pass filtered. When the induction machine is supplied by a hysteresis current-controlled PWM inverter, the undesirable time-harmonic components will be spread over the entire frequency domain and this property complicates the design of the filter. However, when a current-controlled PWM with sine–triangle modulation is used, the harmonic frequencies are fixed at the harmonic frequencies of the triangular wave and thus it is easier to design the required filter than in the previous case. The filter can be a moving-average filter; such a filter can easily be implemented in a microprocessor or by utilizing a shift register and Sample-and-Hold (S/H) circuits.

If the input signal to the moving-average filter is $x(t)$, the output signal is

$$y(t)=\frac{1}{T}\int x(t)\,\mathrm{d}t, \tag{8.1-2}$$

where T is the sampling rate. Thus this filter yields the mean value of the input signal in the time interval $t-T$, t. The integral given by eqn (8.1-2) can be replaced by a summation and, if the input signal is sampled k times in the period T, then the output signal at instant t_i is

$$y(t_i)=\frac{1}{k}\sum x(t_j). \tag{8.1-3}$$

By utilizing eqn (8.1-3), Fig. 8.2 shows an implementation of the moving average filter. The input to the circuit shown in Fig. 8.2 is the signal $x(t)$, which is a line-to-line voltage in the experiment described above. The sample instants are determined by a rotating pulse in a Shift Register (SR). The sampling pulses are $s_1, s_2, s_3, \ldots, s_k$. The frequency of the input pulse train of the shift register is $f=k/T$. Each S/H circuit holds one sample during the period T. The k samples of the line-to-line voltage are added, and are then divided by k to yield the moving-average value.

The accuracy of the test described above can be checked by either comparing the obtained value of the rotor time constant with the value obtained with other tests, or, and this is simpler, by utilizing the measured drive to check the accuracy. For this purpose it is possible to use the relationship between the output torque and the reference torque in the steady state. This will only be linear if the correct value of the rotor time constant is used in the calculation block of the angular rotor slip frequency. Since the rotor resistance can change significantly with temperature during the operation of the drive, the initial value has to be corrected. For this purpose it is also possible to use a simplified thermal model of the machine.

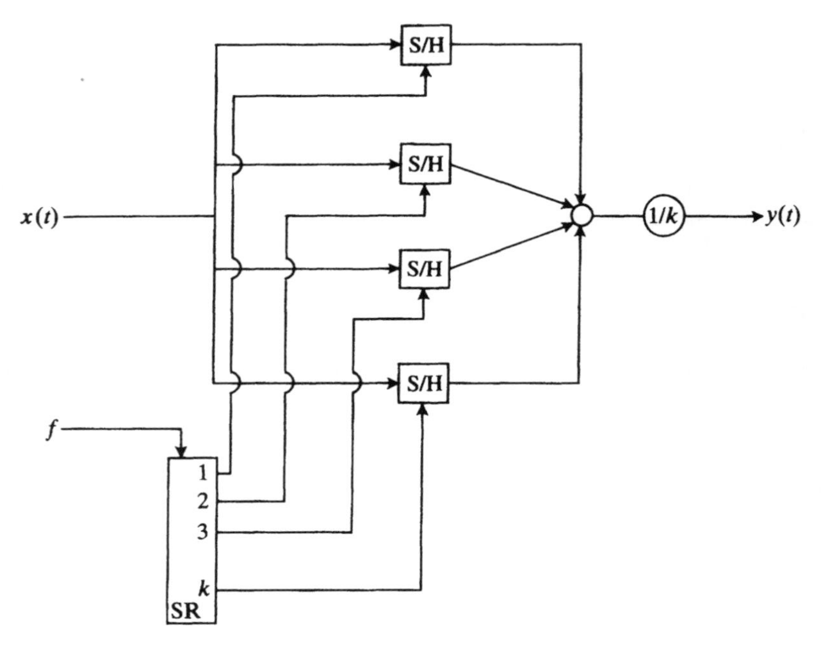

Fig. 8.2. An implementation of the moving average filter.

8.2 Determination of electrical and mechanical parameters of an induction machine with impressed stator voltages and subjected to rotor-flux-oriented control

It is assumed that the induction machine is supplied by a voltage source inverter, as discussed in Section 4.1.1, and rotor-flux-oriented control is employed. It is assumed that all control tasks (inverter control, rotor-flux-oriented control) can be performed by a computer. The stator current in stator phase sA is measured and its instantaneous value is i_{sA}. The electrical parameters of the machine are obtained when the machine is at standstill by utilizing measured values of appropriately formed stator currents and measured terminal voltages. For this purpose the voltage equations of the machine expressed in the stationary reference frame are utilized, but with the two-axis components of the stator currents (i_{sD}, i_{sQ}) and the two axis rotor magnetizing-current components (i_{mrD}, i_{mrQ}) chosen as state-variables.

It follows fron eqns (4.1-34) and (4.1-29) that the space-phasor form of the stator and rotor voltage equations in the stationary reference frame are as follows if the machine is at standstill ($\omega_r = 0$) and if the effects of magnetic saturation are neglected:

$$\bar{u}_s = R_s \bar{i}_s + L'_s \frac{d\bar{i}_s}{dt} + (1-\sigma)L_s \frac{d\bar{i}_{mr}}{dt} \tag{8.2-1}$$

$$\frac{d\bar{i}_{mr}}{dt} = \frac{\bar{i}_s - \bar{i}_{mr}}{T_r}. \tag{8.2-2}$$

In eqn (8.2-1) the derivative or the rotor magnetizing-current space phasor is present. When its expression [eqn (8.2-2)] is substituted into eqn (8.2-1), the stator voltage equation can be expressed as

$$\bar{u}_s = (R_s + R_{rref})\bar{i}_s + L'_s \frac{d\bar{i}_s}{dt} - \frac{(1-\sigma)L_s \bar{i}_{mr}}{T_r}. \tag{8.2-3}$$

By introducing the referred value of the rotor flux-linkage space phasor,

$$\bar{\psi}_{rref} = (1-\sigma)L_s \bar{i}_{mr} = \frac{L_m}{L_r} \bar{\psi}'_r, \tag{8.2-4}$$

where $\bar{\psi}'_r$ is the space phasor of the rotor flux linkages and L_m/L_r is a referring factor (whose role has been described in Section 4.1.1), eqns (8.2-3) and (8.2-2) can be put into the form,

$$\bar{u}_s = (R_s + R_{rref})\bar{i}_s + L'_s \frac{d\bar{i}_s}{dt} - \frac{\bar{\psi}_{rref}}{T_r} \tag{8.2-5}$$

$$\frac{d\bar{\psi}_{rref}}{dt} = R_{rref}\bar{i}_s - \frac{\bar{\psi}_{rref}}{T_r}, \tag{8.2-6}$$

where R_{rref} is the referred value of the rotor resistance,

$$R_{rref} = \left(\frac{L_m}{L_r}\right)^2 R_r \tag{8.2-7}$$

and has been shown in the equivalent circuit given in Figs. 4.10(a) and (c).

Equations (8.2-5)–(8.2-7) are used to obtain several parameters (stator transient time constant, stator resistance, rotor resistance, rotor time constant, etc.) of the induction machine supplied by the voltage-source PWM inverter by employing tests performed at standstill.

8.2.1 DETERMINATION OF THE STATOR TRANSIENT INDUCTANCE

First the stator transient inductance (L'_s) is determined. For this purpose appropriate short voltage impulses are generated by the inverter which supplies the induction machine itself. It follows from eqns (8.2-5)–(8.2-7) that if the direct-axis stator current expressed in the stationary reference frame is zero, $i_{sD}=0$, and the direct-axis component of the referred value of the rotor flux-linkage space phasor expressed in the stationary reference frame is zero, $\psi_{\mathrm{rref}\alpha}=0$, then the direct-axis component of the equations yields

$$L'_s = u_{sD}/(di_{sD}/dt), \tag{8.2-8}$$

where u_{sD} is the direct-axis component of the stator-voltage space phasor in the stationary reference frame. Equation (8.2-8) is used to obtain the stator transient inductance. For this purpose first two appropriate stator terminals (SR, TR) are supplied by the d.c. voltage U_d, using the inverter itself, where U_d is the d.c. link voltage. From the definition of the space phasor of the stator voltages, eqn (2.1-126), $u_{sD}=\frac{2}{3}U_d$ is obtained and this is the value of which has to be used in eqn (8.2-8). Figure 8.3 shows the terminal voltage u_{SR} and the current response in phase sA.

The duration of the pulses in the terminal voltage is selected to be much shorter than the open-circuit rotor time constant (T_r) and it follows from the voltage equations that, with respect to $i_{sA}=i_{sD}$, the induction machine behaves like a first-order delay element and the initial rise of the stator current i_{sD} is determined by the stator transient inductance. In Fig. 8.3 at instant t_1 the voltage U_d appears at the two stator terminals indicated above. Thus the stator current starts to rise and at instant t_2 it reaches its peak value (peak of rated stator current I_s). At this instant the stator terminals are short-circuited with the aid of the inverter valves and thus the stator current starts to decrease. At instant t_3, where the phase current is approximately equal to half the peak value of the rated current, a negative value of the d.c. link voltage is applied and thus the stator current will further decrease and will become negative, as shown in Fig. 8.3. At instant t_4 the stator current reaches its negative peak value and at this instant the stator terminals are again short-circuited. Thus the stator current will start to rise.

In eqn (8.2-8) the derivative di_{sD}/dt is present. This can be computed by considering the monitored values of the stator current shown in Fig. 8.3 in either the interval t_1–t_2 or the interval t_3–t_4. However, since $t_4-t_3>t_2-t_1$ and

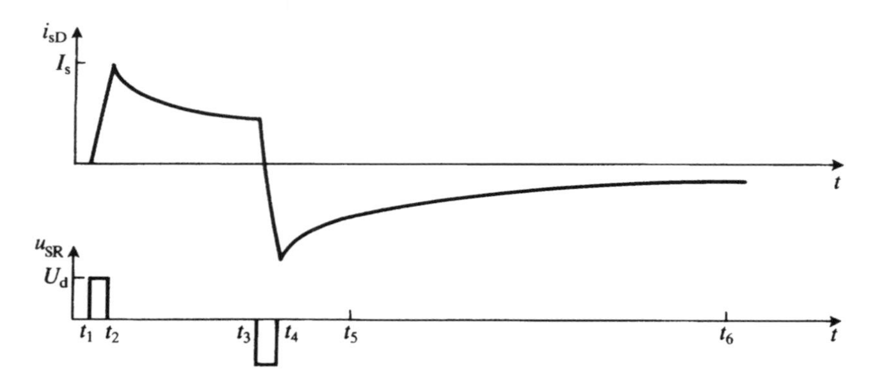

Fig. 8.3. Variation of u_{SR} and i_{sA}.

$i_{sD}(t_3)-i_{sD}(t_4)<i_{sD}(t_2)-i_{sD}(t_1)$, more accurate results are obtained if the values of the current at the instants t_3 and t_4 are used for the calculation of the derivative of the stator current. Furthermore, the average value of the stator current in the second interval is closer to zero than in the first interval and this also leads to a more accurate value of L_s'. It follows that L_s' is computed by considering the following expression,

$$L_s' = \frac{2}{3} U_d \frac{t_4 - t_3}{i_{sD}(t_3) - i_{sD}(t_4)}. \tag{8.2-9}$$

Equation (8.2-9) has been derived by considering the voltage equations, eqns (4.1-33) and (4.1-29), which have been obtained by assuming ideal conditions, e.g. by neglecting the effects of magnetic saturation, eddy currents, and current displacement. Thus when the stator transient inductance is determined by utilizing eqn (8.2-9), the value obtained will be smaller than the actual value. However, an improvement can be obtained if instead of the measured value of the stator current at instant t_4, $i_{sD}(t_4)$, a computed value $i_{sDc}(t_4)$ is used, which is obtained by fitting an exponential current in the interval t_5–t_6 shown in Fig. 8.3 and by using this to compute the value of the stator current at instant t_4. This value, $i_{sDc}(t_4)$, is smaller than the measured value $i_{sD}(t_4)$, and thus $i_{sD}(t_3)-i_{sD}(t_4)>i_{sD}(t_3)-i_{sDc}(t_4)$. The stator transient inductance can be computed from

$$L_s' = \frac{2}{3} U_d \frac{t_4 - t_3}{i_{sD}(t_3) - i_{sDc}(t_4)}. \tag{8.2-10}$$

When the stator transient inductance is determined by using eqn (8.2-10) it will be larger than the value obtained by using eqn (8.2-9). It is possible to use a similar technique to obtain the variation of the other two stator currents and to determine the resulting transient stator inductance. The average value of the results will give a more accurate representation of the actual value of the stator transient inductance.

8.2.2 DETERMINATION OF THE STATOR RESISTANCE

The stator resistance, the referred value of the rotor resistance, and the rotor time constant can be obtained by impressing different values of direct current on the machine. This could be performed by a current controller and a PWM modulator. The current controller can be implemented by software. Figure 8.4 shows the employed stator current and also the output of the current controller, (u_c), the average value of which is proportional to the terminal voltage.

It should be noted that the interval t_1–t_2 in Fig. 8.3 is in the millisecond region, and the interval t_7–t_{10} shown in Fig. 8.4 can be a continuation of the interval t_1–t_6 shown in Fig. 8.3. However, interval t_7–t_{10} is much longer than t_1–t_6, and can be a few seconds long. At instant t_7 a d.c. current is applied to the stator winding sA, which is approximately one third of the peak value of the rated current (I_s). This value is chosen so that it should be less than the magnetizing

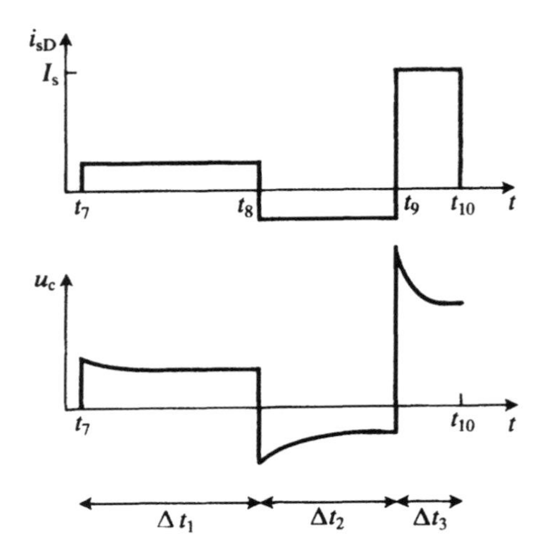

Fig. 8.4. The impressed stator current and the output voltage of the current controller.

current of the machine to ensure that the effects of magnetic saturation do not affect the measured parameters. Since the value of the magnetizing current is unknown prior to the measurements, this current can be reduced to a lower level if it is established that saturated conditions are obtained when the d.c. current is equal to approximately one third of the peak value of the rated current. As a result of the applied d.c. current, a voltage (u_c) will appear on the output of the current controller and will start to decrease until it reaches the steady state, as shown in Fig. 8.4. In the steady state, at instant t_8, a negative value of the previously impressed d.c. current is applied and thus the voltage on the output of the current controller will be negative and will start to increase until the steady state is reached. Thus at instant t_9 the applied d.c. current is equal to the peak of the rated current and a large voltage will arise on the output of the current controller; this will start to decrease and at instant t_{10} the steady state is reached.

It follows from eqns (8.2-5)–(8.2-7) that, in the steady state, the direct-axis stator voltage in the stationary reference frame is

$$u_{sD} = R_s i_{sD}. \tag{8.2-11}$$

As mentioned above, the average value of the terminal voltage is proportional to the voltage on the output of the current controller and can be determined if the d.c. voltage is known. However, the voltage drops due to the semiconductor valves cause a small deviation. To obtain a correct value of the stator resistance, the following expression can be used:

$$R_s = \frac{u_{sD}(\Delta t_3) - u_{sD}(\Delta t_1)}{i_{sD}(\Delta t_3) - i_{sD}(\Delta t_1)}, \tag{8.2-12}$$

where $u_{sD}(\Delta t_3)$ and $u_{sD}(\Delta t_1)$ are the average values of the terminal voltage in intervals (t_9-t_{10}) and (t_7-t_8) respectively, and $i_{sD}(\Delta t_3)$, $i_{sD}(\Delta t_1)$ are the direct-axis stator current components in the same two intervals.

The stator transient inductance obtained in the previous section, together with the stator resistance, can be used for the estimation of the stator transient time constant, $T'_s = L'_s/R_s$.

8.2.3 DETERMINATION OF THE ROTOR TIME CONSTANT AND THE REFERRED VALUE OF THE ROTOR RESISTANCE

It follows from eqns (8.2-5)–(8.2-7) that in interval t_8-t_9 of Fig. 8.4, the direct-axis component of the referred rotor flux $\psi_{rref\alpha}$ (and of course the unreferred rotor flux as well) changes exponentially from $\psi_{rref\alpha}(\Delta t_1) = T_r R_{rref} i_{sD}(\Delta t_1)$ to $\psi_{rref\alpha}(\Delta t_2) = T_r R_{rref} i_{sD}(\Delta t_2)$, and the time constant of the exponential change is equal to the rotor time constant. Since, according to eqn (8.2-5), the stator voltage also contains the change of the rotor flux, the rotor time constant can be obtained from the output voltage of the controller in the interval Δt_2. In practice, when the controller is realized digitally, the output voltage of the controller is not smooth because of the sampling and thus an exponential function is fitted to it in the interval Δt_2, and the rotor time constant is obtained from this fitted curve. It should be noted that the third interval Δt_3 shown in Fig. 8.4 cannot be used for the estimation of the rotor time constant, since in this region the stator current is high (it is equal to the peak rated value) and thus the rotor flux very quickly reaches its saturated value and this would result in too small a value of the rotor time constant.

From eqns (8.2-5)–(8.2-7), in the third interval shown in Fig. 8.4, the initial value of the terminal voltage can be expressed as

$$u_{sDO}(\Delta t_3) = R_s i_{sD}(\Delta t_3) + R_{rref}[i_{sD}(\Delta t_3) - i_{sD}(\Delta t_2)]. \qquad (8.2\text{-}13)$$

Thus it follows that the referred value of the rotor resistance can be obtained as

$$R_{rref} = \frac{u_{sDO}(\Delta t_3) - R_s i_{sD}(\Delta t_3)}{i_{sD}(\Delta t_3) - i_{sD}(\Delta t_2)}. \qquad (8.2\text{-}14)$$

Equation (8.2-14) can be used to obtain R_{rref}, and the stator resistance can be obtained from eqn (8.2-12). Similarly to above, instead of the terminal voltage the output of the controller is used, and thus $u_{cO}(\Delta t_3)$ is used instead of $u_{sDO}(\Delta t_3)$, where $u_{cO}(\Delta t_3)$ is the value of the output voltage of the controller at the beginning of the third interval shown in Fig. 8.4. Furthermore, to avoid the effects of current displacement, a similar technique can be used to that presented above in connection with the determination of the stator transient inductance (where computed values are used instead of measured ones and these are obtained from fitted curves).

8.2.4 DETERMINATION OF THE STATOR SELF-INDUCTANCE AND THE RESULTANT LEAKAGE FACTOR

Since T_r and R_{rref} are known and according to eqn (8.2-7) the referred rotor resistance can be expressed as $R_{rref}=(L_m/L_r)^2R_r$, it follows that $R_{rref}=L_m^2/(L_rT_r)$ and thus

$$L_m^2/L_r=(1-\sigma)L_s=T_rR_{rref}, \qquad (8.2\text{-}15)$$

where σ is the resultant leakage factor, $\sigma=1-L_m^2/(L_sL_r)$, and $\sigma L_s=L_s'$ is the stator transient inductance. It should be noted that the factor L_m^2/L_r is also present in the expression for the electromagnetic torque, eqn (2.1-197).

Since L_s' can be determined from eqn (8.2-10), it follows from eqn (8.2-15) that the stator self-inductance L_s can be estimated by using

$$L_s=L_s'+T_rR_{rref} \qquad (8.2\text{-}16)$$

and, since L_s and L_s' are now known quantities, the leakage factor can be determined by considering

$$\sigma=\frac{L_s'}{L_s}. \qquad (8.2\text{-}17)$$

It should be noted that, in practice, some of the machine parameters can change during the operation of the drive. For example, the rotor resistance can change considerably with temperature. Thus during the operation of the drive some of the machine parameters derived on the basis of the method described above must be corrected. A background computer program can take care of this updating. When the drive is closed down, the required parameters can be saved in RAM for reuse.

8.2.5 DETERMINATION OF THE INERTIA OF THE DRIVE

It is also possible to obtain the inertia of the drive by using on-line measurements, but standstill tests cannot be used. The inertia can also be computed from suitable measurements of the currents and speed of the induction machine during a test run.

For the purpose of obtaining the expression for the inertia as a function of machine currents and rotor speed, the equation of motion is considered. For simplicity a two-pole induction machine is considered. It follows from eqn (2.1-109) that if $D=0$, the equation of motion can be expressed as

$$t_e-t_l=J\frac{d\omega_r}{dt}, \qquad (8.2\text{-}18)$$

where t_e is the developed electromagnetic torque of the machine, t_l is the load torque, J is the inertia, and ω_r is the rotor speed. Since it is assumed that rotor-flux-oriented control is employed, it is useful to express the electromagnetic torque in terms of the torque-producing stator current (i_{sy}) and the modulus of the rotor magnetizing current space phasor $|\bar{i}_{mr}|$ and, by considering eqn (2.1-197), it follows that

$$t_e = \frac{3}{2}\frac{L_m^2}{L_r}|\bar{i}_{mr}|i_{sy}. \tag{8.2-19}$$

It is possible to determine the inertia by considering the response of the drive to changes in the reference speed and to obtain an expression for the inertia which does not contain the load torque. This allows the technique to be used for a large range of drive applications. If the machine runs at a given rotor speed $\omega_r(\Delta t_1)$, approximately 30% of the base speed, during interval Δt_1 of the test run, then the speed reference is changed (interval Δt_2) and in interval Δt_3 the rotor speed is $\omega_r(\Delta t_3)$, and, say, the speed can rise to approximately, 60% of the base speed in this region, then it follows from eqns (8.2-18)–(8.2-19) that if the mean values of the load torque in intervals Δt_1 and Δt_3 are the same, $t_{lm} = t_{lm}(\Delta t_1) = t_{lm}(\Delta t_3)$, then

$$t_{em}(\Delta t_1) - t_{lm} = J\Delta\omega_r(\Delta t_1)/T \tag{8.2-20}$$

and

$$t_{em}(\Delta t_3) - t_{lm} = J\Delta\omega_r(\Delta t_3)/T, \tag{8.2-21}$$

where $t_{em}(\Delta t_1)$ and $t_{em}(\Delta t_3)$ are the mean values of the electromagnetic torque in intervals Δt_1 and Δt_3 respectively and $\Delta\omega_r(\Delta t_1)$ and $\Delta\omega_r(\Delta t_3)$ represent the changes in rotor speed in the intervals Δt_1 and Δt_3 respectively. Furthermore, $T = \Delta t_1 = \Delta t_3$ is the duration of the intervals where the measurement of the speed is performed, together with the measurement of the torque-producing stator current in the same two intervals, as will be discussed below. It follows from eqn (8.2-19) that for a fixed value of the rotor flux, the electromagnetic torque is the product of a constant $c = \frac{3}{2}(L_m^2/L_r)|\bar{i}_{mr}|$ and the quadrature-axis stator current, which is expressed in the rotor-flux-oriented reference frame but is accessible in the drive system. In the two intervals Δt_1 and Δt_3 the mean values of the electromagnetic torque can thus be expressed as $ci_{sym}(\Delta t_1)$ and $ci_{sym}(\Delta t_3)$ respectively, where $i_{sym}(\Delta t_1)$ and $i_{sym}(\Delta t_3)$ are the mean values of the torque-producing stator current component in the same two intervals. Thus

$$ci_{sym}(\Delta t_1) - t_{lm} = J\Delta\omega_r(\Delta t_1)/T \tag{8.2-22}$$

and

$$ci_{sym}(\Delta t_3) - t_{lm} = J\Delta\omega_r(\Delta t_3)/T \tag{8.2-23}$$

are obtained and, when eqn (8.2-23) is subtracted from eqn (8.2-22), finally the following expression is obtained for the inertia of the drive,

$$J = \frac{cT[i_{sym}(\Delta t_3) - i_{sym}(\Delta t_1)]}{\Delta\omega_r(\Delta t_3) - \Delta\omega_r(\Delta t_1)}. \tag{8.2-24}$$

It should be noted that it has been assumed that the mean values of the load torque in the intervals Δt_3 and Δt_1 are equal, and thus eqn (8.2-24) gives an accurate value of the inertia only if this condition is fulfilled. In practice, the load torque can be constant and this condition is automatically satisfied, or it can change with the speed or time. However, even in these two cases, the error in the determination of the inertia will be small if T is chosen to be small.

It should be noted that if prior to starting the drive the parameters of the speed controller are not known, initially these parameters can be set to small values which yield safe operation and during the interval $\Delta t_1+\Delta t_2+\Delta t_3$ these parameters can be held at their initial values. In the next interval the speed-controller parameters can be adjusted by using the known value of the inertia. During the commissioning of the drive the output of the speed controller is monitored in order to identify closed-loop oscillations. In the case of oscillations the parameters of the speed controller are reduced.

It also follows from eqn (8.2-24) that when $i_{sym}(\Delta t_3)-i_{sym}(\Delta t_1)$ is small, i.e. when the acceleration torque is small, an incorrect value of the inertia will be obtained. This can be the case if the change of the reference speed is too small, or if the torque limit is very near to the load torque or if the speed ramp is too slow. If torsionally flexible coupling exists between the machine and the load, then eqn (8.2-24) cannot be used.

8.2.6 IDENDIFICATION OF VARIOUS INDUCTION MACHINE PARAMETERS BY THE RLS TECHNIQUE; IDENTIFICATION OF THE SATURATED L_m

In the present section various induction machine parameters (R_s, R_r, L_r, L_m) are identified by using the recursive least squares (RLS) algorithm. For this purpose the measured stator-current response of the induction machine at standstill is utilized, which is obtained to a single-phase step voltage. First, the effects of magnetic saturation are neglected, and it should be noted that in the previous sections, saturation of the magnetizing inductance was not considered. However, as shown below, it is also possible to identify the non-linear magnetizing inductance of the induction machine at standstill without locking the rotor by using the recursive least squares (RLS) algorithm. The identification technique described below is suitable for self-commissioning purposes.

At standstill the induction machine is described, eqns (8.2-1) and (8.2-2) or by eqns (8.2-5) and (8.2-6), where $\bar{\psi}_{rref}=(L_m/L_r)\bar{\psi}'_r$. It is assumed that there is magnetizing current only in the direct-axis (of the stator reference frame), thus $u_{sQ}=0$, $i_{sQ}=0$ and also there is no rotor current in the quadrature-axis ($i_{rq}=0$). First the effects of magnetic saturation are neglected and it then follows from the machine equations at standstill that

$$u_{sD}=R_s i_{sD}+\frac{d\psi_{sD}}{dt}=R_s i_{sD}+L'_s\frac{di_{sD}}{dt}+\frac{L_m}{L_r}\frac{d\psi_{rd}}{dt} \tag{8.2-25}$$

$$T_r\frac{d\psi_{rd}}{dt}+\psi_{rd}=L_m i_{sD}. \tag{8.2-26}$$

It is also assumed that the stator and rotor leakage inductances are identical. Thus from eqns (8.2-25) and (8.2-26) the discrete transfer function $\hat{i}_{sD}(z)/u_{sD}(z)$ is obtained as

$$\frac{\hat{i}_{sD}(z)}{u_{sD}(z)}=\frac{\hat{b}_1 z^{-1}+\hat{b}_2 z^{-2}}{1+\hat{a}_1 z^{-1}+\hat{a}_2 z^{-2}}, \tag{8.2-27}$$

where the parameters $\hat{a}_1$, $\hat{a}_2$, $\hat{b}_1$, $\hat{b}_2$ are defined as

$$\hat{a}_1=-\exp(\lambda_1 T)-\exp(\lambda_2 T) \tag{8.2-28}$$

$$\hat{a}_2=\exp(\lambda_1 T)\exp(\lambda_2 T) \tag{8.2-29}$$

$$\hat{b}_1=r_1[\exp(\lambda_1 T)-1]+r_2[\exp(\lambda_2 T)-1] \tag{8.2-30}$$

$$\hat{b}_2=r_1\exp(\lambda_2 T)[1-\exp(\lambda_1 T)]+r_2\exp(\lambda_1 T)[1-\exp(\lambda_2 T)] \tag{8.2-31}$$

and are estimated by using the RLS algorithm [Bunte and Grotstollen 1993]. In eqns (8.2-28)–(8.2-31) T is the sampling time, λ_1, λ_2, are eigenvalues of the system and r_1, r_2 are residuals:

$$\lambda_{1,2}=-\tfrac{1}{2}[L_r(R_s+R_r)/(L_r^2-L_m^2)]\pm\tfrac{1}{2}\{[L_r(R_s+R_r)]^2/(L_r^2-L_m^2)^2$$
$$-4R_r R_s/(L_r^2-L_m^2)\}^{1/2} \tag{8.2-32}$$

$$r_1=(1+\lambda_1 T_r)/[R_s(\lambda_1-\lambda_2)] \tag{8.2-33}$$

$$r_2=(1+\lambda_2 T_r)/[R_s(\lambda_2-\lambda_1)], \tag{8.2-34}$$

where T_r is the rotor time constant, $T_r=L_r/R_r$.

For better understanding, the identification steps of the machine parameters are now summarized:

Step 1 The machine is supplied by a single-phase voltage which is a step voltage,

$$u_{sD}(t)=U_{s0}.$$

Step 2 The current response, $i_{sD}(t)$, is measured and the data are stored.

Step 3 The parameter vector $\hat{\mathbf{p}}$ (which contains the four elements $\hat{a}_1$, $\hat{a}_2$, $\hat{b}_1$, $\hat{b}_2$) is estimated by the RLS algorithm.

Step 4 The machine parameters ($\hat{R}_s$, $\hat{R}_r$, $\hat{L}_r$, $\hat{L}_m$) are obtained by solving eqns (8.2-28)–(8.2-34). It should be noted that $L_s=L_r$ has been assumed and magnetic saturation has been neglected.

The parameters obtained by the technique described above cannot be used in the drive scheme if the drive operates in the field-weakening region, since in this case, the saturated magnetizing inductance must be used. However, for this purpose a simple correction to the magnetizing inductance can be made [Bunte and Grotstollen 1993]. This uses the measured steady-state value of the stator current response, $I_s=i_{sD}(t\to\infty)$, due to the applied single-phase step voltage (U_{s0}). Thus the following correction factor is defined:

$$k_f=I_s(1+\hat{a}_1+\hat{a}_2)/[U_{s0}(\hat{b}_1+\hat{b}_2)]. \tag{8.2-35}$$

The correction of the magnetizing inductance is made by as follows: if $k_f<1$, L_m has to be decreased; if $k_f=1$, L_m does not have to be changed; if $k_f>1$, L_m has to be increased. Equation (8.2-35) follows from the fact that in the steady state, when magnetic saturation is neglected, it follows from eqn (8.2-27) that $I_s=U_{s0}(\hat{b}_1+\hat{b}_2)/(1+\hat{a}_1+\hat{a}_2)$. Furthermore, in the identification scheme a linear machine model has been used, thus the estimated parameters in $I_s=U_{s0}(\hat{b}_1+\hat{b}_2)/(1+\hat{a}_1+\hat{a}_2)$ are erroneous. The correction is done by using an iterative method which minimizes the error function:

$$J=\tfrac{1}{2}\sum[i_{sD}(k)-\hat{i}_{sD}(k)], \tag{8.2-36}$$

where $i_{sD}(k)$ is obtained from the measured stator-current response, and $\hat{i}_{sD}(k)$ is obtained from the machine model using the estimated machine parameters $\hat{R}_s$, $\hat{R}_r$, $\hat{L}_r$, $\hat{L}_m$).

It should be noted that it is also possible to identify the saturated magnetizing inductance by using the measured magnetizing curve [$\psi_s(I_m)$ curve], which is obtained point-by-point, and then $L_s(i_s)=\psi_s/I_s$ is obtained and thus $L_m=L_s-L_{sl}$, where L_{sl} is the stator leakage inductance. To obtain a point of this curve, the stator voltages are not monitored, but a stator voltage is first reconstructed from the reference voltage (voltage command applied to the inverter) by also considering the non-linear voltage error $\Delta u_s(i_s)$ of the inverter. Thus the stator voltage is obtained as $u_s=u_{sref}+\Delta u_s(i_s)$. The steps of obtaining each point of the magnetizing curve are:

Step 1 A step of reference voltage from 0 to U_{sref} is applied to the machine, which is at standstill. However, care must be taken to ensure that the voltage does not exceed a maximum value yielding excessive stator current. For this purpose it is possible to use the results of another test which gives the $U_{sref}(I_s)$ static characteristic (by applying different U_{sref} and measuring the stator current in the steady state). This static curve can also be used to obtain $\Delta u_s(I_s)$, since $\Delta u_s(I_s)=U_{sref}(I_s)-R_sI_s$.

Step 2 The response $i_s(t)$ is measured.

Step 3 Flux linkage is obtained by integrating stator voltage applied to machine; thus

$$\psi_s=\int(U_{sref}-\Delta u_s(i_s)-R_si_s)\,dt,$$

where $\Delta u_s(i_s)$ is obtained by using the $\Delta u_s(I_s)$ characteristic (obtained in Step 1).

Step 4 Stator inductance is obtained as $L_S(i_s)=\psi_s/I_s$.

Step 5 Magnetizing inductance is obtained as $L_m=L_s-L_{sl}$.

Bibliography

Borsting, H., Knudsen, M., and Vadstrup, P. (1995). Standstill estimation of electrical parameters in induction machines using an optimal input signal. *EPE, Seville*, pp. 814–19.

Bunte, A. and Grotstollen, H. (1993). Parameter identification of an inverter-fed induction motor at standstill with a correlation method. *EPE, Brighton*, pp. 97–102.

Bunte, A. and Grotstollen, H. (1996). Off-line parameter identification of an inverter-fed induction motor at standstill. *EPE, Seville*, 3.492–3.496.

Du, T., Vas, P., and Stronach, F. (1995). Design and application of extended observers for joint state and parameter estimation in high-performance a.c. drive. *Proc. IEE Pt. B* **142**, 71–8.

Elten, D. and Filbert, D. (1990). Identification of electrical parameters and the speed of small three phase induction motors. *ETZ-Archiv* **12**, 379–83.

Gorter, R. J. A., Duarte, J. L., and Van de Bosch, P. P. J. (1995). Parameter estimation for induction machines. *EPE, Seville*, pp. 627–32.

Grotstollen, H. and Wiesing, J. (1995). Torque capability and control of a saturated induction motor over a wide range of flux weakening. *IEEE Transaction on Ind. Electronics* **42**, 374–81.

Harms, K. and Leonhard, W. (1987). Parameter adaptive control of induction motor based on steady-state model. *2nd European Conference on Power Electronics and Applications, Grenoble*, pp. 2.57–2.62.

Heinemann, G. and Leonhard, W. (1990). Self-tuning field orientated control of an induction motor drive. *IPEC, Tokyo*, pp. 465–72.

Hindmarsh, J. (1977). *Electrical machines and their applications*. Pergamon Press, Oxford.

Holtz, J. and Thimm, T. (1989). Identification of the machine parameters in a vector controlled induction motor drive. *IEEE IAS Meeting*, pp. 601–6.

Kudor, T., Ishihara, K., and Naitoh, H. (1993). Self-commissioning for vector controlled induction motors. *IEEE IAS Meeting*, pp. 528–35.

Loeser, F. and Sattler, P. K. (1985). Identification and compensation of the rotor temperature of a.c. drives by an observer. *IEEE Transactions on Industry Applications* **IA-21**, 1387–93.

Matsuo, T. and Lipo, T. A. (1985). A rotor parameter identification scheme for vector-controlled induction motor drives. *IEEE Transactions on Industry Applications* **IA-21**, 624–32.

Moons, C. and de Moor, B. (1995). Parameter identification of induction motor drives. *Automatica* **31**, 1137–47.

Naunin, D. and Beierke, S. (1992). Rotor time constant online identification of induction cage motors for transputer controlled servo drives. *ICEM, Manchester*, pp. 587–91.

Rasmussen, H., Tonnes, M., and Knudsen, M. (1995). Inverter and motor model adaptation at standstill using reference voltages and measured currents. *EPE, Seville*, 1.367–1.372.

Ruff, M. and Grotstollen, H. (1993). Identification of the saturated mutual inductance of an asynchronous motor at standstill by recursive least squares algorithm. *EPE, Brighton*, pp. 103–9.

Ruff, M., Bunte, A., and Grotstollen, H. (1994). A new self-commissioning scheme for an asynchronous motor drive system. *IEEE IAS Meeting*, pp. 616–23.

Say, M. G. (1976). *Alternating current machines*. Pitman, London.

Schierling, H. (1988). Fast and reliable commissioning of a.c. variable speed drives by self-commissioning. *IEEE IAS Annual Meeting, Pittsburgh*, pp. 489–92.

Stronach, F. and Vas, P. (1995). Variable-speed drives incorporating multi-loop adaptive controllers. *Proc. IEE (Pt. D)* **142**, 411–20.

Tonness, M. and Rasmussen, H. (1993). Self-tuning speed controllers for automatic tuning of torque regultated a.c. drives. *EPE, Brighton*, pp. 168–73.

Vas, P. (1990). *Vector control of a.c. machines*. Oxford University Press.

Vas, P. (1993). *Parameter estimation, condition monitoring a diagnosis of electrical machines*. Oxford University Press.

Wang, C., Novotny, D. W., and Lipo, T. A. (1986). An automated rotor time constant measurement system for indirect field oriented drives. *IEEE IAS Annual Meeting, Denver*, pp. 140–6.

Willis, J. R., Brock, G. J., and Edmonds, J. S. (1989). Derivation of induction motor models from standstill frequency response tests. *IEEE Transactions on Energy Conversion* **4**, 608–13.

Index

Appendix: Texas Instruments' motion control DSPs

In the present book numerous motion control applications of DSPs are described. For this purpose it is very suitable to use the new family of Texas Instruments DSP controllers, the TMS320C24x family. This includes six different members based on the 320C2XIp providing 20 MIPS performance. The 320C24x products support power switching device commutation, command generation, control algorithm processing, data communication and system monitoring functions. All the members offer an ideal single chip solution for digital motor control applications. They include program memory Flash or ROM, RAM for the data memory, Event Manager (EV) for generating pulse width modulation signals, A/D converters, UART and CAN controllers.

The TMS320F240 is suitable for an application requiring a large amount of memory on chip. It includes 16K Word of Flash, two 10-bit A/D converters, an EV including 3 timers, a serial port interface for an easy connection with other peripherals, an UART, a 16-bit data and address bus for connecting external memory and peripherals.

The TMS320C240 is the ROM-based version of the TMS320F240 and is ideal for large volume applications. The EV module is used for operations particularly useful for digital motion control applications. The EV module generates outputs and acquires input signals with a minimum CPU load. Up to 4 input captures and 12 output PWMs are available. The A/D converter module has two 10 bit–10 ns converters for a total of 16 input channels. Two sample-and-holds allow parallel and simultaneous sampling and conversion. Conversion starts by external signal transition, software instruction or EV event. The reference voltage of the A/D module is 0–5 volts; this can be supplied either internally or externally. The device also includes a watchdog timer and a Real Time Interrupt (RTI) module.

The TMS320F241 and TMS320C241 are 8K word versions (Flash and ROM). They include a new 10-bit A/D converter, which can process the conversion of a signal in less than 850 ns. They are also the first DSPs to contain a CAN controller on chip; this makes them ideal for automotive and numerous industrial applications, which require fast and secure communication in noisy environments.

The TMS320C242 is a 4K ROM-based device designed for all motor control applications in the consumer area. The 4K ROM program memory is satisfactory to include a complete sensorless software which will control optimally in space vector mode the speed of an induction motor in a washing machine.

The TMS320F243 is also available for an application that requires CAN controller and extended addressing range capability.

Bibliography

Texas Instruments TMS320C24x manual, 1998.